LIFE

The Science of Biology

TENTH EDITION

LIFE

The Science of Biology
TENTH EDITION

DAVID
SADAVA
The Claremont Colleges

DAVID M.
HILLIS
University of Texas

H. CRAIG
HELLER
Stanford University

MAY R.
BERENBAUM
University of Illinois

SINAUER

MACMILLAN

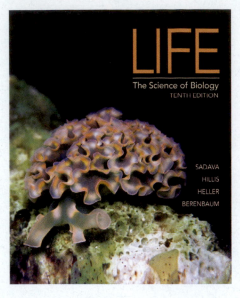

THE COVER
The sea slug *Elysia crispata*. This animal is able to carry out photosynthesis using chloroplasts incorporated from the algae it feeds on (see back cover). Photograph © Alex Mustard/Naturepl.com.

THE FRONTISPIECE
Red-crowned cranes, *Grus japonensis*, gather on a river in Hokkaido, Japan. ©Steve Bloom Images/Alamy.

LIFE: The Science of Biology, Tenth Edition

ADDRESS EDITORIAL CORRESPONDENCE TO:
Sinauer Associates, Inc., 23 Plumtree Road, Sunderland, MA 01375 U.S.A.

www.sinauer.com
publish@sinauer.com

ADDRESS ORDERS TO:
MPS / W. H. Freeman & Co., Order Dept., 16365 James Madison Highway, U.S. Route 15, Gordonsville, VA 22942 U.S.A.

EXAMINATION COPY INFORMATION: 1-800-446-8923

Certified Chain of Custody
Promoting Sustainable Forestry
www.sfiprogram.org
SFI-00712

Planet Friendly Publishing
✔ Made in the United States
✔ Printed on Recycled Paper
Text: 10%   Cover: 10%
Learn more: www.greenedition.org

Courier Corporation, the manufacturer
of this book, owns the *Green Edition* Trademark

Library of Congress Cataloging-in-Publication Data

Life : the science of biology / David Sadava ... [et al.]. -- 10th ed.
    p. cm.
Includes bibliographical references and index.
ISBN 978-1-4292-9864-3 (casebound) — 978-1-4641-4122-5 (pbk. : v. 1) —
ISBN 978-1-4641-4123-2 (pbk. : v. 2) — ISBN 978-1-4641-4124-9 (pbk. : v. 3)
1. Biology--Textbooks. I. Sadava, David E.
QH308.2.L565 2013
570--dc23   2012039164

Printed in U.S.A.
First Printing December 2012
The Courier Companies, Inc.

To all the educators who have worked tirelessly
for quality biology education

The Authors

DAVID HILLIS   MAY BERENBAUM

DAVID SADAVA is the Pritzker Family Foundation Professor of Biology, Emeritus at the Keck Science Center of Claremont McKenna, Pitzer, and Scripps, three of The Claremont Colleges. In addition, he is Adjunct Professor of Cancer Cell Biology at the City of Hope Medical Center in Duarte, California. Twice winner of the Huntoon Award for superior teaching, Dr. Sadava has taught courses on introductory biology, biotechnology, biochemistry, cell biology, molecular biology, plant biology, and cancer biology. In addition to *Life: The Science of Biology* and *Principles of Life*, he is the author or coauthor of books on cell biology and on plants, genes, and crop biotechnology. His research has resulted in many papers coauthored with his students, on topics ranging from plant biochemistry to pharmacology of narcotic analgesics to human genetic diseases. For the past 15 years, he has investigated multidrug resistance in human small-cell lung carcinoma cells with a view to understanding and overcoming this clinical challenge. At the City of Hope, his current work focuses on new anti-cancer agents from plants. He is the featured lecturer in "Understanding Genetics: DNA, Genes and their Real-World Applications," a video course for The Great Courses series.

DAVID M. HILLIS is the Alfred W. Roark Centennial Professor in Integrative Biology and the Director of the Dean's Scholars Program at the University of Texas at Austin, where he also has directed the School of Biological Sciences and the Center for Computational Biology and Bioinformatics. Dr. *Hillis* has taught courses in introductory biology, genetics, evolution, systematics, and biodiversity. He has been elected to the National Academy of Sciences and the American Academy of Arts and Sciences, awarded a John D. and Catherine T. MacArthur fellowship, and has served as President of the Society for the Study of Evolution and of the Society of Systematic Biologists. He served on the National Research Council committee that wrote the report BIO 2010: *Undergraduate Biology Education for Research Biologists*. His research interests span much of evolutionary biology, including studies of viral evolution, empirical studies of natural molecular evolution, applications of phylogenetics, analyses of terrestrial and evolutionary modeling. He is particularly interested of evolutionary teaching and research about the practical applications of evolutionary biology.

H. CRAIG ...logy.
in Biologica...
sity. He has ta...orry I. Lokey/Business Wire Professor
...Human Biology at Stanford Univer-
...biology courses at Stanford since

1972 and served as Director
Chairman of the Biological
ate Dean of Research. Dr.
Association for the Advan
the Walter J. Gores Awar
Kenneth Cuthberson Aw
ford University. His rese
and circadian rhythms, m
of body temperature, the
and the neurobiology of
huge variety of animals a
from sleeping kangaroo
photoperiodic hamsters
extended his enthusias
development of a two—
the middle grades, thro
teractive computer-ba

MAY BERENBAUM is th
Department of Ento
bana-Champaign. Sh
mal biology, entomo
and has received tea
levels from the Ento
the National Acade
Arts and Sciences, a
served as Presiden
ences in 2009 and
AAAS. Her resea
ranges from mol
pacts of herbivor
he practical app
...les, she has ex
...e change
...fosystem
and Priz

*To all the educators who have worked tirelessly
for quality biology education*

The Authors

DAVID HILLIS   MAY BERENBAUM   CRAIG HELLER   DAVID SADAVA

DAVID SADAVA is the Pritzker Family Foundation Professor of Biology, Emeritus at the Keck Science Center of Claremont McKenna, Pitzer, and Scripps, three of The Claremont Colleges. In addition, he is Adjunct Professor of Cancer Cell Biology at the City of Hope Medical Center in Duarte, California. Twice winner of the Huntoon Award for superior teaching, Dr. Sadava has taught courses on introductory biology, biotechnology, biochemistry, cell biology, molecular biology, plant biology, and cancer biology. In addition to *Life: The Science of Biology and Principles of Life*, he is the author or coauthor of books on cell biology and on plants, genes, and crop biotechnology. His research has resulted in many papers coauthored with his students, on topics ranging from plant biochemistry to pharmacology of narcotic analgesics to human genetic diseases. For the past 15 years, he has investigated multidrug resistance in human small-cell lung carcinoma cells with a view to understanding and overcoming this clinical challenge. At the City of Hope, his current work focuses on new anti-cancer agents from plants. He is the featured lecturer in "Understanding Genetics: DNA, Genes and their Real-World Applications," a video course for The Great Courses series.

DAVID M. HILLIS is the Alfred W. Roark Centennial Professor in Integrative Biology and the Director of the Dean's Scholars Program at the University of Texas at Austin, where he also has directed the School of Biological Sciences and the Center for Computational Biology and Bioinformatics. Dr. Hillis has taught courses in introductory biology, genetics, evolution, systematics, and biodiversity. He has been elected to the National Academy of Sciences and the American Academy of Arts and Sciences, awarded a John D. and Catherine T. MacArthur fellowship, and has served as President of the Society for the Study of Evolution and of the Society of Systematic Biologists. He served on the National Research Council committee that wrote the report *BIO 2010: Transforming Undergraduate Biology Education for Research Biologists*. His research interests span much of evolutionary biology, including experimental studies of viral evolution, empirical studies of natural molecular evolution, applications of phylogenetics, analyses of biodiversity, and evolutionary modeling. He is particularly interested in teaching and research about the practical applications of evolutionary biology.

H. CRAIG HELLER is the Lorry I. Lokey/Business Wire Professor in Biological Sciences and Human Biology at Stanford University. He has taught in the core biology courses at Stanford since 1972 and served as Director of the Program in Human Biology, Chairman of the Biological Sciences Department, and Associate Dean of Research. Dr. Heller is a fellow of the American Association for the Advancement of Science and a recipient of the Walter J. Gores Award for excellence in teaching and the Kenneth Cuthberson Award for Exceptional Service to Stanford University. His research is on the neurobiology of sleep and circadian rhythms, mammalian hibernation, the regulation of body temperature, the physiology of human performance, and the neurobiology of learning. He has done research on a huge variety of animals and physiological problems, including from sleeping kangaroo rats, diving seals, hibernating bears, photoperiodic hamsters, and exercising athletes. Dr. Heller has extended his enthusiasm for promoting active learning via the development of a two-year curriculum in human biology for the middle grades, through the production of Virtual Labs—interactive computer-based modules to teach physiology.

MAY BERENBAUM is the Swanlund Professor and Head of the Department of Entomology at the University of Illinois at Urbana-Champaign. She has taught courses in introductory animal biology, entomology, insect ecology, and chemical ecology and has received teaching awards at the regional and national levels from the Entomological Society of America. A fellow of the National Academy of Sciences, the American Academy of Arts and Sciences, and the American Philosophical Society, she served as President of the American Institute for Biological Sciences in 2009 and currently serves on the Board of Directors of AAAS. Her research addresses insect–plant coevolution and ranges from molecular mechanisms of detoxification to impacts of herbivory on community structure. Concerned with the practical application of ecological and evolutionary principles, she has examined impacts of genetic engineering, global climate change, and invasive species on natural and agricultural ecosystems. In recognition of her work, she received the 2011 Tyler Prize for Environmental Achievement. Devoted to fostering science literacy, she has published numerous articles and five books on insects for the general public.

Contents in Brief

PART ONE ■ THE SCIENCE OF LIFE AND ITS CHEMICAL BASIS

1 Studying Life   1
2 Small Molecules and the Chemistry of Life   21
3 Proteins, Carbohydrates, and Lipids   39
4 Nucleic Acids and the Origin of Life   62

PART TWO ■ CELLS

5 Cells: The Working Units of Life   77
6 Cell Membranes   105
7 Cell Communication and Multicellularity   125

PART THREE ■ CELLS AND ENERGY

8 Energy, Enzymes, and Metabolism   144
9 Pathways that Harvest Chemical Energy   165
10 Photosynthesis: Energy from Sunlight   185

PART FOUR ■ GENES AND HEREDITY

11 The Cell Cycle and Cell Division   205
12 Inheritance, Genes, and Chromosomes   232
13 DNA and Its Role in Heredity   259
14 From DNA to Protein: Gene Expression   281
15 Gene Mutation and Molecular Medicine   304
16 Regulation of Gene Expression   328

PART FIVE ■ GENOMES

17 Genomes   352
18 Recombinant DNA and Biotechnology   373
19 Differential Gene Expression in Development   392
20 Genes, Development, and Evolution   412

PART SIX ■ THE PATTERNS AND PROCESSES OF EVOLUTION

21 Mechanisms of Evolution   427
22 Reconstructing and Using Phylogenies   449
23 Speciation   467
24 Evolution of Genes and Genomes   485
25 The History of Life on Earth   505

PART SEVEN ■ THE EVOLUTION OF DIVERSITY

26 Bacteria, Archaea, and Viruses   525
27 The Origin and Diversification of Eukaryotes   549
28 Plants without Seeds: From Water to Land   569

29 The Evolution of Seed Plants   588
30 The Evolution and Diversity of Fungi   608
31 Animal Origins and the Evolution of Body Plans   629
32 Protostome Animals   651
33 Deuterostome Animals   678

PART EIGHT ■ FLOWERING PLANTS: FORM AND FUNCTION

34 The Plant Body   708
35 Transport in Plants   726
36 Plant Nutrition   740
37 Regulation of Plant Growth   756
38 Reproduction in Flowering Plants   778
39 Plant Responses to Environmental Challenges   797

PART NINE ■ ANIMALS: FORM AND FUNCTION

40 Physiology, Homeostasis, and Temperature Regulation   815
41 Animal Hormones   834
42 Immunology: Animal Defense Systems   856
43 Animal Reproduction   880
44 Animal Development   902
45 Neurons, Glia, and Nervous Systems   924
46 Sensory Systems   946
47 The Mammalian Nervous System   967
48 Musculoskeletal Systems   986
49 Gas Exchange   1005
50 Circulatory Systems   1025
51 Nutrition, Digestion, and Absorption   1048
52 Salt and Water Balance and Nitrogen Excretion   1071
53 Animal Behavior   1093

PART TEN ■ ECOLOGY

54 Ecology and the Distribution of Life   1121
55 Population Ecology   1149
56 Species Interactions and Coevolution   1169
57 Community Ecology   1188
58 Ecosystems and Global Ecology   1207
59 Biodiversity and Conservation Biology   1228

Preface

Biology is a constantly changing scientific field. New discoveries about the living world are being made every day, and more than 1 million new research articles in biology are published each year. Beyond the constant need to update the concepts and facts presented in any science textbook, in recent years ideas about how best to educate the upcoming generation of biologists have undergone dynamic and exciting change.

Although we and many of our colleagues had thought about the nature of biological education as individuals, it is only recently that biologists have come together to discuss these issues. Reports from the National Academy of Sciences, Howard Hughes Medical Institute, and College Board AP Biology Program not only express concern about how best to instruct undergraduates in biology, but offer concrete suggestions about how to design the introductory biology course—and by extension, our book. We have followed these discussions closely and have been especially impressed with the report "Vision and Change in Undergraduate Biology Education" (visionandchange.org). As participants in the educational enterprise, we have answered the report's call to action with this textbook and its associated ancillary materials.

The "Vision and Change" report proposes five core concepts for biological literacy:

1. Evolution
2. Structure and function
3. Information flow, exchange, and storage
4. Pathways and transformations of energy and matter
5. Systems

These five concepts have always been recurring themes in Life, but in this Tenth Edition we have brought them even more "front and center."

"Vision and Change" also advocates that students learn and demonstrate core competencies, including the ability to apply the process of science using quantitative reasoning. Life has always emphasized the experimental nature of biology. This edition responds further to these core competency issues with a new working with data feature and the addition of a statistics primer (Appendix B). The authors' multiple educational perspectives and areas of expertise, as well as input from many colleagues and students who used previous editions, have informed the approach to this new edition.

Enduring Features

We remain committed to blending the presentation of core ideas with an emphasis on introducing students to the process of scientific inquiry. Having pioneered the idea of depicting important experiments in unique figures designed to help students understand and appreciate the way scientific investigations work, we continue to develop this approach in the book's 70 **Investigating Life** figures. Each of these figures sets the experiment in perspective and relates it to the accompanying text. As in previous editions, these figures employ the structure Hypothesis, Method, Results, and Conclusion. We have added new information focusing on the individuals who performed these experiments so students can appreciate more fully that science is a human and very personal activity. Each Investigating Life figure has a reference to BioPortal (*yourBioPortal.com*), where discussion and references to follow-up research can be found. A related feature is the **Research Tools** figures, which depict laboratory and field methods used in biology. These, too, have been expanded to provide more useful context for their importance.

Some 15 years ago, *Life's* authors and publishers pioneered the use of **balloon captions** in our figures. We recognized then that many students are visual learners, and this fact is even truer today. *Life's* balloon captions bring the crucial explanations of intricate, complex processes directly into the illustration, allowing students to integrate information without repeatedly going back and forth between the figure, its legend, and the text.

We continue to refine our chapter organization. Our **opening stories** have always provide historical, medical, or social context to intrigue students and show how the subject of each chapter relates to the world around them. In the Tenth Edition, the opening stories all end with a question that is revisited throughout the chapter. At the end of each chapter the answer is presented in the light of material the student encountered in the body of the chapter.

A **chapter outline** asks questions to emphasize scientific inquiry, each of which is answered in a major section of the chapter. A **Recap** summarizes each section's key concepts and poses questions that help the student review and test their mastery of these concepts. The recap questions are similar in form to the learning objectives used in many introductory biology courses. The **Chapter Summaries** highlight each chapter's key figures and defined terms, while restating the major concepts presented in the chapter in a concise and student-friendly manner, with references to specific figures and to the activities and animated tutorials available in BioPortal.

At the end of the book, students will find a much-expanded glossary that continues *Life's* practice of providing Latin or Greek derivations for many of the defined terms. As students become gradually (and painlessly) more familiar with such root words, the mastery of vocabulary as they continue in their biological or medical studies will be easier. In addition, the popular **Tree of Life appendix** (Appendix A) presents the phylogenetic tree of life as a reference tool that allows students to place any group of organisms mentioned in the text into the context of the rest of life. The web-based version of Appendix A provides links to photos, keys, species lists, distribution maps, and other information (via the online database at DiscoverLife.org) to help students explore biodiversity in greater detail.

New Features

The Tenth Edition of *Life* has a different look and feel from its predecessors. The new color palette and more open design will, we hope, be more accessible to students. And, in keeping with our heightened emphasis on scientific inquiry and quantitative analysis, we have added **Working with Data** exercises to almost all chapters. In these innovative exercises, we describe the context and approach of a research paper that provides the basis of the analysis. We then ask questions that require students to analyze data, make calculations, and draw conclusions. Answers (or suggested possible answers) to these questions are included in BioPortal and can be made available to students at the instructor's discretion.

Because many of the questions in the Working with Data exercises require the use of basic statistical methods, we have included a **Statistics Primer** as the book's Appendix B, describing the concepts and some methods of statistical analysis. We hope that the Working with Data exercises and statistics primer will reinforce students' skills and their ability to apply quantitative analysis to biology.

We have added links to **Media Clips** in the body of the text, with at least one per chapter. These brief clips are intended to enlighten and entertain. Recognizing the widespread use of "smart phones" by students, the textbook includes **instant access (QR) codes** that bring the Media Clips, Animated Tutorials, and Interactive Summaries directly to the screen in your hand. If you do not have a smart phone, never fear, we also provide direct web addresses to these features.

As educators, we follow current discussions of pedagogy in biological education. The chapter-ending **Chapter Reviews** now contain multiple levels of questions based on Bloom's taxonomy: Remembering, Understanding and Applying, and Analyzing and Evaluating. Answers to these questions appear at the end of the book.

For a detailed description of the media and supplements available for the Tenth Edition, please turn to "*Life's* Media and Supplements" on page xvii.

The Ten Parts

PART ONE, THE SCIENCE OF LIFE AND ITS CHEMICAL BASIS Chapter 1 introduces the core concepts set forth in the "Vision and Change" report and continues the much-praised approach of focusing on a specific series of experiments that introduces students to biology as an experimentally based and constantly expanding science. Chapter 1 emphasizes the principles of biology that are the foundation for the rest of the book, including the unity of life at the cellular level and how evolution unites the living world. Chapters 2–4 cover the chemical principles and building blocks that underlie life. Chapter 4 also includes a discussion of how life could have evolved from inanimate chemicals.

PART TWO, CELLS The nature of cells and their role as the structural and functional basis of life is foundational to biology. These revised chapters include expanded explanations of how experimental manipulations of living systems have been used to discover cause and effect in biology. Students who are intrigued by the question "Where did the first cells come from?" will appreciate the updated discussion of ideas on the origin of cells and organelles, as well as expanded discussion of the evolution of multicellularity and cell interactions. In response to reviewer comments, the discussion of membrane potential has been moved to Chapter 45, where students may find it to be more relevant.

PART THREE, CELLS AND ENERGY The biochemistry of life and energy transformations are among the most challenging topics for many students. We have worked to clarify such concepts as enzyme inhibition, allosteric enzymes, and the integration of biochemical systems. Revised presentations of glycolysis and the citric acid cycle now focus, in both text and figures, on key concepts and attempt to limit excessive detail. There are also revised discussions of the ecological roles of alternate pathways of photosynthetic carbon fixation, as well as the roles of accessory pigments and reaction center in photosynthesis.

PART FOUR, GENES AND HEREDITY This crucial section of the book is revised to improve clarity, link related concepts, and provide updates from recent research results. Rather than being segregated into separate chapters, material on prokaryotic genetics and molecular medicine are now interwoven into relevant chapters. Chapter 11 on the cell cycle includes a new discussion of how the mechanisms of cell division are altered in cancer cells. Chapter 12 on transmission genetics now includes coverage of this phenomenon in prokaryotes. Chapters 13 and 14 cover gene expression and gene regulation, including new discoveries about the roles of RNA and an expanded discussion of epigenetics. Chapter 15 covers the subject of gene mutations and describes updated applications of medical genetics.

PART FIVE, GENOMES This extensive and up-to-date coverage of genomes expands and reinforces the concepts covered in Part Four. The first chapter of Part Five describes how genomes are analyzed and what they tell us about the biology of prokaryotes and eukaryotes, including humans. Methods of DNA sequencing and genome analysis, familiar to many students in a general way, are rapidly improving, and we discuss these advances as well as how bioinformatics is used. This leads to a chapter describing how our knowledge of molecular biology and genetics underpins biotechnology—the application of this knowledge to practical problems and issues such as stem cell research. Part Five closes with a unique sequence of two chapters that explore the interface of developmental processes with molecular biology (Chapter 19) and with evolution (Chapter 20), providing students with a link between these two crucial topics and a bridge to Part Six.

PART SIX, THE PATTERNS AND PROCESSES OF EVOLUTION   Many students come to the introductory biology course with ideas about evolution already firmly in place. One common view, that evolution is only about Darwin, is firmly put to rest at the start of Chapter 21, which not only illustrates the practical value of fully understanding modern evolutionary biology, but briefly and succinctly traces the history of "Darwin's dangerous idea" through the twentieth century and up to the present syntheses of molecular evolutionary genetics and evolutionary developmental biology—fields of study that uphold and support the principles of evolutionary biology as the basis for comparing and comprehending all other aspects of biology. The remaining sections of Chapter 21 describe the mechanisms of evolution in clear, matter-of-fact terms. Chapter 22 describes phylogenetic trees as a tool not only of classification but also of evolutionary inquiry. The remaining chapters cover speciation and molecular evolution, concluding with an overview of the evolutionary history of life on Earth.

PART SEVEN, THE EVOLUTION OF DIVERSITY   Continuing the theme of how evolution has shaped our world, Part Seven introduces the latest views on biodiversity and the evolutionary relationships among organisms. The chapters have been revised with the aim of making it easier for students to appreciate the major evolutionary changes that have taken place within the different groups of organisms. These chapters emphasize understanding the big picture of organismal diversity—the tree of life—as opposed to memorizing a taxonomic hierarchy and names. Throughout the book, the tree of life is emphasized as a way of understanding and organizing biological information.

PART EIGHT, FLOWERING PLANTS: FORM AND FUNCTION   The emphasis of this modern approach to plant form and function is not only on the basic findings that led to the elucidation of mechanisms for plant growth and reproduction, but also on the use of genetics of model organisms. In response to users of earlier editions, material covering recent discoveries in plant molecular biology and signaling has been reorganized and streamlined to make it more accessible to students. There are also expanded and clearer explanations of such topics as water relations, the plant body plan, and gamete formation and double fertilization.

PART NINE, ANIMALS: FORM AND FUNCTION   This overview of animal physiology begins with a sequence of chapters covering the systems of information—endocrine, immune, and neural. Learning about these information systems provides important groundwork and explains the processes of control and regulation that affect and integrate the individual physiological systems covered in the remaining chapters of the Part. Chapter 45, "Neurons and Nervous Systems," has been rearranged and contains descriptions of exciting new discoveries about glial cells and their role in the vertebrate nervous system. The organization of several other chapters has been revised to reflect recent findings and to allow the student to more readily identify the most important concepts to be mastered.

PART TEN, ECOLOGY   Part Ten continues *Life*'s commitment to presenting the experimental and quantitative aspects of biology, with increased emphasis on how ecologists design and conduct experiments. New exercises provide opportunities for students to see how ecological data are acquired in the laboratory and in the field, how these data are analyzed, and how the results are applied to answer questions. There is also an expanded discussion of aquatic biomes and a more synthetic explanation of how aquatic, terrestrial, and atmospheric components integrate to influence the distribution and abundance of life on Earth. In addition there is an expanded emphasis on examples of successful strategies proposed by ecologists to mitigate human impacts on the environment; rather than an inventory of ways human activity adversely affects natural systems, this revised Tenth Edition provides more examples of ways that ecological principles can be applied to increase the sustainability of these systems.

Exceptional Value Formats

We again provide *Life* both as the full book and as a set of paperback volumes. Thus, instructors who want to use less than the whole book can choose from these split volumes, each of which contains the book's front matter, appendices, glossary, and index.

- Volume I, *The Cell and Heredity*, includes: Part One, The Science of Life and Its Chemical Basis (Chapters 1–4); Part Two, Cells (Chapters 5–7); Part Three, Cells and Energy (Chapters 8–10); Part Four, Genes and Heredity (Chapters 11–16); and Part Five, Genomes (Chapters 17–20).

- Volume II, *Evolution, Diversity, and Ecology*, includes: Chapter 1, Studying Life; Part Six, The Patterns and Processes of Evolution (Chapters 21–25); Part Seven, The Evolution of Diversity (Chapters 26–33); and Part Ten, Ecology (Chapters 54–59).

- Volume III, *Plants and Animals*, includes: Chapter 1, Studying Life; Part Eight, Flowering Plants: Form and Function (Chapters 34–39); and Part Nine, Animals: Form and Function (Chapters 40–53).

Responding to student concerns, there also are two ways to obtain the entire book at a significantly reduced cost. The loose-leaf edition of *Life* is a shrink-wrapped, unbound, three-hole-punched version that fits into a three-ring binder. Students take only what they need to class and can easily integrate instructor handouts and other resources.

Life was the first comprehensive biology text to offer the entire book as a truly robust eBook, and we offer the Tenth Edition in this flexible, interactive format that gives students a different way to read the text and learn the material. The eBook integrates student media resources (animations, activities, interactive summaries, and quizzes) and offers instructors a powerful way to customize the textbook with their own text, images, web links, and, in BioPortal, quizzes, and other materials.

We are proud that our print edition is a greener *Life* that minimizes environmental impact. *Life* was the first introductory biology text to be printed on paper earning the Forest Stewardship Council label, the "gold standard in green paper," and it continues to be manufactured from wood harvested from sustainable forests.

Many People to Thank

One of the wisest pieces of advice ever given to a textbook author is to "be passionate about your subject, but don't put your ego on the page." Considering all the people who looked over our shoulders throughout the process of creating this book, this advice could not be more apt. We are indebted to the many people who help to make this book what it is. First and foremost among these are our colleagues, biologists from over 100 institutions. Before we set pen to paper, we solicited the advice of users of *Life*'s Ninth Edition, as well as users of other books. These reviewers gave detailed suggestions for improvements. Other colleagues acted as reviewers when the book was almost completed, pointing out inaccuracies or lack of clarity. All of these biologists are listed in the reviewer credits, along with the dozens who reviewed all of the revised assessment resources.

Once we began writing, we had the superb advice of a team of experienced, knowledgeable, and patient biologists working as development and line editors. Laura Green of Sinauer Associates headed the team and coordinated her own fine work with that of Jane Murfett, Norma Roche, and Liz Pierson

to produce a polished and professional text. We are especially indebted to Laura for her work on the important Investigating Life and new Working with Data elements. For the tenth time in ten editions, Carol Wigg oversaw the editorial process. Her positive influence pervades the entire book. Artist Elizabeth Morales again translated our crude sketches into beautiful new illustrations. We hope you agree that our art program remains superbly clear and elegant. Johannah Walkowicz effectively coordinated the hundreds of reviews described above. David McIntyre, photo editor extraordinaire, researched and provided us with new photographs, including many of his own, to enrich the book's content and visual statement. Joanne Delphia is responsible for the crisp new design and layout that make this edition of *Life* not just clear and readable but beautiful as well. Christopher Small headed Sinauer's production team and contributed in innumerable ways to bringing *Life* to its final form. Jason Dirks coordinated the creation of our array of media and instructor resources, with Mary Tyler, Mitch Walkowicz, and Carolyn Wetzel serving as editors for our expanded assessment supplements.

W. H. Freeman continues to bring *Life* to a wider audience. Associate Director of Marketing Debbie Clare, the regional specialists, regional managers, and experienced sales force are effective ambassadors and skillful transmitters of the features and unique strengths of our book. We depend on their expertise and energy to keep us in touch with how *Life* is perceived by its users. Thanks also to the Freeman media group for eBook and BioPortal production.

Finally, we thank our friend Andy Sinauer. Like ours, his name is on the cover of the book, and he truly cares deeply about what goes into it.

DAVID SADAVA

DAVID HILLIS

CRAIG HELLER

MAY BERENBAUM

Reviewers for the Tenth Edition

Between Edition Reviewers

Shivanthi Anandan, Drexel University
Brian Bagatto, The University of Akron
Mary Bisson, University at Buffalo, The State University of New York
Meredith Blackwell, Louisiana State University
Randy Brooks, Florida Atlantic University
Heather Caldwell, Kent State University
Jeffrey Carrier, Albion College
David Champlin, University of Southern Maine
Wesley Colgan, Pikes Peak Community College
Emma Creaser, Unity College
Karen Curto, University of Pittsburgh
John Dennehy, Queens College, The City University of New York
Rajinder Dhindsa, McGill University
James A. Doyle, University of California, Davis
Scott Edwards, Harvard University
David Eldridge, Baylor University
Joanne Ellzey, The University of Texas at El Paso
Douglas Gayou, University of Missouri
Stephen Gehnrich, Salisbury University
Arundhati Ghosh, University of Pittsburgh
Nathalia Glickman Holtzman, Queens College, The City University of New York
Elizabeth Good, University of Illinois at Urbana-Champaign
Harry Greene, Cornell University
Alice Heicklen, Columbia University
Albert Herrera, University of Southern California
David Hibbett, Clark University
Mark Holbrook, University of Iowa
Craig Jordan, The University of Texas at San Antonio
Walter Judd, University of Florida

John M. Labavitch, University of California, Davis
Nathan H. Lents, John Jay College of Criminal Justice, The City University of New York
Barry Logan, Bowdoin College
Barbara Lom, Davidson College
David Low, University of California, Davis
Janet Loxterman, Idaho State University
Sharon Lynn, The College of Wooster
Julin Maloof, University of California, Davis
Richard McCarty, Johns Hopkins University
Sheila McCormick, University of California, Berkeley
Marcie Moehnke, Baylor University
Roberta Moldow, Seton Hall University
Tsafrir Mor, Arizona State University
Alexander Motten, Duke University
Barbara Musolf, Clayton State University
Stuart Newfeld, Arizona State University
Bruce Ostrow, Grand Valley State University
Laura K. Palmer, The Pennsylvania State University, Altoona
Robert Pennock, Michigan State University
Kamini Persaud, University of Toronto, Scarborough
Roger Persell, Hunter College, The City University of New York
Matthew Rand, Carleton College
Susan Richardson, Florida Atlantic University
Brian C. Ring, Valdosta State University
Jay Rosenheim, University of California, Davis
Ben Rowley, University of Central Arkansas
Ann Rushing, Baylor University

Mikal Saltveit, University of California, Davis
Joel Schildbach, Johns Hopkins University
Christopher J. Schneider, Boston University
Paul Schulte, University of Nevada, Las Vegas
Leah Sheridan, University of Northern Colorado
Gary Shin, University of California, Los Angeles
Mitchell Singer, University of California, Davis
William Taylor, The University of Toledo
Sharon Thoma, University of Wisconsin, Madison
James F. A. Traniello, Boston University
Terry Trier, Grand Valley State University
Sara Via, University of Maryland
Curt Walker, Dixie State College
Fred Wasserman, Boston University
Alexander J. Werth, Hampden-Sydney College
Elizabeth Willott, University of Arizona

Accuracy Reviewers

Rebecca Rashid Achterman, Western Washington University
Maria Ambrosetti, Emory University
Miriam Ashley-Ross, Wake Forest University
Felicitas Avendaño, Grand View University
David Bailey, St. Norbert College
Chhandak Basu, California State University, Northridge
Jim Bednarz, Arkansas State University
Charlie Garnett Benson, Georgia State University
Katherine Boss-Williams, Emory University
Ben Brammell, Asbury University

Christopher I. Brandon, Jr., Georgia Gwinnett College

Carolyn J. W. Bunde, Idaho State University

Darlene Campbell, Cornell University

Jeffrey Carmichael, University of North Dakota

David J. Carroll, Florida Institute of Technology

Ethan Carver, The University of Tennessee at Chattanooga

Peter Chabora, Queens College, The City University of New York

Heather Cook, Wagner College

Hsini Lin Cox, The University of Texas at El Paso

Douglas Darnowski, Indiana University Southeast

Stephen Devoto, Wesleyan University

Rajinder Dhindsa, McGill University

Jesse Dillon, California State University, Long Beach

James A. Doyle, University of California, Davis

Devin Drown, Indiana University

Richard E. Duhrkopf, Baylor University

Weston Dulaney, Nashville State Community College

David Eldridge, Baylor University

Kenneth Filchak, University of Notre Dame

Kerry Finlay, University of Regina

Kevin Folta, University of Florida

Douglas Gayou, University of Missouri

David T. Glover, Food and Drug Administration

Russ Goddard, Valdosta State University

Elizabeth Godrick, Boston University

Leslie Goertzen, Auburn University

Elizabeth Good, University of Illinois at Urbana-Champaign

Ethan Graf, Amherst College

Eileen Gregory, Rollins College

Julie C. Hagelin, University of Alaska, Fairbanks

Nathalia Glickman Holtzman, Queens College, The City University of New York

Dianne Jennings, Virginia Commonwealth University

Jamie Jensen, Bringham Young University

Glennis E. Julian

Erin Keen-Rhinehart, Susquehanna University

Henrik Kibak, California State University, Monterey Bay

Brandi Brandon Knight, Emory University

Daniel Kueh, Emory University

John G. Latto, University of California, Santa Barbara

Kristen Lennon, Frostburg State University

David Low, University of California, Santa Barbara

Jose-Luis Machado, Swarthmore College

Jay Mager, Ohio Northern University

Stevan Marcus, University of Alabama

Nilo Marin, Broward College

Marlee Marsh, Columbia College South Carolina

Erin Martin, University of South Florida, Sarasota-Manatee

Brad Mehrtens, University of Illinois at Urbana-Champaign

Michael Meighan, University of California, Berkeley

Tsafrir Mor, Arizona State University

Roderick Morgan, Grand Valley State University

Jacalyn Newman, University of Pittsburgh

Alexey Nikitin, Grand Valley State University

Zia Nisani, Antelope Valley College

Laura K. Palmer, The Pennsylvania State University, Altoona

Nancy Pencoe, State University of West Georgia

David P. Puthoff, Frostburg State University

Brett Riddle, University of Nevada, Las Vegas

Leslie Riley, Ohio Northern University

Brian C. Ring, Valdosta State University

Heather Roffey, McGill University

Lori Rose, Hill College

Naomi Rowland, Western Kentucky University

Beth Rueschhoff, Indiana University Southeast

Ann Rushing, Baylor University

Illya Ruvinsky, University of Chicago

Paul Schulte, University of Nevada, Las Vegas

Susan Sharbaugh, University of Alaska, Fairbanks

Jonathan Shenker, Florida Institute of Technology

Gary Shin, California State University, Long Beach

Ken Spitze, University of West Georgia

Bruce Stallsmith, The University of Alabama in Huntsville

Robert M. Steven, The University of Toledo

Zuzana Swigonova, University of Pittsburgh

Rebecca Symula, The University of Mississippi

Mark Taylor, Baylor University

Mark Thogerson, Grand Valley State University

Elethia Tillman, Spelman College

Terry Trier, Grand Valley State University

Michael Troyan, The Pennsylvania State University, University Park

Sebastian Velez, Worcester State University

Sheela Vemu, Northern Illinois University

Andrea Ward, Adelphi University

Katherine Warpeha, University of Illinois at Chicago

Fred Wasserman, Boston University

Michelle Wien, Bryn Mawr College

Robert Wisotzkey, California State University, East Bay

Greg Wray, Duke University

Joanna Wysocka-Diller, Auburn University

Catherine Young, Ohio Northern University

Heping Zhou, Seton Hall University

Assessment Reviewers

Maria Ambrosetti, Georgia State University

Cecile Andraos-Selim, Hampton University

Felicitas Avendaño, Grand View University

David Bailey, St. Norbert College

Jim Bednarz, Arkansas State University

Charlie Garnett Benson, Georgia State University

Katherine Boss-Williams, Emory University

Ben Brammell, Asbury University

Christopher I. Brandon, Jr., Georgia Gwinnett College

Brandi Brandon Knight, Emory University

Ethan Carver, The University of Tennessee, Chattanooga

Heather Cook, Wagner College

Hsini Lin Cox, The University of Texas at El Paso

Douglas Darnowski, Indiana University Southeast

Jesse Dillon, California State University, Long Beach

Devin Drown, Indiana University

Richard E. Duhrkopf, Baylor University

Weston Dulaney, Nashville State Community College

Kenneth Filchak, University of Notre Dame

Elizabeth Godrick, Boston University

Elizabeth Good, University of Illinois at Urbana-Champaign

Susan Hengeveld, Indiana University Bloomington

Nathalia Glickman Holtzman, Queens College, The City College of New York

Glennis E. Julian

Erin Keen-Rhinehart, Susquehanna University

Stephen Kilpatrick, University of Pittsburgh

Daniel Kueh, Emory University

Stevan Marcus, University of Alabama

Nilo Marin, Broward College

Marlee Marsh, Columbia College

Erin Martin, University of South Florida, Sarasota-Manatee

Brad Mehrtens, University of Illinois at Urbana-Champaign

Darlene Mitrano, Christopher Newport University

Anthony Moss, Auburn University

Jacalyn Newman, University of Pittsburgh

Alexey Nikitin, Grand Valley State University

Zia Nisani, Antelope Valley College

Sabiha Rahman, University of Ottawa

Nancy Rice, Western Kentucky University

Brian C. Ring, Valdosta State University

Naomi Rowland, Western Kentucky University

Jonathan Shenker, Florida Institute of Technology

Gary Shin, California State University, Long Beach

Jacob Shreckengost, Emory University

Michael Smith, Western Kentucky University

Ken Spitze, University of West Georgia

Bruce Stallsmith, The University of Alabama in Huntsville

Zuzana Swigonova, University of Pittsburgh

William Taylor, The University of Toledo

Mark Thogerson, Grand Valley State University

Elethia Tillman, Spelman College

Michael Troyan, The Pennsylvania State University

Ximena Valderrama, Ramapo College of New Jersey

Sheela Vemu, Northern Illinois University

Suzanne Wakim, Butte College

Katherine Warpeha, University of Illinois at Chicago

Fred Wasserman, Boston University

Michelle Wien, Bryn Mawr College

Robert Wisotzkey, California State University, East Bay

Heping Zhou, Seton Hall University

LIFE's Media and Supplements

BIO P🌐RTAL

yourBioPortal.com

BioPortal is the online gateway to all of *Life*'s digital resources, including the fully interactive eBook, a wide range of student and instructor media resources, and powerful assessment tools. BioPortal includes the following features and resources:

Life, Tenth Edition eBook
(eBook also available stand-alone)

- Complete online version of the textbook
- Integration of all Media Clips, Activities, Animated Tutorials, and other media resources
- In-text links to all glossary entries, with audio pronunciations
- A flexible notes feature and easy text highlighting
- Searchable glossary and index
- Full-text search

Additional eBook features for instructors:

- *Content Customization*: Instructors can easily hide chapters or sections that they don't cover in their course, re-arrange the order of chapters and sections, and add their own content directly into the eBook.
- *Instructor Notes*: Instructors can annotate the eBook with their own notes and content on any page. Instructor notes can include text, Web links, images, links to BioPortal resources, uploaded documents, and more.

LearningCurve

New for the Tenth Edition, LearningCurve is a powerful adaptive quizzing system with a game-like format that engages students. Rather than simply answering a fixed set of questions, students answer dynamically-selected questions to progress toward a target level of understanding. At any point, students can view a report of how well they are performing in each topic area (with links to eBook sections and media resources), to help them focus on problem areas.

Student BioPortal Resources

DIAGNOSTIC QUIZZING. The pre-built diagnostic quizzes assesses student understanding of each section of each chapter, and generates a Personalized Study Plan to effectively focus student study time. The plan includes links to specific textbook sections, animated tutorials, and activities.

INTERACTIVE SUMMARIES. For each chapter, these dynamic summaries combine a review of important concepts with links to all of the key figures, Activities, and Animated Tutorials.

ANIMATED TUTORIALS. In-depth tutorials that present complex topics in a clear, easy-to-follow format that combines a detailed animation or simulation with an introduction, conclusion, and brief quiz.

MEDIA CLIPS. New for the Tenth Edition, these short, engaging video clips depict fascinating examples of some of the many organisms, processes, and phenomena discussed in the textbook.

ACTIVITIES. A range of interactive activities that help students learn and review key facts and concepts through labeling diagrams, identifying steps in processes, and matching concepts.

LECTURE NOTEBOOK. New for the Tenth Edition, the Lecture Notebook is included online in BioPortal. The Notebook includes all of the textbook's figures and tables, with space for note-taking, and is available as downloadable PDF files.

BIONEWS FROM SCIENTIFIC AMERICAN. BioNews makes it easy for instructors to bring the dynamic nature of the biological sciences and up-to-the minute currency into their course, via an automatically updated news feed.

BIONAVIGATOR. A unique visual way to explore all of the Animated Tutorials and Activities across the various levels of biological inquiry—from the global scale down to the molecular scale.

WORKING WITH DATA. Online versions of the Working with Data exercises that are included in the textbook.

FLASHCARDS AND KEY TERMS. The Flashcards and Key Terms provide an ideal way for students to learn and review the extensive terminology of introductory biology, featuring a review mode and a quiz mode.

INVESTIGATING LIFE LINKS. For each Investigating Life figure in the textbook, BioPortal includes an overview of the experiment featured in the figure with links to the original paper(s), related research or applications that followed, and additional information related to the experiment.

GLOSSARY. The full glossary, with audio pronunciations for all terms.

TREE OF LIFE. An interactive version of the Tree of Life from Appendix A. The online Tree links to a wealth of information on each group listed.

MATH FOR LIFE. A collection of mathematical shortcuts and references to help students with the quantitative skills they need in the biology laboratory.

SURVIVAL SKILLS. A guide to more effective study habits, including time management, note-taking, effective highlighting, and exam preparation.

Instructor BioPortal Resources

Assessment

- LearningCurve and Diagnostic Quizzing reports provide instructors with a wealth of information on student comprehension, by textbook section, along with targeted lecture resources for those areas requiring the most attention.
- Comprehensive question banks include questions from the Test Bank, LearningCurve, Diagnostic Quizzes, Study Guide, and textbook Chapter Review.
- Question filtering allows instructors to select questions based on Bloom's category and/or textbook section, in order to easily select the desired mix of question types.
- Easy-to-use assessment tools allow instructors to create quizzes and many other types of assignments using any combination of publisher-provided questions and those created by the instructor.

Media Resources

(see Instructor's Media Library below for details)

- Videos
- PowerPoint Presentations (Figures & Tables, Lecture, Editable Labels, Layered Art)
- Supplemental Photos
- Active Learning Exercises
- Instructor's Manual
- Lecture Notes
- Answers to Working with Data Exercises
- Course management features
- Complete course customization capabilities
- Custom resources/document posting
- Robust gradebook
- Communication Tools: Announcements, Calendar, Course Email, Discussion Boards

Student Supplements

Life, Tenth Edition Study Guide

(Paper, ISBN 978-1-4641-2365-8)

The *Life* Study Guide offers a variety of study and review resources to accompany each chapter of the textbook. The opening Big Picture section gives students a concise overview of the main concepts covered in the chapter. The Study Strategies section points out common problem areas that students may find more challenging, and suggests strategies for learning the material most effectively. The Key Concept Review section combines a detailed review of each section with questions that help students synthesize and apply what they have learned, including diagram questions, short-answer questions, and more open-ended questions. Each chapter concludes with a Test Yourself section that allows students to test their comprehension. All questions include answers, explanations, and references to textbook sections.

Life Flashcards App

Available for iPhone/iPad and Android, the *Life* Flashcards App is a great way for students to learn and review all the key terminology from the textbook, whenever and wherever they want to study, in an intuitive flashcard interface. Available in the iTunes App Store and Google Play.

CatchUp Math & Stats

Michael Harris, Gordon Taylor, and Jacquelyn Taylor
(ISBN 978-1-4292-0557-3)

Presented in brief, accessible units, this primer will help students quickly brush up on the quantitative skills they need to succeed in biology.

Student Handbook for Writing in Biology, Third Edition

Karen Knisely (ISBN 978-1-4292-3491-7)

This book provides practical advice to students who are learning to write according to the conventions in biology, using the standards of journal publication as a model.

Bioethics and the New Embryology: Springboards for Debate

Scott F. Gilbert, Anna Tyler, and Emily Zackin
(ISBN 978-0-7167-7345-0)

Our ability to alter the course of human development ranks among the most significant changes in modern science and has brought embryology into the public domain. The question that must be asked is: Even if we can do such things, should we?

BioStats Basics: A Student Handbook

James L. Gould and Grant F. Gould (ISBN 978-0-7167-3416-1)

Engaging and informal, *BioStats Basics* provides introductory-level biology students with a practical, accessible introduction to statistical research.

Inquiry Biology: A Laboratory Manual, Volumes 1 and 2

Mary Tyler, Ryan W. Cowan, and Jennifer L. Lockhart (Volume 1 ISBN 978-1-4292-9288-7; Volume 2 ISBN 978-1-4292-9289-4)

This introductory biology laboratory manual is inquiry-based—instructing in the process of science by allowing students to ask their own questions, gather background information, formulate hypotheses, design and carry out experiments, collect and analyze data, and formulate conclusions.

Hayden-McNeil Life Sciences Lab Notebook

(ISBN 978-1-4292-3055-1)

This carbonless laboratory notebook is of the highest quality and durability, allowing students to hand in originals or copies, not entire composition books. Contains Hayden-McNeil's unique white paper carbonless copies and biology-specific reference materials.

Instructor Media & Supplements

Instructor's Media Library

(Available both online via BioPortal and on disc; disc version ISBN 978-1-4641-2364-1)

The *Life,* Tenth Edition Instructor's Media Library includes a wide range of electronic resources to help instructors plan their course, present engaging lectures, and effectively assess their students. The Media Library includes the following resources:

TEXTBOOK FIGURES AND TABLES. Every figure and table from the textbook (including all photos and all un-numbered figures) is provided in both JPEG (high- and low-resolution) and PDF formats, in multiple versions.

UNLABELED FIGURES. Every figure is provided in an unlabeled format, useful for student quizzing and custom presentations.

SUPPLEMENTAL PHOTOS. The supplemental photograph collection contains over 1,500 photographs, giving instructors a wealth of additional imagery to draw upon.

ANIMATIONS. An extensive collection of detailed animations, all built specifically for Life, and viewable in either narrated or step-through mode.

VIDEOS. Featuring many new segments for the Tenth Edition, the wide-ranging collection of video segments help demonstrate the complexity and beauty of life.

POWERPOINT RESOURCES. For each chapter of the textbook, many different PowerPoint presentations are available, providing instructors the flexibility to build presentations in the manner that best suits their needs, including the following:

- Textbook Figures and Tables
- Lecture Presentation
- Figures with Editable Labels
- Layered Art Figures
- Supplemental Photos
- Videos
- Animations
- Active Learning Exercises

INSTRUCTOR'S MANUAL, LECTURE NOTES, and **TEST BANK** are available in Microsoft Word format for easy use in lecture and exam preparation.

MEDIA GUIDE. A PDF version of the Media Guide from the Instructor's Resource Kit, convenient for searching.

ACTIVE LEARNING EXERCISES. Set up for easy integration into lectures, each exercise poses a question or problem for the class to discuss or solve during lecture. Each also includes a multiple-choice element, for easy use with clicker systems.

ANSWERS TO WORKING WITH DATA EXERCISES. Complete answers to all of the Working with Data exercises.

Instructor's Resource Kit

(Binder, ISBN 978-1-4641-4131-7)

The *Life,* Tenth Edition Instructor's Resource Kit includes a wealth of information to help instructors in the planning and teaching of their course. The Kit includes:

INSTRUCTOR'S MANUAL

- *Chapter Overview*: A brief, high-level synopsis of the chapter.
- *What's New*: A guide to the revisions, updates, and new content added to the Tenth Edition.
- *Key Concepts & Learning Objectives*: New for the Tenth Edition, this section includes the major learning goals for the chapter, a detailed set of key concepts, and specific learning objectives for each key concept.
- *Chapter Outline*: All of the chapter's section headings and sub-headings.
- *Key Terms*: All of the important terms introduced in the chapter.

LECTURE NOTES. Detailed lecture outlines for each chapter, including references to relevant figures and media resources.

MEDIA GUIDE. A visual guide to the extensive media resources available with Life, including all animations, activities, videos, and supplemental photos.

Overhead Transparencies

(ISBN 978-1-4641-4127-0)

The set of overheads includes over 1,000 transparencies—including all of the four-color line art and all of the tables from the text—in two convenient binders. All figures have been formatted and color-enhanced for clear projection in a wide range of conditions. Labels and images have been resized for improved readability.

Test File

(Paper, ISBN 978-1-4292-5579-0)

The *Life*, Tenth Edition Test File includes over 5,000 questions and has been revised and reviewed for both accuracy and effectiveness. All questions are referenced to specific textbook headings and categorized according to Bloom's taxonomy. This allows instructors to easily build quizzes and exams with the desired mix of content, coverage, and question types (factual, conceptual, analyzing/applying, etc.). Each chapter includes a wide range of multiple choice and fill-in-the-blank questions, in addition to diagram questions that involve the student in working with illustrations of structures, graphs, steps in processes, and more.

Computerized Test Bank

(CD, ISBN 978-1-4641-4128-7)

The entire Test File, plus the Diagnostic Quizzes, Learning-Curve questions, Study Guide questions, and Textbook End-of-Chapter Review questions are all included in Wimba's easy-to-use Diploma program (software included). Designed for both novice and advanced users, Diploma allows instructors to quickly and easily create or edit questions, create quizzes or exams with a "drag-and-drop" feature (using any combination of publisher-provided and instructor-added questions), publish to online courses, and print paper-based assessments.

Figure Correlation Tool

An invaluable resource for instructors switching to *Life,* Tenth Edition from another textbook or from *Life,* Ninth Edition, this online tool provides correlations between all of the figures in *Life,* Tenth Edition and figures in other majors biology textbooks and *Life,* Ninth Edition.

Course Management System Support

As a service for *Life* adopters using Blackboard, WebCT, AN-GEL, or other course management systems, full electronic course packs are available.

Faculty Lounge for Majors Biology is the first publisher-provided website for the majors biology community that lets instructors freely communicate and share peer-reviewed lecture and teaching resources. The Faculty Lounge offers convenient access to peer-recommended and vetted resources, including the following categories: Images, News, Videos, Labs, Lecture Resources, and Educational Research. **majorsbio.facultylounge.whfreeman.com**

LabPartner is a site designed to facilitate the creation of customized lab manuals. Its database contains a wide selection of experiments published by W. H. Freeman and Hayden-McNeil Publishing. Instructors can preview, choose, and re-order labs, interleave their own original experiments, add carbonless graph paper and a pocket folder, customize the cover both inside and out, and select a binding type. Manuals are printed on-demand. **www.whfreeman.com/labpartner**

The Scientific Teaching Book Series is a collection of practical guides, intended for all science, technology, engineering and mathematics (STEM) faculty who teach undergraduate and graduate students in these disciplines. The purpose of these books is to help faculty become more successful in all aspects of teaching and learning science, including classroom instruction, mentoring students, and professional development. Authored by well-known science educators, the Series provides concise descriptions of best practices and how to implement them in the classroom, the laboratory, or the department. For readers interested in the research results on which these best practices are based, the books also provide a gateway to the key educational literature.

Scientific Teaching

Jo Handelsman, Sarah Miller, and Christine Pfund (ISBN 978-1-4292-0188-9)

Transformations:
Approaches to College Science Teaching

Deborah Allen and Kimberly Tanner (ISBN 978-1-4292-5335-2)

Entering Research: A Facilitator's Manual
Workshops for Students Beginning Research in Science

Janet L. Branchaw, Christine Pfund, and Raelyn Rediske (ISBN 978-1-429-25857-9)

Discipline-Based Science Education Research:
A Scientist's Guide

Stephanie Slater, Tim Slater, and Janelle M. Bailey (ISBN 978-1-4292-6586-7)

Assessment in the College Classroom

Clarissa Dirks, Mary Pat Wenderoth, Michelle Withers (ISBN 978-1-4292-8197-3)

iclicker

Developed for educators by educators, iclicker is a hassle-free radio-frequency classroom response system that makes it easy for instructors to ask questions, record responses, take attendance, and direct students through lectures as active participants. For more information, visit **www.iclicker.com**.

Contents

PART ONE
The Science of Life and Its Chemical Basis

1 Studying Life   1

1.1 What Is Biology?   2

Life arose from non-life via chemical evolution   3

Cellular structure evolved in the common ancestor of life   3

Photosynthesis allows some organisms to capture energy from the sun   4

Biological information is contained in a genetic language common to all organisms   5

Populations of all living organisms evolve   6

Biologists can trace the evolutionary tree of life   6

Cellular specialization and differentiation underlie multicellular life   9

Living organisms interact with one another   9

Nutrients supply energy and are the basis of biosynthesis   10

Living organisms must regulate their internal environment   10

1.2 How Do Biologists Investigate Life?   11

Observing and quantifying are important skills   11

Scientific methods combine observation, experimentation, and logic   11

Good experiments have the potential to falsify hypotheses   12

Statistical methods are essential scientific tools   13

Discoveries in biology can be generalized   14

Not all forms of inquiry are scientific   14

1.3 Why Does Biology Matter? 15

Modern agriculture depends on biology   15

Biology is the basis of medical practice   15

Biology can inform public policy   16

Biology is crucial for understanding ecosystems   17

Biology helps us understand and appreciate biodiversity   17

2 Small Molecules and the Chemistry of Life   21

2.1 How Does Atomic Structure Explain the Properties of Matter?   22

An element consists of only one kind of atom   22

Each element has a unique number of protons   22

The number of neutrons differs among isotopes   22

The behavior of electrons determines chemical bonding and geometry   24

2.2 How Do Atoms Bond to Form Molecules?   26

Covalent bonds consist of shared pairs of electrons   26

Ionic attractions form by electrical attraction   28

Hydrogen bonds may form within or between molecules with polar covalent bonds   30

Hydrophobic interactions bring together nonpolar molecules   30

van der Waals forces involve contacts between atoms   30

2.3 How Do Atoms Change Partners in Chemical Reactions?   31

2.4 What Makes Water So Important for Life?   32

Water has a unique structure and special properties   32

The reactions of life take place in aqueous solutions   33

Aqueous solutions may be acidic or basic   34

3 Proteins, Carbohydrates, and Lipids   39

3.1 What Kinds of Molecules Characterize Living Things?   40

Functional groups give specific properties to biological molecules   40

Isomers have different arrangements of the same atoms   41

The structures of macromolecules reflect their functions   41

Most macromolecules are formed by condensation and broken down by hydrolysis 42

3.2 What Are the Chemical Structures and Functions of Proteins? 42

Amino acids are the building blocks of proteins 43

Peptide linkages form the backbone of a protein 43

The primary structure of a protein is its amino acid sequence 45

The secondary structure of a protein requires hydrogen bonding 45

The tertiary structure of a protein is formed by bending and folding 46

The quaternary structure of a protein consists of subunits 48

Shape and surface chemistry contribute to protein function 48

Environmental conditions affect protein structure 50

Protein shapes can change 50

Molecular chaperones help shape proteins 51

3.3 What Are the Chemical Structures and Functions of Carbohydrates? 51

Monosaccharides are simple sugars 52

Glycosidic linkages bond monosaccharides 53

Polysaccharides store energy and provide structural materials 53

Chemically modified carbohydrates contain additional functional groups 55

3.4 What Are the Chemical Structures and Functions of Lipids? 56

Fats and oils are triglycerides 56

Phospholipids form biological membranes 57

Some lipids have roles in energy conversion, regulation, and protection 57

4 Nucleic Acids and the Origin of Life 62

4.1 What Are the Chemical Structures and Functions of Nucleic Acids? 63

Nucleotides are the building blocks of nucleic acids 63

Base pairing occurs in both DNA and RNA 63

DNA carries information and is expressed through RNA 65

The DNA base sequence reveals evolutionary relationships 66

Nucleotides have other important roles 66

4.2 How and Where Did the Small Molecules of Life Originate? 67

Experiments disproved the spontaneous generation of life 67

Life began in water 68

Life may have come from outside Earth 69

Prebiotic synthesis experiments model early Earth 69

4.3 How Did the Large Molecules of Life Originate? 71

Chemical evolution may have led to polymerization 71

RNA may have been the first biological catalyst 71

4.4 How Did the First Cells Originate? 71

Experiments explore the origin of cells 73

Some ancient cells left a fossil imprint 74

PART TWO   Cells

5 Cells: The Working Units of Life 77

5.1 What Features Make Cells the Fundamental Units of Life? 78

Cell size is limited by the surface area-to-volume ratio 78

Microscopes reveal the features of cells 79

The plasma membrane forms the outer surface of every cell 79

Cells are classified as either prokaryotic or eukaryotic 81

5.2 What Features Characterize Prokaryotic Cells? 82

Prokaryotic cells share certain features 82

Specialized features are found in some prokaryotes 83

5.3 What Features Characterize Eukaryotic Cells? 84

Compartmentalization is the key to eukaryotic cell function 84

Organelles can be studied by microscopy or isolated for chemical analysis 84

Ribosomes are factories for protein synthesis 84

The nucleus contains most of the generic information 85

The endomembrane system is a group of interrelated organelles 88

Some organelles transform energy 91

There are several other membrane-enclosed organelles 93

The cytoskeleton is important in cell structure and movement 94

Biologists can manipulate living systems to establish cause and effect 98

5.4 What Are the Roles of Extracellular Structures? 99

The plant cell wall is an extracellular structure 99

The extracellular matrix supports tissue functions in animals 100

5.5 How Did Eukaryotic Cells Originate? 101

Internal membranes and the nuclear envelope probably came from the plasma membrane 101

Some organelles arose by endosymbiosis 102

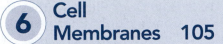

6 Cell Membranes 105

6.1 What Is the Structure of a Biological Membrane? 106

Lipids form the hydrophobic core of the membrane 106

Membrane proteins are asymmetrically distributed 107

Membranes are constantly changing 109

Plasma membrane carbohydrates are recognition sites 109

6.2 How Is the Plasma Membrane Involved in Cell Adhesion and Recognition? 110

Cell recognition and adhesion involve proteins and carbohydrates at the cell surface 111

Three types of cell junctions connect adjacent cells 111

Cell membranes adhere to the extracellular matrix 111

6.3 What Are the Passive Processes of Membrane Transport? 113

Diffusion is the process of random movement toward a state of equilibrium 113

Simple diffusion takes place through the phospholipid bilayer 114

Osmosis is the diffusion of water across membranes 114

Diffusion may be aided by channel proteins 115

Carrier proteins aid diffusion by binding substances 117

6.4 What are the Active Processes of Membrane Transport? 118

Active transport is directional 118

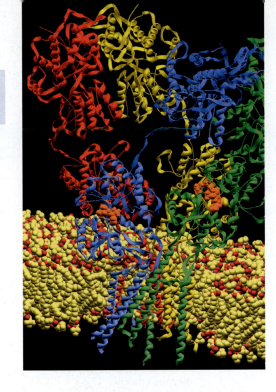

Different energy sources distinguish different active transport systems 118

6.5 How Do Large Molecules Enter and Leave a Cell? 120

Macromolecules and particles enter the cell by endocytosis 120

Receptor-mediated endocytosis is highly specific 121

Exocytosis moves materials out of the cell 122

7 Cell Communication and Multicellularity 125

7.1 What Are Signals, and How Do Cells Respond to Them? 126

Cells receive signals from the physical environment and from other cells 126

A signal transduction pathway involves a signal, a receptor, and responses 126

7.2 How Do Signal Receptors Initiate a Cellular Response? 127

Receptors that recognize chemical signals have specific binding sites 127

Receptors can be classified by location and function 128

Intracellular receptors are located in the cytoplasm or the nucleus 130

7.3 How Is the Response to a Signal Transduced through the Cell? 131

A protein kinase cascade amplifies a response to ligand binding 131

Second messengers can amplify signals between receptors and target molecules 132

Signal transduction is highly regulated 136

7.4 How Do Cells Change in Response to Signals? 137

Ion channels open in response to signals 137

Enzyme activities change in response to signals 138

Signals can initiate DNA transcription 139

7.5 How Do Cells in a Multicellular Organism Communicate Directly? 139

Animal cells communicate through gap junctions 139

Plant cells communicate through plasmodesmata 140

Modern organisms provide clues about the evolution of cell–cell interactions and multicellularity 140

PART THREE
Cells and Energy

8 Energy, Enzymes, and Metabolism 144

8.1 What Physical Principles Underlie Biological Energy Transformations? 145

There are two basic types of energy 145

There are two basic types of metabolism 145

The first law of thermodynamics: Energy is neither created nor destroyed 146

The second law of thermodynamics: Disorder tends to increase 146

Chemical reactions release or consume energy 147

Chemical equilibrium and free energy are related 148

8.2 What Is the Role of ATP in Biochemical Energetics? 149

ATP hydrolysis releases energy 149

ATP couples exergonic and endergonic reactions   150

8.3 What Are Enzymes?   151

To speed up a reaction, an energy barrier must be overcome   151

Enzymes bind specific reactants at their active sites   152

Enzymes lower the energy barrier but do not affect equilibrium   153

8.4 How Do Enzymes Work?   154

Enzymes can orient substrates   154

Enzymes can induce strain in the substrate   154

Enzymes can temporarily add chemical groups to substrates   154

Molecular structure determines enzyme function   155

Some enzymes require other molecules in order to function   155

The substrate concentration affects the reaction rate   156

8.5 How Are Enzyme Activities Regulated?   156

Enzymes can be regulated by inhibitors   157

Allosteric enzymes are controlled via changes in shape   159

Allosteric effects regulate many metabolic pathways   160

Many enzymes are regulated through reversible phosphorylation   161

Enzymes are affected by their environment   161

9 Pathways That Harvest Chemical Energy   165

9.1 How Does Glucose Oxidation Release Chemical Energy?   166

Cells trap free energy while metabolizing glucose   166

Redox reactions transfer electrons and energy   167

The coenzyme NAD^+ is a key electron carrier in redox reactions   167

An overview: Harvesting energy from glucose   168

9.2 What Are the Aerobic Pathways of Glucose Catabolism?   169

In glycolysis, glucose is partially oxidized and some energy is released   169

Pyruvate oxidation links glycolysis and the citric acid cycle   170

The citric acid cycle completes the oxidation of glucose to CO_2 170

Pyruvate oxidation and the citric acid cycle are regulated by the concentrations of starting materials   171

9.3 How Does Oxidative Phosphorylation Form ATP?   171

The respiratory chain transfers electrons and protons, and releases energy   172

Proton diffusion is coupled to ATP synthesis   173

Some microorganisms use non-O_2 electron acceptors   176

9.4 How Is Energy Harvested from Glucose in the Absence of Oxygen?   177

Cellular respiration yields much more energy than fermentation   178

The yield of ATP is reduced by the impermeability of mitochondria to NADH   178

9.5 How Are Metabolic Pathways Interrelated and Regulated?   179

Catabolism and anabolism are linked   179

Catabolism and anabolism are integrated   180

Metabolic pathways are regulated systems   181

10 Photosynthesis: Energy from Sunlight   185

10.1 What Is Photosynthesis   186

Experiments with isotopes show that O_2 comes from H_2O in oxygenic photosynthesis   186

Photosynthesis involves two pathways   188

10.2 How Does Photosynthesis Convert Light Energy into Chemical Energy?   188

Light energy is absorbed by chlorophyll and other pigments   188

Light absorption results in photochemical change   190

Reduction leads to ATP and NADPH formation   191

Chemiosmosis is the source of the ATP produced in photophosphorylation   192

10.3 How Is Chemical Energy Used to Synthesize Carbohydrates?   193

Radioisotope labeling experiments revealed the steps of the Calvin cycle   193

The Calvin cycle is made up of three processes   194

Light stimulates the Calvin cycle   196

10.4 How Have Plants Adapted Photosynthesis to Environmental Conditions?   197

Rubisco catalyzes the reaction of RuBP with O_2 or CO_2 197

C_3 plants undergo photorespiration but C_4 plants do not   198

CAM plants also use PEP carboxylase   200

10.5 How Does Photosynthesis Interact with Other Pathways?   200

PART FOUR
Genes and Heredity

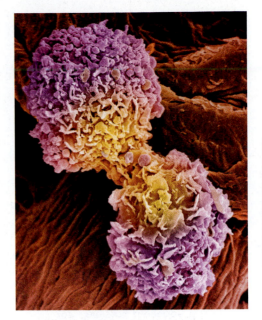

11 **The Cell Cycle and Cell Division 205**

11.1 How Do Prokaryotic and Eukaryotic Cells Divide? 206

Prokaryotes divide by binary fission 206

Eukaryotic cells divide by mitosis or meiosis followed by cytokinesis 207

11.2 How Is Eukaryotic Cell Division Controlled? 208

Specific internal signals trigger events in the cell cycle 208

Growth factors can stimulate cells to divide 211

11.3 What Happens during Mitosis? 211

Prior to mitosis, eukaryotic DNA is packed into very compact chromosomes 211

Overview: Mitosis segregates copies of genetic information 212

The centrosomes determine the plane of cell division 212

The spindle begins to form during prophase 213

Chromosome separation and movement are highly organized 214

Cytokinesis is the division of the cytoplasm 216

11.4 What Role Does Cell Division Play in a Sexual Life Cycle? 217

Asexual reproduction by mitosis results in genetic constancy 217

Sexual reproduction by meiosis results in genetic diversity 218

11.5 What Happens during Meiosis? 219

Meiotic division reduces the chromosome number 219

Chromatid exchanges during meiosis I generate genetic diversity 219

During meiosis homologous chromosomes separate by independent assortment 220

Meiotic errors lead to abnormal chromosome structures and numbers 222

The number, shapes, and sizes of the metaphase chromosomes constitute the karyotype 224

Polyploids have more than two complete sets of chromosomes 224

11.6 In a Living Organism, How Do Cells Die? 225

11.7 How Does Unregulated Cell Division Lead to Cancer? 227

Cancer cells differ from normal cells 227

Cancer cells lose control over the cell cycle and apoptosis 228

Cancer treatments target the cell cycle 228

12 **Inheritance, Genes, and Chromosomes 232**

12.1 What Are the Mendelian Laws of Inheritance? 233

Mendel used the scientific method to test his hypotheses 233

Mendel's first experiments involved monohybrid crosses 234

Mendel's first law states that the two copies of a gene segregate 236

Mendel verified his hypotheses by performing test crosses 237

Mendel's second law states that copies of different genes assort independently 237

Probability can be used to predict inheritance 239

Mendel's laws can be observed in human pedigrees 240

12.2 How Do Alleles Interact? 241

New alleles arise by mutation 241

Many genes have multiple alleles 242

Dominance is not always complete 242

In codominance, both alleles at a locus are expressed 243

Some alleles have multiple phenotypic effects 243

12.3 How Do Genes Interact? 244

Hybrid vigor results from new gene combinations and interactions 244

The environment affects gene action 245

Most complex phenotypes are determined by multiple genes and the environment 246

12.4 What Is the Relationship between Genes and Chromosomes? 247

Genes on the same chromosome are linked 247

Genes can be exchanged between chromatids and mapped 247

Linkage is revealed by studies of the sex chromosomes 249

12.5 What Are the Effects of Genes Outside the Nucleus? 252

12.6 How Do Prokaryotes Transmit Genes? 253

Bacteria exchange genes by conjugation 253

Bacterial conjugation is controlled by plasmids 254

13 **DNA and Its Role in Heredity 259**

13.1 What Is the Evidence that the Gene Is DNA? 260

DNA from one type of bacterium genetically transforms another type 260

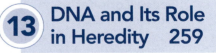

Viral infection experiments confirmed that DNA is the genetic material   261

Eukaryotic cells can also be genetically transformed by DNA   263

13.2 What Is the Structure of DNA?   264

Watson and Crick used modeling to deduce the structure of DNA   264

Four key features define DNA structure   265

The double-helical structure of DNA is essential to its function   266

13.3 How Is DNA Replicated?   267

Three modes of DNA replication appeared possible   267

An elegant experiment demonstrated that DNA replication is semiconservative   268

There are two steps in DNA replication   268

DNA polymerases add nucleotides to the growing chain   269

Many other proteins assist with DNA polymerization   272

The two DNA strands grow differently at the replication fork   272

Telomeres are not fully replicated and are prone to repair   275

13.4 How Are Errors in DNA Repaired?   276

13.5 How Does the Polymerase Chain Reaction Amplify DNA?   277

The polymerase chain reaction makes multiple copies of DNA sequences   277

14 From DNA to Protein: Gene Expression   281

14.1 What Is the Evidence that Genes Code for Proteins?   282

Observations in humans led to the proposal that genes determine enzymes   282

Experiments on bread mold established that genes determine enzymes   282

One gene determines one polypeptide   283

14.2 How Does Information Flow from Genes to Proteins?   284

Three types of RNA have roles in the information flow from DNA to protein   285

In some cases, RNA determines the sequence of DNA   285

14.3 How Is the Information Content in DNA Transcribed to Produce RNA?   286

RNA polymerases share common features   286

Transcription occurs in three steps   286

The information for protein synthesis lies in the genetic code   288

14.4 How Is Eukaryotic DNA Transcribed and the RNA Processed?   290

Many eukaryotic genes are interrupted by noncoding sequences   290

Eukaryotic gene transcripts are processed before translation   291

14.5 How Is RNA Translated into Proteins?   293

Transfer RNAs carry specific amino acids and bind to specific codons   293

Each tRNA is specifically attached to an amino acid   294

The ribosome is the workbench for translation   294

Translation takes place in three steps   295

Polysome formation increases the rate of protein synthesis   297

14.6 What Happens to Polypeptides after Translation?   298

Signal sequences in proteins direct them to their cellular destinations   298

Many proteins are modified after translation   300

15 Gene Mutation and Molecular Medicine   304

15.1 What Are Mutations?   305

Mutations have different phenotypic effects   305

Point mutations are changes in single nucleotides   306

Chromosomal mutations are extensive changes in the genetic material   307

Retroviruses and transposons can cause loss of function mutations or duplications   308

Mutations can be spontaneous or induced   308

Mutagens can be natural or artificial   310

Some base pairs are more vulnerable than others to mutation   310

Mutations have both benefits and costs   310

15.2 What Kinds of Mutations Lead to Genetic Diseases?   311

Genetic mutations may make proteins dysfunctional   311

Disease-causing mutations may involve any number of base pairs   312

Expanding triplet repeats demonstrate the fragility of some human genes   313

Cancer often involves somatic mutations   314

Most diseases are caused by multiple genes and environment 314

15.3 How Are Mutations Detected and Analyzed? 315

Restriction enzymes cleave DNA at specific sequences 315

Gel electrophoresis separates DNA fragments 316

DNA fingerprinting combines PCR with restriction analysis and electrophoresis 317

Reverse genetics can be used to identify mutations that lead to disease 318

Genetic markers can be used to find disease-causing genes 318

The DNA barcode project aims to identify all organisms on Earth 319

15.4 How Is Genetic Screening Used to Detect Diseases? 320

Screening for disease phenotypes involves analysis of proteins and other chemicals 320

DNA testing is the most accurate way to detect abnormal genes 320

Allele-specific oligonucleotide hybridization can detect mutations 321

15.5 How Are Genetic Diseases Treated? 322

Genetic diseases can be treated by modifying the phenotype 322

Gene therapy offers the hope of specific treatments 323

PART FIVE
Genomes

 Genomes 352

17.1 How Are Genomes Sequenced? 353

New methods have been developed to rapidly sequence DNA 353

Genome sequences yield several kinds of information 355

 Regulation of Gene Expression 328

16.1 How Is Gene Expression Regulated in Prokaryotes? 329

Regulating gene transcription conserves energy 329

Operons are units of transcriptional regulation in prokaryotes 330

Operator–repressor interactions control transcription in the *lac* and *trp* operons 330

Protein synthesis can be controlled by increasing promoter efficiency 332

RNA polymerases can be directed to particular classes of promoters 332

16.2 How Is Eukaryotic Gene Transcription Regulated? 333

General transcription factors act at eukaryotic promoters 333

Specific proteins can recognize and bind to DNA sequences and regulate transcription 335

Specific protein–DNA interactions underlie binding 335

The expression of transcription factors underlies cell differentiation 336

The expression of sets of genes can be coordinately regulated by transcription factors 336

16.3 How Do Viruses Regulate Their Gene Expression? 339

17.2 What Have We Learned from Sequencing Prokaryotic Genomes? 356

Prokaryotic genomes are compact 356

The sequencing of prokaryotic and viral genomes has many potential benefits 357

Metagenomics allows us to describe new organisms and ecosystems 357

Some sequences of DNA can move about the genome 358

Many bacteriophages undergo a lytic cycle 339

Some bacteriophages can undergo a lysogenic cycle 340

Eukaryotic viruses can have complex life cycles 341

HIV gene regulation occurs at the level of transcription elongation 341

16.4 How Do Epigenetic Changes Regulate Gene Expression? 343

DNA methylation occurs at promoters and silences transcription 343

Histone protein modifications affect transcription 344

Epigenetic changes can be induced by the environment 344

DNA methylation can result in genomic imprinting 344

Global chromosome changes involve DNA methylation 345

16.5 How Is Eukaryotic Gene Expression Regulated after Transcription? 346

Different mRNAs can be made from the same gene by alternative splicing 346

Small RNAs are important regulators of gene expression 347

Translation of mRNA can be regulated by proteins and riboswitches 348

Will defining the genes required for cellular life lead to artificial life? 359

17.3 What Have We Learned from Sequencing Eukaryotic Genomes? 361

Model organisms reveal many characteristics of eukaryotic genomes 361

Eukaryotes have gene families 363

Eukaryotic genomes contain many repetitive sequences 364

17.4 What Are the Characteristics of the Human Genome? 366

The human genome sequence held some surprises 366

Comparative genomics reveals the evolution of the human genome 366

Human genomics has potential benefits in medicine 367

17.5 What Do the New Disciplines of Proteomics and Metabolomics Reveal? 369

The proteome is more complex than the genome 369

Metabolomics is the study of chemical phenotype 370

18 Recombinant DNA and Biotechnology 373

18.1 What Is Recombinant DNA? 374

18.2 How Are New Genes Inserted into Cells? 375

Genes can be inserted into prokaryotic or eukaryotic cells 376

A variety of methods are used to insert recombinant DNA into host cells 376

Reporter genes help select or identify host cells containing recombinant DNA 377

18.3 What Sources of DNA Are Used in Cloning? 379

Libraries provide collections of DNA fragments 379

cDNA is made from mRNA transcripts 379

Synthetic DNA can be made by PCR or by organic chemistry 380

18.4 What Other Tools Are Used to Study DNA Function? 380

Genes can be expressed in different biological systems 380

DNA mutations can be created in the laboratory 381

Genes can be inactivated by homologous recombination 381

Complementary RNA can prevent the expression of specific genes 382

DNA microarrays reveal RNA expression patterns 382

18.5 What Is Biotechnology? 383

Expression vectors can turn cells into protein factories 384

18.6 How Is Biotechnology Changing Medicine and Agriculture? 384

Medically useful proteins can be made using biotechnology 384

DNA manipulation is changing agriculture 386

There is public concern about biotechnology 388

19 Differential Gene Expression in Development 392

19.1 What Are the Processes of Development? 393

Development involves distinct but overlapping processes 393

Cell fates become progressively more restricted during development 394

19.2 How Is Cell Fate Determined? 395

Cytoplasmic segregation can determine polarity and cell fate 395

Inducers passing from one cell to another can determine cell fates 395

19.3 What Is the Role of Gene Expression in Development? 397

Cell fate determination involves signal transduction pathways that lead to differential gene expression 397

Differential gene transcription is a hallmark of cell differentiation 398

19.4 How Does Gene Expression Determine Pattern Formation? 399

Multiple proteins interact to determine developmental programmed cell death 399

Plants have organ identity genes 400

Morphogen gradients provide positional information 401

A cascade of transcription factors establishes body segmentation in the fruit fly 401

19.5 Is Cell Differentiation Reversible? 405

Plant cells can be totipotent 405

Nuclear transfer allows the cloning of animals 406

Multipotent stem cells differentiate in response to environmental signals 408

Pluripotent stem cells can be obtained in two ways 408

20 Genes, Development, and Evolution 412

20.1 How Can Small Genetic Changes Result in Large Changes in Phenotype? 413

Developmental genes in distantly related organisms are similar 413

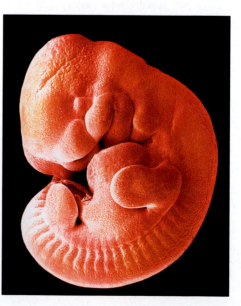

20.2 How Can Mutations with Large Effects Change Only One Part of the Body? 415

Genetic switches govern how the genetic toolkit is used 415

Modularity allows for differences in the patterns of gene expression 416

20.3 How Can Developmental Changes Result in Differences among Species? 418

Differences in Hox gene expression patterns result in major differences in body plans 418

Mutations in developmental genes can produce major morphological changes 418

20.4 How Can the Environment Modulate Development? 420

Temperature can determine sex 420

Dietary information can be a predictor of future conditions 421

A variety of environmental signals influence development 421

20.5 How Do Developmental Genes Constrain Evolution? 423

Evolution usually proceeds by changing what's already there 423

Conserved developmental genes can lead to parallel evolution 423

PART SIX
The Patterns and Processes of Evolution

21 Mechanisms of Evolution 427

21.1 What Is the Relationship between Fact and Theory in Evolution? 428

Darwin and Wallace introduced the idea of evolution by natural selection 428

Evolutionary theory has continued to develop over the past century 430

Genetic variation contributes to phenotypic variation 431

21.2 What Are the Mechanisms of Evolutionary Change? 432

Mutation generates genetic variation 432

Selection acting on genetic variation leads to new phenotypes 432

Gene flow may change allele frequencies 433

Genetic drift may cause large changes in small populations 434

Nonrandom mating can change genotype or allele frequencies 434

21.3 How Do Biologists Measure Evolutionary Change? 436

Evolutionary change can be measured by allele and genotype frequencies 436

Evolution will occur unless certain restrictive conditions exist 437

Deviations from Hardy–Weinberg equilibrium show that evolution is occurring 438

Natural selection acts directly on phenotypes 438

Natural selection can change or stabilize populations 439

21.4 How Is Genetic Variation Distributed and Maintained within Populations? 441

Neutral mutations accumulate in populations 441

Sexual recombination amplifies the number of possible genotypes 441

Frequency-dependent selection maintains genetic variation within populations 441

Heterozygote advantage maintains polymorphic loci 442

Genetic variation within species is maintained in geographically distinct populations 443

21.5 What Are the Constraints on Evolution? 444

Developmental processes constrain evolution 444

Trade-offs constrain evolution 445

Short-term and long-term evolutionary outcomes sometimes differ 446

22 Reconstructing and Using Phylogenies 449

22.1 What Is Phylogeny? 450

All of life is connected through evolutionary history 451

Comparisons among species require an evolutionary perspective 451

22.2 How Are Phylogenetic Trees Constructed? 452

Parsimony provides the simplest explanation for phylogenetic data 454

Phylogenies are reconstructed from many sources of data 454

Mathematical models expand the power of phylogenetic reconstruction 456

The accuracy of phylogenetic methods can be tested 457

22.3 How Do Biologists Use Phylogenetic Trees? 458

Phylogenetic trees can be used to reconstruct past events 458

Phylogenies allow us to compare and contrast living organisms 459

Phylogenies can reveal convergent evolution 459

Ancestral states can be reconstructed 460

Molecular clocks help date evolutionary events 461

22.4 How Does Phylogeny Relate to Classification? 462

Evolutionary history is the basis for modern biological classification 463

Several codes of biological nomenclature govern the use of scientific names 463

23 Speciation 467

23.1 What Are Species? 468

We can recognize many species by their appearance 468

Reproductive isolation is key 468

The lineage approach takes a long-term view 469

The different species concepts are not mutually exclusive 469

23.2 What Is the Genetic Basis of Speciation? 470

Incompatibilities between genes can produce reproductive isolation 470

Reproductive isolation develops with increasing genetic divergence 470

23.3 What Barriers to Gene Flow Result in Speciation? 472

Physical barriers give rise to allopatric speciation 472

Sympatric speciation occurs without physical barriers 473

23.4 What Happens When Newly Formed Species Come into Contact? 475

Prezygotic isolating mechanisms prevent hybridization 476

Postzygotic isolating mechanisms result in selection against hybridization 478

Hybrid zones may form if reproductive isolation is incomplete 478

23.5 Why Do Rates of Speciation Vary? 480

Several ecological and behavioral factors influence speciation rates 480

Rapid speciation can lead to adaptive radiation 481

24 Evolution of Genes and Genomes 485

24.1 How Are Genomes Used to Study Evolution? 486

Evolution of genomes results in biological diversity 486

Genes and proteins are compared through sequence alignment 486

Models of sequence evolution are used to calculate evolutionary divergence 487

Experimental studies examine molecular evolution directly 489

24.2 What Do Genomes Reveal about Evolutionary Processes? 491

Much of evolution is neutral 492

Positive and purifying selection can be detected in the genome 492

Genome size also evolves 494

24.3 How Do Genomes Gain and Maintain Functions? 496

Lateral gene transfer can result in the gain of new functions 496

Most new functions arise following gene duplication 496

Some gene families evolve through concerted evolution 498

24.4 What Are Some Applications of Molecular Evolution? 499

Molecular sequence data are used to determine the evolutionary history of genes 499

Gene evolution is used to study protein function 500

In vitro evolution is used to produce new molecules 500

Molecular evolution is used to study and combat diseases 501

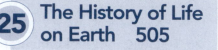

25 The History of Life on Earth 505

25.1 How Do Scientists Date Ancient Events? 506

Radioisotopes provide a way to date fossils and rocks 507

Radiometric dating methods have been expanded and refined 507

Scientists have used several methods to construct a geological time scale 508

25.2 How Have Earth's Continents and Climates Changed over Time? 508

The continents have not always been where they are today 509

Earth's climate has shifted between hot and cold conditions 510

Volcanoes have occasionally changed the history of life 510

Extraterrestrial events have triggered changes on Earth 511

Oxygen concentrations in Earth's atmosphere have changed over time 511

25.3 What Are the Major Events in Life's History? 514

Several processes contribute to the paucity of fossils 514

Precambrian life was small and aquatic 515

Life expanded rapidly during the Cambrian period 516

Many groups of organisms that arose during the Cambrian later diversified 516

Geographic differentiation increased during the Mesozoic era 521

Modern biotas evolved during the Cenozoic era 521

The tree of life is used to reconstruct evolutionary events 522

PART SEVEN
The Evolution of Diversity

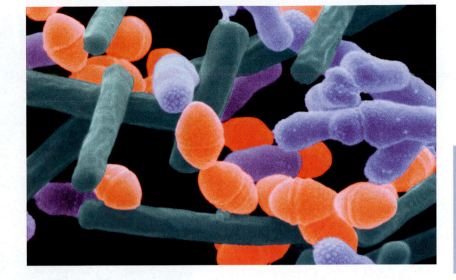

26 Bacteria, Archaea, and Viruses   525

26.1 Where Do Prokaryotes Fit into the Tree of Life?   526

The two prokaryotic domains differ in significant ways   526

The small size of prokaryotes has hindered our study of their evolutionary relationships   527

The nucleotide sequences of prokaryotes reveal their evolutionary relationships   528

Lateral gene transfer can lead to discordant gene trees   529

The great majority of prokaryote species have never been studied   530

26.2 Why Are Prokaryotes So Diverse and Abundant?   530

The low-GC Gram-positives include some of the smallest cellular organisms   530

Some high-GC Gram-positives are valuable sources of antibiotics   532

Hyperthermophilic bacteria live at very high temperatures   532

Hadobacteria live in extreme environments   532

Cyanobacteria were the first photosynthesizers   532

Spirochetes move by means of axial filaments   533

Chlamydias are extremely small parasites   533

The proteobacteria are a large and diverse group   534

Gene sequencing enabled biologists to differentiate the domain Archaea   534

Most crenarchaeotes live in hot or acidic places   536

Euryarchaeotes are found in surprising places   536

Korarchaeotes and nanoarchaeotes are less well known   537

26.3 How Do Prokaryotes Affect Their Environments?   537

Prokaryotes have diverse metabolic pathways   537

Prokaryotes play important roles in element cycling   538

Many prokaryotes form complex communities   539

Prokaryotes live on and in other organisms   539

Microbiomes are critical to human health   539

A small minority of bacteria are pathogens   541

26.4 How Do Viruses Relate to Life's Diversity and Ecology?   543

Many RNA viruses probably represent escaped genomic components of cellular life   544

Some DNA viruses may have evolved from reduced cellular organisms   544

Vertebrate genomes contain endogenous retroviruses   545

Viruses can be used to fight bacterial infections   545

Viruses are found throughout the biosphere   546

27 The Origin and Diversification of Eukaryotes   549

27.1 How Did the Eukaryotic Cell Arise?   550

The modern eukaryotic cell arose in several steps   550

Chloroplasts have been transferred among eukaryotes several times   551

27.2 What Features Account for Protist Diversity?   552

Alveolates have sacs under their plasma membranes   553

Stramenopiles typically have two flagella of unequal length   555

Rhizaria typically have long, thin pseudopods   557

Excavates began to diversify about 1.5 billion years ago   558

Amoebozoans use lobe-shaped pseudopods for locomotion   559

27.3 What Is the Relationship between Sex and Reproduction in Protists?   562

Some protists reproduce without sex and have sex without reproduction   562

Some protist life cycles feature alternation of generations   562

27.4 How Do Protists Affect Their Environments?   563

Phytoplankton are primary producers   563

Some microbial eukaryotes are deadly   563

Some microbial eukaryotes are endosymbionts   564

We rely on the remains of ancient marine protists   565

28 Plants without Seeds: From Water to Land   569

28.1 How Did Photosynthesis Arise in Plants?   570

Several distinct clades of algae were among the first photosynthetic eukaryotes   571

Two groups of green algae are the closest relatives of land plants   572

There are ten major groups of land plants   573

28.2   When and How Did Plants Colonize Land?   574

Adaptations to life on land distinguish land plants from green algae   574

Life cycles of land plants feature alternation of generations   574

Nonvascular land plants live where water is readily available   575

The sporophytes of nonvascular land plants are dependent on the gametophytes   575

Liverworts are the sister clade of the remaining land plants   577

Water and sugar transport mechanisms emerged in the mosses   577

Hornworts have distinctive chloroplasts and stalkless sporophytes   578

28.3   What Features Allowed Land Plants to Diversify in Form?   579

Vascular tissues transport water and dissolved materials   579

Vascular plants allowed herbivores to colonize the land   580

The closest relatives of vascular plants lacked roots   580

The lycophytes are sister to the other vascular plants   581

Horsetails and ferns constitute a clade   581

The vascular plants branched out   582

Heterospory appeared among the vascular plants   584

29   The Evolution of Seed Plants   588

29.1   How Did Seed Plants Become Today's Dominant Vegetation?   589

Features of the seed plant life cycle protect gametes and embryos   589

The seed is a complex, well-protected package   591

A change in stem anatomy enabled seed plants to grow to great heights   591

29.2   What Are the Major Groups of Gymnosperms?   592

There are four major groups of living gymnosperms   592

Conifers have cones and no swimming sperm   593

29.3   How Do Flowers and Fruits Increase the Reproductive Success of Angiosperms?   596

Angiosperms have many shared derived traits   596

The sexual structures of angiosperms are flowers   596

Flower structure has evolved over time   597

Angiosperms have coevolved with animals   598

The angiosperm life cycle produces diploid zygotes nourished by triploid endosperms   600

Fruits aid angiosperm seed dispersal   601

Recent analyses have revealed the phylogenetic relationships of angiosperms   601

29.4   How Do Plants Benefit Human Society?   604

Seed plants have been sources of medicine since ancient times   604

Seed plants are our primary food source   605

30   The Evolution and Diversity of Fungi   608

30.1   What Is a Fungus?   609

Unicellular yeasts absorb nutrients directly   609

Multicellular fungi use hyphae to absorb nutrients   609

Fungi are in intimate contact with their environment   610

30.2   How Do Fungi Interact with Other Organisms?   611

Saprobic fungi are critical to the planetary carbon cycle   611

Some fungi engage in parasitic or predatory interactions   611

Mutualistic fungi engage in relationships that benefit both partners   612

Endophytic fungi protect some plants from pathogens, herbivores, and stress   615

30.3   How Do Major Groups of Fungi Differ in Structure and Life History?   615

Fungi reproduce both sexually and asexually   616

Microsporidia are highly reduced, parasitic fungi   617

Most chytrids are aquatic   617

Some fungal life cycles feature separate fusion of cytoplasms and nuclei   619

Arbuscular mycorrhizal fungi form symbioses with plants   619

The dikaryotic condition is a synapomorphy of sac fungi and club fungi   620

The sexual reproductive structure of sac fungi is the ascus   620

The sexual reproductive structure of club fungi is the basidium   622

30.4   What Are Some Applications of Fungal Biology?   623

Fungi are important in producing food and drink   623

Fungi record and help remediate environmental pollution   624

Lichen diversity and abundance are indicators of air quality   624

Fungi are used as model organisms in laboratory studies   624

Reforestation may depend on mycorrhizal fungi   626

Fungi provide important weapons against diseases and pests 626

31 Animal Origins and the Evolution of Body Plans 629

31.1 What Characteristics Distinguish the Animals? 630

Animal monophyly is supported by gene sequences and morphology 630

A few basic developmental patterns differentiate major animal groups 633

31.2 What Are the Features of Animal Body Plans? 634

Most animals are symmetrical 634

The structure of the body cavity influences movement 635

Segmentation improves control of movement 636

Appendages have many uses 636

Nervous systems coordinate movement and allow sensory processing 637

31.3 How Do Animals Get Their Food? 637

Filter feeders capture small prey 637

Herbivores eat plants 637

Predators and omnivores capture and subdue prey 638

Parasites live in or on other organisms 638

Detritivores live on the remains of other organisms 639

31.4 How Do Life Cycles Differ among Animals? 639

Many animal life cycles feature specialized life stages 639

Most animal life cycles have at least one dispersal stage 640

Parasite life cycles facilitate dispersal and overcome host defenses 640

Some animals form colonies of genetically identical, physiologically integrated individuals 640

No life cycle can maximize all benefits 641

31.5 What Are the Major Groups of Animals? 643

Sponges are loosely organized animals 643

Ctenophores are radially symmetrical and diploblastic 644

Placozoans are abundant but rarely observed 645

Cnidarians are specialized predators 645

Some small groups of parasitic animals may be the closest relatives of bilaterians 648

32 Protostome Animals 651

32.1 What Is a Protostome? 652

Cilia-bearing lophophores and trochophores evolved among the lophotrochozoans 652

Ecdysozoans must shed their cuticles 654

Arrow worms retain some ancestral developmental features 655

32.2 What Features Distinguish the Major Groups of Lophotrochozoans? 656

Most bryozoans and entoprocts live in colonies 656

Flatworms, rotifers, and gastrotrichs are structurally diverse relatives 656

Ribbon worms have a long, protrusible feeding organ 658

Brachiopods and phoronids use lophophores to extract food from the water 658

Annelids have segmented bodies 659

Mollusks have undergone a dramatic evolutionary radiation 662

32.3 What Features Distinguish the Major Groups of Ecdysozoans? 665

Several marine ecdysozoan groups have relatively few species 665

Nematodes and their relatives are abundant and diverse 666

32.4 Why Are Arthropods So Diverse? 667

Arthropod relatives have fleshy, unjointed appendages 667

Jointed appendages appeared in the trilobites 668

Chelicerates have pointed, nonchewing mouthparts 668

Mandibles and antennae characterize the remaining arthropod groups 669

More than half of all described species are insects 671

33 Deuterostome Animals 678

33.1 What Is a Deuterostome? 679

Deuterostomes share early developmental patterns 679

There are three major deuterostome clades 679

Fossils shed light on deuterostome ancestors 679

33.2 What Features Distinguish the Echinoderms, Hemichordates, and Their Relatives? 680

Echinoderms have unique structural features 680

Hemichordates are wormlike marine deuterostomes 682

33.3 What New Features Evolved in the Chordates? 683

Adults of most lancelets and tunicates are sedentary 684

A dorsal supporting structure replaces the notochord in vertebrates 684

The phylogenetic relationships of jawless fishes are uncertain 685

Jaws and teeth improved feeding efficiency 686

Fins and swim bladders improved stability and control over locomotion 686

33.4 How Did Vertebrates Colonize the Land? 689

Jointed limbs enhanced support and locomotion on land   689

Amphibians usually require moist environments   690

Amniotes colonized dry environments   692

Reptiles adapted to life in many habitats   693

Crocodilians and birds share their ancestry with the dinosaurs   693

Feathers allowed birds to fly   695

Mammals radiated after the extinction of non-avian dinosaurs   696

33.5 What Traits Characterize the Primates?   701

Two major lineages of primates split late in the Cretaceous   701

Bipedal locomotion evolved in human ancestors   702

Human brains became larger as jaws became smaller   704

Humans developed complex language and culture   705

PART EIGHT
Flowering Plants: Form and Function

34 The Plant Body   708

34.1 What Is the Basic Body Plan of Plants?   709

Most angiosperms are either monocots or eudicots   709

Plants develop differently than animals   710

Apical–basal polarity and radial symmetry are characteristics of the plant body   711

34.2 What Are the Major Tissues of Plants?   712

The plant body is constructed from three tissue systems   712

Cells of the xylem transport water and dissolved minerals   714

Cells of the phloem transport the products of photosynthesis   714

34.3 How Do Meristems Build a Continuously Growing Plant?   715

Plants increase in size through primary and secondary growth   715

A hierarchy of meristems generates the plant body   715

Indeterminate primary growth originates in apical meristems   715

The root apical meristem gives rise to the root cap and the root primary meristems   716

The products of the root's primary meristems become root tissues   716

The root system anchors the plant and takes up water and dissolved minerals   718

The products of the stem's primary meristems become stem tissues   719

The stem supports leaves and flowers   720

Leaves are determinate organs produced by shoot apical meristems   720

Many eudicot stems and roots undergo secondary growth   721

34.4 How Has Domestication Altered Plant Form?   723

35 Transport in Plants   726

35.1 How Do Plants Take Up Water and Solutes?   727

Water potential differences govern the direction of water movement   727

Water and ions move across the root cell plasma membrane   728

Water and ions pass to the xylem by way of the apoplast and symplast   729

35.2 How Are Water and Minerals Transported in the Xylem?   730

The transpiration–cohesion–tension mechanism accounts for xylem transport   731

35.3 How Do Stomata Control the Loss of Water and the Uptake of CO_2?   732

The guard cells control the size of the stomatal opening   733

Plants can control their total numbers of stomata   734

35.4 How Are Substances Translocated in the Phloem?   734

Sucrose and other solutes are carried in the phloem   734

The pressure flow model appears to account for translocation in the phloem   735

36 Plant Nutrition   740

36.1 What Nutrients Do Plants Require?   741

All plants require specific macronutrients and micronutrients   741

Deficiency symptoms reveal inadequate nutrition   742

Hydroponic experiments identified essential elements   742

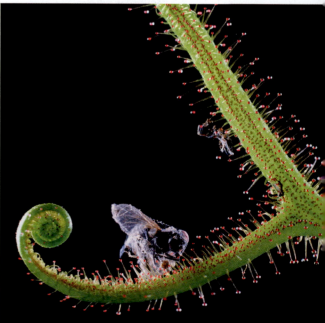

36.2 How Do Plants Acquire Nutrients? 743

Plants rely on growth to find nutrients 743

Nutrient uptake and assimilation are regulated 744

36.3 How Does Soil Structure Affect Plants? 744

Soils are complex in structure 745

Soils form through the weathering of rock 745

Soils are the source of plant nutrition 746

Fertilizers can be used to add nutrients to soil 746

36.4 How Do Fungi and Bacteria Increase Nutrient Uptake by Plant Roots? 747

Plants send signals for colonization 747

Mycorrhizae expand the root system 748

Soil bacteria are essential in getting nitrogen from air to plant cells 749

Nitrogenase catalyzes nitrogen fixation 749

Biological nitrogen fixation does not always meet agricultural needs 750

Plants and bacteria participate in the global nitrogen cycle 750

36.5 How Do Carnivorous and Parasitic Plants Obtain a Balanced Diet? 751

Carnivorous plants supplement their mineral nutrition 751

Parasitic plants take advantage of other plants 752

The plant–parasite relationship is similar to plant–fungus and plant–bacteria associations 753

37 Regulation of Plant Growth 756

37.1 How Does Plant Development Proceed? 757

In early development, the seed germinates and forms a growing seedling 757

Several hormones and photoreceptors help regulate plant growth 758

Genetic screens have increased our understanding of plant signal transduction 759

37.2 What Do Gibberellins and Auxin Do? 760

Gibberellins have many effects on plant growth and development 760

Auxin plays a role in differential plant growth 762

Auxin affects plant growth in several ways 765

At the molecular level, auxin and gibberellins act similarly 767

37.3 What Are the Effects of Cytokinins, Ethylene, and Brassinosteroids? 768

Cytokinins are active from seed to senescence 768

Ethylene is a gaseous hormone that hastens leaf senescence and fruit ripening 769

Brassinosteroids are plant steroid hormones 771

37.4 How Do Photoreceptors Participate in Plant Growth Regulation? 771

Phototropins, cryptochromes, and zeaxanthin are blue-light receptors 771

Phytochromes mediate the effects of red and far-red light 772

Phytochrome stimulates gene transcription 773

Circadian rhythms are entrained by light reception 774

38 Reproduction in Flowering Plants 778

38.1 How Do Angiosperms Reproduce Sexually? 779

The flower is an angiosperm's structure for sexual reproduction 779

Flowering plants have microscopic gametophytes 779

Pollination in the absence of water is an evolutionary adaptation 780

A pollen tube delivers sperm cells to the embryo sac 780

Many flowering plants control pollination or pollen tube growth to prevent inbreeding 782

Angiosperms perform double fertilization 783

Embryos develop within seeds contained in fruits 784

Seed development is under hormonal control 785

38.2 What Determines the Transition from the Vegetative to the Flowering State? 785

Shoot apical meristems can become inflorescence meristems 785

A cascade of gene expression leads to flowering 786

Photoperiodic cues can initiate flowering 787

Plants vary in their responses to photoperiodic cues 787

Night length is a key photoperiodic cue that determines flowering 788

The flowering stimulus originates in a leaf 788

Florigen is a small protein 790

Flowering can be induced by temperature or gibberellin 790

Some plants do not require an environmental cue to flower 792

38.3 How Do Angiosperms Reproduce Asexually? 792

Many forms of asexual reproduction exist 792

Vegetative reproduction has a disadvantage 793

Vegetative reproduction is important in agriculture 793

39 **Plant Responses to Environmental Challenges   797**

39.1 How Do Plants Deal with Pathogens?   798

Physical barriers form constitutive defenses   798

Plants can seal off infected parts to limit damage   798

General and specific immunity both involve multiple responses   799

Specific immunity involves gene-for-gene resistance   800

Specific immunity usually leads to the hypersensitive response   800

Systemic acquired resistance is a form of long-term immunity   801

39.2 How Do Plants Deal with Herbivores?   801

Mechanical defenses against herbivores are widespread   801

Plants produce constitutive chemical defenses against herbivores   802

Some secondary metabolites play multiple roles   803

Plants respond to herbivory with induced defenses   803

Jasmonates trigger a range of responses to wounding and herbivory   805

Why don't plants poison themselves?   805

Plants don't always win the arms race   806

39.3 How Do Plants Deal with Environmental Stresses?   806

Some plants have special adaptations to live in very dry conditions   806

Some plants grow in saturated soils   808

Plants can respond to drought stress   809

Plants can cope with temperature extremes   810

39.4 How Do Plants Deal with Salt and Heavy Metals?   810

Most halophytes accumulate salt   811

Some plants can tolerate heavy metals   811

PART NINE
Animals: Form and Function

40 **Physiology, Homeostasis, and Temperature Regulation   815**

40.1 How Do Multicellular Animals Supply the Needs of Their Cells?   816

An internal environment makes complex multicellular animals possible   816

Physiological systems are regulated to maintain homeostasis   816

40.2 What Are the Relationships between Cells, Tissues, and Organs?   817

Epithelial tissues are sheets of densely packed, tightly connected cells   817

Muscle tissues generate force and movement   818

Connective tissues include bone, blood, and fat   818

Neural tissues include neurons and glial cells   819

Organs consist of multiple tissues   820

40.3 How Does Temperature Affect Living Systems?   820

Q_{10} is a measure of temperature sensitivity   821

Animals acclimatize to seasonal temperatures   821

40.4 How Do Animals Alter Their Heat Exchange with the Environment?   822

Endotherms produce substantial amounts of metabolic heat   822

Ectotherms and endotherms respond differently to changes in environmental temperature   822

Energy budgets reflect adaptations for regulating body temperature   823

Both ectotherms and endotherms control blood flow to the skin   824

Some fish conserve metabolic heat   825

Some ectotherms regulate metabolic heat production   825

40.5 How Do Endotherms Regulate Their Body Temperatures?   826

Basal metabolic rates correlate with body size   826

Endotherms respond to cold by producing heat and adapt to cold by reducing heat loss   827

Evaporation of water can dissipate heat, but at a cost   829

The mammalian thermostat uses feedback information   829

Fever helps the body fight infections   830

Some animals conserve energy by turning down the thermostat   830

41 **Animal Hormones 834**

41.1 What Are Hormones and How Do They Work?   835

Endocrine signaling can act locally or at a distance   835

Hormones can be divided into three chemical groups   836

Hormone action is mediated by receptors on or within their target cells   836

Hormone action depends on the nature of the target cell and its receptors   837

41.2   What Have Experiments Revealed about Hormones and Their Action?   838

The first hormone discovered was the gut hormone secretin   838

Early experiments on insects illuminated hormonal signaling systems   839

Three hormones regulate molting and maturation in arthropods   840

41.3   How Do the Nervous and Endocrine Systems Interact?   842

The pituitary is an interface between the nervous and endocrine systems   842

The anterior pituitary is controlled by hypothalamic neurohormones   844

Negative feedback loops regulate hormone secretion   844

41.4   What Are the Major Endocrine Glands and Hormones?   845

The thyroid gland secretes thyroxine   845

Three hormones regulate blood calcium concentrations   847

PTH lowers blood phosphate levels   848

Insulin and glucagon regulate blood glucose concentrations   848

The adrenal gland is two glands in one   849

Sex steroids are produced by the gonads   850

Melatonin is involved in biological rhythms and photoperiodicity   851

Many chemicals may act as hormones   851

41.5   How Do We Study Mechanisms of Hormone Action?   852

Hormones can be detected and measured with immunoassays   852

A hormone can act through many receptors   853

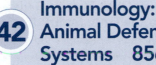

42   Immunology: Animal Defense Systems   856

42.1   What Are the Major Defense Systems of Animals?   857

Blood and lymph tissues play important roles in defense   857

White blood cells play many defensive roles   858

Immune system proteins bind pathogens or signal other cells   858

42.2   What Are the Characteristics of the Innate Defenses?   859

Barriers and local agents defend the body against invaders   859

Cell signaling pathways stimulate the body's defenses   860

Specialized proteins and cells participate in innate immunity   860

Inflammation is a coordinated response to infection or injury   861

Inflammation can cause medical problems   862

42.3   How Does Adaptive Immunity Develop?   862

Adaptive immunity has four key features   862

Two types of adaptive immune responses interact: an overview   863

Adaptive immunity develops as a result of clonal selection   865

Clonal deletion helps the immune system distinguish self from nonself   865

Immunological memory results in a secondary immune response   865

Vaccines are an application of immunological memory   866

42.4   What Is the Humoral Immune Response?   867

Some B cells develop into plasma cells   867

Different antibodies share a common structure   867

There are five classes of immunoglobulins   868

Immunoglobulin diversity results from DNA rearrangements and other mutations   868

The constant region is involved in immunoglobulin class switching   869

Monoclonal antibodies have many uses   871

42.5   What Is the Cellular Immune Response?   871

T cell receptors bind to antigens on cell surfaces   871

MHC proteins present antigen to T cells   872

T-helper cells and MHC II proteins contribute to the humoral immune response   872

Cytotoxic T cells and MHC I proteins contribute to the cellular immune response   874

Regulatory T cells suppress the humoral and cellular immune responses   874

MHC proteins are important in tissue transplants   874

42.6   What Happens When the Immune System Malfunctions?   875

Allergic reactions result from hypersensitivity   875

Autoimmune diseases are caused by reactions against self antigens   876

AIDS is an immune deficiency disorder   876

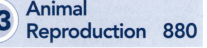

43   Animal Reproduction   880

43.1   How Do Animals Reproduce without Sex?   881

Budding and regeneration produce new individuals by mitosis   881

Parthenogenesis is the development of unfertilized eggs   881

43.2   How Do Animals Reproduce Sexually?   882

Gametogenesis produces eggs and sperm   882

Fertilization is the union of sperm and egg   884

Getting eggs and sperm together   887

Some individuals can function as both male and female   887

The evolution of vertebrate reproductive systems parallels the move to land   888

Animals with internal fertilization are distinguished by where the embryo develops   889

43.3 How Do the Human Male and Female Reproductive Systems Work? 889

Male sex organs produce and deliver semen 889

Male sexual function is controlled by hormones 892

Female sex organs produce eggs, receive sperm, and nurture the embryo 892

The ovarian cycle produces a mature egg 893

The uterine cycle prepares an environment for a fertilized egg 893

Hormones control and coordinate the ovarian and uterine cycles 894

FSH receptors determine which follicle ovulates 895

In pregnancy, hormones from the extraembryonic membranes take over 896

Childbirth is triggered by hormonal and mechanical stimuli 896

43.4 How Can Fertility Be Controlled? 897

Humans use a variety of methods to control fertility 897

Reproductive technologies help solve problems of infertility 897

44 Animal Development 902

44.1 How Does Fertilization Activate Development? 903

The sperm and the egg make different contributions to the zygote 903

Rearrangements of egg cytoplasm set the stage for determination 903

44.2 How Does Mitosis Divide Up the Early Embryo? 904

Cleavage repackages the cytoplasm 904

Early cell divisions in mammals are unique 905

Specific blastomeres generate specific tissues and organs 906

Germ cells are a unique lineage even in species with regulative development 908

44.3 How Does Gastrulation Generate Multiple Tissue Layers? 908

Invagination of the vegetal pole characterizes gastrulation in the sea urchin 908

Gastrulation in the frog begins at the gray crescent 909

The dorsal lip of the blastopore organizes embryo formation 910

Transcription factors and growth factors underlie the organizer's actions 911

The organizer changes its activity as it migrates from the dorsal lip 912

Reptilian and avian gastrulation is an adaptation to yolky eggs 913

The embryos of placental mammals lack yolk 914

44.4 How Do Organs and Organ Systems Develop? 915

The stage is set by the dorsal lip of the blastopore 915

Body segmentation develops during neurulation 916

Hox genes control development along the anterior–posterior axis 916

44.5 How Is the Growing Embryo Sustained? 918

Extraembryonic membranes form with contributions from all germ layers 918

Extraembryonic membranes in mammals form the placenta 919

44.6 What Are the Stages of Human Development? 919

Organ development begins in the first trimester 920

Organ systems grow and mature during the second and third trimesters 920

Developmental changes continue throughout life 920

45 Neurons, Glia, and Nervous Systems 924

45.1 What Cells Are Unique to the Nervous System? 925

The structure of neurons reflects their functions 925

Glia are the "silent partners" of neurons 926

45.2 How Do Neurons Generate and Transmit Electric Signals? 927

Simple electrical concepts underlie neural function 927

Membrane potentials can be measured with electrodes 928

Ion transporters and channels generate membrane potentials 928

Ion channels and their properties can now be studied directly 929

Gated ion channels alter membrane potential 930

Graded changes in membrane potential can integrate information 932

Sudden changes in Na^+ and K^+ channels generate action potentials 932

Action potentials are conducted along axons without loss of signal 934

Action potentials jump along myelinated axons 935

45.3 How Do Neurons Communicate with Other Cells? 936

The neuromuscular junction is a model chemical synapse 936

The arrival of an action potential causes the release of neurotransmitter 936

Synaptic functions involve many proteins 936

The postsynaptic membrane responds to neurotransmitter 936

Synapses can be excitatory or inhibitory 938

The postsynaptic cell sums excitatory and inhibitory input 938

Synapses can be fast or slow 938

Electrical synapses are fast but do not integrate information well 939

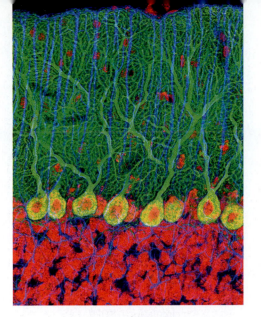

The action of a neurotransmitter depends on the receptor to which it binds   939

To turn off responses, synapses must be cleared of neurotransmitter   940

The diversity of receptors makes drug specificity possible   940

45.4 How Are Neurons and Glia Organized into Information-Processing Systems?   940

Nervous systems range in complexity   940

The knee-jerk reflex is controlled by a simple neural network   941

The vertebrate brain is the seat of behavioral complexity   943

46 Sensory Systems 946

46.1 How Do Sensory Receptor Cells Convert Stimuli into Action Potentials?   947

Sensory transduction involves changes in membrane potentials   947

Sensory receptor proteins act on ion channels   947

Sensation depends on which neurons receive action potentials from sensory cells   947

Many receptors adapt to repeated stimulation   948

46.2 How Do Sensory Systems Detect Chemical Stimuli?   949

Olfaction is the sense of smell   949

Some chemoreceptors detect pheromones   950

The vomeronasal organ contains chemoreceptors   950

Gustation is the sense of taste   951

46.3 How Do Sensory Systems Detect Mechanical Forces?   952

Many different cells respond to touch and pressure   952

Mechanoreceptors are also found in muscles, tendons, and ligaments   952

Hair cells are mechanoreceptors of the auditory and vestibular systems   953

Auditory systems use hair cells to sense sound waves   954

Flexion of the basilar membrane is perceived as sound   955

Various types of damage can result in hearing loss   956

The vestibular system uses hair cells to detect forces of gravity and momentum   956

46.4 How Do Sensory Systems Detect Light?   957

Rhodopsin is a vertebrate visual pigment   957

Invertebrates have a variety of visual systems   958

Image-forming eyes evolved independently in vertebrates and cephalopods   958

The vertebrate retina receives and processes visual information   959

Rod and cone cells are the photoreceptors of the vertebrate retina   960

Information flows through layers of neurons in the retina   962

47 The Mammalian Nervous System: Structure and Higher Functions 967

47.1 How Is the Mammalian Nervous System Organized?   968

Functional organization is based on flow and type of information   968

The anatomical organization of the CNS emerges during development   968

The spinal cord transmits and processes information   969

The brainstem carries out many autonomic functions   969

The core of the forebrain controls physiological drives, instincts, and emotions   970

Regions of the telencephalon interact to control behavior and produce consciousness   970

The size of the human brain is off the curve   973

47.2 How Is Information Processed by Neural Networks?   973

Pathways of the autonomic nervous system control involuntary physiological functions   974

The visual system is an example of information integration by the cerebral cortex   975

Three-dimensional vision results from cortical cells receiving input from both eyes   977

47.3 Can Higher Functions Be Understood in Cellular Terms?   978

Sleep and dreaming are reflected in electrical patterns in the cerebral cortex   978

Language abilities are localized in the left cerebral hemisphere   980

Some learning and memory can be localized to specific brain areas   981

We still cannot answer the question "What is consciousness?"   982

48 Musculoskeletal Systems 986

48.1 How Do Muscles Contract?   987

Sliding filaments cause skeletal muscle to contract   987

Actin–myosin interactions cause filaments to slide   988

Actin–myosin interactions are controlled by calcium ions   989

Cardiac muscle is similar to and different from skeletal muscle   991

Smooth muscle causes slow contractions of many internal organs   993

48.2 What Determines Skeletal Muscle Performance?   994

The strength of a muscle contraction depends on how many fibers are contracting and at what rate   994

Muscle fiber types determine endurance and strength   995

A muscle has an optimal length for generating maximum tension   996

Exercise increases muscle strength and endurance   996

Muscle ATP supply limits performance   997

Insect muscle has the greatest rate of cycling   997

48.3 How Do Skeletal Systems and Muscles Work Together?   999

A hydrostatic skeleton consists of fluid in a muscular cavity   999

Exoskeletons are rigid outer structures   999

Vertebrate endoskeletons consist of cartilage and bone   999

Bones develop from connective tissues   1001

Bones that have a common joint can work as a lever   1001

49 Gas Exchange   1005

49.1 What Physical Factors Govern Respiratory Gas Exchange?   1006

Diffusion of gases is driven by partial pressure differences   1006

Fick's law applies to all systems of gas exchange   1006

Air is a better respiratory medium than water   1007

High temperatures create respiratory problems for aquatic animals   1007

O_2 availability decreases with altitude   1007

CO_2 is lost by diffusion   1008

49.2 What Adaptations Maximize Respiratory Gas Exchange?   1008

Respiratory organs have large surface areas   1008

Ventilation and perfusion of gas exchange surfaces maximize partial pressure gradients   1009

Insects have airways throughout their bodies   1009

Fish gills use countercurrent flow to maximize gas exchange   1009

Birds use unidirectional ventilation to maximize gas exchange   1010

Tidal ventilation produces dead space that limits gas exchange efficiency   1012

49.3 How Do Human Lungs Work?   1013

Respiratory tract secretions aid ventilation   1013

Lungs are ventilated by pressure changes in the thoracic cavity   1015

49.4 How Does Blood Transport Respiratory Gases?   1016

Hemoglobin combines reversibly with O_2 1016

Myoglobin holds an O_2 reserve   1017

Hemoglobin's affinity for O_2 is variable   1017

CO_2 is transported as bicarbonate ions in the blood   1018

49.5 How Is Breathing Regulated?   1019

Breathing is controlled in the brainstem   1019

Regulating breathing requires feedback   1020

50 Circulatory Systems   1025

50.1 Why Do Animals Need a Circulatory System?   1026

Some animals do not have a circulatory system   1026

Circulatory systems can be open or closed   1026

Open circulatory systems move extracellular fluid   1026

Closed circulatory systems circulate blood through a system of blood vessels   1026

50.2 How Have Vertebrate Circulatory Systems Evolved?   1027

Circulation in fish is a single circuit   1028

Lungfish evolved a gas-breathing organ   1028

Amphibians have partial separation of systemic and pulmonary circulation   1029

Reptiles have exquisite control of pulmonary and systemic circulation   1029

Birds and mammals have fully separated pulmonary and systemic circuits   1030

50.3 How Does the Mammalian Heart Function?   1030

Blood flows from right heart to lungs to left heart to body   1030

The heartbeat originates in the cardiac muscle   1032

A conduction system coordinates the contraction of heart muscle   1034

Electrical properties of ventricular muscles sustain heart contraction   1034

The ECG records the electrical activity of the heart   1035

50.4 What Are the Properties of Blood and Blood Vessels?   1037

Red blood cells transport respiratory gases   1038

Platelets are essential for blood clotting   1039

Arteries withstand high pressure, arterioles control blood flow   1039

Materials are exchanged in capillary beds by filtration, osmosis, and diffusion   1039

Blood flows back to the heart through veins   1041

Lymphatic vessels return interstitial fluid to the blood   1042

Vascular disease is a killer   1042

50.5 How Is the Circulatory System Controlled and Regulated?   1043

Autoregulation matches local blood flow to local need   1044

Arterial pressure is regulated by hormonal and neural mechanisms   1044

51 Nutrition, Digestion, and Absorption   1048

51.1 What Do Animals Require from Food?   1049

Energy needs and expenditures can be measured   1049

Sources of energy can be stored in the body   1050

Food provides carbon skeletons for biosynthesis   1051

Animals need mineral elements for a variety of functions   1052

Animals must obtain vitamins from food   1053

Nutrient deficiencies result in diseases   1054

51.2 How Do Animals Ingest and Digest Food?   1054

The food of herbivores is often low in energy and hard to digest   1054

Carnivores must find, capture, and kill prey   1055

Vertebrate species have distinctive teeth   1055

Digestion usually begins in a body cavity   1056

Tubular guts have an opening at each end   1056

Digestive enzymes break down complex food molecules   1057

51.3 How Does the Vertebrate Gastrointestinal System Function?   1058

The vertebrate gut consists of concentric tissue layers   1058

Mechanical activity moves food through the gut and aids digestion   1059

Chemical digestion begins in the mouth and the stomach   1060

The stomach gradually releases its contents to the small intestine   1061

Most chemical digestion occurs in the small intestine   1061

Nutrients are absorbed in the small intestine   1063

Absorbed nutrients go to the liver   1063

Water and ions are absorbed in the large intestine   1063

Herbivores rely on microorganisms to digest cellulose   1063

51.4 How Is the Flow of Nutrients Controlled and Regulated?   1064

Hormones control many digestive functions   1065

The liver directs the traffic of the molecules that fuel metabolism   1065

The brain plays a major role in regulating food intake   1067

52 Salt and Water Balance and Nitrogen Excretion   1071

52.1 How Do Excretory Systems Maintain Homeostasis?   1072

Water enters or leaves cells by osmosis   1072

Excretory systems control extracellular fluid osmolarity and composition   1072

Aquatic invertebrates can conform to or regulate their osmotic and ionic environments   1072

Vertebrates are osmoregulators and ionic regulators   1073

52.2 How Do Animals Excrete Nitrogen?   1074

Animals excrete nitrogen in a number of forms   1074

Most species produce more than one nitrogenous waste   1074

52.3 How Do Invertebrate Excretory Systems Work?   1075

The protonephridia of flatworms excrete water and conserve salts   1075

The metanephridia of annelids process coelomic fluid   1075

Malpighian tubules of insects use active transport to excrete wastes   1076

52.4 How Do Vertebrates Maintain Salt and Water Balance?   1077

Marine fishes must conserve water   1077

Terrestrial amphibians and reptiles must avoid desiccation   1077

Mammals can produce highly concentrated urine   1078

The nephron is the functional unit of the vertebrate kidney   1078

Blood is filtered into Bowman's capsule   1078

The renal tubules convert glomerular filtrate to urine   1079

52.5 How Does the Mammalian Kidney Produce Concentrated Urine?   1079

Kidneys produce urine and the bladder stores it   1080

Nephrons have a regular arrangement in the kidney   1081

Most of the glomerular filtrate is reabsorbed by the proximal convoluted tubule   1082

The loop of Henle creates a concentration gradient in the renal medulla   1082

Water permeability of kidney tubules depends on water channels   1084

The distal convoluted tubule fine-tunes the composition of the urine   1084

Urine is concentrated in the collecting duct   1084

The kidneys help regulate acid–base balance   1084

Kidney failure is treated with dialysis   1085

52.6 How Are Kidney Functions Regulated?   1087

Glomerular filtration rate is regulated   1087

Regulation of GFR uses feedback information from the distal tubule   1087

Blood osmolarity and blood pressure are regulated by ADH   1088

The heart produces a hormone that helps lower blood pressure   1090

53 Animal Behavior 1093

53.1 What Are the Origins of Behavioral Biology?   1094
Conditioned reflexes are a simple behavioral mechanism   1094

Ethologists focused on the behavior of animals in their natural environment   1094

Ethologists probed the causes of behavior   1095

53.2 How Do Genes Influence Behavior?   1096
Breeding experiments can produce behavioral phenotypes   1096

Knockout experiments can reveal the roles of specific genes   1096

Behaviors are controlled by gene cascades   1097

53.3 How Does Behavior Develop?   1098
Hormones can determine behavioral potential and timing   1098

Some behaviors can be acquired only at certain times   1099

Birdsong learning involves genetics, imprinting, and hormonal timing   1099

The timing and expression of birdsong are under hormonal control   1101

53.4 How Does Behavior Evolve?   1102
Animals are faced with many choices   1103

Behaviors have costs and benefits   1103

Territorial behavior carries significant costs   1103

Cost–benefit analysis can be applied to foraging behavior   1104

53.5 What Physiological Mechanisms Underlie Behavior?   1106
Biological rhythms coordinate behavior with environmental cycles   1106

Animals must find their way around their environment   1109

Animals use multiple modalities to communicate   1110

53.6 How Does Social Behavior Evolve?   1113
Mating systems maximize the fitness of both partners   1113

Fitness can include more than your own offspring   1114

Eusociality is the extreme result of kin selection   1115

Group living has benefits and costs   1116

Can the concepts of sociobiology be applied to humans?   1116

PART TEN
Ecology

54 Ecology and the Distribution of Life   1121

54.1 What Is Ecology?   1122
Ecology is not the same as environmentalism   1122

Ecologists study biotic and abiotic components of ecosystems   1122

54.2 Why Do Climates Vary Geographically?   1122
Solar radiation varies over Earth's surface   1123

Solar energy input determines atmospheric circulation patterns   1124

Atmospheric circulation and Earth's rotation result in prevailing winds   1124

Prevailing winds drive ocean currents   1124

Organisms adapt to climatic challenges   1125

54.3 How Is Life Distributed in Terrestrial Environments?   1126
Tundra is found at high latitudes and high elevations   1128

Evergreen trees dominate boreal and temperate evergreen forests   1129

Temperate deciduous forests change with the seasons   1130

Temperate grasslands are widespread   1131

Hot deserts form around 30° latitude   1132

Cold deserts are high and dry   1133

Chaparral has hot, dry summers and wet, cool winters 1134

Thorn forests and tropical savannas have similar climates 1135

Tropical deciduous forests occur in hot lowlands 1136

Tropical rainforests are rich in species 1137

54.4 How Is Life Distributed in Aquatic Environments? 1139

The marine biome can be divided into several life zones 1139

Freshwater biomes may be rich in species 1140

Estuaries have characteristics of both freshwater and marine environments 1141

54.5 What Factors Determine the Boundaries of Biogeographic Regions? 1141

Geological history influences the distribution of organisms 1141

Two scientific advances changed the field of biogeography 1142

Discontinuous distributions may result from vicariant or dispersal events 1143

Humans exert a powerful influence on biogeographic patterns 1145

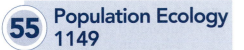

55 Population Ecology 1149

55.1 How Do Ecologists Measure Populations? 1150

Ecologists use a variety of approaches to count and track individuals 1150

Ecologists can estimate population densities from samples 1151

A population's age structure influences its capacity to grow 1151

A population's dispersion pattern reflects how individuals are distributed in space 1152

55.2 How Do Ecologists Study Population Dynamics? 1153

Demographic events determine the size of a population 1153

Life tables track demographic events 1154

Survivorship curves reflect life history strategies 1155

55.3 How Do Environmental Conditions Affect Life Histories? 1156

Survivorship and fecundity determine a population's growth rate 1156

Life history traits vary with environmental conditions 1156

Life history traits are influenced by interspecific interactions 1157

55.4 What Factors Limit Population Densities? 1157

All populations have the potential for exponential growth 1157

Logistic growth occurs as a population approaches its carrying capacity 1158

Population growth can be limited by density-dependent or density-independent factors 1159

Different population regulation factors lead to different life history strategies 1159

Several ecological factors explain species' characteristic population densities 1159

Some newly introduced species reach high population densities 1160

Evolutionary history may explain species abundances 1160

55.5 How Does Habitat Variation Affect Population Dynamics? 1161

Many populations live in separated habitat patches 1161

Corridors may allow subpopulations to persist 1162

55.6 How Can We Use Ecological Principles to Manage Populations? 1163

Management plans must take life history strategies into account 1163

Management plans must be guided by the principles of population dynamics 1163

Human population growth has been exponential 1164

56 Species Interactions and Coevolution 1169

56.1 What Types of Interactions Do Ecologists Study? 1170

Interactions among species can be grouped into several categories 1170

Interaction types are not always clear-cut 1171

Some types of interactions result in coevolution 1171

56.2 How Do Antagonistic Interactions Evolve? 1172

Predator–prey interactions result in a range of adaptations   1172

Herbivory is a widespread interaction   1175

Parasite–host interactions may be pathogenic   1176

56.3 How Do Mutualistic Interactions Evolve?   1177

Some mutualistic partners exchange food for care or transport   1178

Some mutualistic partners exchange food or housing for defense   1178

Plants and pollinators exchange food for pollen transport   1180

Plants and frugivores exchange food for seed transport   1181

56.4 What Are the Outcomes of Competition?   1182

Competition is widespread because all species share resources   1182

Interference competition may restrict habitat use   1183

Exploitation competition may lead to coexistence   1183

Species may compete indirectly for a resource   1184

Competition may determine a species' niche   1184

57   Community Ecology   1188

57.1 What Are Ecological Communities?   1189

Energy enters communities through primary producers   1189

Consumers use diverse sources of energy   1190

Fewer individuals and less biomass can be supported at higher trophic levels   1190

Productivity and species diversity are linked   1192

57.2 How Do Interactions among Species Influence Communities?   1193

Species interactions can cause trophic cascades   1193

Keystone species have disproportionate effects on their communities   1194

57.3 What Patterns of Species Diversity Have Ecologists Observed?   1195

Diversity comprises both the number and the relative abundance of species   1195

Ecologists have observed latitudinal gradients in diversity   1196

The theory of island biogeography suggests that immigration and extinction rates determine diversity on islands   1196

57.4 How Do Disturbances Affect Ecological Communities?   1199

Succession is the predictable pattern of change in a community after a disturbance   1199

Both facilitation and inhibition influence succession   1201

Cyclical succession requires adaptation to periodic disturbances   1201

Heterotrophic succession generates distinctive communities   1202

57.5 How Does Species Richness Influence Community Stability?   1202

Species richness is associated with productivity and stability   1202

Diversity, productivity, and stability differ between natural and managed communities   1202

58   Ecosystems and Global Ecology   1207

58.1 How Does Energy Flow through the Global Ecosystem?   1208

Energy flows and chemicals cycle through ecosystems   1208

The geographic distribution of energy flow is uneven   1208

Human activities modify the flow of energy   1210

58.2 How Do Materials Move through the Global Ecosystem?   1210

Elements move between biotic and abiotic compartments of ecosystems   1211

The atmosphere contains large pools of the gases required by living organisms   1211

The terrestrial surface is influenced by slow geological processes   1213

Water transports elements among compartments   1213

Fire is a major mover of elements   1214

58.3 How Do Specific Nutrients Cycle through the Global Ecosystem?   1214

Water cycles rapidly through the ecosystem   1215

The carbon cycle has been altered by human activities   1216

The nitrogen cycle depends on both biotic and abiotic processes 1218

The burning of fossil fuels affects the sulfur cycle 1219

The global phosphorus cycle lacks a significant atmospheric component 1220

Other biogeochemical cycles are also important 1221

Biogeochemical cycles interact 1221

58.4 What Goods and Services Do Ecosystems Provide? 1223

58.5 How Can Ecosystems Be Sustainably Managed? 1224

59 Biodiversity and Conservation Biology 1228

59.1 What Is Conservation Biology? 1229

Conservation biology aims to protect and manage biodiversity 1229

Biodiversity has great value to human society 1230

59.2 How Do Conservation Biologists Predict Changes in Biodiversity? 1230

Our knowledge of biodiversity is incomplete 1230

We can predict the effects of human activities on biodiversity 1231

59.3 What Human Activities Threaten Species Persistence? 1232

Habitat losses endanger species 1233

Overexploitation has driven many species to extinction 1234

Invasive predators, competitors, and pathogens threaten many species 1235

Rapid climate change can cause species extinctions 1236

59.4 What Strategies Are Used to Protect Biodiversity? 1237

Protected areas preserve habitat and prevent overexploitation 1237

Degraded ecosystems can be restored 1237

Disturbance patterns sometimes need to be restored 1239

Ending trade is crucial to saving some species 1240

Species invasions must be controlled or prevented 1241

Biodiversity has economic value 1241

Changes in human-dominated landscapes can help protect biodiversity 1243

Captive breeding programs can maintain a few species 1244

Earth is not a ship, a spaceship, or an airplane 1244

**APPENDIX A
The Tree of Life 1248**

**APPENDIX B
Statistics Primer 1255**

**APPENDIX C
Some Measurements Used in Biology 1264**

ANSWERS TO CHAPTER REVIEW QUESTIONS A-1

GLOSSARY G-1

ILLUSTRATION CREDITS C-1

INDEX I-1

1

Studying Life

CHAPTEROUTLINE

1.1 What Is Biology?

1.2 How Do Biologists Investigate Life?

1.3 Why Does Biology Matter?

What's Happening to the Frogs? Tyrone Hayes grew up near the great Congaree Swamp in South Carolina collecting turtles, snakes, frogs, and toads. He is now a professor of biology at the University of California at Berkeley. In the laboratory and in the field, he is studying how and why populations of frogs are endangered by agricultural pesticides.

AMPHIBIANS—frogs, salamanders, and wormlike caecilians—have been around so long they watched the dinosaurs come and go. But for the last three decades, amphibian populations around the world have been declining dramatically. Today more than a third of the world's amphibian species are threatened with extinction. Why are these animals disappearing?

Tyrone Hayes, a biologist at the University of California at Berkeley, probed the effects of certain chemicals that are applied to croplands in large quantities and that accumulate in the runoff water from the fields. Hayes focused on the effects on amphibians of atrazine, a weed killer (herbicide) widely used in the United States and some other countries, where it is a common contaminant in fresh water (its use has been banned in the European Union). In the U.S., atrazine is usually applied in the spring, when many amphibians are breeding and thousands of tadpoles swim in the ditches, ponds, and streams that receive runoff from farms.

In his laboratory, Hayes and his associates raised frog tadpoles in water containing no atrazine and also in water with concentrations ranging from 0.01 parts per billion (ppb) up to 25 ppb. Concentrations as low as 0.1 ppb had a dramatic effect on tadpole development: it feminized the males. When these males became adults, their vocal structures—which are used in mating calls and thus are crucial for successful reproduction—were smaller than normal; in some, eggs were growing in the testes; some developed female sex organs. In other studies, normal adult male frogs exposed to 25 ppb had a tenfold reduction in testosterone levels and did not produce sperm. You can imagine the disastrous effects of such developmental and hormonal changes on the capacity of frogs to breed and reproduce.

But these experiments were performed in the laboratory, with a species of frog bred for laboratory use. Would the results be the same in nature? To find out, Hayes and his students traveled from Utah to Iowa, sampling water and collecting frogs. They analyzed the water for atrazine and examined the frogs. The only site where the frogs were normal was one where atrazine was undetectable. At all other sites, male frogs had abnormalities of the sex organs.

Like other biologists, Hayes made observations. He then made predictions based on those observations, and designed and carried out experiments to test his predictions.

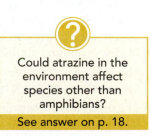

Could atrazine in the environment affect species other than amphibians?

See answer on p. 18.

(A) *Sulfolobus*

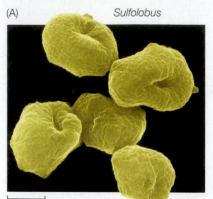

0.5 µm

(B) *Escherichia coli*

0.6 µm

(C) *Coronosphaera mediterranea*

4 µm

(D) *Passiflora quadrangularis* (passion flower)

(E) *Phallus indusiatus* (stinkhorn mushrooms)

(F) *Phymateus morbillosus* (milkweed grasshopper)

(G) *Chelonoidis nigra* (giant tortoise)   *Buteo galapagoensis* (Galápagos hawk)

1.1 The Many Faces of Life   The processes of evolution have led to the millions of diverse organisms living on Earth today. Archaea (A) and bacteria (B) are all single-celled, prokaryotic organisms, as described in Chapter 26. (C) Many protists are unicellular but, as discussed in Chapter 27, their cell structures are more complex than those of the prokaryotes. This protist has manufactured "plates" of calcium carbonate that surround and protect its single cell. (D–G) Most of the visible life on Earth is multicellular. Chapters 28 and 29 cover the green plants (D). The other broad groups of multicellular organisms are the fungi (E), discussed in Chapter 30, and the animals (F, G), covered in Chapters 31–33.

1.1 What Is Biology?

Biology is the scientific study of living things, which we call organisms (**Figure 1.1**). The living organisms we know about are all descended from a common origin of life on Earth that occurred almost 4 billion years ago. Living organisms share many characteristics that allow us to distinguish them from the nonliving world:

- Organisms are made up of a common set of chemical components, including particular carbohydrates, fatty acids, nucleic acids, and amino acids, among others.

- The building blocks of most organisms are **cells**—individual structures enclosed by plasma membranes.

- The cells of living organisms convert molecules obtained from their environment into new biological molecules.

- Cells extract energy from the environment and use it to do biological work.

- Organisms contain genetic information that uses a nearly universal code to specify the assembly of proteins.

- Organisms share similarities among a fundamental set of genes and replicate this genetic information when reproducing themselves.

- Organisms exist in populations that evolve through changes in the frequencies of genetic variants within the populations over time.

- Living organisms self-regulate their internal environments, thus maintaining the conditions that allow them to survive.

Taken together, these characteristics logically lead to the conclusion that all life has a common ancestry, and that the diverse organisms alive today all originated from one life form. If life had multiple origins, we would not expect to see the striking similarities across gene sequences, the nearly universal genetic code, or the common set of amino acids that characterizes every known living organism. Organisms from a separate origin of life—say, on another planet—might be similar in some ways to life on Earth. For example, such life forms would probably possess heritable genetic information that they could pass on to offspring. But we would not expect the details of their genetic code or the fundamental sequences of their genomes to be the same as or even similar to ours.

The list is necessarily simplified, and some forms of life may not display all of the listed characteristics all of the time. For example, the seed of a desert plant may go for many years without extracting energy from the environment, converting molecules, regulating its internal environment, or reproducing; yet the seed is alive. And there are viruses, which are not composed of cells and cannot carry out physiological functions on their own (they parasitize host cells to function for them). Yet viruses contain genetic information, and they mutate and evolve. So even though viruses are not independent cellular organisms, their existence depends on cells. In addition, it is highly probable that viruses evolved from cellular life forms. Thus most biologists consider viruses to be a part of life.

This book will explore the details of the common characteristics of life, how these characteristics arose, and how they work together to enable organisms to survive and reproduce. Not all organisms survive and reproduce with equal success, and it is through differential survival and reproduction that living systems evolve and become adapted to Earth's many environments. The processes of evolution have generated the enormous diversity of life on Earth, and evolution is a central theme of biology.

Life arose from non-life via chemical evolution

Geologists estimate that Earth formed between 4.6 and 4.5 billion years ago. At first the planet was not a very hospitable place. It was some 600 million years or more before the earliest life evolved. If we picture the 4.6-billion-year history of Earth as a 30-day month, life first appeared some time around the end of the first week (**Figure 1.2**).

When we consider how life might have arisen from nonliving matter, we must take into account the properties of the young Earth's atmosphere, oceans, and climate, all of which were very different than they are today. Biologists postulate that complex biological molecules first arose through the random physical association of chemicals in that environment. Experiments simulating the conditions on early Earth have confirmed that the generation of complex molecules under such conditions is possible, even probable. The critical step for the evolution of life, however, was the appearance of **nucleic acids**—molecules that could reproduce themselves and also serve as templates for the

1.2 Life's Timeline Depicting the 4.6 billion years of Earth's history on the scale of a 30-day month provides a sense of the immensity of evolutionary time.

synthesis of **proteins**, large molecules with complex but stable shapes. The variation in the shapes of these proteins enabled them to participate in increasing numbers and kinds of chemical reactions with other molecules. These subjects are covered in Part One of this book.

Cellular structure evolved in the common ancestor of life

Another important step in the history of life was the enclosure of complex proteins and other biological molecules by membranes that contained them in a compact internal environment separate from the surrounding (external) environment. Molecules called fatty acids played a critical role because these molecules do not dissolve in water; rather they form membranous

(A)

(B)

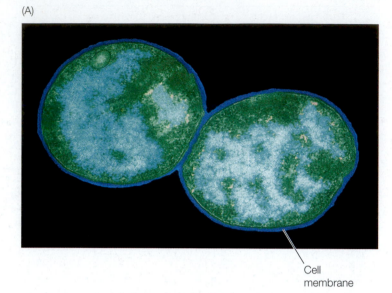

Cell
membrane

Cell
membrane

Membrane
of nucleus

Mitochondria
(membrane-enclosed)

1.3 Cells Are Building Blocks for Life   These photographs of cells were taken with a transmission electron microscope (see Figure 5.3) and enhanced with added color to highlight details. (A) Two pro-karyotic cells of an *Enterococcus* bacterium that lives in the human digestive system. Prokaryotes are unicellular organisms with genetic and biochemical material enclosed inside a single membrane. (B) A human white blood cell (lymphocyte) represents one of the many specialized cell types that make up a multicellular eukaryote. Multi-ple membranes within the cell-enclosing outer membrane segregate the different biochemical processes of eukaryotic cells.

films that, when agitated, can form spherical structures. These membranous structures could have enveloped assemblages of biological molecules. The creation of an internal environment that concentrated the reactants and products of chemical reac-tions opened up the possibility that those reactions could be integrated and controlled within a tiny cell (**Figure 1.3**). Scien-tists postulate that this natural process of membrane formation resulted in the first cells with the ability to reproduce—that is, the evolution of the first cellular organisms.

For the first few billion years of cellular life, all the organ-isms that existed were unicellular and were enclosed by a single outer membrane. Such organisms, like the bacteria that are still abundant on Earth today, are called **prokaryotes**. Two main groups of prokaryotes emerged early in life's history: the **bacteria** and **archaea**. Some representatives of each of these groups began to live in a close, interdependent relationship with one another, and eventually merged to form a third ma-jor lineage of life, the **eukaryotes**. In addition to their outer membranes, the cells of eukaryotes have internal membranes that enclose specialized organelles within their cells. Eukaryote organelles include the nucleus that contains the genetic mate-rial and the mitochondria that power the cell. The structure of prokaryote and eukaryote cells and their membranes are the subjects of Part Two.

At some point, the cells of some eukaryotes failed to sepa-rate after cell division, remaining attached to each other. Such permanent colonial aggregations of cells made it possible for some of the associated cells to specialize in certain functions, such as reproduction, while other cells specialized in other functions, such as absorbing nutrients. This **cellular specializa-tion** enabled multicellular eukaryotes to increase in size and become more efficient at gathering resources and adapting to specific environments.

Photosynthesis allows some organisms to capture energy from the sun

Living cells require energy in order to function, and the bio-chemistry of the fundamental processes of energy conversion that drive life is covered in Part Three.

To fuel their cellular **metabolism** (energy transformations), the earliest prokaryotes took in small molecules directly from their environment and broke them down to their component atoms, thus releasing and using the energy contained in the chemical bonds. Many modern prokaryotes still function this way, and they function very successfully. But about 2.5 bil-lion years ago, the emergence of **photosynthesis** changed the nature of life on Earth.

The chemical reactions of photosynthesis transform the energy of sunlight into a form of biological energy that powers the synthesis of large molecules. These large mol-ecules can then be broken down to provide metabolic en-ergy. Photosynthesis is the basis of much of life on Earth today because its energy-capturing processes provide food for other organisms. Early photosynthetic cells were prob-ably similar to present-day prokaryotes called cyanobacteria (**Figure 1.4**). Over time, photosynthetic prokaryotes became so abundant that vast quantities of oxygen gas (O_2), which is a by-product of photosynthesis, began to accumulate in the atmosphere.

During the early eons of life, there was no O_2 in Earth's at-mosphere. In fact, O_2 was poisonous to many of the prokary-otes living at that time. As O_2 levels increased, however, those

(A)

0.5 cm

(B)

Stromatolites form as small grains of sediment are cemented together by communities of microorganisms, especially cyanobacteria.

10 cm

1.4 Photosynthetic Organisms Changed Earth's Atmosphere (A) Colonies of photosynthetic cyanobacteria and other microorganisms produced structures called stromatolites that were preserved in the ancient fossil record. This section of fossilized stromatolite reveals layers representing centuries of growth. (B) Living stromatolites can still be found in appropriate environments.

organisms that *did* tolerate O_2 were able to proliferate. The abundance of O_2 opened up vast new avenues of evolution because **aerobic metabolism**—a biochemical process that uses O_2 to extract energy from nutrient molecules—is far more efficient than **anaerobic metabolism** (which does not use O_2). Aerobic metabolism allows organisms to grow larger and is used by the majority of organisms today.

Oxygen in the atmosphere also made it possible for life to move onto land. For most of life's history, UV radiation falling on Earth's surface was so intense that it destroyed any organism that was not well shielded by water. But the accumulation of photosynthetically generated O_2 in the atmosphere for more than 2 billion years gradually produced a thick layer of ozone (O_3) in the upper atmosphere. By about 500 million years ago, the ozone layer was sufficiently dense and absorbed enough of the sun's UV radiation to make it possible for organisms to leave the protection of the water and live on land.

Biological information is contained in a genetic language common to all organisms

The information that specifies what an organism will look like and how it will function—its "blueprint" for existence—is contained in the organism's **genome**: the sum total of all the DNA molecules contained in each of its cells. **DNA** (deoxyribonucleic acid) molecules are long sequences of four different subunits called **nucleotides**. The sequence of these four nucleotides contains genetic information. **Genes** are specific segments of DNA that encode the information the cell uses to create amino acids and form them into proteins (**Figure 1.5**). Protein molecules govern the chemical reactions within cells and form much of an organism's structure.

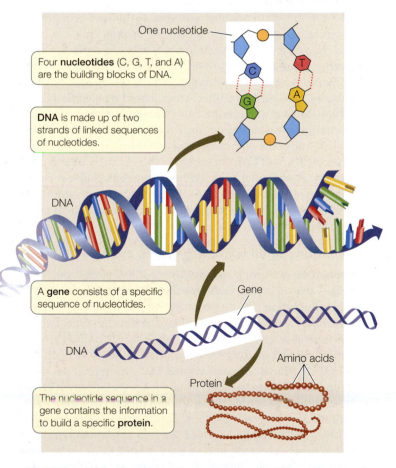

One nucleotide

Four **nucleotides** (C, G, T, and A) are the building blocks of DNA.

DNA is made up of two strands of linked sequences of nucleotides.

DNA

A **gene** consists of a specific sequence of nucleotides.

Gene

DNA

Amino acids

Protein

The nucleotide sequence in a gene contains the information to build a specific **protein**.

1.5 DNA Is Life's Blueprint The instructions for life are contained in the sequences of nucleotides in DNA molecules. Specific DNA nucleotide sequences comprise genes. The average length of a single human gene is 16,000 nucleotides. The information in each gene provides the cell with the information it needs to manufacture molecules of a specific protein.

By analogy with a book, the nucleotides of DNA are like the letters of an alphabet, and protein molecules are sentences. Combinations of proteins that form structures and control biochemical processes are the paragraphs. The structures and processes that are organized into different systems with specific tasks (such as digestion or transport) are the chapters of the book, and the complete book is the organism. If you were to write out your own genome using four letters to represent the four nucleotides, you would write more than 3 billion letters. Using the size type you are reading now, your genome would fill about 1,000 books the size of this one. The mechanisms of evolution are the authors and editors of all the books in the library of life.

All the cells of a multicellular organism contain essentially the same genome, yet different cells have different functions and form different structures—contractile proteins form in muscle cells, hemoglobin in red blood cells, digestive enzymes in gut cells, and so on. Therefore different types of cells in an organism must express different parts of the genome. How cells control gene expression in ways that enable a complex organism to develop and function is a major focus of current biological research.

The genome of an organism consists of thousands of genes. This entire genome must be replicated as new cells are produced. However, the replication process is not perfect, and a few errors, known as mutations, are likely to occur each time the genome is replicated. Mutations occur spontaneously; they can also be induced by outside factors, including chemicals and radiation. Most mutations are either harmful or have no effect, but occasionally a mutation improves the functioning of the organism under the environmental conditions it encounters.

The discovery of DNA in the latter half of the twentieth century and the subsequent elucidation of the remarkable mechanisms by which this material encodes and transmits information transformed biological science. These crucial discoveries are detailed in Parts Four and Five.

Populations of all living organisms evolve

A **population** is a group of individuals of the same type of organism—that is, of the same **species**—that interact with one another. **Evolution** acts on populations; it is the change in the genetic makeup of biological populations through time. Evolution is the major unifying principle of biology. Charles Darwin compiled factual evidence for evolution in his 1859 book *On the Origin of Species*. Darwin argued that differential survival and reproduction among individuals in a population, which he termed **natural selection**, could account for much of the evolution of life.

Although Darwin proposed that all organisms are descended from a common ancestor and therefore are related to one another, he did not have the advantage of understanding the mechanisms of genetic inheritance and mutation. Even so, he observed that offspring resembled their parents; therefore, he surmised, such mechanisms had to exist. Part Six will describe how Darwin's theory of natural selection is both supported and explained by the massive body of molecular genetic data elucidated during the twentieth century, and how these elements coincide and mesh in the modern field of evolutionary biology.

If all the organisms on Earth today are the descendants of a single kind of unicellular organism that lived almost 4 billion years ago, how have they become so different? As mentioned earlier, organisms reproduce by replicating their genomes, and mutations are introduced almost every time a genome is replicated. Some of these mutations give rise to structural and functional changes in organisms. As individuals mate with one another, the genetic variants stemming from mutation can change in frequency within a population, and the population is said to evolve.

Any population of a plant or animal species displays variation, and if you select breeding pairs on the basis of some particular trait, that trait is more likely to be present in their offspring than in the general population. Darwin himself bred pigeons, and was well aware of how pigeon fanciers selected breeding pairs to produce offspring with unusual feather patterns, beak shapes, or body sizes (see Figure 21.5). He realized that if humans could select for specific traits in domesticated animals, the same process could operate in nature; hence the term "natural selection" as opposed to artificial (human-imposed) selection.

How does natural selection function? Darwin postulated that different probabilities of survival and reproductive success would do the job. He reasoned that the reproductive capacity of plants and animals, if unchecked, would result in unlimited growth of populations, but we do not observe such growth in nature; in most species, only a small percentage of an individual's offspring will survive to reproduce. Thus any trait that confers even a small increase in the probability that its possessor will survive and reproduce would spread in the population.

Because organisms with certain traits survive and reproduce best under specific sets of conditions, natural selection leads to **adaptations**: structural, physiological, or behavioral traits that enhance an organism's chances of survival and reproduction in its environment (**Figure 1.6**). In addition to natural selection, evolutionary processes such as sexual selection (for example, selection due to mate choice) and genetic drift (the random fluctuation of gene frequencies in a population due to chance events) contribute to the rise of biodiversity. These processes operating over evolutionary history have led to the remarkable diversity of life on Earth.

Biologists can trace the evolutionary tree of life

As populations become geographically isolated from one another, they evolve differences. As populations diverge from one another, individuals in each population become less likely to reproduce with individuals of the other population. Eventually these differences between populations become so great that the two populations are considered different species. Thus species that share a fairly recent evolutionary history are generally more similar to each other than species

(A) *Dyscophus guineti*

(B) *Xenopus laevis*

(C) *Agalychnis callidryas*

(D) *Rhacophorus nigropalmatus*

1.6 Adaptations to the Environment   The limbs of frogs show adaptations to the different environments of each species. (A) This terrestrial frog walks across the ground using its short legs and peglike digits (toes). (B) Webbed rear feet are evident in this highly aquatic species of frog. (C) This arboreal species has toe pads, which are adaptations for climbing. (D) A different arboreal species has extended webbing between the toes, which increases surface area and allows the frog to glide from tree to tree.

that share an ancestor in the more distant past. By identifying, analyzing, and quantifying similarities and differences between species, biologists can construct **phylogenetic trees** that portray the evolutionary histories of the different groups of organisms.

Tens of millions of species exist on Earth today; many times that number lived in the past but are now extinct. Biologists give each of these species a distinctive scientific name formed from two Latinized names—a **binomial**. The first name identifies the species' **genus** (plural *genera*)—a group of species that share a recent common ancestor. The second is the name of the species. For example, the scientific name for the human species is *Homo sapiens*: *Homo* is our genus, *sapiens* our species. *Homo* is Latin for "man," and *sapiens* is from the Latin word for "wise" or "rational." Our closest relatives in the genus *Homo* are the Neanderthals, *Homo neanderthalensis*. Neanderthals are now extinct and are known only from their fossil remains.

Much of biology is based on comparisons among species, and these comparisons are useful precisely because we can place species in an evolutionary context relative to one another. Our ability to do this has been greatly enhanced in recent decades by our ability to sequence and compare the genomes of different species. Genome sequencing and other molecular techniques have allowed biologists to augment evolutionary knowledge based on the fossil record with a vast array of molecular evidence. The result is the ongoing compilation of phylogenetic trees that document and diagram evolutionary relationships as part of an overarching tree of life, the broadest categories of which are shown in **Figure 1.7** and will be surveyed in more detail in Part Seven. (The tree is expanded in Appendix A, and you can also explore the tree interactively.)

Although many details remain to be clarified, the broad outlines of the tree of life have been determined. Its branching patterns are based on a rich array of evidence from fossils, structures, metabolic processes, behavior, and molecular

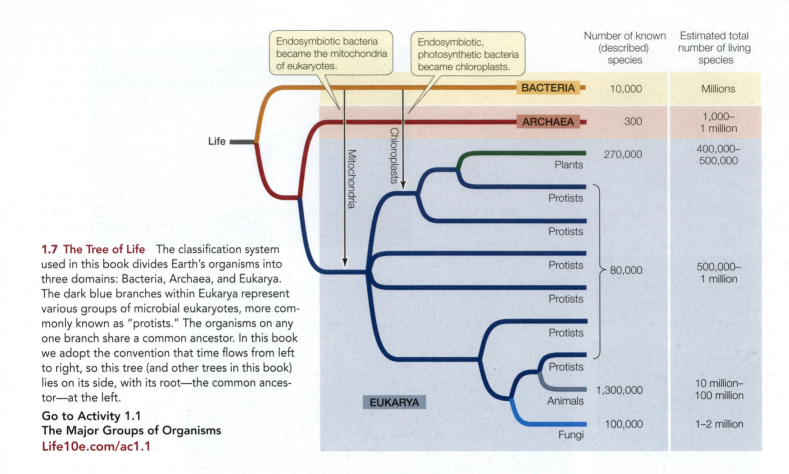

Endosymbiotic bacteria became the mitochondria of eukaryotes.

Endosymbiotic, photosynthetic bacteria became chloroplasts.

	Number of known (described) species	Estimated total number of living species
BACTERIA	10,000	Millions
ARCHAEA	300	1,000– 1 million
Plants	270,000	400,000– 500,000
Protists		
Protists		
Protists	80,000	500,000– 1 million
Protists		
Protists		
Protists		
Animals	1,300,000	10 million– 100 million
Fungi	100,000	1–2 million

Life

Mitochondria

Chloroplasts

EUKARYA

1.7 The Tree of Life The classification system used in this book divides Earth's organisms into three domains: Bacteria, Archaea, and Eukarya. The dark blue branches within Eukarya represent various groups of microbial eukaryotes, more commonly known as "protists." The organisms on any one branch share a common ancestor. In this book we adopt the convention that time flows from left to right, so this tree (and other trees in this book) lies on its side, with its root—the common ancestor—at the left.

Go to Activity 1.1
The Major Groups of Organisms
Life10e.com/ac1.1

analyses of genomes. Two of the three main domains of life—Archaea and Bacteria—are single-celled prokaryotes, as mentioned earlier in this chapter. However, members of these two groups differ so fundamentally in their metabolic processes that they are believed to have separated into distinct evolutionary lineages very early. Species belonging to the third domain—Eukarya—have eukaryotic cells whose mitochondria and chloroplasts originated from endosymbioses of bacteria.

Plants, fungi, and animals are examples of familiar multicellular eukaryotes that evolved independently, from different groups of the unicellular eukaryotes informally known as protists. We know that plants, fungi, and animals had independent origins of multicellularity because each of these three groups is most closely

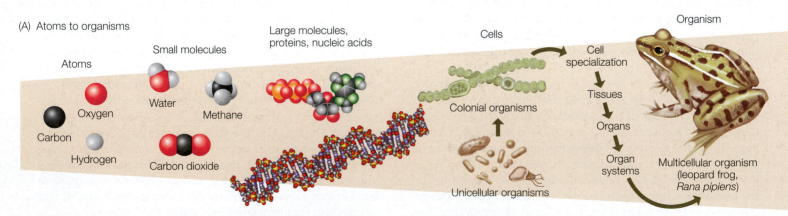

(A) Atoms to organisms

Atoms — Small molecules — Large molecules, proteins, nucleic acids — Cells — Cell specialization — Organism

Oxygen, Water, Methane, Carbon, Hydrogen, Carbon dioxide

Colonial organisms

Unicellular organisms

Tissues → Organs → Organ systems

Multicellular organism (leopard frog, *Rana pipiens*)

1.8 Biology Is Studied at Many Levels of Organization
(A) Life's properties emerge when DNA and other molecules are organized in cells, which form building blocks for organisms. (B) Organisms exist in populations and interact with other populations

to form communities, which interact with the physical environment to make up the many ecosystems of the biosphere.
Go to Activity 1.2 The Hierarchy of Life
Life10e.com/ac1.2

related to different groups of unicellular protists, as can be seen from the branching pattern of Figure 1.7.

Cellular specialization and differentiation underlie multicellular life

Looking back at Figure 1.2, you can see that for more than half of Earth's history, all life was unicellular. Unicellular species remain ubiquitous and highly successful in the present, even though the diverse multicellular organisms, owing to their much larger size, may seem to us to dominate the planet.

With the evolution of cells specialized for different functions within the same organism, these differentiated cells lost many of the functions carried out by single-celled organisms, and a **biological hierarchy** emerged (**Figure 1.8A**). To accomplish their specialized tasks, assemblages of differentiated cells are organized into **tissues**. For example, a single muscle cell cannot generate much force, but when many cells combine to form the tissue of a working muscle, considerable force and movement can be generated. Different tissue types are organized to form **organs** that accomplish specific functions. The heart, brain, and stomach are each constructed of several types of tissues, as are the roots, stems, and leaves of plants. Organs whose functions are interrelated can be grouped into **organ systems**; the esophagus, stomach, and intestines, for example, are all part of the digestive system. The physiology of two major groups of multicellular organisms (land plants and animals) is discussed in detail in Parts Eight and Nine, respectively.

Living organisms interact with one another

Organisms do not live in isolation, and the internal hierarchy of the individual organism is matched by the external hierarchy of the biological world (**Figure 1.8B**). As mentioned earlier in this section, a group of individuals of the same species that interact with one another is a population. The populations of all the species that live and interact in a defined area (areas are defined in different ways and can be small or large) are called a **community**. Communities together with their abiotic (nonliving) environment constitute an **ecosystem**.

Individuals in a population interact in many different ways. Animals eat plants and other animals (usually members of another species) and compete with other species for food and other resources. Some animals prevent other individuals of their own species from exploiting a resource, be it food, nesting sites, or mates. Animals may also cooperate with members of their own species, forming social units such as a termite colony or a flock of birds. Such interactions have resulted in the evolution of social behaviors such as communication and courtship displays.

Plants also interact with their external environment, which includes other plants, fungi, animals, and microorganisms. All terrestrial plants depend on partnerships with fungi, bacteria, and animals. Some of these partnerships are necessary to obtain nutrients, some to produce fertile seeds, and still others to disperse seeds. Plants compete with each other for light and water and have ongoing evolutionary interactions with the animals that eat them. Through time, many adaptations have evolved in plants that protect them from predation (such as thorns) or that help then attract the animals that assist in their reproduction (such as sweet nectar or colorful flowers). The interactions of populations of plant and animal species in a community are major evolutionary forces that produce specialized adaptations.

Communities interacting over a broad geographic area with distinguishing physical features form ecosystems; examples

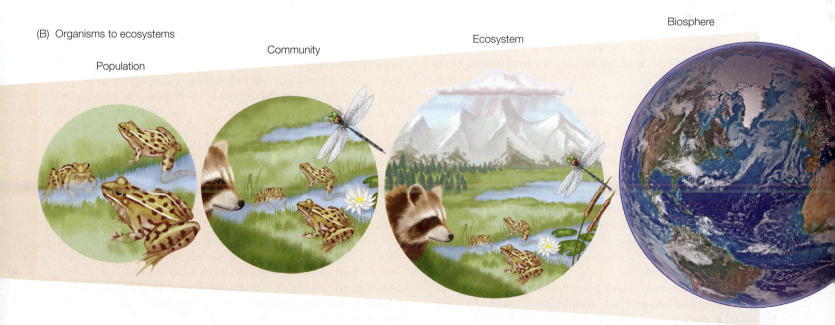

(B) Organisms to ecosystems

Population

Community

Ecosystem

Biosphere

(A) *Propithecus verreauxi*

(B) *Spermophilus parryii*

Go to Media Clip 1.1
Leaping Lemurs
Life10e.com/mc1.1

1.9 Energy Can Be Used Immediately or Stored (A) Animal cells break down food molecules and use the energy contained in the chemical bonds of those molecules to do mechanical work, such as running and jumping. This composite image of a sifaka (a type of lemur from Madagascar) shows the same individual at five stages of a single jump. (B) The cells of this Arctic ground squirrel have broken down the complex carbohydrates in the plants it consumed and converted those molecules into fats. The fats are stored in the animal's body to provide an energy supply for the cold months.

include Arctic tundra, coral reef, and tropical rainforest. The ways in which species interact with one another and with their environment in populations, communities, and ecosystems is the subject of ecology, covered in Part Ten of this book.

Nutrients supply energy and are the basis of biosynthesis

Living organisms acquire nutrients from the environment. Nutrients supply the organism with energy and raw materials for carrying out biochemical reactions. Life depends on thousands of biochemical reactions that occur inside cells. Some of these reactions break down nutrient molecules into smaller chemical units, and in the process some of the energy contained in the chemical bonds of the nutrients is captured by high-energy molecules that can be used to do different kinds of cellular work.

One obvious kind of work cells do is mechanical—moving molecules from one cellular location to another, moving whole cells or tissues, or even moving the organism itself, as muscles do (**Figure 1.9A**). The most basic cellular work is the building, or synthesis, of new complex molecules and structures from smaller chemical units. For example, we are all familiar with the fact that carbohydrates eaten today may be deposited in the body as fat tomorrow (**Figure 1.9B**). Still another kind of work is the electrical work that is the essence of information processing in nervous systems.

The myriad biochemical reactions that take place in cells are integrally linked in that the products of one reaction are the raw materials of the next. These complex networks of reactions must be integrated and precisely controlled; when they are not, the result is malfunction and disease.

Living organisms must regulate their internal environment

The specialized cells, tissues, and organ systems of multicellular organisms exist in and depend on an **internal environment** that is made up of extracellular fluids. Because this environment serves the needs of the cells, its physical and chemical composition must be maintained within a narrow range of physiological conditions that support survival and function. The maintenance of this narrow range of conditions is known as **homeostasis**. A relatively stable internal (but extracellular) environment means that cells can function efficiently even when conditions outside the organism's body become unfavorable for cellular processes.

The organism's regulatory systems obtain information from sensory cells that provide information about both the internal and external conditions the organism is subject to at a given time. The cells of regulatory systems process and integrate this information and send signals to components of physiological systems, which can change in response to these signals so that the organism's internal environment remains reasonably constant.

The concept of homeostasis extends beyond the internal environment of multicellular organisms, however. In both unicellular and multicellular organisms, individual cells must regulate physiological parameters (such as acidity and salinity), maintaining them within a range that allows those cells to survive and function. Individual cells regulate these properties through actions of the plasma membrane that encloses them and are the cell's interface with its environment (either internal or external). Thus self-regulation to maintain a more or less constant internal environment is a general attribute of all living organisms.

RECAP 1.1

All organisms are related by common descent from a single ancestral form. They contain genetic information that encodes how they look and how they function. They also reproduce, extract energy from their environment, and use energy to do biological work, synthesize complex molecules to construct biological structures, regulate their internal environment, and interact with one another.

- Why did the evolution of photosynthesis so radically affect the course of life on Earth? **See pp. 4–5**
- Describe the relationship between evolution by natural selection and the genetic code. **See p. 6**
- What information have biologists used to construct a tree of life? **See pp. 6–8 and Figure 1.7**
- What do we mean by "homeostasis," and why is it crucial to living organisms? **See p. 10**

The preceding section briefly outlined the major features of life—features that will be covered in depth in subsequent chapters of this book. Before going into the details of what we know about life, however, it is important to understand how scientists obtain information and how they use that information in broadening our understanding of Earth's diverse living organisms and putting this understanding to practical use.

1.2 How Do Biologists Investigate Life?

Scientific investigations are based on observation, data, experimentation, and logic. Scientists use many different tools and methods in making observations, collecting data, designing experiments, and applying logic, but they are always guided by established principles that allow us to discover new aspects about the structure, function, evolution, and interactions of organisms.

Observing and quantifying are important skills

Biologists have always observed the world around them, but today our ability to observe is greatly enhanced by technologies such as electron microscopes, rapid genome sequencing, magnetic resonance imaging, and global positioning satellites. These technologies allow us to observe everything from the distribution of molecules in the body to the movement of animals across continents and oceans.

Observation is a basic tool of biology, but as scientists we must also be able to quantify the information, or **data**, we collect as we observe. Whether we are testing a new drug or mapping the migrations of the great whales, applying mathematical and statistical calculations to the data we collect is essential. For example, biologists once classified organisms based entirely on qualitative descriptions of the physical differences among them. There was no way of objectively determining evolutionary relationships of organisms, and biologists had to depend on the fossil record for insight. Today our ability to quantify the molecular and physical differences among species, combined with explicit mathematical models

of the evolutionary process, enables quantitative analyses of evolutionary history. These mathematical calculations, in turn, facilitate comparative investigations of all other aspects of an organism's biology.

Scientific methods combine observation, experimentation, and logic

Textbooks often describe "*the* scientific method," as if there is a single, simple flow chart that all scientists follow. This is an oversimplification. Although flow charts such as the one shown in **Figure 1.10** incorporate much of what scientists do, you should not conclude that scientists necessarily progress through the steps of the process in one prescribed, linear order.

Observations lead to questions, and scientists make additional observations and often do experiments to answer those

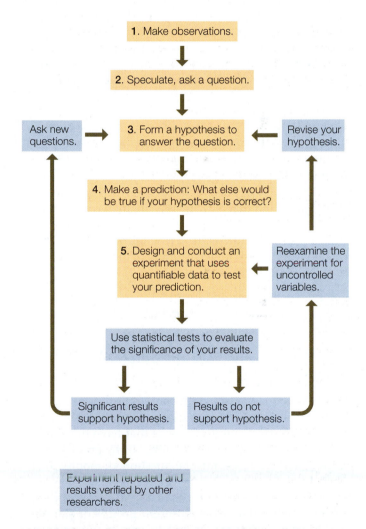

1.10 Scientific Methodology   The process of observation, speculation, hypothesis, prediction, and experimentation is a cornerstone of modern science, although scientists may initiate their research at several different points. Answers gleaned through experimentation lead to new questions, more hypotheses, further experiments, and expanding knowledge.

questions. This hypothesis–prediction approach traditionally has five steps: (1) making observations; (2) asking questions; (3) forming hypotheses, which are tentative answers to the questions; (4) making predictions based on the hypotheses; and (5) testing the predictions by making additional observations or conducting experiments.

After posing a question, a scientist often uses **inductive logic** to propose a tentative answer. Inductive logic involves taking observations or facts and creating a new proposition that is compatible with those observations or facts. Such a tentative proposition is a **hypothesis** (plural *hypotheses*). In formulating a hypothesis, scientists put together the facts and data at their disposal to formulate one or more possible answers to the question. For example, at the opening of this chapter you learned that scientists have observed the rapid decline of amphibian populations worldwide and are asking why. Some scientists have hypothesized that a fungal disease is a cause; other scientists have hypothesized that increased exposure to ultraviolet radiation is a cause. Tyrone Hayes hypothesized that exposure to agricultural chemicals, specifically the widely used herbicide atrazine, could be a cause.

The next step in the scientific method is to apply a different form of logic—**deductive logic**—that starts with a statement believed to be true (the hypothesis) and then goes on to predict what facts would also have to be true to be compatible with that statement. Hayes knew that atrazine is commonly applied in the spring, when amphibians are breeding, and that atrazine is a common contaminant in the waters in which amphibians live as they develop into adults. Thus he predicted that frog tadpoles exposed to atrazine would show adverse effects of the chemical once they reached adulthood.

Go to Animated Tutorial 1.1
Using Scientific Methodology
Life10e.com/at1.1

Good experiments have the potential to falsify hypotheses

Once predictions are made from a hypothesis, experiments can be designed to test those predictions. The most informative experiments are those that have the ability to show that the prediction is wrong. If the prediction is wrong, the hypothesis must be questioned, modified, or rejected.

There are two general types of experiments, both of which compare data from different groups or samples. A **controlled experiment** manipulates one or more of the factors being tested; **comparative experiments** compare unmanipulated data gathered from different sources. As described at the opening of this chapter, Tyrone Hayes and his colleagues conducted both types of experiments to test the prediction that the herbicide atrazine, a contaminant in freshwater ponds and streams throughout the world, affects the development of frogs.

INVESTIGATING**LIFE**

1.11 Controlled Experiments Manipulate a Variable   The Hayes laboratory created controlled environments that differed only in the concentrations of atrazine in the water. Eggs from leopard frogs (*Rana pipiens*) raised specifically for laboratory use were allowed to hatch and the tadpoles were separated into experimental tanks containing water with different concentrations of atrazine.[a]

HYPOTHESIS   Exposure to atrazine during larval development causes abnormalities in the reproductive tissues of male frogs.

Method
1. Establish 9 tanks in which all attributes are held constant except the water's atrazine concentration. Establish 3 atrazine conditions (3 replicate tanks per condition): 0 ppb (control condition), 0.1 ppb, and 25 ppb.
2. Place *Rana pipiens* tadpoles from laboratory-reared eggs in the 9 tanks (30 tadpoles per replicate).
3. When tadpoles have transitioned into adults, sacrifice the animals and evaluate their reproductive tissues.
4. Test for correlation of degree of atrazine exposure with the presence of abnormalities in the gonads (testes) of male frogs.

Results

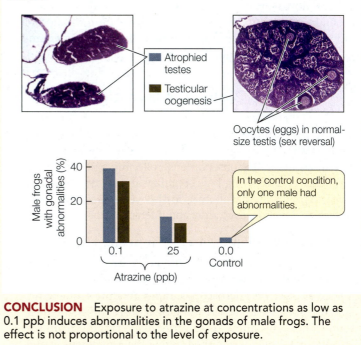

CONCLUSION   Exposure to atrazine at concentrations as low as 0.1 ppb induces abnormalities in the gonads of male frogs. The effect is not proportional to the level of exposure.

Go to **BioPortal** for discussion and relevant links for all INVESTIGATING**LIFE** figures.

[a]Hayes, T. et al. 2003. *Environmental Health Perspectives III*: 568–575.

In a controlled experiment, we start with groups or samples that are as similar as possible. We predict on the basis of our hypothesis that some critical factor, or variable, has an effect on the phenomenon we are investigating. We devise some method to manipulate *only that variable* in an "experimental" group and compare the resulting data with data from an unmanipulated "control" group. If the predicted difference occurs, we then apply statistical tests to ascertain the probability that the manipulation created the difference (as opposed to the difference being the result of random

INVESTIGATING**LIFE**

1.12 Comparative Experiments Look for Differences among Groups To see whether the presence of atrazine correlates with testicular abnormalities in male frogs, the Hayes lab collected frogs and water samples from different locations around the U.S. The analysis that followed was "blind," meaning that the frogs and water samples were coded so that experimenters working with each specimen did not know which site the specimen came from.[a]

HYPOTHESIS Presence of the herbicide atrazine in environmental water correlates with gonadal abnormalities in frog populations.

Method
1. Based on commercial sales of atrazine, select 4 sites (sites 1–4) less likely and 4 sites (sites 5–8) more likely to be contaminated with atrazine.
2. Visit all sites in the spring (i.e., when frogs have transitioned from tadpoles into adults); collect frogs and water samples.
3. In the laboratory, sacrifice frogs and examine their reproductive tissues, documenting abnormalities.
4. Analyze the water samples for atrazine concentration (the sample for site 7 was not tested).
5. Quantify and correlate the incidence of reproductive abnormalities with environmental atrazine concentrations.

Results

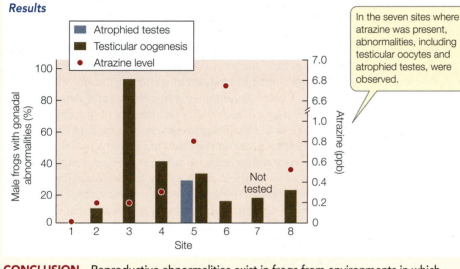

In the seven sites where atrazine was present, abnormalities, including testicular oocytes and atrophied testes, were observed.

CONCLUSION Reproductive abnormalities exist in frogs from environments in which aqueous atrazine concentration is 0.2 ppb or above. The incidence of abnormalities does not appear to be proportional to atrazine concentration at the time of transition to adulthood.

Go to **BioPortal** for discussion and relevant links for all INVESTIGATING**LIFE** figures.

[a]Hayes, T. et al. 2003. *Nature* 419: 895–896.

chance). **Figure 1.11** describes one of the many controlled experiments performed by the Hayes laboratory to quantify the effects of atrazine on male frogs.

The basis of controlled experiments is that one variable is manipulated while all others are held constant. The variable that is manipulated is called the **independent variable**, and the response that is measured is the **dependent variable**. A good controlled experiment is not easy to design because biological variables are so interrelated that it is difficult to alter just one.

A comparative experiment starts with the prediction that there will be a difference between samples or groups based on the hypothesis. In comparative experiments, however, we cannot control the variables; often we cannot even identify all the variables that are present. We are simply gathering and comparing data from different sample groups.

When his controlled experiments indicated that atrazine indeed affects reproductive development in frogs, Hayes and his colleagues performed a comparative experiment. They collected frogs and water samples from eight widely separated sites across the United States and compared the incidence of abnormal frogs from environments with very different levels of atrazine (**Figure 1.12**). Of course, the sample sites differed in many ways besides the level of atrazine present.

The results of experiments frequently reveal that the situation is more complex than the hypothesis anticipated, thus raising new questions. In the Hayes experiments, for example, there was no clear direct relationship between the *amount* of atrazine present and the percentage of abnormal frogs: there were fewer abnormal frogs at the highest concentrations of atrazine than at lower concentrations. There are no "final answers" in science. Investigations consistently reveal more complexity than we expect, so scientists must design systematic approaches to identify, assess, and understand that complexity.

Statistical methods are essential scientific tools

Whether we do comparative or controlled experiments, at the end we have to decide whether there is a difference between the samples, individuals, groups, or populations in the study. How do we decide whether a measured difference is enough to support or falsify a hypothesis? In other words, how do we decide in an unbiased, objective way that the measured difference is significant?

Significance can be measured with statistical methods. Scientists use statistics because they recognize that variation is always present in any set of measurements. Statistical tests calculate the probability that the differences observed in an experiment could be due to random variation. The results of statistical tests are therefore probabilities. A statistical test starts with a **null hypothesis**—the premise that any observed differences are simply the result of random differences that arise from drawing two finite samples from the same population. When quantified observations, or data, are collected, statistical methods are applied to those data to calculate the likelihood that the null hypothesis is correct.

More specifically, statistical methods tell us the probability of obtaining the same results by chance even if the null hypothesis were true. *We need to eliminate, insofar as possible, the chance that any differences showing up in the data are merely the result of random variation in the samples tested.* Scientists generally conclude that the differences they measure are significant if statistical tests show that the probability of error (that is, the probability that a difference as large as the one observed could be obtained by mere chance) is 5 percent or lower, although more stringent levels of significance may be set for some problems. Appendix B of this book is a short primer on statistical methods that you can refer to as you analyze data that will be presented throughout the text.

Discoveries in biology can be generalized

Because all life is related by descent from a common ancestor, shares a genetic code, and consists of similar biochemical building blocks, knowledge gained from investigations of one type of organism can, with thought and care, be generalized to other organisms. Biologists use **model systems** for research, knowing that they can extend their findings from such systems to other organisms. For example, our basic understanding of the chemical reactions in cells came from research on bacteria but is applicable to all cells, including those of humans. Similarly, the biochemistry of photosynthesis—the process by which all green plants use sunlight to produce biological molecules—was largely worked out from experiments on *Chlorella,* a unicellular green alga. Much of what we know about the genes that control plant development is the result of work on *Arabidopsis thaliana,* a relative of the mustard plant. Knowledge about how animals, including humans, develop has come from work on sea urchins, frogs, chickens, roundworms, mice, and fruit flies. Being able to generalize from model systems is a powerful tool in biology.

Not all forms of inquiry are scientific

Science is a unique human endeavor that has certain standards of practice. Other areas of scholarship share with science the practice of making observations and asking questions, but scientists are distinguished by *what they do with their observations* and *how they frame the answers.* Quantifiable data, subjected to appropriate statistical analysis, are critical in evaluating hypotheses (the Working with Data exercises you will find throughout this book are intended to reinforce this way of thinking). In short, scientific observation and evaluation is the most powerful approach humans have devised for learning about the world and how it works.

Scientific explanations for natural processes are objective and reliable because *a hypothesis must be testable* and *a hypothesis must have the potential of being rejected* by direct observations and experiments. Scientists must clearly describe the methods they use to test hypotheses so that other scientists can repeat their results. Not all experiments are repeated, but surprising or controversial results are always subjected to independent verification. Scientists worldwide share this process of testing and rejecting hypotheses, contributing to a common body of scientific knowledge.

If you understand the methods of science, you can distinguish science from non-science. Art, music, and literature all contribute to the quality of human life, but they are not science.

They do not use scientific methods to establish what is fact. Religion is not science, although religions have historically attempted to explain natural events ranging from unusual weather patterns to crop failures to human diseases. Most such phenomena that at one time were mysterious can now be explained in terms of scientific principles. Fundamental tenets of religious faith, such as the existence of a supreme deity or deities, cannot be confirmed or refuted by experimentation and are thus outside the realm of science.

The power of science derives from strict objectivity and absolute dependence on evidence based on *reproducible and quantifiable observations.* A religious or spiritual explanation of a natural phenomenon may be coherent and satisfying for the person holding that view, but it is not testable and therefore it is not science. To invoke a supernatural explanation (such as a "creator" or "intelligent designer" with no known bounds) is to depart from the world of science. Science does not necessarily say that religious beliefs are wrong; they are simply not part of the world of science, and many religious beliefs are untestable using scientific methods.

Science describes how the world works. It is silent on the question of how the world "ought to be." Many scientific advances that contribute to human welfare also raise major ethical issues. Recent developments in genetics and developmental biology may enable us to select the sex of our children, to use stem cells to repair our bodies, and to modify the human genome. Although scientific knowledge allows us to do these things, science cannot tell us whether or not we *should* do so or, if we choose to do them, how we should regulate them. Such issues are as crucial to human society as the science itself, and a responsible scientist does not lose sight of these questions or neglect the contributions of the humanities or social sciences in attempting to come to grips with them.

RECAP **1.2**

Scientific methods of inquiry start with the formulation of hypotheses based on observations and data. Comparative and controlled experiments are carried out to test hypotheses.

- Explain the relationship between a hypothesis and an experiment. **See pp. 11–12 and Figure 1.10**
- What is controlled in a controlled experiment? **See pp. 11–12 and Figure 1.11**
- What features characterize questions that can be answered only by using a comparative approach? **See p. 13 and Figure 1.12**
- Explain why arguments must be supported by quantifiable and reproducible data in order to be considered scientific. **See pp. 13–14**
- Why can the results of biological research on one species often be generalized to very different species? **See p. 14**

The vast body of scientific knowledge accumulated over centuries of human civilization allows us to understand and manipulate aspects of the natural world in ways that no other species can. These abilities present us with challenges, opportunities, and above all, responsibilities.

1.3 Why Does Biology Matter?

Human beings exist in and depend on a world of living organisms. The oxygen in the air we breathe is produced by photosynthesis conducted by countless billions of individual organisms. The food that fuels our bodies comes from the tissues of other living organism. The fuels that drive our cars and power our electric plants are, for the most part, various forms of carbon molecules produced by living organisms—mostly millions of years ago. Inside and out, our bodies are covered in complex communities of living unicellular organisms, most of which help us maintain our health. There are also harmful species that invade our bodies and can cause mild to serious diseases, or even death. These interactions with other species are not limited to humans. Ecosystem function depends on thousands of complex interactions among the millions of species that inhabit Earth. In other words, understanding biological principles is essential to our lives and for maintaining the functioning of Earth as we know it and depend on it.

Modern agriculture depends on biology

Agriculture represents some of the earliest human applications of biological principles. Even in prehistoric times, farmers selected the most productive or otherwise favorable plants and animals to use as seed stock for propagation, and over generations farmers continued and refined these practices. His knowledge of this kind of artificial selection helped Charles Darwin understand the importance of natural selection in evolution across all of life.

In modern times, increasing knowledge of plant biology has transformed agriculture in many ways and has resulted in huge boosts in food production (**Figure 1.13**), which in turn has allowed the planet to support a far larger human population than it once could have. Over the past few decades, detailed knowledge of the genomes of many domestic species and the development of technology for directly recombining genes have allowed biologists to develop new breeds and strains of animals, plants, and fungi of agricultural interest. For example, new strains of crop plants are being developed that are resistant to pests or can tolerate drought. Moreover, understanding evolutionary theory allows biologists to devise strategies for the application of pesticides that minimize the evolution of pest resistance. And better understanding of plant–fungus relationships results in better plant health and higher productivity. These are just a few of the many ways that biology continues to inform and improve agricultural practice.

Biology is the basis of medical practice

People have speculated about the causes of diseases and searched for methods to combat them since ancient times. Long before the microbial causes of many diseases were known, people recognized that infections could be passed from one person to another, and the isolation of infected persons has been practiced as long as written records have been available.

Modern biological research informs us about how living organisms work, and about why they develop the problems and

1.13 A Green Revolution The agricultural advancements of the last 100 years have vastly increased yields and nutritional value of crops such as grains that sustain the expanding human population. In the last 30 years, these advancements have included genetic recombination techniques. Here a researcher with the U.S. Department of Agriculture works with a strain of "supernutritious" rice that provides high levels of the amino acid lysine.

infections that we call disease. In addition to diseases caused by infection of other organisms, we now know that many diseases are genetic—meaning that variants of genes in our genomes cause particular problems in the way we function. Developing appropriate treatments or cures for diseases depends on understanding the origin, basis, and effects of these diseases, as well understanding the consequences of any changes that we make. For example, the recent resurgence of tuberculosis is the result of the evolution of bacteria that are resistant to antibiotics. Dealing with future tuberculosis epidemics requires understanding aspects of molecular biology, physiology, microbial ecology, and evolution—in other words, many of the general principles of modern biology.

Many of the microbial organisms that are periodically epidemic in human populations have short generation times and high mutation rates. For example, we need yearly vaccines for flu because of the high rate of evolution of influenza viruses, the causative agent of flu. Evolutionary principles help us understand how influenza viruses are changing, and can even help us predict which strains of influenza virus are likely to lead to future flu epidemics. This medical understanding—which combines an application of molecular biology, evolutionary

1.14 Medical Applications of Biology Improve Human Health
Vaccination to prevent disease is a biologically based medical
practice that began in the eighteenth century. Today evolutionary
biology and genomics provide the basis for constant updates to
vaccines that protect humans from virus-borne diseases such as flu.
In the developed world, vaccinations have become so common-
place that some are offered on a "drive-through" basis.

theory, and basic principles of ecology—allows medical researchers to develop effective vaccines and other strategies for
the control of major epidemics (**Figure 1.14**).

Biology can inform public policy

Thanks to the deciphering of genomes and our newfound ability to manipulate them, vast new possibilities now exist for
controlling human diseases and increasing agricultural productivity—but these capabilities raise ethical and policy issues. How much and in what ways should we tinker with the
genes of humans and other species? Does it matter whether
the genomes of our crop plants and domesticated animals are
changed by traditional methods of controlled breeding and
crossbreeding or by the biotechnology of gene transfer? What
rules should govern the release of genetically modified organisms into the environment? Science alone cannot provide all
the answers, but wise policy decisions must be based on accurate scientific information.

Biologists are increasingly called on to advise government
agencies concerning the laws, rules, and regulations by which
society deals with the increasing number of challenges that
have a biological basis. As an example of the value of scientific
knowledge for the assessment and formulation of public policy, consider a management problem. Scientists and fishermen
have long known that Atlantic bluefin tuna (*Thunnus thynnus*)
have a western breeding ground in the Gulf of Mexico and
an eastern breeding ground in the Mediterranean Sea (**Figure
1.15**). Overfishing led to declining numbers of bluefin tuna,

(A)

1.15 Bluefin Tuna Do Not Recognize Boundaries   (A) Marine biologist Barbara
Block attaches computerized data-recording tracking tags to a live bluefin tuna
before returning it to the Atlantic Ocean, where its travels will be monitored.
(B) At one time we assumed that bluefins from western- and eastern-breeding
populations also fed on their respective sides of the Atlantic, so separate fishing
quotas for each side (dashed line) in an attempt to speed recovery of the endangered western population. Now, however, tracking data have shown that the
two populations *do not* remain separate after spawning, so in fact the arbitrary
boundary and quotas do not protect the endangered population.

(B)

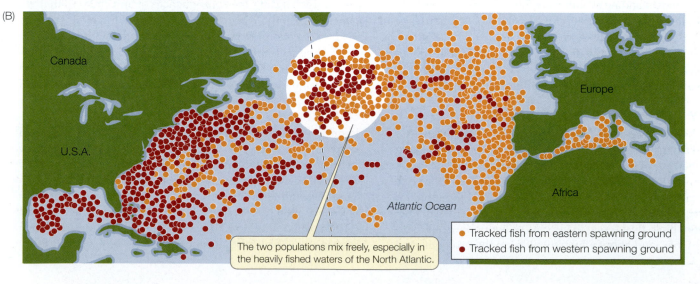

(A) 1941

Riggs
Glacier

Muir
Glacier

(B) 2004

Riggs
Glacier

1.16 A Warmer World Earth's climate has been steadily warming for the last 150 years. The rate of this warming trend has also steadily increased, resulting in the rapid melting of polar ice caps, glaciers, and alpine (mountaintop) snow and ice. This photograph shows the effects of 63 years of climate change on two ancient, longstanding glaciers in Alaska. Over that time, Muir Glacier retreated some 7 kilometers and can no longer be seen from the original vantage point. Understanding how biological populations respond to such change requires integration of biological principles from molecular biology to ecosystem ecology.

especially in the western-breeding populations, to the point of these populations being endangered.

Initially it was assumed by scientists, fishermen, and policy makers alike that the eastern and western populations had geographically separate feeding grounds as well as separate breeding grounds. Acting on this assumption, an international commission drew a line down the middle of the Atlantic Ocean and established strict fishing quotas on the western side of the line, with the intent of allowing the western population to recover. Modern tracking data, however, revealed that in fact the eastern and western bluefin populations mix freely on their feeding grounds across the entire North Atlantic—a swath of ocean that includes the most heavily fished waters in the world. Tuna caught on the eastern side of the line could just as likely be from the western breeding population as the eastern; thus the established policy could not achieve its intended goal.

Policy makers take more things into consideration than scientific knowledge and recommendations. For example, studies on the effects of atrazine on amphibians have led one U.S. group, the Natural Resources Defense Council, to take legal action to have atrazine banned on the basis of the Endangered Species Act. The U.S. Environmental Protection Agency, however, must also consider the potential loss to agriculture that such a ban would create and thus has continued to approve atrazine's use as long as environmental levels do not exceed 30 to 40 ppb—which is 300 to 400 times the levels shown to induce abnormalities in the Hayes studies. Scientific conclusions do not always prevail in the political world. Some scientific conclusions may have more influence than others, however, especially when they indicate a strong possibility of negative effects on humans.

Biology is crucial for understanding ecosystems

The world has been changing since its formation and continues to change with every passing day. Human activity, however, is resulting in an unprecedented *rate* of change in the world's ecosystems. For example, the mining and consumption of fossil fuels is releasing massive quantities of carbon dioxide into Earth's atmosphere. This anthropogenic (human-generated) increase in atmospheric carbon dioxide is largely responsible for the rapid rate of climate warming recorded over the last 50 years (**Figure 1.16**).

Our use of natural resources is putting stress on the ability of Earth's ecosystems to continue to produce the goods and services on which our society depends. Human activities are changing global climates at an unprecedented rate and are leading to the extinctions of large numbers of species (such as the amphibians featured in this chapter). The modern, warmer world is also experiencing the spread of new diseases and the resurgence of old ones. Biological knowledge is vital for determining the causes of these changes and for devising policies to deal with them.

Biology helps us understand and appreciate biodiversity

Beyond issues of policy and pragmatism lies the human "need to know." Humans are fascinated by the richness and diversity of life, and most people want to know more about organisms and how they interact. Human curiosity might even be seen as an adaptive trait—it is possible that such a trait could have been selected for if individuals who were motivated to learn about their surroundings were likely to have survived and reproduced better, on average, than their less curious relatives.

1.17 Discovering Life on Earth   These biologists are collecting insects in the top boughs of a spruce tree in the Carmanah Valley of Vancouver, Canada. Biologists estimate that the number of species discovered to date is only a small percentage of the number of species that inhabit Earth. To fill this gap in our knowledge, biologists around the world are applying thorough sampling techniques and new genetic tools to document and understand the Earth's biodiversity.

Far from ending the process, new discoveries and greater knowledge typically engender questions no one thought to ask before. There are vast numbers of questions for which we do not yet have answers, and the most important motivator of most scientists is curiosity.

Observing the living world motivates many biologists to learn more and to constantly collect new information (**Figure 1.17**). An intimate understanding of the **natural history** of a group of organisms—that is, how those organisms get their food, reproduce, behave, regulate their internal environments, and interact with other organisms—facilitates observations and provides a stronger basis for framing hypotheses about about those observations. The more information biologists have and the more the observer knows about general principles, the more he or she is likely to gain new insights from observing nature.

Most humans engage in activities that depend on biodiversity. You may be an avid birdwatcher, or enjoy gardening, or seek out particular species if you hunt or fish. Some people like to observe or collect butterflies, or mushrooms, or other groups of plants, animals, and fungi. Displays of spring wildflowers bring out throngs of human viewers in many areas of the world. Hiking and camping in natural areas full of diverse species are activities enjoyed by millions. All of these interests support the growing industry of eco-tourism, which depends on the observation of rare or unusual species. Learning about biology greatly increases our enjoyment of these activities.

RECAP 1.3

Biology informs us about the structure, processes, and interactions of the living organisms that make up our world. Informed decisions about food and energy production, health, and our environment depend on biological knowledge. Biology also addresses the human need to understand the world around us, and helps us appreciate the diverse planet we call home.

- Describe an example of how modern biology is applied to agriculture. **See p. 15**

- Why are some antibiotics not as effective for treating bacterial diseases as they were when the drugs were originally introduced? **See p. 15**

- What is an example of a biological problem that is directly related to global climate change? **See p. 17**

This chapter has provided a brief roadmap of the rest of the book. Thinking about the principles outlined here may help you to clarify and make sense of the pages of detailed description to come. At the end of the course you may wish to revisit Chapter 1 and see if you have a different perspective on the world of biology.

?

Could atrazine in the environment affect species other than amphibians?

ANSWER

An important aspect of the scientific process is the replication of experimental results. In some cases the exact same experiment is repeated in another laboratory by other investigators and the results are compared. In other cases the experiment is repeated on other species to test the generality of the findings.

Following the publications by Hayes and his students, other investigators tested the effects of atrazine on other species of amphibians as well as on vertebrates other than amphibians. Feminizing effects of atrazine have now been demonstrated in fish, reptiles, and mammals. These results are not surprising, because as you will learn in Chapters 41 and 43, the hormonal controls of sex development and function are the same, and therefore the effects of atrazine should generalize to other vertebrate species.

Biologists have now studied the molecular mechanisms of the effects of atrazine on the hormonal control of sex and found that very similar responses to atrazine are seen in fish and in cultures of human cells. So atrazine in the environment is increasingly a concern for the health of many other species—and that includes humans.

CHAPTER**SUMMARY** 1

1.1 What Is Biology?

- **Biology** is the scientific study of living organisms, including their characteristics, functions, and interactions.

- All living organisms are related to one another through common descent. Shared features of all living organisms, such as specific chemical building blocks, a nearly universal genetic code, and sequence similarities across fundamental genes, support the common ancestry of life.

- Cells evolved early in the history of life. **Cellular specialization** allowed multicellular organisms to increase in size and diversity. **Review Figure 1.2**

- The instructions for a cell are contained in its **genome**, which consists of **DNA** molecules made up of sequences of **nucleotides**. Specific segments of DNA called **genes** contain the information the cell uses to make **proteins**. **Review Figure 1.5**

- **Photosynthesis** provided a means of capturing energy directly from sunlight and over time changed Earth's atmosphere.

- **Evolution**—change in the genetic makeup of biological **populations** through time—is a fundamental principle of life. Populations evolve through several different processes, including **natural selection**, which is responsible for the diversity of **adaptations** found in living organisms.

- Biologists use fossils, anatomical similarities and differences, and molecular comparisons of genomes to reconstruct the history of life. Three domains—**Bacteria**, **Archea**, and **Eukarya**—represent the major divisions, which were established very early in life's history. **Review Figure 1.7, ACTIVITY 1.1**

- Life can be studied at different levels of organization within a **biological hierarchy**. The specialized cells of multicellular organisms are organized into **tissues**, **organs**, and **organ systems**. Individual organisms form populations and interact with other organisms of their own and other species. The populations that live and interact in a defined area form a **community**, and communities together with their abiotic (nonliving) environment constitute an **ecosystem**. **Review Figure 1.9, ACTIVITY 1.2**

- Living organisms, whether unicellular or multicellular, must regulate their internal environment to maintain **homeostasis**, the range of physical conditions necessary for their survival and function.

1.2 How Do Biologists Investigate Life?

- Scientific methods combine observation, gathering information (**data**), experimentation, and logic to study the natural world. Many scientific investigations involve five steps: making observations, asking questions, forming hypotheses, making predictions, and testing those predictions. **Review Figure 1.10**

- **Hypotheses** are tentative answers to questions. Predictions made on the basis of a hypothesis are tested with additional observations and two kinds of experiments, **comparative** and **controlled experiments**. **Review Figures 1.11, 1.12, ANIMATED TUTORIAL 1.1**

- Quantifiable data are critical in evaluating hypotheses. Statistical methods are applied to quantitative data to establish whether or not the differences observed could be the result of chance. These methods start with the **null hypothesis** that there are no differences. **See Appendix B**

- Biological knowledge obtained from a **model system** may be generalized to other species.

1.3 Why Does Biology Matter?

- Application of biological knowledge is responsible for vastly increased agricultural production.

- Understanding and treatment of human disease requires an integration of a wide range of biological principles, from molecular biology through cell biology, physiology, evolution, and ecology.

- Biologists are often called on to advise government agencies on the solution of important problems that have a biological component.

- Biology is increasing important for understanding how organisms interact in a rapidly changing world.

- Biology helps us understand and appreciate the diverse living world.

Go to the Interactive Summary to review key figures, Animated Tutorials, and Activities
Life10e.com/is1

CHAPTER**REVIEW**

REMEMBERING

1. Which of the following is *not* an attribute common to all living organisms?
   a. They are made up of a common set of chemical components, including particular nucleic and amino acids.
   b. They contain genetic information that uses a nearly universal code to specify the assembly of proteins.
   c. They share sequence similarities among their genes.
   d. They exist in populations that evolve over time.
   e. They extract energy from the sun in a process called photosynthesis.

2. In describing the hierarchy of life, which of the following descriptions of relationships is *not* accurate?
   a. An organ is a structure consisting of different types of cells and tissues.
   b. A population consists of all of the different animals in a particular type of environment.
   c. An ecosystem includes different communities.
   d. A tissue consists of a particular type of cells.
   e. A community consists of populations of different species.

3. Which of the following is a property of a good hypothesis?
   a. It is a statement of facts.
   b. It is general enough to explain a variety of possible experimental outcomes.
   c. It is independent of any observations.
   d. It explains things that are not addressable by experimentation.
   e. It can be falsified by experiments.

4. Which of the following events was most directly responsible for increasing oxygen in Earth's atmosphere?
   a. The cooling of the planet
   b. The origin of eukaryotes
   c. The origin of multicellularity
   d. The origin of photosynthesis
   e. The origin of prokaryotes

5. Which of the following is a reason to use statistics to evaluate data?
   a. It enables you to prove that your hypothesis is correct.
   b. It enables you to exclude data that do not fit your hypothesis.
   c. It makes it possible to exclude the null hypothesis.
   d. It enables you to predict experimental results.
   e. It accounts for variation in scientific measurements.

UNDERSTANDING & APPLYING

6. Why is it important in science to design and perform experiments that are capable of falsifying a hypothesis?

7. What is the significance of the fact that mitochondria and chloroplasts contain the DNA that instructs their form and function?

8. The results in Dr. Hayes's comparative experiments were more variable than the results from his controlled experiments. How would you explain this?

ANALYZING & EVALUATING

9. Biologists can now isolate genes from organisms and decode their DNA. When the nucleotide sequences from the same gene in different species are compared, differences are discovered. How could you use those data to deduce the evolutionary relationships among the organisms in your comparison?

10. Mitochondria are cell organelles that have their own DNA and replicate independently of the cell itself. In most organisms, mitochondria are inherited only from the mother. Based on this observation, when might it be advantageous or disadvantageous to use mitochondrial DNA rather than nuclear DNA for studying evolutionary relationships among populations?

Go to BioPortal at **yourBioPortal.com** for Animated Tutorials, Activities, LearningCurve Quizzes, Flashcards, and many other study and review resources.

2 Small Molecules and the Chemistry of Life

CHAPTER**OUTLINE**

2.1 How Does Atomic Structure Explain the Properties of Matter?

2.2 How Do Atoms Bond to Form Molecules?

2.3 How Do Atoms Change Partners in Chemical Reactions?

2.4 What Makes Water So Important for Life?

Big Teeth   Isotopes in *Camarasaurus* teeth yield clues about the behavior of these huge dinosaurs—150 million years after the last of them disappeared.

"**Y**OU ARE WHAT YOU EAT—and that applies to teeth" is a modification of a famous saying about body chemistry. As we pointed out in Chapter 1, living things are made up of the same kinds of atoms that make up the inanimate universe. One of these atoms is oxygen (O), which is part of water (H_2O). Oxygen has two naturally occurring variants called isotopes; they have the same chemical properties but different weights because their nuclei have different numbers of neutrons. Both isotopes of O are incorporated into the bodies of animals that consume the isotopes in water and food.

The hard surface of teeth, called enamel, is made up largely of calcium phosphate, which has the chemical formula $Ca_3(PO_4)_2$. Calcium phosphate has a lot of oxygen, and the isotopic composition of the oxygen in enamel varies depending on where an animal was living when the enamel was made. When water evaporates from the ocean, it forms clouds that move inland and release rain. Water made up of the heavier isotope of O is heavier, and tends to fall more readily than water containing the lighter isotope. Regions of the world that are closer to the ocean receive rain containing more heavy water than regions further away, and these differences are reflected in the bodies of animals that dwell in these regions.

This property has been used to reveal an astounding fact about dinosaurs that lived in the great basins of southwestern North America about 150 million years ago. *Camarasaurus* was big, really big—up to 75 feet long and weighing up to 50 tons.

Henry Ficke from Colorado College analyzed the oxygen isotopes in the enamel of *Camarasaurus* fossils and found two kinds of teeth: Some had the heavy oxygen content typical of rains and rocks in the basin region. But others, surprisingly, had a lower proportion of heavy oxygen, indicating that the animals had lived at higher elevations 300 km to the west. This indicates for the first time that dinosaurs migrated a long way from west to east. The reason for this migration is not clear. *Camarasaurus* ate a plant-based diet, and perhaps the migration was directed at finding food.

Life millions of years ago, as today, was based on chemistry. Just like the dinosaurs, we are what we eat—including our teeth. Indeed, biologists accept that life is based on chemistry and obeys universal laws of chemistry and physics. This physical–chemical view of life forms much of the basis of this book, and has led to great advances in biological science.

Can isotope analysis of water be used to detect climate change?
See answer on p. 36.

2.1 How Does Atomic Structure Explain the Properties of Matter?

All matter is composed of **atoms**. Atoms are tiny—more than a trillion (10^{12}) of them could fit on top of the period at the end of this sentence. Each atom consists of a dense, positively charged **nucleus**, around which one or more negatively charged **electrons** move (**Figure 2.1**). The nucleus contains one or more positively charged **protons** and may contain one or more **neutrons** with no electric charge. Atoms and their component particles have volume and mass, which are characteristics of all matter. **Mass** is a measure of the quantity of matter present; the greater the mass, the greater the quantity of matter.

The mass of a proton serves as a standard unit of measure called the **dalton** (named after the English chemist John Dalton). A single proton or neutron has a mass of about 1 dalton (Da), which is 1.7×10^{-24} grams (0.0000000000000000000000017 g), but an electron is even tinier at 9×10^{-28} g (0.0005 Da). Because the mass of an electron is negligible compared with the mass of a proton or a neutron, the contribution of electrons to the mass of an atom can usually be ignored when measurements and calculations are made. It is electrons, however, that determine how atoms will combine with other atoms to form stable associations.

Each proton has a positive electric charge, defined as +1 unit of charge. An electron has a negative charge equal and opposite to that of a proton (–1). The neutron, as its name suggests, is electrically neutral, so its charge is 0. Charges that are different (+/−) attract each other, whereas charges that are alike (+/+, −/−) repel each other. Generally, atoms are electrically neutral because the number of electrons in an atom equals the number of protons.

An element consists of only one kind of atom

An **element** is a pure substance that contains only one kind of atom. The element hydrogen consists only of hydrogen atoms; the element iron consists only of iron atoms. The atoms of each element have certain characteristics or properties that distinguish them from the atoms of other elements. These physical and chemical (reactive) properties depend on the numbers of subatomic particles the atoms contain. Such properties include mass and how the atoms interact and associate with other atoms.

There are 94 elements in nature and at least another 24 have been made in physics laboratories. About 98 percent of the mass of every living organism is composed of just six elements:

Carbon (symbol C)	Hydrogen (H)	Nitrogen (N)
Oxygen (O)	Phosphorus (P)	Sulfur (S)

The biological roles of these elements will be our major concern in this book, but other elements are found in living organisms as well. Sodium and potassium, for example, are essential for nerve function; calcium can act as a biological signal; iodine is a component of a vital hormone; and magnesium is bound to chlorophyll in plants.

Each element has a unique number of protons

An element differs from other elements by the number of protons in the nucleus of each of its atoms; the number of protons

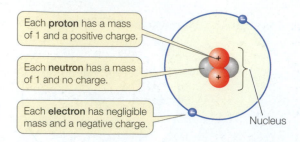

is designated the **atomic number**. This atomic number is unique to each element and does not change. The atomic number of helium is 2, and an atom of helium always has two protons; the atomic number of oxygen is 8, and an atom of oxygen always has eight protons. Since the number of protons (and electrons) determines how an element behaves in chemical reactions, it is possible to arrange the elements in a table such that those with similar chemical properties are grouped together. This is the familiar **periodic table** that is shown in **Figure 2.2**.

2.1 The Helium Atom This representation of a helium atom is called a Bohr model. Although the nucleus accounts for virtually all of the atomic weight, it occupies only 1/10,000 of the atom's volume.

Go to Media Clip 2.1
The Elements Song
Life10e.com/mc2.1

Along with a definitive number of protons, every element except hydrogen has one or more neutrons in its nucleus. The **mass number** of an atom is the total number of protons and neutrons in its nucleus. The nucleus of a carbon atom contains six protons and six neutrons and has a mass number of 12. Oxygen has eight protons and eight neutrons and has a mass number of 16. Since the mass of an electron is negligible, the mass number is essentially the mass of the atom (see below) in daltons.

By convention, we often print the symbol for an element with the atomic number at the lower left and the mass number at the upper left, both immediately preceding the symbol. Thus hydrogen, carbon, and oxygen can be written as $^{1}_{1}\text{H}$, $^{12}_{6}\text{C}$, and $^{16}_{8}\text{O}$, respectively.

The number of neutrons differs among isotopes

In some elements, the number of neutrons in the atomic nucleus is not always the same. Different **isotopes** of the same element have the same number of protons but different numbers of neutrons, as you saw in the opening story of this chapter. Many elements have several isotopes. Generally, isotopes are formed when atoms combine and/or release particles (decay). The isotopes of hydrogen shown below each have special names, but the isotopes of most elements do not have distinct names.

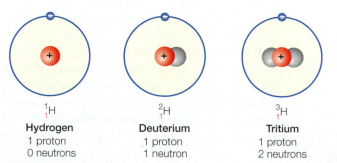

$^{1}_{1}\text{H}$	$^{2}_{1}\text{H}$	$^{3}_{1}\text{H}$
Hydrogen	**Deuterium**	**Tritium**
1 proton	1 proton	1 proton
0 neutrons	1 neutron	2 neutrons

2.2 The Periodic Table The periodic table groups the elements according to their physical and chemical properties. Elements 1–94 occur in nature; elements with atomic numbers above 94 were created in the laboratory.

Atomic number (number of protons) — 2
He

Chemical symbol (for helium)

Atomic weight — 4.003

The six elements highlighted in yellow make up 98% of the mass of most living organisms.

Elements in the same vertical columns have similar properties because they have the same number of electrons in their outermost shell.

Elements highlighted in orange are present in small amounts in many organisms.

Masses in parentheses indicate unstable elements that decay rapidly to form other elements.

Elements without a chemical symbol are as yet unnamed.

1 H 1.0079																	2 He 4.003
3 Li 6.941	4 Be 9.012											5 B 10.81	6 C 12.011	7 N 14.007	8 O 15.999	9 F 18.998	10 Ne 20.179
11 Na 22.990	12 Mg 24.305											13 Al 26.982	14 Si 28.086	15 P 30.974	16 S 32.06	17 Cl 35.453	18 Ar 39.948
19 K 39.098	20 Ca 40.08	21 Sc 44.956	22 Ti 47.88	23 V 50.942	24 Cr 51.996	25 Mn 54.938	26 Fe 55.847	27 Co 58.933	28 Ni 58.69	29 Cu 63.546	30 Zn 65.38	31 Ga 69.72	32 Ge 72.59	33 As 74.922	34 Se 78.96	35 Br 79.909	36 Kr 83.80
37 Rb 85.4778	38 Sr 87.62	39 Y 88.906	40 Zr 91.22	41 Nb 92.906	42 Mo 95.94	43 Tc (99)	44 Ru 101.07	45 Rh 102.906	46 Pd 106.4	47 Ag 107.870	48 Cd 112.41	49 In 114.82	50 Sn 118.69	51 Sb 121.75	52 Te 127.60	53 I 126.904	54 Xe 131.30
55 Cs 132.905	56 Ba 137.34	71 Lu 174.97	72 Hf 178.49	73 Ta 180.948	74 W 183.85	75 Re 186.207	76 Os 190.2	77 Ir 192.2	78 Pt 195.08	79 Au 196.967	80 Hg 200.59	81 Tl 204.37	82 Pb 207.19	83 Bi 208.980	84 Po (209)	85 At (210)	86 Rn (222)
87 Fr (223)	88 Ra 226.025	103 Lr (260)	104 Rf (261)	105 Db (262)	106 Sg (266)	107 Bh (264)	108 Hs (269)	109 Mt (268)	110 Ds (269)	111 Rg (272)	112 Cn (277)	113	114 (285)	115 (289)	116	117	118 (293)

Lanthanide series

57 La 138.906	58 Ce 140.12	59 Pr 140.9077	60 Nd 144.24	61 Pm (145)	62 Sm 150.36	63 Eu 151.96	64 Gd 157.25	65 Tb 158.924	66 Dy 162.50	67 Ho 164.930	68 Er 167.26	69 Tm 168.934	70 Yb 173.04

Actinide series

89 Ac 227.028	90 Th 232.038	91 Pa 231.0359	92 U 238.02	93 Np 237.0482	94 Pu (244)	95 Am (243)	96 Cm (247)	97 Bk (247)	98 Cf (251)	99 Es (252)	100 Fm (257)	101 Md (258)	102 No (259)

The natural isotopes of carbon, for example, are ^{12}C (six neutrons in the nucleus), ^{13}C (seven neutrons), and ^{14}C (eight neutrons). Note that all three (called "carbon-12," "carbon-13," and "carbon-14") have six protons, so they are all carbon. Most carbon atoms are ^{12}C, about 1.1 percent are ^{13}C, and a tiny fraction are ^{14}C. All have virtually the same chemical reactivity, which is an important property for their use in experimental biology and medicine.

An element's **atomic weight** (or relative atomic mass) is equivalent to the average of the mass numbers of a representative sample of atoms of that element, with all the isotopes in their normally occurring proportions. More precisely, an element's atomic weight is defined as the ratio of the average mass per atom of the element to 1/12 of the mass of an atom of ^{12}C. Because it is a ratio, atomic weight is a dimensionless physical quantity—it is not expressed in units. The atomic weight of hydrogen, taking into account all of its isotopes and

their typical abundances, is 1.00794. This number is fractional because it is the average of the contributing masses of all of the isotopes. This definition implies that in any given sample of hydrogen atoms of a particular element found on Earth, the average composition of isotopes will be constant. But as you saw in the opening to this chapter, that is not necessarily so. Some water has more of the heavy isotopes. So chemists are now listing atomic weights as ranges, for example, H: 1.00784–1.00811.

Most isotopes are stable. But some, called **radioisotopes**, are unstable and spontaneously give off energy in the form of α (alpha), β (beta), or γ (gamma) radiation from the atomic nucleus. Known as **radioactive decay**, this release of energy transforms the original atom. The type of transformation varies depending on the radioisotope, but some result in a different number of protons, so that the original atom becomes a different element.

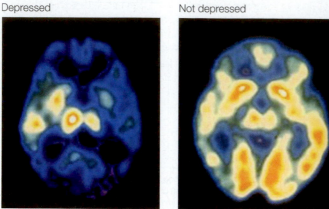

Depressed   Not depressed

2.3 Tagging the Brain   In these images from live persons, a radio-actively labeled sugar is used to detect differences between the brain activity of a depressed person (left) and that of a person who is not depressed. The more active a brain region, the more sugar it takes up (shown as orange areas). The brain of the depressed person (left) shows less activity than the brain of the person who is not depressed.

With sensitive instruments, scientists can use the released radiation to detect the presence of radioisotopes. For instance, if an earthworm is given food containing a radioisotope, its path through the soil can be followed using a simple detector called a Geiger counter. Most atoms in living organisms are organized into stable associations called **molecules**. If a radioisotope is incorporated into a molecule, it acts as a tag or label, allowing a researcher or physician to track the molecule in an experiment or in the body (**Figure 2.3**). Radioisotopes are also used to date fossils, an application described in Section 25.1.

Although radioisotopes are useful in research and in medicine, even a low dose of the radiation they emit has the potential to damage molecules and cells. However, these damaging effects are sometimes used to our advantage; for example, the radiation from ^{60}Co (cobalt-60) is used in medicine to kill cancer cells.

The behavior of electrons determines chemical bonding and geometry

The number of electrons in an atom determines how it will combine with other atoms. Biologists are interested in how chemical changes take place in living cells. When considering atoms, they are concerned primarily with electrons because the behavior of electrons explains how chemical reactions occur. Chemical reactions alter the atomic compositions of substances and thus alter their properties. Reactions usually involve changes in the distribution of electrons between atoms.

The location of a given electron in an atom at any given time is impossible to determine. We can only describe a volume of space within the atom where the electron is likely to be. The region of space where the electron is found at least 90 percent of the time is the electron's **orbital**. Orbitals have characteristic shapes and orientations, and a given orbital can be occupied by a maximum of two electrons (**Figure 2.4**). Thus any atom larger than helium (atomic number 2) must have electrons in two or more orbitals. As we move from lighter to heavier atoms in the periodic table, the orbitals are filled in a specific sequence, in a series of what are known as **electron shells**, or energy levels, around the nucleus.

- *First shell*: The innermost electron shell consists of just one orbital, called an *s* orbital. A hydrogen atom ($_1$H) has one electron in its first shell; helium ($_2$He) has two. Atoms of all other elements have two or more shells to accommodate orbitals for additional electrons.

- *Second shell*: The second shell contains four orbitals (an *s* orbital and three *p* orbitals) and hence holds up to eight electrons. As depicted in Figure 2.4, *s* orbitals have the shape of a sphere, whereas *p* orbitals are oriented at right angles to one another. The orientations of these orbitals in space contribute to the three-dimensional shapes of molecules when atoms link to other atoms.

- *Additional shells*: Elements with more than ten electrons have three or more electron shells. The farther a shell is from the nucleus, the higher the energy level is for an electron occupying that shell.

Go to Activity 2.1 Electron Orbitals   Life10e.com/ac2.1

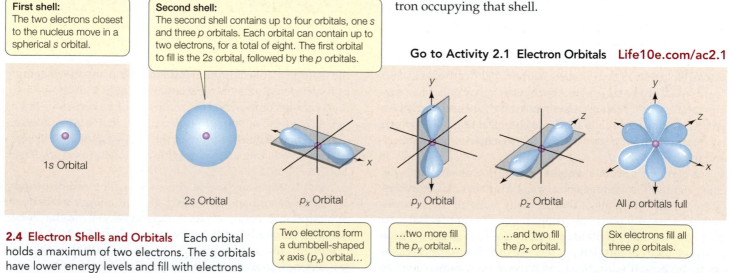

First shell:
The two electrons closest to the nucleus move in a spherical *s* orbital.

Second shell:
The second shell contains up to four orbitals, one *s* and three *p* orbitals. Each orbital can contain up to two electrons, for a total of eight. The first orbital to fill is the 2*s* orbital, followed by the *p* orbitals.

1*s* Orbital

2*s* Orbital

p$_x$ Orbital

p$_y$ Orbital

p$_z$ Orbital

All *p* orbitals full

2.4 Electron Shells and Orbitals   Each orbital holds a maximum of two electrons. The *s* orbitals have lower energy levels and fill with electrons before the *p* orbitals do.

Two electrons form a dumbbell-shaped x axis (*p$_x$*) orbital...

...two more fill the *p$_y$* orbital...

...and two fill the *p$_z$* orbital.

Six electrons fill all three *p* orbitals.

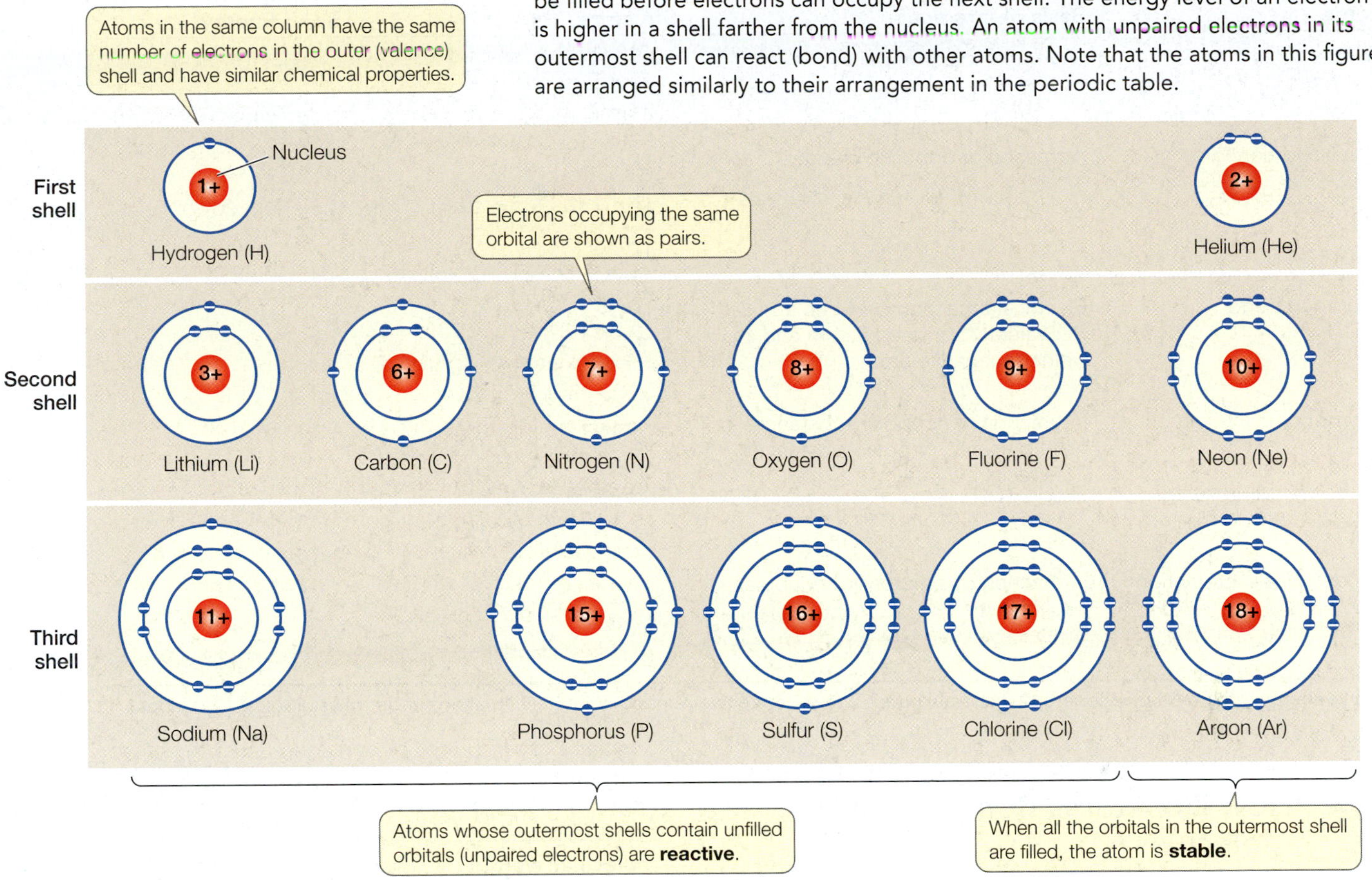

2.5 Electron Shells Determine the Reactivity of Atoms Each shell can hold a specific maximum number of electrons. Going out from the nucleus, each shell must be filled before electrons can occupy the next shell. The energy level of an electron is higher in a shell farther from the nucleus. An atom with unpaired electrons in its outermost shell can react (bond) with other atoms. Note that the atoms in this figure are arranged similarly to their arrangement in the periodic table.

Atoms in the same column have the same number of electrons in the outer (valence) shell and have similar chemical properties.

Electrons occupying the same orbital are shown as pairs.

Nucleus

First shell
Hydrogen (H) 1+
Helium (He) 2+

Second shell
Lithium (Li) 3+
Carbon (C) 6+
Nitrogen (N) 7+
Oxygen (O) 8+
Fluorine (F) 9+
Neon (Ne) 10+

Third shell
Sodium (Na) 11+
Phosphorus (P) 15+
Sulfur (S) 16+
Chlorine (Cl) 17+
Argon (Ar) 18+

Atoms whose outermost shells contain unfilled orbitals (unpaired electrons) are **reactive**.

When all the orbitals in the outermost shell are filled, the atom is **stable**.

The *s* orbitals fill with electrons first, and their electrons have the lowest energy level. Subsequent shells have different numbers of orbitals, but the outermost shells usually hold only eight electrons. In any atom, the outermost electron shell (the **valence shell**) determines how the atom combines with other atoms—that is, how the atom behaves chemically. When a valence shell with four orbitals contains eight electrons, there are no unpaired electrons and the atom is stable—it is least likely to react with other atoms (**Figure 2.5**). Examples of chemically stable elements are helium, neon, and argon. By contrast, atoms that have one or more unpaired electrons in their outer shells are capable of reacting with other atoms.

Atoms with unpaired electrons (i.e., partially filled orbitals) in their outermost electron shells are unstable and will undergo reactions in order to fill their outermost shells. *Reactive atoms can attain stability either by sharing electrons with other atoms or by losing or gaining one or more electrons.* In either case, the atoms involved are bonded together into stable associations called molecules. The tendency of atoms to form stable molecules so that they have eight electrons in their outermost shells is known as the octet rule. Many atoms in biologically important molecules—for example, carbon (C) and nitrogen (N)—follow this rule. An important exception is hydrogen (H), which attains stability when two electrons occupy its single shell (consisting of just one *s* orbital).

RECAP 2.1

Living organisms are composed of the same set of chemical elements as the rest of the universe. An atom consists of a nucleus of protons and neutrons, and a characteristic configuration of electrons in orbitals around the nucleus. This structure determines the atom's chemical properties.

- Describe the arrangement of protons, neutrons, and electrons in an atom. **See Figure 2.1**

- Use the periodic table to identify some of the similarities and differences in atomic structure among oxygen, carbon, and helium. How does the configuration of the valence shell influence the placement of an element in the periodic table? **See p. 23 and Figures 2.2, 2.5**

- How does bonding help a reactive atom achieve stability? **See p. 25 and Figure 2.5**

We have introduced the individual players on the biochemical stage—the atoms. We have shown how the number of unpaired electrons in an atom's valence shell drives its "quest for stability." Next we will describe the different types of chemical bonds that can lead to stability—joining atoms together into molecular structures with hosts of different properties.

TABLE**2.1**

Chemical Bonds and Interactions

Name	Basis of Interaction	Structure	Bond Energy[a]
Covalent bond	Sharing of electron pairs		50–110
Ionic attraction	Attraction of opposite charges		3–7
Hydrogen bond	Electrical attraction between a covalently bonded H atom and an electronegative atom		3–7
Hydrophobic interaction	Interaction of nonpolar substances in the presence of polar substances (especially water)		1–2
van der Waals interaction	Interaction of electrons of nonpolar substances		1

[a]Bond energy is the amount of energy in kcal/mol needed to separate two bonded or interacting atoms under physiological conditions.

2.2 How Do Atoms Bond to Form Molecules?

A **chemical bond** is an attractive force that links two atoms together in a molecule. There are several kinds of chemical bonds (**Table 2.1**). In this section we will begin with covalent bonds, the strong bonds that result from the sharing of electrons. Next we will examine ionic attractions, which form when an atom gains or loses one or more electrons to achieve stability. We will then consider other, weaker kinds of interactions, including hydrogen bonds.

> Go to Animated Tutorial 2.1
> **Chemical Bond Formation**
> Life10e.com/at2.1

Covalent bonds consist of shared pairs of electrons

A **covalent bond** forms when two atoms attain stable electron numbers in their outermost shells by sharing one or more pairs of electrons. Consider two hydrogen atoms coming into close proximity, each with an unpaired electron in its single shell (**Figure 2.6**). When the electrons pair up, a stable association is formed, and this links the two hydrogen atoms in a covalent bond, forming the molecule H_2.

A **compound** is a pure substance made up of two or more different elements bonded together in a fixed ratio. Chemical symbols identify the different elements in a compound, and subscript numbers indicate how many atoms of each

element are present (e.g., H_2O has two atoms of hydrogen bonded to a single oxygen atom). Every compound has a **molecular weight** (relative molecular mass) that is the sum of the atomic weights of all atoms in the molecule. Looking at

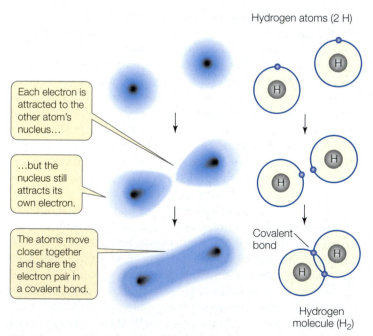

2.6 Electrons Are Shared in Covalent Bonds Two hydrogen atoms can combine to form a hydrogen molecule. A covalent bond forms when the electron orbitals of the two atoms overlap in an energetically stable manner.

2.7 Covalent Bonding Can Form Compounds

(A) Bohr models showing the formation of covalent bonds in methane, whose molecular formula is CH_4. Electrons are shown in shells around the nucleus. (B) Three additional ways of representing the structure of methane. The ball-and-stick model and the space-filling model show the spatial orientations of the bonds. The space-filling model indicates the overall shape and surface of the molecule. In the chapters that follow, different conventions will be used to depict molecules. Bear in mind that these are models to illustrate certain properties, not accurate portrayals of how atoms would actually appear.

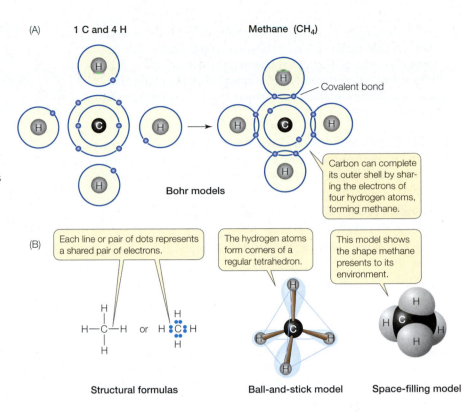

(A) 1 C and 4 H   Methane (CH_4)

Covalent bond

Carbon can complete its outer shell by sharing the electrons of four hydrogen atoms, forming methane.

Bohr models

(B) Each line or pair of dots represents a shared pair of electrons.

The hydrogen atoms form corners of a regular tetrahedron.

This model shows the shape methane presents to its environment.

Structural formulas   Ball-and-stick model   Space-filling model

the periodic table in Figure 2.2, you can calculate the molecular weight of water to be 18.01. (But remember that this value comes from the average atomic weights of hydrogen and oxygen; the molecular weight of the heavy water in our opening story is higher because it is formed from heavier isotopes.) Molecules that make up living organisms can have molecular weights of up to half a billion, and covalent bonds are found in all.

How are the covalent bonds formed in a molecule of methane gas (CH_4)? The carbon atom has six electrons: two electrons fill its inner shell, and four unpaired electrons travel in its outer shell. Because its outer shell can hold up to eight electrons, carbon can share electrons with up to four other atoms—*it can form four covalent bonds* (**Figure 2.7A**). When an atom of carbon reacts with four hydrogen atoms, methane forms. Thanks to electron sharing, the outer shell of methane's carbon atom is now filled with eight electrons, a stable configuration. The outer shell of each of the four hydrogen atoms is also filled. Four covalent bonds—four shared electron pairs—hold methane together. **Figure 2.7B** shows several different ways to represent the molecular structure of methane. **Table 2.2** shows the covalent bonding capacities of some biologically significant elements.

STRENGTH AND STABILITY Covalent bonds are very strong, meaning that it takes a lot of energy to break them. At temperatures where life exists, the covalent bonds of biological

molecules are quite stable, as are their three-dimensional structures. However, this stability does not preclude change, as we will discover.

ORIENTATION For a given pair of elements—for example, carbon bonded to hydrogen—the length of the covalent bond is always the same. And for a given atom within a molecule, the angle of each of its covalent bonds, with respect to the other bonds, is generally the same. This is true regardless of the type of larger molecule that contains the atom. For example, the four filled orbitals around the carbon atom in methane are always distributed in space so that the bonded hydrogen atoms point to the corners of a regular tetrahedron, with carbon in the center (see Figure 2.7B). Even when carbon is bonded to four atoms other than hydrogen, this three-dimensional orientation is more or less maintained. The orientation of covalent bonds in space gives the molecules their three-dimensional geometry, and the shapes of molecules contribute to their biological functions, as we will see in Section 3.1.

Even though the orientations of bonds around each atom are fairly stable, the shapes of molecules can change. Think of a single covalent bond as an axle around which the two atoms, along with their other bonded atoms, can rotate.

Dichloroethane

TABLE 2.2
Covalent Bonding Capabilities of Some Biologically Important Elements

Element	Usual Number of Covalent Bonds
Hydrogen (H)	1
Oxygen (O)	2
Sulfur (S)	2
Nitrogen (N)	3
Carbon (C)	4
Phosphorus (P)	5

This phenomenon has enormous implications for the large molecules that make up living tissues. In long chains of atoms (especially carbons) that can rotate freely, there are many possibilities for the arrangement of atoms within the chain. This allows molecules to alter their structures, for example, to fit other molecules.

MULTIPLE COVALENT BONDS Two atoms can share more than one pair of electrons, forming multiple covalent bonds. These can be represented by lines between the chemical symbols for the linked atoms:

- A single bond involves the sharing of a single pair of electrons (for example, H—H or C—H).
- A double bond involves the sharing of four electrons (two pairs) (C=C).
- Triple bonds—six shared electrons—are rare, but there is one in nitrogen gas (N≡N), which is the major component of the air we breathe.

UNEQUAL SHARING OF ELECTRONS If two atoms of the same element are covalently bonded, there is an equal sharing of the pair(s) of electrons in their outermost shells. However, when the two atoms are of different elements, the sharing is not necessarily equal. One nucleus may exert a greater attractive force on the electron pair than the other nucleus, so that the pair tends to be closer to that atom.

The attractive force that an atomic nucleus exerts on electrons in a covalent bond is called its **electronegativity**. The electronegativity of an atom depends on how many positive charges it has (atoms with more protons are more positive and thus more attractive to electrons) and on the distance between the nucleus and the electrons in the outer (valence) shell (the closer the electrons, the greater the electronegative pull). **Table 2.3** shows the electronegativities (which are calculated to produce dimensionless quantities) of some elements important in biological systems.

If two atoms are close to each other in electronegativity, they will share electrons equally in what is called a **nonpolar covalent bond**. Two oxygen atoms, for example, each with an electronegativity of 3.5, will share electrons equally. So will two hydrogen atoms (each with an electronegativity of 2.1). But when hydrogen bonds with oxygen to form water, the electrons involved are unequally shared; they tend to be nearer to the oxygen nucleus because it is the more electronegative of the two. When electrons are drawn to one nucleus more than to the other, the result is a **polar covalent bond** (**Figure 2.8**).

Because of this unequal sharing of electrons, the oxygen end of the hydrogen–oxygen bond has a slightly negative charge (symbolized by δ− and spoken of as "delta negative," meaning a partial unit of charge), and the hydrogen end has a slightly positive charge (δ+). The bond is **polar** because these opposite charges are separated at the two ends, or poles, of the bond. The partial charges that result from polar covalent bonds produce polar molecules or polar regions of large molecules. Polar

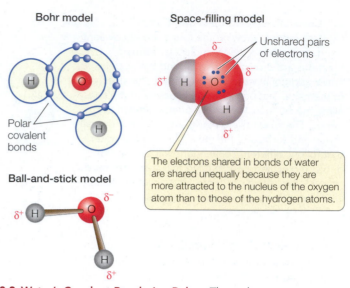

2.8 Water's Covalent Bonds Are Polar These three representations all illustrate polar covalent bonding in water (H_2O). When atoms with different electronegativities, such as oxygen and hydrogen, form a covalent bond, the electrons are drawn to one nucleus more than to the other. A molecule held together by such a polar covalent bond has partial (δ+ and δ-) charges at different surfaces. In water, the shared electrons are displaced toward the oxygen atom's nucleus.

bonds within molecules greatly influence the interactions they have with other polar molecules. Water (H_2O) is a polar compound, and this polarity has significant effects on its physical properties and chemical reactivity, as we will see in later chapters.

Ionic attractions form by electrical attraction

When one interacting atom is much more electronegative than the other, a complete transfer of one or more electrons may take place. Consider sodium (electronegativity 0.9) and chlorine (3.1). A sodium atom has only one electron in its outermost shell; this condition is unstable. A chlorine atom has seven electrons in its outermost shell—another unstable condition. Since the electronegativity of chlorine is so much greater than that of sodium, any electrons involved in bonding will tend to transfer completely

TABLE 2.3

Some Electronegativities

Element	Electronegativity
Oxygen (O)	3.5
Chlorine (Cl)	3.1
Nitrogen (N)	3.0
Carbon (C)	2.5
Phosphorus (P)	2.1
Hydrogen (H)	2.1
Sodium (Na)	0.9
Potassium (K)	0.8

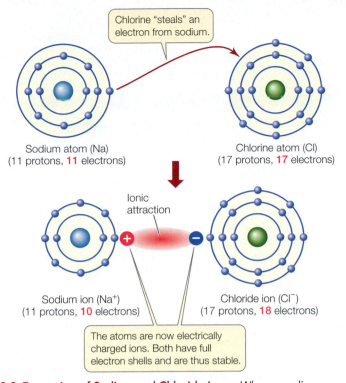

Chlorine "steals" an electron from sodium.

Sodium atom (Na)
(11 protons, **11** electrons)

Chlorine atom (Cl)
(17 protons, **17** electrons)

Ionic attraction

Sodium ion (Na⁺)
(11 protons, **10** electrons)

Chloride ion (Cl⁻)
(17 protons, **18** electrons)

The atoms are now electrically charged ions. Both have full electron shells and are thus stable.

2.9 Formation of Sodium and Chloride Ions When a sodium atom reacts with a chlorine atom, the more electronegative chlorine fills its outermost shell by "stealing" an electron from the sodium. In so doing, the chlorine atom becomes a negatively charged chloride ion (Cl⁻). With one less electron, the sodium atom becomes a positively charged sodium ion (Na⁺).

from sodium's outermost shell to that of chlorine (**Figure 2.9**). This reaction between sodium and chlorine makes the resulting atoms more stable because they both have eight fully paired electrons in their outer shells. The result is two **ions**.

Ions are electrically charged particles that form when atoms gain or lose one or more electrons:

- The sodium ion (Na⁺) in our example has a +1 unit of charge because it has one less electron than it has protons. The outermost electron shell of the sodium ion is full, with eight electrons, so the ion is stable. Positively charged ions are called **cations**.

- The chloride ion (Cl⁻) has a −1 unit of charge because it has one more electron than it has protons. This additional electron gives Cl⁻ a stable outermost shell with eight electrons. Negatively charged ions are called **anions**.

Some elements can form ions with multiple charges by losing or gaining more than one electron. Examples are Ca²⁺ (the calcium ion, a calcium atom that has lost two electrons) and Mg²⁺ (the magnesium ion). Two biologically important elements can each yield more than one stable ion. Iron yields Fe²⁺ (the ferrous ion) and Fe³⁺ (the ferric ion), and copper yields Cu⁺ (the cuprous ion) and Cu²⁺ (the cupric ion). Groups of covalently bonded atoms that carry an electric charge are called **complex ions**; examples include NH₄⁺ (the ammonium ion), SO₄²⁻ (the sulfate ion), and PO₄³⁻ (the phosphate ion). Once formed, ions are usually stable and no more electrons are lost or gained.

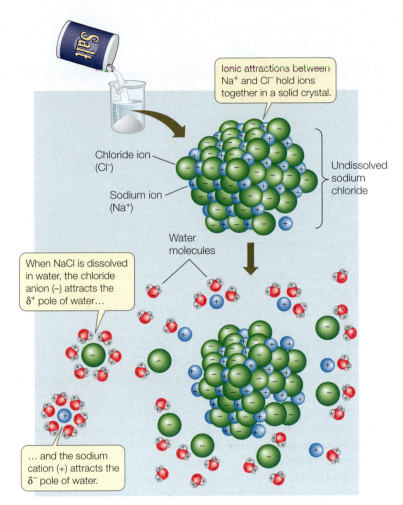

Ionic attractions between Na⁺ and Cl⁻ hold ions together in a solid crystal.

Chloride ion (Cl⁻)

Sodium ion (Na⁺)

Undissolved sodium chloride

Water molecules

When NaCl is dissolved in water, the chloride anion (−) attracts the δ⁺ pole of water...

... and the sodium cation (+) attracts the δ⁻ pole of water.

2.10 Water Molecules Surround Ions When an ionic solid dissolves in water, polar water molecules cluster around the cations and anions, preventing them from reassociating.

Ionic attractions are bonds formed as a result of the electrical attraction between ions bearing opposite charges. Ions can form bonds that result in stable solid compounds, which are referred to by the general term salts. Examples are sodium chloride (NaCl) and potassium phosphate (K₃PO₄). In sodium chloride—familiar to us as table salt—cations and anions are held together by ionic attractions. In solids, the attractions are strong because the ions are close together. However, when ions are dispersed in water, the distances between them can be large; the strength of the attraction is thus greatly reduced. Under the conditions in living cells, an ionic attraction is less strong than a nonpolar covalent bond (see Table 2.1).

Not surprisingly, ions can interact with polar molecules, since both are charged. This interaction results when a solid salt such as NaCl dissolves in water. Water molecules surround the individual ions, separating them (**Figure 2.10**). The negatively charged chloride ions attract the positive poles of the water molecules, while the positively charged sodium ions attract the negative poles of the water molecules. This special property of water (its polarity) is one reason it is such a good biological solvent (see Section 2.4).

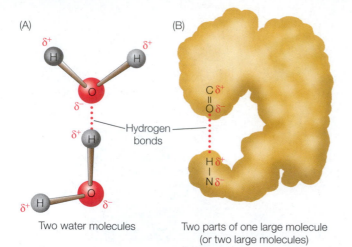

(A) (B)

Two water molecules

Two parts of one large molecule
(or two large molecules)

2.11 Hydrogen Bonds Can Form between or within Molecules (A) A hydrogen bond between two molecules is an attraction between a negative charge on one molecule and the positive charge on a hydrogen atom of the second molecule. (B) Hydrogen bonds can form between different parts of the same large molecule.

Hydrogen bonds may form within or between molecules with polar covalent bonds

In liquid water, the negatively charged oxygen (δ^-) atom of one water molecule is attracted to the positively charged hydrogen (δ^+) atoms of another water molecule (**Figure 2.11A**). The bond resulting from this attraction is called a **hydrogen bond**. Later in this chapter we'll see how hydrogen bonding between water molecules contributes to many of the properties that make water so important for living systems. Hydrogen bonds are not restricted to water molecules. Such a bond can also form between a strongly electronegative atom in one molecule and a hydrogen atom that is involved in a polar covalent bond in another molecule, or another part of the same molecule (**Figure 2.11B**).

A hydrogen bond is weaker than most ionic attractions because its formation is due to partial charges (δ^+ and δ^-). It is much weaker than a covalent bond between a hydrogen atom and an oxygen atom (see Table 2.1). Although individual hydrogen bonds are weak, there can be many of them within a single molecule or between two molecules. In these cases, the hydrogen bonds together have considerable strength and can greatly influence the structure and properties of substances. For example, hydrogen bonds play important roles in determining and maintaining the three-dimensional shapes of giant molecules such as DNA and proteins (see Section 3.2).

Hydrophobic interactions bring together nonpolar molecules

Just as water molecules can interact with one another through hydrogen bonds, any molecule that is polar can interact with other polar molecules through the weak (δ^+ to δ^-) attractions of hydrogen bonds. If a polar molecule interacts with water in this way, it is called **hydrophilic** ("water-loving") (**Figure 2.12A**).

Nonpolar molecules, in contrast, tend to interact with other nonpolar molecules. For example, carbon (electronegativity 2.5) forms nonpolar bonds with hydrogen (electronegativity 2.1), and molecules containing only hydrogen and carbon atoms—called

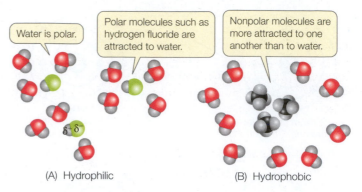

(A) Hydrophilic (B) Hydrophobic

2.12 Hydrophilic and Hydrophobic (A) Molecules with polar covalent bonds are attracted to polar water (they are hydrophilic). (B) Molecules with nonpolar covalent bonds show greater attraction to one another than to water (they are hydrophobic).

hydrocarbon molecules—are nonpolar. In water these molecules tend to aggregate with one another rather than with the polar water molecules. Therefore, nonpolar molecules are known as **hydrophobic** ("water-hating"), and the interactions between them are called **hydrophobic interactions** (**Figure 2.12B**). Of course, hydrophobic substances do not really "hate" water; they can form weak interactions with it, since the electronegativities of carbon and hydrogen are not exactly the same. But these interactions are far weaker than the hydrogen bonds between the water molecules (see Table 2.1), so the nonpolar substances tend to aggregate.

van der Waals forces involve contacts between atoms

The interactions between nonpolar substances are enhanced by **van der Waals forces**, which occur when the atoms of two molecules are in close proximity. These brief interactions result from random variations in the electron distribution in one molecule, which create opposite charge distributions in the adjacent molecule. So there will be a weak, temporary δ^+ to δ^- attraction. Although a single van der Waals interaction is brief and weak, the sum of many such interactions over the entire span of a large nonpolar molecule can result in substantial attraction. This is important when hydrophobic regions of different molecules such as an enzyme and a substrate come together (see Chapter 8).

RECAP 2.2

Some atoms form strong covalent bonds with other atoms by sharing one or more pairs of electrons. Unequal sharing of electrons produces polarity. Other atoms become ions by losing or gaining electrons, and they interact with other ions or polar molecules.

- Why is a covalent bond stronger than an ionic attraction? **See pp. 26–29 and Table 2.1**

- How do variations in electronegativity result in the unequal sharing of electrons in polar molecules? **See p. 28, Table 2.3, and Figure 2.8**

- What is a hydrogen bond and how is it important in biological systems? **See p. 30 and Figure 2.11**

The bonding of atoms into molecules is not necessarily a permanent affair. The dynamic of life involves constant change, even at the molecular level. In the next section we will examine how molecules interact with one another—how they break up, how they find new partners, and what the consequences of those changes can be.

2.3 How Do Atoms Change Partners in Chemical Reactions?

A **chemical reaction** occurs when moving atoms collide with sufficient energy to combine or to change their bonding partners. Consider the combustion reaction that takes place in the flame of a propane stove. When propane (C_3H_8) reacts with oxygen gas (O_2), the carbon atoms become bonded to oxygen atoms instead of hydrogen atoms, and the hydrogen atoms become bonded to oxygen instead of carbon (**Figure 2.13**). As the covalently bonded atoms change partners, the composition of the matter changes; propane and oxygen gas become carbon dioxide and water. This chemical reaction can be represented by the equation

$$C_3H_8 + 5 O_2 \rightarrow 3 CO_2 + 4 H_2O + Energy$$

$$Reactants \rightarrow Products$$

In this equation, the propane and oxygen are the **reactants**, and the carbon dioxide and water are the **products**. In fact, this is a special type of reaction called an oxidation–reduction reaction. Electrons and protons (i.e., hydrogen atoms) are transferred from propane (the reducing agent) to oxygen (the oxidizing agent) to form water. You will see this kind of reaction involving electron/proton transfer many times in later chapters.

The products of a chemical reaction can have very different properties from the reactants. In the case shown in Figure 2.13, the reaction is *complete*: all the propane and oxygen are used up in forming the two products. The arrow symbolizes the direction of the chemical reaction. The numbers preceding the molecular formulas indicate how many molecules are used or produced.

Note that in this and all other chemical reactions, *matter is neither created nor destroyed*. The total number of carbon atoms on the left side of the equation (3) equals the total number of carbon atoms on the right (3). In other words, the equation is *balanced*. However, there is another aspect of this reaction: the heat and light of the stove's flame reveal that the reaction between propane and oxygen releases a great deal of energy.

Energy is defined as the capacity to do work, but in the context of chemical reactions, it can be thought of as the capacity for change. Chemical reactions do not create or destroy energy, but *changes in the form of energy* usually accompany chemical reactions.

In the reaction between propane and oxygen, a large amount of heat energy is released. This energy was present in another form, called potential chemical energy, in the covalent bonds within the propane and oxygen gas molecules. Not all reactions release energy; indeed, many chemical reactions require that

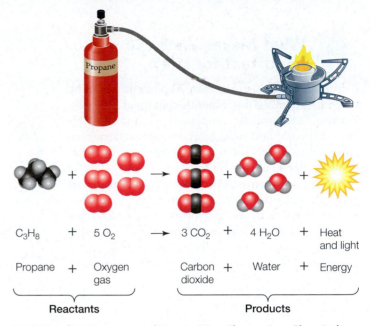

C_3H_8 + 5 O_2 → 3 CO_2 + 4 H_2O + Heat and light

Propane + Oxygen gas → Carbon dioxide + Water + Energy

Reactants — Products

2.13 Bonding Partners and Energy May Change in a Chemical Reaction One molecule of propane (a gas used for cooking) from this burner reacts with five molecules of oxygen gas to give three molecules of carbon dioxide and four molecules of water. This reaction releases energy in the form of heat and light.

energy be supplied from the environment. Some of this energy is then stored as potential chemical energy in the bonds formed in the products. We will see in future chapters how reactions that release energy and reactions that require energy can be linked together.

Many chemical reactions take place in living cells, and some of these have a lot in common with the oxidation–reduction reaction that happens in the combustion of propane. In cells, the reactants are different (they may be sugars or fats), and the reactions proceed by many intermediate steps that permit the released energy to be harvested and put to use by the cells. But the products are the same: carbon dioxide and water. We will discuss energy changes, oxidation–reduction reactions, and several other types of chemical reactions that are prevalent in living systems in Part Three of this book.

RECAP 2.3

In a chemical reaction, a set of reactants is converted to a set of products with different chemical compositions. This is accomplished by breaking old bonds and making new ones. A reaction may release energy or require its input.

- Explain how a chemical equation is balanced. **See p. 31 and Figure 2.13**
- How can the form of energy change during a chemical reaction? **See p.31**

We will return to chemical reactions and how they occur in living systems in Part Three of this book. First, however, we will examine the unique properties of the substance in which most biochemical reactions take place: water.

2.4 What Makes Water So Important for Life?

A human body is more than 70 percent water by weight, excluding the minerals contained in bones. Water is the dominant component of virtually all living organisms, and most biochemical reactions take place in this watery, or aqueous, environment. What makes water so important?

Water is an unusual substance with unusual properties. Under conditions on Earth, water exists in solid, liquid, and gas forms, all of which have relevance to living systems. Water allows chemical reactions to occur inside living organisms, and it is necessary for the formation of certain biological structures. In this section we will explore how the structure and interactions of water molecules make water essential to life.

Water has a unique structure and special properties

The molecule H_2O has unique chemical features. As we have already learned, water is a polar molecule that can form hydrogen bonds. The four pairs of electrons in the outer shell of the oxygen atom repel one another, giving the water molecule a tetrahedral shape:

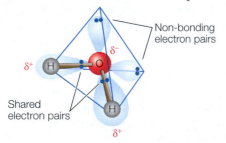

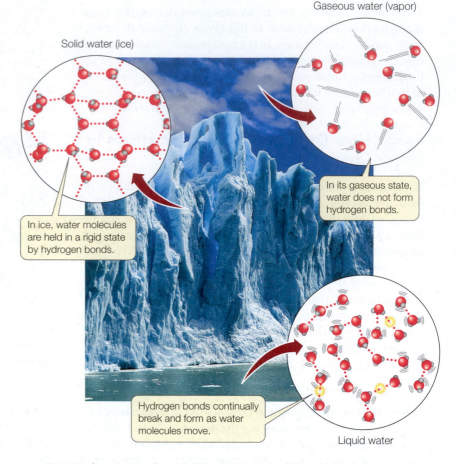

Solid water (ice)

In ice, water molecules are held in a rigid state by hydrogen bonds.

Gaseous water (vapor)

In its gaseous state, water does not form hydrogen bonds.

Hydrogen bonds continually break and form as water molecules move.

Liquid water

2.14 Hydrogen Bonding and the Properties of Water Hydrogen bonding occurs between the molecules of water in both its liquid and solid states. Ice is more structured but less dense than liquid water, which is why ice floats. Water forms a gas when its hydrogen bonds are broken and the molecules move farther apart.

These chemical features explain some of the interesting properties of water, such as the ability of ice to float, the melting and freezing temperatures of water, the ability of water to store heat, the formation of water droplets, water's ability to dissolve many substances, and its inability to dissolve many others.

ICE FLOATS In water's solid state (ice), individual water molecules are held in place by hydrogen bonds. Each molecule is bonded to four other molecules in a rigid, crystalline structure (**Figure 2.14**). Although the molecules are held firmly in place, they are farther apart from one another than they are in liquid water, where the molecules are moving about. In other words, solid water is less dense than liquid water, which is why ice floats.

Think of the biological consequences if ice were to sink in water. A pond would freeze from the bottom up, becoming a solid block of ice in winter and killing most of the organisms living there. Once the whole pond was frozen, its temperature could drop well below the freezing point of water. But in fact ice floats, forming an insulating layer on the top of the pond, and reducing heat flow to the cold air above. Thus fish, plants, and other organisms in the pond are not subjected to temperatures lower than 0°C, which is the freezing point of pure water.

MELTING, FREEZING, AND HEAT CAPACITY Compared with many other substances that have molecules of similar size, ice requires a great deal of heat energy to melt. This is because so many hydrogen bonds must be broken in order for water to change from solid to liquid. In the opposite process—freezing—a great deal of energy is released to the environment.

This property of water contributes to the surprising constancy of the temperatures found in oceans and other large bodies of water throughout the year. The temperature changes of coastal land masses are also moderated by large bodies of water. Indeed, water helps minimize variations in atmospheric temperature across the planet. This moderating ability is a result of the high heat capacity of liquid water, which is in turn a result of its high specific heat.

The **specific heat** of a substance is the amount of heat energy required to raise the temperature of 1 gram of that substance by 1°C. Raising the temperature of liquid water takes a relatively large amount of heat because much of the heat energy is used to break the hydrogen bonds that hold the liquid together. Compared with other small molecules that are liquids, water has a high specific heat. For example, water has twice the specific heat of ethyl alcohol.

Water also has a high **heat of vaporization**, which means that a lot of heat is required to change water from its liquid to its gaseous state (the process of evaporation). Once again, much of the heat energy is used to break the many hydrogen bonds between the water molecules. This heat must be absorbed from the environment in contact with the water. Evaporation thus has a cooling effect on the environment—whether a leaf, a forest, or an entire land mass. This effect explains why sweating cools the human body: as sweat evaporates from the skin, it uses up some of the adjacent body heat (**Figure 2.15A**).

COHESION AND SURFACE TENSION In liquid water, individual molecules are able to move about. The hydrogen bonds between the molecules continually form and break (see Figure 2.14). Chemists estimate that this occurs about a trillion times a minute for a single water molecule, making it a truly dynamic structure.

At any given time, a water molecule will form on average 3.4 hydrogen bonds with other water molecules. These hydrogen bonds explain the cohesive strength of liquid water. This cohesive strength, or **cohesion**, is defined as the capacity of water molecules to resist coming apart from one another when placed under tension. Water's cohesive strength permits narrow columns of liquid water to move from the roots to the leaves of tall trees. When water evaporates from the leaves, the entire column moves upward in response to the pull of the molecules at the top (**Figure 2.15B**).

The surface of liquid water exposed to the air is difficult to puncture because the water molecules at the surface are hydrogen-bonded to other water molecules below them. This surface tension of water permits a container to be filled slightly above its rim without overflowing, and it permits spiders to walk on the surface of a pond (**Figure 2.15C**).

The reactions of life take place in aqueous solutions

A **solution** is produced when a substance (the **solute**) is dissolved in a liquid (the **solvent**). If the solvent is water, then the solution is an aqueous solution. Many of the important molecules in biological systems are polar, and therefore soluble in water. Many important biochemical reactions occur in aqueous solutions within cells.

Biologists who are interested in the biochemical reactions within cells need to identify the reactants and products and to determine their amounts:

- Qualitative analyses deal with the identification of substances involved in chemical reactions. For example, a qualitative analysis would be used to investigate the steps involved and the products formed during respiration, when carbon-containing compounds are broken down to release energy in living tissues.

- Quantitative analyses measure concentrations or amounts of substances. For example, a biochemist would use a quantitative analysis to measure how much of a certain product is formed in a chemical reaction. What follows is a brief introduction to some of the quantitative chemical terms you will see in this book.

High heat of vaporization: Sweating uses evaporation of water to cool the body.

Cohesion: Water's cohesive strength helps it to flow from the roots to the leaves in a tree.

Surface tension: Water molecules stick to one another and help prevent this wolf spider from sinking.

2.15 Water in Biology These three properties of water make it beneficial to organisms.

Fundamental to quantitative thinking in chemistry and biology is the concept of the mole. A **mole** is the amount of a substance (in grams) that is numerically equal to its molecular weight. So a mole of table sugar ($C_{12}H_{22}O_{11}$) weighs about 342 grams; a mole of sodium ion (Na^+) weighs 23 grams; and a mole of hydrogen gas (H_2) weighs 2 grams.

Quantitative analyses do not yield direct counts of molecules. Because the amount of a substance in 1 mole is directly related to its molecular weight, it follows that the number of molecules in 1 mole is constant for all substances. So 1 mole of salt contains the same number of molecules as 1 mole of table sugar. This constant number of molecules in a mole is called **Avogadro's number**, and it is 6.02×10^{23} molecules per mole. Chemists work with moles of substances (which can be weighed in the laboratory) instead of actual molecules, which are too numerous to be counted. Consider 34.2 grams (just over 1 ounce) of table sugar, $C_{12}H_{22}O_{11}$. This is one-tenth of a mole, or as Avogadro puts it, 6.02×10^{23} molecules.

A chemist can dissolve a mole of sugar (342 g) in water to make 1 liter of solution, knowing that the mole contains 6.02×10^{23} individual sugar molecules. This solution—1 mole of a substance dissolved in water to make 1 liter—is called a 1 molar (1M) solution. When a physician injects a certain volume and molar concentration of a drug into the bloodstream of a patient, a rough calculation can be made of the actual number of drug molecules that will interact with the patient's cells.

The many molecules dissolved in the water of living tissues are not present at concentrations anywhere near 1 molar. Most are in the micromolar (millionths of a mole per liter of solution; µM) to millimolar (thousandths of a mole per liter; mM) range. Some, such as hormone molecules, are even less concentrated than that. While these molarities seem to indicate very low concentrations, remember that even a 1 µM solution has 6.02×10^{17} molecules of the solute per liter.

Aqueous solutions may be acidic or basic

When some substances dissolve in water, they release hydrogen ions (H^+), which are actually single, positively charged protons. Hydrogen ions can interact with other molecules and change their properties. For example, the protons in "acid rain" can damage plants, and you probably have experienced the excess of hydrogen ions that we call "acid indigestion."

Here we will examine the properties of **acids** (defined as substances that release H^+) and **bases** (defined as substances that accept H^+). We will distinguish between strong and weak acids and bases, and provide a quantitative means for stating the concentration of H^+ in solutions: the pH scale.

ACIDS RELEASE H⁺   When hydrochloric acid (HCl) is added to water, it dissolves, releasing the ions H^+ and Cl^-:

$$HCl \rightarrow H^+ + Cl^-$$

Because its H^+ concentration has increased, the solution is acidic.

Acids are substances that *release* H^+ ions in solution. HCl is an acid, as is H_2SO_4 (sulfuric acid). One molecule of sulfuric acid will ionize to yield two H^+ and one SO_4^{2-}. Biological compounds that contain —COOH (the carboxyl group) are also acids because the carboxyl group ionizes to —COO$^-$, releasing H^+:

$$—COOH \rightarrow —COO^- + H^+$$

Acids that fully ionize in solution, such as HCl and H_2SO_4 are called strong acids. However, not all acids ionize fully in water. For example, if acetic acid (CH_3COOH) is added to water, some of it will dissociate into two ions (CH_3COO^- and H^+), but some of the original acetic acid will remain as well. Because the reaction is not complete, acetic acid is a weak acid.

BASES ACCEPT H⁺   Bases are substances that *accept* H^+ in solution. As with acids, there are strong and weak bases. If NaOH (sodium hydroxide) is added to water, it dissolves and ionizes, releasing OH^- and Na^+ ions:

$$NaOH \rightarrow Na^+ + OH^-$$

Because OH^- absorbs H^+ to form water, such a solution is basic. This reaction is complete, and so NaOH is a strong base.

Weak bases include the bicarbonate ion (HCO_3^-), which can accept an H^+ ion and become carbonic acid (H_2CO_3), and ammonia (NH_3), which can accept H^+ and become an ammonium ion (NH_4^+). Biological compounds that contain —NH$_2$ (the amino group) are also bases because —NH$_2$ accepts H^+:

$$—NH_2 + H^+ \rightarrow —NH_3^+$$

ACID–BASE REACTIONS MAY BE REVERSIBLE   When acetic acid is dissolved in water, two reactions happen. First, the acetic acid forms its ions:

$$CH_3COOH \rightarrow CH_3COO^- + H^+$$

Then, once the ions are formed, some of them re-form acetic acid:

$$CH_3COO^- + H^+ \rightarrow CH_3COOH$$

This pair of reactions is reversible. A **reversible reaction** can proceed in either direction—left to right or right to left—depending on the relative starting concentrations of the reactants and products. The formula for a reversible reaction can be written using a double arrow:

$$CH_3COOH \rightleftharpoons CH_3COO^- + H^+$$

In terms of acids and bases, there are two types of reactions, depending on the extent of the reversibility:

- The ionization of strong acids and bases in water is virtually irreversible.

- The ionization of weak acids and bases in water is somewhat reversible.

WATER IS A WEAK ACID AND A WEAK BASE   The water molecule has a slight but significant tendency to ionize into a hydroxide ion (OH^-) and a hydrogen ion (H^+). Actually, two water molecules participate in this reaction. One of the two molecules "captures" a hydrogen ion from the other, forming a hydroxide ion and a hydronium ion:

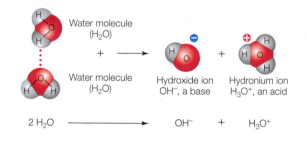

The hydronium ion is, in effect, a hydrogen ion bound to a water molecule. For simplicity, biochemists tend to use a modified representation of the ionization of water:

$$H_2O \rightarrow H^+ + OH^-$$

The ionization of water is important to all living creatures. This fact may seem surprising, since only about 1 water molecule in 500 million is ionized at any given time. But this is less surprising if we focus on the abundance of water in living systems, and the reactive nature of the H^+ ions produced by ionization.

pH: HYDROGEN ION CONCENTRATION As we have seen, compounds can be either acids or bases, and thus solutions can be either acidic or basic. We can measure how acidic or basic a solution is by measuring its concentration of H^+ in moles per liter (its molarity; see p. 34). Here are some examples:

- Pure water has a H^+ concentration of 10^{-7} M.
- A 1 M HCl solution has a H^+ concentration of 1 M (recall that all the HCl dissociates into its ions).
- A 1 M NaOH solution has a H^+ concentration of 10^{-14} M.

This is a very wide range of numbers to work with—think about the decimals! It is easier to work with the logarithm of the H^+ concentration, because logarithms compress this range: the $\log_{10}$ of 100, for example, is 2; and the $\log_{10}$ of 0.01 is –2. Because most H^+ concentrations in living systems are less than 1 M, their $\log_{10}$ values are negative. For convenience, we convert these negative numbers into positive ones by using the *negative* of the logarithm of the H^+ molar concentration. This number is called the **pH** of the solution.

Since the H^+ concentration of pure water is 10^{-7} M, its pH is $-\log(10^{-7}) = -(-7)$, or 7. A smaller negative logarithm means a larger number. In practical terms, a lower pH means a higher H^+ concentration, or greater acidity. In 1 M HCl, the H^+ concentration is 1 M, so the pH is the negative logarithm of 1 ($-\log 10^0$), or 0. The pH of 1 M NaOH is the negative logarithm of 10^{-14}, or 14.

A solution with a pH of less than 7 is acidic—it contains more H^+ ions than OH^- ions. A solution with a pH of 7 is referred to as neutral, and a solution with a pH value greater than 7 is basic. **Figure 2.16** shows the pH values of some common substances.

Why is this discussion of pH so relevant to biology? Many reactions involve the transfer of an ion or charged group from one molecule to another, and the presence of positive or negative ions in the environment can greatly influence the rates of such reactions. Furthermore, pH can influence the shapes of molecules. Many biologically important molecules contain charged groups (e.g., —COO⁻) that can interact with the polar regions of water, and these interactions influence the way such molecules fold up into three-dimensional shapes. If these charged groups combine with H^+ or other ions in their environment to form uncharged groups (e.g., —COOH, see above), they will have a reduced tendency to interact with water. These uncharged (hydrophobic) groups might induce the molecule to fold up differently so that they are no longer in contact with the watery environment. Since the three-dimensional structures of biological molecules greatly affect the way they function, organisms do all they can to minimize changes in the pH of their cells and tissues. An important way to do this is with buffers.

BUFFERS The maintenance of internal constancy—homeostasis—is a hallmark of all living things and extends to pH. If biological molecules lose or gain H^+ ions, their properties can change, thus upsetting homeostasis. Internal constancy is achieved with buffers: solutions that maintain a relatively constant pH even when substantial amounts of acid or base are added. How does this work?

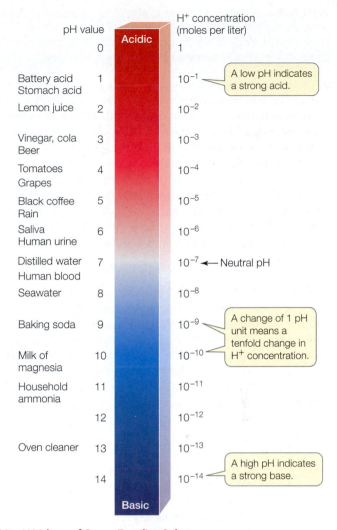

2.16 pH Values of Some Familiar Substances

A **buffer** is a mixture of a weak acid and its corresponding base, or a weak base and its corresponding acid. For example, a weak acid is carbonic acid (H_2CO_3), and its corresponding base is the bicarbonate ion (HCO_3^-). If another acid is added to a solution containing this mixture (a buffered solution), not all the H^+ ions from the acid remain in solution. Instead, many of them combine with the bicarbonate ions to produce more carbonic acid:

$$HCO_3^- + H^+ \rightleftharpoons H_2CO_3$$

This reaction uses up some of the H^+ ions in the solution and decreases the acidifying effect of the added acid. If a base is added, the reaction essentially reverses. Some of the carbonic acid ionizes to produce bicarbonate ions and more H^+, which counteracts some of the added base. In this way, the buffer minimizes the effect that an added acid or base has on pH. The carbonic acid/bicarbonate buffering system is present in the blood, where it is important for preventing significant changes in pH that could disrupt the ability of the blood to carry vital oxygen to tissues. A given amount of acid or base causes a smaller pH change in a buffered solution than in a non-buffered one (**Figure 2.17**).

Buffers illustrate an important chemical principle of reversible reactions, called the **law of mass action**. Addition of a

2.17 Buffers Minimize Changes in pH When a base is added to a solution, the pH of the solution increases. Without a buffer, the change is large and the slope of the pH graph is steep. In the presence of a buffer, however, the slope within the buffering range is shallow.

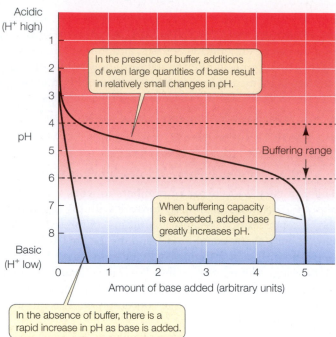

In the presence of buffer, additions of even large quantities of base result in relatively small changes in pH.

Buffering range

When buffering capacity is exceeded, added base greatly increases pH.

In the absence of buffer, there is a rapid increase in pH as base is added.

reactant on one side of a reversible system drives the reaction in the direction that uses up that compound. In the case of buffers, addition of an acid drives the reaction in one direction; addition of a base drives the reaction in the other direction.

We use a buffer to relieve the common problem of indigestion. The lining of the stomach constantly secretes hydrochloric acid, making the stomach contents acidic. But excessive stomach acid inhibits digestion and causes discomfort. We can relieve this discomfort by ingesting a salt such as $NaHCO_3$ (sodium bicarbonate), which acts as a buffer.

RECAP 2.4

Most of the chemistry of life occurs in water, which has unique properties that make it an ideal medium for supporting life. Aqueous solutions can be acidic or basic, depending on the concentration of hydrogen ions. The cells and tissues of organisms are buffered, however, because changes in pH can change the properties of biological molecules.

- What are some biologically important properties of water that arise from its molecular structure? **See pp. 32–33 and Figure 2.14**

- What is a solution, and why do we call water "the medium of life"? **See p. 33**

- What is the relationship among hydrogen ions, acids, and bases? Explain what the pH scale measures. **See p.35 and Figure 2.16**

- How does a buffer work, and why is buffering important to living systems? **See pp. 35–36 and Figure 2.17**

An Overview and a Preview

Now that we have covered the major properties of atoms and molecules, let's review them and see how these properties relate to the major molecules of biological systems.

- *Molecules vary in size.* Some are small, such as those of hydrogen gas (H_2) and methane (CH_4). Others are larger, such as a molecule of table sugar ($C_{12}H_{22}O_{11}$), which has 45 atoms. Still others, especially proteins and nucleic acids, are gigantic, containing tens of thousands or even millions of atoms.

- *Each molecule can have a specific three-dimensional shape.* For example, the orientations of the bonding orbitals around the carbon atom give the methane molecule (CH_4) the shape of a regular tetrahedron (see Figure 2.7B). Larger molecules have complex shapes that result from the numbers and kinds of atoms present, and the ways in which they are linked together. Some large molecules, such as the protein hemoglobin (the oxygen carrier in red blood cells), have compact, ball-like shapes. Others, such as the protein keratin that makes up hair, have long, thin,

ropelike structures. Their shapes relate to the roles these molecules play in living cells.

- *Molecules are characterized by certain chemical properties* that determine their biological roles. Chemists use atomic composition, structure (three-dimensional shape), reactivity, and solubility to distinguish a pure sample of one molecule from a sample of a different molecule. The presence of certain groups of atoms can impart distinctive chemical properties to a molecule.

Between the small molecules discussed in this chapter and the world of the living cell are the macromolecules. We will discuss these larger molecules—proteins, lipids, carbohydrates, and nucleic acids—in the next two chapters.

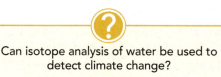

Can isotope analysis of water be used to detect climate change?

ANSWER

Water evaporates in warmer regions at the tropical latitudes on Earth and moves toward the cooler poles. As an air mass moves from a warmer to a cooler region, water vapor condenses and is removed as precipitation. The heavy isotopes of H and O tend to fall as precipitation more readily than the lighter isotopes, so as the water vapor moves toward the poles, it becomes enriched in the lighter isotopes. The ratio of heavy to light isotopes that reach the poles depends on the climate—the cooler the climate, the lower the ratio, because more water precipitates as it moves toward the poles, depleting more of the heavier isotopes. Analyses of polar ice cores show that heavy-to-light isotope ratios vary over geological time scales. This has allowed scientists to reconstruct climate change in the past, and to relate it to fossil organisms that lived at those times.

2.1 How Does Atomic Structure Explain the Properties of Matter?

- Matter is composed of atoms. Each **atom** consists of a positively charged **nucleus** made up of **protons** and **neutrons**, surrounded by **electrons** bearing negative charges. **Review Figure 2.1**

- The number of protons in the nucleus defines an **element**. There are many elements in the universe, but only a few of them make up the bulk of living organisms: C, H, O, P, N, and S. **Review Figure 2.2**

- **Isotopes** of an element differ in their numbers of neutrons. **Radioisotopes** are radioactive, emitting radiation as they break down.

- Electrons are distributed in **electron shells**, which are volumes of space defined by specific numbers of orbitals. Each **orbital** contains a maximum of two electrons. **Review Figures 2.4, 2.5, ACTIVITY 2.1**

- In losing, gaining, or sharing electrons to become more stable, an atom can combine with other atoms to form a **molecule**.

2.2 How Do Atoms Bond to Form Molecules?
See ANIMATED TUTORIAL 2.1

- A **chemical bond** is an attractive force that links two atoms together in a molecule. **Review Table 2.1**

- A **compound** is a substance made up of molecules with two or more different atoms bonded together in a fixed ratio, such as water (H_2O).

- **Covalent bonds** are strong bonds formed when two atoms share one or more pairs of electrons. **Review Figure 2.6**

- When two atoms of unequal electronegativity bond with each other, a **polar covalent bond** is formed. The two ends, or poles, of the bond have partial charges (δ^+ or δ^-). **Review Figure 2.8**

- An **ion** is an electrically charged body that forms when an atom gains or loses one or more electrons in order to form a more stable electron configuration. **Anions** and **cations** are negatively and positively charged ions, respectively. Different charges attract, and like charges repel each other.

- **Ionic attractions** occur between oppositely charged ions. Ionic attractions are strong in solids (salts) but weaken when the ions are separated from one another in solution. **Review Figure 2.9**

- A **hydrogen bond** is a weak electrical attraction that forms between a δ^+ hydrogen atom in one molecule and a δ^- atom in another molecule (or in another part of the same, large molecule). Hydrogen bonds are abundant in water. **Review Figure 2.11**

- Nonpolar molecules interact very little with polar molecules, including water. Nonpolar molecules are attracted to one another by very weak bonds called **van der Waals forces**.

2.3 How Do Atoms Change Partners in Chemical Reactions?

- In **chemical reactions**, atoms combine or change their bonding partners. **Reactants** are converted into **products**.

- Some chemical reactions release **energy** as one of their products; other reactions can occur only if energy is provided to the reactants.

- Neither matter nor energy is created or destroyed in a chemical reaction, but both change form. **Review Figure 2.13**

- Some chemical reactions, especially in biology, are reversible. That is, the products formed may be converted back to the reactants.

- In organisms, chemical reactions take place in multiple steps so that released energy can be harvested for cellular activities.

2.4 What Makes Water So Important for Life?

- Water's molecular structure and its capacity to form hydrogen bonds give it unique properties that are significant for life. **Review Figure 2.14**

- The high **specific heat** of water means that water gains or loses a great deal of heat when it changes state. Water's high **heat of vaporization** ensures effective cooling when water evaporates.

- The **cohesion** of water molecules refers to their capacity to resist coming apart from one another. Hydrogen bonding between the water molecules plays an essential role in this property.

- A **solution** is produced when a solid substance (the **solute**) dissolves in a liquid (the **solvent**). Water is the critically important solvent for life.

Go to the Interactive Summary to review key figures, Animated Tutorials, and Activities
Life10e.com/is2

CHAPTER**REVIEW**

■■■ REMEMBERING

1. The atomic number of an element
   a. equals the number of neutrons in an atom.
   b. equals the number of protons in an atom.
   c. equals the number of protons minus the number of neutrons.
   d. equals the number of neutrons plus the number of protons.
   e. depends on the isotope.

2. The mass number of an element
   a. equals the number of neutrons in an atom.
   b. equals the number of protons in an atom.
   c. equals the number of electrons in an atom.
   d. equals the number of neutrons plus the number of protons.
   e. depends on the relative abundances of its electrons and neutrons.

3. Which of the following statements about the isotopes of an element is *not* true?
   a. They all have the same atomic number.
   b. They all have the same number of protons.
   c. They all have the same number of neutrons.
   d. They all have the same number of electrons.
   e. They all have identical chemical properties.

4. Which of the following statements about covalent bonds is *not* true?
   a. A covalent bond is stronger than a hydrogen bond.
   b. A covalent bond can form between atoms of the same element.
   c. Only a single covalent bond can form between two atoms.
   d. A covalent bond results from the sharing of electrons by two atoms.
   e. A covalent bond can form between atoms of different elements.

5. Which of the following statements about water is *not* true?
   a. It releases a large amount of heat when changing from liquid into vapor.
   b. Its solid form is less dense than its liquid form.
   c. It is the most effective solvent for polar molecules.
   d. It is typically the most abundant substance in a living organism.
   e. It takes part in some important chemical reactions.

6. The reaction $HCl \rightarrow H^+ + Cl^-$ in the human stomach is an example of the
   a. cleavage of a hydrophobic bond.
   b. formation of a hydrogen bond.
   c. elevation of the pH of the stomach.
   d. formation of ions by dissociation of an acid.
   e. formation of polar covalent bonds.

■■■ UNDERSTANDING & APPLYING

7. Using the information in the periodic table (Figure 2.2), draw a Bohr model (see Figures 2.5 and 2.7) of silicon dioxide, showing electrons shared in covalent bonds.

8. Compare a covalent bond between two hydrogen atoms with a hydrogen bond between a hydrogen and an oxygen atom, with regard to the electrons involved, the role of polarity, and the strength of the bond.

9. Use Tables 2.2 and 2.3 to determine for each of the pairs of bonded atoms below:
   a. whether the bond is polar or nonpolar;
   b. if polar, which end is δ^-; and
   c. whether the bond is hydrophilic or hydrophobic.

   C–H   C=O   O–P   C–C

■■■ ANALYZING & EVALUATING

10. Geckos are lizards that are amazing climbers. A gecko can climb up a glass surface and stick to it with a single toe. Professor Kellar Autumn at Lewis and Clark College and his students and collaborators have shown that each toe of a gecko has millions of micrometer-sized hairs, and that each hair splits into hundreds of 200-nanometer tips that provide intimate contact with a surface. Careful measurements show that a million of these tips could easily support the animal, but it has far more. The toes stick well on hydrophilic and hydrophobic surfaces. Bending the hairs allows the gecko to detach. What kind of noncovalent force is involved in gecko sticking?

11. Would you expect the elemental composition of Earth's crust to be the same as that of the human body?

3 Proteins, Carbohydrates, and Lipids

CHAPTER**OUTLINE**

3.1 What Kinds of Molecules Characterize Living Things?

3.2 What Are the Chemical Structures and Functions of Proteins?

3.3 What Are the Chemical Structures and Functions of Carbohydrates?

3.4 What Are the Chemical Structures and Functions of Lipids?

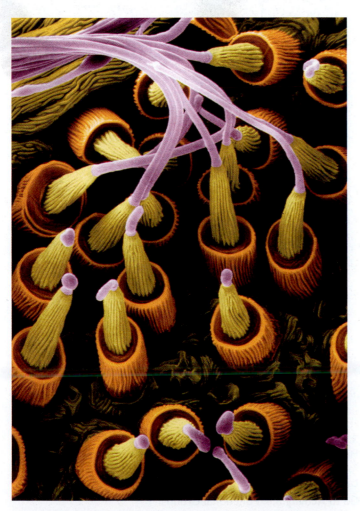

A Complex Macromolecule Spider silk (purple) being spun from a gland by the shiny black spider, *Castercantha*.

A SPIDER WEB is an amazing structure. It is not only beautiful to look at, but it is an architectural wonder that is the spider's home, its mating place, and its way to capture food. Think of a fly that chances to interact with a spider web. The fibers of the web must slow down the fly, but they cannot break, so they need to stretch to dissipate the energy of the fly's movement. The fibers holding the fly cannot stretch too much, however. They must be strong enough to hold the web in place and not let it wobble out of control. Web fibers are far thinner than a human hair, yet they are five times tougher than steel and in some cases more elastic than nylon. The fibers can also be long; for example, the Darwin's bark spider makes strands up to 25 meters long.

Spider silk is composed of variations on a single type of large molecule—a macromolecule called protein. Proteins are polymers: long chains of individual smaller units called amino acids. The proteins in spider silks have characteristic structures and amino acid compositions depending on their particular functions. Proteins in the stretchy web fibers have amino acids that allow them to curl into spirals, and these spirals can slip along one another to change the fiber's length. Another kind of spider silk is the dragline silk, which is less stretchy and used to construct the outline of the web, its spokes, and the lifeline of the spider. The proteins in these strong fibers are made up of amino acids that cause the proteins to fold into flat sheets with ratchets, so that parallel sheets can fit together like Lego blocks. This arrangement makes these fibers hard to pull apart. The relationship between chemical structure and biological function is a recurring theme in biochemistry, as you will see in this and the succeeding chapters.

Proteins are one of the four major kinds of large molecules that characterize living systems. These macromolecules, which also include carbohydrates, lipids, and nucleic acids, differ in several significant ways from the small molecules and ions described in Chapter 2. First—no surprise—they are larger; the molecular masses of some nucleic acids reach billions of daltons. Second, these molecules all contain carbon atoms, and so belong to a group known as organic compounds. Third, the atoms of individual macromolecules are held together mostly by covalent bonds, which gives them structural stability and distinctive three-dimensional geometries. These distinctive shapes are the basis of many of the functions of macromolecules, particularly the proteins.

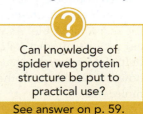

? Can knowledge of spider web protein structure be put to practical use?

See answer on p. 59.

3.1 What Kinds of Molecules Characterize Living Things?

Four kinds of molecules are characteristic of living things: proteins, carbohydrates, lipids, and nucleic acids. With the exception of the lipids, these biological molecules are **polymers** (*poly*, "many"; *mer*, "unit") constructed by the covalent bonding of smaller molecules called **monomers**. Each kind of biological molecule is made up of monomers with similar chemical structures:

- Proteins are formed from different combinations of 20 amino acids, all of which share chemical similarities.

- Carbohydrates can form giant molecules by linking together chemically similar sugar monomers (monosaccharides) to form polysaccharides.

- Nucleic acids are formed from four kinds of nucleotide monomers linked together in long chains.

- Lipids also form large structures from a limited set of smaller molecules, but in this case noncovalent forces maintain the interactions between the lipid monomers.

Polymers with molecular weights exceeding 1,000 are considered to be **macromolecules**. The proteins, carbohydrates, and nucleic acids of living systems certainly fall into this category. Although large lipid structures are not polymers in the strictest sense, it is convenient to treat them as a special type of macromolecule (see Section 3.4).

How the macromolecules function and interact with other molecules depends on the properties of certain chemical groups in their monomers, the functional groups.

Go to Animated Tutorial 3.1
Macromolecules
Life10e.com/at3.1

Functional groups give specific properties to biological molecules

Certain small groups of atoms, called **functional groups**, occur frequently in biological molecules (**Figure 3.1**). Each functional group has specific chemical properties, and when it is attached to a larger molecule, it confers those properties on the larger molecule. One of these properties is polarity. Looking at the structures in Figure 3.1, can you determine which functional groups are the most polar? (Hint: look for C—O, N—H, and P—O bonds.) The consistent chemical behavior of functional groups helps us understand the properties of the molecules that contain them.

Because macromolecules are so large, they contain many different functional groups. A single large protein may contain hydrophobic, polar, and charged functional groups, each of which gives different specific properties to local sites on the macromolecule. As we will see, sometimes these different groups interact within the same macromolecule. They help determine the shape of the macromolecule as well as how it interacts with other macromolecules and with smaller molecules.

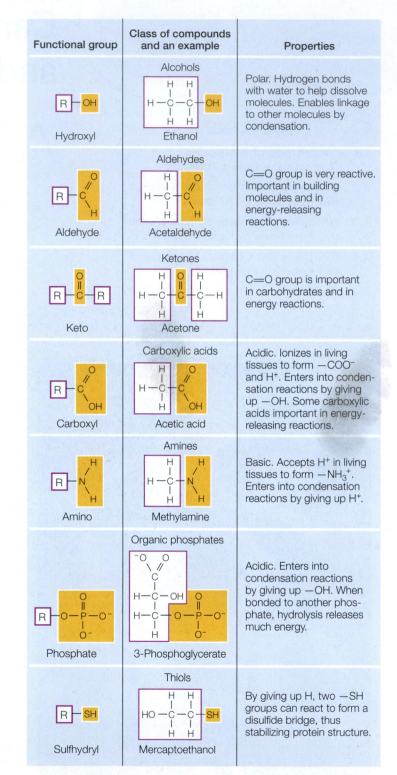

3.1 Some Functional Groups Important to Living Systems
Highlighted here are the seven functional groups most commonly found in biologically important molecules. "R" is a variable chemical grouping.

Go to Activity 3.1 Functional Groups
Life10e.com/ac3.1

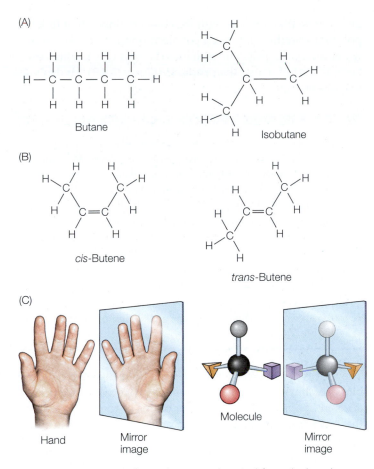

(A) Butane / Isobutane

(B) *cis*-Butene / *trans*-Butene

(C) Hand / Mirror image / Molecule / Mirror image

3.2 Isomers Isomers have the same chemical formula, but the atoms are arranged differently. Pairs of isomers often have different chemical properties.

Isomers have different arrangements of the same atoms

Isomers are molecules that have the same chemical formula—the same kinds and numbers of atoms—but with the atoms arranged differently. (The prefix *iso-*, meaning "same," is encountered in many biological terms.) Of the different kinds of isomers, we will consider three: structural isomers, *cis-trans* isomers, and optical isomers.

Structural isomers differ in how their atoms are joined together. Consider two simple molecules, each composed of four carbon and ten hydrogen atoms bonded covalently, both with the formula C_4H_{10}. These atoms can be linked in two different ways, resulting in different molecules (**Figure 3.2A**).

In biological molecules, **cis-trans isomers** typically involve a double bond between two carbon atoms, where the carbons share two pairs of electrons. When the remaining two bonds of each of these carbons are to two different atoms or groups of atoms (e.g., a hydrogen and a methyl group; **Figure 3.2B**), these can be oriented on the same side or different sides of the double-bonded molecule. If the different atoms or groups of atoms are on the same side, the double bond is called *cis*; if they are on opposite sides, the bond is *trans*. These molecules can have very different properties.

Optical isomers occur when a carbon atom has four different atoms or groups of atoms attached to it. This pattern allows for two different ways of making the attachments, each the mirror image of the other (**Figure 3.2C**). Such a carbon atom is called an asymmetric carbon, and the two resulting molecules are optical isomers of one another. You can envision your right and left hands as optical isomers. Just as a glove is specific for a particular hand, some biochemical molecules that can interact with one optical isomer of a carbon compound are unable to "fit" the other.

The structures of macromolecules reflect their functions

The four kinds of biological macromolecules are present in roughly the same proportions in all living organisms (**Figure 3.3**). Furthermore, a protein that has a certain function in an apple tree probably has a similar function in a human being, because the protein's chemistry is the same wherever it is found. Such biochemical unity reflects the evolution of all life from a common ancestor, by descent with modification. An important advantage of biochemical unity is that some organisms can acquire needed raw materials by eating other organisms. When you eat an apple, the molecules you take in include carbohydrates, lipids, and proteins that can be broken down and rebuilt into the varieties of those molecules needed by humans.

Each type of macromolecule performs one or more functions such as energy storage, structural support, catalysis (speeding up of chemical reactions), transport of other molecules, regulation of other molecules, defense, movement, or information storage. These roles are not necessarily exclusive; for example, both carbohydrates and proteins can play structural roles, supporting and protecting tissues and organs. However, only the nucleic acids specialize in information storage and transmission. These macromolecules function as hereditary material, carrying the traits of both species and individuals from generation to generation.

The functions of macromolecules are directly related to their three-dimensional shapes and to the sequences and chemical properties of their monomers. Some macromolecules fold into

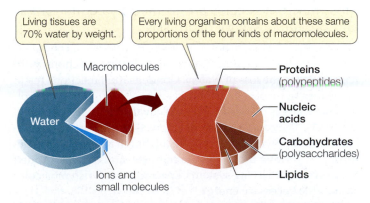

Living tissues are 70% water by weight.

Every living organism contains about these same proportions of the four kinds of macromolecules.

Macromolecules / Water / Ions and small molecules / Proteins (polypeptides) / Nucleic acids / Carbohydrates (polysaccharides) / Lipids

3.3 Substances Found in Living Tissues The substances shown here make up the nonmineral components of living tissues (bone would be an example of a mineral component).

(A) Condensation

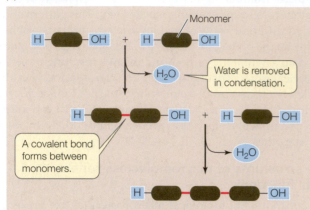

Monomer

Water is removed in condensation.

A covalent bond forms between monomers.

(B) Hydrolysis

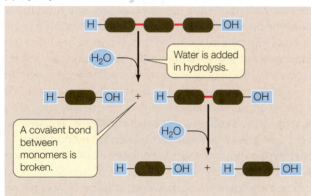

Water is added in hydrolysis.

A covalent bond between monomers is broken.

3.4 Condensation and Hydrolysis of Polymers   (A) Condensation reactions link monomers into polymers and produce water. (B) Hydrolysis reactions break polymers into individual monomers and consume water.

compact forms with surface features that make them water-soluble and capable of intimate interactions with other molecules. Some proteins and carbohydrates form long, fibrous structures (such as those found in hair or spider silk) that provide strength and rigidity to cells and tissues. The long, thin assemblies of proteins in muscles can contract, resulting in movement.

Most macromolecules are formed by condensation and broken down by hydrolysis

Polymers are formed from monomers by a series of **condensation reactions** (sometimes called dehydration reactions; both terms refer to the loss of water). Condensation reactions result in the formation of covalent bonds between monomers. A molecule of water is released with each covalent bond formed (**Figure 3.4A**). The condensation reactions that produce the different kinds of polymers differ in detail, but in all cases polymers form only if water molecules are removed and energy is added to the system. In living systems, specific energy-rich molecules supply the necessary energy.

The reverse of a condensation reaction is a **hydrolysis reaction** (*hydro*, "water"; *lysis*, "break"). Hydrolysis reactions result in the breakdown of polymers into their component

monomers. Water reacts with the covalent bonds that link the polymer together. For each covalent bond that is broken, a water molecule splits into two ions (H^+ and OH^-), which each become part of one of the products (**Figure 3.4B**). Hydrolysis releases energy.

RECAP 3.1

The four kinds of large molecules that distinguish living tissues are proteins, lipids, carbohydrates, and nucleic acids. Most are polymers: chains of linked monomers. Very large polymers are called macromolecules. Biological molecules carry out a variety of life-sustaining functions.

- How do functional groups affect the structures and functions of macromolecules? (Keep this question in mind as you read the rest of this chapter.) **See p. 40 and Figure 3.1**
- What are the differences between structural, *cis-trans*, and optical isomers? **See p. 41 and Figure 3.2**
- How do monomers link up to form polymers, and how do they break down into monomers again? **See p. 42 and Figure 3.4**

The four types of macromolecules can be seen as the building blocks of life. We will cover the unique properties of the nucleic acids in Chapter 4. The remainder of this chapter will describe the structures and functions of the proteins, carbohydrates, and lipids.

3.2 What Are the Chemical Structures and Functions of Proteins?

Proteins have very diverse roles. In virtually every chapter of this book you will study examples of their extensive functions (**Table 3.1**). Among the functions of macromolecules listed in Section

TABLE 3.1
Proteins and Their Functions

Category	Function
Enzymes	Catalyze (speed up) biochemical reactions
Structural proteins	Provide physical stability and movement
Defensive proteins	Recognize and respond to nonself substances (e.g., antibodies)
Signaling proteins	Control physiological processes (e.g., hormones)
Receptor proteins	Receive and respond to chemical signals
Membrane transporters	Regulate passage of substances across cellular membranes
Storage proteins	Store amino acids for later use
Transport proteins	Bind and carry substances within the organism
Gene regulatory proteins	Determine the rate of expression of a gene

3.1, only two—energy storage and information storage—are not usually performed by proteins.

All **proteins** are polymers made up of 20 amino acids in different proportions and sequences. Proteins range in size from small ones such as insulin, which has 51 amino acids and a molecular weight of 5,733, to huge molecules such as the muscle protein titin, with 26,926 amino acids and a molecular weight of 2,993,451. Proteins consist of one or more **polypeptide chains**—unbranched (linear) polymers of covalently linked amino acids. Variation in the sequences of amino acids in the polypeptide chains allows for the vast diversity in protein structure and function. Each chain folds into a particular three-dimensional shape that is specified by the sequence of amino acids present in the chain.

Amino acids are the building blocks of proteins

Each **amino acid** has both a carboxyl functional group and an amino functional group (see Figure 3.1) attached to the same carbon atom, called the α (alpha) carbon. Also attached to the α carbon atom are a hydrogen atom and a **side chain**, or **R group**, designated by the letter R.

The α carbon is asymmetrical because it is bonded to four different atoms or groups of atoms. Therefore, amino acids can exist as optical isomers called D-amino acids and L-amino acids. D and L are abbreviations of the Latin terms for right (*dextro*) and left (*levo*). Only L-amino acids (with the configuration shown above) are commonly found in the proteins of most organisms, and their presence is an important chemical "signature" of life.

At the pH levels typically found in cells (usually about pH 7), both the carboxyl and amino groups of amino acids are ionized: the carboxyl group has lost a hydrogen ion:

$$-COOH \rightarrow -COO^- + H^+$$

and the amino group has gained a hydrogen ion:

$$-NH_2 + H^+ \rightarrow -NH_3^+$$

Thus amino acids are simultaneously acids and bases.

The side chains (or R groups) of amino acids contain functional groups that are important in determining the three-dimensional structure and thus the function of the protein. As **Table 3.2** shows, the 20 amino acids found in living organisms are grouped and distinguished by their side chains:

- Five amino acids have electrically charged (ionized) side chains at pH levels typical of living cells. These side chains attract water (are hydrophilic) and attract oppositely charged ions of all sorts.
- Five amino acids have polar side chains. They are also hydrophilic and attract other polar or charged molecules.

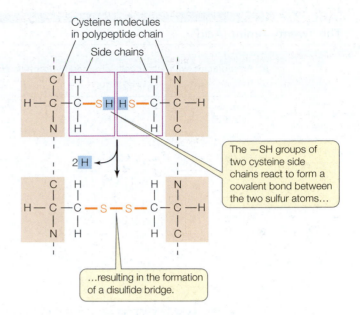

The —SH groups of two cysteine side chains react to form a covalent bond between the two sulfur atoms…

…resulting in the formation of a disulfide bridge.

3.5 A Disulfide Bridge Two cysteine molecules in a polypeptide chain can form a disulfide bridge (—S—S—) by oxidation (removal of H atoms).

- Seven amino acids have side chains that are nonpolar and thus hydrophobic. In the watery environment of the cell, these hydrophobic groups may cluster together in the interior of the protein.

Three amino acids—cysteine, glycine, and proline—are special cases, although the side chains of the latter two are generally hydrophobic.

- The cysteine side chain, which has a terminal —SH group, can react with another cysteine side chain in an oxidation reaction to form a covalent bond (**Figure 3.5**). Such a bond, called a **disulfide bridge** or disulfide bond (—S—S—), helps determine how a polypeptide chain folds.
- The glycine side chain consists of a single hydrogen atom. It is small enough to fit into tight corners in the interiors of protein molecules where larger side chains could not fit.
- Proline possesses a modified amino group that lacks a hydrogen and instead forms a covalent bond with the hydrocarbon side chain, resulting in a ring structure. This limits both its hydrogen-bonding ability and its ability to rotate about the α carbon. Thus proline is often found where a protein bends or loops.

Go to Activity 3.2 **Features of Amino Acids** Life10e.com/ac3.2

Peptide linkages form the backbone of a protein

When amino acids polymerize, the carboxyl and amino groups attached to the α carbon are the reactive groups. The carboxyl group of one amino acid reacts with the amino group of another, undergoing a condensation reaction that forms a **peptide linkage** (also called a peptide bond). **Figure 3.6** gives a simplified description of this reaction.

TABLE 3.2
The Twenty Amino Acids

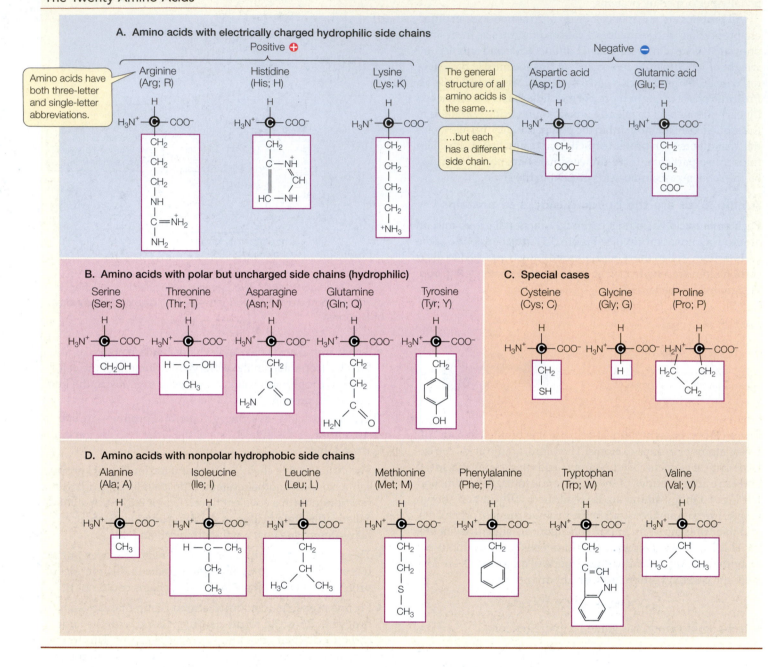

Just as a sentence begins with a capital letter and ends with a period, polypeptide chains have a beginning and an end. The "capital letter" marking the beginning of a polypeptide is the amino group of the first amino acid added to the chain and is known as the N terminus. The "period" is the carboxyl group of the last amino acid added; this is the C terminus.

Two characteristics of the peptide bond are especially important in the three-dimensional structures of proteins:

● In the C—N linkage, the adjacent α carbons (α-C—C—N—α-C) are not free to rotate fully, which limits the folding of the polypeptide chain.

● The oxygen bound to the carbon (C=O) in the carboxyl group carries a slight negative charge (δ^-), whereas the hydrogen bound to the nitrogen (N—H) in the amino group is slightly positive (δ^+). This asymmetry of charge favors hydrogen bonding within the protein molecule itself and between molecules. These bonds contribute to the structures and functions of many proteins.

In addition to these characteristics of the peptide linkage, the particular sequence of amino acids—with their various R groups—in the polypeptide chain also plays a vital role in determining a protein's structure and function.

3.6 Formation of Peptide Linkages In living things, the reaction leading to a peptide linkage (also called a peptide bond) has many intermediate steps, but the reactants and products are the same as those shown in this simplified diagram.

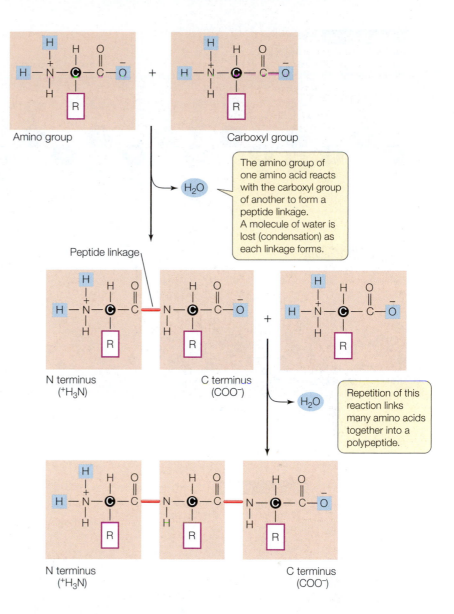

The amino group of one amino acid reacts with the carboxyl group of another to form a peptide linkage. A molecule of water is lost (condensation) as each linkage forms.

Repetition of this reaction links many amino acids together into a polypeptide.

The primary structure of a protein is its amino acid sequence

The precise sequence of amino acids in a polypeptide chain held together by peptide bonds constitutes the **primary structure** of a protein (**Figure 3.7A**). The backbone of the polypeptide chain consists of the repeating sequence —N—C—C— made up of the N atom from the amino group, the α C atom, and the C atom from the carboxyl group in each amino acid.

The single-letter abbreviations for amino acids (see Table 3.2) are used to record the amino acid sequence of a protein. Here, for example, are the first 20 amino acids (out of a total of 124) in the protein ribonuclease from a cow:

KETAAAKFERQHMDSSTSAA

The theoretical number of different proteins is enormous. Since there are 20 different amino acids, there could be $20 \times 20 = 400$ distinct dipeptides (two linked amino acids) and $20 \times 20 \times 20 = 8,000$ different tripeptides (three linked amino acids). Imagine this process of multiplying by 20 extended to a protein made up of 100 amino acids (which would be considered a small protein). There could be 20^{100} (that's approximately 10^{130}) such small proteins, each with its own distinctive primary structure. How large is the number 20^{100}? Physicists tell us that there aren't that many electrons in the entire universe.

The sequence of amino acids in the polypeptide chain(s) determines its final shape. The properties associated with each functional group in the side chains of the amino acids (see Table 3.2) determine how the protein can twist and fold, thus adopting a specific stable structure that distinguishes it from every other protein.

The secondary structure of a protein requires hydrogen bonding

A protein's **secondary structure** consists of regular, repeated spatial patterns in different regions of a polypeptide chain. There are two basic types of secondary structure, both determined by hydrogen bonding between the amino acids that make up the primary structure: the α helix and the β pleated sheet.

THE ALPHA HELIX The α (**alpha**) **helix** is a right-handed coil that turns in the same direction as a standard wood screw (**Figure 3.7B** and **Figure 3.8**). The R groups extend outward from the peptide backbone of the helix. The coiling results from hydrogen bonds that form between the δ⁺ hydrogen of the N—H of one amino acid and the δ⁻ oxygen of the C=O of another. When this pattern of hydrogen bonding is established repeatedly over a segment of the protein, it stabilizes the coil.

THE BETA PLEATED SHEET A β (**beta**) **pleated sheet** is formed from two or more polypeptide chains that are almost completely extended and aligned. The sheet is stabilized by hydrogen bonds between the N—H groups on one chain and the C=O groups on the other (**Figure 3.7C**). A β pleated sheet may form between separate polypeptide chains or between different regions of a single polypeptide chain that is bent back on itself. The ratcheted, stacked sheets in dragline spider silks (see the opening story at the beginning of the chapter) are made up of β pleated sheets. Many proteins contain regions of both α helix and β pleated sheet in the same polypeptide chain.

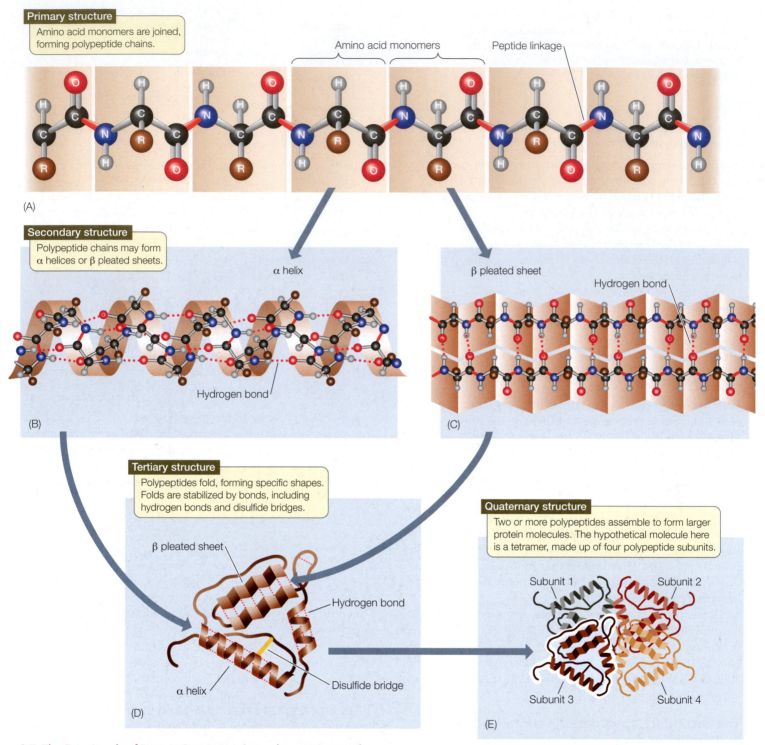

Primary structure
Amino acid monomers are joined, forming polypeptide chains.

Amino acid monomers

Peptide linkage

(A)

Secondary structure
Polypeptide chains may form α helices or β pleated sheets.

α helix

β pleated sheet

Hydrogen bond

Hydrogen bond

(B)

(C)

Tertiary structure
Polypeptides fold, forming specific shapes. Folds are stabilized by bonds, including hydrogen bonds and disulfide bridges.

β pleated sheet

Hydrogen bond

α helix

Disulfide bridge

(D)

Quaternary structure
Two or more polypeptides assemble to form larger protein molecules. The hypothetical molecule here is a tetramer, made up of four polypeptide subunits.

Subunit 1

Subunit 2

Subunit 3

Subunit 4

(E)

3.7 The Four Levels of Protein Structure   Secondary, tertiary, and quaternary structure all arise from the primary structure of the protein.

The tertiary structure of a protein is formed by bending and folding

In many proteins, the polypeptide chain is bent at specific sites and then folded back and forth, resulting in the **tertiary structure** of the protein (**Figure 3.7D**). Although α helices and β pleated sheets contribute to the tertiary structure, usually only portions of the macromolecule have these secondary structures, and large regions consist of tertiary structure unique to a particular protein. For example, the proteins found in stretchy spider silks have repeated amino acid sequences that cause the proteins to fold into structures called right-handed β-spirals. Tertiary structure is a macromolecule's definitive three-dimensional shape, often including a buried interior as well as a surface that is exposed to the environment.

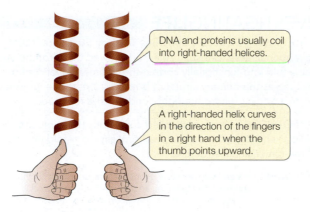

3.8 Left- and Right-Handed Helices A protein will often have one or more right-handed helices as part of its secondary structure.

The protein's exposed outer surfaces present functional groups capable of interacting with other molecules in the cell. These molecules might be other macromolecules, including proteins, nucleic acids, carbohydrates, and lipid structures, or smaller chemical substances.

Whereas hydrogen bonding between the N—H and C=O groups within and between chains is responsible for secondary structure, the interactions between R groups—the amino acid side chains—and between R groups and the environment determine tertiary structure. We described the various strong and weak interactions between atoms in Section 2.2. Many of these interactions are involved in determining and maintaining tertiary structure.

- Covalent disulfide bridges can form between specific cysteine side chains (see Figure 3.5), holding a folded polypeptide in place.

- Hydrogen bonds between side chains also stabilize folds in proteins.

- Hydrophobic side chains can aggregate together in the interior of the protein, away from water, folding the polypeptide in the process. Close interactions between the hydrophobic side chains are stabilized by van der Waals forces.

- Ionic attractions can form between positively and negatively charged side chains, forming salt bridges between amino acids. Salt bridges can be near the surfaces of polypeptides or buried deep within a protein, away from water. These interactions occur between positively and negatively charged amino acids, for example glutamic acid (which has a negatively charged R group) and arginine (which is positively charged) (see Table 3.2):

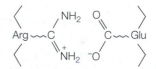

A complete description of a protein's tertiary structure would specify the location of every atom in the molecule in three-dimensional space relative to all the other atoms. **Figure 3.9** shows three models of the structure of the protein lysozyme. The space-filling model might be used to study how other molecules interact with specific sites and R groups on the protein's surface. The stick model emphasizes the sites where bends occur, resulting in

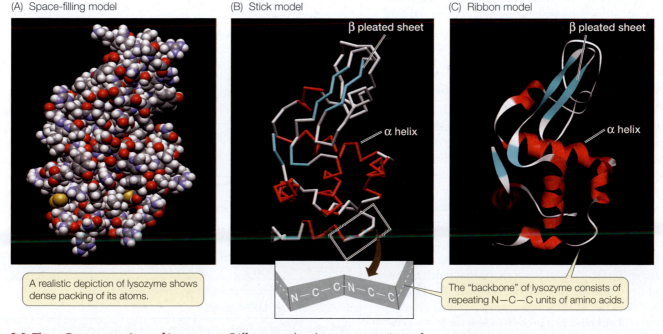

(A) Space-filling model

A realistic depiction of lysozyme shows dense packing of its atoms.

(B) Stick model

β pleated sheet

α helix

The "backbone" of lysozyme consists of repeating N—C—C units of amino acids.

(C) Ribbon model

β pleated sheet

α helix

3.9 Three Representations of Lysozyme Different molecular representations of a protein emphasize different aspects of its tertiary structure: surface features, sites of bends and folds, or sites where alpha or beta structures predominate. These three representations of lysozyme are similarly oriented.

 Go to Media Clip 3.1
Protein Structures in 3D
Life10e.com/mc3.1

folds in the polypeptide chain. The ribbon model, perhaps the most widely used, shows the different types of secondary structure and how they fold into the tertiary structure.

Remember that both secondary and tertiary structure derive from primary structure. If a protein is heated slowly and moderately, the heat energy will disrupt only the weak interactions, causing the secondary and tertiary structure to break down. The protein is then said to be **denatured**. But in some cases the protein can return to its normal tertiary structure when it cools, demonstrating that all the information needed to specify the unique shape of a protein is contained in its primary structure. This was first shown (using chemicals instead of heat to denature the protein) by biochemist Christian Anfinsen for the protein ribonuclease (**Figure 3.10**).

The quaternary structure of a protein consists of subunits

Many functional proteins contain two or more polypeptide chains, called subunits, each of them folded into its own unique tertiary structure. The protein's **quaternary structure** results from the ways in which these subunits bind together and interact (**Figure 3.7E**).

The models of hemoglobin in **Figure 3.11** illustrate quaternary structure. Hydrophobic interactions, van der Waals forces, hydrogen bonds, and ionic attractions all help hold the four subunits together to form a hemoglobin molecule. However, the weak nature of these forces permits small changes in the quaternary structure to aid the protein's function—which is to carry oxygen in red blood cells. As hemoglobin binds one O_2 molecule, the four subunits shift their relative positions slightly, changing the quaternary structure. Ionic attractions are broken, exposing buried side chains that enhance the binding of additional O_2 molecules. The quaternary structure changes back when hemoglobin releases its O_2 molecules to the cells of the body.

Shape and surface chemistry contribute to protein function

The shapes and structures of proteins allow specific sites on their exposed surfaces to bind noncovalently to other molecules, which may be large or small. The binding is usually very specific because only certain compatible chemical groups will bind to one another. The specificity of protein binding depends on two general properties of the protein: its shape, and the chemistry of its exposed surface groups.

- *Shape.* When a small molecule collides with and binds to a much larger protein, it is like a baseball being caught by a catcher's mitt: the mitt has a shape that binds to the ball and fits around it. Just as a hockey puck or a Ping-Pong ball does not fit a baseball catcher's mitt, a given molecule will not bind to a protein unless there is a general "fit" between their three-dimensional shapes.

INVESTIGATING**LIFE**

3.10 Primary Structure Specifies Tertiary Structure Using the protein ribonuclease, Christian Anfinsen showed that proteins spontaneously fold into functionally correct three-dimensional configurations.[a] As long as the primary structure is not disrupted, the information for correct folding (under the right conditions) is retained.

HYPOTHESIS Under controlled conditions that simulate the normal cellular environment, a denatured protein can refold into a functional three-dimensional structure.

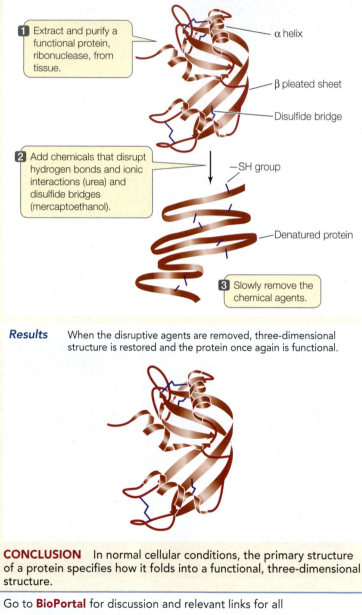

Method Chemically denature a functional ribonuclease so that only its primary structure (i.e., an unfolded polypeptide chain) remains. Once denaturation is complete, remove the disruptive chemicals.

1 Extract and purify a functional protein, ribonuclease, from tissue.

α helix

β pleated sheet

Disulfide bridge

2 Add chemicals that disrupt hydrogen bonds and ionic interactions (urea) and disulfide bridges (mercaptoethanol).

SH group

Denatured protein

3 Slowly remove the chemical agents.

Results When the disruptive agents are removed, three-dimensional structure is restored and the protein once again is functional.

CONCLUSION In normal cellular conditions, the primary structure of a protein specifies how it folds into a functional, three-dimensional structure.

Go to **BioPortal** for discussion and relevant links for all INVESTIGATING**LIFE** figures.

[a]Anfinsen, C. B. et al. 1961. *Proceedings of the National Academy of Sciences USA* 47: 1309–1314.

WORKING WITH **DATA:**

Primary Structure Specifies Tertiary Structure

Original Papers

Anfinsen, C. B., E. Haber, M. Sela, and F. White, Jr. 1961. The kinetics of formation of native ribonuclease during oxidation of the reduced polypeptide chain. *Proceedings of the National Academy of Sciences USA* 47: 1309–1314.

White, Jr., F. 1961. Regeneration of native secondary and tertiary structures by air oxidation of reduced ribonuclease. *Journal of Biological Chemistry* 236: 1353–1360.

Analyze the Data

After the tertiary structures of proteins were shown to be highly specific, the question arose as to how the order of amino acids determined the three-dimensional structure. The second protein whose structure was determined was ribonuclease A (RNase A). This enzyme was readily available from cow pancreases at slaughterhouses and, because it works in the highly acidic environment of the cow stomach, was stable compared with most proteins and easy to purify. RNase A has 124 amino acids. Among these are eight cysteine residues, which form four disulfide bridges. Were these covalent links between cysteines essential for the three-dimensional structure of RNase A? Christian Anfinsen and his colleagues set out to answer this question. They first destroyed these links by reducing the S—S bonds to —SH and —SH. With the links destroyed, they looked at the three-dimensional structure of the protein (the extent of denaturation) and assessed protein function by measuring the loss of enzyme activity. They then removed the reducing agent (mercaptoethanol) and allowed the S—S bonds to re-form. They

found that links between amino acids were indeed essential for tertiary structure and function. Anfinsen was awarded the Nobel Prize in Chemistry in 1973.

QUESTION 1

Initially, the disulfide bonds (S—S) in RNase A were eliminated because the sulfur atoms in cysteine residues were all reduced (—SH). At time zero, reoxidation began; and at various times, the amount of S—S bond re-formation and the activity of the enzyme were measured by chemical methods. The data are shown in **FIGURE A**.

At what time did disulfide bonds begin to form? At what time did enzyme activity begin to appear? Explain the difference between these times.

QUESTION 2

The three-dimensional structure of RNase A was examined by ultraviolet spectroscopy. In this technique, the protein was exposed to different wavelengths of ultraviolet light (measured in nanometers) and the amount of light absorbed by the protein at each wavelength was measured (E). The results are plotted in **FIGURE B**.

Look carefully at the plots. What are the differences between the peak absorbances of native (untreated) and reduced (denatured) RNase A? What happened when reduced RNase A was reoxidized (renatured)? What can you conclude about the structure of RNase A from these experiments?

FIGURE A

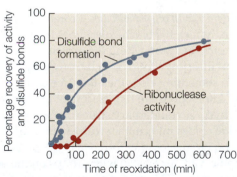

FIGURE B

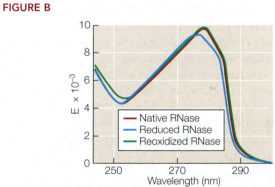

Go to **BioPortal** for all WORKING WITH **DATA** exercises

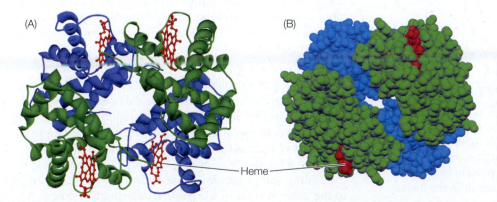

3.11 Quaternary Structure of a Protein Hemoglobin consists of four folded polypeptide subunits that assemble themselves into the quaternary structure represented by the ribbon model (A) and space-filling model (B). In both graphic representations, each type of subunit is a different color (α subunits are blue and β subunits are green). The heme groups (red) contain iron and are the oxygen-carrying sites.

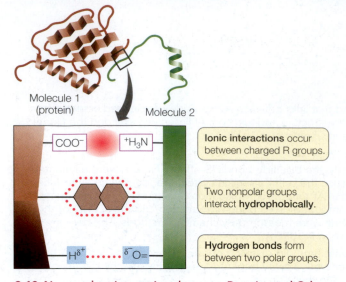

3.12 Noncovalent Interactions between Proteins and Other Molecules Noncovalent interactions (see p. 26) allow a protein (brown) to bind tightly to another molecule (green) with specific properties. Noncovalent interactions also allow regions within the same protein to interact with one another.

- *Chemistry.* The exposed R groups on the surface of a protein permit chemical interactions with other substances (**Figure 3.12**). Three types of interactions may be involved: ionic, hydrophobic, or hydrogen bonding. Many important functions of proteins involve interactions between surface R groups and other molecules.

Environmental conditions affect protein structure

Because they are determined by weak forces, the three-dimensional structures of proteins are influenced by environmental conditions. Conditions that would not break covalent bonds can disrupt the weaker, noncovalent interactions that determine secondary, tertiary, and quaternary structure. Such alterations may affect a protein's shape and thus its function. Various conditions can alter the weak, noncovalent interactions:

- Increases in temperature cause more rapid molecular movements and thus can break hydrogen bonds and hydrophobic interactions.
- Alterations in pH can change the pattern of ionization of exposed carboxyl and amino groups in the R groups of amino acids, thus disrupting the pattern of ionic attractions and repulsions.
- High concentrations of polar substances such as urea can disrupt the hydrogen bonding that is crucial to protein structure. Urea was used in the experiment on reversible protein denaturation shown in Figure 3.10.
- Nonpolar substances may also disrupt normal protein structure in cases where hydrophobic interactions are essential to maintain the structure.

Although denaturation is reversible in many cases (see Figure 3.10), in other cases it can be irreversible, such as when amino acids that were buried in the interior of the protein become exposed at the surface, or vice versa. This can result in the formation of new structures with different properties. Boiling an egg denatures its proteins and is, as you know, not reversible.

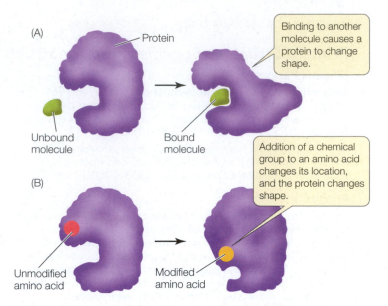

3.13 Protein Structure Can Change Proteins can change their tertiary structure when they bind to other molecules (A) or are modified chemically (B).

Protein shapes can change

As we saw in the case of hemoglobin, which undergoes subtle shape changes when it binds oxygen, the shapes of proteins can change as a result of their interactions with other molecules. Proteins can also change shape if they undergo covalent modifications.

- *Proteins interact with other molecules.* Proteins do not exist in isolation. In fact, if a biochemist "goes fishing" with a particular protein, by attaching the protein to a chemical "hook" and inserting it into cells, the protein will often be attached to something else when it is "reeled in." These molecular interactions are reminiscent of the interactions that make up quaternary structure (see above). If a polypeptide comes into contact with another molecule, R groups on its surface may form weak interactions (e.g., hydrophobic, van der Waals) with groups on the surface of the other molecule. This may disrupt some of the interactions between R groups within the polypeptide, causing it to undergo a change in shape (**Figure 3.13A**). You will see many instances of this in the coming chapters. An important example is an enzyme, which changes shape when it comes into contact with a reactant in a biochemical reaction (see Section 8.4).
- *Proteins undergo covalent modifications.* After it is made, the structure of a protein can be modified by the covalent bonding of a chemical group to the side chain of one or more of its amino acids. The chemical modification of just one amino acid can alter the shape and function of a protein. An example is the addition of a charged phosphate group to a relatively nonpolar R group. This can cause the amino acid to become more hydrophilic and to move to the outer surface of the protein, altering the shape of the protein in the region near the amino acid (**Figure 3.13B**).

Molecular chaperones help shape proteins

Within a living cell, a polypeptide chain is sometimes in danger of binding the wrong substance. There are two major situations when this can occur:

- *Just after a protein is made.* When a protein has not yet folded completely, it can present a surface that binds the wrong molecule.

- *Following denaturation.* Certain conditions, such as moderate heat, can cause some proteins in a living cell to denature without killing the organism. Before the protein can re-fold, it may present a surface that binds the wrong molecule. In these cases, the inappropriate binding may be irreversible. Many cells have a special class of proteins, called **chaperones**, that protect the three-dimensional structures of other proteins. Like the chaperones at a high school dance, they prevent inappropriate interactions and enhance appropriate ones. Typically, a chaperone protein has a cagelike structure that pulls in a polypeptide, causes it to fold into the correct shape, and then releases it (**Figure 3.14**). Tumors make chaperone proteins, possibly to stabilize proteins important in the cancer process, and so chaperone-inhibiting drugs are being designed for use in chemotherapy. In some clinical situations, treatment with these inhibitors results in the inappropriate folding of proteins in tumor cells, causing the tumors to stop growing.

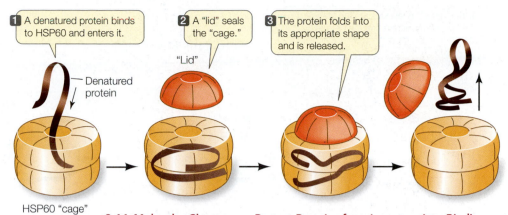

1 A denatured protein binds to HSP60 and enters it.

2 A "lid" seals the "cage."

3 The protein folds into its appropriate shape and is released.

"Lid"

Denatured protein

HSP60 "cage"

3.14 Molecular Chaperones Protect Proteins from Inappropriate Binding Chaperone proteins surround new or denatured proteins and prevent them from binding to the wrong substances. Heat shock proteins such as HSP60, shown here, make up one class of chaperone proteins.

RECAP 3.2

Proteins are polymers of amino acids. The sequence of amino acids in a protein determines its primary structure. Secondary, tertiary, and quaternary structures arise through interactions among the amino acids. A protein's three-dimensional shape and exposed chemical groups establish its binding specificity for other substances.

- What are the attributes of an amino acid's R group that would make it hydrophobic? Hydrophilic? **See p. 43 and Table 3.2**

- Sketch and explain how two amino acids link together to form a peptide linkage. **See pp. 43–45 and Figure 3.6**

- What are the four levels of protein structure, and how are they all ultimately determined by the protein's primary structure (i.e., its amino acid sequence)? **See pp. 45–48 and Figure 3.7**

- How do environmental factors such as temperature and pH affect the weak interactions that give a protein its specific shape and function? **See p. 50**

The seemingly infinite number of protein configurations made possible by the biochemical properties of the 20 amino acids has driven the evolution of life's diversity. The linkage configurations of sugar monomers (monosaccharides) determine the structures of the next group of macromolecules, the carbohydrates, which provide energy for life.

3.3 What Are the Chemical Structures and Functions of Carbohydrates?

Carbohydrates make up a large group of molecules that all have similar atomic compositions but differ greatly in size, chemical properties, and biological functions. Carbohydrates usually have the general formula $C_mH_{2n}O_n$, (where m and n stand for numbers), which makes them appear as hydrates of carbon [associations between water molecules and carbon in the ratio $C_m(H_2O)_n$], hence their name. However, carbohydrates are not really "hydrates" because the water molecules are not intact. Rather, the linked carbon atoms are bonded with hydrogen atoms (—H) and hydroxyl groups (—OH), the components of water. Carbohydrates have three major biochemical roles:

- They are a source of stored energy that can be released in a form usable by organisms.

- They are used to transport stored energy within complex organisms.

- They serve as carbon skeletons that can be rearranged to form new molecules.

Some carbohydrates are relatively small, with molecular weights of less than 100. Others are true macromolecules, with molecular weights in the hundreds of thousands.

There are four categories of biologically important carbohydrate defined by the number of monomers:

- **Monosaccharides** (*mono*, "one"; *saccharide*, "sugar"), such as glucose, are simple sugars. They are the monomers from which the larger carbohydrates are constructed.

- **Disaccharides** (*di*, "two") consist of two monosaccharides linked together by covalent bonds. The most familiar is sucrose, which is made up of covalently bonded glucose and fructose molecules.

- **Oligosaccharides** (*oligo*, "several") are made up of several (3–20) monosaccharides.

- **Polysaccharides** (*poly*, "many"), such as starch, glycogen, and cellulose, are polymers made up of hundreds or thousands of monosaccharides.

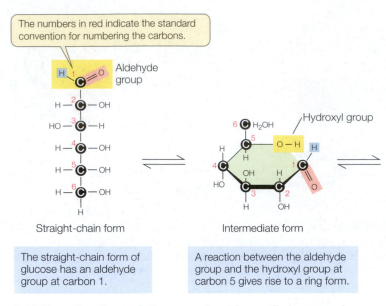

The numbers in red indicate the standard convention for numbering the carbons.

Aldehyde group

Hydroxyl group

The dark line indicates that the edge of the molecule extends toward you; the thin line extends back away from you.

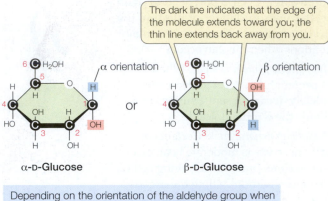

α orientation

or

β orientation

α-D-Glucose

β-D-Glucose

Straight-chain form

The straight-chain form of glucose has an aldehyde group at carbon 1.

Intermediate form

A reaction between the aldehyde group and the hydroxyl group at carbon 5 gives rise to a ring form.

Depending on the orientation of the aldehyde group when the ring closes, either of two molecules—α-D-glucose or β-D-glucose—forms.

3.15 From One Form of Glucose to the Other All glucose molecules have the formula $C_6H_{12}O_6$, but their structures vary. When dissolved in water, the α and β "ring" forms of glucose interconvert. The convention used here for numbering the carbon atoms is standard in biochemistry.

Go to Activity 3.3 Forms of Glucose   Life10e.com/ac3.3

Monosaccharides are simple sugars

All living cells contain the monosaccharide **glucose**; it is the familiar "blood sugar," used to transport energy in humans. Cells use glucose as an energy source, breaking it down through a series of reactions that release stored energy and produce water and carbon dioxide; this is a cellular form of the combustion reaction described in Section 2.3.

Glucose exists in straight chains and in ring forms. The ring forms predominate in virtually all biological circumstances because they are more stable under physiological conditions. There are two versions of the glucose ring, called α- and β-glucose, which differ only in the orientation of the —H and —OH groups attached to carbon 1 (**Figure 3.15**). The α and β forms interconvert and exist in equilibrium when dissolved in water.

Different monosaccharides contain different numbers of carbons. Some monosaccharides are structural isomers, with the same kinds and numbers of atoms but in different arrangements (**Figure 3.16**). Such seemingly small structural changes can significantly alter their properties. Most of the monosaccharides in living systems belong to the D (right-handed) series of optical isomers.

Pentoses (*pente*, "five") are five-carbon sugars. Two pentoses are of particular biological importance: the backbones of the nucleic acids RNA and DNA contain ribose and deoxyribose, respectively (see Section 4.1). These two pentoses are not isomers of each other; rather, one oxygen atom is missing from carbon 2 in deoxyribose (*de-*, "absent"). The absence of this oxygen atom is an important distinction between RNA and DNA.

The **hexoses** (*hex*, "six") shown in Figures 3.15 and 3.16 are a group of structural isomers with the formula $C_6H_{12}O_6$. Common hexoses are glucose, fructose (so named because it was first found in fruits), mannose, and galactose.

Three-carbon sugar

Glyceraldehyde is the smallest monosaccharide and exists only as the straight-chain form.

Glyceraldehyde

Five-carbon sugars (pentoses)

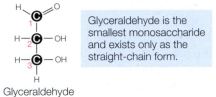

Ribose and deoxyribose each have five carbons, but very different chemical properties and biological roles.

Ribose   Deoxyribose

Six-carbon sugars (hexoses)

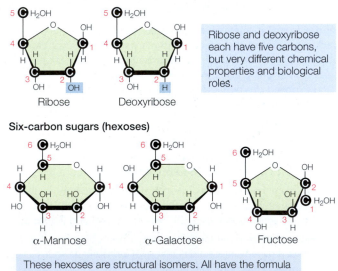

α-Mannose   α-Galactose   Fructose

These hexoses are structural isomers. All have the formula $C_6H_{12}O_6$, but each has distinct biochemical properties.

3.16 Monosaccharides Are Simple Sugars Monosaccharides are made up of varying numbers of carbons. Some hexoses are structural isomers that have the same kind and number of atoms, but the atoms are arranged differently. Fructose, for example, is a hexose but forms a five-membered ring like the pentoses.

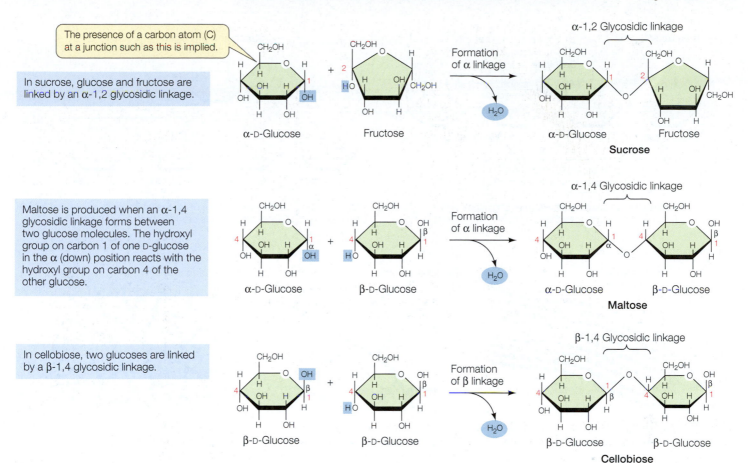

The presence of a carbon atom (C) at a junction such as this is implied.

In sucrose, glucose and fructose are linked by an α-1,2 glycosidic linkage.

α-1,2 Glycosidic linkage

α-D-Glucose + Fructose → (Formation of α linkage) → Sucrose

Maltose is produced when an α-1,4 glycosidic linkage forms between two glucose molecules. The hydroxyl group on carbon 1 of one D-glucose in the α (down) position reacts with the hydroxyl group on carbon 4 of the other glucose.

α-1,4 Glycosidic linkage

α-D-Glucose + β-D-Glucose → (Formation of α linkage) → Maltose

In cellobiose, two glucoses are linked by a β-1,4 glycosidic linkage.

β-1,4 Glycosidic linkage

β-D-Glucose + β-D-Glucose → (Formation of β linkage) → Cellobiose

3.17 Disaccharides Form by Glycosidic Linkages Glycosidic linkages between two monosaccharides can create many different disaccharides. The particular disaccharide formed depends on which monosaccharides are linked, on the site of linkage (i.e., which carbon atoms are involved), and on the form (α or β) of the linkage.

Glycosidic linkages bond monosaccharides

The disaccharides, oligosaccharides, and polysaccharides are all constructed from monosaccharides that are covalently bonded together by condensation reactions that form **glycosidic linkages** (**Figure 3.17**). A single glycosidic linkage between two monosaccharides forms a disaccharide. For example, sucrose—common table sugar in the human diet and a major disaccharide in plants—is formed from a glucose and a fructose molecule. The disaccharides maltose and cellobiose are made from two glucose molecules (see Figure 3.17). Maltose and cellobiose are structural isomers, both having the formula $C_{12}H_{22}O_{11}$. However, they have different chemical properties and are recognized by different enzymes in biological tissues. For example, maltose can be hydrolyzed into its monosaccharides in the human body, whereas cellobiose cannot.

Oligosaccharides contain several monosaccharides bound by glycosidic linkages at various sites. Many oligosaccharides have additional functional groups, which give them special properties. Oligosaccharides are often covalently bonded to proteins and lipids on the outer cell surface, where they serve as recognition signals. The different human blood groups (for example, the ABO blood types) get their specificities from oligosaccharide chains.

Polysaccharides store energy and provide structural materials

Polysaccharides are large (sometimes gigantic) polymers of monosaccharides connected by glycosidic linkages (**Figure 3.18**). In contrast to proteins, polysaccharides are not necessarily linear chains of monomers. Each monomer unit has several sites that are capable of forming glycosidic linkages, and thus branched molecules are possible.

STARCH **Starches** comprise a family of giant molecules of broadly similar structure. While all starches are polysaccharides of glucose with α-glycosidic linkages (α–1,4 and α–1,6 glycosidic bonds; see Figure 3.18A), the different starches can be distinguished by the amount of branching that occurs at carbons 1 and 6 (see Figure 3.18B). Starch is the principal energy storage compound of plants. Some plant starches, such as amylose, are unbranched; others are moderately branched (for example, amylopectin). Starch readily binds water. When the water is removed, however, hydrogen bonds tend to form between the unbranched polysaccharide chains, which then aggregate. Large starch aggregates called starch grains can be observed in the storage tissues of plant seeds (see Figure 3.18C).

GLYCOGEN **Glycogen** is a water-insoluble, highly branched polymer of glucose. It is used to store glucose in the liver and muscles and is thus an energy storage compound for animals, as starch is for plants. Both glycogen and starch are readily

(A) Molecular structure

Cellulose

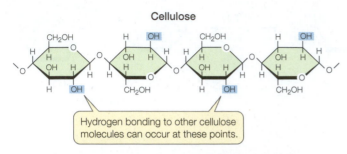

Hydrogen bonding to other cellulose molecules can occur at these points.

Cellulose is an unbranched polymer of glucose with β-1,4 glycosidic linkages that are chemically very stable.

Starch and glycogen

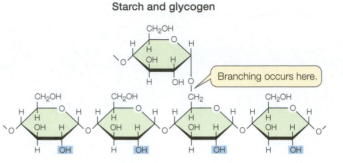

Branching occurs here.

Glycogen and starch are polymers of glucose with α-1,4 glycosidic linkages. α-1,6 Glycosidic linkages produce branching at carbon 6.

(B) Macromolecular structure

Linear (cellulose)

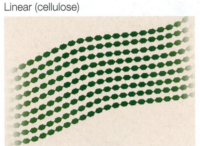

Parallel cellulose molecules form hydrogen bonds, resulting in thin fibrils.

Branched (starch)

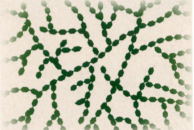

Branching limits the number of hydrogen bonds that can form in starch molecules, making starch less compact than cellulose.

Highly branched (glycogen)

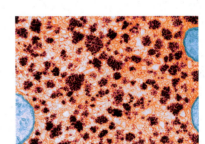

The high amount of branching in glycogen makes its solid deposits more compact than starch.

(C) Polysaccharides in cells

Layers of cellulose fibrils, as seen in this scanning electron micrograph, give plant cell walls great strength.

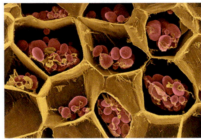

Within these potato cells, starch deposits (colored red in this scanning electron micrograph) have a granular shape.

The dark clumps in this electron micrograph are glycogen deposits.

3.18 Representative Polysaccharides Cellulose, starch, and glycogen have different levels of branching and compaction of the polysaccharides.

hydrolyzed into glucose monomers, which in turn can be broken down to liberate their stored energy.

But if it is glucose that is needed for fuel, why store it in the form of glycogen? The reason is that 1,000 glucose molecules would exert 1,000 times the osmotic pressure of a single glycogen molecule, causing water to enter cells where glucose is stored (see Section 6.3). If it were not for polysaccharides, many organisms would expend a lot of energy expelling excess water from their cells.

CELLULOSE As the predominant component of plant cell walls, **cellulose** is by far the most abundant organic compound on Earth. Like starch and glycogen, cellulose is a polysaccharide of glucose, but its individual monosaccharides are connected by β- rather than by α-glycosidic linkages. Starch is easily degraded by the actions of chemicals or enzymes. Cellulose, however, is chemically more stable because of its β-glycosidic linkages. Thus whereas starch is easily broken down to supply glucose for energy-producing reactions, cellulose is an excellent structural material that can withstand harsh environmental conditions without substantial change.

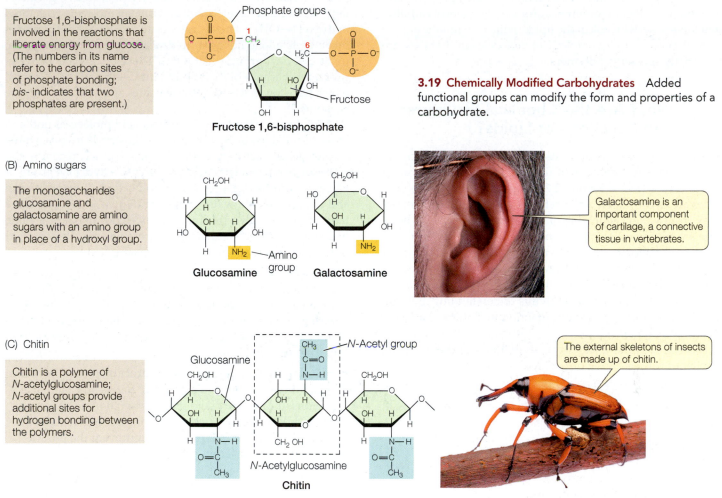

(A) Sugar phosphate

Fructose 1,6-bisphosphate is involved in the reactions that liberate energy from glucose. (The numbers in its name refer to the carbon sites of phosphate bonding; *bis-* indicates that two phosphates are present.)

Phosphate groups

Fructose

Fructose 1,6-bisphosphate

3.19 Chemically Modified Carbohydrates Added functional groups can modify the form and properties of a carbohydrate.

(B) Amino sugars

The monosaccharides glucosamine and galactosamine are amino sugars with an amino group in place of a hydroxyl group.

CH₂OH

Amino group

Glucosamine

CH₂OH

Galactosamine

Galactosamine is an important component of cartilage, a connective tissue in vertebrates.

(C) Chitin

Chitin is a polymer of *N*-acetylglucosamine; *N*-acetyl groups provide additional sites for hydrogen bonding between the polymers.

Glucosamine

N-Acetyl group

CH₂OH

CH₂OH

N-Acetylglucosamine

Chitin

The external skeletons of insects are made up of chitin.

Chemically modified carbohydrates contain additional functional groups

Some carbohydrates are chemically modified by oxidation–reduction reactions, or by the addition of functional groups such as phosphate, amino, or *N*-acetyl groups (**Figure 3.19**). For example, carbon 6 in glucose may be oxidized from —CH₂OH to a carboxyl group (—COOH), producing glucuronic acid. Or a phosphate group may be added to one or more of the —OH sites. Some of the resulting sugar phosphates, such as fructose 1,6-bisphosphate, are important intermediates in cellular energy reactions, which we will discuss in Chapter 9.

When an amino group is substituted for an —OH group, amino sugars, such as glucosamine and galactosamine, are produced. These compounds are important in the extracellular matrix (see Section 5.4), where they form parts of glycoproteins, which are molecules involved in keeping tissues together. Galactosamine is a major component of cartilage, the material that forms caps on the ends of bones and stiffens the ears and nose. A derivative of glucosamine is present in the polymer chitin, the principal structural polysaccharide in the external skeletons of insects and many crustaceans (such as crabs and lobsters),

and a component of the cell walls of fungi. Because these are among the most abundant complex organisms on Earth, chitin rivals cellulose as one of the most abundant substances in the living world.

RECAP 3.3

Carbohydrates are composed of carbon, hydrogen, and oxygen and have the general formula $C_mH_{2n}O_n$. They provide energy and structure to cells and are precursors of numerous important biological molecules. Monosaccharide monomers can be connected by glycosidic linkages to form disaccharides, oligosaccharides, and polysaccharides.

- Draw the chemical structure of a disaccharide formed from two monosaccharides. **See Figure 3.17**

- What qualities of the polysaccharides starch and glycogen make them useful for energy storage? **See pp. 53–54 and Figure 3.18**

- After looking at the cellulose molecule in Figure 3.18A, can you see why a large number of hydrogen bonds are present in the linear structure of cellulose shown in Figure 3.18B? Why is this structure so strong? **See p. 54**

We have seen how amino acid monomers form protein polymers and how sugar monomers form the polymers of carbohydrates. Now we will look at the lipids, which are unique among the four classes of large biological molecules in that they are not, strictly speaking, polymers.

3.4 What Are the Chemical Structures and Functions of Lipids?

Lipids—colloquially called fats—are hydrocarbons that are insoluble in water because of their many nonpolar covalent bonds. As we saw in Section 2.2, nonpolar hydrocarbon molecules are hydrophobic and preferentially aggregate together, away from water, which is polar. When nonpolar hydrocarbons are sufficiently close to one another, weak but additive van der Waals forces help hold them together. The huge macromolecular aggregations that can form are not polymers in a strict chemical sense, because the individual lipid molecules are not covalently bonded. With this understanding, it is still useful to consider aggregations of individual lipids as a different sort of polymer.

There are several different types of lipids, and they play a number of roles in living organisms:

- Fats and oils store energy.
- Phospholipids play important structural roles in cell membranes.
- Carotenoids and chlorophylls help plants capture light energy.
- Steroids and modified fatty acids play regulatory roles as hormones and vitamins.
- Fat in animal bodies serves as thermal insulation.
- A lipid coating around nerves provides electrical insulation.
- Oil or wax on the surfaces of skin, fur, feathers, and leaves repels water and prevents excessive evaporation of water from terrestrial animals and plants.

Fats and oils are triglycerides

Chemically, fats and oils are **triglycerides**, also known as simple lipids. Triglycerides that are solid at room temperature (around 20°C) are called **fats**; those that are liquid at room temperature are called **oils**.

3.20 Synthesis of a Triglyceride In living things, the reaction that forms a triglyceride is more complex, but the end result is the same as shown here.

Triglycerides are composed of two types of building blocks: fatty acids and glycerol. **Glycerol** is a small molecule with three hydroxyl (—OH) groups (thus it is an alcohol). A **fatty acid** is made up of a long nonpolar hydrocarbon chain and an acidic polar carboxyl group (—COOH). These chains are very hydrophobic because of their abundant C—H and C—C bonds, which have low electronegativity values and are nonpolar (see Section 2.2).

A triglyceride contains three fatty acid molecules and one molecule of glycerol. Synthesis of a triglyceride involves three condensation (dehydration) reactions. In each reaction, the carboxyl group of a fatty acid bonds with a hydroxyl group of glycerol, resulting in a covalent bond called an **ester linkage** and the release of a water molecule (**Figure 3.20**). The three fatty acids in a triglyceride molecule need not all have the same hydrocarbon chain length or structure; some may be saturated fatty acids, whereas others may be unsaturated:

- In **saturated fatty acids**, all the bonds between the carbon atoms in the hydrocarbon chain are single bonds—there are no double bonds. That is, all the bonds are saturated with hydrogen atoms (**Figure 3.21A**). These fatty acid molecules are relatively straight, and they pack together tightly, like pencils in a box.

- In **unsaturated fatty acids**, the hydrocarbon chain contains one or more double bonds. Linoleic acid is an example of a polyunsaturated fatty acid that has two double bonds near the middle of the hydrocarbon chain, causing kinks in the molecule (**Figure 3.21B**). Such kinks prevent the unsaturated fat molecules from packing together tightly.

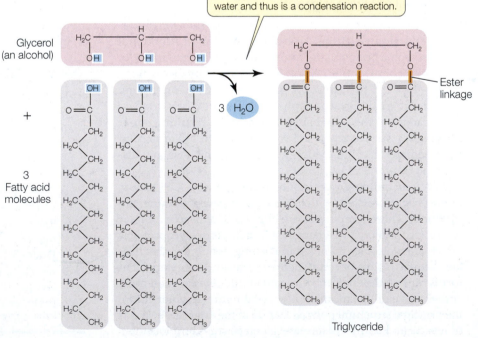

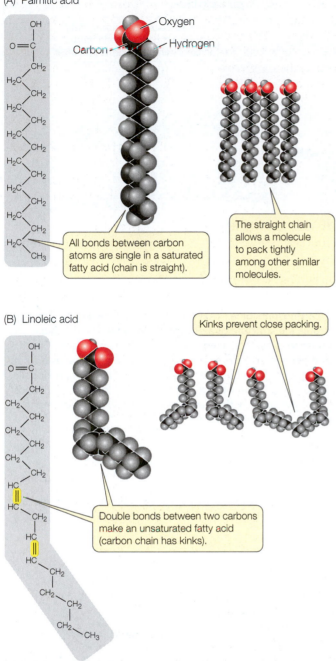

(A) Palmitic acid

Oxygen
Carbon
Hydrogen

All bonds between carbon atoms are single in a saturated fatty acid (chain is straight).

The straight chain allows a molecule to pack tightly among other similar molecules.

(B) Linoleic acid

Kinks prevent close packing.

Double bonds between two carbons make an unsaturated fatty acid (carbon chain has kinks).

3.21 Saturated and Unsaturated Fatty Acids (A) The straight hydrocarbon chain of a saturated fatty acid allows the molecule to pack tightly with other, similar molecules. (B) In unsaturated fatty acids, kinks in the chain prevent close packing. The color convention in the models shown here (gray, H; red, O; black, C) is commonly used.

The kinks in fatty acid molecules are important in determining the fluidity and melting points of lipids. The triglycerides of animal fats tend to have many long-chain saturated fatty acids packed tightly together; these fats are usually solids at room temperature and have high melting points. The triglycerides of plants, such as corn oil, tend to have short or unsaturated fatty acids. Because of their kinks, these fatty acids pack together poorly and have low melting points, and these triglycerides are usually liquids at room temperature.

Fatty acids are excellent storehouses for chemical energy. As you will see in Chapter 9, when the C—H bond is broken, it releases significant energy that an organism can use for its own purposes, such as movement or building up other complex molecules.

Phospholipids form biological membranes

We have mentioned the hydrophobic nature of the many C—C and C—H bonds in fatty acids. But what about the carboxyl functional group at the end of the molecule? When it ionizes and forms COO⁻, it is strongly hydrophilic. So a fatty acid is a molecule with a hydrophilic end and a long hydrophobic tail. It has two opposing chemical properties; the technical term for this is **amphipathic**. When fatty acids are bonded to glycerol, their carboxyl groups are incorporated into the ester bonds, and the resulting triglyceride is hydrophobic.

Like triglycerides, **phospholipids** contain fatty acids bound to glycerol by ester linkages. In phospholipids, however, any one of several phosphate-containing compounds replaces one of the fatty acids, giving phospholipids amphipathic properties (**Figure 3.22A**). The phosphate functional group has a negative electric charge, so this portion of the molecule is hydrophilic, attracting polar water molecules. But the two fatty acids are hydrophobic, so they tend to avoid water and aggregate together or with other hydrophobic substances.

In an aqueous environment, phospholipids line up in such a way that the nonpolar, hydrophobic "tails" pack tightly together and the phosphate-containing "heads" face outward, where they interact with water. The phospholipids thus form a **bilayer**: a sheet two molecules thick, with water excluded from the core (**Figure 3.22B**). Biological membranes have this kind of **phospholipid bilayer** structure, and we will devote Chapter 6 to their biological functions.

Some lipids have roles in energy conversion, regulation, and protection

In the paragraphs above we focused on triglycerides and phospholipids—lipids that are involved in energy storage and cell structure. However, there are other nonpolar and amphipathic lipids that have different structures and roles.

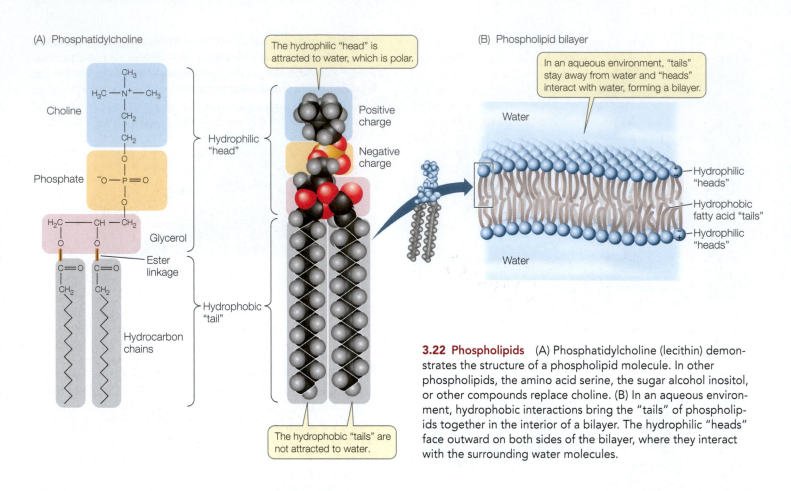

(A) Phosphatidylcholine

Choline

Phosphate

Glycerol

Ester linkage

Hydrocarbon chains

Hydrophilic "head"

Hydrophobic "tail"

The hydrophilic "head" is attracted to water, which is polar.

Positive charge

Negative charge

The hydrophobic "tails" are not attracted to water.

(B) Phospholipid bilayer

In an aqueous environment, "tails" stay away from water and "heads" interact with water, forming a bilayer.

Water

Hydrophilic "heads"

Hydrophobic fatty acid "tails"

Hydrophilic "heads"

Water

3.22 Phospholipids (A) Phosphatidylcholine (lecithin) demonstrates the structure of a phospholipid molecule. In other phospholipids, the amino acid serine, the sugar alcohol inositol, or other compounds replace choline. (B) In an aqueous environment, hydrophobic interactions bring the "tails" of phospholipids together in the interior of a bilayer. The hydrophilic "heads" face outward on both sides of the bilayer, where they interact with the surrounding water molecules.

CAROTENOIDS The carotenoids are a family of light-absorbing pigments found in plants and animals. Beta-carotene (β-carotene) is one of the pigments that traps light energy in leaves during photosynthesis. In humans, a molecule of β-carotene can be broken down into two vitamin A molecules. Vitamin A is used to make the pigment *cis*-retinal, which is required for vision.

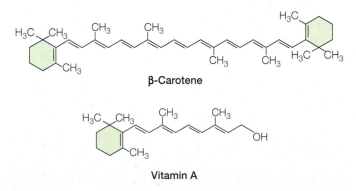

β-Carotene

Vitamin A

Carotenoids are responsible for the colors of carrots, tomatoes, pumpkins, egg yolks, and butter. The brilliant yellows and oranges of autumn leaves are also from carotenoids.

STEROIDS The steroids are a family of organic compounds whose multiple rings are linked through shared carbons. The

steroid cholesterol is an important constituent of membranes, helping maintain membrane integrity (see Section 6.1).

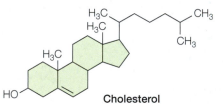

Cholesterol

Other steroids function as hormones: chemical signals that carry messages from one part of the body or in some cases are synthesized in inadequate amounts to another (see Chapter 41). Cholesterol is synthesized in the liver and is the starting material for making steroid hormones such as testosterone and estrogen.

VITAMINS Vitamins are small molecules that are not synthesized by the human body or in some cases are synthesized in inadequate amounts and so must be acquired from the diet (see Chapter 51). For example, vitamin A is formed from the β-carotene found in green and yellow vegetables (see above). In humans, a deficiency of vitamin A leads to dry skin, eyes, and internal body surfaces, retarded growth and development, and night blindness, which is a diagnostic symptom for the deficiency. Vitamins D, E, and K are also lipids.

WAXES Birds and mammals have glands in their skins that secrete waxy coatings onto their hair or feathers. These coatings repel water and help keep the hair and feathers pliable. The shiny leaves of plants such as holly, familiar during winter holidays, also have waxy coatings. Waxy coatings on plants can help them retain water and exclude pathogens. Bees make their honeycombs out of wax. Waxes are substances that are hydrophobic and plastic, or malleable, at room temperature. Each wax molecule consists of a saturated, long-chain fatty acid and a saturated, long-chain alcohol joined by an ester linkage. The result is a very long molecule with 40–60 CH_2 groups.

■ RECAP 3.4

Lipids include both hydrophobic and amphipathic molecules that are largely composed of carbon and hydrogen. They are important in energy storage, light absorption, regulation, and biological structures. A phospholipid is composed of two hydrophobic fatty acids linked to glycerol and a hydrophilic phosphate group. Cell membranes contain phospholipid bilayers.

- Draw the molecular structures of fatty acids and glycerol and show how they are linked to form a triglyceride. **See p. 56 and Figure 3.20**
- What is the difference between fats and oils? **See pp. 56–57 and Figure 3.21**
- How does the polar nature of phospholipids result in their forming a bilayer? **See p. 57 and Figure 3.22**
- Why are steroids and some vitamins classified as lipids? **See p. 58**

In this chapter we discussed three of the classes of macromolecules that are characteristic of living organisms, but a final class of biological macromolecules has special importance to the living world. Nucleic acids transmit life's "blueprint" to each new organism. This chapter illustrated the wonderful biochemical unity of life, implying that all life has a common origin (see Section 1.1). Essential to this origin were the monomeric nucleotides and their polymers, nucleic acids. In the next chapter we will turn to the related topics of nucleic acids and the origin of life.

Can knowledge of spider web protein structure be put to practical use?

ANSWER

Because of its strength, spider silk is much desired for human uses, ranging from surgical sutures in medicine to bulletproof vests in the military. "Farming" live spiders is tedious, costly, and gives a low yield of usable silk for industry. Unlike your hair, which grows continuously, spider silk is synthesized and stored as a liquid precursor solution in silk glands, and then "spun" out into fibers as needed. Recently, biotechnology has been used to genetically engineer silkworms, which produce their own form of silk, to make spider silk instead; even bacteria have been coaxed into making massive amounts of the protein. Moreover, by carefully studying how spiders do it, scientists have successfully spun out usable fibers from these artificial systems.

CHAPTER**SUMMARY** 3

3.1 What Kinds of Molecules Characterize Living Things?
See ANIMATED TUTORIAL 3.1

- **Macromolecules** are **polymers** constructed by the formation of covalent bonds between smaller molecules called **monomers**. Macromolecules in living organisms include polysaccharides, proteins, and nucleic acids. Large lipid structures may also be considered macromolecules.
- **Functional groups** are small groups of atoms that are consistently found together in a variety of different macromolecules. Functional groups have particular chemical properties that they confer on any larger molecule of which they are a part. **Review Figure 3.1, ACTIVITY 3.1**
- Structural, *cis-trans*, and **optical isomers** have the same kinds and numbers of atoms but differ in their structures and properties. **Review Figure 3.2**
- The many functions of macromolecules are directly related to their three-dimensional shapes, which in turn result from the sequences and chemical properties of their monomers.
- Monomers are joined by **condensation reactions**, which release a molecule of water for each bond formed. **Hydrolysis reactions** use water to break polymers into monomers. **Review Figure 3.4**

3.2 What Are the Chemical Structures and Functions of Proteins?

- The functions of **proteins** include support, protection, catalysis, transport, defense, regulation, and movement. **Review Table 3.1**
- Proteins consist of one or more **polypeptide chains**, which are polymers of **amino acids**. Four atoms or groups are attached to a central carbon atom: a hydrogen atom, an amino group, a carboxyl group, and a variable R group. The particular properties of each amino acid depend on its **side chain**, or **R group**, which may be charged, polar, or hydrophobic. **Review Table 3.2, ACTIVITY 3.2**
- **Peptide linkages**, also called peptide bonds, covalently link amino acids into polypeptide chains. These bonds form by condensation reactions between the carboxyl and amino groups. **Review Figure 3.6**
- The **primary structure** of a protein is the sequence of amino acids in the chain. This chain is folded into a **secondary structure**, which in different parts of the protein may form an **α helix** or a **β pleated sheet**. **Review Figure 3.7A–C**
- **Disulfide bridges** and noncovalent interactions between amino acids cause polypeptide chains to fold into three-dimensional **tertiary structures**. Weak, noncovalent interactions allow multiple poly-peptide chains to form **quaternary structures**. **Review Figure 3.7D, 3.7E**

continued

- Heat, alterations in pH, or certain chemicals can all result in a protein becoming **denatured**. This involves the loss of tertiary and/or secondary structure as well as biological function. **Review Figure 3.10**

- The specific shape and structure of a protein allows it to bind noncovalently to other molecules. In addition, amino acids may be modified by the covalent bonding of chemical groups to their side chains. Such binding may result in the protein changing its shape. **Review Figures 3.12, 3.13**

- **Chaperone proteins** enhance correct protein folding and prevent inappropriate binding to other molecules. **Review Figure 3.14**

3.3 What Are the Chemical Structures and Functions of Carbohydrates?

- **Carbohydrates** contain carbon bonded to hydrogen and oxygen atoms and have the general formula $C_mH_{2n}O_n$.

- **Monosaccharides** are the monomers that make up carbohydrates. **Hexoses** such as **glucose** are six-carbon monosaccharides; **pentoses** have five carbons. **Review Figure 3.16, ACTIVITY 3.3**

- **Glycosidic linkages**, which have either an α or a β orientation in space, are covalent bonds between monosaccharides. Two linked monosaccharides are called **disaccharides**; larger units are **oligosaccharides** and **polysaccharides**. **Review Figure 3.17**

- **Starch** is a polymer of glucose that stores energy in plants, and **glycogen** is an analogous polymer in animals. They can be easily broken down to release stored energy. **Review Figure 3.18**

- **Cellulose** is a very stable glucose polymer and is the principal structural component of plant cell walls.

3.4 What Are the Chemical Structures and Functions of Lipids?

- **Lipids** are hydrocarbons that are insoluble in water because of their many nonpolar covalent bonds. They play roles in energy storage, membrane structure, light harvesting, regulation, and protection.

- **Fats** and **oils** are **triglycerides**. A triglyceride is composed of three **fatty acids** covalently bonded to a molecule of glycerol by **ester linkages**. **Review Figure 3.20**

- A **saturated fatty acid** has a hydrocarbon chain with no double bonds. These molecules can pack together tightly. The hydrocarbon chain of an **unsaturated fatty acid** has one or more double bonds that bend the chain, preventing close packing. **Review Figure 3.21**

- A **phospholipid** has a hydrophobic hydrocarbon "tail" and a hydrophilic phosphate "head"; that is, it is **amphipathic**. In water, the interactions of the tails and heads of phospholipids generate a **phospholipid bilayer**. The heads are directed outward, where they interact with the surrounding water. The tails are packed together in the interior of the bilayer, away from water. **Review Figure 3.22**

- Other lipids include vitamins A, D, E, and K, steroids, and plant pigments such as carotenoids.

 Go to the Interactive Summary to review key figures, Animated Tutorials, and Activities Life10e.com/is3

CHAPTER**REVIEW**

REMEMBERING

1. The most abundant molecule in the cell is
   a. a carbohydrate.
   b. a lipid.
   c. a nucleic acid.
   d. a protein.
   e. water.

2. All lipids are
   a. triglycerides.
   b. polar.
   c. hydrophilic.
   d. polymers of fatty acids.
   e. more soluble in nonpolar solvents than in water.

3. All carbohydrates
   a. are polymers.
   b. are simple sugars.
   c. consist of one or more simple sugars.
   d. are found in biological membranes.
   e. are more soluble in nonpolar solvents than in water.

4. Which of the following statements about the primary structure of a protein is *not* true?
   a. It may be branched.
   b. It is held together by covalent bonds.
   c. It is unique to that protein.
   d. It determines the tertiary structure of the protein.
   e. It is the sequence of amino acids in the protein.

5. The amino acid leucine
   a. is found in all proteins.
   b. cannot form peptide linkages.
   c. has a hydrophobic side chain.
   d. has a hydrophilic side chain.
   e. is identical to the amino acid lysine.

6. The amphipathic nature of phospholipids is
   a. determined by the fatty acid composition.
   b. important in membrane structure.
   c. polar but not nonpolar.
   d. shown only if the lipid is in a nonpolar solvent.
   e. important in energy storage by lipids.

UNDERSTANDING & APPLYING

7. A single amino acid change in a protein can change its shape. Normally, at a certain position in a protein is the amino acid glycine (see Table 3.2). If glycine is replaced with either glutamic acid or arginine, the protein shape near that amino acid changes significantly. There are two possible explanations for this:

   a. A small amino acid at that position in the polypeptide is necessary for normal shape.

   b. An uncharged amino acid is necessary for normal shape.

   Further amino acid substitutions are done to distinguish between these possibilities. Replacing glycine with serine or alanine results in normal shape; but replacing glycine with valine changes the shape. Which of the two possible explanations is supported by the observations? Explain your answer.

8. Examine the hexose isomers mannose and galactose below. What makes them structural isomers of one another? Which functional groups do these carbohydrates contain, and what properties do these functional groups give to the molecules?

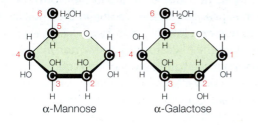

α-Mannose   α-Galactose

9. How does high temperature affect protein structure? When an organism is exposed to high temperature, it often makes a special class of molecular chaperones called heat shock proteins. How do you think these proteins work?

ANALYZING & EVALUATING

10. Suppose that, in a given protein, one lysine is replaced by aspartic acid (see Table 3.2). Does this change occur in the primary structure or in the secondary structure? How might it result in a change in tertiary structure? In quaternary structure?

11. Human hair is composed of the protein keratin. At the hair salon, two techniques are used to modify the three-dimensional shape of hair. Styling involves heat, and a perm involves cleaving and re-forming disulfide bonds. How would you investigate these phenomena in terms of protein structure?

Go to BioPortal at **yourBioPortal.com** for Animated Tutorials, Activities, LearningCurve Quizzes, Flashcards, and many other study and review resources.

4 Nucleic Acids and the Origin of Life

CHAPTEROUTLINE

4.1 **What Are the Chemical Structures and Functions of Nucleic Acids?**

4.2 **How and Where Did the Small Molecules of Life Originate?**

4.3 **How Did the Large Molecules of Life Originate?**

4.4 **How Did the First Cells Originate?**

Fast Cat Cheetahs are among the swiftest of the world's land animals. The 7,000 cheetahs living today have almost identical DNA sequences in their genomes, resulting from an evolutionary event about 10,000 years ago that wiped out all but a few individuals. The history of life is largely written in its DNA.

THE NAMES OF THE CHEETAH, *Acinonyx jubatus*, describe it well. "Cheetah" comes from the Hindi word *chiita*, meaning "spotted," for the small black spots on the animal's yellow fur. *Acinonyx* means "no-move claw" in Greek. Cheetahs cannot fully retract their claws—an advantage for running fast and hunting. In Latin, *jubatus* means "maned"—a characteristic of cheetah cubs.

This sleek, muscular cat is a solitary hunter of mammals such as gazelles and hares. It stalks its prey until it is 10–30 meters away and then chases it at speeds of up to 110 km/h (70 mph). Usually, the chase is over within a minute.

There are only about 7,000 cheetahs in the world today, most of them in Africa. The recent decline in their numbers is mostly due to humans: loss of habitat, and killing by farmers trying to protect their livestock. But—written in the cheetah's DNA—is more to the story of their decline.

Like proteins and polysaccharides, DNA is a macromolecule, in this case composed of a set of four different monomers called nucleotides. The nucleotide sequence of DNA is essential to its function, which is to carry information that determines an organism's characteristics. If you compare the sequence of the billions of nucleotides in your own DNA with that of an unrelated person in your class, the sequences will be about 0.5 percent different. This variation is reflected in the many differences among individual humans.

The genomes of cheetahs have a remarkable degree of similarity, almost as if all cheetahs descended from a single set of parents. The modern cheetah probably evolved about 15 million years ago and was widespread until the last ice age, which ended about 10,000 years ago. At that point many other large mammals (e.g.,the sabre-toothed tiger)died out, but a few cheetahs apparently survived and were the ancestors of the modern animals. So it is presumed that all current cheetahs—and their DNA—derive from the few individuals that survived an event that almost wiped out the species.

DNA belongs to a class of large molecules called nucleic acids. In Chapters 2 and 3 we described molecules that are important for biological structure and function. Here we turn to the nucleic acids, which are involved in perpetuating of life.

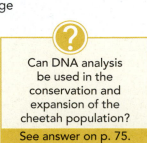

?

Can DNA analysis be used in the conservation and expansion of the cheetah population?

See answer on p. 75.

4.1 What Are the Chemical Structures and Functions of Nucleic Acids?

From medicine to evolution, from agriculture to forensics, the properties of nucleic acids affect our lives every day. It is with nucleic acids that the concept of "information" entered the biological vocabulary. Nucleic acids are uniquely capable of coding for and transmitting biological information.

Nucleic acids are polymers specialized for the storage, transmission, and use of genetic information. There are two types of nucleic acids: **DNA (deoxyribonucleic acid)** and **RNA (ribonucleic acid)**. DNA is a macromolecule that encodes hereditary information and passes it from generation to generation. Through RNA intermediates, the information encoded in DNA is used to specify the amino acid sequences of proteins and control the expression synthesis of other RNAs. During cell division and reproduction, information flows from existing DNA to the newly formed DNA in a new cell or organism. In the nonreproductive activities of the cell, information flows from DNA to RNA to proteins. It is the proteins that ultimately carry out many of life's functions.

Nucleotides are the building blocks of nucleic acids

Nucleic acids are polymers composed of monomers called nucleotides. A **nucleotide** consists of three components: a nitrogen-containing **base**, a pentose sugar, and one to three phosphate groups (**Figure 4.1**). Molecules consisting of a pentose sugar and a nitrogenous base—but no phosphate group—are

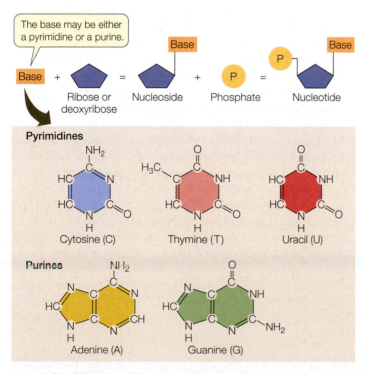

The base may be either a pyrimidine or a purine.

Base + Ribose or deoxyribose = Nucleoside + Phosphate P = Nucleotide

Pyrimidines

Cytosine (C)   Thymine (T)   Uracil (U)

Purines

Adenine (A)   Guanine (G)

4.1 Nucleotides Have Three Components Nucleotide monomers are the building blocks of DNA and RNA polymers.

Go to Activity 4.1 Nucleic Acid Building Blocks
Life10e.com/ac4.1

TABLE 4.1
Distinguishing RNA from DNA

Nucleic Acid	Sugar	Bases	Name of Nucleoside	Strands
RNA	Ribose	Adenine	Adenosine	Single
		Cytosine	Cytidine	
		Guanine	Guanosine	
		Uracil	Uridine	
DNA	Deoxyribose	Adenine	Deoxyadenosine	Double
		Cytosine	Deoxycytidine	
		Guanine	Deoxyguanosine	
		Thymine	Deoxythymidine	

called **nucleosides**. The nucleotides that make up nucleic acids contain just one phosphate group—they are nucleoside *monophosphates*.

The bases of the nucleic acids take one of two chemical forms: a six-membered single-ring structure called a **pyrimidine**, or a fused double-ring structure called a **purine** (see Figure 4.1). In DNA, the pentose sugar is **deoxyribose**, which differs from the **ribose** found in RNA by the absence of one oxygen atom (see Figure 3.16).

During the formation of a nucleic acid, new nucleotides are added to an existing chain one at a time. The pentose sugar in the last nucleotide of the existing chain and the phosphate on the new nucleotide undergo a condensation reaction (see Figure 3.4), and the resulting bond is called a **phosphodiester linkage** (**Figure 4.2**). The phosphate on the new nucleotide is attached to the 5'-carbon atom of its sugar, and the linkage occurs between it and the 3'-carbon on the last sugar of the existing chain. Because each nucleotide is added to the 3'-carbon of the last sugar, nucleic acids are said to *grow in the 5'-to-3' direction*.

As with carbohydrates (see Section 3.3), nucleic acids can range in size. Oligonucleotides are relatively short, with about 20 nucleotide monomers, whereas polynucleotides can be much longer.

- Oligonucleotides include RNA molecules that function as "primers" to begin the duplication of DNA; RNA molecules that regulate the expression of genes; and synthetic DNA molecules used for amplifying and analyzing other, longer nucleotide sequences.

- Polynucleotides, more commonly referred to as nucleic acids, include DNA and most RNA. Polynucleotides can be very long, and indeed are the longest polymers in the living world. Some DNA molecules in humans contain hundreds of millions of nucleotides.

Base pairing occurs in both DNA and RNA

DNA and RNA differ somewhat in their sugar groups, bases, and strand structure (**Table 4.1**). Four bases are found in DNA: **adenine (A)**, **cytosine (C)**, **guanine (G)**, and **thymine (T)**. RNA is also made up of four different monomers, but its nucleotides include **uracil (U)** instead of thymine. The sugar in DNA is deoxyribose, whereas the sugar in RNA is ribose.

4.2 Linking Nucleotides Together
Growth of a nucleic acid (RNA in this figure) from its monomers occurs in the 5' (phosphate) to 3' (hydroxyl) direction.

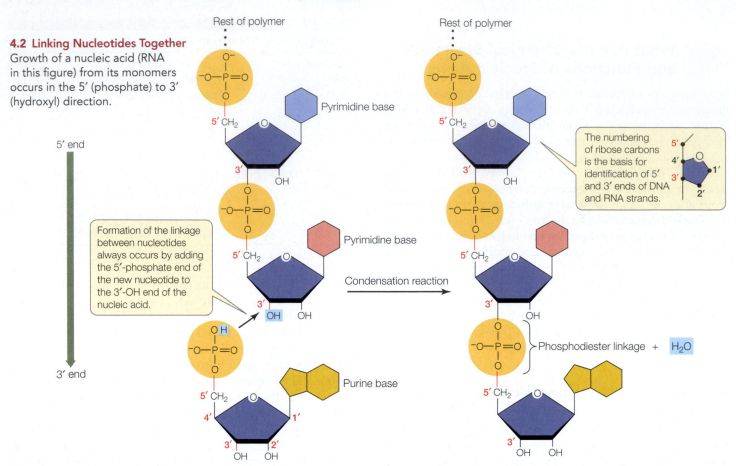

The numbering of ribose carbons is the basis for identification of 5' and 3' ends of DNA and RNA strands.

Formation of the linkage between nucleotides always occurs by adding the 5'-phosphate end of the new nucleotide to the 3'-OH end of the nucleic acid.

Condensation reaction

Phosphodiester linkage + H_2O

The key to understanding the structure and function of nucleic acids is the principle of **complementary base pairing**. In DNA, thymine and adenine always pair (T-A), and cytosine and guanine always pair (C-G). In RNA, the base pairs are A-U and C-G.

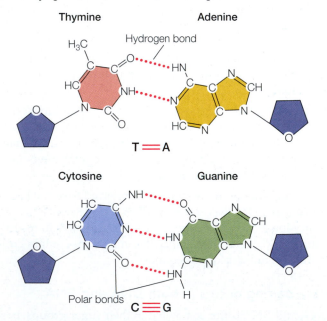

Base pairs are held together primarily by hydrogen bonds. As you can see, there are polar C=O and N—H covalent bonds in the bases; these can form hydrogen bonds between the δ⁻ on an oxygen or nitrogen of one base and the δ⁺ on a hydrogen of another base.

Individual hydrogen bonds are relatively weak, but there are so many of them in a DNA or RNA molecule that collectively they provide a considerable force of attraction, which can bind together two polynucleotide strands, or a single strand that folds back onto itself. This attraction is not as strong as a covalent bond, however. This means that individual base pairs are relatively easy to break with a modest input of energy. As you will see, the breaking and making of hydrogen bonds in nucleic acids is vital to their role in living systems.

RNA Even though RNA is generally single-stranded (**Figure 4.3A**), base pairing can occur between different regions of the molecule. Portions of the single-stranded RNA molecule can fold back and pair with one another (**Figure 4.3B**). Thus complementary hydrogen bonding between ribonucleotides plays an important role in determining the three-dimensional shapes of some RNA molecules. Complementary base pairing can also take place between ribonucleotides and deoxyribonucleotides. Adenine in an RNA strand can pair either with uracil (in another RNA strand) or with thymine (in a DNA strand). Similarly, an adenine in DNA can pair either with thymine (in the complementary DNA strand) or with uracil (in RNA).

DNA Usually, DNA is double-stranded; that is, it consists of two separate polynucleotide strands of the same length that are held together by hydrogen bonds between base pairs (**Figure 4.4A**). In contrast to RNA's diversity in three-dimensional structure, DNA is remarkably uniform. The A-T and G-C base pairs are about

(A)

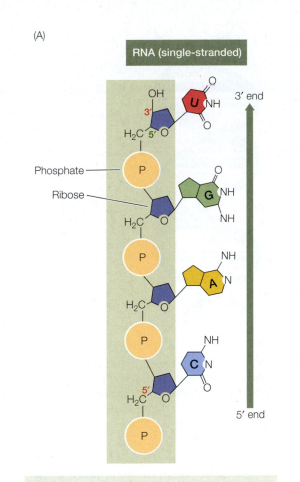

In RNA, the bases are attached to ribose. The bases in RNA are the purines adenine (A) and guanine (G) and the pyrimidines cytosine (C) and uracil (U).

(B)

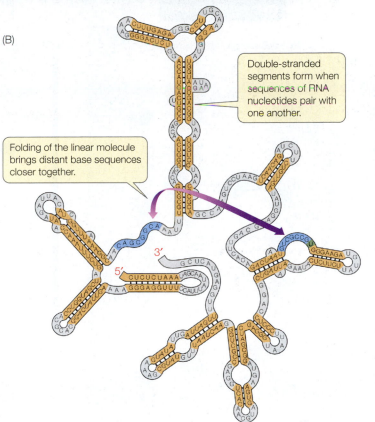

4.3 RNA (A) RNA is usually a single strand. (B) When a single-stranded RNA folds back on itself, hydrogen bonds between complementary sequences can stabilize it into a three-dimensional shape with complex surface characteristics.

the same size (each is a purine paired with a pyrimidine), and the two polynucleotide strands form a "ladder" that twists into a **double helix** (**Figure 4.4B**). The sugar–phosphate groups form the sides of the ladder, and the bases with their hydrogen bonds form the "rungs" on the inside. DNA carries genetic information in its sequence of base pairs rather than in its three-dimensional structure. The key differences among DNA molecules are their different nucleotide base sequences.

Go to Activity 4.2 DNA Structure
Life10e.com/ac4.2

DNA carries information and is expressed through RNA

DNA is a purely informational molecule. The information is encoded in the sequence of bases carried in its strands. For example, the information encoded in the sequence TCAGCA is different from the information in the sequence CCAGCA. DNA transmits information in two ways:

- DNA can be reproduced exactly. This is called **DNA replication**. It is done by polymerization using an existing strand as a base-pairing template.

- Certain DNA sequences can be copied into RNA, in a process called **transcription**. The nucleotide sequence in the RNA can then be used to specify a sequence of amino acids in a

polypeptide chain. This process is called **translation**. The overall process of transcription and translation is called **gene expression**.

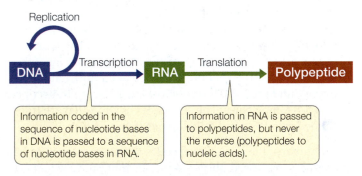

Information coded in the sequence of nucleotide bases in DNA is passed to a sequence of nucleotide bases in RNA.

Information in RNA is passed to polypeptides, but never the reverse (polypeptides to nucleic acids).

The details of these important processes are described in later chapters, but it is important to realize two things at this point:

1. *DNA replication and transcription depend on the base-pairing properties of nucleic acids.* Recall that the hydrogen-bonded base pairs are A-T and G-C in DNA and A-U and G-C in RNA. Consider, for example, this double-stranded DNA region:

$$5'\text{-TCAGCA-}3'$$
$$3'\text{-AGTCGT-}5'$$

Transcription of the lower strand will result in a single strand of RNA with the sequence 5'-UCAGCA-3'. Can you figure out the sequence that the top strand would produce?

(A)

DNA (double-stranded)

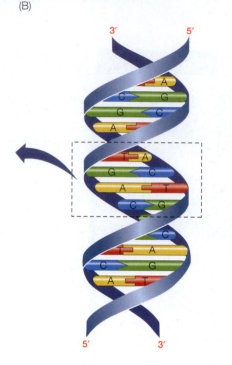

Deoxyribose

Pyrimidine base   Purine base

3′ end

5′ end

Phosphate

5′ end

3′ end

Hydrogen
bond

In DNA, the bases are attached to deoxyribose, and the base thymine (T) is found instead of uracil. Hydrogen bonds between purines and pyrimidines hold the two strands of DNA together.

4.4 DNA (A) DNA usually consists of two strands running in opposite directions that are held together by hydrogen bonds between purines and pyrimidines on the two strands. (B) The two strands in DNA are coiled in a right-handed double helix.

2. *DNA replication usually involves the entire DNA molecule.* Since DNA holds essential information, it must be replicated completely and accurately so that each new cell or new organism receives a complete set of DNA from its parent (**Figure 4.5A**). The complete set of DNA in a living organism is called its **genome**. However, not all of the information in the genome is needed at all times and in all tissues, and only small sections of the DNA are transcribed into RNA molecules. The sequences of DNA that are transcribed into RNA are called **genes** (**Figure 4.5B**).

In humans, the gene that encodes the major protein in hair (keratin) is expressed only in skin cells that produce hair. The genetic information in the keratin-encoding gene is transcribed into RNA and then translated into a keratin polypeptide. In other tissues such as the muscles, the keratin gene is not transcribed, but other genes are—for example, the genes that encode proteins present in muscles but not in skin or hair.

The DNA base sequence reveals evolutionary relationships

DNA carries hereditary information from one generation to the next, gradually accumulating changes in its base sequences over long periods of time. A series of DNA molecules stretches back through the lineage of every organism to the beginning of

biological evolution on Earth, about 4 billion years ago. Therefore closely related living species have more similar base sequences than species that are more distantly related. The same is true for closely related versus distantly related individuals within a species. The details of how scientists use this information are covered in Chapter 24. We described one such analysis, of the cheetah, in the opening story of this chapter.

Remarkable developments in DNA sequencing and computer technology have enabled scientists to determine the entire DNA base sequences—the genome—of many organisms, including humans, whose genome contains about 3 billion base pairs. These studies have confirmed many of the evolutionary relationships that had been inferred previously from more traditional comparisons of body structure, biochemistry, and physiology. For example, traditional comparisons had indicated that the closest living relative of humans (*Homo sapiens*) is the chimpanzee (genus *Pan*). In fact, the chimpanzee genome shares more than 98 percent of its DNA base sequence with the human genome. Increasingly, scientists turn to DNA analyses to elucidate evolutionary relationships when other comparisons are not possible or are not conclusive. For example, DNA studies revealed a close relationship between starlings and mockingbirds that was not expected on the basis of their anatomy or behavior.

Nucleotides have other important roles

Nucleotides are more than just the building blocks of nucleic acids. As we will describe in later chapters, there are several nucleotides (or modified nucleotides) with other functions:

- ATP (adenosine triphosphate) acts as an energy transducer in many biochemical reactions (see Section 8.2).

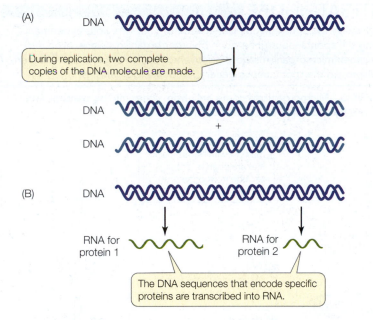

(A) DNA

During replication, two complete copies of the DNA molecule are made.

DNA

+

DNA

(B) DNA

RNA for protein 1

RNA for protein 2

The DNA sequences that encode specific proteins are transcribed into RNA.

4.5 DNA Replication and Transcription DNA is usually completely replicated (A) but only partially transcribed (B). RNA transcripts are produced from genes that code for specific proteins. Transcription of different genes occurs at different times and, in multicellular organisms, in different cells of the body.

- GTP (guanosine triphosphate) serves as an energy source, especially in protein synthesis. It also plays a role in the transfer of information from the environment to cells (see Section 7.2).

- cAMP (cyclic adenosine monophosphate) is a special nucleotide with an additional bond between the sugar and the phosphate group. It is essential in many processes, including the actions of hormones and the transmission of information by the nervous system (see Section 7.3).

- Nucleotides play roles as carriers in the synthesis and breakdown of carbohydrates and lipids.

■ RECAP 4.1

The nucleic acids DNA and RNA are polymers made up of nucleotide monomers. The sequence of nucleotides in DNA carries the information that is used by RNA to specify primary protein structure. The genetic information in DNA is passed from generation to generation and can be used to understand evolutionary relationships.

- List the key differences between DNA and RNA, and between purines and pyrimidines. **See pp. 63–65, Table 4.1, and Figure 4.1**

- How do purines and pyrimidines pair up in complementary base pairing? **See p. 64**

- What are the differences between DNA replication and transcription? **See pp. 65–66 and Figure 4.5**

- How can DNA molecules be very diverse, even though they appear to be structurally similar? **See p. 65**

We have seen that the nucleic acids RNA and DNA carry the blueprint of life, and that the inheritance of these macromolecules reaches back to the beginning of evolutionary time. But when, where, and how did nucleic acids arise on Earth? How did the building blocks of life, such as amino acids and sugars, originally arise?

4.2 How and Where Did the Small Molecules of Life Originate?

Chapter 2 pointed out that living things are composed of the same atomic elements as the inanimate universe. But the arrangements of these atoms into molecules are unique in biological systems. You will not find biological molecules in inanimate matter unless they came from a once-living organism.

It is impossible to know for certain how life on Earth began. But one thing is sure: life (or at least life as we know it) is not constantly being restarted. That is, **spontaneous generation** of life from inanimate nature is not happening repeatedly before our eyes. Now and for many millenia in the past, all life has come from life that existed before. But people, including scientists, did not always believe this.

Experiments disproved the spontaneous generation of life

The idea that life can originate repeatedly from nonliving matter has been common in many cultures and religions. During the European Renaissance (from the fourteenth to seventeenth centuries, a period that witnessed the birth of modern science), most people thought that at least some forms of life arose repeatedly and directly from inanimate or decaying matter by spontaneous generation. Many thought that mice arose from sweaty clothes placed in dim light; that frogs sprang directly from moist soil; and that rotting meat produced flies. Scientists such as the Italian physician and poet Francesco Redi, however, doubted these assumptions. Redi proposed that flies arose not by some mysterious transformation of decaying meat, but from other flies that laid their eggs on the meat. In 1668, Redi performed a scientific experiment—a relatively new concept at the time—to test his hypothesis. He set out several jars containing chunks of meat.

- One jar contained meat exposed to both air and flies.

- A second jar was covered with a fine cloth so that the meat was exposed to air but not to flies.

- The third jar was sealed with a lid so the meat was exposed to neither air nor flies.

No lid   Fine cloth cover   Lid

As he had hypothesized, Redi found maggots, which then hatched into flies, only in the first jar. This finding demonstrated that maggots could occur only where flies were present before. The idea that a complex organism like a fly could appear spontaneously from a nonliving substance in the meat, or from "something in the air," was laid to rest. Well, perhaps not quite to rest.

In the 1660s, newly developed microscopes revealed a vast new biological world. Under microscopic observation, virtually every environment on Earth was found to be teeming with tiny organisms. Some scientists believed these organisms arose spontaneously from their rich chemical environment, by the action of a "life force." But experiments in the nineteenth century by the great French scientist Louis Pasteur showed that microorganisms can arise only from other microorganisms, and that an environment without life remains lifeless (**Figure 4.6**).

Go to Animated Tutorial 4.1
Pasteur's Experiment
Life10e.com/at4.1

Pasteur's and Redi's experiments indicated that living organisms cannot arise from nonliving materials *under the conditions that exist on Earth now*. But their experiments did not prove that spontaneous generation never occurred. Eons ago, conditions on Earth and in the atmosphere above it were vastly different than they are today. Indeed, conditions similar to those found on primitive Earth may have existed, or may exist now, on other bodies in our solar system and elsewhere. This has led scientists to ask whether life has originated on other bodies in space, as it did on Earth.

Life began in water

As we emphasized in Chapter 2, water is an essential component of life as we know it. This is why there was great excitement when remote laboratories sent from Earth detected water ice on Mars. Astronomers believe our solar system began forming about 4.6 billion years ago, when a star exploded and collapsed to form the sun and about 500 bodies called planetesimals. These planetesimals collided with one another to form the inner planets, including Earth and Mars. The first chemical signatures indicating the presence of life on Earth are about 4 billion years old. So it took 600 million years for the chemical conditions on Earth to become just right for life. Key among those conditions was the presence of water.

Ancient Earth probably had a lot of water high in its atmosphere. But the new planet was hot, and the water remained in vapor form and dissipated into space. As Earth cooled, it became possible for water to condense on the planet's surface—but where did that water come from? One current view is that comets (loose agglomerations of dust and ice that have orbited the sun since the planets formed) struck Earth

INVESTIGATINGLIFE

4.6 Disproving the Spontaneous Generation of Life Previous experiments disproving the spontaneous generation of larger organisms were called into question when microorganisms were discovered. Louis Pasteur's classic experiments disproved the spontaneous generation of microorganisms.[a]

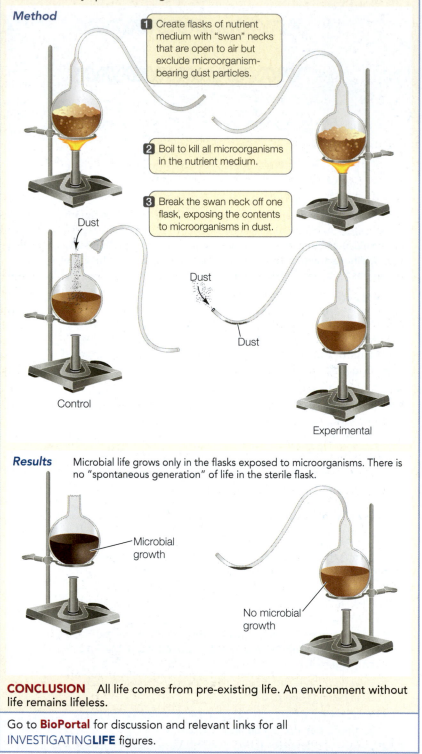

HYPOTHESIS Microorganisms come only from other microorganisms and cannot arise by spontaneous generation.

Method

1 Create flasks of nutrient medium with "swan" necks that are open to air but exclude microorganism-bearing dust particles.

2 Boil to kill all microorganisms in the nutrient medium.

3 Break the swan neck off one flask, exposing the contents to microorganisms in dust.

Dust

Dust

Dust

Control

Experimental

Results Microbial life grows only in the flasks exposed to microorganisms. There is no "spontaneous generation" of life in the sterile flask.

Microbial growth

No microbial growth

CONCLUSION All life comes from pre-existing life. An environment without life remains lifeless.

Go to **BioPortal** for discussion and relevant links for all INVESTIGATING**LIFE** figures.

[a]Pasteur gave a talk on his research at the "Sorbonne Scientific Soirée" on April 7, 1864. This talk has been translated into English: http://rc.usf.edu/~levineat/pasteur.pdf

and Mars repeatedly, bringing to those planets not only water but also other chemical components of life, such as nitrogen.

As the planets cooled and chemicals from their crusts dissolved in the water, simple chemical reactions would have taken place. Some of these reactions might have led to life, but impacts by large comets and rocky meteorites released enough energy to heat the developing oceans almost to boiling, thus destroying any early life that might have existed. On Earth, these large impacts eventually subsided, and some time around 3.8 to 4 billion years ago, life gained a foothold. There has been life on Earth ever since.

Several models have been proposed to explain the origin of life on Earth. The next sections will discuss two alternative theories: that life came from outside Earth, or that life arose on Earth through chemical evolution.

Life may have come from outside Earth

In 1969 a remarkable event led to the discovery that a meteorite from space carried molecules that were characteristic of life on Earth. On September 28 of that year, fragments of a meteorite fell around the town of Murchison, Australia. Using gloves to avoid Earth-derived contamination, scientists immediately shaved off tiny pieces of the rock, put them in test tubes, and extracted them in water (**Figure 4.7**). They found several of the molecules that are unique to life, including purines, pyrimidines, sugars, and ten amino acids.

Go to Media Clip 4.1
DNA Building Blocks from Space
Life10e.com/mc4.1

Were these molecules truly brought from space as part of the meteorite, or did they get there after the rock landed on Earth? There are a number of reasons to believe the molecules were not Earthly contaminants:

- The scientists took great care to avoid contamination. They used gloves and sterile instruments, took pieces from below the rock's surface, and did their work very soon after it landed (they hoped before organisms from Earth could contaminate the samples).

- Amino acids in living organisms on Earth are L-amino acids: they are found in only one of the two possible optical isomeric forms (see Figure 3.2). The amino acids in the meteorite were a mixture of L- and D-isomers, with a slight preponderance of the L form. Thus the amino acids in the meteorite were not likely to have come from a living organism on Earth.

- In the story that opened Chapter 2, we described how the ratio of isotopes in a living organism reflects the ratio of the same isotopes in the environment where the organism lives. The isotope ratios for carbon and hydrogen in the sugars from the meteorite were different from the ratios of those elements found on Earth.

More than 90 meteorites from Mars have been recovered on Earth. Many show signs of water, for example minerals such as carbonates that are precipitated from aqueous solution. Some also contain organic molecules that are the chemical signatures

4.7 The Murchison Meteorite Pieces from a fragment of the meteorite that landed in Australia in 1969 were put into test tubes with water. Soluble molecules present in the rock—including amino acids, nucleotide bases, and sugars—dissolved in the water. Plastic gloves and sterile instruments were used to reduce the possibility of contamination with substances from Earth.

of life. While the presence of such molecules suggests that these rocks once harbored life, it does not prove that there were living organisms in the rocks when they landed on Earth. Many scientists find it hard to believe that an organism could survive thousands of years of traveling through space in a meteorite, followed by intense heat as the meteorite passed through Earth's atmosphere. But there is evidence that the heat at the centers of some meteorites may not have been severe. If this was the case, then a long interplanetary trip by living organisms might have been possible.

Prebiotic synthesis experiments model early Earth

It is clear that other bodies in the solar system have, or once had, water and other simple organic molecules. Possibly, a meteorite was the source of the simple molecules that were the original building blocks for life on Earth. But a second theory for the origin of life on Earth, **chemical evolution**, holds that conditions on primitive Earth led to the formation of these simple molecules (prebiotic synthesis), and that these molecules led to the formation of life forms. Scientists have sought to reconstruct those primitive conditions, both physically (by varying temperature) and chemically (by re-creating the mixes of elements that may have been present).

HOT CHEMISTRY In oxygenated water, some trace metals such as molybdenum and rhenium are soluble, and their presence in sediments under oceans and lakes is directly proportional to the amount of oxygen gas (O_2) that was present in and above the water at the times the rocks were formed. Measurements of dated sedimentary cores indicate that none of these rare metals was present prior to 2.5 billion years ago. This and other lines of evidence suggest that there was little O_2 in Earth's early atmosphere. Oxygen gas is thought to have accumulated about 2.5 billion years ago as the by-product of photosynthesis by single-celled life forms; today 21 percent of our atmosphere is O_2.

In the 1950s Stanley Miller and Harold Urey at the University of Chicago set up an experimental "atmosphere" containing the gases they thought were present in Earth's early atmosphere: hydrogen gas, ammonia, methane gas, and water vapor. They passed an electrical spark through these gases to simulate lightning, a source of energy to drive chemical reactions. Then they cooled the system so the gases would condense and collect in a watery solution, or "ocean" (**Figure 4.8**). After a week of continuous operation, the system contained numerous organic molecules, including a variety of amino acids—the building blocks of proteins.

Go to Animated Tutorial 4.2
Synthesis of Prebiotic Molecules
Life10e.com/at4.2

COLD CHEMISTRY Stanley Miller also performed a long-term experiment in which the electrical spark was not used. In 1972 he filled test tubes with ammonia gas, water vapor, and cyanide (HCN), another molecule that is thought to have formed on primitive Earth. After checking that there were no contaminating substances or organisms that might confound the results, he sealed the tubes and cooled them to –78°C, the temperature of the ice that covers Europa, one of Jupiter's moons. Opening the tubes 27 years later, Miller found amino acids and nucleotide bases. Apparently, pockets of liquid water within the ice had allowed high concentrations of the starting materials to accumulate, thereby speeding up chemical reactions. The important conclusion is that the cold water within ice on ancient Earth, and other celestial bodies such as Mars, Europa, and Enceladus (one of Saturn's moons; satellite photos have revealed geysers of liquid water coming from its interior), may have provided environments for the prebiotic synthesis of molecules required for the subsequent formation of simple living systems.

The results of these experiments were profoundly important in giving weight to speculations about the chemical origin of life on Earth and elsewhere in the universe. Decades of experimental work and critical evaluation followed Miller and Urey's original experiments. In science, an experiment and its results must be repeatable and be reinterpreted and refined as more knowledge accumulates. For example, ideas about Earth's original atmosphere have changed. There is abundant evidence indicating that major volcanic eruptions occurred 4 billion years ago; these would have released carbon dioxide (CO_2), nitrogen (N_2), hydrogen sulfide (H_2S), and sulfur dioxide (SO_2) into the atmosphere. Experiments using these gases in addition to the ones in the original Miller–Urey experiment have produced a more diverse list of organic products:

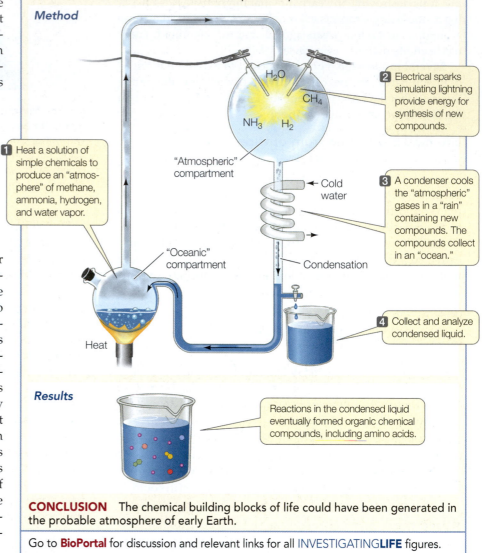

INVESTIGATING**LIFE**

4.8 Miller and Urey Synthesized Prebiotic Molecules in an Experimental Atmosphere With an increased understanding of the atmospheric conditions that existed on primitive Earth, the researchers devised an experiment to see if these conditions could lead to the formation of organic molecules.[a,b]

HYPOTHESIS Organic chemical compounds can be generated under conditions similar to those that existed in the atmosphere of primitive Earth.

Method

1 Heat a solution of simple chemicals to produce an "atmosphere" of methane, ammonia, hydrogen, and water vapor.

H_2O
CH_4
NH_3 H_2
"Atmospheric" compartment

2 Electrical sparks simulating lightning provide energy for synthesis of new compounds.

Cold water

3 A condenser cools the "atmospheric" gases in a "rain" containing new compounds. The compounds collect in an "ocean."

"Oceanic" compartment
Condensation
Heat

4 Collect and analyze condensed liquid.

Results

Reactions in the condensed liquid eventually formed organic chemical compounds, including amino acids.

CONCLUSION The chemical building blocks of life could have been generated in the probable atmosphere of early Earth.

Go to **BioPortal** for discussion and relevant links for all INVESTIGATING**LIFE** figures.

[a]Miller, S. L. 1953. *Science* 117: 528–519.
[b]Miller, S. L. and H. C. Urey. 1959. *Science* 130: 245–251.

- All five bases that are present in DNA and RNA (i.e., A, T, C, G, and U)
- All of the 20 amino acids used in protein synthesis
- Many 3- to 6-carbon sugars
- Certain fatty acids
- Vitamin B_6 (pantothenic acid, a component of coenzyme A)
- Nicotinamide (part of NAD, which is involved in energy metabolism)
- Carboxylic acids such as succinic and lactic acids (also involved in energy metabolism)

Could Biological Molecules Have Been Formed from Chemicals Present in Earth's Early Atmosphere?

Original Papers

Miller, S. L. 1953. A production of amino acids under possible primitive earth conditions. *Science* 117: 528–519.

Miller, S. L. and H. C. Urey. 1959. Organic compound synthesis on the primitive earth. *Science* 130: 245–251.

Analyze the Data

In the 1950s the Nobel Prize–winning chemist Harold Urey proposed that the molecules present in primitive Earth's atmosphere were methane (CH_4), ammonia (NH_3), hydrogen (H_2), and water (H_2O). He suggested that it might be possible to generate the building blocks of life, such as amino acids, in a laboratory simulation of these early conditions. Urey's graduate student Stanley Miller ran the simulation, which consisted of the four gases in an enclosed apparatus, an electric discharge to provide energy, and a cooling condenser to allow any substances that formed to dissolve in a watery "ocean" (see Figure 4.8). After a week, Miller analyzed the water using paper chromatography and found amino acids. This hallmark experiment was the first to demonstrate that organic molecules may have formed on Earth before life appeared.

The data Miller and Urey gave for sources of energy impinging on Earth are shown in the table.

QUESTION 1

Of the total energy from the sun, only a small fraction is in the ultraviolet range, less than 250 nm. What proportion of total solar energy is the energy with wavelengths below 250 nm?

QUESTION 2

The molecules CH_4, H_2O, NH_3, and CO_2 absorb light at wavelengths of less than 200 nm. What fraction of total solar radiation is in this range?

QUESTION 3

Miller and Urey used electric discharges as their energy source. What other sources of energy could be used in similar experiments?

Source	Energy (cal cm^{-2} yr^{-1})
Total radiation from sun	260,000
Ultraviolet light	
Wavelength <250 nm[a]	570
Wavelength <200 nm	85
Wavelength <150 nm	3.5
Electric discharges	4
Cosmic rays	0.0015
Radioactivity	0.8
Volcanoes	0.13

[a]Nanometer, 10^{-9} meters.

Go to **BioPortal** for all WORKING WITH**DATA** exercises

RECAP 4.2

Life does not arise repeatedly through spontaneous generation, but comes from pre-existing life. Water is an essential ingredient for the emergence of life. Meteorites that have landed on Earth provide some evidence for an extraterrestrial origin of life. Chemical synthesis experiments provide support for the idea that life's simple molecules formed in the prebiotic environment on Earth.

- Explain how Redi's and Pasteur's experiments disproved spontaneous generation. **See pp. 67–68 and Figure 4.6**
- What is the evidence that life on Earth came from other bodies in the solar system? **See p. 69**
- What is the significance of the Miller–Urey experiment, what did it find, and what were its limitations? **See pp. 70–71 and Figure 4.8**

Chemistry experiments modeling the conditions of ancient Earth provide clues about the origins of the monomers (such as amino acids) that make up the polymers (such as proteins) that characterize life. How did these polymers develop?

4.3 How Did the Large Molecules of Life Originate?

The Miller–Urey experiment and others that followed provide a plausible scenario for the formation of the building blocks of life under conditions that prevailed on primitive Earth. The next step in forming and supporting a general theory on the origin of life would be an explanation for how polymers formed from these monomers.

Chemical evolution may have led to polymerization

Scientists have used a number of model systems to try to simulate conditions under which polymers might have been made. Each of these systems is based on several observations and speculations:

- Solid mineral surfaces, such as powderlike clays, have large surface areas. Scientists speculate that the silicates in clay may have catalyzed (speeded up) the condensation reactions that resulted in organic polymers.
- Hydrothermal vents deep in the ocean, where hot water emerges from beneath Earth's crust, lack oxygen gas and contain metals such as iron and nickel. In laboratory experiments, these metals have been shown to catalyze the polymerization of amino acids in the absence of oxygen.
- In hot pools at the edges of oceans, evaporation may have concentrated monomers to the point where polymerization was favored (the "primordial soup" hypothesis).

In whatever ways the earliest stages of chemical evolution occurred, they resulted in the emergence of monomers and polymers that have probably remained unchanged in their general structures and functions for several billion years.

RNA may have been the first biological catalyst

A hallmark of living organisms is chemical change, for example, DNA replication and RNA synthesis, which we described earlier in this chapter (see Figure 4.5). There are many other chemical changes that occur in living systems, often involving the hydrolysis or synthesis of macromolecules or conversions among small molecules. As you will see in Chapter 8, these chemical changes can occur spontaneously in aqueous solutions like those that exist in biological systems, but most would occur extremely slowly. The evolution of catalysts—molecules that speed up biochemical conversions—solved this problem. So a key to the origin of life is the appearance of catalysts.

In life today, the main catalysts are proteins called enzymes. The myriad shapes of proteins allow them to bind to diverse substances in solution and speed up chemical reactions. But we know that proteins are made from information in nucleic acids (see p. 65). So we have a problem: If proteins are needed for life, nucleic acids must have appeared first, so that the proteins could be made. But if nucleic acids appeared before proteins, the proteins could not have been the first catalysts. Could nucleic acids be catalysts, in addition to their role as blueprints for protein synthesis? The answer is yes.

Like a protein, the three-dimensional structure of a folded RNA molecule presents a unique surface to the external environment (see Figure 4.3). The surfaces of RNA molecules can be every bit as specific as those of proteins. The three-dimensional shapes and other chemical properties of certain RNA molecules allow them to function as catalysts. Catalytic RNAs, called **ribozymes**, can speed up reactions involving their own nucleotides as well as other cellular substances. Although in retrospect it is not too surprising, the discovery of catalytic RNAs was a major shock to a community of biologists who were convinced that all biological catalysts were proteins (enzymes). It took almost a decade for the work of the scientists involved, Thomas Cech and Sidney Altman, to be fully accepted by other scientists. Later, they were awarded the Nobel Prize.

Given that RNA can be both informational (in its nucleotide sequence) and catalytic (because of its ability to form unique three-dimensional shapes), it has been hypothesized that early life consisted of an "RNA world"—a world before DNA. It is thought that when RNA was first made, it could have acted as a catalyst for its own replication as well as for the synthesis of proteins. DNA could eventually have evolved from RNA (**Figure 4.9**). Several lines of evidence support this scenario:

- In living organisms today, the formation of peptide linkages (see Figure 3.6) is catalyzed by ribozymes.

- In certain viruses called retroviruses, there is an enzyme called reverse transcriptase that catalyzes the synthesis of DNA from RNA.

- When a short, naturally occuring RNA molecule is added to a mixture of nucleotides, RNA polymers are formed at a rate 7 million times greater than the formation of polymers without the added RNA. This indicates that the added RNA is a catalyst, not just a template.

- An artificial ribozyme has been developed that can catalyze the assembly of short RNAs into a longer molecule that is an exact copy of itself (**Figure 4.10**). This may be how nucleic acid replication evolved.

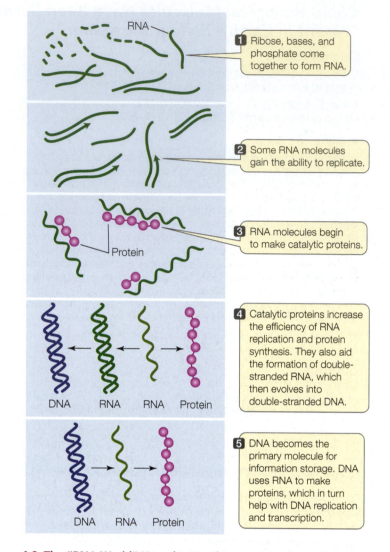

1 Ribose, bases, and phosphate come together to form RNA.

2 Some RNA molecules gain the ability to replicate.

3 RNA molecules begin to make catalytic proteins.

4 Catalytic proteins increase the efficiency of RNA replication and protein synthesis. They also aid the formation of double-stranded RNA, which then evolves into double-stranded DNA.

5 DNA becomes the primary molecule for information storage. DNA uses RNA to make proteins, which in turn help with DNA replication and transcription.

4.9 The "RNA World" Hypothesis This view postulates that in a world before DNA, RNA alone was both the blueprint for protein synthesis and a catalyst for its own replication. Eventually, the information storage molecules of DNA could have evolved from RNA.

RECAP **4.3**

The formation of the large polymers that are characteristic of life may have occurred on the surfaces of clay particles, near hydrothermal vents, or in hot pools at the edges of oceans. RNA may have been the first genetic material and catalyst.

- Why was the discovery of ribozymes important for the development of the "RNA world" hypothesis? See p. 72

- How can RNA self-replicate? See p. 72 and Figure 4.10

The discovery of mechanisms for the formation of small and large molecules is essential to answering questions about the origin of life on Earth. But we also need to understand how organized living systems formed. Such systems display the characteristic properties of life, including reproduction, energy processing, and responsiveness to the environment. These are properties of cells, whose origin we will explore in the next section.

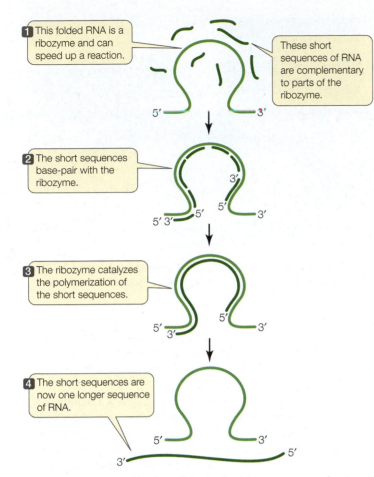

1 This folded RNA is a ribozyme and can speed up a reaction.

These short sequences of RNA are complementary to parts of the ribozyme.

2 The short sequences base-pair with the ribozyme.

3 The ribozyme catalyzes the polymerization of the short sequences.

4 The short sequences are now one longer sequence of RNA.

4.10 An Early Catalyst for Life? In the laboratory, a synthetic ribozyme (a folded RNA molecule) can catalyze the polymerization of shorter RNA strands into a longer molecule that is identical to itself. This may be how the earliest nucleic acids replicated.

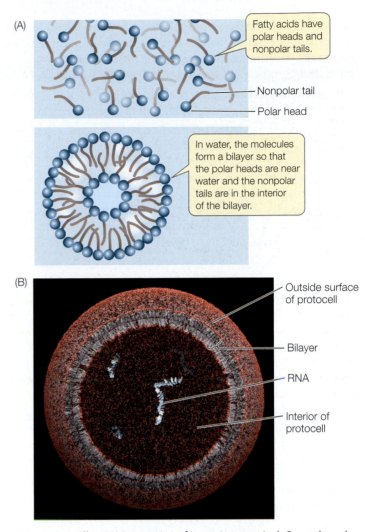

Fatty acids have polar heads and nonpolar tails.

Nonpolar tail

Polar head

In water, the molecules form a bilayer so that the polar heads are near water and the nonpolar tails are in the interior of the bilayer.

Outside surface of protocell

Bilayer

RNA

Interior of protocell

4.11 Protocells (A) In a series of experiments, Jack Szostak and his colleagues mixed fatty acid molecules in water. The molecules formed spherical structures called protocells, with water surrounded by bilayers of fatty acids. (B) A model of a protocell. A portion of the "membrane" has been cut away to reveal the inside of the protocell and the membrane's bilayer structure. Nutrients and nucleotides pass through the "membrane" and enter the protocell, where they copy an already present RNA template. The new copies of RNA remain in the protocell.

4.4 How Did the First Cells Originate?

As you have seen from many of the theories for the origin of life, the evolution of biochemistry occurred under localized conditions. That is, the chemical reactions of life could not occur in a dilute aqueous environment. There had to be a compartment of some sort that brought together and concentrated the compounds involved in these events. Biologists have proposed that initially this compartment may have simply been a tiny droplet of water on the surface of a rock. But another major event in the origin of life was necessary.

Life as we know it is separated from the environment within structurally defined units called **cells**. The internal contents of a cell are separated from the nonbiological environment by a special barrier—a **membrane**. The membrane is not just a barrier; it regulates what goes into and out of the cell, as we will describe in Chapter 6. This role of the surface membrane is very important because it permits the interior of the cell to maintain a chemical composition that is different from its external environment. How did the first cells with membranes come into existence?

Experiments explore the origin of cells

Jack Szostak and his colleagues at Harvard University built a laboratory model that gives insights into the origin of cells. To do this, they first put fatty acids (which can be made in prebiotic

experiments) into water. Recall from Chapter 3 that fatty acids are amphipathic: they have a hydrophilic polar head and a long, nonpolar tail that is hydrophobic (see Figure 3.22). When placed in water, fatty acids will arrange themselves in a round "huddle" much like a football team: the hydrophilic heads point outward to interact with the aqueous environment, and the fatty acid tails point inward, away from the water molecules.

What if some water becomes trapped in the interior of this "huddle"? Now the layer of hydrophobic fatty acid tails is in water, which is an unstable situation. To stabilize this structure, a second layer of fatty acids forms. This **lipid bilayer** has the polar heads of the fatty acids facing both outward and inward, because they are attracted to the polar water molecules present on each side of the double layer. The nonpolar tails form the interior of the bilayer (**Figure 4.11**). These prebiotic, water-filled structures, defined by a lipid bilayer membrane, very much resemble living cells. Scientists refer to these compartments as **protocells**. Examining their properties revealed that

- Large molecules such as DNA or RNA could not pass through the bilayer to enter the protocells, but small molecules such as sugars and individual nucleotides could.

- Nucleic acids inside the protocells could replicate using the nucleotides from outside. When the investigators placed a short nucleic acid strand capable of self-replication inside protocells and added nucleotides to the watery environment outside, the nucleotides crossed the barrier, entered the protocells, and became incorporated into new polynucleotide chains. This replication, which can occur without protein catalysis, may have been the first step toward cell reproduction.

Were these protocells truly cells, and was the lipid bilayer produced in these experiments a true cell membrane? Certainly not. The protocells could not fully reproduce, nor could they carry out all the metabolic reactions that take place in modern cells. The simple lipid bilayer had few of the sophisticated functions of modern cell membranes. Nevertheless, the protocell may be a reasonable facsimile of a cell as it evolved billions of years ago:

- It can act as an organized system of parts, with substances interacting and reacting, in some cases catalytically.

- It includes an interior that is distinct from the exterior environment.

- It is capable of self-replication.

These are all fundamental characteristics of living cells.

Some ancient cells left a fossil imprint

In the 1990s scientists made a rare find: a formation of ancient rocks in Australia that had remained relatively unchanged since the rocks first formed 3.5 billion years ago. In one of these rock samples, geologist J. William Schopf of the University of California, Los Angeles, saw chains and clumps of what looked tantalizingly like contemporary cyanobacteria, or "blue-green" bacteria (**Figure 4.12**). Cyanobacteria are believed to have been among the first organisms because they can perform photosynthesis, converting CO_2 and water into

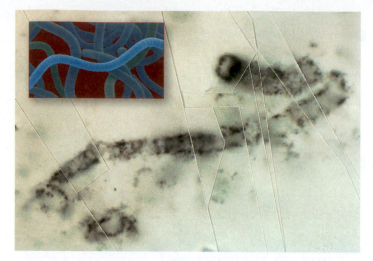

4.12 The Earliest Cells? This fossil from Western Australia is 3.5 billion years old. Its form is similar to that of modern filamentous cyanobacteria (inset).

carbohydrates. Schopf needed to prove that the chains were once alive, not just the results of simple chemical reactions. He and his colleagues looked for chemical evidence of photosynthesis in the rock samples.

The use of carbon dioxide in photosynthesis is a hallmark of life and leaves a unique chemical signature—a specific ratio of carbon isotopes (^{13}C:^{12}C) in the resulting carbohydrates. Schopf showed that the Australian material had this isotope signature. Furthermore, microscopic examination of the chains revealed *internal* substructures that are characteristic of living systems and were not likely to be the result of simple chemical reactions. Schopf's evidence suggests that the Australian sample is indeed the remains of a truly ancient living organism.

In 2011 a different team of scientists, working about 20 miles from Schopf's discovery, found similar-looking microfossil structures in sandstone rocks that were about 3.4 billion years old. In this case, a chemical analysis of the rocks indicated that these cells used sulfur instead of oxygen in the series of cellular reactions that release chemical energy.

Taking geological, chemical, and biological evidence into account, it is plausible that it took about 500 million to a billion years from the formation of Earth until the appearance of the first cells (**Figure 4.13**). Life has been cellular ever since. In the next chapter we will begin our study of cell structure and function.

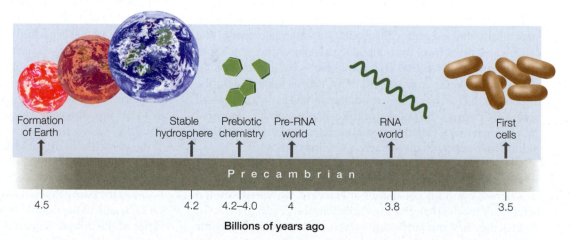

4.13 The Origin of Life
This highly simplified timeline gives a sense of the major events that culminated in the origin of life more than 3.5 billion years ago.

| Formation of Earth | Stable hydrosphere | Prebiotic chemistry | Pre-RNA world | RNA world | First cells |

Precambrian

Billions of years ago

4.5   4.2   4.2–4.0   4   3.8   3.5

RECAP 4.4

The chemical reactions that preceded living organisms probably occurred in specialized compartments, such as water droplets on the surfaces of minerals. Life as we know it did not begin until the emergence of cells. Protocells made in the laboratory have some of the properties of modern cells. Cell-like structures fossilized in ancient rocks date the first cells to about 3.5 billion years ago.

- Explain the importance of the cell membrane to the evolution of living organisms. **See p. 73**
- What is the evidence that ancient rocks contain the fossils of cells? **See p. 74**

Can DNA analysis be used in the conservation and expansion of the cheetah population?

ANSWER

At the Cheetah Conservation Fund in Otjiwarongo, Namibia, a DNA sequencing laboratory analyzes populations of wild cheetahs that live in that area. The aim is to mate pairs of cheetahs that differ the most in their DNA sequences so that their offspring will have the greatest possible genetic diversity. Because of their genetic homogeneity, cheetah males produce a low amount of sperm, and the scientists in Namibia have developed artificial insemination methods to overcome this.

CHAPTER SUMMARY 4

4.1 What Are the Chemical Structures and Functions of Nucleic Acids?

- The unique functions of the **nucleic acids**—**DNA** and **RNA**—are information storage and transfer. DNA is the hereditary material that passes genetic information from one generation to the next, and RNA uses that information to specify the structures of proteins.
- Nucleic acids are polymers of nucleotides. A **nucleotide** consists of a phosphate group, a pentose sugar (**ribose** in RNA and **deoxyribose** in DNA), and a nitrogen-containing **base**. **Review Figure 4.1**
- In DNA, the nucleotide bases are **adenine** (**A**), **guanine** (**G**), **cytosine** (**C**), and **thymine** (**T**). **Uracil** (**U**) replaces thymine in RNA. C, T, and U have single-ring structures and are **pyrimidines**. A and G have double-ring structures and are **purines**.
- The nucleotides in DNA and RNA are joined by **phosphodiester linkages** involving the sugar of one nucleotide and the phosphate of the next, forming a nucleic acid polymer. **Review Figure 4.2, ACTIVITY 4.1**
- **Complementary base pairing** due to hydrogen bonds between A and T, A and U, and G and C occurs in nucleic acids. In RNA, the hydrogen bonds result in a folded molecule. In DNA, the hydrogen bonds connect two strands into a double helix. **Review Figures 4.3, 4.4, ACTIVITY 4.2**
- The information content of DNA and RNA resides in their base sequences.
- DNA is expressed as RNA in **transcription**. RNA can then specify the amino acid sequence of a protein in **translation**. **Review Figure 4.5**

4.2 How and Where Did the Small Molecules of Life Originate?

- Historically, many cultures believed that life originated repeatedly by **spontaneous generation**. This was disproven experimentally. **Review Figure 4.6, ANIMATED TUTORIAL 4.1**
- A prerequisite for life is the presence of water.
- Some meteorites that have landed on Earth contain organic molecules, suggesting that life might have originated extraterrestrially.

- An alternative hypothesis is **chemical evolution**: the idea that organic molecules were formed on Earth before life began.
- Chemical experiments modeling the prebiotic conditions on Earth support the idea of chemical evolution. **Review Figure 4.8, ANIMATED TUTORIAL 4.2**

4.3 How Did the Large Molecules of Life Originate?

- Chemical evolution may have led to the polymerization of small molecules into polymers. This may have occurred on the surfaces of clay particles, in hydrothermal vents, or in hot pools at the edges of oceans.
- A catalyst speeds up a chemical reaction. Today most catalysts are proteins, but some RNA molecules can function as both catalysts and information molecules. A catalytic RNA is called a **ribozyme**.
- The existence of ribozymes supports the idea of an "RNA world"—a world before DNA. On early Earth, RNA may have acted as a catalyst for its own replication as well as for the synthesis of proteins. DNA could eventually have evolved from RNA. **Review Figure 4.9**
- In support of the "RNA world" hypothesis, an artificial self-replicating ribozyme was developed in the laboratory. **Review Figure 4.10**

4.4 How Did the First Cells Originate?

- A key to the emergence of living cells was the prebiotic generation of compartments enclosed by **membranes**. Such enclosed compartments permitted the generation and maintenance of internal chemical conditions that were different from those in the exterior environment.
- In the laboratory, fatty acids assemble into **protocells** that have some of the characteristics of cells. **Review Figure 4.11**
- Ancient rocks (3.5 billion years old) have been found with imprints that are probably fossils of early cells.

 Go to the Interactive Summary to review key figures, Animated Tutorials, and Activities
Life10e.com/is4

CHAPTER**REVIEW**

■ REMEMBERING

1. A nucleotide in DNA is made up of
   a. four bases.
   b. a base plus a ribose sugar.
   c. a base plus a deoxyribose sugar plus a phosphate.
   d. a sugar plus a phosphate.
   e. a sugar and a base.

2. Nucleotides in RNA are connected to one another in the polynucleotide chain by
   a. covalent bonds between bases.
   b. covalent bonds between sugars.
   c. covalent bonds between sugar and phosphate.
   d. hydrogen bonds between purines.
   e. hydrogen bonds between any bases.

3. Which is a difference between DNA and RNA?
   a. DNA is single-stranded and RNA is double-stranded.
   b. DNA is only informational and RNA is only catalytic.
   c. DNA contains deoxyribose and RNA contains ribose.
   d. DNA is transcribed and RNA is replicated.
   e. DNA contains uracil (U) and RNA contains thymine (T).

4. The components in the atmosphere for the Miller–Urey experiment on prebiotic synthesis did not include
   a. H_2.
   b. H_2O.
   c. O_2.
   d. NH_3.
   e. CH_4.

5. The "RNA world" hypothesis proposes that
   a. RNA formed from DNA.
   b. RNA was both a catalyst and genetic material.
   c. RNA was a catalyst only.
   d. RNA formed after proteins.
   e. DNA formed after RNA was broken down.

6. Findings in ancient rocks indicate cells first appeared
   a. about 4.5 billion years ago.
   b. about 3.5 billion years ago.
   c. about 2 billion years ago.
   d. before rocks were formed.
   e. before water arrived on Earth.

■ UNDERSTANDING & APPLYING

7. What conditions existing on Earth today might preclude the origin of life from the prebiotic molecules Miller and Urey used?

8. Applied biologists are trying to develop reagents with specific three-dimensional shapes to bind to target molecules. An active research area in this regard is the development of oligonucleotides of RNA. What properties of oligonucleotide chains cause them to fold into a precise shape? How long do you think a chain would have to be to take a unique shape?

9. Why was the evolution of a self-contained cell essential for life as we know it?

■ ANALYZING & EVALUATING

10. The interpretation of Pasteur's experiment (see Figure 4.6) depended on the inactivation of microorganisms by heat. We now know of microorganisms that can survive extremely high temperatures (see Chapter 26). Does this change the interpretation of Pasteur's experiment? What experiments would you do to inactivate such microbes?

11. The Miller–Urey experiment (see Figure 4.8) showed that it was possible for amino acids to be formed from gases that were hypothesized to have been in Earth's early atmosphere. These amino acids were dissolved in water. Knowing what you do about the polymerization of amino acids into proteins (see Figure 3.6), how would you set up experiments to show that proteins can form under the conditions of early Earth?

Go to BioPortal at **yourBioPortal.com** for Animated Tutorials, Activities, LearningCurve Quizzes, Flashcards, and many other study and review resources.

5 Cells: The Working Units of Life

CHAPTER**OUTLINE**

5.1 What Features Make Cells the Fundamental Units of Life?

5.2 What Features Characterize Prokaryotic Cells?

5.3 What Features Characterize Eukaryotic Cells?

5.4 What Are the Roles of Extracellular Structures?

5.5 How Did Eukaryotic Cells Originate?

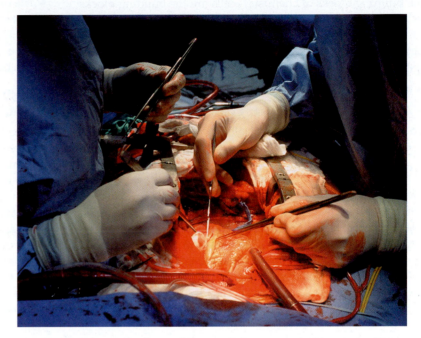

New Heart Tissue Cardiac problems are often treated with surgery. Stem cells that are coaxed to differentiate into heart cells may be used to repair a damaged heart.

IT IS A DAY in the not-too-distant future. Decades of eating fatty foods, combined with an inherited tendency to deposit cholesterol in his arteries, have finally caught up with 70-year-old Don. A blood clot has closed off blood flow to part of his heart, leading to a heart attack and severe damage to that vital organ.

If this had happened today, Don would have been faced with a long period of rehabilitation, taking medications to manage his weakened heart. Instead, his physicians take a pinch of skin tissue from his arm and bring it to a laboratory. After certain DNA sequences are added, Don's skin cells no longer look and act like skin cells: they are undifferentiated (unspecialized) and reproduce continuously in the laboratory dish. These cells are also multipotent stem cells, able to differentiate into almost any type of cell in the body if given the right environment. When they are injected directly into Don's heart, his stem cells soon become heart muscle cells, repairing the damage caused by the heart attack. Don leaves the hospital with full cardiac function and recommendations for a healthy diet.

The potential uses of stem cells in medicine have generated a lot of excitement in recent years. Such widely read periodicals as *Time* have hailed advances in stem cell research as "breakthroughs of the year." Patients with the neurological disorder Parkinson's disease dream of the day when their skin cells can be turned into brain cells to fix their damaged nervous systems. People with diabetes hope for stem cells to repair their pancreas. The list is long.

Behind all of this hope and the research it has inspired is a cornerstone of biological science: the cell theory. As you saw in the last chapter, a key event in the emergence of life was the enclosure of biochemical reactions inside cells, thus concentrating them and separating them from the external environment. The emergence of cells was essential for the evolution of life as we know it—and this is reflected in the first two basic tenets of cell theory: cells are the fundamental units of life; and all living organisms are composed of cells. Don's stem cells contain not just the activities of a living entity, but also the potential to change those activities in new directions. The third (and equally important) tenet of cell theory states that the cell is the unit of reproduction: all cells come from pre-existing cells. Stem cell therapy does not create new cells out of thin air; it coaxes existing ones to differentiate and reproduce along the desired path.

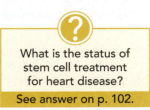

What is the status of stem cell treatment for heart disease?

See answer on p. 102.

5.1 What Features Make Cells the Fundamental Units of Life?

In Chapter 1 we introduced some of the characteristics of life: chemical complexity, growth and reproduction, the ability to refashion substances from the environment, and the ability to move specific substances into and out of the organism. These characteristics are all demonstrated by cells. Just as atoms are the building blocks of chemistry, cells are the building blocks of life.

The **cell theory** is an important unifying principle of biology. There are three critical components of the cell theory:

- Cells are the fundamental units of life.
- All living organisms are composed of cells.
- All cells come from preexisting cells.

To the original cell theory, first stated in 1838, should be added:

- Evolution through natural selection explains the diversity of modern cells.

Cells contain water and the other small and large molecules, which we examined in Chapters 2–4. Each cell contains at least 10,000 different types of molecules, most of them present in many copies. Cells use these molecules to transform matter and energy, to respond to their environments, and to reproduce themselves.

The cell theory has three important implications:

- Studying cell biology is in some sense the same as studying life. The principles that underlie the functions of the single cell of a bacterium are similar to those governing the approximately 60 trillion cells of your body.

- Life is continuous. All those cells in your body came from a single cell, a fertilized egg (zygote). That zygote came from the fusion of two cells, a sperm and an egg, from your parents. The cells of your parents' bodies were all derived from their parents, and so on back through generations and evolution to the first living cell.

- The origin of life on Earth was marked by the origin of the first cells (see Chapter 4).

Even the largest creatures on Earth are composed of cells, but the cells themselves are usually too small for the naked eye to see. Why are cells so small?

Cell size is limited by the surface area-to-volume ratio

Most cells are tiny. In 1665 Robert Hooke estimated that in one square inch of cork, which he examined under his magnifying lens, there were 1,259,712,000 cells! The diameters of cells range from about 1 to 100 micrometers (μm). There are some exceptions: the eggs of birds are single cells that are, relatively speaking, enormous, and individual cells of several types of algae and bacteria are large enough to be viewed with the unaided eye (**Figure 5.1**).

Small cell size is a practical necessity arising from the change in the **surface area-to-volume ratio** of any object as it increases in size. As an object increases in volume, its surface area also increases, but not at the same rate (**Figure 5.2**). This phenomenon has great biological significance. To appreciate this point,

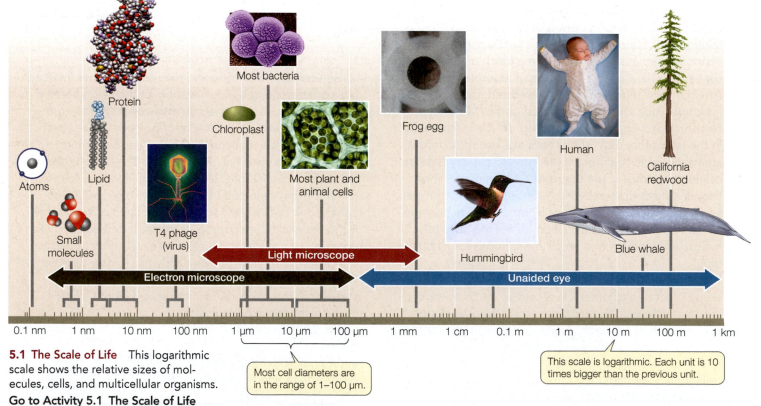

5.1 The Scale of Life This logarithmic scale shows the relative sizes of molecules, cells, and multicellular organisms.

Atoms   Small molecules   Lipid   Protein   T4 phage (virus)   Chloroplast   Most bacteria   Most plant and animal cells   Frog egg   Hummingbird   Human   Blue whale   California redwood

Light microscope

Electron microscope

Unaided eye

0.1 nm   1 nm   10 nm   100 nm   1 μm   10 μm   100 μm   1 mm   1 cm   0.1 m   1 m   10 m   100 m   1 km

Most cell diameters are in the range of 1–100 μm.

This scale is logarithmic. Each unit is 10 times bigger than the previous unit.

Go to Activity 5.1 The Scale of Life
Life10e.com/ac5.1

Diameter	1 μm	2 μm	3 μm
Surface area $4 \pi r^2$	3.14 μm²	12.56 μm²	28.26 μm²
Volume $4/3 \pi r^3$	0.52 μm³	4.19 μm³	14.18 μm³
Surface area-to-volume ratio	6:1	3:1	2:1

5.2 Why Cells Are Small   As an object grows larger, its volume increases more rapidly than its surface area. Cells must maintain a large surface area-to-volume ratio in order to function. This explains why large organisms are composed of many small cells rather than a few huge ones.

let's assume that the amount of chemical activity carried out by a cell is proportional to its volume. The surface area of the cell determines the amount of substances that can enter it from the outside environment, and the amount of waste products that can exit to the environment.

As a living cell grows larger, its chemical activity, and thus its need for resources and its rate of waste production, increases faster than its surface area. (The surface area being two-dimensional, increases in proportion to the square of the radius, whereas the volume being three-dimensional, increases much more—in proportion to the cube of the radius.) In addition, substances must move from one site to another within the cell; the smaller the cell, the more easily this is accomplished. This explains why large organisms must consist of many small cells: cells must be small in volume in order to maintain a large enough surface area-to-volume ratio and an ideal internal volume. The large surface area represented by the many small cells of a multicellular organism enables it to carry out the many different functions required for survival.

Microscopes reveal the features of cells

Microscopes do two different things that allow cells and details within them to be seen by the human eye. They provide the ability to see great detail, which then allows the viewer to magnify the image of interest. The property that allows detail to be seen is called resolution. Formally defined, resolution is the minimum distance two objects can be apart and still be seen as two objects. Resolution for the human eye is about 0.2 mm (200 μm). Most cells are much smaller than 200 μm and thus are invisible to the human eye. Microscopes magnify and increase resolution so that cells and their internal structures can be seen clearly (**Figure 5.3**).

There are two basic types of microscopes—light microscopes and electron microscopes—that use different forms of radiation (see Figure 5.3). While the resolution is better in electron microscopy, only dead cells are visualized because they must be prepared in a vacuum. Light microscopes, by contrast, can be used to visualize living cells (for example, by phase-contrast microscopy; see Figure 5.3).

Before we delve into the details of cell structure, it is useful to consider the many uses of microscopy. An entire branch of medicine, pathology, makes use of many different methods of microscopy to aid in the analysis of cells and the diagnosis of diseases. For instance, a surgeon might remove from a body some tissue suspected of being cancerous. The pathologist might:

- examine the tissue quickly by phase-contrast microscopy or interference-contrast microscopy to determine the size, shape, and spread of the cells;

- stain the tissue with a general dye and examine it by bright-field microscopy to bring out features such as the shapes of the nuclei, or cell division characteristics;

- examine the tissue under the electron microscope to observe internal structures such as the mitochondria or the chromatin (these are described in Section 5.3);

- stain the tissue with a specific dye and examine it by microscopy for the presence of proteins that are diagnostic of particular cancers. The results can influence the choice of therapy.

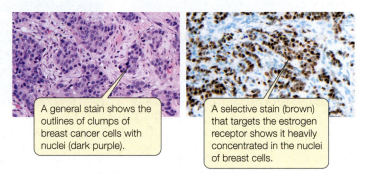

A general stain shows the outlines of clumps of breast cancer cells with nuclei (dark purple).

A selective stain (brown) that targets the estrogen receptor shows it heavily concentrated in the nuclei of breast cells.

The plasma membrane forms the outer surface of every cell

While the structural diversity of cells can often be observed using light microscopy, the **plasma membrane** is best observed with an electron microscope. This very thin structure forms the outer surface of every cell, and it has more or less the same thickness and molecular structure in all cells. We will describe the membrane in more detail in Chapter 6. For now we should keep in mind that it consists of a phospholipid bilayer (see Section 3.4), and that a variety of proteins are embedded within the bilayer. There is much compositional and functional diversity in the proteins associated with the plasma membrane. The membrane has several important roles:

- The plasma membrane acts as a selectively permeable barrier, preventing some substances from crossing it while permitting other substances to enter and leave the cell. For example, macromolecules such as DNA and proteins cannot normally cross the plasma membrane, but some smaller molecules such as oxygen can. In addition to size, other factors (particularly polarity) determine a molecule's ability to cross the plasma membrane. Because the membrane is composed mostly of hydrophobic fatty acids,

5.3 Looking at Cells   The six images on this page show some techniques used in light microscopy. The three images on the following page were created using electron microscopes. All of these images are of a particular type of cultured cell known as HeLa cells. Note that the images in most cases are flat, two-dimensional views. As you look at images of cells, keep in mind that they are three-dimensional structures.

Light microscope

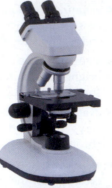

- Ocular lens
- Light beam
- Objective lens
- Specimen
- Condenser lens
- Light source

In a *light microscope*, glass lenses and visible light are used to form an image. The resolution is about 0.2 μm, which is 1,000 times greater than that of the human eye. Light microscopy allows visualization of cell sizes and shapes and some internal cell structures. Internal structures are hard to see under visible light, so cells are often chemically treated and stained with various dyes to make certain structures stand out by increasing contrast.

30 μm

In **bright-field microscopy**, light passes directly through these human cells. Unless natural pigments are present, there is little contrast and details are not distinguished.

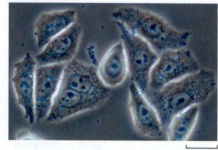

30 μm

In **phase-contrast microscopy**, contrast in the image is increased by emphasizing differences in refractive index (the capacity to bend light), thereby enhancing light and dark regions in the cell.

30 μm

Differential interference-contrast microscopy uses two beams of polarized light. The combined images look as if the cell is casting a shadow on one side.

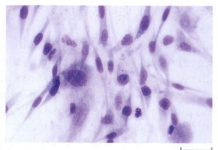

30 μm

In **stained bright-field microscopy**, a stain enhances contrast and reveals details not otherwise visible. Stains differ greatly in their chemistry and their capacity to bind to cell materials, so many choices are available.

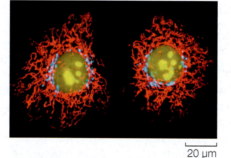

20 μm

In **fluorescence microscopy**, a natural substance in the cell or a fluorescent dye that binds to a specific cell material is stimulated by a beam of light, and the longer-wavelength fluorescent light is observed coming directly from the dye.

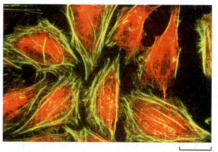

20 μm

Confocal microscopy uses fluorescent materials but adds a system of focusing both the stimulating and emitted light so that a single plane through the cell is seen. The result is a sharper two-dimensional image than with standard fluorescence microscopy.

nonpolar molecules cross it more easily than polar or charged molecules.

- The plasma membrane allows the cell to maintain a more or less constant internal environment. The maintenance of a constant internal environment (known as homeostasis) is a key characteristic of life and will be discussed in detail in Chapter 40. The membrane contributes to homeostasis by actively regulating the transport of substances across it. This dynamic process is distinct from the more passive process of diffusion, which is dependent on only the size of a molecule.

5.3 Looking at Cells *(continued)*   Go to Activity 5.2 Know Your Techniques Life10e.com/ac5.2

Transmission electron microscope

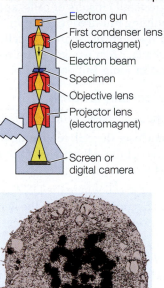

- Electron gun
- First condenser lens (electromagnet)
- Electron beam
- Specimen
- Objective lens
- Projector lens (electromagnet)
- Screen or digital camera

In an *electron microscope*, electromagnets are used to focus an electron beam, much as a light microscope uses glass lenses to focus a beam of light. Since we cannot see electrons, the electron microscope directs them through a vacuum at a fluorescent screen or digital camera to create a visible image. The resolution of electron microscopes is about 2 nm, which is about 100,000 times greater than that of the human eye. This resolution permits the details of many subcellular structures to be distinguished.

|— 10 μm —|

In **transmission electron microscopy** (TEM), a beam of electrons is focused on the object by magnets. Objects appear darker if they absorb the electrons. If the electrons pass through they are detected on a fluorescent screen.

|— 20 μm —|

Scanning electron microscopy (SEM) directs electrons to the surface of the sample, where they cause other electrons to be emitted. These electrons are viewed on a screen. The three-dimensional surface of the sample can be visualized.

|— 0.1 μm —|

In **freeze-fracture microscopy**, cells are frozen and then a knife is used to crack them open. The crack often passes through the interior of plasma and internal membranes. The "bumps" that appear are usually large proteins or aggregates embedded in the interior of the membrane.

- As the cell's boundary with the outside environment, the plasma membrane is important in communicating with adjacent cells and receiving signals from the environment. We will describe this function in Chapter 7.

- The plasma membrane often has proteins protruding from it that are responsible for binding and adhering to adjacent cells. Thus the plasma membrane plays an important structural role and contributes to cell shape.

Cells are classified as either prokaryotic or eukaryotic

As we learned in Section 1.1, biologists classify all living things into three domains: Archaea, Bacteria, and Eukarya. The organisms in Archaea and Bacteria are collectively called **prokaryotes**, and they have in common a prokaryotic cell organization. A prokaryotic cell does not typically have membrane-enclosed internal compartments; in particular, it does not have a nucleus. The first cells were probably similar in organization to those of modern prokaryotes.

Eukaryotic cell organization is found in members of the domain Eukarya (**eukaryotes**), which includes the protists,

plants, fungi, and animals. In contrast to prokaryotic cells, eukaryotic cells contain membrane-enclosed compartments called **organelles**. The most notable organelle is the cell **nucleus**, where most of the cell's DNA is located and where gene expression begins:

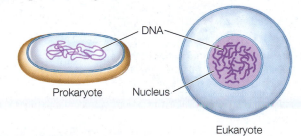

DNA

Prokaryote   Nucleus

Eukaryote

Just as a cell is an enclosed compartment, separating its contents from the surrounding environment, so each organelle provides a compartment that separates molecules and biochemical reactions from the rest of the cell. This "division of labor" provides possibilities for regulation and efficiency that were important in the evolution of complex organisms and helps to explain the complexity of eukaryotic cells relative to prokaryotic cells.

RECAP 5.1

The cell theory is a unifying principle of biology. Cell size is limited in order to maintain a high surface area-to-volume ratio. Both prokaryotic and eukaryotic cells are enclosed within a plasma membrane, but prokaryotic cells lack the membrane-enclosed organelles that are found in eukaryotic cells.

- How does cell biology embody all the principles of life? See **p. 78**
- Why are cells small? See **pp. 78–79** and **Figure 5.2**
- Explain the importance of the plasma membrane and the membranes that surround organelles. See **pp. 79–81**

As we mentioned in this section, there are two structural themes in cell architecture: prokaryotic and eukaryotic. We will now turn to the organization of prokaryotic cells.

5.2 What Features Characterize Prokaryotic Cells?

In terms of sheer numbers, prokaryotes are the most successful organisms on Earth. As we examine prokaryotic cells in this section, bear in mind that there are vast numbers of prokaryotic species, and that the Bacteria and Archaea are distinguished in numerous ways. These differences, and the vast diversity of organisms in these two domains, will be the subject of Chapter 26.

Prokaryotic cells, with diameters or lengths in the range 1–10 μm (micrometers), are generally smaller than eukaryotic cells, whose diameters are usually in the range of 10–100 μm. Each individual prokaryote is a single cell, but many types of prokaryotes form chains or small clusters of cells, and some occur in large clusters containing hundreds of cells. In this section we will first consider the features shared by cells in the domains Bacteria and Archaea. Then we will examine structural features that are found in some, but not all, prokaryotes.

Prokaryotic cells share certain features

All prokaryotic cells have the same basic structure (**Figure 5.4**):

- The plasma membrane encloses the cell, regulating the traffic of materials into and out of the cell, and separating its interior from the external environment.

- The **nucleoid** is a region in the cell where the DNA is located. As we described in Section 4.1, DNA is the hereditary material that controls cell growth, maintenance, and reproduction.

- The rest of the material enclosed in the plasma membrane is called the **cytoplasm**. The cytoplasm consists of a liquid component, the cytosol, and a variety of insoluble filaments and particles, the most abundant of which are ribosomes (see below).

- The **cytosol** consists mostly of water that contains dissolved ions, small molecules, and soluble macromolecules such as proteins.

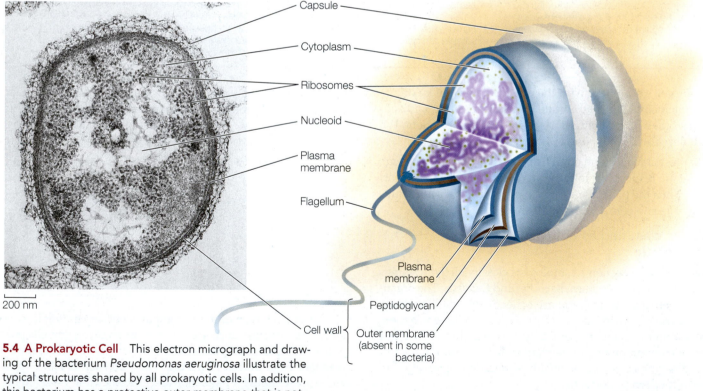

200 nm

5.4 A Prokaryotic Cell This electron micrograph and drawing of the bacterium *Pseudomonas aeruginosa* illustrate the typical structures shared by all prokaryotic cells. In addition, this bacterium has a protective outer membrane that is not present in all prokaryotes. The flagellum and capsule are also structures found in some, but not all, prokaryotic cells.

- **Ribosomes** are complexes of RNA and proteins that are about 25 nm (nanometers) in diameter. They can only be visualized with the electron microscope. They are the sites of protein synthesis, where information coded for in nucleic acids directs the sequential linking of amino acids to form proteins.

The cytoplasm is not a static region. Rather, the substances in this environment are in constant motion. For example, a typical protein moves around the entire cell within a minute, and it collides with many other molecules along the way. This motion helps ensure that biochemical reactions proceed at rates sufficient to meet the needs of the cell. Although they are structurally less complex than eukaryotic cells, prokaryotic cells are functionally complex, carrying out thousands of biochemical reactions.

Specialized features are found in some prokaryotes

As they evolved, some prokaryotes developed specialized structures that gave a selective advantage to those that had them: cells with these structures were better able to survive and reproduce in particular environments than cells lacking them.

CELL WALLS Most prokaryotes have a **cell wall** located outside the plasma membrane. The rigidity of the cell wall supports the cell and determines its shape. The cell walls of most bacteria, but not archaea, contain peptidoglycan, a polymer of amino sugars that is linked at regular intervals to short peptides. Cross-linking among these peptides results in a single giant molecule around the entire cell. In some bacteria, another layer, the **outer membrane** (a polysaccharide-rich phospholipid

membrane), encloses the peptidoglycan layer (see Figure 5.4). Unlike the plasma membrane, this outer membrane is not a major barrier to the movement of molecules across it.

Enclosing the cell wall in some bacteria is a slimy layer composed mostly of polysaccharides and referred to as a **capsule**. In some cases these capsules protect the bacteria from attack by white blood cells in the animals they infect. Capsules also help keep the cells from drying out, and sometimes they help bacteria attach to other cells. Many prokaryotes produce no capsule, and those that do have capsules can survive even if they lose them, so the capsule is not essential to prokaryotic life.

INTERNAL MEMBRANES Some groups of bacteria—including the cyanobacteria—carry out photosynthesis: they use energy from the sun to convert carbon dioxide and water into carbohydrates. These bacteria have an **internal membrane** system that contains molecules needed for photosynthesis. The development of photosynthesis, which requires membranes, was an important event in the early evolution of life on Earth. Other prokaryotes have internal membrane folds that are attached to the plasma membrane. These folds may function in cell division or in various energy-releasing reactions.

FLAGELLA AND PILI Some prokaryotes swim by using appendages called **flagella**, which sometimes look like tiny corkscrews (**Figure 5.5A**). In bacteria, the filament of the flagellum is made of a protein called flagellin. A complex motor protein spins the flagellum on its axis like a propeller, driving the cell along. The motor protein is anchored to the plasma membrane and,

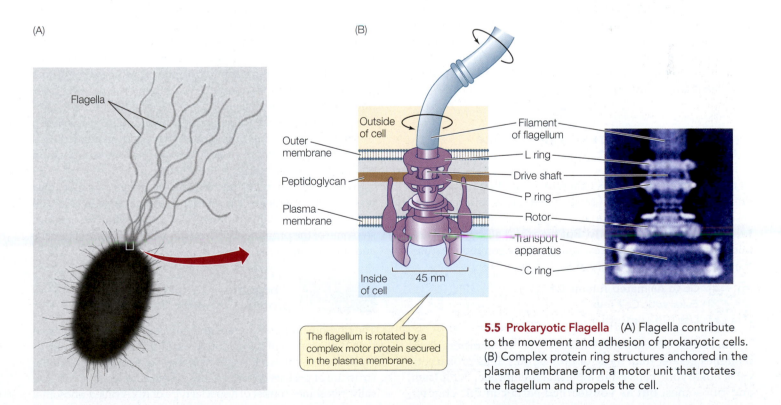

(A)

Flagella

(B)

Outside of cell

Outer membrane

Peptidoglycan

Plasma membrane

Inside of cell

Filament of flagellum

L ring

Drive shaft

P ring

Rotor

Transport apparatus

C ring

45 nm

The flagellum is rotated by a complex motor protein secured in the plasma membrane.

5.5 Prokaryotic Flagella (A) Flagella contribute to the movement and adhesion of prokaryotic cells. (B) Complex protein ring structures anchored in the plasma membrane form a motor unit that rotates the flagellum and propels the cell.

in some bacteria, to the outer membrane of the cell wall (**Figure 5.5B**). We know that the flagella cause the motion of cells because if they are removed, the cells do not move.

Pili are structures made of protein that project from the surfaces of some types of bacterial cells. These hairlike structures are shorter than flagella and are used for adherence. Conjugative pili (sex pili) help bacteria join to one another to exchange genetic material. Fimbriae are composed of the same proteins as pili but are shorter, and help cells adhere to surfaces such as animal cells, for food and protection.

CYTOSKELETON   The **cytoskeleton** is the collective name for protein filaments that play roles in cell division or in maintaining the shapes of cells. One such protein forms a ring structure that constricts during cell division, whereas another forms helical structures that extend down the lengths of rod-shaped cells, helping maintain their shapes. In the past it was thought that only eukaryotic cells had cytoskeletons (see Section 5.3), but more recently, biologists have recognized that cytoskeletal components are also widely distributed among prokaryotes.

RECAP (5.2)

Prokaryotic cells share basic features, including the plasma membrane, the nucleoid, and the cytoplasm, which consists of the liquid cytosol and insoluble filaments and particles, including ribosomes. Other features, such as cell walls, internal membranes, and flagella, are present in some but not all prokaryotes.

- What structures are present in all prokaryotic cells?
  See pp. 82–83 and Figure 5.4
- Describe the structure and function of a specialized prokaryotic cell feature, such as the cell wall, capsule, flagellum, or pilus. **See pp. 83–84 and Figure 5.5**

As we mentioned earlier, the prokaryotic cell is one of two types of cell recognized in cell biology. The other is the eukaryotic cell. Eukaryotic cells are more structurally and functionally complex than prokaryotic cells.

(5.3) What Features Characterize Eukaryotic Cells?

In the opening story of this chapter we saw that human cells arise by the differentiation of stem cells. This differentiation results in hundreds of different cell types, all with specialized functions in the human body. But these and all other eukaryotic cells share many features, and it is these common features that we will discuss in this section.

 Go to Animated Tutorial 5.1
Eukaryotic Cell Tour
Life10e.com/at5.1

Eukaryotic cells generally have lengths or diameters about ten times greater than those of prokaryotes. Like prokaryotic cells, eukaryotic cells have a plasma membrane, cytoplasm, and ribosomes. But as you learned earlier in this chapter,

eukaryotic cells also have compartments within the cytoplasm whose interiors are separated from the cytosol by membranes.

Compartmentalization is the key to eukaryotic cell function

The membranous compartments of eukaryotic cells are called organelles. Each type of organelle has a specific role: some organelles have been characterized as factories that make specific products, whereas others are like power plants that take in energy in one form and convert it to a more useful form. These functional roles are defined by the chemical reactions that occur within the organelles. Eukaryotic cells also have some structures that are analogous to those in prokaryotes. For example, they have a cytoskeleton composed of protein fibers and, outside the cell membrane, an extracellular matrix.

Animal and plant cells have many organelles and structures in common—the most obvious is the cell nucleus. But they also have some differences. For example, many plant cells have chloroplasts that perform photosynthesis.

Organelles can be studied by microscopy or isolated for chemical analysis

Cell organelles and structures were first detected by light and then by electron microscopy. The functions of the organelles could sometimes be inferred by observations and experiments, leading, for example, to the hypothesis (later confirmed) that the nucleus contained the genetic material. Later, the use of stains targeted to specific macromolecules allowed cell biologists to determine the chemical compositions of organelles (see Figure 5.14, which shows three different cytoskeletal proteins in a single cell).

Another way to analyze cells is to take them apart in a process called cell fractionation. This process permits cell organelles and other cytoplasmic structures to be separated from each other and examined using chemical methods. Cell fractionation begins with the destruction of the plasma membrane, which allows the cytoplasmic components to flow out into a test tube. The various organelles can then be separated from one another on the basis of size or density (**Figure 5.6**). Biochemical analyses can then be done on the isolated organelles.

Microscopy and cell fractionation have complemented each other, giving us a more complete picture of the composition and function of each organelle and structure.

Microscopy of plant and animal cells has revealed that many of the organelles are similar in appearance in each cell type (**Figure 5.7**). By comparing Figures 5.7 and 5.4 you can see some of the prominent differences between eukaryotic cells and prokaryotic cells.

 ### Ribosomes are factories for protein synthesis

The ribosomes of prokaryotes and eukaryotes are similar in that both types consist of two different-sized subunits. Eukaryotic ribosomes are somewhat larger than those of prokaryotes, but the structure of prokaryotic ribosomes is better understood. Chemically, ribosomes consist of a special type of RNA called ribosomal

RESEARCH**TOOLS**

5.6 Cell Fractionation Organelles can be separated from one another after cells are broken open and their contents suspended in an aqueous medium. The medium is placed in a tube and spun in a centrifuge, which rotates about an axis at high speed. Centrifugal forces (measured in multiples of gravity, × *g*) cause particles to sediment (form a pellet) at the bottom of the tube, which may be collected for biochemical study. Heavier particles sediment at lower speeds (lower centrifugal forces) than lighter particles. By adjusting the speed of centrifugation, researchers can separate and partially purify cellular organelles and large particles such as ribosomes.

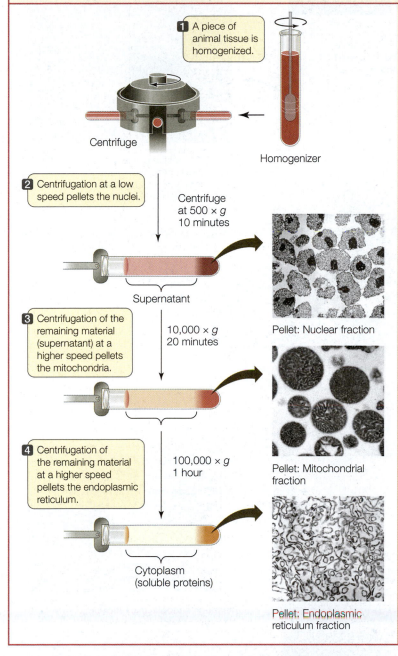

1 A piece of animal tissue is homogenized.

Centrifuge

Homogenizer

2 Centrifugation at a low speed pellets the nuclei.

Centrifuge at 500 × *g* 10 minutes

Supernatant

Pellet: Nuclear fraction

3 Centrifugation of the remaining material (supernatant) at a higher speed pellets the mitochondria.

10,000 × *g* 20 minutes

Pellet: Mitochondrial fraction

4 Centrifugation of the remaining material at a higher speed pellets the endoplasmic reticulum.

100,000 × *g* 1 hour

Cytoplasm (soluble proteins)

Pellet: Endoplasmic reticulum fraction

RNA (rRNA). Ribosomes also contain more than 50 different protein molecules, which are noncovalently bound to the rRNA.

In prokaryotic cells, ribosomes generally float freely in the cytoplasm. In eukaryotic cells they are found in multiple places: in the cytoplasm, where they may be free or attached to the surface of the endoplasmic reticulum (a membrane-bound organelle; see below), inside the mitochondria, and inside the chloroplasts in plant cells. In each of these locations, the

ribosomes are molecular factories where proteins are synthesized. Although they seem small in comparison with the cells that contain them, by molecular standards ribosomes are huge complexes (about 25 nm in diameter), made up of several dozen different molecules.

The nucleus contains most of the genetic information

As we discussed in Chapter 4, hereditary information is stored in the sequence of nucleotides in DNA molecules. Most of the DNA in eukaryotic cells resides in the nucleus (see Figure 5.7). Information encoded in the DNA is translated into proteins at the ribosomes. This process is described in detail in Chapter 14.

Most cells have a single nucleus, which is usually the largest organelle. The nucleus of a typical animal cell is approximately 5 μm (micrometers) in diameter—substantially larger than many prokaryotic cells (**Figure 5.8A**). The nucleus has several functions in the cell:

- It is the location of most of the cell's DNA and the site of DNA replication.
- It is the site where gene transcription is turned on or off.
- A region within the nucleus, the **nucleolus**, is where ribosomes begin to be assembled from RNA and proteins.

The contents of the nucleus, aside from the nucleolus, are referred to as the nucleoplasm. Similar to the cytoplasm, the nucleoplasm consists of the liquid content of the nucleus and the insoluble molecules suspended within it.

The nucleus is surrounded by an integrated structure comprised of two membranes, called the **nuclear envelope**. This structure separates the genetic material from the cytoplasm. Functionally, it separates DNA transcription (which occurs in the nucleus) from translation (which occurs in the cytoplasm) (see Section 4.1). The two membranes of the nuclear envelope are perforated by thousands of nuclear pores, each measuring approximately 9 nm in diameter, which connect the nucleoplasm with the cytoplasm.

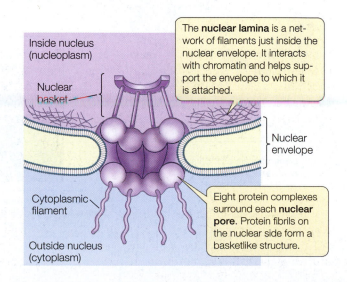

Inside nucleus (nucleoplasm)

The **nuclear lamina** is a network of filaments just inside the nuclear envelope. It interacts with chromatin and helps support the envelope to which it is attached.

Nuclear basket

Nuclear envelope

Cytoplasmic filament

Eight protein complexes surround each **nuclear pore**. Protein fibrils on the nuclear side form a basketlike structure.

Outside nucleus (cytoplasm)

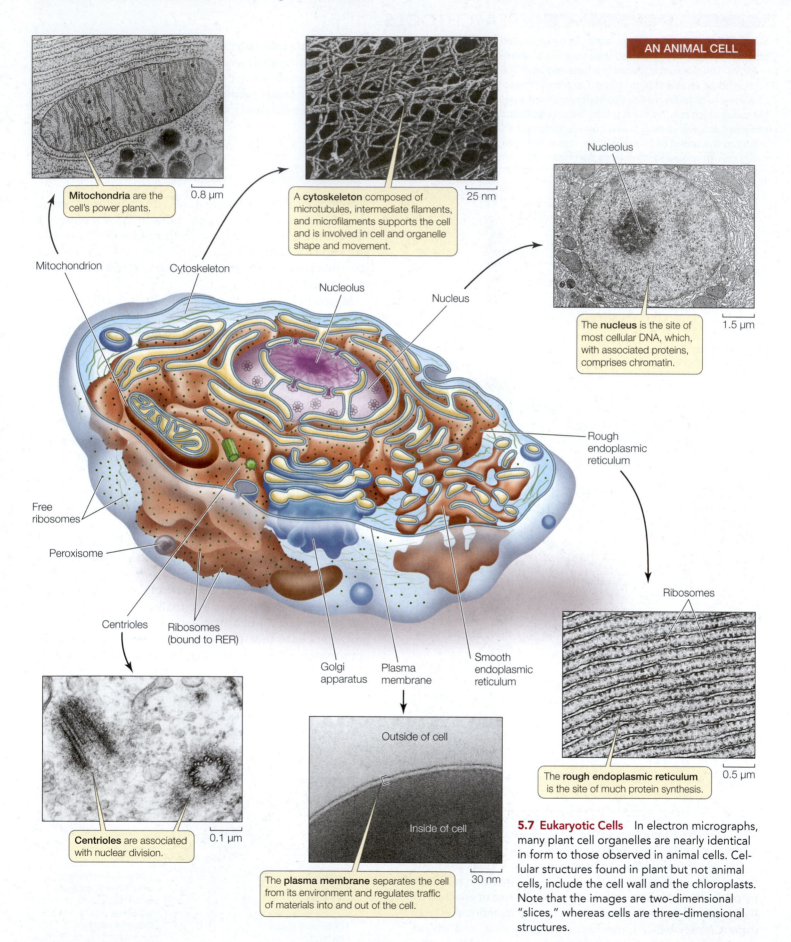

AN ANIMAL CELL

Mitochondria are the cell's power plants.

0.8 μm

Mitochondrion

A **cytoskeleton** composed of microtubules, intermediate filaments, and microfilaments supports the cell and is involved in cell and organelle shape and movement.

25 nm

Cytoskeleton

Nucleolus

Nucleolus

Nucleus

The **nucleus** is the site of most cellular DNA, which, with associated proteins, comprises chromatin.

1.5 μm

Rough endoplasmic reticulum

Free ribosomes

Peroxisome

Centrioles

Ribosomes (bound to RER)

Golgi apparatus

Plasma membrane

Smooth endoplasmic reticulum

Ribosomes

Centrioles are associated with nuclear division.

0.1 μm

Outside of cell

Inside of cell

The **plasma membrane** separates the cell from its environment and regulates traffic of materials into and out of the cell.

30 nm

The **rough endoplasmic reticulum** is the site of much protein synthesis.

0.5 μm

5.7 Eukaryotic Cells In electron micrographs, many plant cell organelles are nearly identical in form to those observed in animal cells. Cellular structures found in plant but not animal cells, include the cell wall and the chloroplasts. Note that the images are two-dimensional "slices," whereas cells are three-dimensional structures.

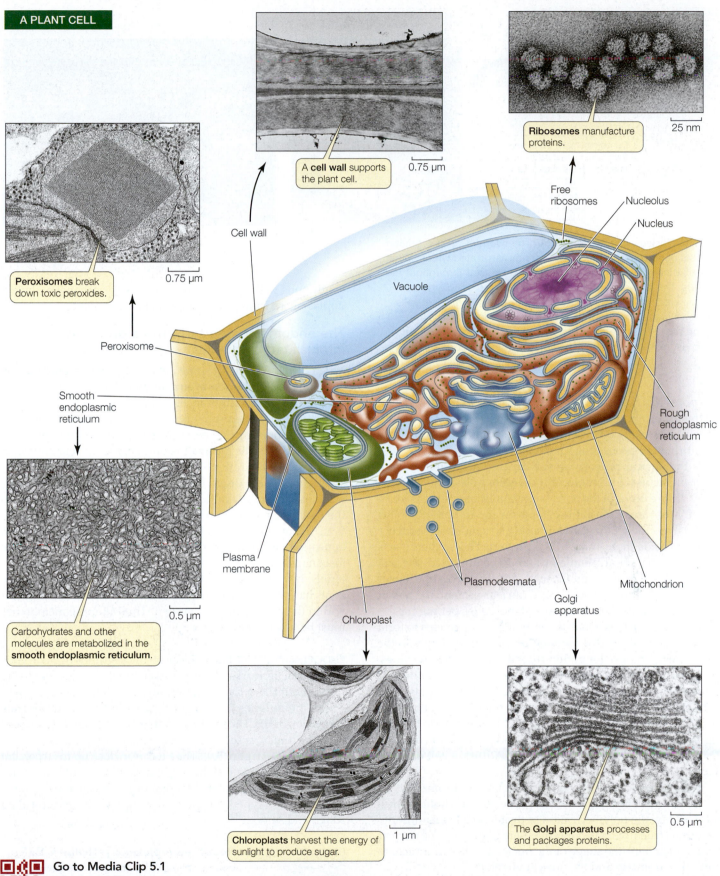

A PLANT CELL

Peroxisomes break down toxic peroxides.

0.75 μm

A **cell wall** supports the plant cell.

0.75 μm

Ribosomes manufacture proteins.

25 nm

Cell wall

Free ribosomes

Nucleolus

Nucleus

Vacuole

Peroxisome

Smooth endoplasmic reticulum

Rough endoplasmic reticulum

Carbohydrates and other molecules are metabolized in the **smooth endoplasmic reticulum**.

0.5 μm

Plasma membrane

Plasmodesmata

Mitochondrion

Golgi apparatus

Chloroplast

Chloroplasts harvest the energy of sunlight to produce sugar.

1 μm

The **Golgi apparatus** processes and packages proteins.

0.5 μm

Go to Media Clip 5.1
The Inner Life of a Cell
Life10e.com/mc5.1

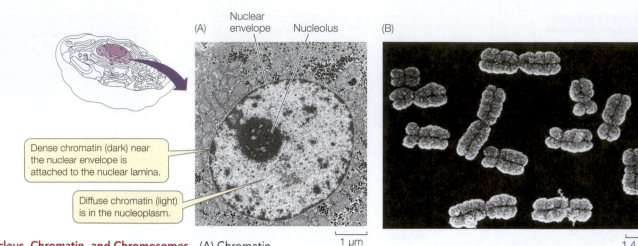

5.8 The Nucleus, Chromatin, and Chromosomes   (A) Chromatin consists of nuclear DNA and the proteins associated with it. When the cell is not dividing, the chromatin is dispersed throughout the nucleus. This two-dimensional image was made using a transmission electron microscope. (B) The chromatin in dividing cells becomes highly condensed so that the individual chromosomes can be seen. This three-dimensional image of isolated metaphase chromosomes was produced using a scanning electron microscope.

The pores regulate the traffic between these two cellular compartments by allowing some molecules to enter or exit the nucleus and blocking others. This allows the nucleus to regulate its information-processing functions.

At the nuclear pore, small substances, including ions and other molecules with molecular weights of less than 10,000, freely diffuse through the pore. Larger molecules, such as many proteins that are made in the cytoplasm and imported into the nucleus, cannot get through without a specific short sequence of amino acids that is part of the protein. This sequence acts as a signal that identifies the protein to be imported. We will describe this sequence and the evidence for its role in Chapter 14 (see Figure 14.19). For a typical nuclear protein, the rate of import into the nucleus is about 100 molecules per minute.

Inside the nucleus, DNA is combined with proteins to form a fibrous complex called **chromatin**. Chromatin occurs in the form of exceedingly long, thin threads called **chromosomes**. Different eukaryotic organisms have different numbers of chromosomes (ranging from two in one kind of Australian ant to hundreds in some plants). Prior to cell division, the chromatin becomes tightly compacted and condensed so that the individual chromosomes are visible under a light microscope. This facilitates distribution of the DNA during cell division (**Figure 5.8B**).

At the interior periphery of the nucleus, the chromatin is attached to a protein meshwork, called the nuclear lamina, which is formed by the polymerization of proteins called lamins into long thin structures called intermediate filaments. The nuclear lamina maintains the shape of the nucleus by its attachment to both the chromatin and the nuclear envelope.

At the exterior of the nucleus, the outer membrane of the nuclear envelope folds outward into the cytoplasm and is continuous with the membrane of another organelle, the endoplasmic reticulum, which we will discuss next.

The endomembrane system is a group of interrelated organelles

Much of the volume of some eukaryotic cells is taken up by an extensive **endomembrane system**. This is an interconnected system of membrane-enclosed compartments that are sometimes flattened into sheets and sometimes have other characteristic shapes (see Figure 5.7). The endomembrane system includes the plasma membrane, the nuclear envelope, the endoplasmic reticulum, the Golgi apparatus, and lysosomes, which are derived from the Golgi. Tiny, membrane-surrounded droplets called vesicles shuttle substances between the various components of the endomembrane system (**Figure 5.9**). In drawings and electron microscope pictures, this system appears static, fixed in space and time. But these depictions are just snapshots; in the living cell, the membranes and the materials they contain are in constant motion. Membrane components have been observed to shift from one organelle to another within the endomembrane system. Thus all these membranes must be functionally related.

ENDOPLASMIC RETICULUM   Electron micrographs of eukaryotic cells reveal networks of interconnected membranes branching throughout the cytoplasm, forming tubes and flattened sacs. These membranes are collectively called the **endoplasmic reticulum**, or **ER**. The interior compartment of the ER, referred to as the lumen, is separate and distinct from the surrounding cytoplasm (see Figure 5.9). The ER can enclose up to 10 percent of the interior volume of the cell, and its folds result in a surface area many times greater than that of the plasma membrane. There are two types of endoplasmic reticulum, the so-called rough and smooth.

Rough endoplasmic reticulum (**RER**) is called "rough" because of the many ribosomes attached to the outer surface of the membrane, giving it a "rough" appearance

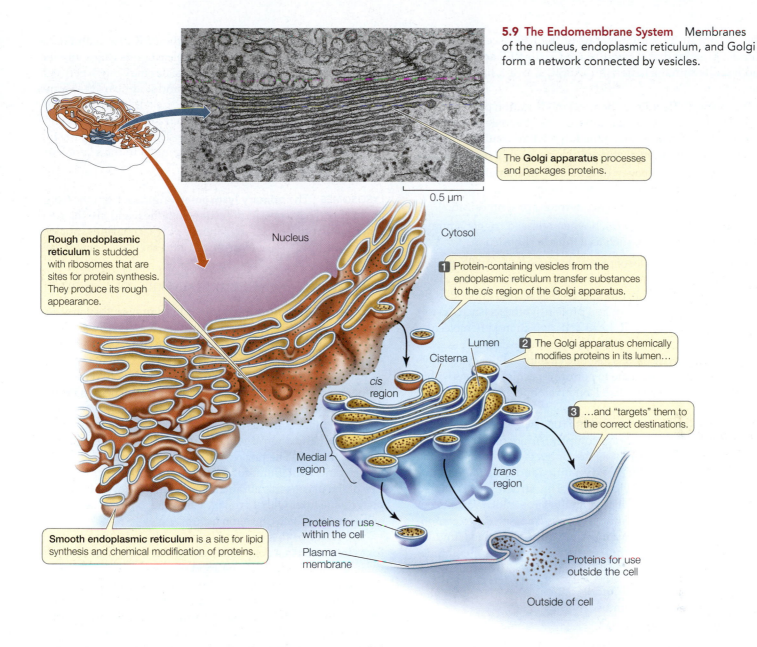

5.9 The Endomembrane System Membranes of the nucleus, endoplasmic reticulum, and Golgi form a network connected by vesicles.

The **Golgi apparatus** processes and packages proteins.

0.5 μm

Nucleus

Cytosol

Rough endoplasmic reticulum is studded with ribosomes that are sites for protein synthesis. They produce its rough appearance.

1 Protein-containing vesicles from the endoplasmic reticulum transfer substances to the *cis* region of the Golgi apparatus.

Lumen

Cisterna

2 The Golgi apparatus chemically modifies proteins in its lumen…

cis region

3 …and "targets" them to the correct destinations.

Medial region

trans region

Smooth endoplasmic reticulum is a site for lipid synthesis and chemical modification of proteins.

Proteins for use within the cell

Plasma membrane

Proteins for use outside the cell

Outside of cell

in electron microscopy (see Figure 5.7). The bound ribosomes are actively involved in protein synthesis, but that is not the entire story:

- The RER receives into its lumen certain newly synthesized proteins (including exported proteins and those destined for lysosomes and the plasma membrane), segregating them away from the cytoplasm. The RER also participates in transporting these proteins to other locations in the cell.

- While inside the RER, proteins can be chemically modified to alter their functions and to "tag" them for delivery to specific cellular destinations.

- Proteins are shipped to destinations elsewhere in the cell enclosed within vesicles that pinch off from the RER.

- Most membrane-bound proteins are made in the RER. A protein enters the lumen of the RER through a pore as it is synthesized. As with a protein passing through a nuclear

pore, this is accomplished via a sequence of amino acids on the protein, which acts as a RER localization signal ("address"; see Section 14.6). Once in the lumen of the RER, proteins undergo several changes, including the formation of disulfide bridges and folding into their tertiary structures (see Figure 3.7).

Some proteins are covalently linked to carbohydrate groups in the RER, thus becoming glycoproteins. In the case of proteins directed to the lysosomes, the carbohydrate groups are part of an "addressing" system that ensures that the right proteins are directed to those organelles. This addressing system is very important because the enzymes within the lysosomes are some of the most destructive the cell makes. Were they not properly addressed and contained, they could destroy the cell.

The **smooth endoplasmic reticulum** (**SER**) lacks ribosomes and is more tubular (and less like flattened sacs) than the RER, but it

shows continuity with portions of the RER (see Figure 5.9). Certain proteins that are synthesized in the RER are chemically modified within the lumen of the SER. The SER has four other important roles:

- It is responsible for the chemical modification of small molecules taken in by the cell that may be toxic to the cell. These modifications make the targeted molecules more polar, so they are more water-soluble and easily removed.

- It is the site for glycogen degradation in animal cells. We discuss this important process in Chapter 9.

- It is the site where lipids and steroids are synthesized.

- It stores calcium ions, which when released trigger a number of cell responses, such as a muscle contraction.

Cells that synthesize a lot of protein for export are usually packed with RER. Examples include glandular cells that secrete digestive enzymes and white blood cells that secrete antibodies. In contrast, cells that carry out less protein synthesis (such as storage cells) contain less RER. Liver cells, which modify molecules (including toxins) that enter the body from the digestive system, have abundant SER. This is just one example of the many ways individual cells differentiate to perform specific functions within a multicellular organism (like Don in our opening story).

 **GOLGI APPARATUS** The **Golgi apparatus** (or Golgi complex), more often referred to merely as the Golgi, is another part of the diverse, dynamic, and extensive endomembrane system (see Figure 5.9). This structure was named after its discoverer, Camillo Golgi. Its appearance varies from species to species, but it almost always consists of two components: flattened membranous sacs called cisternae (singular cisterna) that are piled up like saucers, and small membrane-enclosed vesicles. The entire apparatus is about 1 μm long.

Go to Animated Tutorial 5.2
The Golgi Apparatus
Life10e.com/at5.2

The Golgi has several roles:

- It receives protein-containing vesicles from the RER.

- It modifies, concentrates, packages, and sorts proteins before they are sent to their cellular or extracellular destinations.

- It adds carbohydrates to proteins and modifies other carbohydrates that were attached to proteins in the RER.

- It is where some polysaccharides for the plant cell wall are synthesized.

The cisternae of the Golgi apparatus have three functionally distinct regions: the *cis* region lies nearest to the nucleus or a patch of RER, the *trans* region lies closest to the plasma membrane, and the medial region lies in between (see Figure 5.9). (The terms *cis*, *trans*, and medial derive from Latin words meaning, respectively, "on the same side," "on the opposite side," and "in the middle.") These three parts of the Golgi apparatus contain different enzymes and perform different functions.

Protein-containing vesicles from the RER fuse with the *cis* membrane of the Golgi apparatus, releasing its cargo into the lumen of the Golgi cisterna. Other vesicles may move between the cisternae, transporting proteins, and it appears that some proteins move from one cisterna to the next through tiny channels. Vesicles budding off from the *trans* region carry their contents away from the Golgi apparatus. These vesicles go to the plasma membrane, or to another organelle in the endomembrane system called the lysosome.

LYSOSOMES The **primary lysosomes** originate from the Golgi apparatus. They contain digestive enzymes, and are the sites where macromolecules—proteins, polysaccharides, nucleic acids, and lipids—are hydrolyzed into their monomers (see Figure 3.4).

$$R_1—R_2 \text{ (linked monomers)} + H_2O \rightarrow R_1—OH + R_2—H$$

Lysosomes are about 1 micrometer in diameter; they are surrounded by a single membrane and have a densely staining, featureless interior (**Figure 5.10**). There may be dozens of lysosomes in a cell, depending on its needs.

Lysosomes are sites for the breakdown of food, other cells, or foreign objects that are taken up by the cell. These materials enter the cell by a process called **phagocytosis** (*phago*, "eat"; *cytosis*, "cellular"). In this process, a pocket forms in the plasma membrane and then deepens and encloses material from outside the cell. The pocket becomes a small vesicle called a phagosome, containing food or other material, which breaks free of the plasma membrane to move into the cytoplasm. The phagosome fuses with a primary lysosome to form a **secondary lysosome**, in which digestion occurs.

The effect of this fusion is rather like releasing hungry foxes into a chicken coop: the enzymes in the secondary lysosome quickly hydrolyze the food particles. These reactions are enhanced by the mild acidity of the lysosome's interior, where the pH is lower than in the surrounding cytoplasm. The products of digestion pass through the membrane of the lysosome, providing energy and raw materials for other cellular processes. The "used" secondary lysosome, now containing undigested particles, then moves to the plasma membrane, fuses with it, and releases the undigested contents to the environment.

Phagocytes (see Section 42.1) are specialized cells that have an essential role in taking up and breaking down materials; they are found in nearly all animals and many protists. You will encounter them and their activities again at many places in this book, but at this point one example suffices: in the human liver and spleen, phagocytes digest approximately 10 billion aged or damaged blood cells each day! The digestion products are then used to make new cells to replace those that are digested.

Lysosomes are active even in cells that do not perform phagocytosis. Cells are dynamic systems; some cell components are continually being broken down and replaced by new ones. The programmed destruction of cell components is called **autophagy**, and lysosomes are where the cell breaks down its

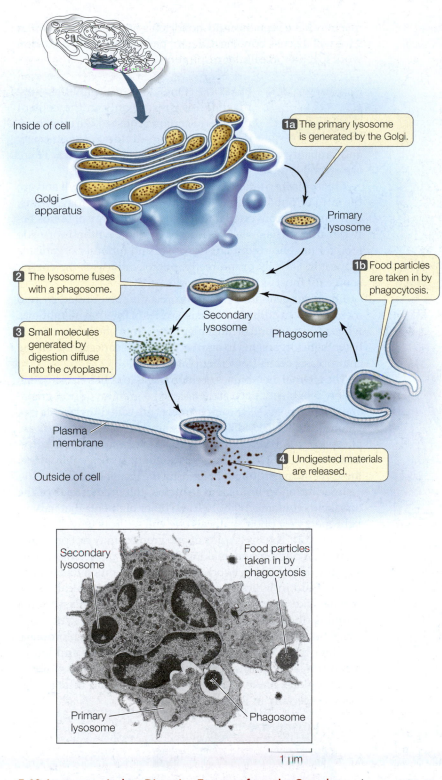

diseases are invariably very harmful or fatal. An example is Tay-Sachs disease, in which a particular lipid called a ganglioside is not broken down in the lysosomes and instead accumulates in brain cells. In the most common form of this disease, a baby starts exhibiting neurological symptoms and becomes blind, deaf, and unable to swallow after six months of age. Death occurs before age 4.

Plant cells do not appear to contain lysosomes, but the vacuole of a plant cell (which we will describe below) may function in an equivalent capacity because it, like lysosomes, contains many digestive enzymes.

Some organelles transform energy

A cell requires energy to make the molecules it needs for activities such as growth, reproduction, responsiveness, and movement. Energy is harvested from fuel molecules in the mitochondria (found in all eukaryotic cells) and from sunlight in the chloroplasts of plant cells. In contrast, energy transformations in prokaryotic cells are associated with enzymes attached to the inner surface of the plasma membrane or to extensions of the plasma membrane that protrude into the cytoplasm.

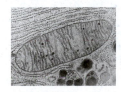

MITOCHONDRIA In eukaryotic cells, the breakdown of fuel molecules such as glucose begins in the cytosol. The molecules that result from this partial degradation enter the **mitochondria** (singular *mitochondrion*), whose primary function is to harvest the chemical energy of those fuel molecules in a form that the cell can use, namely the energy-rich molecule ATP (adenosine triphosphate) (see Section 8.2). The production of ATP in the mitochondria, using fuel molecules and molecular oxygen (O_2), is called **cellular respiration**.

Typical mitochondria are somewhat less than 1.5 μm in diameter and are 2–8 μm in length—about the size of many bacteria. They can divide independently of the central nucleus. The number of mitochondria per cell ranges from one gigantic organelle in some unicellular protists to a few hundred thousand in large egg cells. An average human liver cell contains more than 1,000 mitochondria. Cells that are active in movement and growth require the most chemical energy, and these tend to have the most mitochondria per unit of volume.

Mitochondria have two membranes. The outer membrane is smooth and protective, and it offers little resistance to the movement of substances into and out of the organelle. Immediately inside the outer membrane is an inner membrane, which folds inward in many places and thus has a surface area much

5.10 Lysosomes Isolate Digestive Enzymes from the Cytoplasm Lysosomes are sites for the hydrolysis of material taken into the cell by phagocytosis.

Go to Activity 5.3 Lysosomal Digestion **Life10e.com/ac5.3**

own materials. With the proper signal, lysosomes can engulf entire organelles, hydrolyzing their constituents.

How important is autophagy? An entire class of human diseases called lysosomal storage diseases are caused by the failure of lysosomes to digest specific cellular components; these

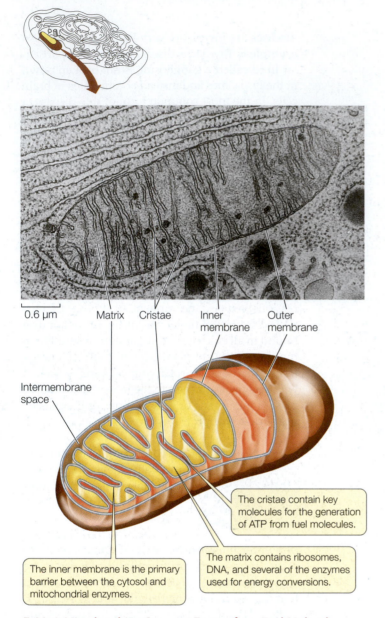

0.6 μm | Matrix | Cristae | Inner membrane | Outer membrane

Intermembrane space

The cristae contain key molecules for the generation of ATP from fuel molecules.

The inner membrane is the primary barrier between the cytosol and mitochondrial enzymes.

The matrix contains ribosomes, DNA, and several of the enzymes used for energy conversions.

5.11 A Mitochondrion Converts Energy from Fuel Molecules into ATP The electron micrograph is a two-dimensional slice through a three-dimensional organelle. As the drawing emphasizes, the cristae are extensions of the inner mitochondrial membrane.

greater than that of the outer membrane (**Figure 5.11**). The folds tend to be quite regular, giving rise to shelflike structures called cristae. The inner membrane exerts much more control over what enters and leaves the space it encloses than does the outer membrane. Embedded in the inner mitochondrial membrane are many large protein complexes that participate in cellular respiration.

The space enclosed by the inner membrane is referred to as the mitochondrial matrix. In addition to many enzymes, the matrix contains ribosomes and DNA that are used to make some of the proteins needed for cellular respiration. As we will discuss later in this chapter, it is likely that this DNA is the remnant of a larger, complete chromosome from a prokaryote

that may have been the mitochondrion's progenitor. In Chapter 9 we will discuss how the different parts of the mitochondrion work together in cellular respiration.

 **PLASTIDS** One class of organelles—the plastids—is present only in the cells of plants and certain protists. Like mitochondria, plastids can divide autonomously and probably evolved from independent prokaryotes. There are several types of plastids, with different functions.

Chloroplasts contain the green pigment chlorophyll and are the sites of photosynthesis (**Figure 5.12**). In photosynthesis, light energy is converted into the chemical energy of bonds between atoms. The molecules formed by photosynthesis provide food for the photosynthetic organism and for other organisms that eat it. Directly or indirectly, photosynthesis is the energy source for most of the living world.

Like a mitochondrion, a chloroplast is surrounded by two membranes. In addition, there is a series of internal membranes whose structure and arrangement vary from one group of photosynthetic organisms to another. Here we concentrate on the chloroplasts of the flowering plants.

The internal membranes of chloroplasts look like stacks of flat, hollow pita bread. Each stack is called a granum (plural grana) and the pita bread–like compartments are called **thylakoids** (see Figure 5.12). Thylakoid lipids are distinctive: only 10 percent are phospholipids, whereas the rest are galactose-substituted diglycerides and sulfolipids. Because of the abundance of chloroplasts, these are the most abundant lipids in the biosphere.

In addition to lipids and proteins, the membranes of the thylakoids contain chlorophyll and other pigments that harvest light energy for photosynthesis (we will see how they do this in Section 10.2). The thylakoids of one granum may be connected to those of other grana, making the interior of the chloroplast a highly developed network of membranes, much like the ER.

The fluid in which the grana are suspended is called the stroma. Like the mitochondrial matrix, the chloroplast stroma contains ribosomes and DNA, which are used to synthesize some, but not all, of the proteins that make up the chloroplast.

Other types of plastids, such as chromoplasts and leucoplasts, have functions different from those of chloroplasts. Chromoplasts make and store red, yellow, and orange pigments, especially in flowers and fruits.

Chromoplast

5 μm

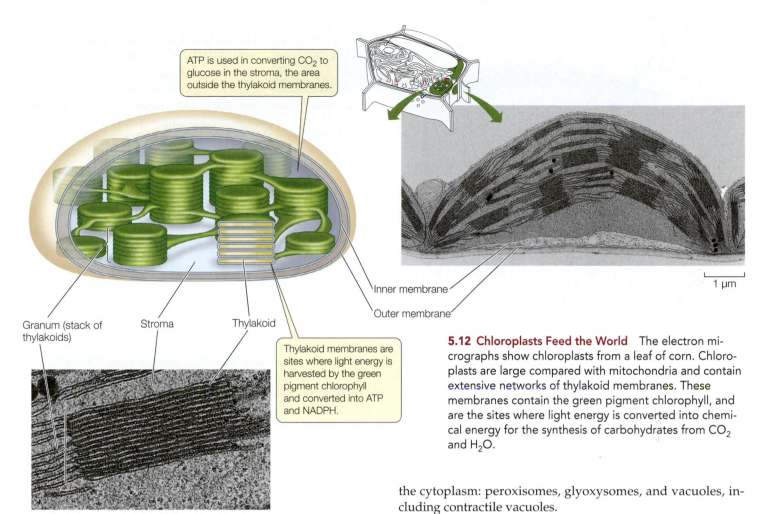

ATP is used in converting CO_2 to glucose in the stroma, the area outside the thylakoid membranes.

Granum (stack of thylakoids) Stroma Thylakoid

Thylakoid membranes are sites where light energy is harvested by the green pigment chlorophyll and converted into ATP and NADPH.

Inner membrane

Outer membrane

0.25 μm

1 μm

5.12 Chloroplasts Feed the World The electron micrographs show chloroplasts from a leaf of corn. Chloroplasts are large compared with mitochondria and contain extensive networks of thylakoid membranes. These membranes contain the green pigment chlorophyll, and are the sites where light energy is converted into chemical energy for the synthesis of carbohydrates from CO_2 and H_2O.

the cytoplasm: peroxisomes, glyoxysomes, and vacuoles, including contractile vacuoles.

Peroxisomes are organelles that accumulate toxic peroxides, such as hydrogen peroxide (H_2O_2), that occur as by-products of some biochemical reactions. These peroxides are safely broken down inside the peroxisomes without mixing with other parts of the cell.

$$RH_2 + O_2 \rightarrow R + H_2O_2 \text{ (cellular reactions)}$$

$$2\,H_2O_2 \rightarrow 2\,H_2O + O_2 \text{ (inside peroxisome)}$$

Peroxisomes are small organelles, about 0.2–1.7 micrometers in diameter. They have a single membrane and a granular interior containing specialized enzymes. Peroxisomes are found in at least some of the cells of almost every eukaryotic species.

As with lysosomes, there are rare inherited diseases in humans that involve peroxisomes. In Zellweger syndrome there is a defect in peroxisome assembly, and affected infants are born without peroxisomes. As you can imagine, a consequence of this is the accumulation of toxic peroxides, and the infants seldom live beyond one year of age.

Glyoxysomes are similar to peroxisomes and are found only in plants. They are most abundant in young plants and are the locations where stored lipids are converted into carbohydrates for transport to growing cells.

Leucoplasts are storage organelles that do not contain pigments. An amyloplast is a leucoplast that stores starch.

Leucoplast

Starch grains

1 μm

There are several other membrane-enclosed organelles

There are several other organelles whose boundary membranes separate their specialized chemical reactions and contents from

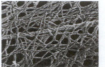

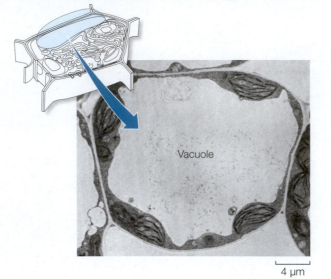

Vacuole

4 µm

5.13 Vacuoles in Plant Cells Are Usually Large   The large central vacuole in this cell is typical of mature plant cells.

Vacuoles occur in many eukaryotic cells but particularly those of plants, fungi, and protists. Plant vacuoles (**Figure 5.13**) have several functions:

- *Storage*: Plant cells produce a number of toxic by-products and waste products, many of which are simply stored within vacuoles. Because they are poisonous or distasteful, these stored materials deter some animals from eating the plants and may thus contribute to plant defense and survival.

- *Structure*: In many plant cells, vacuoles take up more than 90 percent of the cell volume and grow as the cell grows. The presence of dissolved substances in the vacuole causes water to enter it from the cytoplasm, making the vacuole swell like a balloon. A mature plant cell does not swell when the vacuole fills with water, since it has a rigid cell wall. Instead, it stiffens from the increase in water pressure (called turgor), and this supports the plant (see Figure 6.9).

- *Reproduction*: Vacuoles contain some of the pigments (especially blue and pink ones) in the petals and fruits of flowering plants. These pigments—the anthocyanins—are visual cues, which help attract animals that assist in pollination or seed dispersal.

- *Digestion*: In some plants, the vacuoles in seeds contain enzymes that hydrolyze stored proteins into monomers. During seed germination, the monomers are used as food by the developing plant seedlings.

Contractile vacuoles are found in many freshwater protists. Their function is to get rid of the excess water that rushes into the cell because of the imbalance in solute concentration between the interior of the cell and its freshwater environment. The contractile vacuole enlarges as water enters, then abruptly contracts, forcing the water out of the cell through a special pore structure.

So far we have discussed numerous membrane-enclosed organelles. Now we will turn to a group of cytoplasmic structures without membranes.

The cytoskeleton is important in cell structure and movement

From the earliest observations, light microscopy revealed distinctive cell shapes that would sometimes change, and rapid movements within cells. With the advent of electron microscopy, a new world of cellular substructure was revealed, including a meshwork of filaments inside cells. Experimentation showed that this meshwork—called the cytoskeleton—fills several important roles:

- It supports the cell and maintains its shape.
- It holds cell organelles and other particles in position within the cell.
- It moves organelles and other particles around in the cell.
- It is involved with movements of the cytoplasm, called cytoplasmic streaming.
- It interacts with extracellular structures, helping anchor the cell in place.

There are three components of the eukaryotic cytoskeleton: microfilaments (smallest diameter), intermediate filaments, and microtubules (largest diameter). These filaments have very different functions.

MICROFILAMENTS   Microfilaments can exist as single filaments, in bundles, or in networks. They are about 7 nanometers in diameter and up to several micrometers long. Microfilaments have two major roles:

- They help the entire cell or parts of the cell move.
- They determine and stabilize cell shape.

Microfilaments are assembled from monomers of **actin**, a protein that exists in several forms and has many functions, especially in animals. The actin found in microfilaments (which are also known as actin filaments) has distinct ends, designated "plus" and "minus." These ends permit actin monomers to interact with one another to form long, double helical chains (**Figure 5.14A**). Within cells, the polymerization of actin into microfilaments is reversible, and the microfilaments can disappear from cells by breaking down into monomers of free actin. Special actin-binding proteins mediate these processes.

In the muscle cells of animals, actin filaments are associated with another protein, the "motor protein" **myosin**, and the interactions of these two proteins account for the contraction of muscles (described in Section 48.1). In non-muscle cells, actin filaments are associated with localized changes in cell shape. For example, microfilaments are involved in the flowing movement of the cytoplasm called cytoplasmic streaming, in amoeboid movement, and in the "pinching" contractions that divide an animal cell into two daughter cells. Microfilaments are also involved in the formation of cellular extensions called pseudopodia (*pseudo*, "false"; *podia*, "feet") that enable some cells to move (**Figure 5.15**). As you will see in Chapter 42, cells of the immune system must move toward other cells during the immune response.

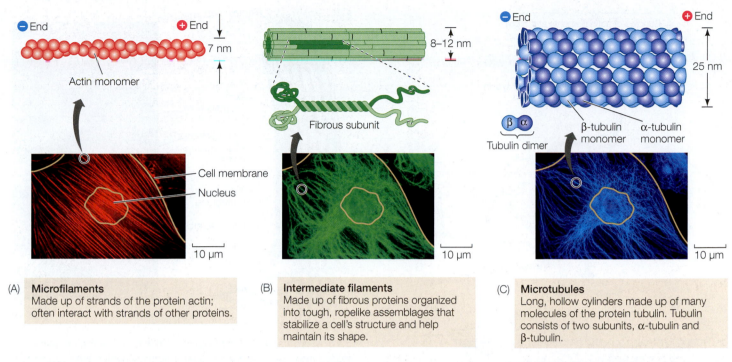

(A) Microfilaments
Made up of strands of the protein actin; often interact with strands of other proteins.

(B) Intermediate filaments
Made up of fibrous proteins organized into tough, ropelike assemblages that stabilize a cell's structure and help maintain its shape.

(C) Microtubules
Long, hollow cylinders made up of many molecules of the protein tubulin. Tubulin consists of two subunits, α-tubulin and β-tubulin.

5.14 The Cytoskeleton Three highly visible and important structural components of the cytoskeleton are shown here in detail. The photographs are all of the same cell, treated with different fluorescent antibodies that detect microfilaments (A), intermediate filaments (B), or microtubules (C). These structures maintain and reinforce cell shape and contribute to cell movement. The position of the cell's nucleus is near the center of the photos.

In some cell types, microfilaments form a meshwork just inside the plasma membrane. Actin-binding proteins then cross-link the microfilaments to form a rigid netlike structure that supports the cell. For example, microfilaments support the tiny microvilli that line the human intestine, giving it a larger surface area through which to absorb nutrients (**Figure 5.16**).

INTERMEDIATE FILAMENTS There are at least 50 different kinds of **intermediate filaments**, many of them specific to a few cell types. They generally fall into six molecular classes (based on amino acid sequence) that share the same general structure. One of these classes consists of fibrous proteins of the keratin family, which also includes the proteins that make up hair and fingernails. Intermediate filaments are tough, ropelike protein assemblages 8–12 nanometers in diameter (**Figure 5.14B**). They are more permanent than the other two types in that they do not continually form and reform, as the microtubules and microfilaments do.

Intermediate filaments have two major structural functions:

- They anchor cell structures in place. In some cells, intermediate filaments radiate from the nuclear envelope and help maintain the positions of the nucleus and other organelles in the cell. The lamins of the nuclear lamina are intermediate filaments. Other kinds of intermediate filaments help hold in place the complex apparatus of microfilaments in the microvilli of intestinal cells (see Figure 5.16).

- They resist tension. For example, they maintain rigidity in body surface tissues by stretching through the cytoplasm and connecting specialized membrane structures called desmosomes (see Figure 6.7).

MICROTUBULES The largest-diameter components of the cytoskeletal system, **microtubules**, are long, hollow, unbranched cylinders about 25 nanometers in diameter and up to several micrometers long. Microtubules have two roles in the cell:

- They form a rigid internal skeleton for some cells.

- They act as a framework along which motor proteins can move structures within the cell.

Microtubules are assembled from dimers of the protein **tubulin**. A dimer is a molecule made up of two monomers. The polypeptide monomers that make up the tubulin dimer are known as α-tubulin and β-tubulin. Thirteen chains of tubulin dimers surround the central cavity of the microtubule (**Figure 5.14C**; see also Figure 5.17B). As in microfilaments, the two ends of a microtubule are different: one is designated the "plus" end and the other the "minus" end. Tubulin dimers can be rapidly added or subtracted, mainly at the plus end, lengthening or shortening the microtubule.

Many microtubules radiate from a region of the cell called the microtubule organizing center. Tubulin polymerization results in a rigid structure (sometimes called an endoskeleton), and tubulin depolymerization leads to its collapse. The capacity to change length rapidly makes microtubules dynamic structures: they are readily adapted for new purposes in the cell. For example, by disassembly and reassembly,

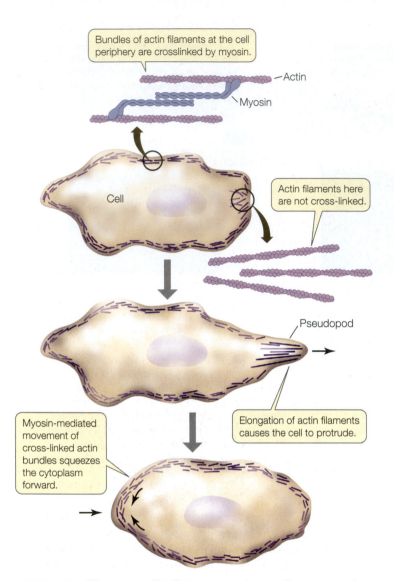

5.15 Microfilaments and Cell Movements   Microfilaments mediate the movement of whole cells (as illustrated here for amoeboid movement), as well as the movement of cytoplasm within a cell.

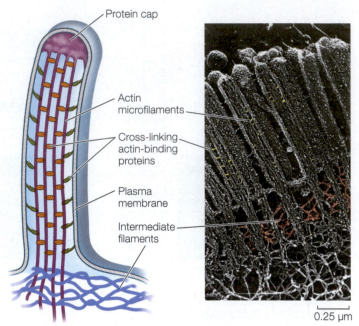

0.25 µm

5.16 Microfilaments for Support   Cells that line the intestine are folded into tiny projections called microvilli, which are supported by microfilaments. The microfilaments interact with intermediate filaments at the base of each microvillus. Microvilli increase the surface area of a cell, facilitating its absorption of small molecules.

microtubules can move to new parts of the cell and assemble new structures needed for cell division. Microtubules from all eukaryotes have this dynamic property, indicating that it is evolutionarily advantageous over a static, unchanging structure.

In plants, microtubules help control the arrangement of the cellulose fibers of the cell wall. Electron micrographs of plants frequently show microtubules lying just inside the plasma membranes of cells that are forming or extending their cell walls. If the orientation of these microtubules is altered experimentally, it leads to a similar change in the cell wall and a new shape for the cell.

Microtubules serve as tracks for **motor proteins**, specialized molecules that use cellular energy to change their shapes and move. Motor proteins bind to and move along the microtubules,

carrying materials from one part of the cell to another. Microtubules are also essential in distributing chromosomes to daughter cells during cell division. Because of this, drugs such as vincristine and taxol, which disrupt microtubule dynamics, also disrupt cell division. These drugs are useful for treating cancer, where cell division is excessive.

CILIA AND FLAGELLA   Microtubules and their associated proteins line the interior of certain movable appendages on eukaryotic cells: the **cilia** (**Figure 5.17A**) and flagella. Many cells have one or the other of these appendages, which form from projections of the plasma membrane:

- Cilia are only 0.25 µm in length. They occur by the hundreds on individual cells and move stiffly to either propel the cell (for example, in protists) or to move fluid over a stationary cell (as in the human respiratory system).

- Flagella are longer—100 to 200 µm—and occur singly or in pairs. They can push or pull a cell through its aqueous environment.

In cross section, a typical cilium or eukaryotic flagellum is surrounded by the plasma membrane and contains a "9 + 2" array of microtubules. As **Figure 5.17B** shows, nine fused pairs of microtubules—called doublets—form an outer cylinder, and one pair of unfused microtubules runs up the center. Each doublet is connected to the center of the structure by a radial spoke. This structure is essential to the bending motion of both cilia and flagella. How does this bending occur?

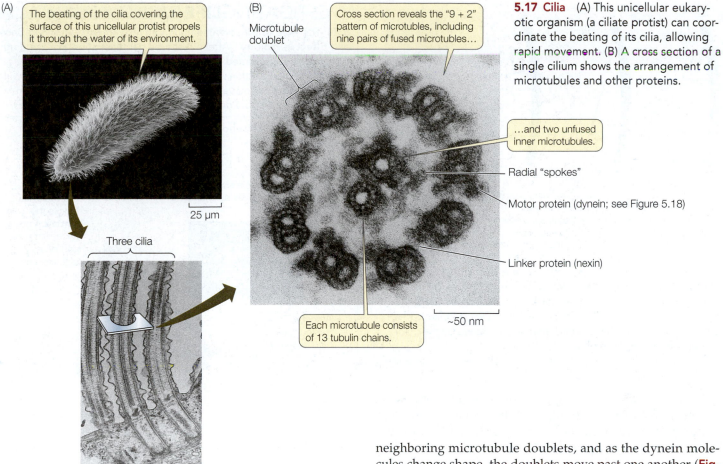

(A)

The beating of the cilia covering the surface of this unicellular protist propels it through the water of its environment.

25 μm

Three cilia

250 nm

(B)

Microtubule doublet

Cross section reveals the "9 + 2" pattern of microtubules, including nine pairs of fused microtubules...

...and two unfused inner microtubules.

Radial "spokes"

Motor protein (dynein; see Figure 5.18)

Linker protein (nexin)

Each microtubule consists of 13 tubulin chains.

~50 nm

5.17 Cilia (A) This unicellular eukaryotic organism (a ciliate protist) can coordinate the beating of its cilia, allowing rapid movement. (B) A cross section of a single cilium shows the arrangement of microtubules and other proteins.

The motion of cilia and flagella results from the sliding of the microtubule doublets past one another. This sliding is driven by the motor protein dynein, which, like other motor proteins, works by undergoing reversible shape changes that require chemical energy. Dynein molecules bind between two neighboring microtubule doublets, and as the dynein molecules change shape, the doublets move past one another (**Figure 5.18**). Another protein, nexin, cross-links the doublets and appears to limit how far the doublets can slide. This causes the cilium or flagellum to bend.

Other motor proteins, including kinesin, carry protein-laden vesicles or other organelles from one part of the cell to another (**Figure 5.19**). These proteins bind to the organelle and "walk" it along a microtubule by a repeated series of shape changes. Recall that microtubules are directional, with a plus

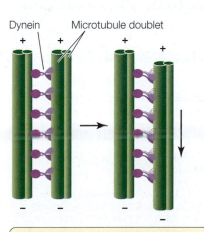

Dynein   Microtubule doublet

In isolated cilia without nexin cross-links, movement of dynein motor proteins causes microtubule doublets to slide past one another.

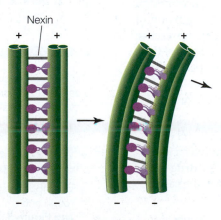

Nexin

When nexin is present to cross-link the doublets, they cannot slide and the force generated by dynein movement causes the cilium to bend.

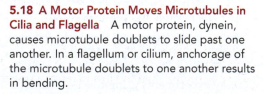

5.18 A Motor Protein Moves Microtubules in Cilia and Flagella A motor protein, dynein, causes microtubule doublets to slide past one another. In a flagellum or cilium, anchorage of the microtubule doublets to one another results in bending.

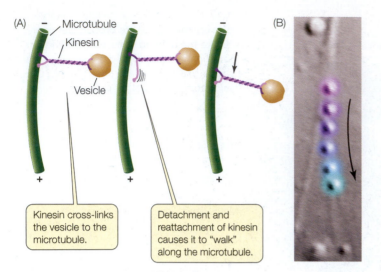

5.19 A Motor Protein Pulls Vesicles along Microtubules
(A) Kinesin delivers vesicles to various parts of the cell by moving along microtubule "railroad tracks." (B) A vesicle is pulled by kinesin along a microtubule in the protist *Dictyostelium*. The time sequence, with half-second intervals, is shown by the color changes from purple to blue.

and a minus end. Cytoplasmic dynein (which has a different role than the one found in cilia and flagella) moves attached organelles toward the minus end, whereas kinesin moves them toward the plus end (see Figure 5.14).

Biologists can manipulate living systems to establish cause and effect

How do we know that the structural fibers of the cytoskeleton can achieve all these dynamic functions? We can observe an individual structure under the microscope, and we can observe the functions of living cells that contain that structure. These observations may suggest that the structure carries out a particular function, but mere correlation does not show cause and effect. For example, light microscopy of living cells reveals that the cytoplasm is actively streaming around the cell, and that cytoplasm flows into an extended portion of an amoeboid cell during movement. The observed presence of cytoskeletal components *suggests, but does not prove,* their role in this process. Science seeks to show the specific links that relate one process, A, to a function, B. In cell biology, two approaches are often used to show that a structure or process A causes function B:

- *Inhibition*: use a drug that inhibits A and see if B still occurs. If it does not, then A is probably a causative factor for B. **Figure 5.20** shows an experiment with such a drug (an inhibitor) that demonstrates cause and effect in the case of the cytoskeleton and cell movement.

- *Mutation*: examine a cell that lacks the gene (or genes) for A and see if B still occurs. If it does not, then A is probably a causative factor for B. Part Four of this book describes many experiments using this genetic approach.

INVESTIGATING**LIFE** ▮

5.20 The Role of Microfilaments in Cell Movement—Showing Cause and Effect in Biology After a test tube demonstration that the drug cytochalasin B prevented microfilament formation from monomeric precursors, the question was asked: Will the drug work like this in living cells and inhibit cell movement in *Amoeba*? Complementary experiments showed that the drug did not poison other cellular processes.

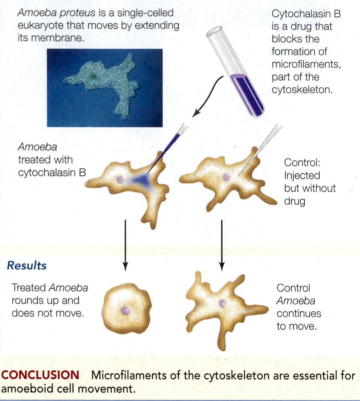

HYPOTHESIS Amoeboid cell movements are caused by the cytoskeleton.

Method

Amoeba proteus is a single-celled eukaryote that moves by extending its membrane.

Cytochalasin B is a drug that blocks the formation of microfilaments, part of the cytoskeleton.

Amoeba treated with cytochalasin B

Control: Injected but without drug

Results

Treated *Amoeba* rounds up and does not move.

Control *Amoeba* continues to move.

CONCLUSION Microfilaments of the cytoskeleton are essential for amoeboid cell movement.

Go to **BioPortal** for discussion and relevant links for all INVESTIGATING**LIFE** figures.

Pollard, T. D. and R. R. Weihing. 1974. *CRC Critical Reviews of Biochemistry* 2: 1–65.

RECAP 5.3

The hallmark of eukaryotic cells is compartmentalization. Membrane-enclosed organelles process information, transform energy, form internal compartments for transporting proteins, and carry out intracellular digestion. An internal cytoskeleton plays several structural roles.

- What are some advantages of organelle compartmentalization? **See p. 84**

- Describe the structural and functional differences between rough and smooth endoplasmic reticulum. **See pp. 88–90 and Figure 5.9**

- Explain how motor proteins and microtubules move materials within the cell. **See pp. 97–98 and Figures 5.18, 5.19**

WORKING WITH**DATA:**

The Role of Microfilaments in Cell Movement

Original Paper

Pollard, T. D. and R. R. Weihing. 1974. Actin and myosin in cell movement. *CRC Critical Reviews of Biochemistry* 2: 1–65.

Analyze the Data

In a search for natural molecules that have effects on cells—particularly cancer cells—a team of chemists and biologists at Imperial Chemical Industries examined extracts of the fungus *Helminthosporium dematiodeium*. When the extracts appeared to inhibit cell division, the scientists purified the active ingredient and called it cytochalasin B (from the Greek *cyto*, "cell" and *chalasis*, "dislocation"). Remarkably, application of cytochalasin B to dividing cells blocked the division of the cytoplasm but not division of the nucleus, so the result was a binucleate cell. In addition, the drug inhibited cell movement and phagocytosis. These dynamic processes were both hypothesized to involve cytoplasmic microfilaments (actin filaments). In the test tube, cytochalasin B blocked the polymerization of actin monomers into actin filaments. This prompted the use of cytochalasin B in cause-and-effect experiments: if a cellular process was inhibited by the drug, that process must involve microfilaments. This is the basis of the experiment shown in Figure 5.20.

Several important controls were done to validate the conclusions of the experiment. The experiment was repeated in the presence of the following drugs: cycloheximide, which inhibits new protein synthesis; dinitrophenol, which inhibits new ATP formation (energy); and colchicine, which inhibits the polymerization of microtubules. The results are shown in the table.

Condition	Rounded cells (%)
No drug	3
Cytochalasin B	95
Colchicine	4
Cycloheximide	3
Cycloheximide + cytochalasin B	94
Dinitrophenol	5
Dinitrophenol + cytochalasin B	85

QUESTION 1

Explain the reasoning behind each experiment. Why were these controls important?

QUESTION 2

Interpret the results of each experiment. What can you conclude about movements in *Amoeba* and the cytoskeleton?

Go to BioPortal for all WORKING WITH**DATA** exercises

All cells interact with their environments. Many eukaryotic cells are parts of multicellular organisms and must closely coordinate their activities with other cells. The plasma membrane plays a crucial role in these interactions, but other structures outside that membrane are involved as well.

5.4 What Are the Roles of Extracellular Structures?

Although the plasma membrane is the functional barrier between the inside and the outside of a cell, many structures are produced by cells and secreted to the outside of the plasma membrane, where they play essential roles in protecting, supporting, or attaching cells to each other. Because they are outside the plasma membrane, these structures are said to be extracellular. The peptidoglycan cell wall of bacteria is an example of an extracellular structure (see Figure 5.4). In eukaryotes, other extracellular structures—the cell walls of plants and the extracellular matrices found between the cells of animals—play similar roles. Each of these structures is made up of two main components: a prominent fibrous macromolecule and a gel-like medium in which the fibers are embedded.

The plant cell wall is an extracellular structure

The plant cell wall performs the same role as skeletal structures in animals. It is a semirigid structure outside the plasma membrane (**Figure 5.21**). We will consider the structure and role of

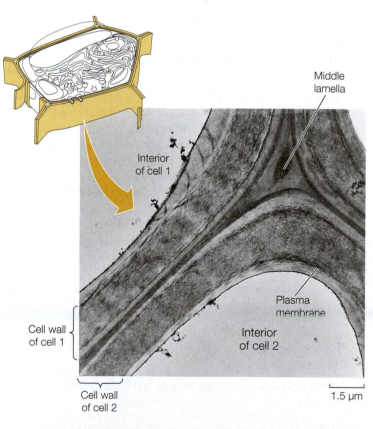

5.21 The Plant Cell Wall The semirigid cell wall provides support for plant cells. It is composed of cellulose fibrils embedded in a matrix of polysaccharides and proteins.

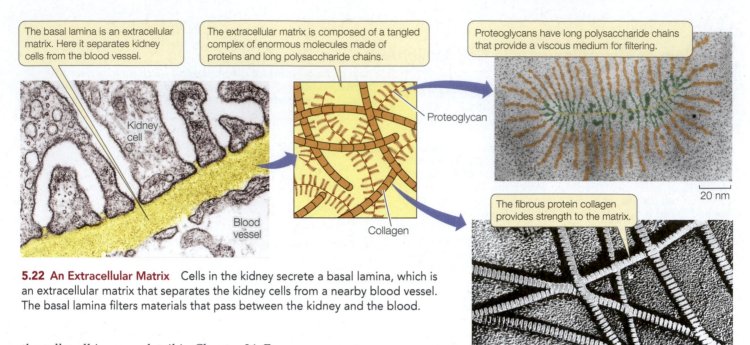

5.22 An Extracellular Matrix   Cells in the kidney secrete a basal lamina, which is an extracellular matrix that separates the kidney cells from a nearby blood vessel. The basal lamina filters materials that pass between the kidney and the blood.

the cell wall in more detail in Chapter 34. For now, we note that it consists of cellulose fibers (see Figure 3.18) embedded in other complex polysaccharides and proteins. The plant cell wall has three major roles:

- It provides support for the cell and plant by remaining rigid. Yet it is flexible enough that it can allow the plant to bend in the wind, for example.

- It acts as a barrier to infection by fungi and other organisms that can cause plant diseases.

- It contributes to plant form by growing as the plant cells expand.

In some cells, such as those in a leaf, the cell wall is porous to allow the passage of molecules into and out of the cell. In other cells, such as those of the plant's vascular system (which transports water and small molecules between organs), the wall is not porous.

Because of their thick cell walls, plant cells viewed under a light microscope appear to be entirely isolated from one another. But electron microscopy reveals that this is not the case. The cytoplasms of adjacent plant cells are connected by numerous plasma membrane–lined channels called **plasmodesmata**, which are about 20–40 nanometers in diameter and extend through the cell walls (see Figure 5.7). Plasmodesmata permit the diffusion of water, ions, small molecules, RNA, and proteins between connected cells, allowing for the use of these substances far from their sites of synthesis.

The extracellular matrix supports tissue functions in animals

Animal cells lack the semirigid wall that is characteristic of plant cells, but many animal cells are surrounded by, or in contact with, an **extracellular matrix**. This matrix is composed of three types of molecules: fibrous proteins such as **collagen** (the most abundant protein in mammals, constituting over 25 percent of the protein in the human body); a matrix of glycoproteins termed **proteoglycans**, consisting primarily of sugars; and a third group of proteins that link the fibrous proteins and the gel-like proteoglycan matrix together (**Figure 5.22**). These proteins and proteoglycans are secreted, along with other substances that are specific to certain body tissues, by cells that are present in or near the matrix.

The functions of the extracellular matrix are many:

- It holds cells together in tissues. In Chapter 6 we will see how there is an intercellular "glue" that is involved in both cell recognition and adhesion.

- It contributes to the physical properties of cartilage, skin, and other tissues. For example, the mineral component of bone is laid down on an organized extracellular matrix.

- It helps filter materials passing between different tissues. This is especially important in the kidney.

- It helps orient cell movements during embryonic development and during tissue repair.

- It plays a role in chemical signaling from one cell to another. Proteins connect the cell's plasma membrane to the extracellular matrix. These proteins (for example, integrin) span the plasma membrane and are involved with transmitting signals to the interior of the cell. This allows communication between the extracellular matrix and the cytoplasm of the cell.

RECAP 5.4

Extracellular structures are produced by cells and secreted outside the plasma membrane. Most consist of a fibrous component in a gel-like medium.

- What are the functions of the cell wall in plants and the extracellular matrix in animals? **See p. 100**

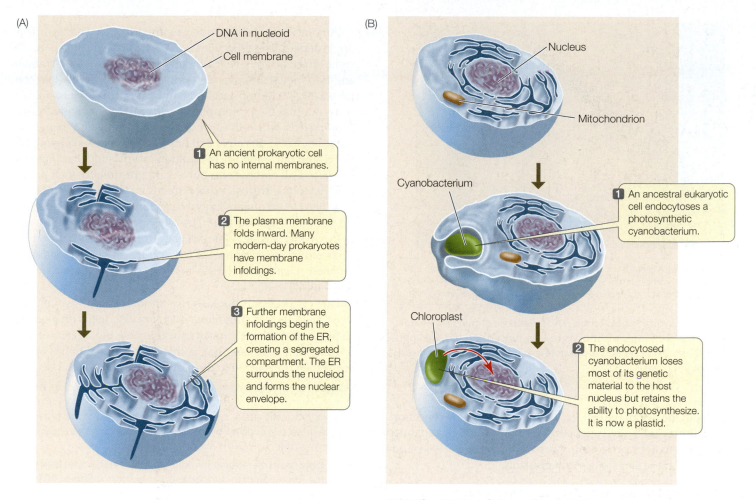

5.23 The Origins of Organelles (A) The endomembrane system and nuclear envelope may have been formed by infolding and then fusion of the plasma membrane. (B) The endosymbiosis theory proposes that some organelles may be descended from prokaryotes that were engulfed by other, larger cells.

We have now discussed the structures and some functions of prokaryotic and eukaryotic cells. Both exemplify the cell theory, showing that cells are the basic units of life and of biological continuity. Much of the rest of this part of the book will deal with these two aspects of cells. There is abundant evidence that the simpler prokaryotic cells are more ancient than eukaryotic cells, and that the first cells were probably prokaryotic. We will now turn to the next step in cellular evolution, the origin of eukaryotic cells.

 ## 5.5 How Did Eukaryotic Cells Originate?

Life on Earth was entirely prokaryotic for about 2 billion years—from the time when prokaryotic cells first appeared until about 1.5 billion years ago, when eukaryotic cells arrived on the scene. The advent of compartmentalization—the hallmark of eukaryotes—was a major event in the history of life. It permitted many more biochemical functions to exist in the same cell than had previously been possible. Compared with a typical eukaryote, a single prokaryotic cell is often biochemically specialized, limited in the resources it can use and the functions it can perform.

What is the origin of compartmentalization? We will describe the evolution of eukaryotic organelles in more detail in Section 27.1. Here we outline two major themes in this process.

Internal membranes and the nuclear envelope probably came from the plasma membrane

We noted earlier that some bacteria contain internal membranes. How could these arise? In electron micrographs, the internal membranes of prokaryotes often appear to be inward folds of the plasma membrane. This has led to a theory that the endomembrane system and the cell nucleus originated by related processes (**Figure 5.23A**). The close relationship between the endoplasmic reticulum (ER) and the nuclear envelope in today's eukaryotes is consistent with this theory.

A bacterium with enclosed compartments would have several evolutionary advantages. Chemicals could be concentrated in particular regions of the cell, allowing chemical reactions to proceed more efficiently. A biochemical process could be segregated within an organelle with, for example, a different

Daughter cells

This cell has a green, photosynthetic plastid acquired from an algal cell.

After cell division, only one of the daughter cells inherits the plastid; the other cell must ingest a new one from the environment.

5.24 Endosymbiosis in Action A *Hatena* cell engulfs an algal cell, which then loses most of its cellular functions other than photosynthesis. This reenacts a possible event in the origin of plastids in eukaryotic cells.

pH from the rest of the cell, creating more favorable conditions for that process. Finally, gene transcription could be separated from translation, providing more opportunities for separate control of these steps in gene expression.

Some organelles arose by endosymbiosis

Symbiosis means "living together," and often refers to two organisms that coexist, each one supplying something that the other needs. Biologists have proposed that some organelles—the mitochondria and the plastids—arose not by an infolding of the plasma membrane but by one cell ingesting (but not digesting) another cell, giving rise to a symbiotic relationship. Eventually, the ingested cell lost its autonomy and some of its functions. In addition, many of the ingested cell's genes were transferred to the host's DNA. Mitochondria and plastids in today's eukaryotic cells are the remnants of these symbionts, retaining some specialized functions that benefit their host cells. This is the essence of the **endosymbiosis theory** for the origin of organelles.

Consider the case of the plastid. About 2.5 billion years ago some prokaryotes (the cyanobacteria) developed photosynthesis (see Figure 1.4). The emergence of these prokaryotes was a key event in the evolution of complex organisms because they increased the O_2 concentration in Earth's atmosphere (see Section 1.1)

According to the endosymbiosis theory, photosynthetic prokaryotes also provided the precursor of the modern-day plastid. Cells without cell walls can engulf relatively large particles by phagocytosis (see Figure 5.10). In some cases, such as that of phagocytes in the human immune system, the engulfed particle can be an entire cell, such as a bacterium. Plastids may have arisen by a similar event involving an ancestral eukaryote and a cyanobacterium (**Figure 5.23B**).

Among the abundant evidence supporting the endosymbiotic origin of plastids (see Section 27.1), perhaps the most remarkable comes from a sandy beach in Japan. In 2006 Noriko Okamoto and Isao Inouye discovered a single-celled eukaryote that contains a large "chloroplast" and named it *Hatena* (**Figure 5.24**). It turns out that the "chloroplast" is the remains of a green alga, *Nephroselmis*, which lives among the *Hatena* cells. When living autonomously, this algal cell has flagella, a cytoskeleton, ER, Golgi, and mitochondria in addition to a plastid. Once ingested by *Hatena*, all of these structures, and presumably their associated functions, are lost. What remains is essentially the plastid.

When *Hatena* divides, only one of the two daughter cells ends up with the "chloroplast." The other cell finds and ingests its own *Nephroselmis* alga—almost like a "replay" of what may have occurred in the evolution of eukaryotic cells. No wonder the Japanese scientists call the host cell *Hatena*: in Japanese, it means "how odd"!

The cover of this book shows another example of chloroplast theft: the sea slug *Elysia chlorotica* also feeds on algae for their photosynthetic capacity. In this case biologists recently found that a gene essential to chloroplast function has apparently been transferred to the slug nucleus.

■ RECAP 5.5

Eukaryotic cells arose long after prokaryotic cells. Some organelles may have evolved by infolding of the plasma membrane, whereas others probably evolved by endosymbiosis.

- How could membrane infolding in a prokaryotic cell lead to the formation of the endomembrane system? **See p. 101 and Figure 5.23A**
- Explain the endosymbiosis theory for the origin of chloroplasts. **See p. 102 and Figure 5.23B**

In this chapter we presented an overview of the components of cells, with some ideas about their structures, functions, and origins. As you now embark on the study of major cellular processes, keep in mind that cellular components do not exist in isolation: they are part of a dynamic, interacting system. In Chapter 6 we will show that the plasma membrane is far from a passive barrier, but instead is a multifunctional system that connects the inside of the cell with its extracellular environment.

?

What is the status of stem cell treatment for heart disease?

ANSWER

Although there are active clinical trials for using stem cell therapy in patients with heart disease in the United States and western Europe, by far the most active use of this treatment occurs in China. Stem cells are typically collected from umbilical cords at childbirth and stored for later use. There have been anecdotal reports of great successes in using such treatments to repair damaged hearts. An examination of treated and functioning hearts indicated that the stem cells were involved in repair of the heart muscle and the blood vessels that supply it.

CHAPTER**SUMMARY** 5

5.1 What Features Make Cells the Fundamental Units of Life?

- The **cell theory** is the unifying theory of cell biology. All living things are composed of cells, and all cells come from preexisting cells.

- A cell is small in order to maintain a large **surface area-to-volume ratio**. This allows it to exchange adequate quantities of materials with its environment. **Review Figures 5.1, 5.2, ACTIVITY 5.1**

- Cell structures can be studied with light and electron microscopes. **Review Figure 5.3, ACTIVITY 5.2**

- All cells are enclosed by a selectively permeable **plasma membrane** that separates their contents from the external environment.

- Whereas certain biochemical processes, molecules, and structures are shared by all kinds of cells, there are two categories of organisms—**prokaryotes** and **eukaryotes**—that can be distinguished by characteristic cell structures.

- Eukaryotic cells are generally larger and more complex than prokaryotic cells. They contain membrane-bound **organelles**, including the **nucleus**.

5.2 What Features Characterize Prokaryotic Cells?

- Prokaryotic cells have no internal compartments but have a **nucleoid** region containing DNA, and a **cytoplasm** containing **cytosol**, **ribosomes**, proteins, and small molecules. Some prokaryotes have additional protective structures, including a **cell wall**, an **outer membrane**, and a **capsule**. **Review Figure 5.4**

- Some prokaryotes have folded **internal membranes** such as those used in photosynthesis, and some have **flagella** or **pili** for motility or attachment. **Review Figure 5.5**

- Filamentous proteins in the cytoplasm make up the **cytoskeleton**, which assists in cell division and the maintenance of cell shape.

5.3 What Features Characterize Eukaryotic Cells?

See ANIMATED TUTORIAL 5.1

- Eukaryotic cells are larger than prokaryotic cells and contain many membrane-enclosed **organelles**. The membranes that envelop organelles ensure compartmentalization of their functions. **Review Figure 5.7**

- The **nucleus** contains most of the cell's DNA and participates in the control of protein synthesis. The DNA and the proteins associated with it form a material called **chromatin**. Each long, thin DNA molecule occurs in a discrete chromatin structure called a **chromosome**. **Review Figure 5.8**

- Within the nucleus is the **nucleolus**, where ribosome assembly begins. After partial assembly, the **ribosomes** are transported to the cytoplasm, where they are completed and function as sites of protein synthesis.

- The **endomembrane system**—consisting of the **endoplasmic reticulum** and the **Golgi apparatus**—is a series of interrelated compartments enclosed by membranes. It segregates proteins and modifies them. **Lysosomes** contain many digestive enzymes. **Review Figures 5.9, 5.10, ACTIVITY 5.3, ANIMATED TUTORIAL 5.2**

- **Mitochondria** and **chloroplasts** are semiautonomous organelles that process energy. Mitochondria are present in most eukaryotic organisms and contain the enzymes needed for **cellular respiration**. The cells of photosynthetic eukaryotes contain chloroplasts that harvest light energy for photosynthesis. **Review Figures 5.11, 5.12**

- Large **vacuoles** are present in many plant cells. A vacuole consists of a membrane-enclosed compartment full of water and dissolved substances.

- The **microfilaments**, **intermediate filaments**, and **microtubules** of the cytoskeleton provide the cell with shape, strength, and movement. **Review Figure 5.14**

- **Motor proteins** use cellular energy to change shape and move. They drive the bending movements of **cilia** and **flagella**, and transport organelles along microtubules within the cell. **Review Figures 5.18, 5.19**

5.4 What Are the Roles of Extracellular Structures?

- The plant **cell wall** consists principally of cellulose. Cell walls are pierced by **plasmodesmata** that join the cytoplasms of adjacent cells.

- In animals, the **extracellular matrix** consists of different kinds of proteins, including **collagen** and **proteoglycans**. **Review Figure 5.22**

5.5 How Did Eukaryotic Cells Originate?

- Infoldings of the plasma membrane could have led to the formation of some membrane-enclosed organelles, such as the endomembrane system and the nucleus. **Review Figure 5.23A**

- **Symbiosis** means "living together." The **endosymbiosis theory** states that mitochondria and chloroplasts originated when larger cells engulfed, but did not digest, smaller cells. Mutual benefits permitted this symbiotic relationship to be maintained, allowing the smaller cells to evolve into the eukaryotic organelles observed today. **Review Figure 5.23B**

 Go to the Interactive Summary to review key figures, Animated Tutorials, and Activities
Life10e.com/is5

CHAPTER**REVIEW**

■■■■ REMEMBERING

1. Which structure is generally present in both prokaryotic cells and eukaryotic plant cells?
   a. Chloroplasts
   b. Cell wall
   c. Nucleus
   d. Mitochondria
   e. Microtubules

2. The major factor limiting cell size is the
   a. concentration of water in the cytoplasm.
   b. need for energy.
   c. presence of membrane-enclosed organelles.
   d. ratio of surface area to volume.
   e. composition of the plasma membrane.

3. Which statement about plastids is true?
   a. They are found in prokaryotes.
   b. They are surrounded by a single membrane.
   c. They are the sites of cellular respiration.
   d. They are found only in fungi.
   e. They may contain various pigments or polysaccharides.

4. Which structure is *not* surrounded by one or more membranes?
   a. Ribosome
   b. Chloroplast
   c. Mitochondrion
   d. Peroxisome
   e. Vacuole

5. The cytoskeleton consists of
   a. cilia, flagella, and microfilaments.
   b. cilia, microtubules, and microfilaments.
   c. internal cell walls.
   d. microtubules, intermediate filaments, and microfilaments.
   e. calcified microtubules.

6. Microfilaments
   a. are composed of polysaccharides.
   b. are composed of actin.
   c. allow cilia and flagella to move.
   d. make up the spindle that aids the movement of chromosomes.
   e. maintain the position of the chloroplast in the cell.

■■■■ UNDERSTANDING & APPLYING

7. If all the lysosomes within a cell suddenly ruptured, what would be the most likely result?
   a. The macromolecules in the cytosol would break down.
   b. More proteins would be made.
   c. The DNA in mitochondria would break down.
   d. The mitochondria and chloroplasts would divide.
   e. There would be no change in cell function.

8. Through how many membranes would a molecule have to pass in moving from the interior (stroma) of a chloroplast to the interior (matrix) of a mitochondrion? From the interior of a lysosome to the outside of a cell? From one ribosome to another?

9. Compare the extracellular matrix of the animal cell with the plant cell wall, with respect to composition of the fibrous and nonfibrous components, rigidity, and connectivity of cells.

■■■■ ANALYZING & EVALUATING

10. The drug vincristine is used to treat many cancers. It apparently works by causing microtubules to depolymerize. Vincristine use has many side effects, including loss of dividing cells and nerve problems. Explain why this might be so.

11. The movements of newly synthesized proteins can be followed through cells using a "pulse–chase" experiment. During synthesis, proteins are tagged with a radioactive isotope (the "pulse"), and then the cells are allowed to process the proteins for varying periods of time. The locations of the radioactive proteins are then determined by isolating cell organelles and quantifying their levels of radioactivity. What results would you expect for (a) a lysosomal enzyme and (b) a protein that is released from the cell?

Go to BioPortal at **yourBioPortal.com** for Animated Tutorials, Activities, LearningCurve Quizzes, Flashcards, and many other study and review resources.

6 Cell Membranes

CHAPTEROUTLINE

6.1 What Is the Structure of a Biological Membrane?

6.2 How Is the Plasma Membrane Involved in Cell Adhesion and Recognition?

6.3 What Are the Passive Processes of Membrane Transport?

6.4 What are the Active Processes of Membrane Transport?

6.5 How Do Large Molecules Enter and Leave a Cell?

Sweating: A Membrane Activity During physical activity, water is transported across the cell membranes of sweat glands and out of the skin by exocytosis.

A FEW DAYS AFTER HE BECAME PRIME MINISTER, as World War II spread across Europe, Winston Churchill told the British Parliament, "I have nothing to offer but blood, toil, tears and sweat." He may not have known that the last two, tears and sweat, are transported across cell membranes inside vesicles. The harder we work, the hotter we get and the more we sweat. As you saw in Chapter 2, sweating is a way to reduce body heat by using excess heat to evaporate water. At peak activity, we may lose as much as 2 liters of water an hour, and if you know anything about the German air attacks on London during the war, you know that the people indeed toiled hard and must have sweated a lot.

The sweat glands lie just below the surface of the skin. They are essentially cell-lined tubes surrounded by extracellular fluid. When sweating is triggered, these tubes fill with water and dissolved substances. To get from the extracellular fluid into the tube, water must go through the cells that line the tube.

A hallmark of living cells is the ability to regulate what enters and leaves their cytoplasms. This is a function of the plasma (or cell) membrane, a hydrophobic lipid bilayer with associated proteins. Because it is insoluble in the aqueous environment both inside and outside cells, the membrane is a physical barrier. But it is also a functional barrier. Whereas water is polar, the interior of the membrane is nonpolar—so water has a natural tendency to avoid the membrane. The rate of movement of water across a lipid bilayer is modest. When you engage in normal activities such as reading this book, the cell membranes enclosing the cells lining the sweat glands do not allow much water to enter or leave. But when you exercise vigorously, tiny membrane-enclosed vesicles inside the cells fill with water and some dissolved salts. In a process called exocytosis, these vesicles fuse with the cell membrane and release their watery contents (sweat) into the tubes. From there the sweat flows to the surface of the skin and evaporates.

Vesicles are not the only way to get polar water across a nonpolar membrane. In the mammalian kidney, and in plant roots, stems, and leaves, special pores called aquaporins occur in the cell membrane. Water can flow through them readily, as the proteins lining the channel have a hydrophilic inner surface. The water-carrying function of aquaporins is so well known that there is a cosmetic moisturizing cream called Aquaporin Active!

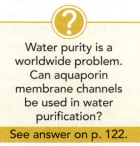

Water purity is a worldwide problem. Can aquaporin membrane channels be used in water purification?

See answer on p. 122.

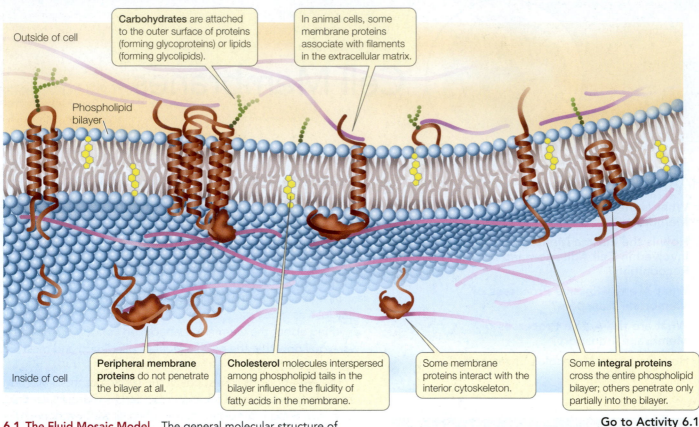

Outside of cell

Carbohydrates are attached to the outer surface of proteins (forming glycoproteins) or lipids (forming glycolipids).

In animal cells, some membrane proteins associate with filaments in the extracellular matrix.

Phospholipid bilayer

Inside of cell

Peripheral membrane proteins do not penetrate the bilayer at all.

Cholesterol molecules interspersed among phospholipid tails in the bilayer influence the fluidity of fatty acids in the membrane.

Some membrane proteins interact with the interior cytoskeleton.

Some **integral proteins** cross the entire phospholipid bilayer; others penetrate only partially into the bilayer.

6.1 The Fluid Mosaic Model The general molecular structure of a biological membrane is a continuous phospholipid bilayer which has proteins embedded in it or associated with it.

Go to Activity 6.1
The Fluid Mosaic Model **Life10e.com/ac6.1**

 ### 6.1 What Is the Structure of a Biological Membrane?

The physical organization and functioning of all biological membranes depend on their constituents: lipids, proteins, and carbohydrates. You are already familiar with these molecules from Chapter 3; it may be useful to review that chapter now. The lipids establish the physical integrity of the membrane and create an effective barrier to the rapid passage of hydrophilic materials such as water and ions. In addition, the phospholipid bilayer serves as a lipid "lake" in which a variety of proteins "float" (**Figure 6.1**). This general design is known as the **fluid mosaic model**. It is *mosaic* because it is made up of many discrete components, and *fluid* because they can move freely.

In the fluid mosaic model for biological membranes, the proteins are noncovalently embedded in the phospholipid bilayer by their hydrophobic regions (or domains), but their hydrophilic domains are exposed to the watery conditions on either side of the bilayer. These membrane proteins have several functions, including moving materials through the membrane and receiving chemical signals from the cell's external environment. Each membrane has a set of proteins suitable for the specialized functions of the cell or organelle it surrounds.

The carbohydrates associated with membranes are attached either to the lipids or to protein molecules. In plasma membranes, carbohydrates are located on the outside of the cell, where they may interact with substances in the external environment. Like some of the membrane proteins, carbohydrates

are crucial in recognizing specific molecules, such as those on the surfaces of adjacent cells.

Although the fluid mosaic model is largely valid for membrane structure, it does not say much about membrane composition. As you read about the various molecules in membranes in the next sections, keep in mind that some membranes have more protein than lipids, others are lipid-rich, others have significant amounts of cholesterol or other sterols, and still others are rich in carbohydrates.

Lipids form the hydrophobic core of the membrane

The lipids in biological membranes are usually phospholipids. Recall from Section 2.2 that some compounds are hydrophilic ("water-loving") and others are hydrophobic ("water-hating"), and from Section 3.4 that a phospholipid molecule has regions of both kinds:

- *Hydrophilic regions*: The phosphorus-containing "head" of the phospholipid is electrically charged and therefore associates with polar water molecules.

- *Hydrophobic regions*: The long, nonpolar fatty acid "tails" of the phospholipid associate with other nonpolar materials; they do not dissolve in water or associate with hydrophilic substances.

Because of these properties, one way in which phospholipids can coexist with water is to form a bilayer, with the fatty acid "tails" of the two layers interacting with each other and the polar

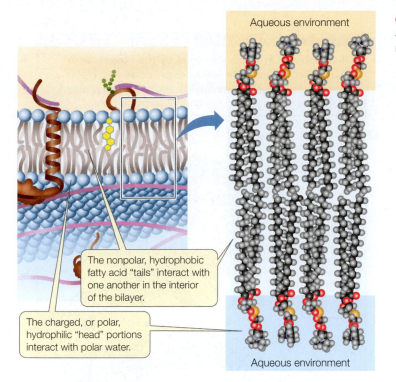

Aqueous environment

The nonpolar, hydrophobic fatty acid "tails" interact with one another in the interior of the bilayer.

The charged, or polar, hydrophilic "head" portions interact with polar water.

Aqueous environment

6.2 A Phospholipid Bilayer The phospholipid bilayer separates two aqueous regions. The eight phospholipid molecules shown on the right represent a small cross section of a membrane bilayer.

"heads" facing the outside aqueous environment (**Figure 6.2**). The thickness of a biological membrane is about 8 nanometers (0.008 μm), which is twice the length of a typical phospholipid—another indication that the membrane consists of a lipid bilayer. A typical sheet of paper is about 8,000 times thicker than this.

In the laboratory, it is easy to make artificial bilayers with the same organization as natural membranes. Small holes in such bilayers seal themselves spontaneously. This capacity of lipids to associate with one another and maintain a bilayer organization helps biological membranes fuse during vesicle formation, phagocytosis, and related processes.

All biological membranes have a similar structure, but they differ in the kinds of proteins and lipids they contain. Membranes from different cells or organelles may differ greatly in their lipid composition. Phospholipids can differ in terms of fatty acid chain length (number of carbon atoms), degree of unsaturation (number of double bonds) in the fatty acids, and the polar groups present (see Chapter 3). The saturated chains allow close packing of fatty acids in the bilayer, whereas the "kinks" in unsaturated fatty acids (see Figure 3.21) make for a less dense, more fluid packing.

Up to 25 percent of the lipid content of an animal cell plasma membrane may be the steroid cholesterol (see Section 3.4). Cholesterol preferentially associates with saturated fatty acids. When present, cholesterol is important for membrane integrity, the cholesterol in your membranes is not hazardous to your health.

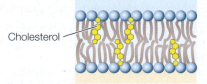

Cholesterol

The fatty acids of the phospholipids make the hydrophobic interior of the membrane somewhat fluid—about as fluid

as lightweight olive oil. This fluidity permits some molecules to move laterally within the plane of the membrane. A given phospholipid molecule in the plasma membrane can travel from one end of the cell to the other in a little more than a second! However, a phospholipid molecule in one half of the bilayer is unlikely to spontaneously flip over to the other side. For that to happen, the polar part of the molecule would have to move through the hydrophobic interior of the membrane. Since spontaneous phospholipid flip-flops are rare, the inner and outer halves of the bilayer may be quite different in the kinds of phospholipids they contain.

Membrane fluidity is affected by several factors, two of which are particularly important:

- *Lipid composition*: Cholesterol and long-chain, saturated fatty acids pack tightly beside one another, with little room for movement. This close packing results in less-fluid membranes. A membrane with shorter-chain fatty acids, unsaturated fatty acids, or less cholesterol is more fluid.

- *Temperature*: Because molecules move more slowly and fluidity decreases at reduced temperatures, cellular processes that take place within the membrane may slow down or stop under cold conditions in organisms that cannot keep their bodies warm. To address this problem, some organisms simply change the lipid composition of their membranes when they get cold, replacing saturated with unsaturated fatty acids and using fatty acids with shorter tails. These changes play a role in the survival of plants, bacteria, and hibernating animals during the winter.

Go to Animated Tutorial 6.1
Lipid Bilayer Composition
Life10e.com/at6.1

Membrane proteins are asymmetrically distributed

All biological membranes contain proteins. Typically, plasma membranes have 1 protein molecule for every 25 lipid molecules. This ratio varies depending on membrane function. In the inner membrane of the mitochondrion, which is specialized for energy processing, there is 1 protein for every 15 lipids. However, myelin—a membrane that encloses portions of some neurons (nerve cells) and acts as an electrical insulator—has only 1 protein for every 70 lipids.

Membrane proteins are very diverse. In fact, about one-fourth of the protein-coding genes in the eukaryotic genome encode membrane proteins. There are two general types of membrane proteins: peripheral proteins and integral proteins.

Peripheral membrane proteins lack exposed hydrophobic groups and are not embedded in the bilayer. Instead, they have polar or charged regions that interact with exposed parts of integral membrane proteins, or with the polar heads of phospholipid molecules (see Figure 6.1).

Integral membrane proteins are at least partly embedded in the phospholipid bilayer (see Figure 6.1). Like phospholipids,

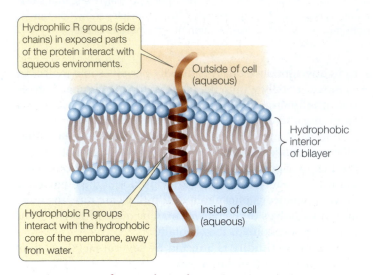

Hydrophilic R groups (side chains) in exposed parts of the protein interact with aqueous environments.

Outside of cell (aqueous)

Hydrophobic interior of bilayer

Hydrophobic R groups interact with the hydrophobic core of the membrane, away from water.

Inside of cell (aqueous)

6.3 Interactions of Integral Membrane Proteins An integral membrane protein is held in the membrane by the distribution of the hydrophilic and hydrophobic side chains on its amino acids. The hydrophilic parts of the protein extend into the aqueous cell exterior and the internal cytoplasm. The hydrophobic side chains interact with the hydrophobic lipid core of the membrane.

these proteins have both hydrophilic and hydrophobic regions (domains) (**Figure 6.3**).

- *Hydrophilic domains*: Stretches of amino acids with hydrophilic side chains (R groups; see Table 3.2) give certain regions of the protein a polar character. These hydrophilic domains interact with water and stick out into the aqueous environment inside or outside the cell.

- *Hydrophobic domains*: Stretches of amino acids with hydrophobic side chains give other regions of the protein a nonpolar character. These domains interact with the fatty acids in the interior of the phospholipid bilayer, away from water.

A special preparation method for electron microscopy, called freeze-fracturing, reveals proteins that are embedded in the phospholipid bilayers of cellular membranes (**Figure 6.4**). When the two lipid leaflets (or layers) that make up the bilayer are separated, the proteins can be seen as bumps that protrude from the interior of each membrane. The bumps are not observed when artificial bilayers of pure lipid are freeze-fractured.

Membrane proteins and lipids generally interact only noncovalently. The polar ends of proteins can interact with the polar ends of lipids, and the nonpolar regions of both molecules can interact hydrophobically. However, some membrane proteins have fatty acids or other lipid groups covalently attached to them. Their hydrophobic lipid components allow these proteins to tether themselves to the phospholipid bilayer.

Proteins are asymmetrically distributed on the inner and outer surfaces of membranes. An integral protein that extends all the way through the phospholipid bilayer and protrudes on both sides is known as a **transmembrane protein**.

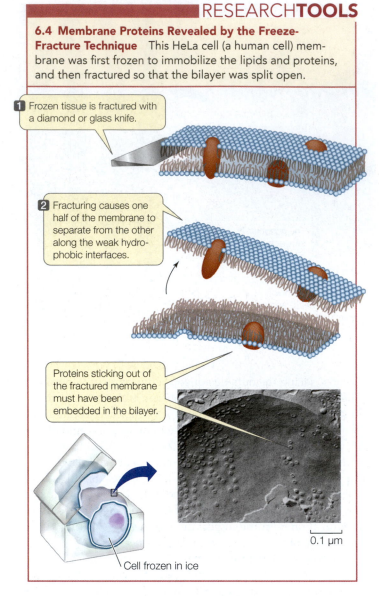

RESEARCHTOOLS

6.4 Membrane Proteins Revealed by the Freeze-Fracture Technique This HeLa cell (a human cell) membrane was first frozen to immobilize the lipids and proteins, and then fractured so that the bilayer was split open.

1 Frozen tissue is fractured with a diamond or glass knife.

2 Fracturing causes one half of the membrane to separate from the other along the weak hydrophobic interfaces.

Proteins sticking out of the fractured membrane must have been embedded in the bilayer.

Cell frozen in ice

0.1 μm

In addition to one or more **transmembrane domains** that extend through the bilayer, such a protein may have domains with other specific functions on the inner and outer sides of the membrane. Peripheral membrane proteins are located on one side of the membrane or the other. This asymmetrical arrangement of membrane proteins gives the two surfaces of the membrane different properties. As we will soon see, these differences have great functional significance.

Like lipids, some membrane proteins move around relatively freely within the phospholipid bilayer. Experiments that involve the technique of cell fusion illustrate this migration dramatically. When two cells are fused, a single continuous membrane forms and surrounds both cells, and some proteins from each cell distribute themselves uniformly around this membrane (**Figure 6.5**).

Although some proteins are free to migrate in the membrane, others are not, but rather appear to be "anchored" to a specific region of the membrane. These membrane regions are

INVESTIGATING**LIFE**

6.5 Rapid Diffusion of Membrane Proteins Two animal cells can be fused together in the laboratory, forming a single large cell (heterokaryon). This phenomenon was used to test whether membrane proteins can diffuse independently in the plane of the plasma membrane.[a]

HYPOTHESIS Proteins embedded in a membrane can diffuse freely within the membrane.

Method

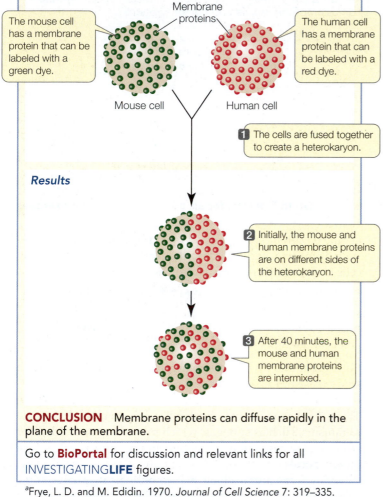

The mouse cell has a membrane protein that can be labeled with a green dye.

Membrane proteins

The human cell has a membrane protein that can be labeled with a red dye.

Mouse cell   Human cell

1 The cells are fused together to create a heterokaryon.

Results

2 Initially, the mouse and human membrane proteins are on different sides of the heterokaryon.

3 After 40 minutes, the mouse and human membrane proteins are intermixed.

CONCLUSION Membrane proteins can diffuse rapidly in the plane of the membrane.

Go to **BioPortal** for discussion and relevant links for all INVESTIGATING**LIFE** figures.

[a]Frye, L. D. and M. Edidin. 1970. *Journal of Cell Science* 7: 319–335.

like a corral of horses on a farm: the horses are free to move around within the fenced area but not outside it. An example is the protein in the plasma membrane of a muscle cell that recognizes a chemical signal from a neuron. This protein is normally found only at the specific region where the neuron meets the muscle cell. How does this happen?

Proteins inside the cell can restrict the movement of proteins within a membrane. The cytoskeleton may have components just below the inner face of the membrane that are attached to membrane proteins protruding into the cytoplasm. The stability of the cytoskeletal components may thus restrict movement of attached membrane proteins.

Membranes are constantly changing

Membranes in eukaryotic cells are constantly forming, transforming from one type to another, fusing with one another, and breaking down. As we discussed in Chapter 5, fragments of membrane move, in the form of vesicles, from the endoplasmic reticulum (ER) to the Golgi, and from the Golgi to the plasma membrane (see Figure 5.9). Secondary lysosomes form when primary lysosomes from the Golgi fuse with phagosomes from the plasma membrane (see Figure 5.10).

Because all membranes appear similar under the electron microscope, and because they interconvert readily, we might expect all subcellular membranes to be chemically identical. However, that is not the case: there are major chemical differences among the membranes of even a single cell. Membranes are changed chemically when they form parts of certain organelles. In the Golgi apparatus, for example, the membranes of the *cis* face closely resemble those of the ER in chemical composition, but those of the *trans* face are more similar to the plasma membrane.

Plasma membrane carbohydrates are recognition sites

In addition to lipids and proteins, the plasma membrane contains carbohydrates (see Figure 6.1). The carbohydrates are located on the outer surface of the plasma membrane and serve as recognition sites for other cells and molecules, as you will see in Section 6.2.

Membrane-associated carbohydrates may be covalently bonded to lipids or to proteins:

- A **glycolipid** consists of a carbohydrate covalently bonded to a lipid. Extending out from the cell surface, the carbohydrate may serve as a recognition signal for interactions between cells. For example, the carbohydrates on some glycolipids change when cells become cancerous. This change may allow white blood cells to target cancer cells for destruction.

- A **glycoprotein** consists of one or more short carbohydrate chains covalently bonded to a protein. The bound carbohydrates are oligosaccharides, usually not exceeding 15 monosaccharide units in length (see Section 3.3). A proteoglycan (see Section 5.4) is a more heavily glycosylated protein: it has more carbohydrate molecules attached to it, and the carbohydrate chains are often longer than they are in glycoproteins. The carbohydrates of glycoproteins and proteoglycans often function in cell recognition and adhesion.

The "alphabet" of monosaccharides on the outer surfaces of membranes can generate a large diversity of messages. Recall from Section 3.3 that monosaccharides are simple carbohydrates, often containing five or six carbons in a ring structure, which can bond with one another in various configurations. They may form linear or branched oligosaccharides with many different three-dimensional shapes. An oligosaccharide of a specific shape on one cell can bind to a complementary shape on an adjacent cell. This binding is the basis of cell–cell adhesion.

WORKING WITH**DATA:**

Rapid Diffusion of Membrane Proteins

Original paper

Frye, L. D. and M. Edidin. 1970. The rapid intermixing of cell surface antigens after formation of mouse-human heterokaryons. *Journal of Cell Science* 7: 319–335.

Analyze the Data

One of the key experiments providing evidence for the fluid mosaic model was performed by Louis Frye and Michael Edidin at Johns Hopkins University. The scientists took advantage of the recently developed technique of cell fusion and used it to show that membrane proteins rapidly diffuse within the plane of the membrane (see Figure 6.5). Under the right conditions, two cells could fuse together, forming a binucleate cell with one continuous membrane. In this case, mouse and human cells were fused, and membrane proteins specific to each cell type were visualized using antibodies—the mouse one coupled to a green dye and the human to a red dye. Immediately after fusion, half of the cell membrane stained red, and half green. Then over time, the colors intermixed, demonstrating that the mouse and human proteins were diffusing within the membrane. The percentage of cells that had red and green colors fully intermixed was calculated over time after cell fusion. The results are shown in the table.

QUESTION 1

Plot the percentage of fully mixed cells over time. How long did it take for complete mixing?

QUESTION 2

What does your answer to Question 1 indicate about the rate of diffusion of the mouse and human proteins?

Time (min)	Cells with fully mixed proteins (%)
5	0
10	3
25	40
40	94
120	100

Go to **BioPortal** for all WORKING WITH**DATA** exercises

RECAP 6.1

The fluid mosaic model applies to the plasma membrane and the membranes of organelles. An integral membrane protein has both hydrophilic and hydrophobic domains, which affect its position and function in the membrane. Carbohydrates that attach to lipids and proteins on the outside of the membrane serve as recognition sites.

- What are some of the features of the fluid mosaic model of biological membranes? **See p. 106**
- Explain how the hydrophobic and hydrophilic regions of phospholipids cause a membrane bilayer to form. **See Figures 6.1 and 6.2**
- What differentiates an integral protein from a peripheral protein? **See pp. 107–108 and Figure 6.1**
- What is the experimental evidence that membrane proteins can diffuse in the plane of the membrane? **See p. 108 and Figure 6.5**

Now that you understand the structure of biological membranes, let's see how their components function. In the next section we'll focus on the membrane that surrounds individual cells: the plasma membrane. We'll then look at how the plasma membrane allows individual cells to be grouped together into multicellular systems of tissues.

6.2 How Is the Plasma Membrane Involved in Cell Adhesion and Recognition?

Often the cells of multicellular organisms exist in specialized groups with similar functions, called tissues. Your body has about 60 trillion cells organized into various kinds of tissues—such as muscle, nerve, and epithelium. Two processes allow cells to arrange themselves in groups:

- **Cell recognition**, in which one cell specifically binds to another cell of a certain type
- **Cell adhesion**, in which the connection between the two cells is strengthened

Both processes involve the plasma membrane. One way to study these processes is to break down a tissue into its individual cells and then allow them to adhere to one another again. This type of experiment is most easily done in relatively simple organisms, such as sponges, which provide good models for studying processes that also occur in the complex tissues of larger species.

Sponges are multicellular marine animals that have only a few distinct cell layers (see Section 31.5). The cells of a sponge adhere to one another but can be separated mechanically by passing the animal several times through a fine wire screen (**Figure 6.6**). Through this process, what was a single animal becomes hundreds of individual cells suspended in seawater. If such cells are stirred gently for a few hours, cell recognition occurs: The cells bump into and recognize one another, sticking together in the same shape and tissue organization as the original sponge. This recognition is species-specific; if disaggregated sponge cells from two different species are placed in the same container and shaken, individual cells will stick only to other cells of the same species. Two different sponges form, just like the ones at the start of the experiment.

Such tissue-specific and species-specific cell recognition and cell adhesion are essential to the formation and maintenance of tissues in multicellular organisms. Think of your own body.

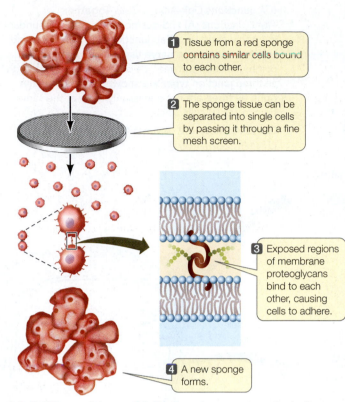

1 Tissue from a red sponge contains similar cells bound to each other.

2 The sponge tissue can be separated into single cells by passing it through a fine mesh screen.

3 Exposed regions of membrane proteoglycans bind to each other, causing cells to adhere.

4 A new sponge forms.

6.6 Cell Recognition and Adhesion In most cases (including the aggregation of animal cells into tissues), the binding between molecules is homotypic (same to same).

What keeps muscle cells bound to muscle cells and skin to skin? Specific cell adhesion is so obvious a characteristic of complex organisms that it is easy to overlook. You will see many examples of specific cell adhesion throughout this book; here we describe its general principles. As you will see, cell recognition and cell adhesion depend on plasma membrane proteins.

Cell recognition and adhesion involve proteins and carbohydrates at the cell surface

The molecules responsible for cell recognition and adhesion in sponges are proteoglycans (often 80% carbohydrate by molecular weight) that carry two kinds of carbohydrates. One kind is relatively small and binds to membrane components, keeping the proteoglycan attached to the cell. The other kind of carbohydrate is a larger, sulfated polysaccharide. If the sulfated polysaccharide from a particular species of sponge is purified and attached to cellulose beads, the beads will aggregate together or with sponge cells—but only with cells of the same species from which the polysaccharide was purified. This demonstrates that the sulfated polysaccharide is responsible for both the specific recognition and adhesion of the sponge cells.

Cell adhesion can result from interactions between the carbohydrates that are parts of glycolipids, glycoproteins, or proteoglycans—as is the case in sponge cells. In other cases, a carbohydrate on one cell interacts with a membrane protein on another cell. Or two proteins can interact directly. As we described in Section 3.2, a protein not only has a specific shape, it

also has specific chemical groups exposed on its surface where they can interact with other substances, including other proteins. Both of these features allow binding to other specific molecules. Cell adhesion occurs in all kinds of multicellular organisms. In plants, cell adhesion may be mediated by both integral membrane proteins and specific carbohydrates in the cell walls.

In most cases, the binding of cells in a tissue is **homotypic**; that is, the same molecule sticks out of both cells, and the exposed surfaces bind to each other. But **heterotypic** binding (between different molecules on different cells) also occurs. In this case, different chemical groups on different surface molecules have an affinity for one another. For example, when the mammalian sperm meets the egg, different proteins on the two types of cells have complementary binding surfaces. Similarly, some algae form male and female reproductive cells (analogous to sperm and eggs) that have flagella to propel them toward each other. Male and female cells can recognize each other by heterotypic glycoproteins on their flagella.

Three types of cell junctions connect adjacent cells

In a complex multicellular organism, cell recognition molecules allow specific types of cells to bind to one another. Often, after the initial binding, both cells contribute material to form additional membrane structures that connect them to one another. These specialized structures, called **cell junctions**, are most evident in electron micrographs of epithelial tissues, which are layers of cells that line body cavities or cover body surfaces. These surfaces often receive stresses or must retain their contents under pressure, so it is particularly important that their cells adhere tightly. We will examine three types of cell junctions that enable animal cells to seal intercellular spaces, reinforce attachments to one another, and communicate with each other. Tight junctions, desmosomes, and gap junctions, respectively, perform these three functions (**Figure 6.7**).

- **Tight junctions** prevent substances from moving through the spaces between cells. For example, cells lining the bladder have tight junctions so urine cannot leak out into the body cavity. Another important function of tight junctions is to maintain distinct faces of a cell within a tissue by restricting the migration of membrane proteins over the cell surface from one face to the other.

- **Desmosomes** hold neighboring cells firmly together, acting like spot welds or rivets. Materials can still move around in the extracellular matrix. This provides mechanical stability for tissues such as skin that receive physical stress.

- **Gap junctions** are channels that run between membrane pores in adjacent cells, allowing substances to pass between cells. In the heart, for example, gap junctions allow the rapid spread of electric current (mediated by ions) so the heart muscle cells beat in unison.

Cell membranes adhere to the extracellular matrix

In Section 5.4 we described the extracellular matrices of animal cells, which are composed of collagen protein arranged in

(A)

Plasma membranes

Intercellular space

Junctional proteins (interlocking)

The proteins of **tight junctions** form a "quilted" seal, barring the movement of dissolved materials through the space between epithelial cells.

6.7 Junctions Link Animal Cells Together
Tight junctions (A) and desmosomes (B) are abundant in epithelial tissues. Gap junctions (C) are also found in some muscle and nerve tissues, in which rapid communication between cells is important. Although all three junction types are shown in the cell at the right, all three are not necessarily seen at the same time in actual cells.

Go to Activity 6.2 Animal Cell Junctions
Life10e.com/ac6.2

(B)

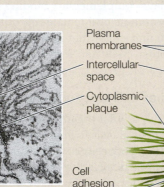

Plasma membranes

Intercellular space

Cytoplasmic plaque

Cell adhesion molecules

Keratin fiber (cytoskeleton filaments)

Desmosomes link adjacent cells tightly but permit materials to move around them in the intercellular space.

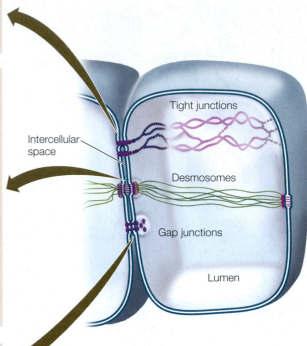

Tight junctions

Intercellular space

Desmosomes

Gap junctions

Lumen

(C)

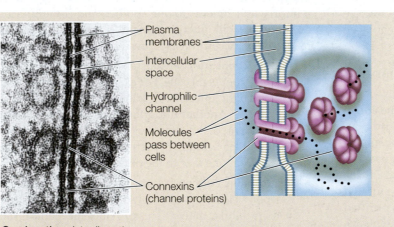

Plasma membranes

Intercellular space

Hydrophilic channel

Molecules pass between cells

Connexins (channel proteins)

Gap junctions let adjacent cells communicate.

(**Figure 6.8A**). More than 24 different integrins have been described in human cells. All of them bind to a protein in the extracellular matrix outside the cell, and to actin filaments, which are part of the cytoskeleton, inside the cell. So in addition to adhesion, integrin has a role in maintaining cell structure via its interaction with the cytoskeleton.

The binding of integrin to the extracellular matrix is noncovalent and reversible. When a cell moves its location within a tissue or organism, one side of the cell detaches from the extracellular matrix while the other side extends in the direction of movement, forming new attachments in that direction (**Figure 6.8B**). The integrin at the "back" of the cell (away from the direction of movement) is brought into the cytoplasm by endocytosis (see Section 6.5) so that it can be recycled and used for new attachments at the "front" of the cell. These events are important for cell movement within the developing embryo, and for the spread of cancer cells.

fibers in a gelatinous matrix of proteoglycans. The attachment of a cell to the extracellular matrix is important in maintaining the integrity of a tissue. In addition, some cells can detach from their neighbors, move, and attach to other cells; this is often mediated by interactions with the extracellular matrix.

A transmembrane protein called **integrin** often mediates the attachment of epithelial cells to the extracellular matrix

(A)

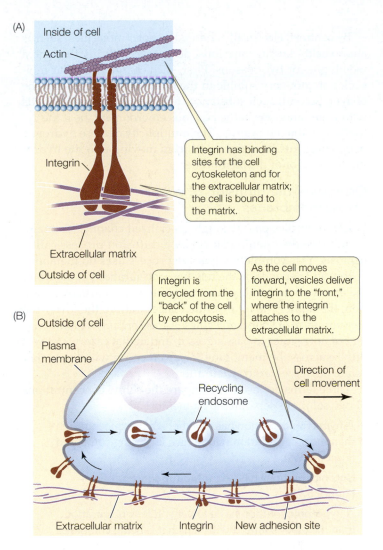

6.8 Integrins and the Extracellular Matrix (A) Integrins mediate the attachment of cells to the extracellular matrix. (B) Cell movements are mediated by integrin attachment.

RECAP 6.2

In multicellular organisms, cells arrange themselves into tissues via the processes of cell recognition and cell adhesion. These processes are mediated by membrane-associated proteins and carbohydrates. Cell membrane proteins also interact with the extracellular matrix. Cell junctions assist in strengthening tissues and allow cells to communicate with one another.

- Describe the difference between cell recognition and cell adhesion. See p. 110

- How do the three types of cell junctions regulate the passage of materials between cells and through the intercellular space? See p. 111 and Figure 6.7

We have just examined how the plasma membrane and molecules associated with it accommodate binding between cells and the maintenance of cell adhesion. We'll turn now to another major function of membranes: regulating the substances that enter or leave a cell or organelle.

6.3 What Are the Passive Processes of Membrane Transport?

As you have already learned, biological membranes have many functions, and control of the cell's internal composition is one of the most important. Biological membranes allow some substances, but not others, to pass through them. This characteristic of membranes is called **selective permeability**. Selective permeability allows the membrane to determine what substances enter or leave a cell or organelle.

There are two fundamentally different processes by which substances cross biological membranes:

- The processes of **passive transport** do not require the input of chemical energy to drive them.

- The processes of **active transport** require the input of chemical energy (metabolic energy).

This section focuses on the passive processes by which substances cross membranes. The energy for the passive transport of a substance comes from the difference between its concentration on one side of the membrane and its concentration on the other—its **concentration gradient**. Passive transport can involve either of two types of diffusion: simple diffusion through the phospholipid bilayer, or facilitated diffusion via channel proteins or carrier proteins.

▶ Go to Animated Tutorial 6.2
Passive Transport
Life10e.com/at6.2

Diffusion is the process of random movement toward a state of equilibrium

In a solution, there is a tendency for all of the components to be evenly distributed. You can see this when a drop of ink is allowed to fall into a gelatin suspension. Initially the pigment molecules are very concentrated, but they will move about at random, slowly spreading until the intensity of the color is exactly the same throughout the gelatin.

A solution in which the solute particles are uniformly distributed is said to be at equilibrium because there will be no future net change in their concentration. Being at equilibrium does not mean that the particles have stopped moving; it just means that they are moving in such a way that their overall distribution does not change.

Diffusion is the process of random movement toward a state of equilibrium. Although the motion of each individual particle is absolutely random, the net movement of particles is directional until equilibrium is reached. Diffusion is thus a net movement from regions of greater concentration to regions of lesser concentration.

In a complex solution (one with many different solutes), the diffusion of each solute is independent of those of the others. How fast a substance diffuses depends on three factors:

- The *diameter* of the molecules or ions: smaller molecules diffuse faster.

- The *temperature* of the solution: higher temperatures lead to faster diffusion because ions or molecules have more energy, and thus move more rapidly, at higher temperatures.

- The *concentration gradient* in the system—that is, the change in solute concentration with distance in a given direction: the greater the concentration gradient, the more rapidly a substance diffuses.

DIFFUSION WITHIN CELLS AND TISSUES In a small volume such as that inside a cell, solutes distribute themselves rapidly by diffusion. Small molecules and ions may move from one end of an organelle to another in a millisecond (10^{-3} s, or one-thousandth of a second). However, the usefulness of diffusion as a transport mechanism declines drastically as distances become greater. In the absence of mechanical stirring, diffusion across more than a centimeter may take an hour or more, and diffusion across meters may take years! Diffusion would not be adequate to distribute materials over the length of a human body, much less that of a larger organism. But within our cells or across layers of one or two cells, diffusion is rapid enough to distribute small molecules and ions almost instantaneously.

DIFFUSION ACROSS MEMBRANES In a solution without barriers, all the solutes diffuse at rates determined by temperature, their physical properties, and their concentration gradients. If a biological membrane divides the solution into separate compartments, then the movement of the different solutes can be affected by the properties of the membrane. The membrane is said to be permeable to solutes that can cross it more or less easily, but impermeable to substances that cannot move across it.

Molecules to which the membrane is impermeable remain in separate compartments, and their concentrations may be different on the two sides of the membrane. Molecules to which the membrane is permeable diffuse from one compartment to the other until their concentrations are equal on both sides of the membrane, and equilibrium is reached. After that point, individual molecules will continue to pass through the membrane, but *there will be no net change in concentration.*

Simple diffusion takes place through the phospholipid bilayer

In **simple diffusion**, small molecules pass through the phospholipid bilayer of the membrane. A molecule that is itself hydrophobic, and is therefore soluble in lipids, enters the membrane readily and is able to pass through it. The more lipid-soluble the molecule is, the more rapidly it diffuses through the membrane bilayer. This statement holds true over a wide range of molecular weights.

By contrast, electrically charged or polar molecules, such as amino acids, sugars, and ions, do not pass readily through a membrane for two reasons. First, such charged or polar molecules are not very soluble in the hydrophobic interior of the bilayer. Second, such substances form many hydrogen bonds with water and ions in the aqueous environment, be it the cytoplasm or the cell exterior. The multiplicity of these hydrogen bonds prevents the substances from moving into the hydrophobic interior of the membrane.

Osmosis is the diffusion of water across membranes

Water molecules pass through specialized channels in membranes (see the opening story) by a diffusion process called **osmosis**. This completely passive process uses no metabolic energy and depends on the relative concentrations of the water molecules on each side of the membrane. In a particular solution, the higher the total solute concentration, the lower the concentration of water molecules. A membrane may allow water but not solutes to pass across it, and in that case, water will diffuse across the membrane toward the side with the higher solute (lower water) concentration.

Three terms are used to compare the solute concentrations of two solutions separated by a membrane:

- A **hypertonic** solution has a higher solute concentration than the other solution with which it is being compared (**Figure 6.9A**).

- **Isotonic** solutions have equal solute concentrations (**Figure 6.9B**).

- A **hypotonic** solution has a lower solute concentration than the other solution with which it is being compared (**Figure 6.9C**).

Water moves from a hypotonic solution across a membrane to a hypertonic solution.

When we say that "water moves," bear in mind that we are referring to the net movement of water. Since it is so abundant, water is constantly moving through protein channels across the plasma membrane into and out of cells. What concerns us here is whether the overall movement is greater in one direction or the other.

The concentration of solutes in the environment determines the direction of osmosis in all animal cells. A red blood cell takes up water from a solution that is hypotonic to the cell's contents. The cell bursts because its plasma membrane cannot withstand the pressure created by the water entry and the resultant swelling. Conversely, the cell shrinks if the solution surrounding it is hypertonic to its contents. The integrity of red and white blood cells is absolutely dependent on the maintenance of a constant solute concentration in the blood plasma: the plasma must be isotonic to the blood cells if the cells are not to burst or shrink. Regulation of the solute concentration of body fluids is thus an important process for organisms without cell walls.

In contrast to animal cells, the cells of plants, archaea, bacteria, fungi, and some protists have cell walls that limit their volumes and keep them from bursting. Cells with sturdy walls

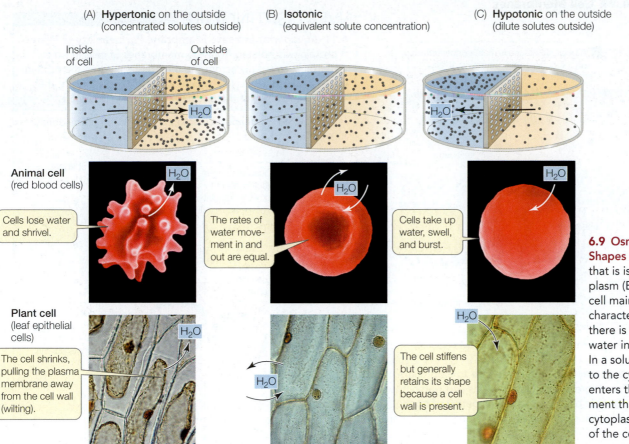

(A) Hypertonic on the outside (concentrated solutes outside)

(B) Isotonic (equivalent solute concentration)

(C) Hypotonic on the outside (dilute solutes outside)

Inside of cell

Outside of cell

H_2O

Animal cell (red blood cells)

H_2O

Cells lose water and shrivel.

H_2O

The rates of water movement in and out are equal.

H_2O

Cells take up water, swell, and burst.

Plant cell (leaf epithelial cells)

H_2O

The cell shrinks, pulling the plasma membrane away from the cell wall (wilting).

H_2O

H_2O

H_2O

The cell stiffens but generally retains its shape because a cell wall is present.

6.9 Osmosis Can Modify the Shapes of Cells In a solution that is isotonic with the cytoplasm (B), a plant or animal cell maintains a consistent, characteristic shape because there is no net movement of water into or out of the cell. In a solution that is hypotonic to the cytoplasm (C), water enters the cell. An environment that is hypertonic to the cytoplasm (A) draws water out of the cell.

take up a limited amount of water, and in so doing they build up internal pressure against the cell wall, which prevents further water from entering. This pressure within the cell is called **turgor pressure**. Turgor pressure keeps plants upright (and lettuce crisp) and is the driving force for the enlargement of plant cells. It is a normal and essential component of plant growth. If enough water leaves the cells, turgor pressure drops and the plant wilts. Turgor pressure reaches about 100 pounds per square inch (0.7 kg/cm²)—several times greater than the pressure in automobile tires. This pressure is so great that the cells would change shape and detach from one another were it not for adhesive molecules in the plant cell walls.

Diffusion may be aided by channel proteins

As we saw earlier, polar or charged substances such as water, amino acids, sugars, and ions do not readily diffuse across membranes. But they can cross the hydrophobic phospholipid bilayer passively (that is, without the input of energy) in one of two ways, depending on the substance:

- **Channel proteins** are integral membrane proteins that form channels across the membrane through which certain substances can pass.
- **Carrier proteins** bind substances and speed up their diffusion through the phospholipid bilayer.

Both of these processes are forms of **facilitated diffusion**. That is, the substances diffuse according to their concentration gradients, but their diffusion is facilitated by protein channels or carriers.

ION CHANNELS The best-studied channel proteins are the **ion channels**. As you will see in later chapters, the movement of ions across membranes is important in many biological processes such as respiration within the mitochondria, the electrical activity of the nervous system, and the opening of pores in leaves to allow gas exchange with the environment. Several types of ion channel have been identified—each specific for a particular ion. All show the same basic structure of a hydrophilic pore that allows a particular ion to move through its center (**Figure 6.10**).

Just as a fence may have a gate that can be opened or closed, most ion channels are gated: they can be opened or closed to ion passage. A **gated channel** opens when a stimulus causes a change in the three-dimensional shape of the channel. In some cases, this stimulus is the binding of a chemical signal, or **ligand** (see Figure 6.10). Channels controlled in this way are called ligand-gated channels. In contrast, a voltage-gated channel is stimulated to open or close by a change in the voltage (electric charge difference) across the membrane. As you will see in Section 45.2, voltage-gated channels are important for nerve cell function.

THE SPECIFICITY OF ION CHANNELS How does an ion channel allow one ion, but not another, to pass through? It is not simply a matter of charge and size of the ion. For example, a sodium ion (Na⁺), with a radius of 0.095 nanometers (nm), is smaller than K⁺ (0.130 nm), and both carry the same positive charge. Yet the potassium channel lets only K⁺ pass through the membrane, and not the smaller Na⁺. Nobel laureate Roderick MacKinnon at The Rockefeller University found an elegant explanation for this when he deciphered the structure of a potassium channel from a bacterium.

Being charged, both Na⁺ and K⁺ are attracted to water molecules. They are surrounded by water "shells" in solution, held by the attraction of their positive charges to the negatively

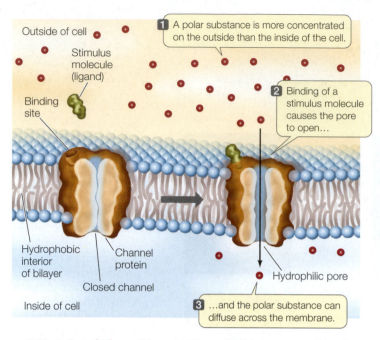

Outside of cell

Stimulus molecule (ligand)

Binding site

1 A polar substance is more concentrated on the outside than the inside of the cell.

2 Binding of a stimulus molecule causes the pore to open…

Hydrophobic interior of bilayer

Channel protein

Closed channel

Hydrophilic pore

Inside of cell

3 …and the polar substance can diffuse across the membrane.

6.10 A Gated Channel Protein Opens in Response to a Stimulus
The channel protein has a pore of polar amino acids and water. It is anchored in the hydrophobic bilayer by its outer coating of nonpolar R groups. The protein changes its three-dimensional shape when a stimulus molecule (ligand) binds to it, opening the pore so that specific hydrophilic substances can pass through. Other (voltage) gated channels open in response to an electrical potential (voltage).

charged oxygen atoms on the water molecules (see Figure 2.10). The potassium channel contains highly polar oxygen atoms at its opening. The gap enclosed by these atoms is exactly the right size so that when a K^+ ion approaches the opening, it is more strongly attracted to the oxygen atoms there than to those of the water molecules in its shell. It sheds its water shell and passes through the channel. The smaller Na^+ ion, however, is kept a bit more distant from the oxygen atoms at the opening of the channel because extra water molecules can fit between the ion (with its shell) and the oxygen atoms at the opening. This hydration prevents entry of Na^+ into the potassium channel.

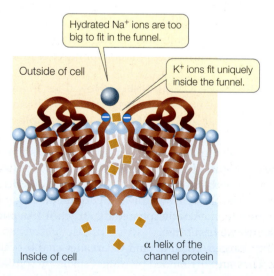

Hydrated Na^+ ions are too big to fit in the funnel.

Outside of cell

K^+ ions fit uniquely inside the funnel.

α helix of the channel protein

Inside of cell

INVESTIGATINGLIFE

6.11 Aquaporins Increase Membrane Permeability to Water
A protein was isolated from the membranes of cells in which water diffuses rapidly across the membranes. When the protein was inserted into oocytes, which do not normally have it, the water permeability of the oocytes was greatly increased.[a]

HYPOTHESIS Aquaporin increases membrane permeability to water.

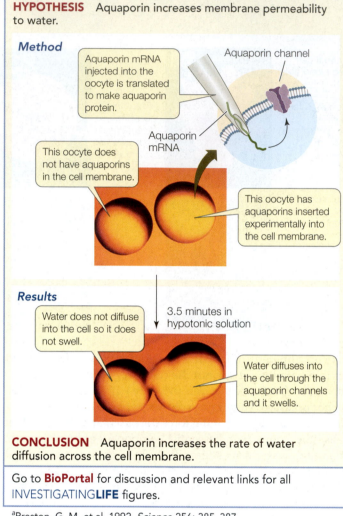

Method

Aquaporin mRNA injected into the oocyte is translated to make aquaporin protein.

Aquaporin channel

This oocyte does not have aquaporins in the cell membrane.

Aquaporin mRNA

This oocyte has aquaporins inserted experimentally into the cell membrane.

Results

Water does not diffuse into the cell so it does not swell.

3.5 minutes in hypotonic solution

Water diffuses into the cell through the aquaporin channels and it swells.

CONCLUSION Aquaporin increases the rate of water diffusion across the cell membrane.

Go to **BioPortal** for discussion and relevant links for all INVESTIGATINGLIFE figures.

[a]Preston, G. M. et al. 1992. *Science* 256: 385–387.

Such a potassium channel is voltage-gated. The mechanism for opening and closing it depends on interactions between positively charged arginine residues on the protein and negative charges on membrane phospholipids.

AQUAPORINS FOR WATER As you saw in the opening story of this chapter, water can cross membranes through protein channels called **aquaporins**. These channels function as a cellular plumbing system for moving water. Like the K^+ channel, the aquaporin channel is highly specific. Water molecules move in single file through the channel, which excludes ions so that the electrical properties of the cell are maintained. Aquaporins were first identified when a protein from red blood cell membranes was inserted into frog oocytes (immature egg cells). The membranes of these cells are normally impermeable to water, but the membranes of the cells treated with aquaporins became much more permeable (**Figure 6.11**).

(A)

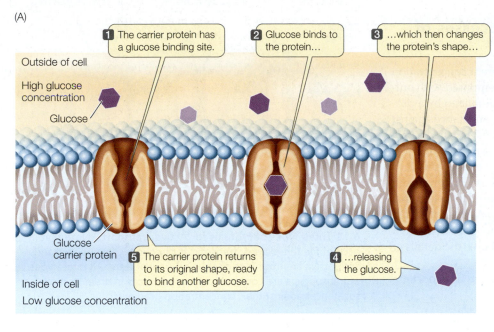

1 The carrier protein has a glucose binding site.

2 Glucose binds to the protein...

3 ...which then changes the protein's shape...

Outside of cell

High glucose concentration

Glucose

Glucose carrier protein

5 The carrier protein returns to its original shape, ready to bind another glucose.

4 ...releasing the glucose.

Inside of cell

Low glucose concentration

(B)

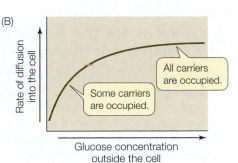

Rate of diffusion into the cell

All carriers are occupied.

Some carriers are occupied.

Glucose concentration outside the cell

6.12 A Carrier Protein Facilitates Diffusion
The glucose transporter is a carrier protein that allows glucose to enter the cell at a faster rate than would be possible by simple diffusion. (A) The transporter binds to glucose, brings it into the membrane interior, then changes shape, releasing glucose into the cell cytoplasm. (B) The graph shows the rate of glucose entry via a carrier versus the concentration of glucose outside the cell. As the glucose concentration increases, the rate of diffusion increases until the point at which all the available transporters are being used (the system is saturated).

Carrier proteins aid diffusion by binding substances

As we mentioned earlier, another type of facilitated diffusion involves the binding of the transported substance to a membrane protein called a carrier protein. Like channel proteins, carrier proteins facilitate the passive diffusion of substances into or out of cells or organelles. Carrier proteins transport polar molecules such as sugars and amino acids.

Glucose is the major energy source for most cells, and living systems require a great deal of it. Glucose is polar and cannot readily diffuse across membranes. Eukaryotic cell membranes contain a carrier protein—the glucose transporter—that facilitates glucose uptake into the cell. Binding of glucose to a specific three-dimensional site on one side of the transporter protein causes the protein to change its shape and release glucose on the other side of the membrane (**Figure 6.12A**). Since glucose is either broken down or otherwise removed almost as soon as it enters a cell, there is almost always a strong concentration gradient favoring glucose entry (that is, a higher concentration outside the cell than inside).

Transport by carrier proteins is different from simple diffusion. In simple diffusion, the rate of movement depends on the concentration gradient across the membrane. This is also true for carrier-mediated transport, up to a point. In carrier-mediated transport, as the concentration gradient increases, the diffusion rate also increases, but its *rate* of increase slows, and a point is reached at which the diffusion rate becomes constant. At this point, the facilitated diffusion system is said to be *saturated* (**Figure 6.12B**). A particular cell has a specific number of carrier protein molecules in its plasma membrane. The rate of diffusion reaches a maximum when all the carrier molecules are fully loaded with solute molecules. Think of waiting for the elevator on the ground floor of a hotel with 50 other people. You can't all get in

the elevator (carrier) at once, so the rate of transport (say, ten people at a time) is at its maximum, and the transport system is "saturated." As a consequence, cells that require large amounts of energy, such as muscle cells, have high concentrations of glucose transporters in their membranes so that the maximum rate of facilitated diffusion is greater. Likewise, the human brain has high glucose needs, and the blood vessels that nourish it have high concentrations of glucose transporters.

■ **RECAP** 6.3

Diffusion is the movement of ions or molecules from a region of greater concentration to a region of lesser concentration. Osmosis is the diffusion of water through a selectively permeable cell membrane. Channel proteins and carrier proteins can facilitate the diffusion of charged and polar substances, including water, across cell membranes.

- What properties of a substance determine whether, and how fast, it will diffuse across a membrane? **See p. 114**
- Describe osmosis and explain the terms hypertonic, hypotonic, and isotonic. **See p. 114 and Figure 6.9**
- How does a channel protein facilitate diffusion? **See pp. 115–116 and Figure 6.10**

The process of diffusion tends to equalize the concentrations of substances outside and inside cells. However, one hallmark of a living cell is that it can have an internal composition quite different from that of its environment. To achieve this, a cell must sometimes move substances against their concentration gradients. This process requires an input of energy and is known as active transport.

6.4 What are the Active Processes of Membrane Transport?

In many biological situations, there is a different concentration of a particular ion or small molecule inside compared with outside a cell. In these cases, the imbalance is maintained by a protein in the plasma membrane that moves the substance against its concentration and/or electrical gradient. This is called active transport, and because it is acting "against the normal flow," it requires the expenditure of energy. Often the energy source is adenosine triphosphate (ATP). In eukaryotes, ATP is produced in the mitochondria. It has chemical energy stored in its terminal phosphate bond. This energy is released when ATP is converted to adenosine diphosphate (ADP) in a hydrolysis reaction that breaks the terminal phosphate bond. We will give the details of how ATP provides energy to cells in Section 8.2.

The differences between diffusion and active transport are summarized in Table 6.1.

Go to Animated Tutorial 6.3
Active Transport
Life10e.com/at6.3

Active transport is directional

In many cases of simple and facilitated diffusion, ions or molecules can move down their concentration gradients in either direction across the cell membrane. In contrast, active transport is directional, and moves a substance either into or out of the cell or organelle, depending on need. There are three kinds of membrane proteins that carry out active transport (Figure 6.13):

- A **uniporter** moves a single substance in one direction. For example, a calcium-binding protein found in the plasma membrane and the ER membranes of many cells actively transports Ca^{2+} to locations where it is more highly concentrated, either outside the cell or inside the lumen of the ER.

- A **symporter** moves two substances in the same direction. For example, a symporter in the cells that line the intestine must bind Na^+ in addition to an amino acid in order to absorb amino acids from the intestine.

- An **antiporter** moves two substances in opposite directions, one into the cell (or organelle) and the other out of the cell (or organelle). For example, many cells have a sodium–potassium pump that moves Na^+ out of the cell and K^+ into it.

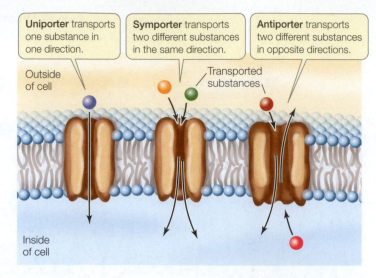

6.13 Three Types of Proteins for Active Transport   Note that in each of the three cases, transport is directional. Symporters and antiporters are examples of coupled transporters. All three types of transporters are coupled to energy sources in order to move substances against their concentration gradients.

Symporters and antiporters are also known as **coupled transporters** because they move two substances at once.

Different energy sources distinguish different active transport systems

There are two basic types of active transport:

- **Primary active transport** involves the direct hydrolysis of ATP, which provides the energy required for transport.

- **Secondary active transport** does not use ATP directly. Instead, its energy is supplied by an ion concentration gradient established by primary (ATP-driven) active transport. Secondary active transport uses the energy of ATP indirectly in the form of the gradient.

In primary active transport, energy released by the hydrolysis of ATP drives the movement of specific ions against their concentration gradients. For example, we mentioned earlier that concentrations of potassium ions (K^+) inside a cell are often much higher than in the fluid bathing the cell. However, the concentration of sodium ions (Na^+) is often much higher outside the cell. A protein in the plasma membrane pumps Na^+ out of the cell and K^+ into the cell against these concentration gradients, ensuring that the gradients are maintained (Figure 6.14). This **sodium–potassium (Na^+–K^+) pump** is found in all animal cells. The pump is an integral membrane glycoprotein.

TABLE 6.1

Membrane Transport Mechanisms

	Simple Diffusion	Facilitated Diffusion (through Channel or Carrier)	Active Transport
Cellular energy required?	No	No	Yes
Driving force	Concentration gradient	Concentration gradient	ATP hydrolysis (against concentration gradient)
Membrane protein required?	No	Yes	Yes
Specificity	No	Yes	Yes

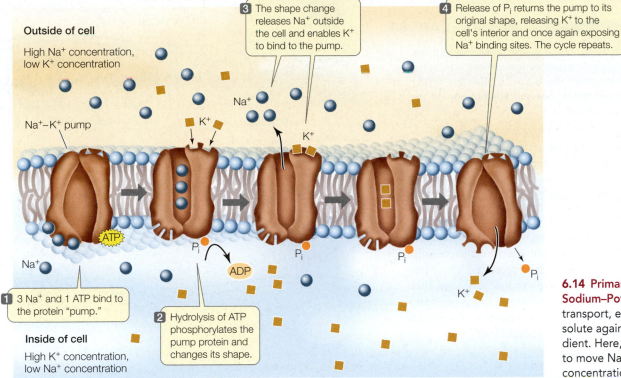

1 3 Na$^+$ and 1 ATP bind to the protein "pump."

2 Hydrolysis of ATP phosphorylates the pump protein and changes its shape.

3 The shape change releases Na$^+$ outside the cell and enables K$^+$ to bind to the pump.

4 Release of P$_i$ returns the pump to its original shape, releasing K$^+$ to the cell's interior and once again exposing Na$^+$ binding sites. The cycle repeats.

Outside of cell

High Na$^+$ concentration, low K$^+$ concentration

Na$^+$–K$^+$ pump

Inside of cell

High K$^+$ concentration, low Na$^+$ concentration

6.14 Primary Active Transport: The Sodium–Potassium Pump In active transport, energy is used to move a solute against its concentration gradient. Here, energy from ATP is used to move Na$^+$ and K$^+$ against their concentration gradients.

It breaks down a molecule of ATP to ADP and a free phosphate ion (P$_i$) and uses the energy released to bring two K$^+$ ions into the cell and export three Na$^+$ ions. The Na$^+$–K$^+$ pump is thus an antiporter because it moves two substances in different directions.

In secondary active transport, the movement of a substance against its concentration gradient is accomplished using energy "regained" by letting ions move across the membrane with their concentration gradients. For example, once the sodium–potassium pump establishes a concentration gradient of sodium ions, the passive diffusion of some Na$^+$ back into the cell can provide energy for the secondary active transport of glucose into the cell (**Figure 6.15**). This occurs when glucose is absorbed into the bloodstream from the digestive tract. Secondary active transport aids in the uptake of amino acids and sugars, which are essential raw materials for cell maintenance and growth. Both types of coupled transport proteins—symporters and antiporters—are used for secondary active transport.

Primary active transport
The Na$^+$–K$^+$ pump moves Na$^+$, using the energy of ATP hydrolysis to establish a concentration gradient of Na$^+$.

Secondary active transport
Na$^+$, moving with the concentration gradient established by the Na$^+$–K$^+$ pump, drives the transport of glucose against its concentration gradient.

Outside of cell

High Na$^+$ concentration, low K$^+$ concentration

Na$^+$–K$^+$ pump (antiporter)

Glucose (lower concentration)

Glucose (higher concentration)

Inside of cell

High K$^+$ concentration, low Na$^+$ concentration

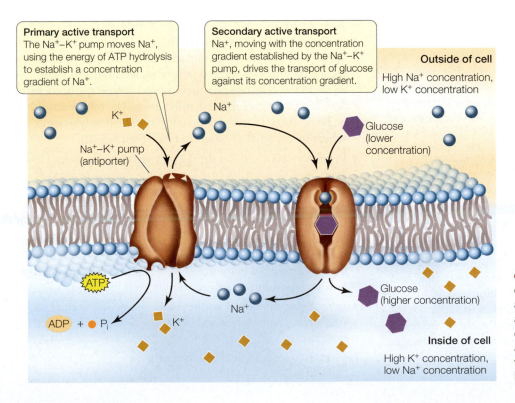

6.15 Secondary Active Transport The Na$^+$ concentration gradient established by primary active transport (left) powers the secondary active transport of glucose (right). A symporter protein couples the movement of glucose across the membrane against its concentration gradient to the passive movement of Na$^+$ into the cell.

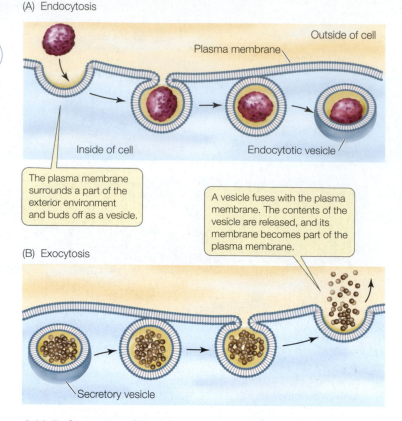

6.16 Endocytosis and Exocytosis   Endocytosis (A) and exocytosis (B) are used by eukaryotic cells to take up and release fluids, large molecules, and particles. Smaller cells, such as invading bacteria, can be taken up by endocytosis.

Go to Media Clip 6.1
An Amoeba Eats by Phagocytosis
Life10e.com/mc6.1

RECAP 6.4

Active transport across a membrane is directional and requires an input of energy to move substances against their concentration gradients. Active transport allows a cell to maintain small molecules and ions at concentrations very different from those in the surrounding environment.

- Why is energy required for active transport? **See p. 118**
- Why is the sodium–potassium (Na⁺–K⁺) pump classified as an antiporter? **See p. 118–119 and Figure 6.14**
- Explain the difference between primary active transport and secondary active transport. **See p. 118 and Figure 6.15**

We have examined a number of passive and active ways in which ions and small molecules can enter and leave cells. But what about large molecules such as proteins? Many proteins are so large that they diffuse very slowly, and their bulk makes it difficult for them to pass through the phospholipid bilayer. A completely different mechanism is needed to move intact large molecules across membranes.

6.5 How Do Large Molecules Enter and Leave a Cell?

Macromolecules such as proteins, polysaccharides, and nucleic acids are simply too large and too charged or polar to pass through biological membranes. This is actually a fortunate property—think of the consequences if such molecules diffused out of cells. A red blood cell would not retain its hemoglobin! As we discussed in Chapter 5, the development of a selectively permeable membrane was essential for the functioning of the first cells when life on Earth began. The interior of a cell can be maintained as a separate compartment with a different composition from that of the exterior environment, which is subject to abrupt changes. However, cells must sometimes take up or secrete (release to the external environment) intact large molecules. In Section 5.3 we described phagocytosis, the mechanism by which solid particles can be brought into the cell by means of vesicles that pinch off from the plasma membrane. The general terms for the mechanisms by which substances enter and leave the cell via membrane vesicles are endocytosis and exocytosis.

 Go to Animated Tutorial 6.4
Endocytosis and Exocytosis
Life10e.com/at6.4

Macromolecules and particles enter the cell by endocytosis

Endocytosis is a general term for a group of processes that bring small molecules, macromolecules, large particles, and even small cells into the eukaryotic cell (**Figure 6.16A**). We described an example of endocytosis earlier in the chapter, involving integrins (see Figure 6.8). Generally, there are three types of endocytosis: phagocytosis, pinocytosis, and receptor-mediated endocytosis. In all three, the plasma membrane

invaginates (folds inward), forming a small pocket around materials from the environment. The pocket deepens, forming a vesicle. This vesicle separates from the plasma membrane and migrates with its contents to the cell's interior.

- In **phagocytosis** ("cellular eating"), part of the plasma membrane engulfs large particles or even entire cells. Unicellular protists use phagocytosis for feeding, and some white blood cells use phagocytosis to defend the body by engulfing foreign cells and substances. The food vacuole or phagosome that forms usually fuses with a lysosome, where its contents are digested (see Figure 5.10).

- In **pinocytosis** ("cellular drinking"), vesicles also form. However, these vesicles are smaller, and the process operates to bring fluids and dissolved substances into the cell. Like phagocytosis, pinocytosis can be relatively nonspecific regarding what it brings into the cell. For example, pinocytosis occurs constantly in the endothelium, the single layer of cells that separates a tiny blood capillary from the surrounding tissue. Pinocytosis allows cells of the endothelium to rapidly acquire fluids and dissolved solutes from the blood.

- In **receptor-mediated endocytosis**, molecules at the cell surface recognize and trigger the uptake of specific materials.

Let's take a closer look at this last process.

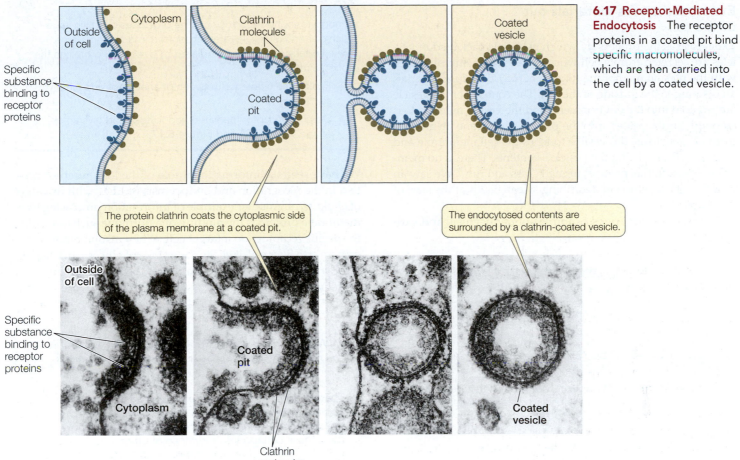

6.17 Receptor-Mediated Endocytosis The receptor proteins in a coated pit bind specific macromolecules, which are then carried into the cell by a coated vesicle.

The protein clathrin coats the cytoplasmic side of the plasma membrane at a coated pit.

The endocytosed contents are surrounded by a clathrin-coated vesicle.

Receptor-mediated endocytosis is highly specific

Receptor-mediated endocytosis is used by animal cells to capture specific macromolecules from the cell's environment. This process depends on **receptor proteins**, which are proteins that can bind to specific molecules within the cell or in the cell's external environment. In receptor-mediated endocytosis, the receptors are integral membrane proteins located at particular regions on the extracellular surface of the plasma membrane. These membrane regions are called coated pits because they form slight depressions in the plasma membrane and their cytoplasmic surfaces are coated by other proteins, such as clathrin. The uptake process is similar to that in phagocytosis.

When a receptor protein binds to its specific ligand (in this case, the macromolecule to be taken into the cell), its coated pit invaginates and forms a coated vesicle around the bound macromolecule. The clathrin molecules strengthen and stabilize the vesicle, which carries the macromolecule away from the plasma membrane and into the cytoplasm (**Figure 6.17**). Once inside, the vesicle loses its clathrin coat and may fuse with a lysosome, where the engulfed material is digested (by the hydrolysis of polymers to monomers) and the products released into the cytoplasm. Because of its specificity for particular macromolecules, receptor-mediated endocytosis is an efficient method of taking up substances that may exist at low concentrations in the cell's environment.

Receptor-mediated endocytosis is the method by which cholesterol is taken up by most mammalian cells. Water-insoluble cholesterol and triglycerides are packaged by liver cells into lipoprotein particles. Most of the cholesterol is packaged into a type of lipoprotein particle called low-density lipoprotein, or LDL, which is circulated via the bloodstream. When a particular cell requires cholesterol, it produces specific LDL receptors, which are inserted into the plasma membrane in clathrin-coated pits. Binding of LDLs to the receptor proteins triggers the uptake of the LDLs via receptor-mediated endocytosis. Within the resulting vesicle, the LDL particles are freed from the receptors. The receptors segregate to a region that buds off and forms a new vesicle, which is recycled to the plasma membrane. The freed LDL particles remain in the original vesicle, which fuses with a lysosome. There, the LDLs are digested and the cholesterol made available for cell use.

In healthy individuals, the liver takes up unused LDLs for recycling. People with the inherited disease familial hypercholesterolemia have a deficient LDL receptor in their livers. This prevents receptor-mediated endocytosis of LDLs, resulting in dangerously high levels of cholesterol in the blood. The cholesterol builds up in the arteries that nourish the heart and causes heart attacks. In extreme cases where only the deficient receptor is present, children and teenagers can have severe cardiovascular disease.

Exocytosis moves materials out of the cell

Exocytosis is the process by which materials packaged in vesicles are secreted from a cell when the vesicle membrane fuses with the plasma membrane (**Figure 6.16B**). This fusing makes an opening to the outside of the cell. The contents of the vesicle are released into the environment, and the vesicle membrane is smoothly incorporated into the plasma membrane. In another form of exocytosis, the vesicle touches the cell membrane and a pore forms, releasing the vesicle's contents. There is no membrane fusion in this process, termed "kiss and run." We saw an example of exocytosis in describing sweat glands at the start of the chapter.

Table 6.2 summarizes examples of endocytosis and exocytosis.

TABLE 6.2
Endocytosis and Exocytosis

Type of Process	Example
Endocytosis	
Receptor-mediated endocytosis	Specific uptake of large molecules, e.g., LDL
Pinocytosis	Nonspecific uptake of extracellular fluid, e.g., fluids and dissolved substances from blood
Phagocytosis	Nonspecific uptake of large undissolved particles, e.g., invading bacteria by cells of the immune system
Exocytosis	
Release of large molecules	Vesicle fusion with cell membrane, e.g., digestive enzymes in the pancreas
Release of small molecules	Vesicle fusion with cell membrane, e.g., neurotransmitters at the synapse

RECAP 6.5

Endocytosis and exocytosis are the processes by which large particles and molecules are transported into and out of the cell. Endocytosis may be mediated by a receptor protein in the plasma membrane.

- Explain the difference between phagocytosis and pinocytosis. **See p. 120**
- Describe receptor-mediated endocytosis and give an example. **See p. 121 and Figure 6.17**

We have seen the informational role of the LDL receptor protein in the recognition and endocytosis of LDL, with its cargo of cholesterol. Another example of information processing by a membrane protein is the binding of a hormone such as insulin to specific receptors on a target cell. When insulin binds to receptors on a liver cell, it elicits the uptake of glucose. In Chapter 7 we will see many other examples of the roles of membrane proteins in information processing.

?

Water purity is a worldwide problem. Can aquaporin membrane channels be used in water purification?

ANSWER

There are more than ten genes encoding aquaporins in humans, each channel having its particular location in the body (for example, salivary glands and the kidney). All of these aquaporins, and those of other organisms, including plants, share a common structure that spans the plasma membrane and has a channel through which water molecules pass in single file. Efforts are underway to insert aquaporins into synthetic membranes used for industrial applications. Because aquaporins allow only water to pass through (no solutes), such membranes could be used to purify contaminated fresh water or to desalinate seawater for drinking.

CHAPTER SUMMARY 6

6.1 What Is the Structure of a Biological Membrane?

- Biological membranes consist of lipids, proteins, and carbohydrates. The **fluid mosaic model** of membrane structure describes a phospholipid bilayer in which proteins can move about within the plane of the membrane. **Review ACTIVITY 6.1**

- The two layers of a membrane may have different properties because of their different lipid compositions. Animal cell membranes may contain high concentrations (up to 25%) of cholesterol. **Review ANIMATED TUTORIAL 6.1**

- The properties of membranes also depend on the **integral membrane proteins** and **peripheral membrane proteins** associated with them. Some proteins, called **transmembrane proteins**, span the membrane. **Review Figure 6.1**

- Carbohydrates, attached to proteins in **glycoproteins** or to phospholipids in **glycolipids**, project from the external surface of the plasma membrane and function as recognition signals.

- Membranes are not static structures, but are constantly forming, exchanging components, and breaking down.

6.2 How Is the Plasma Membrane Involved in Cell Adhesion and Recognition?

- In order for cells to assemble into tissues, they must recognize and adhere to one another. **Cell recognition** and **cell adhesion** depend on membrane-associated proteins and carbohydrates. **Review Figure 6.6**

- Adhesion can involve binding between identical (**homotypic**) or different (**heterotypic**) molecules on adjacent cells.

- **Cell junctions** connect adjacent cells. **Tight junctions** prevent the passage of molecules through the intercellular spaces between cells, and they restrict the migration of membrane proteins over the cell surface. **Desmosomes** cause cells to adhere firmly to one another. **Gap junctions** provide channels for communication between adjacent cells. **Review Figure 6.7, ACTIVITY 6.2**

- **Integrins** mediate the attachment of animal cells to the extracellular matrix. Detachment and recycling of integrins allow cells to move. **Review Figure 6.8**

continued

6.3 What Are the Passive Processes of Membrane Transport?
See ANIMATED TUTORIAL 6.2

- Membranes exhibit **selective permeability**, regulating which substances pass through them. Substances can cross the membrane by either **passive transport**, which requires no input of chemical energy, or **active transport**, which uses chemical energy.

- **Diffusion** is the movement of a solute from a region of higher concentration to a region of lower concentration. Equilibrium is reached when there is no further net change in concentration.

- In **osmosis**, water diffuses across a membrane from a region of higher water concentration to a region of lower water concentration.

- Most cells are in an **isotonic** environment, where total solute concentrations on both sides of the plasma membrane are equal. If the solution surrounding a cell is **hypotonic** to the cell interior, more water enters the cell than leaves it, causing it to swell. In plant cells, this contributes to **turgor pressure**. In a **hypertonic** solution, more water leaves the cell than enters it, causing it to shrivel. **Review Figure 6.9**

- A substance can diffuse passively across a membrane by either **simple diffusion** or **facilitated diffusion**, via a **channel protein** or a carrier protein.

- **Ion channels** are membrane proteins that allow the rapid facilitated diffusion of ions through membranes. **Gated channels** can be opened or closed by either chemical **ligands** or changes in membrane voltage. **Review Figure 6.10**

- **Aquaporins** are water channels. **Review Figure 6.11**

- **Carrier proteins** bind to polar molecules such as sugars and amino acids and transport them across the membrane. The maximum rate of this type of facilitated diffusion is limited by the number of carrier (transporter) proteins in the membrane. **Review Figure 6.12**

6.4 What Are the Active Processes of Membrane Transport?
See ANIMATED TUTORIAL 6.3

- **Active transport** requires the use of chemical energy to move substances across membranes against their concentration or electrical gradients. Active transport proteins may be **uniporters**, **symporters**, or **antiporters**. **Review Figure 6.13**

- In **primary active transport**, energy from the hydrolysis of ATP is used to move ions into or out of cells. The **sodium–potassium pump** is an important example. **Review Figure 6.14**

- **Secondary active transport** couples the passive movement of one substance down its concentration gradient to the movement of another substance against its concentration gradient. Energy from ATP is used indirectly to establish the concentration gradient that results in the movement of the first substance. **Review Figure 6.15**

6.5 How Do Large Molecules Enter and Leave a Cell?
See ANIMATED TUTORIAL 6.4

- **Endocytosis** is the transport of macromolecules, large particles, and small cells into eukaryotic cells via the invagination of the plasma membrane and the formation of vesicles. **Phagocytosis** and **pinocytosis** are types of endocytosis. **Review Figure 6.16A**

- In **exocytosis**, materials in vesicles are secreted from the cell when the vesicles fuse with the plasma membrane. **Review Figure 6.16B**

- In **receptor-mediated endocytosis**, a specific **receptor protein** on the plasma membrane binds to a particular macromolecule. **Review Figure 6.17**

Go to the Interactive Summary to review key figures, Animated Tutorials, and Activities
Life10e.com/is6

CHAPTER**REVIEW**

REMEMBERING

1. When a hormone molecule binds to a specific protein on the plasma membrane, the protein it binds to is called a
   a. ligand.
   b. clathrin.
   c. receptor protein.
   d. hydrophobic protein.
   e. cell adhesion molecule.

2. Which statement about membrane proteins is *not* true?
   a. They all extend from one side of the membrane to the other.
   b. Some serve as channels for ions to cross the membrane.
   c. Many are free to migrate laterally within the membrane.
   d. Their position in the membrane is determined by their structure.
   e. Some have both hydrophobic and hydrophilic regions.

3. Which statement about membrane carbohydrates is *not* true?
   a. Some are bound to proteins.
   b. Some are bound to lipids.
   c. They are added to proteins in the Golgi apparatus.
   d. They show little diversity.
   e. They are important in recognition reactions at the cell surface.

4. Which statement about ion channels is *not* true?
   a. They form pores in the membrane.
   b. They are proteins.
   c. All ions pass through the same type of channel.
   d. Movement through them is from regions of high concentration to regions of low concentration.
   e. Movement through them is by diffusion.

5. Facilitated diffusion and active transport both
   a. require ATP.
   b. require the use of proteins as carriers or channels.
   c. carry ions and not small molecules.
   d. increase without limit as the concentration gradient increases.
   e. depend on the solubility of the solute in lipids.

6. Primary and secondary active transport both
   a. generate ATP.
   b. are based on passive movement of Na⁺ ions.
   c. include the passive movement of glucose molecules.
   d. use ATP directly.
   e. can move solutes against their concentration gradients.

UNDERSTANDING & APPLYING

7. You are studying how the protein transferrin enters cells. When you examine cells that have taken up transferrin, you find it inside clathrin-coated vesicles. Therefore the most likely mechanism for uptake of transferrin is
   a. facilitated diffusion.
   b. an antiporter.
   c. receptor-mediated endocytosis.
   d. gap junctions.
   e. ion channels.

8. Muscle function requires calcium ions (Ca^{2+}) to be pumped into a subcellular compartment against a concentration gradient. What types of molecules are required for this to happen?

9. Section 27.2 will describe the diatoms, which are protists that have complex glassy structures in their cell walls (see Figure 27.8). These structures form within the Golgi apparatus. How do these structures reach the cell wall without having to pass through a membrane?

Eight diatom cells inside their ornate cell walls

ANALYZING & EVALUATING

10. Organisms that live in fresh water are almost always hypertonic to their environment. In what way is this a serious problem? How could some organisms cope with this problem?

11. When a normal lung cell becomes a lung cancer cell, there are several important changes in plasma membrane properties. How would you investigate the following phenomena? (a) The cancer cell membrane is more fluid, with more rapid diffusion in the plane of the membrane of both lipids and proteins. (b) The cancer cell has altered cell adhesion properties, binding to other tissues in addition to lung cells.

Go to BioPortal at **yourBioPortal.com** for Animated Tutorials, Activities, LearningCurve Quizzes, Flashcards, and many other study and review resources.

7 Cell Communication and Multicellularity

CHAPTER**OUTLINE**

7.1 What Are Signals, and How Do Cells Respond to Them?

7.2 How Do Signal Receptors Initiate a Cellular Response?

7.3 How Is the Response to a Signal Transduced through the Cell?

7.4 How Do Cells Change in Response to Signals?

7.5 How Do Cells in a Multicellular Organism Communicate Directly?

PRAIRIE VOLES (*Microtus ochrogaster*) are small rodents that live in temperate climates, where they dig tunnels in fields. When a male prairie vole encounters a female, mating often ensues. After mating (which can take as long as a day) the pair stays together, building a nest and raising their pups together. The bond between the two voles is so strong that they stay together for life. Contrast this behavior with that of the montane vole (*M. montanus*), which is closely related to the prairie vole and lives in the hills not far away. In this species, mating is quick, and afterward the pair separates. The male looks for new mates and the female abandons her young soon after they are born.

The explanation for these dramatic behavioral differences lies in the brains of these two species. When prairie voles mate, the brains of both males and females release specific peptides consisting of nine amino acids. In females, the peptide is oxytocin; in males, it is vasopressin. The peptides are circulated in the bloodstream and reach all tissues in the body, but they bind to only a few cell types. These cells have surface proteins called receptors, to which the peptides specifically bind, like a key inserting into a lock.

The interaction of peptide and receptor causes the receptor, which extends across the plasma membrane, to change shape. Within the cytoplasm, this change sets off a series of events that ultimately result in changes in behavior. The receptors for oxytocin and

Voles Prairie voles display extensive bonding behaviors after mating. These behaviors are mediated by peptides acting as intercellular signals.

vasopressin in prairie voles are most concentrated in the regions of the brain that are responsible for behaviors such as bonding and caring for the young. In montane voles, there are far fewer receptors for these peptides, and as a result, fewer bonding and caring behaviors. Clearly, oxytocin and vasopressin are signals that induce these behaviors.

Intercellular signaling is a hallmark of multicellular organisms. Even in the simplest such organisms, the differentiation of a group of cells into a specialized tissue (e.g., reproductive cells) must be integrated into the organism as a whole. A cell's response to a signal molecule takes place in three sequential steps. First, the signal binds to a receptor in the cell, often embedded in the outside surface of the plasma membrane. Second, signal binding conveys a message to the cell. Third, the cell changes its activity in response to the signal. In a multicellular organism, these steps lead to changes in that organism's functioning.

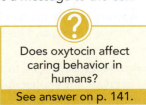

Does oxytocin affect caring behavior in humans?

See answer on p. 141.

7.1 What Are Signals, and How Do Cells Respond to Them?

Both prokaryotic and eukaryotic cells process information from their environments. This information can be in the form of a physical stimulus, such as the light reaching your eyes as you read this book, or chemicals that bathe a cell, such as lactose in a bacterial growth medium. It may come from outside the organism, such as the scent of a female moth seeking a mate in the dark, or from a neighboring cell within the organism, such as in the heart, where thousands of muscle cells contract in unison by transmitting signals to one another.

Of course, the mere presence of a signal does not mean that a cell will respond to it, just as you do not pay close attention to every stimulus in your environment as you study. To respond to a signal, a cell must have a specific receptor that can detect it and a way to use that information to influence cellular processes. A **signal transduction pathway** is a sequence of molecular events and chemical reactions that lead to a cell's response to a signal. Signal transduction pathways vary greatly in their details, but every such pathway involves a signal, a receptor, and a response. In this section we will discuss signals and provide a brief overview of signal transduction. We will consider receptors in Section 7.2, and other aspects of signal transduction in Sections 7.3 and 7.4.

Cells receive signals from the physical environment and from other cells

The physical environment is full of signals. For example, our sense organs allow us to respond to light (a physical signal), or odors and tastes (chemical signals) from our environment. Bacteria and protists can respond to minute chemical changes in their surroundings. Plants respond to light as a signal as well as an energy source, for example, by growing toward the source of light. However, a cell deep inside a large multicellular organism is far away from the exterior environment—its signals come from neighboring cells and the surrounding extracellular fluids. In such organisms, chemical signals are often made in one part of the body and arrive at target cells by local diffusion or by circulation in the blood or the plant vascular system. These signals are usually present in tiny concentrations (as low as 10^{-10} M) (see Chapter 2 for an explanation of molar concentrations) and differ in their sources and mode of delivery (**Figure 7.1**):

- **Autocrine** signals diffuse to and affect the cells that make them. For example, many tumor cells reproduce uncontrollably because they both make, and respond to, signals that stimulate cell division.
- **Juxtacrine** signals affect only cells adjacent to the cell producing the signal. This is especially common during development.
- **Paracrine** signals diffuse to and affect nearby cells. An example is a neurotransmitter made by one nerve cell that diffuses to a nearby cell and stimulates it (see Section 45.3).

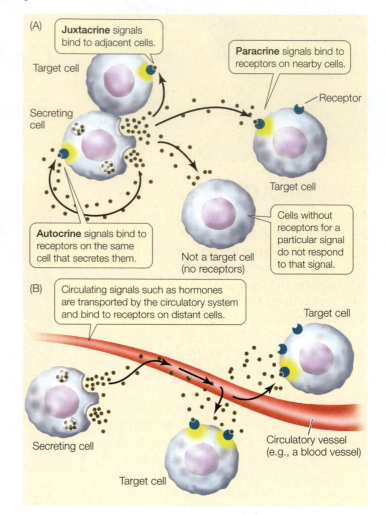

7.1 Chemical Signaling Systems (A) A signal molecule can diffuse to and act on the cell that produces it, an adjacent cell, or a nearby cell. (B) Many signals act on distant cells and must be transported by the organism's circulatory system.

Go to Activity 7.1 Chemical Signaling Systems
Life10e.com/ac7.1

- Signals that travel through the circulatory systems of animals or the vascular systems of plants are generally called **hormones**.

A signal transduction pathway involves a signal, a receptor, and responses

As we said earlier, the basic elements of any signal transduction pathway are a signal, a receptor, and a response (**Figure 7.2**). For the information from a signal to be transmitted to a cell, the target cell must be able to receive the signal and respond to it. This is the job of receptors. In a mammal, all cells may be exposed to a chemical signal that is circulated in the blood, such as oxytocin or vasopressin (see the opening of this chapter). However, most body cells are not capable of responding to these signals. *Only cells with the appropriate receptors can respond.*

The "response" can involve enzymes, which catalyze biochemical reactions, and transcription factors, which are proteins that turn the expression of particular genes on and off. An important feature of signal transduction is that the activities of

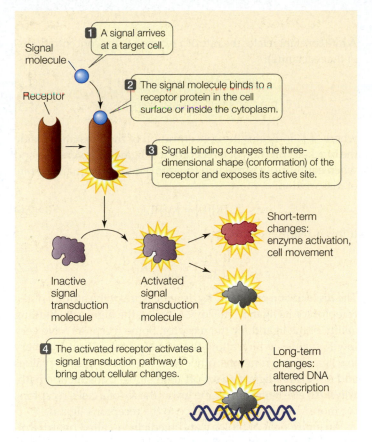

1 A signal arrives at a target cell.

Signal molecule

Receptor

2 The signal molecule binds to a receptor protein in the cell surface or inside the cytoplasm.

3 Signal binding changes the three-dimensional shape (conformation) of the receptor and exposes its active site.

Inactive signal transduction molecule

Activated signal transduction molecule

Short-term changes: enzyme activation, cell movement

4 The activated receptor activates a signal transduction pathway to bring about cellular changes.

Long-term changes: altered DNA transcription

7.2 A Signal Transduction Pathway This general pathway is common to many cells and situations. The ultimate effects on the cell are either short-term or long-term molecular changes, or both.

specific enzymes and transcription factors are regulated: they are either activated or inactivated to bring about cellular changes (see Figure 7.2). For example, an enzyme may be activated by the addition of a phosphate group (phosphorylation) to a particular site on the protein, thereby changing the enzyme's shape (see Figure 3.13B) and exposing its active site. The activity of a protein can also be regulated by mechanisms that control its location in the cell. For example, a transcription factor located in the cytoplasm is inactive because it is separated from the genetic material in the nucleus; a signal transduction pathway may result in the transport of the factor to the nucleus, where it can affect gene expression. There are many other ways in which enzymes and transcription factors are regulated. We will encounter some of these mechanisms in Sections 7.3 and 7.4.

In this chapter we consider signal transduction pathways in isolation from one another. But life is not that simple. In fact, there is a great deal of **crosstalk**: interactions between different signal transduction pathways. For example, signal transduction pathways often branch: a single activated protein (receptor or enzyme) might activate enzymes or transcription factors in multiple pathways, leading to multiple responses to a single stimulus. Multiple signal transduction pathways might converge on a single transcription factor, allowing the transcription of a single gene to be adjusted in response to several different signals. Crosstalk can also result in the activation of one pathway and the inhibition of another. This phenomenon inside the cell is analogous to the "crosstalk" that occurs at the level of the whole

body. For example, in your limbs you have opposing muscles. When you bend your elbow, you contract one set of muscles and relax the opposing muscles, so that the arm will bend. Because of crosstalk, biologists often refer to "signaling networks" rather than signal transduction pathways, reflecting the high degree of complexity in cellular signaling.

The general features of signal transduction pathways described in this section will recur in more detail throughout the chapter. First let's consider more closely the nature of the receptors that bind signal molecules.

7.2 How Do Signal Receptors Initiate a Cellular Response?

Any given cell in a multicellular organism is bombarded with many signals. However, it responds to only some of them, because no cell makes receptors for all signals. A **receptor** protein recognizes its signal very specifically, in much the same way that a membrane transport protein recognizes and binds to the substance it transports. This specificity ensures that only those cells that make a specific receptor will respond to a given signal.

Receptors that recognize chemical signals have specific binding sites

A specific chemical signal molecule fits into a three-dimensional site on its protein receptor (**Figure 7.3A**). A molecule that binds to a receptor site on another molecule in this way is called a **ligand**. Binding of the signaling ligand causes the receptor protein to change its three-dimensional shape, and that conformational change initiates a cellular response. The ligand does not contribute further to this response. In fact, the ligand is usually not metabolized into a useful product; its role is purely to "knock on the door." (This is in sharp contrast to the enzyme–substrate interaction, which we will describe in Chapter 8. The entire purpose of that interaction is to convert the substrate into a useful product.)

The sensitivity of a cell to a signal is determined in part by the affinity of the cell's receptors for the signal ligand—the likelihood that the receptor will bind to the ligand at any given

(A)

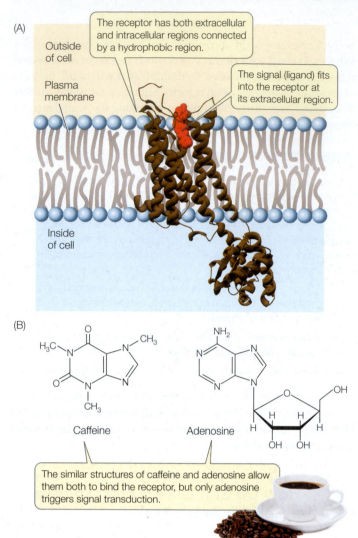

The receptor has both extracellular and intracellular regions connected by a hydrophobic region.

Outside of cell

The signal (ligand) fits into the receptor at its extracellular region.

Plasma membrane

Inside of cell

(B)

Caffeine   Adenosine

The similar structures of caffeine and adenosine allow them both to bind the receptor, but only adenosine triggers signal transduction.

7.3 A Signal and Its Receptor (A) The adenosine 2A receptor occurs in the human brain, where it is involved in inhibiting arousal. (B) Adenosine is the normal ligand for the receptor. Caffeine has a similar structure to that of adenosine and can act as an antagonist that binds the receptor and prevents its normal functioning.

ligand concentration. Receptors (R) bind to their ligands (L) according to chemistry's law of mass action. This means that the binding is reversible:

$$R + L \rightleftharpoons RL \qquad (7.1)$$

For most ligand–receptor complexes (RL), the equilibrium point is far to the right—that is, binding is favored. Reversibility is important, however, because if the ligand were never released, the receptor would be continuously stimulated and the cell would never stop responding.

As with any reversible chemical reaction, the binding and dissociation processes each have a rate constant, here designated K_1 and K_2:

$$\text{Binding:} \qquad R + L \xrightarrow{K_1} RL \qquad (7.2)$$

$$\text{Dissociation:} \qquad RL \xrightarrow{K_2} R + L \qquad (7.3)$$

A rate constant relates the rate of a reaction to the concentration(s) of the reactant(s):

$$\text{Rate of binding} = K_1[R][L] \qquad (7.4)$$

$$\text{Rate of dissociation} = K_2[RL] \qquad (7.5)$$

where "[ ]" indicates the concentration of the substance inside the brackets. Binding of a receptor to a ligand is reversible, and when equilibrium is reached the rate of binding equals the rate of dissociation:

$$K_1[R][L] = K_2[RL] \qquad (7.6)$$

If this is rearranged, we get:

$$\frac{[R][L]}{[RL]} = \frac{K_2}{K_1} = K_D \qquad (7.7)$$

The **dissociation constant**, K_D, is a measure of the affinity of the receptor for its ligand. The lower the K_D, the higher the binding ability of the ligand for the receptor. Some receptors have very low K_D values, which allow them to bind their ligands at very low ligand concentrations; other receptors have higher K_D values and need more ligand to set off their signal transduction pathways. All else being equal, the lower the K_D of a cell's receptors, the more sensitive the cell will be to the receptor's ligand.

An entire field of biology and medicine—called pharmacology—is devoted to the study of drugs. Drugs function as ligands that bind specific receptors. In the discovery and design of new drugs, it is helpful to know the specific receptor that the drug will bind, because then it is possible to determine the K_D value of its binding. This is one factor that can be taken into consideration when determining dosage levels. Of course, many drugs have side effects, and these are also dosage-dependent!

An inhibitor (or antagonist) can also bind to a receptor protein, instead of the normal ligand. There are both natural and artificial antagonists of receptor binding. Many substances that alter human behavior bind to specific receptors in the brain, and prevent the binding of the receptors' specific ligands. One example is caffeine, which is probably the world's most widely consumed stimulant. In the brain, the nucleoside adenosine acts as a ligand that binds to a receptor on nerve cells, initiating a signal transduction pathway that reduces brain activity, especially arousal. Because caffeine has a similar molecular structure to that of adenosine, it also binds to the adenosine receptor (**Figure 7.3B**). But in this case binding does not initiate a signal transduction pathway. Rather, it "ties up" the receptor, preventing adenosine binding and thereby allowing continued nerve cell activity and arousal.

Receptors can be classified by location and function

The chemistry of ligand signals is quite variable: some ligands can diffuse through membranes whereas others cannot. Physical signals such as light also vary in their ability to penetrate cells and tissues. Correspondingly, a receptor can be classified by its location in the cell, which largely depends on the nature of its signal (**Figure 7.4**):

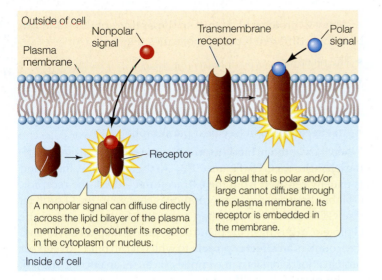

7.4 Two Locations for Receptors Receptors can be located inside the cell (in the cytoplasm or nucleus) or in the plasma membrane.

- *Membrane receptors*: Large or polar ligands cannot cross the lipid bilayer. Insulin, for example, is a protein hormone that cannot diffuse through the plasma membrane; instead, it binds to a transmembrane receptor with an extracellular binding domain.

- *Intracellular receptors*: Small or nonpolar ligands can diffuse across the nonpolar phospholipid bilayer of the plasma membrane and enter the cell. Estrogen, for example, is a lipid-soluble steroid hormone that can easily diffuse across the plasma membrane; it binds to a receptor inside the cell. Light of certain wavelengths can penetrate the cells in a plant leaf quite easily, and many plant light receptors (photoreceptors) are also intracellular.

In complex eukaryotes such as mammals and higher plants, there are three well-studied categories of plasma membrane receptors that are grouped according to their functions: ion channels, protein kinase receptors, and G protein-linked receptors.

ION CHANNEL RECEPTORS As described in Section 6.3, the plasma membranes of many types of cells contain gated **ion channels** that allow ions such as Na^+, K^+, Ca^{2+}, or Cl^- to enter or leave the cell. The gate-opening mechanism is an alteration in the three-dimensional shape of the channel protein upon interaction with a signal; thus these proteins function as receptors. Each type of ion channel responds to a specific signal, including sensory stimuli such as light, sound, and electric charge differences across the plasma membrane, as well as chemical ligands such as hormones and neurotransmitters.

The acetylcholine receptor, which is located in the plasma membrane of skeletal muscle cells, is an example of a gated ion channel. This receptor protein is a sodium channel that binds the ligand acetylcholine, which is a neurotransmitter—a chemical signal released from neurons (nerve cells) (**Figure 7.5**). When two molecules of acetylcholine bind to the receptor, it opens for about a thousandth of a second. That

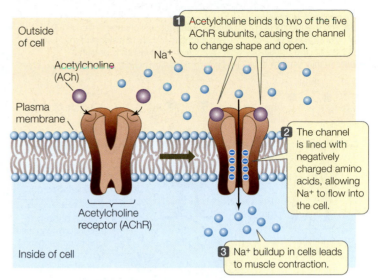

7.5 A Gated Ion Channel The acetylcholine receptor (AChR) is a ligand-gated ion channel for sodium ions. It is made up of five polypeptide subunits. When acetylcholine molecules (ACh) bind to two of the subunits, the gate opens and Na^+ flows into the cell. This channel helps regulate membrane polarity (see Chapter 45).

is enough time for Na^+, which is more concentrated outside the cell than inside, to rush into the cell, moving in response to both concentration and electrical potential gradients. The change in Na^+ concentration in the cell initiates a series of events that result in muscle contraction.

PROTEIN KINASE RECEPTORS Some eukaryotic receptor proteins are **protein kinases**. When they are activated, they catalyze the phosphorylation of themselves and/or other proteins, thus changing their shapes and therefore their functions.

$$\text{Target protein} + \text{ATP} \xrightarrow{\text{Protein kinase}} \text{protein-} \textcircled{P} + \text{ADP}$$
$$\text{(altered shape and function)}$$

The receptor for insulin is an example of a protein kinase receptor. Insulin is a protein hormone made by the mammalian pancreas. Its receptor has two copies each of two different polypeptide subunits called α and β (**Figure 7.6**). When insulin binds to the receptor, the receptor becomes activated and is able to phosphorylate itself and certain cytoplasmic proteins that are appropriately called insulin response substrates. These proteins then initiate many cellular responses, including the insertion of glucose transporters (see Figure 6.12) into the plasma membrane.

G PROTEIN-LINKED RECEPTORS A third category of eukaryotic plasma membrane receptors is the **G protein-linked receptors**, also referred to as the seven-transmembrane domain receptors. These receptors have many physiological roles, including light detection in the mammalian retina (photoreceptors), detection of odors (olfactory receptors), and the regulation of mood and behavior. For example, the receptors that bind the hormones oxytocin and vasopressin, which affect mating behavior in voles (see the opening story), are G protein-linked receptors.

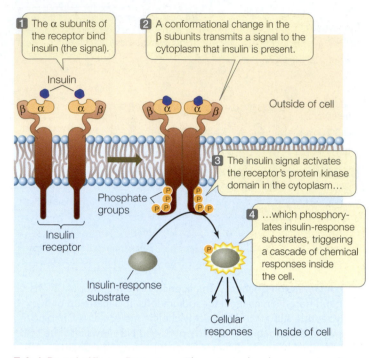

1 The α subunits of the receptor bind insulin (the signal).

2 A conformational change in the β subunits transmits a signal to the cytoplasm that insulin is present.

Insulin

Outside of cell

3 The insulin signal activates the receptor's protein kinase domain in the cytoplasm…

Phosphate groups

Insulin receptor

4 …which phosphorylates insulin-response substrates, triggering a cascade of chemical responses inside the cell.

Insulin-response substrate

Cellular responses   Inside of cell

7.6 A Protein Kinase Receptor   The mammalian hormone insulin binds to a receptor on the outside surface of the cell and initiates a response.

The descriptive name identifies a fascinating group of receptors, each of which is composed of a single protein with seven transmembrane domains. These seven domains pass through the phospholipid bilayer and are separated by short loops that extend either outside or inside the cell. Ligand binding on the extracellular side of the receptor changes the shape of its cytoplasmic region, exposing a site that binds to a mobile membrane protein called a **G protein**. The G protein is partially inserted into the lipid bilayer and partially exposed on the cytoplasmic surface of the membrane.

Many G proteins have three polypeptide subunits and can bind three different types of molecules (**Figure 7.7A**):

- The receptor
- GDP and GTP (guanosine diphosphate and triphosphate, respectively; these are nucleoside phosphates like ADP and ATP)
- An effector protein (see next paragraph)

When the G protein binds to an activated receptor protein, GDP is exchanged for GTP (**Figure 7.7B**). At the same time, the ligand is usually released from the extracellular side of the receptor. GTP binding causes a conformational change in the G protein. The GTP-bound subunit then separates from the rest of the G protein, diffusing in the plane of the phospholipid bilayer until it encounters an **effector protein** to which it can bind. An effector protein is just what its name implies: it causes an effect in the cell. The binding of the GTP-bearing G protein subunit activates the effector—which may be an enzyme or an ion channel—thereby causing changes in cell function (**Figure 7.7C**).

After activation of the effector protein, the GTP bound to the G protein is hydrolyzed to GDP. The now inactive G protein subunit separates from the effector protein and diffuses in the membrane to collide with and bind to the other two G protein subunits. When the three components of the G protein are reassembled, the protein is capable of binding again to an activated receptor. After binding, the activated receptor exchanges the GDP on the G protein for a GTP, and the cycle begins again.

Intracellular receptors are located in the cytoplasm or the nucleus

Intracellular receptors are located inside the cell and respond to physical signals such as light (for example, some photoreceptors in plants) or chemical signals that can diffuse across the plasma membrane (for example, steroid hormones in animals). Many intracellular receptors are transcription factors. Some are

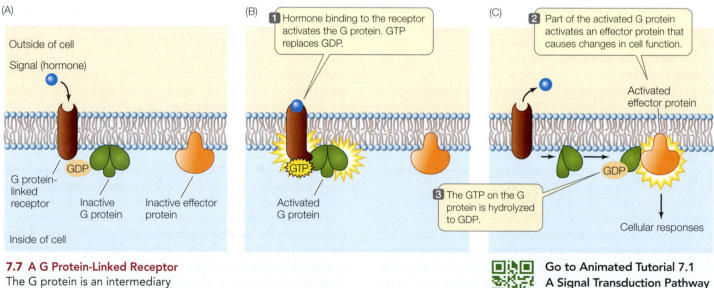

(A)

Outside of cell

Signal (hormone)

G protein-linked receptor   Inactive G protein   Inactive effector protein

Inside of cell

(B)

1 Hormone binding to the receptor activates the G protein. GTP replaces GDP.

Activated G protein

(C)

2 Part of the activated G protein activates an effector protein that causes changes in cell function.

Activated effector protein

3 The GTP on the G protein is hydrolyzed to GDP.

Cellular responses

7.7 A G Protein-Linked Receptor
The G protein is an intermediary between the receptor and its effector.

Go to Animated Tutorial 7.1
A Signal Transduction Pathway
Life10e.com/at7.1

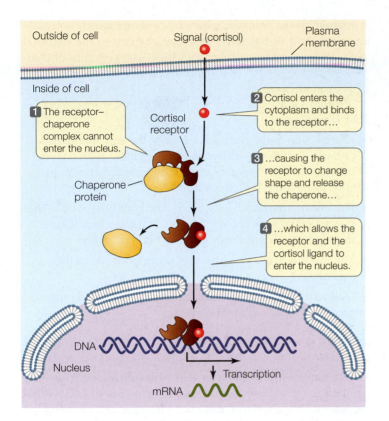

Outside of cell
Signal (cortisol)
Plasma membrane

Inside of cell

1 The receptor–chaperone complex cannot enter the nucleus.

Cortisol receptor

2 Cortisol enters the cytoplasm and binds to the receptor…

Chaperone protein

3 …causing the receptor to change shape and release the chaperone…

4 …which allows the receptor and the cortisol ligand to enter the nucleus.

DNA

Nucleus

Transcription

mRNA

7.8 An Intracellular Receptor The receptor for cortisol is in the cytoplasm bound to a chaperone protein that is released when cortisol binds to the receptor.

located in the cytoplasm until they are activated; after binding their ligands, these transcription factors move to the nucleus where they bind to DNA and alter the expression of specific genes. A typical example is the receptor for the steroid hormone cortisol. This receptor is normally bound to a chaperone protein that blocks it from entering the nucleus. Binding of the hormone causes the receptor to change its shape so that the chaperone is released (**Figure 7.8**). This release allows the receptor to enter the nucleus, where it affects DNA transcription. Another group of intracellular receptors are always located in the nucleus, and their ligands must enter the nucleus before binding.

RECAP 7.2

Receptors are proteins that bind, or are changed by, specific ligands or physical signals. The changed receptor initiates a response in the cell. These receptors are located in the plasma membrane or inside the cell.

- What is the nature and importance of specificity in the binding of a receptor to its particular ligand? **See p. 127**
- How is a dissociation constant calculated, and what is its relevance to ligand–receptor binding? **See p. 128**
- Describe three important categories of plasma membrane receptors that are seen in complex eukaryotes. **See pp. 129–130 and Figures 7.5, 7.6, 7.7**

Now that we have discussed signals and receptors, let's examine the characteristics of the molecules (transducers) that mediate the cellular response.

7.3 How Is the Response to a Signal Transduced through the Cell?

As we have seen, there are many different kinds of signals and receptors. Not surprisingly, the ways that signals are transduced, and the resulting cellular responses, are also highly varied. Some signal transduction pathways are quite simple and direct, whereas others involve multiple steps. As we mentioned in Section 7.1, signal transduction pathways can involve enzymes and transcription factors. In addition, small nonprotein molecules called **second messengers** can diffuse throughout the cytoplasm and mediate further steps in pathways.

In many cases, the signal can initiate a chain (cascade) of events, in which proteins interact with other proteins, which interact with still other proteins until the final responses are achieved. Through such a cascade, an initial signal can be both amplified and distributed to cause several different responses in the target cell. In this section we will examine the kinds of molecules that transduce signals and look at several different signal transduction pathways.

A protein kinase cascade amplifies a response to ligand binding

We have seen that when a signal binds to a protein kinase receptor, the receptor's conformation changes, exposing a protein kinase active site on the receptor's cytoplasmic domain. The protein kinase then catalyzes the phosphorylation of target proteins. Protein kinase receptors are important in binding signals called growth factors that stimulate cell division in both plants and animals.

Scientists worked out the signal transduction pathway for one growth factor by studying a cell that went wrong. Many human bladder cancers contain an abnormal form of a protein called Ras (so named because a similar protein was previously isolated from a *rat* sarcoma tumor). Investigations of these bladder cancers showed that Ras was a G protein and that the abnormal form was always active because it was permanently bound to GTP and thus caused continuous cell division (**Figure 7.9**). If this abnormal form of Ras was inhibited, the cells stopped dividing. This discovery has led to a major effort to develop specific Ras inhibitors for cancer treatment.

 Go to Animated Tutorial 7.2
Signal Transduction and Cancer
Life10e.com/at7.2

Other cancer cells have abnormalities in different parts of the same signal transduction pathway. Biologists compared the defects in these cells with the normal signaling process in non-cancer cells and thus worked out the entire signaling pathway (**Figure 7.10**). In Section 7.2 we discussed G protein-linked receptors, but other kinds of receptors can

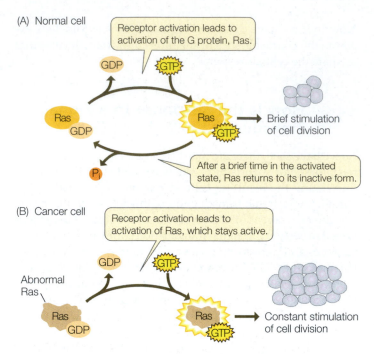

7.9 Signal Transduction and Cancer (A) Ras is a G protein that regulates cell division. (B) In some tumors, the Ras protein is permanently active, resulting in uncontrolled cell division.

interact with G proteins as well. The G protein Ras mediates a response after activation by the protein kinase receptor. The resulting signal transduction pathway is an example of a **protein kinase cascade**, where one protein kinase activates the next, and so on. Such cascades are key to the external regulation of many cellular activities. The genomes of complex eukaryotes, such as humans, typically encode hundreds of protein kinases.

Protein kinase cascades are useful signal transducers for four reasons:

- At each step in the cascade of events, the signal is *amplified*, because each newly activated protein kinase is an enzyme that can catalyze the phosphorylation of many target proteins (see Figure 7.10, steps 5 and 6).

- The information from a signal that originally arrived at the plasma membrane is *communicated* to the nucleus where the expression of multiple genes is often modified.

- The multitude of steps provides some *specificity* to the process.

- Different target proteins at each step in the cascade can provide *variation* in the response.

Second messengers can amplify signals between receptors and target molecules

Often, there is a small, nonprotein molecule intermediary between the activated receptor and the cascade of events that ensues. Earl Sutherland and his colleagues at Case Western Reserve University discovered one such molecule when they were investigating the activation of the liver enzyme glycogen phosphorylase by the hormone epinephrine. The

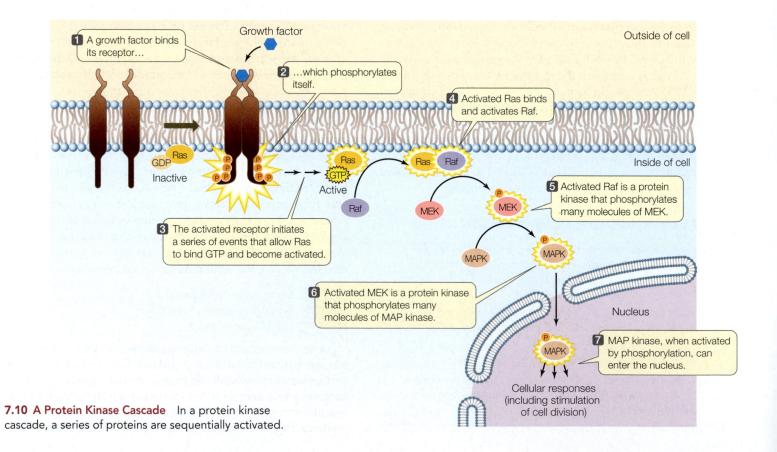

7.10 A Protein Kinase Cascade In a protein kinase cascade, a series of proteins are sequentially activated.

INVESTIGATING**LIFE**

7.11 The Discovery of a Second Messenger Glycogen phosphorylase is activated in liver cells after epinephrine binds to a membrane receptor. Sutherland and his colleagues observed that this activation could occur in vitro only if fragments of the plasma membrane were present. They designed experiments to show that a second messenger caused the activation of glycogen phosphorylase.[a]

HYPOTHESIS A second messenger mediates between receptor activation at the plasma membrane and enzyme activation in the cytoplasm.

Method

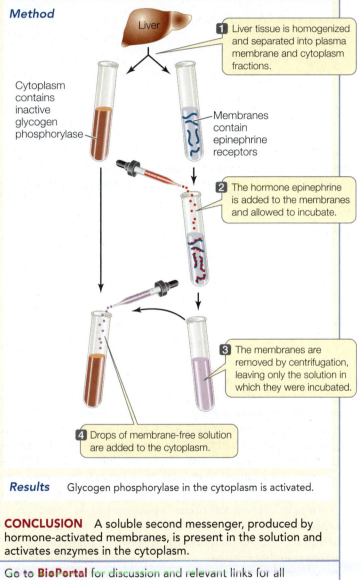

Liver

Cytoplasm contains inactive glycogen phosphorylase

Membranes contain epinephrine receptors

1 Liver tissue is homogenized and separated into plasma membrane and cytoplasm fractions.

2 The hormone epinephrine is added to the membranes and allowed to incubate.

3 The membranes are removed by centrifugation, leaving only the solution in which they were incubated.

4 Drops of membrane-free solution are added to the cytoplasm.

Results Glycogen phosphorylase in the cytoplasm is activated.

CONCLUSION A soluble second messenger, produced by hormone-activated membranes, is present in the solution and activates enzymes in the cytoplasm.

Go to **BioPortal** for discussion and relevant links for all INVESTIGATING**LIFE** figures.

[a]Rall, T. W. et al. 1957. *Journal of Biological Chemistry* 224: 463.

hormone is released when an animal faces life-threatening conditions and needs energy fast for the fight-or-flight response (see Figure 41.3). Glycogen phosphorylase catalyzes the breakdown of glycogen stored in the liver so that the resulting glucose molecules can be released to the blood. The

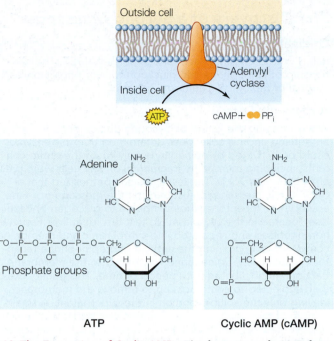

Outside cell

Adenylyl cyclase

Inside cell

ATP

cAMP + PP$_i$

Adenine

Phosphate groups

ATP

Cyclic AMP (cAMP)

7.12 The Formation of Cyclic AMP The formation of cAMP from ATP is catalyzed by adenylyl cyclase, an enzyme that is activated by G proteins.

enzyme is present in the liver cell cytoplasm but is inactive unless the liver cells are exposed to epinephrine:

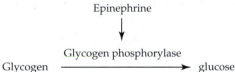

Epinephrine

Glycogen phosphorylase

Glycogen ⟶ glucose

The researchers found that epinephrine could activate glycogen phosphorylase in liver cells that had been broken open, but only if the entire cell contents, including plasma membrane fragments, were present. Under these conditions epinephrine was bound to the plasma membrane fragments (the location of its receptor), but the active phosphorylase was present in the solution. Adding epinephrine to just cytoplasm with inactive phosphorylase did not result in activation. They hypothesized that there must be a "second messenger" that transmits the epinephrine signal (epinephrine being the "first messenger"). They investigated this by separating plasma membrane fragments from the cytoplasmic fractions of broken liver cells and following the sequence of steps described in **Figure 7.11**. The experiments confirmed the existence of a second messenger, later identified as **cyclic AMP** (**cAMP**), which is produced from ATP by the enzyme adenylyl cyclase (**Figure 7.12**). Adenylyl cyclase is activated via a G protein-linked epinephrine receptor (see the first steps in Figure 7.18).

In contrast to the specificity of receptor binding, second messengers such as cAMP allow a cell to respond to a single event at the plasma membrane with many events inside the cell. Thus second messengers serve to rapidly amplify and distribute the signal—for example, binding of a single epinephrine molecule leads to the production of many molecules of cAMP, which

WORKING WITHDATA:

The Discovery of a Second Messenger

Original Paper

Rall, T. W., E. W. Sutherland, and J. Berthet. 1957. The relationship of epinephrine and glucagon to liver phosphorylase. *Journal of Biological Chemistry* 224: 463.

Analyze the Data

While studying the action of glycogen phosphorylase, Earl Sutherland and his colleagues determined that this enzyme could be activated by epinephrine only when the entire contents, including membrane fragments, of liver cells were present (see Figure 7.11). The researchers hypothesized that a cytoplasmic messenger must transmit the message from the epinephrine receptor at the membrane to glycogen phosphorylase, located in the cytoplasm. To test this idea, liver tissue was homogenized and separated into cytoplasmic and membrane components, containing the enzyme and epinephrine receptors, respectively. Epinephrine was added to the membrane fraction and incubated for a period of time. This fraction was then subjected to centrifugation to remove the membranes, leaving only the soluble portion in the supernatant. A small sample of the membrane-free solution was added to the cytoplasmic fraction, which was then assayed for the presence of glycogen phosphorylase activity. The assay showed that active glycogen phosphorylase was indeed present in the cytoplasmic fraction. These results confirmed the hypothesis that a soluble second messenger was produced in response to epinephrine binding to its receptor in the membrane, and then diffused into the cytoplasm to activate the enzyme. Later research by Sutherland identified cAMP as the second messenger involved in the mechanism of action of epinephrine, as well as many other hormones. Sutherland's research was highly regarded in the scientific community, and in 1971 he won the Nobel Prize in Physiology or Medicine for his discoveries concerning "the mechanisms of the action of hormones."

QUESTION 1

As part of Sutherland's research, the activity of glycogen phosphorylase was measured in various liver cell fractions, with or without incubation with epinephrine. The table shows the results. Explain how these data support the hypothesis that there is a soluble second messenger that activates the enzyme.

Condition	Enzyme activity (units)
Homogenate	0.4
Homogenate + epinephrine	2.5
Cytoplasm	0.2
Cytoplasm + epinephrine	0.4
Membranes + epinephrine	0.4
Cytoplasm + membranes + epinephrine	2.0

QUESTION 2

Propose an experiment to test whether the factor (second messenger) that activates the enzyme is stable on heating (and therefore probably not a protein), and give predicted data.

QUESTION 3

The second messenger, cAMP, was purified from the hormone-treated membrane fraction. Propose experiments to show that cAMP could replace the membrane fraction and hormone treatment in the activation of glycogen phosphorylase, and create a table to show possible results.

*Go to **BioPortal** for all WORKING WITHDATA exercises*

---

then activate many enzyme targets by binding to them noncovalently. In the case of epinephrine and the liver cell, glycogen phosphorylase is just one of several enzymes that are activated.

In addition, second messengers are often involved in crosstalk between different signaling pathways. Activation of the epinephrine receptor is not the only way for a cell to produce cAMP; and as noted, there are multiple targets of cAMP in the cell, and these targets are parts of other pathways.

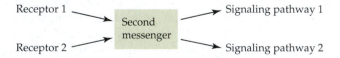

Several other classes of second messengers have since been identified, including lipid-derived second messengers, calcium ions, and nitric oxide.

LIPID-DERIVED SECOND MESSENGERS In addition to their role as structural components of the plasma membrane, phospholipids are also involved in signal transduction. When certain phospholipids are hydrolyzed into their component parts by enzymes called phospholipases, second messengers are formed.

The best-studied examples of lipid-derived second messengers come from the hydrolysis of the phospholipid **phosphatidyl inositol-bisphosphate** (**PIP₂**). Like all phospholipids, PIP₂ has a hydrophobic portion embedded in the plasma membrane: two fatty acid tails attached to a molecule of glycerol, which together form **diacylglycerol**, or **DAG**. The hydrophilic portion of PIP₂ is **inositol trisphosphate**, or **IP₃**, which projects into the cytoplasm.

As with cAMP, the receptors involved in this second-messenger system are often G protein-linked receptors. A G protein subunit is activated by the receptor, then diffuses within the plasma membrane and activates phospholipase C, an enzyme that is also located in the membrane. This enzyme cleaves off the IP_3 from PIP_2, leaving the diacylglycerol (DAG) in the phospholipid bilayer:

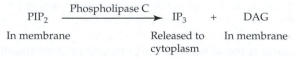

IP_3 and DAG are both second messengers; they have different modes of action that build on each other to activate protein kinase C (PKC) (**Figure 7.13**). PKC refers to a family of protein

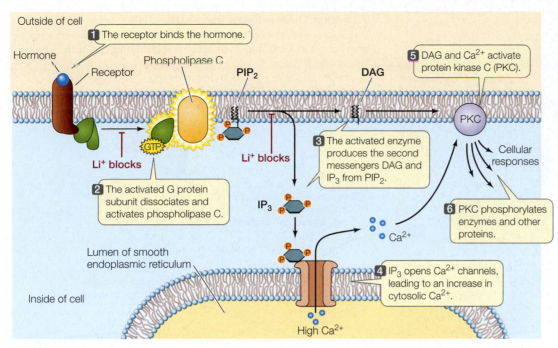

7.13 The IP$_3$/DAG Second-Messenger System Phospholipase C hydrolyzes the phospholipid PIP$_2$ into its components, IP$_3$ and DAG, both of which are second messengers. Lithium ions (Li$^+$) block this pathway and are used to treat bipolar disorder (red type).

kinases that can phosphorylate a wide variety of target proteins, leading to a multiplicity of cellular responses that vary depending on the tissue or cell type.

The IP$_3$/DAG pathway is apparently a target for the ion lithium (Li$^+$), which has been used for many years as a psychoactive drug to treat bipolar (manic-depressive) disorder. This serious illness occurs in about 1 in every 100 people. In these patients, an overactive IP$_3$/DAG signal transduction pathway in the brain leads to excessive brain activity in certain regions. Lithium "tones down" this pathway in two ways, as indicated by the red notations in Figure 7.13: it inhibits G protein activation of phospholipase C, as well as the synthesis of IP$_3$. The overall result is that brain activity returns to normal.

CALCIUM IONS Calcium ions (Ca^{2+}) are scarce inside most cells, which have cytosolic Ca^{2+} concentrations of only about 0.1 μM. Ca^{2+} concentrations outside cells and within the endoplasmic reticulum are usually much higher. Active transport proteins in the plasma and ER membranes maintain this concentration difference by pumping Ca^{2+} out of the cytosol. In contrast to cAMP and the lipid-derived second messengers, Ca^{2+} cannot be synthesized to increase the intracellular Ca^{2+} concentration. Instead, Ca^{2+} ion levels are regulated via the opening and closing of ion channels, and the action of membrane pumps.

There are many signals that can cause calcium channels to open, including IP$_3$ (see Figure 7.13). The entry of a sperm into an egg is a very important signal that causes a massive opening of calcium channels, resulting in numerous and dramatic changes that prepare the now fertilized egg for cell division and development (**Figure 7.14**). Whatever the initial signal that causes calcium channels to open, their opening results in a dramatic increase in cytosolic Ca^{2+} concentration, which can increase up to 100-fold within a fraction of a second. As we saw earlier, this increase activates protein kinase C. In addition,

Ca^{2+} controls other ion channels and stimulates secretion by exocytosis in many cell types.

NITRIC OXIDE Most signaling molecules and second messengers are solutes that remain dissolved in either the aqueous or hydrophobic components of cells. It was a great surprise to find that a gas could also be active in signal transduction. Nitric oxide (NO) is a second messenger in the signal transduction pathway between the neurotransmitter acetylcholine

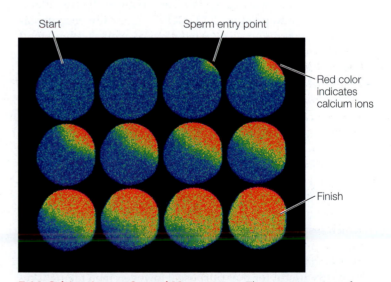

7.14 Calcium Ions as Second Messengers The concentration of Ca^{2+} can be measured using a dye that fluoresces when it binds the ion. Here, fertilization in a starfish egg causes a rush of Ca^{2+} from the environment into the cytoplasm. Areas of high Ca^{2+} concentration are indicated by the red color, and the events are photographed at 5-second intervals. Calcium signaling occurs in virtually all animal groups and triggers cell division in fertilized eggs, initiating the development of new individuals.

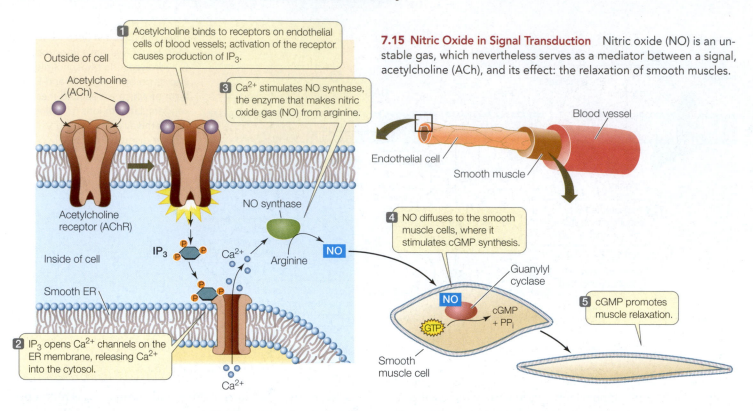

1 Acetylcholine binds to receptors on endothelial cells of blood vessels; activation of the receptor causes production of IP$_3$.

Outside of cell

Acetylcholine (ACh)

3 Ca^{2+} stimulates NO synthase, the enzyme that makes nitric oxide gas (NO) from arginine.

Acetylcholine receptor (AChR)

NO synthase

Inside of cell

IP$_3$ Ca^{2+} Arginine   NO

Smooth ER

2 IP$_3$ opens Ca^{2+} channels on the ER membrane, releasing Ca^{2+} into the cytosol.

Ca^{2+}

7.15 Nitric Oxide in Signal Transduction Nitric oxide (NO) is an unstable gas, which nevertheless serves as a mediator between a signal, acetylcholine (ACh), and its effect: the relaxation of smooth muscles.

Blood vessel

Endothelial cell   Smooth muscle

4 NO diffuses to the smooth muscle cells, where it stimulates cGMP synthesis.

Guanylyl cyclase

NO   cGMP + PP$_i$

GTP

5 cGMP promotes muscle relaxation.

Smooth muscle cell

(see Section 7.2) and the relaxation of smooth muscles lining blood vessels, which allows more blood flow (**Figure 7.15**). In the body, NO is made from the amino acid arginine by the enzyme NO synthase. When the acetylcholine receptor on the surface of an endothelial cell is activated, IP$_3$ is released from the membrane (via the pathway shown in Figure 7.13), causing a calcium channel in the ER membrane to open and a subsequent increase in cytosolic Ca^{2+}. The Ca^{2+} then activates NO synthase to produce NO. NO is chemically very unstable, readily reacting with oxygen gas as well as other small molecules. Although NO diffuses readily, it does not get far. Conveniently, the endothelial cells are close to the underlying smooth muscle cells, where NO activates an enzyme called guanylyl cyclase (a close relative of adenylyl cyclase). This enzyme catalyzes the formation of cyclic GMP (cGMP): yet another second messenger that contributes to the relaxation of muscle cells.

The discovery of NO as a participant in signal transduction explained the action of nitroglycerin, a drug that has been used for more than a century to treat angina, the chest pain caused by insufficient blood flow to the heart. Nitroglycerin releases NO, which results in relaxation of the blood vessels and increased blood flow. The drug sildenafil (Viagra) was developed to treat angina via the NO signal transduction pathway but was only modestly useful for that purpose. However, men taking it reported more pronounced penile erections. During sexual stimulation, NO acts as a signal, causing an increase in cGMP and a subsequent relaxation of the smooth muscles surrounding the arteries in the corpus cavernosum of the penis. As a result of this signal, the penis fills with blood, producing an erection. Sildenafil acts by inhibiting an enzyme (a phosphodiesterase) that breaks down cGMP—resulting in more cGMP and better erections.

Signal transduction is highly regulated

There are several ways in which cells can regulate the activity of a transducer molecule. The concentration of NO, which breaks down quickly, can be regulated only by how much of it is made. By contrast, membrane pumps and ion channels regulate the cytosolic concentration of Ca^{2+}, as we have seen. To regulate protein kinase cascades, G proteins, and cAMP, there are enzymes that inactivate the activated transducer (**Figure 7.16**).

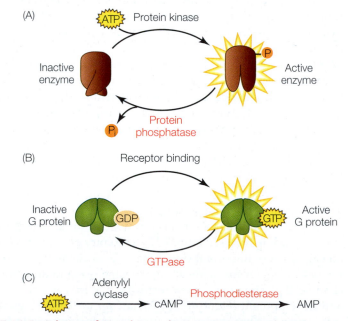

(A)   ATP   Protein kinase

Inactive enzyme   P   Active enzyme

P$_i$ Protein phosphatase

(B)   Receptor binding

Inactive G protein   GDP   GTP   Active G protein

GTPase

(C)   Adenylyl cyclase   Phosphodiesterase

ATP → cAMP → AMP

7.16 Regulation of Signal Transduction Some signals lead to the production of active transducers such as (A) protein kinases, (B) G proteins, and (C) cAMP. Other enzymes (shown in red) inactivate or remove these transducers.

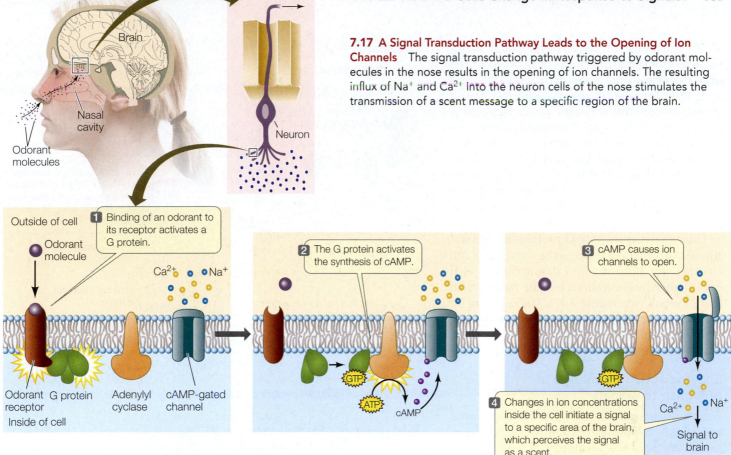

7.17 A Signal Transduction Pathway Leads to the Opening of Ion Channels The signal transduction pathway triggered by odorant molecules in the nose results in the opening of ion channels. The resulting influx of Na⁺ and Ca²⁺ into the neuron cells of the nose stimulates the transmission of a scent message to a specific region of the brain.

The balance between the activities of enzymes that activate and inactivate transducers determines the ultimate cellular response to a signal. Cells can alter this balance in several ways:

- *Synthesis or breakdown of the enzymes.* For example, synthesis of adenylyl cyclase (which synthesizes cAMP) and breakdown of phosphodiesterase (which breaks down cAMP) would tilt the balance in favor of more cAMP in the cell.

- *Activation or inhibition of the enzymes by other molecules.* An example is the inhibition of phosphodiesterase by sildenafil.

Because cell signaling is so important in diseases such as cancer, a search is under way for new drugs that can modulate the activities of enzymes that participate in signal transduction pathways.

RECAP 7.3

Signal transduction is the series of steps between the binding of a signal to a receptor and the ultimate cellular response. A protein kinase cascade amplifies a signal through a series of protein phosphorylation reactions. In many cases, a second messenger serves to amplify and distribute the downstream effects of the signal.

- How does a protein kinase cascade amplify a signal's message inside the cell? **See p. 132 and Figure 7.10**
- What is the role of cAMP as a second messenger? **See p. 133 and Figure 7.11**
- How are signal transduction cascades regulated? **See pp. 136–137 and Figure 7.16**

We have seen how the binding of a signal to its receptor initiates the response of a cell to the signal, and how signal transduction pathways amplify the signal and distribute its effects to numerous targets in the cell. In the next section we will consider the third step in the signal transduction process, the actual effects of the signal on cell function.

7.4 How Do Cells Change in Response to Signals?

The effects of a signal on cell function take three primary forms: the opening of ion channels, changes in the activities of enzymes, or differential gene expression. These events set the cell on a path for further and sometimes dramatic changes in form and function.

Ion channels open in response to signals

We have seen that ion channels can function as receptors in cell signaling, and as components of more complex signal transduction pathways: for example, the calcium ion channel in the pathway shown in Figure 7.13. In some cases, the opening of an ion channel is itself the cellular response to a signal. For example, the opening of ion channels is a key response in the nervous system. In sense organs, specialized cells have receptors that respond to external stimuli such as light, sound, taste, odor, or pressure. The alteration of the receptor results in the opening of ion channels. We will focus here on one such signal transduction pathway, that for the sense of smell, which responds to gaseous molecules in the environment (**Figure 7.17**).

The sense of smell is well developed in mammals. Each of the thousands of neurons in the nose expresses one of many different odorant receptors. The identification of which chemical signal, or odorant, activates which receptor is just getting under way. Humans have the genetic capacity to make about 950 different odorant receptor proteins, but very few people express more than 400 of them. Some express far fewer, which may explain why you are able to smell certain things that your roommate cannot, or vice versa.

Odorant receptors are G protein-linked, and signal transduction leads to the opening of ion channels for sodium and calcium ions, which have higher concentrations outside the cell than in the cytosol (see Figure 7.17). The resulting influx of Na^+ and Ca^{2+} causes the neuron to become stimulated so that it sends a signal to the brain that a particular odor is present.

Enzyme activities change in response to signals

Enzymes are often modified during signal transduction—either covalently or noncovalently. We have seen examples of both types of protein modification in earlier sections of this chapter. For example, addition of a phosphate group to an enzyme by a protein kinase is a covalent change; cAMP binding is noncovalent. Both types of modification change the enzyme's shape, activating or inhibiting its function. In the case of activation, the shape change exposes a previously inaccessible active site, and the target enzyme goes on to perform a new cellular role.

The G protein-mediated protein kinase cascade that is stimulated by epinephrine in liver cells results in the activation by cAMP of a key signaling molecule, protein kinase A. In turn, protein kinase A phosphorylates two other enzymes, with opposite effects:

- *Inhibition*: Glycogen synthase, which catalyzes the joining of glucose molecules to synthesize the energy-storing molecule glycogen, is inactivated when a phosphate group is added to it by protein kinase A. Thus the epinephrine signal *prevents glucose from being stored* in the form of glycogen (**Figure 7.18, step 1**).

- *Activation*: Phosphorylase kinase is activated when a phosphate group is added to it. It is part of a protein kinase cascade that ultimately leads to the activation of glycogen phosphorylase, another key enzyme in glucose metabolism. This enzyme results in the *liberation of glucose molecules* from glycogen (**Figure 7.18, steps 2 and 3**).

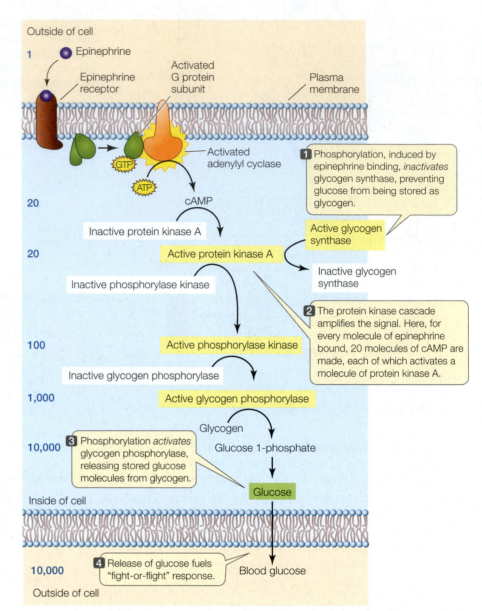

7.18 A Cascade of Reactions Leads to Altered Enzyme Activity   Liver cells respond to epinephrine by activating G proteins, which in turn activate the synthesis of the second messenger cAMP. Cyclic AMP initiates a protein kinase cascade, greatly amplifying the epinephrine signal, as indicated by the blue numbers. The cascade both inhibits the conversion of glucose to glycogen and stimulates the release of previously stored glucose.

The amplification of the signal in this pathway is impressive; as detailed in Figure 7.18, each molecule of epinephrine that arrives at the plasma membrane ultimately results in the release of 10,000 molecules of glucose into the bloodstream:

1	molecule of epinephrine bound to the membrane leads to
20	molecules of cAMP, which activate
20	molecules of protein kinase A, which activate
100	molecules of phosphorylase kinase, which activate
1,000	molecules of glycogen phosphorylase, which produce
10,000	molecules of glucose 1-phosphate, which produce
10,000	molecules of blood glucose

Signals can initiate DNA transcription

As we introduced in Section 4.1, the genetic material, DNA, is expressed by transcription as RNA, which is then translated into a protein whose amino acid sequence is specified by the original DNA sequence. Proteins are important in all cellular functions, so a key way to regulate specific functions in a cell is to regulate which proteins are made, and therefore which DNA sequences are transcribed.

Signal transduction plays an important role in determining which DNA sequences are transcribed. Common targets of signal transduction are proteins called transcription factors, which bind to specific DNA sequences in the cell nucleus and activate or inactivate transcription of the adjacent DNA regions. For example, the Ras signaling pathway (see Figure 7.10) ends in the nucleus. The final protein kinase in the Ras signaling cascade, MAPK (mitogen-activated protein kinase; a mitogen is a type of signal that stimulates cell division), enters the nucleus and phosphorylates a protein that stimulates the expression of several genes involved in cell proliferation.

In this chapter we have concentrated on signaling pathways that occur in animal cells. However, signal transduction pathways play equally important roles in other organisms, including plants, as you will see in Part Eight of this book.

▆ RECAP (7.4)

Cells respond to signal transduction by opening membrane channels, activating or inactivating enzymes, and stimulating or inhibiting gene transcription.

- What role does cAMP play in the sense of smell? **See pp. 137–138 and Figure 7.17**
- How does amplification of a signal occur, and why is it important in a cell's response to changes in its environment? **See p. 138 and Figure 7.18**

We have described how signals from a cell's environment can influence the cell. But the environment of a cell in a multicellular organism is more than the extracellular medium—it includes neighboring cells as well. In the next section we'll see how specialized junctions between cells allow them to pass signals from one to another.

(7.5) How Do Cells in a Multicellular Organism Communicate Directly?

The hallmark of multicellular organisms is their ability to have specialized functions in subsets of cells within their bodies. How do these cells communicate with one another so that they can work together for the good of the entire organism? As we learned in Section 7.1, some intercellular signals travel through the circulatory system to reach their target cells. But cells also have more direct ways of communicating. Cells that are packed together within a tissue can communicate directly with their neighbors via specialized intercellular junctions: gap junctions in animals (see Figure 6.7) and plasmodesmata in plants (see Figure 5.7).

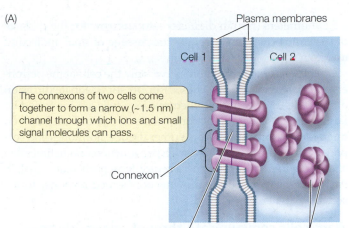

(A)

Plasma membranes

Cell 1　　Cell 2

The connexons of two cells come together to form a narrow (~1.5 nm) channel through which ions and small signal molecules can pass.

Connexon

Space between cells ("gap"; ~2 nm)　Connexins

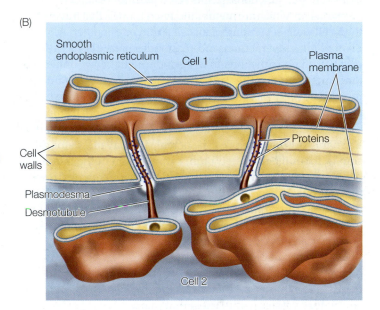

(B)

Smooth endoplasmic reticulum　Cell 1　Plasma membrane

Proteins

Cell walls

Plasmodesma

Desmotubule

Cell 2

7.19 Communicating Junctions (A) An animal cell may contain hundreds of gap junctions connecting it to neighboring cells. The pores of gap junctions allow small molecules to pass from cell to cell, ensuring similar concentrations of important signaling molecules in adjacent cells so that the cells can coordinate their activities. (B) Plasmodesmata connect plant cells. The desmotubule, derived from the smooth endoplasmic reticulum, fills up most of the space inside a plasmodesma, leaving a tiny gap through which small metabolites and ions can pass.

Animal cells communicate through gap junctions

Gap junctions are channels between adjacent cells that occur in many animals, occupying up to 25 percent of the area of the plasma membrane (**Figure 7.19A**). Gap junctions traverse the narrow space between the plasma membranes of two adjacent cells (the "gap") by means of channel structures called **connexons**. The walls of a connexon are composed of six subunits of the integral membrane protein connexin. In adjacent cells, two connexons come together to form a gap junction that links the cytoplasms of the two cells. There may be hundreds of these channels between a cell and its neighbors. The channel pores are about

1.5 nanometers (nm) in diameter—too narrow for the passage of large molecules but adequate for passage of small molecules and ions.

In the lens of the mammalian eye, only the cells at the periphery are close enough to the blood supply for adequate diffusion of nutrients and wastes. But because lens cells are connected by large numbers of gap junctions, material can diffuse between them rapidly and efficiently. In other tissues, hormones and second messengers can move through gap junctions. Sometimes just a few cells in a tissue have the receptor for a particular signal; in such cases, gap junctions allow a coordinated response to the signal by all the cells in the tissue.

Plant cells communicate through plasmodesmata

Instead of gap junctions, plants have **plasmodesmata** (singular *plasmodesma*), which are membrane-lined tunnels that traverse the thick cell walls separating plant cells from one another. A typical plant cell has several thousand plasmodesmata. Plasmodesmata differ from gap junctions in one fundamental way: unlike gap junctions, in which the wall of the channel is made of integral membrane proteins from the adjacent plasma membranes, plasmodesmata are lined by the fused plasma membranes themselves.

The diameter of a plasmodesma is about 6 nm, far larger than a gap junction channel. But the actual space available for diffusion is about the same—1.5 nm. Examination of the interior of the plasmodesma by transmission electron microscopy reveals that a tubule called the **desmotubule**, apparently derived from the endoplasmic reticulum, fills up most of the opening of the plasmodesma (**Figure 7.19B**). Typically, only small metabolites and ions can move between plant cells.

Plasmadesmata play an important role in plant physiology because the bulk transport system in plants, the vascular system, lacks the tiny circulatory vessels (capillaries) that many animals have for bringing gases and nutrients to every cell. For example, diffusion from cell to cell across plasma membranes is probably inadequate to account for the movement of a plant hormone from the site of production to the site of action. Instead, plants rely on more rapid diffusion through plasmodesmata to ensure that all cells of a tissue respond to a signal at the same time. There are cases in which larger molecules or particles can pass between cells via plasmodesmata. For example, some viruses can move through plasmodesmata by using "movement proteins" to assist their passage.

Modern organisms provide clues about the evolution of cell–cell interactions and multicellularity

Even though single-celled organisms continue to be highly successful on Earth, over time complex multicellular organisms evolved, along with their division of biological labor among specialized cells. The transition from single-celled to multicellular life took a long time. Indeed, while there is evidence that single-celled organisms arose about 500 million to a billion years after the formation of Earth (see Chapter 4), the first evidence of true multicellular organisms dates from more than a billion years later.

Studying the evolutionary origin of multicellularity is a challenge because it happened so long ago. The closest unicellular relatives of most modern animals and plants probably existed hundreds of millions of years ago. The evolution from single-celled to multicellular organisms may have occurred in several steps:

- Aggregation of cells into a cluster
- Intercellular communication within the cluster
- Specialization of some cells within the cluster
- Organization of specialized cells into groups (tissues)

A key event would have been the evolution of intercellular communication, which is necessary to coordinate the activities of different cells within a multicellular organism.

We can visualize how the evolution of multicellularity might have occurred by looking at the "Volvocine line" of aquatic green algae (Chlorophyta). These plants range from single cells to complex multicellular organisms with differentiated cell clusters (**Figure 7.20**). Included in this range are a single-celled organism (*Chlamydomonas*); an organism that occurs

Chlamydomonas is single-celled.

Gonium is a cluster of cells.

Pandorina is a larger cluster of 16 cells.

Eudorina is a still larger cluster of cells.

Pleodorina has some cells specialized for reproduction.

Volvox is larger, with internal specialized reproductive cells.

7.20 Multicellularity The evolution of intercellular interactions in a multicellular organism can be inferred from these green algae.

in small cell clusters (*Gonium*); species with larger cell clusters (*Pandorina* and *Eudorina*); a colony of somatic and reproductive cells (*Pleodorina*); and a larger, 1,000-celled alga with somatic and reproductive cells organized into separate tissues (*Volvox*).

Chlamydomonas is the single-celled member of this group. It has two cellular phases: a swimming phase, when the cells have flagella and move about, and a non-swimming phase, when the flagella are reabsorbed (disaggregated) and the cell undergoes cell division (reproduction). Compare this with *Volvox*: most of the cells of this multicellular, spherical organism are on the surface; the beating of their flagella gives the organism a rolling motion as it swims toward light, where it can perform photosynthesis. But some *Volvox* cells are larger and located inside the sphere. These cells are specialized for reproduction: they lose their flagella and then divide to form offspring.

The separation of somatic and reproductive functions in *Volvox* is possible because of a key intercellular signaling mechanism that coordinates the activities of the separate tissues within the organism. *Volvox* has a gene whose protein product is produced by the outer, motile cells and travels to the reproductive cells, causing them to lose their flagella and divide. This gene is not active in species such as *Gonium* and *Pandorina*, which show cell aggregation but no cell specialization.

Go to Media Clip 7.1
Social Amoebas Aggregate on Cue
Life10e.com/mc7.1

RECAP 7.5

Cells can communicate with their neighbors through specialized cell junctions. In animals, these structures are gap junctions; in plants, they are plasmodesmata. The evolution of intercellular communication and tissue formation can be inferred from existing organisms, such as certain related green algae.

- What are the roles that gap junctions and plasmodesmata play in cell signaling? **See pp. 139–140 and Figure 7.19**
- How does the Volvocine line of green algae show possible steps in the evolution of cell communication and tissue formation? **See pp. 140–141 and Figure 7.20**

Does oxytocin affect caring behavior in humans?

ANSWER

Oxytocin is released during sexual activity in humans, and the release results in bonding behaviors just as it does in voles. Other behaviors are affected by oxytocin as well. Human volunteers given a nasal spray of oxytocin show more trust when investing money with a stranger than do volunteers given an inert spray. This kind of experiment has opened up a new field called neuroeconomics.

CHAPTER**SUMMARY** 7

7.1 What Are Signals, and How Do Cells Respond to Them?

- Cells receive many signals from the physical environment and from other cells. Chemical signals are often at very low concentrations. **Autocrine** signals affect the cells that make them; **juxtacrine** signals affect adjacent cells; **paracrine** signals diffuse to and affect nearby cells; and **hormones** are carried through the circulatory systems of animals or the vascular systems of plants. **Review Figure 7.1, ACTIVITY 7.1**

- A **signal transduction pathway** involves the interaction of a signal molecule with a receptor; the transduction of the signal via a series of steps within the cell; and effects on the function of the cell. **Review Figure 7.2**

- Signal transduction pathways involve regulation of enzymes and transcription factors. A great deal of **crosstalk** occurs between pathways.

7.2 How Do Signal Receptors Initiate a Cellular Response?

- Cells respond to signals only if they have specific **receptor** proteins that can recognize those signals.

- Binding of a signal **ligand** to its receptor obeys the chemical law of mass action. A key measurement of the strength of binding is the **dissociation constant (K_D)**.

- Depending on the nature of its signal or ligand, a receptor may be located in the plasma membrane or inside the target cell. **Review Figure 7.4**

- Receptors located in the plasma membrane include **ion channels**, **protein kinases**, and **G protein-linked receptors**.

- Ion channel receptors are "gated." The gate "opens" when the three-dimensional structure of the channel protein is altered by ligand binding. **Review Figure 7.5**

- Protein kinase receptors catalyze the phosphorylation of themselves and/or other proteins. **Review Figure 7.6**

- A **G protein** has three important binding sites, which bind a G protein-linked receptor, GDP or GTP, and an **effector protein**. A G protein can either activate or inhibit an effector protein. **Review Figure 7.7, ANIMATED TUTORIAL 7.1**

- **Intracellular receptors** include certain photoreceptors in plants and steroid hormone receptors in animals. A lipid-soluble ligand such as a steroid hormone may enter the cytoplasm or the nucleus before binding. Many intracellular receptors are transcription factors. **Review Figure 7.8**

7.3 How Is the Response to a Signal Transduced through the Cell?

- A **protein kinase cascade** amplifies the response to receptor binding. **Review Figure 7.10, ANIMATED TUTORIAL 7.2**

- Second messengers include **cyclic AMP (cAMP)**, **inositol trisphosphate (IP_3)**, **diacylglycerol (DAG)**, and **calcium** ions. IP_3 and DAG are derived from the phospholipid **phosphatidyl inositol-bisphosphate (PIP_2)**.

- The gas nitric oxide (NO) is involved in signal transduction in human smooth muscle cells. **Review Figure 7.15**

continued

- Signal transduction can be regulated in several ways. The balance between activating and inactivating the molecules involved determines the ultimate cellular response to a signal. **Review Figure 7.16**

7.4 How Do Cells Change in Response to Signals?

- The cellular responses to signals may include the opening of ion channels, the alteration of enzyme activities, or changes in gene expression. **Review Figure 7.17**
- Activated enzymes may activate other enzymes in a signal transduction pathway, leading to impressive amplification of a signal. **Review Figure 7.18**
- Protein kinases covalently add phosphate groups to target proteins; cAMP binds target proteins noncovalently. Both kinds of binding change the target protein's conformation to expose or hide its active site.

7.5 How Do Cells in a Multicellular Organism Communicate Directly?

- Many adjacent animal cells can communicate with one another directly through small pores in their plasma membranes called **gap junctions**. Protein structures called **connexons** form thin channels between two adjacent cells through which small signal molecules and ions can pass. **Review Figure 7.19A**
- Plant cells are connected by somewhat larger pores called **plasmodesmata**, which traverse both plasma membranes and cell walls. The **desmotubule** narrows the opening of the plasmodesma. **Review Figure 7.19B**
- The evolution of cell communication and tissue formation can be inferred from existing organisms, such as certain green algae. **Review Figure 7.20**

See ACTIVITY 7.2 for a concept review of this chapter.

Go to the Interactive Summary to review key figures, Animated Tutorials, and Activities
Life10e.com/is7

CHAPTER**REVIEW**

▬▬ REMEMBERING

1. What is the correct order for the following events in the interaction of a cell with a signal? (1) Alteration of cell function; (2) signal binds to receptor; (3) signal released from source; (4) signal transduction.
   - a. 1234
   - b. 2314
   - c. 3214
   - d. 3241
   - e. 3421

2. Steroid hormones such as estrogen act on target cells by
   - a. initiating second-messenger activity.
   - b. binding to membrane proteins.
   - c. initiating gene expression.
   - d. activating enzymes.
   - e. binding to membrane lipids.

3. Which of the following is *not* a consequence of a signal binding to a receptor?
   - a. Activation of receptor enzyme activity
   - b. Diffusion of the receptor in the plasma membrane
   - c. Change in conformation of the receptor protein
   - d. Breakdown of the receptor to amino acids
   - e. Release of the signal from the receptor

4. A nonpolar ligand such as a steroid hormone usually binds to a/an
   - a. intracellular receptor.
   - b. protein kinase.
   - c. ion channel.
   - d. phospholipid.
   - e. second messenger.

5. Which of the following is *not* true of a protein kinase cascade?
   - a. The signal is amplified.
   - b. A second messenger is formed.
   - c. Target proteins are phosphorylated.
   - d. The cascade ends up at the mitochondrion.
   - e. The cascade begins at the plasma membrane.

6. Plasmodesmata and gap junctions
   - a. allow small molecules and ions to pass rapidly between cells.
   - b. are both membrane-lined channels.
   - c. are channels about 1 millimeter in diameter.
   - d. are present only once per cell.
   - e. are involved in cell recognition.

▬▬ UNDERSTANDING & APPLYING

7. Why do some signals ("first messengers") trigger "second messengers" to activate target cells?
   - a. The first messenger requires activation by ATP.
   - b. The first messenger is not water-soluble.
   - c. The first messenger binds to many types of cells.
   - d. The first messenger cannot cross the plasma membrane.
   - e. There are no receptors for the first messenger.

8. The major difference between a cell that responds to a signal and one that does not is the presence of a
   - a. DNA sequence that binds to the signal.
   - b. nearby blood vessel.
   - c. receptor.
   - d. second messenger.
   - e. transduction pathway.

9. Cyclic AMP is a second messenger in many different responses. How can the same messenger act in different ways in different cells?

10. Compare direct communication via plasmodesmata or gap junctions with receptor-mediated communication between cells. What are the advantages of one method over the other?

ANALYZING & EVALUATING

11. Like the Ras protein itself, the various components of the Ras signaling pathway are changed in cancer cells. What might be the biochemical consequences of mutations in the genes coding for (*a*) Raf and (*b*) MAP kinase that result in rapid cell division?

12. The tiny invertebrate *Hydra* has an apical region with tentacles and a long, slender body. *Hydra* can reproduce asexually when cells on the body wall differentiate and form a bud, which then breaks off as a new organism. Buds form only at certain distances from the apex, leading to the idea that the apex releases a signal molecule that diffuses down the body and, at high concentrations (i.e., near the apex), inhibits bud formation. *Hydra* lacks a circulatory system, so this inhibitor must diffuse from cell to cell. If you had an antibody that binds to connexons and plugs up the gap junctions, how would you test the hypothesis that *Hydra*'s inhibitory factor passes through these junctions?

Apex

Bud

Go to BioPortal at **yourBioPortal.com** for Animated Tutorials, Activities, LearningCurve Quizzes, Flashcards, and many other study and review resources.

8 Energy, Enzymes, and Metabolism

CHAPTEROUTLINE

8.1 What Physical Principles Underlie Biological Energy Transformations?

8.2 What Is the Role of ATP in Biochemical Energetics?

8.3 What Are Enzymes?

8.4 How Do Enzymes Work?

8.5 How Are Enzyme Activities Regulated?

Cleaning Aids   Enzymes are an important component of detergents. By hydrolyzing the macromolecules that make up stains, enzymes help the stains dissolve in wash water.

"CLEAN THE DISHES!" probably gives you the same reaction as "Clean your clothes!" For most people, washing things has never been a favorite activity. In chemical terms, the problem is that nonpolar substances, such as food and dirt, stick to nonpolar surfaces on dishes and clothes. Water alone doesn't do a very good job of unsticking them. Perhaps the best-known cleaning aid is soap, which has polar and nonpolar regions that help separate nonpolar substances from one another. The polar regions allow the soap to dissolve in water, and the nonpolar regions solubilize the targeted stuff on your dishes and clothes. Even though synthetic detergents have now improved on traditional soaps, the TV ads will tell you that's still not enough. The "really tough stains" need something more.

A century ago, German chemist Otto Rohm came up with the idea of using enzymes to make the stains on clothes and dishes more water-soluble. Stains from meat or blood contain insoluble polymers such as protein. Rohm's idea was to mimic the mammalian digestive system, where insoluble polymers are hydrolyzed to soluble monomers by enzymes—proteins that speed up chemical reactions. Rohm isolated the digestive enzyme trypsin, which speeds up the hydrolysis of proteins, and added it to a detergent. The result was a dramatic improvement in the removal of stains. Nevertheless, it was many decades before this approach was used widely, because there were several challenges that had to be surmounted. First, the supply of enzymes from Rohm's original source (the pancreas) was limited. To solve this, biologists used bacteria, yeast, and fungi to make enzymes in huge amounts. Second, enzymes are proteins, and their three-dimensional structure is vital to their activity. The detergents and ions used for washing tend to destroy enzyme structure. To solve this, scientists screened many organisms for enzymes that would work well under these conditions and in a wide range of temperatures. They also performed genetic manipulations to improve on nature. Finally, not all stains are proteins. The spaghetti sauce you might eat stains your shirt with fats and starch. To solve this, scientists added lipases that hydrolyze lipids, and amylases that hydrolyze starch.

The result of all this chemistry is a range of modern cleaning products that attack dirt and grime on many fronts. The enzymes in these products perform specific chemical transformations and are active in washing conditions.

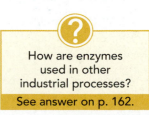

How are enzymes used in other industrial processes?

See answer on p. 162.

8.1 What Physical Principles Underlie Biological Energy Transformations?

A **chemical reaction** occurs when atoms have sufficient energy to combine or change their bonding partners. Consider the hydrolysis of the disaccharide sucrose to its component monomers, glucose and fructose (see p. 53 for the chemical structures of these sugars). We can express this reaction by a chemical equation:

$$\underset{(C_{12}H_{22}O_{11})}{\text{Sucrose}} + H_2O \rightarrow \underset{(C_6H_{12}O_6)}{\text{glucose}} + \underset{(C_6H_{12}O_6)}{\text{fructose}}$$

In this equation, sucrose and water are the **reactants**, and glucose and fructose are the **products**. During the reaction, some of the bonds in sucrose and water are broken and new bonds are formed, resulting in products with chemical properties that are very different from those of the reactants. The sum total of all the chemical reactions occurring in a biological system at a given time is called **metabolism**. Metabolic reactions involve energy changes; for example, the energy contained in the chemical bonds of sucrose (reactants) is greater than the energy in the bonds of the two products, glucose and fructose.

What is energy? Physicists define it as the capacity to do work, which occurs when a force operates on an object over a distance. In biochemistry, it is more useful to consider energy as *the capacity for change*. In biochemical reactions, energy changes are usually associated with changes in the chemical compositions and properties of molecules.

There are two basic types of energy

Energy comes in many forms: chemical, electrical, heat, light, and mechanical (**Table 8.1**). But all forms of energy can be considered as one of two basic types (**Figure 8.1**):

- **Potential energy** is the energy of state or position—that is, stored energy. It can be stored in many forms: in chemical bonds, as a concentration gradient, or even as an electric charge imbalance.

Potential chemical energy is converted into kinetic mechanical energy when the cat leaps.

Potential chemical energy is stored in the muscles of the cat.

8.1 Energy Conversions and Work A leaping cat illustrates both the conversion between potential and kinetic energy and the conversion of energy from one form (chemical) to another (mechanical).

- **Kinetic energy** is the energy of movement—that is, the type of energy that does work, that makes things change. For example, heat causes molecular motions and can even break chemical bonds.

Potential energy can be converted into kinetic energy and vice versa, and the form that the energy takes can also be converted. Think of reading this book: light energy is converted to chemical energy in your eyes, and then is converted to electrical energy in the nerve cells that carry messages to your brain. When you decide to turn a page, the electrical and chemical energy of nerves and muscles are converted to kinetic energy for movement of your hand and arm.

There are two basic types of metabolism

Energy changes in living systems usually occur as chemical changes, in which energy is stored in, or released from, chemical bonds.

Anabolic reactions (collectively anabolism) link simple molecules to form more complex molecules (for example, the synthesis of sucrose from glucose and fructose). Anabolic reactions require an input of energy. Energy is captured in the chemical bonds that are formed (for example, the glycosidic bond between the two monosaccharides). This captured energy is stored in the chemical bonds as potential energy.

TABLE 8.1
Energy in Biology

Form of Energy	Example in Biology
Chemical: Stored in bonds	Chemical energy is released during the hydrolysis of polymers
Electrical: Separation of charges	Electrical gradients across cell membranes help drive the movement of ions through channels
Heat: Transfer due to temperature difference	Heat can be released by chemical reactions
Light: Electromagnetic radiation stored as photons	Light energy is captured by pigments in the eye
Mechanical: Energy of motion	Mechanical energy is used in muscle movements

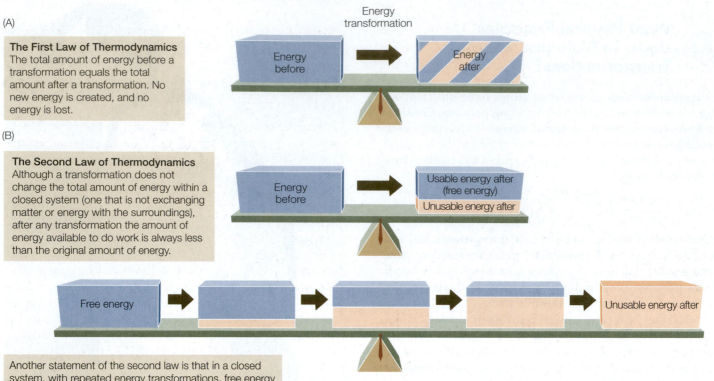

8.2 The Laws of Thermodynamics (A) The first law states that energy cannot be created or destroyed. (B) The second law states that after energy transformations, some energy becomes unavailable to do work.

Catabolic reactions (collectively catabolism) break down complex molecules into simpler ones and release the energy stored in the chemical bonds. For example, when sucrose is hydrolyzed, energy is released. In a biological system the released energy may be recaptured in new chemical bonds, or it may be used as kinetic energy—moving atoms, molecules, cells, or the whole organism.

Catabolic and anabolic reactions are often linked. The energy released in catabolic reactions is often used to drive anabolic reactions—that is, to do biological work. For example, the energy released by the breakdown of glucose (catabolism) is used to drive anabolic reactions such as the synthesis of triglycerides. This is why you accumulate fat if you eat food in excess of your energy requirements.

The **laws of thermodynamics** (*thermo*, "energy"; *dynamics*, "change") were derived from studies of the fundamental physical properties of energy, and the ways it interacts with matter. The laws apply to all matter and all energy transformations in the universe. Their application to living systems helps us understand how organisms and cells harvest and transform energy to sustain life.

The first law of thermodynamics: Energy is neither created nor destroyed

The first law of thermodynamics states that in any energy conversion, energy is neither created nor destroyed. In other words, during any conversion of energy, the total energy before and after the conversion is the same (**Figure 8.2A**). As you will see in the next two chapters, the potential energy present in the chemical bonds of carbohydrates and lipids can be converted to

potential energy in the form of adenosine triphosphate (ATP). This can then be converted into kinetic energy to do mechanical work (such as in muscle contractions) or biochemical work (such as protein synthesis).

The second law of thermodynamics: Disorder tends to increase

Although energy cannot be created or destroyed, the second law of thermodynamics states that when energy is converted from one form to another, some of that energy becomes unavailable for doing work (**Figure 8.2B**). In other words, no physical process or chemical reaction is 100 percent efficient; some of the released energy is lost to a form associated with disorder. Think of disorder as a kind of randomness that is due to the thermal motion of particles; this energy is of such a low value and so dispersed that it is unusable. **Entropy** is a measure of the disorder in a system.

It takes energy to impose order on a system. Unless energy is applied to a system, it will be randomly arranged or disordered. The second law applies to all energy transformations, but we will focus here on chemical reactions in living systems.

NOT ALL ENERGY CAN BE USED In any system, the total energy includes the usable energy that can do work and the unusable energy that is lost to disorder:

$$\text{Total energy} = \text{usable energy} + \text{unusable energy}$$

In biological systems, the total energy is called **enthalpy** (*H*). The usable energy that can do work is called **free energy** (*G*). Free energy is what cells require for all the chemical reactions involved in growth, cell division, and maintenance. The unusable energy is represented by entropy (S) multiplied by the absolute temperature (T). Thus we can rewrite the word equation above more precisely as:

$$H = G + TS \tag{8.1}$$

Because we are interested in usable energy, we rearrange Equation 8.1:

$$G = H - TS \tag{8.2}$$

Although we cannot measure G, H, or S absolutely, we can determine the change in each at a constant temperature. Such energy changes are measured in calories (cal) or joules (J).* A change in energy is represented by the Greek letter delta (Δ). The change in free energy (ΔG) of any chemical reaction is equal to the difference in free energy between the products and the reactants:

$$\Delta G_{reaction} = G_{products} - G_{reactants} \tag{8.3}$$

Such a change can be either positive or negative; that is, the free energy of the products can be more or less than the free energy of the reactants. If the products have more free energy than the reactants, then there must have been some input of energy into the reaction. (Remember that energy cannot be created, so some energy must have been added from an external source.)

At a constant temperature, ΔG is defined in terms of the change in total energy (ΔH) and the change in entropy (ΔS):

$$\Delta G = \Delta H - T\Delta S \tag{8.4}$$

Equation 8.4 tells us whether free energy is released or consumed by a chemical reaction:

- If ΔG is negative (ΔG < 0), free energy is released.

- If ΔG is positive (ΔG > 0), free energy is required (consumed).

If the necessary free energy is not available, the reaction does not occur. The sign and magnitude of ΔG depend on the two factors on the right side of the equation:

- ΔH: In a chemical reaction, ΔH is the total amount of energy added to the system (ΔH > 0) or released (ΔH < 0).

- ΔS: Depending on the sign and magnitude of ΔS, the entire term, TΔS, may be negative or positive, large or small. In other words, in living systems at a constant temperature (no change in T), the magnitude and sign of ΔG can depend a lot on changes in entropy.

If a chemical reaction increases entropy, its products are more disordered or random than its reactants. If there are more products than reactants, as in the hydrolysis of a protein to its amino acids, the products have considerable freedom to move around. The disorder in a solution of amino acids will be large compared

with that in the protein, in which peptide bonds and other forces prevent free movement. So in hydrolysis, the change in entropy (ΔS) will be positive. Conversely, if there are fewer products and they are more restrained in their movements than the reactants (as for amino acids being joined in a protein), ΔS will be negative.

DISORDER TENDS TO INCREASE The second law of thermodynamics also predicts that, as a result of energy transformations, disorder tends to increase; some energy is always lost to random thermal motion (entropy). Chemical changes, physical changes, and biological processes all tend to increase entropy (see Figure 8.2B), and this tendency gives direction to these processes. It explains why some reactions proceed in one direction rather than another.

How does the second law apply to organisms? Consider the human body, with its highly organized tissues and organs composed of large, complex molecules. This level of complexity appears to be in conflict with the second law but is not for two reasons. First, the construction of complexity is coupled to the generation of disorder. Constructing 1 kg of a human body requires the catabolism of about 10 kg of highly ordered biological materials (our food), which are converted into CO_2, H_2O, and other simple molecules that move independently and randomly. So metabolism creates far more disorder (more energy is lost to entropy) than the amount of order (total energy; enthalpy) stored in 1 kg of flesh. Second, life requires a constant input of energy to maintain order. Without this energy, the complex structures of living systems would break down. Because energy is used to generate and maintain order, there is no conflict with the second law of thermodynamics.

Having seen that the laws of thermodynamics apply to living things, we will now turn to a consideration of how these laws apply to biochemical reactions.

Chemical reactions release or consume energy

As we saw earlier, anabolic reactions link simple molecules to form more complex molecules, so they tend to increase complexity (order) in the cell. By contrast, catabolic reactions break down complex molecules into simpler ones, so they tend to decrease complexity (generate disorder).

- Catabolic reactions may break down an ordered reactant into smaller, more randomly distributed products. Reactions that release free energy (–ΔG) are called **exergonic** (or exothermic) reactions (**Figure 8.3A**). For example:

  Complex molecules → free energy + small molecules

- Anabolic reactions may make a single product (a highly ordered substance) out of many smaller reactants (less ordered). Reactions that require or consume free energy (+ΔG) are called **endergonic** (or endothermic) reactions (**Figure 8.3B**). For example:

  Free energy + small molecules → complex molecules

In principle, chemical reactions are reversible and can run both forward and backward. For example, if compound A can be converted into compound B (A → B), then B, in principle,

---

*A calorie is the amount of heat energy needed to raise the temperature of 1 gram of pure water from 14.5°C to 15.5°C. In the SI system, energy is measured in joules. 1 J = 0.239 cal; conversely, 1 cal = 4.184 J. Thus, for example, 486 cal = 2,033 J, or 2.033 kJ. Although they are defined here in terms of heat, the calorie and the joule are measures of mechanical, electrical, or chemical energy. When you compare data on energy, always compare joules with joules and calories with calories.

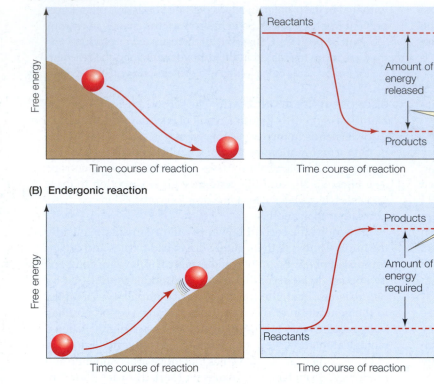

(A) Exergonic reaction

Free energy

Time course of reaction

Reactants

Amount of energy released

Products

Time course of reaction

In an exergonic reaction, *energy is released* as the reactants form lower-energy products. ΔG is negative.

(B) Endergonic reaction

Free energy

Time course of reaction

Products

Amount of energy required

Reactants

Time course of reaction

Energy must be added for an endergonic reaction, in which reactants are converted to products with a higher energy level. ΔG is positive.

8.3 Exergonic and Endergonic Reactions (A) In an exergonic reaction, the reactants behave like a ball rolling down a hill, and energy is released. (B) A ball will not roll uphill by itself. Driving an endergonic reaction, like moving a ball uphill, requires the addition of free energy.

can be converted into A (B → A), although *the concentrations of A and B determine which of these directions will be favored.* Think of the overall reaction as resulting from competition between the forward and reverse reactions (A ⇌ B). According to the law of mass action, increasing the concentration of A makes the forward reaction happen more often relative to the reverse reaction, just as B favors the reverse reaction.

There are concentrations of A and B at which the forward and reverse reactions take place at the same rate. At these concentrations, no further net change in the system is observable, although individual molecules are still forming and breaking apart. This balance between forward and reverse reactions is known as **chemical equilibrium**. Chemical equilibrium is a state of no net change, and a state in which ΔG = 0.

Chemical equilibrium and free energy are related

Every chemical reaction proceeds to a certain extent, but not necessarily to completion (all reactants converted into products). Each reaction has a specific equilibrium point, which is related to

the free energy released by the reaction under specified conditions. To understand the principle of equilibrium, consider the following example.

Most cells contain glucose 1-phosphate, which is converted into glucose 6-phosphate.

Glucose 1-phosphate ⇌ glucose 6-phosphate

Imagine that we start out with an aqueous solution of glucose 1-phosphate that has a concentration of 0.02 M. (M stands for molar concentration; see Section 2.4.) The solution is maintained under constant environmental conditions (25°C and pH 7). As the reaction proceeds to equilibrium, the concentration of the product, glucose 6-phosphate, rises from 0 to 0.019 M, while the concentration of the reactant, glucose 1-phosphate, falls to 0.001 M. At this point, equilibrium is reached (**Figure 8.4**). At equilibrium, the reverse reaction, from glucose 6-phosphate to glucose 1-phosphate, progresses at the same rate as the forward reaction.

At equilibrium, then, this reaction has a product-to-reactant ratio of 19:1 (0.019/0.001), so the forward reaction has gone 95 percent of the way to completion ("to the right," as written above). This result is obtained every time the experiment is run under the same conditions.

The change in free energy (ΔG) for any reaction is related directly to its point of equilibrium. The further toward completion the point of equilibrium lies, the more free energy is released. In an exergonic reaction, ΔG is a negative number. The value of ΔG also depends on the beginning concentrations of the reactants and products and other conditions such as temperature,

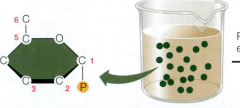
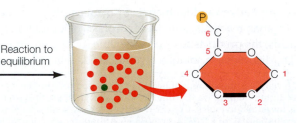

8.4 Chemical Reactions Run to Equilibrium No matter what quantities of glucose 1-phosphate and glucose 6-phosphate are dissolved in water, when equilibrium is attained, there will always be 95 percent glucose 6-phosphate and 5 percent glucose 1-phosphate.

Initial condition:
100% Glucose 1-phosphate
(0.02 M concentration)

Reaction to equilibrium

At equilibrium:
95% Glucose 6-phosphate (0.019 M concentration)
5% Glucose 1-phosphate (0.001 M concentration)

pressure, and pH of the solution. Biochemists often calculate ΔG using standard laboratory conditions: 25°C, one atmosphere pressure, one molar (1M) concentrations of the solutes, and pH 7. The standard free energy change calculated using these conditions is designated $\Delta G^{0'}$. In our example of the conversion of glucose 1-phosphate to glucose 6-phosphate, $\Delta G^{0'} = -1.7$ kcal/mol, or -7.1 kJ/mol.

A large, positive ΔG for a reaction means that it proceeds hardly at all to the right (A → B). If the concentration of B is initially high relative to that of A, such a reaction runs "to the left" (A ← B), and at equilibrium nearly all of B is converted into A. A ΔG value near zero is characteristic of a readily reversible reaction: reactants and products have almost the same free energies.

In Chapters 9 and 10 we will examine the metabolic reactions that harvest energy from food and light. In turn, this energy is used to synthesize carbohydrates, lipids, and proteins. All of the chemical reactions carried out by living organisms are governed by the principles of thermodynamics and equilibrium.

RECAP 8.1

Two laws of thermodynamics govern energy transformations in biological systems. A biochemical reaction can release or consume energy, and it may not run to completion, but instead end up at a point of equilibrium.

- What is the difference between potential energy and kinetic energy? Between anabolism and catabolism? **See pp. 145–146**

- What are the laws of thermodynamics? How do they relate to biology? **See pp. 146–147 and Figure 8.2**

- What is the difference between endergonic and exergonic reactions, and what is the importance of ΔG? **See p. 147 and Figure 8.3**

The principles of thermodynamics that we have been discussing apply to all energy transformations in the universe, so they are very powerful and useful. Next we'll apply them to reactions in cells that involve the currency of biological energy, ATP.

8.2 What Is the Role of ATP in Biochemical Energetics?

Cells rely on adenosine triphosphate (ATP) for the capture and transfer of the free energy they require to do chemical work. ATP operates as a kind of "energy currency." Just as it is more effective, efficient, and convenient for you to trade money for a lunch than to trade your actual labor, it is useful for cells to have a single currency for transferring energy between different reactions and cell processes. So some of the free energy that is released by exergonic reactions is captured in the formation of ATP from adenosine diphosphate (ADP) and inorganic phosphate (HPO_4^{2-}, which is commonly abbreviated to P_i). The ATP can then be hydrolyzed at other sites in the cell to release free energy to drive endergonic reactions. [In some reactions, guanosine triphosphate (GTP) is used as the energy transfer molecule instead of ATP, but we will focus on ATP here.]

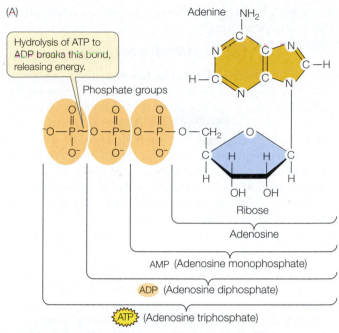

(A)

(B) *Photuris pennsylvanica*

8.5 ATP (A) ATP is richer in energy than its relatives ADP and AMP. (B) Fireflies use ATP to initiate the oxidation of luciferin. This process converts chemical energy into light energy, emitting rhythmic flashes that signal the insect's readiness to mate.

ATP has another important role in the cell beyond its use as an energy currency: it can be converted into a building block for nucleic acids (see Chapter 4). The structure of ATP is similar to those of other nucleoside triphosphates, but two things about ATP make it especially useful to cells.

- ATP releases a relatively large amount of energy when hydrolyzed to ADP and P_i.

- ATP can phosphorylate (donate a phosphate group to) many different molecules, which gain some of the energy that was stored in the ATP.

ATP hydrolysis releases energy

An ATP molecule consists of the nitrogenous base adenine bonded to ribose (a sugar), which is attached to a sequence of three phosphate groups (**Figure 8.5A**). The hydrolysis of a molecule of ATP yields free energy, as well as ADP and an inorganic phosphate ion (P_i). Thus:

$$ATP + H_2O \rightarrow ADP + P_i + \text{free energy}$$

The important property of this reaction is that it is exergonic, releasing free energy. Under standard laboratory conditions, the change in free energy for this reaction (ΔG) is about −7.3 kcal/mol (−30 kJ/mol). However, under cellular conditions, ΔG can be as much as −14 kcal/mol. We give both values here because you will encounter them both and you should be aware of their origins. Both are correct, but in different conditions.

A molecule of ATP can be hydrolyzed either to ADP and P_i, or to adenosine monophosphate (AMP) and a pyrophosphate ion ($P_2O_7^{4-}$; commonly abbreviated as PP_i). Two characteristics of ATP account for the free energy released by the loss of one or two of its phosphate groups:

- Because phosphate groups are negatively charged and so repel each other, it takes energy to get two phosphates near enough to each other to make the covalent bond that links them together. Some of this energy is stored as potential energy in the P~O bonds between the phosphates in ATP (the wavy line indicates a high energy bond).

- The free energy of this P~O bond (called a phosphoric acid anhydride bond) is much higher than the energy of the O—H bond that forms as a result of hydrolysis. So some usable energy is released by hydrolysis.

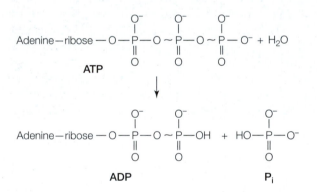

Cells use the energy released by ATP hydrolysis to fuel endergonic reactions (such as the biosynthesis of complex molecules), for active transport, and for movement. Another interesting example of the use of ATP involves converting its chemical energy into light energy.

BIOLUMINESCENCE The production of light by living organisms is referred to as **bioluminesence** (**Figure 8.5B**). It is an example of an endergonic reaction driven by ATP hydrolysis that involves an interconversion of energy forms (chemical to light). The chemical that becomes luminescent is called luciferin (after the light-bearing fallen angel, Lucifer):

$$\text{Luciferin} + O_2 + ATP \xrightarrow{\text{Luciferase}} \text{oxyluciferin} + AMP + PP_i + \text{light}$$

This reaction and the enzyme that catalyzes it (luciferase) occur in a wide variety of organisms in addition to the familiar firefly. These include a variety of marine organisms, microorganisms,

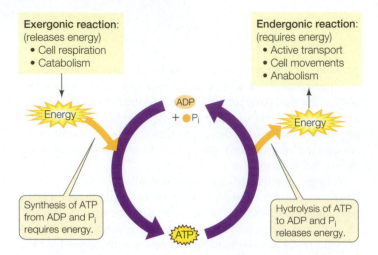

8.6 Coupling of Reactions Exergonic cellular reactions release the energy needed to make ATP from ADP. The energy released from the conversion of ATP back to ADP can be used to fuel endergonic reactions.

Go to Activity 8.1 ATP and Coupled Reactions
Life10e.com/ac8.1

worms, and mushrooms. The light is generally used to avoid predators or to attract potential mates.

Go to Media Clip 8.1
Bioluminescence in the Deep Sea
Life10e.com/mc8.1

ATP couples exergonic and endergonic reactions

As we have just seen, the hydrolysis of ATP is exergonic and yields ADP, P_i, and free energy (or AMP, PP_i, and free energy). The reverse reaction, the formation of ATP from ADP and P_i, is endergonic and consumes as much free energy as is released by the hydrolysis of ATP:

$$ADP + P_i + \text{free energy} \rightarrow ATP + H_2O$$

Many different exergonic reactions in the cell can provide the energy to convert ADP into ATP. For eukaryotes and many prokaryotes, the most important of these reactions is cellular respiration, in which some of the energy released from fuel molecules is captured in ATP. The formation and hydrolysis of ATP constitute what might be called an "energy-coupling cycle," in which ADP picks up energy from exergonic reactions to become ATP, which then donates energy to endergonic reactions. ATP is the common component of these reactions and is the agent of coupling, as illustrated in **Figure 8.6**.

Coupling of exergonic and endergonic reactions is very common in metabolism. Free energy is captured and retained in the P~O bonds of ATP. ATP then diffuses to another site in the cell, where its hydrolysis releases the free energy to drive an endergonic reaction. For example, the formation of glucose 6-phosphate from glucose (**Figure 8.7**), which has a positive ΔG (is endergonic), will not proceed without the input of free energy from ATP hydrolysis, which has a negative ΔG (is exergonic). The overall ΔG for the coupled reactions (when the two ΔGs are added together) is negative. Hence the reactions proceed exergonically when they are coupled, and glucose 6-phosphate is synthesized. As you will see in Chapter 9, this is the initial reaction in the catabolism of glucose.

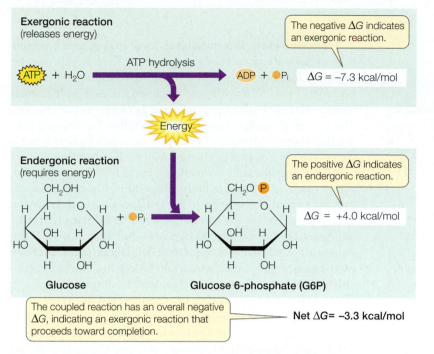

Exergonic reaction
(releases energy)

The negative ΔG indicates an exergonic reaction.

ATP hydrolysis

ATP + H_2O → ADP + Pi   ΔG = –7.3 kcal/mol

Energy

Endergonic reaction
(requires energy)

The positive ΔG indicates an endergonic reaction.

CH₂OH + Pi → CH₂O-P   ΔG = +4.0 kcal/mol

Glucose   Glucose 6-phosphate (G6P)

The coupled reaction has an overall negative ΔG, indicating an exergonic reaction that proceeds toward completion.

Net ΔG= –3.3 kcal/mol

8.7 Coupling of ATP Hydrolysis to an Endergonic Reaction The addition of phosphate derived from the hydrolysis of ATP to glucose forms the molecule glucose 6-phosphate (in a reaction catalyzed by hexokinase). ATP hydrolysis is exergonic and the energy released drives the second reaction, which is endergonic.

An active cell requires the production of millions of molecules of ATP per second to drive its biochemical machinery. You are already familiar with some of the activities in the cell that require energy from the hydrolysis of ATP:

- Active transport across a membrane (see Figure 6.14)
- Condensation reactions that use enzymes to form polymers (see Figure 3.4A)
- Modifications of cell signaling proteins by protein kinases (see Figure 7.16)
- Motor proteins that move vesicles along microtubules (see Figure 5.19)

An ATP molecule is typically consumed within a second of its formation. At rest, an average person produces and hydrolyzes about 40 kg of ATP per day—as much as some people weigh. This means that each ATP molecule undergoes about 10,000 cycles of synthesis and hydrolysis every day!

RECAP 8.2

ATP is the "energy currency" of cells. Some of the free energy released by exergonic reactions can be captured in the form of ATP. This energy can then be released by ATP hydrolysis and used to drive endergonic reactions.

- How does ATP store energy? **See pp. 149–150**
- What are coupled reactions? **See pp. 150–151 and Figure 8.7**

ATP is synthesized and used up very rapidly. But these biochemical reactions—and nearly all the others that take place inside a cell—could not proceed so rapidly without the help of enzymes.

8.3 What Are Enzymes?

When we know the change in free energy (ΔG) of a reaction, we know where the equilibrium point of the reaction lies: the more negative the ΔG value is, the further the reaction proceeds toward completion. However, ΔG tells us nothing about the *rate* of a reaction—the speed at which it moves toward equilibrium. The reactions that cells depend on have spontaneous rates that are so slow that the cells would not survive without a way to speed up the reactions. That is the role of catalysts: substances that speed up reactions without themselves being permanently altered. A catalyst does not cause a reaction to occur that would not proceed without it, *but merely increases the rate of the reaction*, allowing equilibrium to be approached more rapidly. This is an important point: *no catalyst makes a reaction occur that cannot otherwise occur.*

Most biological catalysts are proteins called enzymes. Although we will focus here on proteins, some catalysts are RNA molecules called ribozymes (see Section 4.3). A biological catalyst, whether protein or RNA, is a framework or scaffold within which chemical catalysis takes place. This molecular framework binds the reactants and sometimes participates in the reaction itself; however, such participation does not permanently change the enzyme. The catalyst ends up in exactly the same chemical condition after a reaction as before it. Although there is considerable evidence that the first enzymes to evolve were ribozymes, cells now use proteins rather than RNA to catalyze most biochemical reactions. Compared with RNA, proteins show greater diversity in their three-dimensional structures, and in the chemical functions provided by their functional groups (see Table 3.2).

In this section we will discuss the energy barrier that controls the rate of a chemical reaction. Then we will focus on the roles of enzymes: how they interact with specific reactants, how they lower the energy barrier, and how they permit reactions to proceed more quickly.

To speed up a reaction, an energy barrier must be overcome

An exergonic reaction may release free energy, but without a catalyst it will take place very slowly. This is because there is an energy barrier between reactants and products. Think about the hydrolysis of sucrose, which we described in Section 8.1:

Sucrose + H_2O → glucose + fructose

In humans, this reaction is part of the process of digestion. Even if water is abundant, the sucrose molecule will only very rarely bind the H atom and the —OH group of water at the appropriate locations to break the covalent bond between glucose and fructose *unless there is an input of energy to initiate the reaction.* Such an input of energy will place the sucrose into a reactive mode called the **transition state**. The energy input required for sucrose to reach this state is called the **activation energy** (E_a).

(A)

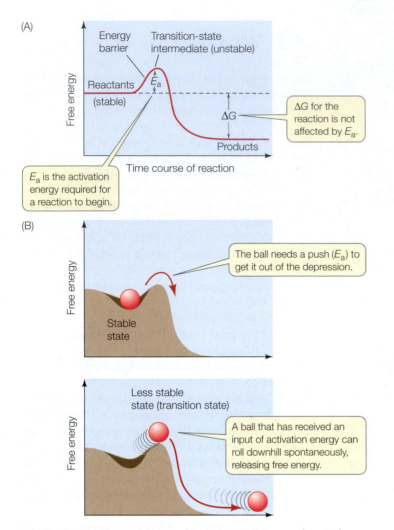

8.8 Activation Energy Initiates Reactions (A) In any chemical reaction, an initial stable state must become less stable before change is possible. (B) A ball on a hillside provides a physical analogy to the biochemical principle graphed in (A).

The following example will help illustrate the ideas of activation energy and transition state:

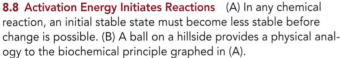

Activation energy in the form of a spark is needed to excite the molecules in the fireworks so they will react with oxygen in the air. Once the transition state is reached, the reaction occurs.

In general, exergonic reactions proceed only after the reactants are pushed over the energy barrier by some added energy. The energy barrier thus represents the amount of energy needed to start the reaction, the activation energy (**Figure 8.8A**). Recall the ball rolling down the hill in Figure 8.3A. The ball has a lot of potential energy at the top of the hill. However, if it is stuck in a small depression, it will not roll down the hill, even though that action is exergonic. To start the ball rolling, a small amount of energy (activation energy) is needed to push it out of the depression (**Figure 8.8B**). In a chemical reaction, the activation energy is the energy needed to change the reactants into unstable molecular forms called transition-state intermediates.

Transition-state intermediates have higher free energies than either the reactants or the products. Their bonds may be stretched and therefore unstable. Although the amount of activation energy needed for different reactions varies, it is often small compared with the change in free energy of the overall reaction. The activation energy put in to start a reaction is recovered during the ensuing "downhill" phase of the reaction, so it is not a part of the net free energy change, ΔG (see Figure 8.8A).

Where does the activation energy come from? In any collection of reactants at room or body temperature, the molecules are moving around. A few are moving fast enough that their kinetic energy can overcome the energy barrier, enter the transition state, and react. So the reaction takes place—but very slowly. If the system is heated, all the reactant molecules move faster and have more kinetic energy, and the reaction speeds up. You have probably used this technique in the chemistry laboratory.

However, adding enough heat to increase the average kinetic energy of the molecules would not work in living systems. Such a nonspecific approach would accelerate all reactions, including destructive ones such as the denaturation of proteins (see Chapter 3). A more effective way to speed up a reaction in a living system is to lower the energy barrier by bringing the reactants close together. In living cells, enzymes and ribozymes accomplish this task.

Enzymes bind specific reactants at their active sites

Catalysts increase the rates of chemical reactions. Most nonbiological catalysts are nonspecific. For example, powdered platinum catalyzes virtually any reaction in which molecular hydrogen (H_2) is a reactant. In contrast, most biological catalysts are highly specific. An enzyme or ribozyme usually recognizes and binds to only one or a few closely related reactants, and it catalyzes only a single chemical reaction. In the discussion that follows, we focus on enzymes, but remember that similar rules of chemical behavior apply to ribozymes as well.

In an enzyme-catalyzed reaction, the reactants are called **substrates**. Substrate molecules bind to a particular site on the enzyme, called the **active site**, where catalysis takes place (**Figure 8.9**). The specificity of an enzyme results from the exact three-dimensional shape and structure of its active site, into which only a narrow range of substrates can fit. Other molecules—with different shapes, different functional groups, and different properties—cannot fit properly and bind to the active site. This specificity is comparable to the specific binding of a membrane transport protein or receptor protein to its specific ligand, as described in Chapters 6 and 7.

The names of enzymes often reflect their functions and end with the suffix "ase." For example the enzyme sucrase catalyzes the hydrolysis of sucrose, which we have referred to above:

$$\text{Sucrose} + H_2O \xrightarrow{\text{Sucrase}} \text{glucose} + \text{fructose}$$

And as we saw in the opening story, lipases and amylases catalyze the hydrolysis of lipids and starch, respectively (starch is sometimes referred to as amylum).

The binding of a substrate to the active site of an enzyme produces an **enzyme–substrate complex (ES)** that is held

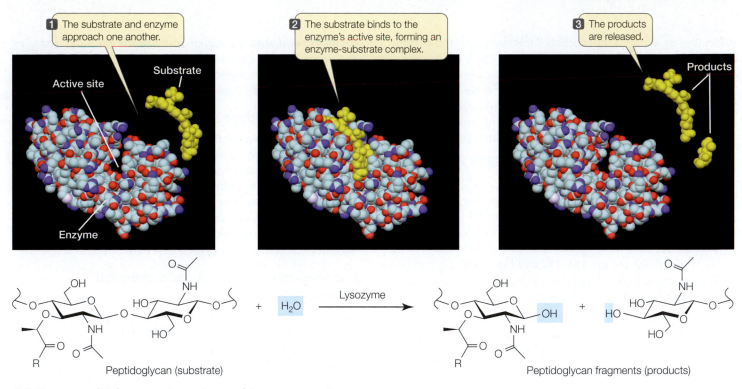

1 The substrate and enzyme approach one another.

2 The substrate binds to the enzyme's active site, forming an enzyme-substrate complex.

3 The products are released.

Active site

Substrate

Enzyme

Products

Peptidoglycan (substrate)

$+ H_2O$ Lysozyme

Peptidoglycan fragments (products)

8.9 Enzyme and Substrate A reaction involving an enzyme is illustrated by lysozyme. Lysozyme catalyzes breakage of bonds in the peptidoglycans of bacterial cell walls. (See Section 5.2 for a description of peptidoglycans.)

together by several interactions, such as hydrogen bonding, electrical attraction, or temporary covalent bonding. The enzyme–substrate complex gives rise to product and free enzyme:

$$E + S \rightarrow ES \rightarrow E + P$$

where E is the enzyme, S is the substrate, P is the product, and ES is the enzyme–substrate complex. The free enzyme (E) is in the same chemical form at the end of the reaction as at the beginning. While bound to the substrate, it may change chemically, but by the end of the reaction it has been restored to its initial form and is ready to bind more substrate.

The equation that forms ES may be familiar to you from Chapter 7: it is the same as the binding of a receptor and ligand (RL) in cell signaling. In that chapter, we defined the dissociation constant (K_D) as a measure of the affinity of the receptor for its ligand. The lower the K_D, the tighter the binding. For enzymes and their substrates, K_D values are often in the range 10^{-5} to 10^{-6} M. This favors the formation of ES. In practical terms, it means that the binding between enzyme and substrate is somewhat reversible; the substrate can be released before the reaction. However, this is counteracted by the fact that ES is short-lived and the product(s) form quickly.

Enzymes lower the energy barrier but do not affect equilibrium

When reactants are bound to the enzyme, forming an enzyme–substrate complex, they require less activation energy than the transition-state intermediates in the corresponding uncatalyzed reaction (**Figure 8.10**). Thus the enzyme lowers the energy

barrier for the reaction—it offers the reaction an easier path, speeding it up. When an enzyme lowers the energy barrier, both the forward and the reverse reactions speed up, so the enzyme-catalyzed reaction proceeds toward equilibrium more rapidly than the uncatalyzed reaction. *The final equilibrium is the same with or without the enzyme.* Similarly, adding an enzyme to a reaction does not change the difference in free energy (ΔG) between the reactants and the products (see Figure 8.10).

Enzymes can change the rate of a reaction substantially. For example, if a particular protein that has arginine as its terminal amino acid just sits in solution, the protein molecules tend toward disorder and the terminal peptide bonds break, releasing the arginine residues (ΔS increases). Without an enzyme this is a very slow reaction—it takes about 7 years for half of the protein

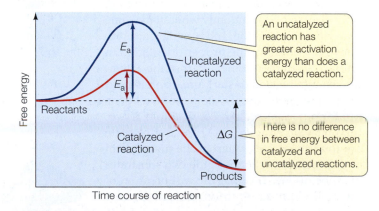

An uncatalyzed reaction has greater activation energy than does a catalyzed reaction.

E_a

Uncatalyzed reaction

E_a

Free energy

Reactants

Catalyzed reaction

ΔG

There is no difference in free energy between catalyzed and uncatalyzed reactions.

Products

Time course of reaction

8.10 Enzymes Lower the Energy Barrier Although the activation energy is lower in an enzyme-catalyzed reaction than in an uncatalyzed reaction, the energy released is the same with or without catalysis. In other words, E_a is lower, but ΔG is unchanged. Lower activation energy means the reaction will take place at a faster rate.

Go to Activity 8.2 Free Energy Changes   Life10e.com/ac8.2

molecules to undergo the reaction. However, with the enzyme carboxypeptidase A catalyzing the reaction, half the arginines are released in less than a second! Rate enhancement by enzymes varies from 1 million times to an amazing 10^{17} times for the champion enzyme orotidine monophosphate decarboxylase! The consequence of catalysis for living cells is not difficult to imagine. Such increased reaction rates make new realities possible.

> ### RECAP 8.3
>
> A chemical reaction requires a "push" over the energy barrier to get started. An enzyme reduces the activation energy needed to start a reaction by binding the reactants (substrates). This speeds up the reaction.
>
> - What is activation energy, and how is it supplied in the chemistry lab? **See pp. 151–152 and Figure 8.8**
> - Explain how the structure of an enzyme makes that enzyme specific. **See p.152 and Figure 8.9**
> - How does an enzyme speed up a reaction? **See Figure 8.10**
> - What is the relationship between an enzyme and the equilibrium point of a reaction? **See p. 153**

Now that you have a general understanding of the structures, functions, and specificities of enzymes, let's look more closely at how they work.

8.4 How Do Enzymes Work?

During and after the formation of the enzyme–substrate complex, chemical interactions occur. These interactions contribute directly to the breaking of old bonds and the formation of new ones. In catalyzing a reaction, an enzyme may use one or more mechanisms.

Enzymes can orient substrates

When free in solution, substrates are moving from place to place randomly while at the same time vibrating, rotating, and tumbling around. They may not have the proper orientation to interact when they collide. Part of the activation energy needed to start a reaction is used to bring together specific atoms so that bonds can form (**Figure 8.11A**). For example, if acetyl coenzyme A (acetyl CoA) and oxaloacetate are to form citrate (a step in the metabolism of glucose; see Section 9.2), the two substrates must be oriented so that the carbon atom of the methyl group of acetyl CoA can form a

covalent bond with the carbon atom of the carbonyl group of oxaloacetate. The active site of the enzyme citrate synthase has just the right shape to bind these two molecules so that these atoms are adjacent.

Enzymes can induce strain in the substrate

Once a substrate has bound to its active site, an enzyme can cause bonds in the substrate to stretch, putting it in an unstable transition state (**Figure 8.11B**). For example, lysozyme is a protective enzyme abundant in tears and saliva that destroys invading bacteria by cleaving peptidoglycans in their cell walls (see Figure 8.9). Lysozyme's active site "stretches" the bonds between the glycan monomers, rendering the bonds unstable and more reactive to lysozyme's other substrate, water.

Enzymes can temporarily add chemical groups to substrates

The side chains (R groups) of an enzyme's amino acids may be direct participants in making its substrates more chemically reactive (**Figure 8.11C**).

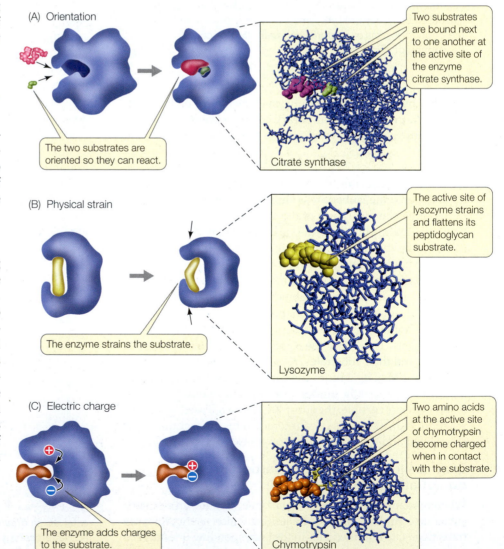

(A) Orientation

The two substrates are oriented so they can react.

Two substrates are bound next to one another at the active site of the enzyme citrate synthase.

Citrate synthase

(B) Physical strain

The enzyme strains the substrate.

The active site of lysozyme strains and flattens its peptidoglycan substrate.

Lysozyme

(C) Electric charge

The enzyme adds charges to the substrate.

Two amino acids at the active site of chymotrypsin become charged when in contact with the substrate.

Chymotrypsin

8.11 Life at the Active Site Enzymes have several ways of causing their substrates to enter the transition state: (A) orientation, (B) physical strain, and (C) chemical charge.

- In *acid–base catalysis*, the acidic or basic side chains of the amino acids in the active site transfer H⁺ to or from the substrate, destabilizing a covalent bond in the substrate, and permitting it to break.

- In *covalent catalysis*, a functional group in a side chain forms a temporary covalent bond with a portion of the substrate.

- In *metal ion catalysis*, metal ions such as copper, iron, and manganese, which are often firmly bound to side chains of enzymes, can lose or gain electrons without detaching from the enzymes. This ability makes them important participants in oxidation–reduction reactions, which involve the loss or gain of electrons.

Molecular structure determines enzyme function

Most enzymes are much larger than their substrates. An enzyme is typically a protein containing hundreds of amino acids. It may consist of a single folded polypeptide chain or of several subunits (see Section 3.2). Its substrate is generally a small molecule or a small part of a large molecule. The active site of the enzyme is usually quite small, not more than 6 to 12 amino acids. Two questions arise from these observations:

- What features of the active site allow it to recognize and bind the substrate?

- What is the role of the rest of the huge protein?

THE ACTIVE SITE IS SPECIFIC TO THE SUBSTRATE(S) The remarkable ability of an enzyme to select exactly the right substrate(s) depends on a precise interlocking of molecular shapes and interactions of chemical groups at the active site. The binding of a substrate to the active site depends on the same kinds of forces that maintain the tertiary structure of the enzyme: hydrogen bonds, the attraction and repulsion of electrically charged groups, and hydrophobic interactions. The specific fit of the substrate in the active site of lysozyme is illustrated in Figures 8.9 and 8.11B.

AN ENZYME CHANGES SHAPE WHEN IT BINDS A SUBSTRATE Just as a membrane receptor protein may undergo precise changes in conformation upon binding to its ligand (see Chapter 7), some enzymes change their shapes when they bind their substrate(s). These shape changes, which are called **induced fit**, alter the shape of the active site(s) of the enzyme.

An example of induced fit can be seen in the enzyme hexokinase (see Figure 8.7), which catalyzes the reaction

$$\text{Glucose} + \text{ATP} \rightarrow \text{glucose 6-phosphate} + \text{ADP}$$

Induced fit brings reactive side chains from the hexokinase active site into alignment with the substrates (**Figure 8.12**), facilitating its catalytic mechanisms. Equally important, the folding of hexokinase to fit around the substrates (glucose and ATP) excludes water from the active site. This is essential, because if water were present, the ATP could be hydrolyzed to ADP and P_i. But since water is absent, the transfer of a phosphate from ATP to glucose is favored.

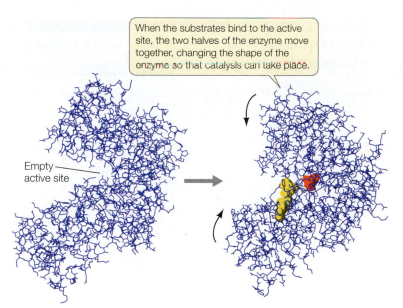

When the substrates bind to the active site, the two halves of the enzyme move together, changing the shape of the enzyme so that catalysis can take place.

Empty active site

8.12 Some Enzymes Change Shape When Substrate Binds to Them Shape changes result in an induced fit between enzyme and substrate, improving the catalytic ability of the enzyme. Induced fit can be observed in the enzyme hexokinase, seen here with and without its substrates, glucose (red) and ATP (yellow).

Induced fit at least partly explains why enzymes are so large. The rest of the macromolecule may have one or more of the following three roles:

- It provides a framework so that the amino acids of the active site are properly positioned in relation to the substrate(s).

- It participates in significant changes in protein shape and structure that result in induced fit.

- It provides binding sites for regulatory molecules (see Section 8.5).

Some enzymes require other molecules in order to function

As large and complex as enzymes are, many of them require the presence of nonprotein chemical "partners" in order to function (**Table 8.2**):

- *Prosthetic groups* are distinct, non–amino acid atoms or molecular groupings that are permanently bound to their enzymes. An example is flavin adenine dinucleotide (FAD), which is bound to succinate dehydrogenase, an important enzyme in cellular respiration (see Section 9.3).

- *Inorganic cofactors* include ions such as copper, zinc, and iron that are permanently bound to certain enzymes. For example, the enzyme alcohol dehydrogenase contains the cofactor zinc.

- A *coenzyme* is a nonprotein carbon-containing molecule that is required for the action of one or more enzymes. It is usually relatively small compared with the enzyme to which it temporarily binds.

A coenzyme moves from enzyme to enzyme, adding or removing chemical groups from the substrate. A coenzyme is like a

TABLE**8.2**

Some Examples of Nonprotein "Partners" of Enzymes

Type of Molecule	Role in Catalyzed Reactions
Prosthetic groups	
Heme	Binds ions, O_2, and electrons
FAD	Carries electrons/protons
Retinal	Converts light energy
Inorganic cofactors	
Iron (Fe^{2+} or Fe^{3+})	Oxidation/reduction
Copper (Cu^+ or Cu^{2+})	Oxidation/reduction
Zinc (Zn^{2+})	Stabilizes DNA binding structure
Coenzymes	
Biotin	Carries —COO^-
Coenzyme A	Carries —CO—CH_3
NAD	Carries electrons/protons
ATP	Provides/extracts energy

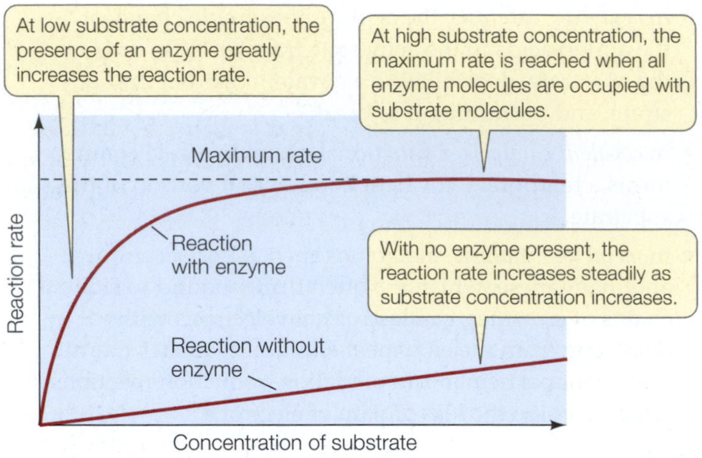

At low substrate concentration, the presence of an enzyme greatly increases the reaction rate.

At high substrate concentration, the maximum rate is reached when all enzyme molecules are occupied with substrate molecules.

With no enzyme present, the reaction rate increases steadily as substrate concentration increases.

8.13 Catalyzed Reactions Reach a Maximum Rate   Because there is usually less enzyme than substrate present, the reaction rate levels off when the enzyme becomes saturated.

substrate in that it does not permanently bind to the enzyme: it binds to the active site, changes chemically during the reaction, and then separates from the enzyme to participate in other reactions. There is actually no clear distinction between the functions of coenzymes and some substrates. For example, ATP and ADP have been described as coenzymes, even though they are really substrates that gain or lose phosphate groups during chemical reactions. The term coenzyme was coined before the functions of these molecules were fully understood. Biochemists continue to use the term, and for consistency with the field, we use it in this book.

In the next chapter we will encounter other coenzymes that function in energy-harvesting reactions by accepting or donating electrons or hydrogen atoms. In animals, some coenzymes are produced from vitamins—substances that must be obtained from food because they cannot be synthesized by the body. For example, the B vitamin niacin is used to make the coenzyme nicotinamide adenine dinucleotide (NAD).

The substrate concentration affects the reaction rate

For a reaction of the type A → B, the rate of the uncatalyzed reaction is directly proportional to the concentration of A. The higher the concentration of substrate, the faster the rate of the reaction. The appropriate enzyme not only speeds up the reaction; it also changes the shape of a plot of rate versus substrate concentration (**Figure 8.13**). For a given concentration of enzyme, the rate of the enzyme-catalyzed reaction initially increases as the substrate concentration increases from zero, but then it levels off. At some point, further increases in the substrate concentration do not significantly increase the reaction rate—the maximum rate has been reached.

Since the concentration of an enzyme is usually much lower than that of its substrate and does not change as substrate concentration changes, what we see is a saturation phenomenon like the one that occurs in facilitated diffusion (see Figure 6.12). When all the enzyme molecules are bound to substrate

molecules, the enzyme is working as fast as it can—at its maximum rate. Nothing is gained by adding more substrate, because no free enzyme molecules are left to act as catalysts. Under these conditions the active sites are said to be saturated.

The maximum rate of a catalyzed reaction can be used to measure how efficient the enzyme is. The turnover number is the maximum number of substrate molecules that one enzyme molecule can convert to product per unit of time. This number ranges from 1 molecule every 2 seconds for lysozyme to an amazing 40 million molecules per second for the liver enzyme catalase.

■ **RECAP** **8.4**

Enzymes orient their substrates to bring together specific atoms so that bonds can form. An enzyme can participate in the reaction it catalyzes by temporarily changing shape or destabilizing the enzyme–substrate complex. Some enzymes require prosthetic groups, inorganic cofactors, or coenzymes in order to function.

- What are three mechanisms of enzyme catalysis? **See p.154 and Figure 8.11**
- What are the chemical roles of coenzymes in enzymatic reactions? **See pp. 155–156 and Table 8.2**

We've seen in this section how individual enzymes work on their substrates. However, enzymes inside organisms don't operate in isolation—there may be thousands of different enzymes within a given cell. Let's see how all these different enzymes work together in a complex organism.

8.5 **How Are Enzyme Activities Regulated?**

A major characteristic of life is homeostasis—the maintenance of stable internal conditions (see Chapter 40). How does a cell maintain a relatively constant internal environment while thousands of chemical reactions are going on? These chemical reactions operate within metabolic pathways in which the product of one reaction is a reactant for the next. These pathways do not exist in isolation, but interact extensively, and each reaction in each pathway is catalyzed by a specific enzyme.

Within a cell or organism, the presence and activity of enzymes determine the "flow" of chemicals through different metabolic pathways. The amount of enzyme activity, in turn, is controlled in part via the regulation of gene expression. Many signal transduction pathways (described in Chapter 7) end with changes in gene expression, and often the genes that are switched on or off encode enzymes. But the simple presence of an enzyme does not ensure that it is functioning. Another way cells can control which pathways are active at a particular time is by the activation or inactivation of existing enzymes. If one enzyme in the pathway is inactive, that step and all subsequent steps shut down. Thus some enzymes are target points for the regulation of entire sequences of chemical reactions.

Regulation of the rates at which thousands of different enzymes operate contributes to homeostasis within an organism. Such control permits cells to make orderly changes in their functions in response to changes in the external environment. In Chapter 7 we described a number of enzymes that become activated in signal transduction pathways, illustrating how enzyme activation can dramatically alter cell functions. (For example, see the activation of glycogen phosphorylase in Figure 7.18.)

The flow of chemicals such as carbon atoms through interacting metabolic pathways can be studied, but this process becomes complicated quickly, because each pathway influences the others. Computer algorithms are used to model these pathways and show how they mesh in an interdependent system (**Figure 8.14**). Such models can help predict what will happen if the concentration of one molecule or another is altered. This new field of biology is called **systems biology**, and it has numerous applications.

In this section we will investigate the roles of enzymes in organizing and regulating metabolic pathways. We will also examine how the environment—particularly temperature and pH—affects enzyme activity.

Enzymes can be regulated by inhibitors

Various chemical inhibitors can bind to enzymes, slowing down the rates of the reactions they catalyze. Some inhibitors occur naturally in cells; others are artificial. Naturally occurring inhibitors regulate metabolism; artificial ones can be used to treat disease, to kill pests, or to study how enzymes work. In some cases the inhibitor binds the enzyme irreversibly, and the enzyme becomes permanently inactivated. In other cases the inhibitor has reversible effects; it can separate from the enzyme, allowing the enzyme to function fully as before. The removal of a natural reversible inhibitor increases an enzyme's rate of catalysis.

IRREVERSIBLE INHIBITION If an inhibitor covalently binds to certain side chains at the active site of an enzyme, it will permanently inactivate the enzyme by destroying its capacity

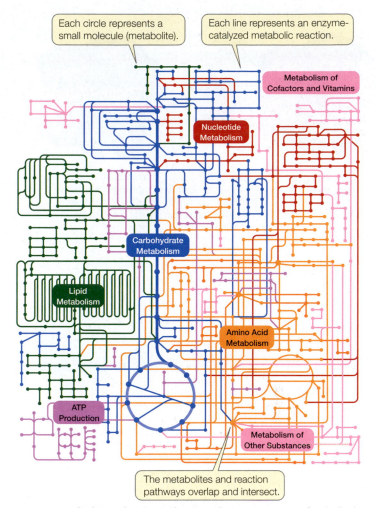

Each circle represents a small molecule (metabolite).

Each line represents an enzyme-catalyzed metabolic reaction.

Metabolism of Cofactors and Vitamins

Nucleotide Metabolism

Carbohydrate Metabolism

Lipid Metabolism

Amino Acid Metabolism

ATP Production

Metabolism of Other Substances

The metabolites and reaction pathways overlap and intersect.

8.14 Metabolic Pathways The complex interactions of metabolic pathways can be modeled by the tools of systems biology. In cells, the main elements controlling these pathways are enzymes.

to interact with its normal substrate. An example of an irreversible inhibitor is DIPF (diisopropyl phosphorofluoridate), which reacts with serine (**Figure 8.15**). DIPF is an irreversible inhibitor of acetylcholinesterase, whose operation is essential for the normal functioning of the nervous system. Because of their effect on acetylcholinesterase, DIPF and other similar compounds are classified as nerve gases, and were developed for biological warfare. One of these compounds, Sarin, was used in an attack on the Tokyo subway in 1995, resulting in a dozen deaths and the hospitalization of hundreds more. The widely used insecticide malathion is a derivative of DIPF that inhibits only insect acetylcholinesterase, not the mammalian enzyme. The irreversible inhibition of enzymes is of practical use to humans, but this form of regulation is not common in the cell, because the enzyme is permanently inactivated and cannot be recycled. Instead, cells use reversible inhibition.

REVERSIBLE INHIBITION In some cases an inhibitor is similar enough to a particular enzyme's natural substrate to bind noncovalently to its active site, yet different enough that the

Acetylcholinesterase

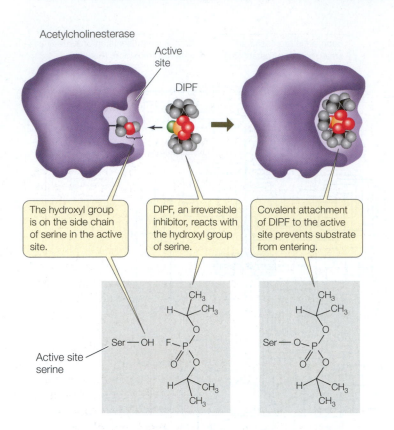

The hydroxyl group is on the side chain of serine in the active site.

DIPF, an irreversible inhibitor, reacts with the hydroxyl group of serine.

Covalent attachment of DIPF to the active site prevents substrate from entering.

Active site serine

8.15 Irreversible Inhibition DIPF forms a stable covalent bond with the side chain of the amino acid serine at the active site of the enzyme acetylcholinesterase, thus irreversibly disabling the enzyme.

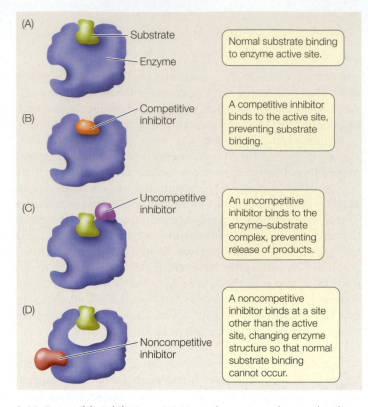

8.16 Reversible Inhibition (A) Normal enzyme–substrate binding. (B) Competitive inhibition. (C) Uncompetitive inhibition. (D) Noncompetitive inhibition.

Go to Animated Tutorial 8.1
Enzyme Catalysis
Life10e.com/at8.1

enzyme catalyzes no chemical reaction. While such a molecule is bound to the enzyme, the natural substrate cannot enter the active site and the enzyme is unable to function. Such a molecule is called a **competitive inhibitor** because it competes with the natural substrate for the active site (**Figure 8.16A, B**). In this case, the degree of inhibition depends on the relative concentrations of the substrate and the inhibitor: if the inhibitor concentration is higher, it is more likely to bind the active site of the enzyme than the substrate, and vice versa. The inhibition is reversible because if the concentration of substrate is increased or if the concentration of inhibitor is reduced, the substrate is more likely to bind, and the enzyme is active again.

An example of a competitive inhibitor is the drug methotrexate. An important coenzyme in the formation of purines (components of nucleic acids) is tetrahydrofolate, which is formed from dihydrofolate in a reaction catalyzed by dihydrofolate reductase (DHFR):

$$\text{Dihydrofolate} \xrightarrow{\text{DHFR}} \text{tetrahydrofolate}$$

When cancer cells reproduce, they need to replicate their DNA, and so they need to produce purines. This makes DHFR an ideal target for an anticancer drug. A team led by Sidney Farber at Harvard Medical School first showed that an analog of dihydrofolate could treat leukemia. The drug used by Farber

(aminopterin) has since been replaced by a similar analog, methotrexate:

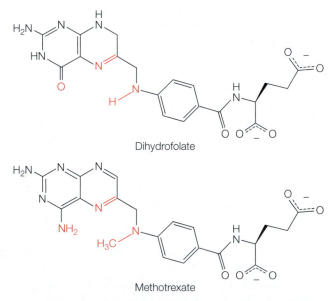

Dihydrofolate

Methotrexate

This successful drug is used to treat inflammatory diseases such as psoriasis and rheumatoid arthritis, as well as cancer.

An **uncompetitive inhibitor** (**Figure 8.16C**) binds to the enzyme–substrate complex, preventing the complex from releasing products. Unlike competitive inhibition, it cannot be overcome by adding more substrate.

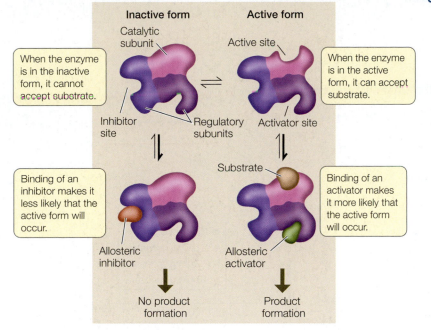

8.17 Allosteric Regulation of Enzymes Active and inactive forms of an enzyme can be interconverted, depending on the binding of effector molecules at sites other than the active site. Binding an inhibitor stabilizes the inactive form, and binding an activator stabilizes the active form.

Go to Animated Tutorial 8.2
Allosteric Regulation of Enzymes
Life10e.com/at8.2

A **noncompetitive inhibitor** binds to an enzyme at a site distinct from the active site. This binding causes a change in the shape of the enzyme that alters its activity (**Figure 8.16D**). The active site may no longer bind the substrate, or if it does, the rate of product formation may be reduced. Like competitive inhibitors, noncompetitive inhibitors can become unbound, so their effects are reversible.

Allosteric enzymes are controlled via changes in shape

The change in enzyme shape that is due to noncompetitive inhibitor binding is an example of allostery (*allo*, "different"; *stereos*, "shape"). **Allosteric regulation** occurs when an effector molecule binds to a site other than the active site of an enzyme, *inducing the enzyme to change its shape*. The change in shape alters the affinity of the active site for the substrate, and so the rate of the reaction is changed.

Often, an enzyme will exist in the cell in more than one possible shape (**Figure 8.17**):

- The *active form* of the enzyme has the proper shape for substrate binding.
- The *inactive form* of the enzyme has a shape that cannot bind the substrate.

Other molecules, collectively referred to as effectors, can influence which form the enzyme takes:

- Binding of an inhibitor to a site other than the active site can stabilize the inactive form of the enzyme, making it less likely to convert to the active form.

- The active form can be stabilized by the binding of an activator to another site on the enzyme.

Like substrate binding, the binding of inhibitors and activators to their regulatory sites (also called allosteric sites) is highly specific. Most (but not all) enzymes that are allosterically regulated are proteins with quaternary structure; that is, they are made up of multiple polypeptide subunits. The polypeptide that has the active site is called the catalytic subunit. The allosteric sites are often located on different polypeptides, called the regulatory subunits (see Figure 8.17).

Some enzymes have multiple subunits containing active sites, and the binding of substrate to one of the active sites causes allosteric effects. When substrate binds to one subunit, there is a slight change in protein structure that influences the adjacent subunit. The slight change to the second subunit makes its active site more likely to bind to the substrate. So the reaction speeds up as the sites become sequentially activated.

As a result, an allosteric enzyme with multiple active sites and a nonallosteric enzyme with a single active site differ greatly in their reaction rates when the substrate concentration is low. Graphs of reaction rates plotted against substrate concentrations show this relationship. For a nonallosteric enzyme, the plot is hyperbolic:

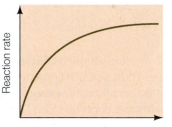

The reaction rate first increases sharply with increasing substrate concentration, then tapers off to a constant maximum rate as the supply of enzyme becomes saturated.

For a multisubunit allosteric enzyme, the graph looks different, having a sigmoid (S-shaped) appearance:

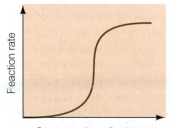

At low substrate concentrations, the reaction rate increases only gradually as substrate concentration increases. After the substrate binds to the first active site of the enzyme (the slowly increasing part of the curve), there is a change in the quaternary structure such that the other sites become more likely to bind substrate, so the reaction speeds up (the rapidly increasing part of

WORKING WITH**DATA:**

How Does an Herbicide Work?

Original Paper

Boocock, M. R. and J. R. Coggins. 1983. Kinetics of 5-enol-pyruvylshikimate-3-phosphate synthase inhibition by glyphosate. *FEBS Letters* 154: 127–133.

Analyze the Data

Glyphosate (also called Roundup) is used to kill weeds growing in farmer's fields or backyard gardens. Glyphosate kills plants by inhibiting an enzyme (5-enolpyruvylshikimate-3-phosphate synthase; EPSP synthase) in the metabolic pathway used to synthesize the amino acids phenylalanine, tyrosine, and tryptophan. Plants treated with glyphosate can't make these amino acids and die.

To investigate how glyphosate inhibits EPSP synthase, Boocock and Coggins isolated the enzyme from *Neurospora crassa* (a fungus, which was a more convenient source than plants). They measured the rate of the EPSP synthase reaction in the presence of different concentrations of glyphosate and of one of EPSP synthase's substrates, phosphoenolpyruvate (PEP). Their results are shown in the graph (presented in a different format than in the original paper).

QUESTION 1

At about what substrate concentration does EPSP synthase become saturated when no glyphosate is present? How much substrate is needed to saturate EPSP synthase in the presence of 18 μM glyphosate? In each case, what is the reaction rate at saturation?

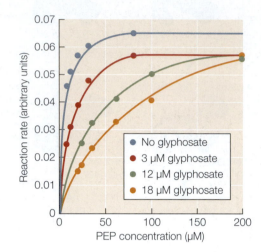

QUESTION 2

Looking at the curve for the reaction rate without inhibitor, is EPSP synthase a multi-subunit allosteric enzyme? Explain your answer.

QUESTION 3

Based on these data, what is the most likely mechanism for glyphosate inhibition of EPSP synthase: competitive, noncompetitive, or uncompetitive? Why?

Go to BioPortal for all WORKING WITH**DATA** exercises

the curve). Once all sites are saturated with substrate, the reaction rate reaches a plateau (the upper, flat part of the curve). Within a certain range, the reaction rate is extremely sensitive to relatively small changes in substrate concentration. In addition, allosteric enzymes are very sensitive to low concentrations of inhibitors. Because of this sensitivity, allosteric enzymes are important in regulating entire metabolic pathways.

Allosteric effects regulate many metabolic pathways

Metabolic pathways typically involve a starting material, various intermediate products, and an end product that is used for some purpose by the cell. In each pathway there are a number of reactions, each forming an intermediate product and each catalyzed by a different enzyme. The first step in a pathway is called the commitment step, meaning that once this enzyme-catalyzed reaction occurs, the "ball is rolling," and the other reactions happen in sequence, leading to the end product. But what if the cell has no requirement for that product—for example, if that product is available from its environment in adequate amounts? It would be energetically wasteful for the cell to continue making something it does not need.

One way to avoid this problem is to shut down the metabolic pathway by having the

final product inhibit the enzyme that catalyzes the commitment step (**Figure 8.18**). Often this inhibition occurs allosterically. When the end product is present at a high concentration, some of it binds to an allosteric site on the commitment step enzyme, thereby causing it to become inactive. Thus the final product acts as a noncompetitive inhibitor (described earlier in this section)

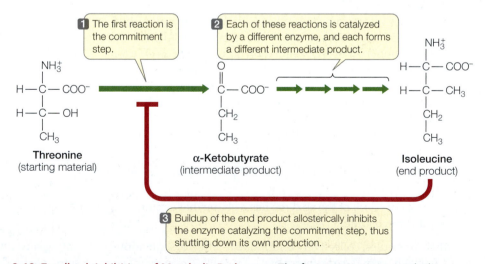

8.18 Feedback Inhibition of Metabolic Pathways The first reaction in a metabolic pathway is referred to as the commitment step. It is often catalyzed by an enzyme that can be allosterically inhibited by the end product of the pathway. The specific pathway shown here is the synthesis of isoleucine from threonine in bacteria. It is typical of many enzyme-catalyzed biosynthetic pathways.

of the first enzyme in the pathway. This mechanism is known as feedback inhibition or end-product inhibition. We will describe many other examples of such inhibition in later chapters.

Many enzymes are regulated through reversible phosphorylation

As we saw in Chapter 7, many enzymes involved in signal transduction are regulated via reversible phosphorylation (see Figure 7.16A). An enzyme can be activated by a protein kinase, which adds a phosphate to one or more specific amino acids. This results in a change in the shape of the enzyme, making it active. Such activation is reversible because another enzyme called a protein phosphatase can remove the phosphate groups, so that the enzyme becomes inactive again. In addition to the enzymes involved in signal transduction, many other enzymes and proteins in the cell (such as ion channels) are regulated via reversible phosphorylation. Reflecting the important role of protein phosphorylation in cell functions, the human genome contains about 500 protein kinase genes: about 2 percent of all the protein-coding genes we have.

Enzymes are affected by their environment

Enzymes enable cells to perform chemical reactions and carry out complex processes rapidly without using the extremes of temperature and pH employed by chemists in the laboratory. However, because of their three-dimensional structures and the chemistry of the side chains in their active sites, enzymes (and their substrates) are highly sensitive to changes in temperature and pH. This was one of the obstacles to the use of enzymes in laundry detergent (see the opening story). In Section 3.2 we described the general effects of these environmental factors on proteins. Here we will examine their effects on enzyme function (which, of course, depends on enzyme structure and chemistry).

pH AFFECTS ENZYME ACTIVITY The rates of most enzyme-catalyzed reactions depend on the pH of the solution in which they occur. While the water inside cells is generally at a neutral pH of 7, the presence of acids, bases, and buffers can alter this. Each enzyme is most active at a particular pH; its activity decreases as the solution is made more acidic or more basic than the ideal (optimal) pH (**Figure 8.19**). As an example, consider the human digestive system (see Section 51.3). The pH inside the human stomach is highly acidic, around pH 1.5. However, many enzymes that hydrolyze macromolecules in the intestines, such as proteases, have pH optima in the neutral range. So when food enters the small intestine, a buffer (bicarbonate) is secreted into the intestine to raise the pH to 6.5. This allows the hydrolytic enzymes to be active and digest the food.

An important factor in the effect of pH on enzyme function is ionization of the carboxyl, amino, and other groups on either the substrate or the enzyme. In neutral or basic solutions, carboxyl groups (—COOH) release H^+ to become negatively charged carboxylate groups (—COO$^-$). However, in neutral or acidic solutions, amino groups (—NH$_2$) accept H^+ to become positively charged —NH$_3^+$ groups (see the discussion of acids and bases in Section 2.4). Thus in a neutral solution, an amino group is electrically attracted to a carboxyl group on another molecule

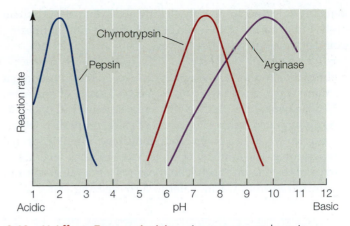

8.19 pH Affects Enzyme Activity An enzyme catalyzes its reaction at a maximum rate. The activity curve for each enzyme peaks at its optimal pH. For example, pepsin is active in the acidic environment of the stomach, whereas chymotrypsin is active in the small intestine.

or another part of the same molecule, because both groups are ionized and have opposite charges. If the pH changes, however, the ionization of these groups may change. For example, at a low pH (high H^+ concentration, such as the stomach contents where the enzyme pepsin is active), the excess H^+ may react with —COO$^-$ to form —COOH. If this happens, the group is no longer negatively charged and can no longer interact with positively charged groups in the protein, so the folding of the protein may be altered. If such a change occurs at the active site of an enzyme, the enzyme may no longer be able to bind to its substrate.

TEMPERATURE AFFECTS ENZYME ACTIVITY In general, warming increases the rate of a chemical reaction because a greater proportion of the reactant molecules have enough kinetic energy to provide the activation energy for the reaction. Enzyme-catalyzed reactions are no different (**Figure 8.20**). However, temperatures that are too high inactivate enzymes, because at high temperatures enzyme molecules vibrate and twist so rapidly that some of their noncovalent bonds break. When an enzyme's tertiary structure is changed by heat it loses its function. Some enzymes

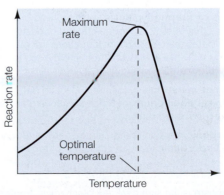

8.20 Temperature Affects Enzyme Activity Each enzyme is most active at a particular optimal temperature. At higher temperatures the enzyme becomes denatured and inactive; this explains why the activity curve falls off abruptly at temperatures above the optimal.

denature at temperatures only slightly above that of the human body, but a few are stable even at the boiling point (or freezing point) of water. All enzymes, however, have an optimal temperature for activity.

Individual organisms adapt to changes in the environment in many ways, one of which is based on groups of enzymes, called **isozymes**, that catalyze the same reaction but have different amino acid compositions and physical properties. Different isozymes within a given group may have different optimal temperatures. The rainbow trout, for example, has several isozymes of the enzyme acetylcholinesterase. If a rainbow trout is transferred from warm water to near-freezing water (2°C),

the fish produces an isozyme of acetylcholinesterase that is different from the one it produces at the higher temperature. The new isozyme has a lower optimal temperature, allowing the fish's nervous system to perform normally in the colder water.

In general, enzymes adapted to warm temperatures do not denature at those temperatures, because their tertiary structures are held together largely by covalent bonds, such as charge interactions or disulfide bridges, instead of the more heat-sensitive weak chemical interactions. Most enzymes in humans are more stable at high temperatures than those of the bacteria that infect us, so that a moderate fever tends to denature bacterial enzymes, but not our own.

RECAP 8.5

The rates of most enzyme-catalyzed reactions are affected by interacting molecules (such as inhibitors and activators) and by environmental factors (such as temperature and pH). Reversible phosphorylation is another important mechanism for regulating enzyme activity.

- What is the difference between reversible and irreversible enzyme inhibition? **See p. 157**

- How are allosteric enzymes regulated? **See pp. 159–160 and Figure 8.17**

- Explain the concept of feedback inhibition. How might the reactions shown in Figure 8.18 fit into a systems diagram such as the one shown in Figure 8.14? **See p. 160**

How are enzymes used in other industrial processes?

ANSWER

The commercial application of purified enzymes is a multibillion-dollar industry. Examples in the food industry include pectinase, an enzyme that hydrolyzes plant cell wall components and is used in clarifying fruit juices; and glucoamylase, which hydrolyzes starch and is used to make the widely used sweetener, high-fructose corn syrup. Many enzymes used in industry come from microbes such as bacteria and fungi. One reason is that microbes are relatively easy to grow in large quantities in a controlled industrial setting. Another reason is that microbes use enzymes to hydrolyze molecules in their environment, so they often either secrete the enzymes or carry them exposed on the cell surface. This makes extracting and purifying these enzymes relatively straightforward.

CHAPTER SUMMARY 8

8.1 What Physical Principles Underlie Biological Energy Transformations?

- Energy is the capacity to do work. In a biological system, the usable energy is called **free energy** (G). The unusable energy is **entropy** (S), a measure of the disorder in the system.

- **Potential energy** is the energy of state or position; it includes the energy stored in chemical bonds. **Kinetic energy** is the energy of motion; it is the type of energy that can do work.

- The **laws of thermodynamics** apply to living organisms. The first law states that energy cannot be created or destroyed. The second law states that energy transformations decrease the amount of energy available to do work (free energy) and increase disorder. **Review Figure 8.2**

- The change in free energy (ΔG) of a reaction determines its point of **chemical equilibrium**, at which the forward and reverse reactions proceed at the same rate.

- An **exergonic** reaction releases free energy and has a negative ΔG. An **endergonic** reaction consumes or requires free energy and has a positive ΔG. Endergonic reactions proceed only if free energy is provided. **Review Figure 8.3**

- **Metabolism** is the sum of all the biochemical (metabolic) reactions in an organism. **Catabolic reactions** are associated with the breakdown of complex molecules and release energy (are exergonic). **Anabolic reactions** build complexity in the cell and are endergonic.

8.2 What Is the Role of ATP in Biochemical Energetics?

- Adenosine triphosphate (ATP) serves as an energy currency in cells. Hydrolysis of ATP releases a relatively large amount of free energy.

- The ATP cycle couples exergonic and endergonic reactions, harvesting free energy from exergonic reactions, and providing free energy for endergonic reactions. **Review Figure 8.6, ACTIVITY 8.1**

8.3 What Are Enzymes?

- The rate of a chemical reaction is independent of ΔG but is determined by the energy barrier. **Review Figure 8.8**

- Enzymes are protein catalysts that affect the rates of biological reactions by lowering the energy barrier, supplying the **activation energy** (E_a) needed to initiate reactions. **Review Figure 8.10, ACTIVITY 8.2**

- A **substrate** binds to the enzyme's active site—the site of catalysis—forming an **enzyme–substrate (ES) complex**. Enzymes are highly specific for their substrates. **Review Figure 8.9**

continued

8.4 How Do Enzymes Work?

- At the active site, a substrate can be oriented correctly, chemically modified, or strained. As a result, the substrate readily forms its **transition state**, and the reaction proceeds. **Review Figure 8.11**

- Binding substrate causes many enzymes to change shape, exposing their active site(s) and allowing catalysis. The change in enzyme shape caused by substrate binding is known as **induced fit**. **Review Figure 8.12**

- Some enzymes require other substances, known as cofactors, to carry out catalysis. Prosthetic groups are permanently bound to enzymes; coenzymes are not. A coenzyme can be considered a substrate, as it is changed by the reaction and then released from the enzyme.

- Substrate concentration affects the rate of an enzyme-catalyzed reaction.

8.5 How Are Enzyme Activities Regulated?

- Metabolism is organized into pathways in which the product of one reaction is a reactant for the next reaction. Each reaction in the pathway is catalyzed by a different enzyme.

- Enzyme activity is subject to regulation. Some inhibitors bind irreversibly to enzymes. Others bind reversibly. **Review Figures 8.15, 8.16, ANIMATED TUTORIAL 8.1**

- An allosteric effector binds to a site other than the active site and stabilizes the active or inactive form of an enzyme. **Review Figure 8.17, ANIMATED TUTORIAL 8.2**

- The end product of a metabolic pathway may inhibit an enzyme that catalyzes the commitment step of that pathway. **Review Figure 8.18**

- Reversible phosphorylation is another important mechanism for regulating enzyme activity.

- Enzymes are sensitive to their environments. Both pH and temperature affect enzyme activity. **Review Figures 8.19, 8.20**

 Go to the Interactive Summary to review key figures, Animated Tutorials, and Activities Life10e.com/is8

CHAPTER**REVIEW**

REMEMBERING

1. Coenzymes differ from enzymes in that coenzymes are
   a. only active outside the cell.
   b. polymers of amino acids.
   c. smaller molecules, such as vitamins.
   d. specific for one reaction.
   e. always carriers of high-energy phosphate.

2. Which statement about thermodynamics is *true*?
   a. Free energy is used up in an exergonic reaction.
   b. Free energy cannot be used to do work.
   c. The total amount of energy can change after a chemical transformation.
   d. Free energy can be kinetic but not potential energy.
   e. Entropy has a tendency to increase.

3. The active site of an enzyme
   a. never changes shape.
   b. forms no chemical bonds with substrates.
   c. determines, by its structure, the specificity of the enzyme.
   d. looks like a lump projecting from the surface of the enzyme.
   e. changes the ΔG of the reaction.

4. The molecule ATP is
   a. a component of most proteins.
   b. high in energy because of the presence of adenine.
   c. required for many energy-transforming biochemical reactions.
   d. a catalyst.
   e. used in some exergonic reactions to provide energy.

5. In an enzyme-catalyzed reaction,
   a. a substrate does not change.
   b. the rate decreases as substrate concentration increases.
   c. the enzyme can be permanently changed.
   d. strain may be added to a substrate.
   e. the rate is not affected by substrate concentration.

6. Which statement about enzyme inhibitors is *not* true?
   a. A competitive inhibitor binds to the active site of the enzyme.
   b. An allosteric inhibitor binds to a site on the active form of the enzyme.
   c. A noncompetitive inhibitor binds to 2-3,10-1118-19,20-21
   d. a site other than the active site.
   e. Noncompetitive inhibition cannot be completely overcome by the addition of more substrate.
   f. Competitive inhibition can be completely overcome by the addition of more substrate.

UNDERSTANDING & APPLYING

7. What makes it possible for endergonic reactions to proceed in organisms?

8. Consider two proteins: one is an enzyme dissolved in the cytosol of a cell, the other is an ion channel in its plasma membrane. Contrast the structures of the two proteins, indicating at least two important differences.

9. Plot free energy versus the time course of an endergonic reaction, and the same for an exergonic reaction. Include the activation energy on both plots. Label E_a and ΔG on both graphs.

10. When potatoes are peeled, the enzyme polyphenol oxidase causes discoloration by catalyzing the oxidation of certain molecules, using O_2 as a substrate. Explain the following observations:

    a. If potatoes are peeled under water and kept there, browning is reduced.

    b. Potatoes that have been boiled at 100°C and then sliced do not turn brown.

    c. If lemon juice (pH 3) is applied to newly peeled potatoes, they do not brown.

ANALYZING & EVALUATING

11. Consider an enzyme that is subject to allosteric regulation. If a competitive inhibitor (not an allosteric inhibitor) is added to a solution containing such an enzyme, the ratio of enzyme molecules in the active form to those in the inactive form increases. Explain this observation.

12. In humans, hydrogen peroxide (H_2O_2) is a dangerous toxin produced as a by-product of several metabolic pathways. The accumulation of H_2O_2 is prevented by its conversion to harmless H_2O, a reaction catalyzed by the appropriately named enzyme catalase. Air pollutants can inhibit this enzyme and leave individuals susceptible to tissue damage by H_2O_2. How would you investigate whether catalase has an allosteric or a nonallosteric mechanism, and whether the pollutants are acting as competitive or noncompetitive inhibitors?

$$2\,H_2O_2 \xrightarrow{\text{Catalase}} O_2 + 2\,H_2O$$

Go to BioPortal at **yourBioPortal.com** for Animated Tutorials, Activities, LearningCurve Quizzes, Flashcards, and many other study and review resources.

9 Pathways That Harvest Chemical Energy

CHAPTER**OUTLINE**

9.1 How Does Glucose Oxidation Release Chemical Energy?

9.2 What Are the Aerobic Pathways of Glucose Catabolism?

9.3 How Does Oxidative Phosphorylation Form ATP?

9.4 How Is Energy Harvested from Glucose in the Absence of Oxygen?

9.5 How Are Metabolic Pathways Interrelated and Regulated?

THE "OBESITY EPIDEMIC" is probably the biggest recent story about the health of people in the developed world. There have been so many declarations by physicians, public health experts, and politicians that it might be easy to think it is all exaggerated. However, with 15 percent of children and 30 percent of adults in the United States described as obese, we face dramatic increases in diseases associated with obesity such as diabetes, heart disease, and cancer. The human and financial toll of this will be huge. What can be done?

Most people agree that in the majority of cases, obesity can be prevented or reduced if we eat less and exercise more. It is all a matter of energy: if we eat more energy-yielding molecules than we need to build up our bodies and to fuel activities such as brain functions and physical activity, we will store the unneeded energy as fat. Evolutionarily, this is usually advantageous, because fat stores energy in its C—C and C—H bonds for later use when food is scarce. But excess fat, as noted above, has adverse consequences.

Not all fat tissues, or adipose tissues, are the same. White adipose tissue (sometimes referred to simply as "white fat") is used primarily to store energy. But some adipose tissue is brown because of its high concentration of mitochondria, which have iron-containing pigments. When energy-rich molecules in brown fat are catabolized, the stored energy is released not as chemical energy but as heat. The cells in brown fat make a protein called UCP1 (uncoupling protein 1) that inserts into the inner membranes of mitochondria, making them permeable to protons (H^+). As you will see,

Built-in Heater A newborn baby has abundant brown fat stores, which can be oxidized to release heat instead of chemical energy.

the general impermeability of these membranes to H^+ is key to coupling the catabolism of molecules such as fats to the release of their stored energy in chemical form (to make ATP). If the membranes become permeable to H^+, this coupling is lost and the stored energy is released as heat.

Human infants are born with a lot of brown fat in their back and shoulder regions—it comprises about 5 percent of their body weight. Because infants have a high surface area-to-volume ratio, they tend to lose a lot of heat. One way that they keep warm is to produce heat in their brown fat tissues. As a child grows up, the brown fat content of the body is reduced. Adults have mostly white fat, which has less UCP1 and generates less heat when the fat is catabolized.

It used to be thought that brown fat was not present in adult humans. But recently it was confirmed that adults do have some brown fat, in the upper chest and neck area. Interestingly, obese people have less brown fat than lean people, leading to the suggestion that brown fat may be somehow associated with a tendency to remain lean. A tantalizing possibility is that recent discoveries about brown fat and UCP1 may lead to new methods for weight loss.

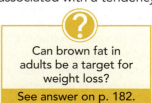

Can brown fat in adults be a target for weight loss?

See answer on p. 182.

9.1 How Does Glucose Oxidation Release Chemical Energy?

Energy is stored in the covalent bonds of fuels, and it can be released and transformed. Wood burning in a campfire releases energy as heat and light. In cells, fuel molecules release chemical energy that is used to make ATP, which in turn drives endergonic reactions. ATP is central to the energy transformations of all living organisms. Photosynthetic cells and organisms use energy from sunlight to synthesize their own fuels, as we will describe in Chapter 10. In nonphotosynthetic cells, the most common chemical fuel is the sugar glucose ($C_6H_{12}O_6$). Other molecules, including other carbohydrates, fats, and proteins, can supply energy to the whole organism. However, to release their energy they must be converted into glucose or intermediate compounds that can enter into the various pathways of glucose metabolism.

We should note here that some prokaryotes (bacteria and archaea) can harvest chemical energy from inorganic sources such as metal ions, hydrogen sulfide, and ammonia. In addition, prokaryotes use a variety of metabolic pathways to convert chemical energy (both organic and inorganic) into a usable form. These pathways are forms of anaerobic respiration or fermentation, which we will discuss in Sections 9.3 and 9.4.

In this section we explore how cells obtain energy from glucose by the chemical process of oxidation, which is carried out through a series of metabolic pathways. Five principles govern metabolic pathways:

- A complex chemical transformation occurs in a series of separate reactions that form a metabolic pathway.

- Each reaction is catalyzed by a specific enzyme.

- Many metabolic pathways are similar in all organisms, from bacteria to humans.

- In eukaryotes, many metabolic pathways are compartmentalized, with certain reactions occurring inside specific organelles.

- Some key enzymes in each metabolic pathway can be inhibited or activated to alter the rate of the pathway.

Cells trap free energy while metabolizing glucose

As we saw in Section 2.3, the familiar process of combustion (burning) is very similar to the chemical processes that release energy in cells. If glucose is burned in a flame, it reacts with oxygen gas (O_2), forming carbon dioxide and water and releasing energy in the form of heat. The balanced equation for the complete reaction is

$$C_6H_{12}O_6 + 6\ O_2 \rightarrow 6\ CO_2 + 6\ H_2O + \text{free energy}$$
$$(\Delta G = -686\ \text{kcal/mol})$$

This is an oxidation–reduction reaction (see next page), in which glucose loses electrons (becomes oxidized) and oxygen gains them (becomes reduced). The energy that is released can be used to do work. The same equation applies to the overall catabolism of glucose in cells. However, in contrast to combustion, the catabolism of glucose is a multistep pathway. Each step is catalyzed by an enzyme, and the process is compartmentalized.

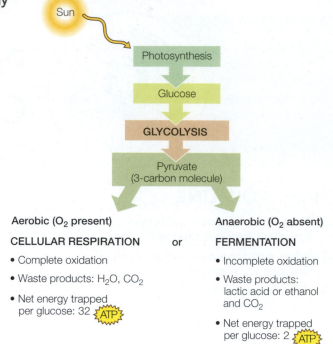

9.1 Energy for Life Many prokaryotes and all eukaryotes obtain their energy from the food compounds produced by photosynthesis. They convert these compounds into glucose, which they metabolize to trap energy in ATP.

Unlike combustion, glucose catabolism is tightly regulated and occurs at temperatures compatible with life.

The glucose catabolism pathway "extracts" the energy stored in the covalent bonds of glucose and stores it instead in ATP molecules via the phosphorylation reaction:

$$ADP + P_i + \text{free energy} \rightarrow ATP$$

ATP is the energy currency of cells (see Chapter 8). The energy trapped in ATP can be used to do cellular work—such as movement of muscles or active transport across membranes—just as the energy captured from combustion can be used to do mechanical work.

The standard free energy change resulting from the complete conversion of glucose and O_2 to CO_2 and water, whether by combustion or by metabolism, is −686 kcal/mol (−2,870 kJ/mol). Thus the overall reaction is highly exergonic and can drive the endergonic formation of a great deal of ATP from ADP and phosphate. Note that in the discussion that follows, "energy" means free energy.

Three catabolic processes harvest the energy in the chemical bonds of glucose: glycolysis, cellular respiration, and fermentation (**Figure 9.1**). All three processes involve pathways made up of many distinct chemical reactions.

- **Glycolysis** begins glucose catabolism. Through a series of chemical rearrangements, glucose is converted to two molecules of the three-carbon product **pyruvate**, and a small amount of energy is captured in usable forms. Glycolysis is an **anaerobic** process because it does not require O_2.

- **Cellular respiration** uses O_2 from the environment, and thus it is **aerobic**. Each pyruvate molecule is completely converted into three molecules of CO_2 through a set of catabolic pathways including pyruvate oxidation, the citric acid cycle, and an electron transport system (the respiratory chain). In the process, a great deal of the energy stored in the covalent bonds of pyruvate is captured to form ATP.

- **Fermentation** does not involve O_2 (it is anaerobic). Fermentation converts pyruvate into lactic acid or ethyl alcohol (ethanol), which are still relatively energy-rich molecules. Because the breakdown of glucose is incomplete, much less energy is released when glycolysis is coupled to fermentation than when it is coupled to cellular respiration.

Redox reactions transfer electrons and energy

As we discussed in Section 8.2, the addition of a phosphate group to ADP to make ATP is an endergonic reaction (see Figure 8.6). It is achieved by coupling an exergonic reaction to ATP production: the energy released in the exergonic reaction is used to drive ATP synthesis. Electrons are transferred in the exergonic reaction. A reaction in which one substance transfers one or more electrons to another substance is called an **oxidation–reduction**, or **redox**, **reaction**.

- **Reduction** is the gain of one or more electrons by an atom, ion, or molecule.
- **Oxidation** is the loss of one or more electrons.

Oxidation and reduction *always occur together*: as one chemical is oxidized, the electrons it loses are transferred to another chemical, reducing it. In a redox reaction, we call the reactant that becomes reduced an oxidizing agent and the one that becomes oxidized a reducing agent:

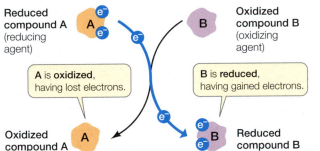

In the metabolism of glucose, glucose is the reducing agent (electron donor) and O_2 is the oxidizing agent (electron acceptor).

Although oxidation and reduction are always defined in terms of electron traffic, it is often simpler to think in terms of the gain or loss of hydrogen atoms. The transfer of electrons is often associated with the transfer of hydrogen ions ($H = H^+ + e^-$). So when a molecule loses hydrogen atoms, it becomes oxidized.

In general, the more reduced a molecule is, the more energy is stored in its covalent bonds (**Figure 9.2**). In a redox reaction, some energy is transferred from the reducing agent to the reduced product. The rest remains in the reducing agent or is lost to entropy. As we will see, some of the key reactions of glycolysis and cellular respiration are highly exergonic redox reactions.

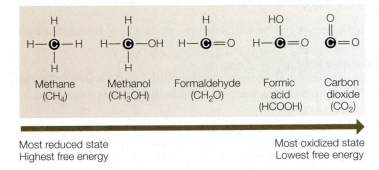

Most reduced state
Highest free energy

Most oxidized state
Lowest free energy

9.2 Oxidation, Reduction, and Energy The more oxidized a carbon atom in a molecule is, the less its stored free energy.

The coenzyme NAD⁺ is a key electron carrier in redox reactions

Section 8.4 describes the role of coenzymes, small molecules that assist in enzyme-catalyzed reactions. ADP acts as a coenzyme when it picks up energy released in an exergonic reaction and packages it to form ATP. The coenzyme nicotinamide adenine dinucleotide (NAD^+) acts as an electron carrier in redox reactions:

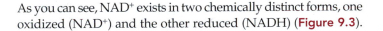

As you can see, NAD^+ exists in two chemically distinct forms, one oxidized (NAD^+) and the other reduced (NADH) (**Figure 9.3**).

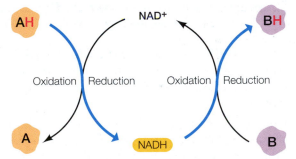

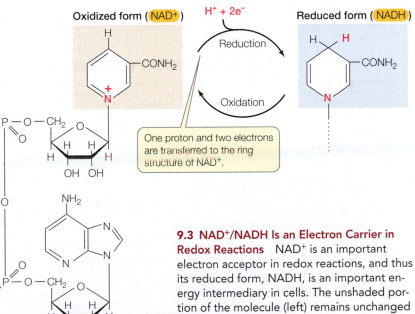

9.3 NAD⁺/NADH Is an Electron Carrier in Redox Reactions NAD^+ is an important electron acceptor in redox reactions, and thus its reduced form, NADH, is an important energy intermediary in cells. The unshaded portion of the molecule (left) remains unchanged by the redox reaction.

Both forms participate in redox reactions. The reduction reaction

$$NAD^+ + H^+ + 2\,e^- \rightarrow NADH$$

is actually the transfer of a proton (the hydrogen ion, H^+) and two electrons, which are released by the accompanying oxidization reaction.

The electrons do not remain with the coenzyme. Oxygen is highly electronegative and readily accepts electrons from NADH. The oxidation of NADH by O_2 (which occurs in several steps)

$$NADH + H^+ + \tfrac{1}{2}\,O_2 \rightarrow NAD^+ + H_2O$$

is highly exergonic, with a standard free energy change at pH 7 (ΔG) of –52.4 kcal/mol (–219 kJ/mol). Note that the oxidizing agent appears here as "$\tfrac{1}{2}\,O_2$" instead of "O." This notation emphasizes that it is molecular oxygen, O_2, that acts as the oxidizing agent.

Just as ATP can be thought of as a package of 7.3 kcal/mol (30.5 kJ/mol) of free energy, NADH can be thought of as a larger package of free energy (52.4 kcal/mol, see above). NAD^+ is a common electron carrier in cells, but not the only one. Another carrier, flavin adenine dinucleotide (FAD), also transfers electrons during glucose metabolism.

An overview: Harvesting energy from glucose

Both eukaryotic and prokaryotic cells can harvest energy from glucose using different combinations of the following metabolic pathways:

- Under aerobic conditions, when O_2 is available as the final electron acceptor, four pathways operate (**Figure 9.4A**). Glycolysis is followed by the three pathways of cellular respiration: pyruvate oxidation, the citric acid cycle (also called the Krebs cycle or the tricarboxylic acid cycle), and electron transport/ATP synthesis (also called the respiratory chain).

- In eukaryotes and many prokaryotes, pyruvate oxidation, the citric acid cycle, and the respiratory chain do not function under anaerobic conditions. The pyruvate produced by

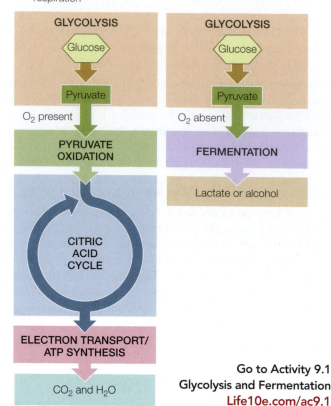

(A) Glycolysis and cellular respiration   (B) Glycolysis and fermentation

Go to Activity 9.1
Glycolysis and Fermentation
Life10e.com/ac9.1

9.4 Energy-Yielding Metabolic Pathways Energy-yielding reactions can be grouped into five metabolic pathways: glycolysis, pyruvate oxidation, the citric acid cycle, the respiratory chain/ATP synthesis, and fermentation. (A) The three lower pathways occur only in the presence of O_2 and are collectively referred to as cellular respiration. (B) When O_2 is unavailable, glycolysis is followed by fermentation.

glycolysis is further metabolized by fermentation (**Figure 9.4B**). Some prokaryotes, however, are able to harvest energy in pathways involving respiratory chains even in the absence of oxygen (anaerobic respiration; see Section 9.3).

The five pathways shown in Figure 9.4 occur in different locations in the cell (**Table 9.1**).

TABLE 9.1

Cellular Locations for Major Energy Pathways in Eukaryotes and Prokaryotes

	Eukaryotes	Prokaryotes
	In cytoplasm	In cytoplasm
	Glycolysis	Glycolysis
	Fermentation	Fermentation
		Citric acid cycle
	Inside mitochondrion	On plasma membrane
	Matrix	Pyruvate oxidation
	Citric acid cycle	Respiratory chain
	Pyruvate oxidation	
	Inner membrane	
	Respiratory chain	

Go to Activity 9.2
Energy Pathways in Cells
Life10e.com/ac9.2

Now that you have an overview of the metabolic pathways that harvest energy from glucose, let's take a closer look at the three pathways involved in aerobic glucose catabolism: glycolysis, pyruvate oxidation, and the citric acid cycle.

9.2 What Are the Aerobic Pathways of Glucose Catabolism?

The aerobic pathways of glucose catabolism oxidize glucose completely to CO_2 and H_2O. Initially, the glycolysis reactions convert the six-carbon glucose molecule to two three-carbon pyruvate molecules (**Figure 9.5**). Pyruvate is then converted to CO_2 in a second series of reactions beginning with pyruvate oxidation and followed by the citric acid cycle. In addition to generating CO_2, the oxidation events are coupled with the reduction of electron carriers, mostly NAD^+. Thus much of the chemical energy in the C—C and C—H bonds of glucose is transferred to NAD^+ to form NADH. Ultimately this energy will be transferred to ATP, but this comes in a separate series of reactions involving electron transport, called the respiratory chain. In the respiratory chain, redox reactions result in the oxidative phosphorylation of ADP by ATP synthase. We will begin our consideration of the catabolism of glucose with a closer look at glycolysis.

In glycolysis, glucose is partially oxidized and some energy is released

Glycolysis takes place in the cytosol and involves ten enzyme-catalyzed reactions. During glycolysis, some of the covalent bonds between carbon and hydrogen atoms in the glucose molecule are oxidized, releasing some of the stored energy. The products are two molecules of pyruvate (pyruvic acid), two molecules of ATP, and two molecules of NADH. Glycolysis can be divided into two stages: the initial energy-investing reactions that consume ATP, and the energy-harvesting reactions that produce ATP and NADH (see Figure 9.5).

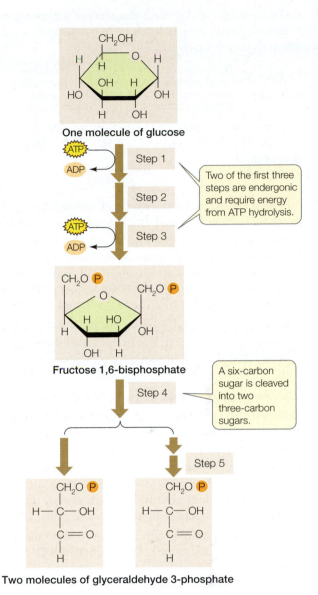

One molecule of glucose

Step 1

Step 2

Step 3

Two of the first three steps are endergonic and require energy from ATP hydrolysis.

Fructose 1,6-bisphosphate

Step 4

A six-carbon sugar is cleaved into two three-carbon sugars.

Step 5

Two molecules of glyceraldehyde 3-phosphate

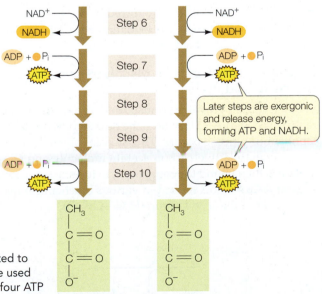

Step 6

Step 7

Step 8

Step 9

Step 10

Later steps are exergonic and release energy, forming ATP and NADH.

Two molecules of pyruvate

9.5 Glycolysis Converts Glucose into Pyruvate Glucose is converted to pyruvate in ten enzyme-catalyzed steps. Along the way, two ATP are used (Steps 1 and 3), two NAD^+ are reduced to two NADH (Step 6), and four ATP are produced (Steps 7 and 10).

To help you understand the process without getting into extensive detail, we will focus on two consecutive reactions in this pathway (Steps 6 and 7 in Figure 9.5).

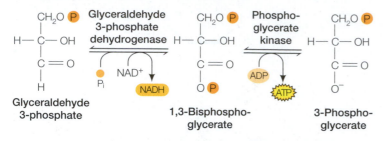

These are examples of two types of reaction that occur repeatedly in glycolysis and in many other metabolic pathways:

1. *Oxidation–reduction*: The first reaction is exergonic—more than 50 kcal/mol of energy are released in the oxidation of glyceraldehyde 3-phosphate. (Look at the bottom carbon atom, where an H is replaced by an O.) The energy is trapped via the reduction of NAD$^+$ to NADH.

2. *Substrate-level phosphorylation*: The second reaction in this series is also exergonic, but in this case less energy is released. It is enough to transfer a phosphate directly from the substrate to ADP, forming ATP.

The end product of glycolysis, pyruvate, is somewhat more oxidized than glucose. In the presence of O$_2$, further oxidation can occur. In prokaryotes these subsequent reactions take place in the cytosol, but in eukaryotes they take place in the mitochondrial matrix.

To summarize:

- The initial steps of glycolysis use the energy of hydrolysis of two ATP molecules per glucose molecule.

- The remaining steps produce four ATP molecules per glucose molecule, so the net production of ATP is two molecules.

- Glycolysis produces two molecules of NADH.

If O$_2$ is present, glycolysis is followed by the three stages of cellular respiration: pyruvate oxidation, the citric acid cycle, and the respiratory chain/ATP synthesis.

Pyruvate oxidation links glycolysis and the citric acid cycle

In eukaryotes, pyruvate is transported into the mitochondrial matrix (see Figure 5.11), where the next step in the aerobic catabolism of glucose occurs. This step involves the oxidation of pyruvate to a two-carbon acetate molecule and CO$_2$. The acetate is then bound to coenzyme A to form **acetyl coenzyme A (acetyl CoA)**; CoA is used in various biochemical reactions as a carrier of the acetyl group (H$_3$C—C=O).

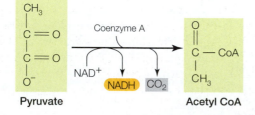

This is the link between glycolysis and further oxidative reactions (see Figure 9.4).

The formation of acetyl CoA is a multistep reaction catalyzed by the pyruvate dehydrogenase complex, which contains 60 individual proteins and 5 different coenzymes. The overall reaction is exergonic, and one molecule of NAD$^+$ is reduced to NADH. But the main role of acetyl CoA is to donate its acetyl group to the four-carbon compound oxaloacetate, forming the six-carbon molecule citrate. This initiates the citric acid cycle, one of life's most important energy-harvesting pathways.

The citric acid cycle completes the oxidation of glucose to CO$_2$

Acetyl CoA is the starting point for the citric acid cycle. This pathway of eight reactions completely oxidizes the two-carbon acetyl group to two molecules of CO$_2$. The free energy released from these reactions is captured by GDP and the electron carriers NAD$^+$ and FAD (**Figure 9.6**). [Remember from Section 7.2 that GDP (guanosine diphosphate) is a nucleoside diphosphate

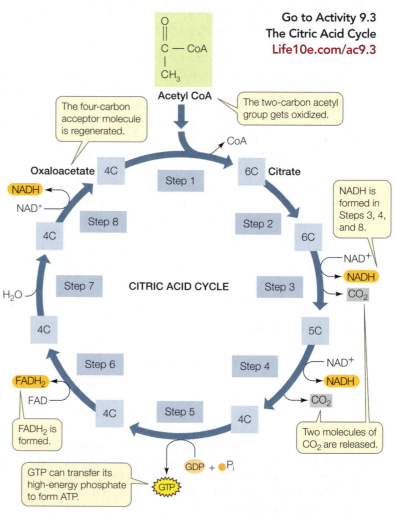

9.6 The Citric Acid Cycle   The citric acid cycle involves eight steps; in the last step, the starting material acceptor, oxaloacetate, is regenerated. Energy is released and captured by reducing NAD$^+$ or FAD, or by producing GTP. "6C," "5C," and so on indicate the number of carbon atoms in each intermediate in the cycle.

like ADP.] This is a cycle because the starting material, oxaloacetate, is regenerated in the last step and is ready to accept another acetate group from acetyl CoA. The citric acid cycle operates twice for each glucose molecule that enters glycolysis (once for each pyruvate that enters the mitochondrion).

Let's focus on the final reaction of the cycle (Step 8 in Figure 9.6), as an example of the kind of reaction that occurs:

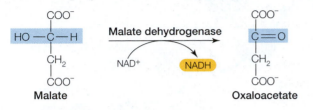

This oxidation reaction (see the carbon atom highlighted in blue) is exergonic, and the released energy is trapped by NAD$^+$, forming NADH. With four such reactions (the FADH$_2$ produced in Step 6 is a reduced coenzyme similar to NADH), the citric acid cycle harvests a great deal of chemical energy from the oxidation of acetyl CoA.

To summarize:

- The inputs to the citric acid cycle are acetate (in the form of acetyl CoA), water, GDP, and the oxidized electron carriers NAD$^+$ and FAD.

- The outputs are carbon dioxide, reduced electron carriers (NADH and FADH$_2$), and a small amount of GTP. The energy in the terminal phosphate of GTP is transferred to ATP:

$$GTP + ADP \rightarrow ATP + GDP$$

Thus the citric acid cycle releases two carbons as CO_2 and produces four reduced electron carrier molecules.

Overall, for each molecule of glucose that is oxidized, two molecules of pyruvate are produced during glycolysis, and after oxidation these feed two turns of the citric acid cycle. So the oxidation of one glucose molecule yields:

- Six CO_2

- Ten NADH (two in glycolysis, two in pyruvate oxidation, and six in the citric acid cycle)

- Two FADH$_2$

- Four ATP

Pyruvate oxidation and the citric acid cycle are regulated by the concentrations of starting materials

We have seen how pyruvate, a three-carbon molecule, is completely oxidized to CO_2 by pyruvate dehydrogenase and the citric acid cycle. For the cycle to continue, the starting molecules—acetyl CoA and oxidized electron carriers—must all be replenished. The electron carriers are reduced during the citric acid cycle and in Step 6 of glycolysis (see Figure 9.5) and they must be reoxidized:

$$NADH \rightarrow NAD^+ + H^+ + 2\ e^-$$
$$FADH_2 \rightarrow FAD + 2\ H^+ + 2\ e^-$$

These oxidation reactions occur as coupled redox reactions in which other molecules get reduced. When it is present, O_2 is

the molecule that eventually accepts these electrons, and it is reduced to form H_2O.

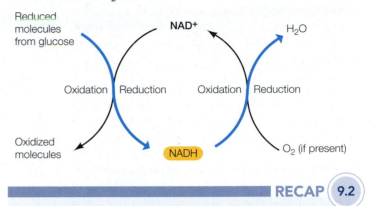

■ **RECAP** (9.2)

The oxidation of glucose in the presence of O_2 involves glycolysis, pyruvate oxidation, and the citric acid cycle. In glycolysis, glucose is converted to pyruvate with some energy capture. Pyruvate is oxidized first to acetyl CoA by pyruvate dehydrogenase, then all the way to CO_2 by the citric acid cycle, releasing energy that is captured in the form of reduced electron carriers.

- What is the net energy yield of glycolysis in terms of energy invested and energy harvested? **See p. 169 and Figure 9.5**

- What role does pyruvate oxidation play in relation to the citric acid cycle? **See p. 170**

- Explain why reoxidation of NADH is crucial for the continuation of the citric acid cycle. **See p. 171 and Figure 9.6**

Pyruvate oxidation and the citric acid cycle cannot continue operating unless O_2 is available to receive electrons during the reoxidation of reduced electron carriers. However, these electrons are not passed directly to O_2, as you will learn next.

(9.3) How Does Oxidative Phosphorylation Form ATP?

The overall process of ATP synthesis resulting from the reoxidation of electron carriers in the presence of O_2 is called **oxidative phosphorylation**. In this section we describe oxidative phosphorylation as it occurs in mitochondria, but the same process occurs in prokaryotes on the plasma membrane (see Table 9.1). Two components of the process can be distinguished:

1. *Electron transport.* The electrons from NADH and FADH$_2$ pass through the **respiratory chain**, a series of membrane-associated electron carriers. The flow of electrons along this pathway results in the active transport of protons out of the mitochondrial matrix and across the inner mitochondrial membrane, creating a proton concentration gradient.

2. *Chemiosmosis.* The protons diffuse back into the mitochondrial matrix through a channel protein, **ATP synthase**, which couples this diffusion to the synthesis of ATP. As we mentioned in the opening story, the inner mitochondrial membrane is normally impermeable to protons, so the only way for them to follow their concentration gradient is through the channel.

Before we proceed with the details of these pathways, let's consider an important question: Why should the respiratory chain be such a complex process? Why don't cells use the following single step?

$$2\,NADH + 2\,H^+ + O_2 \rightarrow 2\,NAD^+ + 2\,H_2O$$

The answer is that this reaction would be untamable. It is extremely exergonic—and oxidizing NADH this way would be rather like setting off a stick of dynamite in the cell. There is no biochemical way to harvest that burst of energy efficiently and put it to physiological use (that is, no single metabolic reaction is so endergonic as to consume a significant fraction of that energy in a single step). To control the release of energy during the oxidation of glucose, cells have evolved a lengthy respiratory chain: a series of reactions, each of which releases a small amount of energy, one step at a time.

The respiratory chain transfers electrons and protons, and releases energy

The respiratory chain is located in the inner mitochondrial membrane. Because of the extensive folding of the membrane, there is much more space for the proteins involved in the chain than there would be in a membrane with less surface area. There are several interacting components, including large integral proteins, a small peripheral protein, and a small lipid molecule. **Figure 9.7** shows a plot of the free energy released as electrons are passed between the carriers.

- Four large protein complexes (I, II, III, and IV) contain electron carriers and associated enzymes. In eukaryotes they are integral proteins of the inner mitochondrial membrane (see Figure 5.11), and three are transmembrane proteins.

- Cytochrome *c* is a small peripheral protein that lies in the intermembrane space. It is loosely attached to the outer surface of the inner mitochondrial membrane.

- Ubiquinone (often referred to as coenzyme Q10; abbreviated Q) is a small, nonpolar, lipid molecule that moves freely within the hydrophobic interior of the phospholipid bilayer of the inner mitochondrial membrane.

As illustrated in Figure 9.7, NADH passes electrons to protein complex I (called NADH-Q reductase), which in turn passes the electrons to Q. This electron transfer is accompanied by a large drop in free energy. Complex II (succinate dehydrogenase) passes electrons to Q from FADH₂, which was generated in Step 6 of the citric acid cycle (see Figure 9.6). These electrons enter the chain later than those from NADH and will ultimately produce less ATP.

Complex III (cytochrome *c* reductase) receives electrons from Q and passes them to cytochrome *c*. Complex IV (cytochrome *c* oxidase) receives electrons from cytochrome *c* and

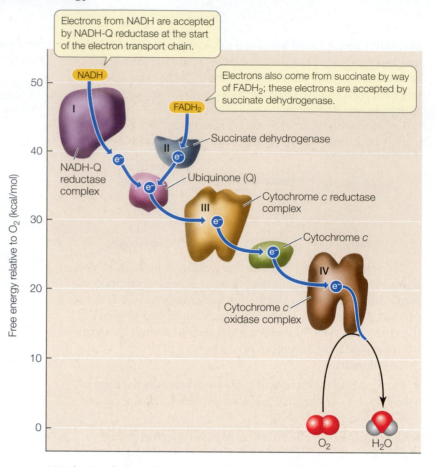

9.7 The Oxidation of NADH and FADH₂ in the Respiratory Chain
Electrons from NADH and FADH₂ are passed along the respiratory chain, a series of protein complexes in the inner mitochondrial membrane containing electron carriers and enzymes. The carriers gain free energy when they become reduced and release free energy when they are oxidized. This illustration shows the standard free energy changes along the respiratory chain.
Go to Activity 9.4 Respiratory Chain   Life10e.com/ac9.4

passes them to oxygen. Finally the reduction of oxygen to H₂O occurs:

$$O_2 + 4\,H^+ + 4\,e^- \rightarrow 2\,H_2O$$

Notice that four protons (H⁺) are also consumed in this reaction. This contributes to the proton concentration gradient across the inner mitochondrial membrane by reducing the proton concentration in the mitochondrial matrix.

During electron transport, protons are also actively transported across the membrane: electron transport within each of the three transmembrane complexes (I, III, and IV) results in the transfer of protons from the matrix to the intermembrane space (**Figure 9.8**). The mechanisms by which the three complexes transfer protons across the membrane are still not well understood but are likely to involve Q. The transfer sets up an imbalance of protons, with the impermeable inner mitochondrial membrane as a barrier. The concentration of H⁺ in the intermembrane space is higher than in the matrix, and this gradient represents a source of potential energy. The diffusion

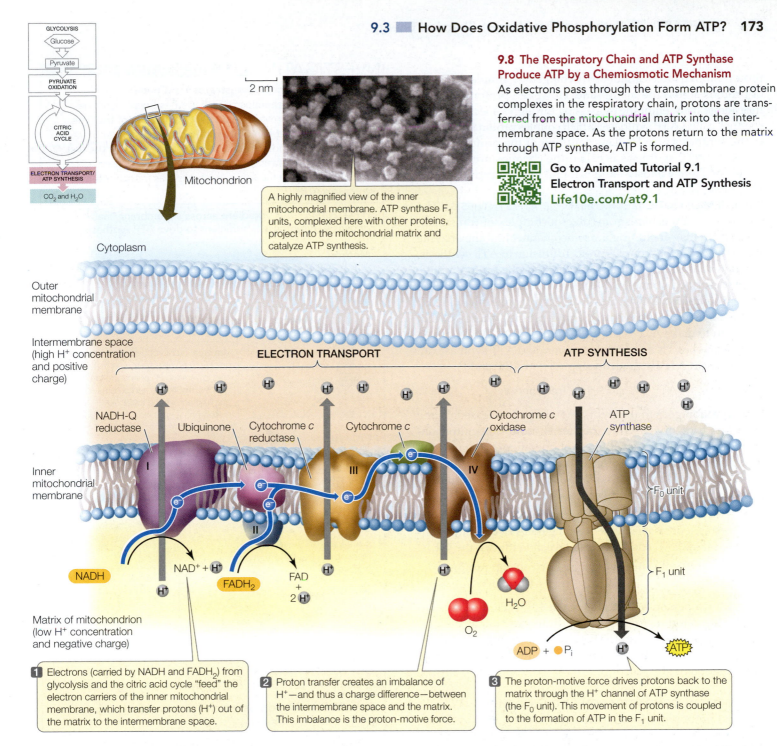

GLYCOLYSIS
Glucose
Pyruvate
PYRUVATE OXIDATION
CITRIC ACID CYCLE
ELECTRON TRANSPORT/ ATP SYNTHESIS
CO_2 and H_2O

2 nm

Mitochondrion

A highly magnified view of the inner mitochondrial membrane. ATP synthase F_1 units, complexed here with other proteins, project into the mitochondrial matrix and catalyze ATP synthesis.

9.8 The Respiratory Chain and ATP Synthase Produce ATP by a Chemiosmotic Mechanism
As electrons pass through the transmembrane protein complexes in the respiratory chain, protons are transferred from the mitochondrial matrix into the intermembrane space. As the protons return to the matrix through ATP synthase, ATP is formed.

Go to Animated Tutorial 9.1
Electron Transport and ATP Synthesis
Life10e.com/at9.1

Cytoplasm

Outer mitochondrial membrane

Intermembrane space (high H^+ concentration and positive charge)

ELECTRON TRANSPORT

ATP SYNTHESIS

H^+ ... (many H^+ labels)

NADH-Q reductase
Ubiquinone
Cytochrome c reductase
Cytochrome c
Cytochrome c oxidase
ATP synthase

Inner mitochondrial membrane

F_0 unit

F_1 unit

NADH
$NAD^+ + H^+$
$FADH_2$
FAD + 2 H^+
H_2O
O_2
ADP + P_i
H^+
ATP

Matrix of mitochondrion (low H^+ concentration and negative charge)

1 Electrons (carried by NADH and $FADH_2$) from glycolysis and the citric acid cycle "feed" the electron carriers of the inner mitochondrial membrane, which transfer protons (H^+) out of the matrix to the intermembrane space.

2 Proton transfer creates an imbalance of H^+—and thus a charge difference—between the intermembrane space and the matrix. This imbalance is the proton-motive force.

3 The proton-motive force drives protons back to the matrix through the H^+ channel of ATP synthase (the F_0 unit). This movement of protons is coupled to the formation of ATP in the F_1 unit.

of those protons across the membrane is coupled with the formation of ATP. Thus the energy originally contained in glucose and other fuel molecules is finally captured in the cellular energy currency, ATP. For each pair of electrons passed along the chain from NADH to oxygen, about 2.5 molecules of ATP are formed. $FADH_2$ oxidation produces about 1.5 ATP molecules.

Proton diffusion is coupled to ATP synthesis

All the electron carriers and enzymes of the respiratory chain, except cytochrome c, are embedded in the inner mitochondrial membrane. As we have just seen, the operation of the respiratory chain results in the active transport of protons from the mitochondrial matrix to the intermembrane space. The transmembrane protein complexes (I, III, and IV) act as proton carriers,

and as a result, the intermembrane space is more acidic than the matrix.

Because of the positive charge carried by a proton (H^+), the respiratory chain creates not only a concentration gradient but also a difference in electric charge across the inner mitochondrial membrane, making the mitochondrial matrix more negative than the intermembrane space. Together, the proton concentration gradient and the electric charge difference constitute a source of potential energy called the **proton-motive force**. This force tends to drive the protons back across the membrane, just as the charge on a battery drives the flow of electrons to discharge the battery.

The hydrophobic interior of the lipid bilayer is essentially impermeable to protons, so the potential energy of the proton-motive force cannot be discharged by simple diffusion of protons across the membrane. However, protons can diffuse across the membrane by passing through a specific proton channel, called ATP synthase, which couples proton movement to the synthesis of ATP. This coupling of the proton-motive force and ATP synthesis is called the chemiosmotic mechanism—or **chemiosmosis**—and is found in all respiring cells.

THE CHEMIOSMOTIC MECHANISM FOR ATP SYNTHESIS The chemiosmotic mechanism involves a complex of transmembrane proteins—including a proton channel and the enzyme ATP synthase—that couples proton diffusion to ATP synthesis. The potential energy of the H^+ gradient, or the proton-motive force (described above), is harnessed by ATP synthase. This protein complex has two roles: it acts as a channel allowing H^+ to diffuse back into the matrix, and it uses the energy of that diffusion to make ATP from ADP and P_i.

ATP synthesis is a reversible reaction, and ATP synthase can also act as an ATPase, hydrolyzing ATP to ADP and P_i:

$$ATP \rightleftharpoons ADP + P_i + \text{free energy}$$

If the reaction goes to the right, free energy is released. In the mitochondrion, it is used to transfer H^+ out of the mitochondrial matrix—not the usual mode of operation. If the reaction goes to the left, it uses the free energy from H^+ diffusion into the matrix to make ATP. What makes it prefer ATP synthesis? There are two answers to this question:

- ATP leaves the mitochondrial matrix for use elsewhere in the cell as soon as it is made, keeping the ATP concentration in the matrix low, and driving the reaction toward the left.
- The H^+ gradient is maintained by electron and proton transport.

Every day a person hydrolyzes about 10^{25} ATP molecules to ADP. This amounts to 9 kg, a significant fraction of the person's entire body weight! The vast majority of this ADP is "recycled"—converted back to ATP—using free energy from the oxidation of glucose.

EXPERIMENTAL DEMONSTRATION OF CHEMIOSMOSIS When it was first proposed, the idea that a proton gradient was the energy intermediate linking electron transport to ATP synthesis was a departure from the conventional thinking of the time. Scientists assumed that the energy for ATP synthesis came directly from the substrate donating the phosphate group to form ATP (i.e., from substrate-level phosphorylation; see Section 9.2). However, no intermediate could be found for this reaction in the mitochondrion, leading to the new hypothesis that chemiosmosis was the mechanism of oxidative phosphorylation. Experimental evidence was needed to support this hypothesis. The key experiment that demonstrated that a proton (H^+) gradient across a membrane could drive ATP synthesis was first performed using chloroplasts, the organelles in plants that convert the energy in sunlight into chemical energy (photosynthesis; see Chapter 10) (**Figure 9.9**). Soon thereafter, the same mechanism was shown to work in mitochondria.

INVESTIGATING**LIFE**

9.9 An Experiment Demonstrates the Chemiosmotic Mechanism The chemiosmosis hypothesis was a bold departure from the conventional scientific thinking of the time. It required an intact compartment enclosed by a membrane. Could a proton gradient drive the synthesis of ATP? The first experiments to answer this question used chloroplasts, plant organelles that use the same mechanism as mitochondria to synthesize ATP.[a]

HYPOTHESIS A H^+ gradient across a membrane that contains ATP synthase is sufficient to drive ATP synthesis.

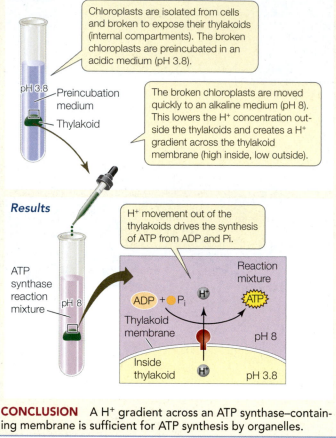

Method

Chloroplasts are isolated from cells and broken to expose their thylakoids (internal compartments). The broken chloroplasts are preincubated in an acidic medium (pH 3.8).

pH 3.8 —Preincubation medium
—Thylakoid

The broken chloroplasts are moved quickly to an alkaline medium (pH 8). This lowers the H^+ concentration outside the thylakoids and creates a H^+ gradient across the thylakoid membrane (high inside, low outside).

Results

H^+ movement out of the thylakoids drives the synthesis of ATP from ADP and Pi.

ATP synthase reaction mixture
pH 8

Reaction mixture
ADP + Pi
H^+
ATP
Thylakoid membrane
pH 8
Inside thylakoid
H^+
pH 3.8

CONCLUSION A H^+ gradient across an ATP synthase–containing membrane is sufficient for ATP synthesis by organelles.

Go to **BioPortal** for discussion and relevant links for all INVESTIGATING**LIFE** figures.

[a]Jagendorf, A. T. and E. Uribe. 1966. *Proceedings of the National Academy of Sciences USA* 55: 170–177.

Go to Animated Tutorial 9.2
Two Experiments Demonstrate the Chemiosmotic Mechanism Life10e.com/at9.2

As we have seen, the coupling of electron transport (which generates the proton gradient) with chemiosmosis is vital for the capture of free energy in the form of ATP. Uncoupling protein 1 (UCP1), which is found in the mitochondria of brown fat cells (see the opening story), demonstrates the importance of this coupling. By disrupting the gradient, UCP1 allows the energy released during electron transport to be in the form of heat, rather than chemical energy trapped in ATP.

Experimental Demonstration of the Chemiosmotic Mechanism

Original Paper

Jagendorf, A. T. and E. Uribe. 1966. ATP formation caused by acid-base transition of spinach chloroplasts. *Proceedings of the National Academy of Sciences USA* 55: 170–177.

Analyze the Data

Previous research into the force driving ATP synthesis in mitochondria led to the postulation of the chemiosmotic mechanism by Peter Mitchell. A key prediction of his hypothesis was that a chemical potential across a membrane, in this case driven by an imbalance of protons (H^+), could provide energy for the synthesis of ATP from ADP and P_i. Working with chloroplasts—which as you will see in Chapter 10 also use the chemiosmotic mechanism to make ATP—André Jagendorf and Ernest Uribe at Johns Hopkins University performed a clever experiment, using rapid changes in pH to create a proton gradient across a membrane and to drive ATP synthesis (see Figure 9.9). They isolated chloroplasts from plant cells, then broke the chloroplasts' outer membranes to expose the thylakoids, internal membrane-enclosed compartments that contain the chloroplast ATP synthase (see Figure 10.10). The broken chloroplasts were placed in a medium at a relatively high H^+ concentration (acid pH) until equilibrium was reached (the interior of the thylakoids became acidic). The broken chloroplasts were then moved to a medium with a relatively low H^+ concentration (alkaline pH). The H^+ ions moved down the concentration gradient (out of the thylakoids), driving the synthesis of ATP. Although chloroplasts normally make ATP during photosynthesis in the light, in these experiments light was not required—just the H^+ gradient across the thylakoid membrane. Mitchell was a brilliant theoretician; Jagendorf, Uribe, and others provided experimental confirmation for his ideas, which won Mitchell the Nobel Prize in 1978.

In Jagendorf and Uribe's experiment, thylakoids were preincubated in a solution at pH 3.8 (or in one experiment, pH 7) and then quickly transferred to an ATP synthase reaction mixture containing ADP and inorganic phosphate (P_i) at pH 8. ATP formation was measured in two ways. The first used the enzyme luciferase, which catalyzes the formation of a luminescent molecule if ATP is present. The second used molybdate to measure phosphorylation directly. The table shows the data from the paper.

Preincubation pH	ATP synthase reaction mixture (pH 8)	ATP formation[a]	
		Luciferase assay	Molybdate assay
3.8	Complete	141	166
7.0	Complete	12	3
3.8	– P_i[b]	12	3
3.8	– ADP	4	3
3.8	– thylakoids	7	3

[a]In nmoles ATP per mg chlorophyll (a pigment in thylakoids).
[b]A minus sign indicates a component that was omitted from the reaction mixture.

QUESTION 1

Which two experiments show that a proton gradient is necessary and sufficient to stimulate ATP formation? Explain your reasoning.

QUESTION 2

Why was there less ATP production in the absence of Pi?

Go to **BioPortal** for all WORKING WITH**DATA** exercises

HOW ATP SYNTHASE WORKS: A MOLECULAR MOTOR Now that we have established that the H^+ gradient is needed for ATP synthesis, a question remains: How does the enzyme actually make ATP from ADP and P_i? This is certainly a fundamental question in biology, as it underlies energy harvesting in most cells. The structure and mechanism of ATP synthase, illustrated in **Figure 9.10A**, are shared by living organisms as diverse as bacteria and humans. ATP synthase is a molecular motor composed of two parts: the F_0 unit, a transmembrane region that is the H^+ channel; and the F_1 unit, which contains the active sites for ATP synthesis. F_1 consists of six subunits (three each of two polypeptide chains), arranged like the segments of an orange around a central polypeptide that interacts with the membrane-embedded F_0. Electron transport sets up an H^+ gradient across the membrane. This gradient has potential energy, and when H^+ diffuses through the channel, the potential energy is converted to kinetic energy, causing the central polypeptide to rotate. The energy is transmitted to the catalytic subunits of F_1, resulting in ATP synthesis. These molecular motors make ATP at rates up to 100 molecules per second.

An ingenious experiment confirmed this rotary motor mechanism (**Figure 9.10B**). Masasuke Yoshida and his colleagues at the Tokyo Institute of Technology isolated the F_1 portion of the ATP

synthase and attached it to a glass slide. Fluorescently labeled microfilaments were attached to the central peptide, and the slide was incubated in a solution containing ATP. In this case there was no proton gradient to drive the molecular motor in the direction of ATP synthesis. Instead, ATP was hydrolyzed to ADP and P_i. This caused the motor to spin, and the rotation of the labeled microfilament was visible under a microscope. The result was amazing, with the labeled filaments clearly rotating like propellers.

ELECTRON TRANSPORT CAN BE TOXIC Oxygen is an excellent electron acceptor, forming water as a result:

$$O_2 + 4\,H^+ + 4\,e^- \rightarrow 2\,H_2O \qquad (9.1)$$

Unfortunately, the consequences of an incomplete transfer of electrons are far from harmless because the intermediates are toxic:

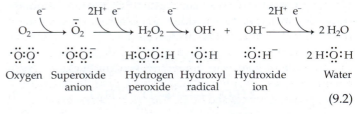

(9.2)

(A)

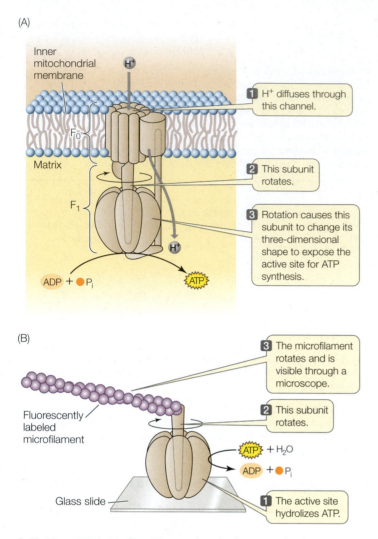

(B)

9.10 How ATP Is Made (A) Mitochondrial ATP synthase is a rotary motor. (B) A clever experiment was used to visualize the rotary motor.

Go to Media Clip 9.1
ATP Synthase in Motion
Life10e.com/mc9.1

Because they have unpaired electrons, superoxide and the hydroxyl radical are highly reactive and "seek out" other molecules to react with to gain or lose electrons and achieve a more stable state. These molecules can include lipids and nucleic acids, whose properties are changed dramatically by these reactions. Superoxide damage has been implicated in many human conditions, ranging from lung damage in emphysema to renal failure, stroke, and even aging. Cells have enzymes that "scavenge" these oxidizing substances before they can do harm, reducing them to harmless substances. Superoxide dismutase converts superoxide to peroxide, and catalase (in peroxisomes; see Section 5.3) converts peroxide to water:

$$O_2^- + 2\,H^+ \xrightarrow{\text{Superoxide dismutase}} O_2 + H_2O_2 \qquad (9.3)$$

$$2\,H_2O_2 \xrightarrow{\text{Catalase}} O_2 + 2\,H_2O \qquad (9.4)$$

TABLE 9.2

Electron Acceptors Used in the Respiratory Chain of Anaerobic Microbes

Terminal Electron Acceptor	Product Formed	Organism
SO_4^{-2}	H_2S	*Desulfovibrio desulfuricans*
Fe^{3+}	Fe^{2+}	*Geobacter metallireducens*
NO_3^-	NO_2^-	*Escherichia coli*
CO_2	CH_4	*Methanosarcina barkeri*
CO_2	CH_3COO^-	*Clostridium aceticum*
Fumarate	Succinate	*Wolinella succinogenes*

Antioxidant vitamins such as vitamin E act in a similar way by reducing these harmful molecules.

Some microorganisms use non-O₂ electron acceptors

A more general way to describe the last reaction in electron transport (Equation 9.1, above) is:

$$X_{\text{oxidized}} + e^- \rightarrow X_{\text{reduced}} \qquad (9.5)$$

Part of the amazing success of bacteria and archaea is that they have evolved biochemical pathways that allow them to exist in environments where O_2 is scarce or absent. As you will see in the next section, for most animals and plants, the anaerobic (no O_2) catabolism of glucose generally yields much less energy than aerobic catabolism. However, many bacteria and archaea exploit their environments to use alternative electron acceptors. This allows them to complete the electron transport chain and produce ATP even in the absence of O_2. **Table 9.2** summarizes some of these pathways, which are referred to as **anaerobic respiration**. Note that some of these microbes use ions as electron acceptors while others use small molecules.

■■■■■■■■ **RECAP 9.3**

The oxidation of reduced electron carriers in the respiratory chain drives the active transport of protons across the inner mitochondrial membrane, generating a proton-motive force. Diffusion of protons down their electrochemical gradient through ATP synthase is coupled to the synthesis of ATP. Electron transport can form toxic intermediates. Some bacteria and archaea can respire using alternative electron acceptors instead of O_2.

- What are the roles of oxidation and reduction in the respiratory chain? **See Figures 9.7 and 9.8**
- What is the proton-motive force, and how does it drive chemiosmosis? **See pp. 173–174**
- Explain how the experiment described in Figure 9.9 demonstrates the chemiosmotic mechanism. **See pp. 174–175 and Figure 9.9**

Oxidative phosphorylation captures a great deal of energy in ATP. But it does not occur if O_2 is absent. We will turn now to the metabolism of glucose in anaerobic conditions.

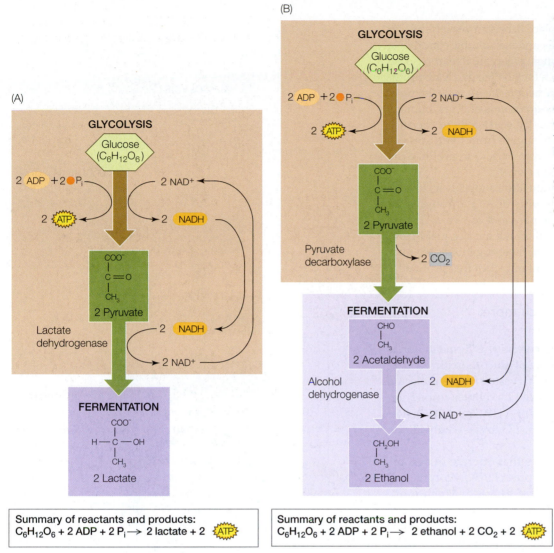

9.11 Fermentation Glycolysis produces pyruvate, ATP, and NADH from glucose. (A) Lactic acid fermentation uses NADH as a reducing agent to reduce pyruvate to lactate, thus regenerating NAD^+ to keep glycolysis operating. (B) In alcoholic fermentation, pyruvate from glycolysis is converted into acetaldehyde, and CO_2 is released. NADH from glycolysis is used to reduce acetaldehyde to ethanol, thus regenerating NAD^+ to keep glycolysis operating.

Summary of reactants and products:
$C_6H_{12}O_6 + 2\ ADP + 2\ P_i \longrightarrow$ 2 lactate + 2 ATP

Summary of reactants and products:
$C_6H_{12}O_6 + 2\ ADP + 2\ P_i \longrightarrow$ 2 ethanol + 2 CO_2 + 2 ATP

9.4 How Is Energy Harvested from Glucose in the Absence of Oxygen?

In eukaryotes, in the absence of O_2 (anaerobic conditions), a small amount of ATP can be produced by glycolysis and fermentation. Like glycolysis, fermentation pathways occur in the cytoplasm. There are many different types of fermentation, but they all operate to regenerate NAD^+ so that the NAD^+-requiring reactions of glycolysis can continue. The two best-understood fermentation pathways are found in a wide variety of organisms, including eukaryotes:

- Lactic acid fermentation, the end product of which is lactic acid (lactate)
- Alcoholic fermentation, the end product of which is ethyl alcohol (ethanol)

In **lactic acid fermentation**, pyruvate serves as the electron acceptor and lactate is the product (**Figure 9.11A**). This process takes place in many microorganisms and complex organisms, including higher plants and vertebrates. A notable example

of lactic acid fermentation occurs in vertebrate muscle tissue. Usually, vertebrates get their energy for muscle contraction aerobically, with the circulatory system supplying O_2 to muscles. In small vertebrates, this is almost always adequate: for example, birds can fly long distances without resting. But in larger vertebrates such as humans, the circulatory system is not up to the task of delivering enough O_2 when the need is great, such as during intense activity. At this point, the muscle cells break down glycogen (a stored polysaccharide; see Figure 3.18) and undergo lactic acid fermentation.

Lactate buildup becomes a problem after prolonged periods of intense exercise because it is associated with an increase in the H^+ concentration in the cell, lowering the pH. This affects cellular activities and causes muscle cramps, resulting in muscle pain, which abates upon resting. Lactate dehydrogenase, the enzyme that catalyzes the fermentation reaction, works in both directions. That is, when O_2 is available it can catalyze the oxidation of lactate to form pyruvate, which is then catabolized to CO_2 with concomitant energy release to form ATP. When lactate levels are decreased, muscle activity can resume.

Alcoholic fermentation takes place in certain yeasts (eukaryotic microbes) and some plant cells under anaerobic conditions. This process requires two enzymes, pyruvate decarboxylase and alcohol dehydrogenase, which metabolize pyruvate to ethanol (**Figure 9.11B**). As with lactic acid fermentation, the reactions are essentially reversible. For thousands of years, humans have used anaerobic fermentation by yeast cells to produce alcoholic beverages. The cells use sugars from plant sources (glucose from grapes or maltose from barley) to produce the end product, ethanol, in wine and beer.

By recycling NAD+, fermentation allows glycolysis to continue, thus producing small amounts of ATP through substrate-level phosphorylation. The net yield of two ATPs per glucose molecule is much lower than the energy yield from cellular respiration. For this reason, most organisms existing in anaerobic environments are small microbes that grow relatively slowly.

Cellular respiration yields much more energy than fermentation

The total net energy yield from glycolysis plus fermentation is two molecules of ATP per molecule of glucose oxidized. The maximum yield of ATP that can be harvested from a molecule of glucose through glycolysis followed by cellular respiration is much greater—about 32 molecules of ATP (**Figure 9.12**). (Review Figures 9.5, 9.6, and 9.8, and p. 173 to see where all the ATP molecules come from.)

Why do the metabolic pathways that operate in aerobic environments produce so much more ATP? Glycolysis and fermentation only partially oxidize glucose. Much more energy remains in the end products of fermentation (lactic acid and ethanol) than in CO_2, the end product of cellular respiration. In cellular respiration, carriers (mostly NAD+) are reduced during pyruvate oxidation and the citric acid cycle. Then the reduced carriers are oxidized by the respiratory chain, with the accompanying production of ATP by chemiosmosis (about 2.5 ATP for each NADH and 1.5 ATP for each $FADH_2$). In an aerobic environment, a cell or organism capable of aerobic metabolism will have the advantage over one that is limited to fermentation, in terms of its ability to harvest chemical energy. Two key events in the evolution of multicellular organisms were the rise in atmospheric O_2 levels (see Chapter 1) and the development of metabolic pathways to use that O_2.

The yield of ATP is reduced by the impermeability of mitochondria to NADH

The total gross yield of ATP from the oxidation of one molecule of glucose to CO_2 is about 32. However, in many eukaryotes the inner mitochondrial membrane is impermeable to NADH, and a "toll" of one ATP must be paid for each NADH molecule that is produced in glycolysis and must be "shuttled" into the mitochondrial matrix. So in these organisms, the net yield of ATP is 30.

NADH shuttle systems transfer the electrons captured by glycolysis onto substrates that are capable of movement across the mitochondrial membranes. In muscle and liver tissues (and

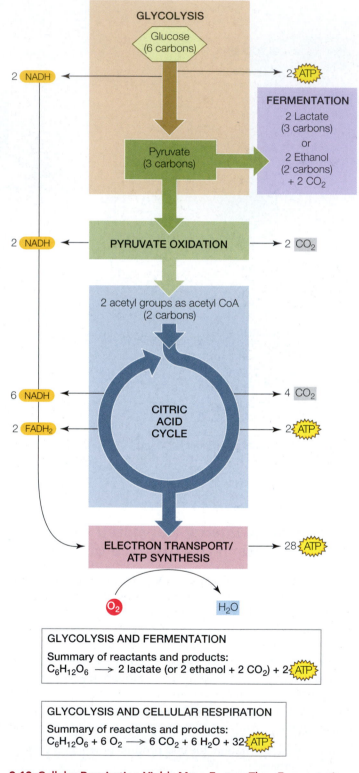

9.12 Cellular Respiration Yields More Energy Than Fermentation Electron carriers are reduced in pyruvate oxidation and the citric acid cycle, then oxidized by the respiratory chain. These reactions produce ATP via chemiosmosis.

Go to Activity 9.5 Energy Levels
Life10e.com/ac9.5

the brown fat in the opening story), an important shuttle involves glycerol 3-phosphate. In the cytosol,

NADH (from glycolysis) + dihydroxyacetone phosphate (DHAP) → NAD$^+$ + glycerol 3-phosphate

Glycerol 3-phosphate is transferred to the outer surface of the inner mitochondrial membrane. At that surface,

FAD + glycerol 3-phosphate → FADH$_2$ + DHAP

The electrons flow from FADH$_2$ into the electron transport chain via ubiquinone (Q) (see Section 9.3). DHAP is able to move back to the cytosol, where it is available to repeat the process. Note that the reducing electrons are transferred from NADH to FADH$_2$. As you know from Figure 9.8, the energy yield in terms of ATP from FADH$_2$ is lower than that from NADH. This lowers the overall energy yield.

RECAP 9.4

In the absence of O$_2$, fermentation pathways use NADH formed by glycolysis to reduce pyruvate and regenerate NAD$^+$. The energy yield of glycolysis coupled to fermentation is low because glucose is only partially oxidized. When O$_2$ is present, the electron carriers of cellular respiration allow for the full oxidation of glucose, so the energy yield from glucose is much higher.

- Why is replenishing NAD$^+$ crucial to cellular metabolism? **See pp. 177–178**
- How does fermentation replenish NAD$^+$? **Review Figure 9.11**
- What is the total energy yield from glucose in human cells in the presence versus the absence of O$_2$? **See p. 178 and Figure 9.12**

Now that you've seen how cells harvest energy, let's see how that energy moves through other metabolic pathways in the cell.

9.5 How Are Metabolic Pathways Interrelated and Regulated?

Glycolysis and the pathways of cellular respiration do not operate in isolation. Rather, there is an interchange of molecules into and out of these pathways, to and from the metabolic pathways for the synthesis and breakdown of amino acids, nucleotides, fatty acids, and other building blocks of life (see Figure 8.14). Carbon skeletons (i.e., the carbon backbones of organic molecules) can enter the catabolic pathways and be broken down to release their energy, or they can enter anabolic pathways to be used in the formation of the macromolecules that are the major constituents of the cell. These relationships are summarized in **Figure 9.13**. In this section we will explore how pathways are interrelated by the sharing of intermediate molecules, and we will see how pathways are regulated by the inhibitors of key enzymes.

Catabolism and anabolism are linked

A hamburger or veggie burger on a bun contains three major sources of carbon skeletons: carbohydrates, mostly in the form of starch (a polysaccharide); lipids, mostly as triglycerides

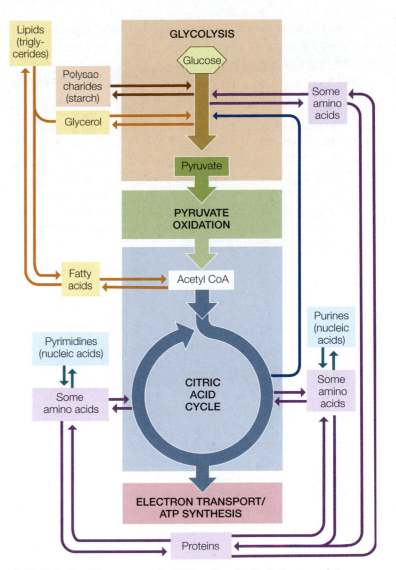

9.13 Relationships among the Major Metabolic Pathways of the Cell Note the central positions of glycolysis and the citric acid cycle in this network of metabolic pathways. Also note that many of the pathways can operate essentially in reverse.

(three fatty acids attached to glycerol); and proteins (polymers of amino acids). Look at Figure 9.13 to see how each of these three types of macromolecules can be hydrolyzed and used in catabolism or anabolism.

CATABOLIC INTERCONVERSIONS Polysaccharides, lipids, and proteins can all be broken down to provide energy:

- *Polysaccharides* are hydrolyzed to glucose. Glucose then passes through glycolysis and cellular respiration, where its energy is captured in ATP.
- *Lipids* are broken down into their constituents, glycerol and fatty acids. Glycerol is converted into dihydroxyacetone phosphate (DHAP), an intermediate in glycolysis. Fatty acids are highly reduced molecules that are converted to acetyl CoA inside the mitochondrion by a series of oxidation enzymes, in a process known as β-oxidation. For example,

the β-oxidation of a 16-carbon (C_{16}) fatty acid occurs in several steps:

$$C_{16} \text{ fatty acid} + \text{CoA} \rightarrow C_{16} \text{ fatty acyl CoA}$$
$$C_{16} \text{ fatty acyl CoA} + \text{CoA} \rightarrow C_{14} \text{ fatty acyl CoA} + \text{acetyl CoA}$$
$$\text{Repeat 6 times} \rightarrow 8 \text{ acetyl CoA}$$

The acetyl CoA can then enter the citric acid cycle and be catabolized to CO_2.

- *Proteins* are hydrolyzed to their amino acid building blocks. The 20 different amino acids feed into glycolysis or the citric acid cycle at different points determined by their structures. For example, the amino acid glutamate is converted into α-ketoglutarate, an intermediate in the citric acid cycle (the five-carbon molecule in Figure 9.6).

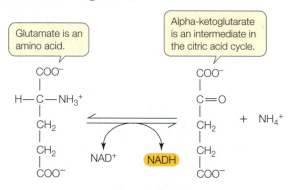

Glutamate is an amino acid.

Alpha-ketoglutarate is an intermediate in the citric acid cycle.

ANABOLIC INTERCONVERSIONS Many catabolic pathways can operate essentially in reverse, with some modifications. Glycolytic and citric acid cycle intermediates, instead of being oxidized to form CO_2, can be reduced and used to form glucose in a process called **gluconeogenesis** (which means "new formation of glucose"). Likewise, acetyl CoA can be used to form

fatty acids. The most common fatty acids have even numbers of carbons: 14, 16, or 18. These are formed by the addition of two-carbon acetyl CoA "units" one at a time until the appropriate chain length is reached. Acetyl CoA is also a building block for various pigments, plant growth substances, rubber, steroid hormones, and other molecules.

Some intermediates in the citric acid cycle are reactants in pathways that synthesize important components of nucleic acids. For example, α-ketoglutarate is a starting point for purines, and oxaloacetate for pyrimidines. In addition, α-ketoglutarate is a starting point for the synthesis of chlorophyll (used in photosynthesis; see Chapter 10) and the amino acid glutamate (used in protein synthesis).

Catabolism and anabolism are integrated

A carbon atom from a protein in your burger can end up in DNA, fat, or CO_2, among other fates. How does the organism "decide" which metabolic pathways to follow, in which cells? With all of the possible interconversions, you might expect that cellular concentrations of various biochemical molecules would vary widely. Remarkably, the levels of these substances in what is called the metabolic pool—the sum total of all the biochemical molecules in a cell—are quite constant. Organisms regulate the enzymes in various cells in order to maintain a balance between catabolism and anabolism. For example, let's look at how glucose levels are maintained in the body during exercise (**Figure 9.14**).

When a person is walking or jogging, there are two major organs where energy is most needed: the leg muscles to power movement and the heart muscles to circulate blood. The energy comes from the catabolism of glucose by glycolysis, the citric acid cycle, and oxidative phosphorylation. These muscles therefore need a lot of glucose. In the leg muscles, there is some

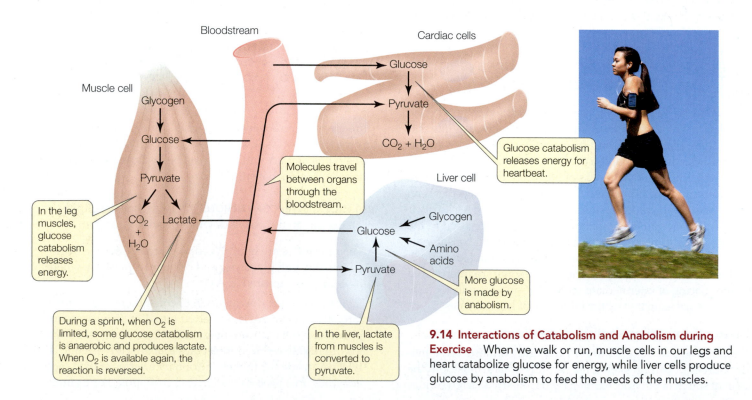

In the leg muscles, glucose catabolism releases energy.

During a sprint, when O_2 is limited, some glucose catabolism is anaerobic and produces lactate. When O_2 is available again, the reaction is reversed.

Molecules travel between organs through the bloodstream.

In the liver, lactate from muscles is converted to pyruvate.

Glucose catabolism releases energy for heartbeat.

More glucose is made by anabolism.

9.14 Interactions of Catabolism and Anabolism during Exercise When we walk or run, muscle cells in our legs and heart catabolize glucose for energy, while liver cells produce glucose by anabolism to feed the needs of the muscles.

stored glycogen, which is hydrolyzed to glucose monomers. In addition, glycogen is hydrolyzed in the liver, which releases the glucose into the blood. Both the heart and leg muscles receive glucose this way. As glucose is used up by the working muscles, more glucose is made in the liver by anabolism from amino acids and pyruvate. Some of the pyruvate used in this anabolism comes from lactate that was made by fermentation in the leg muscles and then transported back to the liver in the blood. Of course, all glucose comes originally from food, and is stored in the liver and the muscles in the form of glycogen. Glucose is catabolized in all cells at all times to provide them with energy for basic cellular activities.

The integration of catabolism and anabolism shown here implies that there are control points in the biochemical pathways. For example, how does the liver "know" that it should be making glucose rather than catabolizing it or storing it?

Metabolic pathways are regulated systems

The regulation of interconnecting biochemical pathways is a problem of systems biology, which seeks to understand how biochemical pathways interact (see Section 8.5). It is a bit like trying to predict traffic patterns in a city: if an accident blocks traffic on a major road, drivers take alternate routes, where the traffic volume consequently changes.

Consider what happens to the starch in your burger bun. In the digestive system, starch is hydrolyzed to glucose, which enters the blood for distribution to the rest of the body. But before the glucose is distributed, a regulatory check must be made: If there is already enough glucose in the blood to supply the body's needs, the excess glucose is converted into glycogen and stored in the liver and muscles. If not enough glucose is supplied by food, glycogen is broken down, or other molecules are used to make glucose by gluconeogenesis. The end result is that the level of glucose in the blood is remarkably constant. How does the body accomplish this?

Glycolysis, the citric acid cycle, and the respiratory chain are subject to allosteric regulation (see Section 8.5) of key enzymes involved. In a metabolic pathway, a high concentration of the final product can inhibit the action of an enzyme that catalyzes an earlier reaction (see Figure 8.18). Furthermore, an excess of the product of one pathway can activate an enzyme in another pathway, speeding up its reactions and diverting raw materials away from synthesis of the first product (**Figure 9.15**). These negative and positive feedback mechanisms are used at many points in the energy-harvesting pathways and are summarized in **Figure 9.16**.

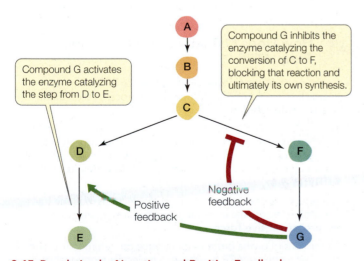

9.15 Regulation by Negative and Positive Feedback
Allosteric feedback regulation plays an important role in metabolic pathways. The accumulation of some products can shut down their synthesis, or can stimulate other pathways that require the same raw materials.

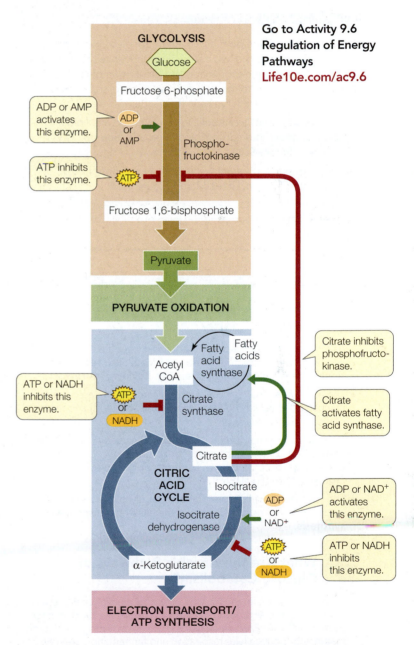

9.16 Allosteric Regulation of Glycolysis and the Citric Acid Cycle
Allosteric regulation controls glycolysis and the citric acid cycle at crucial early steps, increasing their efficiency and preventing the excessive buildup of intermediates.

- The main control point in glycolysis is the glycolytic enzyme phosphofructokinase, which catalyzes Step 3 in Figure 9.5. This enzyme is allosterically inhibited by ATP or citrate, and activated by ADP or AMP. Under anaerobic conditions, fermentation yields a relatively small amount of ATP, and phosphofructokinase operates at a high rate. However, when conditions are aerobic, respiration makes 16 times more ATP than fermentation does, and the abundant ATP allosterically inhibits phosphofructokinase. So glycolysis slows down.

- The main control point in the citric acid cycle is the enzyme *isocitrate dehydrogenase* (catalyzes Step 3 in Figure 9.6). This

enzyme is activated by increases in substrate concentrations (ADP, NAD⁺, and isocitrate) and is inhibited by products of the citric acid cycle: ATP and NADH. If too much ATP or NADH accumulates, the reaction is slowed, and the citric acid cycle shuts down.

- Another control point involves *acetyl CoA*. If the level of ATP is high and the citric acid cycle shuts down, the accumulation of citrate activates fatty acid synthase, diverting acetyl CoA to the synthesis of fatty acids for storage. That is one reason why people who eat too much accumulate fat. These fatty acids may be metabolized later to produce more acetyl CoA.

RECAP 9.5

Glucose can be made from intermediates in glycolysis and the citric acid cycle by a process called gluconeogenesis. The metabolic pathways for the production and breakdown of lipids and amino acids are tied to those of glucose metabolism. Reaction products regulate key enzymes in the various pathways.

- Give examples of a catabolic interconversion of a lipid and of an anabolic interconversion of a protein. **See pp. 179–180 and Figure 9.13**

- How does phosphofructokinase serve as a control point for glycolysis? **See pp.181–182 and Figure 9.16**

- Describe what would happen if there were no allosteric mechanism for modulating the level of acetyl CoA. **See p. 182**

Can brown fat in adults be a target for weight loss?

ANSWER

Yuxiang Sun and her colleagues at the Baylor College of Medicine in Houston, Texas, have bred mice that make more UCP1 as adults. Like normal mice, they have brown fat for the generation of heat as newborns. But these genetically altered mice continue to use UCP1 for heat generation as adults. They eat as much and exercise as much as the normal mice, but they stay thinner by breaking down fat to release heat instead of chemical energy. By targeting the mitochondrial protein UCP1, scientists may be able to turn up brown fat's furnace in humans.

CHAPTER SUMMARY 9

9.1 How Does Glucose Oxidation Release Chemical Energy?

- As a material is **oxidized**, the electrons it loses are transferred to another material, which is thereby **reduced**. Such **redox reactions** transfer large amounts of energy. **Review Figure 9.2**

- The coenzyme NAD⁺ is a key electron carrier in biological redox reactions. It exists in two forms, one oxidized (NAD⁺) and the other reduced (NADH).

- **Glycolysis** does not use O₂. Under **aerobic** conditions, **cellular respiration** continues the process of breaking down glucose. Under **anaerobic** conditions, **fermentation** occurs. **Review Figure 9.4, ACTIVITIES 9.1, 9.2**

- The pathways of cellular respiration after glycolysis are pyruvate oxidation, the citric acid cycle, and the electron transport/**ATP synthesis**.

9.2 What Are the Aerobic Pathways of Glucose Catabolism?

- Glycolysis consists of ten enzyme-catalyzed reactions that occur in the cell cytoplasm. Two **pyruvate** molecules are produced for each partially oxidized molecule of glucose, providing the starting material for both cellular respiration and fermentation. **Review Figure 9.5**

- Pyruvate oxidation follows glycolysis and links glycolysis to the citric acid cycle. This pathway converts pyruvate into **acetyl CoA**.

- Acetyl CoA is the starting point of the citric acid cycle. It reacts with oxaloacetate to produce citrate. A series of eight enzyme-catalyzed reactions oxidize citrate and regenerate oxaloacetate, continuing the cycle. **Review Figure 9.6, ACTIVITY 9.3**

9.3 How Does Oxidative Phosphorylation Form ATP?

- Oxidation of electron carriers in the presence of O₂ releases energy that can be used to form ATP in a process called **oxidative phosphorylation**.

- The NADH and FADH₂ produced in glycolysis, pyruvate oxidation, and the citric acid cycle are oxidized by the **respiratory chain**, regenerating NAD⁺ and FAD. Oxygen (O₂) is the final acceptor of electrons and protons, forming water (H₂O). **Review Figure 9.7, ACTIVITY 9.4**

- The respiratory chain not only transports electrons, but also transfers protons across the inner mitochondrial membrane, creating the **proton-motive force**.

- Protons driven by the proton-motive force can return to the mitochondrial matrix via **ATP synthase**, a molecular motor that couples this movement of protons to the synthesis of ATP. This process is called **chemiosmosis**. **Review Figure 9.8, ANIMATED TUTORIALS 9.1, 9.2**

continued

- Incomplete transfer of electrons can result in the formation of toxic superoxides and hydroxyl radicals. Special enzymes remove them to protect the cell from damage.

9.4 How Is Energy Harvested from Glucose in the Absence of Oxygen?

- In the absence of O_2, glycolysis is followed by fermentation. Together, these pathways partially oxidize pyruvate and generate end products such as lactic acid or ethanol. In the process, NAD^+ is regenerated from NADH so that glycolysis can continue, thus generating a small amount of ATP. **Review Figure 9.11**

- For each molecule of glucose used, glycolysis plus fermentation yields two molecules of ATP. In contrast, glycolysis operating with pyruvate oxidation, the citric acid cycle, and the respiratory chain/ATP synthase yields up to 32 molecules of ATP per molecule of glucose. **Review Figure 9.12, ACTIVITY 9.5**

9.5 How Are Metabolic Pathways Interrelated and Regulated?

- The catabolic pathways for the breakdown of carbohydrates, fats, and proteins feed into the energy-harvesting metabolic pathways. **Review Figure 9.13**

- Anabolic pathways use intermediate components of the energy-harvesting pathways to synthesize fats, amino acids, and other essential building blocks.

- The formation of glucose from intermediates of glycolysis and the citric acid cycle is called **gluconeogenesis**.

- The rates of glycolysis and the citric acid cycle are controlled by allosteric regulation and by the diversion of excess acetyl CoA into fatty acid synthesis. Key regulated enzymes include phospho-fructokinase, citrate synthase, isocitrate dehydrogenase, and fatty acid synthase. **See Figure 9.16, ACTIVITY 9.6**

Go to the Interactive Summary to review key figures, Animated Tutorials, and Activities
Life10e.com/is9

CHAPTER**REVIEW**

REMEMBERING

1. The role of oxygen gas in our cells is to
   a. catalyze reactions in glycolysis.
   b. produce CO_2.
   c. form ATP.
   d. accept electrons from the respiratory chain.
   e. react with glucose to split water.

2. Oxidation and reduction
   a. entail the gain or loss of proteins.
   b. are defined as the loss of electrons.
   c. are both endergonic reactions.
   d. always occur together.
   e. proceed only under aerobic conditions.

3. Glycolysis
   a. takes place in the mitochondrion.
   b. produces no ATP.
   c. is part of the respiratory chain.
   d. is the same thing as fermentation.
   e. reduces two molecules of NAD^+ for every glucose molecule processed.

4. Fermentation
   a. takes place in the mitochondrion.
   b. takes place in all animal cells.
   c. does not require O_2.
   d. requires lactic acid.
   e. prevents glycolysis.

5. The respiratory chain
   a. is located in the mitochondrial matrix.
   b. includes only peripheral membrane proteins.
   c. always produces ATP.
   d. reoxidizes reduced coenzymes.
   e. operates simultaneously with fermentation.

6. Compared with glycolysis coupled with fermentation, the aerobic pathways of glucose metabolism produce
   a. more ATP.
   b. pyruvate.
   c. fewer protons for transfer in the mitochondria.
   d. less CO_2.
   e. more oxidized coenzymes.

UNDERSTANDING & APPLYING

7. Trace the sequence of chemical changes that occurs in mammalian tissue when the oxygen supply is cut off. The first change is that all of the cytochrome *c* becomes reduced, because electrons can still flow from cytochrome *c*, but there is no oxygen to accept electrons from cytochrome *c* oxidase. What are the remaining changes?

8. You eat a burger that contains polysaccharides, proteins, and lipids. Using what you know of the integration of biochemical pathways, explain how the amino acids in the proteins and the glucose in the polysaccharides can end up as fats.

9. The following reaction occurs in the citric acid cycle:

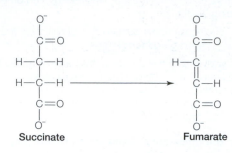

Succinate   Fumarate

Answer each of the following questions, and explain your answers:

a. Is this reaction an oxidation or reduction?

b. Is the reaction exergonic or endergonic?

c. This reaction requires a coenzyme. What kind of coenzyme?

d. What happens to the fumarate after the reaction is completed?

e. What happens to the coenzyme after the reaction is completed?

■■■■ **ANALYZING & EVALUATING**

10. Some cells that use the aerobic pathways of glucose metabolism can also thrive by using fermentation under anaerobic conditions. Given the lower yield of ATP (per molecule of glucose) in fermentation, how can these cells function so efficiently under anaerobic conditions?

11. The drug antimycin A blocks electron transport in mitochondria. Explain what would happen if the experiment in Figure 9.9 were repeated in the presence of this drug.

Go to BioPortal at **yourBioPortal.com** for Animated Tutorials, Activities, LearningCurve Quizzes, Flashcards, and many other study and review resources.

10 Photosynthesis: Energy from Sunlight

CHAPTEROUTLINE

10.1 What Is Photosynthesis?

10.2 How Does Photosynthesis Convert Light Energy into Chemical Energy?

10.3 How Is Chemical Energy Used to Synthesize Carbohydrates?

10.4 How Have Plants Adapted Photosynthesis to Environmental Conditions?

10.5 How Does Photosynthesis Interact with Other Pathways?

I F ALL THE CARBOHYDRATES produced by photosynthesis in a year were in the form of sugar cubes, there would be 300 quadrillion of them. As you may have learned from previous courses, photosynthetic organisms use atmospheric carbon dioxide (CO_2) to produce carbohydrates. The simplified equation says it all:

$$CO_2 + H_2O \xrightarrow{\text{Sunlight}} \text{carbohydrates} + O_2$$

Given the role of CO_2, will the photosynthesis rate change with increasing levels of atmospheric CO_2? Over the past 200 years, the concentration of atmospheric CO_2 has increased—from 280 parts per million (ppm) in 1800 to 395 ppm in 2011—and the increase will probably continue for some time. Carbon dioxide is a "greenhouse gas" that traps heat in the atmosphere, and the rising CO_2 level is predicted by climate scientists to result in global climate change. Plant biologists are being asked two questions about the rise in CO_2: will it lead to an increased rate of photosynthesis, and if so, will it lead to increased plant growth?

To answer these questions under realistic conditions, scientists developed a way to expose plants to high levels of CO_2 in the field. *Free-air concentration enrichment* (FACE) uses rings of pipes to release CO_2 into the air surrounding plants in fields or forests. Wind speed and direction are monitored by a computer, which constantly controls which pipes release CO_2. Data from these experiments confirm that photosynthetic rates increase as the

FACE Free-air concentration enrichment uses pipes to release CO_2 around plants in a field, to estimate the effects of rising atmospheric CO_2 on photosynthesis and plant growth.

concentration of CO_2 rises, and these measurements indicate that as atmospheric CO_2 rises globally, there will be an increase in photosynthesis.

Does this increase in photosynthesis result in an increase in plant growth? Keep in mind that plants, like all organisms, use carbohydrates as an energy source. They perform cellular respiration with the general equation:

$$\text{Carbohydrates} + O_2 \rightarrow CO_2 + H_2O + \text{energy}$$

The challenge facing plant biologists is to determine the balance between photosynthesis and respiration and how this affects the rate of plant growth. The FACE experiments indicate that plant growth and crop yields increase under higher CO_2 concentrations, suggesting that the overall increase in photosynthesis is greater than the increase in respiration.

What possible effects will increased atmospheric CO_2 have on global food production?

See answer on p. 202.

10.1 What Is Photosynthesis?

The energy released by catabolic pathways in almost all organisms (with the exception of those living near deep sea vents) ultimately comes from the sun. **Photosynthesis** (literally, "synthesis from light") is an anabolic process by which the energy of sunlight is captured and used to convert carbon dioxide (CO_2) into more complex carbon-containing compounds. Plants, algae, and cyanobacteria live in aerobic environments and carry out oxygenic photosynthesis: the conversion of CO_2 and water (H_2O) into carbohydrates (which we will represent as a six-carbon sugar, $C_6H_{12}O_6$) and oxygen gas (O_2) (**Figure 10.1**).

$$6\ CO_2 + 6\ H_2O \rightarrow C_6H_{12}O_6 + 6\ O_2 \qquad (10.1)$$

Some kinds of bacteria live in anaerobic environments and carry out anoxygenic photosynthesis, in which energy from the sun is used to convert CO_2 to more complex molecules without the production of O_2. We will describe this process in a little more detail below, but first let's look at oxygenic photosynthesis, summarized in Equation 10.1 above.

Equation 10.1 shows a very endergonic reaction. And while it is essentially correct, it is too general for a real understanding of the processes involved. Several questions arise: What are the precise chemical reactions of photosynthesis? What role does light play in these reactions? How do carbons become linked to form carbohydrates? What carbohydrates are formed? And where does the oxygen gas come from: CO_2 or H_2O?

Experiments with isotopes show that O_2 comes from H_2O in oxygenic photosynthesis

In 1941 Samuel Ruben and Martin Kamen at the University of California, Berkeley performed experiments using the isotopes ^{18}O and ^{16}O to identify the source of the O_2 produced during photosynthesis (**Figure 10.2**). Their results showed that all the oxygen gas produced during photosynthesis comes from water, as is reflected in the revised balanced equation:

$$6\ CO_2 + 12\ H_2O \rightarrow C_6H_{12}O_6 + 6\ O_2 + 6\ H_2O \qquad (10.2)$$

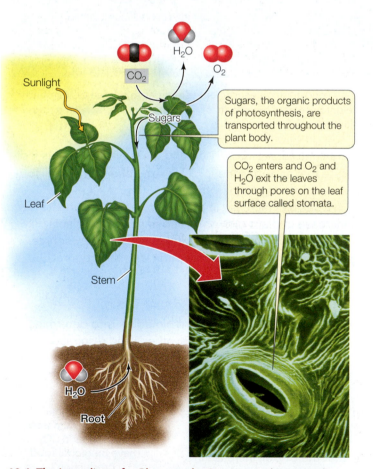

10.1 The Ingredients for Photosynthesis A typical terrestrial plant uses light from the sun, water from the soil, and carbon dioxide from the atmosphere to form organic compounds by photosynthesis.

Sunlight

CO_2

H_2O

O_2

Sugars

Sugars, the organic products of photosynthesis, are transported throughout the plant body.

CO_2 enters and O_2 and H_2O exit the leaves through pores on the leaf surface called stomata.

Leaf

Stem

H_2O

Root

INVESTIGATING**LIFE**

10.2 The Source of the Oxygen Produced by Photosynthesis
Although it was clear that O_2 was made during photosynthesis, its molecular source was not known. Two possibilities were the reactants, CO_2 and H_2O. In two separate experiments, Samuel Ruben and Martin Kamen labeled the oxygen in these molecules with the isotope ^{18}O, then tested the O_2 produced by a green plant to find out which molecules contributed the oxygen.[a]

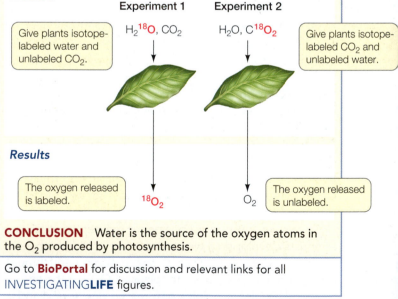

HYPOTHESIS The oxygen released by photosynthesis comes from water rather than CO_2.

Method

	Experiment 1	Experiment 2	

Give plants isotope-labeled water and unlabeled CO_2.

$H_2{}^{18}O$, CO_2 H_2O, $C^{18}O_2$

Give plants isotope-labeled CO_2 and unlabeled water.

Results

The oxygen released is labeled.

$^{18}O_2$ O_2

The oxygen released is unlabeled.

CONCLUSION Water is the source of the oxygen atoms in the O_2 produced by photosynthesis.

Go to **BioPortal** for discussion and relevant links for all INVESTIGATING**LIFE** figures.

[a]Ruben, S., et al. 1941. *Journal of the American Chemical Society* 63: 877–879.

Go to Animated Tutorial 10.1
The Source of the Oxygen Produced by Photosynthesis
Life10e.com/at10.1

Water appears on both sides of the equation because it is both used as a reactant (the 12 molecules on the left) and released as a product (the 6 new ones on the right). This revised equation accounts for all the water molecules needed for all the oxygen gas produced.

The realization that water was the source of photosynthetic O_2 led to an understanding of photosynthesis in terms of oxidation and reduction. As we described in Chapter 9, oxidation–reduction (redox) reactions are coupled: when one molecule becomes oxidized in a reaction, another gets reduced. In this case, oxygen atoms in the reduced state in H_2O get oxidized to O_2:

$$12\ H_2O \rightarrow 24\ H^+ + 24\ e^- + 6\ O_2 \qquad (10.3)$$

while carbon atoms in the oxidized state in CO_2 get reduced to carbohydrate, with the simultaneous production of water:

$$6\ CO_2 + 24\ H^+ + 24\ e^- \rightarrow C_6H_{12}O_6 + 6\ H_2O \qquad (10.4)$$

Adding Equations 10.3 and 10.4 (chemistry students will recognize them as half-cell reactions) gives the overall Equation 10.2 shown above.

As we have just seen, water is the donor of protons and electrons in oxygenic photosynthesis. In anoxygenic photosynthesis, other molecules are used as electron donors in the reduction of CO_2 to carbohydrates. For example, purple sulfur bacteria use hydrogen sulfide (H_2S) as the electron donor:

$$12\ H_2S + 6\ CO_2 + light \rightarrow C_6H_{12}O_6 + 6\ H_2O + 12\ S \quad (10.5)$$

Green sulfur bacteria use sulfide ions, hydrogen, or ferrous iron as electron donors, whereas another group of bacteria use compounds derived from arsenic. The remainder of this chapter will focus on oxygenic photosynthesis, which produces the vast majority of the organic carbon used by life on Earth today and replenishes the O_2 in our atmosphere.

WORKING WITH**DATA:**

Water Is the Source of the Oxygen Produced by Photosynthesis

Original Paper
Ruben, S., M. Randall, M. D. Kamen, and J. L. Hyde. 1941. Heavy oxygen (^{18}O) as a tracer in the study of photosynthesis. *Journal of the American Chemical Society* 63(3): 877–879.

Analyze the Data

In the 1930s, Cornelius van Niel (1897–1985), then a graduate student at Stanford University, was the first to propose that the oxygen released during photosynthesis is not actually derived from carbon dioxide, but rather from the water molecules consumed in the reaction. This hypothesis was formed on the basis of the discovery that the anaerobic purple sulfur bacteria do not release oxygen during photosynthesis. Instead, these organisms convert hydrogen sulfide (H_2S) into elemental sulfur in their photosynthetic pathway (see Equation 10.5).

Given the similarities between this photosynthetic process and that performed by green plants, van Niel proposed that during aerobic photosynthesis, hydrogen is extracted from water and incorporated into glucose, leaving the oxygen to be released as elemental oxygen. This hypothesis was later confirmed by Samuel Ruben and Martin Kamen, who used the "heavy" isotope of oxygen, ^{18}O, to trace the flow of oxygen in plants.

Their results showed that water is the source of the oxygen produced by photosynthesis (see Figure 10.2).

As part of the expanding research on radioisotopes during World War II, the U.S. government set up a radiation laboratory at the University of California, Berkeley. Out of this lab came key experiments that described the light-dependent and light-independent pathways of photosynthesis. In this set of experiments, *Chlorella* algal cells were exposed to water and CO_2, the latter coming from potassium carbonate (K_2CO_3) and potassium bicarbonate ($KHCO_3^-$), which dissolve in water to form CO_2. In Experiment 1 the water contained more ^{18}O than ^{16}O, whereas in Experiment 2 the CO_2 contained more ^{18}O than ^{16}O. A mass spectrometer was used to measure the isotopic contents of the reactants and the O_2 produced, and the data were presented as the isotopic ratio ($^{18}O/^{16}O$). These data are shown in the table.

QUESTION 1
In Experiment 1, was the isotopic ratio of O_2 similar to that of H_2O or to that of CO_2? What about in Experiment 2?

QUESTION 2
What can you conclude from these data?

	Time before start of O_2 collection (min)	Time at end of O_2 collection (min)	$^{18}O/^{16}O$ (percent ^{18}O in compound)		
			H_2O	$HCO_3^- + CO_3^{2-}$ (CO_2 sources)	O_2
Experiment 1:	0		0.85	0.20	
0.09 M $KHCO_3$ + 0.09 M	45	110	0.85	0.41	0.84
K_2CO_3 (^{18}O in H_2O)	110	223	0.85	0.55	0.85
	225	350	0.85	0.61	0.86
Experiment 2:	0		0.20		
0.14 M $KHCO_3$ + 0.06 M	40	110	0.20	0.50	0.20
K_2CO_3 (^{18}O in CO_2)	110	185	0.20	0.40	0.20

Go to **BioPortal** for all WORKING WITH**DATA** exercises

Photosynthesis involves two pathways

Equation 10.2 above summarizes the overall process of oxygenic photosynthesis, but not the stages by which it is completed. Water serves as the electron donor, but there is an intermediary carrier of the H^+ and electrons between the oxidation and reduction reactions. The carrier is the coenzyme nicotinamide adenine dinucleotide phosphate ($NADP^+$).

Like glycolysis and the other metabolic pathways that harvest energy in cells, photosynthesis is a process consisting of many reactions. These reactions are commonly divided into two main pathways:

- The **light reactions** convert light energy into chemical energy in the form of ATP and the reduced electron carrier NADPH. This molecule is similar to NADH (see Section 9.1) but with an additional phosphate group attached to the sugar of its adenosine. In general, NADPH acts as a reducing agent in photosynthesis and other anabolic reactions.

- The **light-independent reactions** (carbon-fixation reactions) do not use light directly, but instead use ATP, NADPH (made by the light reactions), and CO_2 to produce carbohydrate.

The light-independent reactions are sometimes called the dark reactions because they do not directly require light energy. They are also called the carbon-fixation reactions. However, in most plants the light reactions and the light-independent reactions *stop in the dark* because ATP synthesis and $NADP^+$ reduction require light. The reactions of both pathways proceed within the chloroplast, but they occur in different parts of that organelle (**Figure 10.3**).

As we describe these two series of reactions in more detail, you will see that they conform to the principles of biochemistry that we discussed in Chapters 8 and 9: energy transformations, oxidation–reduction, and the stepwise nature of biochemical pathways.

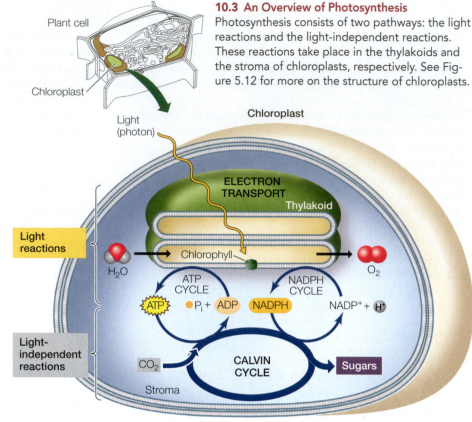

10.3 An Overview of Photosynthesis
Photosynthesis consists of two pathways: the light reactions and the light-independent reactions. These reactions take place in the thylakoids and the stroma of chloroplasts, respectively. See Figure 5.12 for more on the structure of chloroplasts.

We will describe the light reactions and the light-independent reactions separately. In the next section we will begin by discussing the physical nature of light and the specific photosynthetic molecules that capture its energy.

10.2 How Does Photosynthesis Convert Light Energy into Chemical Energy?

Light is a form of energy, and it can be converted to other forms of energy such as heat or chemical energy. Our focus here will be on light as the source of energy to drive the formation of ATP (from ADP and P_i) and NADPH (from $NADP^+$ and H^+).

Light energy is absorbed by chlorophyll and other pigments

It is helpful here to discuss light in terms of its photochemistry and photobiology.

PHOTOCHEMISTRY Light is a form of **electromagnetic radiation**. Electromagnetic radiation is propagated in waves, and the amount of energy in the radiation is inversely proportional to its **wavelength**—the shorter the wavelength, the greater the energy. The visible portion of the electromagnetic spectrum (**Figure 10.4**) encompasses a wide range of wavelengths and energy levels. In addition to traveling in waves, light also behaves as particles, called **photons**, which have no mass. In plants and other photosynthetic organisms, receptive molecules absorb photons in order to harvest their energy for biological processes. These receptive molecules absorb only

---

RECAP 10.1

The light reactions of photosynthesis convert light energy into chemical energy. The light-independent reactions use that chemical energy to reduce CO_2 to carbohydrates. While most photosynthetic organisms use water as the electron donor for reduction of CO_2, some use other molecules, such as hydrogen sulfide (H_2S).

- What is the experimental evidence that water is the source of the O_2 produced during photosynthesis? **See p. 186 and Figure 10.2**

- What is the relationship between the light reactions and the light-independent reactions of photosynthesis? **See p. 188 and Figure 10.3**

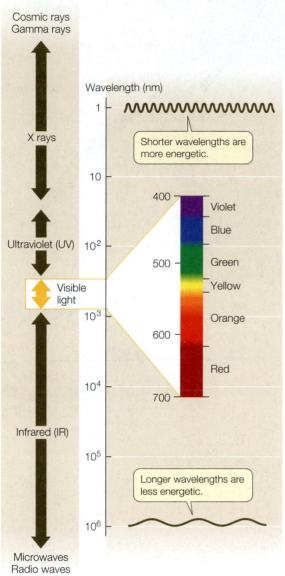

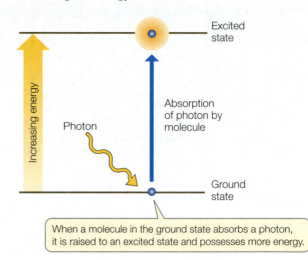

raised from a ground state (with lower energy) to an excited state (with higher energy):

When a molecule in the ground state absorbs a photon, it is raised to an excited state and possesses more energy.

The difference in free energy between the molecule's excited state and its ground state is approximately equal to the free energy of the absorbed photon (a small amount of energy is lost to entropy, according to the second law of thermodynamics). The increase in energy boosts one of the electrons in the molecule into a shell farther from its nucleus; this electron is now held less firmly, making the molecule unstable and more chemically reactive.

PHOTOBIOLOGY The specific wavelengths absorbed by a particular molecule are characteristic of that type of molecule. Molecules that absorb wavelengths in the visible spectrum are called **pigments**.

When a beam of white light (containing all the wavelengths of visible light) falls on a pigment, certain wavelengths are absorbed. The remaining wavelengths are scattered or transmitted and make the pigment appear to us as colored. For example, the pigment **chlorophyll** absorbs both blue and red light, and we see the remaining light, which is primarily green. If we plot light absorbed by a purified pigment against wavelength, the result is an **absorption spectrum** for that pigment.

In contrast to the absorption spectrum, an **action spectrum** is a plot of the rate of photosynthesis carried out by an organism against the wavelengths of light to which it is exposed. An action spectrum for photosynthesis can be determined as follows:

1. Place the organism in a closed container.
2. Expose it to light of a certain wavelength for a period of time.
3. Measure the rate of photosynthesis by the amount of O_2 released.
4. Repeat with light of other wavelengths.

Figure 10.5 shows the absorption spectra of pigments in *Anacharis*, a common aquarium plant, and the action spectrum for photosynthesis by that plant. The two spectra can be compared to see which pigments in *Anacharis* contribute most to light harvesting for photosynthesis.

10.4 The Electromagnetic Spectrum The portion of the electromagnetic spectrum that is visible to humans as light is shown in detail at the right.

specific wavelengths of light—photons with specific amounts of energy.

When a photon meets a molecule, one of three things can happen:

- The photon may bounce off the molecule—it may be scattered or reflected.
- The photon may pass through the molecule—it may be transmitted.
- The photon may be absorbed by the molecule, adding energy to the molecule.

Neither of the first two outcomes causes any change in the molecule. However, in the case of absorption, the photon disappears and its energy is absorbed by the molecule. The photon's energy cannot disappear, because according to the first law of thermodynamics, energy is neither created nor destroyed. When the molecule acquires the energy of the photon, it is

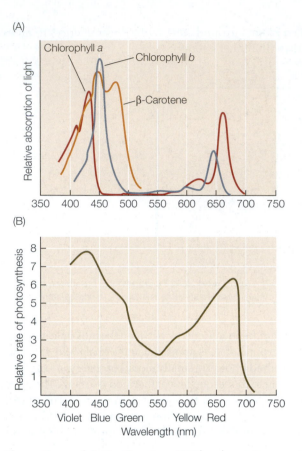

(A)

(B)

10.5 Absorption and Action Spectra (A) The absorption spectra of purified pigments from the common aquarium plant *Anacharis*. (B) The action spectrum for photosynthesis of that plant.

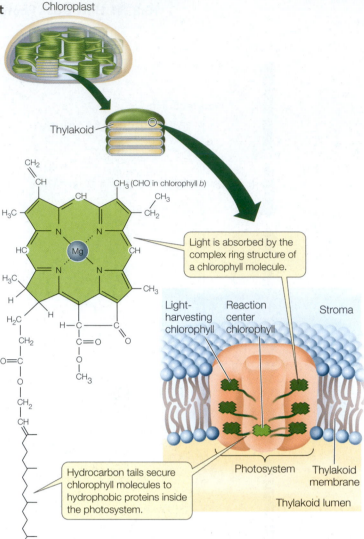

10.6 The Molecular Structure of Chlorophyll *a* Chlorophyll *a* consists of a complex ring structure (green area) with a magnesium atom at the center, plus a hydrocarbon "tail." The tail anchors the molecule to integral membrane proteins in the thylakoid membrane. Chlorophyll *b* is identical except for the replacement of a methyl group (—CH_3) with an aldehyde group (—CHO) at the upper right. These pigments are components of multi-protein complexes called photosystems, which span the thylakoid membrane.

The major pigment used to drive the light reactions of oxygenic photosynthesis is chlorophyll *a* (**Figure 10.6**). (In Figure 10.5 you can see that the wavelengths at which photosynthesis is highest in *Anacharis* are the same wavelengths at which chlorophyll *a* absorbs the most light.) Chlorophyll *a* has a complex ring structure, similar to that of the heme group of hemoglobin, with a magnesium ion at the center. A long hydrocarbon "tail" anchors the molecule to proteins within a large multi-protein complex called a **photosystem**, which spans the thylakoid membrane. Molecules of chlorophyll *a* and various accessory pigments (such as chlorophylls *b* and *c*, carotenoids, and phycobilins; see below) are arranged into **light-harvesting complexes**, also called antenna systems. Multiple light-harvesting complexes surround a single **reaction center** within the photosystem. Light energy is captured by the light-harvesting complexes and transferred to the reaction center, where chlorophyll *a* molecules participate in redox reactions that result in the conversion of the light energy to chemical energy.

As mentioned above, chlorophyll absorbs blue and red light, which are near the two ends of the visible spectrum (see Figure 10.4). The various accessory pigments absorb light in other parts of the spectrum and thus function to broaden the range of wavelengths that can be used for photosynthesis. You can see this in the action spectrum in Figure 10.5B: *Anacharis* is able to photosynthesize at wavelengths of light that chlorophyll *a* doesn't

absorb but that other pigments absorb (e. g., 500 nm). Different photosynthetic organisms have different combinations of accessory pigments. Higher plants and green algae have chlorophyll *b* (with a structure and absorption spectrum very similar to that of chlorophyll *a*) and carotenoids such as β-carotene, which absorb photons in the blue and blue-green wavelengths. Phycobilins, which are found in red algae and cyanobacteria, absorb various yellow-green, yellow, and orange wavelengths.

Light absorption results in photochemical change

When a pigment molecule absorbs light, it enters an excited state. This is an unstable situation, and the molecule rapidly returns to its ground state, releasing most of the absorbed energy. This is an extremely rapid process—measured in picoseconds (trillionths of a second). Within a light-harvesting complex

(A)

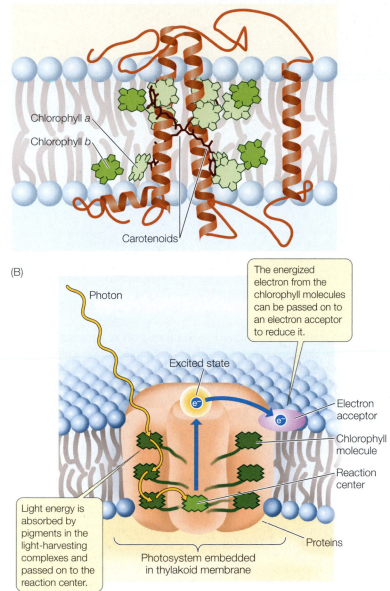

(B)

The energized electron from the chlorophyll molecules can be passed on to an electron acceptor to reduce it.

Photon

Excited state

Electron acceptor

Chlorophyll molecule

Reaction center

Proteins

Light energy is absorbed by pigments in the light-harvesting complexes and passed on to the reaction center.

Photosystem embedded in thylakoid membrane

10.7 Energy Transfer and Electron Transport (A) The molecular structure of a single light-harvesting complex shows the polypeptide in brown with three helices that span the thylakoid membrane. Pigment molecules (carotenoids and chlorophylls *a* and *b*) are bound to the polypeptide. (B) This simplified illustration of the entire photosystem uses chlorophyll molecules to represent the light-harvesting complexes. Energy from a photon is transferred from one pigment molecule to another, until it reaches a chlorophyll *a* molecule in the reaction center. The chlorophyll *a* molecule can give up its excited electron to an electron acceptor.

(**Figure 10.7A**), the energy released by a pigment molecule (for example, chlorophyll *b*) is absorbed by other, adjacent pigment molecules. The energy (not as electrons but in the form of chemical energy called resonance) is passed from molecule to molecule until it reaches a chlorophyll *a* molecule at the reaction center of the photosystem (**Figure 10.7B**).

A ground-state chlorophyll *a* molecule at the reaction center (symbolized by Chl) absorbs the energy from the adjacent chlorophylls and becomes excited (Chl*), but to return to the ground state this chlorophyll does not pass the energy to

another pigment molecule—something very different occurs. *The reaction center converts the absorbed light energy into chemical energy.* The chlorophyll molecule in the reaction center absorbs sufficient energy that it actually *gives up its excited electron to a chemical acceptor*:

$$\text{Chl* + acceptor} \rightarrow \text{Chl}^+ + \text{acceptor}^- \qquad (10.6)$$

This, then, is the first consequence of light absorption by chlorophyll: *the reaction center chlorophyll (Chl*) loses its excited electron in a redox reaction and becomes Chl*.* As a result of this transfer of an electron, the chlorophyll gets oxidized, while the acceptor molecule is reduced.

Reduction leads to ATP and NADPH formation

The electron acceptor that is reduced by Chl* is the first in a chain of electron carriers in the thylakoid membrane. Electrons are passed from one carrier to another in an energetically "downhill" series of reductions and oxidations. Thus the thylakoid membrane has an electron transport system similar to the respiratory chain of mitochondria (see Section 9.3). The final electron acceptor is $NADP^+$, which gets reduced:

$$\text{NADP}^+ + \text{H}^+ + 2\,\text{e}^- \rightarrow \text{NADPH} \qquad (10.7)$$

As in mitochondria, ATP is produced chemiosmotically during the process of electron transport. **Figure 10.8** shows the series of **noncyclic electron transport** reactions that use the energy from light to generate NADPH and ATP. There are two photosystems, each with its own reaction center:

- **Photosystem I** (containing the "P_{700}" chlorophylls at its reaction center) absorbs light energy best at 700 nanometers (nm) and passes its excited electrons (via intermediate molecules) to $NADP^+$, reducing it to NADPH.

- **Photosystem II** (with "P_{680}" chlorophylls at its reaction center) absorbs light energy best at 680 nm, oxidizes water molecules, and passes its energized electrons through a series of carriers to produce ATP.

Let's look in more detail at these photosystems, beginning with photosystem II.

PHOTOSYSTEM II After an excited chlorophyll in the reaction center (Chl*) gives up its energetic electron to reduce a chemical acceptor molecule, the chlorophyll lacks an electron and is very unstable. It has a strong tendency to "grab" an electron from another molecule to replace the one it lost—in chemical terms, it is a strong oxidizing agent. The replenishing electrons come from water, splitting the H—O—H bonds:

$$\text{H}_2\text{O} \rightarrow \tfrac{1}{2}\,\text{O}_2 + 2\,\text{H}^+ + 2\,\text{e}^- \qquad (10.8)$$

$$2\,\text{e}^- + 2\,\text{Chl}^+ \rightarrow 2\,\text{Chl} \qquad (10.9)$$

$$\text{Overall: } 2\,\text{Chl}^+ + \text{H}_2\text{O} \rightarrow 2\,\text{Chl} + 2\,\text{H}^+ + \tfrac{1}{2}\,\text{O}_2 \quad (10.10)$$

Notice that the source of O_2 in photosynthesis is H_2O.

Back to the electron acceptor in the electron transport system: the energetic electrons are passed through a series of membrane-bound carriers to a final acceptor at a lower energy

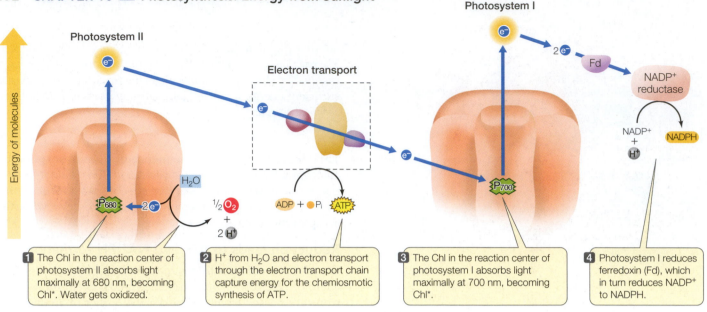

Photosystem II

Electron transport

Photosystem I

1 The Chl in the reaction center of photosystem II absorbs light maximally at 680 nm, becoming Chl*. Water gets oxidized.

2 H+ from H_2O and electron transport through the electron transport chain capture energy for the chemiosmotic synthesis of ATP.

3 The Chl in the reaction center of photosystem I absorbs light maximally at 700 nm, becoming Chl*.

4 Photosystem I reduces ferredoxin (Fd), which in turn reduces NADP+ to NADPH.

10.8 Noncyclic Electron Transport Uses Two Photosystems
Absorption of light energy by chlorophyll molecules in the reaction centers of photosystems I and II allows them to pass electrons into a series of redox reactions.

level. As in the mitochondrion, a proton gradient is generated and is used by ATP synthase to make ATP (see below).

PHOTOSYSTEM I In photosystem I, an excited electron from the Chl* at the reaction center reduces an acceptor. The oxidized chlorophyll (Chl+) now "grabs" an electron, but in this case the electron comes from the last carrier in the electron transport system. This links the two photosystems chemically. They are also linked spatially, with the two photosystems adjacent to one another in the thylakoid membrane. The energetic electrons from photosystem I pass through several molecules and end up reducing NADP+ to NADPH.

Next in the process of harvesting light energy to produce carbohydrates is the series of carbon-fixation reactions. These reactions require more ATP than NADPH. If the pathway we just described—the linear or noncyclic pathway—were the only set of light reactions operating, there might not be sufficient ATP for carbon fixation. **Cyclic electron transport** makes up for this imbalance. This pathway uses photosystem I and the electron transport system to produce ATP but not NADPH; it is cyclic because an electron is passed from an excited chlorophyll and recycles back to the same chlorophyll (**Figure 10.9**).

Chemiosmosis is the source of the ATP produced in photophosphorylation

In Chapter 9 we described the chemiosmotic mechanism for ATP formation in the mitochondrion. A similar mechanism, called **photophosphorylation**, operates in the chloroplast, where

electron transport is coupled to the transport of protons (H+) across the thylakoid membrane, resulting in a proton gradient across the membrane (**Figure 10.10**).

The electron carriers in the thylakoid membrane are oriented so that protons are transferred from the stroma into the lumen of the thylakoid. Thus the lumen becomes acidic with respect to the stroma, resulting in an electrochemical gradient across the thylakoid membrane, whose bilayer is not permeable to H+. Water oxidation creates more H+ in the thylakoid lumen and NADP+ reduction removes H+ in the stroma. Both reactions contribute to the H+ gradient. The high concentration of H+ in the thylakoid space drives the movement of H+ back into the stroma through protein channels in the membrane. These channels are enzymes—ATP synthases—that couple the movement of

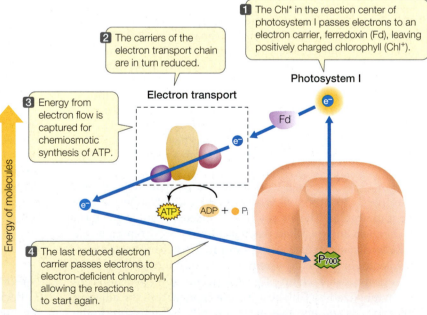

1 The Chl* in the reaction center of photosystem I passes electrons to an electron carrier, ferredoxin (Fd), leaving positively charged chlorophyll (Chl+).

2 The carriers of the electron transport chain are in turn reduced.

3 Energy from electron flow is captured for chemiosmotic synthesis of ATP.

4 The last reduced electron carrier passes electrons to electron-deficient chlorophyll, allowing the reactions to start again.

Electron transport

Photosystem I

10.9 Cyclic Electron Transport Traps Light Energy as ATP
Cyclic electron transport produces ATP but no NADPH.

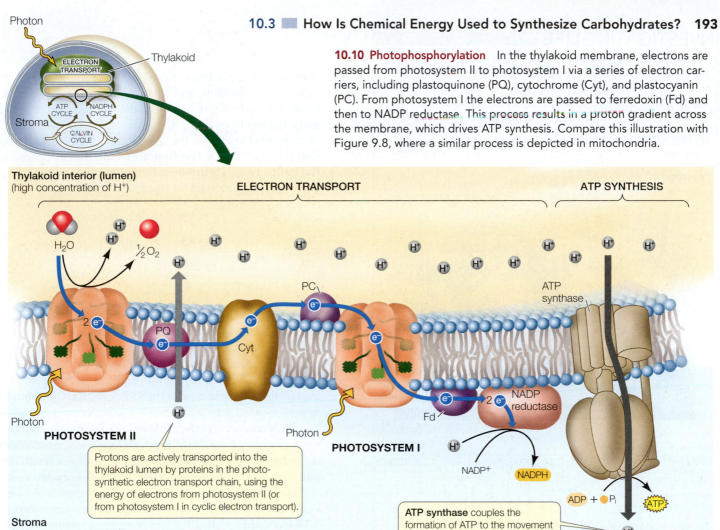

10.10 Photophosphorylation In the thylakoid membrane, electrons are passed from photosystem II to photosystem I via a series of electron carriers, including plastoquinone (PQ), cytochrome (Cyt), and plastocyanin (PC). From photosystem I the electrons are passed to ferredoxin (Fd) and then to NADP reductase. This process results in a proton gradient across the membrane, which drives ATP synthesis. Compare this illustration with Figure 9.8, where a similar process is depicted in mitochondria.

Protons are actively transported into the thylakoid lumen by proteins in the photosynthetic electron transport chain, using the energy of electrons from photosystem II (or from photosystem I in cyclic electron transport).

ATP synthase couples the formation of ATP to the movement of protons back into the stroma.

protons to the formation of ATP, as they do in mitochondria (see Figure 9.8). Indeed, chloroplast ATP synthase is about 60 percent identical to human mitochondrial ATP synthase—a remarkable similarity, given that plants and animals had their most recent common ancestor more than a billion years ago. This is testimony to the evolutionary unity of life.

The mechanisms of the two enzymes are similar, but their orientations differ. In chloroplasts, protons flow through the ATP synthase out of the thylakoid lumen into the stroma (where the ATP is synthesized). In mitochondria the protons flow out of the intermembrane space into the mitochondrial matrix.

RECAP 10.2

Conversion of light energy into chemical energy occurs when pigments absorb photons. Light energy is used to drive a series of protein-associated redox reactions in the thylakoid membranes of the chloroplast.

- How does chlorophyll absorb and transfer light energy? **See pp. 190–191 and Figure 10.7**
- How are electrons produced in photosystem II, and how do they flow to photosystem I? **See pp. 191–192 and Figure 10.8**
- How does cyclic electron transport in photosystem I result in the production of ATP? **See p. 192 and Figure 10.9**

Go to Animated Tutorial 10.2
Photophosphorylation
Life10e.com/at10.2

We have seen how light energy drives the synthesis of ATP and NADPH in the stroma of chloroplasts. We will now turn to the light-independent reactions of photosynthesis, which use energy-rich ATP and NADPH to reduce CO_2 and form carbohydrates.

10.3 How Is Chemical Energy Used to Synthesize Carbohydrates?

Most of the enzymes that catalyze the reactions of CO_2 fixation are dissolved in the stroma of the chloroplast, where those reactions take place. These enzymes use the energy in ATP and NADPH to reduce CO_2 to carbohydrates. Therefore, with some exceptions, CO_2 fixation occurs only in the light, when ATP and NADPH are being generated.

Radioisotope labeling experiments revealed the steps of the Calvin cycle

To identify the reactions by which the carbon from CO_2 ends up in carbohydrates, scientists found a way to label CO_2 so that they could isolate and identify the compounds formed

10.11 Tracing the Pathway of CO₂ How is CO_2 incorporated into carbohydrate during photosynthesis? What is the first stable covalent linkage that forms with the carbon of CO_2? Melvin Calvin and his colleagues used short exposures to $^{14}CO_2$ to identify the first compound formed from CO_2.[a]

HYPOTHESIS The first product of CO_2 fixation is a 3-carbon molecule.

Method

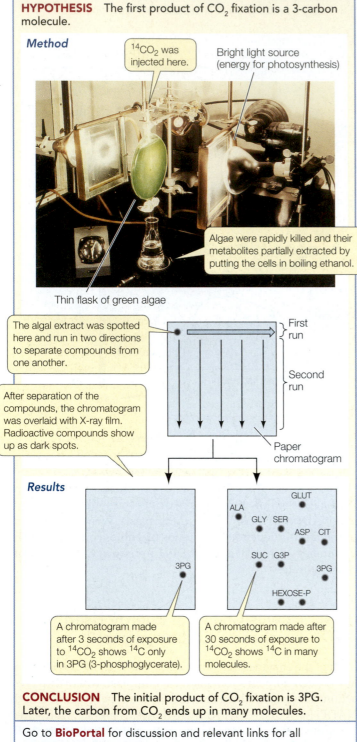

$^{14}CO_2$ was injected here.

Bright light source (energy for photosynthesis)

Algae were rapidly killed and their metabolites partially extracted by putting the cells in boiling ethanol.

Thin flask of green algae

The algal extract was spotted here and run in two directions to separate compounds from one another.

First run

Second run

After separation of the compounds, the chromatogram was overlaid with X-ray film. Radioactive compounds show up as dark spots.

Paper chromatogram

Results

GLUT
ALA
GLY SER
ASP CIT
SUC G3P
3PG
3PG
HEXOSE-P

A chromatogram made after 3 seconds of exposure to $^{14}CO_2$ shows ^{14}C only in 3PG (3-phosphoglycerate).

A chromatogram made after 30 seconds of exposure to $^{14}CO_2$ shows ^{14}C in many molecules.

CONCLUSION The initial product of CO_2 fixation is 3PG. Later, the carbon from CO_2 ends up in many molecules.

Go to **BioPortal** for discussion and relevant links for all INVESTIGATINGLIFE figures.

[a]Benson, A. A., et al. 1950. *Journal of the American Chemical Society* 72: 1710–1718.

Go to Animated Tutorial 10.3
Tracing the Pathway of CO₂
Life10e.com/at10.3

from it during photosynthesis. In the 1950s, Melvin Calvin, Andrew Benson, and their colleagues used radioactively labeled CO_2 in which some of the carbon atoms were the radioisotope ^{14}C rather than the normal ^{12}C. They were able to trace the chemical pathway of CO_2 fixation (**Figure 10.11**). The first molecule that appeared in the pathway was a three-carbon sugar phosphate called 3-phosphoglycerate (3PG) (the ^{14}C is shown in red):

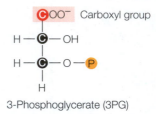

Carboxyl group

3-Phosphoglycerate (3PG)

Using successively longer exposures to $^{14}CO_2$, Calvin and his colleagues were able to trace the route of ^{14}C as it moved through a series of compounds, including monosaccharides and amino acids. It turned out that the pathway the ^{14}C moved through was a cycle. In this cycle, the CO_2 initially bonds covalently to a five-carbon acceptor molecule. The resulting six-carbon intermediate quickly breaks into two three-carbon molecules. As the cycle repeats, a carbohydrate is produced and the initial CO_2 acceptor is regenerated. This pathway was appropriately named the **Calvin cycle**.

The initial reaction in the Calvin cycle adds the one-carbon CO_2 to the five-carbon acceptor molecule ribulose 1,5-bisphosphate (RuBP). The product is an intermediate six-carbon compound, which quickly breaks down and forms two molecules of 3PG (**Figure 10.12**). The intermediate compound is broken down so rapidly that Calvin did not observe radioactive label appearing in it first. But the enzyme that catalyzes its formation, **ribulose bisphosphate carboxylase/oxygenase** (**rubisco**), is the most abundant protein in the world! It constitutes up to 50 percent of all the protein in every plant leaf.

The Calvin cycle is made up of three processes

The Calvin cycle uses the ATP and NADPH made in the light to reduce CO_2 in the stroma to a carbohydrate. Like all biochemical pathways, each reaction is catalyzed by a specific enzyme. The cycle is composed of three distinct processes (**Figure 10.13**):

- *Fixation* of CO_2. As we have seen, this reaction is catalyzed by rubisco, and its stable product is 3PG.

- *Reduction* of 3PG to form glyceraldehyde 3-phosphate (G3P). This series of reactions involves a phosphorylation (using the ATP made in the light reactions) and a reduction (coupled to the oxidation of NADPH made in the light reactions).

- *Regeneration* of the CO_2 acceptor, RuBP. Most of the G3P ends up as ribulose monophosphate (RuMP), and ATP is used to convert this compound into RuBP. So for every "turn" of the cycle, one CO_2 is fixed and one CO_2 acceptor is regenerated.

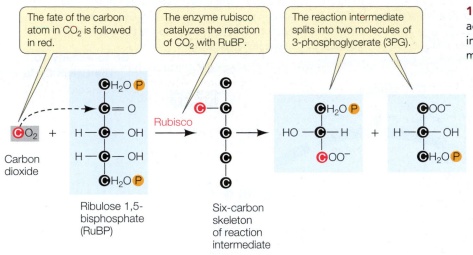

The fate of the carbon atom in CO_2 is followed in red.

The enzyme rubisco catalyzes the reaction of CO_2 with RuBP.

The reaction intermediate splits into two molecules of 3-phosphoglycerate (3PG).

Carbon dioxide

Ribulose 1,5-bisphosphate (RuBP)

Six-carbon skeleton of reaction intermediate

10.12 RuBP Is the Carbon Dioxide Acceptor CO_2 is added to a five-carbon compound, RuBP. The resulting six-carbon compound immediately splits into two molecules of 3PG.

The product of this cycle is **glyceraldehyde 3-phosphate (G3P)**, which is a three-carbon sugar phosphate, also called triose phosphate:

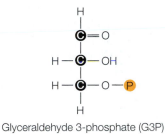

Glyceraldehyde 3-phosphate (G3P)

In a typical leaf, five-sixths of the G3P is recycled into RuBP. There are two fates for the remaining G3P, depending on the time of day and the needs of different parts of the plant:

- Some of the G3P is exported out of the chloroplast to the cytosol, where it is converted to hexoses (glucose and fructose). These molecules may be used in glycolysis and mitochondrial respiration to power the activities of photosynthetic cells (see Chapter 9) or they may be converted into the disaccharide sucrose, which is transported out of the leaf to other organs in the plant. There the sucrose is hydrolyzed to its constituent monosaccharides, which can be used as sources of energy or as building blocks for other molecules.

- Some of the G3P is used to synthesize glucose inside the chloroplast. As the day wears on, glucose molecules accumulate and are linked together to form the polysaccharide starch. This stored carbohydrate can then be drawn upon during the night so that the photosynthetic tissues can continue to export sucrose to the rest of the plant, even when photosynthesis is not taking place. In addition, starch is

WORKING WITH**DATA:**

Tracing the Pathway of CO_2

Original Papers

Calvin and his colleagues described their experiments in a series of 26 papers titled "The Path of Carbon in Photosynthesis." Perhaps the most important was one that showed how paper chromatography and labeled CO_2 could be used as tracers:

Benson, A. A., J. A. Bassham, M. Calvin, T. C. Goodale, V. A. Haas, and W. Stepka. 1950. The path of carbon in photosynthesis. V. Paper chromatography and radioautography of the products. *Journal of the American Chemical Society* 72: 1710–1718.

Analyze the Data

To elucidate the sequence of reactions that allow carbon fixation, Melvin Calvin and colleagues exposed suspensions of the green alga *Chlorella* to $^{14}CO_2$ (see Figure 10.11). After 3 seconds of photosynthesis, the ^{14}C from $^{14}CO_2$ was found only in 3-phosphoglycerate (3PG), but after 30 seconds many compounds were radioactive. Calvin and his colleagues then expanded on these results and were able to determine the exact sequence of reactions and reaction intermediates in the Calvin cycle by exposing the cells to $^{14}CO_2$ for various periods of time.

The first reaction in CO_2 fixation can occur in the dark. To show this, Calvin and his colleagues exposed *Chlorella* cells to $^{14}CO_2$ under bright lights for 20 minutes and harvested the cells.

Then they repeated the experiment, but with the addition of various periods of darkness (30 sec, 2 min, and 5 min) following the 20 minutes of light. They used the harvested cells to make chromatograms and identified the labeled compounds, using a radioactivity detector to quantify the amount of ^{14}C in each compound. The data are shown in the table.

Compound	20 min light	Relative amount of radioactivity after:		
		20 min light + 30 sec dark	20 min light + 2 min dark	20 min light + 5 min dark
3PG	5,500	10,100	10,000	5,200
RuBP	4,900	680	1,850	1,800
Sucrose	13,000	13,500	15,000	14,750

QUESTION 1

Using the data in the table, plot radioactivity in 3PG versus time. What do the data show?

QUESTION 2

Why did the amount of radioactively labeled RuBP go down after 30 seconds in the dark?

Go to BioPortal for all WORKING WITH**DATA** exercises

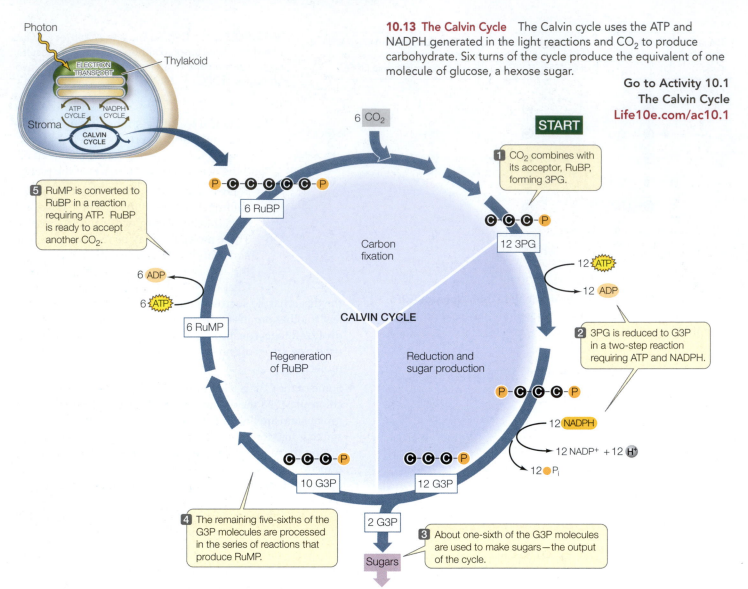

10.13 The Calvin Cycle The Calvin cycle uses the ATP and NADPH generated in the light reactions and CO_2 to produce carbohydrate. Six turns of the cycle produce the equivalent of one molecule of glucose, a hexose sugar.

Go to Activity 10.1
The Calvin Cycle
Life10e.com/ac10.1

1 CO_2 combines with its acceptor, RuBP, forming 3PG.

5 RuMP is converted to RuBP in a reaction requiring ATP. RuBP is ready to accept another CO_2.

Carbon fixation

12 3PG

12 ATP
12 ADP

6 ADP
6 ATP

6 RuMP

CALVIN CYCLE

2 3PG is reduced to G3P in a two-step reaction requiring ATP and NADPH.

Regeneration of RuBP

Reduction and sugar production

12 NADPH
12 NADP⁺ + 12 H⁺
12 Pᵢ

10 G3P

12 G3P

4 The remaining five-sixths of the G3P molecules are processed in the series of reactions that produce RuMP.

2 G3P

Sugars

3 About one-sixth of the G3P molecules are used to make sugars—the output of the cycle.

Other carbon compounds (e.g., starch)

abundant in nonphotosynthetic organs such as roots, underground stems, and seeds, where it provides a ready supply of glucose to fuel cellular activities, including plant growth.

The plant uses the carbohydrates produced in photosynthesis to make other compounds. The carbon molecules are incorporated into amino acids, lipids, and the building blocks of nucleic acids—in fact all the organic molecules in the plant.

The products of the Calvin cycle are of crucial importance to Earth's entire biosphere. For the majority of living organisms on Earth, the C—C and C—H covalent bonds generated by the cycle provide almost all of the energy for life. Photosynthetic organisms, which are also called **autotrophs** ("self-feeders"), release most of this energy by glycolysis and cellular respiration, and use it to support their own growth, development, and reproduction. But plants are also the source of energy for other organisms. Much plant matter ends up being consumed by **heterotrophs** ("other-feeders"), such as animals, which cannot photosynthesize. Heterotrophs depend on autotrophs for both raw materials

and energy. Free energy is released from food by glycolysis and cellular respiration in the cells of heterotrophs.

Light stimulates the Calvin cycle

As we have seen, the Calvin cycle uses NADPH and ATP, which are generated using energy from light. Two other processes connect the light reactions with this CO_2 fixation pathway. Both connections are indirect but significant:

• Light-induced pH changes in the stroma activate some Calvin cycle enzymes. Proton transfer from the stroma into the thylakoid lumen causes an increase in the pH of the stroma from 7 to 8 (a tenfold decrease in H⁺ concentration). This favors the activation of rubisco.

• Light-induced electron transport reduces disulfide bridges in four of the Calvin cycle enzymes, thereby activating them (**Figure 10.14**). When ferredoxin is reduced in photosystem I (see Figure 10.8), it passes some electrons to a small, soluble protein called thioredoxin, and this protein

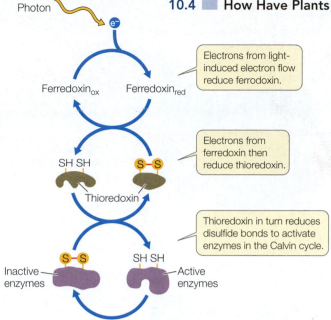

10.14 The Photochemical Reactions Stimulate the Calvin Cycle
By reducing (breaking) disulfide bridges, electrons from the light reactions activate enzymes in CO_2 fixation.

passes electrons to four enzymes in the CO_2 fixation pathway. Reduction of the sulfurs in the disulfide bridges of these enzymes (see Figure 3.5) forms SH groups and breaks the bridges. The resulting changes in their three-dimensional shapes activate the enzymes and increase the rate at which the Calvin cycle operates.

RECAP 10.3

ATP and NADPH produced in the light reactions power the synthesis of carbohydrates by the Calvin cycle. This cycle fixes CO_2, reduces it, and regenerates the acceptor, RuBP, for further fixation.

- Describe the experiments that led to the identification of 3PG as the initial product of carbon fixation. **See p. 194 and Figure 10.11**
- What are the three processes of the Calvin cycle? **See p. 194 and Figure 10.13**
- In what ways does light stimulate the Calvin cycle? **See pp. 196–197 and Figure 10.14**

Although all green plants carry out the Calvin cycle, some plants have evolved variations on, or additional steps in, the light-independent reactions. These variations and additions have permitted plants to adapt to and thrive in certain environmental conditions. Let's look at these environmental limitations and the metabolic bypasses that have evolved to circumvent them.

10.4 How Have Plants Adapted Photosynthesis to Environmental Conditions?

In addition to fixing CO_2 during photosynthesis, rubisco can react with O_2. This reaction, which leads to a process called photorespiration, lowers the overall rate of CO_2 fixation in some plants. After examining this problem, we'll look at some biochemical pathways and features of plant anatomy that compensate for the limitations of rubisco.

Rubisco catalyzes the reaction of RuBP with O_2 or CO_2

As its full name indicates, rubisco (ribulose bisphosphate carboxylase/oxygenase) is an oxygenase as well as a carboxylase—it can add O_2 to the acceptor molecule RuBP instead of CO_2. The affinity of rubisco for CO_2 is about ten times stronger than its affinity for O_2. This means that inside a leaf with a normal exchange of air with the outside, CO_2 fixation is favored even though the concentration of CO_2 in the air is far less than that of O_2. But if there is an even higher concentration of O_2 in the leaf, then the O_2 competes with the CO_2, and rubisco combines RuBP with O_2 rather than CO_2. This reduces the overall amount of CO_2 that is converted into carbohydrates, and may play a role in limiting plant growth.

When O_2 is added to RuBP, one of the products is a two-carbon compound, phosphoglycolate:

$$RuBP + O_2 \rightarrow \text{phosphoglycolate} + \text{3-phosphoglycerate (3PG)} \quad (10.11)$$

The 3PG formed by rubisco's oxygenase activity enters the Calvin cycle, but the phosphoglycolate does not. Plants have evolved a metabolic pathway that can partially recover the carbon in phosphoglycolate. The phosphoglycolate is hydrolyzed to glycolate, which diffuses into peroxisomes (**Figure 10.15**). There, a series of reactions converts it into the amino acid glycine:

$$\text{Glycolate} + O_2 \rightarrow \text{glycine} \quad (10.12)$$

The glycine then diffuses into a mitochondrion, where two glycine molecules are converted in a series of reactions into the amino acid serine, releasing CO_2:

$$2\ \text{Glycine} \rightarrow \text{serine} + CO_2 \quad (10.13)$$

The serine moves into the peroxisome, where it is converted to glycerate. The glycerate then moves into the chloroplast, where it is phosphorylated to make 3PG, which enters the Calvin cycle. Note that it takes two phosphoglycolate molecules from Equation 10.11 to produce the two glycines used in Equation 10.13. So overall:

$$2\ \text{Phosphoglycolate (4 carbons)} + O_2 \rightarrow \text{3PG (3 carbons)} + CO_2 \quad (10.14)$$

This pathway thus reclaims 75 percent of the carbons from phosphoglycolate for the Calvin cycle. In other words, the reaction of RuBP with O_2 instead of CO_2 reduces the net carbon fixed by the Calvin cycle by 25 percent. The pathway is called **photorespiration** because it consumes O_2 and releases CO_2 and because it occurs only in the light (because of the same enzyme activation processes that were mentioned above with regard to the Calvin cycle).

(A)

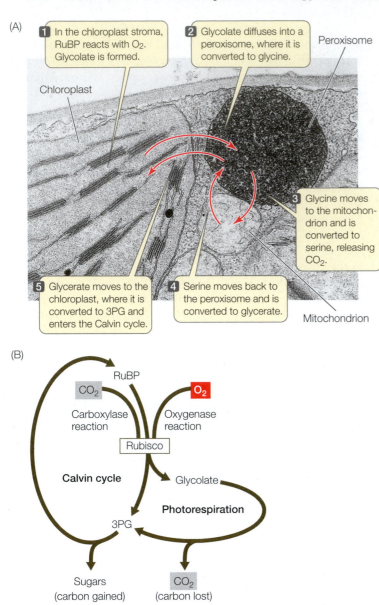

1 In the chloroplast stroma, RuBP reacts with O_2. Glycolate is formed.

2 Glycolate diffuses into a peroxisome, where it is converted to glycine.

Peroxisome

Chloroplast

3 Glycine moves to the mitochondrion and is converted to serine, releasing CO_2.

5 Glycerate moves to the chloroplast, where it is converted to 3PG and enters the Calvin cycle.

4 Serine moves back to the peroxisome and is converted to glycerate.

Mitochondrion

(B)

RuBP

CO_2 O_2

Carboxylase reaction   Oxygenase reaction

Rubisco

Calvin cycle   Glycolate

Photorespiration

3PG

Sugars (carbon gained) CO_2 (carbon lost)

10.15 Organelles of Photorespiration (A) The reactions of photorespiration take place in the chloroplasts, peroxisomes, and mitochondria. (B) Overall, photorespiration consumes O_2 and releases CO_2.

Go to Media Clip 10.1
Chloroplasts in Close-Up
Life10e.com/mc10.1

Why does rubisco act as an oxygenase as well as a carboxylase? Several factors are involved: active site affinities, concentrations of CO_2 and O_2, and temperature:

- As noted above, rubisco has a ten times higher affinity for CO_2 than for O_2, and this favors CO_2 fixation.
- In the leaf, the relative concentrations of CO_2 and O_2 vary. If O_2 is relatively abundant, rubisco acts as an oxygenase and photorespiration ensues. If CO_2 predominates, rubisco fixes it for the Calvin cycle.
- Photorespiration is more likely at high temperatures. On a hot, dry day, small pores in the leaf surface called **stomata** close to prevent water from evaporating from the leaf (see

(A) Arrangement of cells in a C_3 leaf

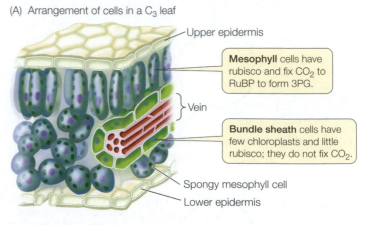

Upper epidermis

Mesophyll cells have rubisco and fix CO_2 to RuBP to form 3PG.

Vein

Bundle sheath cells have few chloroplasts and little rubisco; they do not fix CO_2.

Spongy mesophyll cell

Lower epidermis

(B) Arrangement of cells in a C_4 leaf

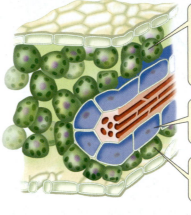

Mesophyll cells have the enzyme PEP carboxylase, which catalyzes the reaction of CO_2 and PEP to form the 4-carbon molecule oxaloacetate, which is converted to malate.

Bundle sheath cells have modified chloroplasts that concentrate CO_2 around rubisco.

Close proximity permits CO_2 "pumping" from mesophyll cells to bundle sheath cells.

10.16 Leaf Anatomy of C_3 and C_4 Plants Carbon dioxide fixation occurs in different organelles and cells of the leaves in (A) C_3 plants and (B) C_4 plants. Cells that are tinted blue have rubisco.

Go to Activity 10.2 C_3 and C_4 Leaf Anatomy
Life10e.com/ac10.2

Figure 10.1). But this also prevents gases from entering and leaving the leaf. The CO_2 concentration in the leaf falls as CO_2 is used up in photosynthetic reactions, and the O_2 concentration rises because of these same reactions. As the ratio of CO_2 to O_2 falls, the oxygenase activity of rubisco is favored, and photorespiration proceeds.

C_3 plants undergo photorespiration but C_4 plants do not

Plants differ in how they fix CO_2, and can be distinguished as C_3 or C_4 plants, based on whether the first product of CO_2 fixation is a three- or four-carbon molecule. In **C_3 plants** such as roses, wheat, and rice, the first product is the three-carbon molecule 3PG—as we have just described for the Calvin cycle. In these plants the cells of the mesophyll, which makes up the main body of the leaf, are full of chloroplasts containing rubisco (**Figure 10.16A**). On a hot day, these leaves close their stomata to conserve water, and as a result, rubisco acts as an oxygenase as well as a carboxylase, and photorespiration occurs.

C_4 plants, which include corn, sugarcane, and tropical grasses, make the four-carbon molecule **oxaloacetate** as the first product of CO_2 fixation (**Figure 10.16B**). On a hot day, they partially close their stomata to conserve water, but their rate of photosynthesis does not fall. What do they do differently?

C_4 plants have evolved a mechanism that increases the concentration of CO_2 around the rubisco enzyme while at the same time isolating the rubisco from atmospheric O_2. Thus in these plants the carboxylase reaction is favored over the oxygenase reaction; the Calvin cycle operates, but photorespiration does not occur. This mechanism involves the initial fixation of CO_2 in the mesophyll cells and then the transfer of the fixed carbon (as a four-carbon molecule) to the **bundle sheath cells**, where the fixed CO_2 is released for use in the Calvin cycle (**Figure 10.17**). The bundle sheath cells are located in the interior of the leaf where less atmospheric O_2 can reach them than reaches cells near the surface of the leaf.

The first enzyme in this C_4 carbon fixation process, called **PEP carboxylase**, is present in the cytosols of mesophyll cells near the leaf's surface. This enzyme fixes CO_2 to a three-carbon acceptor compound, **phosphoenolpyruvate (PEP)**, to produce the four-carbon fixation product, oxaloacetate. PEP carboxylase has two advantages over rubisco:

- It does not have oxygenase activity.
- It fixes CO_2 even at very low CO_2 levels.

So even on a hot day when the stomata are partially closed and the ratio of O_2 to CO_2 rises, PEP carboxylase just keeps on fixing CO_2.

Oxaloacetate is converted to malate, which diffuses out of the mesophyll cells and into the bundle sheath cells (see Figure 10.16B). (Some C_4 plants convert the oxaloacetate to aspartate instead of malate, but we will only discuss the malate pathway here.) The bundle sheath cells contain modified chloroplasts that are designed to concentrate CO_2 around the rubisco. There, the four-carbon malate loses one carbon (is decarboxylated), forming CO_2 and pyruvate. The latter moves back to the mesophyll cells where the three-carbon acceptor compound, PEP, is regenerated at the expense of ATP. So the "expenditure" of ATP in the mesophyll cell "pumps up" the CO_2 concentration around rubisco in the bundle sheath cell, so that it acts as a carboxylase and begins the Calvin cycle.

Under relatively cool or cloudy conditions, C_3 plants have an advantage over C_4 plants in that they don't expend energy to "pump up" the concentration of CO_2 near rubisco. But this advantage begins to be outweighed under conditions that favor photorespiration, such as warmer seasons and climates. Under these conditions C_4 plants have the advantage, especially if there is ample light to supply the extra ATP required for C_4 photosynthesis. For example, Kentucky bluegrass is a C_3 plant that thrives on lawns in April and May. But in the heat of summer it does not do as well, and Bermuda grass, a C_4 plant, takes over the lawn. The same is true on a global scale for crops: C_3 plants such as soybean, wheat, and barley have been adapted for human food production in temperate climates, whereas C_4 plants such as corn and sugarcane originated and are grown mainly in the tropics.

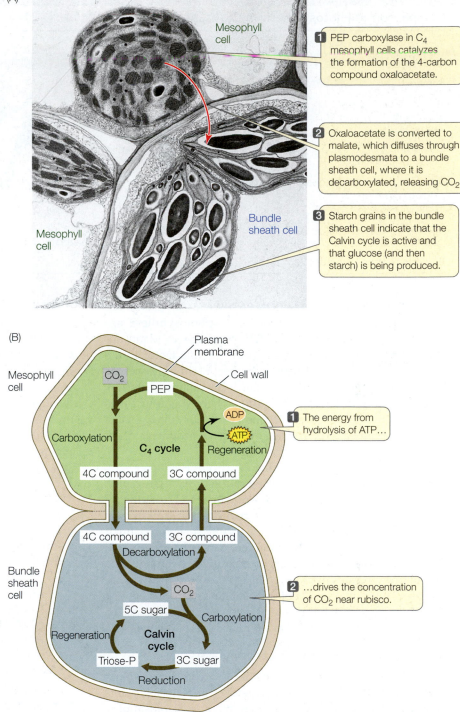

(A)

Mesophyll cell

1 PEP carboxylase in C_4 mesophyll cells catalyzes the formation of the 4-carbon compound oxaloacetate.

2 Oxaloacetate is converted to malate, which diffuses through plasmodesmata to a bundle sheath cell, where it is decarboxylated, releasing CO_2.

Bundle sheath cell

Mesophyll cell

3 Starch grains in the bundle sheath cell indicate that the Calvin cycle is active and that glucose (and then starch) is being produced.

(B)

Plasma membrane

Mesophyll cell

CO_2

PEP

Cell wall

Carboxylation

ADP

ATP

C_4 cycle

Regeneration

1 The energy from hydrolysis of ATP…

4C compound

3C compound

4C compound

3C compound

Decarboxylation

Bundle sheath cell

CO_2

2 …drives the concentration of CO_2 near rubisco.

5C sugar

Carboxylation

Regeneration

Calvin cycle

Triose-P

3C sugar

Reduction

10.17 The Anatomy and Biochemistry of C_4 Carbon Fixation (A) Carbon dioxide is fixed initially in the mesophyll cells but enters the Calvin cycle in the bundle sheath cells. (B) The two cell types share an interconnected biochemical pathway for CO_2 assimilation.

TABLE 10.1

Comparison of Photosynthesis in C_3, C_4, and CAM Plants

	C_3 plants	C_4 plants	CAM plants
Calvin cycle used?	Yes	Yes	Yes
Primary CO_2 acceptor	RuBP	PEP	PEP
CO_2-fixing enzyme	Rubisco	PEP carboxylase	PEP carboxylase
First product of CO_2 fixation	3PG (3-carbon)	Oxaloacetate (4-carbon)	Oxaloacetate (4-carbon)
Affinity of carboxylase for CO_2	Moderate	High	High
Photosynthetic cells of leaf	Mesophyll	Mesophyll and bundle sheath	Mesophyll with large vacuoles
Photorespiration	Extensive	Minimal	Minimal

THE EVOLUTION OF CO_2 FIXATION PATHWAYS C_3 plants are more ancient than C_4 plants. Whereas C_3 photosynthesis appears to have begun about 2.5 billion years ago, C_4 plants appeared about 12 million years ago. A possible factor in the emergence of the C_4 pathway is the decline in atmospheric CO_2. When dinosaurs dominated Earth 100 million years ago, the concentration of CO_2 in the atmosphere was four times what it is now. As CO_2 levels declined thereafter, the C_4 plants would have gained an advantage over their C_3 counterparts in high-temperature, high-light environments.

As we described in the opening essay of this chapter, CO_2 levels have been increasing over the past 200 years. Currently, the level of CO_2 is not enough for maximal CO_2 fixation by rubisco, so photorespiration occurs, reducing the growth rates of C_3 plants. Under hot conditions, C_4 plants are favored. But if CO_2 levels in the atmosphere continue to rise, the reverse will occur and C_3 plants will have a comparative advantage. The overall growth rates of crops such as rice and wheat should increase. This may or may not translate into more food, given that other effects of the human-spurred CO_2 increase (such as global climate change) will also alter Earth's ecosystems.

CAM plants also use PEP carboxylase

Other plants besides the C_4 plants use PEP carboxylase to fix and accumulate CO_2. They include some water-storing plants (succulents) of the family Crassulaceae, many cacti, pineapples, and several other kinds of flowering plants. The CO_2 metabolism of these plants is called **crassulacean acid metabolism**, or **CAM**, after the family of succulents in which it was discovered. CAM is much like the metabolism of C_4 plants in that CO_2 is initially fixed into a four-carbon compound. But in CAM plants the initial CO_2 fixation and the Calvin cycle are *separated in time rather than space*.

- At night, when it is cooler and water loss is minimized, the stomata open. CO_2 is fixed in mesophyll cells to form the four-carbon compound oxaloacetate, which is converted into malate and stored in the vacuole.

- During the day, when the stomata close to reduce water loss, the accumulated malate is shipped from the vacuole

to the chloroplasts, where its decarboxylation supplies the CO_2 for the Calvin cycle and the light reactions supply the necessary ATP and NADPH.

CAM benefits the plant by allowing it to close its stomata during the day. As you will learn in Chapter 35, plants lose most of the water that they take up in their roots by evaporation through the leaves (transpiration). In dry climates, closing stomata is a key to water conservation and survival.

Table 10.1 compares photosynthesis in C_3, C_4, and CAM plants.

RECAP 10.4

Rubisco catalyzes the carboxylation of RuBP to form two 3PG, and the oxygenation of RuBP to form one 3PG and one phosphoglycolate. The diversion of rubisco to its oxygenase function decreases net CO_2 fixation. C_4 photosynthesis and CAM allow plants to fix CO_2 under warm, dry conditions when stomata are closed and CO_2 entry into the leaf is limited.

- Explain how photorespiration recovers some of the carbon that is channeled away from the Calvin cycle. See p. 197 and Figure 10.15

- How do C_4 plants keep the concentration of CO_2 around rubisco high, and why? See pp. 198–199 and Figure 10.17

- What is the pathway for CO_2 fixation in CAM plants? See p. 200

Now that we understand how photosynthesis produces carbohydrates, let's see how the pathways of photosynthesis are connected to other metabolic pathways.

10.5 How Does Photosynthesis Interact with Other Pathways?

Green plants are autotrophs and can synthesize all the molecules they need from simple starting materials: CO_2, H_2O, phosphate, sulfate, ammonium ions (NH_4^+), and small quantities of other mineral nutrients. The NH_4^+ is needed to synthesize amino acids and nucleotides, and it comes from

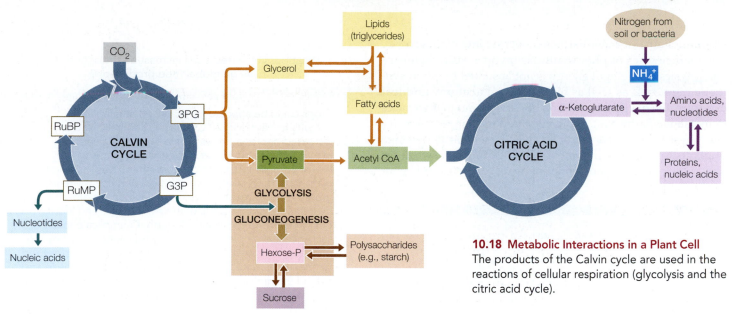

10.18 Metabolic Interactions in a Plant Cell
The products of the Calvin cycle are used in the reactions of cellular respiration (glycolysis and the citric acid cycle).

nitrogen-containing molecules in soil water or from N_2 gas fixed from the atmosphere by bacteria, as we will see in Chapter 36.

Plants use the carbohydrates generated in photosynthesis to provide energy for processes such as active transport and anabolism. Both cellular respiration and fermentation can occur in plants, although the former is far more common. Unlike photosynthesis, plant cellular respiration occurs all the time in both the light and the dark.

Photosynthesis and respiration are closely linked through the Calvin cycle (**Figure 10.18**). The partitioning of G3P is particularly important:

- Some G3P from the Calvin cycle enters glycolysis and is converted into pyruvate in the cytosol. This pyruvate can be used in cellular respiration for energy, or its carbon skeletons can be used in anabolic reactions to make lipids, proteins, and other carbohydrates (see Figure 9.13).

- Some G3P can enter a pathway that is the reverse of glycolysis (gluconeogenesis; see Section 9.5). In this case, hexose phosphates (hexose P) and then sucrose are formed and transported to the nonphotosynthetic tissues of the plant (such as the root).

Energy flows from sunlight to reduced carbon in photosynthesis, then to ATP in respiration. Energy can also be stored in the bonds of macromolecules such as polysaccharides, lipids, and proteins. For a plant to grow, energy

storage (as body structures) must exceed energy release; that is, overall carbon fixation by photosynthesis must exceed respiration. This principle is the basis of the ecological food chain, as we will see in later chapters.

Photosynthesis provides most of the energy that we need for life. Given the uncertainties about the future of photosynthesis (because of changes in CO_2 levels and climate change), it would be wise to seek ways to improve photosynthetic efficiency. **Figure 10.19** shows the various ways in which solar energy is used by plants or lost. In essence, only about 5 percent of

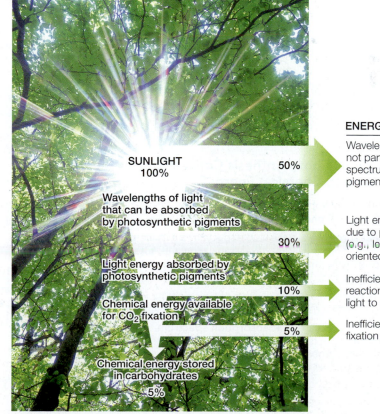

SUNLIGHT 100%

Wavelengths of light that can be absorbed by photosynthetic pigments

50%

Light energy absorbed by photosynthetic pigments

30%

Chemical energy available for CO_2 fixation

10%

Chemical energy stored in carbohydrates

5%

5%

ENERGY LOSS

Wavelengths of light not part of absorption spectrum of photosynthetic pigments (e.g., green light)

Light energy not absorbed due to plant structure (e.g., leaves not properly oriented to sun)

Inefficiency of light reactions converting light to chemical energy

Inefficiency of CO_2 fixation pathways

10.19 Energy Losses in Photosynthesis
Photosynthetic pathways preserve at most about 5 percent of the sun's energy input as chemical energy in carbohydrates.

the sunlight that reaches Earth is converted into plant growth. The inefficiencies of photosynthesis involve basic chemistry and physics (some light energy is not absorbed by photosynthetic pigments) as well as biology (plant anatomy and leaf exposure, the oxygenase reaction of rubisco, and inefficiencies in metabolic pathways). While it is hard to change chemistry and physics, biologists might be able to use their knowledge of plants to improve on the basic biology of photosynthesis. This could result in a more efficient use of resources and better food production.

RECAP 10.5

The products of photosynthesis are used in glycolysis and the citric acid cycle, as well as in the synthesis of lipids, proteins, and other large molecules.

- How do common intermediates link the pathways of glycolysis, the citric acid cycle, and photosynthesis? **See p. 201 and Figure 10.18**
- Why is only 5 percent of the solar radiation that reaches Earth captured by plants? **See pp. 201–202 and Figure 10.19**

What possible effects will increased atmospheric CO_2 have on global food production?

ANSWER

Crops could be affected by increased atmospheric CO_2 in multiple ways. Higher CO_2 levels generally lead to increased photosynthesis. This is especially true for C_3 plants, which are more sensitive than C_4 plants to CO_2 levels. Because increased photosynthesis leads to greater plant growth, C_3 crops such as wheat and rice will tend to grow more. However, it is unclear whether this growth will be in the vegetative parts of the plant (stems and leaves) or in the part we eat (grain). To further complicate matters, such increases in plant growth may be counteracted by the effects of increased CO_2 on climate. For example, increased temperatures would increase the rate of photosynthesis and extend the growing season, but might alter rainfall patterns. In some areas of the world there might be less rain, and this could limit plant growth. Overall, while there will be significant regional variations, biologists estimate that increased CO_2 will result in moderately increased food production.

CHAPTER**SUMMARY** 10

10.1 What Is Photosynthesis?

- In the process of **photosynthesis**, the energy of sunlight is captured and used to convert CO_2 into more complex carbon-containing compounds. **See ANIMATED TUTORIAL 10.1**
- Plants, algae, and cyanobacteria live in aerobic environments and carry out oxygenic photosynthesis: the conversion of CO_2 and H_2O into carbohydrates.
- Some bacteria that live in anaerobic environments carry out anoxygenic photosynthesis, in which energy from the sun is used to fix CO_2 without the use of H_2O and the production of O_2.
- The **light reactions** of photosynthesis convert light energy into chemical energy. They produce ATP and reduce NADP⁺ to NADPH. **Review Figure 10.3**
- The **light-independent reactions** do not use light directly but instead use ATP and NADPH to reduce CO_2, forming carbohydrates.

10.2 How Does Photosynthesis Convert Light Energy into Chemical Energy?

- Light is a form of electromagnetic radiation. It is emitted in particle-like packets called **photons** but has wavelike properties.
- Molecules that absorb light in the visible spectrum are called **pigments**. Photosynthetic organisms have several pigments, most notably **chlorophylls**, but also accessory pigments such as carotenoids and phycobilins.
- Absorption of a photon 5puts an electron of a pigment molecule in an excited state that has more energy than its ground state.
- Each pigment has a characteristic **absorption spectrum**. An **action spectrum** reflects the rate of photosynthesis carried out by a photosynthetic organism at a given wavelength of light. **Review Figure 10.5**

- The pigments in photosynthetic organisms are arranged into **light-harvesting complexes** that absorb energy from light and funnel this energy to chlorophyll *a* molecules in the reaction center of the **photosystem**. Chlorophyll can act as a reducing agent, transferring excited electrons to other molecules. **Review Figure 10.7**
- **Noncyclic electron transport** uses photosystems I and II to produce ATP, NADPH, and O_2. **Cyclic electron transport** uses only photosystem I and produces only ATP. Both systems generate ATP via the electron transport chain. **Review Figures 10.8, 10.9**
- Chemiosmosis is the mechanism of ATP production in **photophosphorylation**. **Review Figure 10.10, ANIMATED TUTORIAL 10.2**

10.3 How Is Chemical Energy Used to Synthesize Carbohydrates?

- The **Calvin cycle** makes carbohydrates from CO_2. The cycle consists of three processes: fixation of CO_2, reduction and carbohydrate production, and regeneration of RuBP. **See ANIMATED TUTORIAL 10.3**
- RuBP is the initial CO_2 acceptor, and 3PG is the first stable product of CO_2 fixation. The enzyme **rubisco** catalyzes the reaction of CO_2 and RuBP to form 3PG. **Review Figure 10.12**
- ATP and NADPH formed by the light reactions are used in the reduction of 3PG to form **G3P**. **Review Figure 10.13, ACTIVITY 10.1**
- Light stimulates enzymes in the Calvin cycle, further integrating the light-dependent and light-independent pathways. **Review Figure 10.14**

continued

10.4 How Have Plants Adapted Photosynthesis to Environmental Conditions?

- Rubisco can catalyze a reaction between O_2 and RuBP in addition to the reaction between CO_2 and RuBP. At high temperatures and low CO_2 concentrations, the oxygenase function of rubisco is favored over its carboxylase function.

- The oxygenase reaction catalyzed by rubisco significantly reduces the efficiency of photosynthesis. The subsequent reactions of **photorespiration** recover some of the fixed carbon that otherwise would be lost. **Review Figure 10.15**

- In **C_4 plants**, CO_2 reacts with **phosphoenolpyruvate (PEP)** to form a four-carbon intermediate in mesophyll cells. The four-carbon product releases its CO_2 to rubisco in the **bundle sheath cells** in the interior of the leaf. **Review Figures 10.16, 10.17, ACTIVITY 10.2**

- **CAM** plants operate much like C_4 plants, but their initial CO_2 fixation by PEP carboxylase is temporally separated from the Calvin cycle, rather than spatially separated as in C_4 plants.

10.5 How Does Photosynthesis Interact with Other Pathways?

- Photosynthesis and cellular respiration are linked through the **Calvin cycle**, the citric acid cycle, and glycolysis. **Review Figure 10.18**

- To survive, a plant must photosynthesize more than it respires.

- Photosynthesis uses only a small portion of the energy of sunlight. **Review Figure 10.19**

Go to the Interactive Summary to review key figures, Animated Tutorials, and Activities
Life10e.com/is10

CHAPTER**REVIEW**

REMEMBERING

1. In noncyclic photosynthetic electron transport, water is used to
   a. excite chlorophyll.
   b. hydrolyze ATP.
   c. reduce P_i.
   d. oxidize NADPH.
   e. reduce chlorophyll.

2. In cyclic electron transport,
   a. oxygen gas is released.
   b. ATP is formed.
   c. water donates electrons and protons.
   d. NADPH forms.
   e. CO_2 reacts with RuBP.

3. In chloroplasts,
   a. light leads to the flow of protons out of the thylakoids.
   b. ATP is formed when protons flow into the thylakoid lumen.
   c. light causes the thylakoid lumen to become less acidic than the stroma.
   d. protons return actively to the stroma through protein channels.
   e. proton transfer requires ATP.

4. Which statement about chlorophylls is *not* true?
   a. Chlorophylls absorb light near both ends of the visible spectrum.
   b. Chlorophylls can accept energy from other pigments, such as carotenoids.
   c. Excited chlorophyll can either reduce another substance or release light energy.
   d. Excited chlorophyll cannot be an oxidizing agent.
   e. Chlorophylls contain magnesium.

5. Which statement about the Calvin cycle is *not* true?
   a. CO_2 reacts with RuBP to form 3PG.
   b. RuBP forms by the metabolism of 3PG.
   c. ATP and NADPH form when 3PG is reduced.
   d. The concentration of 3PG rises if the light is switched off.
   e. Rubisco catalyzes the reaction of CO_2 and RuBP.

6. Photosynthesis in green plants occurs only during the day. Respiration in plants occurs
   a. only at night.
   b. only when there is enough ATP.
   c. only during the day.
   d. all the time.
   e. in the chloroplast after photosynthesis.

UNDERSTANDING & APPLYING

7. Both photosynthetic electron transport and the Calvin cycle stop in the dark. Which specific reaction stops first? Which stops next? Continue answering the question "Which stops next?" until you have explained why both pathways have stopped.

8. Differentiate between cyclic and noncyclic electron transport in terms of (*a*) the products and (*b*) the source of electrons for the reduction of oxidized chlorophyll.

9. Trace the pathway of carbon fixed by CO_2 in photosynthesis to a carbon atom in a protein.

ANALYZING & EVALUATING

10. If water labeled with ^{18}O is added to a suspension of photosynthesizing chloroplasts, which of the following compounds will first become labeled with ^{18}O: ATP, NADPH, O_2, or 3PG? If water labeled with ^3H is added, which of the same compounds will first become radioactive? Which will be first if CO_2 labeled with ^{14}C is added?

11. The Viking I Lander arrived on Mars in 1976 to detect signs of life. Explain the rationale behind the following experiments this unmanned probe performed:

   a. A scoop of dirt was inserted into a container and $^{14}CO_2$ was added. After a while during the Martian day, the $^{14}CO_2$ was removed and the dirt was heated to a high temperature. Scientists monitoring the experiment back on Earth looked for the release of $^{14}CO_2$ as a sign of life.

   b. The same experiment was performed, except that the dirt was heated to a high temperature for 30 minutes and then allowed to cool to Martian temperature right after scooping, and before the $^{14}CO_2$ was added.

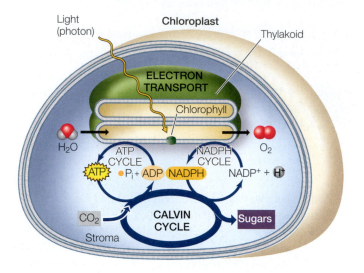

Go to BioPortal at **yourBioPortal.com** for Animated Tutorials, Activities, LearningCurve Quizzes, Flashcards, and many other study and review resources.

11

The Cell Cycle and Cell Division

CHAPTEROUTLINE

11.1 How Do Prokaryotic and Eukaryotic Cells Divide?

11.2 How Is Eukaryotic Cell Division Controlled?

11.3 What Happens during Mitosis?

11.4 What Role Does Cell Division Play in a Sexual Life Cycle?

11.5 What Happens during Meiosis?

11.6 In a Living Organism, How Do Cells Die?

11.7 How Does Unregulated Cell Division Lead to Cancer?

Henrietta Lacks and HeLa Cells
Lacks' tumor cells have far outlived her sad demise from cancer. They reproduce rapidly on the surface of a solid medium. They have been cultured in a laboratory since her death in 1951.

ON JANUARY 29, 1951, 30-year-old Henrietta Lacks visited the nearby Johns Hopkins Hospital in Baltimore, Maryland, because she had been bleeding abnormally after the birth of her last child. The physician found the reason for the bleeding: a tumor the size of a quarter on her cervix. A piece of the tumor was sent to a pathologist in a clinical laboratory, who reported that the tumor was malignant.

A week later Lacks was back at the hospital, where physicians treated her tumor with radiation to try to kill it. But before the treatment began, they took a small sample of cells and sent them to the research lab of George and Margaret Gey, two scientists at the hospital who had been trying for 20 years to coax human cells to live and multiply outside the body. If they could do so, they thought, they might find a cure for cancer. The Geys hit pay dirt with Lacks' cells; they grew more vigorously than any cells they had ever seen. Unfortunately, they also grew fast in Lacks' body, and in a few months they had spread to almost all of her organs. On the day she died, October 4, 1951, Dr. George Gey appeared on national television with a test tube of her cells, which he called HeLa cells. "It is possible that, from a fundamental study such as this, we will ... learn a way by which cancer can be completely wiped out," he said.

Because of their robust ability to reproduce, HeLa cells quickly became a staple of cell biology research. In controlled settings they could be infected with viruses, and they were instrumental in developing the supply of polioviruses that led to the first vaccine against that dread disease. HeLa cells have been used for important basic and applied research ever since. Although Lacks had never been outside Virginia and Maryland, her cells have traveled all over the world and even into space on the space shuttle. Over the past 60 years, tens of thousands of research articles have been published using information obtained from Lacks' cells. Her cells grow so well in the lab that they have sometimes contaminated and taken over cell cultures of other cell types. If they aren't careful, researchers who think they are studying, say, kidney cells may be studying HeLa cells instead.

Understanding the cell division cycle and its control is clearly an important subject for understanding cancer. But cell division is not just important in medicine. It underlies the growth, development, and reproduction of all organisms.

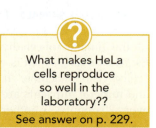

What makes HeLa cells reproduce so well in the laboratory??

See answer on p. 229.

11.1 Important Consequences of Cell Division Cell division is the basis for (A) reproduction, (B) growth, and (C) repair and regeneration of tissues.

(A) Reproduction

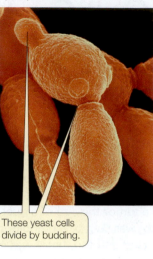

These yeast cells divide by budding.

(B) Growth

Cell division contributes to the growth of this root tissue.

(C) Regeneration

Cell division contributes to the regeneration of a lizard's tail.

11.1 How Do Prokaryotic and Eukaryotic Cells Divide?

The life cycle of an organism, from birth to death, is intimately linked to cell division. Cell division plays important roles in the growth and repair of tissues in multicellular organisms, as well as in the reproduction of all organisms (**Figure 11.1**).

In order for any cell to divide, four events must occur:

- There must be a **reproductive signal**. This signal initiates cell division and may originate from either inside or outside the cell.
- **Replication** of DNA (the genetic material) must occur so that each of the two new cells will have a complete, identical set of genes.
- The cell must distribute the replicated DNA to each of the two new cells. This process is called **segregation**.
- Enzymes and organelles for the new cells must be synthesized, and new material must be added to the plasma membrane (and the cell wall, in organisms that have one), in order to separate the two new cells by a process called **cytokinesis**.

These four events proceed somewhat differently in prokaryotes and eukaryotes.

Prokaryotes divide by binary fission

In prokaryotes, cell division results in the reproduction of the entire single-celled organism. The cell grows in size, replicates its DNA, and then separates the cytoplasm and DNA into two new cells by a process called **binary fission**.

REPRODUCTIVE SIGNALS External factors such as environmental conditions and nutrient concentrations are common signals for the initiation of cell division in prokaryotes. The bacterium *Escherichia coli*, which is widely used in genetic studies, is a "cell division machine." If abundant sources of carbohydrates and mineral nutrients are available, it can divide as often as every 20 minutes. Another bacterium, *Bacillus subtilis*, not only slows its growth when nutrient levels are low but also stops dividing altogether, and then resumes dividing when conditions improve.

REPLICATION OF DNA As we saw in Section 5.3, a **chromosome** consists of a long, thin DNA molecule with proteins attached to it. When a cell divides, all of its chromosomes, which contain the genetic information for the organism, must be replicated, and one copy of each chromosome must find its way into each of the two new cells.

Most prokaryotes have just one main chromosome—a single long DNA molecule with its associated proteins. In *E. coli*, the ends of the DNA molecule are joined to create a circular chromosome. If the *E. coli* DNA were spread out into an actual circle, it would be about 500 micrometers (μm) in diameter. The bacterium itself is only about 2 μm long and 1μm in diameter. Thus if the bacterial DNA were fully extended, it would form a circle more than 200 times larger than the cell! To fit into the cell, bacterial DNA must be compacted. The DNA folds in on itself, and positively charged (basic) proteins bound to the negatively charged (acidic) DNA contribute to this folding.

Two regions of the prokaryotic chromosome play functional roles in cell reproduction:

- *ori*: the site where replication of the circular chromosome starts (the *ori*gin of replication)
- *ter*: the site where replication ends (the *ter*minus of replication)

Chromosome replication takes place as the DNA is threaded through a "replication complex" of proteins near the center of the cell. Replication begins at the *ori* site and moves toward the *ter* site. While the DNA replicates, anabolic metabolism is active and the cell grows. When replication is complete, the two daughter DNA molecules separate and segregate from one another at opposite ends of the cell. In rapidly dividing prokaryotes, DNA replication occupies the entire time between cell divisions.

11.2 Prokaryotic Cell Division (A) The process of cell division in a bacterium. (B) These two cells of the bacterium *Pseudomonas aeruginosa* have almost completed cytokinesis.

SEGREGATION OF DNA MOLECULES Replication begins near the center of the cell, and as it proceeds, the *ori* regions move toward opposite ends of the cell (**Figure 11.2A**). DNA sequences adjacent to the *ori* region bind proteins that are essential for this segregation. This is an active process, since the binding proteins hydrolyze ATP.

CYTOKINESIS Immediately after chromosome replication is finished, cytokinesis begins. At first, the plasma membrane pinches in to form a ring of fibers similar to a purse string. The major component of these fibers is a protein that is related to eukaryotic tubulin (which makes up microtubules; see Figure 5.14). As the membrane pinches in, new cell wall materials are deposited, which finally separate the two cells (**Figure 11.2B**).

Eukaryotic cells divide by mitosis or meiosis followed by cytokinesis

As in prokaryotes, cell reproduction in eukaryotes entails reproductive signals, DNA replication, segregation, and cytokinesis. The details, however, are quite different:

- *Reproductive signal*: Unlike prokaryotes, eukaryotic cells do not constantly divide whenever environmental conditions are adequate. In fact, most eukaryotic cells that are part of a multicellular organism and have become specialized seldom divide. In a eukaryotic organism, the signals for cell division are usually not related to the environment of a single cell, but to the function of the entire organism.

- *Replication*: Whereas most prokaryotes have a single main chromosome, eukaryotes usually have many (humans have 46). Consequently the processes of replication and segregation are more intricate in eukaryotes than in prokaryotes. In eukaryotes, DNA replication is usually limited to a portion of the period between cell divisions.

- *Segregation*: In eukaryotes, the newly replicated chromosomes are closely associated with each other (thus they are known as **sister chromatids**), and a mechanism called **mitosis** segregates them into two new nuclei.

- *Cytokinesis*: Cytokinesis proceeds differently in plant cells (which have a cell wall) than in animal cells (which do not).

The cells resulting from mitosis are identical to the parent cell in the amount and kind of DNA they contain. This contrasts with the second mechanism of nuclear division, meiosis.

Meiosis is the process of nuclear division that occurs in cells involved with sexual reproduction. Whereas the two products of mitosis are genetically identical to the cell that produced them—they both have the same DNA—the products of meiosis are not. As we will see in Section 11.5, meiosis generates diversity by shuffling the genetic material, resulting in new gene combinations. Meiosis plays a key role in the sexual life cycle, which we will discuss in Section 11.4.

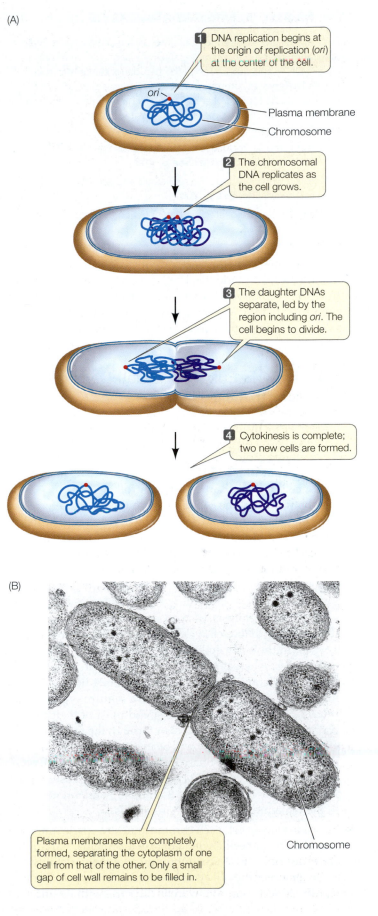

(A)

1 DNA replication begins at the origin of replication (*ori*) at the center of the cell.

ori
Plasma membrane
Chromosome

2 The chromosomal DNA replicates as the cell grows.

3 The daughter DNAs separate, led by the region including *ori*. The cell begins to divide.

4 Cytokinesis is complete; two new cells are formed.

(B)

Plasma membranes have completely formed, separating the cytoplasm of one cell from that of the other. Only a small gap of cell wall remains to be filled in.

Chromosome

What determines whether a eukaryotic cell will divide? How does mitosis lead to identical cells, and meiosis to diversity? Why do most eukaryotic organisms reproduce sexually? In the sections that follow, we will describe the details of mitosis and meiosis and discuss their roles in development and evolution.

11.2 How Is Eukaryotic Cell Division Controlled?

As you will see throughout this book, different cells have different rates of cell division. Some cells, such as those in an early embryo, divide rapidly and continuously. Others, such as neurons in the brain, don't divide at all. Clearly, the signaling pathways for cells to divide are highly controlled.

The period from one cell division to the next is referred to as the **cell cycle**. The cell cycle can be divided into mitosis/cytokinesis and interphase. During **interphase**, the cell nucleus is visible and typical cell functions occur, including DNA replication. This phase of the cell cycle begins when cytokinesis is completed and ends when mitosis (M) begins (**Figure 11.3**). In this section we will describe the events of interphase, especially those that trigger mitosis.

The duration of the cell cycle varies considerably in different cell types. In the early embryo the cell cycle may be as short as 30 minutes, whereas rapidly dividing cells in an adult human typically complete the cycle in about 24 hours. And as mentioned above, many cell types in a mature organism do not divide at all. In general, cells spend most of their time in interphase. So if we take a snapshot through the microscope of a cell population, only a few will be in mitosis or cytokinesis at any given moment. Interphase has three subphases called G1, S, and G2. In a cell cycle of 24 hours, these subphases would typically last for 11 hours (G1), 8 hours (S), and 4 hours (G2), with the remaining 1 hour spent in mitosis.

- *G1 phase.* During G1, each chromosome is a single, unreplicated DNA molecule with associated proteins. Variations in the duration of G1 account for most of the variability in the length of the cell cycle in different cell types. Some rapidly dividing embryonic cells dispense with it entirely,

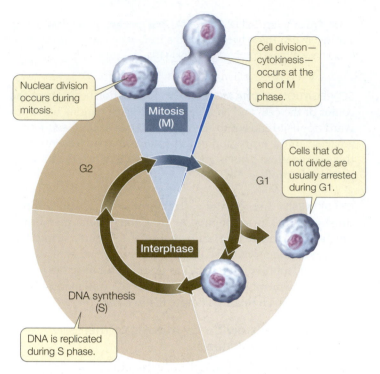

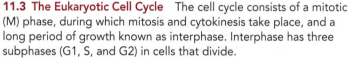

11.3 The Eukaryotic Cell Cycle The cell cycle consists of a mitotic (M) phase, during which mitosis and cytokinesis take place, and a long period of growth known as interphase. Interphase has three subphases (G1, S, and G2) in cells that divide.

whereas other cells may remain in G1 for weeks or even years. In many cases these cells enter a resting phase called G0. Special internal and external signals are needed to prompt a cell to leave G0 and reenter the cell cycle at G1.

- *The G1-to-S transition.* At the **G1-to-S transition** the commitment is made to DNA replication and subsequent cell division.

- *S phase.* DNA replication occurs during **S phase** (see Section 13.3 for a detailed description of DNA replication). Each chromosome is duplicated and thereafter consists of two sister chromatids (the products of DNA replication). The sister chromatids remain joined together until mitosis, when they segregate into two daughter cells.

- *G2 phase.* During **G2 phase**, the cell makes preparations for mitosis—for example, by synthesizing and assembling the structures that move the chromatids to opposite ends of the dividing cell.

The initiation, termination, and operations of these phases are regulated by specific signals.

Specific internal signals trigger events in the cell cycle

Cell fusion experiments were used to reveal the existence of internal signals that control the transitions between stages of the cell cycle. For example, an experiment involving the fusion of HeLa cells (the cells described in the opening story) at different phases of the cell cycle showed that a cell in S phase produces

INVESTIGATING**LIFE**

11.4 Regulation of the Cell Cycle Nuclei in G1 do not undergo DNA replication, but nuclei in S phase do. To determine if there is some signal in the S cells that stimulates G1 cells to replicate their DNA, Rao and Johnson fused together cells in G1 and S phases, creating cells with both G1 and S properties.[a]

HYPOTHESIS A cell in S phase contains an activator of DNA replication.

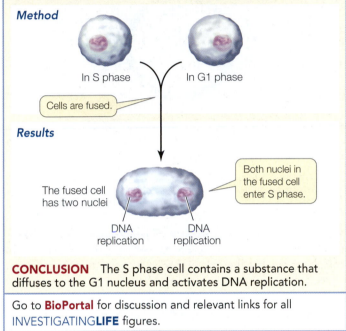

Method

In S phase   In G1 phase

Cells are fused.

Results

The fused cell has two nuclei

Both nuclei in the fused cell enter S phase.

DNA replication   DNA replication

CONCLUSION The S phase cell contains a substance that diffuses to the G1 nucleus and activates DNA replication.

Go to **BioPortal** for discussion and relevant links for all INVESTIGATING**LIFE** figures.

[a]Rao, P. N. and R. T. Johnson. 1970. *Nature* 225: 159–164.

a substance that activates DNA replication (**Figure 11.4**). Similar experiments pointed to the existence of signals controlling entry into M phase. As you will see, the signals that control progress through the cell cycle act through protein kinases.

Progress through the cell cycle depends on the activities of **cyclin-dependent kinases**, or **Cdk's**. Recall from Section 7.2 that a protein kinase is an enzyme that catalyzes the transfer of a phosphate group from ATP to a target protein; this phosphate transfer is called phosphorylation.

$$\text{Protein} + \text{ATP} \xrightarrow{\text{Protein kinase}} \text{protein-P} + \text{ADP}$$
$$(\text{Inactive}) \qquad\qquad\qquad (\text{Active})$$

By catalyzing the phosphorylation of certain target proteins, Cdk's play important roles at various points in the cell cycle. The discovery that Cdk's induce cell division is a beautiful example of how research on different organisms and different cell types can converge on a single mechanism. One group of scientists, led by James Maller at the University of Colorado, was studying immature sea urchin eggs, trying to find out how they are stimulated to divide and eventually form mature eggs. A protein called maturation promoting factor was purified from maturing eggs, which by itself prodded immature egg cells to divide.

WORKING WITH**DATA:**

Regulation of the Cell Cycle

Original Paper

Rao, P. N. and R. T. Johnson. 1970. Mammalian cell fusion: Studies on the regulation of DNA synthesis and mitosis. *Nature* 225: 159–164.

Analyze the Data

The fusion of cellular membranes is a natural process; it occurs during endocytosis and exocytosis, and in fertilization (the fusion of gametes). Membrane fusion also occurs when membrane-enclosed viruses infect their host cells. Occasionally, these viruses also induce the fusion of adjacent host cells, creating a multinucleate cell. This observation led to the use of Sendai virus, a membrane-enclosed mouse respiratory virus, as a tool in the laboratory to fuse cells experimentally. Rao and Johnson used this strategy to study the regulation of the cell cycle (see Figure 11.4).

In their experiment, Rao and Johnson used HeLa cells, which divide continuously (see the opening story of this chapter). They first synchronized separate populations of the cells in G1 or S phase. Before fusion, the cells in S phase were exposed to a radioactively labeled component of DNA (thymidine). The radioactivity was incorporated into these cells' newly replicated DNA, labeling their nuclei. The S and G1 cells were then fused using Sendai virus (resulting in G1/S fusions) and again exposed to labeled thymidine. At various times after fusion, the scientists calculated the percent of previously unlabeled (G1) nuclei that had incorporated new label (i.e., had replicated their DNA) (**FIGURE A**). In a second series of experiments, S and G2 cells were fused in various combinations and then the numbers of cells in mitosis were counted and expressed as a percent of all cells in the population (**FIGURE B**).

QUESTION 1

According to Figure A, how long did it take for all the G1 nuclei in the G1/S cells to become labeled?

QUESTION 2

Examine the data for fused G1/G1 cells and unfused G1 cells in Figure A. Explain why these were appropriate controls for the experiments. When did these nuclei become labeled? Compare these times with each other and with the G1/S nuclei and discuss.

QUESTION 3

Examine the data in Figure B. Why did it take longer for the cells in S phase to begin mitosis than it did for the cells in G2?

QUESTION 4

According to Figure B, did fusion with G2 cells alter the timing of mitosis in the S cell nuclei? Explain what this means in terms of control of the cell cycle.

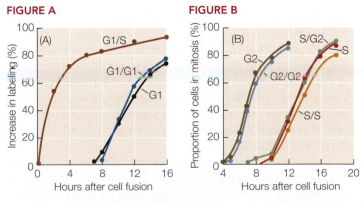

FIGURE A

FIGURE B

Go to **BioPortal** for all WORKING WITH**DATA** exercises

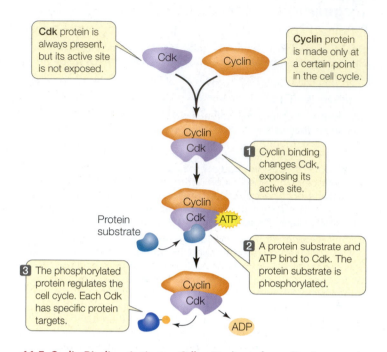

11.5 Cyclin Binding Activates Cdk Binding of a cyclin changes the three-dimensional structure of an inactive Cdk, making it an active protein kinase. Each cyclin–Cdk complex phosphorylates a specific target protein in the cell cycle.

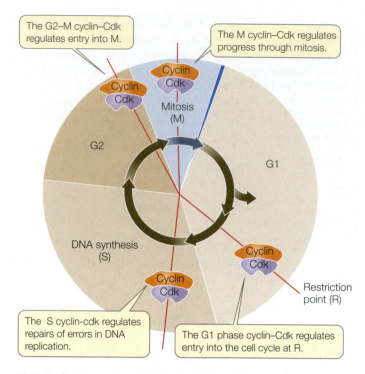

11.6 Cyclin-Dependent Kinases Regulate Progress through the Cell Cycle By acting at checkpoints (red lines), different cyclin–Cdk complexes regulate the orderly sequence of events in the cell cycle.

Meanwhile, Leland Hartwell at the University of Washington was studying the cell cycle in yeast (a single-celled eukaryote; see Figure 11.1A) and found a strain that was stalled at the G1–S boundary because it lacked a Cdk. It turned out that this yeast Cdk and the sea urchin maturation promoting factor had similar properties, and further work confirmed that the sea urchin protein was indeed a Cdk. Similar Cdk's were soon found to control the G1-to-S transition in many other organisms, including humans. This control point in the cell cycle is now called the **restriction (R) point**. Other Cdk's were found to control other parts of the cell cycle.

Cdk's are not active on their own. As their name implies, cyclin-dependent kinases need to be activated by binding to a second type of protein, called **cyclin**. This binding—an example of allosteric regulation (see Section 8.5)—activates the Cdk by altering the shape of its active site (**Figure 11.5**).

The cyclin–Cdk that controls passage from G1 to S phase is not the only such complex involved in regulating the eukaryotic cell cycle. There are different cyclin–Cdk complexes, composed of various cyclins and Cdk's, that act at different stages of the cycle (**Figure 11.6**). The details of how these complexes form and function vary among eukaryotic organisms, but we will focus here on the complexes found in mammalian cells. Let's take a closer look at the cyclin–Cdk complex that controls the G1-to-S transition.

Cyclin–Cdk catalyzes the phosphorylation of a protein called retinoblastoma protein (RB, named because of its role in cancer; see Section 11.7). In many cells, RB or a protein like it acts as an inhibitor of the cell cycle at the R point. To begin

S phase, a cell must get by the RB block. Here is where cyclin–Cdk comes in: it catalyzes the phosphorylation at multiple sites on the RB molecule. This causes a change in the three-dimensional structure of RB, thereby inactivating it. With RB out of the way, the cell cycle can proceed. To summarize:

$$RB \xrightarrow{\text{Cyclin–Cdk}} RB\text{-}P$$
(Active—blocks cell cycle)   (Inactive—allows cell cycle)

Progress through the cell cycle is regulated by the activities of Cdk's, and so regulating *them* is a key to regulating cell division. An effective way to regulate Cdk's is to regulate the presence or absence of cyclins (**Figure 11.7**). Simply put, if a cyclin is not present, its partner Cdk is not active. As their name suggests, cyclins are present cyclically: they are made only at certain times in the cell cycle.

The different cyclin–Cdk's act at **cell cycle checkpoints**, signaling pathways that regulate the cell cycle's progress. For example, if a cell's DNA is substantially damaged by radiation or toxic chemicals, the cell may be prevented from successfully completing a cell cycle. During interphase, there are three checkpoints, with a fourth during mitosis (see Figure 11.6; **Table 11.1**). The table lists the triggers that will cause the cell cycle to pause at each point.

Let's consider the G1 checkpoint (R). If DNA is damaged by radiation during G1, a signaling pathway results in the production of a protein called p21. (The *p* stands for "protein" and the *21* stands for its molecular weight—about 21,000.) The p21

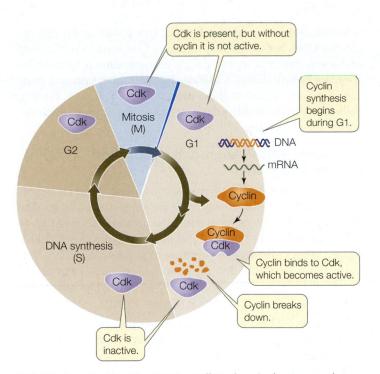

11.7 Cyclins Are Transient in the Cell Cycle Cyclins are made at a particular time and then break down. In this case, the cyclin is present during G1 and activates a Cdk at that time.

protein can bind to the G1–S Cdk, preventing its activation by cyclin. So the cell cycle stops while repairs are made to the DNA (you will learn more about DNA repair in Section 13.4). When the DNA damage pathway is no longer operating, p21 breaks down, allowing the cyclin–Cdk's to function, and the cell cycle proceeds. If DNA damage is severe and it cannot be repaired, the cell will undergo programmed cell death (apoptosis, which we will discuss later in this chapter). Such controls prevent defective cells from proliferating and potentially harming an organism.

Growth factors can stimulate cells to divide

Cyclin–Cdk's provide cells with internal controls of their progress through the cell cycle, but the cell cycle is also influenced by external signals. Not all cells in an organism go through the cell cycle on a regular basis. Some cells either no longer go through the cell cycle and enter G0, or go through it slowly and divide infrequently. If such cells are to divide,

TABLE 11.1

Cell Cycle Checkpoints

Cell Cycle Phase	Checkpoint Trigger
G1	DNA damage
S	Incomplete replication or DNA damage
G2	DNA damage
M	Chromosome unattached to spindle

they must be stimulated by external chemical signals called **growth factors**:

- If you cut yourself and bleed, specialized cell fragments called platelets gather at the wound to initiate blood clotting. The platelets produce and release a protein called platelet-derived growth factor that diffuses to the adjacent cells in the skin and stimulates them to divide and heal the wound.

- Red and white blood cells have limited lifetimes and must be replaced through the division of immature, unspecialized blood cell precursors in the bone marrow. Two types of growth factors, interleukins and erythropoietin, stimulate the division and specialization, respectively, of precursor cells of white blood cells and red blood cells.

In these and other examples, growth factors bind to specific receptors on target cells and activate signal transduction pathways that end with cyclin synthesis, thereby activating Cdk's and the cell cycle.

RECAP 11.2

The eukaryotic cell cycle is under both external and internal control. Cdk's control the eukaryotic cell cycle and are themselves controlled by cyclins. External signals such as growth factors can initiate the cell cycle.

- Draw a cell cycle diagram showing the various stages of interphase. **See p. 208 and Figure 11.3**
- How do cyclin–Cdk's control the progress of the cell cycle? **See pp. 209–210 and Figure 11.6**
- What are growth factors, and how do they act to control the cell cycle? **See p. 211**

11.3 What Happens during Mitosis?

Segregation of the replicated DNA occurs during mitosis. Prior to segregation, the DNA molecules and their associated proteins in each chromosome become condensed into more compact structures. After segregation by mitosis, cytokinesis separates the two cells. Let's now look at these steps more closely.

Go to Animated Tutorial 11.1
Mitosis
Life10e.com/at11.1

Prior to mitosis, eukaryotic DNA is packed into very compact chromosomes

A eukaryotic chromosome consists of one or two long, linear, double-stranded DNA molecules bound with many proteins (the complex of DNA and proteins is referred to as **chromatin**). Before S phase, each chromosome contains only one double-stranded DNA molecule. After it replicates during S phase, however, there are two double-stranded DNA molecules: the sister chromatids (**Figure 11.8**). Throughout G2 the sister chromatids are held together along most of their length by a protein complex called

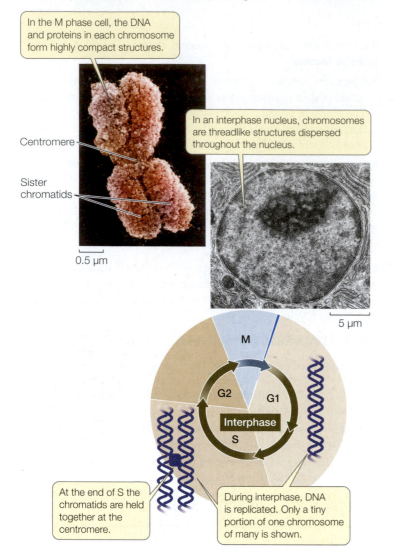

In the M phase cell, the DNA and proteins in each chromosome form highly compact structures.

Centromere

Sister chromatids

0.5 μm

In an interphase nucleus, chromosomes are threadlike structures dispersed throughout the nucleus.

5 μm

M

G2   G1

Interphase

S

At the end of S the chromatids are held together at the centromere.

During interphase, DNA is replicated. Only a tiny portion of one chromosome of many is shown.

11.8 Chromosomes, Chromatids, and Chromatin   DNA in the interphase nucleus is diffuse and becomes compacted as mitosis begins.

cohesin. At mitosis most of the cohesin is removed, except in a region called the **centromere**, where the chromatids remain held together. At the end of G2 and the beginning of mitosis, a second group of proteins called condensins coat the DNA molecules and make them more compact.

If all of the DNA in a typical human cell were put end to end, it would be nearly 2 meters long. Yet the nucleus is only 5 μm (0.000005 meters) in diameter. So eukaryotic DNA is extensively packaged in a highly organized way (**Figure 11.9**). This packing is achieved largely by proteins call histones (*histos*, "web" or "loom"), which are positively charged at cellular pH because of their high content of the basic amino acids lysine and arginine. The charged R groups on these amino acids bind to the negatively charged phosphate groups on DNA by ionic attractions. These DNA–histone interactions, as well as histone–histone interactions, result in the formation of beadlike units called **nucleosomes** (see Figure 11.9).

During interphase, the chromatin that makes up each chromosome is less densely packaged and consists of single DNA molecules running around vast numbers of nucleosomes that

resemble beads on a string. During this phase of the cell cycle, the DNA is accessible to proteins involved in replication and transcription. Once a mitotic chromosome is formed, its compact nature makes it inaccessible to replication and transcription factors, and so these processes cannot occur. Further coiling of the chromatin continues up to the time at which the chromatids begin to move apart.

Overview: Mitosis segregates copies of genetic information

In mitosis, a single nucleus gives rise to two nuclei that are genetically identical to each other and to the parent nucleus. Mitosis (the M phase of the cell cycle) ensures the accurate segregation of the eukaryotic cell's multiple chromosomes into the daughter nuclei. While mitosis is a continuous process in which each event flows smoothly into the next, it is convenient to subdivide it into a series of stages: prophase, prometaphase, metaphase, anaphase, and telophase (**Figure 11.10**).

- **Prophase**. At the beginning of mitosis the chromatin becomes condensed and the separate chromatids become visible through a light microscope.

- **Prometaphase**. The nuclear envelope breaks down and the compacted chromosomes, each consisting of two chromatids, attach to the spindle apparatus (see below).

- **Metaphase**. The chromosomes line up at the midline of the cell (equatorial position).

- **Anaphase**. The chromatids separate and move away from each other toward opposite poles.

- **Telophase**. A nuclear envelope forms around each set of chromosomes, nucleoli appear, and the chromosomes become less compact. The spindle disappears. As a result, there are two new nuclei in a single cell.

Let's take a closer look at two cellular structures that contribute to the orderly segregation of the chromosomes during mitosis—the centrosomes and the spindle.

The centrosomes determine the plane of cell division

The **spindle apparatus** (also called the *mitotic spindle* or simply the *spindle*) is the dynamic structure that moves sister chromatids apart during mitosis. It is made up of microtubules. Before the spindle can form, its orientation must be determined. This is accomplished by the **centrosome** ("central body"), an organelle in the cytoplasm near the nucleus. In many organisms the centrosome consists of a pair of **centrioles**, each one a hollow tube formed by nine microtubule triplets. The two tubes are at right angles to each other. During S phase the centrosome doubles, and at the beginning of prophase the two centrosomes separate from one another, moving to opposite ends of the nuclear envelope. These identify the "poles" toward which chromosomes move during anaphase. The cells of plants and fungi lack centrosomes, but distinct microtubule organizing centers at each end of the cell play the same role.

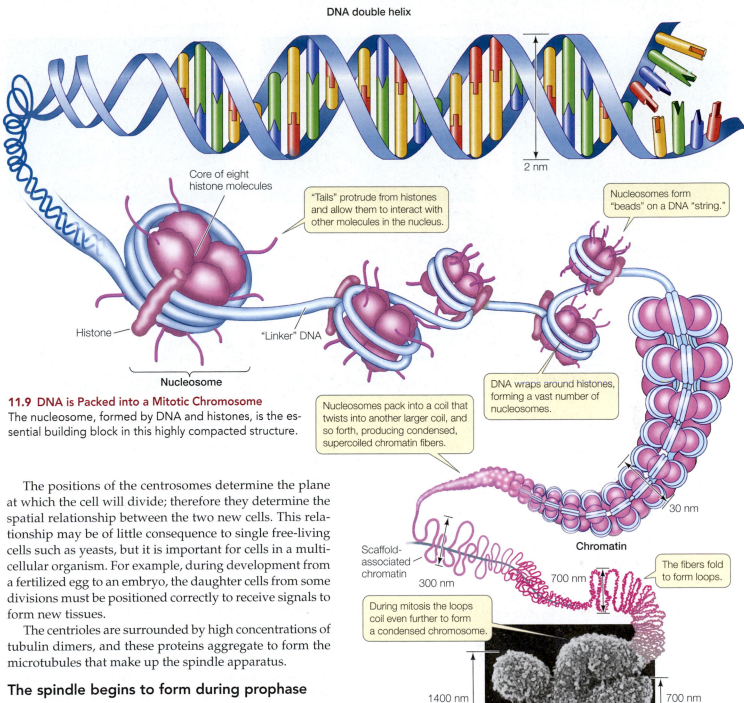

DNA double helix

Core of eight histone molecules

"Tails" protrude from histones and allow them to interact with other molecules in the nucleus.

Nucleosomes form "beads" on a DNA "string."

2 nm

Histone

"Linker" DNA

Nucleosome

DNA wraps around histones, forming a vast number of nucleosomes.

Nucleosomes pack into a coil that twists into another larger coil, and so forth, producing condensed, supercoiled chromatin fibers.

30 nm

Chromatin

Scaffold-associated chromatin

The fibers fold to form loops.

300 nm

700 nm

During mitosis the loops coil even further to form a condensed chromosome.

1400 nm

700 nm

Mitotic chromosome

11.9 DNA is Packed into a Mitotic Chromosome
The nucleosome, formed by DNA and histones, is the essential building block in this highly compacted structure.

The positions of the centrosomes determine the plane at which the cell will divide; therefore they determine the spatial relationship between the two new cells. This relationship may be of little consequence to single free-living cells such as yeasts, but it is important for cells in a multicellular organism. For example, during development from a fertilized egg to an embryo, the daughter cells from some divisions must be positioned correctly to receive signals to form new tissues.

The centrioles are surrounded by high concentrations of tubulin dimers, and these proteins aggregate to form the microtubules that make up the spindle apparatus.

The spindle begins to form during prophase

During interphase, only the nuclear envelope, the nucleoli (see Section 5.3), and a barely discernible tangle of chromatin are visible under the light microscope. The appearance of the nucleus changes as the cell enters prophase. At this stage, most of the cohesin that has held the two products of DNA replication together since S phase is removed, so the individual chromatids become visible. They are still held together by a small amount of cohesin at the centromere. Late in prophase, specialized three-layered structures called **kinetochores** develop in the centromere region, one on each chromatid. These structures will be important for chromosome movement.

Each of the two centrosomes, now on opposite sides of the nucleus, serves as a mitotic center, or pole, toward which the

chromosomes will move (**Figure 11.11A**). During prophase and prometaphase, microtubules form between the poles and the chromosomes to make up the spindle. The spindle serves as a structure to which the chromosomes attach and as a framework keeping the two poles apart. Each half of the spindle develops as tubulin dimers aggregate from around the centrioles and form long fibers that extend into the middle region of the cell. The

Interphase

Prophase

Prometaphase

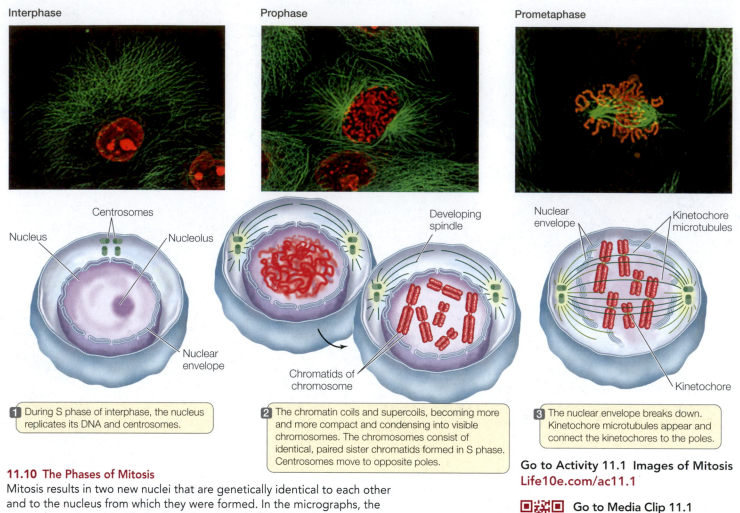

Centrosomes

Nucleus

Nucleolus

Nuclear
envelope

Developing
spindle

Chromatids of
chromosome

Nuclear
envelope

Kinetochore
microtubules

Kinetochore

1 During S phase of interphase, the nucleus replicates its DNA and centrosomes.

2 The chromatin coils and supercoils, becoming more and more compact and condensing into visible chromosomes. The chromosomes consist of identical, paired sister chromatids formed in S phase. Centrosomes move to opposite poles.

3 The nuclear envelope breaks down. Kinetochore microtubules appear and connect the kinetochores to the poles.

11.10 The Phases of Mitosis
Mitosis results in two new nuclei that are genetically identical to each other and to the nucleus from which they were formed. In the micrographs, the green dye stains microtubules (and thus the spindle); the red dye stains the chromosomes. The chromosomes in the diagrams are stylized to emphasize the fates of the individual chromatids.

Go to Activity 11.1 Images of Mitosis
Life10e.com/ac11.1

Go to Media Clip 11.1
Mitosis: Live and Up Close
Life10e.com/mc11.1

microtubules are initially unstable, constantly forming and falling apart, until they contact kinetochores or microtubules from the other half-spindle and become more stable.

There are two types of microtubule in the spindle:

- Polar microtubules form the framework of the spindle and run from one pole to the other.

- Kinetochore microtubules, which form later, attach to the kinetochores on the chromosomes. The two sister chromatids in each chromosome pair become attached to kinetochore microtubules in opposite halves of the spindle (**Figure 11.11B**). This ensures that the two chromatids will eventually move to opposite poles.

Movement of the chromatids is the central feature of mitosis. It accomplishes the segregation that is needed for cell division and completion of the cell cycle. Prophase prepares for this movement, and the actual segregation takes place in the next three phases of mitosis.

Chromosome separation and movement are highly organized

During the next three phases of mitosis—prometaphase, metaphase, and anaphase—dramatic changes take place in the cell and the chromosomes (see above and Figure 11.10). During prometaphase, the nuclear envelope breaks down and spindle formation is completed. During metaphase, the chromosomes line up at the equatorial position of the cell. Now let's consider two key processes of anaphase: separation of the chromatids, and the mechanism of their actual movement toward the poles.

CHROMATID SEPARATION The separation of chromatids occurs at the beginning of anaphase. It is controlled by an M phase cyclin–Cdk (see Figure 11.6), which activates another protein complex called the anaphase-promoting complex (APC). Separation occurs because one subunit of the cohesin protein holding the sister chromatids together is hydrolyzed

Metaphase

Anaphase

Telophase

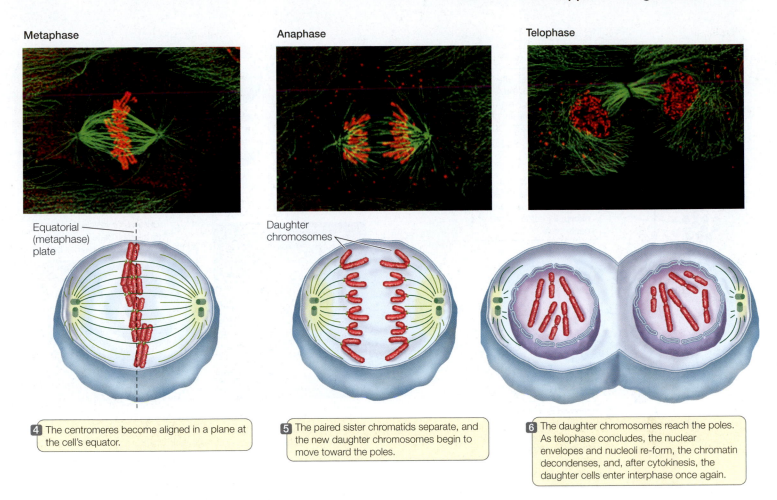

Equatorial
(metaphase)
plate

Daughter
chromosomes

4 The centromeres become aligned in a plane at the cell's equator.

5 The paired sister chromatids separate, and the new daughter chromosomes begin to move toward the poles.

6 The daughter chromosomes reach the poles. As telophase concludes, the nuclear envelopes and nucleoli re-form, the chromatin decondenses, and, after cytokinesis, the daughter cells enter interphase once again.

by a specific protease, appropriately called separase (**Figure 11.12**). A cell cycle checkpoint, often called the spindle assembly checkpoint, occurs at the end of metaphase to inhibit the APC if one of chromosomes is not attached properly to the spindle. When all chromosomes are attached, the APC is activated and the chromatids separate. After separation the chromatids are called **daughter chromosomes**. Note the *difference between chromatids and chromosomes*:

- Chromatids share a centromere.

- Chromosomes have their own centromere.

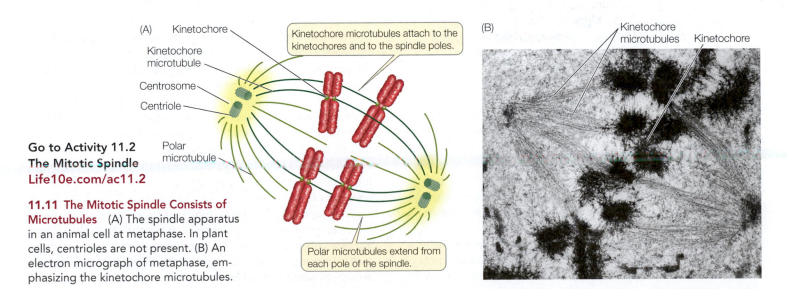

(A) Kinetochore

Kinetochore microtubule

Centrosome

Centriole

Kinetochore microtubules attach to the kinetochores and to the spindle poles.

Polar microtubule

Polar microtubules extend from each pole of the spindle.

(B)

Kinetochore microtubules   Kinetochore

**Go to Activity 11.2
The Mitotic Spindle**
Life10e.com/ac11.2

11.11 The Mitotic Spindle Consists of Microtubules (A) The spindle apparatus in an animal cell at metaphase. In plant cells, centrioles are not present. (B) An electron micrograph of metaphase, emphasizing the kinetochore microtubules.

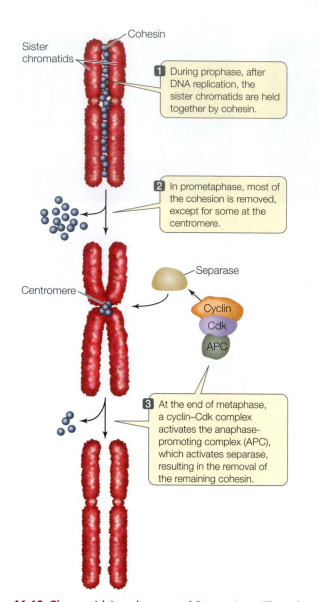

11.12 Chromatid Attachment and Separation The cohesin protein complex holds sister chromatids together at the centromere. The enzyme separase hydrolyzes cohesin at the end of metaphase, allowing the chromatids to separate into daughter chromosomes.

CHROMOSOME MOVEMENT The migration of the two sets of daughter chromosomes to the poles of the cell is a highly organized, active process. Two mechanisms operate to move the chromosomes along. First, the kinetochores contain molecular motor proteins, including kinesins and cytoplasmic dynein (see Figures 5.18 and 5.19), which use energy from ATP hydrolysis to do the work of moving the chromosomes along the microtubules. Second, the kinetochore microtubules shorten from the poles, drawing the chromosomes toward them.

The last stage of mitosis is telophase, when the spindle disappears and a nuclear envelope forms around each set of chromosomes (see Figure 11.10). Finally, the cytoplasms of the two daughter cells separate during cytokinesis: the last stage of cell division.

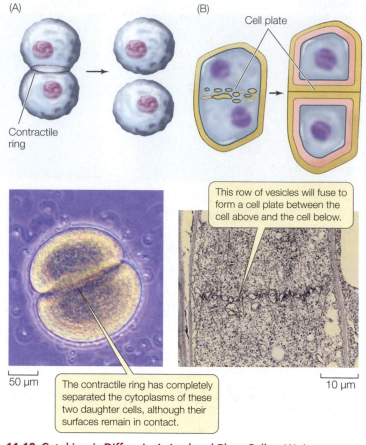

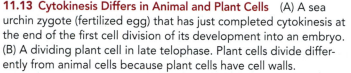

11.13 Cytokinesis Differs in Animal and Plant Cells (A) A sea urchin zygote (fertilized egg) that has just completed cytokinesis at the end of the first cell division of its development into an embryo. (B) A dividing plant cell in late telophase. Plant cells divide differently from animal cells because plant cells have cell walls.

Cytokinesis is the division of the cytoplasm

Cytokinesis divides the cell's cytoplasm after nuclear division in mitosis. There are substantial differences between the process in animal and plant cells. In animal cells, cytokinesis begins with a furrowing of the plasma membrane, as if an invisible thread were cinching the cytoplasm between the two nuclei (**Figure 11.13A**). This contractile ring is composed of microfilaments of actin and myosin (see Figure 5.15), which form a ring on the cytoplasmic surface of the plasma membrane. These two proteins interact to produce a contraction, pinching the cell in two. The microfilaments assemble rapidly from actin monomers that are present in the interphase cytoskeleton. Their assembly is under the control of calcium ions that are released from storage sites in the center of the cell.

The plant cell cytoplasm divides differently because plants have rigid cell walls. In plant cells, as the spindle breaks down after mitosis, membranous vesicles derived from the Golgi apparatus appear along the plane of cell division, roughly midway between the two daughter nuclei. The vesicles are propelled along microtubules by the motor protein kinesin, and fuse to form a new plasma membrane. At the same time they

TABLE 11.2
Summary of Cell Cycle Events

Phase	Events
Interphase:	
G1	Growth; restriction point at end
S	DNA replication
G2	Spindle synthesis begins; preparation for mitosis
Mitosis:	
Prophase	Condensation of chromosomes; spindle assembly
Prometaphase	Nuclear envelope breakdown; chromosome attachment to spindle
Metaphase	Alignment of chromosomes at equatorial plate
Anaphase	Separation of chromatids; migration to poles
Telophase	Chromosomes decondense; nuclear envelope re-forms
Cytokinesis	Cell separation; cell membrane and/or wall formation

contribute their contents to a cell plate, which is the beginning of a new cell wall (**Figure 11.13B**).

Following cytokinesis, each daughter cell contains all the components of a complete cell. A precise distribution of chromosomes is ensured by mitosis. In contrast, organelles such as ribosomes, mitochondria, and chloroplasts do not need to be distributed equally between daughter cells, as long as some of each are present in each cell. Accordingly, there is no mechanism with a precision comparable to that of mitosis to provide for their equal allocation to daughter cells. As we will see in Chapter 19, the unequal distribution of cytoplasmic components during development can have functional significance for the two new cells.

Table 11.2 summarizes the major events of the cell cycle.

■ **RECAP 11.3**

Mitosis is the ordered division of a eukaryotic cell nucleus into two nuclei with identical sets of chromosomes. The process of mitosis, while continuous, can be viewed as a series of events (prophase, prometaphase, metaphase, anaphase, and telophase). Once two nuclei have formed, the cell divides into two cells by cytokinesis.

- What is the difference between a chromosome, a chromatid, and a daughter chromosome? **See Figures 11.8, 11.10**
- What are the various levels of "packing" by which the genetic information contained in linear DNA is condensed during prophase? **See p. 212 and Figure 11.9**
- Describe how chromosomes move during mitosis. **See p. 216 and Figure 11.10**
- What are the differences in cytokinesis between plant and animal cells? **See pp. 216–217 and Figure 11.13**

The intricate process of mitosis results in two cells that are genetically identical. But as we mentioned earlier, there is another eukaryotic cell division process, called meiosis, that results in genetic diversity. In the next section we will discuss the roles of mitosis and meiosis in sexual reproduction, and then turn to the details of meiosis in Section 11.5.

11.4 What Role Does Cell Division Play in a Sexual Life Cycle?

The mitotic cell cycle repeats itself, and by this process a single cell can give rise to many cells with identical nuclear DNA. Meiosis, by contrast, produces just four daughter cells. Mitosis and meiosis are both involved in reproduction but in different ways: asexual reproduction involves only mitosis, whereas the sexual reproduction cycle involves both mitosis and meiosis.

Asexual reproduction by mitosis results in genetic constancy

Asexual reproduction, sometimes called vegetative reproduction, is based on the mitotic division of the nucleus. An organism that reproduces asexually may be single-celled like yeast, reproducing itself with each cell cycle, or it may be multicellular like an aspen in a forest in the Wasatch Mountains of Utah (**Figure 11.14**). Aspen can reproduce sexually, with male and female plants, but in many aspen stands all the trees are the same sex, and DNA analyses have shown that they are **clones** of a single parent organism; the offspring are genetically identical to the parent. In such stands, an extensive root system spreads through the soil and at intervals new stems sprout and grow into trees. Any genetic variation among the trees is most likely due to small environmentally caused changes in the DNA. As you will see, this small amount of variation contrasts with the extensive variation possible in sexually reproducing organisms.

11.14 Asexual Reproduction on a Large Scale In this forest, the aspen trees arose from a single tree by asexual reproduction. They are virtually identical genetically.

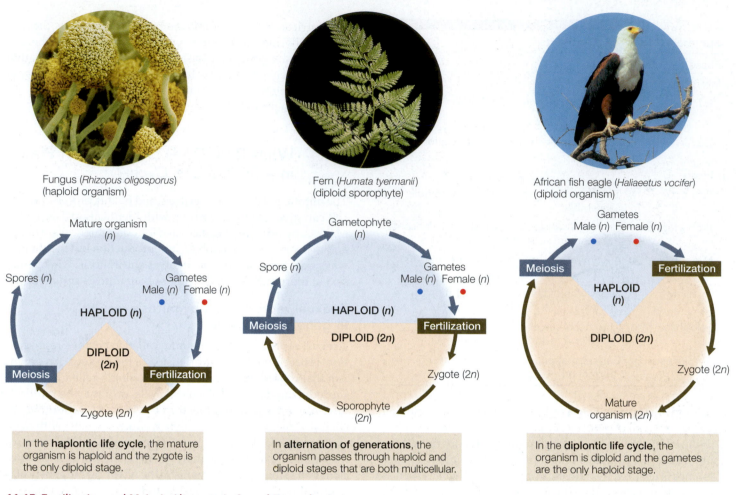

Fungus (*Rhizopus oligosporus*)
(haploid organism)

Fern (*Humata tyermanii*)
(diploid sporophyte)

African fish eagle (*Haliaeetus vocifer*)
(diploid organism)

In the **haplontic life cycle**, the mature organism is haploid and the zygote is the only diploid stage.

In **alternation of generations**, the organism passes through haploid and diploid stages that are both multicellular.

In the **diplontic life cycle**, the organism is diploid and the gametes are the only haploid stage.

11.15 Fertilization and Meiosis Alternate in Sexual Reproduction
In sexual reproduction, haploid (*n*) cells or organisms alternate with diploid (2*n*) cells or organisms.

Go to Activity 11.3 Sexual Life Cycle   Life10e.com/ac11.3

Sexual reproduction by meiosis results in genetic diversity

Unlike asexual reproduction, **sexual reproduction** results in an organism that is not identical to its parents. Sexual reproduction requires **gametes** created by meiosis; two parents each contribute one gamete to each of their offspring. Meiosis can produce gametes—and thus offspring—that differ genetically from each other and from the parents. Because of this genetic variation, some offspring may be better adapted than others to survive and reproduce in a particular environment. Meiosis thus generates the genetic diversity that is the raw material for natural selection and evolution.

In most multicellular organisms, the body cells that are *not* specialized for reproduction, called **somatic cells**, each contain two sets of chromosomes, which are found in pairs. One chromosome of each pair comes from each of the organism's two parents; for example, in humans with 46 chromosomes, 23 come from the mother and 23 from the father. The members of such a **homologous pair** are similar in size and appearance. The exception is the sex chromosomes, which are found in some species. The two chromosomes in a homologous pair (called **homologs**) bear corresponding, though not identical, genetic information. For example, a homologous pair of chromosomes in a plant may carry different versions of a gene that controls seed shape. One homolog may carry the version for wrinkled seeds while the other may carry the version for smooth seeds.

There is no simple relationship between the size of an organism and its chromosome number. A housefly has 5 chromosome pairs and a horse has 32, but the smaller carp (a fish) has 52 pairs. Probably the highest number of chromosomes in any organism is in the fern *Ophioglossum reticulatum*, which has 1,260 (630 pairs)!

In contrast to somatic cells, gametes contain only a single set of chromosomes—that is, one homolog from each pair. The number of chromosomes in a gamete is denoted by *n*, and the cell is said to be **haploid**. During reproduction, two haploid gametes fuse to form a **zygote** in a process called **fertilization**. The zygote thus has two sets of chromosomes, just as the somatic cells do. Its chromosome number is denoted by 2*n*, and the zygote is said to be **diploid**. Depending on the organism, the zygote may divide by either meiosis or mitosis. Either way, a new mature organism develops that is capable of sexual reproduction.

All sexual life cycles involve meiosis to produce haploid cells. **Figure 11.15** presents three types. In some life cycles the products of meiosis undergo cell division, resulting in a mature organism with haploid cells. Specialized cells in these organisms become gametes. In other life cycles the gametes form directly from the products of meiosis. In all cases, the gametes fuse to produce a zygote, beginning the diploid stage of the life cycle. Since the origin of sexual reproduction, evolution has generated many different versions of the sexual life cycle.

The essence of sexual reproduction is the *random selection of half of the diploid chromosome set* to make a haploid gamete, followed by fusion of two haploid gametes to produce a diploid cell. Both of these steps contribute to a shuffling of genetic information in the population, so that no two individuals have exactly the same genetic makeup (unless they are identical twins). The diversity provided by sexual reproduction opens up enormous opportunities for evolution.

RECAP 11.4

Meiosis is necessary for sexual reproduction, in which haploid gametes fuse to produce a diploid zygote. Sexual reproduction increases genetic diversity, the raw material of evolution.

- What is the difference, in terms of genetics, between asexual and sexual reproduction? **See pp. 217–218**
- How does fertilization produce a diploid organism? **See p. 218**
- What general features do all sexual life cycles have in common? **See pp. 218–219 and Figure 11.15**

Meiosis, unlike mitosis, results in daughter cells that differ genetically from the parent cell. We will now look at the details of meiosis to see how this genetic shuffling occurs.

11.5 What Happens during Meiosis?

In the last section we described the role and importance of meiosis in sexual reproduction. Now we will see how meiosis accomplishes the orderly and precise generation of haploid cells.

Meiosis consists of *two* nuclear divisions that together reduce the number of chromosomes to the haploid number, in preparation for sexual reproduction. *Although the nucleus divides twice during meiosis, the DNA is replicated only once.* Unlike the products of mitosis, the products of meiosis are genetically different from one another and from the parent cell. To understand the process of meiosis and its specific details, it is useful to keep in mind the overall functions of meiosis:

- To reduce the chromosome number from diploid to haploid
- To ensure that each of the haploid products has a complete set of chromosomes
- To generate genetic diversity among the products

The events of meiosis are illustrated in **Figure 11.16**. In this section we will discuss some of the key features that distinguish meiosis from mitosis.

Go to Animated Tutorial 11.2
Meiosis
Life10e.com/at11.2

Meiotic division reduces the chromosome number

As noted above, meiosis consists of two nuclear divisions, meiosis I and meiosis II. Two unique features characterize **meiosis I**.

- *Homologous chromosomes come together to pair* along their entire lengths. No such pairing occurs in mitosis.
- *The homologous chromosome pairs separate,* but the individual chromosomes, each consisting of two sister chromatids, remain intact. (The chromatids will separate during meiosis II.)

Like mitosis, meiosis I is preceded by an interphase with an S phase, during which each chromosome is replicated. As a result, each chromosome consists of two sister chromatids, held together by cohesin proteins. At the end of meiosis I, two nuclei form, each with half of the original chromosomes (one member of each homologous pair). Since the centromeres did not separate, these chromosomes are still composed of two sister chromatids. The sister chromatids are separated during **meiosis II**, which is *not* preceded by DNA replication. As a result, the products of meiosis I and II are four cells, each containing the haploid number of chromosomes. But these four cells are not genetically identical.

Chromatid exchanges during meiosis I generate genetic diversity

Meiosis I begins with a long prophase I (the first three panels of Figure 11.16), during which the chromosomes change markedly. The homologous chromosomes pair by adhering along their lengths in a process called **synapsis**. (This does not usually happen in mitosis.) This pairing process lasts from prophase I to the end of metaphase I. The four chromatids of each pair of homologous chromosomes form a **tetrad**, or bivalent. For example, in a human cell at the end of prophase I there are 23 tetrads, each consisting of four chromatids. The four chromatids come from the two partners in each homologous pair of chromosomes.

Throughout prophase I and metaphase I, the chromatin continues to coil and compact, so that the chromosomes appear ever thicker. At a certain point, the homologous chromosomes appear to repel each other, especially near the centromeres, but they remain held together by physical attachments mediated by cohesins. Later in prophase, regions having these attachments take on an X-shaped appearance (**Figure 11.17**) and are called **chiasmata** (singular *chiasma*, "cross").

A chiasma reflects an *exchange of genetic material* between nonsister chromatids on homologous chromosomes—what geneticists call **crossing over** (**Figure 11.18**). The chromosomes usually begin exchanging material shortly after synapsis begins, but chiasmata do not become visible until later, when the homologs are repelling each other. Crossing over results in **recombinant chromatids**, and it increases genetic variation among the products of meiosis by shuffling genetic information among the homologous pairs. In Chapter 12 we will explore further the genetic consequences of crossing over. Mitosis seldom takes more than an hour or two, but meiosis can take much longer. In human males, the cells in

MEIOSIS I

Early prophase I

Mid-prophase I

Late prophase I–Prometaphase

Centrosomes

Pairs of homologs

Tetrad

Chiasma

1 The chromatin begins to condense following interphase.

2 Synapsis aligns homologs, and chromosomes condense further.

3 The chromosomes continue to coil and shorten. The chiasmata reflect crossing over, the exchange of genetic material between nonsister chromatids in a homologous pair. In prometaphase the nuclear envelope breaks down.

MEIOSIS II

Prophase II

Metaphase II

Anaphase II

Equatorial plate

7 The chromosomes condense again, following a brief interphase in which DNA does not replicate.

8 The centromeres of the paired chromatids line up across the equatorial plates of each cell.

9 The chromatids finally separate, becoming chromosomes in their own right, and move to opposite poles. Because of crossing over and independent assortment, each new cell will have a different genetic makeup.

the testis that undergo meiosis take about a week for prophase I and about a month for the entire meiotic cycle. In females, prophase I begins long before a woman's birth, during her early fetal development, and ends as much as decades later, during the monthly ovarian cycle (see Chapter 43).

During meiosis homologous chromosomes separate by independent assortment

A diploid organism has two sets of chromosomes (*2n*): one set derived from its male parent, and the other from its female parent. As the organism grows and develops, its cells undergo

Metaphase I

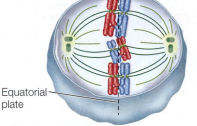

Equatorial plate

4 The homologous pairs line up on the equatorial (metaphase) plate.

Anaphase I

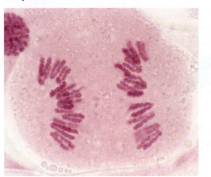

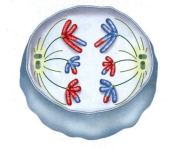

5 The homologous chromosomes (each with two chromatids) move to opposite poles of the cell.

Telophase I

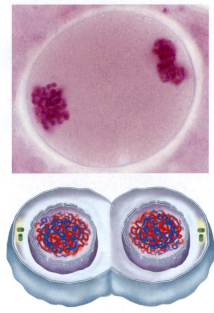

6 The chromosomes gather into nuclei, and the original cell divides.

Telophase II

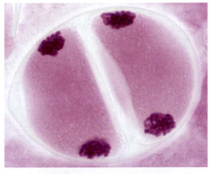

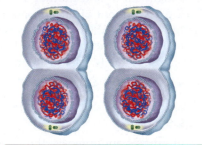

10 The chromosomes gather into nuclei, and the cells divide.

Products

11 Each of the four cells has a nucleus with a haploid number of chromosomes.

11.16 Meiosis: Generating Haploid Cells
In meiosis, four daughter nuclei are produced, each of which has half as many chromosomes as the original cell. Four haploid cells are the result of two successive nuclear divisions. The micrographs show meiosis in the male reproductive organ of a lily; the diagrams show the corresponding phases in an animal cell. (For instructional purposes, the chromosomes from one parent are colored blue and those from the other parent are red.)

Go to Activity 11.4 Images of Meiosis
Life10e.com/ac11.4

mitotic divisions. In mitosis, each chromosome behaves independently of its homolog, and its two chromatids are sent to opposite poles during anaphase. Each daughter nucleus ends up with *2n* chromosomes. In meiosis, things are very different. **Figure 11.19** compares the two processes.

In meiosis I, chromosomes of maternal origin pair with their paternal homologs during synapsis. *This pairing does not occur in mitosis.* Segregation of the homologs during meiotic anaphase I ensures that each newly formed cell receives one member of each homologous pair (see steps 4–6 of Figure 11.16). For example, at the end of meiosis I in humans, each daughter nucleus contains 23 of the original 46 chromosomes. In this way, the chromosome number is decreased from diploid to haploid. Furthermore, meiosis I guarantees that each daughter nucleus gets one full set of chromosomes.

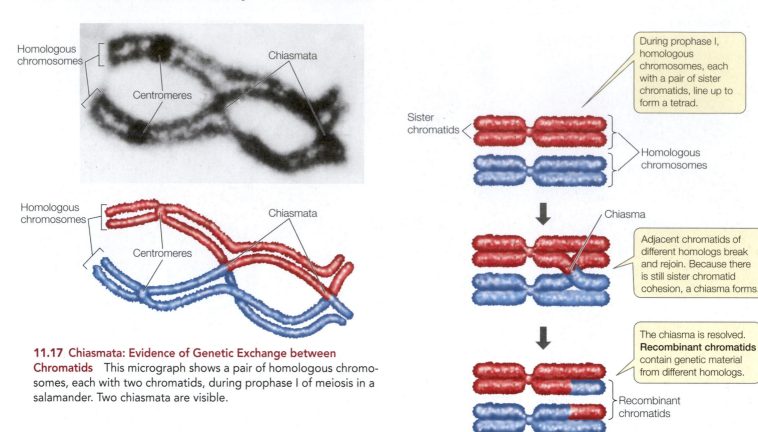

11.17 Chiasmata: Evidence of Genetic Exchange between Chromatids This micrograph shows a pair of homologous chromosomes, each with two chromatids, during prophase I of meiosis in a salamander. Two chiasmata are visible.

Crossing over is one reason for the genetic diversity among the products of meiosis; another is **independent assortment**. It is a matter of chance which member of a homologous pair goes to which daughter cell at anaphase I. For example, imagine there are two homologous pairs of chromosomes in the diploid parent nucleus. A particular daughter nucleus could receive the paternal chromosome 1 and the maternal chromosome 2. Or it could get paternal 2 and maternal 1, or both maternal, or both paternal. It all depends on the way in which the homologous pairs line up at metaphase I. This phenomenon is termed independent assortment.

Note that of the four possible chromosome combinations just described, only two produce daughter nuclei with full complements of either maternal or paternal chromosome sets (apart from the material exchanged by crossing over). *The greater the number of chromosomes, the less probable it is that the original parental combinations will be reestablished, and the greater the potential for genetic diversity.* Most species of diploid organisms have more than two pairs of chromosomes. In humans, with 23 chromosome pairs, 2^{23} (8,388,608) different combinations can be produced just by the mechanism of independent assortment. Taking the extra genetic shuffling afforded by crossing over into account, the number of possible combinations is very large indeed. Crossing over and independent assortment, along with the processes that result in mutations, provide the genetic diversity needed for the differential survival and reproduction of diverse individuals—the basis of evolution by natural selection.

We have seen how meiosis I is fundamentally different from mitosis. In contrast, meiosis II is similar to mitosis

11.18 Crossing Over Forms Genetically Diverse Chromosomes The exchange of genetic material by crossing over results in new combinations of genetic information on the recombinant chromosomes. The two different colors distinguish the chromosomes contributed by the male and female parents.

in that it involves the separation of sister chromatids into daughter nuclei (see steps 7–11 in Figure 11.16). However, because of crossing over during meiosis I, the sister chromatids are not necessarily identical to one another as they would be in mitosis. Chance assortment of the chromatids during meiosis II contributes further to the genetic diversity of the meiotic products. The final products of meiosis I and meiosis II are four haploid daughter cells, each with one set (*n*) of chromosomes.

Meiotic errors lead to abnormal chromosome structures and numbers

In the complex processes of mitosis and meiosis, things occasionally go wrong. In meiosis I, a pair of homologous chromosomes may fail to separate, and in mitosis or meiosis II, sister chromatids may fail to separate. Conversely, homologous chromosomes may fail to remain together during metaphase I of meiosis, and then both may migrate to the same pole in anaphase I. These are all examples of **nondisjunction**, which results in the production of aneuploid cells. **Aneuploidy** is a condition in which one or more chromosomes are either lacking or present in excess. If nondisjunction occurs

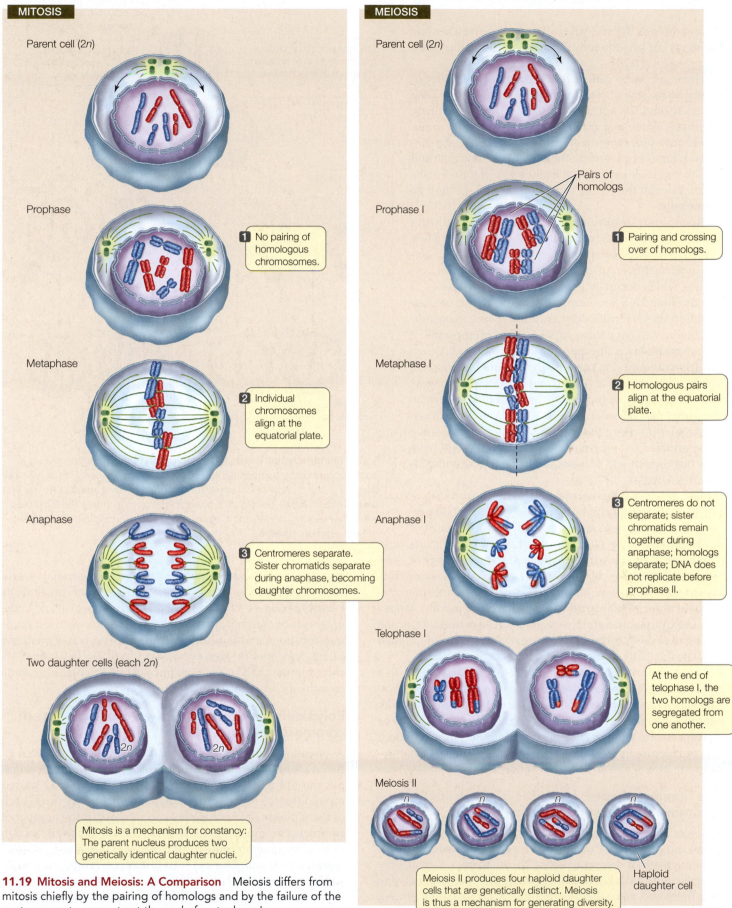

MITOSIS

Parent cell (2n)

Prophase

1 No pairing of homologous chromosomes.

Metaphase

2 Individual chromosomes align at the equatorial plate.

Anaphase

3 Centromeres separate. Sister chromatids separate during anaphase, becoming daughter chromosomes.

Two daughter cells (each 2n)

2n   2n

Mitosis is a mechanism for constancy: The parent nucleus produces two genetically identical daughter nuclei.

MEIOSIS

Parent cell (2n)

Prophase I

Pairs of homologs

1 Pairing and crossing over of homologs.

Metaphase I

2 Homologous pairs align at the equatorial plate.

Anaphase I

3 Centromeres do not separate; sister chromatids remain together during anaphase; homologs separate; DNA does not replicate before prophase II.

Telophase I

At the end of telophase I, the two homologs are segregated from one another.

Meiosis II

n   n   n   n

Meiosis II produces four haploid daughter cells that are genetically distinct. Meiosis is thus a mechanism for generating diversity.

Haploid daughter cell

11.19 Mitosis and Meiosis: A Comparison   Meiosis differs from mitosis chiefly by the pairing of homologs and by the failure of the centromeres to separate at the end of metaphase I.

during meiosis, it can lead to offspring with either one too many, or one too few, chromosomes in all of their cells (**Figure 11.20**).

There are many different causes of aneuploidy, but one cause may be a breakdown in the cohesins that keep sister chromatids and tetrads joined together during prophase. These and other proteins ensure that when the chromosomes line up at the equatorial plate at metaphase I, for example, one homolog will face one pole and the other homolog will face the other pole. If the cohesins break down at the wrong time, both homologs may go to one pole.

Aneuploidy resulting from nondisjunction during meiosis is often lethal for the affected offspring. In a few cases, the affected offspring survive but may have certain abnormalities. An example in humans is Down syndrome, which occurs when a gamete has two copies of chromosome 21. If, for example, an egg with two of these chromosomes is fertilized by a normal sperm, the resulting zygote will have three copies of the chromosome: it will be **trisomic** for chromosome 21. A child with Down syndrome has mild to moderately impaired intellectual ability; characteristic abnormalities of the hands, tongue, and eyelids; and an increased susceptibility to cardiac abnormalities. About 1 child in 800 is born with Down syndrome. If an egg that did not receive chromosome 21 is fertilized by a normal sperm, the zygote will have only one copy: it will be **monosomic** for chromosome 21, and this is lethal.

Trisomies and the corresponding monosomies are surprisingly common in human zygotes, with 10–30 percent of all conceptions showing aneuploidy. But most of the embryos that develop from such zygotes do not survive to birth, and those that do often die before the age of 1 year (trisomies for chromosome 21 are the viable exception). At least one-fifth of all recognized pregnancies are spontaneously terminated (miscarried) during the first 2 months, largely because of trisomies and monosomies. The actual proportion of spontaneously terminated pregnancies is certainly higher, because the earliest ones often go unrecognized.

Other abnormal chromosomal events can also occur. In a process called **translocation**, a piece of a chromosome may break away and become attached to another chromosome. For example, a particular large part of one chromosome 21 may be translocated to another chromosome. Individuals who inherit this translocated piece along with two normal copies of chromosome 21 will also have Down syndrome.

The number, shapes, and sizes of the metaphase chromosomes constitute the karyotype

When cells are in metaphase of mitosis, it is often possible to count and characterize their individual chromosomes. If a photomicrograph of the entire set of chromosomes is made, the images of the individual chromosomes can be manipulated to pair them and place them in an orderly arrangement. Such a

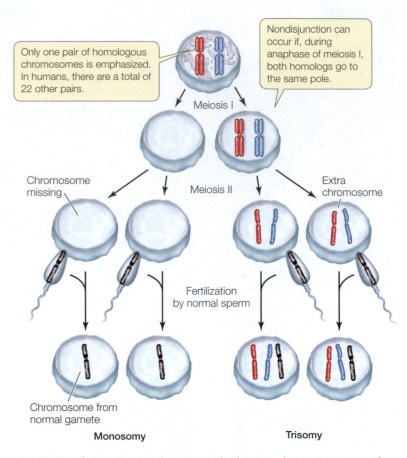

11.20 Nondisjunction Leads to Aneuploidy Nondisjunction occurs if homologous chromosomes fail to separate during meiosis I, as illustrated here, or if chromatids fail to separate during mitosis or meiosis II. The first case is shown here. The result is aneuploidy: one or more chromosomes are either lacking or present in excess. Generally, aneuploidy is lethal to the developing embryo.

rearranged photomicrograph reveals the number, shapes, and sizes of the chromosomes in a cell, which together constitute its **karyotype** (**Figure 11.21**). In humans, karyotypes can aid in the diagnosis of chromosomal abnormalities such as trisomies or translocations, and this has led to an entire branch of medicine called cytogenetics. However, as you will see in Chapter 15, chromosome analysis with the microscope is replaced in some cases by direct analysis of DNA.

Polyploids have more than two complete sets of chromosomes

As mentioned in Section 11.4, mature organisms are often either diploid (for example, most animals) or haploid (for example, most fungi). Under some circumstances, triploid ($3n$), tetraploid ($4n$), or higher-order **polyploid** nuclei may form. Each of these ploidy levels represents an increase in the number of complete chromosome sets present. If there is nondisjunction of all of the chromosomes during meiosis I (see above), diploid gametes will form. This can lead to *auto*polyploidy after fertilization. Autotriploids and autotetraploids have been important in some cases in species formation. A diploid nucleus can

(A)

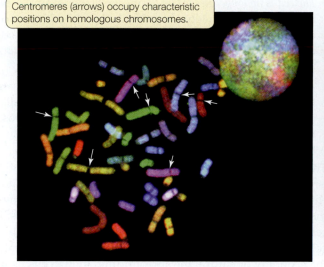

Centromeres (arrows) occupy characteristic positions on homologous chromosomes.

(B)

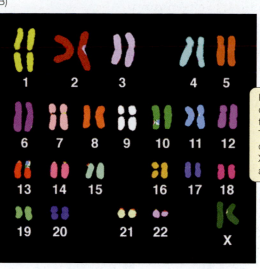

Humans have 23 pairs of chromosomes, including the sex chromosomes. This female's sex chromosomes are X and X; a male would have X and Y chromosomes.

11.21 The Human Karyotype (A) Chromosomes from a human cell in metaphase. The DNA of each chromosome pair has a specific nucleotide sequence that is stained by a particular colored dye, so that the chromosomes in a homologous pair share a distinctive color. Each chromosome at this stage is composed of two chromatids, but they cannot be distinguished. At the upper right is an interphase nucleus. (B) This karyogram, produced by computerized analysis of the image on the left, shows homologous pairs lined up together and numbered, clearly revealing the individual's karyotype.

undergo normal meiosis because there are two sets of chromosomes to make up homologous pairs, which separate during anaphase I. Similarly, a tetraploid nucleus has an even number of each kind of chromosome, so each chromosome can pair with its homolog. However, a triploid nucleus cannot undergo normal meiosis because one-third of the chromosomes would lack partners. Triploid individuals are usually sterile.

Because polyploid nuclei have more chromosome sets, their cells tend to be larger. This has led to the use of polyploid plants in agriculture. Diploid bananas are smaller and produce inedible seeds; triploid bananas are larger and seedless. A similar phenomenon is seen in triploid seedless watermelon. Perhaps the best known, and certainly the most important, polyploid crop plant is wheat. In this case, hybridization occurred between species, forming new *allo*polyploid species:

- Haploid gametes from two species (A and B) mated to form a diploid zygote (chromosomes AB).
- Nondisjunction of all chromosomes occurred during mitosis in the fertilized egg, resulting in a tetraploid (AABB).
- The tetraploid grew up to be a fertile adult.

Modern bread wheat is the result of two such events, which occurred around 8,000 to 10,000 years ago, resulting in a hexaploid (**Figure 11.22**). Wheat's properties of grain formation and environmental adaptation thus come from three different ancestral species. Other allopolyploid crops include cotton, oats, and sugarcane.

RECAP 11.5

Meiosis produces four daughter cells in which the chromosome number is reduced from diploid to haploid. Because of the independent assortment of chromosomes and the crossing over of homologous chromatids, the four products of meiosis are not genetically identical. Meiotic errors, such as the failure of a homologous chromosome pair to separate, can lead to abnormal numbers of chromosomes. Several important crop plants, such as wheat, are polyploid.

- How do crossing over and independent assortment result in unique daughter nuclei? **See pp. 219–222 and Figures 11.16, 11.18**
- What are the differences between meiosis and mitosis? **See pp. 221–222 and Figure 11.19**
- What is aneuploidy, and how can it arise from nondisjunction during meiosis? **See pp. 222–224 and Figure 11.20**

An essential role of cell division in complex eukaryotes is to replace dead cells. What causes cells to die?

11.6 In a Living Organism, How Do Cells Die?

Cells within a living organism can die in one of two ways. The first type of cell death, **necrosis**, occurs when cells and tissues are damaged by mechanical means or toxins, or are starved of oxygen or nutrients. These cells often swell up and burst, releasing their contents into the extracellular environment. This process often results in inflammation (see Section 42.2).

Another kind of cell death occurs during normal developmental processes. **Apoptosis** (Greek, "falling apart") is a programmed series of events that results in cell death. Why would a cell initiate apoptosis, which is essentially cell suicide? In animals, there are two possible reasons:

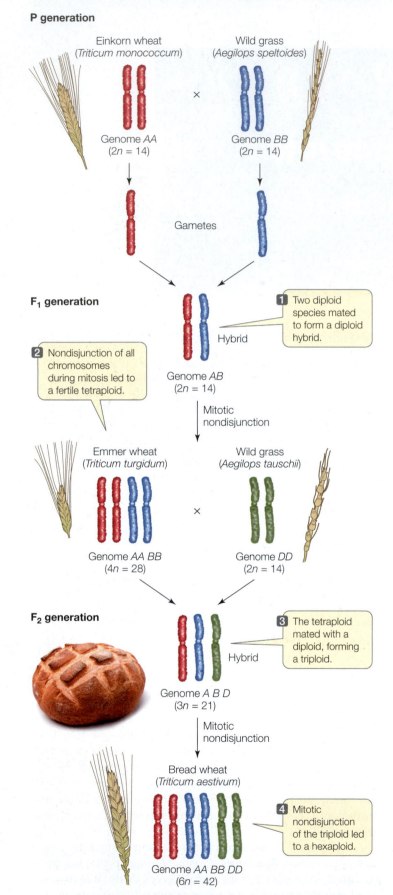

P generation

Einkorn wheat
(*Triticum monococcum*)

Wild grass
(*Aegilops speltoides*)

Genome *AA*
(2*n* = 14)

×

Genome *BB*
(2*n* = 14)

Gametes

F₁ generation

Hybrid

1 Two diploid species mated to form a diploid hybrid.

2 Nondisjunction of all chromosomes during mitosis led to a fertile tetraploid.

Genome *AB*
(2*n* = 14)

Mitotic nondisjunction

Emmer wheat
(*Triticum turgidum*)

Wild grass
(*Aegilops tauschii*)

Genome *AA BB*
(4*n* = 28)

×

Genome *DD*
(2*n* = 14)

F₂ generation

Hybrid

3 The tetraploid mated with a diploid, forming a triploid.

Genome *A B D*
(3*n* = 21)

Mitotic nondisjunction

Bread wheat
(*Triticum aestivum*)

4 Mitotic nondisjunction of the triploid led to a hexaploid.

Genome *AA BB DD*
(6*n* = 42)

11.22 Polyploidy and the Origin of Bread Wheat Three species and two events of mitotic nondisjunction led to modern wheat.

- *The cell is no longer needed by the organism.* For example, before birth, a human fetus has weblike hands, with connective tissue between the fingers. As development proceeds, this unneeded tissue disappears as the cells undergo apoptosis in response to specific signals.

- *The longer cells live, the more prone they are to genetic damage that could lead to cancer.* This is especially true of epithelial cells on the surface of an organism, which may be exposed to radiation or toxic substances. Such cells normally die after only days or weeks and are replaced by new cells.

The outward events of apoptosis are similar in many organisms. The cell becomes detached from its neighbors and its chromatin is digested by enzymes that cut the DNA (between the nucleosomes) into fragments of about 180 base pairs. The cell forms membranous lobes, or "blebs," that break up into cell fragments (**Figure 11.23A**). In a remarkable example of the economy of nature, the surrounding, living cells usually ingest the remains of the dead cell by phagocytosis. Neighboring cells digest the apoptotic cell contents in their lysosomes, and the digested components are recycled.

Apoptosis is also used by plant cells, in an important defense mechanism called the hypersensitive response. Plants can protect themselves from disease by undergoing apoptosis at the site of infection by a fungus or bacterium. With no living tissue to grow in, the invading organism is not able to spread to other parts of the plant. Because of their rigid cell walls, plant cells do not form blebs the way animal cells do. Instead, they digest their own cell contents in the vacuole and then release the digested components into the vascular system.

Despite these differences between plant and animal cells, they share many of the signal transduction pathways that lead to apoptosis. A variety of signals, either external or from inside the cell, can lead to programmed cell death (**Figure 11.23B**). Such signals include hormones, growth factors, viral infection, certain toxins, or extensive DNA damage. These signals activate specific receptors, which in turn activate signal transduction pathways leading to apoptosis. Some apoptotic pathways target the mitochondria, for example, by increasing the permeability of mitochondrial membranes. The cell quickly dies if its mitochondria can't carry out cellular respiration. An important class of enzymes called **caspases** are activated during apoptosis. These enzymes are proteases that hydrolyze target molecules in a cascade of events. The cell dies as the caspases hydrolyze proteins of the nuclear envelope, nucleosomes, and plasma membrane.

■ **RECAP** 11.6

Cell death can occur either by necrosis or by apoptosis. Apoptosis is governed by precise molecular controls.

- What are some differences between apoptosis and necrosis? **See p. 225**
- In what situations is apoptosis necessary? **See p. 226**
- How is apoptosis regulated? **See Figure 11.23**

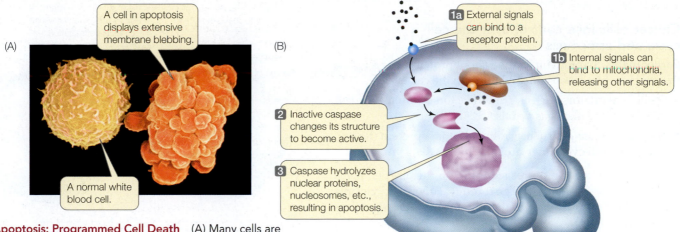

11.23 Apoptosis: Programmed Cell Death (A) Many cells are programmed to "self-destruct" when they are no longer needed, or when they have lived long enough to accumulate a burden of DNA damage that might harm the organism. (B) Both external and internal signals stimulate caspases, the enzymes that break down specific cell constituents, resulting in apoptosis.

Mitosis adds cells to organisms, and apoptosis removes them. Under normal circumstances these processes are balanced so as to benefit the organism as a whole. In the next section we will examine what happens when this balance is disturbed and cell production runs out of control.

11.7 How Does Unregulated Cell Division Lead to Cancer?

Perhaps no malady affecting people in the industrialized world instills more fear than cancer, and most people realize that it involves an inappropriate increase in cell numbers. One in three Americans will have some form of cancer in his or her lifetime, and at present, one in four will die of it. With 1.5 million new cases and half a million deaths in the United States annually, cancer ranks second only to heart disease as a killer.

Cancer cells differ from normal cells

Cancer cells differ from the normal cells from which they originate in two ways:

- Cancer cells lose control over cell division.
- Cancer cells can migrate to other locations in the body.

Most cells in the body divide only if they are exposed to extracellular signals such as growth factors. Cancer cells do not respond to these controls, and instead divide more or less continuously, ultimately forming **tumors** (large masses of cells). By the time a physician can feel a tumor or see one on an X-ray film or CAT scan, it already contains millions of cells. Tumors can be benign or malignant.

Benign tumors resemble the tissue they came from, grow slowly, and remain localized where they develop. For example, a lipoma is a benign tumor of fat cells that may arise in the armpit and remain there. Benign tumors are not cancers, but they must be removed if they impinge on an organ, obstructing its function.

Malignant tumors do not look like their parent tissue at all. A flat, specialized epithelial cell in the lung wall may turn into a relatively featureless, round, malignant lung cancer cell (**Figure 11.24**). Malignant cells often have irregular structures, such as variable nucleus sizes and shapes. This characteristic was used to identify the cells in Henrietta Lacks' tumor as malignant (see the opening story of this chapter).

The second and most fearsome characteristic of cancer cells is their ability to invade surrounding tissues and spread to other parts of the body by traveling through the bloodstream or lymphatic ducts. When malignant cells become lodged in some distant part of the body, they go on dividing and growing, establishing a tumor at that new site. This spreading, called **metastasis**, results in organ failures and makes the cancer very hard to treat.

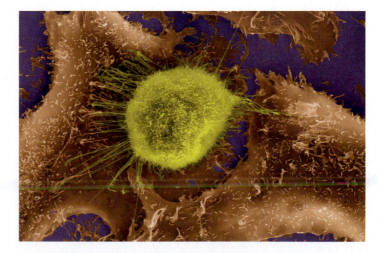

11.24 A Cancer Cell with its Normal Neighbors This lung cancer cell (yellow-green) is quite different from the normal lung cells surrounding it. The cancer cell can divide more rapidly than its normal counterparts, and it can spread to other organs. This form of small-cell cancer is lethal, with a 5-year survival rate of only 10 percent. Most cases are caused by tobacco smoking.

Cancer cells lose control over the cell cycle and apoptosis

Earlier in this chapter you learned about proteins that regulate the progress of a eukaryotic cell through the cell cycle:

- Positive regulators such as growth factors stimulate the cell cycle: they are like "gas pedals."
- Negative regulators such as retinoblastoma protein (RB) inhibit the cell cycle: they are like "brakes."

Just as driving a car requires stepping on the gas pedal *and* releasing the brakes, a cell will go through a division cycle only if the positive regulators are active and the negative regulators are inactive.

In most cells, the two regulatory systems ensure that the cells divide only when needed. In cancer cells, these two processes are abnormal.

- **Oncogene** proteins are positive regulators in cancer cells. They are derived from normal positive regulators that have become mutated to be overly active or that are present in excess, and they stimulate the cancer cells to divide more often. Oncogene products could be growth factors, their receptors, or other components in the signal transduction pathways (see Chapter 7) that stimulate cell division. An example of an oncogene protein is a growth factor receptor in a breast cancer cell (**Figure 11.25A**). Normal breast cells have relatively low numbers of the human epidermal growth factor receptor HER2. So breast cells are not usually stimulated to multiply in the presence of this growth factor. In about 25 percent of breast cancers, a DNA change results in the increased production of the HER2 receptor. This results in positive stimulation of the cell cycle, and a rapid proliferation of cells with the altered DNA.
- **Tumor suppressors** are negative regulators in both cancer and normal cells, but in cancer cells they are inactive. An example is the RB protein, which when active acts at R (the restriction point) in G1 to block the cell cycle (see Figure 11.6). In some cancer cells RB is inactive, allowing the cell cycle to proceed. Some viral proteins can inactivate tumor suppressors. For example, the human papillomavirus infects cells of the cervix and produces a protein called E7. E7 binds to the RB protein and prevents it from inhibiting the cell cycle (**Figure 11.25B**). Another important tumor suppressor is p53, a transcription factor that is involved in cell cycle checkpoint pathways and apoptosis. The importance of p53 as a tumor suppressor is illustrated by the fact that more than 50 percent of human tumors have mutations in the gene that encodes p53.

The discovery of apoptosis (see Section 11.6) changed the way biologists think about cancer. In a population of cells, the net increase in cell numbers over time (the growth rate) is a function of cells added (the rate of cell division) and cells lost (the rate of apoptosis):

$$\begin{array}{c} \text{Growth rate of} \\ \text{cell population} \end{array} = \begin{array}{c} \text{rate of cell} \\ \text{division} \end{array} - \begin{array}{c} \text{rate of} \\ \text{apoptosis} \end{array}$$

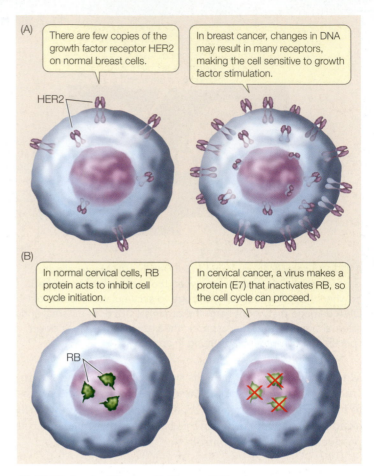

11.25 Molecular Changes in Cancer Cells   In cancer, oncogene proteins become active (A) and tumor suppressor proteins become inactive (B).

In normal nongrowing tissues, the rate of cell division equals the rate of apoptosis, so the cell population as a whole does not grow. Cancer cells are defective in their regulation of the cell cycle, resulting in increased rates of cell division. In addition, cancer cells can lose the ability to respond to positive regulators of apoptosis (see Figure 11.23), and this results in lowered rates of cell death. Both of these defects favor an increased growth rate of the cancer cell population.

Cancer treatments target the cell cycle

The most successful and widely used treatment for cancer is surgery. While physically removing a tumor is optimal, it is often difficult for a surgeon to get all of the tumor cells. (A tumor about 1 centimeter in diameter already has a billion cells!) Tumors are generally embedded in normal tissues. Added to this is the probability that cells of the tumor may have broken off and spread to other organs. This makes it unlikely that localized surgery will be curative. So other approaches are taken to treat or cure cancer, and these generally target the cell cycle (**Figure 11.26**). The goal is to decrease the rate of cell division and/or increase the rate of apoptosis so that the cancer cell population decreases.

An example of a cancer drug that targets the cell cycle is 5-fluorouracil, which blocks the synthesis of thymine, one of the four bases in DNA. The drug paclitaxel prevents the functioning of microtubules in the mitotic spindle. Both drugs inhibit the cell cycle, and apoptosis causes tumor shrinkage. More dramatic is

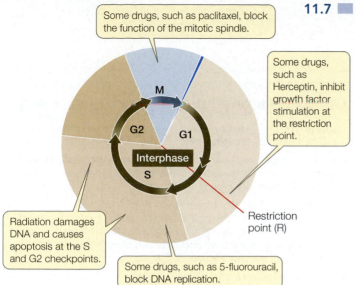

Some drugs, such as paclitaxel, block the function of the mitotic spindle.

Some drugs, such as Herceptin, inhibit growth factor stimulation at the restriction point.

Radiation damages DNA and causes apoptosis at the S and G2 checkpoints.

Restriction point (R)

Some drugs, such as 5-fluorouracil, block DNA replication.

11.26 Cancer Treatment and the Cell Cycle To prevent cancer cells from dividing, physicians use combinations of therapies that attack the cell cycle at different points.

radiation treatment, in which a beam of high-energy radiation is focused on the tumor. DNA damage is extensive, and the cell cycle checkpoint for DNA repair is overwhelmed. As a result, the cell undergoes apoptosis. A major problem with these treatments is that they target normal cells as well as the tumor cells. These treatments are toxic to tissues with large populations of normal dividing cells such as those in the intestine, skin, and bone marrow (where blood cells are produced).

A major effort in cancer research is to find treatments that target only cancer cells. A promising recent example is Herceptin, which targets the HER2 growth factor receptor that occurs at high levels on the surfaces of some breast cancer cells (see Figure 11.25A). Herceptin binds specifically to the HER2 receptor but does not stimulate it. This prevents the natural growth factor from binding, and so the cells are not stimulated to divide. As a result, the tumor shrinks because the apoptosis rate remains the same. More such treatments are on the way.

RECAP 11.7

Cancer cells differ from normal cells in terms of their rapid cell division and their ability to spread (metastasis). Many proteins regulate the cell cycle, either positively (oncogenes) or negatively (tumor suppressors). In cancer, one or another of these proteins is altered in some way, making its activity abnormal. Radiation and many cancer drugs target proteins involved in the cell cycle.

- How are oncogene proteins and tumor suppressor proteins involved in cell cycle control in normal and cancer cells? **Review p. 228 and Figure 11.25**
- How does cancer treatment target the cell cycle? **Review pp. 228–229 and Figure 11.26**

We have now looked at the cell cycle and at cell division by binary fission, mitosis, and meiosis. We have described the normal cell cycle and how its regulation is disrupted in cancer. We have seen how meiosis produces haploid cells in sexual life cycles. In the coming chapters we will examine heredity, genes, and DNA.

?

What makes HeLa cells reproduce so well in the laboratory?

ANSWER

In normal tissues, the rate of cell division is offset by the rate of cell death. Unlike most normal cells, HeLa cells keep growing because they have a genetic imbalance that heavily favors cell reproduction over cell death. Henrietta Lacks was infected with the human papillomavirus, which stimulates cell division in cervical cells. In addition, an enzyme called telomerase, which keeps DNA intact and prevents cell death, is overexpressed in HeLa cells. This combination of traits—increased cell division plus decreased apoptosis—leads to the extraordinary growth rate of HeLa cells.

CHAPTER**SUMMARY** 11

11.1 How Do Prokaryotic and Eukaryotic Cells Divide?

- Cell division is necessary for the reproduction, growth, and repair of organisms.
- Cell division must be initiated by a **reproductive signal**. Before a cell can divide, the genetic material (DNA) must be **replicated** and **segregated** to separate portions of the cell. **Cytokinesis** then divides the cytoplasm into two cells.
- In prokaryotes, most cellular DNA is a single molecule, usually in the form of a circular **chromosome**. Prokaryotes reproduce by **binary fission**. **Review Figure 11.2**
- In eukaryotes, cells divide by either **mitosis** or **meiosis**. Eukaryotic cell division follows the same general pattern as binary fission, but with significant differences. For example, a eukaryotic cell has a distinct nucleus whose chromosomes must be replicated prior to separating the two daughter cells.

11.2 How Is Eukaryotic Cell Division Controlled?

- The eukaryotic cell cycle has two main phases: **interphase**, during which cells are not dividing and the DNA is replicating, and mitosis or M phase, when the cells are dividing.
- During most of the eukaryotic cell cycle, the cell is in interphase, which is divided into three subphases: S, G1, and G2. DNA is replicated during **S phase**. Mitosis (M phase) and cytokinesis follow. **Review Figure 11.3**
- **Cyclin–Cdk** complexes regulate the passage of cells through checkpoints in the cell cycle. Retinoblastoma protein (RB) inhibits the cell cycle at the **restriction point**. The cyclin–Cdk functions by inactivating RB and allows the cell cycle to progress. **Review Figures 11.5, 11.6**
- External controls such as **growth factors** can stimulate the cell to begin a division cycle.

continued

11.3 What Happens during Mitosis?
See ANIMATED TUTORIAL 11.1

- In mitosis, a single nucleus gives rise to two nuclei that are genetically identical to each other and to the parent nucleus.

- DNA is wrapped around proteins called histones, forming bead-like units called **nucleosomes**. A eukaryotic chromosome contains strings of nucleosomes bound to proteins in a complex called **chromatin**. **Review Figure 11.9**

- At mitosis, the replicated chromosomes (**sister chromatids**) are held together at the **centromere**. Each chromatid consists of one double-stranded DNA molecule. During mitosis sister chromatids, attached by cohesin, line up at the equatorial plate and attach to the **spindle apparatus**. The chromatids separate (becoming **daughter chromosomes**) and migrate to opposite ends of the cell. **Review Figures 11.10, 11.11, ACTIVITIES 11.1, 11.2**

- Mitosis can be divided into several phases called **prophase**, **prometaphase**, **metaphase**, **anaphase**, and **telophase**.

- Nuclear division is usually followed by cytokinesis. Animal cell cytoplasms divide via a contractile ring made up of actin microfilaments and myosin. In plant cells, cytokinesis is accomplished by vesicles that fuse to form a cell plate. **Review Figure 11.13**

11.4 What Role Does Cell Division Play in a Sexual Life Cycle?

- **Asexual reproduction** produces **clones**, new organisms that are genetically identical to the parent. Any genetic variation is the result of changes in genes.

- In **sexual reproduction**, two **haploid** gametes—one from each parent—unite in **fertilization** to form a genetically unique, **diploid zygote**. Sexual life cycles can be haplontic, diplontic, or involve alternation of generations. **Review Figure 11.15, ACTIVITY 11.3**

- In non-haplontic sexually reproducing organisms, certain cells in the adult undergo meiosis, a process by which a diploid cell produces haploid gametes.

- Each gamete contains one of each **homologous pair** of chromosomes from the parent.

11.5 What Happens during Meiosis?
See ANIMATED TUTORIAL 11.2

- Meiosis consists of two nuclear divisions, **meiosis I** and **meiosis II**, that together reduce the chromosome number from diploid to haploid. Meiosis ensures that each haploid cell contains one member of each chromosome pair, and results in four genetically diverse haploid cells, usually gametes. **Review Figure 11.16, ACTIVITY 11.4**

- In meiosis I, entire chromosomes, each with two chromatids, migrate to the poles. In meiosis II, the sister chromatids separate.

- During prophase I, homologous chromosomes undergo **synapsis** to form pairs in a **tetrad**. Chromatids can form junctions called **chiasmata**, and genetic material may be exchanged between the two homologs by **crossing over**. **Review Figures 11.17, 11.18**

- Both crossing over during prophase I and **independent assortment** of the homologs as they separate during anaphase I ensure that the gametes are genetically diverse.

- In **nondisjunction**, two members of a homologous pair of chromosomes go to the same pole during meiosis I, or two chromatids go to the same pole during meiosis II or mitosis. This leads to one gamete having an extra chromosome and another lacking that chromosome. **Review Figure 11.20**

- The union between a gamete with an abnormal chromosome number and a normal haploid gamete results in **aneuploidy**. Such genetic abnormalities can be harmful or lethal to the organism.

- The numbers, shapes, and sizes of the metaphase chromosomes constitute the **karyotype** of an organism.

- **Polyploids** have more than two sets of haploid chromosomes. Sometimes these sets come from different species. **Review Figure 11.22**

11.6 In a Living Organism, How Do Cells Die?

- A cell may die by **necrosis**, or it may self-destruct by **apoptosis**, a genetically programmed series of events that includes the fragmentation of its nuclear DNA.

- Apoptosis is regulated by external and internal signals. These signals result in activation of a class of enzymes called **caspases** that hydrolyze proteins in the cell. **Review Figure 11.23**

11.7 How Does Unregulated Cell Division Lead to Cancer?

- Cancer cells divide more rapidly than normal cells and can be **metastatic**, spreading to distant organs in the body.

- Cancer can result from changes in either of two types of proteins that regulate the cell cycle. **Oncogene** proteins stimulate cell division and are activated in cancer. **Tumor suppressor** proteins normally inhibit the cell cycle, but in cancer they are inactive. **Review Figure 11.25**

- Cancer treatment often targets the cell cycle in tumor cells. **Review Figure 11.26**

 Go to the Interactive Summary to review key figures, Animated Tutorials, and Activities.
Life10e.com/is11

CHAPTER**REVIEW**

▬▬ REMEMBERING

1. Which statement about eukaryotic chromosomes is *not* true?
   a. They sometimes consist of two chromatids.
   b. They sometimes consist only of a single chromatid.
   c. They normally possess a single centromere.
   d. They consist only of proteins.
   e. During metaphase they are visible under the light microscope.

2. Which statement about the cell cycle is *not* true?
   a. It consists of interphase, mitosis, and cytokinesis.
   b. The cell's DNA replicates during G1.
   c. A cell can remain in G1 for weeks or much longer.
   d. DNA is not replicated during G2.
   e. Cells enter the cell cycle as a result of internal or external signals.

3. Which statement about mitosis is *not* true?
   a. A single nucleus gives rise to two identical daughter nuclei.
   b. The daughter nuclei are genetically identical to the parent nucleus.
   c. The centromeres separate at the onset of anaphase.
   d. Homologous chromosomes synapse in prophase.
   e. The centrosomes organize the microtubules of the spindle fibers.

4. Apoptosis
   a. occurs in all cells.
   b. involves the formation of the plasma membrane.
   c. does not occur in an embryo.
   d. is a series of programmed events resulting in cell death.
   e. is the same as necrosis.

5. In meiosis,
   a. meiosis II reduces the chromosome number from diploid to haploid.
   b. DNA replicates between meiosis I and meiosis II.
   c. the chromatids that make up a chromosome in meiosis II are identical.
   d. each chromosome in prophase I consists of four chromatids.
   e. homologous chromosomes separate from one another in anaphase I.

6. Which statement about cytokinesis is true?
   a. In animals, a cell plate forms.
   b. In plants, it is initiated by furrowing at the membrane.
   c. It immediately follows the G1 phase of the cell cycle.
   d. In animal cells, actin and myosin play an important role.
   e. It is the division of the nucleus.

▬▬ UNDERSTANDING & APPLYING

7. An animal has a diploid chromosome number of 12. An egg cell of that animal has 5 chromosomes. The most probable explanation is
   a. normal mitosis.
   b. normal meiosis.
   c. nondisjunction in meiosis I.
   d. nondisjunction in meiosis I or II.
   e. nondisjunction in mitosis.

8. The number of daughter chromosomes in a human cell (diploid number 46) in anaphase II of meiosis is
   a. 2.
   b. 23.
   c. 46.
   d. 69.
   e. 92.

9. Contrast mitotic prophase and prophase I of meiosis. Contrast mitotic anaphase and anaphase I of meiosis.

10. The tumor suppressor p53 normally acts by stimulating the expression of p21. Explain why loss of p53 function would lead to uncontrolled cell division.

▬▬ ANALYZING & EVALUATING

11. Cancer-fighting drugs are usually given in combination to target several different stages of the cell cycle. Why might this be a better approach than a single drug?

12. Studying the cell cycle is much easier if the cells under investigation are synchronous—all in the same stage of the cell cycle. Some populations of cells are naturally synchronous. The anther (male sex organ) of a lily plant contains cells that become pollen grains (male gametes). As anthers develop in the flower, their lengths correlate precisely with the stage of the meiotic cycle in those cells. An anther that is 1.5 millimeters long, for example, contains cells in early prophase I. How would you use lily anthers to investigate the roles of cyclins and Cdk's in the meiotic cell cycle?

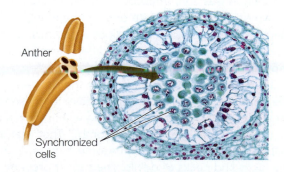

Anther

Synchronized cells

Go to BioPortal at **yourBioPortal.com** for Animated Tutorials, Activities, LearningCurve Quizzes, Flashcards, and many other study and review resources.

12

Inheritance, Genes, and Chromosomes

CHAPTEROUTLINE

12.1 What Are the Mendelian Laws of Inheritance?

12.2 How Do Alleles Interact?

12.3 How Do Genes Interact?

12.4 What Is the Relationship between Genes and Chromosomes?

12.5 What Are the Effects of Genes Outside the Nucleus?

12.6 How Do Prokaryotes Transmit Genes?

Devil Genetics   The Tasmanian devil (*Sarcophilus harrisii*) is in danger of extinction because its genetic makeup has made it susceptible to cancer.

THE TASMANIAN DEVIL (*Sarcophilus harrisii*) does not deserve the reputation that comes with its nickname and certainly does not deserve what has happened to its gene pool. A marsupial (a type of mammal that gives birth to immature young) native to the Australian continent, the devil resembles a small, black dog with patches of white fur. Generally kind and playful, it eats a diet of various dead animals. It is harmless to humans and, cartoon depictions to the contrary, does not eat live rabbits. When startled, a devil often bares its teeth and shrieks to scare off a possible predator; this is how European settlers gave the creature its common name.

Over the years, the devil population was reduced because of hunting and disease; more recently, it has increased again from the few individuals that survived. The close relatedness of individuals within the current population was confirmed by sequencing the genomes of two devils from opposite ends of the island of Tasmania. Of the 3 billion nucleotides in the devil genome, there were fewer than a million differences; humans show about four times as much variation.

Thanks to conservation efforts in the last century, the population of wild Tasmanian devils had risen to about 100,000 by 1996. But lurking within their chromosomes was a "genetic time bomb." That year, biologists noticed a particularly gruesome tumor affecting the faces and necks of devils. The cancer prevents them from feeding and is invariably lethal. It has led to a drastic reduction in the devil population, and at the current rate of spread, the species could be extinct in a few years.

At first, biologists suspected a cancer-causing infectious virus, but they could not find one. Instead, they found that the cancer cells themselves pass from an affected devil to an unaffected one when they bite each other during mating or feeding; the tumor then grows in the second animal. This is highly unusual. If you accidentally got cancer cells from an unrelated person, the genetic differences between the two of you would result in your immune system recognizing the incoming cells as "nonself," and they would be rejected and destroyed. But devils are so closely related genetically that this rejection does not occur.

A massive effort to preserve the remaining Tasmanian devil population is underway. But the lessons from genetics are clear: Related individuals share most of their genes, and genetic diversity is important in disease resistance.

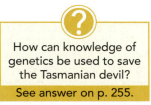

?

How can knowledge of genetics be used to save the Tasmanian devil?

See answer on p. 255.

12.1 What Are the Mendelian Laws of Inheritance?

Genetics, the field of biology concerned with inheritance, has a long history. There is good evidence that people were deliberately breeding animals (horses) and plants (the date palm tree) for desirable characteristics as long as 5,000 years ago. The general idea was to examine the natural variation among the individuals of a species and "breed the best to the best and hope for the best." This was a hit-or-miss method—sometimes the resulting offspring had all the good characteristics of the parents, but often they did not.

By the mid-nineteenth century, two theories had emerged to explain the results of breeding experiments:

- The theory of *blending inheritance* proposed that gametes contained hereditary determinants (what we now call genes) that blended when the gametes fused during fertilization. Like inks of different colors, the two different determinants lost their individuality after blending and could never be separated. For example, if a plant that made smooth, round seeds was crossed (bred) with a plant that made wrinkled seeds, the offspring would be intermediate between the two and the determinants for the two parental characteristics would be lost.

- The theory of *particulate inheritance* proposed that each determinant had a physically distinct nature; when gametes fused in fertilization, the determinants remained intact. According to this theory, if a plant that made round seeds was crossed with a plant that made wrinkled seeds, the offspring (no matter the shape of their seeds) would still contain the determinants for the two characteristics.

The story of how these competing theories were tested provides a great example of how the scientific method can be used to support one theory and reject another. In the following sections we will look in detail at experiments performed in the 1860s by an Austrian monk and scientist, Gregor Mendel, whose work clearly supported the particulate theory.

Mendel used the scientific method to test his hypotheses

After entering the priesthood at a monastery in Brno, in what is now the Czech Republic, Gregor Mendel was sent to the University of Vienna, where he studied biology, physics, and mathematics. He returned to the monastery in 1853 to teach. The abbot in charge had set up a small plot of land to do experiments with plants and encouraged Mendel to continue with them (**Figure 12.1**). Over seven years, Mendel made crosses with many thousands of plants. Analysis of his meticulously gathered data suggested to him that inheritance was due to particulate factors.

Mendel presented his theories in two public lectures in 1865 and a detailed written publication in 1866, but his work was ignored by mainstream scientists until 1900. By that time, the discovery of chromosomes had suggested to biologists that genes might be carried on chromosomes. When they read Mendel's work on particulate inheritance, the biologists connected the dots between genes and chromosomes.

Mendel chose to study the common garden pea because of its ease of cultivation and the feasibility of making controlled crosses. Pea flowers have both male and female sex organs: stamens and pistils, which produce gametes that are contained within the pollen and ovules, respectively.

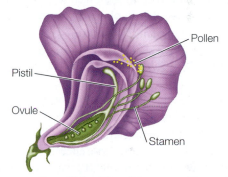

Pea flowers normally self-fertilize. However, the male organs can be removed from a flower so that it can be manually fertilized with pollen from a different flower.

Go to Media Clip 12.1
Mendel's Discoveries
Life10e.com/mc12.1

12.1 Gregor Mendel and His Garden The Austrian monk Gregor Mendel (left) did his groundbreaking genetics experiments in a garden at the monastery at Brno, in what is now the Czech Republic.

There are many varieties of pea plants with easily recognizable characteristics. A **character** is an observable physical feature, such as seed shape. A **trait** is a particular form of a character, such as round or wrinkled seeds. Mendel worked with seven pairs of varieties with contrasting traits for characters such as seed shape, seed color, and flower color. These varieties were true-breeding: that is, when he crossed a plant that produced wrinkled seeds with another of the same variety, all of the offspring plants produced wrinkled seeds.

As we will see, Mendel developed a set of hypotheses to explain the inheritance of particular pea traits, and then designed crossing experiments to test his hypotheses. He performed his crosses in the following manner:

- He removed the stamens from (emasculated) flowers of one parental variety so that it couldn't self-fertilize. Then he collected pollen from another parental variety and placed it on the pistils of the emasculated flowers. The plants providing and receiving the pollen were the **parental generation**, designated **P**.

- In due course, seeds formed and were planted. The seeds and the resulting new plants constituted the **first filial generation**, or **F₁**. (The word "filial" refers to the relationship between offspring and parents, from the Latin *filius*, "son.") Mendel examined each F₁ plant to see which traits it bore and then recorded the number of F₁ plants expressing each trait.

- In some experiments the F₁ plants were allowed to self-pollinate and produce a **second filial generation**, or **F₂**. Again, each F₂ plant was characterized and counted.

Mendel's first experiments involved monohybrid crosses

The term "hybrid" refers to the offspring of crosses between organisms differing in one or more characters. In Mendel's first experiments, he crossed parental (P) varieties with contrasting traits for a single character, producing monohybrids (from the Greek *monos*, "single") in the F₁ generation. He subsequently planted the F₁ seeds and allowed the resulting plants to self-pollinate to produce the F₂ generation. This technique is referred to as a **monohybrid cross**.

Mendel performed the same experiment for seven pea characters. His method is illustrated in **Figure 12.2**, using seed shape as an example. When he crossed a strain that made round seeds with one that made wrinkled seeds, all of the F₁ seeds were round—it was as if the wrinkled seed trait had disappeared completely. However, when F₁ plants were allowed to self-pollinate to produce F₂ seeds, about one-fourth of the seeds were wrinkled. These observations were key to distinguishing the two theories noted above:

- The F₁ offspring were not a blend of the two traits of the parents. Only one of the traits was present (in this case, round seeds).

- Some F₂ offspring had wrinkled seeds. The trait had not disappeared because of blending.

These observations led to a rejection of the blending theory of inheritance and provided support for the particulate theory.

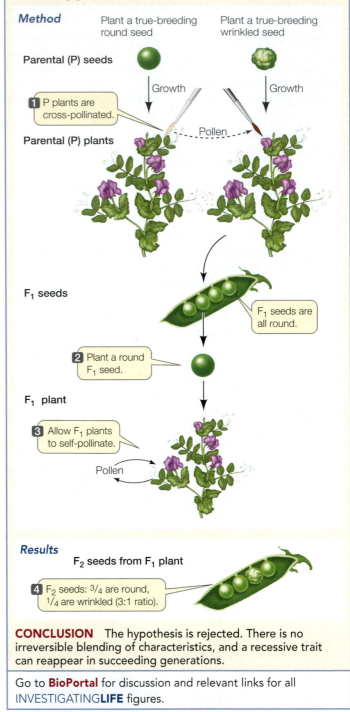

INVESTIGATINGLIFE

12.2 Mendel's Monohybrid Experiments   Mendel performed crosses with pea plants and carefully analyzed the outcomes to show that genetic determinants are particulate.[a]

HYPOTHESIS   When two strains of peas with contrasting traits are bred, their characteristics are irreversibly blended in succeeding generations.

Method   Plant a true-breeding   Plant a true-breeding
              round seed   wrinkled seed

Parental (P) seeds

1 P plants are cross-pollinated.   Growth   Growth

Parental (P) plants   Pollen

F₁ seeds   F₁ seeds are all round.

2 Plant a round F₁ seed.

F₁ plant

3 Allow F₁ plants to self-pollinate.

Pollen

Results   F₂ seeds from F₁ plant

4 F₂ seeds: ³/₄ are round, ¹/₄ are wrinkled (3:1 ratio).

CONCLUSION   The hypothesis is rejected. There is no irreversible blending of characteristics, and a recessive trait can reappear in succeeding generations.

Go to **BioPortal** for discussion and relevant links for all INVESTIGATING**LIFE** figures.

[a]See www.mendelweb.org/Mendel.plain.html

We now know that hereditary determinants are not actually "particulate," but they *are* physically distinct entities: sequences of DNA carried on chromosomes.

WORKING WITHDATA:

Mendel's Monohybrid Experiments

Original Paper

The original German version of Mendel's paper, *Versuche uber Pflanzen-Hybriden*, with an English translation and extensive explanatory notes, is available online: www.mendelweb.org/Mendel.plain.html.

Analyze the Data

Mendel's monohybrid crosses were key to his rejection of the theory of blending inheritance (see Figure 12.2). One of his monohybrid crosses was between true-breeding green-seeded and yellow-seeded pea plants. He observed that all of the pea plants' F_1 generation had yellow seeds. Mendel then allowed the F_1 plants to self-pollinate, and the seed colors of the resulting F_2 generation were analyzed. The table shows actual data from individual plants in the F_2 generation as reported in Mendel's paper. Mendel made mathematical calculations, and in a later part of the paper he showed the overall ratios for these two traits. However, he did not perform a statistical analysis to determine whether the variations in the data reflected a general pattern of inheritance or were simply due to chance.

QUESTION 1

Use the hypothesis that the ratio of yellow to green seeds in the F_2 generation would be 3:1 and perform a chi-square test to analyze the results for each plant in the table (refer to Appendix B for information about the chi-square test). What

can you conclude about this hypothesis from the individual plants? How many crosses have P-values > 0.05?

QUESTION 2

Now total the data from all the plants and rerun the chi-square analysis. What can you conclude? What does your analysis indicate about the need for using a large number of organisms in studies of genetics?

| Plant | Seed color | |
	Yellow	Green
1	25	11
2	32	7
3	14	5
4	70	27
5	24	13
6	20	6
7	32	13
8	44	9
9	50	14
10	44	18

— Go to **BioPortal** for all WORKING WITHDATA exercises

All seven crosses between varieties with contrasting traits gave the same kind of data (**Table 12.1**). In the F_1 generation only one of the two traits was seen, but the other trait reappeared in about one-fourth of the offspring in the F_2 generation. Mendel called the trait that appeared in the F_1 and was more abundant in the F_2 the **dominant** trait, and the other trait

recessive. In the F_2 generation, the *ratio* of dominant to recessive traits was about 3:1. (To calculate the ratios shown in Table 12.1, divide the number of F_2 plants with the dominant trait by the number with the recessive trait.)

You can see in Table 12.1 that for each character, Mendel counted hundreds or even thousands of F_2 seeds or plants to

TABLE12.1

Mendel's Results from Monohybrid Crosses

| Parental generation phenotypes | | | F_2 Generation phenotypes | | | |
Dominant	Recessive		Dominant	Recessive	Total	Ratio
	Round seeds × Wrinkled seeds		5,474	1,850	7,324	2.96:1
	Yellow seeds × Green seeds		6,022	2,001	8,023	3.01:1
	Purple flowers × White flowers		705	224	929	3.15:1
	Inflated pods × Constricted pods		882	299	1,181	2.95:1
	Green pods × Yellow pods		428	152	580	2.82:1
	Axial flowers × Terminal flowers		651	207	858	3.14:1
	Tall stems × Dwarf stems (1 m) (0.3 m)		 787	277	1,064	2.84:1

see how many carried each trait. As we will discuss in more detail below, the probability of a plant inheriting a particular trait is independent of the probability of another plant inheriting the same trait. If Mendel had looked at only a few F_2 progeny from the "round × wrinkled" cross, he might, by chance, have found only round seeds. Or he might have found a higher proportion of wrinkled seeds than he did. In order to discover recurring patterns and to develop his laws of inheritance, Mendel used very large numbers of plants.

Mendel went on to expand on the particulate theory. He proposed that hereditary determinants—we will call them **genes** here, though Mendel did not use that term—occur in pairs and segregate (separate) from one another during the formation of gametes. He concluded that each pea plant has two genes for each character (such as seed shape), one inherited from each parent. We now use the term **diploid** to describe the state of having two copies of each gene; **haploids** have just a single copy.

Mendel concluded that while each gamete contains one copy of each gene, the resulting zygote contains two copies, because it is produced by the fusion of two gametes. Furthermore, different traits arise from different forms of a gene (now called **alleles**) for a particular character. For example, Mendel studied two alleles for seed shape: one that caused round seeds and the other causing wrinkled seeds.

An organism that is **homozygous** for a gene has two alleles that are the same (for example, two copies of the allele for round seeds). An organism that is **heterozygous** for a gene has two different alleles (for example, one allele for round seeds and one allele for wrinkled seeds). In a heterozygote, one of the two alleles may be dominant (such as round, R) and the other recessive (wrinkled, r). By convention, dominant alleles are designated with uppercase letters and recessive alleles with lowercase letters.

The physical appearance of an organism is its **phenotype**. Mendel proposed that the phenotype is the result of the **genotype**, or genetic constitution, of the organism showing the phenotype. Round seeds and wrinkled seeds are two phenotypes resulting from three possible genotypes: the wrinkled seed phenotype is produced by the genotype rr, whereas the round seed phenotype is produced by either of the genotypes RR or Rr (because the R allele is dominant to the r allele).

Mendel's first law states that the two copies of a gene segregate

How do Mendel's theories explain the proportions of traits seen in the F_1 and F_2 generations of his monohybrid crosses? Mendel's first law—the **law of segregation**—states that *when any individual produces gametes, the two copies of a gene separate, so that each gamete receives only one copy.* Thus gametes from a parent with the RR genotype will all carry R; gametes from an rr parent will all carry r; and the progeny derived from a cross between these parents will all be Rr, producing seeds with a round phenotype (**Figure 12.3**).

Now let's consider the composition of the F_2 generation. Because the alleles segregate, half of the gametes produced by the F_1 generation will have the R allele and the other half will have the r allele. What genotypes are produced when these gametes fuse to form the next (F_2) generation?

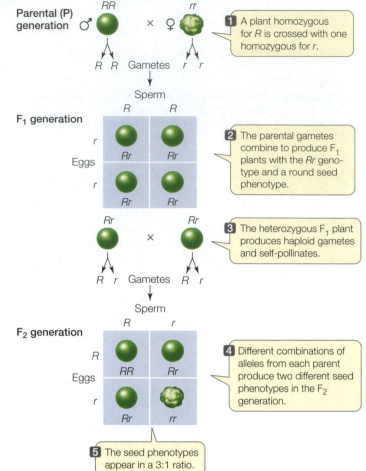

1 A plant homozygous for R is crossed with one homozygous for r.

2 The parental gametes combine to produce F_1 plants with the Rr genotype and a round seed phenotype.

3 The heterozygous F_1 plant produces haploid gametes and self-pollinates.

4 Different combinations of alleles from each parent produce two different seed phenotypes in the F_2 generation.

5 The seed phenotypes appear in a 3:1 ratio.

12.3 Mendel's Explanation of Inheritance Mendel concluded that inheritance depends on discrete factors from each parent that do not blend in the offspring.

The allele combinations that will result from a cross can be predicted using a **Punnett square**, a method devised in 1905 by the British geneticist Reginald Punnett. This device ensures that we consider all possible combinations of gametes when calculating expected genotype frequencies of the resulting offspring. A Punnett square looks like this:

It is a simple grid with all possible male gamete (haploid sperm) genotypes shown along the top and all possible female gamete (haploid egg) genotypes along the left side. The grid is completed by filling in each square with the diploid genotype that can be generated from each combination of gametes. In this example, to fill in the top right square, we put in the R from the female gamete (the egg cell) and the r from the male gamete (a sperm cell in the pollen tube), yielding Rr.

Once the Punnett square is filled in, we can readily see that there are four possible combinations of alleles in the F_2 generation: RR, Rr, rR, and rr (see Figure 12.3). Since R is dominant, there are three ways to get round seeds in the F_2 generation

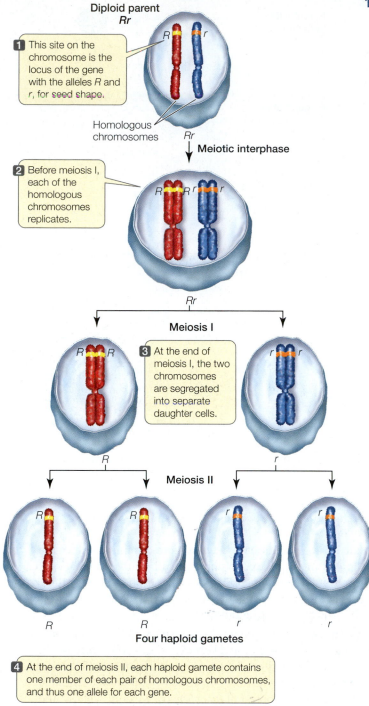

Diploid parent
Rr

1 This site on the chromosome is the locus of the gene with the alleles *R* and *r*, for seed shape.

Homologous chromosomes

Rr
Meiotic interphase

2 Before meiosis I, each of the homologous chromosomes replicates.

Rr

Meiosis I

3 At the end of meiosis I, the two chromosomes are segregated into separate daughter cells.

Meiosis II

Four haploid gametes

4 At the end of meiosis II, each haploid gamete contains one member of each pair of homologous chromosomes, and thus one allele for each gene.

12.4 Meiosis Accounts for the Segregation of Alleles   Although Mendel had no knowledge of chromosomes or meiosis, we now know that a pair of alleles resides on homologous chromosomes, and that those alleles segregate during meiosis.

(genotype *RR*, *Rr*, or *rR*), but only one way to get wrinkled seeds (genotype *rr*). Therefore we predict a 3:1 ratio of these phenotypes in the F$_2$ generation, remarkably close to the ratios Mendel found experimentally for all the traits he compared (see Table 12.1).

Mendel did not live to see his theories placed on a sound physical footing with the discoveries of chromosomes and DNA. Genes are now known to be relatively short sequences of DNA (a few thousand base pairs in length) found on the much longer DNA molecules that make up chromosomes (which are often

millions of base pairs long). Today we can picture the different alleles of a gene segregating as chromosomes separate during meiosis I (**Figure 12.4**).

We also know now that genes determine phenotypes mostly by producing proteins with particular functions, such as enzymes. In many cases a dominant gene is expressed (transcribed and translated) to produce a functional protein, while a recessive gene is mutated so that it is no longer expressed, or it encodes a mutant protein that is nonfunctional. For example, the wrinkled seed phenotype of *rr* peas is caused by the absence of an enzyme called starch branching enzyme 1 (SBE1), which is essential for starch synthesis. With less starch, the developing seed has more sucrose and this causes an inflow of water by osmosis. When the seed matures and dries out, the water is lost, leaving a shrunken seed. A single copy of the *R* allele produces enough functional SBE1 to prevent the wrinkled phenotype, which accounts for the dominance of *R* over *r*.

Mendel verified his hypotheses by performing test crosses

As mentioned above, Mendel arrived at his laws of inheritance by developing a series of hypotheses and then designing experiments to test them. One such hypothesis was that there are two possible allele combinations (*RR* or *Rr*) for seeds with the round phenotype. Mendel verified this hypothesis by performing test crosses with F$_1$ seeds derived from a variety of other crosses. A **test cross** is used to determine whether an individual showing a dominant trait is homozygous or heterozygous. The individual in question is crossed with an individual that is homozygous for the recessive trait—easy to identify, because all individuals with the recessive phenotype are homozygous for that trait.

The recessive homozygote for the seed shape gene has wrinkled seeds and the genotype *rr*. The individual being tested may be described initially as *R_* because we do not yet know the identity of the second allele. We can predict two possible results:

- If the individual being tested is homozygous dominant (*RR*), all offspring of the test cross will be *Rr* and show the dominant trait (round seeds) (**Figure 12.5, left**).

- If the individual being tested is heterozygous (*Rr*), then approximately half the offspring of the test cross will be heterozygous and show the dominant trait (*Rr*), and the other half will be homozygous for the recessive trait (*rr*) (**Figure 12.5, right**).

Mendel obtained results consistent with both of these predictions; thus his hypothesis accurately predicted the results of his test crosses.

Mendel's second law states that copies of different genes assort independently

Consider an organism that is heterozygous for two genes (*RrYy*). In this example, the dominant *R* and *Y* alleles came from one parent, and the recessive *r* and *y* alleles came from the other parent. When this organism produces gametes, do the *R* and *Y* alleles always go together in one gamete, and *r* and *y*

INVESTIGATINGLIFE

12.5 Homozygous or Heterozygous? An individual with a dominant phenotype may have either a homozygous or a heterozygous genotype. The test cross determines which.[a]

HYPOTHESIS The progeny of a test cross can reveal whether an organism is homozygous or heterozygous.

Method

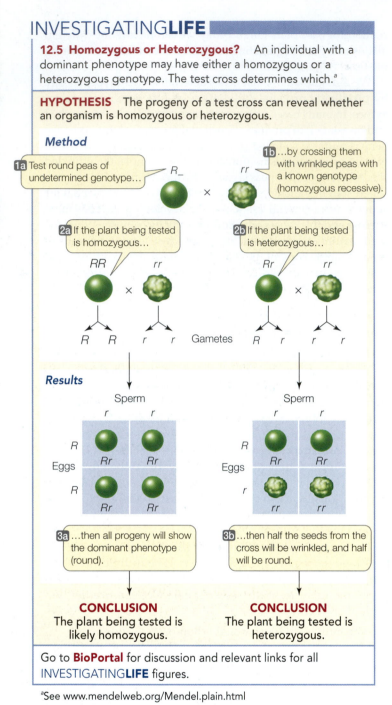

1a Test round peas of undetermined genotype…

1b …by crossing them with wrinkled peas with a known genotype (homozygous recessive).

$R\_$ × rr

2a If the plant being tested is homozygous…

RR × rr

2b If the plant being tested is heterozygous…

Rr × rr

Gametes: R R r r | R r r r

Results

Sperm

	r	r
R	Rr	Rr
R	Rr	Rr

Eggs

Sperm

	r	r
R	Rr	Rr
r	rr	rr

Eggs

3a …then all progeny will show the dominant phenotype (round).

3b …then half the seeds from the cross will be wrinkled, and half will be round.

CONCLUSION
The plant being tested is likely homozygous.

CONCLUSION
The plant being tested is heterozygous.

Go to **BioPortal** for discussion and relevant links for all INVESTIGATINGLIFE figures.

[a]See www.mendelweb.org/Mendel.plain.html

Go to Activity 12.1 Homozygous or Heterozygous?
Life10e.com/ac12.1

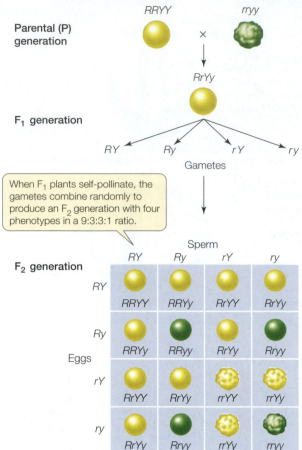

Parental (P) generation

$RRYY$ × $rryy$

F_1 generation

$RrYy$

Gametes: RY Ry rY ry

When F_1 plants self-pollinate, the gametes combine randomly to produce an F_2 generation with four phenotypes in a 9:3:3:1 ratio.

F_2 generation

Sperm

	RY	Ry	rY	ry
RY	$RRYY$	$RRYy$	$RrYY$	$RrYy$
Ry	$RRYy$	$RRyy$	$RrYy$	$Rryy$
rY	$RrYY$	$RrYy$	$rrYY$	$rrYy$
ry	$RrYy$	$Rryy$	$rrYy$	$rryy$

Eggs

12.6 Independent Assortment The 16 possible combinations of gametes in this dihybrid cross result in nine different genotypes. Because R and Y are dominant over r and y, respectively, the nine genotypes result in four phenotypes in a ratio of 9:3:3:1. These results show that the two genes segregate independently.

that are identical double heterozygotes. In this case, he simply allowed the F_1 plants, which were all double heterozygotes, to self-pollinate. Depending on whether the alleles of the two genes are inherited together or separately, there are two possible outcomes, as Mendel saw:

1. *The alleles could maintain the associations they had in the parental generation—they could be linked.* If this were the case, the F_1 plants would produce two types of gametes (RY and ry). The F_2 progeny resulting from self-pollination of these F_1 plants would consist of *two phenotypes*: round yellow and wrinkled green in the ratio of 3:1, just as in the monohybrid cross.

2. *The segregation of R from r could be independent of the segregation of Y from y—the two genes could be unlinked.* In this case, four kinds of gametes would be produced in equal numbers: RY, Ry, rY, and ry. When these gametes combine at random, they should produce an F_2 generation with *four phenotypes* (round yellow, round green, wrinkled yellow, wrinkled green). Putting these possibilities into a Punnett square, we can predict that these four phenotypes would occur in a ratio of 9:3:3:1.

Mendel's dihybrid crosses supported the second prediction: four different phenotypes appeared in the F_2 generation in a ratio of about 9:3:3:1 (**Figure 12.6**). On the basis of such experiments, Mendel proposed his second law—the **law of**

---

alleles in another? Or can a single gamete receive one recessive and one dominant allele (R and y or r and Y)?

Mendel performed another series of experiments to answer these questions. He began with peas that differed in *two* characters: seed shape and seed color. One parental variety produced only round, yellow seeds ($RRYY$), and the other produced only wrinkled, green ones ($rryy$). A cross between these two varieties produced an F_1 generation in which all the plants were $RrYy$. Because the R and Y alleles were dominant, the F_1 seeds were all round and yellow.

Mendel continued this experiment into the F_2 generation by performing a **dihybrid cross**—a cross between individuals

independent assortment: *alleles of different genes assort independently of one another during gamete formation.* In the example above, the segregation of the *R* and *r* alleles is independent of the segregation of the *Y* and *y* alleles. As you will see later in this chapter, this is not as universal as the law of segregation because it does not apply to genes located near one another on the same chromosome. However, it is correct to say that *chromosomes segregate independently* during the formation of gametes, and so do any two genes located on separate chromosome pairs (**Figure 12.7**).

Probability can be used to predict inheritance

One key to Mendel's success was his use of large sample sizes. By counting many progeny from each cross, he observed clear patterns that allowed him to formulate his theories. After his work became widely recognized, geneticists began using simple probability calculations to predict the ratios of genotypes and phenotypes in the progeny of a given cross or mating. They use statistics to determine whether the actual results match the prediction (see Working with Data, p. 235).

You can think of probabilities by considering a coin toss. The basic conventions of probability are simple:

- If an event is absolutely certain to happen, its probability is 1.
- If it cannot possibly happen, its probability is 0.
- All other events have a probability between 0 and 1.

There are two possible outcomes of a coin toss, and both are equally likely, so the probability of heads is ½—as is the probability of tails.

If two coins (say a penny and a dime) are tossed, each acts independently of the other (**Figure 12.8**). What is the probability of both coins coming up heads? In half of the tosses, the penny comes up heads, and in half of that fraction, the dime comes up heads. The probability of both coins coming up heads is ½ × ½ = ¼. In general, *the probability of two independent outcomes occurring together is found by multiplying the two individual probabilities.* This can be applied to a monohybrid cross (see Figure 12.3). After the self-pollination of an *Rr* F₁ plant, the probability that an F₂ plant will have the genotype *RR* is ½ × ½ = ¼, because the chance that the sperm will have the genotype *R* is ½, and the chance that the egg will have the genotype *R* is also ½. Similarly, the probability of *rr* offspring is also ¼.

Probability can also be used to predict the proportions of phenotypes in a dihybrid cross. Let's see how this works for the experiment shown in Figure 12.6. Using the principles described above, we can calculate the probability of an F₂ seed being round. This is found by adding the probability of an *Rr* heterozygote (½) to the probability of an *RR* homozygote (¼): a total of ¾. By the same reasoning, the probability that a seed will be yellow is also ¾. The two characters are determined by separate genes and are independent of each other, so:

- The joint probability for both round and yellow is ¾ × ¾ = 9/16.

What is the probability of F₂ seeds being both wrinkled and yellow? The probability

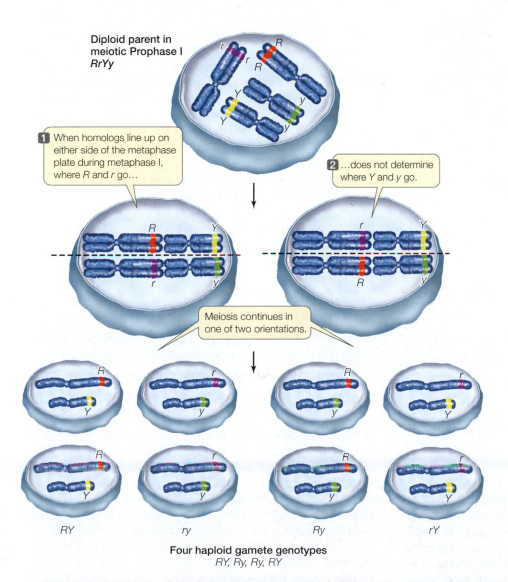

Diploid parent in meiotic Prophase I
RrYy

1 When homologs line up on either side of the metaphase plate during metaphase I, where *R* and *r* go...

2 ...does not determine where *Y* and *y* go.

Meiosis continues in one of two orientations.

RY   ry   Ry   rY

Four haploid gamete genotypes
RY, Ry, Ry, rY

12.7 Meiosis Accounts for Independent Assortment of Alleles We now know that copies of genes on different chromosomes are segregated independently during metaphase I of meiosis. Thus a parent of genotype *RrYy* can form gametes with four different genotypes.

 Go to Animated Tutorial 12.1
Independent Assortment of Alleles
Life10e.com/at12.1

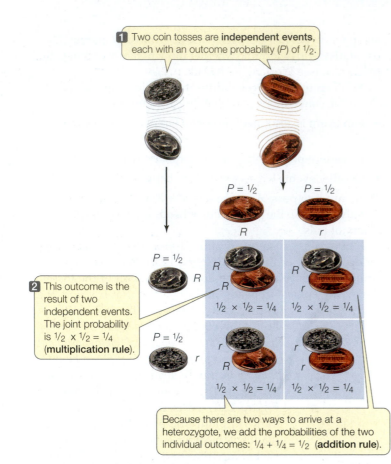

1 Two coin tosses are **independent events**, each with an outcome probability (*P*) of ½.

P = ½   *P* = ½

R   r

P = ½

R   R   R

2 This outcome is the result of two independent events. The joint probability is ½ × ½ = ¼ (**multiplication rule**).

½ × ½ = ¼   ½ × ½ = ¼

P = ½

r   R   r   r

½ × ½ = ¼   ½ × ½ = ¼

Because there are two ways to arrive at a heterozygote, we add the probabilities of the two individual outcomes: ¼ + ¼ = ½ (**addition rule**).

12.8 Using Probability Calculations in Genetics Like the results of a coin toss, the probability of any given combination of alleles appearing in the offspring of a cross can be obtained by multiplying the probabilities of each event. Since a heterozygote can be formed in two ways, these two probabilities are added together.

of being yellow is again ¾; the probability of being wrinkled is ½ × ½ = ¼, so:

- The joint probability for both wrinkled and yellow is ¼ × ¾ = ³⁄₁₆.

By the same reasoning:

- The joint probability for both round (¾) and green (¼) is also ³⁄₁₆.

Finally:

- The joint probability for both wrinkled and green is ¼ × ¼ = ¹⁄₁₆.

Looking at all four phenotypes, we see that they are expected to occur in the ratio of 9:3:3:1.

A Punnett square or these simple probability calculations can be used to determine the *expected* proportions of offspring with particular phenotypes. In the dihybrid cross discussed above, about one-sixteenth of the F_2 seeds are expected to be wrinkled and green. But this does not mean that among 16 F_2 seeds there will always be exactly one wrinkled, green seed. For any toss of a coin, the probability of heads is independent of what happened in all the previous tosses. Even if you get

three heads in a row, the chance of a head in the next toss is still ½, and it is quite possible to toss a coin four times and get four heads. But if you toss the coin many times, you are highly likely to get heads in about half of the tosses. If Mendel had examined only a few progeny in each of his crosses, it is unlikely that he would have observed the phenotypic ratios that he did observe. It was his large sample sizes that allowed him to identify the underlying patterns of inheritance.

Mendel's laws can be observed in human pedigrees

How are Mendel's laws of inheritance applied to humans? Mendel worked out his laws by performing many planned crosses and counting many offspring. Neither of these approaches is possible with humans, so human geneticists rely on **pedigrees**: family trees that show the occurrence of phenotypes (and alleles) in several generations of related individuals.

Because humans have such small numbers of offspring, human pedigrees do not show the clear proportions of phenotypes that Mendel saw in his pea plants. For example, when a man and a woman who are both heterozygous for a recessive allele (say, *Aa*) have children together, each child has a ¼ probability of being a recessive homozygote (*aa*). If this couple were to have dozens of children, about one-fourth of them would be recessive homozygotes. But the offspring of a single couple are likely to be too few to show the exact one-fourth proportion. In a family with only two children, for example, both could easily be *aa* (or *Aa*, or *AA*).

Human geneticists may wish to know whether a particular rare allele that causes an abnormal phenotype is dominant or recessive. **Figure 12.9A** is a pedigree showing the pattern of inheritance of a rare dominant allele. The following are the key features to look for in such a pedigree:

- Every affected person has an affected parent.
- About half of the offspring of an affected parent are also affected. (This is easiest to observe among the twelve cousins in generation III.)

Compare this pattern with the one shown in **Figure 12.9B**, which is typical for the inheritance of a rare recessive allele:

- Affected people can have two parents who are not affected.
- Only a small proportion of people are affected: about one-fourth of children whose parents are both heterozygotes.

In the families of individuals who have a rare recessive phenotype, it is not uncommon to find a marriage of two relatives. This observation is a result of the rarity of recessive alleles that give rise to abnormal phenotypes. For two phenotypically normal parents to have an affected child (*aa*), the parents must both be heterozygous (*Aa*). If a particular recessive allele is rare in the general population, the chance of two people marrying who are both carrying that allele is quite low. However, if that allele is present in a family, two cousins might share it (see Figure 12.9B). For this reason, studies on populations that are isolated either culturally (by religion, as with the Amish in the United States) or geographically (as on islands) have been extremely valuable

12.9 Pedigree Analysis and Inheritance (A) This pedigree represents a family affected by Huntington's disease, which results from a rare dominant allele. Everyone who inherits this allele is affected. (B) The family in this pedigree carries the allele for albinism, a recessive trait. Because the trait is recessive, heterozygotes do not have the albino phenotype, but they can pass the allele on to their offspring. In this family, in generation III the heterozygous parents are cousins; however, the same result could occur if the parents were unrelated but heterozygous.

Go to Animated Tutorial 12.2
Pedigree Analysis
Life10e.com/at12.2

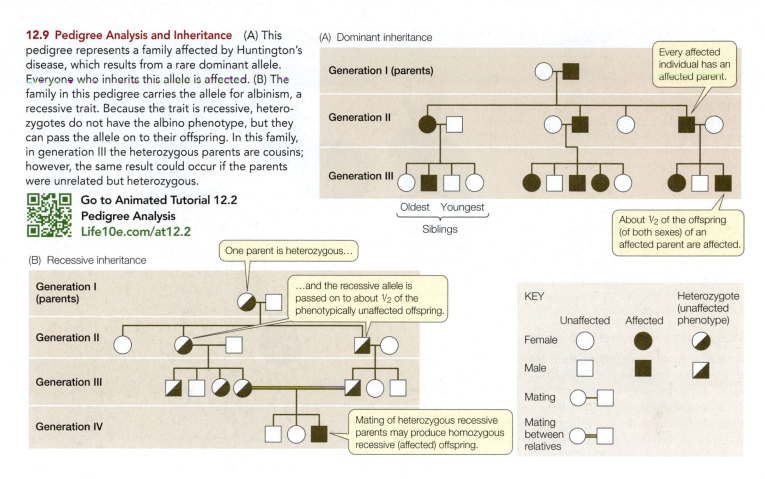

to human geneticists. People in these groups are more likely to marry relatives who may carry the same rare recessive alleles.

RECAP 12.1

Mendel showed that genetic determinants are particulate and do not "blend" when the genes from two gametes combine. Mendel's first law of inheritance states that two copies of a gene segregate during gamete formation. His second law states that genes assort independently during gamete formation. The frequencies with which different allele combinations will be expressed in offspring can be calculated with a Punnett square or using probability theory.

- What results seen in the F_1 and F_2 generations of Mendel's monohybrid cross experiments refuted the blending theory of inheritance? **See p. 234, Figures 12.2, 12.3, and Table 12.1**

- How do events in meiosis explain Mendel's monohybrid cross results? **See p. 237 and Figure 12.4**

- How do events in meiosis explain the independent assortment of alleles in Mendel's dihybrid cross experiments? **See p. 239 and Figures 12.6, 12.7**

- Draw human pedigrees for dominant and recessive inheritance. **See p. 240 and Figure 12.9**

Mendel's laws of inheritance remain valid today; his discoveries laid the groundwork for all future studies of genetics. Inevitably, however, we have learned that things are more complicated than they seemed at first. Let's take a look at some of these complications, beginning with the interactions between alleles.

12.2 How Do Alleles Interact?

Over time genes accumulate changes, giving rise to new alleles. Thus there can be many alleles for a single character. In addition, alleles do not always show simple dominant–recessive relationships. Furthermore, a single allele may have multiple phenotypic effects.

New alleles arise by mutation

Genes are subject to **mutations**, which are stable, inherited changes in the genetic material. In other words, an allele can mutate to become a different allele. For example, you can imagine that at one time all pea plants were tall and had the height allele T. A mutation occurred in that allele that resulted in a new allele, t (conferring a short phenotype). If this mutation was in a cell that underwent meiosis to form gametes, some of the resulting gametes would carry the t allele, and some offspring of this pea plant would carry the t allele. Mutation will be discussed in more detail in Chapter 15. By creating variety, mutations provide the raw material for evolution.

Geneticists usually define one particular allele of a gene as the **wild type**; this allele is the one that is present in most individuals in nature ("the wild") and gives rise to an expected trait or phenotype. Other alleles of that gene, often called mutant

Possible genotypes	CC, Cc^{chd}, Cc^h, Cc	$c^{chd}c^{chd}, c^{chd}c$	c^hc^h, c^hc	cc
Phenotype	Dark gray	Chinchilla	Point restricted	Albino

12.10 Multiple Alleles for Coat Color in Rabbits
These photographs show the phenotypes conferred by four alleles of the *C* gene for coat color in rabbits. Different combinations of two alleles give different coat colors and pigment distributions.

alleles, may produce a different phenotype. The wild-type and mutant alleles reside at the same genetic **locus**, which is their specific position on a chromosome. A genetic locus with a wild-type allele that is present less than 99 percent of the time (the rest of the alleles being mutant) is said to be **polymorphic** (Greek *poly*, "many"; *morph*, "form").

Many genes have multiple alleles

Because of random mutations, more than two alleles of a given gene may exist in a group of individuals. Any one individual has only two alleles—one from its mother and one from its father. But among multiple individuals there may be several different alleles. In fact, there are many examples of such multiple alleles, and they often show a hierarchy of dominance. An example is coat color in rabbits, determined by multiple alleles of the *C* gene:

- *C* determines dark gray
- c^{chd} determines chinchilla, a lighter gray
- c^h determines Himalayan, where pigment is restricted to the extremities (point restricted)
- *c* determines albino

The hierarchy of dominance for these alleles is $C > C^{chd}, C^h > c$. Any rabbit with the *C* allele (paired with itself or any other allele) is dark gray, and a *cc* rabbit is albino. Intermediate colors result from different allele combinations, as shown in **Figure 12.10**. As this example illustrates, multiple alleles can increase the number of possible phenotypes.

Dominance is not always complete

In the pairs of alleles studied by Mendel, dominance is complete in heterozygous individuals. That is, an *Rr* individual always expresses the *R* phenotype. However, many genes have alleles that are not dominant or recessive to one another. Instead, the heterozygotes show an intermediate phenotype—at first glance, like that predicted by the old blending theory of inheritance. For example, if a true-breeding eggplant that produces the familiar purple fruit is crossed with a true-breeding white eggplant, all the F_1 plants produce violet fruit, an intermediate between the two parents. However, further crosses indicate that this apparent blending phenomenon can still be explained in terms of Mendelian genetics (**Figure 12.11**). The purple and white alleles have not disappeared, as those colors reappear when the F_1 plants are interbred.

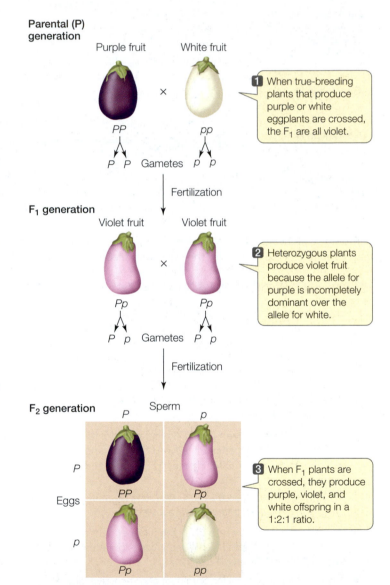

12.11 Incomplete Dominance Follows Mendel's Laws An intermediate phenotype can occur in heterozygotes when neither allele is dominant. The heterozygous phenotype (here, violet fruit) may give the appearance of a blended trait, but the traits of the parental generation reappear in their original forms in succeeding generations, as predicted by Mendel's laws of inheritance.

12.12 ABO Blood Reactions Are Important in Transfusions This table shows the results of mixing red blood cells of types A, B, AB, and O with serum containing anti-A or anti-B antibodies. As you look down the columns, note that each of the types, when mixed separately with anti-A and with anti-B, gives a unique pair of results; this is the basic method by which blood is typed. People with type O blood are good blood donors because O cells do not react with either anti-A or anti-B antibodies. People with type AB blood are good recipients, since they make neither type of antibody. When blood transfusions are incompatible, the reaction (clumping of red blood cells) can have severely adverse consequences for the recipient.

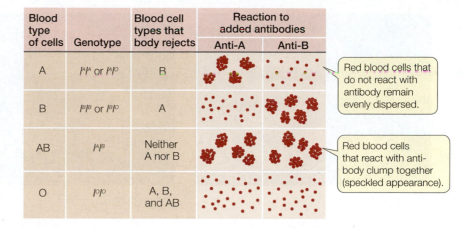

Blood type of cells	Genotype	Blood cell types that body rejects	Reaction to added antibodies	
			Anti-A	Anti-B
A	$I^A I^A$ or $I^A I^O$	B		
B	$I^B I^B$ or $I^B I^O$	A		
AB	$I^A I^B$	Neither A nor B		
O	$I^O I^O$	A, B, and AB		

Red blood cells that do not react with antibody remain evenly dispersed.

Red blood cells that react with antibody clump together (speckled appearance).

When heterozygotes show a phenotype that is intermediate between those of the two homozygotes, the gene is said to be governed by **incomplete dominance**. In other words, neither of the two alleles is dominant. Incomplete dominance is common in nature; in fact, Mendel's study of pea-plant traits is unusual in that all the traits happened to be characterized by complete dominance.

In codominance, both alleles at a locus are expressed

Sometimes the two alleles at a locus produce two different phenotypes that *both* appear in heterozygotes, a phenomenon called **codominance**. Note that this is different from incomplete dominance, where the phenotype of a heterozygote is a blend of the phenotypes of the parents. A good example of codominance is seen in the ABO blood group system in humans (this is also an example of multiple alleles).

Early attempts at blood transfusion frequently killed the patient. Around 1900, the Austrian scientist Karl Landsteiner mixed blood cells and serum (blood from which cells have been removed) from different individuals. He found that only certain combinations of blood and serum are compatible. In other combinations, the red blood cells from one individual form clumps in the presence of serum from the other individual. This discovery led to our ability to administer compatible blood transfusions that do not kill the recipient.

The formation of clumps occurs because of the immune system, which protects the body from invasion by "nonself" molecules or organisms. (This ability to recognize nonself cells is why people do not "catch" cancer like the Tasmanian devils in the opening story.) People make specific proteins in the serum, called antibodies, which react with foreign molecules or particles. The specific part of a molecule that is recognized by an antibody is called an antigen. Oligosaccharides on the surfaces of red blood cells can function as antigens. For example, people with the A blood group make the A antigen, and those with the B group make the B antigen: these antigens are specific oligosaccharides on the surfaces of their red blood cells. If a person with the A blood group is given a transfusion of blood from a person with the B group, their immune system will recognize the B antigen as nonself and make antibodies against it (**Figure 12.12**). Likewise, a person with the B blood group will make

antibodies against the A antigen. However, someone with the codominant AB blood group makes both the A and the B antigens and will not make antibodies against either antigen. A person with the O blood group makes neither the A nor the B antigen (see Figure 12.12).

The oligosaccharides on the surfaces of red blood cells are made by enzymes that catalyze the formation of bonds between specific sugars. The ABO genetic locus encodes one such enzyme and has three alleles, I^A, I^B, and I^O, each producing a different version of the enzyme. The product of the I^A allele adds N-acetylgalactosamine to the end of a pre-existing oligosaccharide chain, resulting in the A antigen. The product of the I^B allele adds galactose to the same chain, making the B antigen. The I^O allele encodes a protein that has no enzymatic activity:

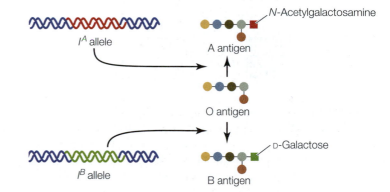

I^A allele — A antigen — N-Acetylgalactosamine

O antigen

I^B allele — B antigen — D-Galactose

Since people inherit one allele from each parent, they may have any combination of these alleles: $I^A I^B$, $I^A I^O$, $I^A I^A$, and so on. The I^A and I^B alleles are codominant because a person with both alleles makes both the A and the B antigens, and both kinds of oligosaccharide occur on their red blood cells. We will learn much more about the functions of antibodies and antigens in Chapter 42.

Some alleles have multiple phenotypic effects

Mendel's principles were further extended when it was discovered that a single allele can influence more than one phenotype. In such a case, we say that the allele is **pleiotropic**. A classic example is the heritable human disease phenylketonuria, which causes mental retardation and reduced hair and skin pigmentation. The disease occurs in people who have a mutation in the gene for a liver enzyme that converts the amino acid

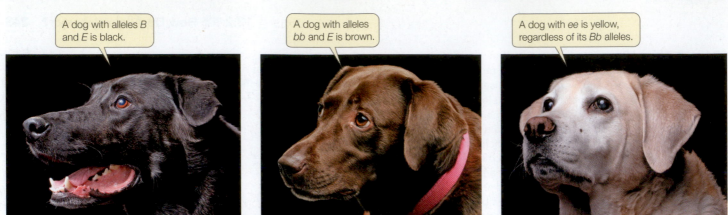

A dog with alleles *B* and *E* is black.

A dog with alleles *bb* and *E* is brown.

A dog with *ee* is yellow, regardless of its *Bb* alleles.

Black (*B_E_*)   Chocolate (*bbE_*)   Yellow (*_ _ee*)

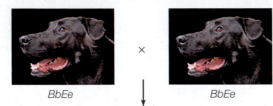

BbEe   ×   BbEe

Sperm

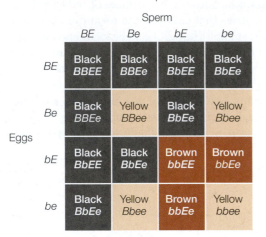

	BE	Be	bE	be
BE	Black *BBEE*	Black *BBEe*	Black *BbEE*	Black *BbEe*
Be	Black *BBEe*	Yellow *BBee*	Black *BbEe*	Yellow *Bbee*
bE	Black *BbEE*	Black *BbEe*	Brown *bbEE*	Brown *bbEe*
be	Black *BbEe*	Yellow *Bbee*	Brown *bbEe*	Yellow *bbee*

Eggs

12.13 Genes May Interact Epistatically Epistasis occurs when one gene alters the phenotypic effect of another gene. In Labrador retrievers, the *Ee* gene determines the expression of the *Bb* gene.

phenylalanine to tyrosine. Without a functional form of this enzyme, phenylalanine builds up in the body to toxic levels, and this affects development in a variety of ways. Given what we now know about genes and their functions, it is not surprising that a gene with such an important metabolic role should have pleiotropic effects. Other examples of pleiotropy include plant and animal genes whose products affect hormone levels, since many hormones play multiple roles in the body.

RECAP 12.2

Genes are subject to random mutations that give rise to new alleles; thus many genes have more than two alleles within a population. Dominance is not necessarily an all-or-nothing phenomenon. Some genes have multiple effects on phenotype.

- How do the crosses in Figure 12.11 demonstrate incomplete dominance? **See pp. 242–243**
- Explain how blood type AB results from codominance. **See p. 243 and Figure 12.12**

Thus far we have discussed phenotypic characters that are affected by single genes. In many cases, however, several genes

interact to determine a phenotype. To complicate things further, the physical environment may interact with the genetic constitution of an individual in determining the phenotype.

12.3 How Do Genes Interact?

We have just seen how two alleles of the same gene can interact to produce a phenotype. If you consider most complex phenotypes, such as human height, you will realize that they are influenced by the products of many genes. We now turn to the genetics of such gene interactions.

Epistasis ("to stand upon") occurs when the phenotypic expression of one gene is affected by another gene. For example, two genes (*B* and *E*) encode proteins that determine coat color in Labrador retrievers:

- Allele *B* (black pigment) is dominant to *b* (brown).
- Allele *E* (pigment deposition in hair) is dominant to *e* (no deposition, so hair is yellow).

An *EE* or *Ee* dog with *BB* or *Bb* is black, and one with *bb* is brown. An *ee* dog is yellow regardless of the *B* gene alleles present (**Figure 12.13**). So the product of the *E* allele is needed for the expression of both the *B* and the *b* alleles, and *E* is said to be epistatic to *B*.

Hybrid vigor results from new gene combinations and interactions

In 1876, Charles Darwin reported that when he crossed two different true-breeding, homozygous genetic strains of corn, the offspring were 25 percent taller than either of the parent strains. Darwin's observation was largely ignored for the next 30 years. In 1908, George Shull "rediscovered" this idea, reporting that not just plant height but the weight of the corn grain produced was dramatically higher in the offspring (**Figure 12.14**). Agricultural scientists took note, and Shull's paper had a lasting impact on the field of applied genetics.

Farmers have known for centuries that matings among close relatives (known as **inbreeding**) can result in offspring

12.14 Hybrid Vigor in Corn Two homozygous parent lines of corn, B73 and Mo17, were crossed to produce the more vigorous hybrid line.

The temperature of the extremities is lower and allows expression of the black coat color gene.

The temperature of most of the body is too high for the expression of the black coat color gene.

12.15 The Environment Influences Gene Expression The rabbit and cat express a coat pattern called "point restriction." Their genotypes specify dark hair/fur, but the enzyme for dark color is inactive at normal body temperature, so only the extremities—the coolest regions of the body—express the phenotype.

of lower quality than matings between unrelated individuals. Agricultural scientists call this inbreeding depression. (This is another of the concerns of conservationists working with the Tasmanian devils in the opening story.) The problems with inbreeding arise because close relatives tend to have the same recessive alleles, some of which may be harmful. The "hybrid vigor" after crossing inbred lines is called **heterosis** (short for heterozygosis). The cultivation of hybrid corn spread rapidly in the United States and all over the world, quadrupling grain production. The practice of hybridization has spread to many other crops and animals used in agriculture. For example, beef cattle that are crossbred are larger and live longer than cattle bred within their own genetic strain.

There has been much controversy over the mechanism by which heterosis works. The dominance hypothesis assumes that all of the extra growth in hybrids can be explained by the lack of inbreeding depression, since hybrids are unlikely to be homozygous for deleterious, recessive alleles. The overdominance hypothesis postulates that in hybrids, new combinations of alleles from the parental strains interact with one another, resulting in superior traits that cannot occur in the parental lines. Many of the characters in question are controlled by multiple genes (see below), and from recent studies it appears that both dominance and overdominance can contribute to heterosis in specific characters.

The environment affects gene action

The phenotype of an individual does not result from its genotype alone. *Genotype and environment interact to determine the phenotype of an organism.* This is especially important to remember in the era of genome sequencing (see Chapter 17). When the sequence of the human genome was completed in 2003, it was hailed as the "book of life," and public expectations of the benefits gained from this knowledge were (and are) high.

But this kind of "genetic determinism" is wrong. Common knowledge tells us that environmental variables such as light, temperature, and nutrition can affect the phenotypic expression of a genotype.

A familiar example of this phenomenon involves "point restriction" coat patterns found in Siamese cats and certain rabbit breeds (**Figure 12.15**). These animals carry a mutant allele of a gene that controls the growth of dark fur all over the body. As a result of this mutation, the enzyme encoded by the gene is inactive at temperatures above a certain point (usually around 35°C). The animals maintain a body temperature above this point, and so their fur is mostly light. However, the extremities—feet, ears, nose, and tail—are cooler, about 25°C, so the fur on these regions is dark. These animals are all white when they are born, because the extremities were kept warm in the mother's womb.

A simple experiment shows that the dark fur is temperature-dependent. If a patch of white fur on a point-restricted rabbit's back is removed and an ice pack is placed on the skin where the patch was, the fur that grows back will be dark. This indicates that although the gene for dark fur was expressed all along, the environment inhibited the activity of the mutant enzyme.

Two parameters describe the effects of genes and environment on the phenotype:

- **Penetrance** is the proportion of individuals in a group with a given genotype that actually show the expected phenotype. For example, many people who inherit a mutant allele of the gene *BRCA1* develop breast cancer. But for some reason, some people with the mutation do not. So the *BRCA1* allele is said to be incompletely penetrant.

12.16 Quantitative Variation Quantitative variation is produced by the interaction of genes at multiple loci and the environment. These students (women [in white] are shorter; men [in blue] are taller) show continuous variation in height that is the result of interactions between many genes and the environment.

- **Expressivity** is the degree to which a genotype is expressed in an individual. For example, a woman with the mutant *BRCA1* allele may develop both breast and ovarian cancer as part of the phenotype, but another woman with the same mutation may only develop breast cancer. So the mutant allele is said to have variable expressivity.

Most complex phenotypes are determined by multiple genes and the environment

Certain simple characters, such as those that Mendel studied in peas, differ in discrete, **qualitative** ways. Mendel used true-breeding parental pea plants that were either short or tall, had purple or white flowers, or had round or wrinkled seeds. But for most complex characters, such as height in humans, the phenotype varies more or less continuously over a range. Some people are short, others are tall, and many are in between the two extremes. Such variation within a population is called **quantitative**, or continuous, variation (**Figure 12.16**).

Sometimes this variation results largely from the alleles that an individual possesses. For instance, much of human eye color is the result of a number of genes controlling the synthesis and distribution of dark melanin pigment. Dark eyes have a lot of it, brown eyes less, and green, gray, and blue eyes even less. In the latter cases, the distribution of other pigments in the eye is what determines light reflection and color.

In most cases, however, *quantitative variation is due to both genes and environment*. Height in humans certainly falls into this category. If you look at families, you often see that parents and their offspring all tend to be tall or short. However, nutrition also plays a role in height: American 18-year-olds today are about 20 percent taller than their great-grandparents were at the same age. Three generations are not enough time for mutations that would exert such a dramatic effect to spread throughout the general population, so the height difference must be due to environmental factors.

Geneticists call the genes that together determine such complex characters **quantitative trait loci**. Identifying these loci is a major challenge, and an important one. For example, the amount of grain that a variety of rice produces in a growing season is determined by many interacting genetic factors. Crop plant breeders have worked hard to decipher these factors in order to breed higher-yielding rice strains. In a similar way, human characteristics such as disease susceptibility and behavior are caused in part by quantitative trait loci. Recently, one of the many genes involved with human height was identified. The gene, *HMGA2*, has an allele that apparently has the potential to add 4 millimeters to human height.

RECAP 12.3

In epistasis, one gene affects the expression of another. Perhaps the most challenging problem for genetics is the explanation of complex phenotypes that are caused by many interacting genes and the environment.

- Explain the difference between penetrance and expressivity. See pp. 245–246
- How is quantitative variation different from qualitative variation? See p. 246

In the next section we'll see how the discovery that genes occupy specific positions on chromosomes enabled Mendel's successors to provide a physical explanation for his model of inheritance, and to provide an explanation for those cases where Mendel's second law does not apply.

12.4 What Is the Relationship between Genes and Chromosomes?

There are far more genes than chromosomes. Studies of different genes that are physically linked on the same chromosome reveal inheritance patterns that are not Mendelian. These patterns have been useful for identifying genes that are linked to one another, and for determining how far apart they are on the chromosome.

Genetic linkage was first discovered in the fruit fly *Drosophila melanogaster*. Its small size, the ease with which it can be bred, its few chromosomes, and its short generation time make this animal an attractive experimental subject. Beginning in 1909, Thomas Hunt Morgan and his students at Columbia University pioneered the study of *Drosophila*, and today it remains a very important model organism for studies of genetics.

 Go to Animated Tutorial 12.3
Alleles That Do Not Assort Independently
Life10e.com/at12.3

Genes on the same chromosome are linked

Some of the crosses Morgan performed with fruit flies yielded phenotypic ratios that were not in accordance with those predicted by Mendel's law of independent assortment. Morgan crossed *Drosophila* with known genotypes at two loci, *B* and *Vg*:

- *B* (wild-type gray body) is dominant over *b* (black body).
- *Vg* (wild-type wing) is dominant over *vg* (vestigial, a very small wing).

Morgan first made an F_1 generation by crossing homozygous dominant *BBVgVg* flies with homozygous recessives (*bbvgvg*). He then performed a test cross with the F_1 flies: *BbVgvg × bbvgvg*.* Morgan expected to see four phenotypes in a ratio of 1:1:1:1, but that is not what he observed. The body color gene and the wing size gene were not assorting independently; rather, they were usually inherited together (**Figure 12.17**).

These results became understandable to Morgan when he considered the possibility that the two loci are on the same chromosome—that is, that they might be linked. Suppose that the *B* and *Vg* loci are indeed located on the same chromosome. Why didn't all of Morgan's F_1 flies have the parental phenotypes—that is, why didn't his cross result in

* Do you recognize this type of cross? It is a test cross for the two gene pairs; see Figure 12.5.

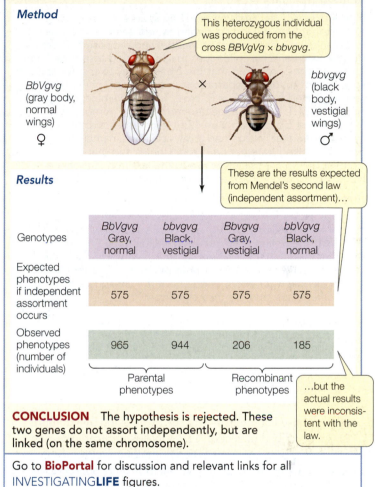

INVESTIGATING**LIFE**

12.17 Some Alleles Do Not Assort Independently

Morgan's studies showed that the genes for body color and wing size in *Drosophila* are linked, so that their alleles do not assort independently.[a]

HYPOTHESIS Alleles for different characteristics always assort independently.

Method

This heterozygous individual was produced from the cross *BBVgVg × bbvgvg*.

BbVgvg (gray body, normal wings) ♀ × *bbvgvg* (black body, vestigial wings) ♂

Results

These are the results expected from Mendel's second law (independent assortment)…

Genotypes	*BbVgvg* Gray, normal	*bbvgvg* Black, vestigial	*Bbvgvg* Gray, vestigial	*bbVgvg* Black, normal
Expected phenotypes if independent assortment occurs	575	575	575	575
Observed phenotypes (number of individuals)	965	944	206	185

Parental phenotypes — Recombinant phenotypes

…but the actual results were inconsistent with the law.

CONCLUSION The hypothesis is rejected. These two genes do not assort independently, but are linked (on the same chromosome).

Go to **BioPortal** for discussion and relevant links for all INVESTIGATING**LIFE** figures.

[a]Morgan, T. H. 1912. *Science* 36: 719–720.

gray flies with normal wings and black flies with vestigial wings, in a 1:1 ratio? If linkage were absolute—that is, if chromosomes always remained intact and unchanged—we would expect to see just those two types of progeny. However, this does not always happen.

Genes can be exchanged between chromatids and mapped

If linkage were absolute, Mendel's law of independent assortment would apply only to loci on different chromosomes. Instead, genes at different loci on the same chromosome *do* sometimes separate from one another during meiosis. Genes may recombine when two homologous chromosomes physically exchange corresponding segments

WORKING WITH DATA:

Some Alleles Do Not Assort Independently

Original Paper

Morgan, T. H. 1912. Complete linkage in the second chromosome of *Drosophila. Science* 36: 719–720.

Analyze the Data

Mendel's work was "rediscovered" 40 years after its publication. At that time, biologists began to find some exceptions to the rules of inheritance that Mendel had proposed. Thomas Hunt Morgan and his colleagues made dihybrid test crosses in fruit flies. They proposed that the clearest way to test for linkage was not to look at aberrations in the 9:3:3:1 phenotypic ratio expected from an $F_1 \times F_1$ cross, but to examine aberrations in the 1:1:1:1 ratio expected from an $F_1 \times$ homozygous recessive test cross (see Figure 12.5). Morgan's group then hypothesized that linkage had a physical basis, namely that genes are linked together on chromosomes and that rare crossing over during meiosis gives rise to the less frequent phenotypes. Examination of actual chromosomal events confirmed this.

QUESTION 1

Morgan first performed a dihybrid cross between black, normal-winged flies (*bbVgVg*) and gray, vestigial-winged flies (*BBvgvg*). The F_1 flies were interbred, yielding the F_2 phenotypes shown in the table (Experiment 1). Compare these data with the expected data in a 9:3:3:1 ratio by using the chi-square test (see Appendix B for information about the chi-square test). Are there differences, and are they significant?

QUESTION 2

To quantify linkage, Morgan crossed homozygous black, normal-winged females with homozygous gray, vestigial-winged

males. He then crossed the F_1 females with black, vestigial-winged males. (You should note that this is not the same test cross as the one shown in Figure 12.17. In that case, the original parents were *BBVgVg* and *bbvgvg*.) The results of this test cross are shown in the table (Experiment 2). Are these genes linked? If they are linked, what is the map distance between the genes? Explain why these data are so different from the data shown in Figure 12.17.

QUESTION 3

In a third experiment, Morgan crossed two genetic strains of flies that were homozygous for the body color and wing genes. The F_1 flies were all gray and normal-winged, and these were interbred. The results are shown in the table (Experiment 3). What were the genotypes and phenotypes of the original parents that produced the F_1?

	Number of progeny showing each phenotype			
Experiment	Gray, normal	Black, normal	Gray, vestigial	Black, vestigial
1	2,316	1,146	737	0
2	578	1,413	1,117	307
3	246	9	65	18

*Go to **BioPortal** for all WORKING WITH DATA exercises*

during prophase I of meiosis—that is, by crossing over (**Figure 12.18**). As described in Section 11.2, DNA is replicated during S phase, so that by prophase I, when homologous chromosome pairs come together to form tetrads, each chromosome consists of two chromatids.

Note that the exchange event involves *only two of the four chromatids* in a tetrad, one from each member of the homologous pair, and can occur at any point along the length of the chromosome. The chromosome segments involved are exchanged reciprocally, so both chromatids involved in crossing over become recombinant (that is, each chromatid ends up with genes from both of the organism's parents). Usually several exchange events occur along the length of each homologous pair.

When crossing over takes place between two linked genes, not all the progeny of a cross have the parental phenotypes. Instead, recombinant offspring appear as well, as they did in Morgan's test cross (see Figure 12.17). They appear in proportions called **recombinant frequencies**, which are calculated by dividing the number of recombinant progeny by the total number of progeny (**Figure 12.19**). Recombinant frequencies will be *greater for loci that are farther apart* on the

chromosome than for loci that are closer together because an exchange event is more likely to occur between genes that are far apart.

By calculating recombination frequencies, geneticists can infer the locations of genes along a chromosome and generate a genetic map. Below is a map showing five genes on a fruit fly chromosome constructed using the recombination frequencies generated by test crosses involving various pairs of genes:

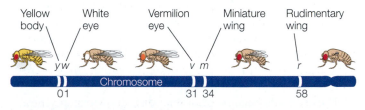

In the chromosome shown above, the recombination frequency between y and w is low, so they are close together on the map. Recombination between y and v is more frequent, so they are farther apart. The recombination frequencies are converted to map units (also called centimorgans, cM); one map unit is equivalent to an average recombination frequency of 0.01 (1 percent).

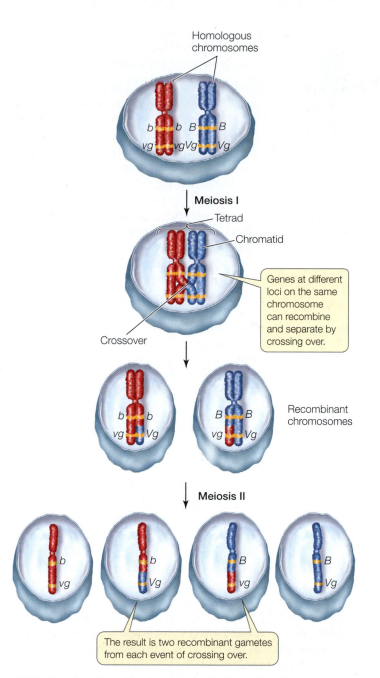

Homologous chromosomes

Meiosis I

Tetrad

Chromatid

Genes at different loci on the same chromosome can recombine and separate by crossing over.

Crossover

Recombinant chromosomes

Meiosis II

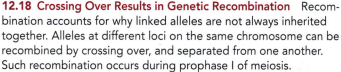

The result is two recombinant gametes from each event of crossing over.

12.18 Crossing Over Results in Genetic Recombination Recombination accounts for why linked alleles are not always inherited together. Alleles at different loci on the same chromosome can be recombined by crossing over, and separated from one another. Such recombination occurs during prophase I of meiosis.

The era of gene sequencing has made mapping less important in some areas of genetics research. However, mapping it still a way to verify that a particular DNA sequence corresponds with a particular phenotype. Linkage has allowed biologists to isolate genes and to identify genetic markers linked to important genes. This is important in breeding new crops and animals for agriculture, and for identifying people carrying medically significant mutations.

Linkage is revealed by studies of the sex chromosomes

In Mendel's work, reciprocal crosses always gave similar results; it did not matter whether a dominant allele was contributed by the female parent or the male parent. But in some cases, the parental origin of a chromosome does matter. For example, human males inherit a bleeding disorder called hemophilia from their mothers, not from their fathers. To understand the types of inheritance in which the parental origin of an allele is important, we must consider the ways in which sex is determined in different species.

SEX DETERMINATION BY CHROMOSOMES In corn, every diploid adult has both male and female reproductive structures. The tissues in these two types of structure are genetically identical, just as roots and leaves are genetically identical. Organisms such as corn, in which the same individual produces both male and female gametes, are said to be **monoecious** (Greek, "one house"). Other organisms, such as date palms, oak trees, and most animals, are **dioecious** ("two houses"), meaning that some individuals can produce only male gametes and others can produce only female gametes. In other words, in dioecious organisms the different sexes are different individuals.

In mammals and birds, sex is determined by differences in the chromosomes, but such determination operates in different ways in different groups of organisms. For example, in many animals, including mammals, sex is determined by a single pair of **sex chromosomes**, which differ from one another. The remaining chromosomes, called **autosomes**, occur in pairs in males and females. For example, in humans there are 22 pairs of autosomes in males and females, and 1 pair of sex chromosomes. The chromosomal bases for sex determination in various groups of animals are summarized in **Table 12.2**.

The sex chromosomes of female mammals consist of a pair of X chromosomes. Male mammals, by contrast, have one X chromosome and a sex chromosome that is not found in females, the Y chromosome. Females may be represented as XX and males as XY.

MALE MAMMALS PRODUCE TWO KINDS OF GAMETES Each gamete produced by a male mammal has a complete set of autosomes, but half the gametes carry an X chromosome and the

TABLE 12.2

Sex Determination in Animals

Animal Group	Mechanism
Bees	Males are haploid, females are diploid
Fruit flies	Fly is female if ratio of X chromosomes to sets of autosomes is 1 or more
Birds	Males ZZ (homogametic), females WZ (heterogametic)
Mammals	Males XY (heterogametic), females XX (homogametic)

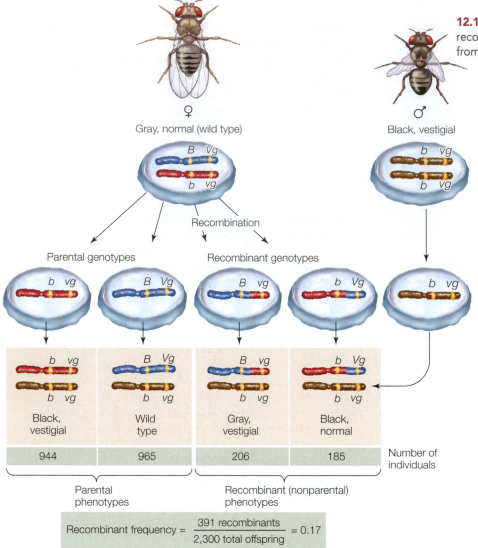

♀
Gray, normal (wild type)

♂
Black, vestigial

12.19 Recombinant Frequencies The frequency of recombinant offspring (those with a phenotype different from either parent) can be calculated.

Recombination

Parental genotypes   Recombinant genotypes

Black, vestigial	Wild type	Gray, vestigial	Black, normal
944	965	206	185

Number of individuals

Parental phenotypes   Recombinant (nonparental) phenotypes

$$\text{Recombinant frequency} = \frac{391 \text{ recombinants}}{2{,}300 \text{ total offspring}} = 0.17$$

other half carry a Y. When an X-bearing sperm fertilizes an egg, the resulting XX zygote is female; when a Y-bearing sperm fertilizes an egg, the resulting XY zygote is male.

SEX CHROMOSOME ABNORMALITIES REVEALED THE GENE THAT DETERMINES SEX Can we determine which chromosome, X or Y, carries the sex-determining gene, and can the gene be identified? One way to determine cause (e.g., the presence of a gene on the Y chromosome) and effect (e.g., maleness) is to look at cases of biological error, in which the expected outcome does not happen.

We can learn something about the functions of X and Y chromosomes from abnormal sex chromosome arrangements resulting from nondisjunction during meiosis or mitosis (see Section 11.5). As you will recall, nondisjunction occurs when a pair of homologous chromosomes (in meiosis I) or sister chromatids (in mitosis or meiosis II) fail to separate. As a result, a gamete may have one too few or one too many chromosomes. If this gamete fuses with another gamete that has the full haploid chromosome set, the resulting offspring will be aneuploid, with fewer or more chromosomes than normal.

In humans, XO individuals sometimes appear. The O indicates that a chromosome is missing—that is, individuals that are XO have only one sex chromosome (an X). Human XO individuals are females who are moderately abnormal physically but normal mentally; usually they are also sterile. The XO condition in humans is called Turner syndrome. It is the only known case in which a person can survive with only one member of a chromosome pair (here, the XY pair), although most XO conceptions are spontaneously terminated early in development. XXY individuals also occur; this condition, which affects males, is called Klinefelter syndrome and results in overlong limbs and sterility.

These observations suggest that the gene controlling maleness is located on the Y chromosome. Observations of people with other types of chromosomal abnormalities helped researchers pinpoint the location of that gene:

- Some women are genetically XY but lack a small portion of the Y chromosome.

- Some men are genetically XX but have a small piece of the Y chromosome attached to another chromosome.

The Y fragments that are respectively missing and present in these two cases are the same and contain the maleness-determining gene, which was named *SRY* (sex-determining region on the Y chromosome).

The *SRY* gene encodes a protein involved in **primary sex determination**—that is, the determination of the kinds of gametes that an individual will produce and the organs that will make them (the male and female gonads). In the presence of the functional SRY protein, an embryo develops sperm-producing testes. (Notice that *italic type* is used for the name of a gene, but roman type is used for the name of a protein.) If the embryo has no Y chromosome, the *SRY* gene is absent, and thus the SRY protein is not made. In the absence of the SRY protein, the embryo develops egg-producing ovaries. In this case, a gene on the X chromosome called *DAX1* produces an anti-testis factor. So the role of SRY in a male is to inhibit the maleness inhibitor encoded by *DAX1*. The SRY protein does this in male cells, but since it is not present in females, DAX1 can act to inhibit maleness.

One function of the gonads is to produce hormones (such as testosterone and estrogen) that send signals to the rest of the body and control the development of **secondary sex characteristics**. These are outward manifestations of maleness and femaleness, such as differences in body type, breast development, body hair, and voice. Secondary sex characteristics distinguish males and females but are not directly part of the reproductive system.

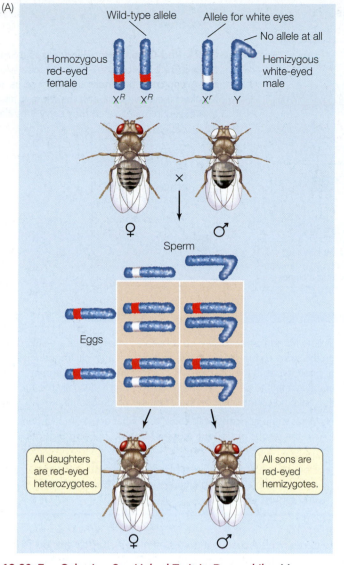

(A)

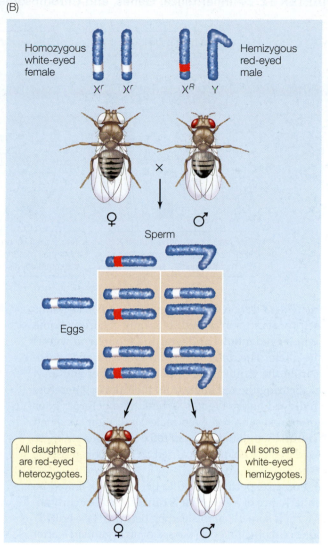

(B)

12.20 Eye Color Is a Sex-Linked Trait in *Drosophila* Morgan demonstrated that a mutant allele that causes white eyes in *Drosophila* is carried on the X chromosome. Note that in this case, the reciprocal crosses do not have the same results.

SEX-LINKED INHERITANCE IN FRUIT FLIES As noted in Table 12.2, sex determination in fruit flies is based on the *proportions* of sex chromosomes, because the numbers of these chromosomes can vary. But most commonly, the fruit fly genome has four pairs of chromosomes: three pairs of autosomes (shown in blue) and (as in humans) a pair of sex chromosomes (green) that differ in size:

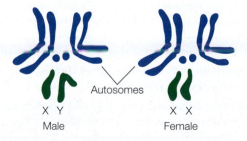

In this case, the female fly has two X chromosomes and the male has only one, the other being the Y chromosome—so the female is XX and the male is XY. As in other organisms, the X and Y chromosomes are not true homologs of one another: *many genes on the X chromosome are not present on the Y*. The X

chromosome of *Drosophila* was one of the first to have specific genes assigned to it.

Thomas Morgan identified a **sex-linked** gene that controls eye color in *Drosophila*. The wild-type allele of the gene confers red eyes, whereas a recessive mutant allele confers white eyes. Morgan's experimental crosses demonstrated that this eye color locus is on the X chromosome. If we abbreviate the eye color alleles as *R* (red eyes) and *r* (white eyes), the presence of the alleles on the X chromosome is designated by X^R and X^r.

Morgan crossed a homozygous red-eyed female (X^RX^R) with a white-eyed male. The male is designated X^rY because the Y chromosome does not carry any allele for this gene. (Any gene that is present as a single copy in a diploid organism is called **hemizygous**.) All the sons and daughters from this cross had red eyes, because the red phenotype is dominant over white and all the progeny had inherited a wild-type X chromosome (X^R) from their mother (**Figure 12.20A**). This phenotypic outcome would have occurred even if the *R* gene had been present on an autosome rather than a sex chromosome. In that case, the male would have been homozygous recessive—*rr*.

When Morgan performed the reciprocal cross, in which a white-eyed female (X^rX^r) was mated with a red-eyed male (X^RY), the results were unexpected: *all the sons were white-eyed*

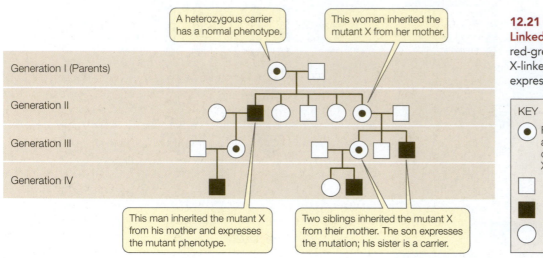

A heterozygous carrier has a normal phenotype.

This woman inherited the mutant X from her mother.

Generation I (Parents)

Generation II

Generation III

Generation IV

This man inherited the mutant X from his mother and expresses the mutant phenotype.

Two siblings inherited the mutant X from their mother. The son expresses the mutation; his sister is a carrier.

12.21 Red-Green Color Blindness Is a Sex-Linked Trait in Humans The mutant allele for red-green color blindness is expressed as an X-linked recessive trait, and therefore is always expressed in males when they carry that allele.

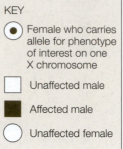

KEY

⊙ Female who carries allele for phenotype of interest on one X chromosome

☐ Unaffected male

■ Affected male

◯ Unaffected female

and all the daughters were red-eyed (**Figure 12.20B**). The sons from the reciprocal cross inherited their only X chromosome from their white-eyed mother and were therefore hemizygous for the white allele. The daughters, however, got an X chromosome bearing the white allele from their mother and an X chromosome bearing the red allele from their father; therefore they were red-eyed heterozygotes. When these heterozygous females were mated with red-eyed males, half their sons had white eyes but all their daughters had red eyes. Together, these results showed that eye color was carried on the X chromosome and not on the Y.

These and other experiments led to the term **sex-linked inheritance**: inheritance of a gene that is carried on a sex chromosome. (This term is somewhat misleading because "sex-linked" inheritance is not really linked to the sex of an organism—after all, both males and females carry X chromosomes.) In mammals, the X chromosome is larger and carries more genes than the Y. For this reason, most examples of sex-linked inheritance involve genes that are carried on the X chromosome.

Many sexually reproducing species, including humans, have sex chromosomes. As in most fruit flies, human males are XY, females are XX, and relatively few of the genes that are present on the X chromosome are present on the Y. Pedigree analyses of X-linked recessive phenotypes like the one in **Figure 12.21** reveal the following patterns (compare with the pedigrees of non–X-linked phenotypes in Figure 12.9):

- The phenotype appears much more often in males than in females, because only one copy of the rare allele is needed for its expression in males, whereas two copies must be present in females.

- A male with the mutation can pass it on only to his daughters; all his sons get his Y chromosome.

- Daughters who receive one X-linked mutation are heterozygous carriers. They are phenotypically normal, but they can pass the mutant allele to their sons or daughters. On average, half their children will inherit the mutant allele since half of their X chromosomes carry the normal allele.

- The mutant phenotype can skip a generation if the mutation passes from a male to his daughter (who will be phenotypically normal) and then to her son.

RECAP 12.4

Simple Mendelian ratios are not observed when genes are linked on the same chromosome. Linkage results in atypical frequencies of phenotypes in the offspring from a test cross. Sex linkage in humans refers to genes on one sex chromosome (usually the X) that have no counterpart on the other sex chromosome.

- What is the concept of linkage, and what are its implications for the results of genetic crosses? **See p. 247 and Figures 12.17, 12.19**

- How does a sex-linked gene behave differently in genetic crosses from a gene on an autosome? **See pp. 251–252 and Figure 12.20**

The genes we've discussed so far in this chapter are all in the cell nucleus. But other organelles, including mitochondria and plastids, also carry genes. What are these genes, and how are they inherited?

12.5 What Are the Effects of Genes Outside the Nucleus?

The nucleus is not the only organelle in a eukaryotic cell that carries genetic material. As described in Section 5.5, mitochondria and plastids contain small numbers of genes, which are remnants of the entire genomes of endosymbiotic prokaryotes that eventually gave rise to these organelles. For example, in humans there are about 21,000 genes coding for proteins in the nuclear genome and 37 in the mitochondrial genome. Plastid genomes are about five times larger than those of mitochondria.

The inheritance of organelle genes differs from that of nuclear genes for several reasons:

- In most organisms, mitochondria and plastids are inherited only from the mother. As you will learn in Chapter 43, eggs contain abundant cytoplasm and organelles, but the only part of the sperm that survives to take part in the union of haploid gametes is the nucleus. So you have inherited your mother's mitochondria (with their genes), but not your father's.

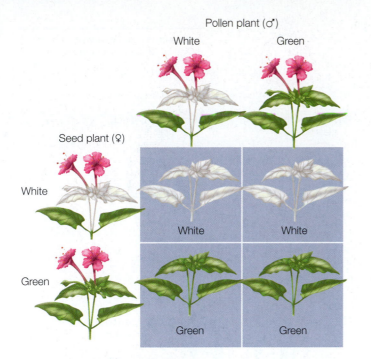

Pollen plant (♂)

Seed plant (♀)

12.22 Cytoplasmic Inheritance In four o'clock plants, leaf color is inherited through the female plant only. The white leaf color is caused by a chloroplast mutation that occurs during the life of the parent plant; the leaves that form before the mutation occurs are green. The mutation is passed on to the germ cells, and the offspring that inherit the mutation are entirely white.

- There may be hundreds of mitochondria or plastids in a cell. So a cell is not diploid for organelle genes.
- Organelle genes tend to mutate at much faster rates than nuclear genes, so organelle genes often have multiple alleles.

Several of the genes carried by cytoplasmic organelles are important for organelle assembly and function, and mutations of these genes can have profound effects on the organism. The phenotypes resulting from such mutations reflect the organelles' roles. For example, in plants and some photosynthetic protists, certain plastid gene mutations affect the proteins that assemble chlorophyll molecules into photosystems. These mutations result in a phenotype that is essentially white instead of green. The inheritance of this phenotype follows a non-Mendelian, maternal pattern (**Figure 12.22**). Mitochondrial gene mutations that affect the respiratory chain result in less ATP production. In animals, these mutations have particularly noticeable effects in tissues with high energy requirements, such as the nervous system, muscles, and kidneys.

RECAP 12.5

Genes in the genomes of organelles, specifically plastids and mitochondria, do not behave in a Mendelian fashion.

- Why are genes carried in organelles usually inherited only from the mother? See p. 252

Mendel and those who followed him focused on eukaryotes, with diploid organisms and haploid gametes. A half-century after the rediscovery of Mendel's work, a process that allows genetic recombination was discovered in prokaryotes as well. We will now turn to that process.

12.6 How Do Prokaryotes Transmit Genes?

As described in Chapter 5, prokaryotic cells lack nuclei; they contain their genetic material mostly as single chromosomes in central regions of their cells. Prokaryotes reproduce asexually by binary fission, a process that gives rise to progeny that are virtually identical genetically. That is, the offspring of cell reproduction in prokaryotes constitute a clone (see Chapter 11). You might expect, therefore, that there is no way for individuals of these organisms to exchange genes, as in sexual reproduction.

How then do prokaryotes evolve? Mutations occur in prokaryotes just as they do in eukaryotes, and the resulting new alleles increase genetic diversity. In addition, it turns out that prokaryotes do have a sexual process for transferring genes between cells. Along with mutation, this process provides for genetic diversity among prokaryotes.

Bacteria exchange genes by conjugation

To illustrate the kind of experiment that led to the discovery of bacterial DNA transfer, let's consider two strains of the bacterium *E. coli* with different alleles for each of six genes. One stain carries the dominant (wild-type) alleles for three of the genes and the recessive (mutant) alleles for the other genes. This situation is reversed in the other strain. Simply put, the two strains have the following genotypes (remember that bacteria are haploid):

$$ABCdef \quad \text{and} \quad abcDEF$$

where capital letters indicate wild-type alleles and lowercase letters indicate mutant alleles.

When the two strains are grown together in the laboratory, most of the cells produce clones. That is, almost all of the cells that grow have the original genotypes. However, out of millions of bacteria, a few occur that have the genotype

$$ABCDEF$$

How could these completely wild-type bacteria arise? One possibility is mutation: in the *abcDEF* bacteria, the *a* allele could have mutated to *A*, the *b* allele to *B*, and the *c* allele to *C*. The problem with this explanation is that a mutation at any particular point in an organism's DNA sequence is a very rare event. The probability of all three events occurring in the same cell is extremely low—much lower than the actual rate of appearance of cells with the *ABCDEF* genotype. So the mutant cells must have acquired wild-type genes some other way—and this turns out to be the transfer of DNA between cells.

Electron microscopy shows that genetic transfers between bacteria can happen via physical contact between the cells (**Figure 12.23A**). Contact is initiated by a thin projection called a **sex pilus** (plural *pili*), which extends from one cell (the donor), attaches to another (the recipient), and draws the two cells together. Genetic material can then pass from the donor cell to the recipient through a thin cytoplasmic bridge called a conjugation tube. There is no reciprocal transfer of DNA from the recipient to the donor. This process is referred to as **bacterial conjugation**.

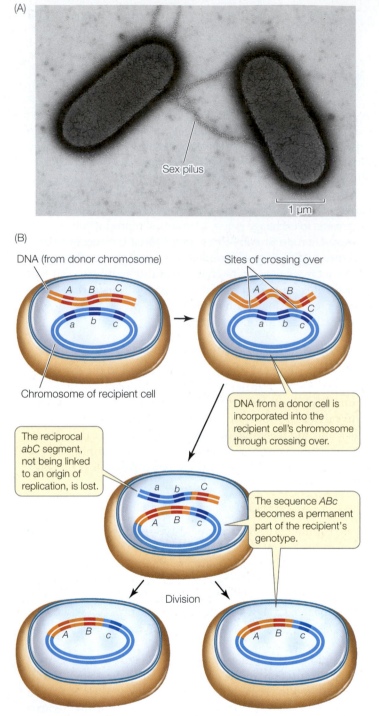

(A)

Sex pilus

1 µm

(B)

DNA (from donor chromosome)

Sites of crossing over

A B C

a b c

Chromosome of recipient cell

A B

C

a b c

DNA from a donor cell is incorporated into the recipient cell's chromosome through crossing over.

The reciprocal *abC* segment, not being linked to an origin of replication, is lost.

a b C

A B c

The sequence *ABc* becomes a permanent part of the recipient's genotype.

Division

A B c

A B c

12.23 Bacterial Conjugation and Recombination (A) Sex pili draw two bacteria into close contact, so that a cytoplasmic conjugation tube can form. DNA is transferred from one cell to the other via the conjugation tube. (B) DNA from a donor cell can become incorporated into a recipient cell's chromosome through crossing over.

Once the donor DNA is inside the recipient cell, it can recombine with the recipient cell's genome. In much the same way that chromosomes pair up, gene for gene, in prophase I of meiosis, the donor DNA can line up beside its homologous genes in the recipient, and crossing over can occur. Some of the genes from the donor can become integrated into the genome of the recipient, thus changing the recipient's genetic constitution (**Figure 12.23B**). When the recipient cells proliferate, the integrated donor genes are passed on to all progeny cells.

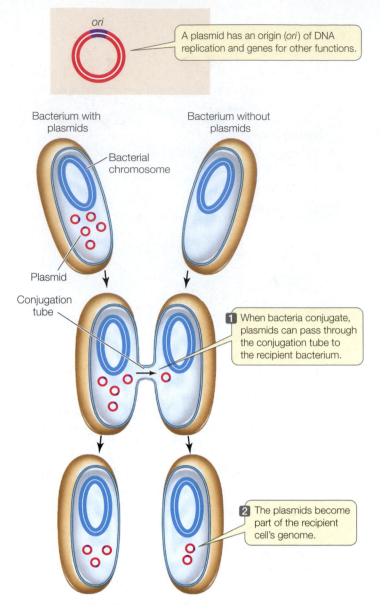

ori

A plasmid has an origin (*ori*) of DNA replication and genes for other functions.

Bacterium with plasmids

Bacterium without plasmids

Bacterial chromosome

Plasmid

Conjugation tube

1 When bacteria conjugate, plasmids can pass through the conjugation tube to the recipient bacterium.

2 The plasmids become part of the recipient cell's genome.

12.24 Gene Transfer by Plasmids When plasmids enter a cell via conjugation, their genes can be expressed in the recipient cell.

Bacterial conjugation is controlled by plasmids

In addition to their main chromosome, many bacteria harbor additional smaller, circular DNA molecules called **plasmids**, which replicate independently of the main chromosome. Plasmids typically contain at most a few dozen genes, which may fall into one of several categories:

- *Genes for unusual metabolic capacities.* For example, bacteria carrying plasmids that confer the ability to break down hydrocarbons can be used to clean up oil spills.

- *Genes for antibiotic resistance.* Plasmids carrying such genes are called R factors, and since they can be transferred between bacteria via conjugation, they are a major threat to human health.

- *Genes that confer the ability to make a sex pilus.*

During bacterial conjugation it is usually plasmids that are transferred from one bacterium to another (**Figure 12.24**). A single strand of the donor plasmid is transferred to the recipient, and then synthesis of a complementary DNA strand results in two complete copies of the plasmid, one in the donor and one in the recipient. Plasmids can replicate independently of the main chromosome, but sometimes they become integrated into the chromosome. When this happens, the genes for conjugation are still expressed, and the plasmid DNA can be transferred between cells. But because the plasmid is part of the main chromosome, DNA from the chromosome is transferred along with the plasmid DNA. The amount of chromosomal DNA that can be transferred this way depends on how long the bacterial cells are in contact; for example, it takes about 100 minutes for the entire *E. coli* chromosome to be transferred by conjugation.

RECAP 12.6

Although they are haploid and reproduce asexually, prokaryotes have the ability to transfer genes from one cell to another. These genes are usually carried on small, circular DNA molecules called plasmids, but chromosomal DNA is sometimes transferred as well.

- How were prokaryotic gene transfer and recombination discovered? **See p. 253**
- What are the differences between recombination after conjugation in prokaryotes and recombination during meiosis in eukaryotes?

?

How can knowledge of genetics be used to save the Tasmanian devil?

ANSWER

The Tasmanian Devil Genome Project, based in the United States but involving biologists from Australia and other countries, is dedicated to preserving the species by determining the genetic diversity of as many devils as possible. Matings are planned between the most genetically diverse individuals in the wild to maximize heterozygosity. (This is not as easy to do as it is to write!) In addition, a captive "insurance population" from an area where devils are cancer-free is being developed. These animals, too, are being genotyped, in the hope of generating a larger uninfected population.

CHAPTER**SUMMARY** 12

12.1 What Are the Mendelian Laws of Inheritance?

- Physical features of organisms, or **characters**, can exist in different forms, or **traits**. A heritable trait is one that can be passed from parent to offspring. A **phenotype** is the physical appearance of an organism; a **genotype** is the genetic constitution of the organism.

- The different forms of a **gene** are called **alleles**. Organisms that have two identical alleles for a trait are called **homozygous**; organisms that have two different alleles for a trait are called **heterozygous**. A gene resides at a particular site on a chromosome called a **locus**.

- Mendel's experiments included reciprocal crosses and **monohybrid crosses** between true-breeding pea plants. Analysis of his meticulously tabulated data led Mendel to propose a particulate theory of inheritance stating that discrete units (now called genes) are responsible for the inheritance of specific traits, to which both parents contribute equally.

- Mendel's first law, the **law of segregation**, states that when any individual produces gametes, the two copies of a gene separate, so that each gamete receives only one member of the pair. Thus every individual in the F_1 inherits one copy from each parent. **Review Figures 12.3, 12.4**

- Mendel used a **test cross** to find out whether an individual showing a dominant phenotype was homozygous or heterozygous. **Review Figure 12.5, ACTIVITY 12.1**

- Mendel's use of **dihybrid crosses** to study the inheritance of two characters led to his second law: the **law of independent assortment**. The independent assortment of genes in meiosis leads to nonparental combinations of phenotypes in the offspring of a dihybrid cross. **Review Figures 12.6, 12.7, ANIMATED TUTORIAL 12.1**

- Probability calculations and **pedigrees** help geneticists trace Mendelian inheritance patterns. **Review Figures 12.8, 12.9, ANIMATED TUTORIAL 12.2**

12.2 How Do Alleles Interact?

- New alleles arise by random **mutation**. Many genes have multiple alleles. A wild-type allele gives rise to the predominant form of a trait. When the wild-type allele is present at a locus less than 99 percent of the time, the locus is said to be **polymorphic**. **Review Figure 12.10**

- In **incomplete dominance**, neither of two alleles is dominant. The heterozygous phenotype is intermediate between the homozygous phenotypes. **Review Figure 12.11**

- **Codominance** exists when two alleles at a locus produce two different phenotypes that both appear in heterozygotes.

- An allele that affects more than one trait is said to be **pleiotropic**.

12.3 How Do Genes Interact?

- In **epistasis**, one gene affects the expression of another. **Review Figure 12.13**

- Environmental conditions can affect the expression of a genotype.

- **Penetrance** is the proportion of individuals in a group with a given genotype that show the expected phenotype. **Expressivity** is the degree to which a genotype is expressed in an individual.

- Variations in phenotypes can be **qualitative** (discrete) or **quantitative** (graduated, continuous). Most quantitative traits result from the effects of several genes and the environment. Genes that together determine quantitative characters are called **quantitative trait loci**.

continued

12.4 **What Is the Relationship between Genes and Chromosomes?**
See ANIMATED TUTORIAL 12.3

- Each chromosome carries many genes. Genes on the same chromosome are referred to as a linkage group.

- Genes on the same chromosome can recombine by crossing over. The resulting recombinant chromosomes have new combinations of alleles. **Review Figures 12.18, 12.19**

- **Sex chromosomes** carry genes that determine whether the organism will produce male or female gametes. All other chromosomes are called **autosomes**. The specific functions of sex chromosomes differ among different groups of organisms.

- **Primary sex determination** in mammals is usually a function of the presence or absence of the *SRY* gene. **Secondary sex characteristics** are the outward manifestations of maleness and femaleness.

- In fruit flies and mammals, the X chromosome carries many genes, but the Y chromosome has only a few. Males have only one allele (are **hemizygous**) for X-linked genes, so recessive **sex-linked** mutations are expressed phenotypically more often in males than in females. Females may be unaffected carriers of such alleles. **Review Figure 12.21**

12.5 **What Are the Effects of Genes Outside the Nucleus?**

- Cytoplasmic organelles such as plastids and mitochondria contain small numbers of genes. In many organisms, cytoplasmic genes are inherited only from the mother because the male gamete contributes only its nucleus (i.e., no cytoplasm) to the zygote at fertilization. **Review Figure 12.22**

12.6 **How Do Prokaryotes Transmit Genes?**

- Prokaryotes reproduce primarily asexually but can exchange genes in a sexual process called **conjugation**. **Review Figure 12.23**

- **Plasmids** are small, extra chromosomes in bacteria that carry genes involved in important metabolic processes and that can be transmitted from one cell to another. **Review Figure 12.24**

See ACTIVITIES 12.2, 12.3 for a concept review of this chapter

Go to the Interactive Summary to review key figures, Animated Tutorials, and Activities
Life10e.com/is12

CHAPTER**REVIEW**

■■■ **REMEMBERING**

1. In a simple Mendelian monohybrid cross, true-breeding tall and short plants are crossed, and the F_1 plants, which are all tall, are allowed to self-pollinate. What fraction of the F_2 generation are both tall and heterozygous?
   a. 1/8
   b. 1/4
   c. 1/3
   d. 2/3
   e. 1/2

2. The phenotype of an individual
   a. depends at least in part on the genotype.
   b. is either homozygous or heterozygous.
   c. determines the individual's genotype.
   d. is the genetic constitution of the organism.
   e. is either monohybrid or dihybrid.

3. Which statement about an individual that is homozygous for an allele is *not* true?
   a. Each of its cells possesses two copies of that allele.
   b. Each of its gametes contains one copy of that allele.
   c. It is true-breeding with respect to that allele.
   d. Its parents were necessarily homozygous for that allele.
   e. It can pass that allele to its offspring.

4. Which statement about a monohybrid test cross is *not* true?
   a. It tests whether an individual of unknown genotype is homozygous or heterozygous.
   b. The test individual is crossed with a homozygous recessive individual.
   c. If the test individual is heterozygous, the progeny will have a 1:1 phenotypic ratio.
   d. If the test individual is homozygous, the progeny will have a 3:1 phenotypic ratio.
   e. Test cross results are consistent with Mendel's model of inheritance.

5. Linked genes
   a. must be immediately adjacent to one another on a chromosome.
   b. have alleles that assort independently of one another.
   c. never show crossing over.
   d. are on the same chromosome.
   e. always have multiple alleles.

6. The genetic sex of a human is determined by
   a. ploidy, with the male being haploid.
   b. the Y chromosome.
   c. X and Y chromosomes, the male being XX.
   d. the number of X chromosomes, the male being XO.
   e. Z and W chromosomes, the male being ZZ.

7. In epistasis
   a. the phenotype does not change from generation to generation.
   b. one gene affects the expression of another.
   c. a portion of a chromosome is deleted.
   d. a portion of a chromosome is inverted.
   e. the behavior of two genes is entirely independent.

UNDERSTANDING & APPLYING

8. The ABO blood groups in humans are determined by a multiple-allele system in which I^A and I^B are codominant and are both dominant to I^O. A newborn infant is type A. The mother is type O. Possible phenotypes of the father are
   a. A, B, or AB.
   b. A, B, or O.
   c. O only.
   d. A or AB.
   e. A or O.

9. In humans, spotted teeth are caused by a dominant sex-linked gene. A man with spotted teeth whose father had normal teeth marries a woman with normal teeth. Therefore
   a. all of their daughters will have normal teeth.
   b. all of their daughters will have spotted teeth.
   c. all of their children will have spotted teeth.
   d. half of their sons will have spotted teeth.
   e. all of their sons will have spotted teeth.

10. In guinea pigs, black body color (B) is completely dominant over albino (b). For the crosses below, give the genotypes of the parents:

Parental phenotypes	Black offspring	Albino offspring	Parental genotypes?
Black × albino	12	0	
Albino × albino	0	12	
Black × albino	5	7	
Black × black	9	3	

11. In the genetic cross $AaBbCcDdEE \times AaBBCcDdEe$, what fraction of the offspring will be heterozygous for all of these genes ($AaBbCcDdEe$)? Assume all genes are unlinked.

12. The pedigree below shows the inheritance of a rare mutant phenotype in humans, congenital cataracts (black symbols).

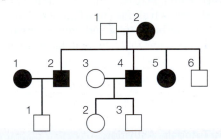

   a. Are cataracts inherited as an autosomal dominant trait? Autosomal recessive? X-linked dominant? X-linked recessive?
   b. Person #5 in the second generation marries a man who does not have cataracts. Two of their four children, a boy and a girl, develop cataracts. What is the chance that their next child will be a girl with cataracts?

13. In cats, black coat (B) is codominant with yellow (b). The coat color gene is on the X chromosome. Calico cats, which have coats with black and yellow patches, are heterozygous for the coat color alleles.
   a. Why are most calico cats females?
   b. A calico female, Pickle, had a litter with one yellow male, two black males, two yellow females, and three calico females. What were the genotype and phenotype of the father?

14. In *Drosophila*, three autosomal genes have alleles as follows:
   Gray body color (G) is dominant over black (g).
   Normal wings (A) is dominant over vestigial (a).
   Red eye (R) is dominant over sepia (r).

   Two crosses were performed, with the following results:

   Cross I: Parents: heterozygous
                red, normal × sepia, vestigial

                Offspring: 131 red, normal
                           120 sepia, vestigial
                           122 red, vestigial
                           127 sepia, normal

   Cross II: Parents: heterozygous
                gray, normal × black, vestigial

                Offspring: 236 gray, normal
                           253 black, vestigial
                           50 gray, vestigial
                           61 black, normal

   Are any of the three genes linked on the same chromosome? If so, what is the distance between the linked genes (in map units)?

15. In a particular plant species, two alleles control flower color, which can be yellow, blue, or white. Crosses of these plants produce the following offspring:

Parental phenotypes	Offspring phenotypes (ratio)
Yellow × yellow	All yellow
Blue × yellow	Blue or yellow (1:1)
Blue × white	Blue or white (1:1)
White × white	All white

   What will be the phenotype, and ratio, of the offspring of a cross of blue × blue?

16. In *Drosophila* the recessive allele p, when homozygous, determines pink eyes. Pp or PP results in wild-type eye color. Another gene on a different chromosome has a recessive allele, sw, that produces short wings when homozygous. Consider a cross between females of genotype PPSwSw and males of genotype ppswsw. Describe the phenotypes and genotypes of the F_1 generation and of the F_2 generation, produced by allowing the F_1 progeny to mate with one another.

17. On the same chromosome of *Drosophila* that carries the *p* (pink eyes) locus, there is another locus that affects the wings. Homozygous recessives, *byby*, have blistery wings, while the dominant allele *By* produces wild-type wings. The *P* and *By* loci are very close together on the chromosome; that is, the two loci are tightly linked. In answering Questions 17a and 17b, assume that no crossing over occurs, and that the F_2 generation is produced by interbreeding the F_1 progeny.

    a. For the cross *PPByBy* × *ppbyby*, give the phenotypes and genotypes of the F_1 and F_2 generations.

    b. For the cross *PPbyby* × *ppByBy*, give the phenotypes and genotypes of the F_1 and F_2 generations.

    c. For the cross of Question 17b, what further phenotype(s) would appear in the F_2 generation if crossing over occurred?

    d. Draw a nucleus undergoing meiosis at the stage in which the crossing over (Question 17c) occurred. In which generation (P, F_1, or F_2) did this crossing over take place?

18. In chickens, when the dominant alleles of the genes for rose comb (*R*) and pea comb (*A*) are present together (*R_A_*), the result is a bird with a walnut comb. Chickens that are homozygous recessive for both genes produce a single comb. A rose-combed bird mated with a walnut-combed bird and produced offspring in the proportion: ⅜ walnut : ⅜ rose : ⅛ pea : ⅛ single.

    What were the genotypes of the parents?

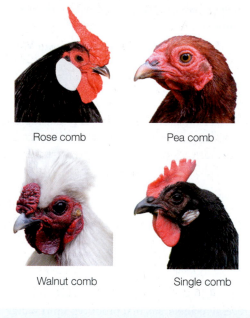

Rose comb

Pea comb

Walnut comb

Single comb

19. In *Drosophila*, white (*w*), eosin (w^e), and wild-type red (w^+) are multiple alleles at a single locus for eye color. This locus is on the X chromosome. A female that has eosin (pale orange) eyes is crossed with a male that has wild-type eyes. All the female progeny are red-eyed; half the male progeny have eosin eyes, and half have white eyes. Assume the female has two X chromosomes and the male has one X and one Y.

    a. What is the order of dominance of these alleles?

    b. What are the genotypes of the parents and progeny?

20. In humans, red-green color blindness is determined by an X-linked recessive allele (*a*), whereas eye color is determined by an autosomal gene, where brown (*B*) is dominant over blue (*b*).

    a. What gametes can be formed with respect to these genes by a heterozygous, brown-eyed, color-blind male?

    b. If a blue-eyed mother with normal vision has a brown-eyed, color-blind son and a blue-eyed, color-blind daughter, what are the genotypes of both parents and children?

21. If the dominant allele *A* is necessary for hearing in humans, and another allele, *B*, located on a different chromosome, results in deafness no matter what other genes are present, what percentage of the offspring of the marriage of *aaBb* × *Aabb* will be deaf?

22. The disease Leber's optic neuropathy is caused by a mutation in a gene carried on mitochondrial DNA. What would be the phenotype of their first child if a man with this disease married a woman who did not have the disease? What would be the result if the wife had the disease and the husband did not?

■■■ **ANALYZING & EVALUATING**

23. Sometimes scientists get lucky. Consider Mendel's dihybrid cross shown in Figure 12.6. Peas have a haploid number of seven chromosomes, so many of their genes are linked. What would Mendel's results have been if the genes for seed color and seed shape were linked with a map distance of 10 units? Now, consider Morgan's fruit flies (see Figure 12.19). Suppose that the genes for body color and wing shape were not linked? What results would Morgan have obtained?

Go to BioPortal at **yourBioPortal.com** for Animated Tutorials, Activities, LearningCurve Quizzes, Flashcards, and many other study and review resources.

13

DNA and Its Role in Heredity

CHAPTEROUTLINE

13.1 What Is the Evidence that the Gene Is DNA?

13.2 What Is the Structure of DNA?

13.3 How Is DNA Replicated?

13.4 How Are Errors in DNA Repaired?

13.5 How Does the Polymerase Chain Reaction Amplify DNA?

Observation, Insight, and Discovery Louis Pasteur's observations and "prepared mind" led to unique insights and scientific breakthroughs in the late nineteenth century. This pattern of research continues to advance the scientific enterprise, resulting in discoveries such as the drug cisplatin.

OVER A CENTURY AGO, the great French microbiologist Louis Pasteur said "chance favors only the prepared mind." He meant that great discoveries come not just from a flash of insight, but from careful observations made by people whose background has made them ready to interpret their observations in a new way. Such "prepared minds" have sparked the development of many new cancer drugs.

Testicular cancer occurs without warning with rapid cell divisions of germ cells. It often spreads to other organs, such as the lungs and brain. With a lifetime risk of about 1 in 250 males, it typically strikes men in their twenties and is the most common cancer in young men. Despite its potential lethality, testicular cancer is one of the few tumors of adults that is highly curable. The cure is primarily due to a drug called cisplatin.

Dr. Barnett Rosenberg, a scientist at Michigan State University, was curious about how electric fields might affect cells. He put bacteria into a growth medium with platinum electrodes connected to a battery. The result was striking: the bacteria stopped dividing. Thinking he was on the road to a major discovery about electromagnetism and cells, Rosenberg tried the experiment again, this time using copper and zinc electrodes. (You are probably familiar with the Cu/Zn system if you've taken a chemistry course.) This time the bacteria kept dividing, with no adverse effects. Only platinum electrodes inhibited cell division.

In light of the data, Rosenberg revised his hypothesis to propose that something leaked out of the platinum electrodes into the medium, and that this "something" blocked cell division. He confirmed his hypothesis by treating bacteria with the medium in which the platinum electrodes had been inserted; the bacteria did not divide. Realizing that cancer cells have uncontrolled cell division, he duplicated his experiments with tumor cells in a laboratory dish. This led to the isolation and development of cisplatin. The drug was so successful with testicular cancer that it has also been used with some success on other tumors.

An essential event for cell division is the complete and precise duplication of the genetic material, DNA. The two strands of DNA unwind and separate, each strand acting as a template for the building of a new strand. Strand separation is possible because the two strands are held together by weak forces, including hydrogen bonds. Cisplatin forms covalent bonds with nucleotides on opposite strands of the DNA, irreversibly cross-linking the two strands together. As a result, the DNA strands cannot separate for replication or expression. With such severe damage to its DNA, the cell then undergoes programmed cell death.

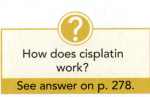

How does cisplatin work?

See answer on p. 278.

13.1 Genetic Transformation Griffith's experiments demonstrated that something in the virulent S strain of pneumococcus could transform nonvirulent R strain bacteria into a lethal form, even when the S strain bacteria had been killed by high temperatures.[a]

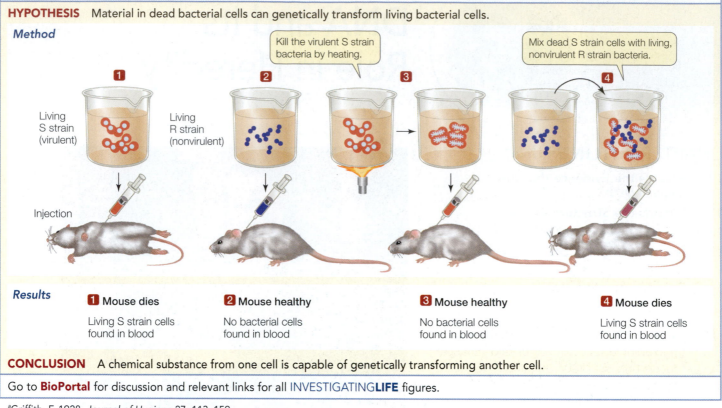

HYPOTHESIS Material in dead bacterial cells can genetically transform living bacterial cells.

Method

Kill the virulent S strain bacteria by heating.

Mix dead S strain cells with living, nonvirulent R strain bacteria.

Living S strain (virulent)

Living R strain (nonvirulent)

Injection

Results

1 Mouse dies
Living S strain cells found in blood

2 Mouse healthy
No bacterial cells found in blood

3 Mouse healthy
No bacterial cells found in blood

4 Mouse dies
Living S strain cells found in blood

CONCLUSION A chemical substance from one cell is capable of genetically transforming another cell.

Go to **BioPortal** for discussion and relevant links for all INVESTIGATING**LIFE** figures.

[a]Griffith, F. 1928. *Journal of Hygiene* 27: 113–159.

 13.1

What Is the Evidence that the Gene Is DNA?

By the early twentieth century, geneticists had associated the presence of genes with chromosomes. Research began to focus on exactly which chemical component of chromosomes comprised this genetic material.

By the 1920s, scientists knew that chromosomes were made up of DNA and proteins. At this time a new dye was developed by Robert Feulgen that could bind specifically to DNA and that stained cell nuclei red in direct proportion to the amount of DNA present in the cell. This technique provided circumstantial evidence that DNA was the genetic material:

- *DNA was in the right place.* DNA was confirmed to be an important component of the nucleus and the chromosomes, which were known to carry genes.

- *DNA was present in the right amounts.* The amount of DNA in somatic cells (body cells not specialized for reproduction) was twice that in reproductive cells (eggs or sperm)—as might be expected for diploid and haploid cells, respectively.

- *DNA varied among species.* When cells from different species were stained with the dye and their color intensity measured, each species appeared to have its own specific amount of nuclear DNA.

But circumstantial evidence is *not* a scientific demonstration of cause and effect. After all, proteins are also present in cell nuclei. Science relies on experiments to test hypotheses. The convincing demonstration that DNA is the genetic material came from two sets of experiments, one with bacteria and the other with viruses.

DNA from one type of bacterium genetically transforms another type

In science, research on one specific topic often contributes to another, apparently unrelated area. Such a case of serendipity is seen in the work of Frederick Griffith, an English physician. In the 1920s Griffith was studying the bacterium *Streptococcus pneumoniae*, or pneumococcus, one of the agents that cause pneumonia in humans. He was trying to develop a vaccine against this devastating illness (antibiotics had not yet been discovered). Griffith was working with two strains of pneumococcus:

- Cells of the S strain produced colonies that looked smooth (S). Covered by a polysaccharide capsule, these cells were protected from attack by a host's immune system. When S cells were injected into mice, they reproduced and caused pneumonia (the strain was virulent).

- Cells of the R strain produced colonies that looked rough (R), lacked the protective capsule, and were *not* virulent.

When Griffith inoculated mice with heat-killed S-type pneumococcus cells, the cells did not produce infection. However, when he inoculated other mice with a mixture of living R-type cells and heat-killed S-type cells, to his astonishment, the mice died of pneumonia (**Figure 13.1**). When he examined blood from these mice, he found it full of living bacteria—many of them with characteristics of the virulent S strain! Griffith concluded that in the presence of the dead S-type pneumococcus cells, some of the living R-type cells had been transformed into virulent S cells. These cells were able to grow in the bodies of the mice, causing pneumonia and multiplying in the blood. The fact that these S-type cells reproduced to make more S-type cells showed that the change from R-type to S-type was genetic.

Did this transformation of the bacteria depend on something that happened in the mouse's body? No. The same transformation could be achieved in a test tube by mixing living R cells with heat-killed S cells, or even with a cell-free extract of the heat-killed S cells. (A cell-free extract contains all the contents of ruptured cells, but no intact cells.) These results demonstrated that some substance from the dead S pneumococcus cells could cause a heritable change in the affected R cells.

Oswald Avery and his colleagues at what is now The Rockefeller University identified the substance causing bacterial transformation in two ways:

- *Eliminating other possibilities.* Cell-free extracts containing the transforming substance were treated with enzymes that destroyed candidates for the genetic material, such as proteins, RNA, and DNA. When the treated samples were tested, the ones treated with RNase and protease (which destroy RNA and proteins, respectively) were still able to transform R-type bacteria into the S-type. But the transforming activity was lost in the extract treated with DNase (which destroys DNA) (**Figure 13.2**).

- *Positive experiment.* They isolated virtually pure DNA from a cell-free extract containing the transforming substance. The DNA alone caused bacterial transformation.

We now know that the gene for the enzyme that catalyzes the synthesis of the polysaccharide capsule, which makes the bacterial colony look "smooth," was transferred into the R cells during transformation.

Viral infection experiments confirmed that DNA is the genetic material

Even with the bacterial transformation experiments, many biologists were still not convinced that DNA is the genetic material. One problem was that DNA, being made up of only four simple nucleotides (see Section 4.1), seemed too uniform a substance to be able to confer all the functions and variety of life. The possibility remained that proteins, with all their chemical and structural diversity, fulfilled that role. Experiments with a virus were designed to distinguish between these alternatives.

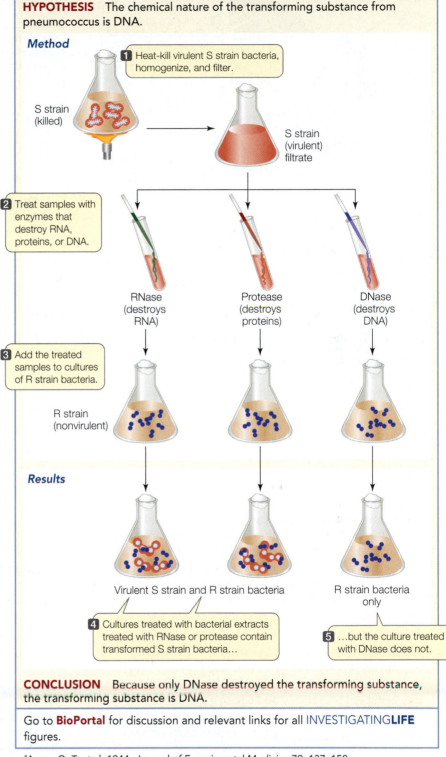

INVESTIGATINGLIFE

13.2 Genetic Transformation by DNA Experiments by Avery and his colleagues showed that DNA from the virulent S strain of pneumococcus was responsible for the transformation in Griffith's experiments (see Figure 13.1).[a]

HYPOTHESIS The chemical nature of the transforming substance from pneumococcus is DNA.

Method

1 Heat-kill virulent S strain bacteria, homogenize, and filter.

S strain (killed)

S strain (virulent) filtrate

2 Treat samples with enzymes that destroy RNA, proteins, or DNA.

RNase (destroys RNA) — Protease (destroys proteins) — DNase (destroys DNA)

3 Add the treated samples to cultures of R strain bacteria.

R strain (nonvirulent)

Results

Virulent S strain and R strain bacteria — R strain bacteria only

4 Cultures treated with bacterial extracts treated with RNase or protease contain transformed S strain bacteria…

5 …but the culture treated with DNase does not.

CONCLUSION Because only DNase destroyed the transforming substance, the transforming substance is DNA.

Go to **BioPortal** for discussion and relevant links for all INVESTIGATINGLIFE figures.

[a]Avery, O. T. et al. 1944. *Journal of Experimental Medicine* 79: 137–158.

Alfred Hershey and Martha Chase of the Cold Spring Harbor Laboratory in New York studied bacteriophage T2 (phage T2), which infects the bacterium *Escherichia coli*. T2 phage consists of a DNA core packed inside a protein coat (**Figure 13.3**). When it attacks a bacterium, part (but not all) of the virus enters

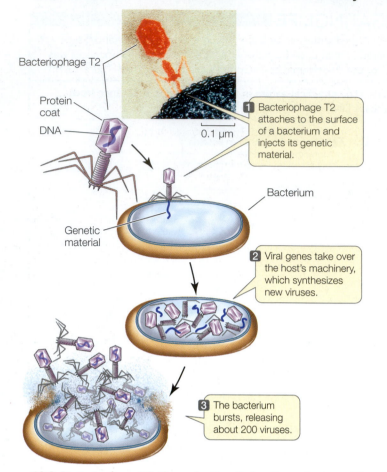

13.3 Bacteriophage T2: Reproduction Cycle Bacteriophage T2 is parasitic on *E. coli*, depending on the bacterium to produce new viruses. The external structures of bacteriophage T2 consist entirely of protein, and the DNA is contained within the protein coat. When the virus infects an *E. coli* cell, its genetic material is injected into the host bacterium.

the bacterial cell. About 20 minutes later, the cell bursts, releasing dozens of particles that are virtually identical to the infecting virus particle. Clearly the virus is somehow able to convert the host cell from its own genetic program into a viral replication machine. Hershey and Chase set out to determine what part of the virus—DNA or protein—enters the host cell to bring about this genetic change. To trace the two components of the virus over its life cycle, the scientists labeled each component with a specific radioisotope:

- *Proteins were labeled with radioactive sulfur*. Proteins contain some sulfur (in the amino acids cysteine and methionine), but DNA does not. Sulfur has a radioactive isotope, ^{35}S. Hershey and Chase grew bacteriophage T2 in a bacterial culture in the presence of ^{35}S, so the proteins of the resulting viruses were labeled with (contained) the radioisotope.

- *DNA was labeled with radioactive phosphorus*. DNA contains a lot of phosphorus (in the deoxyribose–phosphate backbone—see Figure 4.4), whereas proteins contain little or none. Phosphorus also has a radioisotope, ^{32}P. The researchers grew another batch of T2 in a bacterial culture in the presence of ^{32}P, thus labeling the viral DNA with ^{32}P.

INVESTIGATINGLIFE

13.4 The Hershey–Chase Experiment When Hershey and Chase infected bacterial cells with radioactively labeled T2 bacteriophage, only labeled DNA was found in the bacteria.[a] The infected cells were agitated to remove the viral coats from the bacteria and were then centrifuged to pellet the bacteria. The labeled protein remained in the supernatant. This showed that DNA, not protein, is the genetic material.

HYPOTHESIS Either component of a bacteriophage—DNA or protein—might be the hereditary material that enters a bacterial cell to direct the assembly of new viruses.

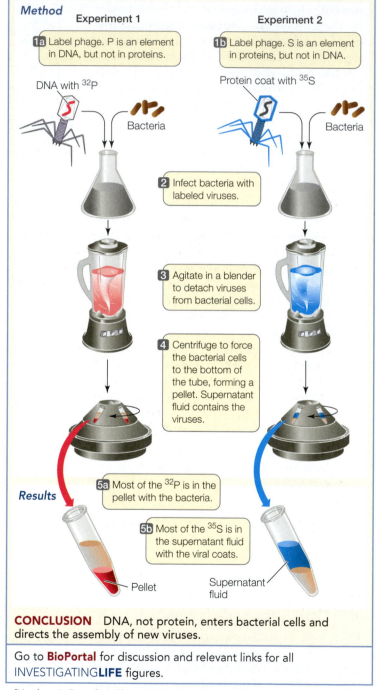

CONCLUSION DNA, not protein, enters bacterial cells and directs the assembly of new viruses.

Go to **BioPortal** for discussion and relevant links for all INVESTIGATINGLIFE figures.

[a]Hershey, A. D. and M. Chase. 1952. *The Journal of General Physiology* 36: 39–56.

Hershey and Chase used these radioactively labeled viruses in their experiments (**Figure 13.4**). In one experiment, they allowed ^{32}P-labeled bacteriophage to infect bacteria; in the other, the bacteria were infected with ^{35}S-labeled bacteriophage. After a few minutes they agitated each mixture vigorously in a kitchen blender, stripping away the parts of the viruses that had not penetrated the bacteria, without bursting the bacteria. Then they separated the bacteria from the rest of the material (the remains of the viruses) in a centrifuge. The result was that the bacterial cells in the centrifuge pellet contained most of the [32]P (and thus the viral DNA), and the supernatant fluid with the viral remains contained most of the [35]S (and thus the viral protein). These results indicated that it was the DNA that had been transferred into the bacteria, and that DNA was the molecule responsible for redirecting the genetic program of the bacterial cell.

 Go to Animated Tutorial 13.1
The Hershey–Chase Experiment
Life10e.com/at13.1

Eukaryotic cells can also be genetically transformed by DNA

The transformation of eukaryotic cells by DNA is often called **transfection**. This can be demonstrated using a **genetic marker**, a gene whose presence in the recipient cells confers an observable phenotype. When transforming both prokaryotes and eukaryotes, researchers often use an antibiotic resistance or nutritional marker gene, which permits the growth of transformed cells but not of nontransformed cells. A common marker in mammalian transfection experiments is a gene that confers resistance to the antibiotic neomycin (**Figure 13.5**). There are many methods for transfection, including chemical treatments that allow the DNA to be taken up by the cells. Any cell can be transfected, even an egg cell. In this case, a whole new genetically transformed organism can result; such an organism is referred to as transgenic. Transformation in eukaryotes is the final line of evidence for DNA as the genetic material.

RECAP 13.1

Experiments on bacteria and on viruses demonstrated that DNA is the genetic material.

- At the time of Griffith's experiments in the 1920s, what circumstantial evidence suggested to scientists that DNA might be the genetic material? **See p. 260**

- How did the experiments of Avery and his colleagues provide further evidence that DNA was the genetic material? **See p. 261 and Figure 13.2**

- What attributes of bacteriophage T2 were key to the Hershey–Chase experiments demonstrating that DNA, rather than protein, is the genetic material? **See pp. 261–262 and Figure 13.4**

The transformation and viral infection experiments convinced biologists that the genetic material is DNA. It had been known for several decades that chemically, DNA is a polymer of nucleotides. Next, scientists turned to DNA's precise three-dimensional structure.

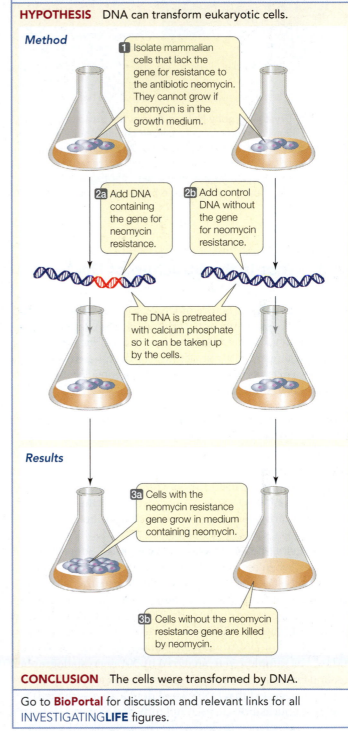

INVESTIGATINGLIFE

13.5 Transfection in Eukaryotic Cells A DNA molecule can be treated chemically so that it is taken up from a solution by mammalian cells.[a] The inclusion of a marker gene shows that the cells have been genetically transformed by the DNA.

HYPOTHESIS DNA can transform eukaryotic cells.

Method

1 Isolate mammalian cells that lack the gene for resistance to the antibiotic neomycin. They cannot grow if neomycin is in the growth medium.

2a Add DNA containing the gene for neomycin resistance.

2b Add control DNA without the gene for neomycin resistance.

The DNA is pretreated with calcium phosphate so it can be taken up by the cells.

Results

3a Cells with the neomycin resistance gene grow in medium containing neomycin.

3b Cells without the neomycin resistance gene are killed by neomycin.

CONCLUSION The cells were transformed by DNA.

Go to **BioPortal** for discussion and relevant links for all INVESTIGATINGLIFE figures.

[a]Bacchetti, S. and F. L. Graham. 1977. *Proceedings of the National Academy of Sciences USA* 74: 1590–1594.

(A)   (B)

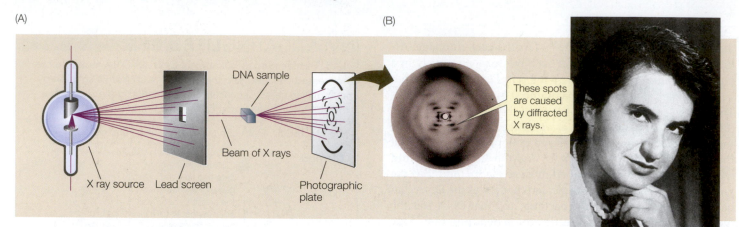

13.6 X-Ray Crystallography Helped Reveal the Structure of DNA
(A) The positions of atoms in a crystallized chemical substance can be inferred by the pattern of diffraction of X rays passed through it. The pattern of DNA is both highly regular and repetitive. (B) Rosalind Franklin's crystallography and her "photograph 51" (shown) helped scientists visualize the helical structure of the DNA molecule.

 ## 13.2 What Is the Structure of DNA?

In determining the structure of DNA, scientists hoped to find the answers to two questions: (1) how is DNA replicated between cell divisions, and (2) how does it direct the synthesis of specific proteins? DNA's structure was deciphered only after many types of experimental evidence were considered together in a theoretical framework.

Watson and Crick used modeling to deduce the structure of DNA

Once pure DNA fibers could be isolated, biophysicists and biochemists examined the DNA for hints about its structure. The evidence eventually used to solve DNA's structure included crucial data obtained using X-ray crystallography and a thorough characterization of the chemical composition of DNA.

PHYSICAL EVIDENCE FROM X-RAY DIFFRACTION Some chemical substances, when they are isolated and purified, can be made to form crystals. The positions of atoms in a crystallized substance can be inferred from the diffraction pattern of X rays passing through the substance. In the early 1950s the New Zealand-born biophysicist Maurice Wilkins discovered a way to make highly ordered fibers of DNA that were suitable for X-ray diffraction studies. His samples were analyzed by X-ray crystallographer Rosalind Franklin of Kings College, London (**Figure 13.6**). Franklin's data suggested that DNA was a double (two-stranded) helix with ten nucleotides in each full turn, and that each full turn was 3.4 nanometers (nm) in length. The molecule's diameter of 2 nm suggested that the sugar–phosphate backbone of each DNA strand must be on the outside of the helix.

CHEMICAL EVIDENCE FROM BASE COMPOSITION Biochemists knew that DNA was a polymer of nucleotides. Each nucleotide

consists of a molecule of the sugar deoxyribose, a phosphate group, and a nitrogen-containing base (see Figure 4.1). The only differences among the four nucleotides of DNA are their nitrogenous bases: the purines **adenine** (**A**) and **guanine** (**G**), and the pyrimidines **cytosine** (**C**) and **thymine** (**T**).

In the early 1950s, biochemist Erwin Chargaff and his colleagues at Columbia University reported that DNA from many different species—and from different sources within a single organism—exhibits certain regularities. This led to the following rule: In any DNA sample, the amount of adenine equals the amount of thymine (A = T), and the amount of guanine equals the amount of cytosine (G = C). As a result, the total abundance of purines (A + G) equals the total abundance of pyrimidines (T + C):

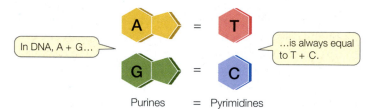

This provided an important clue about the way the bases are arranged in a DNA double helix. Chargaff and colleagues found that this rule held for every organism they examined. However, the relative abundances of A + T versus G + C vary slightly among organisms; in humans the percentages are A= T = 30 percent and G = C = 20 percent.

WATSON AND CRICK'S MODEL If you have taken chemistry courses, you may be familiar with model building, where balls (atoms) and sticks (bonds) are used to put together molecules based on known physical and chemical properties and bond angles. A physicist, Francis Crick, and a geneticist, James D. Watson (**Figure 13.7A**), who were then at the Cavendish Laboratory of Cambridge University, used model building to solve the structure of DNA. They used the physical and chemical evidence we just described:

- To be consistent with Franklin's X-ray diffraction images, Watson and Crick's model had the nucleotide bases on the interior of the two strands, with a sugar–phosphate

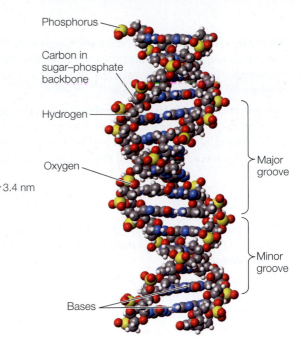

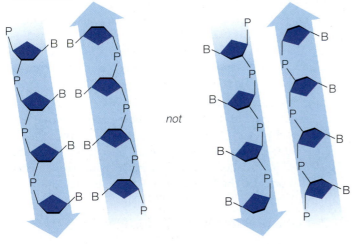

(B) The blue bands represent the two sugar–phosphate backbones, which run in opposite directions:

3.4 nm

13.7 DNA Is a Double Helix (A) James Watson (left) and Francis Crick (right) proposed that the DNA molecule has a double-helical structure. (B) Biochemists can now pinpoint the position of every atom in a DNA molecule. To see that the essential features of the original Watson–Crick model have been verified, follow with your eyes the double-helical chains of sugar–phosphate groups and note the horizontal rungs of the bases.

Go to Media Clip 13.1
Discovery of the Double Helix
Life10e.com/mc13.1

"backbone" on the outside. In addition, the two DNA strands ran in opposite directions, that is, they were **antiparallel**. The two strands would not fit together otherwise:

- To satisfy Chargaff's rule (A = T and G = C), Watson and Crick's model always paired a purine on one strand with a pyrimidine on the opposite strand. These **base pairs** (A-T and G-C) have the same width down the double helix, a uniformity that was also confirmed by X-ray diffraction:

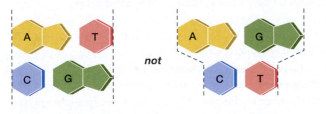

Crick and Watson built their tin model of DNA in late February 1953. This structure explained all the known chemical properties of DNA, and it opened the door to understanding its biological functions. There have been minor amendments to that first published structure, but its principal features remain unchanged.

Four key features define DNA structure

Four features summarize the molecular architecture of the DNA molecule (**Figure 13.7B**):

- DNA is a *double-stranded helix*, with a sugar–phosphate backbone on the outside and base pairs lined up on the inside.
- DNA is usually a *right-handed helix*. If you curl the fingers of your right hand and point your thumb upward, the curve of the helix follows the direction of your fingers, and it winds upward in the direction of your thumb.
- DNA is *antiparallel* (the two strands run in opposite directions).
- DNA has *major and minor grooves* in which the outer edges of the nitrogenous bases are exposed.

THE HELIX The sugar–phosphate backbones of the polynucleotide chains form a coil around the outside of the helix, and the nitrogenous bases point toward the center. The chains are held together by two chemical forces:

1. *Hydrogen bonding* between specifically paired bases. Consistent with Chargaff's rule, adenine (A) pairs with thymine (T) by forming two hydrogen bonds, and guanine (G) pairs with cytosine (C) by forming three hydrogen bonds:

Every base pair consists of one purine (A or G) and one pyrimidine (T or C). This pattern is known as **complementary base pairing**.

2. *Van der Waals* forces between adjacent bases on the same strand. When the base rings come near one another, they tend to stack like poker chips because of these weak attractions.

ANTIPARALLEL STRANDS The backbone of each DNA strand contains repeating units of the five-carbon monosaccharide deoxyribose:

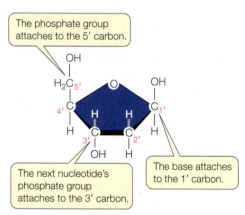

The phosphate group attaches to the 5' carbon.

The base attaches to the 1' carbon.

The next nucleotide's phosphate group attaches to the 3' carbon.

The number followed by a prime (') designates the position of a carbon atom in this sugar molecule. In the sugar–phosphate backbone of DNA, the phosphate groups are connected to the 3' carbon of one deoxyribose molecule and the 5' carbon of the next, linking successive sugars together.

Thus the two ends of a polynucleotide chain differ. At one end of a chain is a free (not connected to another nucleotide) 5' phosphate group (—OPO_3^-); this is called the **5' end**. At the other end is a free 3' hydroxyl group (—OH); this is called the **3' end**. In a DNA double helix, the 5' end of one strand is paired with the 3' end of the other strand, and vice versa. In other words, if you drew an arrow for each strand running from 5' to 3', the arrows would point in opposite directions (see also Figure 4.4A).

BASE EXPOSURE IN THE GROOVES Look back at Figure 13.7B and note the major and minor grooves in the helix. These grooves exist because the backbones of the two strands are closer together on one side of the double helix (forming the

minor groove) than on the other side (forming the major groove). **Figure 13.8** shows the four possible configurations of the flat, hydrogen-bonded base pairs in the major and minor grooves. The exposed outer edges of the base pairs are accessible for additional hydrogen bonding. Note that the arrangements of unpaired atoms and groups differ in A-T and G-C pairs. Thus, the *surfaces of the A-T and C-G base pairs are chemically distinct*, allowing other molecules such as proteins to recognize specific base pair sequences and bind to them. *The binding of proteins to specific base pair sequences is the key to protein–DNA interactions*, which are necessary for the replication and expression of the genetic information in DNA.

The double-helical structure of DNA is essential to its function

The genetic material performs four important functions, and the DNA structure proposed by Watson and Crick was elegantly suited to three of them.

- *The genetic material stores an organism's genetic information.* With its millions of nucleotides, the base sequence of a DNA molecule can encode and store an enormous amount of information. Variations in DNA sequences can account for species and individual differences. DNA fits this role nicely.

- *The genetic material is susceptible to mutations* (permanent changes) *in the information it encodes.* For DNA, mutations might be simple changes in the linear sequence of base pairs.

- *The genetic material is precisely replicated in the cell division cycle.* Replication could be accomplished by complementary base pairing, A with T and G with C. In the original publication of their findings in 1953, Watson and Crick coyly pointed out, "It has not escaped our notice that the specific pairing we have postulated immediately suggests a possible copying mechanism for the genetic material."

- *The genetic material (the coded information in DNA) is expressed as the phenotype.* This function is not obvious in the structure of DNA. However, as we will see in the next chapter, the nucleotide sequence of DNA is copied into RNA, which uses the coded information to specify a linear sequence of amino acids—a protein. The folded forms of proteins determine many of the phenotypes of an organism.

RECAP 13.2

DNA is a double helix made up of two antiparallel polynucleotide chains. The two chains are joined by hydrogen bonds between the nucleotide bases, which pair specifically: A with T, and G with C. Chemical groups on the bases that are exposed in the grooves of the helix are available for hydrogen bonding with other molecules, such as proteins. These molecules can recognize specific sequences of nucleotide bases.

- Describe the evidence that Watson and Crick used to come up with the double helix model for DNA. **See pp. 264–265**

- How does the double-helical structure of DNA relate to its function? **See p. 266 and Figures 13.7, 13.8**

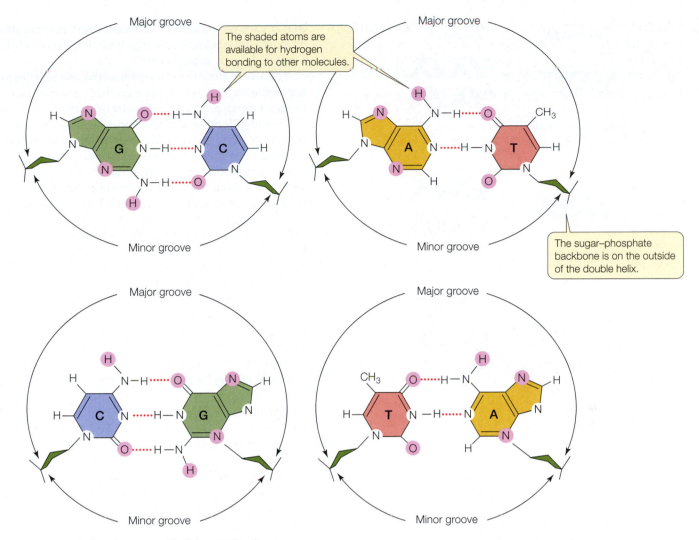

13.8 Base pairs in DNA Can Interact with Other Molecules
These diagrams show the four possible configurations of base pairs within the double helix. Atoms shaded in purple are available for hydrogen bonding with other molecules, such as proteins.

Once the structure of DNA was understood, it was possible to investigate how DNA replicates itself. Next we will examine the experiments that taught us how this elegant process works.

13.3 How Is DNA Replicated?

The mechanism of DNA replication that suggested itself to Watson and Crick was soon confirmed. First, researchers showed that DNA could be replicated in a test tube containing simple substrates and an enzyme. A subsequent study showed that each of the two strands of the double helix can serve as a template for a new strand of DNA.

 Go to Animated Tutorial 13.2
Replication and DNA Polymerization
Life10e.com/at13.2

Three modes of DNA replication appeared possible

New DNA molecules with the same base sequence as an original molecule can be synthesized in a test tube containing the following substances:

- The deoxyribonucleoside triphosphates dATP, dCTP, dGTP, and dTTP. These are the monomers from which the DNA polymers are formed.
- DNA molecules of a particular sequence that serve as **templates** to direct the sequence of nucleotides in the new molecules.
- A **DNA polymerase** enzyme to catalyze the polymerization reaction.
- Salts and a pH buffer, to create an appropriate chemical environment for the DNA polymerase.

The fact that DNA could be synthesized in a test tube confirmed that a DNA molecule contains the information needed

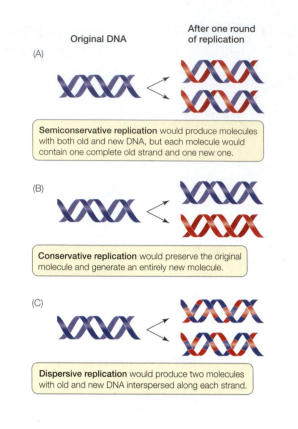

13.9 Three Models for DNA Replication In each model, the original DNA is shown in blue and the newly synthesized DNA is in red.

for its own replication. The next challenge was to determine which of three possible replication patterns occurs during DNA replication:

- *Semiconservative replication*, in which each parent strand serves as a template for a new strand, and the two new DNA molecules each have one old and one new strand (**Figure 13.9A**)

- *Conservative replication*, in which the original double helix serves as a template for, but does not contribute to, a new double helix (**Figure 13.9B**)

- *Dispersive replication*, in which fragments of the original DNA molecule serve as templates for assembling two new molecules, each containing old and new parts, perhaps at random (**Figure 13.9C**)

Watson and Crick's original paper suggested that DNA replication was semiconservative, but the test tube demonstration described above did not provide a basis for choosing among these three models.

An elegant experiment demonstrated that DNA replication is semiconservative

In 1958 Matthew Meselson and Franklin Stahl at the California Institute of Technology convinced the scientific community

that DNA is reproduced by **semiconservative replication**. They used density labeling to distinguish between parent strands of DNA and newly copied ones.

The key to their experiment was the use of a "heavy" isotope of nitrogen. Heavy nitrogen (^{15}N) is a rare, nonradioactive isotope that makes molecules containing it denser than chemically identical molecules containing the common isotope ^{14}N. Two cultures of the bacterium *E. coli* were grown for many generations, one in a medium containing ^{15}N and the other in a medium containing ^{14}N. When DNA extracts from the two cultures were combined and centrifuged in a solution of cesium chloride, which forms a density gradient under centrifugation, two separate bands of DNA formed in the centrifugation tube. The DNA from the ^{15}N culture was heavier than the DNA from the ^{14}N culture, so it formed a band at a different position in the density gradient.

Next, Meselson and Stahl grew another *E. coli* culture in ^{15}N medium, then transferred the bacteria to normal ^{14}N medium and allowed them to continue growing (**Figure 13.10**). The cells replicated their DNA and divided every 20 minutes. Meselson and Stahl collected some of the bacteria at time intervals and extracted DNA from the samples. You can follow their results for the first two generations in Figure 13.10. The results can be explained only by the semiconservative model of DNA replication. The crucial observations demonstrating this model were that all the DNA at the end of the first generation was of intermediate density, while at the end of the second generation there were two discrete bands: one of intermediate and one of light DNA. If the conservative model had been true, there would have been no intermediate density DNA. If the dispersive model were correct, then the DNA would all have been intermediate for the first few generations, with the single intermediate band becoming progressively lighter.

There are two steps in DNA replication

Semiconservative DNA replication in the cell involves several different enzymes and other proteins. It takes place in two general steps:

- The DNA double helix is unwound to separate the two template strands and make them available for new base pairing (this is the step that cisplatin prevents; see the opening story).

- As new nucleotides form complementary base pairs with template DNA, they are covalently linked together by phosphodiester bonds, forming a polymer whose base sequence is complementary to the bases in the template strand.

The nucleotides that make up DNA are deoxyribonucleoside monophosphates because they each contain deoxyribose and one phosphate group (see Figure 4.1). The free monomers that are brought together to form DNA have three phosphate groups. They are the deoxyribonucleoside triphosphates dATP, dTTP, dCTP, and dGTP, collectively referred to as dNTPs. The

INVESTIGATINGLIFE

13.10 The Meselson–Stahl Experiment A centrifuge was used to separate DNA molecules labeled with isotopes of different densities. This experiment revealed a pattern that supports the semiconservative model of DNA replication.[a]

HYPOTHESIS DNA replicates semiconservatively.

Method

Grow bacteria in ^{15}N (heavy) medium.

Transfer some bacteria to ^{14}N (light) medium; bacterial growth continues.

Samples are taken after 0 minutes, 20 minutes (after one round of replication), and 40 minutes (two rounds of replication).

Sample at 0 minutes

Sample after 20 minutes

Sample after 40 minutes

Results

^{14}N/^{14}N (light) DNA

^{14}N/^{15}N (intermediate) DNA

^{15}N/^{15}N (heavy) DNA

Parent (all heavy)

First generation (all intermediate)

Second generation (half intermediate, half light)

Interpretation

Before the bacteria reproduce for the first time in the light medium (at 0 minutes), all DNA (parental) is heavy.

After two generations, half the DNA was intermediate and half was light; there was no heavy DNA.

Parent strand ^{15}N   New strand ^{14}N

CONCLUSION This pattern could only have been observed if each DNA molecule contains a template strand from the parental DNA; thus DNA replication is semiconservative.

Go to **BioPortal** for discussion and relevant links for all INVESTIGATINGLIFE figures.

[a]Meselson, M. and F. Stahl. 1958. *Proceedings of the National Academy of Sciences USA* 44: 671–682.

Go to Animated Tutorial 13.3
The Meselson–Stahl Experiment
Life10e.com/at13.3

three phosphate groups are attached to the 5′ carbon on the deoxyribose sugar (see p. 266).

A key observation regarding DNA replication is that *nucleotides are added to the growing new strand at the 3′ end*—the end at which the DNA strand has a free hydroxyl (—OH) group on the 3′ carbon of its terminal deoxyribose. In the formation of the phosphodiester linkage (a condensation reaction), two of the phosphate groups on an incoming dNTP are removed [as pyrophosphate (PP$_i$)], and the remaining phosphate is bonded to the 3′ carbon on the terminal deoxyribose (**Figure 13.11**; see also Figure 4.2). Just as energy is released when ATP is hydrolyzed to AMP (with subsequent hydrolysis of PP$_i$ to two phosphates), energy is released by the hydrolysis of the dNTP, and this energy is used to drive the condensation reaction.

DNA polymerases add nucleotides to the growing chain

DNA replication begins with the binding of a large protein complex (the pre-replication complex) to a specific site on the DNA molecule. This complex contains several different proteins, including the enzyme DNA polymerase, which catalyzes the addition of nucleotides as the new DNA chain grows. All chromosomes have at least one region called the **origin of replication (*ori*)**, to which the pre-replication complex binds. Binding occurs when proteins in the complex recognize specific DNA sequences within the *ori*.

ORIGINS OF REPLICATION The single circular chromosome of the bacterium *E. coli* has 4×10^6 base pairs (bp) of DNA.

The Meselson–Stahl Experiment

Original Paper

Meselson, M. and F. Stahl. 1958. The replication of DNA in *Escherichia coli. Proceedings of the National Academy of Sciences USA* 44: 671–682.

Analyze the Data

The Meselson–Stahl experiment has been called "the most beautiful experiment in biology" because of its essential simplicity. Meselson and Stahl used density gradients to examine how DNA molecules replicate (see Figure 13.10). The key experimental method was the separation of DNA that contained ^{14}N ("light" DNA) from DNA that contained ^{15}N ("heavy" DNA), using an ultracentrifuge to create a density gradient of cesium chloride. DNA molecules suspended in the cesium chloride formed bands within the gradient, at their own density (**FIGURE A**).

FIGURE B shows the results of the experiment. Each sample contained the same number of bacteria, so the total amount of DNA in each panel was the same. The photos show the DNA bands in the tubes after centrifugation, and the plots show quantitative analysis of the bands, where height indicates amount of DNA.

QUESTION 1

Use the heights of the peaks to estimate the percent of total DNA that was heavy, intermediate, and light at each generational stage. Create a table summarizing these calculations and discuss whether they support the authors' conclusions.

QUESTION 2

What would the data look like if the bacteria had been allowed to divide for three more generations?

QUESTION 3

If Meselson and Stahl had done their experiment starting with light DNA and then added ^{15}N for succeeding generations, what would the bands look like? Draw them alongside the actual data, above.

QUESTION 4

What would the data look like if conservative replication were the correct model? What would the data look like if dispersive replication were correct? Draw these alongside the actual data above.

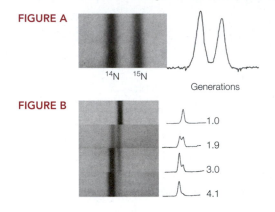

FIGURE A

^{14}N ^{15}N Generations

FIGURE B

1.0
1.9
3.0
4.1

Go to BioPortal for all WORKING WITH **DATA** exercises

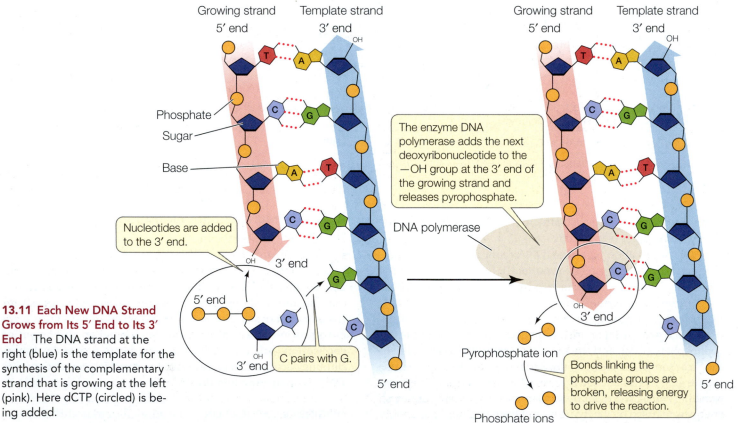

Growing strand
5′ end

Template strand
3′ end

Phosphate
Sugar
Base

Nucleotides are added
to the 3′ end.

5′ end

3′ end

OH
3′ end

C pairs with G.

13.11 Each New DNA Strand Grows from Its 5′ End to Its 3′ End   The DNA strand at the right (blue) is the template for the synthesis of the complementary strand that is growing at the left (pink). Here dCTP (circled) is being added.

Growing strand
5′ end

Template strand
3′ end

The enzyme DNA polymerase adds the next deoxyribonucleotide to the —OH group at the 3′ end of the growing strand and releases pyrophosphate.

DNA polymerase

OH
3′ end

Pyrophosphate ion

Phosphate ions

Bonds linking the phosphate groups are broken, releasing energy to drive the reaction.

5′ end

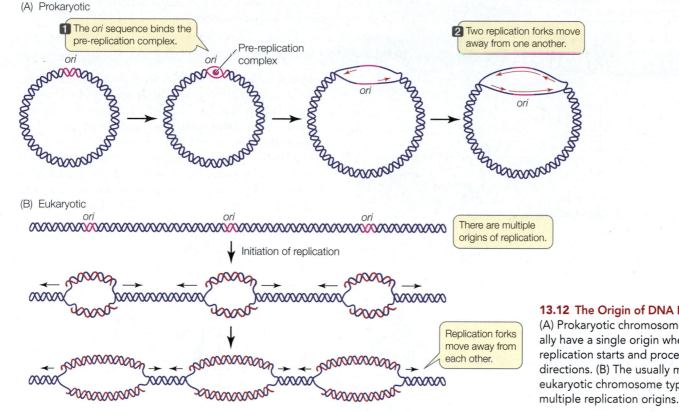

(A) Prokaryotic

1 The *ori* sequence binds the pre-replication complex.

ori

Pre-replication complex

ori

ori

2 Two replication forks move away from one another.

ori

(B) Eukaryotic

ori   *ori*   *ori*

There are multiple origins of replication.

Initiation of replication

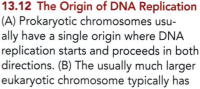

Replication forks move away from each other.

13.12 The Origin of DNA Replication
(A) Prokaryotic chromosomes usually have a single origin where DNA replication starts and proceeds in both directions. (B) The usually much larger eukaryotic chromosome typically has multiple replication origins.

The 245 bp *ori* sequence is at a particular location on the chromosome. Once the pre-replication complex binds to it, the DNA is unwound and replication proceeds in both directions around the circle, forming two **replication forks** (Figure 13.12A). The replication rate in *E. coli* is approximately 1,000 bp per second, so it takes about 40 minutes to fully replicate the chromosome (with two replication forks). Rapidly dividing *E. coli* cells divide every 20 minutes. In these cells, new rounds of replication begin at the *ori* of each new chromosome before the first chromosome has fully replicated. In this way the cells can divide more frequently than the time needed to finish replicating the original chromosome.

Eukaryotic chromosomes are typically much longer than those of prokaryotes—up to a billion bp—and are linear, not circular. If replication occurred from a single *ori* with two forks growing away from each other, it would take weeks to fully replicate a chromosome. So eukaryotic chromosomes have multiple origins of replication, scattered at intervals of 10,000 to 40,000 bp (Figure 13.12B).

DNA REPLICATION BEGINS WITH A PRIMER A DNA polymerase elongates a polynucleotide strand by covalently linking new nucleotides to a preexisting strand. However, it cannot start this process without a short "starter" strand, called a **primer**. In most organisms this primer is a short single strand of RNA (Figure 13.13), but in some organisms it is DNA.

The primer is complementary to the DNA template and is synthesized one nucleotide at a time by an enzyme called a **primase**. The DNA polymerase then adds nucleotides to the 3' end of the primer and continues until the replication of that section of DNA has been completed. Then the RNA primer is degraded, DNA is added in its place, and the resulting DNA fragments are connected by the action of other enzymes. When DNA replication is complete, each new strand consists only of DNA.

DNA POLYMERASES ARE LARGE DNA polymerases are much larger than their substrates (the dNTPs) and the template DNA, which is very thin. Molecular models of the enzyme–substrate–template complex from bacteria show that the enzyme is shaped like an open right hand with a palm, a thumb, and fingers (Figure 13.14). Within the "palm" is the active site of the enzyme, which brings together each dNTP substrate and the template. The "finger" regions have precise shapes that can recognize the different shapes of the four nucleotide bases. They bind to the bases by hydrogen bonding and rotate inward. Most cells contain more than one kind of DNA polymerase, but only one of them is responsible for chromosomal DNA replication. The others are involved in primer removal and DNA repair. Fifteen DNA polymerases have been identified in humans, whereas the bacterium *E. coli* has five DNA polymerases.

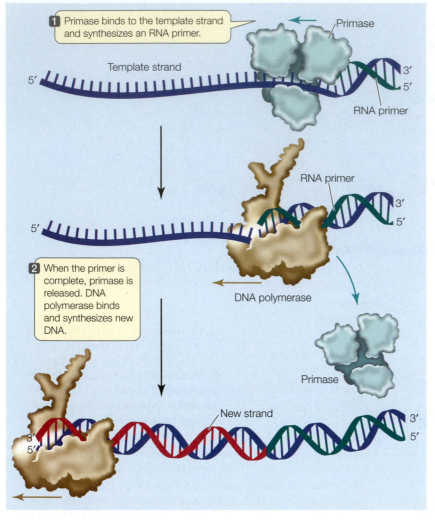

1 Primase binds to the template strand and synthesizes an RNA primer.

Primase

Template strand

RNA primer

2 When the primer is complete, primase is released. DNA polymerase binds and synthesizes new DNA.

RNA primer

DNA polymerase

Primase

New strand

13.13 DNA Forms with a Primer DNA polymerases require a primer—a "starter" strand of DNA or RNA to which they can add new nucleotides.

(denaturation) of the DNA strands. As we discussed in Section 13.2, the two strands are held together by hydrogen bonds and Van der Waals forces. An enzyme called **DNA helicase** uses energy from ATP hydrolysis to unwind and separate the strands, and **single-strand binding proteins** bind to the unwound strands to keep them from reassociating into a double helix. This process makes each of the two template strands available for complementary base pairing.

The two DNA strands grow differently at the replication fork

The DNA at the replication fork—the site where DNA unwinds to expose the bases so that they can act as templates—opens up like a zipper in one direction. Study **Figure 13.16** and try to imagine what is happening over a short period of time. Remember that the two DNA strands are antiparallel; that is, the 3′ end of one strand is paired with the 5′ end of the other.

- One newly replicating strand (the **leading strand**) is oriented so that it can grow continuously at its 3′ end as the fork opens up.

- The other new strand (the **lagging strand**) is oriented so that as the fork opens up, its exposed 3′ end gets farther and farther away from the fork, and an unreplicated gap is formed. This gap would get bigger and bigger if there were not a special mechanism to overcome this problem.

Synthesis of the lagging strand requires the synthesis of relatively small, discontinuous stretches of DNA (100–200 nucleotides in eukaryotes; 1,000–2,000 nucleotides in prokaryotes). These discontinuous stretches are synthesized just as the

Many other proteins assist with DNA polymerization

Various other proteins play roles in other replication tasks; some of these are shown in **Figure 13.15**. The first event at the origin of replication is the localized unwinding and separation

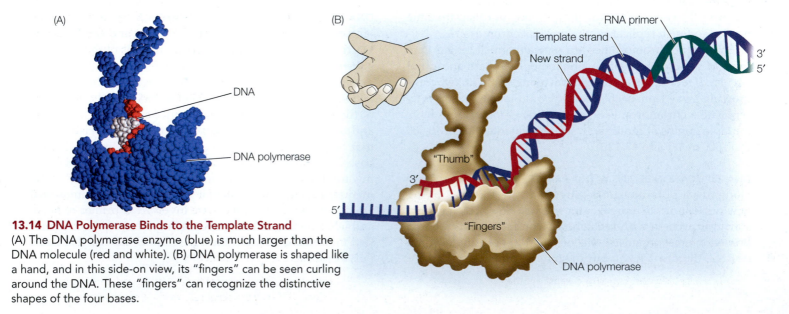

(A)

DNA

DNA polymerase

(B)

RNA primer

Template strand

New strand

"Thumb"

"Fingers"

DNA polymerase

13.14 DNA Polymerase Binds to the Template Strand
(A) The DNA polymerase enzyme (blue) is much larger than the DNA molecule (red and white). (B) DNA polymerase is shaped like a hand, and in this side-on view, its "fingers" can be seen curling around the DNA. These "fingers" can recognize the distinctive shapes of the four bases.

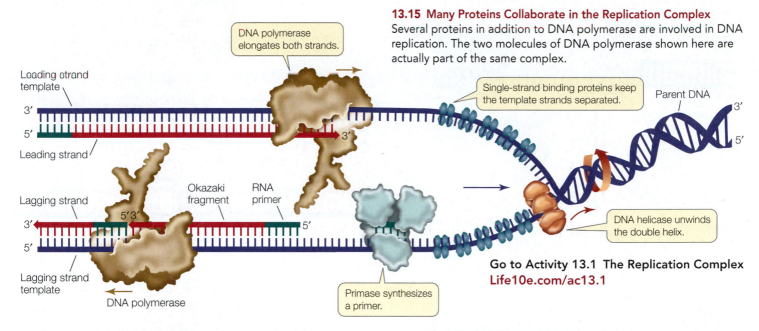

13.15 Many Proteins Collaborate in the Replication Complex Several proteins in addition to DNA polymerase are involved in DNA replication. The two molecules of DNA polymerase shown here are actually part of the same complex.

- DNA polymerase elongates both strands.
- Single-strand binding proteins keep the template strands separated.
- Parent DNA
- Leading strand template
- Leading strand
- Lagging strand
- Okazaki fragment
- RNA primer
- DNA helicase unwinds the double helix.
- Lagging strand template
- DNA polymerase
- Primase synthesizes a primer.

Go to Activity 13.1 **The Replication Complex**
Life10e.com/ac13.1

leading strand is, by the addition of new nucleotides one at a time to the 3′ end of the new strand, but the synthesis of this new strand moves in the direction opposite to that in which the replication fork is moving. These stretches of new DNA are called **Okazaki fragments** (after their discoverer, the Japanese biochemist Reiji Okazaki). While the leading strand grows continuously "forward," the lagging strand grows in shorter, "backward" stretches with gaps between them.

A single primer is needed for synthesis of the leading strand, but each Okazaki fragment requires its own primer to be synthesized by the primase. In bacteria, DNA polymerase III then synthesizes an Okazaki fragment by adding nucleotides to one primer until it reaches the primer of the previous fragment. At this point, DNA polymerase I removes the old primer and replaces it with DNA. Left behind is a tiny nick—the final phosphodiester linkage between the adjacent Okazaki fragments is missing. The enzyme **DNA ligase** catalyzes the formation of that bond, linking the fragments and making the lagging strand whole (**Figure 13.17**).

Working together, DNA helicase, the two DNA polymerases, primase, DNA ligase, and the other proteins of the pre-replication complex do the job of DNA synthesis with a speed and accuracy that are almost unimaginable. In *E. coli*, the replication complex makes new DNA at a rate in excess of 1,000 base pairs per second, committing errors in fewer than one base in a million.

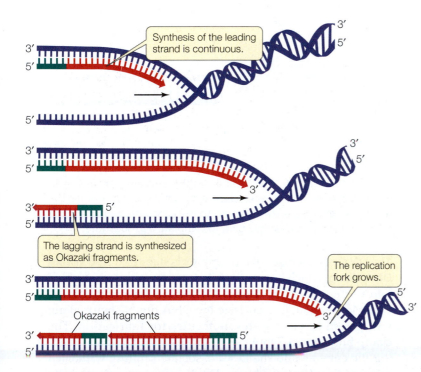

- Synthesis of the leading strand is continuous.
- The lagging strand is synthesized as Okazaki fragments.
- The replication fork grows.
- Okazaki fragments

13.16 The Two New Strands Form in Different Ways As the parent DNA unwinds, both new strands are synthesized in the 5′-to-3′ direction, although their template strands are antiparallel. The leading strand grows continuously forward, but the lagging strand grows in short discontinuous stretches called Okazaki fragments. Eukaryotic Okazaki fragments are hundreds of nucleotides long, with gaps between them.

Go to Animated Tutorial 13.4
Leading and Lagging Strand Synthesis
Life10e.com/at13.4

A SLIDING CLAMP INCREASES THE RATE OF DNA REPLICATION How do DNA polymerases work so fast? We saw in Section 8.3 that an enzyme catalyzes a chemical reaction:

Substrate binds to enzyme → one product is formed →
enzyme is released → cycle repeats

DNA replication would not proceed as rapidly as it does if it went through such a cycle for each nucleotide. Instead, DNA polymerases are **processive**—that is, they catalyze the formation of many phosphodiester linkages each time they bind to a DNA molecule:

Substrates bind to enzyme → many products are
formed → enzyme is released → cycle repeats

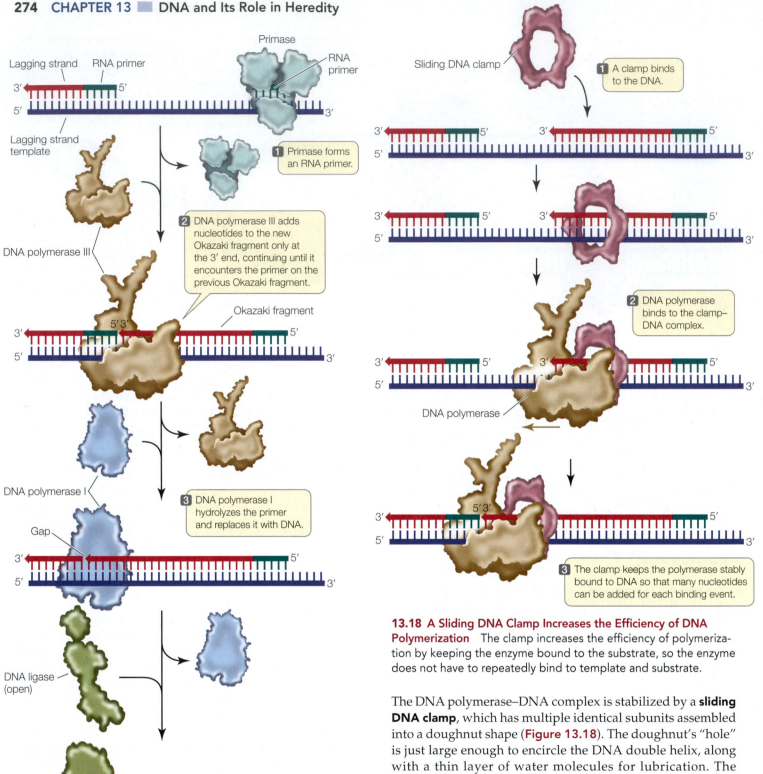

13.17 The Lagging Strand Story In bacteria, DNA polymerase I and DNA ligase cooperate with DNA polymerase III to complete the complex task of synthesizing the lagging strand.

13.18 A Sliding DNA Clamp Increases the Efficiency of DNA Polymerization The clamp increases the efficiency of polymerization by keeping the enzyme bound to the substrate, so the enzyme does not have to repeatedly bind to template and substrate.

The DNA polymerase–DNA complex is stabilized by a **sliding DNA clamp**, which has multiple identical subunits assembled into a doughnut shape (**Figure 13.18**). The doughnut's "hole" is just large enough to encircle the DNA double helix, along with a thin layer of water molecules for lubrication. The clamp binds to the DNA polymerase–DNA complex, keeping the enzyme and the DNA associated tightly with each other. If the clamp is absent, DNA polymerase dissociates from DNA after forming 20 to 100 phosphodiester linkages. With the clamp, it can polymerize up to 50,000 nucleotides before it detaches.

DNA IS THREADED THROUGH A REPLICATION COMPLEX Until recently, DNA replication was always depicted to look like a locomotive (the replication complex) moving along a railroad track (the DNA). While this does occur in some organisms,

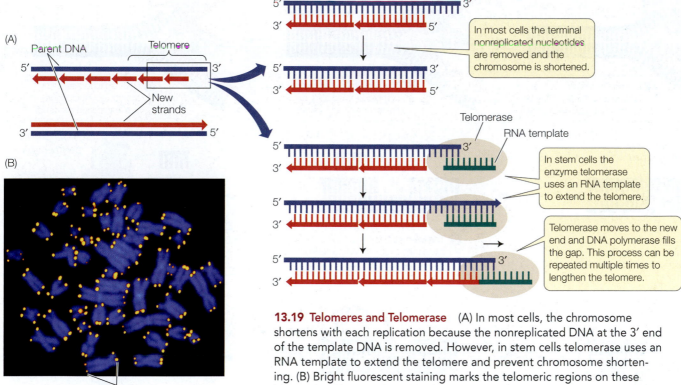

13.19 Telomeres and Telomerase (A) In most cells, the chromosome shortens with each replication because the nonreplicated DNA at the 3' end of the template DNA is removed. However, in stem cells telomerase uses an RNA template to extend the telomere and prevent chromosome shortening. (B) Bright fluorescent staining marks the telomeric regions on these blue-stained human chromosomes.

most commonly in eukaryotes the replication complexes seem to be stationary, attached at specific positions on the nuclear matrix: a dynamic network of protein fibers within the nucleus. It is the DNA that moves, essentially sliding into the replication complex as one double-stranded molecule and emerging as two double-stranded molecules.

Telomeres are not fully replicated and are prone to repair

As we will discuss in Section 13.4, DNA may be damaged by radiation or chemicals. When this happens DNA repair mechanisms are activated, and breaks in DNA are rejoined via a combination of DNA synthesis and DNA ligase activity. So the ends of chromosomes are a potential problem: the DNA repair system might recognize the ends as breaks, and join two chromosomes together. This would create havoc with genomic integrity.

In many eukaryotes, there are repetitive sequences at the ends of chromosomes called **telomeres**. In humans, the telomere sequence is TTAGGG, and it is repeated about 2,500 times at each chromosome end. These repeats bind special proteins that prevent the DNA repair system from recognizing the chromosome ends as breaks. In addition, the repeats may form loops that have a similar protective role.

But there is another problem with chromosome ends. As we have discussed, replication of the lagging strand occurs by the addition of Okazaki fragments to RNA primers. When the terminal RNA primer is removed, no DNA can

be synthesized to replace it because there is no 3' end to extend. In most cells, the short piece of single stranded DNA at each end of the chromosome is removed. Thus the chromosome becomes slightly shorter with each cell division (**Figure 13.19**).

Each human chromosome can lose 50 to 200 base pairs of telomeric DNA after each round of DNA replication and cell division. After many cell divisions, the genes near the ends of the chromosomes can be lost, and the cell dies. This phenomenon explains, in part, why many cell lineages do not last the entire lifetime of the organism: their telomeres are lost. Continuously dividing cells, such as bone marrow stem cells and gamete-producing cells, have a special mechanism for maintaining their telomeric DNA. An enzyme called **telomerase** catalyzes the addition of any lost telomeric sequences in these cells (see Figure 13.19). Telomerase contains an RNA sequence that acts as a template for the telomeric DNA repeat sequence.

Telomerase is expressed in more than 90 percent of human cancers, and may be an important factor in the ability of cancer cells to divide continuously. Since most normal cells do not have this ability, telomerase is an attractive target for drugs designed to attack tumors specifically.

There is also interest in telomerase and aging. When cultured human cells are transformed with a telomerase gene that is expressed at high levels, their telomeres do not shorten. Instead of living 20 to 30 cell generations and then dying, the cells become immortal. It remains to be seen how this finding relates to the aging of a whole organism.

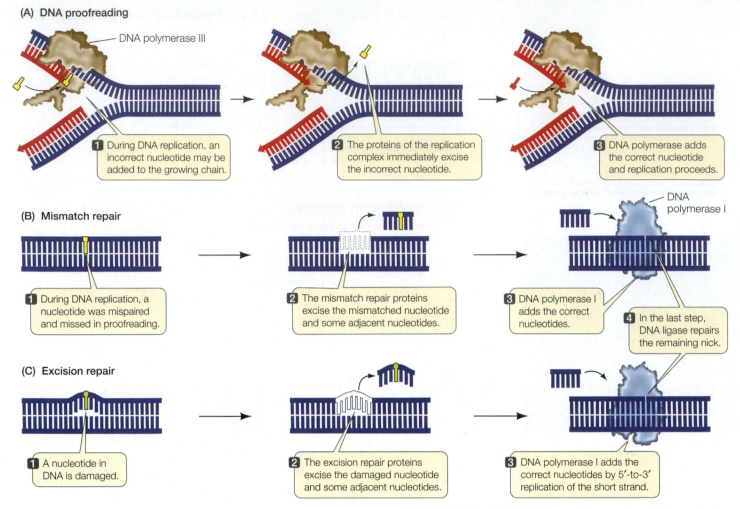

(A) DNA proofreading

DNA polymerase III

1 During DNA replication, an incorrect nucleotide may be added to the growing chain.

2 The proteins of the replication complex immediately excise the incorrect nucleotide.

3 DNA polymerase adds the correct nucleotide and replication proceeds.

(B) Mismatch repair

DNA polymerase I

1 During DNA replication, a nucleotide was mispaired and missed in proofreading.

2 The mismatch repair proteins excise the mismatched nucleotide and some adjacent nucleotides.

3 DNA polymerase I adds the correct nucleotides.

4 In the last step, DNA ligase repairs the remaining nick.

(C) Excision repair

1 A nucleotide in DNA is damaged.

2 The excision repair proteins excise the damaged nucleotide and some adjacent nucleotides.

3 DNA polymerase I adds the correct nucleotides by 5′-to-3′ replication of the short strand.

13.20 DNA Repair Mechanisms The proteins of the replication complex function in DNA repair mechanisms, reducing the rate of errors in the replicated DNA. Another mechanism (excision repair) repairs damage to existing DNA molecules.

RECAP 13.3

Meselson and Stahl showed that DNA replication is semiconservative: each parent DNA strand serves as a template for a new strand. A complex of proteins, most notably DNA polymerase, is involved in replication. New DNA is polymerized in one direction only, and since the two strands are antiparallel, one strand is made continuously and the other is synthesized in short Okazaki fragments that are eventually joined.

- How did the Meselson–Stahl experiment differentiate between the three models for DNA replication? **See p. 268 and Figures 13.9, 13.10**
- Name five enzymes needed for DNA replication. What are their roles? **See pp. 271–274 and Figures 13.13–13.17**
- Why is the leading strand of DNA replicated continuously while the lagging strand must be replicated in fragments? **See pp. 272–273 and Figure 13.16**

The complex process of DNA replication is amazingly accurate, but it is not perfect. What happens when things go wrong?

13.4 How Are Errors in DNA Repaired?

DNA must be accurately replicated and faithfully maintained. This is essential for the proper functioning of every cell, whether a prokaryote or a cell in a complex, multicellular organism. Yet the replication of DNA is not perfectly accurate, and DNA is subject to damage by chemicals and other environmental agents. In the face of these threats, how has life gone on for so long?

DNA repair mechanisms help preserve life. DNA polymerases initially make significant numbers of mistakes in assembling polynucleotide strands. Without DNA repair, the observed error rate of one for every 10^5 bases replicated would result in about 60,000 mutations every time a human cell divided. Fortunately, our cells can repair damaged nucleotides and correct DNA replication errors. Cells have at least three DNA repair mechanisms at their disposal:

- A **proofreading** mechanism corrects errors in replication as DNA polymerase makes them.
- A **mismatch repair** mechanism scans DNA immediately after it has been replicated and corrects any base-pairing mismatches.
- An **excision repair** mechanism removes abnormal bases that have formed because of chemical damage and replaces them with functional bases.

Most DNA polymerases perform a proofreading function each time they introduce a new nucleotide into a growing DNA strand (Figure 13.20A). When a DNA polymerase recognizes a mispairing of bases, it removes the improperly introduced nucleotide and tries again. (Other proteins in the replication complex also play roles in proofreading.) The error rate for this process is only about 1 in 10,000 repaired base pairs, and it lowers the overall error rate for replication to about one error in every 10^{10} bases replicated.

After the DNA has been replicated, a second set of proteins surveys the newly replicated molecule and looks for mismatched base pairs that were missed in proofreading (Figure 13.20B). For example, this mismatch repair mechanism might detect an A-C base pair instead of an A-T pair. The repair system can make the correct choice out of two options: remove the C and replace it with T, or remove the A and replace it with G. When mismatch repair fails, DNA sequences are altered. One form of colon cancer arises in part from a failure of mismatch repair.

DNA molecules can also be damaged during the life of a cell (for example, when it is in G1). High-energy radiation, chemicals from the environment, and random spontaneous chemical reactions can all damage DNA. For example, when adjacent thymines on the same DNA strand absorb ultraviolet light (at about 260 nm), they form a covalent bond between the bases, making a thymine dimer. These dimers interfere with base pairing during replication, leading to mutations. This is the primary cause of skin cancer in humans. Excision repair mechanisms deal with these kinds of damage (Figure 13.20C). Individuals who suffer from a condition known as xeroderma pigmentosum lack an excision repair mechanism that normally corrects the damage caused by ultraviolet radiation. They can develop skin cancers after even a brief exposure to sunlight.

RECAP 13.4

DNA replication is not perfect. In addition, DNA may be altered or damaged by environmental factors. Repair mechanisms detect and repair mismatched or damaged DNA.

- Explain the roles of DNA proofreading, mismatch repair, and excision repair. See Figure 13.20

Understanding how DNA is replicated and repaired has allowed scientists to develop techniques for studying genes. We'll look at just one of those techniques next.

13.5 How Does the Polymerase Chain Reaction Amplify DNA?

The principles underlying DNA replication in cells have been used to develop an important laboratory technique that has been vital in analyzing genes and genomes. This technique allows researchers to make multiple copies of short DNA sequences.

Go to Animated Tutorial 13.5
Polymerase Chain Reaction
Life10e.com/at13.5

The polymerase chain reaction makes multiple copies of DNA sequences

In order to study DNA and perform genetic manipulations, it is often necessary to make multiple copies of a DNA sequence. This is necessary because the amount of DNA isolated from a biological sample is often too small to work with. The **polymerase chain reaction (PCR)** technique essentially automates this replication process by copying a short region of DNA many times in a test tube. This process is referred to as DNA amplification.

The PCR reaction mixture contains:

- a sample of double-stranded DNA from a biological sample, to act as the template;
- two short, artificially synthesized primers that are complementary to the ends of the sequence to be amplified;
- the four dNTPs (dATP, dTTP, dCTP, and dGTP);
- a DNA polymerase that can tolerate high temperatures without becoming degraded; and
- salts and a buffer to maintain a near-neutral pH.

The PCR amplification is a cyclic process in which a sequence of steps is repeated over and over again (Figure 13.21):

- The first step involves heating the reaction mixture to near boiling point, to separate (denature) the two strands of the DNA template.
- The reaction is then cooled to allow the primers to bind (or anneal) to the template strands.
- Next, the reaction is warmed to an optimum temperature for the DNA polymerase to catalyze the production of the complementary new strands.

A single cycle takes a few minutes to produce two copies of the target DNA sequence, leaving the new DNA in the double-stranded state. Repeating the cycle many times leads to an exponential increase in the number of copies of the DNA sequence.

The PCR technique requires that the base sequences at the 3' end of each strand of the target DNA sequence be known, so that complementary primers, usually 15 to 30 bases long, can be made in the laboratory. Because of the uniqueness of DNA sequences, a pair of primers this length will usually bind to only a single region of DNA in an organism's genome. This specificity, despite the incredible diversity of DNA sequences, is a key to the power of PCR.

One initial problem with PCR was its temperature requirements. To denature DNA, it must be heated to more than 90°C—a temperature that destroys most DNA polymerases. When the PCR technique was first being developed, new enzyme had to be added after denaturation in each cycle, which made the technique impractical for widespread use.

This problem was solved by nature: in the hot springs at Yellowstone National Park, as well as in other high-temperature locations, there lives a bacterium called, appropriately, *Thermus aquaticus* ("hot water"). The means by which this organism survives temperatures of up to 95°C was investigated by Thomas Brock and his colleagues at the University of Wisconsin, Madison. They discovered that *T. aquaticus* has an entire metabolic

13.21 The Polymerase Chain Reaction The steps in this cyclic process are repeated many times to produce multiple identical copies of a DNA fragment. This makes enough DNA for chemical analysis and genetic manipulations.

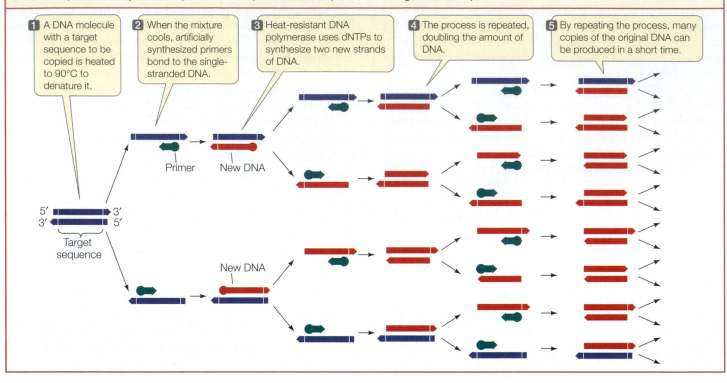

1 A DNA molecule with a target sequence to be copied is heated to 90°C to denature it.

2 When the mixture cools, artificially synthesized primers bond to the single-stranded DNA.

3 Heat-resistant DNA polymerase uses dNTPs to synthesize two new strands of DNA.

4 The process is repeated, doubling the amount of DNA.

5 By repeating the process, many copies of the original DNA can be produced in a short time.

machinery that is heat-resistant, including a DNA polymerase that does not denature at these high temperatures.

Scientists pondering the problem of copying DNA by PCR read Brock's basic research articles and got a clever idea: why not use *T. aquaticus* DNA polymerase in the PCR technique? It could withstand the 90°C denaturation temperature and would not have to be added during each cycle. The idea worked, and it earned biochemist Kary Mullis a Nobel prize. PCR has had an enormous impact on genetic research. Some of its most striking applications will be described in Chapters 15–18. These applications range from amplifying DNA in order to identify an individual person or organism, to detection of diseases.

RECAP 13.5

Knowledge of the mechanisms of DNA replication led to the development of a technique for making multiple copies of DNA sequences.

- What is the role of primers in PCR? **See p. 277 and Figure 13.21**

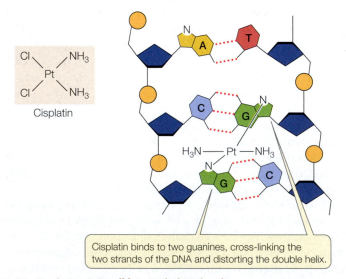

Cisplatin

Cisplatin binds to two guanines, cross-linking the two strands of the DNA and distorting the double helix.

13.22 Cisplatin: A Small but Lethal Molecule

How does cisplatin work?

ANSWER

Cisplatin contains a platinum atom bonded to two amino groups and two chlorine atoms (**Figure 13.22**). In Rosenberg's experiments, this compound was formed when the platinum electrode reacted with salts in the surrounding solution. The bonds between the platinum atom and the chlorine atoms are weak, and the latter can be displaced by electron-rich substances (you may know from chemistry that these are called nucleophiles). In DNA, one of the nitrogen atoms of guanine (see Figure 4.1) displaces one of the chlorines, forming a strong covalent bond. If there is a nearby guanine on the opposite DNA strand, it displaces the other chlorine and the DNA becomes cross-linked. This type of DNA lesion is not repaired by any of the cell's usual DNA repair mechanisms.

13.1 What Is the Evidence that the Gene Is DNA?

- Griffith's experiments in the 1920s demonstrated that some substance in cells can cause heritable changes in other cells. **Review Figure 13.1**

- The location and quantity of DNA in the cell suggested that DNA might be the genetic material. Avery and his colleagues isolated the transforming principle from bacteria and identified it as DNA. **Review Figure 13.2**

- The Hershey–Chase experiments established conclusively that DNA (and not protein) is the genetic material, by tracing the DNA of radioactively labeled viruses, with which they infected bacterial cells. **Review Figure 13.4, ANIMATED TUTORIAL 13.1**

- Genetic transformation of eukaryotic cells is often called **transfection**. Transformation and transfection can be studied with the aid of a **genetic marker** gene that confers a known and observable phenotype. **Review Figure 13.5**

13.2 What Is the Structure of DNA?

- Chargaff's rule states that the amount of **adenine** in DNA is equal to the amount of **thymine**, and that the amount of **guanine** is equal to the amount of **cytosine**; thus the total abundance of purines (A + G) equals the total abundance of pyrimidines (T + C).

- X-ray crystallography showed that the DNA molecule is a double helix. Watson and Crick proposed that the two strands in DNA are **antiparallel**. **Review Figure 13.7**

- **Complementary base pairing** between A and T and between G and C accounts for Chargaff's rule. The bases are held together by hydrogen bonding.

- Reactive groups are exposed in the paired bases, allowing for recognition by other molecules such as proteins. **Review Figure 13.8**

13.3 How Is DNA Replicated?
See ANIMATED TUTORIAL 13.2

- Meselson and Stahl showed that DNA undergoes **semiconservative replication**. Each parent strand acts as a **template** for the synthesis of a new strand; thus the two replicated DNA molecules each contain one parent strand and one newly synthesized strand. **Review Figure 13.10, ANIMATED TUTORIAL 13.3**

- In DNA replication, the enzyme **DNA polymerase** catalyzes the addition of nucleotides to the 3' end of each strand. Which nucleotides are added is determined by complementary base pairing with the template strand. **Review Figure 13.11**

- The pre-replication complex is a huge protein complex that attaches to the chromosome at the **origin of replication** (*ori*).

- Replication proceeds from the origin of replication on both strands in the 5'-to-3' direction, forming a **replication fork**. **Review Figure 13.12**

- **Primase** catalyzes the synthesis of a short RNA **primer** to which nucleotides are added by DNA polymerase. **Review Figure 13.13**

- Many proteins assist in DNA replication. **DNA helicase** separates the strands, and **single-strand binding proteins** keep the strands from reassociating. **Review Figure 13.15, ACTIVITY 13.1**

- The **leading strand** is synthesized continuously and the **lagging strand** in pieces called **Okazaki fragments**. The fragments are joined together by **DNA ligase**. **Review Figures 13.16, 13.17, ANIMATED TUTORIAL 13.4**

- The speed with which DNA polymerization proceeds is attributed to the **processive** nature of DNA polymerases, which can catalyze many polymerizations at a time. A **sliding DNA clamp** helps ensure the stability of this process. **Review Figure 13.18**

- At the ends of eukaryotic chromosomes are regions of repetitive DNA sequence called **telomeres**. Unless the enzyme **telomerase** is present, a short segment at the end of each telomere is lost each time the DNA is replicated. After multiple cell cycles, the telomeres shorten enough to cause chromosome instability and cell death. **Review Figure 13.19**

13.4 How Are Errors in DNA repaired?

- DNA polymerases make about one error in 10^5 bases replicated. DNA is also subject to natural alterations and chemical damage. DNA can be repaired by at least three different mechanisms, including **proofreading**, **mismatch repair**, and **excision repair**. **Review Figure 13.20**

13.5 How Does the Polymerase Chain Reaction Amplify DNA?

- The **polymerase chain reaction** technique uses DNA polymerase to make multiple copies of DNA in the laboratory. **Review Figure 13.21, ANIMATED TUTORIAL 13.5**

 Go to the Interactive Summary to review key figures, Animated Tutorials, and Activities Life10e.com/is13

CHAPTER**REVIEW**

■ REMEMBERING

1. In the Hershey–Chase experiment,
   a. DNA from parent bacteriophages appeared in progeny bacteriophages.
   b. most of the phage DNA never entered the bacteria.
   c. more than three-fourths of the phage protein appeared in progeny phages.
   d. DNA was labeled with radioactive sulfur.
   e. DNA formed the coat of the bacteriophages.

2. Which statement about complementary base pairing is not true?
   a. Complementary base pairing plays a role in DNA replication.
   b. In DNA, T pairs with A.
   c. Purines pair with purines, and pyrimidines pair with pyrimidines.
   d. In DNA, C pairs with G.
   e. The base pairs are of equal length.

3. In semiconservative replication of DNA,
   a. the original double helix remains intact and a new double helix forms.
   b. the strands of the double helix separate and act as templates for new strands.
   c. polymerization is catalyzed by RNA polymerase.
   d. polymerization is catalyzed by a double-helical enzyme.
   e. DNA is synthesized from amino acids.

4. The primer used for DNA replication
   a. is a short strand of RNA added to the 3′ end.
   b. is needed only once on a leading strand.
   c. remains on the DNA after replication.
   d. ensures that there will be a free 5′ end to which nucleotides can be added.
   e. is added to only one of the two template strands.

5. The role of DNA ligase in DNA replication is to
   a. add more nucleotides to the growing strand one at a time.
   b. open up the two DNA strands to expose template strands.
   c. ligate base to sugar to phosphate in a nucleotide.
   d. bond Okazaki fragments to one another.
   e. remove incorrectly paired bases.

6. What is the correct order for the following events in excision repair of DNA? (1) DNA polymerase I adds correct nucleotides by 5′-to-3′ replication; (2) damaged nucleotides are recognized; (3) DNA ligase seals the new strand to existing DNA; (4) part of a single strand is excised.
   a. 1, 2, 3, 4
   b. 2, 1, 3, 4
   c. 2, 4, 1, 3
   d. 3, 4, 2, 1
   e. 4, 2, 3, 1

■ UNDERSTANDING & APPLYING

7. One strand of DNA has the sequence 5′-ATTCCG-3′ The complementary strand for this is
   a. 5′-TAAGGC-3′
   b. 5′-ATTCCG-3′
   c. 5′-ACCTTA-3′
   d. 5′-CGGAAT-3′
   e. 5′-GCCTTA-3′

8. Using the following information, calculate the number of origins of DNA replication on a human chromosome: DNA polymerase adds nucleotides at 3,000 base pairs per minute in one direction; replication is bidirectional; S phase lasts 300 minutes; there are 120 million base pairs per chromosome. In a typical chromosome 3 micrometers (μm) long, how many origins are there per micrometer?

9. The drug dideoxycytidine, used to treat certain viral infections, is a nucleotide made with 2′,3′-dideoxyribose. This sugar lacks —OH groups at both the 2′ and the 3′ positions. Explain why this drug stops the growth of a DNA chain when added to DNA.

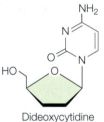

Dideoxycytidine

■ ANALYZING & EVALUATING

10. Suppose that Meselson and Stahl had continued their experiment on DNA replication for another ten bacterial generations. Would there still have been any ^{14}N–^{15}N hybrid DNA present? Would it still have appeared in the centrifuge tube? Explain.

11. Outline a series of experiments using radioactive isotopes (such as ^{32}P and ^{35}S) to show that during bacterial conjugation it is DNA and not protein that moves from the donor cell to the recipient cell and is responsible for bacterial transformation.

Go to BioPortal at **yourBioPortal.com** for Animated Tutorials, Activities, LearningCurve Quizzes, Flashcards, and many other study and review resources.

14 From DNA to Protein: Gene Expression

CHAPTER**OUTLINE**

14.1 What Is the Evidence that Genes Code for Proteins?

14.2 How Does Information Flow from Genes to Proteins?

14.3 How Is the Information Content in DNA Transcribed to Produce RNA?

14.4 How Is Eukaryotic DNA Transcribed and the RNA Processed?

14.5 How Is RNA Translated into Proteins?

14.6 What Happens to Polypeptides after Translation?

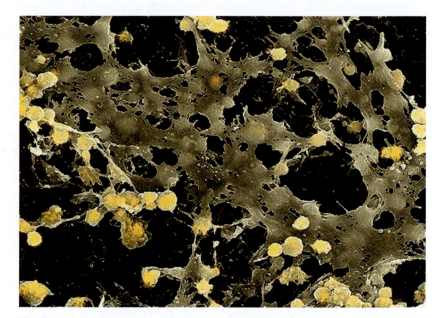

MRSA Methicillin-resistant *Staphylococcus aureus*, a major cause of serious illness and death in the United States and Europe, is treated with antibiotics that target its gene expression.

AT AGE 87, Janet enjoyed her life at the Sunshine Senior Citizen's Rest Home. Her only medical problem was periodic loss of bladder control, but this was alleviated when her nurse placed a plastic tube called a catheter into her bladder with an unobtrusive external bag. Things were going well until one day, unbeknown to Janet, some *Staphylococcus aureus* bacteria from the environment got into the bag and catheter. Finding a hospitable environment on the inner surface of the plastic, the bacteria attached to it via extracellular polysaccharides. Gradually, they divided and formed a colony, recruiting other free-living *S. aureus* cells that happened by. Within a few weeks, a slimy bacteria-laden coating called a biofilm (similar to dental plaque) had formed.

In time, some of the bacteria from the biofilm entered Janet's body and began to reproduce. Her advanced age and weak immune system permitted a significant infection to develop in her bladder and lungs. Fever, chills, shortness of breath, and the beginnings of kidney failure raged in her body. Racing for a treatment, her doctor first got a sample of the bacteria that Janet coughed up and sent it to a pathology lab for testing with antibiotics. The mainstay of treatment of "Staph" has been a penicillin-related antibiotic, methicillin, that binds to a bacterial protein needed to make new cell walls after cell division. Unfortunately, the bacteria had a mutation that made the target protein resistant to the antibiotic.

The lab made a diagnosis of MRSA: methicillin-resistant *S. aureus*. These "superbugs" are now common in hospitals and nursing homes and cause about 20,000 deaths a year in the United States.

Finally, Janet's doctor tried the antibiotic tetracycline. This molecule prevents gene expression in *S. aureus* and many other bacteria, but does not affect gene expression in eukaryotic cells. It has this specificity because it binds to a bacteria-specific protein in the ribosome, preventing the attachment of transfer RNA that carries amino acids to the ribosome for protein synthesis. Eukaryotic ribosomes do not have a binding site for tetracycline. Fortunately, the antibiotic killed the MRSA that infected Janet's bladder and lungs, and she is happy and healthy. But there are strains of MRSA that have developed resistance to tetracycline and other antibiotics as well. Scientists are now working to find new ways to combat this very serious problem.

Can new treatments focused on gene expression control MRSA?

See answer on p. 301.

14.1 What Is the Evidence that Genes Code for Proteins?

Chapter 4 introduced DNA and its role in gene expression. Chapter 13 described how DNA is the carrier of genetic information and how DNA is replicated. Here we will focus on the evidence for proteins as major products of gene expression, and we will describe how a gene is expressed as protein.

The molecular basis of phenotypes was actually discovered before it was known that DNA was the genetic material. Scientists had studied the chemical differences between individuals carrying wild-type and mutant alleles in organisms as diverse as humans and bread molds. They found that the major phenotypic differences resulted from differences in specific proteins.

Observations in humans led to the proposal that genes determine enzymes

The identification of a gene product as a protein began with a mutation. In the early twentieth century, the English physician Archibald Garrod saw several children with a rare disease. One symptom of the disease was that the urine turned dark brown when exposed to air. This was especially noticeable on the infants' diapers. The disease was given the descriptive name alkaptonuria ("black urine").

Garrod noticed that the disease was most common in children whose parents were first cousins. Mendelian genetics had just been "rediscovered," and Garrod realized that because first cousins share on average 1/8 of their alleles, the children of first cousins might inherit a rare mutant allele from both parents. He proposed that alkaptonuria was a phenotype caused by a recessive, mutant allele.

Garrod took the analysis further by identifying the biochemical abnormality in the affected children. He isolated from them an unusual substance, homogentisic acid, which accumulated in blood, joints (where it crystallized and caused severe pain), and urine (where it turned black). The chemical structure of homogentisic acid is similar to that of the amino acid tyrosine:

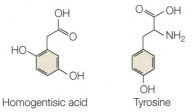

Homogentisic acid   Tyrosine

Enzymes as biological catalysts had just been discovered. Garrod proposed that homogentisic acid was a breakdown product of tyrosine. Normally, homogentisic acid would be converted to a harmless product. According to Garrod, there was a normal human allele that determined the synthesis of an enzyme that catalyzed this conversion:

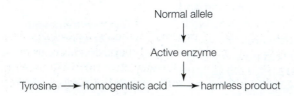

When the allele was mutated, the enzyme was inactive and homogentisic acid accumulated instead:

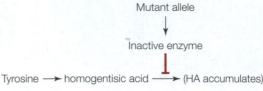

Therefore Garrod correlated *one gene to one enzyme* and coined the term "inborn error of metabolism" to describe this genetically determined biochemical disease. But his hypothesis needed direct confirmation by the identification of the specific enzyme and specific gene mutation involved. This did not occur until the enzyme, homogentisic acid oxidase, was described as active in healthy people and inactive in alkaptonuria patients in 1958, and the specific DNA mutation was described in 1996.

To relate genes and enzymes more generally, biologists turned to simpler organisms that could be manipulated in the laboratory.

Experiments on bread mold established that genes determine enzymes

As they work to explain the principles that govern life, biologists often turn to organisms that are easy to manipulate experimentally. Such **model organisms** have certain characteristics that make them attractive experimental subjects. For example:

- They are easy to grow in the laboratory or greenhouse.
- They have short generation times.
- They are easy to manipulate genetically, by crossing or by other methods.
- They often produce large numbers of progeny.

Biologists have used model organisms to develop principles of genetics that can then be applied more generally to other organisms. You have seen some of these organisms in previous chapters:

- Pea plants (*Pisum sativum*) were used by Mendel in his genetics experiments.
- Fruit flies (*Drosophila*) were used by Morgan in his genetics experiments.
- *Escherichia coli* was used by Meselson and Stahl to study DNA replication.

To this list we now add the bread mold *Neurospora*. *Neurospora* is an ascomycete fungus (see Chapter 30). This mold is haploid for most of its life, so there are no dominant or recessive alleles: all alleles are expressed phenotypically and are not masked by a heterozygous condition. *Neurospora* is simple to grow in the laboratory. Biologists at Stanford University led by George Beadle and Edward Tatum undertook studies to biochemically define the phenotypes in *Neurospora*.

Like Garrod, Beadle and Tatum hypothesized that the expression of a specific gene results in the activity of a specific enzyme. Now, they set out to test this hypothesis directly. They grew *Neurospora* on a nutritional medium containing sucrose, minerals, and biotin, which is the only vitamin that wild-type *Neurospora* cannot synthesize itself. Using this minimal medium, the enzymes of wild-type *Neurospora* could catalyze all the metabolic reactions needed for growth.

The scientists then treated the wild-type *Neurospora* with X rays, which function as a mutagen. A **mutagen** is something that damages DNA, causing **mutations**: heritable alterations in the DNA sequence. After the X-ray treatment, some *Neurospora* strains could no longer grow on the minimal medium. These mutant strains grew only if they were supplied with specific additional nutrients, such as particular vitamins. Beadle and Tatum hypothesized that these genetic strains had mutations in the genes that code for production of enzymes needed to synthesize the additional nutrients. For each mutant strain, the scientists were able to find a single compound that, when added to the minimal medium, supported the growth of that strain. These results suggested that mutations have simple effects, and that each mutation causes a defect in only one enzyme in a metabolic pathway. These conclusions confirmed Garrod's **one-gene, one-enzyme hypothesis**.

Mutations provide a powerful way to determine cause and effect in biology. Nowhere has this been more evident than in the elucidation of biochemical pathways. Such pathways consist of sequential events (chemical reactions) in which each event is dependent on the occurrence of the preceding event. The general reasoning is as follows:

- *Observation.* A particular gene (*a*) is present and a particular reaction catalyzed by a particular enzyme (A) occurs; the two are correlated.

- *Hypothesis.* Gene *a* determines the synthesis of the enzyme A.

- *Test of hypothesis.* Mutate gene *a*. Prediction: no functional enzyme is made, and the reaction does not occur.

Two colleagues of Beadle and Tatum, Adrian Srb and Norman Horowitz, used this experimental approach to isolate *Neurospora* mutants that could not survive without the amino acid arginine in their growth medium. By adding particular compounds to the medium, Srb and Horowitz were able to identify a series of steps in the biochemical pathway leading to the synthesis of arginine (**Figure 14.1**).

One gene determines one polypeptide

The one-gene, one-enzyme relationship has undergone several modifications in light of our current knowledge of

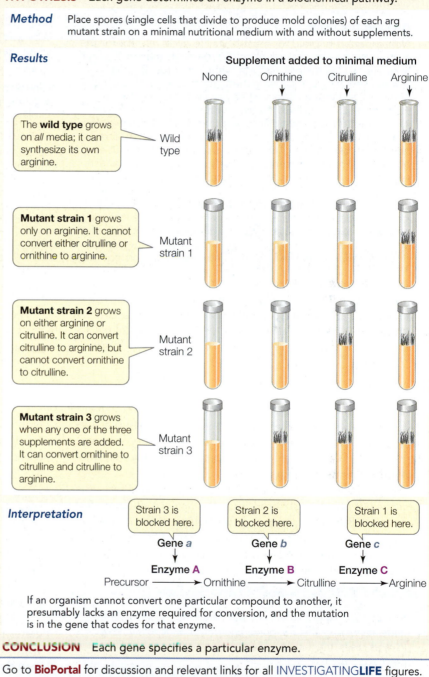

INVESTIGATINGLIFE

14.1 One Gene, One Enzyme Srb and Horowitz had several mutant strains of *Neurospora* that could not make arginine (*arg*). Several compounds are needed for arginine synthesis. By testing these compounds in the growth media for the mutant strains, the researchers deduced that each mutant strain was deficient in one enzyme along a biochemical pathway.[a]

HYPOTHESIS Each gene determines an enzyme in a biochemical pathway.

Method Place spores (single cells that divide to produce mold colonies) of each arg mutant strain on a minimal nutritional medium with and without supplements.

Results

Supplement added to minimal medium

None · Ornithine · Citrulline · Arginine

The **wild type** grows on *all* media; it can synthesize its own arginine. → Wild type

Mutant strain 1 grows only on arginine. It cannot convert either citrulline or ornithine to arginine. → Mutant strain 1

Mutant strain 2 grows on either arginine or citrulline. It can convert citrulline to arginine, but cannot convert ornithine to citrulline. → Mutant strain 2

Mutant strain 3 grows when any one of the three supplements are added. It can convert ornithine to citrulline and citrulline to arginine. → Mutant strain 3

Interpretation

Strain 3 is blocked here. — Gene *a* → Enzyme A

Strain 2 is blocked here. — Gene *b* → Enzyme B

Strain 1 is blocked here. — Gene *c* → Enzyme C

Precursor → Ornithine → Citrulline → Arginine

If an organism cannot convert one particular compound to another, it presumably lacks an enzyme required for conversion, and the mutation is in the gene that codes for that enzyme.

CONCLUSION Each gene specifies a particular enzyme.

Go to **BioPortal** for discussion and relevant links for all **INVESTIGATINGLIFE** figures.

[a]Srb, A. M. and N. H. Horowitz. 1944. *Journal of Biological Chemistry* 154: 129–139.

molecular biology. Many proteins, including many enzymes, are composed of more than one polypeptide chain, or subunit (that is, they have a quaternary structure; see Section 3.2). Look at the illustration of hemoglobin in Figure 3.11.

WORKING WITH**DATA:**

One Gene, One Enzyme

Original Paper

Srb, A. M. and N. H. Horowitz. 1944. The ornithine cycle in *Neurospora* and its genetic control. *Journal of Biological Chemistry* 154: 129–139.

Analyze the Data

Neurospora is haploid for most of its life cycle, except for the formation of a diploid cell when it undergoes mating; the cell then undergoes meiosis to form haploid spores. Beadle and Tatum used X rays to cause genetic mutations in *Neurospora*. They isolated mutant strains that were unable to grow on minimal medium, but were able to grow if the medium was supplemented with particular compounds. Their colleagues Adrian Srb and Norman Horowitz analyzed 15 mutant strains (the *arg* mutants) that could not synthesize arginine, but could grow on medium supplemented with arginine. The scientists tested various compounds and found two, ornithine and citrulline, that could be used instead of arginine to support the growth of some of the mutant strains (see Figure 14.1). Srb and Horowitz concluded that ornithine and citrulline are intermediates in the biochemical pathway leading to the synthesis of arginine. The various *arg* strains had mutations in genes that encode enzymes that catalyzed different steps in this pathway. The results of this and similar experiments supported the "one -gene, one-enzyme" hypothesis.

Fifteen mutant strains that required arginine for growth were tested for growth in the presence of the other substances. The results for three of the strains are shown in the first three rows of the table, with growth expressed as dry weight of fungal material after growth for 5 days.

QUESTION 1

Based on the biochemical pathway for arginine synthesis shown in Figure 14.1, which enzyme (A, B, or C) was mutated in each strain?

QUESTION 2

Why was there some growth in strains 34105 and 33442 even when there were no additions to the growth medium?

QUESTION 3

Nineteen other amino acids were tested as substitutes for arginine in the three strains. In all cases, there was no growth. Explain these results.

QUESTION 4

Sexual reproduction in *Neurospora* was used to create double mutants, which carried the mutations from both parental strains. A double mutant derived from strains 33442 and 36703 had the growth characteristics shown in the last row of the table. Explain these data in terms of the genes, mutations, and biochemical pathway.

Strain	Arginine added	Citrulline added	Ornithine added	No addition
34105	33.2	30.0	25.5	1.1
33442	43.8	42.7	2.5	2.3
36703	20.4	0.0	0.0	0.0
Double mutant	22.0	0.0	0.0	0.0

Go to **BioPortal** for all WORKING WITH**DATA** exercises

This protein has four polypeptides—two α and two β subunits, and the different subunits are encoded by separate genes. Thus it is more correct to speak of a **one-gene, one-polypeptide relationship**.

So far we have seen that in terms of protein synthesis, the *function of a gene is to prescribe the production of a single, specific polypeptide*. But not all genes code for polypeptides. As we will see below and in Chapter 16, there are many DNA sequences that are transcribed to RNA molecules that are not translated into polypeptides, but instead have other functions.

RECAP (14.1)

Studies of mutations in humans and bread molds led to our understanding of the one-gene, one-polypeptide relationship. In most cases, the function of a gene is to code for a specific polypeptide.

- What is a model organism, and why is *Neurospora* a good model for studying biochemical genetics? **See p. 282**

- How were the experiments on *Neurospora* set up to determine the order of steps in a biochemical pathway? **See pp. 282–283 and Figure 14.1**

- Explain the distinction between the phrases "one-gene, one-enzyme" and "one-gene, one-polypeptide." **See pp. 283–284**

Now that we have established the one-gene, one-polypeptide relationship, how does it work? That is, how is the information encoded in DNA used to produce a particular polypeptide?

14.2 How Does Information Flow from Genes to Proteins?

As we discussed in Chapter 13 and Section 14.1, two of the greatest biological discoveries of the twentieth century were that DNA is the hereditary material, and that DNA codes for proteins. We now know that the human genome contains about 21,000 protein-coding genes along with thousands of other genes that are transcribed into noncoding RNA molecules, which themselves have various functions in the cell. In the remainder of this chapter we will focus on the processes that occur when a protein-coding gene is expressed. You will see that certain noncoding RNAs have important roles in this process. Gene expression was briefly outlined in Section 4.1. To review, this process occurs in two major steps:

- During **transcription**, the information in a DNA sequence (a gene) is copied into a complementary RNA sequence.

- During **translation**, this RNA sequence is used to create the amino acid sequence of a polypeptide.

DNA → RNA → Polypeptide
Transcription Translation

Francis Crick and James Watson deciphered the structure of DNA in 1953 (see Section 13.2), and it was they who first proposed this model for gene expression. They took the concept further by suggesting that gene expression can only go in one direction: DNA can be used to create a protein, but a protein can never be used to create DNA. At the time, Crick called this "the **central dogma** of molecular biology."

Three types of RNA have roles in the information flow from DNA to protein

There are numerous types of RNA. Three of them have vital roles in gene expression.

MESSENGER RNA AND TRANSCRIPTION When a particular gene is expressed, one of the two DNA strands in the gene is transcribed to produce a complementary RNA strand, which is then processed to produce **messenger RNA** (**mRNA**). In eukaryotic cells, the mRNA travels from the nucleus to the cytoplasm, where it is translated into a polypeptide (**Figure 14.2**). The nucleotide sequence of the mRNA determines the ordered sequence of amino acids in the polypeptide chain, which is built by a ribosome.

RIBOSOMAL RNA AND TRANSLATION The ribosome is essentially a protein synthesis factory composed of multiple proteins and several **ribosomal RNAs** (**rRNAs**). One of the rRNAs catalyzes peptide bond formation between amino acids, to form a polypeptide.

TRANSFER RNA MEDIATES BETWEEN mRNA AND PROTEIN Another RNA, called **transfer RNA** (**tRNA**), can both bind a specific amino acid and recognize specific sequences of nucleotides in mRNA. It is the tRNA that recognizes which amino acid should be added next to a growing polypeptide chain (see Figure 14.2).

 Go to Media Clip 14.1
Protein Synthesis: An Epic on a Cellular Level
Life10e.com/mc14.1

In some cases, RNA determines the sequence of DNA

Certain viruses present exceptions to the general process of gene expression outlined above. As we saw in Section 13.1, a virus is a non-cellular infectious particle that reproduces inside cells. Many viruses, such as the tobacco mosaic virus, influenza viruses, and poliovirus, have RNA rather than DNA as their genetic material. With its nucleotide sequence, RNA could potentially act as an information carrier and be expressed as a protein. But if RNA is usually single-stranded, how do these viruses replicate? They generally solve this problem by transcribing from RNA to RNA, making an RNA strand that is complementary to their genomes. This "opposite" strand is then used to make multiple copies of the viral genome by transcription:

RNA → RNA → Polypeptide

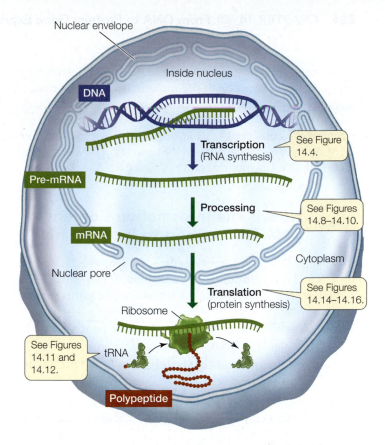

14.2 From Gene to Protein This diagram summarizes the processes of gene expression in eukaryotes.

Go to Activity 14.1 Eukaryotic Gene Expression
Life10e.com/ac14.1

Human immunodeficiency viruses (HIV) and certain rare tumor viruses also have RNA as their genomes, but do not replicate by transcribing from RNA to RNA. Instead, after infecting a host cell, such a virus makes a DNA copy of its genome, which becomes incorporated into the host's genome. The virus relies on the host cell's transcription machinery to make more RNA. This RNA can be either translated to produce viral proteins, or incorporated as the viral genome into new viral particles. Synthesis of DNA from RNA is called **reverse transcription**, and not surprisingly, such viruses are called **retroviruses**.

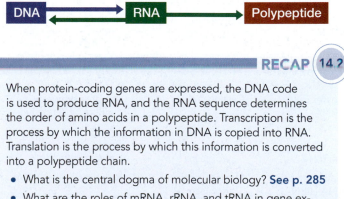

DNA ⇄ RNA → Polypeptide

RECAP 14.2

When protein-coding genes are expressed, the DNA code is used to produce RNA, and the RNA sequence determines the order of amino acids in a polypeptide. Transcription is the process by which the information in DNA is copied into RNA. Translation is the process by which this information is converted into a polypeptide chain.

- What is the central dogma of molecular biology? **See p. 285**
- What are the roles of mRNA, rRNA, and tRNA in gene expression? **See p. 285 and Figure 14.2**

Understanding gene expression is essential for understanding how organisms function at the molecular level. This understanding is key to the application of biology to human welfare, in areas such as agriculture and medicine. Much of the remainder of this book will in one way or another involve DNA and proteins. Let's begin by describing how the information in DNA is transcribed to produce RNA.

14.3 How Is the Information Content in DNA Transcribed to Produce RNA?

In prokaryotic and eukaryotic cells, RNA synthesis is directed by DNA. Transcription—the formation of a specific RNA sequence from a specific DNA sequence—requires several components:

- A DNA template for complementary base pairing; one of the two strands of DNA
- The four ribonucleoside triphosphates ATP, GTP, CTP, and UTP, to act as substrates
- An RNA polymerase enzyme
- Salts and a pH buffer to create an appropriate chemical environment for RNA polymerase if done in a test tube

mRNA is not the only molecule produced by transcription. The same process is responsible for the synthesis of tRNA and rRNA, whose important roles in protein synthesis will be described below. Like polypeptides, these RNAs are encoded by specific genes. Eukaryotes also make many kinds of small RNAs, including small nuclear RNA (snRNA), microRNA (miRNA), and small interfering RNA (siRNA), which are also transcribed. **Table 14.1** summarizes some of the RNAs found in eukaryotic cells. We will discuss the roles of miRNA and siRNA in Chapter 16.

RNA polymerases share common features

RNA polymerases from both prokaryotes and eukaryotes catalyze the synthesis of RNA from the DNA template. There is only one kind of RNA polymerase in bacteria, whereas there

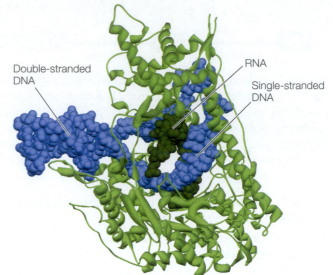

Double-stranded DNA

RNA

Single-stranded DNA

14.3 RNA Polymerase   This enzyme from bacteriophage T7 is similar to most other RNA polymerases. Note the size relationship between enzyme and DNA. See Figure 14.4 for details.

are several kinds in eukaryotes; however, they all share a common structure (**Figure 14.3**). Like DNA polymerases, RNA polymerases catalyze the addition of nucleotides in a 5′-to-3′ direction and are processive; that is, a single enzyme–template binding event results in the polymerization of hundreds of RNA bases. But unlike DNA polymerases (see Figure 13.13), RNA polymerases *do not require a primer*.

Transcription occurs in three steps

Transcription can be divided into three distinct processes: initiation, elongation, and termination. You can follow these processes in **Figure 14.4**.

INITIATION   Transcription begins when RNA polymerase binds to a special sequence of DNA called a **promoter** (see Figure 14.4A). Eukaryotic genes generally have one promoter each, whereas in prokaryotes and viruses, several genes often share one promoter. Promoters are important control sequences that "tell" the RNA polymerase two things:

- Where to start transcription
- Which strand of DNA to transcribe

A promoter reads in a particular direction, so it orients the RNA polymerase and thus "aims" it at the correct strand to use as a template. Part of each promoter is the **initiation site**, where transcription begins. Groups of nucleotides lying "upstream" from the initiation site (5′ on the non-template strand and 3′ on the template strand) help the RNA polymerase bind. Other proteins, which can bind to specific DNA sequences and to RNA polymerase, help direct the polymerase onto the promoter. These proteins, called **sigma factors** and **transcription factors**, help determine which specific genes are expressed at a particular time in the cell.

Although every gene has a promoter, not all promoters are identical. Some are more effective at transcription initiation than others. Furthermore, there are differences between transcription initiation in prokaryotes and in eukaryotes. We will discuss promoters and their roles in the regulation of gene expression in Chapter 16.

TABLE 14.1
Some RNAs in Eukaryotic Cells

RNA Type	Location of Activity	Role
Ribosomal RNA (rRNA)	Cytoplasm (ribosome)	Binding of mRNA and tRNA and protein synthesis
Messenger RNA (mRNA)	Cytoplasm	Carrier of gene sequence
Transfer RNA (tRNA)	Cytoplasm	Adaptor between mRNA and protein sequences
MicroRNA (miRNA)	Nucleus and cytoplasm	Regulates transcription and translation
Small interfering RNA (siRNA)	Nucleus and cytoplasm	Regulates other RNAs
Small nuclear RNA (snRNA)	Nucleus	Mediates mRNA processing

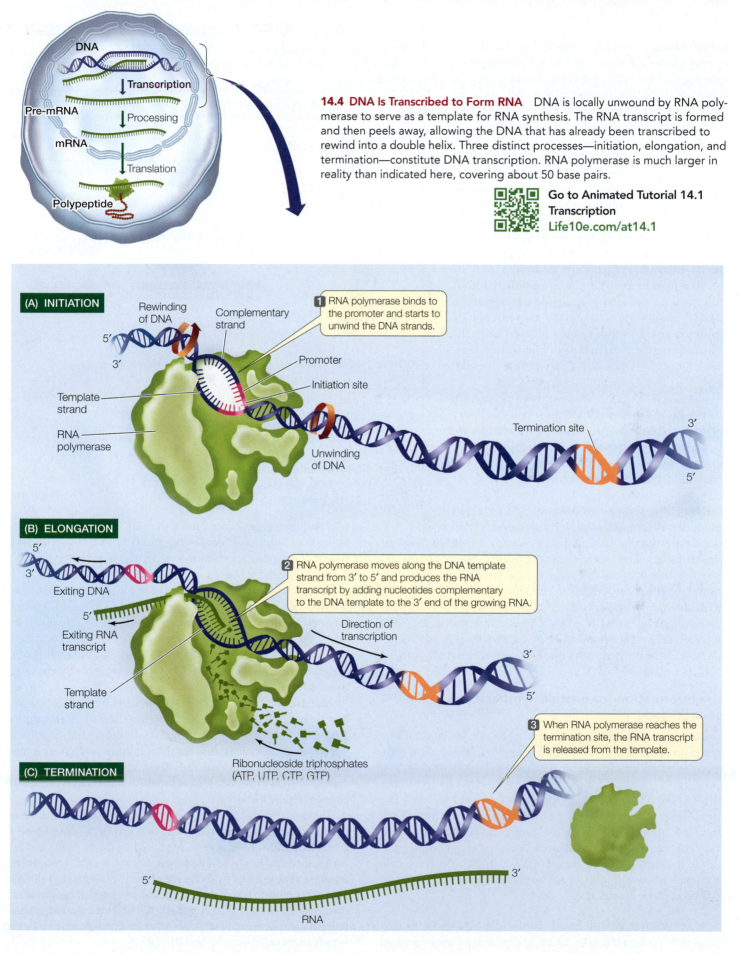

14.4 DNA Is Transcribed to Form RNA DNA is locally unwound by RNA polymerase to serve as a template for RNA synthesis. The RNA transcript is formed and then peels away, allowing the DNA that has already been transcribed to rewind into a double helix. Three distinct processes—initiation, elongation, and termination—constitute DNA transcription. RNA polymerase is much larger in reality than indicated here, covering about 50 base pairs.

Go to Animated Tutorial 14.1
Transcription
Life10e.com/at14.1

DNA
↓ Transcription
Pre-mRNA
↓ Processing
mRNA
↓ Translation
Polypeptide

(A) INITIATION

Rewinding of DNA
Complementary strand

1 RNA polymerase binds to the promoter and starts to unwind the DNA strands.

5′
3′
Promoter
Initiation site
Template strand
RNA polymerase
Unwinding of DNA
Termination site
3′
5′

(B) ELONGATION

5′
3′
Exiting DNA

2 RNA polymerase moves along the DNA template strand from 3′ to 5′ and produces the RNA transcript by adding nucleotides complementary to the DNA template to the 3′ end of the growing RNA.

5′
Exiting RNA transcript
Direction of transcription
3′
Template strand
5′

3 When RNA polymerase reaches the termination site, the RNA transcript is released from the template.

Ribonucleoside triphosphates (ATP, UTP, CTP, GTP)

(C) TERMINATION

5′
3′
RNA

ELONGATION After RNA polymerase has bound to the promoter, it begins the process of **elongation** (see Figure 14.4B). RNA polymerase unwinds the DNA about 10 base pairs at a time and reads the template strand in the 3′-to-5′ direction. The first nucleotide in the new RNA forms its 5′ end, and subsequent nucleotides complementary to the DNA template are added to its 3′ end. Thus the RNA transcript is antiparallel to the DNA template strand.

Recall from Section 13.3 that DNA polymerase uses dNTPs (deoxyribonucleoside triphosphates) as substrates, and forms covalent bonds between each incoming dNTP and the 3′ end of the growing polynucleotide chain (see Figure 13.11). Energy released by the removal of two phosphate groups from the dNTP is used to drive the reaction. Similarly, RNA polymerase uses (ribo)nucleoside triphosphates (NTPs) as substrates, removing two phosphate groups from each substrate molecule and using the released energy to drive the polymerization reaction. But unlike DNA, RNA polymerase does not require a primer to get this process started.

Because RNA polymerases do not proofread, transcription errors occur at a rate of one for every 10^4 to 10^5 bases. Because many copies of RNA are made, however, and because they often have only a relatively short life span, these errors are not as potentially harmful as mutations in DNA.

TERMINATION Just as initiation sites in the DNA template strand specify the starting point for transcription, particular base sequences specify its **termination** (see Figure 14.4C). The mechanisms controlling transcription termination in eukaryotes are not well understood. There are two mechanisms for transcription termination in bacteria. For some genes, the newly formed transcript forms a loop that causes the transcript to fall away from the DNA template and the RNA polymerase. In other cases, a helper protein binds to specific sequences on the transcript and causes the RNA to detach from the DNA template.

The information for protein synthesis lies in the genetic code

The **genetic code** relates genes (DNA) to mRNA, and mRNA to the amino acids that make up proteins. The genetic code specifies which amino acids will be used to build a protein. You can think of the genetic information in an mRNA molecule as a series of sequential, nonoverlapping three-letter "words." The three "letters" are three adjacent nucleotide bases in the mRNA polynucleotide. Each three-letter "word" is called a **codon**, and it specifies a particular amino acid. Each codon is complementary to the corresponding triplet of bases in the DNA molecule from which it was transcribed. The genetic code relates codons to their specific amino acids.

CHARACTERISTICS OF THE CODE Molecular biologists "broke" the genetic code in the early 1960s. The problem they addressed

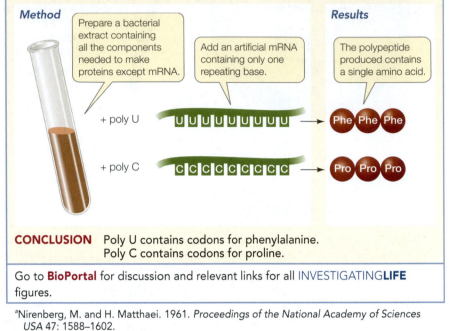

INVESTIGATINGLIFE

14.5 Deciphering the Genetic Code Nirenberg and Matthaei used a test tube protein synthesis system to determine the amino acids specified by synthetic mRNAs of known compositions.[a]

HYPOTHESIS An artificial mRNA containing only one repeating base will direct the synthesis of a protein containing only one repeating amino acid.

Method

Prepare a bacterial extract containing all the components needed to make proteins except mRNA.

Add an artificial mRNA containing only one repeating base.

Results

The polypeptide produced contains a single amino acid.

+ poly U   U U U U U U U U U U → Phe Phe Phe

+ poly C   C C C C C C C C C C → Pro Pro Pro

CONCLUSION Poly U contains codons for phenylalanine.
Poly C contains codons for proline.

Go to **BioPortal** for discussion and relevant links for all INVESTIGATINGLIFE figures.

[a]Nirenberg, M. and H. Matthaei. 1961. *Proceedings of the National Academy of Sciences USA* 47: 1588–1602.

Go to Animated Tutorial 14.2
Deciphering the Genetic Code
Life10e.com/at14.2

was perplexing: how could more than 20 "code words" be written with an "alphabet" consisting of only four "letters"? In other words, how could four bases (A, U, G, and C) code for 20 different amino acids?

A triplet code, based on three-letter codons, was considered likely. Since there are only four letters (A, G, C, and U), a one-letter code clearly could not unambiguously encode 20 amino acids; it could encode only four of them. A two-letter code could have only $4 \times 4 = 16$ unambiguous codons—still not enough. But a triplet code could have $4 \times 4 \times 4 = 64$ codons, more than enough to encode the 20 amino acids.

Marshall W. Nirenberg and J. H. Matthaei, at the U.S. National Institutes of Health, made the first decoding breakthrough in 1961 when they realized that they could use a simple artificial polynucleotide instead of a complex natural mRNA as a messenger. They could then identify the polypeptide that the artificial messenger encoded. This led to the identification of the first codons (**Figure 14.5**).

Other scientists later found that artificial mRNAs only three nucleotides long—each amounting to one codon—could bind to a ribosome, and that the resulting complex could then bind to a corresponding tRNA carrying a specific amino acid. Thus, for example, a simple UUU mRNA caused the tRNA carrying phenylalanine to bind to the ribosome. To discover which amino acid a codon represented, the scientists simply repeated the experiment using a sample of artificial mRNA for that codon, and observed which amino acid became bound to it.

Second letter

First letter	U	C	A	G	Third letter
U	UUU, UUC Phenyl-alanine; UUA, UUG Leucine	UCU, UCC, UCA, UCG Serine	UAU, UAC Tyrosine; UAA Stop codon, UAG Stop codon	UGU, UGC Cysteine; UGA Stop codon; UGG Tryptophan	U C A G
C	CUU, CUC, CUA, CUG Leucine	CCU, CCC, CCA, CCG Proline	CAU, CAC Histidine; CAA, CAG Glutamine	CGU, CGC, CGA, CGG Arginine	U C A G
A	AUU, AUC, AUA Isoleucine; AUG Methionine; start codon	ACU, ACC, ACA, ACG Threonine	AAU, AAC Asparagine; AAA, AAG Lysine	AGU, AGC Serine; AGA, AGG Arginine	U C A G
G	GUU, GUC, GUA, GUG Valine	GCU, GCC, GCA, GCG Alanine	GAU, GAC Aspartic acid; GAA, GAG Glutamic acid	GGU, GGC, GGA, GGG Glycine	U C A G

14.6 The Genetic Code Genetic information is encoded in mRNA in three-letter units—codons—made up of nucleoside monophosphates with the bases uracil (U), cytosine (C), adenine (A), and guanine (G) and is read in a 5′-to-3′ direction on mRNA. To decode a codon, find its first letter in the left column, then read across the top to its second letter, then read down the right column to its third letter. The amino acid the codon specifies is given in the corresponding row. For example, AUG codes for methionine, and GUA codes for valine.

Go to Activity 14.2 The Genetic Code
Life10e.com/ac14.2

The complete genetic code is shown in **Figure 14.6**. Notice that there are many more codons than there are different amino acids in proteins. All possible combinations of the four available "letters" (the bases) give 64 (4^3) different three-letter codons, yet these codons determine only 20 amino acids. AUG, which codes for methionine, is also the **start codon**, the initiation signal for translation. Three of the codons (UAA, UAG, UGA) are **stop codons**, or termination signals for translation. When the translation machinery reaches one of these codons, translation stops, and the polypeptide is released from the translation complex.

THE GENETIC CODE IS REDUNDANT BUT NOT AMBIGUOUS The 60 codons that are not start or stop codons are far more than enough to code for the other 19 amino acids. Indeed, there is more than one codon for almost all amino acids. Thus we say that the genetic code is redundant (or degenerate). For example, leucine is represented by six different codons (see Figure 14.6). Only methionine and tryptophan are represented by just one codon each.

A *redundant* code should not be confused with an *ambiguous* code. If the code were ambiguous, a single codon could specify either of two (or more) different amino acids, and there would be doubt about which amino acid should be incorporated into a growing polypeptide chain. Redundancy in the code simply means that there is more than one clear way to say "Put leucine here." The genetic code is not ambiguous: a given amino acid may be encoded by more than one codon, but a codon can code for only one amino acid.

THE GENETIC CODE IS (NEARLY) UNIVERSAL The same basic genetic code is used by all the species on our planet. Thus the code must be an ancient one that has been maintained intact throughout the evolution of living organisms. Exceptions are known: for example, within mitochondria and chloroplasts, the code differs slightly from that in prokaryotes and in the nuclei of eukaryotic cells; and in one group of protists, UAA and UAG code for glutamine rather than functioning as stop codons. The significance of these differences is not yet clear. What is clear is that the exceptions are few.

The common genetic code means that there is also a common language for evolution. Natural selection acts on phenotypic

variations that result from genetic variation. The genetic code probably originated early in the evolution of life. As we saw in Chapter 4, simulation experiments indicate the plausibility of individual nucleotides and nucleotide polymers arising spontaneously on the primeval Earth. The common code also has profound implications for genetic engineering, as we will see in Chapter 18, since it means that the code for a human gene is the same as that for a bacterial gene. It is therefore impressive, but not surprising, that a human gene can be expressed in *E. coli* via laboratory manipulations, since these cells speak the same "molecular language."

The codons in Figure 14.6 are mRNA codons. The base sequence of the DNA strand that is transcribed to produce the mRNA is complementary and antiparallel to these codons. Thus, for example,

- 3′-AAA-5′ in the template DNA strand corresponds to phenylalanine (which is encoded by the mRNA codon 5′-UUU-3′)
- 3′-ACC-5′ in the template DNA corresponds to tryptophan (which is encoded by the mRNA codon 5′-UGG-3′)

The non-template strand of DNA has the same sequence as the mRNA (but with T's instead of U's), and is often referred to as the "coding strand." By convention, DNA sequences are usually shown beginning with the 5′ end of the coding sequence.

RECAP 14.3

Transcription, which is catalyzed by an RNA polymerase, proceeds in three steps: initiation, elongation, and termination. The genetic code relates the information in mRNA (as a linear sequence of codons) to protein (a linear sequence of amino acids).

- What are the steps of gene transcription that produce mRNA? See pp. 286–288 and Figure 14.4
- How do RNA polymerases work? See pp. 287–288
- How was the genetic code elucidated? See p. 288 and Figure 14.5

The general features of transcription that we have described were first elucidated in model prokaryotes, such as *E. coli*. Biologists then used the same methods to analyze this process in eukaryotes, and although the basics are the same, there are some notable (and important) differences. We will now turn to a more detailed description of eukaryotic gene expression.

14.4 How Is Eukaryotic DNA Transcribed and the RNA Processed?

Since the genetic code is the same, you might expect the process of gene expression to be the same in eukaryotes as it is in prokaryotes. And basically it is. However, there are significant differences in gene structure between prokaryotes and eukaryotes, that is, there are differences in the organization of the nucleotide sequences in the genes. In addition, in eukaryotes but not prokaryotes, a nucleus separates transcription and translation (**Table 14.2**).

Thus far we have examined features of gene transcription that are common to prokaryotes and eukaryotes. We will look in more detail at prokaryotic gene structure in Chapter 16, where we describe how transcription is controlled in prokaryotes. Let's look now at the distinctive structures of eukaryotic genes and their transcription.

Many eukaryotic genes are interrupted by noncoding sequences

The sequence of an mRNA that reaches the ribosome is complementary to the sequence of a gene that is part of DNA. One way to show this is by the technique of **nucleic acid hybridization**, shown in **Figure 14.7A**. This technique involves two steps:

1. A sample of chromosomal DNA containing the gene is denatured to break the hydrogen bonds between the base pairs and separate the two strands.

2. The single-stranded mRNA (called a **probe**) is incubated with the denatured DNA. If the probe has a base sequence complementary to the target DNA, a probe–target double helix forms by hydrogen bonding between the bases. Because the two strands are from different sources, the resulting double-stranded region is called a hybrid.

Hybridization experiments can be performed with various combinations of DNA and RNA (RNA as target and DNA as probe; DNA as both target and probe, etc.). In many hybridization experiments, the probe is labeled in some way so that its binding to a specific target sequence can be detected. The double-stranded hybrids can also be viewed by electron microscopy.

Now let's examine what happens when mRNA probes from prokaryotes and eukaryotes are incubated with their respective chromosomal DNAs and then viewed under an electron microscope.

(A)

1 Upon being slowly heated or placed in a basic solution, the two strands of a DNA molecule denature (separate).

2 If a probe with a complementary base sequence is added to the denatured DNA…

3 …it binds the target DNA strand, forming a *double-stranded* hybrid molecule.

3′ 5′ — Target DNA

Denaturation → → Probe → Hybridization

3′ 5′

3′ 5′   3′ 5′   3′ 5′   3′ 5′

5′ 3′ 5′   3′   3′   5′

(B)

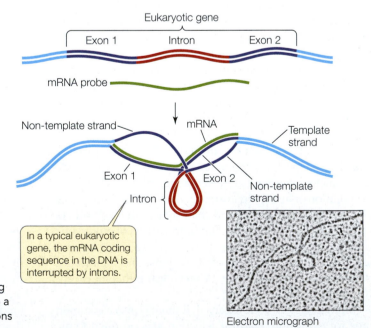

Double-stranded DNA — Prokaryotic gene

mRNA probe

Eukaryotic gene — Exon 1 | Intron | Exon 2

mRNA probe

mRNA

Non-template strand

In a typical prokaryotic gene, the mRNA and DNA are colinear, with full complimentarity.

Non-template strand — mRNA — Template strand

Exon 1 — Exon 2 — Non-template strand

Intron {

In a typical eukaryotic gene, the mRNA coding sequence in the DNA is interrupted by introns.

14.7 Nucleic Acid Hybridization and Introns (A) Base pairing permits the detection of a sequence that is complementary to a probe. (B) Hybridization experiments show that there are introns in eukaryotic genes but generally not in prokaryotic genes.

Electron micrograph of mRNA–DNA hybrid

TABLE14.2

Differences between Prokaryotic and Eukaryotic Gene Expression

Characteristic	Prokaryotes	Eukaryotes
Transcription and translation occurrence	At the same time in the cytoplasm	Transcription in the nucleus, then translation in the cytoplasm
Gene structure	DNA sequence is read in the same order as the amino acid sequence	Noncoding introns within coding sequence
Modification of mRNA after initial transcription but before translation	Usually none	Introns spliced out; 5′ cap and 3′ poly A tail added

- In prokaryotes (**Figure 14.7B, left**), there is usually a 1:1 linear complementarity between the base sequence of the mRNA at the ribosome and that of the chromosomal DNA.

- In eukaryotes (**Figure 14.7B, right**), one or more DNA loops are often observed within the mRNA–DNA hybrid, indicating that there are stretches of DNA sequence that do not have a complementary sequence in the mRNA that is translated at the ribosome.

One question that immediately arises from the eukaryotic result is, does this "extra" DNA actually get transcribed and then removed at the RNA level before mRNA arrives at the ribosome, or does transcription just "skip" these "extra" non-translated sequences? The answer comes from an experiment in which the initial mRNA transcript in the cell nucleus—the **precursor mRNA**, or **pre-mRNA**—is hybridized with chromosomal DNA. In this case, there is full, linear, loop-free hybridization. So the intervening regions (**introns**) of DNA actually get transcribed and then spliced out of the pre-mRNA in the nucleus, leaving expressed sequences (**exons**) in the mRNA that reaches the ribosome (**Figure 14.8**).

Introns *interrupt, but do not scramble*, the DNA sequence of a gene. The base sequences of the exons in the template strand, if joined and taken in order, form a continuous sequence that is complementary to that of the mature mRNA. In some cases,

the separated exons often encode different functional regions, or **domains**, of the protein. For example, the globin polypeptides that make up hemoglobin each have two domains: one for binding to a nonprotein pigment called heme, and another for binding to the other globin subunits. These two domains are encoded by different exons in the globin genes. Most (but not all) eukaryotic genes contain introns, and in rare cases, introns are also found in prokaryotes. The largest human gene encodes a muscle protein called titin; it has 363 exons, which together code for 38,138 amino acids.

Eukaryotic gene transcripts are processed before translation

The primary transcript of a eukaryotic gene is modified in several ways before it leaves the nucleus: both ends of the pre-mRNA are modified, and the introns are removed.

MODIFICATION AT BOTH ENDS Two steps in the processing of pre-mRNA take place in the nucleus, one at each end of the molecule (**Figure 14.9**):

- A **5′ cap** is added to the 5′ end of the pre-mRNA as it is transcribed. The 5′ cap is a chemically modified molecule of guanosine triphosphate (GTP). It facilitates the binding of mRNA to the ribosome for translation, and it protects

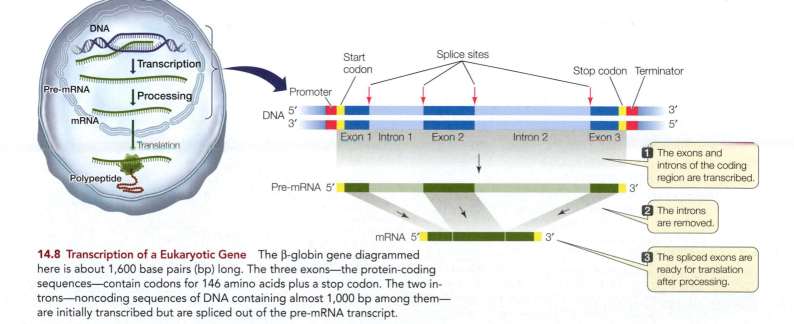

14.8 Transcription of a Eukaryotic Gene The β-globin gene diagrammed here is about 1,600 base pairs (bp) long. The three exons—the protein-coding sequences—contain codons for 146 amino acids plus a stop codon. The two introns—noncoding sequences of DNA containing almost 1,000 bp among them—are initially transcribed but are spliced out of the pre-mRNA transcript.

14.9 Processing the Ends of Eukaryotic Pre-mRNA Modifications at each end of the pre-mRNA transcript—the 5' cap and the poly A tail—are important for mRNA function.

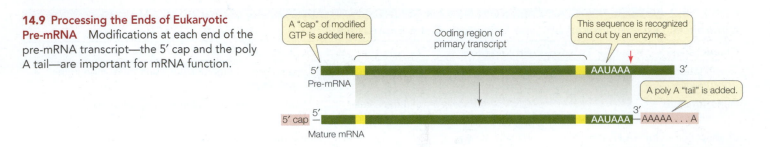

the mRNA from being digested by ribonucleases that break down RNAs.

- A **poly A tail** is added to the 3' end of the pre-mRNA at the end of transcription. Transcription ends downstream of the termination codon: along the coding strand of the DNA in the 3' direction. In eukaryotes, there is usually a "polyadenylation" sequence (AAUAAA) near the 3' end of the pre-mRNA, after the last codon. This sequence acts as a signal for an enzyme to cut the pre-mRNA. Immediately after this cleavage, another enzyme adds 100 to 300 adenine nucleotides (the poly A tail) to the 3' end of the pre-mRNA. This tail may assist in the export of mature mRNA from the nucleus and is important for mRNA stability.

SPLICING TO REMOVE INTRONS The next step in the processing of eukaryotic pre-mRNA within the nucleus is removal of the introns. If these RNA sequences were not removed, a very different amino acid sequence, and possibly a nonfunctional protein, would result. A process called **RNA splicing** removes the introns and splices the exons together.

As soon as the pre-mRNA is transcribed, several **small nuclear ribonucleoprotein particles (snRNPs)** bind to the ends of each intron. Each snRNP has an RNA component—the snRNA (see Table 14.1). There are several types of these RNA–protein particles in the nucleus.

At the boundaries between introns and exons are **consensus sequences**—short stretches of DNA that appear, with little variation, in many different genes. (The polyadenylation sequence mentioned above is another example of a consensus sequence. Most promoters also contain consensus sequences for the binding of transcription factors and RNA polymerases.) The RNA in

14.10 The Spliceosome: An RNA Splicing Machine The binding of snRNPs to consensus sequences bordering the introns on the pre-mRNA results in a series of proteins binding and forming a large complex called a spliceosome. This structure determines the exact position of each cut in the pre-mRNA with great precision.

Go to Animated Tutorial 14.3
RNA Splicing
Life10e.com/at14.3

one of the snRNPs has a stretch of bases that is complementary to the consensus sequence at the 5' exon–intron boundary, and it binds to the pre-mRNA by complementary base pairing. Another snRNP binds to the pre-mRNA near the 3' intron–exon boundary (**Figure 14.10**).

Next, using energy from adenosine triphosphate (ATP), proteins are added to form a large RNA–protein complex called a **spliceosome**. This complex cuts the pre-mRNA at exon–intron boundaries, releases the introns, and joins the ends of the exons together to produce mature mRNA.

Molecular studies of human genetic diseases have provided insights into intron consensus sequences and splicing machinery. For example, people with the genetic disease β-thalassemia have a defect in the production of one of the hemoglobin subunits. These people suffer from severe anemia because they have an inadequate supply of red blood cells. In some cases, the genetic mutation that causes the disease occurs at an intron consensus sequence in the β-globin gene. Consequently, β-globin pre-mRNA cannot be spliced

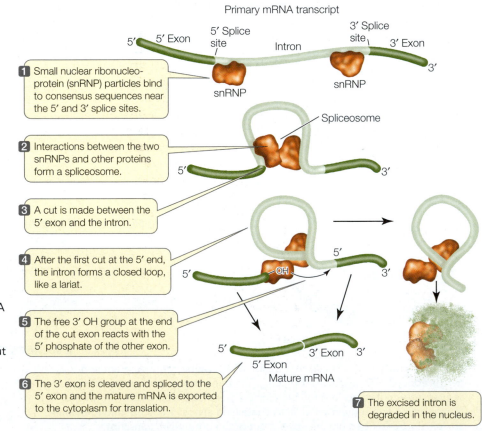

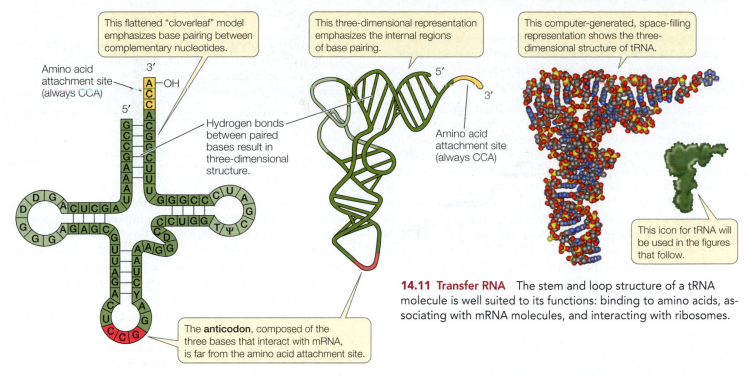

This flattened "cloverleaf" model emphasizes base pairing between complementary nucleotides.

Amino acid attachment site (always CCA)

Hydrogen bonds between paired bases result in three-dimensional structure.

The **anticodon**, composed of the three bases that interact with mRNA, is far from the amino acid attachment site.

This three-dimensional representation emphasizes the internal regions of base pairing.

Amino acid attachment site (always CCA)

This computer-generated, space-filling representation shows the three-dimensional structure of tRNA.

This icon for tRNA will be used in the figures that follow.

14.11 Transfer RNA The stem and loop structure of a tRNA molecule is well suited to its functions: binding to amino acids, associating with mRNA molecules, and interacting with ribosomes.

correctly, and β-globin mRNA that encodes a nonfunctional polypeptide is made. This finding offers another example of how biologists can use mutations to elucidate biological processes.

After processing is completed in the nucleus, the mature mRNA moves out into the cytoplasm through the nuclear pores. In the nucleus, a protein complex called the cap-binding complex binds to the 5' cap of the processed mRNA. This complex is recognized by a receptor at the nuclear pore. Together, these proteins lead the mRNA through the pore. Unprocessed or incompletely processed pre-mRNAs remain in the nucleus.

RECAP 14.4

Most eukaryotic genes contain noncoding sequences called introns, which are removed from the pre-mRNA transcript.

- Describe the method of nucleic acid hybridization and how it shows introns in eukaryotic genes. **See pp. 290–291 and Figure 14.7**

- How is the pre-mRNA transcript modified at the 5' and 3' ends? **See pp. 291–292 and Figure 14.9**

- How does RNA splicing happen? What are the consequences if it does not happen correctly? **See p. 292 and Figure 14.10**

Transcription and posttranscriptional events produce an mRNA that is ready to be translated into a sequence of amino acids in a polypeptide. We will turn now to the events of translation.

14.5 How Is RNA Translated into Proteins?

The translation of mRNA into proteins requires a molecule that links the information contained in mRNA codons with specific amino acids in proteins. This function is performed by transfer

RNA (tRNA). Two key events must take place to ensure that the protein made is the one specified by the mRNA:

- The tRNAs must read mRNA codons correctly.

- The tRNAs must deliver the amino acids that correspond to each mRNA codon.

Once the tRNAs "decode" the mRNA and deliver the appropriate amino acids, components of the ribosome catalyze the formation of peptide bonds between amino acids. We will now turn to these two steps.

Go to Animated Tutorial 14.4
Translation
Life10e.com/at14.4

Transfer RNAs carry specific amino acids and bind to specific codons

There is at least one specific tRNA molecule for each of the 20 amino acids. Each tRNA has three functions that are fulfilled by its structure and base sequence (**Figure 14.11**):

- *tRNAs bind to particular amino acids.* Each tRNA binds to a specific enzyme that attaches it to only 1 of the 20 amino acids. This covalent attachment is at the 3' end of the tRNA. We will describe the details of this vital process in the next section. When it is carrying an amino acid, the tRNA is said to be "charged."

- *tRNAs bind to mRNA.* At about the midpoint on the tRNA polynucleotide chain there is a triplet of bases called the **anticodon**, which is complementary to the mRNA codon for the particular amino acid that the tRNA carries. Like the two strands of DNA, the codon and anticodon bind together via noncovalent hydrogen bonds. For example, the mRNA codon for arginine is 5'-CGG-3', and the complementary tRNA anticodon is 3'-GCC-5'.

• *tRNAs interact with ribosomes*. The ribosome has several sites on its surface that just fit the three-dimensional structure of a tRNA molecule. Interaction between the ribosome and the tRNA is noncovalent.

Recall that 61 different codons encode the 20 amino acids in proteins (see Figure 14.6). Does this mean that the cell must produce 61 different tRNA species, each with a different anticodon? No. The cell gets by with about two-thirds of that number of tRNA species because the specificity for the base at the 3′ end of the codon (and the 5′ end of the anticodon) is not always strictly observed. This phenomenon is called wobble, and it is possible because in some cases unusual or modified nucleotide bases occur in the 5′ position of the anticodon. One such unusual base is inosine (I), which can pair with A, C, and U. For example, the presence of inosine in the tRNA with the anticodon 3′-CGI-5′ allows it to recognize and bind to three of the alanine codons: GCA, GCC, and GCU. Wobble occurs in some matches but not in others; of most importance, it does not allow the genetic code to be ambiguous. That is, *each mRNA codon binds to just one tRNA species, carrying a specific amino acid*.

Each tRNA is specifically attached to an amino acid

The charging of each tRNA with its correct amino acid is achieved by a family of enzymes known as aminoacyl-tRNA synthetases. Each enzyme is specific for one amino acid and for its corresponding tRNA. The reaction uses ATP, forming a high-energy bond between the amino acid and the tRNA (**Figure 14.12**). The energy in this bond is later used in the formation of peptide bonds between amino acids in a growing polypeptide chain.

Clearly, the specificity between the tRNA and its corresponding amino acid is extremely important. These reactions, for example, are highly specific:

$$\text{Cysteine} + \text{tRNAcys} \xrightarrow{\text{Cys-tRNA synthetase}} \text{cys-tRNA}$$
(anticodon ACA)

$$\text{Alanine} + \text{tRNAala} \xrightarrow{\text{Ala-tRNA synthetase}} \text{ala-tRNA}$$
(anticodon CGA)

A clever experiment by Seymour Benzer and his colleagues at Purdue University demonstrated the importance of this specificity. They took the cys-tRNA molecule (see above) and chemically modified the cysteine, converting it into alanine. Which component—the amino acid or the tRNA—would be recognized when this hybrid charged tRNA was put into a protein synthesizing system? The answer was the tRNA. Everywhere in the synthesized protein where cysteine was supposed to be, alanine appeared instead. The cysteine-specific tRNA had delivered its cargo (alanine) to every mRNA codon for cysteine. This experiment showed that the protein synthesis machinery recognizes the anticodon of the charged tRNA, not the amino acid attached to it.

The ribosome is the workbench for translation

The **ribosome** is the molecular workbench where the task of translation is accomplished. Its structure enables it to hold mRNA and charged tRNAs in the correct positions, thus allowing a polypeptide chain to be assembled efficiently. A given ribosome does not specifically produce just one kind

14.12 Charging a tRNA Molecule The amino-acyl-tRNA synthase activates a specific amino acid and charges a specific tRNA with that amino acid.

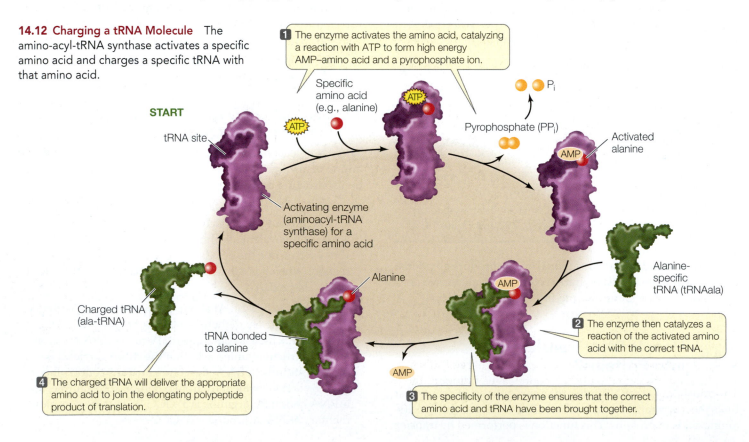

1 The enzyme activates the amino acid, catalyzing a reaction with ATP to form high energy AMP–amino acid and a pyrophosphate ion.

START

tRNA site

Specific amino acid (e.g., alanine)

ATP

P_i

Pyrophosphate (PP_i)

AMP

Activated alanine

Activating enzyme (aminoacyl-tRNA synthase) for a specific amino acid

Alanine-specific tRNA (tRNAala)

Charged tRNA (ala-tRNA)

Alanine

AMP

AMP

tRNA bonded to alanine

2 The enzyme then catalyzes a reaction of the activated amino acid with the correct tRNA.

4 The charged tRNA will deliver the appropriate amino acid to join the elongating polypeptide product of translation.

3 The specificity of the enzyme ensures that the correct amino acid and tRNA have been brought together.

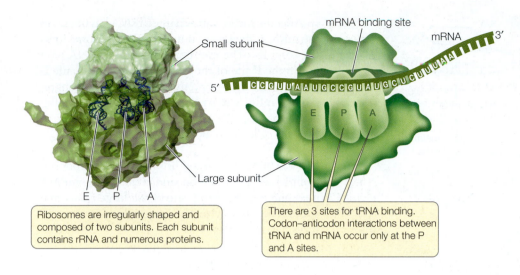

14.13 Ribosome Structure Each ribosome consists of a large and a small subunit. The subunits remain separate when they are not in use for protein synthesis.

Ribosomes are irregularly shaped and composed of two subunits. Each subunit contains rRNA and numerous proteins.

There are 3 sites for tRNA binding. Codon–anticodon interactions between tRNA and mRNA occur only at the P and A sites.

of protein. A ribosome can use any mRNA and all species of charged tRNAs, and thus can be used to make many different polypeptide products. Ribosomes can be used over and over again, and there are thousands of them in a typical cell.

Although ribosomes are small in contrast to other cellular structures, their mass of several million daltons makes them large in comparison with charged tRNAs. Each ribosome consists of two subunits, a large one and a small one (**Figure 14.13**). The two subunits and several dozen other molecules interact noncovalently. In fact, when hydrophobic interactions between the proteins and RNAs are disrupted, the ribosome falls apart. The two subunits separate and all the RNAs and proteins separate from one another. If the disrupting agent is removed, the complex structure self-assembles perfectly! When not active in the translation of mRNA, the ribosomes exist as two separate subunits.

In eukaryotes, the large subunit consists of three different molecules of ribosomal RNA (rRNA) and 49 different protein molecules, arranged in a precise configuration. The small subunit consists of 1 rRNA molecule and 33 different protein molecules. The ribosomes of prokaryotes are somewhat smaller than those of eukaryotes, and their ribosomal proteins and RNAs are different. These differences explain why antibiotics that target the prokaryotic ribosome (such as the tetracycline used to cure Janet's bladder infection in the opening story) can be used to kill bacteria without harming a patient's cells. Mitochondria and chloroplasts also contain ribosomes, some of which are similar to those of prokaryotes (see Chapter 5).

On the large subunit of the ribosome there are three sites to which a tRNA can bind, and these are designated A, P, and E (see Figure 14.13). The mRNA and ribosome move in relation to one another, and as they do so, a charged tRNA traverses these three sites in order:

- The *A (aminoacyl tRNA) site* is where the charged tRNA anticodon binds to the mRNA codon, thus lining up the correct amino acid to be added to the growing polypeptide chain.

- The *P (peptidyl tRNA) site* is where the tRNA adds its amino acid to the polypeptide chain.

- The *E (exit) site* is where the tRNA, having given up its amino acid, resides before being released from the ribosome and going back to the cytosol to pick up another amino acid and begin the process again.

The ribosome has a fidelity function that ensures that the mRNA–tRNA interactions are accurate; that is, that a charged tRNA with the correct anticodon (e.g., 3'-UAC-5') binds to the appropriate codon in mRNA (e.g., 5'-AUG-3'). When proper binding occurs, hydrogen bonds form between the paired bases. The rRNA of the small ribosomal subunit plays a role in validating the three-base-pair match. If hydrogen bonds have not formed between all three base pairs, the tRNA must be the wrong one for that mRNA codon, and the incorrect tRNA is ejected from the ribosome.

Translation takes place in three steps

Translation is the process by which the information in mRNA (derived from DNA) is used to specify and link a specific sequence of amino acids, producing a polypeptide. Like transcription, translation occurs in three steps: initiation, elongation, and termination.

INITIATION The translation of mRNA begins with the formation of an **initiation complex**, which consists of a charged tRNA and a small ribosomal subunit, both bound to the mRNA (**Figure 14.14**).

In prokaryotes, the rRNA of the small ribosomal subunit first binds to a complementary ribosome binding site (AGGAGG; known as the Shine–Dalgarno sequence) on the mRNA. This sequence is less than 10 bases upstream of the actual start codon but lines up the start codon so that it is adjacent to the P site of the large subunit:

mRNA 5' . . . A G G A G G . . . (start codon) . . . 3'
rRNA 3' . . . U C C U C C . . .   (P site)   . . . 5'

14.14 The Initiation of Translation

Translation begins with the formation of an initiation complex. In prokaryotes, the small ribosomal subunit binds to the Shine–Dalgarno sequence to begin the process, whereas in eukaryotes, it binds to the 5' cap.

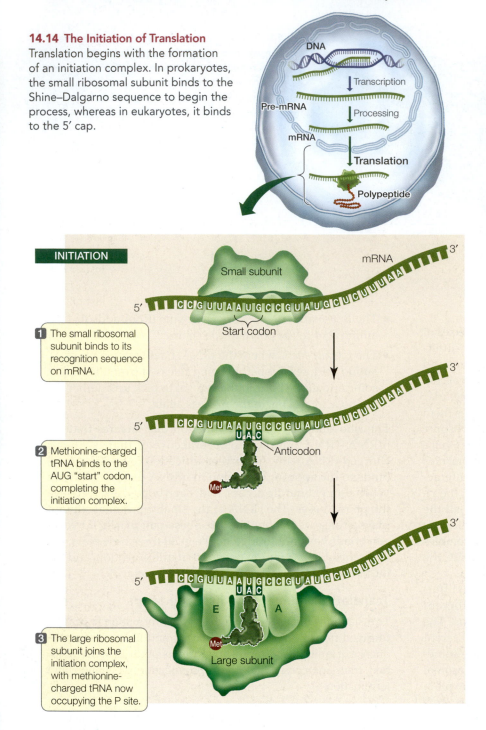

INITIATION

1 The small ribosomal subunit binds to its recognition sequence on mRNA.

2 Methionine-charged tRNA binds to the AUG "start" codon, completing the initiation complex.

3 The large ribosomal subunit joins the initiation complex, with methionine-charged tRNA now occupying the P site.

After the methionine-charged tRNA has bound to the mRNA, the large subunit of the ribosome joins the complex. The methionine-charged tRNA now lies in the P site of the ribosome, and the A site is aligned with the second mRNA codon. These ingredients—mRNA, two ribosomal subunits, and methionine-charged tRNA—are assembled by a group of proteins called initiation factors.

ELONGATION A charged tRNA whose anticodon is complementary to the second codon of the mRNA now enters the open A site of the large ribosomal subunit **Figure 14.15**. The large subunit then catalyzes two reactions:

- It breaks the bond between the tRNA and its amino acid in the P site.
- It catalyzes the formation of a peptide bond between that amino acid and the one attached to the tRNA in the A site.

Because the large ribosomal subunit performs these two actions, it is said to have **peptidyl transferase** activity. In this way, methionine (the amino acid in the P site) becomes the N terminus of the new protein. The second amino acid is now bound to methionine but remains attached to its tRNA at the A site.

How does the large ribosomal subunit catalyze peptide bond formation? Harry Noller and his colleagues at the University of California at Santa Cruz did a series of experiments and found that:

- If they removed almost all of the proteins from the large subunit, it still catalyzed peptide bond formation.
- If the rRNA was extensively modified, peptidyl transferase activity was destroyed.

Thus *rRNA is the catalyst*. The purification and crystallization of ribosomes has allowed scientists to examine their structure in detail, and the catalytic role of rRNA in peptidyl transferase activity has been confirmed. This supports the hypothesis that RNA, and catalytic RNA in particular, evolved before DNA (see Section 4.3).

After the first tRNA releases its methionine, it moves to the E site and is then dissociated from the ribosome, returning to the cytosol to become charged with another methionine. The second tRNA, now bearing a dipeptide (a two-amino acid chain), is shifted to the P site as the ribosome moves one codon along the mRNA in the 5'-to-3' direction.

The elongation process continues, and the polypeptide chain grows, as these steps are repeated. Follow the process in Figure 14.15. All these steps are assisted by ribosomal proteins called elongation factors.

TERMINATION The elongation cycle ends, and translation is terminated, when a stop codon—UAA, UAG, or UGA—enters

Eukaryotes load the mRNA onto the ribosome somewhat differently: the small ribosomal subunit binds to the 5' cap on the mRNA and then moves along the mRNA until it reaches the start codon.

Recall that the mRNA start codon in the genetic code is AUG (see Figure 14.6). The anticodon (UAG) of a methionine-charged tRNA binds to this start codon by complementary base pairing to complete the initiation complex. Thus the first amino acid in a polypeptide chain is always methionine. However, not all mature proteins have methionine as their N-terminal amino acid. In many cases, the initial methionine is removed by an enzyme after translation.

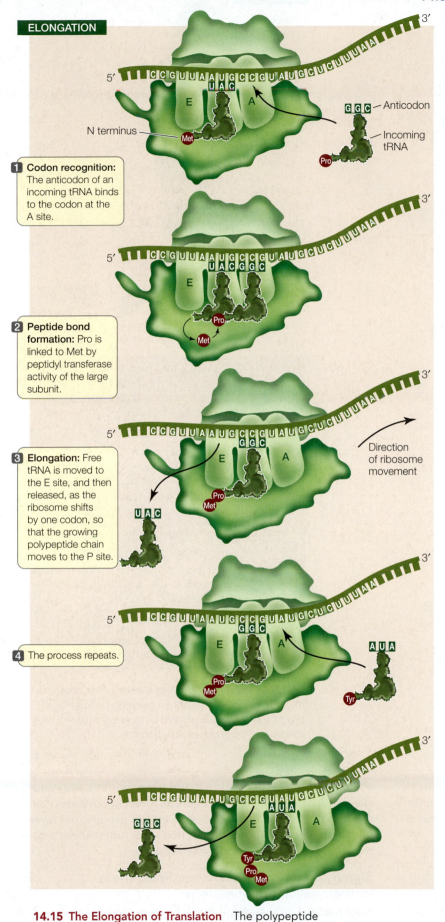

ELONGATION

1 Codon recognition: The anticodon of an incoming tRNA binds to the codon at the A site.

2 Peptide bond formation: Pro is linked to Met by peptidyl transferase activity of the large subunit.

3 Elongation: Free tRNA is moved to the E site, and then released, as the ribosome shifts by one codon, so that the growing polypeptide chain moves to the P site.

4 The process repeats.

N terminus

Anticodon

Incoming tRNA

Direction of ribosome movement

14.15 The Elongation of Translation The polypeptide chain elongates as the mRNA is translated.

TABLE 14.3
Signals that Start and Stop Transcription and Translation

	Transcription	Translation
Initiation	Promoter DNA	AUG start codon in the mRNA
Termination	Terminator DNA	UAA, UAG, or UGA in the mRNA

the A site (**Figure 14.16**). These codons do not correspond with any amino acids, nor do they bind any tRNAs. Rather, they bind a protein release factor, which allows hydrolysis of the bond between the polypeptide chain and the tRNA in the P site. The newly completed polypeptide thereupon separates from the ribosome. Its C terminus is the last amino acid to join the chain. Its N terminus, at least initially, is methionine, as a consequence of the AUG start codon. In its amino acid sequence, it contains information specifying its conformation, as well as its ultimate cellular destination.

Table 14.3 summarizes the nucleic acid signals for initiation and termination of transcription and translation.

Polysome formation increases the rate of protein synthesis

Several ribosomes can work simultaneously at translating a single mRNA molecule, producing multiple polypeptides at the same time. As soon as the first ribosome has moved far enough from the site of translation initiation, a second initiation complex can form, then a third, and so on. An assemblage consisting of a strand of mRNA with its beadlike ribosomes and their growing polypeptide chains is called a **polyribosome**, or **polysome** (**Figure 14.17**). Cells that are actively synthesizing proteins contain large numbers of polysomes and few free ribosomes or ribosomal subunits.

■ RECAP 14.5

A key step in protein synthesis is the attachment of an amino acid to its proper tRNA. This attachment is carried out by an activating enzyme. Translation of the genetic information from mRNA into protein occurs at the ribosome. Multiple ribosomes may act on a single mRNA to make multiple copies of the protein that it encodes.

- How is an amino acid attached to a specific tRNA, and why is specificity in this process so important? **See p. 294 and Figure 14.12**

- Describe the events of initiation, elongation, and termination of translation. **See pp. 295–297 and Figures 14.14–14.16**

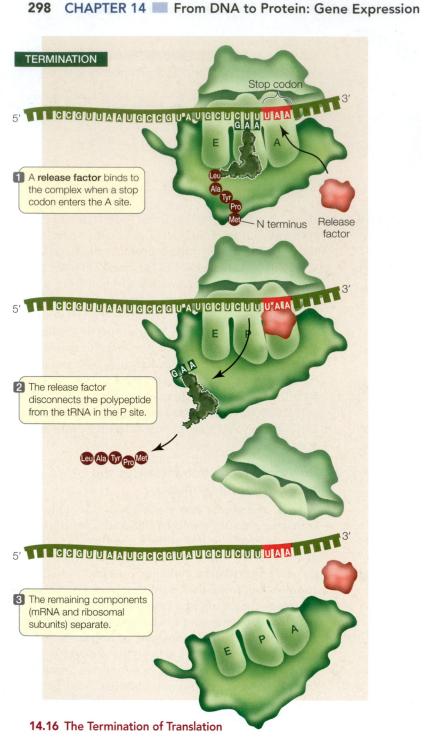

TERMINATION

1 A **release factor** binds to the complex when a stop codon enters the A site.

2 The release factor disconnects the polypeptide from the tRNA in the P site.

3 The remaining components (mRNA and ribosomal subunits) separate.

14.16 The Termination of Translation
Translation terminates when the A site of the ribosome encounters a stop codon on the mRNA.

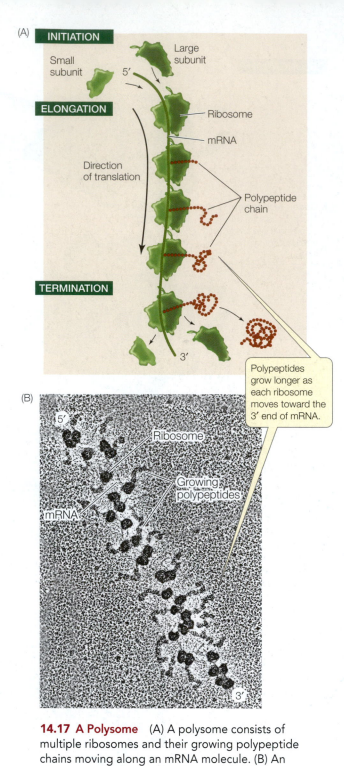

14.17 A Polysome (A) A polysome consists of multiple ribosomes and their growing polypeptide chains moving along an mRNA molecule. (B) An electron micrograph of a polysome.

The polypeptide chain that is released from the ribosome is not necessarily a functional protein. Let's look at some of the posttranslational changes that can affect the fate and function of a polypeptide.

14.6   What Happens to Polypeptides after Translation?

The site of a polypeptide's function may be far away from its point of synthesis in the cytoplasm. This is especially true for eukaryotes. The polypeptide may be moved into an organelle, or even out of the cell. In addition, polypeptides are often modified by the addition of new chemical groups that have functional significance. In this section we examine these posttranslational aspects of protein synthesis.

Signal sequences in proteins direct them to their cellular destinations

As a polypeptide chain emerges from the ribosome it may simply fold into its three-dimensional shape and perform its cellular role. However, a newly formed polypeptide may

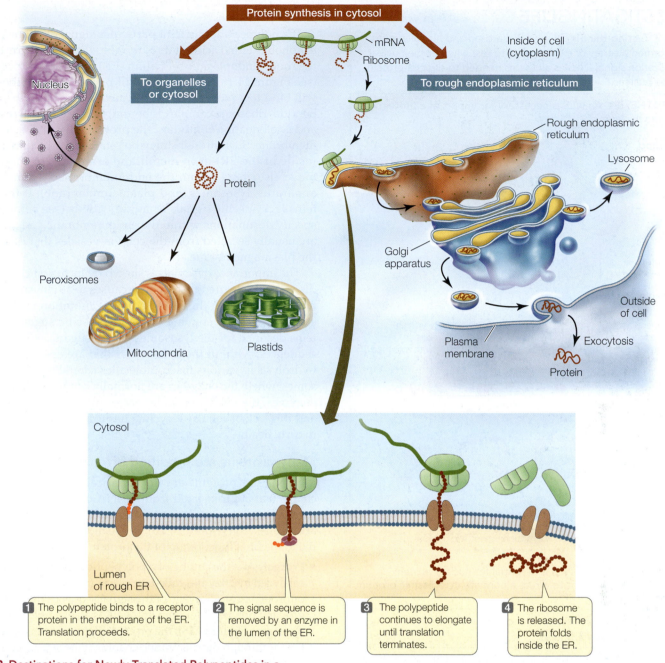

14.18 Destinations for Newly Translated Polypeptides in a Eukaryotic Cell Signal sequences on newly synthesized polypeptides bind to specific receptor proteins on the outer membranes of the organelles to which they are "addressed." Once the protein has bound to it, the receptor forms a channel in the membrane, and the protein enters the organelle.

contain a **signal sequence** (or **signal peptide**)—a short stretch of amino acids that indicates where in the cell the polypeptide belongs. Proteins destined for different locations have different signals.

Protein synthesis always begins on free ribosomes, and the "default" location for a protein is the cytosol. In the absence of a signal sequence, the protein will remain in the same cellular compartment where it was synthesized. Some proteins contain signal sequences that "target" them to the nucleus, mitochondria, plastids, or peroxisomes (**Figure 14.18**). A signal sequence binds to a specific receptor protein at the surface

of the organelle. Once it has bound, a channel forms in the organelle membrane, allowing the targeted protein to move into the organelle. For example, here is a nuclear localization signal (NLS):

-Pro-Pro-Lys-Lys-Lys-Arg-Lys-Val-

How do we know this? The function of this peptide was established using experiments like the one illustrated in **Figure 14.19**. Proteins were made in the laboratory with or without the peptide, and then tested by injecting them into cells. Only proteins with the NLS were found in the nucleus.

INVESTIGATING**LIFE**

14.19 Testing the Signal   A. Richardson and his colleagues performed a series of experiments to test whether the nuclear localization signal (NLS) is all that is needed to direct a protein to the nucleus.[a]

HYPOTHESIS   An NLS is necessary for import of a protein into the nucleus.

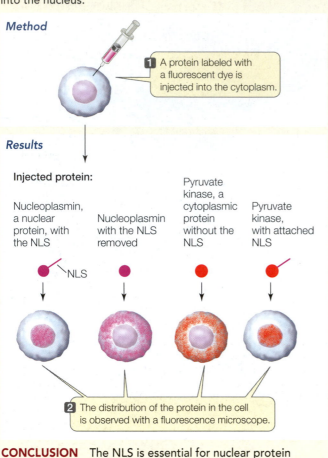

Method

1 A protein labeled with a fluorescent dye is injected into the cytoplasm.

Results

Injected protein:

| Nucleoplasmin, a nuclear protein, with the NLS | Nucleoplasmin with the NLS removed | Pyruvate kinase, a cytoplasmic protein without the NLS | Pyruvate kinase, with attached NLS |

NLS

2 The distribution of the protein in the cell is observed with a fluorescence microscope.

CONCLUSION   The NLS is essential for nuclear protein import and is sufficient to direct a normally cytoplasmic protein to the nucleus.

Go to **BioPortal** for discussion and relevant links for all INVESTIGATING**LIFE** figures.

[a]Dingwall, C. et al. 1988. *The Journal of Cell Biology* 107: 841–849.

If a polypeptide carries a particular signal of about 20 hydrophobic amino acids at its N terminus, it will be directed to the rough endoplasmic reticulum (ER) for further processing (see Figure 14.18). Translation will pause, and the ribosome will bind to a receptor at the ER membrane. Once the polypeptide–ribosome complex is bound, translation will resume, and as elongation continues, the protein will traverse the ER membrane. Such proteins may be retained in the lumen (the inside of the ER) or in membrane of the ER, or they may move elsewhere within the endomembrane system (Golgi apparatus, lysosomes, and plasma membrane). If the proteins lack specific signals or modifications (see below) that specify destinations within the endomembrane system, they are usually secreted from the cell via vesicles that fuse with the plasma membrane.

The importance of signals is shown by Inclusion-cell (I-cell) disease, an inherited disease that causes death in early childhood. People with this disease have a mutation in the gene encoding a Golgi enzyme that adds specific sugars to proteins destined for the lysosomes. These sugars act like signal sequences; without them, enzymes that are essential for the hydrolysis of various macromolecules cannot reach the lysosomes, where the enzymes are normally active. Without these enzymes, the macromolecules accumulate in the lysosomes, and this lack of cellular recycling has drastic effects, resulting in early death.

Many proteins are modified after translation

Most mature proteins are not identical to the polypeptide chains that are translated from mRNA on the ribosomes. Instead, most polypeptides are modified in any of several ways after translation (**Figure 14.20**). These modifications are essential to the final functioning of the protein.

- **Proteolysis** is the cutting of a polypeptide chain, a reaction catalyzed by enzymes called proteases (also called peptidases or proteinases). Cleavage of the signal sequence from the growing polypeptide chain in the ER is an example of proteolysis (see Figure 14.18); the protein might move back out of the ER through the membrane channel if the signal sequence were not cut off. Some proteins are actually made from polyproteins (long polypeptides) that are cut into final products

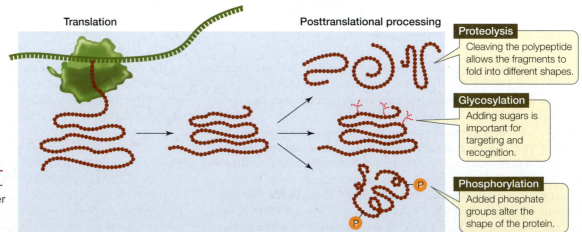

14.20 Posttranslational Modifications of Proteins   Most polypeptides must be modified after translation in order to become functional proteins.

Translation

Posttranslational processing

Proteolysis
Cleaving the polypeptide allows the fragments to fold into different shapes.

Glycosylation
Adding sugars is important for targeting and recognition.

Phosphorylation
Added phosphate groups alter the shape of the protein.

by proteases. These proteases are essential to some viruses, including human immunodeficiency virus (HIV), because the large viral polyprotein cannot fold properly unless it is cut. Certain drugs used to treat acquired immune deficiency syndrome (AIDS) work by inhibiting the HIV protease, thereby preventing the formation of proteins needed for viral reproduction.

- **Glycosylation** is the addition of sugars to proteins to form glycoproteins. In both the ER and the Golgi apparatus, resident enzymes catalyze the addition of various sugars or short sugar chains to certain amino acid R groups on proteins. One such type of "sugar coating" is essential for directing proteins to lysosomes, as mentioned above. Other types are important in the conformation of proteins and their recognition functions at the cell surface (see Section 6.2). Other attached sugars help stabilize extracellular proteins, or proteins stored in vacuoles in plant seeds.

- **Phosphorylation** is the addition of phosphate groups to proteins, and is catalyzed by protein kinases. The charged phosphate groups change the conformation of a protein, often exposing the active site of an enzyme or the binding

site for another protein. We have seen the important role of phosphorylation in cell signaling (see Chapter 7) and the cell cycle (see Chapter 11).

RECAP 14.6

Signal sequences in polypeptides direct them to their appropriate destinations inside or outside the cell. Many polypeptides are modified after translation.

- How do signal sequences determine where a protein will go after it is made? **See pp. 299–300 and Figure 14.18**
- What are some ways in which posttranslational modifications alter protein structure and function? **See pp. 300–301 and Figure 14.20**

All of the processes we have just described result in a functional protein, but only if the amino acid sequence of that protein is correct. If the sequence is not correct, cellular dysfunction may result. Changes in the DNA—mutations—are a major source of errors in amino acid sequences. This is the subject of the next chapter.

?

Can new treatments focused on gene expression control MRSA?

ANSWER

Help may be on the way in the form of a different type of antibiotic that is being developed by Paul Dunman at the University of Rochester, New York. Dr. Dunman and his colleagues have described an important step in gene expression in bacteria, namely the breakdown of mRNA. Bacteria often exist in rapidly changing environments, and so they must adapt quickly by expressing new sets of genes appropriate for the new conditions. They are able to do this in part because they break down

mRNA as soon as it is used, so that its monomers can be used in new mRNA molecules. The scientists screened a library of commercially available small molecules and identified one that targets a protein in this RNA breakdown machinery. Bacteria treated with this molecule cannot recycle old mRNA molecules to provide the monomers needed for new ones. Without the ability to adapt, the bacteria soon die. MRSA are particularly sensitive to this antibiotic.

CHAPTERSUMMARY 14

14.1 What Is the Evidence that Genes Code for Proteins?

- Experiments on metabolic enzymes in the bread mold *Neurospora* led to the **one-gene, one-enzyme hypothesis**. We now know that there is a **one-gene, one-polypeptide relationship**. Review Figure 14.1

14.2 How Does Information Flow from Genes to Proteins?

- The **central dogma** of molecular biology states that DNA encodes RNA, and RNA encodes proteins. Proteins do not encode proteins, RNA, or DNA.

- The process by which the information in DNA is copied to RNA is called **transcription**. The process by which a protein is built from the information in RNA is called **translation**. Review Figure 14.2, ACTIVITY 14.1

- A product of transcription is **messenger RNA (mRNA)**. **Transfer RNAs (tRNAs)** translate the genetic information in the mRNA

into a corresponding sequence of amino acids to produce a polypeptide.

- Certain RNA viruses are exceptions to the central dogma. For example, **retroviruses** synthesize DNA from RNA in a process called **reverse transcription**.

14.3 How Is the Information Content in DNA Transcribed to Produce RNA?

- In a given gene, only one of the two strands of DNA (the template strand) acts as a template for transcription. **RNA polymerase** is the catalyst for transcription.

- RNA transcription from DNA proceeds in three steps: **initiation**, **elongation**, and **termination**. Review Figure 14.4, ANIMATED TUTORIAL 14.1

- Initiation requires a **promoter**, to which RNA polymerase binds. Part of each promoter is the **initiation site**, where transcription begins.

- Elongation of the RNA molecule proceeds from the 5' to 3' end.

continued

- Particular base sequences specify termination, at which point transcription ends and the RNA transcript separates from the DNA template.

- The **genetic code** is a "language" of triplets of mRNA nucleotide bases (**codons**) corresponding to 20 specific amino acids; there are **start** and **stop codons** as well. The code is redundant (an amino acid may be represented by more than one codon) but not ambiguous (no single codon represents more than one amino acid). **Review Figures 14.5, 14.6, ANIMATED TUTORIAL 14.2, ACTIVITY 14.2**

14.4 How Is Eukaryotic DNA Transcribed and the RNA Processed?

- Unlike prokaryotes, where transcription and translation occur in the cytoplasm and are coupled, in eukaryotes transcription occurs in the nucleus and translation occurs later in the cytoplasm.

- Eukaryotic genes contain **introns**, which are noncoding sequences within the transcribed regions of genes. **Review Figures 14.7B, 14.8**

- The initial transcript of a eukaryotic protein-coding gene is modified with a **5′ cap** and a 3′ **poly A tail**. **Review Figure 14.9**

- Pre-mRNA introns are removed in the nucleus via **RNA splicing** by the **small nuclear ribonucleoprotein particles**. Then the mRNA passes through a nuclear pore into the cytoplasm, where it is translated through **ribosomes**. **Review Figure 14.10, ANIMATED TUTORIAL 14.3**

14.5 How Is RNA Translated into Proteins?
See **ANIMATED TUTORIAL 14.4**

- During translation, amino acids are linked together in the order specified by the codons in the mRNA. This task is achieved by tRNAs, which bind to (are charged with) specific amino acids.

- Each tRNA species has an amino acid attachment site as well as an **anticodon** complementary to a specific mRNA codon. A specific activating enzyme charges each tRNA with its specific amino acid. **Review Figures 14.11, 14.12**

- The **ribosome** is the molecular workbench where translation takes place. It has one large and one small subunit, both made of **ribosomal RNA** and proteins.

- Three sites on the large subunit of the ribosome interact with tRNA anticodons. The A site is where the charged tRNA anticodon binds to the mRNA codon; the P site is where the tRNA adds its amino acid to the growing polypeptide chain; and the E site is where the tRNA is released. **Review Figure 14.13**

- Translation occurs in three steps: **initiation**, **elongation**, and **termination**. The **initiation complex** consists of tRNA bearing the first amino acid, the small ribosomal subunit, and mRNA. A specific complementary sequence on the small subunit rRNA binds to the transcription initiation site on the mRNA. **Review Figure 14.14**

- The growing polypeptide chain is elongated by the formation of peptide bonds between amino acids, catalyzed by the rRNA. **Review Figure 14.15**

- When a stop codon reaches the A site, it terminates translation by binding a release factor. **Review Figure 14.16**

- In a **polysome**, more than one ribosome moves along a strand of mRNA at one time. **Review Figure 14.17**

14.6 What Happens to Polypeptides after Translation?

- **Signal sequences** of amino acids direct polypeptides to their cellular destinations. **Review Figures 14.18, 14.19**

- Destinations in the cytoplasm include organelles, which proteins enter after being recognized and bound by surface receptors.

- Proteins "addressed" to the ER bind to a receptor protein in the ER membrane. **Review Figure 14.18**

- Posttranslational modifications of polypeptides include **proteolysis**, in which a polypeptide is cut into smaller fragments; **glycosylation**, in which sugars are added; and **phosphorylation**, in which phosphate groups are added. **Review Figure 14.20**

Go to the Interactive Summary to review key figures, Animated Tutorials, and Activities
Life10e.com/is14

CHAPTER**REVIEW**

REMEMBERING

1. The adapters that allow translation of the four-letter nucleic acid language into the 20-letter protein language are called
   a. aminoacyl-tRNA synthase.
   b. transfer RNAs.
   c. ribosomal RNAs.
   d. messenger RNAs.
   e. ribosomes.

2. Which of the following does *not* occur after eukaryotic mRNA is transcribed?
   a. Binding of a sigma factor to the promoter
   b. Capping of the 5′ end
   c. Addition of a poly A tail to the 3′ end
   d. Splicing out of the introns
   e. Transport to the cytosol

3. Transcription
   a. produces only mRNA.
   b. requires ribosomes.
   c. requires tRNAs.
   d. produces RNA growing from the 5′ end to the 3′ end.
   e. takes place only in eukaryotes.

4. Which statement about translation is *not* true?
   a. Translation is RNA-directed polypeptide synthesis.
   b. An mRNA molecule can be translated by only one ribosome at a time.
   c. The same genetic code operates in almost all organisms and organelles.
   d. Energy is used in the formation of the bond between a tRNA and an amino acid.
   e. There are both start and stop codons.

5. Which statement about RNA is *not* true?
   a. Transfer RNA functions in translation.
   b. Ribosomal RNA functions in translation.
   c. RNAs are produced by transcription.
   d. Messenger RNAs are produced on ribosomes.
   e. DNA codes for mRNA, tRNA, and rRNA.

6. The genetic code
   a. is different for prokaryotes and eukaryotes.
   b. has changed during the course of recent evolution.
   c. has 64 codons that code for amino acids.
   d. has more than one codon for many amino acids.
   e. is ambiguous.

UNDERSTANDING & APPLYING

7. Normally, *Neurospora* can synthesize all 20 amino acids. A certain strain of this mold cannot grow in minimal nutritional medium, but grows only when the amino acid leucine is added to the medium. This strain
   a. is dependent on leucine for energy.
   b. has a mutation affecting a biochemical pathway leading to the synthesis of carbohydrates.
   c. has a mutation affecting the biochemical pathways leading to the synthesis of all 20 amino acids.
   d. has a mutation affecting the biochemical pathway leading to the synthesis of leucine.
   e. has a mutation affecting the biochemical pathways leading to the syntheses of 19 of the 20 amino acids.

8. An mRNA has the sequence 5′-AUGAAAUCCUAG-3′. What is the template DNA strand for this sequence?
   a. 5′-TACTTTAGGATC-3′
   b. 5′-ATGAAATCCTAG-3′
   c. 5′-GATCCTAAAGTA-3′
   d. 5′-TACAAATCCTAG-3′
   e. 5′-CTAGGATTTCAT-3′

9. In rats, a gene 1,440 base pairs long codes for an enzyme made up of 192 amino acids. Discuss this apparent discrepancy.

10. Errors in transcription occur about 100,000 times as often as errors in DNA replication. Why can this high rate be tolerated in RNA synthesis but not in DNA synthesis?

ANALYZING & EVALUATING

11. Har Gobind Khorana at the University of Wisconsin synthesized artificial mRNAs such as poly CA (CACA …) and poly CAA (CAACAACAA …). He found that poly CA codes for a polypeptide consisting of alternating threonine (Thr) and histidine (His) residues. There are two possible codons in poly CA: CAC and ACA. One of these must encode histidine and the other threonine—but which is which? The answer comes from results with poly CAA, which produces three different polypeptides: poly Thr, poly Gln (glutamine), and poly Asn (asparagine). (An artificial mRNA can be read, inefficiently, beginning at any point in the chain; there is no specific initiation signal. Thus poly CAA can be read as a polymer of CAA, of ACA, or of AAC.) Compare the results of the poly CA and poly CAA experiments, and determine which codon corresponds with threonine and which with histidine.

12. Beadle and Tatum's experiments showed that a biochemical pathway could be deduced from mutant strains. In bacteria, the biosynthesis of the amino acid tryptophan (T) from the precursor chorismate (C) involves four intermediate chemical compounds, which we will call D, E, F, and G. Here are the phenotypes of various mutant strains. Each strain has a mutation in a gene for a different enzyme; + means growth with the indicated compound added to the medium, and 0 means no growth. Based on these data, order the compounds (C, D, E, F, G, and T) and enzymes (1, 2, 3, 4, and 5) in a biochemical pathway.

| | Addition to medium | | | | | |
Mutant strain	C	D	E	F	G	T
1	0	0	0	0	+	+
2	0	+	+	0	+	+
3	0	+	0	0	+	+
4	0	+	+	+	+	+
5	0	0	0	0	0	+

Go to BioPortal at **yourBioPortal.com** for Animated Tutorials, Activities, LearningCurve Quizzes, Flashcards, and many other study and review resources.

15 Gene Mutation and Molecular Medicine

CHAPTEROUTLINE

15.1 What Are Mutations?

15.2 What Kinds of Mutations Lead to Genetic Diseases?

15.3 How are Mutations Detected and Analyzed?

15.4 How Is Genetic Screening Used to Detect Diseases?

15.5 How Are Genetic Diseases Treated?

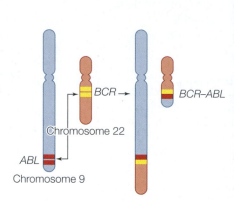

Shuffling the Genetic Deck in Cancer A mutation in a bone marrow cell involves a swap of the ends of two chromosomes. The result is a new gene, encoding a new protein that stimulates cell division. This mutation can cause leukemia.

KAREEM ABDUL-JABBAR had a very successful career as a basketball player. He led his university team (UCLA) to three national championships, and later he led his professional teams to six championships. But in late 2008, with persistent hot flashes and sweats, he went to see his doctor, who sent his blood for testing. A blood smear, where cells are examined in the microscope, showed a high proportion of white blood cells (normally scarce compared with red blood cells). A few days later, there was a diagnosis: chronic myelogenous leukemia (CML), a type of cancer in which immature white blood cells divide continuously. If untreated, the white blood cells crowd out the red blood cells (causing anemia), the platelets (causing clotting disorders), and other types of white blood cells that are part of the immune system (causing infections).

If Abdul-Jabbar had been diagnosed ten years earlier, his CML would have been treated with toxic drugs to stop cancer cell division. He might have expected to live about five years after diagnosis. But a revolutionary treatment has dramatically improved survival in CML. Like many other patients with this cancer, Abdul-Jabbar now has no detectable tumor cells and can expect to live a normal life span. The new treatment came from knowledge of a mutation involved in CML, the abnormal protein produced by that mutation, and a drug targeted to that protein.

The mutation that causes CML occurs in a developing white blood cell in the bone marrow, in which parts of chromosomes 9 and 22 are swapped (translocated). As a result, part of a gene on chromosome 9 (*ABL*) becomes fused with part of a gene on chromosome 22 (*BCR*). The swap usually occurs within the coding regions of the two genes, and the altered chromosome 22 can end up with a new gene made up of the first part of *BCR* and the last part of *ABL*.

The *BCR–ABL* gene fusion happens to be an oncogene; it makes a protein product that stimulates cell division and causes cancer. Immature white blood cells containing the gene fusion divide continuously instead of differentiating into mature white blood cells. The BCR–ABL protein has been crystallized and its structure determined. Its unique structure allowed scientists to design and screen through millions of chemicals to find one that would specifically bind to and inhibit the protein. After a promising chemical was identified, it was modified to improve its binding ability, resulting in the drug imatinib (sold as Gleevec). This drug has been a highly successful targeted therapy for CML and has saved the lives of many patients, including Abdul-Jabbar.

This approach—describing a mutation and its protein phenotype, and then designing a drug for that protein—is at the heart of a new branch of healing called molecular medicine.

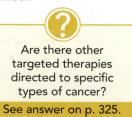

?

Are there other targeted therapies directed to specific types of cancer?

See answer on p. 325.

15.1 What Are Mutations?

In Chapter 12 we described mutations as inherited changes in genes, and we saw that different alleles may produce different phenotypes (short pea plants versus tall, for example). In the following two chapters we described the chemical nature of genes as DNA sequences, and how they are expressed as phenotypes (in particular, proteins). As we mentioned in Section 14.1, a **mutation** is a change in the nucleotide sequence of DNA that can be passed on from one cell, or organism, to another.

As an example of just one cause of mutations, recall from Chapter 13 that DNA polymerases make errors. Repair systems such as proofreading are in place to correct them. But some errors escape being corrected and are passed on to the daughter cells.

Mutations in multicellular organisms can be divided into two types:

- **Somatic mutations** are those that occur in somatic (body) cells. These mutations are passed on to the daughter cells during mitosis, and to the offspring of those cells in turn, but are not passed on to sexually produced offspring. For example, a mutation in a single human skin cell could result in a patch of skin cells that all have the same mutation, but it would not be passed on to the person's children. The mutation that results in chronic myelogenous leukemia (CML; see the opening story) is a somatic mutation in white blood cells.

- **Germ line mutations** are those that occur in the cells of the germ line—the specialized cells that give rise to gametes. A gamete with the mutation passes it on to a new organism at fertilization. The new organism will have the mutation in every cell of its body and will pass the mutation on to all of its progeny.

In either case, the mutations may or may not have phenotypic effects.

Mutations have different phenotypic effects

Phenotypically, we can understand mutations in terms of their effects on proteins and their function (**Figure 15.1**).

- A **silent mutation** does not usually affect protein function (see Figure 15.1B). It can be a mutation in a region of DNA that does not encode a protein, or it can be in the coding region of a gene but not affect the amino acid sequence. Because of the redundancy of the genetic code, a base change in a coding region will not always cause a change in the amino acid sequence when the altered mRNA is translated (see Figure 15.2). Silent mutations are common, and they usually result in genetic diversity that is not expressed as phenotypic differences. We say "usually" here because some silent mutations within coding regions actually do affect protein function. A silent mutation can affect mRNA stability, or it can affect the rate of translation because of differences in the abundance of specific tRNAs. In either case, the abundance of the protein can be affected, and this in turn can affect phenotype.

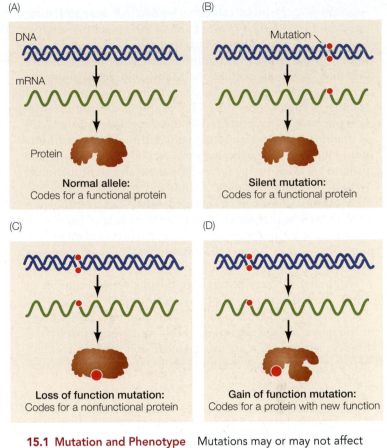

15.1 Mutation and Phenotype Mutations may or may not affect the protein phenotype.

 Go to Animated Tutorial 15.1
Gene Mutations
Life10e.com/at15.1

- A **loss of function mutation** affects protein function (see Figure 15.1C). Such a mutation may cause a gene to not be expressed at all, or the gene may be expressed but produce a protein that no longer plays its cellular role, such as its catalytic function if it is an enzyme. Loss of function mutations almost always show recessive inheritance in diploid organisms, because the presence of one wild-type allele will usually result in sufficient functional protein for the cell. For example, recall from Section 12.1 that the familiar wrinkled-seed allele in pea plants, originally studied by Mendel, is due to a recessive loss of function mutation in the *SBE1* (*starch branching enzyme*). Normally the protein made by this gene catalyzes the branching of starch as seeds develop. In the mutant, the SBE1 protein is not functional, and that leads to osmotic changes, causing the wrinkled appearance.

- A **gain of function mutation** leads to a protein with an altered function (see Figure 15.1D). This kind of mutation usually shows dominant inheritance, because the presence of the wild-type allele does not prevent the mutant allele from functioning. This type of mutation is common in cancer. For example, as we saw in the opening story, the *BCR–ABL* gene fusion is a gain of function mutation that causes uncontrolled cell division.

In addition to the broad categories shown in Figure 15.1, there are mutations that have more subtle effects on phenotype. For example, a **conditional mutation** affects phenotype only under certain restrictive conditions and is not detectable under other, permissive conditions. Many conditional mutations are temperature-sensitive, resulting in proteins with reduced stability at high temperatures. For example, the point restriction phenotype in rabbits and Siamese cats (see Figure 12.15) is due to a temperature-sensitive (conditional, loss of function) mutation in a coat color gene. At body temperature, the protein encoded by the gene is unstable and nonfunctional, so that the animal has dark fur only in its cooler extremities.

Most mutations can be reversed—they can be mutated a second time so that the DNA reverts to its original sequence or to a coding sequence that results in the non-mutant phenotype. These are called **reversion mutations**. When this happens within a gene, it causes the phenotype to go back to wild type.

All mutations are alterations in the nucleotide sequence of DNA (or RNA in the case of viruses with an RNA genome). They can be small-scale mutations that alter only one or a few nucleotides, or they can be large-scale mutations in which entire segments of DNA are rearranged, duplicated, or irretrievably lost. Next we will consider small-scale mutations, in particular, point mutations.

Point mutations are changes in single nucleotides

A **point mutation** is the addition or subtraction of a single nucleotide, or the substitution of one nucleotide base for another. There are two kinds of base substitution:

- A **transition** is the substitution of one purine for the other purine, or one pyrimidine for the other:

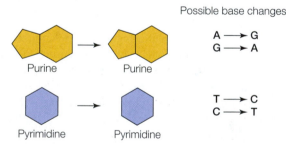

Possible base changes

A ⟶ G
G ⟶ A

Purine   Purine

T ⟶ C
C ⟶ T

Pyrimidine   Pyrimidine

- A **transversion** is the substitution of a purine for a pyrimidine, or vice versa:

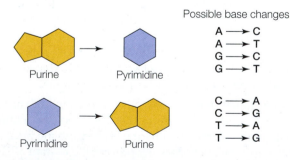

Possible base changes

A ⟶ C
A ⟶ T
G ⟶ C
G ⟶ T

Purine   Pyrimidine

C ⟶ A
C ⟶ G
T ⟶ A
T ⟶ G

Pyrimidine   Purine

A point mutation in the coding region of a gene will result in an alteration in the mRNA sequence, but a change in the mRNA may or may not result in a change in the protein. As we have already discussed, a silent mutation has no effect on the amino acid sequence of an encoded polypeptide. By contrast, missense, nonsense, and frameshift mutations result in changes in the protein, some of them drastic (**Figure 15.2**).

MISSENSE MUTATIONS   Some base substitutions change the genetic code such that one amino acid substitutes for another in a protein. These are called **missense mutations** (see Figure 15.2C). A specific example is the mutation that causes sickle-cell disease, a serious heritable blood disorder. The disease occurs in people who carry two copies of the sickle allele of the gene for β-globin—a subunit of hemoglobin, the protein in human blood that carries oxygen. The sickle allele differs from the normal allele by one base pair, resulting in a polypeptide that differs by one amino acid from the normal protein. Individuals who are homozygous for this recessive allele have defective, sickle-shaped red blood cells (**Figure 15.3**).

A missense mutation may result in a defective protein, but often it has no effect on the protein's function. For example, a hydrophilic amino acid may be substituted for another hydrophilic amino acid, so that the shape of the protein is unchanged. Or a missense mutation might reduce the functional efficiency of a protein rather than completely inactivating it. Therefore individuals homozygous for a missense mutation in a protein essential for life may survive if enough of the protein's function is retained.

In some cases, a gain of function missense mutation occurs. An example is a mutation in the human *TP53* gene, which codes for a tumor suppressor—a protein that inhibits the cell cycle (see Section 11.7). Certain mutations of the *TP53* gene cause this protein to no longer inhibit cell division, but to promote it and prevent programmed cell death. So like the BCR–ABL fusion protein described in the opening story, this kind of mutation results in a TP53 protein that has gained an oncogenic (cancer-causing) function.

NONSENSE MUTATIONS   A **nonsense mutation** involves a base substitution that causes a stop codon (for translation) to form somewhere in the mRNA (see Figure 15.2D). A nonsense mutation results in a shortened protein, since translation does not proceed beyond the point where the mutation occurred. For example, a common mutation causing thalassemia (another blood disorder affecting hemoglobin) in Mediterranean populations is a nonsense mutation that drastically shortens the β-globin subunit. Shortened proteins are usually not functional; however, if the nonsense mutation occurs near the 3' end of the gene, it may have no effect on function.

FRAME-SHIFT MUTATIONS   Not all point mutations are base substitutions. One or two nucleotides may be inserted into, or deleted from, a sequence of DNA. Such mutations in coding sequences are known as **frame-shift mutations** because they alter the reading frame in which the three-base codons are read during translation (see Figure 15.2E). Think again of codons as three-letter words, each corresponding to a particular amino acid. Translation proceeds codon by codon; if a nucleotide is added to the mRNA or subtracted from it, then the three-letter

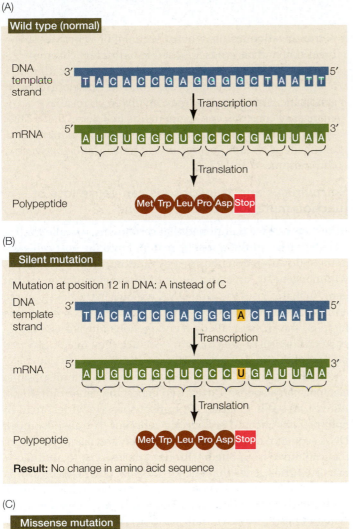

(A) Wild type (normal)

DNA template strand
3′ T A C A C C G A G G G G C T A A T T 5′

↓ Transcription

mRNA
5′ A U G U G G C U C C C C G A U U A A 3′

↓ Translation

Polypeptide
Met Trp Leu Pro Asp Stop

(B) Silent mutation

Mutation at position 12 in DNA: A instead of C

DNA template strand
3′ T A C A C C G A G G G A C T A A T T 5′

↓ Transcription

mRNA
5′ A U G U G G C U C C C U G A U U A A 3′

↓ Translation

Polypeptide
Met Trp Leu Pro Asp Stop

Result: No change in amino acid sequence

(C) Missense mutation

Mutation at position 14 in DNA: A instead of T

DNA template strand
3′ T A C A C C G A G G G C C A A A T T 5′

↓ Transcription

mRNA
5′ A U G U G G C U C C C G G U U A A 3′

↓ Translation

Polypeptide
Met Trp Leu Pro Val Stop

Result: Amino acid change at position 5; Val instead of Asp

(D) Nonsense mutation

Mutation at position 5 in DNA: T instead of C

DNA template strand
3′ T A C A T C G A G G G C C T A A T T 5′

↓ Transcription

mRNA
5′ A U G U A G C U C C C G G A U U A A 3′

↓ Translation

Polypeptide
Met Stop

Result: Only one amino acid translated; no protein made

(E) Frame-shift mutation

Mutation by insertion of T between bases 6 and 7 in DNA

Mutant DNA template strand
3′ T A C A C C T G A G G G C C T A A T T 5′

↓ Transcription

mRNA
5′ A U G U G G A C U C C C G G A U U A A 3′

↓ Translation

Polypeptide
Met Trp Thr Pro Gly Leu →

Result: All amino acids changed beyond the point of insertion

15.2 Point Mutations When they occur in the coding regions of proteins, single base changes can cause silent, missense, nonsense, or frame-shift mutations.

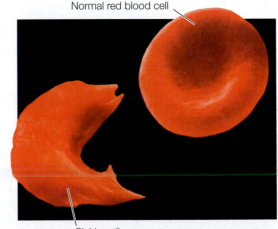

Normal red blood cell

Sickle cell

15.3 Sickle and Normal Red Blood Cells The misshapen red blood cell on the left is caused by a missense mutation and an incorrect amino acid in one of the two polypeptides of hemoglobin.

Go to Media Clip 15.1
Sickle Cells: Deformed by a Mutation
Life10e.com/mc15.1

"words" are altered as translation proceeds beyond that point, and the result is a completely different amino acid sequence. Frame-shift mutations almost always lead to the production of nonfunctional proteins.

Chromosomal mutations are extensive changes in the genetic material

Changes in single nucleotides are not the most dramatic changes that can occur in the genetic material. Whole DNA molecules can break and rejoin, grossly disrupting the sequence of genetic information. There are four types of such

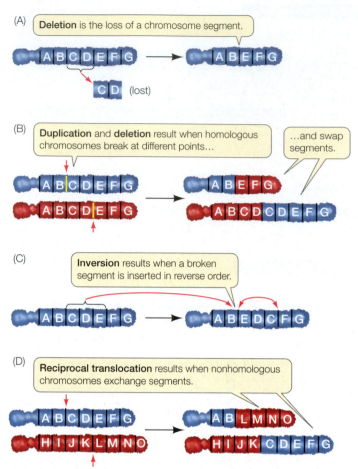

(A) **Deletion** is the loss of a chromosome segment.

A B C D E F G → A B E F G

C D (lost)

(B) **Duplication** and **deletion** result when homologous chromosomes break at different points...

...and swap segments.

A B C D E F G → A B E F G
A B C D E F G → A B C D C D E F G

(C) **Inversion** results when a broken segment is inserted in reverse order.

A B C D E F G → A B E D C F G

(D) **Reciprocal translocation** results when nonhomologous chromosomes exchange segments.

A B C D E F G → A B L M N O
H I J K L M N O → H I J K C D E F G

15.4 Chromosomal Mutations Chromosomes may break during replication, and parts of chromosomes may then rejoin incorrectly. This can result in deletions (A and B), duplications (B), inversions (C), or reciprocal translocations (D). Note that the letters on these illustrations represent large segments of the chromosomes. Because chromosomes contain regions of noncoding DNA, each segment may include anywhere from zero to hundreds or thousands of genes.

chromosomal mutations: deletions, duplications, inversions, and translocations. These mutations can be caused by severe damage to chromosomes resulting from mutagens or by drastic errors in chromosome replication.

- A **deletion** occurs by the removal of part of the genetic material and can happen if a chromosome breaks at two points and then rejoins, leaving out the DNA between the breaks (**Figure 15.4A**).

- A **duplication** can be produced at the same time as a deletion and can occur if homologous chromosomes break at different positions and then reconnect to the wrong partners (**Figure 15.4B**). One of the two chromosomes ends up with a deleted segment, and the other has two copies (a duplication) of the same segment.

- An **inversion** can also result from the breaking and rejoining of a chromosome, and can occur if a segment of DNA becomes "flipped," so that it runs in the opposite direction from its original orientation (**Figure 15.4C**).

- A **translocation** results when a segment of a chromosome breaks off and becomes attached to a different chromosome. As we mentioned in Section 11.5, a translocation of a large segment of chromosome 21 is one cause of Down syndrome. Translocations may involve reciprocal exchanges of chromosome segments, as in **Figure 15.4D**. The opening story of this chapter described the formation of an oncogene as a result of a reciprocal translocation between chromosomes 9 and 22.

Retroviruses and transposons can cause loss of function mutations or duplications

In Section 14.2 we mentioned that certain viruses called retroviruses can insert their genetic material into the host cell's genome. Such insertions happen at random, and if one occurs within a gene, it can cause a loss of function mutation in that gene. In many cases the viral DNA remains in the host genome and is passed on from one generation to the next. When this happens the virus is called an endogenous retrovirus. Endogenous retroviruses are common—in fact, they make up 5 to 8 percent of the human genome!

Another form of DNA, called a transposon or transposable element, can also insert itself into genes and cause mutations. As we will see in Chapter 17, transposons are widespread in both prokaryotic and eukaryotic genomes. A transposon is a DNA sequence of a few hundred to a few thousand base pairs that can move from one position in the genome to another. It usually carries genes that encode the enzymes needed for this movement. Some transposons cut themselves out from their positions in the genome and then insert into other sites (the "cut and paste" mode of transposition). These transposons do not always excise cleanly, but leave behind short sequences of a few base pairs that become permanent mutations in the affected genes. Other transposons first replicate themselves, and then the new copies are inserted into new sites in the genome (the "copy and paste" mode). A sequence of genomic DNA is sometimes carried along with the transposon DNA when it moves, and this results in gene duplication. As we will note below, these gene duplication events play an important role in evolution.

Mutations can be spontaneous or induced

It is useful to distinguish between mutations that are spontaneous or induced, based on their causes. **Spontaneous mutations** are permanent changes in the genetic material that occur without any outside influence. The movement of transposons is an example of spontaneous mutation. Spontaneous mutations can also occur because cellular processes are imperfect, and may occur by several mechanisms:

- *The four nucleotide bases of DNA can have different structures, leading to mistakes during replication.* Each base can exist in two different forms (called tautomers), one of which is common and one rare. When a base temporarily forms its rare tautomer, it can pair with the wrong base. For example, C normally pairs with G, but if C is in its rare tautomer at the time of DNA replication, it pairs with (and DNA

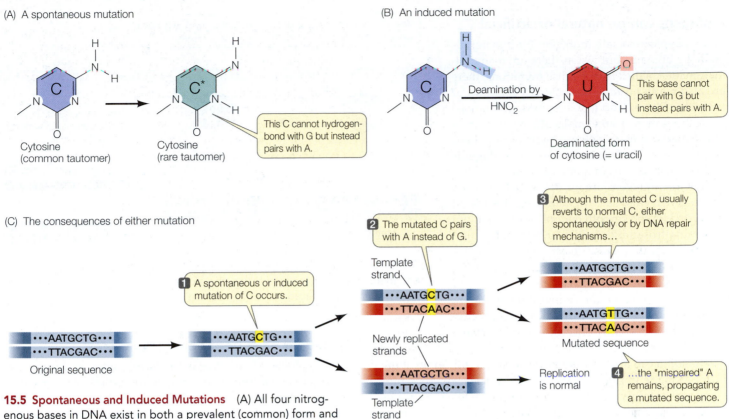

15.5 Spontaneous and Induced Mutations (A) All four nitrogenous bases in DNA exist in both a prevalent (common) form and a rare form. When a base spontaneously forms its rare tautomer, it can pair with a different base. (B) Mutagens such as nitrous acid (HNO₂) can induce changes in the bases. (C) The results of both spontaneous and induced mutations are permanent changes in the DNA sequence following replication.

polymerase will insert) an A. The result is a point mutation: G → A (**Figure 15.5A and C**).

- *Bases in DNA may change because of a chemical reaction*—for example, loss of an amino group in cytosine (a reaction called deamination). If this occurs in a DNA molecule, the error will usually be repaired. However, since the repair mechanism is not perfect, the altered nucleotide will sometimes remain during replication. In these cases, DNA polymerase will add an A (which base-pairs with U) instead of G (which normally pairs with C).

- *DNA polymerase can make errors in replication* (see Section 13.4)—for example, by inserting a T opposite a G. Most of these errors are repaired by the proofreading function of the replication complex, but some errors escape detection and become permanent.

- *Meiosis is not perfect.* Nondisjunction—the failure of homologous chromosomes to separate during meiosis—can occur, leading to one too many chromosomes or one too few (see Figure 11.20). Random chromosome breakage and rejoining can produce deletions, duplications, inversions, or translocations.

Induced mutations occur when some agent from outside the cell—a **mutagen**—causes a permanent change in DNA. As we

mentioned above, retroviruses can function as mutagens. In addition, certain chemicals and radiation can cause mutations:

- *Some chemicals can alter nucleotide bases.* For example, nitrous acid (HNO₂) and similar molecules can react with cytosine and convert it to uracil by deamination. More specifically, they convert an amino group on the cytosine (—NH₂) into a keto group (—C═O) (**Figure 15.5B**). This alteration has the same result as spontaneous deamination: instead of a G, DNA polymerase inserts an A (see Figure 15.5C).

- *Some chemicals add groups to the bases.* For instance, benzopyrene, a component of cigarette smoke, adds a large chemical group to guanine, making it unavailable for base pairing. When DNA polymerase reaches such a modified guanine, it inserts any one of the four bases at random. Three-fourths of the time the inserted base is not cytosine, and a mutation results.

- *Radiation damages the genetic material.* Radiation can damage DNA in two ways. First, ionizing radiation (including X rays, gamma rays, and radiation from unstable isotopes) produces highly reactive chemicals called free radicals. Free radicals can change bases in DNA to forms that are not recognized by DNA polymerase. Ionizing radiation can also break the sugar–phosphate backbone of DNA, causing chromosomal abnormalities. Second, ultraviolet radiation (from the sun or a tanning lamp) is absorbed by thymine, causing it to form covalent bonds with adjacent bases. This, too, plays havoc with DNA replication by distorting the double helix.

Mutagens can be natural or artificial

Many people associate mutagens with materials made by humans, but just as there are many human-made chemicals that cause mutations, there are also many mutagenic substances that occur naturally. Plants (and to a lesser extent animals) make thousands of small molecules with a variety of functions, such as defense against pathogens (see Chapter 39). Some of these are mutagenic and potentially carcinogenic. Examples of human-made mutagens are nitrites, which are used to preserve meats. Once in mammals, nitrites get converted by the smooth endoplasmic reticulum (SER) to nitrosamines, which are strongly mutagenic because they cause deamination of cytosine (see above). An example of a naturally occurring mutagen is aflatoxin, which is made by the mold *Aspergillus*. When mammals ingest the mold, the aflatoxin is converted by the ER into a product that, like benzopyrene from cigarette smoke, binds to guanine; this also causes mutations.

Radiation can also be human-made or natural. Some of the isotopes made in nuclear reactors and nuclear bomb explosions are certainly harmful. For example, extensive studies have shown increased mutations in the survivors of the atom bombs dropped on Japan in 1945. As previously mentioned, natural ultraviolet radiation in sunlight also causes mutations, in this case by affecting thymine and, to a lesser extent, other bases in DNA.

Biochemists have estimated how much DNA damage occurs in the human genome under normal circumstances: among the genome's 3.2 billion base pairs, there are about 16,000 DNA-damaging events per cell per day, of which 80 percent are repaired.

Some base pairs are more vulnerable than others to mutation

In certain regions of DNA, many of the cytosine residues have methyl groups added at their 5' positions, forming 5'–methylcytosine. This methylation plays an important role in gene regulation (see Section 16.4). DNA sequencing has revealed that mutation "hot spots" are often located where cytosines have been methylated.

As we discussed above, unmethylated cytosine can lose its amino group, either spontaneously or because of a chemical mutagen, to form uracil (see Figure 15.5). This type of error is usually detected by the cell and repaired, because uracil is recognized as inappropriate for DNA (it occurs only in RNA). However, when 5'–methylcytosine loses its amino group, the product is thymine, a natural base for DNA (**Figure 15.6**). The DNA repair mechanism ignores this thymine. During replication, however, the mismatch repair mechanism recognizes that G-T is a mismatched pair, although it cannot tell which base is incorrect. Half of the time it matches a new C to the G, but the other half of the time it matches a new A to the T, resulting in a mutation.

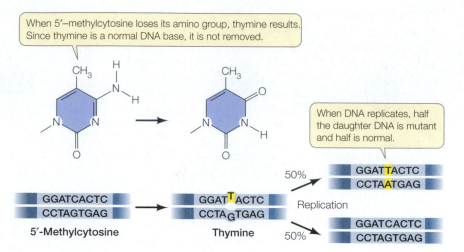

When 5'–methylcytosine loses its amino group, thymine results. Since thymine is a normal DNA base, it is not removed.

When DNA replicates, half the daughter DNA is mutant and half is normal.

15.6 5'-Methylcytosine in DNA Is a "Hot Spot" for Mutations If cytosine has been methylated to 5'-methylcytosine, the mutation is unlikely to be repaired and a C-G base pair is replaced with a T-A pair.

Mutations have both benefits and costs

As we will see in Part Seven of this book, mutations are the raw material of evolution: they provide the genetic diversity that makes natural selection possible. This diversity can be beneficial in two ways. First, a mutation in a somatic cell may benefit the organism immediately. Second, a mutation in a germ line cell may have no immediate selective advantage to the organism, but it may cause a phenotypic change in the organism's offspring. If the environment changes in a later generation, this mutation may be advantageous, and thus selected for, under the new conditions.

We have seen that gene duplication may arise through either chromosomal rearrangements or through the movements of transposons. Gene duplication is not always harmful and is an important source of genetic variation. In a duplicated pair of genes, one gene may continue to play its original role in the cell, while the other may acquire a gain of function mutation that produces a new phenotype. As for any other mutation, this may be of immediate benefit for the organism, or it may provide a later generation with a selective advantage.

By contrast, mutations in genes whose products are needed for normal cellular processes are often deleterious, especially if they occur in germ line cells. In such cases, some offspring can inherit harmful recessive alleles in the homozygous condition. In their extreme form, such mutations produce phenotypes that are lethal, killing the organism during early development.

Mutations in somatic cells can also have dramatic harmful consequences for an organism. In Chapter 11 we described how mutations in oncogenes can result in uncontrolled cell division, whereas loss of function mutations in tumor suppressor genes can prevent the inhibition of cell division in cells that normally do not divide. Both kinds of mutations can lead to cancer, and they can be either spontaneous or induced. While spontaneous mutagenesis is not in our control, we can certainly try to avoid mutagenic substances and radiation. Not surprisingly, many things that cause cancer (carcinogens) are also mutagens. A good example is benzopyrene (discussed above), which is found in coal tar, car exhaust fumes, and charbroiled foods, as well as in cigarette smoke.

A major public policy goal is to reduce the effects of both human-made and natural mutagens on human health. Here are two examples:

- The Montreal Protocol is the only international environmental agreement signed and adhered to by all members of the United Nations. It bans chlorofluorocarbons and other substances that cause depletion of the ozone layer in the upper atmosphere of Earth. Such depletion can result in increased ultraviolet radiation reaching Earth's surface. This would cause more somatic mutations that lead to skin cancer.

- Bans on cigarette smoking have rapidly spread throughout the world. Cigarette smoking causes cancer because of increased exposure of somatic cells in the lungs and throat to benzopyrene and other carcinogens.

RECAP 15.1

Mutations are alterations in the nucleotide sequence of DNA. They may be changes in single nucleotides or extensive rearrangements of chromosomes. If they occur in somatic cells, they will be passed on to daughter cells; if they occur in germ line cells, they will be passed on to offspring.

- What are the various kinds of point mutations? **See p. 306 and Figure 15.2**
- What distinguishes the various kinds of chromosomal mutations: deletions, duplications, inversions, and translocations? **See p. 308 and Figure 15.4**
- Explain the difference between spontaneous and induced mutagenesis. Give an example of each. **See pp. 308–309 and Figure 15.5**
- Why do many mutations involve G-C base pairs? **See p. 310 and Figure 15.6**

We have seen that there are many different ways in which DNA can be altered, in terms of both the types of changes and the mechanisms by which they occur. We will turn now to the ways in which mutations can cause disease.

15.2 What Kinds of Mutations Lead to Genetic Diseases?

The biochemistry that relates genotype (DNA) and phenotype (proteins) has been most completely described for model organisms, such as the prokaryote *E. coli* and the eukaryotes yeast and *Drosophila*. While the details vary, there is great similarity in the fundamental processes among these forms of life. These similarities have permitted the application of knowledge and methods discovered using these model organisms to the study of human biochemical genetics. Our focus in this chapter is mutations that affect human phenotypes, leading to diseases.

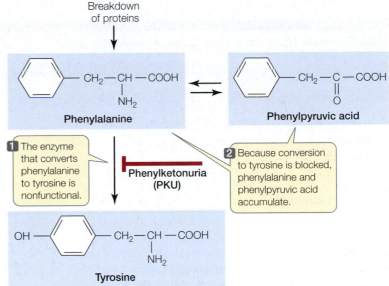

15.7 One Gene, One Enzyme Phenylketonuria is caused by an abnormality in a specific enzyme that metabolizes the amino acid phenylalanine. Knowing the molecular causes of such single-gene, single-enzyme metabolic diseases can aid researchers in developing screening tests as well as treatments.

Genetic mutations may make proteins dysfunctional

Genetic mutations are often expressed phenotypically as proteins that differ from normal (wild-type) proteins. Abnormalities in enzymes, receptor proteins, transport proteins, structural proteins, and most of the other functional classes of proteins have all been implicated in genetic diseases.

LOSS OF ENZYME FUNCTION In 1934, the urine of two mentally retarded young siblings was found to contain phenylpyruvic acid, an unusual by-product of the metabolism of the amino acid phenylalanine. It was not until two decades later that the complex clinical phenotype of the disease that afflicted these children, called phenylketonuria (PKU), was traced back to its molecular cause. The disease results from an abnormality in a single enzyme, phenylalanine hydroxylase (PAH), which catalyzes the conversion of dietary phenylalanine to tyrosine (**Figure 15.7**). This enzyme is not active in the livers of PKU patients, leading to excesses of phenylalanine and phenylpyruvic acid in the blood. Since then, the nucleotide sequence of the *PAH* gene has been compared between healthy people and those with the PKU disease, and more than 400 different disease-causing mutations have been found. A common one is a missense mutation that results in tryptophan instead of arginine at position 408 in the polypeptide chain (**Table 15.1**). As is often the case with loss of function mutations, the mutant alleles are recessive, because one functional allele is all that is needed to produce enough functional PAH to prevent the disease.

Hundreds of human genetic diseases that result from enzyme abnormalities have been discovered, some of which lead

TABLE 15.1
Two Common Mutations That Cause Phenylketonuria

	Codon 408 (20% of PKU Cases)		Codon 280 (2% of PKU Cases)	
	Normal	Mutant	Normal	Mutant
Length of PAH protein	452 amino acids	452 amino acids	452 amino acids	452 amino acids
DNA at codon	...CGG...	...TGG...	...GAA...	...AAA...
	...GCC...	...ACC...	...CTT...	...TTT...
mRNA at codon	...CGG...	...UGG...	...GAA...	...AAA...
Amino acid at codon	Arginine	Tryptophan	Glutamic acid	Lysine
Active PAH enzyme?	Yes	No	Yes	No

to mental retardation and premature death. Most of these diseases are rare; PKU, for example, shows up in 1 out of every 12,000 newborns. But these diseases are just the tip of the mutation iceberg. Some mutations result in amino acid changes that have no obvious clinical effects (see Figure 15.1). As we have just seen, there can be numerous alleles of a gene. Some produce proteins that function normally, whereas others produce variants that cause disease.

ABNORMAL HEMOGLOBIN As we mentioned in Section 15.1, sickle-cell disease is caused by a recessive, missense mutation. This blood disorder most often afflicts people whose ancestors came from the tropics or from the Mediterranean region. About 1 in 655 African-Americans is homozygous for the sickle allele and has the disease.

Recall that human hemoglobin contains four globin subunits—two α chains and two β chains—as well as the pigment heme (see Figure 3.11). In sickle-cell disease, one of the 146 amino acids in the β-globin polypeptide chain is abnormal: at position 6, glutamic acid is replaced by valine. This replacement changes the charge of the protein (glutamic acid is negatively charged and valine is neutral), causing it to form long, needlelike aggregates in the red blood cells. The phenotypic result is sickle-shaped red blood cells (see Figure 15.3) and an impaired ability of the blood to carry oxygen. The sickled cells tend to block narrow blood capillaries, resulting in tissue damage and eventually death by organ failure.

Because hemoglobin is easy to isolate and study, its variations in the human population have been extensively documented (**Figure 15.8**). Hundreds of single amino acid alterations in β-globin have been reported. For example, at the same position that is mutated in sickle-cell disease (resulting in hemoglobin S), glutamic acid may be replaced by lysine, causing hemoglobin C disease. In this case, the resulting anemia is usually not severe. Many alterations of hemoglobin do not affect the protein's function. In fact, about 5 percent of all humans carry at least one missense point mutation in a β-globin allele.

Some of the more common examples of inherited diseases caused by specific protein defects are listed in **Table 15.2**. These mutations can be dominant, codominant, or recessive, and some are sex-linked.

Disease-causing mutations may involve any number of base pairs

Disease-causing mutations may involve a single base pair, a long stretch of DNA, multiple segments of DNA, or even entire chromosomes (as we saw for Down syndrome in Section 11.5).

POINT MUTATIONS There are many examples of point mutations in human genetic diseases. In some cases, all of the people with the disease have the same genetic mutation. This is the case with sickle-cell anemia. In other cases, many different loss of function point mutations in one gene can lead to the same disease, as we saw above for PKU. This makes sense if you think about the three-dimensional structure of an enzyme protein and the many amino acid changes that could affect its activity.

LARGE DELETIONS Larger mutations may involve many base pairs of DNA. For example, deletions in the X chromosome that

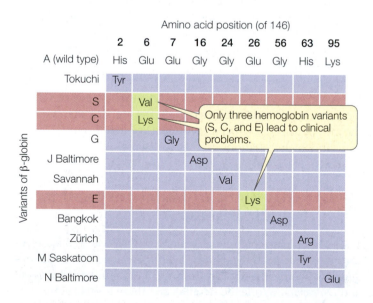

15.8 Hemoglobin Polymorphism Each of these mutant alleles codes for a protein with a single amino acid change in the 146-amino acid chain of β-globin. Only three of the hundreds of known variants of β-globin, shown on the left, are known to lead to clinical abnormalities. "S" is the sickle-cell anemia allele.

TABLE 15.2
Some Human Genetic Diseases

Disease Name	Inheritance Pattern; Births Frequency	Gene Mutated; Protein Product	Clinical Phenotype
Familial hypercholesterolemia	Autosomal codominant; 1 in 500 heterozygous	*LDLR*; low-density lipoprotein receptor	High blood cholesterol, heart disease
Cystic fibrosis	Autosomal recessive; 1 in 4,000	*CFTR*; chloride ion channel in membrane	Immune, digestive, and respiratory illness
Duchenne muscular dystrophy	Sex-linked recessive; 1 in 3,500 males	*DMD*; the muscle membrane protein dystrophin	Muscle weakness
Hemophilia A	Sex-linked recessive; 1 in 5,000 males	*HEMA*; factor VIII blood clotting protein	Inability to clot blood after injury, hemorrhage

include the gene for the protein dystrophin result in Duchenne muscular dystrophy. Dystrophin is important in organizing the structure of muscles, and people who have only the abnormal form have severe muscle weakness. Sometimes only part of the dystrophin gene is missing, leading to an incomplete but partly functional protein and a mild form of the disease. In other cases, deletions span the entire sequence of the gene, so that the protein is missing entirely, resulting in a severe form of the disease. In yet other cases, deletions involve millions of base pairs and cover not only the dystrophin gene but adjacent genes as well; the result may be several diseases in the same person.

CHROMOSOMAL ABNORMALITIES Chromosomal abnormalities also cause human diseases. Such abnormalities result from the gain or loss of complete chromosomes (aneuploidy) (see Figure 11.20), or from the gain or loss of chromosomal segments (see Figure 15.4). About 1 newborn in 200 has a chromosomal abnormality. This may be inherited from a parent who also has the abnormality, or it may result from an error in meiosis during the formation of gametes in one of the parents. One example is the fragile-X syndrome, which is a restriction in the tip of the X chromosome that can result in mental retardation (**Figure 15.9**). About 1 male in 1,500 and 1 female in 2,000 are affected. Although the basic pattern of inheritance is that of an X-linked recessive trait, there are departures from this pattern. Not all people with the fragile-X chromosomal abnormality are mentally retarded, as we will see.

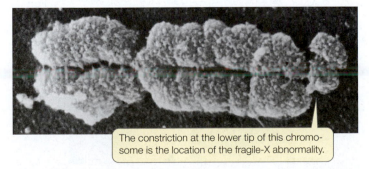

The constriction at the lower tip of this chromosome is the location of the fragile-X abnormality.

15.9 A Fragile-X Chromosome at Metaphase The chromosomal abnormality associated with fragile-X syndrome shows up under the microscope as a constriction in the chromosome. This occurs during preparation of the chromosome for microscopy.

Expanding triplet repeats demonstrate the fragility of some human genes

About one-fifth of all males who have the fragile-X chromosomal abnormality are phenotypically normal, as are most of their daughters. But many of those daughters' sons are mentally retarded. In a family in which the fragile-X syndrome appears, later generations tend to show earlier onset and more severe symptoms of the disease. It is almost as if the abnormal allele itself is changing—and getting worse. And that's exactly what is happening.

The gene responsible for fragile-X syndrome (*FMR1*) contains a repeated triplet, CGG, at a certain point in the promoter region (**Figure 15.10**). In normal people, this triplet is repeated 6 to 54 times (the average is 29). In mentally retarded people with fragile-X syndrome, the CGG sequence is repeated 200 to 2,000 times.

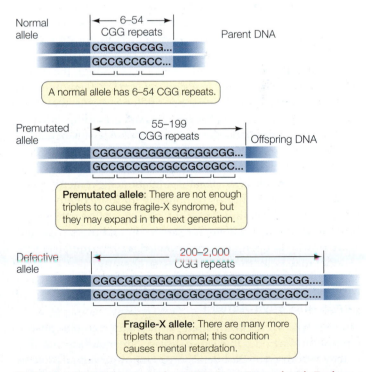

15.10 The CGG Repeats in the *FMR1* Gene Expand with Each Generation The genetic defect in fragile-X syndrome is caused by 200 or more repeats of the CGG triplet.

Males carrying a moderate number of repeats (55–199) show no symptoms and are called premutated. These repeats become more numerous as the daughters of these men pass the chromosome on to their children. With 200 or more repeats, increased methylation of the cytosines in the CGG triplets is likely, which inhibits transcription of the *FMR1* gene. The normal role of the protein product of this gene is to bind to mRNAs involved in neuron function and to regulate their translation at the ribosome. When the FMR1 protein is not made in adequate amounts, these mRNAs are not properly translated, and nerve cells die. Their loss often results in mental retardation.

This phenomenon of **expanding triplet repeats** has been found in more than a dozen other diseases, such as myotonic dystrophy (involving repeated CTG triplets) and Huntington's disease (in which CAG is repeated). Such repeats, which may be found within a protein-coding region or outside it, appear to be present in many other genes without causing harm. How the repeats expand is not known; one hypothesis is that DNA polymerase may slip after copying a repeat and then fall back to copy it again.

Cancer often involves somatic mutations

As we saw in the opening story of this chapter and discussed in Chapter 11, a mutation in a somatic cell can result in cancer. Many chromosomal and point mutations have been described in cancer cells. Such mutations affect oncogenes, whose products stimulate cell division, or tumor suppressor genes, whose products inhibit cell division.

More than two gene mutations are usually needed for full-blown cancer. Because colon cancer progresses to full malignancy slowly, it has been possible to identify the gene mutations that lead to each stage. **Figure 15.11** outlines the "molecular biography" of this form of cancer. At least three tumor suppressor genes and one oncogene must be mutated in sequence in a cell in the colon lining for cancer to develop. Although the occurrence of all of these events in a single cell might seem unlikely, remember that the colon lining has millions of cells, that these cells arise from stem cells that are constantly dividing, and that these changes take place over many years of exposure to natural and synthetic substances in foods, which may act as mutagens.

The description of somatic mutations (as well as the rarer germ line mutations) in cancer is a major achievement of molecular medicine. Cancer is responsible for more deaths in the Western world that any other disease except heart disease. Mutational analysis allows for screening and diagnosis, and as we saw in the opening story, knowledge of the mutated gene product may point the way to targeted therapy.

Most diseases are caused by multiple genes and environment

The example of cancer illustrates how many common phenotypes, including ones that cause disease, are **multifactorial**; that is, they are caused by the interactions of many genes and proteins with one or more factors in the environment. When studying genetics, we tend to call individuals either normal (wild type) or abnormal (mutant); however, in reality every individual contains thousands or millions of genetic variations that arose through

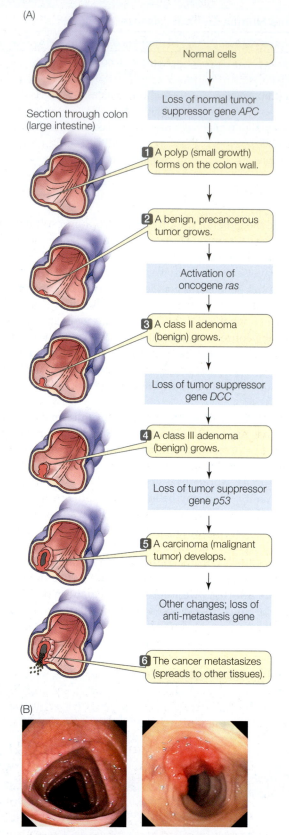

15.11 Multiple Somatic Mutations Transform a Normal Colon Epithelial Cell into a Cancer Cell   (A) At least five genes must be mutated in a single cell to produce colon cancer. (B) These images from a screening test reveal a normal colon (left) and colon cancer (right).

mutations. Our susceptibility to disease is often determined by complex interactions between these genotypes and factors in the environment, such as the foods we eat or the pathogens we encounter. For example, a complex set of genotypes determines who among us can eat a high-fat diet and not experience a heart attack, or who will succumb to disease when exposed to infectious bacteria. Estimates suggest that up to 60 percent of all people are affected by diseases that are genetically influenced. Identifying these genetic influences is another major task of molecular medicine and human genome sequencing.

RECAP 15.2

Many genetic mutations are expressed as nonfunctional enzymes, structural proteins, or membrane proteins. Human genetic diseases may be inherited in dominant, codominant, or recessive patterns, and they may be sex-linked.

- Describe an example of an abnormal protein in humans that results from a genetic mutation and causes a disease. **See pp. 311–313**
- Describe an example of an abnormal protein in humans that results from a genetic mutation and does *not* cause a disease. **See p. 312**
- How do expanding repeats cause genetic diseases? **See pp. 313–314 and Figure 15.10**
- How do somatic mutations cause cancer? **See p. 314 and Figure 15.11**

In the previous section we described the ways in which mutations can lead to human disease. We will turn now to the ways that biologists detect mutations in DNA.

15.3 How Are Mutations Detected and Analyzed?

A challenge for biologists studying mutations is to precisely describe the DNA changes that lead to specific protein changes—an area of research called molecular genetics. Of course, the most direct and comprehensive way to analyze DNA is to determine its sequence of bases. DNA sequencing technologies are continually improving, and the entire genomes of many organisms have now been sequenced completely. Furthermore, the genomes of closely related organisms have been compared in order to identify mutations. We will discuss sequencing technology in Chapter 17. In this section we will look at some of the techniques that are used in combination with DNA sequencing to study DNA, and to identify mutations that cause disease.

Restriction enzymes cleave DNA at specific sequences

All organisms, including bacteria, must have ways of dealing with their enemies. As we saw in Section 13.1, bacteria can be attacked by viruses called bacteriophages. These viruses inject their genetic material into the host cell and turn it into a virus-producing factory, eventually killing the cell. Some bacteria defend themselves against such invasions by producing **restriction enzymes** (also known as restriction endonucleases), which

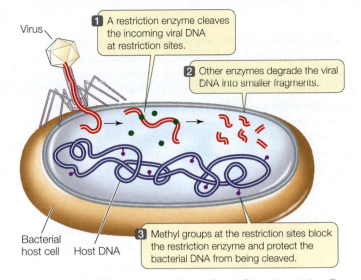

1 A restriction enzyme cleaves the incoming viral DNA at restriction sites.

2 Other enzymes degrade the viral DNA into smaller fragments.

3 Methyl groups at the restriction sites block the restriction enzyme and protect the bacterial DNA from being cleaved.

Virus

Bacterial host cell   Host DNA

15.12 Bacteria Fight Invading Viruses by Making Restriction Enzymes

cut double-stranded DNA molecules—such as those injected by bacteriophages—into smaller, noninfectious fragments (**Figure 15.12**). These enzymes break the bonds of the DNA backbone between the 3' hydroxyl group of one nucleotide and the 5' phosphate group of the next nucleotide. This cutting process is called **restriction digestion**.

There are many such restriction enzymes, each of which cleaves DNA at a specific sequence of bases called a **recognition sequence** or a **restriction site**. Most recognition sequences are four to six base pairs long. Because each sequence of bases has a unique structure (see Section 13.2), it can be specifically recognized by a particular restriction enzyme. Cells protect themselves from being digested by their own enzymes by modifying their DNA, often with methyl groups, to prevent binding by the restriction enzymes.

Restriction enzymes can be isolated from the cells that make them and used as biochemical reagents in the laboratory to give information about the nucleotide sequences of DNA molecules from other organisms. If DNA from any organism is incubated in a test tube with a restriction enzyme (along with buffers and salts that help the enzyme function), that DNA will be cut wherever the restriction site occurs. A specific sequence of bases defines each restriction site. For example, the enzyme *Eco*RI (named after its source strain of the bacterium *E. coli*) cuts DNA only where it encounters the following paired sequence in the DNA double helix:

$$5'\ldots GAATTC\ldots 3'$$
$$3'\ldots CTTAAG\ldots 5'$$

Note that this sequence is palindromic, like the word "mom." This means that both strands have the same sequence when they are read from their 5' (or their 3') ends. The *Eco*RI enzyme has two identical active sites on its two subunits, which cleave the two strands simultaneously between the G and the A of each strand:

$5'\ldots GAATTC\ldots 3'$ $5'\ldots G$ $AATTC\ldots 3'$
$3'\ldots CTTAAG\ldots 5'$ ⟶ $3'\ldots CTTAA$ $G\ldots 5'$

The *Eco*RI recognition sequence occurs, on average, about once in every 4,000 base pairs in a typical prokaryotic genome, or about once per four prokaryotic genes. So *Eco*RI can chop a large piece of DNA into smaller pieces containing, on average, just a few genes. Using *Eco*RI in the laboratory to cut small genomes, such as those of viruses that have tens of thousands of base pairs, may result in just a few fragments. For a huge eukaryotic chromosome with tens of millions of base pairs, a very large number of fragments will be created.

Of course, "on average" does not mean that the enzyme cuts all stretches of DNA at regular intervals. For example, the *Eco*RI recognition sequence does not occur even once in the 40,000 base pairs of the T7 bacteriophage genome—a fact that is crucial to the survival of this virus, since its host is *E. coli*. Fortunately for *E. coli*, the *Eco*RI recognition sequence does appear in the DNA of other bacteriophages.

Gel electrophoresis separates DNA fragments

Restriction enzyme digestion is used to manipulate DNA in the laboratory and to identify and analyze mutations. After a laboratory sample of DNA has been cut with a restriction enzyme, the DNA is in fragments, which must be separated to identify (map) where the cuts were made. Because the recognition sequence does not occur at regular intervals, the fragments are not all the same size, and this property provides a way to separate them from one another. Separating the fragments is necessary to determine the number and molecular sizes (in base pairs) of the fragments produced, or to identify and purify an individual fragment for further analysis or for use in an experiment.

A convenient way to separate or purify DNA fragments is by **gel electrophoresis**. Samples containing the fragments are placed in wells at one end of a semi-solid gel (usually made of agarose or polyacrylamide), and an electric field is applied to the gel (**Figure 15.13**). Because of its phosphate groups, DNA is negatively charged at neutral pH; therefore, because opposite charges attract, the DNA fragments move through the gel toward the positive end of the field. Because the spaces between the polymers of the gel are small, small DNA molecules can move through the gel faster than larger ones. Thus DNA fragments of different sizes separate from one another, forming bands that can be detected with a dye. This provides three types of information:

- *The number of fragments*. The number of fragments produced by digestion of a DNA sample with a given restriction enzyme depends on how many times that enzyme's recognition sequence occurs in the sample. Thus gel electrophoresis can provide some information about the presence of specific DNA sequences (the restriction sites) in the DNA sample.

15.13 Separating Fragments of DNA by Gel Electrophoresis
A mixture of DNA fragments is placed in a gel, and an electric field is applied across the gel. The negatively charged DNA moves toward the positive end of the field, with smaller molecules moving faster than larger ones. After minutes to hours for separation, the electric power is shut off and the separated fragments can be analyzed.

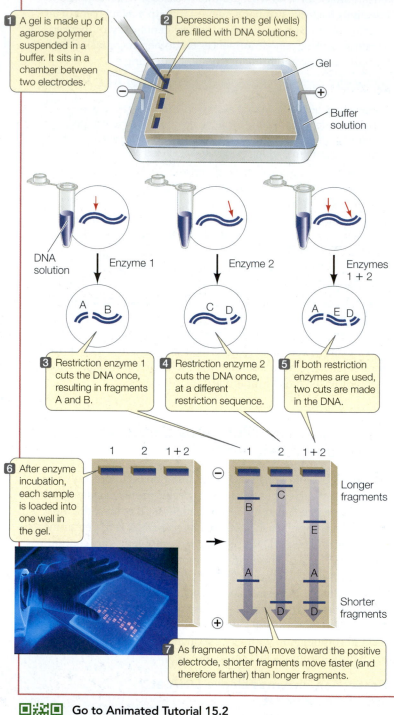

1 A gel is made up of agarose polymer suspended in a buffer. It sits in a chamber between two electrodes.

2 Depressions in the gel (wells) are filled with DNA solutions.

Gel

Buffer solution

DNA solution

Enzyme 1

Enzyme 2

Enzymes 1 + 2

3 Restriction enzyme 1 cuts the DNA once, resulting in fragments A and B.

4 Restriction enzyme 2 cuts the DNA once, at a different restriction sequence.

5 If both restriction enzymes are used, two cuts are made in the DNA.

6 After enzyme incubation, each sample is loaded into one well in the gel.

Longer fragments

Shorter fragments

7 As fragments of DNA move toward the positive electrode, shorter fragments move faster (and therefore farther) than longer fragments.

Go to Animated Tutorial 15.2
Gel Electrophoresis
Life10e.com/at15.2

- *The sizes of the fragments*. DNA fragments of known size (size markers) are often placed in one well of the gel to provide a standard for comparison. The size markers are used to determine the sizes of the DNA fragments in samples in the other wells. By comparing the fragment sizes obtained with two or more restriction enzymes, the locations of their recognition sites relative to one another can be worked out (mapped).

- *The relative abundance of a fragment.* In many experiments, the investigator is interested in how much DNA is present. The relative intensity of a band produced by a specific fragment can indicate the amount of that fragment.

DNA fingerprinting combines PCR with restriction analysis and electrophoresis

The methods we have just described are used in **DNA fingerprinting**, which identifies individuals based on differences in their DNA sequences. DNA fingerprinting works best with sequences that are highly polymorphic—that is, sequences that have multiple alleles (because of many point mutations during the evolution of the organism) and are therefore likely to be different in different individuals. Two types of polymorphisms are especially informative:

- **Single nucleotide polymorphisms** (**SNPs**; pronounced "snips") are inherited variations involving a single nucleotide base—they are point mutations. These polymorphisms have been mapped for many organisms. If one parent is homozygous for the base A at a certain point on the genome, and the other parent is homozygous for a G at that point, the offspring will be heterozygous: one chromosome will have A at that point and the other will have G. If a SNP occurs in a restriction enzyme recognition site, such that one variant is recognized by the enzyme and the other isn't, then individuals can be distinguished from one another very easily using the polymerase chain reaction (PCR) (see Section 13.5). A fragment containing the polymorphic sequence is amplified by PCR from samples of total DNA isolated from each individual. The fragments are then cut with the restriction enzyme and analyzed by gel electrophoresis.

Go to Activity 15.1 Allele-Specific Cleavage
Life10e.com/ac15.1

- **Short tandem repeats** (**STRs**) are short, repetitive DNA sequences that occur side by side on the chromosomes, usually in the noncoding regions. These repeat patterns, which contain one to five base pairs, are also inherited. For example, at a particular locus on chromosome 15 there may be an STR of "AGG." An individual may inherit an allele with six copies of the repeat (AGGAGGAGGAGGAGGAGG) from her mother and an allele with two copies (AGGAGG) from her father. Again, PCR is used to amplify DNA fragments containing these repeated sequences, and then the amplified fragments, which have different sizes because of the different lengths of the repeats, are distinguished by gel electrophoresis (**Figure 15.14A**).

The method of DNA fingerprinting used most commonly today involves STR analysis. The Federal Bureau of Investigation

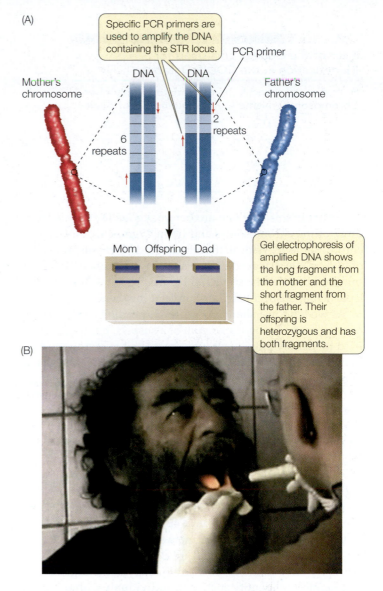

15.14 DNA Fingerprinting with Short Tandem Repeats (A) A particular STR locus can be analyzed to determine the number of repeat sequences that were inherited by an individual from each parent. The two alleles can be identified in an electrophoresis gel on the basis of their sizes. When several STR loci are analyzed, the pattern can constitute a definitive identification of an individual. (B) When the dictator Saddam Hussein was captured in Iraq, a sample of his cheek epithelial cells was taken for DNA fingerprinting. A comparison with DNA fingerprints of relatives provided military scientists with evidence that the man in question was indeed Saddam Hussein.

in the United States uses 13 STR loci in its Combined DNA Index System (CODIS) database (**Table 15.3**). An analysis of these loci in your DNA would reveal your particular DNA fingerprint. Looking at Table 15.3, you might inherit:

From your mother: allele 72 from chromosome 4; allele 23 from chromosome 7; allele 14 from chromosome 11; and allele 12 from chromosome 18

From your father: allele 56 from chromosome 4; allele 22 from chromosome 7; allele 16 from chromosome 11; and allele 12 from chromosome 18

TABLE 15.3
Four of the Genetic Loci Used for Identification in the CODIS Database

Human Chromosome	Locus Name	Repeated Sequence	Number of Alleles
4	FGA	CTTT	80
7	D7S820	GATA	30
11	TH01	TCAT	20
18	S18S51	AGAA	51

Note that in this case you are heterozygous for the alleles on three of the chromosomes and homozygous for the allele on chromosome 18. With all the alleles and 13 loci, the probability of two people sharing the same alleles is very small. So a DNA sample from a crime scene can be used to determine whether a particular suspect left that sample at the scene.

As we have just shown, DNA fingerprinting can be used to help prove the innocence or guilt of a suspect, but it can also be used to identify individuals who are related to one another. On May 2, 2011, Osama bin Laden was killed by U.S. soldiers at his home in Pakistan. He was identified at the scene by comparison with photographs, a wife who pointed him out, and instant analysis using a digital camera with facial recognition software. In addition, DNA fingerprinting was used. Bin Laden's son Khalid was also killed in the raid, and a sister had previously died in the U.S. Analyses of their DNA along with that of Osama indicated that the three shared many polymorphisms and were highly likely to be closely related. The same methods were also used to identify Saddam Hussein, who was captured in 2003 and later executed in Iraq (**Figure 15.14B**).

DNA analysis with genetic markers such as SNPs and STRs has applications throughout all areas of biological research. For example, these markers are used to analyze the organization of genomes, to identify species or individuals within species, to compare species or organisms to see how closely related they are, and to analyze particular genes and the phenotypes associated with them. In the remainder of this chapter we will focus on the use of these markers and other technologies that are used to study and treat genetic diseases.

Reverse genetics can be used to identify mutations that lead to disease

We have seen for diseases such as PKU and sickle-cell anemia that the clinical phenotypes of inherited diseases could be traced to individual proteins, and that the genes could then be identified. With the advent of new ways to identify DNA variations, a new pattern of human genetic analysis has emerged. In these cases, the clinical phenotype is first related to a DNA variation, and then the protein involved is identified. This pattern of discovery is called **reverse genetics**, because it proceeds in the opposite direction to genetic analyses done before the mid-1980s. For example, in sickle-cell anemia, the protein abnormality in hemoglobin was described first (a single amino acid change), and then the gene for β-globin was isolated and the DNA mutation was pinpointed.

Clinical phenotype → protein phenotype → gene

By contrast, for cystic fibrosis (see Table 15.2), a mutant version of the gene *CFTR* was isolated first, and then the protein was characterized:

Clinical phenotype → gene → protein phenotype

Whichever approach is used, final identification of the protein(s) involved in a disease is important in designing specific therapies.

Genetic markers can be used to find disease-causing genes

To identify a mutant gene by reverse genetics, close linkage to a marker sequence is used, in a process called linkage analysis. To understand this type of analysis, imagine an astronaut looking down from space, trying to find her son on a park bench on Chicago's North Shore. The astronaut first picks out reference points—landmarks that will lead her to the park. She recognizes the shape of North America, then moves to Lake Michigan, then the Willis Tower, and so on. Once she has zeroed in on the North Shore Park, she can use advanced optical instruments to find her son.

The reference points for gene isolation are **genetic markers**. Genetic markers such as STRs and SNPs can be used as landmarks to find a gene of interest, if the latter also has multiple alleles (for example, normal and disease-causing alleles). The key to this method is the well-established observation that if two genes are located near each other on the same chromosome, they are usually passed on together from parent to offspring (see Section 12.4). The same holds true for any pair of DNA genetic markers. In the case of linkage analysis, the idea is to find markers that are progressively closer to the gene of interest.

As noted above, SNPs and STRs are widespread in eukaryotic genomes. There is roughly one SNP for every 1,330 base pairs in the human genome, and many regions of the genome also contain repetitive DNA sequences such as those found in STRs. SNPs and STRs can be analyzed using the PCR techniques mentioned above. SNPs can also be detected using sophisticated chemical methods such as mass spectrometry.

To narrow down the location of a gene, a scientist must find a genetic marker that is *always inherited with the gene*. To do this, family medical histories are taken and pedigrees are constructed. If a genetic marker and a genetic disease are inherited together in many families, then they must be near each other on the same chromosome (**Figure 15.15**). This narrows down the location of the gene to a few hundred thousand base pairs.

Once a linked DNA region is identified, many methods are available to identify the actual gene responsible for a genetic disease. The complete sequence of the region can be searched for candidate genes, using information available from databases of genome sequences. With luck a scientist can make an educated guess, based on biochemical or physiological information about the disease, along with information about the functions of candidate genes, as to which gene is responsible

The DNA barcode project aims to identify all organisms on Earth

One of the most exciting aspects of DNA technology for biologists is its potential to identify species, varieties, and even individual organisms from their DNA. In order to repeat experiments and report scientific results, it is essential that biologists know exactly what species or varieties they are studying. However, different organisms can sometimes look very much alike in nature. About 1.8 million species have been named and described, but about ten times that number probably have yet to be identified. A proposal to use DNA technology to identify known species and detect the unknown ones has been endorsed by a large group of scientific organizations known as the Consortium for the Barcode of Life (CBOL).

Evolutionary biologist Paul Hebert at the University of Guelph in Ontario, Canada, was walking down the aisle of a supermarket in 1998 when he noticed the barcodes on all the packaged foods. This gave him an idea to identify each species with a "DNA barcode" that is based on a short sequence from a single gene. The gene he chose is the cytochrome oxidase gene, a component of the respiratory chain that is present in most organisms. Because this gene mutates readily, there are many allelic differences among species. A fragment of 650 to 750 base pairs in this gene is being sequenced for all organisms, and so far sufficient variation has been detected to make it diagnostic for each species (**Figure 15.16**).

Once the DNA of the targeted gene fragment has been sequenced for all known species, a simple device can be used in the field to analyze DNA from an organism and identify its species. The barcode project has the potential to advance biological research on evolution, to track species diversity in ecologically significant areas, to help identify new species, and even to detect undesirable microbes or bioterrorism agents.

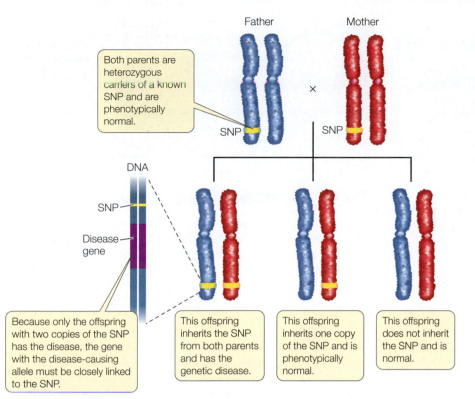

15.15 DNA Linkage Analysis Linkage of an SNP to a disease-causing allele in many families narrows down the location of the defective gene, making its isolation and identification possible. In the example shown here, the disease-causing allele is recessive.

Both parents are heterozygous carriers of a known SNP and are phenotypically normal.

Because only the offspring with two copies of the SNP has the disease, the gene with the disease-causing allele must be closely linked to the SNP.

This offspring inherits the SNP from both parents and has the genetic disease.

This offspring inherits one copy of the SNP and is phenotypically normal.

This offspring does not inherit the SNP and is normal.

for the disease. The identification of DNA polymorphisms within candidate genes that correlate with the presence or absence of disease can also help narrow down the search. A variety of techniques, such as analyzing mRNA levels of candidate genes in diseased and healthy individuals, are used to confirm that the correct gene has been identified.

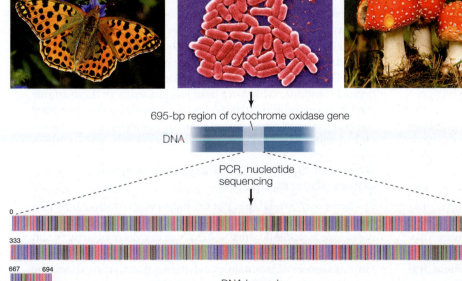

695-bp region of cytochrome oxidase gene

PCR, nucleotide sequencing

DNA barcode

Go to Media Clip 15.2
Barcode of Life
Life10e.com/mc15.2

15.16 A DNA Barcode A 650- to 750-base-pair region of the cytochrome oxidase gene can be amplified by PCR from any organism and then sequenced. This knowledge is used to make a bar code in which each of the four DNA bases is represented as a different color. Such a species barcode permits accurate and rapid identification of a particular species for experimental, ecological, or evolutionary studies.

RECAP 15.3

Large DNA molecules can be cut into smaller pieces by restriction digestion and then sorted by gel electrophoresis. PCR is used to amplify sequences of interest from complex samples. These techniques are used in DNA fingerprinting to analyze DNA polymorphisms for the purpose of identifying individuals. Genes involved in disease can be identified by first detecting the abnormal DNA sequence and then the protein that the wild-type allele encodes. Scientists hope to be able to identify all species using DNA analysis.

- How does a restriction enzyme recognize a restriction site on DNA? **See p. 315**

- How does gel electrophoresis separate DNA fragments? **See p. 316 and Figure 15.13**

- What are STRs, and how are they used to identify individuals? **See pp. 317–318 and Figure 15.14**

- How can a gene mutation that causes a disease be mapped and detected before its protein product is known? **See p. 318 and Figure 15.15**

The determination of the precise molecular phenotypes and genotypes of various human genetic diseases has made it possible to diagnose these diseases even before symptoms appear. Let's take a detailed look at some of these genetic screening techniques.

15.4 How Is Genetic Screening Used to Detect Diseases?

Genetic screening is the use of a test to identify people who have, are predisposed to, or are carriers of a genetic disease. It can be done at many times of life and used for many purposes.

- *Prenatal screening* can be used to identify an embryo or fetus with a disease so that medical intervention can be applied or decisions can be made about whether or not to continue the pregnancy.

- *Newborn babies* can be screened so that proper medical intervention can be initiated quickly for those babies who need it.

- *Asymptomatic people* who have relatives with a genetic disease can be screened to determine whether they are carriers of the disease-associated allele or are likely to develop the disease themselves.

Genetic screening can be done at the level of either the phenotype or the genotype.

Screening for disease phenotypes involves analysis of proteins and other chemicals

Genetic screening can involve examining a protein or other chemical that is relevant to a phenotype associated with a particular disease. Perhaps the best example is the test for phenylketonuria (PKU), which has made it possible to identify the disease in newborns, so that treatment can be started immediately. It is very likely that you were screened as a newborn for PKU.

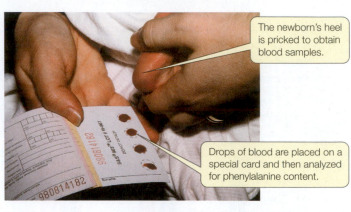

The newborn's heel is pricked to obtain blood samples.

Drops of blood are placed on a special card and then analyzed for phenylalanine content.

15.17 Genetic Screening of Newborns for Phenylketonuria
A blood test is used to screen newborns for phenylketonuria. Small samples of blood are taken from a newborn's heel. The samples are placed in a machine that measures the phenylalanine concentration in the blood. Early detection means that the symptoms of the condition can be prevented by putting the baby on a therapeutic diet.

Initially, babies born with PKU have a normal phenotype because excess phenylalanine in their blood before birth diffuses across the placenta to the mother's circulatory system. Since the mother is almost always heterozygous, and therefore has adequate phenylalanine hydroxylase activity, her body metabolizes the excess phenylalanine from the fetus. After birth, however, the baby begins to consume protein-rich food (milk) and to break down some of his or her own proteins. Phenylalanine begins to accumulate in the blood. After a few days, the phenylalanine level in the baby's blood may be ten times higher than normal. Within days, the developing brain is damaged, and untreated children with PKU become severely mentally retarded. But if detected early, PKU can be treated with a special diet low in phenylalanine to avoid the brain damage that would otherwise result. Thus early detection is imperative.

Newborn screening for PKU and other diseases began in 1963 with the development of a simple, rapid test for the presence of excess phenylalanine in blood serum (**Figure 15.17**). This method uses dried blood spots from newborn babies and can be automated so that a screening laboratory can process many samples in a day. Newborn babies' blood is now screened for up to 35 genetic diseases. Some are common, such as congenital hypothyroidism, which occurs about once in 4,000 births and causes reduced growth and mental retardation because of low levels of thyroid hormone. With early intervention, many of these infants can be successfully treated. So it is not surprising that newborn screening is legally mandatory in many countries, including the United States and Canada.

DNA testing is the most accurate way to detect abnormal genes

The level of phenylalanine in the blood is an indirect measure of phenylalanine hydroxylase activity in the liver. But how can we screen for genetic diseases that are not detectable by blood tests? What if blood is difficult to obtain, as it is in a fetus? How are genetic abnormalities in heterozygotes, who express the normal protein at some level, identified?

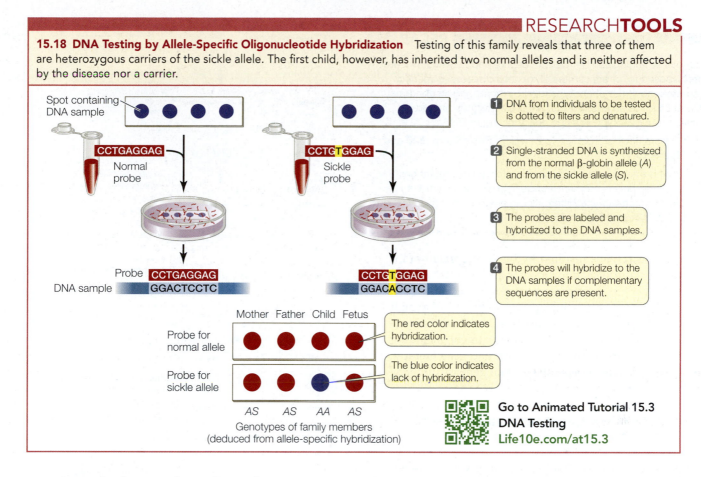

RESEARCHTOOLS

15.18 DNA Testing by Allele-Specific Oligonucleotide Hybridization Testing of this family reveals that three of them are heterozygous carriers of the sickle allele. The first child, however, has inherited two normal alleles and is neither affected by the disease nor a carrier.

Spot containing DNA sample

CCTGAGGAG
Normal probe

CCTGTGGAG
Sickle probe

1 DNA from individuals to be tested is dotted to filters and denatured.

2 Single-stranded DNA is synthesized from the normal β-globin allele (*A*) and from the sickle allele (*S*).

3 The probes are labeled and hybridized to the DNA samples.

Probe CCTGAGGAG
DNA sample GGACTCCTC

Probe CCTGTGGAG
DNA sample GGACACCTC

4 The probes will hybridize to the DNA samples if complementary sequences are present.

Mother Father Child Fetus

Probe for normal allele

The red color indicates hybridization.

Probe for sickle allele

The blue color indicates lack of hybridization.

AS   AS   AA   AS
Genotypes of family members
(deduced from allele-specific hybridization)

Go to Animated Tutorial 15.3
DNA Testing
Life10e.com/at15.3

DNA testing is the direct analysis of DNA for a mutation, and it offers the most direct and accurate way of detecting an abnormal allele. Now that the mutations responsible for many human diseases have been identified, any cell in the body can be examined at any time of life for mutations. The amplification power of PCR means that only one or a few cells are needed for testing. These methods work best for diseases caused by only one or a few different mutations.

Consider, for example, two parents who are both heterozygous for the cystic fibrosis allele but who want to have a normal child. If treated with the appropriate hormones, the mother can be induced to "superovulate," releasing several eggs. An egg can be injected with a single sperm from her husband and the resulting zygote allowed to divide to the eight-cell stage. If one of these embryonic cells is removed, it can be tested for the presence of the cystic fibrosis allele. If the test is negative, the remaining seven-cell embryo can be implanted in the mother's womb where with luck, it will develop normally.

Such preimplantation screening is performed only rarely. More typical are analyses of fetal cells after normal fertilization and implantation in the womb. Fetal cells can be analyzed at about the tenth week of pregnancy by chorionic villus sampling, or during the thirteenth to seventeenth weeks by amniocentesis. In either case, only a few fetal cells are necessary to perform DNA testing. Recently, very sensitive methods were developed so that DNA testing can be done with the few fetal cells that are released into the mother's blood. A 10-milliliter blood sample from a pregnant woman has enough fetal cells for the analysis of

many disorders, including Down syndrome and cystic fibrosis. This relatively noninvasive procedure could replace amniocentesis and chorionic villus sampling—which both carry a slight risk of causing a miscarriage—in the near future.

DNA testing can also be performed with newborns. The blood samples used for screening for PKU and other disorders contain enough of the baby's blood cells to permit DNA analysis using PCR-based techniques. DNA analysis is now being used to screen for sickle-cell disease and cystic fibrosis; similar tests for other diseases will surely follow. Of the numerous methods of DNA testing available, we will describe DNA hybridization, using sickle-cell anemia as an example.

Allele-specific oligonucleotide hybridization can detect mutations

Nucleic acid hybridization (see Figure 14.7) can be used to detect the presence of a specific DNA sequence, such as a sequence containing a particular mutation. Samples of DNA are collected from people who may or may not carry the mutation, and PCR is used to amplify the region of DNA where the mutation may occur. Short synthetic DNA strands called oligonucleotide probes are hybridized with the denatured PCR products. The probe is labeled in some way (e.g., with radioactivity or a fluorescent dye) so that hybridization can be readily detected (**Figure 15.18**).

Detection of a mutation by DNA screening can be used for diagnosis of a genetic disease, so that appropriate treatment can begin. In addition, DNA screening provides a person with important information about his or her genome.

RECAP 15.4

Genetic screening can be used to identify people who have, are predisposed to, or are carriers of, genetic diseases. Screening can be done at the phenotype level by identifying an abnormal protein such as an enzyme with altered activity. It can also be done at the genotype level by direct testing of DNA.

- How are newborn babies screened for PKU? **See p. 320 and Figure 15.17**
- What is the advantage of screening for genetic mutations by allele-specific oligonucleotide hybridization relative to screening phenotype differences in enzyme activity? **See p. 321 and Figure 15.18**

Ongoing research has resulted in the development of increasingly accurate diagnostic tests and a better understanding of various genetic diseases at the molecular level. This knowledge is now being applied to the development of new treatments for genetic diseases. In the next section we will survey various approaches to treatment, including modifications of the mutant phenotype and gene therapy, in which the normal version of a mutant gene is supplied.

15.5 How Are Genetic Diseases Treated?

Most treatments for genetic diseases simply try to alleviate the patient's symptoms. But to effectively treat these diseases—whether they affect all cells, as in inherited disorders such as PKU, or only somatic cells, as in cancer—physicians must be able to diagnose the disease accurately, understand how the disease works at the molecular level, and intervene early, before the disease ravages or kills the individual. There are two main approaches to treating genetic diseases:

- modifying the disease phenotype
- replacing the defective gene (modifying the genotype)

Genetic diseases can be treated by modifying the phenotype

Altering the phenotype of a genetic disease so that it no longer harms an individual is commonly done in one of three ways: by restricting the substrate of a deficient enzyme, by inhibiting a harmful metabolic reaction, or by supplying a missing protein product (**Figure 15.19**).

RESTRICTING THE SUBSTRATE Restricting the substrate of a deficient enzyme is the approach taken when a newborn is diagnosed with PKU. In this case, the deficient enzyme is phenylalanine hydroxylase, and the substrate is phenylalanine (see Figure 15.7). The infant's inability to break down phenylalanine in food leads to a buildup of the substrate, which causes the clinical symptoms. So the infant is immediately put on a special diet that contains only enough phenylalanine for immediate use. Lofenelac, a milk-based product that is low in

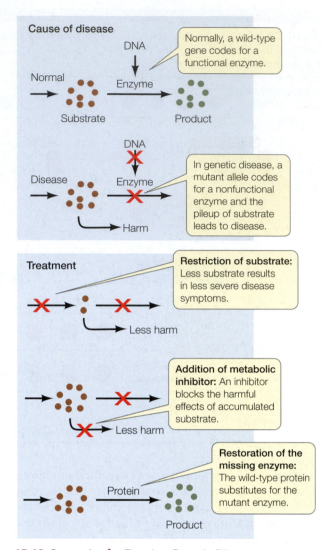

15.19 Strategies for Treating Genetic Diseases

phenylalanine, is fed to these infants just like formula. Later, certain fruits, vegetables, cereals, and noodles low in phenylalanine can be added to the diet. Meat, fish, eggs, dairy products, and bread, which contain high amounts of phenylalanine, must be avoided, especially during childhood, when brain development is most rapid. The artificial sweetener aspartame must also be avoided because it is made of two amino acids, one of which is phenylalanine.

People with PKU are generally advised to stay on a low-phenylalanine diet for life. Although maintaining these dietary restrictions may be difficult, it is effective. Numerous follow-up studies since newborn screening was initiated have shown that people with PKU who stay on the diet are no different from the rest of the population in terms of mental ability. This is an impressive achievement in public health, given the severity of mental retardation in untreated patients.

METABOLIC INHIBITORS In Section 11.7 we described how drugs that are inhibitors of various cell cycle processes are used to treat cancer. Drugs are also used to treat the symptoms of many genetic diseases. As biologists have gained insight into the molecular characteristics of these diseases and the specific proteins involved, a more specific approach to treatment is taking shape. This approach—called molecular medicine—was used to develop the inhibitor used to treat Kareem Abdul-Jabar's chronic myelogenous leukemia (see the opening story).

SUPPLYING THE MISSING PROTEIN An obvious way to treat a disease caused by the lack of a functional protein is to supply that protein. This approach is used to treat hemophilia A, a disease in which blood factor VIII is missing and blood clotting is impaired (see Table 15.2). At first the missing protein was obtained from blood and was sometimes contaminated with viruses (e.g., HIV) or other pathogens that could harm the recipient. Now, however, human clotting proteins are produced by recombinant DNA technology (see Chapter 18), making it possible to provide the protein in a much purer form.

Unfortunately, the phenotypes of many diseases caused by genetic mutations are very complex. In these cases, simple interventions like those we have just described do not work. Indeed, a recent survey of 351 diseases caused by single-gene mutations showed that current therapies increased patients' life spans by an average of only 15 percent.

Gene therapy offers the hope of specific treatments

If a cell lacks an allele that encodes a functional product, an optimal treatment would be to provide a functional allele. The objective of **gene therapy** is to add a new gene that will be expressed in appropriate cells in a patient. What cells should be targeted? There are two approaches:

- Germ line gene therapy, in which the new gene is inserted into a gamete (usually an egg) or the fertilized egg. In this case, all cells of the adult will carry the new gene. Ethical considerations preclude its use in humans.

- Somatic cell gene therapy, in which the new gene is inserted into somatic cells involved in the disease. This method is being tried for numerous diseases, ranging from inherited genetic disorders to cancer.

There are two approaches to somatic cell gene therapy:

- In *ex vivo* gene therapy, target cells are removed from the patient, given the new gene, and then reinserted into the patient. This approach is being used, for example, for diseases caused by defects in genes that are expressed in white blood cells.

- In *in vivo* gene therapy, the gene is actually inserted directly into a patient, targeted to the appropriate cells. An example is a treatment for lung cancer in which a solution with a therapeutic gene is actually squirted onto a tumor.

Armed with knowledge of how genes are expressed (see Chapter 14) and regulated (see Chapter 16), physicians can design a therapeutic gene that contains not only a normal protein-coding sequence but also other sequences—such as an appropriate promotor—required for the gene's expression in targeted cells.

INVESTIGATING**LIFE**

15.20 Gene Therapy Andrew Feigin and his colleagues showed that a virus can be used to insert a therapeutic gene into the brains of patients with Parkinson's disease.[a] The gene they used encodes glutamate decarboxylase, the enzyme that catalyzes the synthesis of the neurotransmitter GABA.

HYPOTHESIS Adding a gene for glutamate decarboxylase to brain cells can alleviate symptoms caused by GABA deficiency.

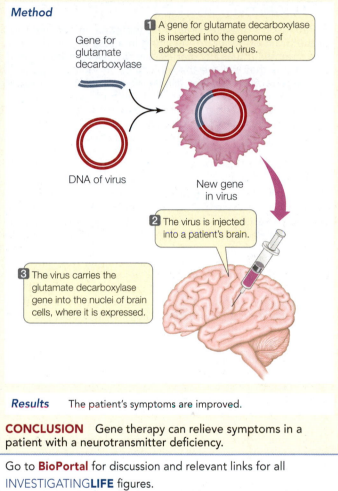

Method

Gene for glutamate decarboxylase

DNA of virus

1 A gene for glutamate decarboxylase is inserted into the genome of adeno-associated virus.

New gene in virus

2 The virus is injected into a patient's brain.

3 The virus carries the glutamate decarboxylase gene into the nuclei of brain cells, where it is expressed.

Results The patient's symptoms are improved.

CONCLUSION Gene therapy can relieve symptoms in a patient with a neurotransmitter deficiency.

Go to **BioPortal** for discussion and relevant links for all INVESTIGATING**LIFE** figures.

[a]LeWitt, P. A. et al. 2011. *Lancet Neurology* 10: 309–319.

A major challenge has been getting the therapeutic gene into cells. Uptake of DNA into eukaryotic cells is a rare event, and once the DNA is inside a cell, its entry into the nucleus and expression are rarer still. One solution to these problems is to insert the gene into a carrier virus that can infect human cells but has been altered genetically to prevent replication. An example is the DNA virus called **adeno-associated virus**, which has been widely used in human gene therapy clinical trials. This virus has a small genome that allows splicing in of a human gene; infects most human cells, including nondividing cells such as neurons; is harmless to humans; does not provoke rejection by the immune system; and enters the cell nucleus, where its DNA with the new gene can be expressed.

A recent success in human gene therapy using adeno-associated virus is shown in **Figure 15.20**. Parkinson's disease is a neurological condition that affects about 1 in 200 persons

WORKING WITH**DATA:**

Gene Therapy for Parkinson's Disease

Original Paper

LeWitt, P. A. and 20 additional authors. 2011. AAV2-GAD gene therapy for advanced Parkinson's disease: a double-blind, sham-surgery controlled randomized trial. *Lancet Neurology* 10: 309–319.

Analyze the Data

In Parkinson's disease, brain cells (neurons) in specific regions of the brain degenerate, resulting in a loss of control of movements. In its early stages, the disease can be treated with drugs that replace the molecules no longer supplied by the degenerated neurons. However, as the disease progresses, these drugs become less effective and have unacceptable side effects. An alternate approach to treatment in such patients is gene therapy to supply an enzyme whose activity produces a molecule that can reduce symptoms. In this case, the enzyme is glutamate decarboxylase and the molecule produced is the neurotransmitter γ-aminobutyric acid (GABA).

Feigin and his colleagues inserted the glutamate decarboxylase gene into a virus that normally infects human cells, in this case, adeno-associated virus (see Figure 15.20). This virus was then injected directly into the region of the brain involved in Parkinson's disease. Sixteen patients received a dose of gene therapy: an injection of a saline solution containing the virus. Twenty-one other patients, with similar ages (average 60) and times since diagnosis (10 years), were in a control group that got a "sham" operation—an injection of saline solution without the virus. Neither the investigators doing the work nor the patients knew who was in the gene therapy and control groups, so this was a "double-blind" study.

At several times after injection, the patients in both groups were examined for changes in motor function. They were each given a score using the unified Parkinson's disease rating scale (UPDRS), which combines both the patient's monitoring of daily activities and a physician's evaluation of motor functions such as walking. A reduction in the score indicates an improvement of function. **Figure A** shows the results, plotted as means with error bars that represent +/– one standard deviation (see Appendix B).

QUESTION 1

Compare the control (sham) group with the gene therapy group. Were there any differences in their UPDRS scores? If so, when were the differences initially apparent? Were the differences statistically significant? How can you tell?

FIGURE A

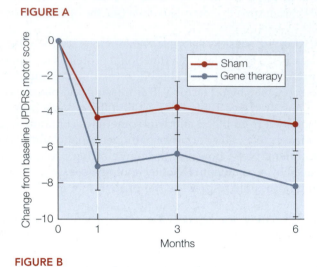

FIGURE B

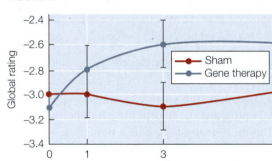

QUESTION 2

Why do you think the score for the control group changed after the sham treatment?

QUESTION 3

A second way to assess symptoms is the global rating of Parkinsonism, in which the patients are asked to evaluate their own symptoms. In this case, an increase in the score indicates improvement. The results are shown in **Figure B**. How did the two groups of patients compare in their own evaluation of their symptoms? How does the fact that this was a double-blind study influence the strength of your conclusions?

Go to **BioPortal** for all WORKING WITH**DATA** exercises

---

in their 60s, and 1 in 25 persons in their 80s. Its symptoms include muscle stiffness and shaking and—in about half of the patients—dementia. In patients with Parkinson's disease, there is inadequate production of the neurotransmitter γ-amino butyric acid (GABA; see Section 45.3) in a particular part of the brain, because of degeneration of the neurons that normally produce GABA. This results in poor coordination of movement. Synthesis of GABA is catalyzed by the enzyme glutamate decarboxylase. A team led by Dr. Andrew Feigin

at the Feinstein Institute in New York has successfully treated patients with Parkinson's disease by using a glutamate decarboxylase gene packaged into adeno-associated virus and injected into the affected part of the brain. After six months, the patients who received the virus-encapsulated gene had increased levels of GABA and significant improvement in their disease symptoms. Other diseases being treated with gene therapy using this virus include cystic fibrosis, hemophilia, and muscular dystrophy.

RECAP 15.5

Treatment of a human genetic disease may involve an attempt to modify the abnormal phenotype by restricting the substrate of a deficient enzyme, inhibiting a harmful metabolic reaction, or supplying a missing protein. By contrast, gene therapy aims to address a genetic defect by inserting a normal allele into a patient's cells.

- How do metabolic inhibitors used in chemotherapy function in treating cancer? **See pp. 322–323 and Figure 15.19**

- How does *in vivo* gene therapy work? Can you give an example? **See pp. 323–324 and Figure 15.20**

In this chapter we dealt with mutations in general, focusing on DNA changes that affect phenotypes through specific protein products. But there is much more to molecular genetics than the sequences of genes and proteins. Determining which genes will be expressed when and where is a major function of the genome. In Chapter 16 we will turn to gene regulation.

(?)

Are there other targeted therapies directed to specific types of cancer?

ANSWER

As the somatic mutations in various cancers are described, specific therapies are being developed. For example, some breast cancer cells overexpress the receptor for the steroid hormone estrogen. This makes the cancer cells abnormally sensitive to estrogen, which stimulates them to divide. Tamoxifen, an analog of estrogen that binds but does not activate the estrogen receptor, is being used to treat breast cancers of this type. Other cancer cells express the receptor for epidermal growth factor, another signal that stimulates cell division. In these cases, treatment with erlotinib, a drug targeted to that receptor, may be effective in slowing cancer growth.

CHAPTERSUMMARY 15

15.1 What Are Mutations?

- **Mutations** are heritable changes in DNA. **Somatic mutations** are passed on to daughter cells, but only **germ line mutations** are passed on to sexually produced offspring. **Review ANIMATED TUTORIAL 15.1**

- **Point mutations** result from alterations in single base pairs of DNA. **Silent mutations** can occur in noncoding DNA or in coding regions of genes and do not affect the amino acid sequences of proteins. **Missense**, **nonsense**, and **frame-shift** mutations all cause changes in protein sequences. **Review Figure 15.2**

- Chromosomal mutations (**deletions**, **duplications**, **inversions**, and **translocations**) involve large regions of chromosomes. **Review Figure 15.4**

- **Spontaneous mutations** occur because of instabilities in DNA or chromosomes. **Induced mutations** occur when a mutagen damages DNA. **Review Figure 15.5**

- Mutations can occur in hot spots where cytosine has been methylated to 5′-methylcytosine. **Review Figure 15.6**

- Mutations, although often detrimental to an individual organism, are the raw material of evolution.

15.2 What Kinds of Mutations Lead to Genetic Diseases?

- Abnormalities in proteins have been implicated in genetic diseases.

- While a single amino acid difference can be the cause of disease, amino acid variations have been detected in many functional proteins. **Review Figures 15.7, 15.8**

- **Point mutations**, **deletions**, and chromosome abnormalities are associated with genetic diseases. **Review Figure 15.9**

- The effects of fragile-X syndrome worsen with each generation. This pattern is the result of an **expanding triplet repeat**. **Review Figure 15.10**

- A series of genetic mutations can lead to colon cancer. **Review Figure 15.11**

- **Multifactorial** diseases are caused by the interactions of many genes and proteins with the environment. They are much more common than diseases caused by mutations in a single gene.

15.3 How Are Mutations Detected and Analyzed?

- **Restriction enzymes**, which are made by microorganisms as a defense against viruses, bind to and cut DNA at specific **recognition sequences** (also called **restriction sites**). These enzymes can be used to produce small fragments of DNA for study, a technique known as restriction digestion. **Review Figure 15.12**

- DNA fragments can be separated by size using **gel electrophoresis**. **Review Figure 15.13, ANIMATED TUTORIAL 15.2**

- **DNA fingerprinting** is used to distinguish among specific individuals or to reveal which individuals are most closely related to one another. It involves the detection of DNA polymorphisms, including **single nucleotide polymorphisms (SNPs)** and **short tandem repeats (STRs)**. **Review Figure 15.14, ACTIVITY 15.1**

- It is possible to isolate both the mutant genes and the abnormal proteins responsible for human diseases. **Review Figure 15.15**

- The goal of the DNA barcoding project is to sequence a single region of DNA in all species for identification purposes.

15.4 How Is Genetic Screening Used to Detect Diseases?

- **Genetic screening** is used to detect human genetic diseases, alleles predisposing people to those diseases, or carriers of those disease alleles.

- Genetic screening can be done by looking for abnormal protein expression. **Review Figure 15.17**

- **DNA testing** is the direct identification of mutant alleles. Any cell can be tested at any time in the life cycle. **Review Figure 15.18, ANIMATED TUTORIAL 15.3**

continued

15.5 **How Are Genetic Diseases Treated?**

- There are three ways to modify the phenotype of a genetic disease: restrict the substrate of a deficient enzyme, inhibit a harmful metabolic reaction, or supply a missing protein. **Review Figure 15.19**
- Cancer sometimes can be treated with treated with metabolic inhibitors.

- In **gene therapy**, a mutant gene is replaced with a normal gene. Both *ex vivo* and *in vivo* therapies are being developed. **Review Figure 15.20**

 Go to the Interactive Summary to review key figures, Animated Tutorials, and Activities
Life10e.com/is15

CHAPTER**REVIEW**

■■■ REMEMBERING

1. Phenylketonuria is an example of a genetic disease in which
   a. a single enzyme is not functional.
   b. inheritance is sex-linked.
   c. two parents without the disease cannot have a child with the disease.
   d. mental retardation always occurs, regardless of treatment.
   e. a transport protein does not work properly.

2. Mutations of the gene for β-globin
   a. are usually lethal.
   b. occur only at amino acid position 6.
   c. number in the hundreds.
   d. always result in sickling of red blood cells.
   e. can always be detected by gel electrophoresis.

3. Multifactorial (complex) diseases
   a. are less common than single-gene diseases.
   b. involve the interaction of many genes with the environment.
   c. affect less than 1 percent of humans.
   d. involve the interactions of several mRNAs.
   e. are exemplified by sickle-cell disease.

4. Mutational "hot spots" in human DNA
   a. always occur in genes that are transcribed.
   b. are common at cytosines that have been modified to 5′–methylcytosine.
   c. involve long stretches of nucleotides.
   d. occur only where there are long repeats.
   e. are very rare in genes that code for proteins.

5. Genetic diagnosis by DNA testing
   a. detects only mutant and not normal alleles.
   b. can be done only on eggs or sperm.
   c. involves hybridization to rRNA.
   d. often uses restriction enzymes and a polymorphic restriction site.
   e. cannot be done with PCR.

6. Current treatments for human genetic diseases include all of the following *except*
   a. restricting a dietary substrate.
   b. replacing the mutant gene in all cells.
   c. alleviating the patient's symptoms.
   d. inhibiting a harmful metabolic reaction.
   e. supplying a protein that is missing.

■■■ UNDERSTANDING & APPLYING

7. Compare the following:
   a. Loss of function mutation versus gain of function mutation
   b. Missense mutation versus nonsense mutation
   c. Spontaneous versus induced mutation

8. In the past, it was common for people with phenylketonuria (PKU) who were placed on a low-phenylalanine diet after birth to be allowed to return to a normal diet during their teenage years. Although the levels of phenylalanine in their blood were high, their brains were thought to be beyond the stage when they could be harmed. If a woman with PKU becomes pregnant, however, a problem arises. Typically, the fetus is heterozygous but is unable, at early stages of development, to metabolize the high levels of phenylalanine that arrive from the mother's blood. (a) Why is the fetus likely to be heterozygous? (b) What do you think would happen to the fetus during this "maternal PKU" situation? (c) What would be your advice to a woman with PKU who wants to have a child?

ANALYZING & EVALUATING

9. Cystic fibrosis is an autosomal recessive disease in which thick mucus is produced in the lungs and airways. The gene responsible for this disease encodes a protein composed of 1,480 amino acids. In most patients with cystic fibrosis, the protein has 1,479 amino acids: a phenylalanine is missing at position 508. A baby is born with cystic fibrosis. He has an older brother who is not affected. How would you test the DNA of the older brother to determine whether he is a carrier for cystic fibrosis? How would you design a gene therapy protocol to "cure" the cells in the younger brother's lungs and airways?

10. Several efforts are under way to identify human genetic polymorphisms that correlate with multifactorial diseases such as diabetes, heart disease, and cancer. What would be the uses of such information? What concerns do you think are being raised about this kind of genetic testing?

11. Tay-Sachs disease is caused by a recessively inherited mutation in the gene coding for the enzyme hexosaminidase A (HEXA), which normally breaks down a lipid called GM2 ganglioside. Accumulation of this lipid in the brain leads to progressive deterioration of the nervous system and death, usually by age 4. HEXA activity in blood serum is 0 to 6 percent in homozygous recessives and 7 to 35 percent in heterozygous carriers, compared with non-carriers (100 percent). The most common mutation in the *HEXA* gene is an insertion of four base pairs, which presumably leads to a premature stop codon. How would you do genetic screening for carriers of this disease by enzyme testing and by DNA testing? What are the advantages of DNA testing? How would you investigate the premature stop codon hypothesis?

12. Alkaptonuria is an inborn error of metabolism, caused by defects in an enzyme in the pathway that breaks down tyrosine (see Section 14.1). Humans who are homozygous for one of these mutations make nonfunctional enzyme and accumulate the enzyme's substrate, homogentisic acid, which causes their disease symptoms. In 1996, researchers in Spain cloned and sequenced the gene for the enzyme, and characterized several mutant alleles. Here is the wild-type coding strand sequence for part of the gene, with the amino acid sequence of the protein below:

… TTG ATA CCC ATT GCC …

… Leu   Ile   Pro   Ile   Ala …

Here is the sequence for one of the mutant alleles:

… TTG ATA TCC ATT GCC …

a. What is the amino acid sequence produced by the mutant allele? What type of mutation is this: silent, nonsense, missense, or frame-shift?

b. Why is this mutation likely to affect the function of the enzyme? (Hint: see Section 3.2, especially Table 3.2.)

c. The mutation creates a new recognition site for the restriction enzyme *Eco*RV (5′-GATATC-3′). How would you take advantage of this new site to screen individuals for the mutant allele?

Go to BioPortal at **yourBioPortal.com** for Animated Tutorials, Activities, LearningCurve Quizzes, Flashcards, and many other study and review resources.

16 Regulation of Gene Expression

CHAPTER**OUTLINE**

16.1 How Is Gene Expression Regulated in Prokaryotes?

16.2 How Is Eukaryotic Gene Transcription Regulated?

16.3 How Do Viruses Regulate Their Gene Expression?

16.4 How Do Epigenetic Changes Regulate Gene Expression?

16.5 How Is Eukaryotic Gene Expression Regulated after Transcription?

A NY TEENAGER OR PARENT OF A TEENAGER will tell you that those years can be rough. But some young people have a rougher time than others. Teens whose mothers were extremely stressed during pregnancy often have more behavioral problems than those whose mothers had calmer pregnancies. A recent study led by psychologist Thomas Elbert and evolutionary biologist Axel Meyer at the University of Konstanz in Germany implicates the regulation of gene expression in these behavioral differences. The study examined a gene for the glucocorticoid receptor, which is involved in regulating hormonal responses to stress. The researchers found that teenagers whose mothers had suffered physical abuse during pregnancy had higher rates of methylation in the promoter of this gene than teenagers whose mothers had not suffered such abuse.

A major control point for gene expression is the promoter, a sequence of DNA adjacent to the coding region of a gene where proteins bind and control the rate of transcription. The ability of these transcription factors to bind to the promoter is affected by the level of DNA methylation in the promoter. As we mentioned in Section 15.1, in certain regions of DNA many of the cytosine residues have methyl groups added at their 5′ positions, forming 5′-methylcytosine. If a gene promoter has a high degree of methylation, some transcription factors can't bind to it. Instead, specific repressor proteins bind to the methylated DNA and prevent expression of the gene. DNA methylation plays an important role in gene regulation and is a normal part of development. But the level of methylation can change over time and can vary among individuals, as was the case for the

Stress Extreme physical or emotional stress during pregnancy can have negative effects on a child many years later. A little relaxation, on the other hand, never hurt anyone.

teenagers in Elbert and Meyer's study. Their finding is interesting because it shows a correlation between maternal stress in humans and DNA methylation in their offspring. However, a lot more work is needed to show how the mother's stress brought about this DNA change, and how this change affects behavior.

Other studies show a link between early life experiences, promoter methylation in the glucocorticoid receptor gene, and behavior. People who are abused as children have altered hormonal stress responses and an increased risk of suicide. One study showed that suicide victims who were abused as children had more methylation of the glucocorticoid receptor gene promotor than in people who were not abused, and this led to lower expression of the gene.

Such studies have spawned the new field of behavioral epigenetics. Epigenetics is the study of heritable changes in gene expression that do not involve changes in the DNA sequence. A deeper understanding of the regulation of gene expression may lead to better treatments of important social problems.

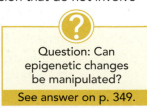

Question: Can epigenetic changes be manipulated?

See answer on p. 349.

16.1 How Is Gene Expression Regulated in Prokaryotes?

Prokaryotes conserve energy and resources by making certain proteins only when they are needed. The protein content of a bacterium can change rapidly when conditions warrant. There are several ways in which a prokaryotic cell could shut off the supply of an unneeded protein. The cell could:

- downregulate the transcription of mRNA for that protein;
- hydrolyze the mRNA after it is made, thereby preventing translation;
- prevent translation of the mRNA at the ribosome;
- hydrolyze the protein after it is made; or
- inhibit the function of the protein.

Whichever mechanism is used, it must be both responsive to environmental signals and efficient. The earlier the cell intervenes in the process of protein synthesis, the less energy it wastes. Selective blocking of transcription is far more efficient than transcribing the gene, translating the message, and then degrading or inhibiting the protein. While all five mechanisms for regulating protein levels are found in nature, prokaryotes generally use the most efficient one: transcriptional regulation.

As we described in Chapter 14, gene expression begins at the promoter, where RNA polymerase binds to initiate transcription. In a genome with many genes, not all promoters are active at a given time—there is selective gene transcription. The "decision" regarding which genes to activate involves two types of regulatory proteins that bind to DNA: repressor proteins and activator proteins. In both cases, these proteins bind to the promoter to regulate the gene (**Figure 16.1**):

- In **negative regulation**, binding of a repressor protein prevents transcription.
- In **positive regulation**, an activator protein binds DNA to stimulate transcription.

You will see examples of these mechanisms, or combinations of them, as we examine regulation in prokaryotes, eukaryotes, and viruses.

Regulating gene transcription conserves energy

As a normal inhabitant of the human intestine, *E. coli* must be able to adjust to sudden changes in its chemical environment. Its host may present it with one foodstuff one hour (e.g., glucose in fruit juice) and another the next (e.g., lactose in milk). Such changes in nutrients present the bacterium with a metabolic challenge. Glucose is its preferred energy source, and is the easiest sugar to metabolize. Lactose is a β-galactoside—a disaccharide containing galactose β-linked to glucose (see Section 3.3). Three proteins are involved in the initial uptake and metabolism of lactose by *E. coli*:

- β-Galactoside permease is a carrier protein in the bacterial plasma membrane that moves the sugar into the cell.

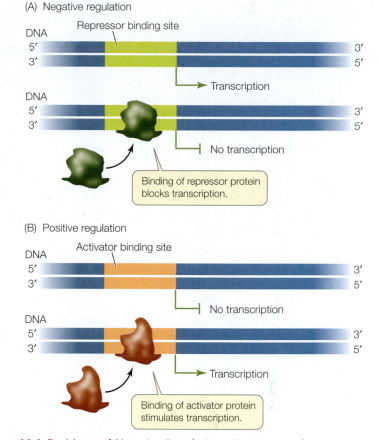

(A) Negative regulation

Repressor binding site

Binding of repressor protein blocks transcription.

(B) Positive regulation

Activator binding site

Binding of activator protein stimulates transcription.

16.1 Positive and Negative Regulation   Proteins regulate gene expression by binding to DNA and preventing or allowing RNA polymerase to bind DNA at the promotor region to control transcription of the gene.

- β-Galactosidase is an enzyme that hydrolyses lactose to glucose and galactose.
- β-Galactoside transacetylase transfers acetyl groups from acetyl CoA to certain β-galactosides. Its role in the metabolism of lactose is not clear.

When *E. coli* is grown on a medium that contains glucose but no lactose or other β-galactosides, the levels of these three proteins are extremely low—the cell does not waste energy and materials making the unneeded enzymes. But if the environment changes such that lactose is the predominant sugar available and very little glucose is present, the bacterium promptly begins making all three enzymes after a short lag period. There are only a few molecules of β-galactosidase present in an *E. coli* cell when glucose is present in the medium. But when glucose is absent, the addition of lactose can induce the synthesis of about 1,500 times more molecules of β-galactosidase per cell (**Figure 16.2A**)!

What is behind this dramatic increase? An important clue comes from measuring the amount of mRNA for β-galactosidase. The mRNA level dramatically increases during the lag period after lactose is added to the medium, and this mRNA is presumably translated into protein (**Figure 16.2B**). Moreover, the high mRNA level depends on the presence of

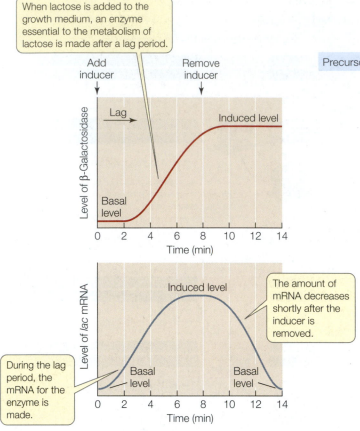

When lactose is added to the growth medium, an enzyme essential to the metabolism of lactose is made after a lag period.

Add inducer | Remove inducer

Lag | Induced level

Basal level

Time (min)

The amount of mRNA decreases shortly after the inducer is removed.

Induced level

During the lag period, the mRNA for the enzyme is made.

Basal level | Basal level

Time (min)

16.2 An Inducer Stimulates the Expression of a Gene for an Enzyme (A) When lactose is added to the growth medium for the bacterium *E. coli*, the synthesis of β-galactosidase begins only after an initial lag. (B) There is a lag because the mRNA for β-galactosidase has to be made before the protein can be made. The amount of mRNA decreases rapidly after the lactose is removed, indicating that transcription is no longer occurring. These changes in mRNA levels indicate that the mechanism of induction by lactose is transcriptional regulation.

lactose, because if the lactose is removed, the mRNA level goes down. The response of the bacterial cell to lactose is clearly at the level of transcription.

Compounds that, like lactose, stimulate the synthesis of a protein are called **inducers**. The proteins that are produced are called **inducible proteins**, whereas proteins that are made all the time at a constant rate are called **constitutive proteins**. (Think of the constitution of a country, a document that does not change under normal circumstances.)

We have now seen two basic ways of regulating the rate of a metabolic pathway. In Section 8.5 we described the allosteric regulation of enzyme activity, which allows the rapid fine-tuning of metabolism. Regulation of protein synthesis—that is, regulation of the concentration of enzymes—is slower but results in greater savings of energy and resources. Protein synthesis is a highly endergonic process, since assembling mRNA, charging tRNA, and moving the ribosomes along mRNA all require the hydrolysis of nucleoside triphosphates such as ATP. **Figure 16.3** compares these two modes of regulation.

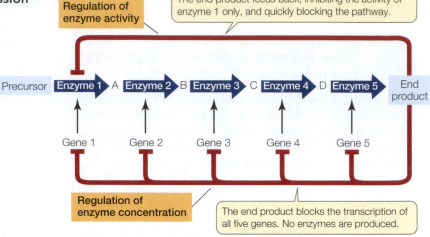

Regulation of enzyme activity

The end product feeds back, inhibiting the activity of enzyme 1 only, and quickly blocking the pathway.

Precursor | Enzyme 1 | A | Enzyme 2 | B | Enzyme 3 | C | Enzyme 4 | D | Enzyme 5 | End product

Gene 1 | Gene 2 | Gene 3 | Gene 4 | Gene 5

Regulation of enzyme concentration

The end product blocks the transcription of all five genes. No enzymes are produced.

16.3 Two Ways to Regulate a Metabolic Pathway Feedback from the end product of a metabolic pathway can block enzyme activity (allosteric regulation), or it can stop the transcription of genes that code for the enzymes in the pathway (transcriptional regulation).

Operons are units of transcriptional regulation in prokaryotes

The genes that encode the three enzymes for processing lactose in *E. coli* are **structural genes**; they specify the primary structures (the amino acid sequences) of protein molecules. Structural genes are genes that can be transcribed into mRNA. The three genes lie adjacent to one another on the *E. coli* chromosome. This arrangement is no coincidence: the genes share a single promoter, and their DNA is transcribed into a single, continuous molecule of mRNA. Because this particular mRNA governs the synthesis of all three lactose-metabolizing enzymes, either all or none of these enzymes are made, depending on whether their common message—their mRNA—is present in the cell.

A cluster of genes with a single promoter is called an **operon**, and the operon that encodes the three lactose-metabolizing enzymes in *E. coli* is called the *lac* operon. The *lac* operon promoter can be very efficient (the maximum rate of mRNA synthesis can be high), but mRNA synthesis can be shut down when the enzymes are not needed. In addition to the promoter, an operon has other **regulatory sequences** that are not transcribed. A typical operon consists of a promoter, an operator, and two or more structural genes (**Figure 16.4**). The **operator** is a short stretch of DNA that lies between the promoter and the structural genes. It can bind very tightly with regulatory proteins that either activate or repress transcription.

There are numerous mechanisms to control the transcription of operons; here we will focus on three examples:

- An inducible operon regulated by a repressor protein
- A repressible operon regulated by a repressor protein
- An operon regulated by an activator protein

Operator–repressor interactions control transcription in the *lac* and *trp* operons

The *lac* operon contains a promoter, to which RNA polymerase binds to initiate transcription, and an operator, to which a **repressor** protein can bind. The gene that encodes this repressor is located near the *lac* operon on the *E. coli* chromosome.

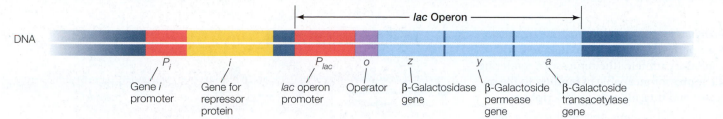

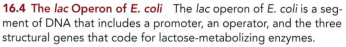

16.4 The *lac* Operon of *E. coli* The *lac* operon of *E. coli* is a segment of DNA that includes a promoter, an operator, and the three structural genes that code for lactose-metabolizing enzymes.

When the repressor is bound, transcription of the operon is blocked. This example of negative regulation was elegantly worked out by Nobel Prize winners François Jacob and Jacques Monod.

The repressor protein has two binding sites: one for the operator and the other for the inducer. The environmental signal that induces the *lac* operon (for example, in the human digestive tract) is lactose, but the actual inducer is allolactose, a molecule that forms from lactose once it enters the cell. In the absence of the inducer, the repressor protein fits into the major groove of the operator DNA and recognizes and binds to a specific nucleotide base sequence. This prevents the binding of RNA polymerase to the promoter, and the operon is not transcribed (**Figure 16.5A**). When the inducer is present, it binds to the repressor and changes the shape of the repressor. This change in three-dimensional structure (conformation) prevents the repressor from binding to the operator. As a result, RNA polymerase can bind to the promoter and start transcribing the structural genes of the *lac* operon (**Figure 16.5B**).

You can see from this example that a key to transcriptional control of gene expression is the presence of regulatory sequences that do not code for proteins, but are binding sites for regulatory proteins and other proteins involved in transcription.

In contrast to the inducible system of the *lac* operon, other operons in *E. coli* are repressible; that is, they are repressed only under specific conditions. In such a system, the repressor is not normally bound to the operator. But if another molecule called a **co-repressor** binds to the repressor, the repressor changes shape and binds to the operator, thereby inhibiting transcription. An example is the *trp* operon, whose structural genes catalyze the synthesis of the amino acid tryptophan:

5 Enzyme-catalyzed reactions

Precursor molecules → → → → → tryptophan

When tryptophan is present in the cell in adequate concentrations, it is advantageous to stop making the enzymes for tryptophan synthesis. To do this, the cell uses a repressor that binds to an operator in the *trp* operon. But the repressor of the *trp* operon is not normally bound to the operator; it only binds when its shape is changed by binding to tryptophan, the co-repressor.

To summarize the differences between these two types of operons:

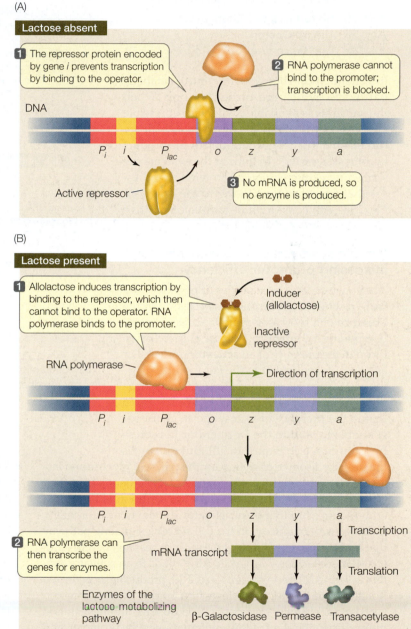

16.5 The *lac* Operon: An Inducible System (A) When lactose is absent, the synthesis of enzymes for its metabolism is inhibited. (B) Lactose (the inducer) leads to synthesis of the enzymes in the lactose-metabolizing pathway by binding to the repressor protein and preventing its binding to the operator.

 Go to Animated Tutorial 16.1
The *lac* Operon
Life10e.com/at16.1

- In *inducible* systems, the substrate of a metabolic pathway (the inducer) interacts with a regulatory protein (the repressor), rendering the repressor incapable of binding to the operator and thus allowing transcription.

- In *repressible* systems, the product of a metabolic pathway (the co-repressor) binds to a regulatory protein, which is then able to bind to the operator and block transcription.

Go to Animated Tutorial 16.2
The *trp* Operon
Life10e.com/at16.2

In general, inducible systems control catabolic pathways (which are turned on only when the substrate is available), whereas repressible systems control anabolic pathways (which are turned on until the concentration of the product becomes excessive). In both of the systems described here, the regulatory protein is a repressor that functions by binding to the operator. Next we will consider an example of positive control involving an activator.

Protein synthesis can be controlled by increasing promoter efficiency

Above we described examples of negative control, in which transcription is *decreased* in the presence of a repressor protein. *E. coli* can also use positive control to *increase* transcription through the presence of an **activator** protein. For an example we return to the *lac* operon, where the relative levels of glucose and lactose determine the amount of transcription. We have seen that in the presence of lactose the *lac* repressor is unable to bind to the *lac* operator to repress transcription (see Figure 16.5B). But glucose is the preferred source of energy for the cell, so if glucose and lactose levels are both high, the *lac* operon is still not transcribed efficiently. This is because efficient transcription of the *lac* operon requires binding of an activator protein to its promoter.

Low levels of glucose in the cell set off a signaling pathway that leads to increased levels of the second messenger cyclic AMP (cAMP) (see Section 7.3). Cyclic AMP binds to an activator protein called *cAMP receptor protein* (CRP), producing a conformational change in CRP that allows it to bind to the *lac* promoter. CRP is an activator of transcription, because its binding results in more efficient binding of RNA polymerase to the promoter, and thus increased transcription of the structural genes (**Figure 16.6**). In the presence of abundant glucose, cAMP levels are low, CRP does not bind to the promoter, and the efficiency of transcription of the *lac* operon is reduced. This is an example of **catabolite repression**, a system of gene regulation in which the presence of the preferred energy source represses other catabolic pathways. The mechanisms controlling positive and negative regulation of the *lac* operon are summarized in **Table 16.1**.

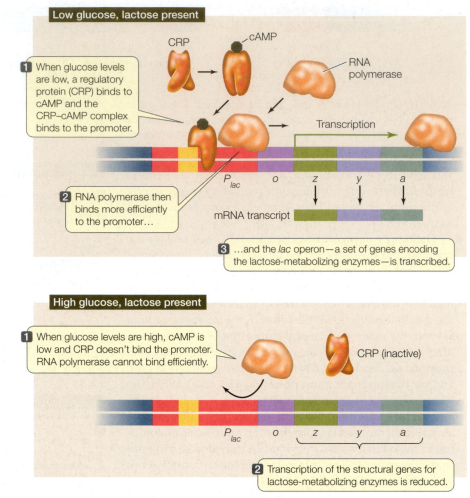

16.6 Catabolite Repression Regulates the *lac* Operon The promoter for the *lac* operon does not function efficiently in the absence of cAMP, as occurs when glucose levels are high. High glucose levels thus repress the enzymes that metabolize lactose.

RNA polymerases can be directed to particular classes of promoters

Thus far we have described a promoter as a specific DNA sequence located upstream of a transcription initiation site. It binds and orients RNA polymerase so that the polymerase transcribes the correct DNA strand. As we mentioned in Chapter 14, not all promoters are identical. However, they share **consensus sequences** that allow them to be recognized by the RNA polymerase and other proteins. (Recall from Section 14.4 that a consensus sequence is a short stretch of DNA that appears, with little variation, in many different genes.)

Prokaryotic promoters generally have two sites for consensus sequences, which begin 10 and 35 base pairs upstream of the transcription start site (the –10 element and the –35 element). Different classes of promoters have different consensus sequences at these two sites. The largest class consists of promoters for "housekeeping genes," which are all the genes that are normally expressed in actively growing cells. In these genes, the consensus –10 element is 5'-TATAAT-3', and the

TABLE 16.1
Positive and Negative Regulation in the *lac* Operon

Glucose	cAMP Levels	RNA Polymerase Binding to Promoter	Lactose	*lac* Repressor	Transcription of *lac* Genes?	Lactose Used by Cells?
Present	Low	Absent	Absent	Active and bound to operator	No	No
Present	Low	Present, not efficient	Present	Inactive and not bound to operator	Low level	No
Absent	High	Present, very efficient	Present	Inactive and not bound to operator	High level	Yes
Absent	High	Absent	Absent	Active and bound to operator	No	No

consensus –35 element is 5′-TTGACAT-3′ (N stands for any nucleotide):

```
5′-NNNTTGACATNNNNNNNNNNNNNNNNNNNNTATAATNNNN*NNNNNNNNNNNNN-3′
3′-NNNAACTGTANNNNNNNNNNNNNNNNNNNNATATTANNNN*NNNNNNNNNNNNN-5′
    –35 Element                    –10 Element   Transcription start site
```

Other classes of genes have different consensus sequences at their –10 and –35 sites. The different classes of consensus sequences are recognized by regulatory proteins called sigma factors.

Sigma factors are proteins in prokaryotic cells that bind to RNA polymerase and direct it to specific classes of promoters. The RNA polymerase must be bound to a sigma factor before it can recognize a promoter and begin transcription. Genes that encode proteins with related functions may be at different locations in the genome, but because they share consensus sequences in their promoters, they are recognized by a particular sigma factor. The sigma-70 factor is active most of the time and binds to the consensus sequences of housekeeping genes; other sigma factors are activated only under specific conditions. For example, when *E. coli* cells experience stress conditions such as DNA damage or osmotic stress, the sigma-38 factor is activated, and it directs RNA polymerase to the promoters of various genes that are expressed under stress conditions. *E. coli* has seven sigma factors; this number varies in other prokaryotes.

As we will see in the next section, this form of global gene regulation by proteins binding to RNA polymerase is also common in eukaryotes.

RECAP 16.1

Gene expression in prokaryotes is most commonly regulated through control of transcription. An operon consists of a set of closely linked structural genes and the DNA sequences (promoter and operator) that control their transcription. Operons can be regulated by both negative and positive controls. Sigma factors control the expression of specific classes of prokaryotic genes that share consensus sequences in their promoters.

- What is the difference between positive and negative regulation of gene expression? **See Figure 16.1 and Table 16.1**
- Describe the molecular conditions at the *lac* operon promoter in the presence versus absence of lactose. **See p. 331 and Figure 16.5**
- What are the key differences between an inducible system and a repressible system? **See p. 332**
- How do sigma factors and consensus sequences act to affect the expressions of classes of genes? **See pp. 332–333**

Studies of bacteria have provided a basic understanding of mechanisms that regulate gene expression and of the roles of regulatory proteins in both positive and negative regulation. We will now turn to the transcriptional control of gene expression in eukaryotes.

16.2 How Is Eukaryotic Gene Transcription Regulated?

For the normal development of a multicellular organism from fertilized egg to adult, and for each cell to acquire and maintain its proper specialized function, certain proteins must be made at just the right times and in just the right cells; these proteins must not be made at other times in other cells. Here are two examples from humans:

- In human pancreatic exocrine cells, the digestive enzyme procarboxypeptidase A makes up 7.6 percent of all the protein in the cell; in other cell types it is usually undetectable.
- In human breast duct cells, alpha-lactalbumin, a protein in breast milk, is made only late in pregnancy and during lactation. Alpha-lactalbumin is not made in any other cell types.

Clearly the expression of eukaryotic genes must be precisely regulated.

As in prokaryotes, eukaryotic gene expression can be regulated at several different points in the process of transcribing and translating the gene into a protein (**Figure 16.7**). In this section we will describe the mechanisms that result in the selective transcription of specific genes. The mechanisms for regulating gene expression in eukaryotes have similar themes as in those of prokaryotes. Both types of cells use DNA–protein interactions and negative and positive control. However, there are many differences, some of them dictated by the presence of a nucleus, which physically separates transcription and translation (**Table 16.2**).

General transcription factors act at eukaryotic promoters

As in prokaryotes, a promoter in a eukaryotic gene is a sequence of DNA near the 5′ end of the coding region, where RNA polymerase binds and initiates transcription. Although eukaryotic promoters are much more diverse than those of prokaryotes, many contain an element similar to the –10 element in prokaryotic promoters. This element is usually located close to the transcription start site and is called the **TATA box** because it

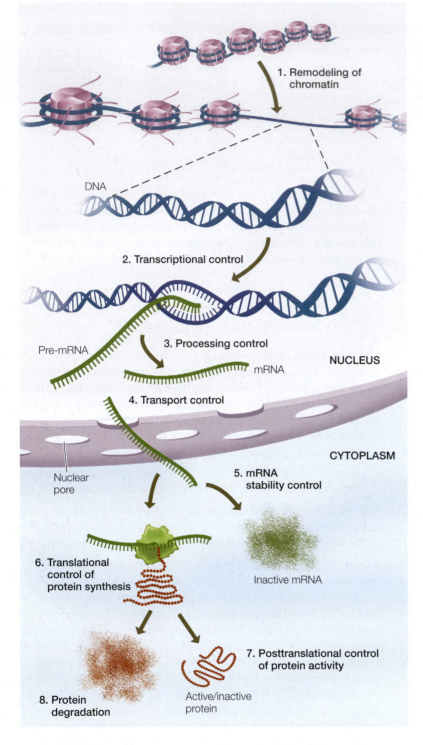

16.7 Potential Points for the Regulation of Gene Expression
Gene expression can be regulated before transcription (1), during transcription (2, 3), after transcription but before translation (4, 5), at translation (6), or after translation (7).

Go to Activity 16.1 Eukaryotic Gene Expression Control Points
Life10e.com/ac16.1

is rich in AT base pairs. The TATA box is the site where DNA begins to denature so that the template strand can be exposed. In addition to the TATA box, eukaryotic promoters typically include multiple regulatory sequences that are recognized and

TABLE 16.2		
Transcription in Prokaryotes and Eukaryotes		
	Prokaryotes	Eukaryotes
Locations of functionally related genes	Often clustered in operons	Often distant from one another with separate promoters
RNA polymerases	One	Three: I transcribes rRNA II transcribes mRNA III transcribes tRNA and small RNAs
Promoters and other regulatory sequences	Few	Many
Initiation of transcription	Binding of RNA polymerase to promoter	Binding of many proteins, including RNA polymerase

bound by **transcription factors**: regulatory proteins that help control transcription.

Like the prokaryotic RNA polymerase, eukaryotic RNA polymerase II cannot simply bind to the promoter and initiate transcription. Rather, it does so only after various **general transcription factors** have assembled on the chromosome (**Figure 16.8**). First, the protein TFIID ("TF" stands for transcription factor) binds to the TATA box. Binding of TFIID changes both its own shape and that of the DNA, presenting a new surface that attracts the binding of other general transcription factors to form a transcription initiation complex. RNA polymerase II binds only after several other proteins have bound to this complex.

Each general transcription factor has a role in gene expression:

- TFIIB binds both DNA polymerase and TFIID, and helps identify the transcription initiation site.

- TFIIF prevents nonspecific binding of the complex to DNA and helps recruit RNA polymerase to the complex; it is similar in function to a bacterial sigma factor.

- TFIIE binds to the promoter and stabilizes the denaturation of the DNA.

- TFIIH opens up the DNA for transcription.

Some regulatory DNA sequences, such as the TATA box, are common to the promoters of many eukaryotic genes and are recognized by general transcription factors that are found in all the cells of an organism. Other regulatory sequences are present in only a few genes and are recognized by specific transcription factors. These factors may be found only in certain types of cells or at certain stages of the cell cycle, or they may be activated by signaling pathways in response to cellular or environmental signals (see Chapter 7). Specific transcription factors play an important role in cell differentiation—the structural and functional specialization of cells during development.

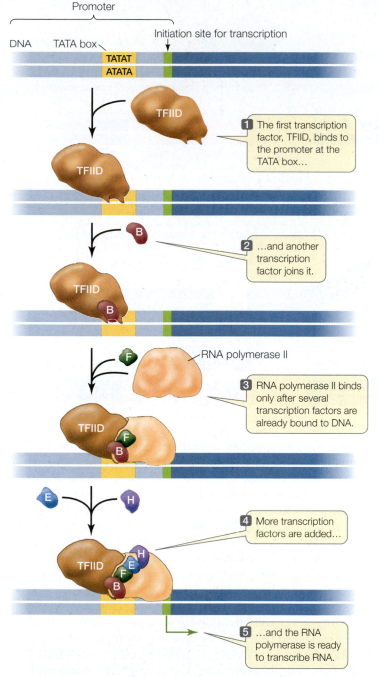

1 The first transcription factor, TFIID, binds to the promoter at the TATA box…

2 …and another transcription factor joins it.

RNA polymerase II

3 RNA polymerase II binds only after several transcription factors are already bound to DNA.

4 More transcription factors are added…

5 …and the RNA polymerase is ready to transcribe RNA.

16.8 The Initiation of Transcription in Eukaryotes Apart from TFIID, which binds to the TATA box, each general transcription factor in this transcription complex has binding sites only for the other proteins in the complex, and does not bind directly to DNA. B, E, F, and H are general transcription factors.

Go to Animated Tutorial 16.3
Initiation of Transcription
Life10e.com/at16.3

Specific proteins can recognize and bind to DNA sequences and regulate transcription

Some regulatory DNA sequences are positive elements termed **enhancers**. They bind transcription factors that either activate transcription or increase the rate of transcription. Other regulatory elements are **silencers**: they bind factors that repress

transcription. Most of the regulatory elements needed for correct expression of a gene are found within a few hundred base pairs of the transcription start site. For example, the mouse albumin gene promoter contains all the information needed for liver cell–specific expression within 170 base pairs upstream of the transcription start site. Other regulatory elements may be located thousands of base pairs away (as far as 20,000), and they may affect the expression of several nearby genes. When transcription factors bind to these elements, they interact with the RNA polymerase complex, causing the DNA to bend (**Figure 16.9**).

Often many transcription factors are involved, and *the combination of factors present determines the rate of transcription*. For example, the immature red blood cells in bone marrow make large amounts of β-globin. At least 13 different transcription factors are involved in regulating transcription of the β-globin gene in these cells. Not all of these factors are present or active in other cells, such as the immature white blood cells produced by the same bone marrow. As a result, the β-globin gene is not transcribed in those cells. So although the same genes are present in all cells, the fate of the cell is determined by which of its genes are expressed. How do transcription factors recognize specific DNA sequences?

Specific protein–DNA interactions underlie binding

As we have seen, transcription factors with specific DNA-binding domains are involved in the activation and inactivation of specific genes. There are several common structural themes in the protein domains that bind to DNA. These themes, or **structural motifs**, consist of different combinations of structural elements (protein conformations) and may include special components such as zinc. One of the common structural motifs is the helix-turn-helix, in which two α helices are connected via a nonhelical turn. The interior-facing "recognition" helix interacts with the bases inside the DNA. The exterior-facing helix sits on the sugar–phosphate backbone, ensuring that the interior helix is presented to the bases in the correct configuration:

α helices

"Turn"

Helix-turn-helix motif   Motif bound to DNA

How does a protein recognize a sequence in DNA? As pointed out in Section 13.2, the complementary bases in DNA not only form hydrogen bonds with each other, but also can form additional hydrogen bonds with proteins, particularly at points exposed in the major and minor grooves. In this way, an intact DNA double helix can be recognized by a protein motif whose structure:

- fits into the major or minor groove;
- has amino acids that can project into the interior of the double helix; and
- has amino acids that can form hydrogen bonds with the interior bases.

16.9 Transcription Factors and Transcription Initiation
The actions of many proteins determine whether and where RNA polymerase II will transcribe DNA.

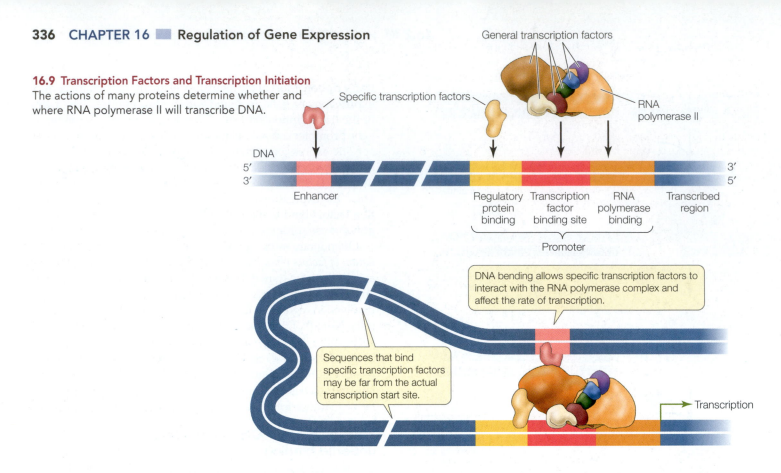

The helix-turn-helix motif fits these three criteria. Many repressor proteins, including the bacterial *lac* repressor, have this helix-turn-helix motif in their structure:

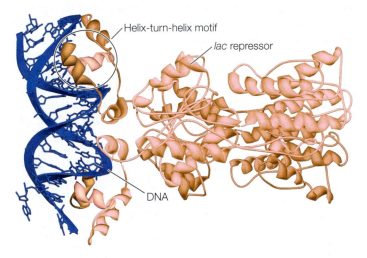

These proteins often form dimers, as shown here (one molecule is shown in white, the other in gray).

Repressors can inhibit transcription in several different ways. They can prevent the binding of transcriptional activators to DNA, or they can interact with other DNA-binding proteins to decrease the rate of transcription.

The expression of transcription factors underlies cell differentiation

During the development of a complex organism from fertilized egg to adult, cells become more and more differentiated (specialized). Differentiation is mediated in many cases by changes in gene expression, resulting from the activation (and inactivation) of various transcription factors. We will discuss this topic in more detail in Chapter 19. For now, remember that all differentiated cells contain the entire genome, and that their specific characteristics arise from differential gene expression.

Currently there is great interest in cellular therapy: providing new, functional cells to patients who have diseases that involve the degeneration of certain cell types. An example is Alzheimer's disease, which involves the degeneration of neurons in the brain. Because of the possibility of immune system rejection (see Chapter 42), it would be optimal if patients could receive their own cells, modified in some way to be functional. Since specialized functions are mediated by transcription factors, turning readily available cells into a particular desired cell type might be achieved by altering transcription factor expression. Marius Wernig and his colleagues at Stanford University have made important progress toward this goal (**Figure 16.10**). They took skin fibroblasts from mice and manipulated the expression of transcription factors in the cells to change them into neurons. By repeating their experiments on human fibroblasts, they have brought cellular therapy closer to reality.

The expression of sets of genes can be coordinately regulated by transcription factors

How do eukaryotic cells coordinate the regulation of several genes whose transcription must be turned on at the same time? Prokaryotes solve this problem by arranging multiple genes in an operon that is controlled by a single promoter, and by using sigma factors to recognize particular classes of promoters. Most eukaryotic genes have their own separate promoters, and genes that are coordinately regulated may be far apart. In these

16.10 Expression of Specific Transcription Factors Turns Fibro-blasts into Neurons Fibroblasts are cells that secrete abundant extracellular matrix and contribute to the structural integrity of organs. Neurons are highly specialized cells in the nervous system. Marius Wernig and his colleagues performed a series of experiments to find out whether expressing neuronal transcription factors in fibroblasts would be sufficient to cause the fibroblasts to become neurons.[a]

HYPOTHESIS Expression of neuron-specific transcription factors in fibroblasts will turn the latter into neurons.

Method

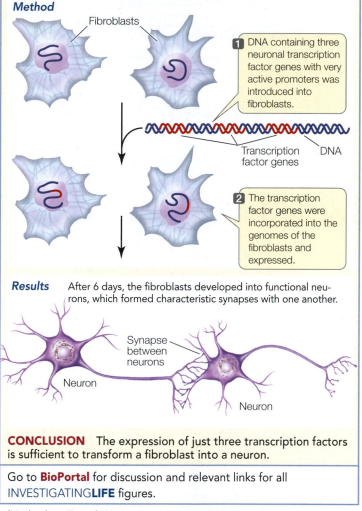

Fibroblasts

1 DNA containing three neuronal transcription factor genes with very active promoters was introduced into fibroblasts.

Transcription factor genes DNA

2 The transcription factor genes were incorporated into the genomes of the fibroblasts and expressed.

Results After 6 days, the fibroblasts developed into functional neurons, which formed characteristic synapses with one another.

Synapse between neurons

Neuron

Neuron

CONCLUSION The expression of just three transcription factors is sufficient to transform a fibroblast into a neuron.

Go to **BioPortal** for discussion and relevant links for all INVESTIGATINGLIFE figures.

[a]Vierbuchen, T. et al. 2010. *Nature* 463: 1035–1041.

cases, the expression of genes can be coordinated if they share regulatory sequences that bind the same transcription factors.

This type of coordination is used by organisms to respond to stress—for example, by plants in response to drought. Under conditions of drought stress, a plant must simultaneously synthesize several proteins whose genes are scattered throughout the genome. The synthesis of these proteins comprises the stress response. To coordinate expression, each of these genes has a specific regulatory sequence near its promoter called the stress response element (SRE). A transcription factor binds to this element and stimulates mRNA synthesis (**Figure 16.11**). The stress response proteins not only help the plant conserve

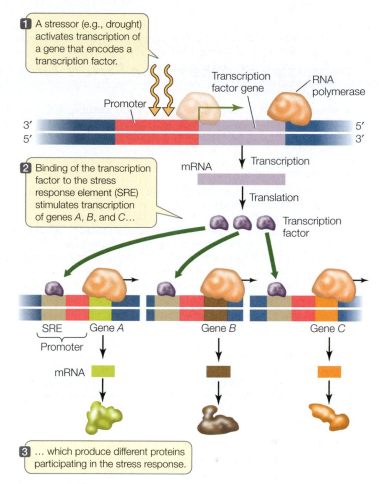

1 A stressor (e.g., drought) activates transcription of a gene that encodes a transcription factor.

Transcription factor gene RNA polymerase

Promoter

3' 5'
5' 3'

mRNA Transcription

2 Binding of the transcription factor to the stress response element (SRE) stimulates transcription of genes A, B, and C…

Translation

Transcription factor

SRE Gene A Gene B Gene C
Promoter

mRNA

3 … which produce different proteins participating in the stress response.

16.11 Coordinating Gene Expression A single environmental signal, such as drought stress, causes the synthesis of a transcription factor that acts on many genes.

water, but also protect the plant against excess salt in the soil and freezing. This finding has considerable importance for agriculture because crops are often grown under less than optimal conditions or are affected by weather.

RECAP 16.2

A number of general transcription factors must bind to a eukaryotic promoter before RNA polymerase will bind to it and begin transcription. Other, specific transcription factors bind to regulatory DNA sequences and interact with the RNA polymerase complex to control differential gene expression. This provides several ways to increase or decrease transcription.

- Describe some of the different ways in which transcription factors regulate gene transcription. See pp. 334–336 and Figure 16.9
- How can more than one gene be regulated at the same time? See pp. 336–337 and Figure 16.11

We have seen how prokaryotes and eukaryotes regulate the transcription of their genes and operons. In the next section we will see how prokaryotic and eukaryotic viruses are able to hijack transcription mechanisms in order to complete their life cycles.

WORKING WITH**DATA:**

Expression of Transcription Factors Turns Fibroblasts into Neurons

Original Paper

Vierbuchen, T., A. Ostermeier, Z. P. Pang, Y. Kokubu, T. C. Südhof, and M. Wernig. 2010. Direct conversion of fibroblasts to functional neurons by defined factors. *Nature* 463: 1035–1041.

Analyze the Data

Different cell types in an organism are usually distinguished by differences in the expression of their genes. When the regulatory regions of differentially expressed genes are analyzed, they are found to bind different transcription factors. Would expression of a specific set of transcription factors be sufficient to change one cell type into another? Marius Wernig and his colleagues at Stanford University set out to answer this question by attempting to change fibroblasts into neurons (see Figure 16.10).

Fibroblasts occur widely in the body. They secrete extracellular matrix materials and infiltrate various organs, contributing to the organs' structural integrity. These cells divide actively when cultured in the laboratory. Neurons have highly specialized roles in the nervous system (see Chapter 45) and do not divide in culture. Wernig and his colleagues found 19 transcription factors that were strongly expressed in mouse neurons and not in fibroblasts. When five of these transcription factors—Ascl1, Brn2, Mytl1, Zic1, and Olig2—were introduced into fibroblasts and expressed from very strong promoters, the fibroblasts became neurons. Subsequent experiments showed that a combination just three of these—Ascl1, Brn2, and Mytl1—was sufficient to cause efficient transformation. In a further study, the researchers were able to repeat these experiments with human cells, using four transcription factors to convert fibroblasts into neurons. Since fibroblasts are easily isolated from human skin, the possibility of creating neurons from them that can be used for cellular therapy in diseases where neurons degenerate (e.g., Alzheimer's disease) is exciting.

Three main criteria were used to determine that the transformed cells were neurons: morphology, electrical excitability, and lack of cell division. The transformed cells clearly showed the distinctive cellular architecture of neurons (see Figure 45.1). Wernig and his colleagues conducted additional experiments to test for the other two properties of neurons in their transformed cells.

QUESTION 1:

As you will see in Chapter 45, neurons respond to electrical stimulation by generating an action potential. The electrical activity of a stimulated transformed fibroblast cell is shown in **FIGURE A**—8, 12, and 20 days after addition of the transcription factors. What is the magnitude of the action potential of the transformed cell in millivolts (mv)? Look up Figure 45.10. How does this compare?

QUESTION 2:

The rate of cell division in the population of transformed cells was measured by the incorporation of the labeled nucleotide BrdU into their DNA. The percentage of labeled—and hence dividing—cells is shown in **FIGURE B**. Did cell division stop in the transformed cells? Explain your answer.

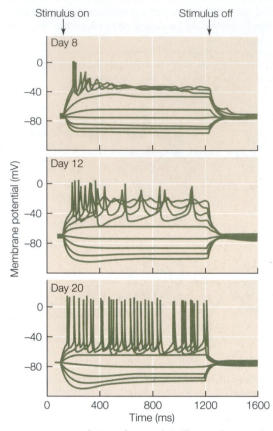

FIGURE A Response of Transformed Cells to Electrical Stimulation The different traces show the cell's response to different amounts of stimulation, some of which were too small to trigger the production of action potentials (upward spikes).

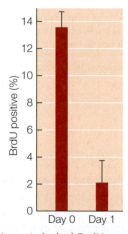

FIGURE B Cell Division Labeled BrdU was added to the transformed cells at the time transcription factors were added (Day 0) or one day later (Day 1). The number of labeled cells was assessed 13 days after the transcription factors were added.

Go to **BioPortal** for all WORKING WITH**DATA** exercises

16.12 Bacteriophage and Host (A) *E. coli* cells (viewed here from the side) are the host for bacteriophage T2. (B) Bacteriophages have attached to this *E. coli* cell, and the reproductive cycle is underway, producing new phage particles. The cell is viewed in transverse section.

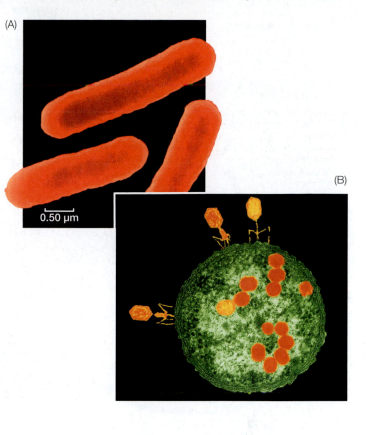

(A)

(B)

0.50 µm

16.3 How Do Viruses Regulate Their Gene Expression?

"A virus is a piece of bad news wrapped in protein." This quote from immunologist Sir Peter Medawar is certainly true for the cells that viruses infect. As we described in Chapter 13, a bacterial virus (**bacteriophage**) injects its genetic material into a host cell and turns that cell into a virus factory (see Figure 13.3). Other viruses enter cells intact and then shed their coats and take over the cell's replication machinery. Viral life cycles can be very efficient. An example is the poliovirus: a single poliovirus infecting a mammalian cell can produce more than 100,000 new virus particles!

Viruses are small infectious agents that infect cellular organisms and that cannot reproduce outside their host cells. Most virus particles, called **virions**, consist of only two or three components: the genetic material made up of DNA or RNA, a protein coat that protects the genetic material, and in some cases, an envelope of lipids that surrounds the protein coat. As we will see in this section, viral genomes include sequences that encode regulatory proteins. These proteins "hijack" the host cells' transcriptional machinery, allowing the viruses to complete their life cycles.

Many bacteriophages undergo a lytic cycle

The Hershey–Chase experiment (see Figure 13.4) involved a typical **lytic** viral reproductive cycle, so named because soon after infection, the host cell bursts (lyses), releasing progeny viruses. In this cycle, the viral genetic material takes over the host's synthetic machinery for its own reproduction immediately after infection. In the case of some bacteriophages, the process is extremely rapid—within 15 minutes, new phage particles appear in the bacterial cell (**Figure 16.12**). Ten minutes later, the "game is over," and these particles are released from the lysed cell. What happens during this rapid life cycle?

At the molecular level, the reproductive cycle of a typical lytic virus has two stages: early and late, as illustrated in **Figure 16.13**. The reproductive cycle involves both positive

16.13 The Lytic Cycle: A Strategy for Viral Reproduction
In a host cell infected with a virus, the viral genome uses its early genes to shut down host transcription while it replicates itself. Once the viral genome is replicated, its late genes produce capsid proteins that package the genome and other proteins that lyse the host cell.

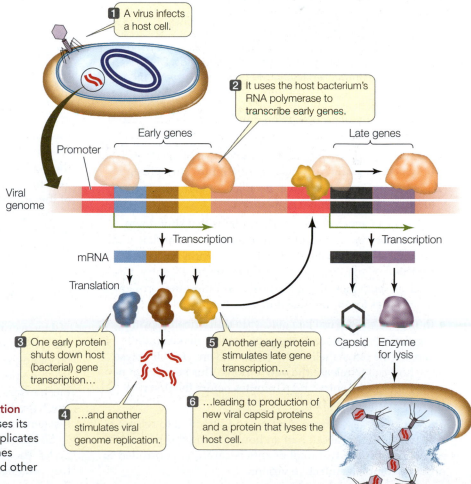

1 A virus infects a host cell.

2 It uses the host bacterium's RNA polymerase to transcribe early genes.

Early genes

Late genes

Promoter

Viral genome

Transcription

Transcription

mRNA

Translation

3 One early protein shuts down host (bacterial) gene transcription…

4 …and another stimulates viral genome replication.

5 Another early protein stimulates late gene transcription…

6 …leading to production of new viral capsid proteins and a protein that lyses the host cell.

Capsid Enzyme for lysis

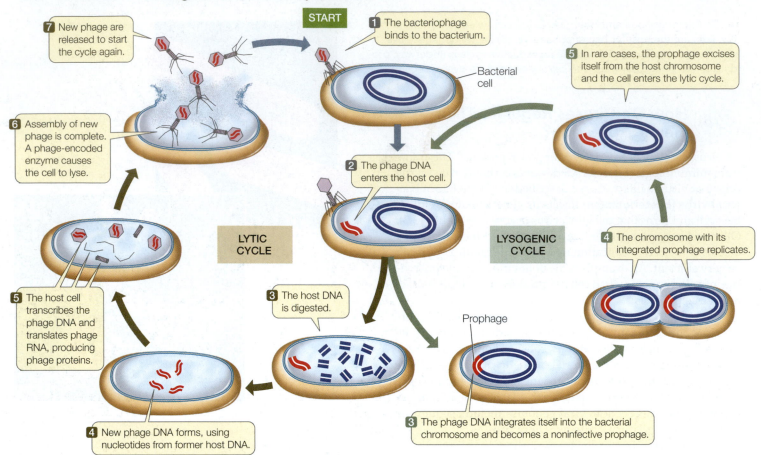

16.14 The Lytic and Lysogenic Cycles of Bacteriophages   In the lytic cycle, infection of a bacterium by viral DNA leads directly to multiplication of the virus and lysis of the host cell. In the lysogenic cycle, an inactive prophage is integrated into the host DNA where it is replicated during the bacterial life cycle.

and negative regulation, which stimulate and inhibit, respectively, gene expression:

- The viral genome contains a promoter that binds host RNA polymerase. In the early stage (1–2 minutes after phage DNA entry), viral genes that lie adjacent to this promoter are transcribed (positive regulation).

- These early genes often encode proteins that shut down host transcription (negative regulation) and stimulate viral genome replication and transcription of viral late genes (positive regulation). Three minutes after DNA entry, viral nuclease enzymes digest the host's chromosome, providing nucleotides for the synthesis of viral genomes.

- In the late stage, viral late genes are transcribed (positive regulation); they encode the proteins that make up the **capsid** (the outer shell of the virus) and enzymes that lyse the host cell to release the new virions. This begins 9 minutes after DNA entry and 6 minutes before the first new phage particles appear.

The entire process—from binding and infection to release of new phage—takes about half an hour. During this period, the sequence of transcriptional events is carefully controlled to produce complete, infective virions.

Some bacteriophages can undergo a lysogenic cycle

Like all nucleic acid genomes, those of viruses can mutate and evolve by natural selection. Some viruses have evolved an advantageous process called **lysogeny** that postpones the lytic cycle. In lysogeny, the viral DNA becomes integrated into the host DNA and becomes a **prophage** (**Figure 16.14**). As the host cell divides, the viral DNA gets replicated along with that of the host. The prophage can remain inactive within the bacterial genome for thousands of generations, producing many copies of the original viral DNA.

However, if the host cell is not growing well, the virus "cuts its losses." It switches to a lytic cycle, in which the prophage excises itself from the host chromosome and reproduces. In other words, the virus is able to enhance its chances of multiplication and survival by inserting its DNA into the host chromosome, where it sits as a silent passenger until conditions are right for lysis.

Uncovering the regulation of gene expression that underlies the lysis/lysogeny switch was a major achievement of molecular biologists. Here we present just an outline of the process to give you an idea of the positive and negative regulatory mechanisms involved (**Figure 16.15**). The model virus bacteriophage λ (lambda) has been used extensively to study the lysogenic mechanism.

So how does a prophage "know" when to switch to the lytic cycle? The virus genome includes genes that encode the regulatory proteins cI and Cro, which compete for specific promoters

16.15 Control of Bacteriophage λ Lysis and Lysogeny
Two regulatory proteins, Cro and cI, compete to control expression of one another and genes for viral lysis and lysogeny.

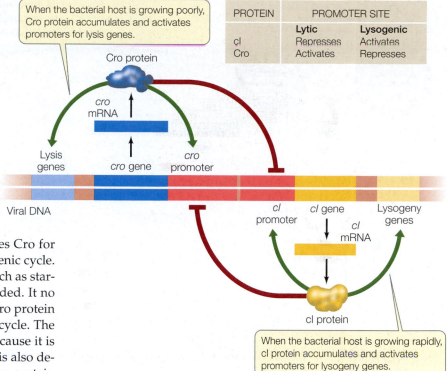

When the bacterial host is growing poorly, Cro protein accumulates and activates promoters for lysis genes.

Cro protein

PROTEIN	PROMOTER SITE	
	Lytic	**Lysogenic**
cI	Represses	Activates
Cro	Activates	Represses

cro mRNA

Lysis genes — *cro* gene — *cro* promoter

Viral DNA

cI promoter — *cI* gene — Lysogeny genes

cI mRNA

cI protein

When the bacterial host is growing rapidly, cI protein accumulates and activates promoters for lysogeny genes.

on the viral DNA—including their own promoters (see Figure 16.15). Cro and cI have opposite effects on each promoter: cI represses the expression of genes involved in the lytic cycle and promotes expression of genes involved in lysogeny, whereas Cro has the opposite effect.

The outcome of a bacteriophage infection depends on the relative abundance of these two regulatory proteins. Under favorable bacterial growth conditions, cI accumulates and outcompetes Cro for DNA binding, and the bacteriophage enters a lysogenic cycle. But if the host cell undergoes stressful conditions such as starvation, exposure to toxins, or radiation, cI is degraded. It no longer represses transcription of the *Cro* gene, the Cro protein is produced, and the bacteriophage enters the lytic cycle. The cI protein is degraded under stressful conditions because it is structurally similar to an *E. coli* protein (LexA) that is also degraded under such conditions. LexA is a regulatory protein that represses DNA repair mechanisms under normal conditions, but it is degraded by other proteins when the cell becomes stressed.

The reproductive cycle of bacteriophage λ is a paradigm for our understanding of viral life cycles in general. This system has served as a model to help us understand how the complicated reproductive cycles of other viruses, including HIV, are controlled.

Eukaryotic viruses can have complex life cycles

Eukaryotes are susceptible to infection by various kinds of viruses whose genomes may consist of RNA or DNA. A subgroup of RNA viruses are called retroviruses (see also Section 26.4):

- *DNA viruses*. Many viral particles contain double-stranded DNA. However, some contain single-stranded DNA, and a complementary strand is made after the viral genome enters the host cell. Like some bacteriophages, DNA viruses that infect eukaryotes are capable of undergoing both lytic and lysogenic life cycles. Examples include the herpes viruses and papillomaviruses (which cause warts).

- *RNA viruses*. Some viral genomes are made up of RNA that is usually, but not always, single-stranded. The RNA is translated by the host's machinery to produce viral proteins, some of which are involved in replication of the RNA genome. The influenza virus has an RNA genome.

- *Retroviruses*. As we described in Section 14.2, a **retrovirus** is an RNA virus that carries a gene for **reverse transcriptase**, a protein that synthesizes DNA from an RNA template. The retrovirus uses this protein to make a DNA copy of its genome, which then becomes integrated into the host

genome. The integrated DNA acts as a template for both mRNA and new viral genomes. Human immunodeficiency virus (HIV) is a retrovirus that infects cells of the immune system and causes acquired immune deficiency syndrome (AIDS).

HIV gene regulation occurs at the level of transcription elongation

As we have discussed so far, many instances of gene regulation occur at the level of transcription *initiation*, involving both activator and repressor proteins that bind to the promoters of genes. However, studies of HIV and other viruses have revealed that transcription can also be controlled at the *elongation* stage.

HIV is an **enveloped virus**; it is enclosed within a phospholipid membrane derived from its host cells (a specific type of immune system cell) (**Figure 16.16**). During infection, proteins in this membrane interact with proteins on the host cell surface, and the viral envelope fuses with the host plasma membrane. After the virus enters the cell, its capsid is broken down. The viral reverse transcriptase then uses the virus's RNA template to produce a complementary DNA (cDNA) strand, while at the same time degrading the viral RNA. The enzyme then makes a complementary copy of the cDNA, and the resulting double-stranded DNA is inserted into the host's chromosome by the enzyme integrase. The integrated DNA is referred to as the **provirus**. Both the reverse transcriptase and the integrase are carried inside the HIV virion, along with other proteins needed at the very early stages of infection.

The provirus resides permanently in the host chromosome, and can remain in a latent (inactive) state for many years. During this time transcription of the viral DNA is initiated,

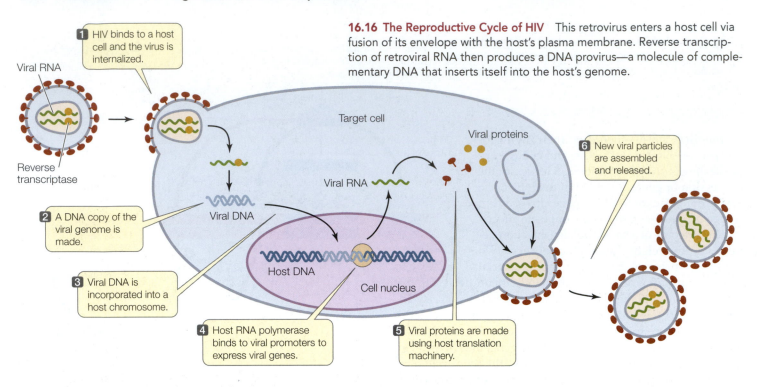

16.16 The Reproductive Cycle of HIV This retrovirus enters a host cell via fusion of its envelope with the host's plasma membrane. Reverse transcription of retroviral RNA then produces a DNA provirus—a molecule of complementary DNA that inserts itself into the host's genome.

1. HIV binds to a host cell and the virus is internalized.

Viral RNA

Reverse transcriptase

2. A DNA copy of the viral genome is made.

Viral DNA

Target cell

Viral proteins

Viral RNA

6. New viral particles are assembled and released.

3. Viral DNA is incorporated into a host chromosome.

Host DNA

Cell nucleus

4. Host RNA polymerase binds to viral promoters to express viral genes.

5. Viral proteins are made using host translation machinery.

but host cell proteins prevent the RNA from elongating, and transcription is terminated prematurely (**Figure 16.17A**). Under some circumstances, such as when the host immune cell is activated, the level of transcription initiation increases and some viral RNA is made. One of these viral genes encodes a protein called tat (*trans*activator of *t*ranscription), which binds to a stem-and-loop structure at the 5' end of the viral RNA. As a result of tat binding, the production of full-length viral RNA is dramatically increased (**Figure 16.17B**), and the rest of the viral reproductive cycle is able to proceed. It was only after the discovery of this mechanism in HIV and similar viruses that

researchers found that many eukaryotic genes are regulated at the level of transcription elongation.

Almost every step in the reproductive cycle of HIV is, in principle, a potential target for drugs to treat AIDS. The classes of anti-HIV drugs currently in use include:

- Reverse transcriptase inhibitors that block viral DNA synthesis from RNA (step 2 in Figure 16.16)

- Integrase inhibitors that block the incorporation of viral DNA into the host chromosome (step 3)

- Protease inhibitors that block the posttranslational processing of viral proteins (step 5)

Combinations of drugs from these classes have been spectacularly successful in treating HIV infection. But a problem with HIV treatment is the rapid emergence of drug-resistant strains,

(A)

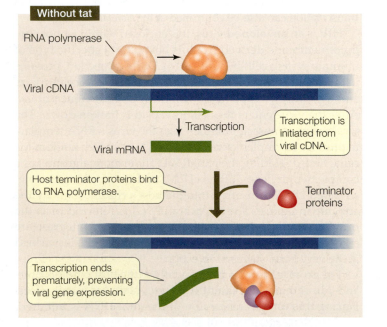

Without tat

RNA polymerase

Viral cDNA

Transcription

Transcription is initiated from viral cDNA.

Viral mRNA

Host terminator proteins bind to RNA polymerase.

Terminator proteins

Transcription ends prematurely, preventing viral gene expression.

(B)

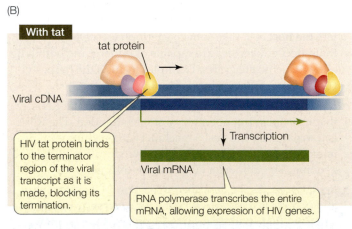

With tat

tat protein

Viral cDNA

Transcription

Viral mRNA

HIV tat protein binds to the terminator region of the viral transcript as it is made, blocking its termination.

RNA polymerase transcribes the entire mRNA, allowing expression of HIV genes.

16.17 Regulation of Transcription by HIV The tat protein acts as an antiterminator, allowing transcription of the HIV genome.

and drugs targeted to other steps in the HIV reproductive cycle are being developed. These include drugs that interact with the binding of viral particles to host cells, and drugs that interfere with tat activity.

RECAP 16.3

A virus consists of nucleic acids, a few proteins, and in some cases, a lipid envelope. Viruses require host cells to reproduce. Viral life cycles can include lytic and lysogenic stages. Like its prokaryotic host, bacteriophage λ uses both positive and negative regulators of transcription initiation. Studies of HIV revealed a new mechanism for gene regulation: the regulation of transcription elongation.

- What are the lytic and lysogenic cycles of bacteriophages? See pp. 339–341 and Figure 16.14
- Describe positive and negative regulation of gene expression in the bacteriophage λ and HIV life cycles. See pp. 340–342 and Figures 16.15, 16.17

So far we have discussed mechanisms that cells and viruses use to control gene transcription. These mechanisms usually involve the interaction of regulatory proteins with specific DNA sequences. However, there are other mechanisms for controlling gene expression that do not depend on specific DNA sequences. We will discuss these mechanisms in the next section.

16.4 How Do Epigenetic Changes Regulate Gene Expression?

In the mid-twentieth century, the great developmental biologist Conrad Hal Waddington coined the term "epigenetics" and defined it as "that branch of biology which studies the causal interactions between genes and their products which bring the phenotype into being." Today **epigenetics** is defined more specifically, referring to the study of changes in gene expression that occur without changes in the DNA sequence. These changes are reversible but sometimes are stable and heritable. You saw an example of this phenomenon in the opening story of this chapter. Epigenetic changes include two processes: DNA **methylation** and chromosomal protein alterations.

DNA methylation occurs at promoters and silences transcription

Depending on the organism, from 1 to 5 percent of cytosine residues in the DNA are chemically modified by the addition of a methyl group ($-CH_3$) to the 5′ carbon, to form 5′-methylcytosine (**Figure 16.18**). This covalent addition is catalyzed by the enzyme **DNA methyltransferase** and, in mammals, usually occurs in C residues that are adjacent to G residues. DNA regions rich in these doublets are called **CpG islands**, and are especially abundant in promoters.

This covalent change in DNA is heritable: when DNA is replicated, a **maintenance methylase** catalyzes the formation of 5′-methylcytosine in the new DNA strand. However, the pattern of cytosine methylation can also be altered, because

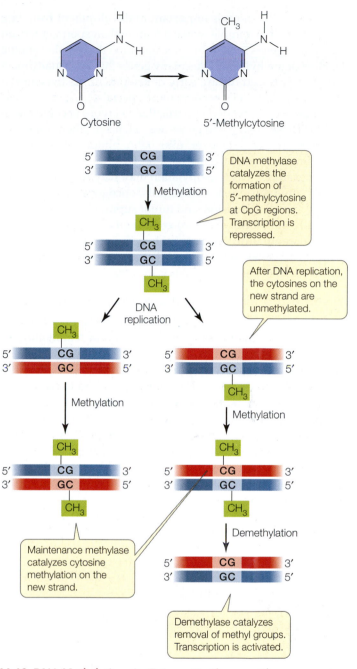

16.18 DNA Methylation: An Epigenetic Change The reversible formation of 5′-methylcytosine in DNA can alter the rate of transcription.

methylation is reversible: a third enzyme, appropriately called **demethylase**, catalyzes the removal of the methyl group from cytosine (see Figure 16.18).

What is the effect of DNA methylation? During replication and transcription, 5′-methylcytosine behaves just like plain cytosine: it base-pairs with guanine. But extra methyl groups in a promoter attract proteins that bind methylated DNA. These proteins are generally involved in the repression of gene transcription; thus heavily methylated genes tend to be inactive. This form of genetic regulation is epigenetic because it affects gene expression patterns without altering the DNA sequence.

DNA methylation is important in development from egg to embryo. For example, when a mammalian sperm enters an egg, many genes in first the male and then the female genome become demethylated. Thus many genes that are usually inactive are expressed during early development. As the embryo develops and its cells become more specialized, genes whose products are not needed in particular cell types become methylated. These methylated genes are "silenced"; their transcription is repressed. However, unusual or abnormal events can sometimes turn silent genes back on.

For example, DNA methylation may play roles in the genesis of some cancers. In cancer cells, oncogenes get activated and promote cell division, and tumor suppressor genes (which normally inhibit cell division) are turned off (see Chapter 11). This misregulation can occur when the promoters of oncogenes become demethylated while those of tumor suppressor genes become methylated. This is the case in colorectal cancer.

Histone protein modifications affect transcription

Another mechanism for epigenetic gene regulation is the alteration of chromatin structure, or chromatin remodeling. As we saw in Chapter 11, DNA is packaged with histone proteins into nucleosomes (see Figure 11.9), which can make DNA physically inaccessible to RNA polymerase and the rest of the transcription apparatus. Each histone protein has a "tail" of approximately 20 amino acids at its N terminus that sticks out of the compact structure and contains certain positively charged amino acids (notably lysine). Ordinarily there is strong ionic attraction between the positively charged histone proteins and DNA, which is negatively charged because of its phosphate groups. However, enzymes called histone acetyltransferases can add acetyl groups to these positively charged amino acids, thus changing their charges:

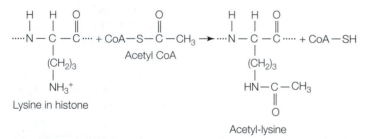

Lysine in histone

Acetyl CoA

Acetyl-lysine

Reducing the positive charges of the histone tails reduces the affinity of the histones for DNA, opening up the compact nucleosome (**Figure 16.19**). Additional chromatin remodeling proteins can bind to the loosened nucleosome–DNA complex, opening up the DNA for gene expression. Histone acetyltransferases can thus activate transcription.

Another kind of chromatin remodeling protein, histone deacetylase, can remove the acetyl groups from histones and thereby repress transcription. Histone deacetylases are targets for drug development to treat some forms of cancer. As noted above, certain genes block cell division in normal specialized tissues. In some cancers these genes are less active than in normal cells, and the histones near them show excessive levels of deacetylation. Theoretically, a drug acting as a histone deacetylase inhibitor could tip the balance toward

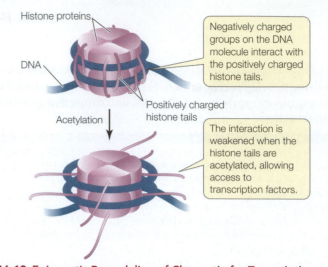

16.19 Epigenetic Remodeling of Chromatin for Transcription Initiation of transcription requires that nucleosomes change their structure, becoming less compact. This chromatin remodeling makes DNA accessible to the transcription initiation complex (see Figure 16.8).

acetylation, and this might activate genes that normally inhibit cell division.

Other types of histone modification can affect gene activation and repression. For example, histone methylation (not to be confused with DNA methylation) is associated with gene inactivation, and histone phosphorylation also affects gene expression, the specific effects depending on which amino acids are modified. All of these effects are reversible, and so the activity of a eukaryotic gene may be determined by very complex patterns of histone modification.

Epigenetic changes can be induced by the environment

Despite the fact that they are reversible, many epigenetic changes such as DNA methylation and histone modification can permanently alter gene expression patterns in a cell. If the cell is a germ line cell that forms gametes, the epigenetic changes can be passed on to the next generation. But what determines these epigenetic changes? A clue comes from a recent study of monozygotic (identical) twins.

Monozygotic twins come from a single fertilized egg that divides to produce two separate cells; each of these goes on to develop a separate individual. Monozygotic twins thus have identical genomes. But are they identical in their epigenomes? A comparison of DNA in hundreds of such twin pairs shows that in tissues of three-year-olds, the DNA methylation patterns are virtually the same. But by age 50, by which time the twins have usually been living apart in different environments for decades, the patterns are quite different. This indicates that the *environment plays an important role in epigenetic modifications*, and thus in the regulation of genes that these modifications affect.

 Go to Media Clip 16.1
The Surprising Epigenetics of Identical Twins
Life10e.com/mc16.1

DNA methylation can result in genomic imprinting

In mammals specific patterns of methylation develop for each sex during gamete formation. This happens in two stages: first, the existing methyl groups are removed from the 5'-methyl-cytosines by a demethylase, and then a DNA methylase adds methyl groups to a new set of cytosines. When the gametes form, they carry this new pattern of methylation.

The DNA methylation pattern in male gametes (sperm) differs from that in female gametes (eggs) at about 200 genes in the mammalian genome. That is, a given gene in this group may be methylated in eggs but unmethylated in sperm (Figure 16.20). In this case the offspring would inherit a maternal gene that is transcriptionally inactive (methylated) and a paternal gene that is transcriptionally active (demethylated). This is called **genomic imprinting**.

An example of imprinting is found in a region on human chromosome 15 called 15q11. This region is imprinted differently during the formation of male and female gametes, and offspring normally inherit both the paternally and maternally derived patterns. In rare cases there is a chromosome deletion in one of the gametes, and the newborn baby inherits just the male or the female imprinting pattern in this particular chromosome region:

- If the male pattern is the only one present (female region deleted), the baby develops Angelman syndrome, characterized by epilepsy, tremors, and constant smiling.

- If the female pattern is the only one present (male region deleted), the baby develops a quite different phenotype called Prader-Willi syndrome, marked by muscle weakness and obesity.

Note that the gene sequences are the same in both cases: it is the epigenetic patterns that are different.

Imprinting of specific genes occurs primarily in mammals and flowering plants. Most imprinted genes are involved with embryonic development. An embryo must have both the paternally and maternally imprinted gene patterns to develop properly. In fact, attempts to make an embryo that has chromosomes from only one sex (for example, by chemically treating an egg cell to double its chromosomes) usually fail. So imprinting has an important lesson for genetics: *males and females may be the same genetically (except for the X and Y chromosomes), but they differ epigenetically.*

Global chromosome changes involve DNA methylation

Like single genes, large regions of chromosomes or even entire chromosomes can have distinct patterns of DNA methylation. Under a microscope, two kinds of chromatin can be distinguished in the stained interphase nucleus: **euchromatin** and **heterochromatin**. The euchromatin appears diffuse and stains lightly; it contains the DNA that is transcribed into mRNA. Heterochromatin is condensed and stains darkly; any genes it contains are generally not transcribed.

Perhaps the most dramatic example of heterochromatin is the inactive X chromosome of mammals. A normal female mammal has two X chromosomes; a normal male has an X and a Y (see Section 12.4). The X and Y chromosomes probably arose from a pair of autosomes (non–sex chromosomes) about 300 million years ago. Over time, mutations in the Y chromosome resulted in maleness-determining genes, and the Y chromosome gradually lost most of the genes it once shared with

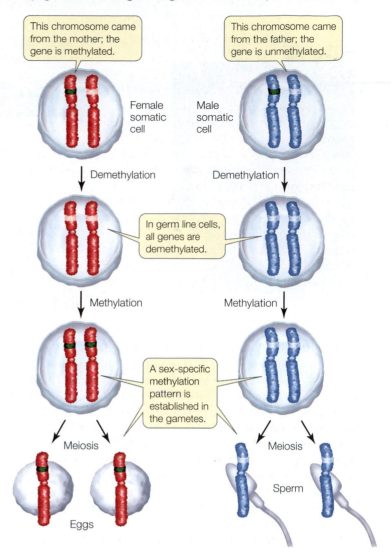

16.20 Genomic Imprinting For some genes, epigenetic DNA methylation differs in male and female gametes. As a result, an individual might inherit an allele from the female parent that is transcriptionally silenced; but the same allele from the male parent would be expressed.

its X homolog. As a result, females and males differ greatly in the "dosage" of X-linked genes. Each female cell has two copies of each gene on the X chromosome and therefore has the potential to produce twice as much of each protein product. Nevertheless, for 75 percent of the genes on the X chromosome, transcription is generally the same in males and in females. How does this happen?

During early embryonic development, one of the X chromosomes in each cell of a female is largely inactivated with regard to transcription. The same X chromosome remains inactive in all that cell's descendants. In a given embryonic cell, the "choice" of which X in the pair to inactivate is random. Recall that one X in a female comes from her father and one from her mother. Thus in one embryonic cell the paternal X might be the one remaining transcriptionally active, but in a neighboring cell the maternal X might be active.

The inactivated X chromosome is identifiable within the nucleus because it is very compact, even during interphase. Typically, a nuclear structure called a Barr body (after its discoverer,

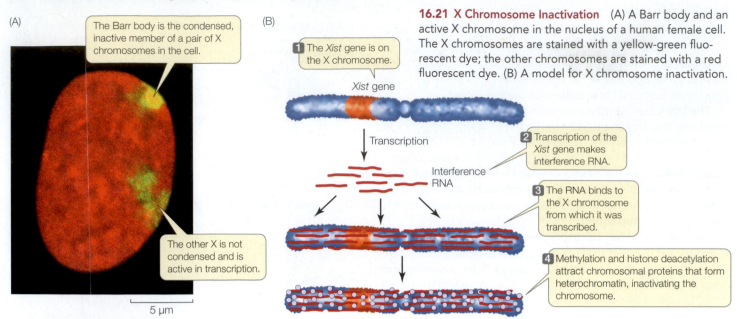

16.21 X Chromosome Inactivation   (A) A Barr body and an active X chromosome in the nucleus of a human female cell. The X chromosomes are stained with a yellow-green fluorescent dye; the other chromosomes are stained with a red fluorescent dye. (B) A model for X chromosome inactivation.

Murray Barr) can be seen in human female cells under the light microscope (**Figure 16.21A**). This clump of heterochromatin, which is not present in normal males, is the inactivated X chromosome, and it consists of heavily methylated DNA. A female with the normal two X chromosomes will have one Barr body, whereas a rare female with three Xs will have two, and an XXXX female will have three. Males that are XXY will have one. These observations suggest that the interphase cells of each person, male or female, have a single active X chromosome, and thus a constant dosage of expressed X chromosome genes.

Condensation of the inactive X chromosome makes its DNA sequences physically unavailable to the transcriptional machinery. Most of the genes of the inactive X are heavily methylated. However, one gene, *Xist* (for *X* i*nactivation-specific transcript*), is only lightly methylated and is transcriptionally active. On the active X chromosome, *Xist* is heavily methylated and not transcribed. The RNA transcribed from *Xist* binds to the X chromosome from which it is transcribed, and this binding leads to a spreading of inactivation along the chromosome. The *Xist* RNA transcript is an example of **interference RNA** (**Figure 16.21B**).

■ **RECAP** **16.4**

Epigenetics describes stable changes in gene expression that do not involve changes in DNA sequences. These changes involve modifications of DNA (cytosine methylation) or of histone proteins bound to DNA. Epigenetic changes can be affected by the environment, and can also result in genome imprinting, in which expression of some genes depends on their parental origin.

- How are DNA methylation patterns established, and how do they affect gene expression? **See p. 343 and Figure 16.18**
- How do histone modifications affect transcription? **See p. 344 and Figure 16.19**
- Why and how does X chromosome inactivation occur? **See pp. 345–346**

Gene expression involves transcription and then translation. So far we have described how gene expression is regulated at the transcriptional level. But as Figure 16.7 shows, there are many points at which regulation can occur after the initial gene transcript is made.

16.5 **How Is Eukaryotic Gene Expression Regulated after Transcription?**

Eukaryotic gene expression can be regulated both in the nucleus prior to mRNA export, and after the mRNA leaves the nucleus. Posttranscriptional control mechanisms include alternative splicing of pre-mRNA, gene silencing by miRNAs and siRNAs, repression of translation, and regulation of protein breakdown in the proteasome.

Different mRNAs can be made from the same gene by alternative splicing

Most primary mRNA transcripts contain several introns (see Figure 14.7). Before the RNA is exported from the nucleus, a splicing mechanism recognizes the boundaries between exons and introns and converts pre-mRNA, which has the introns, into mature mRNA, which does not:

Pre-mRNA ———— Splicing ————→ mRNA
(introns and all exons)   (exons only)

For many genes, alternative splicing can occur, where some exons are spliced out along with the introns (**Figure 16.22**). This mechanism generates a family of different proteins, with different functions, from a single gene. Before the human genome was sequenced, most scientists estimated that there were 80,000 to 150,000 protein-coding genes. You can imagine their surprise when the actual sequence revealed only about 21,000! In fact, there are many more human mRNAs than there are human genes, and most of this variation comes from alternative

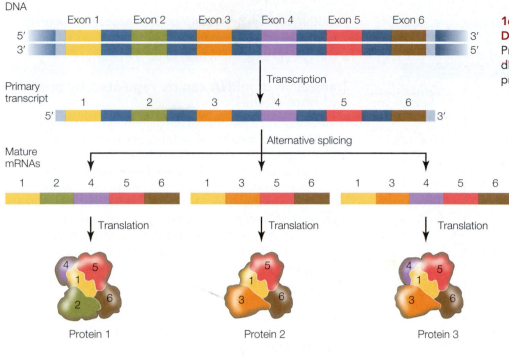

16.22 Alternative Splicing Results in Different Mature mRNAs and Proteins Pre-mRNA can be spliced differently in different tissues, resulting in different proteins.

splicing. Indeed, recent surveys show that about half of all human genes are alternatively spliced. Alternative splicing may be a key to the differences in levels of complexity among organisms. For example, although humans and chimpanzees have similar-sized genomes, there is more alternative splicing in the human brain than in the brain of a chimpanzee.

Alternative RNA splicing is controlled both by regulatory elements in the RNA sequence that bind specific proteins (similar to regulatory sequences in DNA) and by secondary RNA structures that form by hybridization between nucleotides in the single-stranded RNA molecule.

Small RNAs are important regulators of gene expression

As we will discuss in Chapter 17, less than 5 percent of the genome in most plants and animals codes for proteins. Some of the genome encodes ribosomal RNA and transfer RNAs, but until recently biologists thought that the rest of the genome was not transcribed; some even called it "junk." Recent investigations, however, have shown that some of these noncoding regions are transcribed. The RNAs produced from these regions are often very small and therefore difficult to detect. In both prokaryotes and eukaryotes, these tiny RNA molecules are called **microRNA (miRNA)**.

The first miRNA sequences were found in the worm *Caenorhabditis elegans*. This model organism, which has been studied extensively by developmental biologists, goes through several larval stages. Victor Ambros at the University of Massachusetts found mutations in two genes that had different effects on progress through these stages:

- *lin-14* mutations (named for abnormal cell *lin*eage) caused the larvae to skip the first stage and go straight to the second stage. Thus the gene's normal role is to facilitate events in the first larval stage.

- *lin-4* mutations caused certain cells in later larval stages to repeat a pattern of development normally shown in the first larval stage. It was as if the cells were stuck in that stage. So the normal role of this gene is to negatively regulate *lin-14*, turning its expression off so the cells can progress to the next stage.

Not surprisingly, further investigation showed that *lin-14* encodes a transcription factor that affects the transcription of genes involved in larval cell progression. It was originally expected that *lin-4*, the negative regulator, would encode a protein that downregulates genes activated by the lin-14 protein. But this turned out to be incorrect. Instead, *lin-4* encodes a 22-base miRNA that inhibits *lin-14* expression posttranscriptionally, by binding to its mRNA.

More than 5,000 miRNAs have now been described in eukaryotes. The human genome has about 1,000 miRNA-encoding regions. Each miRNA is about 22 bases long and usually has dozens of mRNA targets because the base pairing between the miRNA and the target mRNA doesn't have to be perfect. MicroRNAs are transcribed as longer precursors that fold into double-stranded RNA molecules and are then processed through a series of steps into single-stranded miRNAs. A protein complex guides the miRNA to its target mRNA, where translation is inhibited (**Figure 16.23A**). The remarkable conservation of the miRNA gene-silencing mechanism indicates that it is evolutionarily ancient and biologically important.

In addition to miRNAs, the eukaryotic RNA silencing mechanism also recognizes a similar class of molecules called **small interfering RNAs (siRNAs)**. These often arise from viral infections, when two complementary strands of a viral genome are transcribed. Large double-stranded RNAs are formed, and as with miRNAs, these are converted into shorter single-stranded sequences; these bind to the target RNA and cause its

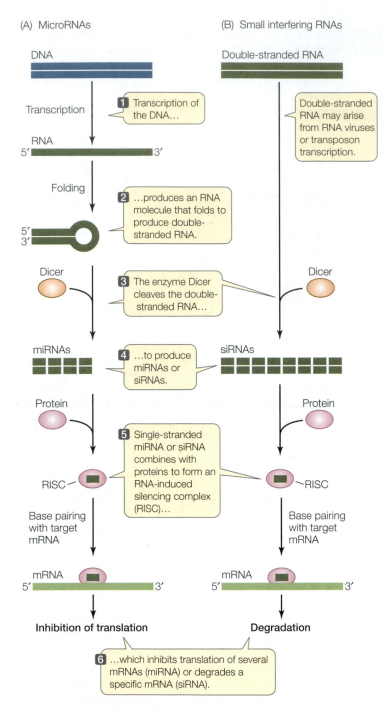

(A) MicroRNAs

DNA

Transcription

1 Transcription of the DNA…

RNA

5' ━━━━━━ 3'

Folding

2 …produces an RNA molecule that folds to produce double-stranded RNA.

5'
3'

Dicer

3 The enzyme Dicer cleaves the double-stranded RNA…

miRNAs

4 …to produce miRNAs or siRNAs.

Protein

5 Single-stranded miRNA or siRNA combines with proteins to form an RNA-induced silencing complex (RISC)…

RISC

Base pairing with target mRNA

mRNA

5' ━━━━━━━ 3'

Inhibition of translation

(B) Small interfering RNAs

Double-stranded RNA

Double-stranded RNA may arise from RNA viruses or transposon transcription.

Dicer

siRNAs

Protein

RISC

Base pairing with target mRNA

mRNA

5' ━━━━━━━ 3'

Degradation

6 …which inhibits translation of several mRNAs (miRNA) or degrades a specific mRNA (siRNA).

16.23 mRNA Inhibition by RNAs MicroRNAs and small interfering RNAs can inhibit translation by binding to target mRNAs.

degradation (**Figure 16.23B**). Small interfering RNAs are also derived from transposon sequences, which are widespread in eukaryotic genomes (see Section 15.1). Therefore it is likely that gene silencing involving siRNAs evolved as a defense mechanism to prevent the translation of viral and transposon sequences. MicroRNAs and siRNAs are similar molecules that are processed by the same cellular enzymes. A major difference between them is that:

- miRNAs are synthesized from DNA sequences separate from their target, whereas
- siRNAs are targeted to their sequence of origin.

Translation of mRNA can be regulated by proteins and riboswitches

Is the amount of a protein in a cell determined by the amount of its mRNA? Scientists have examined the relationship between mRNA abundance and protein abundance in yeast cells.

For about one-third of the many genes surveyed, there was a clear correlation between mRNA and protein: more of one led to more of the other. But for two-thirds of the proteins, there was no apparent relationship between the two: sometimes there was lots of mRNA and little or no protein, or lots of protein and little mRNA. The concentrations of these proteins must therefore have been determined by factors acting after the mRNA was made. Cells have two major ways to control the amount of a protein after transcription: (1) they can regulate translation of the protein's mRNA and (2) they can regulate how long a newly synthesized protein persists in the cell (protein longevity).

REGULATION OF TRANSLATION There are a variety of ways in which the translation of mRNA can be regulated. One way, as we saw in the previous section, is to inhibit translation with siRNAs and miRNAs. A second way involves modification of the guanosine triphosphate cap on the 5' end of the mRNA (see Section 14.4). An mRNA that is capped with an unmodified GTP molecule is not translated. For example, stored mRNAs in the egg cells of the tobacco hornworm moth are capped with unmodified GTP molecules and are not translated. After the egg is fertilized, however, the caps are modified, allowing the mRNA to be translated to produce the proteins needed for early embryonic development.

In another system, repressor proteins directly block translation. For example, in mammalian cells the protein ferritin binds free iron ions (Fe^{2+}). When iron is present in excess, ferritin synthesis rises dramatically, but the amount of ferritin mRNA remains constant, indicating that the increase in ferritin synthesis is due to an increased rate of mRNA translation. Indeed, when the iron level in the cell is low, a translational repressor protein binds to the 5' noncoding region of ferritin mRNA and prevents its translation by blocking its attachment to a ribosome. When the iron level rises, some of the excess Fe^{2+} ions bind to the repressor and alter its three-dimensional structure, causing the repressor to detach from the mRNA and allowing translation to proceed (**Figure 16.24**).

The binding site for the translational repressor on mRNA is a stem-and-loop region with sufficient three-dimensional structure for recognition by a protein or metabolite. This regulatory mechanism occurs widely, and the RNA region that is bound is called a **riboswitch**.

REGULATION OF PROTEIN LONGEVITY The protein content of a cell at any given time is a function of both protein synthesis and protein degradation. Certain proteins can be targeted for destruction in a chain of events that begins when an enzyme attaches a 76-amino acid protein called **ubiquitin** (so named because it is ubiquitous, or widespread) to a lysine residue of the protein to be destroyed. Other ubiquitins then attach to the primary one, forming a polyubiquitin chain. The protein–polyubiquitin complex then binds to a huge protein complex called a **proteasome** (from *protease* and *soma*, "body") (**Figure 16.25**). Upon entering the proteasome, the polyubiquitin is removed and ATP energy is used to unfold the target protein. Three different proteases then digest the protein into small peptides and amino acids.

You may recall from Section 11.2 that cyclins are proteins that regulate the activities of key enzymes at specific points in the cell cycle. Cyclins must be broken down at just the right time, and this is done by attaching ubiquitin to them and degrading them in the proteasomes. Viruses can hijack this system. For example, some strains of the human papillomavirus (HPV) add ubiquitin to the p53 and retinoblastoma proteins, targeting them for proteasomal degradation. These proteins normally inhibit the cell cycle, so the result of this HPV activity is unregulated cell division (cancer).

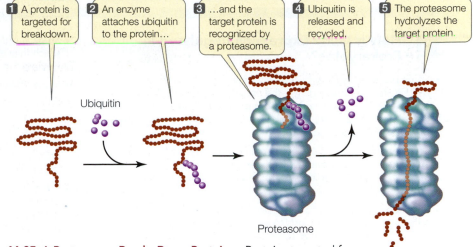

1 A protein is targeted for breakdown.

2 An enzyme attaches ubiquitin to the protein…

3 …and the target protein is recognized by a proteasome.

4 Ubiquitin is released and recycled.

5 The proteasome hydrolyzes the target protein.

Ubiquitin

Proteasome

16.25 A Proteasome Breaks Down Proteins Proteins targeted for degradation are bound to ubiquitin, which then binds the targeted protein to a proteasome. The proteasome is a complex structure where proteins are digested by several powerful proteases.

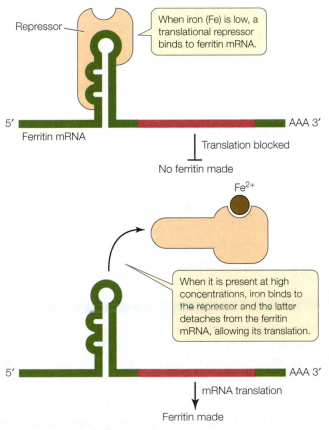

Repressor

When iron (Fe) is low, a translational repressor binds to ferritin mRNA.

5′

Ferritin mRNA

AAA 3′

Translation blocked

No ferritin made

Fe²⁺

When it is present at high concentrations, iron binds to the repressor and the latter detaches from the ferritin mRNA, allowing its translation.

5′

AAA 3′

mRNA translation

Ferritin made

16.24 Translational Repressor Can Repress Translation Binding of a protein to a target mRNA can inhibit its translation.

RECAP 16.5

One of the most important means of posttranscriptional regulation is alternative RNA splicing, which allows more than one protein to be made from a single gene. The stability of mRNA in the cytoplasm can also be regulated. MicroRNAs, siRNAs, mRNA modifications, and translational repressors can prevent mRNA translation. Proteins in the cell can be targeted for breakdown by ubiquitin and then hydrolyzed in proteasomes.

- How can a single pre-mRNA sequence encode several different proteins? **See pp. 346–347 and Figure 16.22**
- How do miRNAs and siRNAs regulate gene expression? **See pp. 347–348 and Figure 16.23**
- What is a riboswitch? **See p. 348 and Figure 16.24**
- Explain the role of the proteasome. **See p. 349 and Figure 16.25**

?

Can epigenetic changes be manipulated?

ANSWER

Epigenetic changes often involve the addition of methyl (–CH₃) groups. Nutrients in the diet such as folic acid and SAM-e (*S*-adenosyl methinione) contain methyl groups and participate in reactions that modify DNA. Experiments with mice have shown that feeding young animals a diet enriched with these nutrients causes changes in epigenetic patterns and gene expression that remain throughout life. It remains to be seen what the effects of such a diet on human infants would be. But the results of these experiments raise the possibility of altering gene expression by diet.

16.1 How Is Gene Expression Regulated in Prokaryotes?

- Some proteins are synthesized only when they are needed. Proteins that are made only in the presence of a particular compound—an **inducer**—are **inducible proteins**. Proteins that are made at a constant rate regardless of conditions are **constitutive proteins**. Review Figure 16.2

- An **operon** consists of a promoter, an **operator**, and two or more **structural genes**. Promoters and operators do not code for proteins, but serve as binding sites for regulatory proteins. **Review Figure 16.4**

- Regulatory genes code for regulatory proteins, such as **repressors**. When a repressor binds to an operator, transcription of the structural gene is inhibited. **Review Figure 16.5, ANIMATED TUTORIALS 16.1, 16.2**

- The *lac* operon is an example of an inducible system, in which the presence of an inducer (lactose) keeps the repressor from binding the operator, allowing the transcription of structural genes for lactose metabolism.

- Transcription can be enhanced by the binding of an **activator** protein to the promoter. **Review Figure 16.6**

- **Catabolite repression** is the inhibition of a catabolic pathway for one energy source by a different, preferred energy source.

16.2 How Is Eukaryotic Gene Transcription Regulated?

- Eukaryotic gene expression is regulated both during and after transcription. **Review Figure 16.7, ACTIVITY 16.1**

- **Transcription factors** and other proteins bind to DNA and affect the rate of initiation of transcription at the promoter. **Review Figures 16.8, 16.9, ANIMATED TUTORIAL 16.3**

- The interactions of these proteins with DNA are highly specific and depend on protein domains and DNA sequences.

- Genes at distant locations from one another can be coordinately regulated by transcription factors and promoter elements. **Review Figure 16.11**

16.3 How Do Viruses Regulate Their Gene Expression?

- **Viruses** are not cells, and rely on host cells to reproduce.

- The basic unit of a virus is a **virion**, which consists of a nucleic acid genome (DNA or RNA) and a protein coat, called a **capsid**.

- **Bacteriophages** are viruses that infect bacteria.

- Viruses undergo a **lytic** cycle, which causes the host cell to burst, releasing new virions.

- Some viruses have promoters that bind host RNA polymerase, which they use to transcribe their own genes and proteins. **Review Figure 16.13**

- Some viruses can also undergo **lysogeny**, in which a molecule of their DNA, called a **prophage**, is inserted into the host chromosome, where it replicates for generations. **Review Figure 16.14**

- The cellular environment determines whether a phage undergoes a lytic or a lysogenic cycle. Regulatory proteins that compete for promoters on phage DNA control the switch between the two life cycles. **Review Figure 16.15**

- A **retrovirus** uses **reverse transcriptase** to generate a cDNA **provirus** from its RNA genome. The provirus is incorporated into the host's DNA and can be activated to produce new virions. **Review Figure 16.16**

16.4 How Do Epigenetic Changes Regulate Gene Expression?

- **Epigenetics** refers to changes in gene expression that do not involve changes in DNA sequences.

- **Methylation** of cytosine residues generally inhibits transcription. **Review Figure 16.18**

- Modifications of histone proteins in nucleosomes make transcription either easier or more difficult. **Review Figure 16.19**

- Epigenetic changes can occur because of the environment.

- DNA methylation can explain **genomic imprinting**, where the expression of a gene depends on its parental origin. **Review Figure 16.20**

16.5 How Is Eukaryotic Gene Expression Regulated after Transcription?

- Alternative splicing of pre-mRNA can produce different proteins. **Review Figure 16.22**

- Small RNAs do not code for proteins but regulate the translation and longevity of mRNA. **Review Figure 16.23**

- The translation of mRNA to proteins can be regulated by translational repressors.

- The **proteasome** can break down proteins, thus affecting protein longevity. **Review Figure 16.25**

See ACTIVITY 16.2 for a concept review of this chapter.

 Go to the Interactive Summary to review key figures, Animated Tutorials, and Activities
Life10e.com/is16

CHAPTER**REVIEW**

REMEMBERING

1. Which of the following statements about the *lac* operon is *not* true?
   a. When the inducer binds to the repressor, the repressor can no longer bind to the operator.
   b. When the inducer binds to the operator, transcription is stimulated.
   c. When the repressor binds to the operator, transcription is inhibited.
   d. When the inducer binds to the repressor, the shape of the repressor is changed.
   e. The repressor has binding sites for both DNA and the inducer.

2. In the lysogenic cycle of bacteriophage λ,
   a. a repressor, cI, blocks the lytic cycle.
   b. the bacteriophage carries DNA between bacterial cells.
   c. both early and late viral genes are transcribed.
   d. the viral genome is made into RNA, which stays in the host cell.
   e. many new viruses are made immediately, regardless of host health.

3. An operon is
   a. a molecule that can turn genes on and off.
   b. an inducer bound to a repressor.
   c. a series of regulatory sequences controlling transcription of protein-coding genes.
   d. any long sequence of DNA.
   e. a promoter, an operator, and a group of linked structural genes.

4. Which of the following is true of both positive and negative gene regulation?
   a. They directly reduce the rate of transcription of certain genes.
   b. They involve transcription factors (or RNA) binding to DNA.
   c. They involve transcription of all genes in the genome.
   d. They are not both active in the same organism or virus.
   e. They act away from the promoter.

5. Which statement about selective gene transcription in eukaryotes is *not* true?
   a. Transcription factors can bind at a site on DNA distant from the promoter.
   b. Transcription requires transcription factors.
   c. Genes are usually transcribed as groups called operons.
   d. Both positive and negative regulation occur.
   e. Many proteins bind at the promoter.

6. Control of gene expression in eukaryotes includes all of the following except
   a. alternative RNA splicing.
   b. binding of proteins to DNA.
   c. transcription factors.
   d. stabilization of mRNA by miRNA.
   e. DNA methylation.

UNDERSTANDING & APPLYING

7. Compare the life cycles of a lysogenic bacteriophage and HIV (Figures 16.14 and 16.16) with respect to:
   a. how the virus enters the cell.
   b. how the new viruses are released from the cell.
   c. how the viral genome is replicated.
   d. how new viruses are produced.

8. Compare the roles of proteins and DNA sequences in the initiation of transcription in a prokaryote gene and a eukaryotic gene.

9. A protein-coding gene in a eukaryote has three introns. How many different proteins could be made by alternative splicing of the pre-mRNA from this gene?

ANALYZING & EVALUATING

10. The repressor protein that acts on the *lac* operon of *E. coli* is encoded by a regulatory gene. The repressor is made in small quantities and at a constant rate. Would you surmise that the promoter for this repressor protein is efficient or inefficient? Is synthesis of the repressor constitutive, or is it inducible and under environmental control?

11. In colorectal cancer, tumor suppressor genes are not active. This important factor results in uncontrolled cell division. Two possible explanations for the inactive genes are (a) mutations in the coding regions, resulting in inactive proteins, or (b) epigenetic silencing at the promoters of the genes, resulting in reduced transcription. How would you investigate these two possibilities?

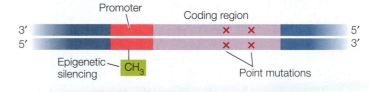

Go to BioPortal at **yourBioPortal.com** for Animated Tutorials, Activities, LearningCurve Quizzes, Flashcards, and many other study and review resources.

17 Genomes

CHAPTEROUTLINE

17.1 How Are Genomes Sequenced?

17.2 What Have We Learned from Sequencing Prokaryotic Genomes?

17.3 What Have We Learned from Sequencing Eukaryotic Genomes?

17.4 What Are the Characteristics of the Human Genome?

17.5 What Do the New Disciplines of Proteomics and Metabolomics Reveal?

Variation in Dogs   The Chihuahua (bottom) and the Brazilian mastiff (top) are the same species, *Canis lupus familiaris*, yet show great variation in size. Genome sequencing has revealed insights into how size is controlled by genes.

*C*ANIS LUPUS FAMILIARIS, the dog, was domesticated by humans from the gray wolf thousands of years ago. While there are many kinds of wolves, they all look about the same. Not so with "man's best friend." The American Kennel Club recognizes about 155 different breeds, which not only look different, but also vary greatly in size. For example, an adult Chihuahua weighs just 1.5 kg, whereas a Scottish deerhound weighs 70 kg. No other mammal shows such large phenotypic variation. Also, there are hundreds of genetic diseases in dogs, many of which have counterparts in humans. To find out about the genes behind the phenotypic variation, and to elucidate the relationships between genes and diseases, the Dog Genome Project was started in the late 1990s.

Two dogs—a boxer and a poodle—were the first to have their entire genomes sequenced. The dog genome contains 2.8 billion base pairs of DNA in 39 pairs of chromosomes. There are 19,000 protein-coding genes, most of them with close counterparts in other mammals, including humans. The whole genome sequence made it easy to create a map of genetic markers—specific nucleotides or short sequences of DNA at particular locations on the genome that differ among individual dogs or breeds.

Genetic markers are being used to map the locations of (and thus identify) genes that control particular traits. For example, Dr. Elaine Ostrander and her colleagues at the National Institutes of Health studied Portuguese water dogs to identify genes that control size. Taking samples of cells for DNA isolation was relatively easy: a cotton swab was swept over the inside of the cheek. As Dr. Ostrander said, the dogs "didn't care, especially if they thought they were going to get a treat or if there was a tennis ball in our other hand." It turned out that the gene for *insulin-like growth factor 1* (IGF-1) is important in determining size: large breeds have an allele that codes for an active IGF-1, and small breeds have a different allele that codes for a less active IGF-1.

Inevitably, some scientists have set up companies to test dogs for genetic variations, using DNA supplied by anxious owners and breeders. Some traditional breeders disapprove, but others say it will improve the breeds and give more joy (and prestige) to owners. In other words, the issues surrounding the Dog Genome Project are not unlike those arising from the Human Genome Project.

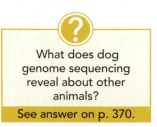

?

What does dog genome sequencing reveal about other animals?

See answer on p. 370.

17.1 How Are Genomes Sequenced?

Genome sequencing involves determining the nucleotide base sequence of the entire genome of an organism. For a prokaryotic organism with a single chromosome, the genome sequence is one continuous series of base pairs (bp). In the case of a diploid, sexually reproducing species with multiple autosomes and a pair of sex chromosomes (see Section 12.4), the "sequenced genome" usually means the sequence of all the bases in one set of autosomes and in each of the two sex chromosomes. With advances in the technology for DNA sequencing, there has been an explosion of genetic information that scientists can use in a variety of ways:

- The genomes of different species can be compared to find out how they differ at the DNA level, and this can be used to trace evolutionary relationships.

- The sequences of individuals within a species can be compared to identify mutations that affect particular phenotypes.

- The sequence information can be used to identify genes for particular traits, such as genes associated with diseases.

The notion of sequencing the entire genome of a complex organism was not contemplated until 1986. The Nobel laureate Renato Dulbecco and others proposed at that time that the world scientific community be mobilized to undertake the sequencing of the entire human genome. One motive was to detect DNA damage in people who had survived the atomic bomb attacks and been exposed to radiation in Japan during World War II. But in order to detect changes in the human genome, scientists first needed to know its normal sequence.

The result was the publicly funded **Human Genome Project**, an enormous undertaking that was successfully completed in 2003. This effort was aided and complemented by privately funded groups. The project benefited from the development of many new methods that were first developed to sequence smaller genomes—those of prokaryotes and simple eukaryotes, the model organisms you have encountered in earlier chapters of this book. Many of these methods are still applied widely, and powerful new methods for sequencing genomes have emerged. These methods are complemented by new ways to examine phenotypic diversity in a cell's proteins and in the metabolic products of the cell's enzymes.

New methods have been developed to rapidly sequence DNA

Many prokaryotes have a single chromosome, whereas eukaryotes have many. Because of their differing sizes, chromosomes can be separated from one another, identified, and experimentally manipulated. It might seem that the most straightforward way to sequence a chromosome would be to start at one end and simply sequence the DNA molecule one nucleotide at a time. The task is somewhat simplified because only one of the two strands needs to be sequenced, the other being complementary. However, this large-polymer approach is not practical, since at most only several hundred bp can be sequenced at a time using current methods.

As you will see, the key to determining genome sequences is to perform many sequencing reactions simultaneously, after first breaking the DNA up into millions of small, overlapping fragments.

In the 1970s Frederick Sanger and his colleagues invented a way to sequence DNA by using chemically modified nucleotides that were originally developed to stop cell division in cancer. This method, or a variation of it, was used to obtain the first human genome sequence as well as those of several model organisms. However, it was relatively slow, expensive, and labor-intensive. The first decade of the new millennium saw the development of faster and less expensive methods, often referred to under the general term **high-throughput sequencing**. These methods use miniaturization techniques first developed for the electronics industry, as well as the principles of DNA replication, often in combination with the polymerase chain reaction (PCR).

High-throughput sequencing methods are rapidly evolving. Just one of the many approaches is outlined here and illustrated in **Figure 17.1**. First the DNA is prepared for sequencing by attaching it to a solid surface and amplifying the DNA by PCR (see Figure 17.1A):

1. A large molecule of DNA is cut into small fragments of about 100 bp each. This can be done physically, using mechanical forces to shear (break up) the DNA, or by using enzymes that hydrolyze the phosphodiester bonds between nucleotides at intervals in the DNA backbone.

2. The DNA is denatured by heat, breaking the hydrogen bonds that hold the two strands together. Each single strand acts as a template for the synthesis of new, complementary DNA.

3. Short, synthetic oligonucleotides are attached to each end of each fragment, and these are attached to a solid support. The support can be a microbead or a flat surface.

4. The DNA is amplified by PCR (see Section 13.5) using primers complementary to the synthetic oligonucleotides attached to the ends of each DNA. The multiple (approximately 1,000) copies of the DNA at a single location allow for easy detection of added nucleotides during the sequencing steps.

Once the DNA has been attached to a solid substrate and amplified, it is ready for sequencing (see Figure 17.1B):

1. At the beginning of each sequencing cycle, the fragments are heated to denature them. A solution containing a universal primer (complementary to one of the same synthetic oligonucleotides used for the PCR amplification step), DNA polymerase, and the four deoxyribonucleoside triphosphates (dNTPs: dATP, dGTP, dCTP, and dTTP) is then added to the DNA. Recall that dNTPs are the substrates that the DNA polymerase uses in DNA synthesis (see Section 13.3). Each of the four kinds of dNTP is tagged with a different colored fluorescent dye.

2. The DNA synthesis reaction is set up so that only one nucleotide is added to the new DNA strand in each sequencing cycle. After each addition, the unincorporated dNTPs are removed.

17.1 DNA Sequencing High-throughput sequencing involves (A) the chemical amplification of DNA fragments and (B) the synthesis of complementary strands using fluorescently labeled nucleotides.

Go to Animated Tutorial 17.1
Sequencing the Genome
Life10e.com/at17.1

Go to Animated Tutorial 17.2
High-Throughput Sequencing
Life10e.com/at17.2

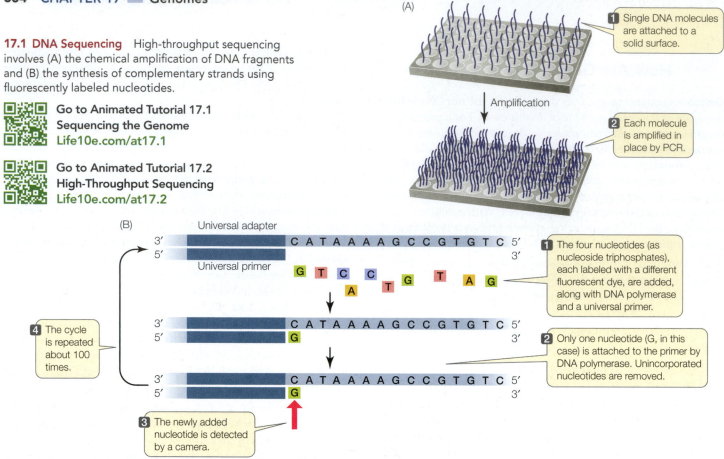

(A)
1 Single DNA molecules are attached to a solid surface.

Amplification

2 Each molecule is amplified in place by PCR.

(B)
Universal adapter
Universal primer

1 The four nucleotides (as nucleoside triphosphates), each labeled with a different fluorescent dye, are added, along with DNA polymerase and a universal primer.

2 Only one nucleotide (G, in this case) is attached to the primer by DNA polymerase. Unincorporated nucleotides are removed.

4 The cycle is repeated about 100 times.

3 The newly added nucleotide is detected by a camera.

3. The fluorescence of the new nucleotide at each location is detected with a camera. The color of the fluorescence indicates which of the four nucleotides was added.

4. The fluorescent tag is removed from the nucleotide that is already attached, and then the DNA synthesis cycle is repeated. Images are captured after each nucleotide is added. The series of colors at each location indicate the sequence of nucleotides in the growing DNA strand at that location.

The power of this method derives from the fact that:

• It is fully automated and miniaturized.

• Millions of different fragments are sequenced at the same time.

• It is an inexpensive way to sequence large genomes. For example, at the time of this writing, a complete human genome could be sequenced in a few days for several thousand dollars. This is in contrast to the Human Genome Project, which took 13 years and $2.7 billion to sequence one genome!

The technology used to sequence millions of short DNA fragments is only half the story, however. Once these sequences have been determined, the problem becomes how to put them together. In other words, how are they arranged in the chromosomes from which they came? Imagine if you cut out every word in this book (there are more than half a million of them), put them on a table, and tried to arrange them in

their original order! The enormous task of determining DNA sequences is possible because the original DNA fragments are overlapping.

Let's illustrate the process using a single 10-bp DNA molecule. (This is a double-stranded molecule, but for convenience we show only the sequence of the noncoding strand.) The molecule is cut three ways (for example, using three different restriction enzymes). Cutting with the first enzyme generates the fragments:

TG, ATG, and CCTAC

Cutting the same molecule with the second enzyme generates the fragments:

AT, GCC, and TACTG

Cutting with the third enzyme results in:

CTG, CTA, and ATGC

Can you put the fragments in the correct order? (The answer is ATGCCTACTG.) For genome sequencing, the sequence fragments are called "reads" (**Figure 17.2**). Of course, the problem of ordering 2.5 million fragments from human chromosome 1 (246 million bp) is more challenging than our 10-bp example above! The field of **bioinformatics** was developed to analyze DNA sequences using sophisticated mathematics and computer programs to handle the large amounts of data generated in genome sequencing.

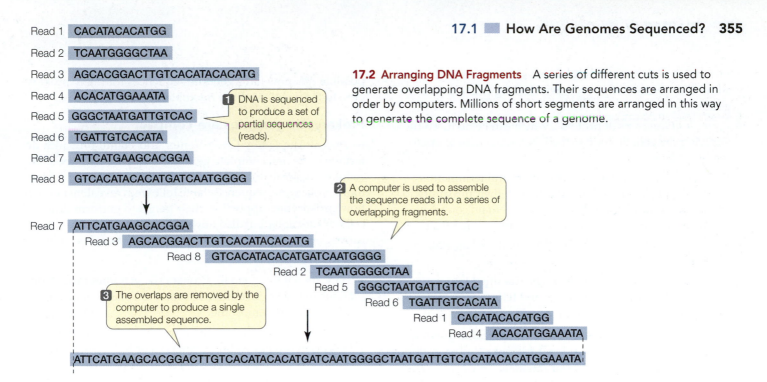

17.2 Arranging DNA Fragments A series of different cuts is used to generate overlapping DNA fragments. Their sequences are arranged in order by computers. Millions of short segments are arranged in this way to generate the complete sequence of a genome.

Genome sequences yield several kinds of information

New genome sequences are being published at an accelerating pace, creating a torrent of biological information (**Figure 17.3**). This information is used in two related fields of research, both focused on studying genomes. In **functional genomics**,

biologists use sequence information to identify the functions of various parts of genomes. These parts include:

- *Open reading frames*, which are the coding regions of genes. For protein-coding genes, these regions can be recognized by the start and stop codons for translation, and by consensus sequences that indicate the locations of

17.3 The Genomic Book of Life Genome sequences contain many features, some of which are summarized in this overview. Sifting through all the information contained in a genome sequence can help us understand how an organism functions and what its evolutionary history might be.

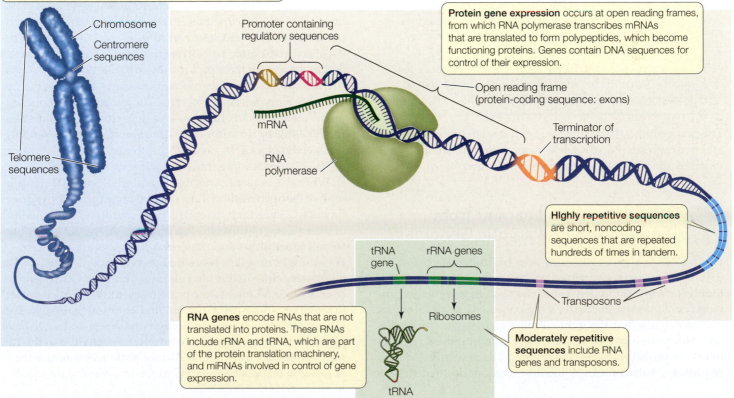

introns. A major goal of functional genomics is to understand the function of every open reading frame in each genome.

- *Amino acid sequences* of proteins, which can be deduced by applying the genetic code to the DNA sequences of open reading frames.

- *Regulatory sequences*, such as promoters and terminators for transcription. These are identified by their proximity to open reading frames and because they contain consensus sequences for the binding of specific transcription factors.

- *RNA genes*, including rRNA, tRNA, small nuclear RNA, and microRNA genes.

- *Other noncoding sequences* that can be classified into various categories, including centromeric and telomeric regions, transposons, and other repetitive sequences.

Sequence information is also used in **comparative genomics**: the comparison of a newly sequenced genome (or parts thereof) with sequences from other organisms. This can provide further information about the functions of sequences and can be used to trace evolutionary relationships among different organisms.

RECAP 17.1

The sequencing of genomes involves cutting large chromosomes into fragments, sequencing the fragments, and then assembling the fragment sequences into continuous sequences for entire chromosomes. Current sequencing methods use automation and powerful computers. They use labeled nucleotides that are detected at the ends of growing polynucleotide chains.

- Describe one high-throughput method for DNA sequencing. **See pp. 353–354 and Figure 17.1**

- For sequencing genomes, why are overlapping sequences obtained, and how are they arranged to give the final sequence? **See p. 354 and Figure 17.2**

- How are open reading frames recognized in a genomic sequence? What kind of information can be derived from an open reading frame? **See pp. 355–356**

The first genomes to be fully sequenced were those of viruses and prokaryotes. Next we will discuss the information provided by the relatively simple prokaryotic genomes.

17.2 What Have We Learned from Sequencing Prokaryotic Genomes?

When DNA sequencing became possible in the late 1970s, the first life forms to be sequenced were the simplest viruses with their relatively small genomes. The sequences quickly provided new information on how these viruses infect their hosts and reproduce. But the manual sequencing techniques used on viruses were not up to the task of studying the larger genomes of prokaryotes and eukaryotes. The newer, automated sequencing techniques we just described made such studies

possible. We now have genome sequences for many prokaryotes, to the great benefit of microbiology and medicine.

Prokaryotic genomes are compact

In 1995 a team led by Craig Venter and Hamilton Smith determined the first complete genomic sequence of a free-living cellular organism, the bacterium *Haemophilus influenzae*. Many more prokaryotic sequences have followed, revealing not only how prokaryotes apportion their genes to perform different cellular functions, but also how their specialized functions are carried out. There are several notable features of bacterial and archaeal genomes:

- They are relatively small. Prokaryotic genomes range from about 160,000 to 12 million bp and are usually organized into a single chromosome.

- They are compact. Typically, more than 85 percent of the DNA is in protein-coding sequences or RNA genes, with only short sequences between genes.

- The genes usually do not have introns. An exception is the rRNA and tRNA genes of archaea, which frequently contain introns.

- In addition to the main chromosome, prokaryotes often have smaller, circular molecules of DNA called plasmids, which may be transferred between cells (see Chapter 12).

Beyond these similarities, there is great diversity among these single-celled organisms, reflecting the huge variety of environments in which they are found (see Chapter 26).

Let's examine prokaryotic genomes in terms of functional and comparative genomics.

FUNCTIONAL GENOMICS As described above, functional genomics is the biological discipline that assigns functions to the products of genes. This field, less than 20 years old, is now a major occupation of biologists. You can see the various functions encoded by the genomes of three prokaryotes in **Table 17.1**.

The only host for *H. influenzae* is humans. It lives in the upper respiratory tract and can cause ear infections or, more seriously, meningitis in children. Its single circular chromosome has 1,830,138 bp. In addition to its origin of replication and the genes coding for rRNAs and tRNAs, this bacterial chromosome has 1,727 open reading frames with promoters nearby.

When this sequence was first announced, only 1,007 (58 percent) of the open reading frames coded for proteins with known functions. Since then, scientists have identified the functions of many more of the encoded proteins. All of the major biochemical pathways and molecular functions are represented. For example, there are genes that encode enzymes involved in glycolysis, fermentation, and electron transport. Other gene sequences code for membrane proteins, including those involved in active transport. An important finding was that highly infective strains of *H. influenzae*, but not noninfective strains, have genes for surface proteins that attach the bacterium to the human respiratory tract. These surface proteins are now a focus of research on possible treatments for *H. influenzae* infections.

TABLE 17.1
Gene Functions in Three Bacteria

Category	Number of Genes in:		
	E. coli	H. influenzae	M. genitalium
Total protein-coding genes	4,288	1,727	482
Biosynthesis of amino acids	131	68	1
Biosynthesis of cofactors	103	54	5
Biosynthesis of nucleotides	58	53	19
Cell envelope proteins	237	84	17
Energy metabolism	243	112	31
Intermediary metabolism	188	30	6
Lipid metabolism	48	25	6
DNA replication, recombination, and repair	115	87	32
Protein folding	9	6	7
Regulatory proteins	178	64	7
Transcription	55	27	12
Translation	182	141	101
Uptake of molecules from the environment	427	123	34

COMPARATIVE GENOMICS Soon after the sequence of *H. influenzae* was announced, smaller (*Mycoplasma genitalium*: 580,073 bp) and larger (*E. coli*: 4,639,221 bp) prokaryotic sequences were completed. Thus began the era of comparative genomics. Scientists can identify genes that are present in one bacterium and missing in another, allowing them to relate these genes to bacterial function.

M. genitalium, for example, lacks enzymes needed to synthesize amino acids, whereas *E. coli* and *H. influenzae* both possess such enzymes. This finding reveals that *M. genitalium* must obtain all its amino acids from its environment (usually the human urogenital tract). Furthermore, *E. coli* has 55 genes that encode transcriptional activators, whereas *M. genitalium* has only 7. This relative lack of control over gene expression suggests that the biochemical flexibility of *M. genitalium* must be limited compared with that of *E. coli*.

The sequencing of prokaryotic and viral genomes has many potential benefits

Prokaryotic genome sequencing is providing insights into microorganisms that are important for agriculture and medicine. Scientists who analyze the sequences have discovered previously unknown genes and proteins that can be targeted for isolation and functional study. They have also discovered surprising relationships between some organisms, suggesting that genes may be transferred between different species.

● *Rhizobium* species are bacteria that form symbiotic associations with plants, living inside the roots of legumes such as beans, peas, and clover. The bacteria fix atmospheric nitrogen from the air and convert it into forms usable by the plants, reducing the need for nitrogen-containing

fertilizers. Genome sequences from several *Rhizobium* species have been used to identify the genes involved in successful symbiosis, and this knowledge is being used both to improve the efficiency of this process and to broaden the range of plants that can form these beneficial associations.

● *E. coli* strain O157:H7 causes illness (sometimes severe) in at least 70,000 people a year in the United States. Its genome has 5,416 genes, of which 1,387 are different from those in the familiar (and harmless) laboratory strains of this bacterium. Many of these unique genes are also present in other pathogenic bacteria, such as *Salmonella* and *Shigella*. This finding suggests that there is extensive genetic exchange among these species, and that "superbugs" that have acquired multiple genes for antibiotic resistance may be on the horizon.

● *Severe acute respiratory syndrome* (*SARS*) was first detected in southern China in 2002 and rapidly spread in 2003. There is no effective treatment, and 10 percent of infected people die. Isolation of the causative agent, a virus, and the rapid sequencing of its genome revealed several novel proteins that are possible targets for antiviral drugs or vaccines. Research is underway on both fronts, since further outbreaks are anticipated.

Genome sequencing also provides insights into organisms involved in global ecological cycles (see Chapter 58). In addition to carbon dioxide, another important gas that contributes to global warming is methane (CH_4; see Figure 2.7). Some bacteria, such as *Methanococcus*, produce methane in the stomachs of cows. Others, such as *Methylococcus*, remove methane from the air and use it as an energy source. The genomes of both of these bacteria have been sequenced. Understanding the genes involved in methane production and consumption may help us slow the progress of global warming.

Metagenomics allows us to describe new organisms and ecosystems

If you take a microbiology laboratory course you will learn how to identify various prokaryotes on the basis of their growth on particular artificial media. For example, staphylococci are a group of bacteria that infect skin and nasal passages. When grown on a medium called blood agar, they form round, raised colonies. Microorganisms can also be identified by their nutritional requirements or the conditions under which they will grow (for example, aerobic versus anaerobic). Such culture methods have been the mainstay of microbial identification for more than a century and are still useful and important. However, scientists can now use PCR and DNA sequencing to identify microbes without culturing them in the laboratory.

In 1985 Norman Pace, then at Indiana University, came up with the idea of isolating DNA directly from environmental samples. He used PCR to amplify specific sequences from the samples to determine whether particular microbes were present. The PCR products were sequenced to explore their

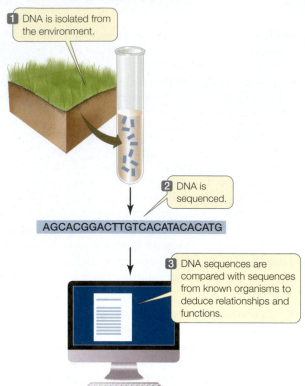

1 DNA is isolated from the environment.

2 DNA is sequenced.

AGCACGGACTTGTCACATACACATG

3 DNA sequences are compared with sequences from known organisms to deduce relationships and functions.

17.4 Metagenomics Microbial DNA extracted from the environment can be sequenced and analyzed. This has led to the description of many new genes and species.

diversity. The term **metagenomics** was coined to describe this approach of analyzing genes without isolating the intact organism. It is now possible to sequence DNA samples from almost any environment. The sequences can be used to detect the presence of both known microbes and heretofore unidentified organisms (**Figure 17.4**). For example:

- Sequencing of DNA from 200 liters of seawater indicated that it contained 5,000 different viruses and 2,000 different bacteria, many of which had not been described previously.

- One kilogram of marine sediment contained 1 million different viruses, most of them new.

- Water runoff from a mine contained many new species of prokaryotes thriving in this apparently inhospitable environment. Some of these organisms exhibited metabolic pathways that were previously unknown to biologists. These organisms and their capabilities may be useful in cleaning up pollutants from the water.

These and other discoveries are truly extraordinary and potentially very important. It is estimated that 90 percent of the microbial world has been invisible to biologists and is only now being revealed by metagenomics. Entirely new ecosystems of bacteria and viruses are being discovered in which, for example, one species produces a molecule that another metabolizes. It is hard to overemphasize the

importance of such an increase in our knowledge of the hidden world of microbes. This knowledge will help us understand natural ecological processes, and has the potential to help us find better ways to manage environmental catastrophes such as oil spills, or to remove toxic heavy metals from soil.

Some sequences of DNA can move about the genome

As we mentioned in Section 15.1, **transposable elements** (or **transposons**) are segments of DNA that can move from place to place in the genome. Genome sequencing has allowed scientists to study these elements more broadly, and they are now known to be widespread in both prokaryotes and eukaryotes. Prokaryotic transposable elements are often short sequences of 1,000 to 2,000 bp, and they can be found in both chromosomes and in plasmids. A transposable element might be at one location in the genome of one *E. coli* strain, and at a different location in another strain. The insertion of this movable DNA sequence from elsewhere in the genome into the middle of a protein-coding gene disrupts that gene (**Figure 17.5A**). Any mRNA expressed from the disrupted gene will contain the extra sequence, and the protein it encodes will be altered and almost certainly nonfunctional. So transposable elements can produce significant phenotypic effects. Sometimes, two transposable elements located near one another (within a few thousand bp) will transpose together and carry the intervening DNA sequence with them. These are referred to as **composite transposons** (**Figure 17.5B**). Genes for

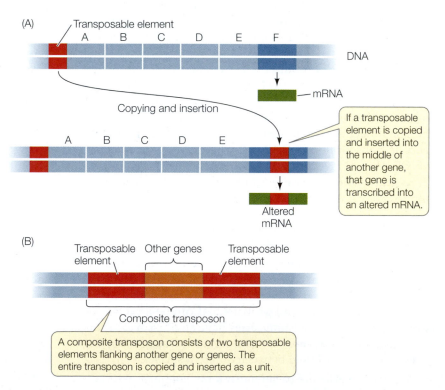

(A) Transposable element

A B C D E F

DNA

mRNA

Copying and insertion

A B C D E

If a transposable element is copied and inserted into the middle of another gene, that gene is transcribed into an altered mRNA.

Altered mRNA

(B)

Transposable element   Other genes   Transposable element

Composite transposon

A composite transposon consists of two transposable elements flanking another gene or genes. The entire transposon is copied and inserted as a unit.

17.5 DNA Sequences That Move Transposable elements are DNA sequences that move from one location to another. (A) In one method of transposition ("copy and paste"), the DNA sequence is replicated and the copy inserts elsewhere in the genome. (B) Composite transposons contain additional genes flanked by two transposable elements.

17.6 Using Transposon Mutagenesis to Determine the Minimal Genome

Mycoplasma genitalium has one of the smallest known genomes of any prokaryote. But are all of its genes essential to life? By inactivating the genes one by one, scientists determined which of them are essential for the cell's survival. This research may lead to the construction of artificial cells with customized genomes, designed to perform functions such as degrading oil and making plastics.[a]

HYPOTHESIS Only some of the genes in a bacterial genome are essential for cell survival.

Method

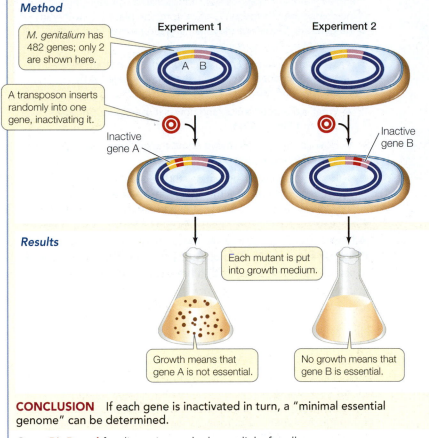

M. genitalium has 482 genes; only 2 are shown here.

A transposon inserts randomly into one gene, inactivating it.

Experiment 1

Experiment 2

A B

Inactive gene A

Inactive gene B

Results

Each mutant is put into growth medium.

Growth means that gene A is not essential.

No growth means that gene B is essential.

CONCLUSION If each gene is inactivated in turn, a "minimal essential genome" can be determined.

Go to **BioPortal** for discussion and relevant links for all INVESTIGATING LIFE figures.

[a]Hutchison, C. et al. 1999. *Science* 286: 2165–2169, and Glass, J. I. et al. 2006. *Proceedings of the National Academy of Sciences USA* 103: 425–430.

all organisms (universal genes). Not surprisingly, these include genes whose products are involved in DNA replication, transcription, and RNA translation to form proteins. There are also some (nearly) universal gene segments that are present in many genes in many organisms; for example, the sequence that codes for an ATP binding site in a protein. These findings suggest that there is some ancient, minimal set of DNA sequences that is common to all cells. One way to identify these sequences is to look for them in computer analyses of sequenced genomes.

Another way to define the minimal genome is to take an organism with a simple genome and deliberately mutate one gene at a time to see what happens. *M. genitalium* has one of the smallest known genomes—only 482 protein-coding genes. Even so, some of its genes are dispensable under some circumstances. For example, *M. genitalium* has genes for metabolizing both glucose and fructose, but it can survive in the laboratory on a medium containing only one of these sugars. Under such conditions it doesn't need the genes for metabolizing the other sugar.

What about other genes? A team led by Craig Venter has addressed this question with experiments involving the use of transposons as mutagens. When transposons in the bacterium are activated, they insert themselves into genes at random, mutating and inactivating them (**Figure 17.6**). The mutated bacteria are tested for growth and survival, and DNA from interesting mutants is sequenced to find out which genes contain transposons. The astonishing result of these studies is that *M. genitalium* can survive in the laboratory with a minimal genome of only 382 protein-coding genes!

One goal of this research is to design new life forms with specific purposes, such as bacteria that will clean up oil spills. The next step toward this goal is to create an artificial genome and insert it into bacterial cells. Venter's team recently synthesized the entire genome of *Mycoplasma mycoides*, using a computer-directed blueprint that included some extra sequences that could be used to track the survival of the genome in multiplying cells. They then transplanted this genome into empty cells of a different species, *Mycoplasma capricolum*, whose own DNA had been hydrolyzed. The new DNA directed the cell to perform all the biochemical functions of life, including reproduction (**Figure 17.7**). Since the new cell's genome had some extra sequences, it was an entirely new organism, called *Mycoplasma mycoides* JCV1-syn.1.0.

At the time of DNA transfer, the recipient cell still had its original small molecules and proteins, but after approximately 30 cell divisions, all of the proteins in the new cell colony had been made using the synthetic genome's sequences. The cells had used substances in their environment to make their own small and large molecules. They were new individuals of a new organism.

antibiotic resistance can be multiplied and transferred between bacteria in this way: composite transposons carrying genes for antibiotic resistance can insert into a plasmid that then moves between bacteria by conjugation (see Section 12.6).

As we mentioned in Chapter 15, the mechanisms that allow transposable elements to move vary. For example, a transposable element may be replicated, and then the copy can be inserted into another site in the genome (the "copy and paste" mode). Or the element might splice out of one location and move to another location ("cut and paste"). The elements usually carry genes for enzymes such as transposases, which catalyze the reactions needed for transposition. Often the elements are flanked by inverted repeat DNA sequences that are recognized by these enzymes.

Will defining the genes required for cellular life lead to artificial life?

When the genomes of prokaryotes and eukaryotes are compared, a striking conclusion arises: certain genes are present in

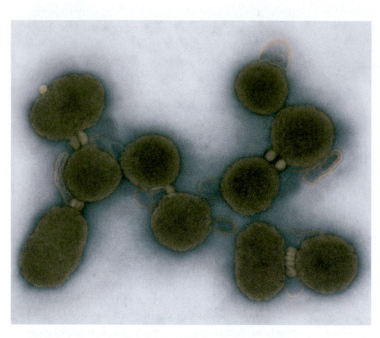

17.7 Synthetic Cells   Cells of *Mycoplasma mycoides JCVI-syn 1.0*, the first synthetic organism, are shown in this false-colored micrograph.

RECAP 17.2

DNA sequencing is used to study the genomes of prokaryotes that are important to humans and to ecosystems. Functional genomics uses gene sequences to determine the functions of the gene products. Comparative genomics compares gene sequences from different organisms to help identify their functions and evolutionary relationships. Transposable elements, including composite transposons, move from one place to another in the genome. Studies of the minimal genome may lead to the creation of artificial species.

- Give some examples of prokaryotic genomes that have been sequenced. What have the sequences shown? **See pp. 356–357**

- What is metagenomics, and how is it used? **See pp. 357–358 and Figure 17.4**

- How do transposons move about the genome? **See pp. 358–359 and Figure 17.5**

- How are selective inactivation studies being used to determine the minimal genome, and what are possible practical applications of this? **See p. 359 and Figures 17.6 and 17.7**

Advances in DNA sequencing and sequence analysis have led to the rapid sequencing of eukaryotic genomes. We will now look at some of the new insights that have come from these studies.

WORKING WITH**DATA:**

Using Transposon Mutagenesis to Determine the Minimal Genome

Original Papers

Hutchison, C., S. N. Peterson, S. R. Gill, R. T. Cline, O. White, C. M. Fraser, H. O. Smith, and J. C. Venter. 1999. Global transposon mutagenesis and a minimal *Mycoplasma* genome. *Science* 286: 2165–2169.

Glass, J. I., N. Assad-Garcia, N. Alperovich, S. Yooseph, M. R. Lewis, M. Maruf, C. A. Hutchison III, H. O. Smith, and J. C. Venter. 2006. Essential genes of a minimal bacterium. *Proceedings of the National Academy of Sciences USA* 103: 425–430.

Analyze the Data

Mycoplasma genitalium is a tiny parasitic bacterium. Its genome was the second bacterial genome to be sequenced, and the genome's size (580,073 bp) and number of protein-coding genes (482) made it the smallest known bacterial genome at the time. For this reason, it was a useful organism for exploring the question of how many genes are necessary for life. An approach to answering this question is transposon mutagenesis. A transposon is a movable sequence of DNA that can insert within the coding region of a gene. If the gene is essential to life, a cell harboring the transposon will not survive. A team led by Craig Venter used this approach to identify 382 *M. genitalium* genes that are essential for its survival in the laboratory (see Figure 17.6). They went on to chemically synthesize all of these genes, put them together to make a chromosome, and put the chromosome into a *Mycoplasma* cell

whose DNA had been destroyed. They thereby created a new cell with just the synthetic minimal genome (see Figure 17.7).

The growth of *M. genitalium* strains with insertions in genes (intragenic regions) was compared with the growth of strains with insertions in noncoding (intergenic) regions of the genome. The results are shown in the table.

QUESTION 1

Explain these data in terms of genes essential for growth and survival. Are all of the genes in *M. genitalium* essential for growth? If not, how many are essential? Why did some of the insertions in intergenic regions prevent growth?

QUESTION 2

If a transposon inserts into the following regions of a gene, there might be no effect on the phenotype. Explain in each case:

  a. near the 3′ end of a coding region

  b. within a gene coding for rRNA (see Section 17.3)

How does this affect your answer to Question 1?

Type of insertion	Number of different genes/regions	Number that grew
Intragenic	482	100
Intergenic	199	184

Go to BioPortal for all WORKING WITH**DATA** exercises

17.3 What Have We Learned from Sequencing Eukaryotic Genomes?

As genomes have been sequenced and described, a number of major differences have emerged between eukaryotic and prokaryotic genomes. For example, in **Table 17.2** compare the bacterial genomes with those of yeasts, plants, and animals—all eukaryotes. Key differences include the following:

- *Eukaryotic genomes are larger than those of prokaryotes*, and they have more protein-coding genes. This difference is not surprising given that multicellular organisms have many cell types with specialized functions. As we saw above, the simple prokaryote *Mycoplasma* has several hundred protein-coding genes in a genome of 0.58 million bp. A rice plant, in contrast, has almost 43,000 genes!

- *Eukaryotic genomes have more regulatory sequences*—and many more regulatory proteins—than prokaryotic genomes. The greater complexity of eukaryotes requires much more regulation, which is evident in the many points of control associated with the expression of eukaryotic genes (see Figure 16.13).

- *Much of eukaryotic DNA is noncoding*. Distributed throughout many eukaryotic genomes are various kinds of DNA sequences that are not transcribed into mRNA, most notably introns and gene control sequences. As we discussed in Chapter 16, some noncoding sequences are transcribed into

microRNAs. In addition, eukaryotic genomes contain various kinds of repeated sequences. These features are rare in prokaryotes.

- *Eukaryotes have multiple chromosomes*, whereas prokaryotes often have a single, circular chromosome. As we have described in previous chapters, eukaryotic chromosomes have multiple origins of replication, a centromere region that holds the replicated chromosomes together before mitosis, and a telomeric sequence at each end of the chromosome that maintains chromosome integrity.

Model organisms reveal many characteristics of eukaryotic genomes

Most of our information about eukaryotic genomes has come from model organisms that have been studied extensively: the yeast *Saccharomyces cerevisiae*, the nematode (roundworm) *Caenorhabditis elegans*, the fruit fly *Drosophila melanogaster*, and the thale cress plant (*Arabidopsis thaliana*). Model organisms have been chosen because they are relatively easy to grow and study in a laboratory, their genetics are well studied, and they exhibit characteristics that represent a larger group of organisms.

YEAST: THE BASIC EUKARYOTIC MODEL Yeasts are single-celled eukaryotes. Like most eukaryotes, they have membrane-enclosed organelles, such as the nucleus and endoplasmic reticulum, and a life cycle that alternates between haploid and diploid generations (see Figure 11.15).

TABLE 17.2
Representative Sequenced Genomes

Organism	Haploid Genome Size (Mb)	Number of Protein-coding Genes	Percent of Genome That Codes for Proteins
Bacteria			
Mycoplasma genitalium	0.58	482	88
Haemophilus influenzae	1.8	1,727	89
Escherichia coli	4.6	4,288	88
Yeasts			
Saccharomyces cerevisiae	12.2	6,275	70
Schizosaccharomyces pombe	13.8	4,824	60
Plants			
Arabidopsis thaliana	125	25,498	25
Oryza sativa (rice)	420	42,653	12
Animals			
Caenorhabditis elegans (nematode)	100	20,470	25
Drosophila melanogaster (fruit fly)	140	15,016	13
Tetraodon nigroviridis (pufferfish)	340	27,918	10
Gallus gallus (chicken)	1,060	~20,000	3
Homo sapiens (human)	3,200	~21,000	1.2

Mb = millions of base pairs

TABLE 17.3

Comparison of the Genomes of *E. coli* and *S. cerevisiae*

	E. coli	Yeast
Genome length (base pairs)	4,640,000	12,157,000
Number of protein-coding genes	4,288	6,275
Proteins with roles in:		
Metabolism	650	650
Energy production/storage	240	175
Membrane transport	280	250
DNA replication/repair/recombination	115	175
Transcription	230	400
Translation	182	350
Protein targeting/secretion	35	430
Cell structure	180	250

Whereas the prokaryote *E. coli* has a single circular chromosome with about 4.6 million bp and 4,288 protein-coding genes, budding yeast (*Saccharomyces cerevisiae*) has 16 linear chromosomes and a haploid content of about 12.2 million bp, with 6,275 protein-coding genes. Gene inactivation studies similar to those carried out for *M. genitalium* (see Figure 17.6) indicate that fewer than 20 percent of the yeast's genes are essential to survival.

The most striking difference between the yeast genome and that of *E. coli* is the number of genes for targeting proteins to organelles (Table 17.3). Both of these single-celled organisms appear to use about the same number of genes to perform the basic functions of cell survival. It is the compartmentalization of the eukaryotic yeast cell into organelles that requires it to have many more genes. This finding is direct, quantitative confirmation of something we have known for a century: the eukaryotic cell is structurally more complex than the prokaryotic cell.

THE NEMATODE: UNDERSTANDING EUKARYOTIC DEVELOPMENT   In 1965 Sydney Brenner was fresh from being part of the team that first isolated mRNA. He was looking for a simple organism in which to study multicellularity, and settled on *Caenorhabditis elegans*, a 1-millimeter-long nematode (roundworm) that normally lives in the soil. It can also live in the laboratory, where it has become a favorite model organism of developmental biologists (see Section 19.4). The nematode has a transparent body that develops over 3 days from a fertilized egg to an adult worm made up of nearly 1,000 cells. In spite of its small number of cells, the nematode has a nervous system, digests food, reproduces sexually, and ages. So it is not surprising that an intense effort was made to sequence the genome of this model organism.

The *C. elegans* genome (100 million bp) is eight times larger than that of the yeast *Saccharomyces cerevisiae* and has 3.3 times as many protein-coding genes (20,470). Gene inactivation studies have shown that the worm can survive in laboratory cultures with only 10 percent of these genes. So the minimal genome of a worm is about twice the size of that of the yeast, which in turn is about three times the size of the minimal genome for *Mycoplasma*. What do these extra genes do?

All cells must have genes for survival, growth, and division. In addition, the cells of multicellular organisms must have genes for holding cells together to form tissues, for cell differentiation, and for intercellular communication. Looking at Table 17.4, you will recognize functions that we discussed in earlier chapters, including gene regulation (see Chapter 16) and cell communication (see Chapter 7).

***DROSOPHILA MELANOGASTER*: RELATING GENETICS TO GENOMICS**   The fruit fly *Drosophila melanogaster* is a famous model organism. Studies of fruit flies resulted in the formulation of many basic principles of genetics (see Section 12.4). More than 2,500 mutations of *D. melanogaster* had been described by the 1990s when genome sequencing began, and this fact alone was a good reason for sequencing the fruit fly's DNA. The fruit fly is a much larger organism than *C. elegans*, both in size (it has ten times more cells) and complexity, and it undergoes complicated developmental transformations from egg to larva to pupa to adult.

Not surprisingly, the fly's genome (about 140 million bp) is larger than that of *C. elegans*. But as we mentioned earlier, genome size does not necessarily correlate with the number of genes encoded. In this case, the larger fruit fly genome contains fewer genes (15,016) than the smaller nematode genome. Figure 17.8 summarizes the functions of the *Drosophila* genes that have been characterized so far; this distribution is typical of complex eukaryotes.

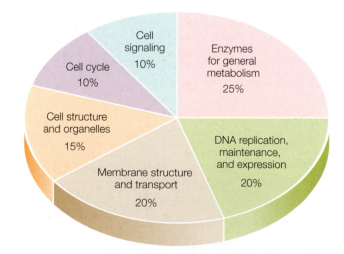

17.8 Functions of the Eukaryotic Genome   The distribution of gene functions in *Drosophila melanogaster* shows a pattern that is typical of many complex organisms.

TABLE 17.4
C. elegans Genes Essential to Multicellularity

Function	Protein/Domain	Number of Genes
Transcription control	Zinc finger; homeobox	540
RNA processing	RNA binding domains	100
Nerve impulse transmission	Gated ion channels	80
Tissue formation	Collagens	170
Cell interactions	Extracellular domains; glycotransferases	330
Cell–cell signaling	G protein–linked receptors; protein kinases; protein phosphatases	1,290

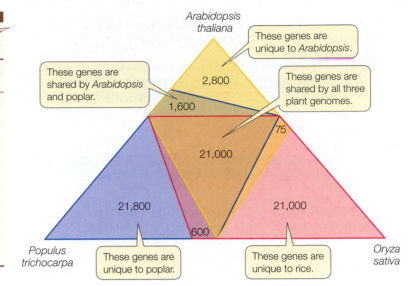

17.9 Plant Genomes Three plant genomes share a common set of approximately 21,000 genes that appear to comprise the "minimal" plant genome.

ARABIDOPSIS: STUDYING THE GENOMES OF PLANTS About 250,000 species of flowering plants dominate the land and fresh water. But in the context of the history of life, the flowering plants are fairly young, having evolved only about 200 million years ago. The genomes of some plants are huge—for example, the genome of corn is about 3 billion bp, and that of wheat is 16 billion bp. So although we are naturally most interested in the genomes of plants we use as food and fiber, it is not surprising that scientists first chose to sequence a simpler flowering plant.

Arabidopsis thaliana, thale cress, is a member of the mustard family and has long been a favorite model organism of plant biologists. It is small (hundreds could grow and reproduce in the space occupied by this page), it is easy to manipulate, and it has a relatively small genome of 125 million bp.

The *Arabidopsis* genome has an estimated 25,498 protein-coding genes, but remarkably, many of these genes are duplicates and probably originated by chromosomal rearrangements. When these duplicate genes are subtracted from the total, about 15,000 unique genes are left—similar to the number of genes found in fruit flies. Indeed, many of the genes found in these animals have homologs (related genes) in *Arabidopsis* and other plants, suggesting that plants and animals have a common ancestor.

But *Arabidopsis* has some genes that distinguish it as a plant (**Table 17.5**). These include genes involved in photosynthesis, in the transport of water into the root and throughout the plant, in the assembly of the cell wall, in the uptake and metabolism of

TABLE 17.5
Arabidopsis Genes Unique to Plants

Function		Number of Genes
Cell wall and growth		42
Water channels		300
Photosynthesis		139
Defense and metabolism		94

inorganic substances from the environment, and in the synthesis of specific molecules used for defense against microbes and herbivores (organisms that eat plants). Plants cannot escape their enemies or other adverse conditions as animals can, and so they must cope with the situation where they are. The ability to make tens of thousands of unusual molecules helps them fight their enemies and adapt to the environment (see Chapter 39).

The plant-specific genes in *Arabidopsis* are also found in the genomes of other plants, including rice (*Oryza sativa*), the first major crop plant whose sequence has been determined. Rice is the world's most important crop; it is a staple in the diet of 3 billion people. Despite its larger genome, rice has a set of genes remarkably similar to those of *Arabidopsis*. More recently the genome of the poplar tree *Populus trichocarpa* was sequenced. This rapidly growing tree is widely used for manufacturing paper and is a potential source of fixed carbon for making fuel. A comparison of the three genomes shows many genes in common, comprising the basic minimal plant genome (**Figure 17.9**).

Eukaryotes have gene families

About half of all eukaryotic protein-coding genes exist as only one copy in the haploid genome (two copies in somatic cells). The rest are present in multiple copies, which arose from gene duplications. Over evolutionary time, different copies of genes have undergone separate mutations, giving rise to groups of closely related genes called **gene families**. Some gene families, such as those encoding the globin proteins that make up hemoglobin, contain only a few members. Other families, such as the genes encoding the immunoglobulins that make up antibodies, have hundreds of members. In the human genome, there are about 21,000 protein-coding genes, but 16,000 of these are members of gene families. So only about one-third of the human genes are unique.

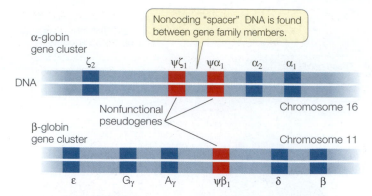

α-globin
gene cluster

Noncoding "spacer" DNA is found
between gene family members.

17.10 The Globin Gene Family The α-globin and β-globin clusters of the human globin gene family are located on different chromosomes. The genes of each cluster are separated by noncoding "spacer" DNA. The nonfunctional pseudogenes are indicated by the Greek letter psi (ψ). The γ gene has two variants, A_γ and G_γ.

The DNA sequences in a gene family are usually different from one another. As long as at least one member encodes a functional protein, the other members may mutate in ways that change the functions of the proteins they encode. During evolution, the availability of multiple copies of a gene allows for selection of mutations that provide advantages under certain circumstances. If a mutated gene is useful, it may be selected for in succeeding generations. If the mutated gene is a total loss, the functional copy is still there to carry out its role.

The family of genes that encode globins is a good example of a gene family in vertebrates. The globins are components of hemoglobin and myoglobin (an oxygen-binding protein present in muscle). The globin genes all arose long ago from a single common ancestral gene. In humans there are three functional members of the α-globin cluster and five in the β-globin cluster (**Figure 17.10**). In adults, each hemoglobin molecule is a tetramer containing two identical α-globin subunits, two identical β-globin subunits, and four heme pigments (see Figure 3.11).

During human development, different members of the globin gene cluster are expressed at different times and in different tissues. This differential gene expression has great physiological significance. For example, hemoglobin that contains γ-globin, a subunit found in the hemoglobin of the human fetus, binds O_2 more tightly than adult hemoglobin does. This specialized form of hemoglobin ensures that in the placenta, O_2 will be transferred from the mother's blood to the developing fetus's blood. Just before birth the liver stops synthesizing fetal hemoglobin and the bone marrow cells take over, making the adult form (2 α and 2 β). Thus hemoglobins with different binding affinities for O_2 are provided at different stages of human development.

In addition to genes that encode proteins, many gene families include nonfunctional **pseudogenes**, which are designated with the Greek letter psi (ψ) (see Figure 17.10). These pseudogenes result from mutations that cause a loss of function rather than an enhanced or new function. The DNA sequence of a pseudogene may not differ greatly from that of other family members.

It may simply lack a promoter, for example, and thus fail to be transcribed. Or it may lack a recognition site needed for the removal of an intron, so that the transcript it makes is not correctly processed into a useful mature mRNA. In some gene families pseudogenes outnumber functional genes. In such cases, there appears to be no evolutionary advantage for the deletion of the pseudogenes, even though they have no apparent function.

Eukaryotic genomes contain many repetitive sequences

Eukaryotic genomes contain numerous repetitive DNA sequences that do not code for polypeptides. These include highly repetitive sequences, moderately repetitive sequences, and transposons.

Highly repetitive sequences are short (less than 100 bp) sequences that are repeated thousands of times in tandem (side-by-side) arrangements in the genome. They are not transcribed. Their proportion in eukaryotic genomes varies, from 10 percent in humans to about half the genome in some species of fruit flies. Often they are associated with heterochromatin, the densely packed, transcriptionally inactive part of the genome. Other highly repetitive sequences are scattered around the genome. For example, short tandem repeats (STRs) of 1 to 5 bp can be repeated up to 100 times at a particular chromosomal location. The copy number of an STR at a particular location varies among individuals and is inherited. In Chapter 15 we described how STRs can be used in the identification of individuals (DNA fingerprinting).

Moderately repetitive sequences are repeated 10 to 1,000 times in the eukaryotic genome. These sequences include the genes that are transcribed to produce tRNAs and rRNAs, which are used in protein synthesis. The cell makes tRNAs and rRNAs constantly, but even at the maximum rate of transcription, single copies of the tRNA and rRNA genes would be inadequate to supply the large amounts of these molecules needed by most cells. Thus the genome has multiple copies of these genes.

In mammals, four different rRNA molecules make up the ribosome: the 18S, 5.8S, 28S, and 5S rRNAs. (The S stands for Svedberg unit, which is a measure of size.) The 18S, 5.8S, and 28S rRNAs are transcribed together as a single precursor RNA molecule (**Figure 17.11**). As a result of several posttranscriptional steps, the precursor is cut into the final three rRNA products, and the noncoding "spacer" RNA is discarded. The sequence encoding these RNAs is moderately repetitive in humans: a total of 280 copies of the sequence are located in clusters on five different chromosomes.

Apart from the RNA genes, most moderately repetitive sequences are transposons, which, like the prokaryotic transposons we discussed earlier, can move about in the genome. Transposons make up more than 40 percent of the human genome and about 50 percent of the corn genome, although the percentage is smaller (3–10 percent) in many other eukaryotes. Table 17.6 summarizes the four main types of transposons in eukaryotes. Three types—long terminal repeats (LTRs), long interspersed elements (LINEs), and short interspersed elements (SINEs)—are retrotransposons. Retrotransposons move about

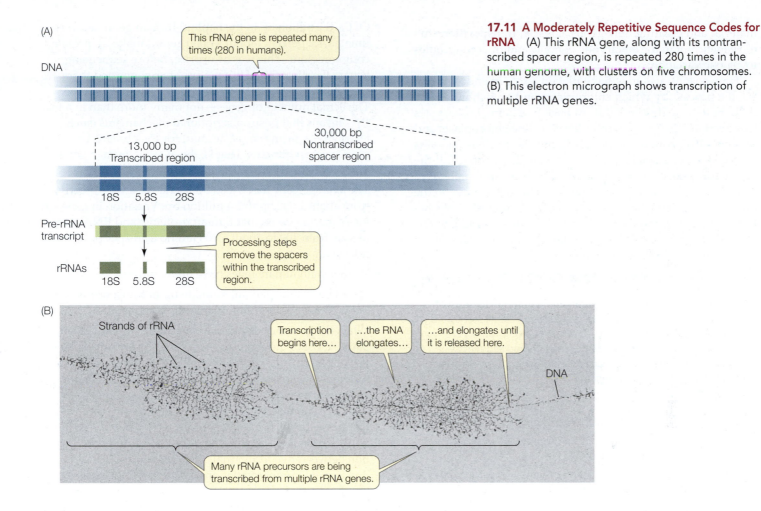

17.11 A Moderately Repetitive Sequence Codes for rRNA (A) This rRNA gene, along with its nontranscribed spacer region, is repeated 280 times in the human genome, with clusters on five chromosomes. (B) This electron micrograph shows transcription of multiple rRNA genes.

the genome in a distinctive way: they are transcribed into RNA, which then acts as a template for new DNA. The new DNA becomes inserted at a new location in the genome. This "copy and paste" mechanism results in two copies of the transposon: one at the original location and the other at a new location. A single type of SINE retrotransposon, the 300-bp Alu element, accounts for 11 percent of the human genome; it is present in a million copies!

Transposons of the fourth type, the DNA transposons, do not use RNA intermediates. Like some prokaryotic transposable elements, they are excised from the original location and become inserted at a new location without being replicated (a "cut and paste" mechanism).

What role do these moving sequences play in the cell? The best answer so far seems to be that transposons are simply cellular parasites that can be replicated. The insertion of a transposon at a new location can have important consequences. For example, the insertion of a transposon into the coding region of a gene results in a mutation (see Figure 17.5). This phenomenon accounts for rare forms of several genetic diseases in humans, including hemophilia and muscular dystrophy. If the insertion of a transposon takes place in the germ line, a gamete with a new mutation results. If the insertion takes place in a somatic cell, cancer may result.

Sometimes an adjacent gene can be replicated along with a transposon, resulting in a gene duplication. A transposon can carry a gene, or a part of it, to a new location in the genome, shuffling the genetic

TABLE 17.6

Major Transposable Element Groups in the Eukaryotic Genome

Class	Element	Length (bp)	Number in Human Genome	Percent of Human Genome
Retrotransposon	LTRs (long terminal repeats)	100–5,000	450,000	8
	LINEs (long, interspersed elements)	6,000–8,000	850,000	17
	SINEs (short interspersed elements)	<300	1,500,000	15
DNA transposon		2,000–3,000	300,000	3

material and creating new genes. Clearly, transposition stirs the genetic pot in the eukaryotic genome and thus contributes to genetic variation.

Transposons also may have played a role in endosymbiosis, the process by which chloroplasts and mitochondria are thought to have descended from once free-living prokaryotes (see Section 5.5). In modern eukaryotes the chloroplasts and mitochondria contain some DNA, but the nucleus contains most of the genes that encode the organelles' proteins. If the organelles were once independent, they must originally have contained all of those genes. How did the genes move to the nucleus? They may have done so by DNA transpositions between organelles and the nucleus, which still occur today. The DNA that remains in the organelles may be the remnants of more complete prokaryotic genomes.

RECAP 17.3

The genomes of eukaryotes contain more genes than those of prokaryotes. Some of these genes encode functions associated with the compartmentalization of eukaryotic cells; others are needed to support multicellularity. The genome sequences of model organisms have been used to identify common features of the eukaryotic genome, including the presence of abundant regulatory sequences, repetitive sequences, and noncoding DNA. Some eukaryotic genes are in families, which may include members that are mutated and nonfunctional.

- What are the major differences between prokaryotic and eukaryotic genomes? **See p. 361**
- Describe one class of proteins found in *C. elegans* that has few counterparts in yeasts. **See p. 362 and Table 17.4**
- What is the evolutionary role of eukaryotic gene families? **See pp. 363–364 and Figure 17.10**
- Why are there multiple copies of sequences coding for rRNA in the mammalian genome? **See p. 364**
- What effects can transposons have on a genome? **See pp. 365–366**

The analysis of eukaryotic genomes has resulted in an enormous amount of useful information, as we have seen. In the next section we will look more closely at the human genome.

17.4 What Are the Characteristics of the Human Genome?

Since the first human genome sequence was completed early in the first decade of this millennium, the haploid genomes of numerous other individuals have been sequenced and published. With the rapid development of sequencing technologies, the time is approaching when a human genome can be sequenced for less than $1,000.

The human genome sequence held some surprises

The following are just some of the interesting facts that we have learned about the human genome:

- Of the 3.2 billion bp in the haploid human genome, an estimated 1.2 percent (about 21,000 genes) make up protein-coding regions. This was a surprise. Before sequencing began, the diversity of human proteins suggested there might be 80,000 to 150,000 genes in the human genome. The actual number of genes—not many more than in a fruit fly—means that posttranscriptional mechanisms (such as alternative splicing) must account for the observed number of proteins in humans. That is, the average human gene must code for several different proteins.

- The average gene has 27,000 bp. Gene sizes vary greatly, from about 1,000 bp to 2.4 million bp. Variation in gene size was expected given that human proteins (and RNAs) vary in size, from 100 to about 5,000 amino acids per polypeptide chain.

- Virtually all human genes have many introns.

- About half of the genome is made up of transposons (see Table 17.6) and other highly repetitive sequences.

- When the genomes of two unrelated individuals are compared, most of the sequence—about 99.5 percent—is identical. Despite this apparent homogeneity, there are many differences, and as more genomes are sequenced, more variants are found. Current estimates suggest that each haploid genome contains about 3.3 million single nucleotide polymorphisms (SNPs; pronounced "snips") (see Section 15.3), so these account for about one-fifth of the variation between two individuals. The remaining four-fifths are due to copy number variation: differences in sequence copy number that have arisen through chromosomal deletions, duplications, or translocations (see Figure 15.4) or through duplications caused by transposons.

- Genes are not evenly distributed over the genome. Chromosome 19 is packed densely with genes, whereas chromosome 8 has long stretches without coding regions. The Y chromosome has the fewest genes (about 230), and chromosome 1 has the most (about 3,000).

Go to Media Clip 17.1
A Big Surprise from Genomics
Life10e.com/mc17.1

Comparative genomics reveals the evolution of the human genome

Comparisons between sequenced genomes from prokaryotes and eukaryotes have revealed some of the evolutionary relationships between genes. Some genes are present in both prokaryotes and eukaryotes, others are only in eukaryotes, and still others are only in animals or only in vertebrates (**Figure 17.12**).

The genomes of various other primates, including all of the great apes, have now been sequenced. The search is on for a set of human genes that differ from those found in other primates and that make us unique. Chimpanzees are our closest living relatives, sharing nearly 99 percent of our DNA sequence. Gorillas and orangutans are next closest, with genomes that are about 98 percent and 97 percent similar to ours, respectively.

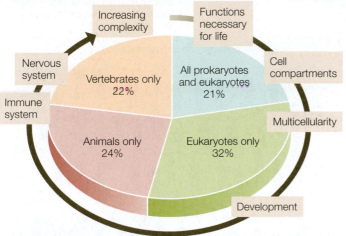

17.12 Evolution of the Genome A comparison of the human and other genomes has revealed how genes with new functions have been added over the course of evolution. Each percentage number refers to genes in the human genome. Thus 21 percent of human genes have homologs in prokaryotes and other eukaryotes, 32 percent of human genes occur only in other eukaryotes, and so on.

The gorilla sequence was completed in 2012 and was the last of the great ape genomes to be sequenced. The international team of researchers who published the sequence have identified about 500 protein-coding genes that have undergone accelerated evolution in humans, chimpanzees, and gorillas, including genes involved in hearing and brain development. Further analyses of these sequences may reveal genes that distinguish us from other apes, and that "make humans human."

Other clues about "human" genes have come from sequencing the genomes of ancient human relatives. An international team of scientists led by Svante Pääbo at the Max Planck Institute has extracted and sequenced DNA from the bones of Neanderthals, who lived in Europe up to 50,000 years ago. The entire Neanderthal genome has been sequenced. It is more than 99 percent identical to our human DNA, justifying the classification of Neanderthals as part of the same genus, *Homo*.

Comparisons of humans and Neanderthals with regard to specific genes and mutations are ongoing and have already revealed several interesting facts:

- The gene *MC1R* is involved in skin and hair pigmentation. A point mutation found in Neanderthals but not humans caused lower activity of the MC1R protein when it was introduced into cell cultures. Such lower activity of MC1R is known to result in fair skin and red hair in humans. So it appears that at least some Neanderthals may have had pale skin and red hair (**Figure 17.13**).

- The gene *FOXP2* is involved in vocalization in many organisms, including birds and mammals. Mutations in this gene result in severe speech impairment in humans. The Neanderthal *FOXP2* gene is identical to that of humans, whereas that of chimpanzees is slightly different. This has led to speculation that Neanderthals were capable of speech.

- While the human and Neanderthal genome sequences are very similar, there are differences in many point mutations and larger chromosomal arrangements. There are

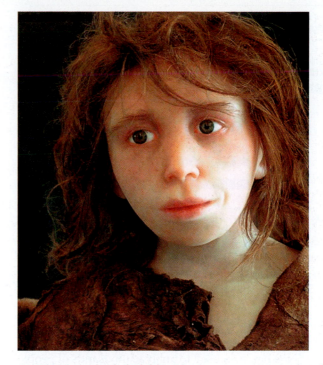

17.13 A Neanderthal Child Genome sequencing and analyses have led to this reconstruction of a Neanderthal child who lived about 60,000 years ago.

distinctive "human" DNA sequences, and also distinctive "Neanderthal" sequences. There is some mixture of the two, indicating that humans and Neanderthals interbred, with transfer of DNA between the two.

Human genomics has potential benefits in medicine

Most complex phenotypes are determined not by single genes but by multiple genes interacting with the environment. The single-allele explanations of phenylketonuria and sickle-cell anemia (see Chapter 15) do not apply to such common disorders as diabetes, heart disease, and Alzheimer's disease. To understand the genetic bases of these diseases, biologists are now using rapid genotyping technologies to create "haplotype maps," which are used to identify SNPs that are linked to genes involved in disease.

HAPLOTYPE MAPPING The SNPs that differ among individuals are not inherited as independent alleles. Rather, a set of SNPs that are present on a segment of chromosome are usually inherited as a unit. This linked piece of a chromosome is called a **haplotype**. You can think of the chromosome as a sentence, the haplotype as a word, and the SNP as a letter in the word. Analyzing SNPs is faster and less expensive than sequencing whole genomes, so haplotype mapping provides a shortcut for identifying the locations of genes and mutations involved in particular diseases (see Section 15.3). By comparing the haplotypes of individuals with and without a particular genetic disease, the genetic loci associated with the disease can be identified (**Figure 17.14**).

New technologies are continually being developed to analyze thousands or millions of SNPs in the genomes of individuals. Such technologies include rapid sequencing methods and

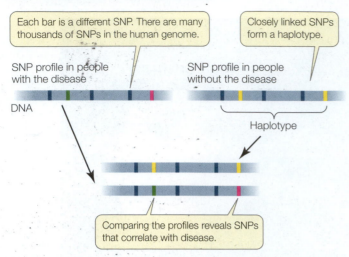

17.14 SNP Genotyping and Disease   Scanning the genomes of people with and without particular diseases reveals correlations between SNPs and complex diseases.

DNA microarrays (see Chapter 18) that depend on DNA hybridization to identify specific SNPs. For example, a microarray of 500,000 SNPs has been used to analyze thousands of people to find out which SNPs are associated with specific diseases. The amount of data is prodigious: 500,000 SNPs, thousands of people, thousands of medical records. With so much natural variation, statistical measures of association between a haplotype and a disease need to be very rigorous.

GENOTYPING TECHNOLOGY AND PERSONAL GENOMICS   Association tests like the one described above have revealed particular haplotypes or alleles that are associated with modestly increased risks for such diseases as breast cancer, diabetes, arthritis, obesity, and coronary heart disease (**Table 17.7**). Private companies can now scan a human genome for these variants—and the price for this service keeps getting lower. However, at this point it is unclear what a person without symptoms should do with the information, since multiple genes, environmental influences, and epigenetic effects all contribute to the development of these diseases.

Of course, the most comprehensive way to analyze a person's genome is by actually sequencing it. Once the cost of genome sequencing is within an affordable range, SNP testing may be superseded.

TABLE 17.7 ▆

SNP Human Genome Scans and Diseases

Disease	Location of SNP (Chromosome Number)	Percent of Increased Risk	
		Heterozygotes	Homozygotes
Breast cancer	8	20	63
Coronary heart disease	9	20	56
Heart attack	9	25	64
Obesity	16	32	67
Diabetes	10	65	277
Prostate cancer	8	26	58

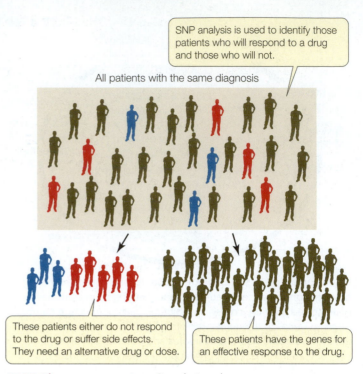

17.15 Pharmacogenomics   Correlations between genotypes and responses to drugs will help physicians develop personalized medical care. The different colors indicate individuals with different SNPs.

PHARMACOGENOMICS   Genetic variation can affect how an individual responds to a particular drug. For example, a drug may be chemically modified in the liver to make it more or less active. Consider an enzyme that catalyzes the following reaction:

$$\text{Active drug} \rightarrow \text{less active drug}$$

A mutation in the gene that encodes this enzyme may make the enzyme less active. For a given dose of the drug, a person with the mutation would have more active drug in the bloodstream than a person without the mutation. So the effective dose of the drug would be lower in the person with the mutation.

Now consider a different case, in which the liver enzyme is needed to make the drug active:

$$\text{Inactive drug} \rightarrow \text{active drug}$$

A person carrying a mutation in the gene encoding this liver enzyme would not be affected by the drug, since the activating enzyme is not present.

The study of how an individual's genome affects his or her response to drugs or other outside agents is called **pharmacogenomics**. Just as it is possible to identify haplotypes or SNPs that are associated with particular disease susceptibilities, it is also possible to identify SNPs that are associated with specific drug responses. This type of analysis makes it possible to predict whether a drug will be effective. The objective is to personalize drug treatment so that a physician can know in advance whether an individual will benefit from a particular drug (**Figure 17.15**). This approach might also be used to reduce the incidence of adverse drug reactions by identifying individuals who will metabolize a drug slowly, which can lead to a dangerously high level of the drug in the body.

Genome sequencing has advanced our understanding of biology enormously. High-throughput technologies are now being applied to other components of the cell: proteins and metabolites. We will now turn to the results of these studies.

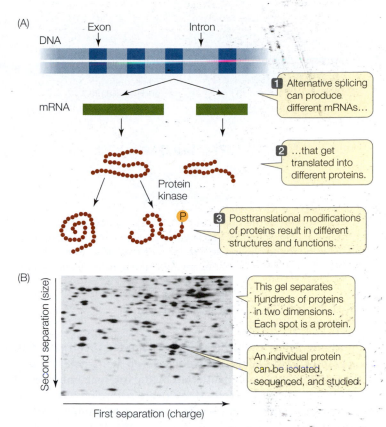

17.16 Proteomics (A) A single gene can code for multiple proteins. (B) A cell's proteins can be separated on the basis of charge and size by two-dimensional gel electrophoresis. The two separations can distinguish most proteins from one another.

17.5 What Do the New Disciplines of Proteomics and Metabolomics Reveal?

"The human genome is the book of life." Statements like this were common at the time the human genome sequence was first revealed. They reflect "genetic determinism," that a person's phenotype is determined by his or her genotype. But is an organism just a product of gene expression? We know that it is not. The proteins and small molecules present in any cell at a given point in time reflect not just gene expression but also modifications by the intracellular and extracellular environment. Two new fields have emerged to complement genomics and take a more complete snapshot of a cell and organism: proteomics and metabolomics.

The proteome is more complex than the genome

As mentioned above, many genes encode more than a single protein (**Figure 17.16A**). Alternative splicing leads to different combinations of exons in the mature mRNAs transcribed from a single gene (see Figure 16.22). Posttranslational modifications also increase the number of protein variants that can be derived from one gene (see Figure 14.20). However, in a multicellular organism many proteins are produced only by certain cells under specific conditions. Even single-celled organisms express only a subset of their genes at any particular time. The **proteome** is the sum total of the proteins produced by a cell, tissue, or organism at a given time, under defined conditions. Two methods are commonly used to analyze the proteome:

- Because of their unique amino acid compositions (primary structures), most proteins have unique combinations of electric charge and size. On the basis of these two properties, they can be separated by two-dimensional gel electrophoresis. Thus isolated, individual proteins can be analyzed, sequenced, and studied (**Figure 17.16B**).

- Mass spectrometry uses electromagnets to identify proteins by the masses of their atoms. Each protein has a unique pattern of these atomic masses.

The ultimate aim of proteomics is just as ambitious as that of genomics. Whereas genomics seeks to describe the genome and its expression, proteomics seeks to identify and characterize all of the expressed proteins.

Comparisons of the proteomes of humans and other eukaryotic organisms have revealed a common set of proteins that can be categorized into groups with similar amino acid sequences and similar functions. When considered on a whole organism basis, 46 percent of the yeast proteome, 43 percent of the nematode proteome, and 61 percent of the fruit fly proteome are shared by the human proteome. Functional analyses indicate that this set of 1,300 proteins provides the basic metabolic functions of a eukaryotic cell, such as glycolysis, the citric acid cycle, membrane transport, protein synthesis, DNA replication, and so on (**Figure 17.17**).

Of course, these are not the only human proteins. There are many more, which presumably distinguish us as *human* eukaryotic organisms. As we have mentioned before, proteins have different functional regions called domains (for example, a domain for binding a substrate, or a domain for spanning a membrane). While a particular organism may have many unique proteins, those proteins are often just unique combinations of domains that exist in other organisms. *This reshuffling of the genetic deck is a key to evolution.*

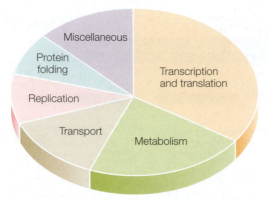

17.17 Proteins of the Eukaryotic Proteome   About 1,300 proteins are common to all eukaryotes and fall into these categories. Although their amino acid sequences may differ to a limited extent, they perform the same essential functions in all eukaryotes.

Metabolomics is the study of chemical phenotype

Studying genes and proteins gives a limited picture of what is going on in a cell. But as we have seen, both gene function and protein function are affected by the internal and external environments of the cell. Many proteins are enzymes, and their activities affect the concentrations of their substrates and products. So as the proteome changes, so will the abundances of small molecules called metabolites. The **metabolome** is the complete set of small molecules present in a cell, tissue, or organism under defined conditions. These include:

- *Primary metabolites* involved in normal processes, such as intermediates in pathways such as glycolysis. This category also includes hormones and other signaling molecules.

- *Secondary metabolites*, which are often unique to particular organisms or groups of organisms. They are often involved in special responses to the environment. Examples are antibiotics made by microbes, and the many chemicals made by plants that are used in defense against pathogens and herbivores.

Not surprisingly, measuring metabolites involves sophisticated analytical instruments. If you have studied organic or analytical chemistry, you may be familiar with gas chromatography and high-performance liquid chromatography, which separate molecules, and mass spectrometry and nuclear magnetic resonance spectroscopy, which are used to identify them. These measurements result in "chemical snapshots" of cells or organisms, which can be related to physiological states.

There has been some progress in defining the human metabolome. A database created by David Wishart and colleagues at the University of Alberta contains more than 6,500 metabolite entries. The challenge now is to relate levels of these substances to physiology. For example, you probably know that high levels of glucose in the blood are associated with diabetes. But what about the early stages of heart disease? There may be a pattern of metabolites that is diagnostic of this disease. This could aid in early diagnosis and treatment.

Plant biologists are far ahead of medical researchers in the field of metabolomics. Over the years, tens of thousands of secondary metabolites have been identified in plants, many of them made in response to environmental challenges. Some of these will be discussed in Chapter 39. The metabolome of the model organism *Arabidopsis thaliana* is being described, and will give insight into how a plant copes with stresses such as drought or pathogen attack. This knowledge could be helpful in optimizing plant growth for agriculture.

RECAP 17.5

The proteome is the total of all proteins produced by a cell, tissue, or organism under specific conditions. The metabolome is the total content of small molecules such as intermediates in primary metabolism, hormones, and secondary metabolites. The proteome and the metabolome can be analyzed using chemical methods that separate and identify molecules.

- How is the proteome analyzed? **See p. 369 and Figure 17.17**

- Explain the differences between genome, protoeome, and metabolome.

?

What does dog genome sequencing reveal about other animals?

ANSWER

Some breeds of dogs have leg muscles that are highly developed and permit them to run very fast. Genome analyses have shown that the gene for myostatin, a protein that inhibits muscle growth, is mutated and leads to an inactive protein in these dogs. The lack of myostatin leads to bulkier muscles (**Figure 17.18**). Comparative genomics reveals that the myostatin gene is also mutated in a cattle breed that has overdeveloped muscles. In humans there is potential for manipulating myostatin to treat muscle-wasting diseases such as muscular dystrophy. As might be expected, athletes anxious to have bulkier muscles are also very interested in this gene and its protein product.

17.18 Muscular Gene   These dogs are both whippets, but the muscle-bound dog (right) has a mutation in the myostatin gene.

━━━━━━━━━━ **CHAPTERSUMMARY** (17)

(17.1) How Are Genomes Sequenced?

- New methods of DNA sequencing involve miniaturization and computerized analysis. **Review Figure 17.1, ANIMATED TUTORIALS 17.1, 17.2**

- Genomes are sequenced in overlapping fragments, and then the fragments are lined up to give the final sequence. **Review Figure 17.2**

- The analysis of genome sequences gives information about protein-coding and noncoding regions. **Review Figure 17.3**

(17.2) What Have We Learned from Sequencing Prokaryotic Genomes?

- DNA sequencing is used to study the genomes of prokaryotes that are important to humans and ecosystems.

- **Metagenomics** is the identification of DNA sequences without first isolating, growing, and identifying the organisms present in an environmental sample. Many of these sequences are from prokaryotes that were heretofore unknown to biologists. **Review Figure 17.4**

- **Transposable elements** and **composite transposons** can move about the genome. **Review Figure 17.5**

- Transposon mutagenesis can be used to inactivate genes one by one. The mutated organism can be tested for survival and an artificial genome created based on a minimal set of essential genes. **Review Figures 17.6, 17.7**

(17.3) What Have We Learned from Sequencing Eukaryotic Genomes?

- Genome sequences from model organisms have demonstrated some common features of the eukaryotic genome. In addition, there are specialized genes for cellular compartmentalization, development, and features unique to plants. **Review Tables 17.2–17.5, Figures 17.8, 17.9**

- Some eukaryotic genes exist as members of **gene families**. Proteins may be made from these closely related genes at different times and in different tissues. Some members of gene families may be nonfunctional **pseudogenes**. **Review Figure 17.10**

- Repeated sequences are present in the eukaryotic genome. **Review Table 17.6**

- **Moderately repetitive sequences** include those coding for rRNA and transposons. **Review Figure 17.11**

(17.4) What Are the Characteristics of the Human Genome?

- The haploid human genome has 3.2 billion bp.

- Only about 1.2 percent of the genome codes for proteins; the rest consists of repeated sequences and noncoding DNA.

- Virtually all human genes have introns, and alternative splicing leads to the production of more than one protein per gene.

- SNP genotyping (haplotype mapping) correlates variations in the genome with diseases or drug sensitivity. It may lead to personalized medicine. **Review Figure 17.14**

- **Pharmacogenomics** is the analysis of genetics as applied to drug metabolism. **Review Figure 17.15**

(17.5) What Do the New Disciplines of Proteomics and Metabolomics Reveal?

- The **proteome** is the total protein content of an organism.

- There are more proteins than there are protein-coding genes in the genome.

- The proteome can be analyzed using chemical methods that separate and identify proteins. These include two-dimensional electrophoresis and mass spectrometry. **See Figure 17.16**

- The **metabolome** is the total content of small molecules, such as intermediates in primary metabolism, hormones, and secondary metabolites.

See ACTIVITY 17.1 for a concept review of this chapter

Go to the Interactive Summary to review key figures, Animated Tutorials, and Activities
Life10e.com/is17

CHAPTER**REVIEW**

━━━ **REMEMBERING**

1. Genome sequencing is done
   a. one chromosome at a time.
   b. on entire chromosomal DNA molecules.
   c. on overlapping fragments.
   d. manually, with large electrophoresis gels.
   e. using radioactive isotopes to label DNA.

2. The minimal genome of *Mycoplasma genitalium*
   a. has 100 genes.
   b. has been used to create new species.
   c. is made of RNA.
   d. is larger than the genome of *E. coli*.
   e. was derived from point mutations in each gene.

3. Which is *not* true of metagenomics?
   a. It has been done with bacteria.
   b. It has revealed many new species.
   c. It has revealed many new metabolic capacities.
   d. It involves extracting DNA from the environment.
   e. It cannot be done on seawater.

4. Transposons
   a. always use RNA for replication.
   b. are approximately 50 bp long.
   c. are made up of either DNA or RNA.
   d. do not contain genes coding for proteins.
   e. make up more than 40 percent of the human genome.

5. Vertebrate gene families
   a. contain only active genes.
   b. include the globins.
   c. are not produced by gene duplications.
   d. increase the number of unique genes in the genome.
   e. are not transcribed.

6. The human genome
   a. contains very few repeated sequences.
   b. has 3.2 billion bp.
   c. is 10 percent protein-coding sequences.
   d. has genes evenly distributed along chromosomes.
   e. has few genes with introns.

UNDERSTANDING & APPLYING

7. Eukaryotic protein-coding genes differ from their prokaryotic counterparts in that eukaryotic genes
   a. are double-stranded.
   b. are present in only a single copy.
   c. contain introns.
   d. have promoters.
   e. are transcribed into mRNA.

8. A comparison of the genomes of yeasts and bacteria shows that only yeasts have many genes for
   a. energy metabolism.
   b. cell wall synthesis.
   c. intracellular protein targeting.
   d. DNA-binding proteins.
   e. RNA polymerase.

9. The genomes of the fruit fly and the nematode are similar to those of yeasts, except that the former organisms have many genes for
   a. intercellular signaling.
   b. synthesis of polysaccharides.
   c. cell cycle regulation.
   d. intracellular protein targeting.
   e. transposable elements.

10. Why is identifying an organism's proteome and metabolome more complex than defining its genome?

ANALYZING & EVALUATING

11. The genomes of rice, wheat, and corn are similar to one another and to that of *Arabidopsis* in many ways. Discuss how these plants might nevertheless have very different proteins.

12. It is the year 2025. You are taking care of a patient who is concerned about having an early stage of kidney cancer. His mother died from this disease.
    a. Assume that the SNPs linked to genes involved in the development of this type of cancer have been identified. How would you determine if this man has a genetic predisposition for developing kidney cancer? Explain how you would do the analysis.
    b. How might you develop a metabolomic profile for kidney cancer and then use it to determine whether your patient has kidney cancer?
    c. If the patient was diagnosed with cancer by the methods in (a) and (b), how would you use pharmacogenomics to choose the right medications to treat the tumor in this patient?

Go to BioPortal at **yourBioPortal.com** for Animated Tutorials, Activities, LearningCurve Quizzes, Flashcards, and many other study and review resources.

18 Recombinant DNA and Biotechnology

CHAPTEROUTLINE

18.1 What Is Recombinant DNA?

18.2 How Are New Genes Inserted into Cells?

18.3 What Sources of DNA Are Used in Cloning?

18.4 What Other Tools Are Used to Study DNA Function?

18.5 What Is Biotechnology?

18.6 How Is Biotechnology Changing Medicine and Agriculture?

I N 2010 THE DEEPWATER HORIZON oil rig in the Gulf of Mexico exploded, and crude oil began gushing out of the well below it. The oil flowed unabated for three months until the well-head was finally capped; a total of 210 million gallons were released. The oil slick was visible from space—it damaged marine habitats and killed wildlife, washed up on beaches, and shut down fishing and tourism in the area. Efforts to remove the giant slick included collecting the oil, burning it, and dispersing it chemically. In addition, scientists identified bacteria living in the Gulf waters that were able to digest and break down components of the crude oil. Despite the clean-up efforts and the actions of these bacteria, much of the oil remains in the Gulf today.

Bioremediation—the use of microorganisms to remove pollutants in the environment—is an attractive option for cleaning up these spills. One approach is to encourage the growth of natural microorganisms that digest components of the crude oil. After the oil tanker *Exxon Valdez* ran aground near the Alaskan shore in 1989, nitrogen fertilizers were applied to nearby beaches to encourage the growth of oil-eating bacteria. Similar approaches have been tried in Kuwait, where destruction of wells during the 1991 Gulf War led to a massive release of oil. But the success of such methods has been limited because the naturally occurring bacteria digest only certain components of the crude oil, and because of technical difficulties in bringing the organisms into contact with the oil. Within the scientific community, there is great interest in producing genetically modified organisms that can clean up oil more rapidly and effectively.

Oil on the Beach A massive oil spill from a drilling rig in the Gulf of Mexico in 2010 released oil that reached the seashore. On this beach in Louisiana, huge piles of contaminated sand await cleaning. Naturally occurring bacteria broke down a lot of the oil, but biotechnology may be able to improve the process.

The possibility of genetically changing bacteria for bioremediation began in 1971. Ananda Chakrabarty, at the General Electric Research Center in New York, used genetic crosses to develop a strain of the bacterium *Pseudomonas* with multiple genes for the breakdown of various hydrocarbons in oil. He and his company applied for a patent to legally protect their discovery and profit from it. In a landmark case, the U.S. Supreme Court ruled in 1980 that "a live, human-made microorganism is patentable" under the U.S. Constitution.

The biotechnology industry has flourished since then. Many genetically modified organisms (GMOs) have been developed and approved for use in agriculture, medicine, and other industries. At the same time, concerns about the safety of GMOs have been raised. In particular, there are well-founded fears concerning the environmental consequences of releasing genetically modified "superbugs" into the environment. Such approaches to environmental cleanup remain at the experimental stage.

? Are there other uses for microorganisms in environmental cleanup?

See answer on p. 389.

18.1 What Is Recombinant DNA?

Recombinant DNA is a DNA molecule that has been made in the laboratory using at least two different sources of DNA. In Chapter 15 we discussed the use of restriction enzymes (restriction endonucleases) to detect mutations. These bacterial enzymes are also used in the laboratory to cut up (cleave) DNA. In the late 1960s scientists discovered other enzymes that act on DNA. One such enzyme is **DNA ligase**, which catalyzes the joining of DNA fragments; one of its functions is to join Okazaki fragments during DNA replication (see Section 13.3).

With these enzymes in hand, scientists could cut up DNA molecules and then splice the fragments together in new combinations ("recombine" the DNA fragments). In 1973 Stanley Cohen at Stanford University, Herbert Boyer at the University of California, and their colleagues did just that. They isolated two different plasmids (circular DNA molecules that replicate independently in bacterial cells; see Figure 12.24) from *Escherichia coli*. These plasmids contained different antibiotic resistance genes. The scientists cut the two plasmids with restriction enzymes, mixed the fragments together, and then used DNA ligase to rejoin the fragments. The products of this ligation reaction were inserted into new *E. coli* cells, and the cells were grown on solid medium containing both antibiotics. (See Section 18.2 for a discussion of how DNA is inserted into cells.) A few of the bacteria had been transformed with a recombinant plasmid containing both of the antibiotic resistance genes and could grow on the medium (**Figure 18.1**). With this experiment, **recombinant DNA technology** was born.

Let's look more closely at what happens when DNA is cut with a restriction enzyme and then rejoined with DNA ligase. Some restriction enzymes recognize palindromic DNA sequences—sequences that read the same way in both directions. For example, you can read the DNA recognition sequence for the restriction enzyme *Eco*R1 from 5′ to 3′ as GAATTC on both strands:

5′.......GAATTC......3′

3′.......CTTAAG......5′

Some restriction enzymes cut the DNA straight through the middle of the palindrome, generating "blunt-ended" fragments. Others, such as *Eco*RI, make staggered cuts—they cut the phosphodiester bond of one strand of the double helix several bases away from where they cut the other (**Figure 18.2**). After *Eco*RI makes its two cuts in the complementary strands, the ends of the strands are held together only by the hydrogen bonds between four base pairs. At warm temperatures (above room temperature) these hydrogen bonds are too weak to hold the two strands together, so the DNA separates into fragments when it is warmed. Each fragment carries a single-stranded "overhang" at the location of each cut. These overhangs are called **sticky ends** because they have specific base sequences that can bind by base pairing with complementary sticky ends.

Any two sticky ends that are complementary can form hydrogen bonds with one another. The sticky ends of the original

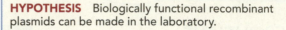
18.1 Recombinant DNA With the discovery of restriction enzymes and DNA ligase, it became possible to combine DNA fragments from different sources in the laboratory. But would such "recombinant DNA" be functional when inserted into a living cell? The results of this experiment completely changed the scope of genetic research, increasing our knowledge of gene structure and function, and ushered in the new field of biotechnology.[a]

HYPOTHESIS Biologically functional recombinant plasmids can be made in the laboratory.

Method E. coli plasmids carrying a gene for resistance to either the antibiotic kanamycin (K) or tetracycline (T) are cut with a restriction enzyme.

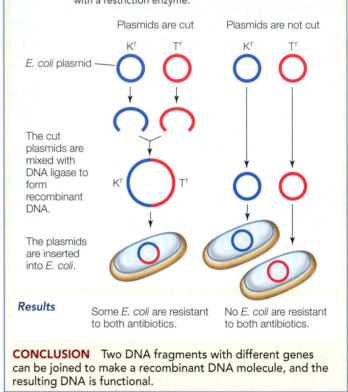

Results

Some *E. coli* are resistant to both antibiotics.

No *E. coli* are resistant to both antibiotics.

CONCLUSION Two DNA fragments with different genes can be joined to make a recombinant DNA molecule, and the resulting DNA is functional.

Go to **BioPortal** for discussion and relevant links for all INVESTIGATING**LIFE** figures.

[a]Cohen, S. N. et al. 1973. *Proceedings of the National Academy of Sciences USA* 70: 3240–3244.

 Go to Media Clip 18.1
Striking Views of Recombinant DNA Being Made
Life10e.com/mc18.1

molecule may rejoin, or two different fragments may join (see Figure 18.2). Indeed, a fragment from one source, such as a human, can be joined to a fragment from another source, such as a bacterium. Initially the fragments are held together only by weak hydrogen bonds, but the enzyme ligase catalyzes the formation of covalent bonds between adjacent nucleotides at the ends of the fragments, joining them to form a single, larger molecule.

Besides *Eco*RI, there are hundreds of other restriction enzymes, each with its own recognition sequence. With these tools—restriction enzymes and DNA ligase—scientists can cut and rejoin different DNA molecules from any and all sources,

WORKING WITH **DATA:**

Recombinant DNA

Original Paper

Cohen, S. N., A. C. Y. Chang, H. W. Boyer, and R. B. Helling. 1973. Construction of biologically functional bacterial plasmids *in vitro. Proceedings of the National Academy of Sciences USA* 70: 3240–3244.

Analyze the Data

In the 1960s scientists discovered restriction enzymes and DNA ligase. In 1973 Stanley Cohen, Hebert Boyer, and colleagues used these two enzymes as chemical reagents to show that biologically functional recombinant plasmids can be constructed in the laboratory (see Figure 18.1). Specifically, they used the restriction enzyme *EcoRI* to cut two *E. coli* plasmids, one containing a resistance gene for kanamycin and the other containing a resistance gene for tetracycline. The fragments were mixed together, and DNA ligase was used to join them in random combinations. The ligated molecules were then inserted back into *E. coli*. Some of the transformed cells were resistant to both antibiotics, indicating that they had been transformed with a recombinant plasmid that contained both the tetracycline and kanamycin resistance genes. For this landmark experiment on recombinant DNA, Cohen was awarded the Nobel Prize in 1986.

QUESTION 1

Two plasmids were used in this study. One, pSC101, had a gene for tetracycline resistance, and the other, pSC102, had a gene for kanamycin resistance. In one experiment, some pSC101 was cut with the restriction enzyme *EcoRI* but not sealed up with DNA ligase. Cut or intact pSC101 were used to transform *E. coli* cells sensitive to antibiotics. The transformed cells were plated on media containing tetracycline, kanamycin, or chloramphenicol (another antibiotic). The results are shown in **TABLE A**. What can you conclude from this experiment?

QUESTION 2

In another experiment, pSC101 and pSC102 were mixed and treated in various ways. The mixtures were used to transform antibiotic-sensitive *E. coli*. The three treatments were: intact

("None"), cut with *EcoRI*, and cut with *EcoRI* then sealed with DNA ligase. The results are shown in **TABLE B**. Did treatment with DNA ligase improve the efficiency of genetic transformation by the cut plasmids? What is the quantitative evidence for your statement?

QUESTION 3

How did the antibiotic-resistant bacteria arise in the "None" DNA treatment?

QUESTION 4

Did the *EcoRI* + DNA ligase treatment result in an increase in doubly-resistant bacteria over controls? What data provide evidence for your statement?

QUESTION 5

For the *EcoRI* + DNA ligase treatment, compare the number of transformants that were resistant to either tetracycline or kanamycin alone to the number that were doubly resistant. What accounts for the large difference? (Hint: look at Figure 18.2.)

TABLE A

Plasmid	Transformants per μg DNA	
	Tetracycline	Kanamycin
Intact pSC101	3×10^5	None
EcoRI-treated pSC101	2.8×10^4	None

TABLE B

Treatment of DNA	Transformants per μg DNA		
	Tetracycline	Kanamycin	Tetracycline and Kanamycin
None	2×10^5	1×10^5	2×10^2
EcoRI	1×10^4	1.1×10^3	7×10^1
EcoRI + DNA ligase	1.2×10^4	1.3×10^3	5.7×10^2

Go to **BioPortal** for all WORKING WITH **DATA** exercises

including artificially synthesized DNA sequences. Recently, new tools have been developed for making recombinant DNA. Methods based on the polymerase chain reaction (PCR) allow the joining of any two DNA molecules without the need for conveniently placed restriction enzyme sites. There are even methods for the construction of multiple artificial genes using programmable microchips. Despite these advances, restriction enzymes and DNA ligase are still used routinely for recombinant DNA construction in biology labs.

RECAP 18.1

DNA fragments from different sources can be linked together to make recombinant DNA.

- How did Cohen and Boyer make the first recombinant DNA? **See Figure 18.1**
- How does a staggered cut in DNA create a "sticky end"? **See p. 374 and Figure 18.2**

Recombinant DNA has no biological significance until it is inserted inside a living cell, which can replicate and transcribe the transplanted genetic information. How can recombinant DNA made in the laboratory be inserted and expressed in living cells?

18.2 How Are New Genes Inserted into Cells?

One goal of recombinant DNA technology is to **clone**—that is, to produce many identical copies of—a particular gene. We have seen the term "clone" used in the context of whole cells or organisms (see Chapters 11 and 12) that are genetically identical to one another. A gene can be cloned by inserting it into a bacterial cell such as *E. coli*. The bacterium is allowed to reproduce and multiply into millions of identical cells, all carrying copies of the gene. Cloning might be done for analysis,

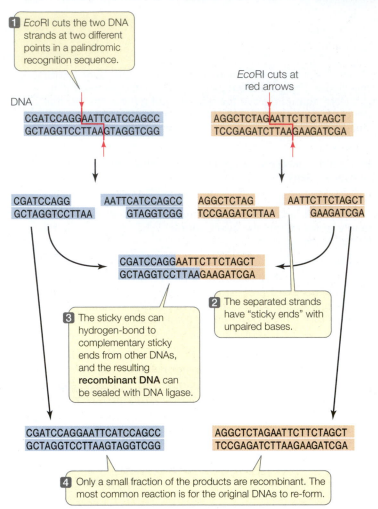

1 *Eco*RI cuts the two DNA strands at two different points in a palindromic recognition sequence.

DNA

*Eco*RI cuts at red arrows

CGATCCAGGAATTCATCCAGCC
GCTAGGTCCTTAAGTAGGTCGG

AGGCTCTAGAATTCTTCTAGCT
TCCGAGATCTTAAGAAGATCGA

CGATCCAGG
GCTAGGTCCTTAA

AATTCATCCAGCC
GTAGGTCGG

AGGCTCTAG
TCCGAGATCTTAA

AATTCTTCTAGCT
GAAGATCGA

CGATCCAGGAATTCTTCTAGCT
GCTAGGTCCTTAAGAAGATCGA

3 The sticky ends can hydrogen-bond to complementary sticky ends from other DNAs, and the resulting **recombinant DNA** can be sealed with DNA ligase.

2 The separated strands have "sticky ends" with unpaired bases.

CGATCCAGGAATTCATCCAGCC
GCTAGGTCCTTAAGTAGGTCGG

AGGCTCTAGAATTCTTCTAGCT
TCCGAGATCTTAAGAAGATCGA

4 Only a small fraction of the products are recombinant. The most common reaction is for the original DNAs to re-form.

18.2 Cutting, Splicing, and Joining DNA Some restriction enzymes (*Eco*RI is shown here) make staggered cuts in DNA. *Eco*RI can be used to cut two different DNA molecules (blue and orange). The exposed bases can hydrogen-bond with complementary exposed bases on other DNA fragments, forming recombinant DNA. DNA ligase stabilizes the recombinant molecule by forming covalent bonds in the DNA backbone.

to produce a protein product in quantity, or as a step toward creating an organism with a new phenotype.

Recombinant DNA is cloned by inserting it into host cells in a process known as **transformation**—or **transfection** if the host cells are derived from an animal. A host cell or organism that contains recombinant DNA is referred to as a **transgenic** cell or organism, and the non-native DNA is called a transgene. Later in this chapter we will encounter many examples of transgenic organisms, including yeast, mice, wheat plants, and even cows.

Various methods are used to create transgenic cells. Generally these methods are inefficient in that only a few of the cells that are exposed to the recombinant DNA actually become transformed with it. In order to select for the transgenic cells, **selectable marker** genes, such as genes that confer resistance to antibiotics, are often included as part of the recombinant DNA molecule. When an antibiotic resistance gene is used as the selectable marker, the cells from the transformation experiment are grown in the presence of the antibiotic; the antibiotic

kills all nontransgenic cells, leaving only the transgenic cells. Antibiotic resistance genes were the markers used in Cohen and Boyer's experiment (see Figure 18.1).

Genes can be inserted into prokaryotic or eukaryotic cells

In theory, any cell or organism can act as a host for the introduction of recombinant DNA. Most research has been done using model organisms:

- *Bacteria* are easily grown and manipulated in the laboratory. Much of their molecular biology is known, especially for well-studied bacteria such as *E. coli*. Furthermore, bacteria contain plasmids, which are easily manipulated to carry recombinant DNA into the cell. But since the processes of transcription, translation, and posttranslational modification proceed differently in prokaryotes than they do in eukaryotes, bacteria might not be suitable as hosts to express eukaryotic genes.

- *Yeasts* such as *Saccharomyces cerevisiae* are commonly used as eukaryotic hosts for recombinant DNA studies. The advantages of using yeasts include rapid cell division (a life cycle completed in 2–4 hours), ease of growth in the laboratory, and a relatively small genome size (see Table 17.2). In addition, yeast cells have most of the characteristics of other eukaryotic cells, with a notable exception being those associated with multicellularity.

- *Plant cells* are good hosts because of their ability to make stem cells (unspecialized totipotent cells) from mature plant tissues. The unspecialized cells can be transformed with recombinant DNA and then studied in culture, or grown into new plants. There are also methods for making whole transgenic plants without going through the cell culture step. These methods result in plants that carry the recombinant DNA in all their cells, including the germ line cells.

- *Cultured animal cells* can be used to study expression of human or animal genes, for example for medical purposes. Whole transgenic animals can also be created.

A variety of methods are used to insert recombinant DNA into host cells

Methods for inserting DNA into host cells vary. The cells may be chemically treated to make their outer membranes more permeable, and then mixed with the DNA so that it can diffuse into the cells. Another approach is called electroporation: a short electric shock is used to create temporary pores in the membranes through which the DNA can enter. Viruses can be altered so that they carry recombinant DNA into cells. A common method for transforming plants involves a specific bacterium that inserts DNA into plant cells. Transgenic animals can be produced by injecting recombinant DNA into the nuclei of fertilized eggs. There are even "gene guns" that "shoot" the host cells with tiny metal particles coated with the DNA.

The challenge of inserting new DNA into a cell lies not only in getting it into the host cell, but also in getting it to replicate as the host cell divides. DNA polymerase does not bind to and copy just any sequence. If the new DNA is to be replicated, it must become part of a segment of DNA that contains an origin of replication (see Figure 13.12). Such a DNA molecule is called a **replicon**, or replication unit.

There are two general ways in which the newly introduced DNA can become part of a replicon within the host cell:

- It may be inserted into a host chromosome. Although the site of insertion is usually random, this is nevertheless a common method of integrating new genes into host cells.

- It can enter the host cell as part of a carrier DNA sequence, called a **vector**, and can then either integrate into a host chromosome or be replicated from its own origin of replication.

Several types of vectors are used to get DNA into cells, some of which we will now describe in more detail.

PLASMIDS AS VECTORS As we described in Chapter 12, plasmids are small, circular DNA molecules that replicate autonomously in many prokaryotic cells. Several characteristics make plasmids useful as transformation vectors:

- Plasmids are relatively small (an *E. coli* plasmid usually has 2,000–6,000 base pairs) and are therefore easy to manipulate in the laboratory.

- A typical plasmid has one or more restriction enzyme recognition sequences that each occur only once in the plasmid sequence. These sites make it easy to insert additional DNA into the plasmid before it is used to transform host cells.

- Many plasmids contain genes that confer resistance to antibiotics, which can serve as selectable markers.

- Plasmids have a bacterial origin of replication (*ori*) and can replicate independently of the host chromosome. It is not uncommon for a bacterial cell to contain hundreds of copies of a recombinant plasmid. For this reason, the power of bacterial transformation to amplify a gene is extraordinary. A one-liter culture of bacteria harboring the human β-globin gene in a typical plasmid has as many copies of that gene as there are cells in a typical adult human (10^{14}).

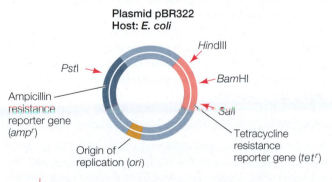

Plasmid pBR322
Host: *E. coli*

*Hind*III
*Pst*I
*Bam*HI
Ampicillin resistance reporter gene (*amp^r*)
*Sal*I
Origin of replication (*ori*)
Tetracycline resistance reporter gene (*tet^r*)

→ Recognition sites for restriction enzymes

The plasmids used as vectors in the laboratory have been extensively altered to include convenient features: multiple cloning sites, often with 20 or more unique restriction enzyme sites for cloning purposes; origins of replication that will function in a variety of host cells; and various kinds of reporter genes (see below), such as selectable marker genes.

PLASMID VECTORS FOR PLANTS An important vector for carrying new DNA into many types of plants is a plasmid found in the bacterium *Agrobacterium tumefaciens*. This bacterium lives in the soil, infects plants, and causes a disease called crown gall, which is characterized by the presence of growths (or tumors) on the plant. *Agrobacterium* contains a plasmid called Ti (for tumor-inducing). When the bacterium infects a plant cell, a region of the Ti plasmid called the T DNA is inserted into the cell, where it becomes incorporated into one of the plant's chromosomes. The Ti plasmid carries the genes needed for this transfer and incorporation of the T DNA. The T DNA carries genes that are expressed by the host cell, causing the growth of tumors and the production of specific sugars that the bacterium uses as sources of energy. Scientists have exploited this remarkable natural "genetic engineer" to insert foreign DNA into the genomes of plants.

When used as a vector for plant transformation, the tumor-inducing and sugar-producing genes on the T DNA are removed and replaced with foreign DNA. The recombinant Ti plasmids are first used to transform *Agrobacterium* cells from which the original Ti plasmids have been removed. Then the *Agrobacterium* cells are used to infect plant cells.

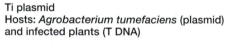

Ti plasmid
Hosts: *Agrobacterium tumefaciens* (plasmid) and infected plants (T DNA)

T DNA
Sites for several restriction enzymes
ori

VIRUSES AS VECTORS Constraints on plasmid replication limit the size of the new DNA that can be inserted into a plasmid to about 10,000 base pairs. Although many prokaryotic genes may be smaller than this, most eukaryotic genes—with their introns and extensive flanking sequences—are bigger. A vector that accommodates larger DNA inserts is needed for these genes.

Both prokaryotic and eukaryotic viruses are used as vectors for eukaryotic DNA. Bacteriophage λ, which infects *E. coli*, has a DNA genome of about 45,000 base pairs. About 20,000 base pairs are not necessary for the bacteriophage to complete its life cycle (see Figure 16.14). These 20,000 base pairs can be deleted and replaced with DNA from another organism, which then gets replicated along with the virus DNA. Because viruses infect cells naturally, they offer a great advantage over plasmids, which often require artificial means to coax them to enter host cells.

Reporter genes help select or identify host cells containing recombinant DNA

Even when a population of host cells is exposed to an appropriate vector, only a small proportion of the cells actually take up the

RESEARCH**TOOLS**

18.3 Selection for Recombinant DNA Selectable marker (reporter) genes are used by biologists to select for bacteria that have taken up a plasmid. In a typical experiment, most of the bacteria will not take up any DNA. Of those that do, only a small fraction will take up recombinant DNA.

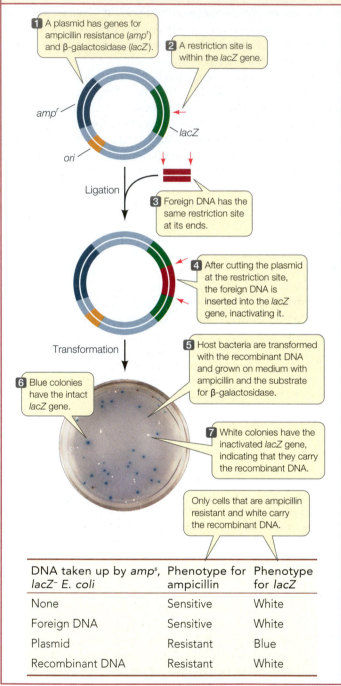

1 A plasmid has genes for ampicillin resistance (*amp*ʳ) and β-galactosidase (*lacZ*).

2 A restriction site is within the *lacZ* gene.

*amp*ʳ

lacZ

ori

Ligation

3 Foreign DNA has the same restriction site at its ends.

4 After cutting the plasmid at the restriction site, the foreign DNA is inserted into the *lacZ* gene, inactivating it.

Transformation

5 Host bacteria are transformed with the recombinant DNA and grown on medium with ampicillin and the substrate for β-galactosidase.

6 Blue colonies have the intact *lacZ* gene.

7 White colonies have the inactivated *lacZ* gene, indicating that they carry the recombinant DNA.

Only cells that are ampicillin resistant and white carry the recombinant DNA.

DNA taken up by *amp*ˢ, *lacZ*⁻ E. coli	Phenotype for ampicillin	Phenotype for *lacZ*
None	Sensitive	White
Foreign DNA	Sensitive	White
Plasmid	Resistant	Blue
Recombinant DNA	Resistant	White

vector. Furthermore, the process of making recombinant DNA is far from perfect. During a ligation reaction, DNA molecules can combine in various ways, many of which do not produce the desired recombinant molecule (see Figures 18.1 and 18.2). Methods have been developed to improve the chance that a desired combination will occur. A simple approach is to cut the vector with two different restriction enzymes that have sites near each other. This produces a molecule with incompatible sticky ends, which is much less likely to simply recircularize during the ligation

reaction. The desired insert molecule is cut with the same two enzymes, so that theoretically a functional circular plasmid can only be produced if the vector and insert ligate with one another. Even so, it is often the case that only a small proportion of the ligation products have the desired recombinant sequence.

How can we identify or select the host cells that contain the desired sequence? One way is to use selectable markers, such as those used in early experiments with recombinant DNA (see Figure 18.1). Selectable markers are one type of **reporter gene**, which is any gene whose expression is easily observed. There are several types of reporter genes:

- An antibiotic resistance gene in a plasmid or other vector allows the selection of transformed host cells. If the host cells are normally sensitive to the antibiotic, they will grow on medium containing the antibiotic only if they have been transformed by the vector. This approach is used in the selection of transgenic prokaryotic and eukaryotic cells, including those of plants and animals.

- The β-galactosidase (*lacZ*) gene in the *E. coli lac* operon (see Figure 16.4) codes for an enzyme that can convert the artificial substrate X-Gal into a bright blue product. Many plasmids contain the *lacZ* gene with a multiple cloning site (i.e., multiple unique restriction enzyme sites) within its sequence. These plasmids also carry genes for antibiotic resistance. To clone a specific DNA fragment, the foreign DNA and the plasmid are each cut with one of the restriction enzymes that can cut within the cloning site. The vector and the foreign DNA are then combined in a ligation reaction, and the ligation products are used to transform bacteria. Bacterial colonies containing the plasmid are selected on a solid medium containing the antibiotic. X-Gal is also included in the medium. If a bacterial colony contains a recombinant plasmid that carries the foreign DNA inserted into the *lacZ* gene, it will not make β-galactosidase and the colony will be white. Clones that contain the original plasmid with no insert express the *lacZ* gene and make blue colonies (**Figure 18.3**).

- Green fluorescent protein (GFP), which normally occurs in the jellyfish *Aequorea victoria*, emits visible green light when exposed to ultraviolet light. The gene for this protein has been isolated and incorporated into vectors. GFP is now widely used as a reporter gene (**Figure 18.4**). GFP has also been modified to emit other colors when exposed to ultraviolet light, and these new variants are widely used by molecular biologists.

■ **RECAP** 18.2

Recombinant DNA can be cloned by using a vector to insert it into a suitable host cell. The vector often has a selectable marker or other reporter gene that gives the host cell a phenotype by which transgenic cells can be identified.

- List the characteristics of a plasmid that make it suitable for introducing new DNA into a host cell. **See p. 377**

- How are cells harboring a vector that carries recombinant DNA identified? **See p. 378 and Figure 18.3**

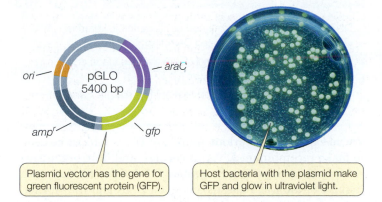

Plasmid vector has the gene for green fluorescent protein (GFP).

Host bacteria with the plasmid make GFP and glow in ultraviolet light.

18.4 Green Fluorescent Protein as a Reporter The presence of a plasmid with the gene for green fluorescent protein (GFP) is readily apparent in transgenic cells because the GFP that the cells produce glows under ultraviolet light. This allows the identification of cells carrying a plasmid without the use of selection on antibiotics. That is, no cells are killed during the selection process.

We have described how DNA can be cut with restriction enzymes, inserted into a vector, and introduced into host cells. We have also seen how host cells carrying recombinant DNA can be identified. Now let's consider where the genes or DNA fragments used in these procedures come from.

18.3 What Sources of DNA Are Used in Cloning?

A major goal of molecular cloning experiments is to learn the functions of the DNA sequences and the proteins they encode. The DNA fragments used in cloning procedures are obtained from several sources. They include random fragments of chromosomes that are maintained as gene libraries, complementary DNA obtained by reverse transcription from mRNA, products of the polymerase chain reaction (PCR), and artificially synthesized or mutated DNA.

Libraries provide collections of DNA fragments

A **genomic library** is a collection of DNA fragments that together comprise the genome of an organism. Restriction enzyme digestion or other means, such as mechanical shearing, can be used to break chromosomes into smaller pieces. These smaller DNA fragments still constitute a genome (**Figure 18.5A**), but the information is now in many smaller "volumes." Each fragment is inserted into a vector, which is then taken up by a host cell. Proliferation of a single transformed cell produces a colony of cells, each of which harbors many copies of the same fragment of DNA.

When plasmids are used as vectors, about 700,000 separate fragments are required to make a library of the human genome. By using bacteriophage λ, which can carry about four times as much DNA as a plasmid, the number of "volumes" in the library can be reduced to about 160,000. Although this still seems like a large number, a single petri plate can hold thousands of phage colonies, or plaques, which are easily screened for the presence of a particular DNA sequence by hybridization to an appropriate DNA probe.

cDNA is made from mRNA transcripts

A much smaller DNA library—one that includes only the genes transcribed in a particular tissue—can be made from **complementary DNA**, or **cDNA** (**Figure 18.5B**). This involves isolating mRNA from cells, then making cDNA copies of that mRNA by complementary base pairing. An enzyme, **reverse transcriptase**, catalyzes this reaction.

A collection of cDNAs from a particular tissue at a particular time in the life cycle of an organism is called a **cDNA**

RESEARCH**TOOLS**

18.5 Constructing Libraries Intact genomic DNA is too large to be introduced into host cells. (A) A genomic library can be made by breaking the DNA into small fragments, incorporating the fragments into a vector, and then transforming host cells with the recombinant vectors. Each colony of cells contains many copies of a small part of the genome. (B) Similarly, there are many mRNAs in a cell. These can be copied into cDNAs and a library made from them. The DNA in these colonies can then be isolated for analysis.

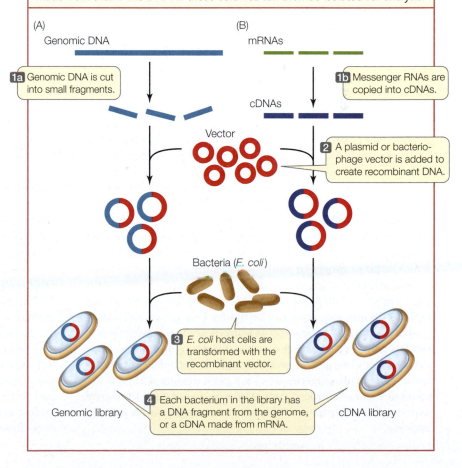

(A) Genomic DNA

(B) mRNAs

1a Genomic DNA is cut into small fragments.

1b Messenger RNAs are copied into cDNAs.

cDNAs

Vector

2 A plasmid or bacteriophage vector is added to create recombinant DNA.

Bacteria (*E. coli*)

3 *E. coli* host cells are transformed with the recombinant vector.

4 Each bacterium in the library has a DNA fragment from the genome, or a cDNA made from mRNA.

Genomic library

cDNA library

library. Because mRNAs do not last long in the cytoplasm, the types and amounts of mRNAs present in a cell are a good representation of the transcription rates of all the genes in that cell. Thus a cDNA library is a "snapshot" that captures the transcription pattern of a set of cells at a given point in time. cDNA libraries have been invaluable for comparing gene expression in different tissues at different stages of development. For example, researchers have found that up to one-third of all the genes of an animal are expressed only during development. cDNA is also a good starting point for cloning eukaryotic genes, because the clones contain only the coding sequences of the genes (the introns having been spliced out; see Figure 14.8). Also, if a eukaryotic gene is highly expressed in a particular tissue, a cDNA library made from that tissue will be enriched for the gene, making it easier to identify and clone the gene.

Reverse transcriptase along with PCR (see below) can also be used to create and amplify a specific cDNA sequence without the need to construct a library. In this case, RNA is isolated from an organism or tissue, and reverse transcriptase is used to make cDNA from the RNA. Then PCR is used to amplify a specific sequence directly from the sample of cDNA. This procedure, called **RT-PCR**, has become an invaluable tool for studying the expression of particular genes in cells and organisms.

Synthetic DNA can be made by PCR or by organic chemistry

PCR is a method of amplifying DNA in a test tube, based on the same processes used to replicate DNA in a cell (see Figure 13.21). PCR can begin with as little as 10^{-12} g of DNA (a picogram). Any fragment of DNA can be amplified by PCR as long as appropriate primers are available. You will recall that DNA replication (by PCR or in a cell) requires not just a template onto which DNA polymerase adds complementary nucleotides, but also a short oligonucleotide primer where replication begins (see Figure 13.13). If the appropriate primers (two are needed—one for each strand of DNA) are added to template DNA in a PCR reaction, millions of copies of the DNA region between the primers can be produced in just a few hours. This amplified DNA can then be inserted into a vector to create recombinant DNA and be cloned in host cells.

Artificial DNA can be synthesized using organic chemistry to link nucleotides together in a specified sequence. This process is now fully automated, and a service laboratory can make large numbers of short- to medium-length sequences overnight. PCR primers, for example, are made in this way. The synthesis of artificial DNA does not require a template, so DNA with any sequence can be made. This flexibility is useful for creating DNA fragments with desirable characteristics, such as convenient restriction sites or specific mutations. Longer synthetic sequences can be pieced together to construct completely artificial genes that have been designed for specific purposes. For example, a gene might be designed to be highly expressed in a particular cell type, or to encode a highly active enzyme.

RECAP 18.3

DNA for cloning can be obtained using genomic libraries, cDNA, or artificially synthesized DNA fragments.

• How are genomic DNA and cDNA libraries made and used? **See pp. 379–380 and Figure 18.5**

• Explain how RT-PCR is used to amplify a specific gene sequence. **See p. 380**

We've explored the various sources of DNA that can be used to make recombinant DNA molecules, and how organisms are transformed with recombinant DNA. We will now turn to some of the ways that recombinant DNA and transformation methods can be used to study the functions of genes and proteins.

18.4 What Other Tools Are Used to Study DNA Function?

So far in this chapter we have seen how recombinant DNA is made and how organisms are transformed with recombinant DNA. In this section we will examine several additional techniques for studying DNA. These approaches include expression of genes in different biological systems, mutagenesis, methods to block gene expression, and DNA microarrays to analyze large numbers of nucleotide sequences.

Genes can be expressed in different biological systems

One way to study a gene and its protein product is to express it in cells that do not normally express that gene. Often a scientist will want to express a gene derived from one kind of organism in another, very different organism—for example, a human gene in a bacterium, or a bacterial gene in a plant. For successful expression, the gene of interest must be combined with a promoter and other regulatory sequences from the host organism: a bacterial promoter will not function in a plant cell, for example. The coding region of the gene of interest is inserted between a promoter and a transcription termination sequence derived from the host organism, or from an organism that uses similar mechanisms for gene regulation.

Another way to study a gene is to overexpress it in cells where it is normally expressed at lower levels, so that much more of its protein product is made. To achieve this, a copy of the coding region is inserted downstream of a different, stronger promoter, and cells are transformed with the recombinant DNA.

There are many thousands of examples of how such experiments have shed light on the functions of genes and their protein products. One example involves a genetic system that probably evolved to prevent inbreeding in plants. Most plants produce flowers with both male and female parts, but many plant species are self-incompatible; they cannot self-pollinate (see Sections 22.3 and 38.1). Their flowers produce a protein that recognizes "self" pollen, and prevents the pollen from fertilizing egg cells in the same flower (see Figure 38.5). Genetic crosses suggested that a particular multi-allelic gene, called the *S* gene, was responsible for self-incompatibility. Definitive proof was obtained when plants were transformed with recombinant DNA containing an *S* allele different from their own *S* alleles. The transgenic plants rejected not only their own pollen but also pollen from flowers that naturally carried the foreign *S* allele.

DNA mutations can be created in the laboratory

Mutations that occur in nature have been important in demonstrating cause-and-effect relationships for a specific gene. However, mutations in nature are rare events. Recombinant DNA technology allows us to ask "what if" questions by creating mutations artificially. Because synthetic DNA can be made with any desired sequence, it can be manipulated to create specific mutations, the consequences of which can be observed when the mutant DNA is expressed in host cells.

Genes can be inactivated by homologous recombination

As we have just discussed, a gene can be studied by expressing it in cells where it is not normally expressed or by expressing it at higher than normal levels. Another way to study a gene is to inactivate it, so that it is not transcribed and translated into a functional protein. An example of this approach is the use of transposon mutagenesis to define the minimal genome (see Section 17.2). In animals, such a manipulation is called a **knockout** experiment. Methods have been developed to knock out genes in model organisms such as mice, fruit flies, and nematodes. We will focus here on the technique used in mice.

In mice, **homologous recombination** can be used to knock out a specific gene (**Figure 18.6**). As we saw in Chapter 11, homologous recombination occurs when a pair of homologous chromosomes line up during meiosis (see Figures 11.17 and 11.18). Homologous recombination also occurs in cells that are not undergoing meiosis, as part of their DNA repair processes. A key feature of homologous recombination is that it involves an exchange of DNA between molecules with identical, or nearly identical, sequences.

To create a knockout mouse, the normal allele of a gene of interest is inserted into a plasmid. Restriction enzymes are then used to insert a fragment containing a selectable marker gene into the middle of the mouse gene. This mutated version of the gene will eventually replace the wild-type gene in a living mouse. The extra DNA interferes with transcription and translation of the targeted gene, making it nonfunctional.

Once the recombinant plasmid has been made, it is used to transfect mouse embryonic stem cells. (A **stem cell** is an unspecialized cell that divides and differentiates into specialized cells.) Much of the targeted gene is still present in the plasmid (although in two separated regions), and these sequences tend

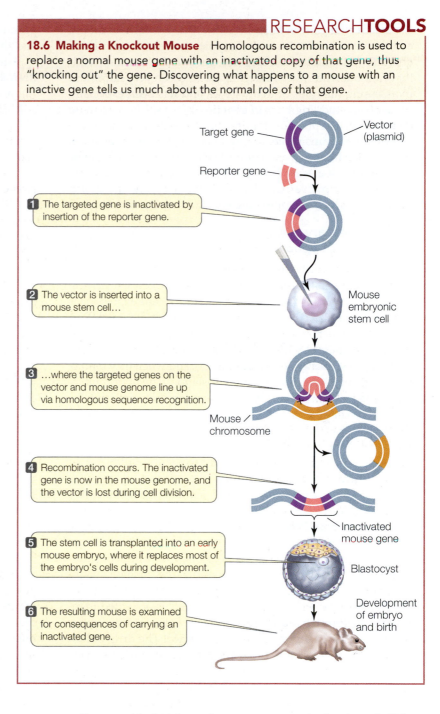

RESEARCH**TOOLS**

18.6 Making a Knockout Mouse Homologous recombination is used to replace a normal mouse gene with an inactivated copy of that gene, thus "knocking out" the gene. Discovering what happens to a mouse with an inactive gene tells us much about the normal role of that gene.

Target gene — Vector (plasmid)

Reporter gene —

1 The targeted gene is inactivated by insertion of the reporter gene.

2 The vector is inserted into a mouse stem cell...

Mouse embryonic stem cell

3 ...where the targeted genes on the vector and mouse genome line up via homologous sequence recognition.

Mouse chromosome

4 Recombination occurs. The inactivated gene is now in the mouse genome, and the vector is lost during cell division.

Inactivated mouse gene

5 The stem cell is transplanted into an early mouse embryo, where it replaces most of the embryo's cells during development.

Blastocyst

Development of embryo and birth

6 The resulting mouse is examined for consequences of carrying an inactivated gene.

to line up with their homologous sequences in the normal allele on the mouse chromosome. Sometimes recombination occurs, and the plasmid's nonfunctional allele is "swapped" with the functional allele in the host cell. The nonfunctional allele is inserted permanently into the host cell's genome and the normal allele is lost, because the plasmid cannot replicate in mouse cells. The active marker gene in the insert is used to select those stem cells carrying the nonfunctional allele.

A transfected stem cell is now transplanted into a mouse embryo at an early stage of development. If the mouse that develops from this embryo has the mutant allele in its germ line cells, its progeny will have the allele in every cell in their bodies. Such mice are inbred to create knockout mice carrying the inactivated gene in homozygous form. The mutant mouse can then be observed for phenotypic changes, to find clues about the function of the targeted gene in the normal (wild-type) animal. The knockout technique has been very important in assessing the roles of many genes and has been especially valuable in studying human genetic diseases. Many of these diseases, such as phenylketonuria, have knockout mouse models—mouse strains produced by homologous recombination that are affected by an analogous disease. These models can be used to study a disease and to test potential treatments. Mario Capecchi, Martin Evans, and Oliver Smithies shared the Nobel Prize in 2007 for developing the knockout mouse technique.

Complementary RNA can prevent the expression of specific genes

Another way to study the expression of a specific gene is to block the translation of its mRNA. This is an example of scientists imitating nature. As we described in Section 16.5, gene expression is sometimes regulated by the production of double-stranded RNA molecules that are processed to produce short, single-stranded microRNA (miRNA). These miRNAs are complementary to specific mRNA sequences (see Figure 16.23), and when they bind to their target mRNAs, they inhibit translation. The hybrid molecules tend to break down rapidly in the cytoplasm. Although the target gene continues to be transcribed, translation does not take place. Scientists have used the idea of hybrid RNAs to develop methods for blocking the expression of particular genes (**Figure 18.7**). An organism can be transformed to produce mRNAs that are complementary to and thus bind to specific endogenous mRNAs. Or cells can be injected with synthetic complementary sequences (see Figure 18.7B). Such a complementary molecule is called **antisense RNA** because it binds by base pairing to the "sense" bases on the mRNA.

There is much interest in the possibility of producing antisense drugs to treat cancer. For example, the gene *bcl2* codes for a protein that blocks apoptosis (programmed cell death; see Section 11.6). In some forms of cancer, *bcl2* is activated inappropriately through mutation. These cells fail to undergo apoptosis, continue to divide, and form a tumor. The drug oblimersen is an antisense RNA that binds to *bcl2* mRNA, prevents

production of the protein, and leads to apoptosis of tumor cells and shrinkage of the tumor.

A technique related to antisense RNA takes advantage of **RNA interference** (**RNAi**), a natural mechanism for inhibiting mRNA translation that involves small interfering RNA (siRNA; see Figure 16.23). siRNAs bind to complementary regions on their target mRNAs, which are then degraded. Since the discovery of RNAi in the late 1990s, scientists have synthesized artificial double-stranded siRNAs to inhibit the expression of known genes (see Figure 18.7C). Because these double-stranded siRNAs are more stable than antisense RNAs, the use of siRNAs is the preferred approach for blocking translation. An RNAi-based therapy is being developed to treat macular degeneration, an eye disease that results in near-blindness when blood vessels proliferate in the eye. The signaling molecule that stimulates vessel proliferation is a growth factor. An siRNA that targets this growth factor's mRNA shows promise in stopping and even reversing the progress of the disease.

Although medical applications for RNAi are still at the experimental stage, antisense RNA and RNAi have been widely used to test cause-and-effect relationships in biological research.

DNA microarrays reveal RNA expression patterns

The science of genomics faces two major quantitative challenges. First, there are very large numbers of genes in eukaryotic genomes. Second, there are myriad distinct patterns of gene expression in different tissues at different times. For example, the cells of a skin cancer at its early stage may have a unique mRNA "fingerprint" that differs from those of normal skin cells and cells from a more advanced skin cancer. In such a case, the pattern of gene expression could provide invaluable information to a clinician trying to characterize a patient's tumor (see below).

To find such patterns, scientists could isolate mRNA from cells and measure the amount of each gene's mRNA one gene at a time by hybridization or RT-PCR. But that would involve many steps and take a very long time. It is far simpler to do these hybridizations all in one step. This is possible with **DNA microarray** technology, which provides large arrays of sequences for hybridization experiments.

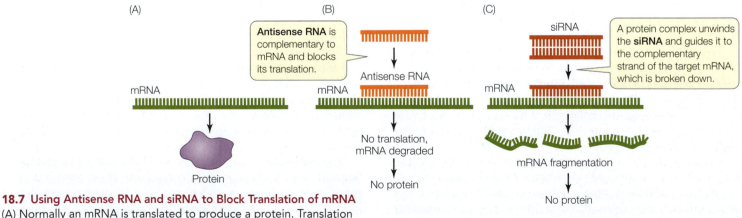

18.7 Using Antisense RNA and siRNA to Block Translation of mRNA
(A) Normally an mRNA is translated to produce a protein. Translation of a target mRNA can be prevented with (B) an antisense RNA or (C) a small interfering RNA (siRNA) that is complementary to the target mRNA.

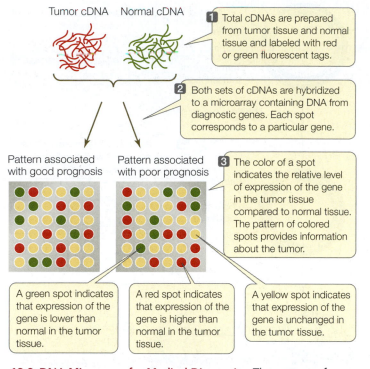

Tumor cDNA   Normal cDNA

1 Total cDNAs are prepared from tumor tissue and normal tissue and labeled with red or green fluorescent tags.

2 Both sets of cDNAs are hybridized to a microarray containing DNA from diagnostic genes. Each spot corresponds to a particular gene.

Pattern associated with good prognosis

Pattern associated with poor prognosis

3 The color of a spot indicates the relative level of expression of the gene in the tumor tissue compared to normal tissue. The pattern of colored spots provides information about the tumor.

A green spot indicates that expression of the gene is lower than normal in the tumor tissue.

A red spot indicates that expression of the gene is higher than normal in the tumor tissue.

A yellow spot indicates that expression of the gene is unchanged in the tumor tissue.

18.8 DNA Microarray for Medical Diagnosis The pattern of expression of 70 genes in tumor tissues indicates whether breast cancer is likely to recur. Actual arrays have more spots than shown here.

Go to Animated Tutorial 18.1
DNA Microarray Technology
Life10e.com/at18.1

DNA microarrays ("gene chips") contain a series of DNA sequences attached to a glass slide in a precise order. The slide is divided into a grid of microscopic spots, or "wells." Each spot contains thousands of copies of a particular oligonucleotide of 20 or more bases. A computer controls the addition of these oligonucleotide sequences in a predetermined pattern. Each oligonucleotide can hybridize with only one DNA or RNA sequence and thus is a unique identifier of a gene. Many thousands of different oligonucleotides can be placed on a single microarray.

Microarrays can be used to examine patterns of gene expression in different tissues and under different conditions, and they can be used to identify individual organisms with particular mutations. You can see the concept of microarray analysis by following the example illustrated in **Figure 18.8**. Most women with breast cancer are treated with surgery to remove the tumor, and then treated with radiation soon afterward to kill cancer cells that the surgery may have missed. But a few cancer cells may still survive in some patients, and these eventually form tumors in the breast or elsewhere in the body. The challenge for physicians is to develop criteria to identify patients likely to have surviving cancer cells so that they can be treated aggressively with tumor-killing chemotherapy. Scientists at the Netherlands Cancer Institute used medical records to identify patients whose cancer did or did not recur. Then they extracted mRNA from the patients'

tumors, which had been stored after surgery, and made cDNA from the samples. The cDNAs were hybridized to microarrays containing sequences derived from 1,000 human genes. The researchers found 70 genes whose expression differed dramatically between tumors from patients whose cancers recurred and tumors from patients whose cancers did not recur. From this information the Dutch group identified gene expression signatures that are useful in clinical decision-making: patients with a good prognosis can avoid unnecessary chemotherapy, whereas those with a poor prognosis can receive more aggressive treatment.

RECAP 18.4

Researchers can study the function of a gene by expressing it in cells where it is not normally expressed, by overexpressing it, or by knocking out the gene in a living organism. Antisense RNAs and siRNAs prevent gene expression by selectively blocking mRNA translation. DNA microarrays allow the simultaneous analysis of many different mRNA transcripts.

- How is a gene "knocked out" in a living organism? **See p. 381 and Figure 18.6**
- How do antisense RNA and siRNA molecules affect gene expression? **See p. 382 and Figure 18.7**

Now that you've seen how DNA can be fragmented, recombined, manipulated, and put back into living organisms, let's see some examples of how these techniques are used to make useful products.

18.5 What Is Biotechnology?

Biotechnology is the use of cells or whole living organisms to produce materials useful to people, such as foods, medicines, and chemicals. People have been using various forms of biotechnology for a long time. For example, the use of yeasts to brew beer and wine dates back at least 8,000 years, and the use of bacterial cultures to make cheese and yogurt is a technique many centuries old. For a long time people exploited these biochemical transformations without being aware of the organisms and genes involved.

About 100 years ago it became clear that specific bacteria, yeasts, and other microbes could be used as biological converters to make certain products. Alexander Fleming's discovery that the mold *Penicillium* makes the antibiotic penicillin led to the large-scale commercial culture of microbes to produce antibiotics as well as other useful chemicals. Today microbes are grown in vast quantities to make much of the industrial-grade alcohol, glycerol, butyric acid, and citric acid that are used as is or as starting materials in the manufacture of other products.

In contrast, the commercial harvesting of proteins, including hormones and enzymes, was until recently limited by the (often) minuscule amounts that could be extracted from organisms that produce them naturally. Yields were low, and purification was difficult and costly. Gene cloning has changed all this. Now that almost any gene can be inserted into bacteria or yeasts and the cells can be induced to make and export the gene product in large amounts, these microbes have become versatile factories for important products. Today there is interest in producing nutritional supplements and pharmaceuticals in whole transgenic animals and harvesting them in large quantities from, for example, the milk of cows or the eggs of chickens. Key to this boom in biotechnology has been the development of specialized vectors that not only carry genes into cells, but also make those cells express the genes at high levels.

Expression vectors can turn cells into protein factories

If a eukaryotic gene is inserted into a typical plasmid and used to transform *E. coli*, little if any of the gene product will be made unless other key prokaryotic DNA sequences are included with the gene. A bacterial promoter, a signal for transcription termination, and a special sequence that is necessary for binding of bacterial ribosomes to the mRNA must all be included in the transformation vector if the gene is to be expressed in a bacterial cell.

To solve this kind of problem, scientists make **expression vectors** that have all the characteristics of cloning vectors, as well as the extra sequences needed for the foreign gene (also called a transgene) to be expressed in the host cell. For bacterial hosts, these additional sequences include the elements named above (**Figure 18.9**). For eukaryotes, they include the poly A–addition sequence and a promoter that contains all the elements needed for expression in a eukaryotic cell. An expression vector can be designed to deliver transgenes to any class of prokaryotic or eukaryotic host and may include additional features:

- An *inducible promoter*, which responds to a specific signal, can be included. For example, a promoter that responds to hormonal stimulation can be used so that the transgene will be expressed at high levels only when the hormone is added.

- A *tissue-specific promoter*, which is expressed only in a certain tissue at a certain time, can be used if localized expression is desired. For example, many seed proteins are expressed only in the plant embryo. Coupling a transgene to a seed-specific promoter will allow the gene to be expressed only in seeds.

- *Signal sequences* can be added so that the gene product is directed to an appropriate destination. For example, when a protein is made by yeast or bacterial cells in a liquid medium, it is economical to include a signal directing the protein to be secreted into the extracellular medium for easier recovery.

RECAP 18.5

Expression vectors maximize the expression of transgenes inserted into host cells.

- How do expression vectors work? **See p. 384 and Figure 18.9**

This chapter has introduced many of the methods that are used in biotechnology. Let's turn now to the ways biotechnology is being applied to meet some specific human needs.

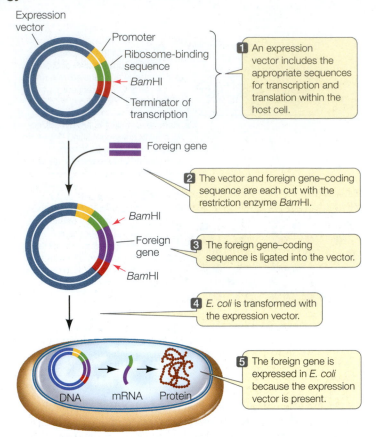

18.9 Expression of a Transgene in a Host Cell Produces Large Amounts of Its Protein Product To be expressed in *E. coli*, a gene derived from a eukaryote requires bacterial sequences for transcription initiation (promoter), transcription termination, and ribosome binding. Expression vectors contain these additional sequences, enabling the eukaryotic protein to be synthesized in the prokaryotic cell.

Go to Activity 18.1 Expression Vectors
Life10e.com/ac18.1

18.6 How Is Biotechnology Changing Medicine and Agriculture?

Huge potential for improvements in health and agriculture derive from recent developments in biotechnology. We now have the ability to make virtually any protein by recombinant DNA technology and to insert transgenes into many kinds of host cells. With these revolutionary developments, concerns have been raised about ethics and safety. We will now turn to the promises and problems of biotechnology that uses DNA manipulation.

Medically useful proteins can be made using biotechnology

Many medically useful products are being made using biotechnology (**Table 18.1**), and more are in various stages of development. A good illustration of a medical application of biotechnology is the manufacture of tissue plasminogen activator (TPA).

When a wound begins bleeding, a blood clot soon forms to stop the flow. Later, as the wound heals, the clot dissolves.

TABLE **18.1**

Some Medically Useful Products of Biotechnology

Product	Use
Colony-stimulating factor	Stimulates production of white blood cells in patients with cancer and AIDS
Erythropoietin	Prevents anemia in patients undergoing kidney dialysis and cancer therapy
Factor VIII	Replaces clotting factor missing in patients with hemophilia A
Growth hormone	Replaces missing hormone in people of short stature
Insulin	Stimulates glucose uptake from blood in people with insulin-dependent (Type I) diabetes
Platelet-derived growth factor	Stimulates wound healing
Tissue plasminogen activator	Dissolves blood clots after heart attacks and strokes
Vaccine proteins: Hepatitis B, herpes, influenza, Lyme disease, meningitis, pertussis, etc.	Prevent and treat infectious diseases

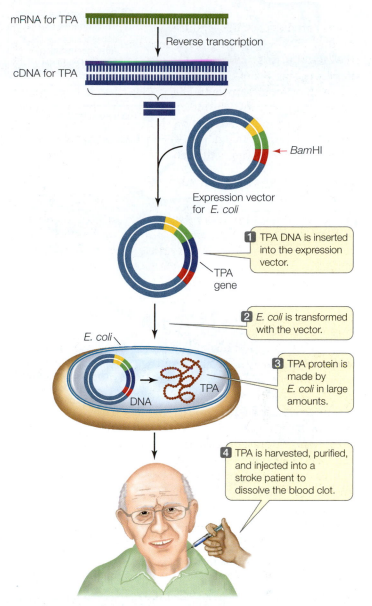

18.10 Tissue Plasminogen Activator TPA is a naturally occurring human protein that dissolves blood clots. It is used to treat patients suffering from blood clotting in heart attacks or strokes, and is manufactured using recombinant DNA technology.

How does the blood perform these conflicting functions at the right times? Mammalian blood contains an enzyme called plasminogen. When activated, it is converted into plasmin and catalyzes the dissolution of the clotting proteins. The conversion of plasminogen into plasmin is catalyzed by the enzyme TPA, which is produced by cells lining the blood vessels:

$$\text{Plasminogen} \xrightarrow{\text{TPA}} \text{plasmin}$$
$$\text{(Inactive)} \qquad \text{(Active)}$$

Heart attacks and strokes can be caused by blood clots that form in major blood vessels leading to the heart or the brain, respectively. In the 1970s a bacterial enzyme called streptokinase was found to stimulate the dissolution of the clots in some patients. Treatment with this enzyme saved lives, but being a foreign protein, it triggered the body's immune system to react against it. Worse, the drug sometimes prevented clotting throughout the entire circulatory system, leading to a dangerous situation in which blood could not clot where needed; uncontrolled bleeding was a serious concern.

When TPA was discovered, it had many advantages over streptokinase: it bound specifically to clots, and it did not provoke an immune reaction because it is an enzyme that is normally found in the body. But the amounts of TPA that could be harvested from human tissues were tiny, certainly not enough to inject at the site of a clot in the emergency room.

Recombinant DNA technology solved this problem. TPA mRNA was isolated and used to make cDNA, which was then inserted into an expression vector and used to transform *E. coli* (**Figure 18.10**). The transgenic bacteria made the protein in quantity, and it soon became available commercially. This drug has had considerable success in dissolving blood clots in people experiencing strokes and heart attacks.

Another way of making medically useful products in large amounts is **pharming**: the production of pharmaceuticals in farm animals or plants. For example, a gene encoding a useful protein might be placed next to the promoter of the gene that encodes lactoglobulin, an abundant milk protein. Transgenic animals carrying this recombinant DNA will secrete large amounts of the foreign protein into their milk. These natural "bioreactors" can produce abundant supplies of the protein, which can be separated easily from the other components of the milk (**Figure 18.11**).

Pharming is being used to produce human growth hormone (hGH), a protein made in the pituitary gland in the brain that has

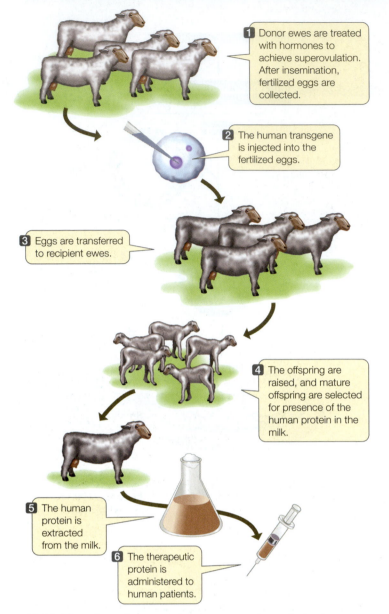

1 Donor ewes are treated with hormones to achieve superovulation. After insemination, fertilized eggs are collected.

2 The human transgene is injected into the fertilized eggs.

3 Eggs are transferred to recipient ewes.

4 The offspring are raised, and mature offspring are selected for presence of the human protein in the milk.

5 The human protein is extracted from the milk.

6 The therapeutic protein is administered to human patients.

18.11 Pharming An expression vector carrying a desired gene can be put into an animal egg, which is implanted into a surrogate mother. The transgenic offspring produce the new protein in their milk. The milk is easily harvested and the protein isolated, purified, and made clinically available to patients.

many effects, especially in growing children (see Chapter 41). Children with growth hormone deficiency have short stature as well as other abnormalities. In the past they were treated with hGH isolated from the pituitary glands of organ donors, but the supply was too limited to meet demand. Recombinant DNA technology was used to coax bacteria to make the protein, but the cost of treatment was high ($30,000 a year per person). In 2004 a team led by Daniel Salamone at the University of Buenos Aires made a transgenic cow that secretes hGH in her milk. The yield is prodigious: only 15 such cows are needed to meet the needs worldwide of children suffering from this type of dwarfism.

DNA manipulation is changing agriculture

The cultivation of plants and the husbanding of animals provide the world's oldest examples of biotechnology, dating back more than 10,000 years. Over the centuries, people have adapted crops and farm animals to their needs, producing organisms with desirable characteristics such as large seeds, high fat content in milk, or resistance to disease.

Until recently, the most common way to improve crop plants and farm animals was to identify individuals with desirable phenotypes that existed as a result of natural variation. Through many deliberate crosses—a process called selective breeding—the genes responsible for the desirable trait could be introduced into a widely used variety or breed of that organism.

Despite some spectacular successes, among them the breeding of high-yielding varieties of wheat, rice, and hybrid corn, such deliberate crossing can be a hit-or-miss affair. Many desirable traits are controlled by multiple genes, and it is hard to predict the results of a cross or to maintain a prized combination as a pure-breeding variety year after year. In sexual reproduction, combinations of desirable genes are quickly separated by meiosis. Furthermore, traditional breeding takes a long time: many plants and animals take years to reach maturity and then can reproduce only once or twice a year—a far cry from the rapid reproduction of bacteria.

Modern recombinant DNA technology has several advantages over traditional methods of breeding (**Figure 18.12**):

- *The ability to identify specific genes.* The development of genetic markers allows breeders to select for specific desirable genes, making the breeding process more precise and rapid.

- *The ability to introduce any gene from any organism into a plant or animal species.* This ability, combined with mutagenesis techniques, vastly expands the range of possible new traits.

- *The ability to generate new organisms quickly.* Manipulating cells in the laboratory and regenerating a whole plant or animal by cloning is much faster than traditional breeding.

Consequently, recombinant DNA technology has found many applications in agriculture (**Table 18.2**). We will describe a few examples to demonstrate the approaches that plant scientists have used to improve crop plants.

PLANTS THAT MAKE THEIR OWN INSECTICIDES Plants are subject to infections by viruses, bacteria, and fungi, but probably the

TABLE 18.2

Agricultural Applications of Biotechnology under Development

Goal	Technology/Genes
Improving the environmental adaptations of plants	Genes for drought tolerance, salt tolerance
Improving nutritional traits	High-lysine seeds; β-carotene in rice
Improving crops after harvest	Delay of fruit ripening; sweeter vegetables
Using plants as bioreactors	Plastics, oils, and drugs produced in plants

Conventional breeding
Many generations; only gene(s) from same species can be used

Biotechnology
One generation; gene(s) from any organism can be used

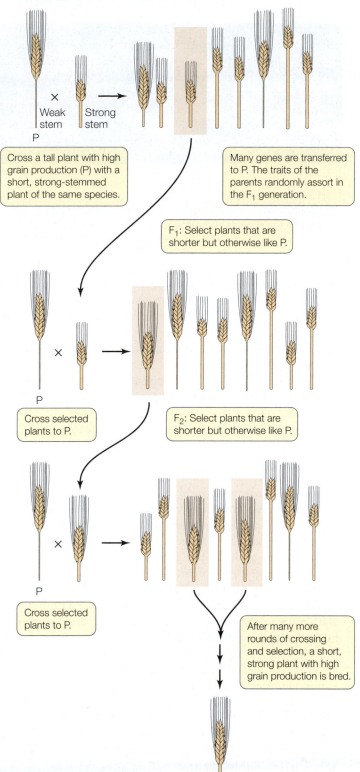

Weak stem × Strong stem
P

Cross a tall plant with high grain production (P) with a short, strong-stemmed plant of the same species.

Many genes are transferred to P. The traits of the parents randomly assort in the F₁ generation.

F_1: Select plants that are shorter but otherwise like P.

P ×

P

Cross selected plants to P.

F_2: Select plants that are shorter but otherwise like P.

× P

P

Cross selected plants to P.

After many more rounds of crossing and selection, a short, strong plant with high grain production is bred.

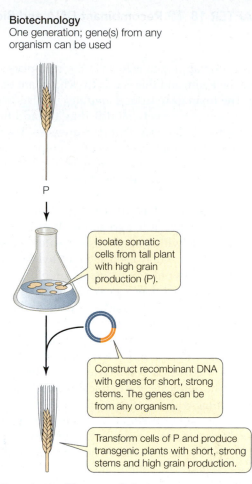

P

Isolate somatic cells from tall plant with high grain production (P).

Construct recombinant DNA with genes for short, strong stems. The genes can be from any organism.

Transform cells of P and produce transgenic plants with short, strong stems and high grain production.

18.12 Genetic Modification of Plants versus Conventional Plant Breeding Plant biotechnology offers many potential advantages over conventional breeding. In the example here, the objective is to transfer gene(s) for short, strong stems into a wheat plant that has high grain production but a tall, weak stem.

including people. What's more, many insecticides persist in the environment for a long time.

Some bacteria protect themselves by producing proteins that can kill insects. For example, the bacterium *Bacillus thuringiensis* produces a protein that is toxic to the insect larvae that prey on it.

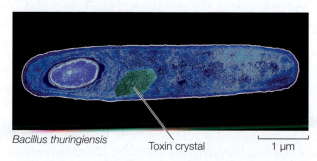

Bacillus thuringiensis　　　Toxin crystal　　　1 μm

most important crop pests are herbivorous insects. From the locusts of biblical (and modern) times to the cotton boll weevil, insects have continually eaten the crops people grow.

The development of insecticides has improved the situation, but insecticides have their own problems. Many, including the organophosphates, are relatively nonspecific and kill beneficial insects in the broader ecosystem as well as crop pests. Some pesticides even have toxic effects on other groups of organisms,

The toxicity of this protein is 80,000 times greater than that of a typical commercial insecticide. When a hapless larva eats the bacteria, the toxin becomes activated and binds specifically to the insect's gut, producing holes and killing the insect. Dried preparations of *B. thuringiensis* have been sold for decades as safe insecticides that break down rapidly in the environment. But the biodegradation of these preparations is their limitation, because it means that the dried bacteria must be applied repeatedly during the growing season.

A longer-acting approach is to have the crop plants themselves make the toxin, and this is exactly what plant scientists have done. The toxin gene from *B. thuringiensis* has been isolated, cloned, and extensively modified by the addition of a plant promoter and other regulatory sequences. Transgenic corn, cotton, soybeans, tomatoes, and other crops are now being grown successfully with this added gene. Farmers growing these transgenic crops use far less of other pesticides.

CROPS THAT ARE RESISTANT TO HERBICIDES   Herbivorous insects are not the only threat to agriculture. Weeds may grow in fields and compete with crop plants for water and soil nutrients. Glyphosate is a widely used and effective herbicide, or weed killer, that works only on plants. It inhibits an enzyme system in the chloroplast that is involved in the synthesis of amino acids. Glyphosate is a broad-spectrum herbicide that kills most weeds, but unfortunately it also kills crop plants. One solution to this problem is to use it to rid a field of weeds before the crop plants start to grow. But as any gardener knows, when the crop begins to grow, the weeds reappear. If the crop were not affected by the herbicide, the herbicide could be applied to the field at any time without harming the crop.

Scientists have used expression vectors to make plants that synthesize a different form of the target enzyme for glyphosate that is unaffected by the herbicide. The gene for this enzyme has been inserted into corn, cotton, and soybean plants, making them resistant to glyphosate. This technology has expanded rapidly, and a large proportion of cotton and soybean plants now carry this gene.

GRAINS WITH IMPROVED NUTRITIONAL CHARACTERISTICS   To remain healthy, humans must consume adequate amounts of β-carotene, which the body converts into vitamin A. About 400 million people worldwide suffer from vitamin A deficiency, which makes them susceptible to infections and blindness. One reason is that rice grains, which do not contain β-carotene, make up a large part of their diets. Rice grains lack the two-enzyme biochemical pathway that synthesizes β-carotene.

Plant biologists Ingo Potrykus and Peter Beyer isolated one of the genes for the β-carotene pathway from the bacterium *Erwinia uredovora* and the other from daffodil plants. They added a promoter and other signals for expression in the developing rice grain and then transformed rice plants with the two genes. The resulting rice plants produce grains that look yellow because of their high β-carotene content. A newer variety with a corn gene replacing the one from daffodils makes even more β-carotene and is golden in color (**Figure 18.13**). A daily intake of about 150 grams of this cooked rice can supply all the β-carotene a person needs. This new transgenic strain has been crossed with strains adapted for various local environments, in the hope of improving the diets of millions of people.

CROPS THAT ADAPT TO THE ENVIRONMENT   Agriculture depends on ecological management—tailoring the environment to the needs of crop plants and animals. A farm field is an unnatural, human-designed system that must be carefully

Wild type   Golden rice 1   Golden rice 2

18.13 Transgenic Rice Rich in β-Carotene   Middle and right: The grains from these transgenic rice strains are colored because they make the pigment β-carotene, which is converted to vitamin A in the human body. Left: Normal rice grains do not contain β-carotene.

managed to maintain optimal conditions for crop growth. For example, excessive irrigation can cause increases in soil salinity. The Fertile Crescent, the region between the Tigris and Euphrates rivers in the Middle East where agriculture probably originated 10,000 years ago, is no longer fertile. It is now a desert, largely because the soil has a high salt concentration. Few plants can grow on salty soils, partly because of osmotic effects that result in wilting, and partly because excess salt ions are toxic to plant cells.

Some plants can tolerate salty soils because they have a protein that transports Na^+ ions out of the cytoplasm and into the vacuole, where the ions can accumulate without harming plant growth (see Section 5.3 for a description of the plant vacuole). Scientists developed a highly active version of this gene and used it to transform crop plants that are less tolerant to salt, including rapeseed, wheat, and tomatoes. When this gene was added to tomato plants, they grew in water that was four times as salty as the typical lethal level (**Figure 18.14**). This finding raises the prospect of growing crops on what were previously unproductive soils.

This example illustrates what could become a fundamental shift in the relationship between crop plants and the environment. *Instead of manipulating the environment to suit the plant, biotechnology may allow us to adapt the plant to the environment.* As a result, some of the negative effects of agriculture, such as water pollution, could be lessened.

There is public concern about biotechnology

Concerns have been raised about the safety and wisdom of genetically modifying crops and other organisms. These concerns are centered on three claims:

- Genetic manipulation is an unnatural interference with nature.
- Genetically altered foods are unsafe to eat.
- Genetically altered crop plants are dangerous to the environment.

Advocates of biotechnology tend to agree with the first claim. However, they point out that all crops are unnatural in the sense that they come from artificially bred plants growing in

(A)

18.14 Salt-Tolerant Tomato Plants Transgenic plants containing a gene for salt tolerance thrive in salty water (A), whereas plants without the transgene die (B). This technology may allow crops to be grown on salty soils.

(B)

a manipulated environment (a farmer's field). Recombinant DNA technology just adds another level of sophistication to these technologies.

To counter the concern about whether genetically engineered crops are safe for human consumption, biotechnology advocates point out that only single genes are added and that these genes are specific for plant function. For example, the *B. thuringiensis* toxin produced by transgenic plants has no effect on people. However, as plant biotechnology moves from adding genes that improve plant growth to adding genes that affect human nutrition, such concerns will become more pressing.

Various negative environmental impacts have been envisaged. There is concern about the possible "escape" of transgenes from crops to other species. If the gene for herbicide resistance, for example, were inadvertently transferred from a crop plant to a closely related weed, that weed could thrive in herbicide-treated areas. Another negative impact is the possibility that increased use of an herbicide will select for weeds with naturally occurring mutations that make them resistant to that herbicide. This is indeed occurring. Widespread use of glyphosate on fields of glyphosate-resistant crops has resulted in the selection of rare mutations in weeds that make them resistant to glyphosate. More than ten resistant weed species have appeared in the U.S.

As we mentioned in the opening story, scientists have developed microorganisms that are able to break down components of crude oil. These have not been released into the environment because of the unknown effects that such organisms might have on natural ecosystems. However, these organisms might provide a way to rapidly clean up catastrophic oil spills.

Because of the potential benefits of biotechnology, scientists believe that it is wise to proceed, albeit with caution.

RECAP 18.6

Biotechnology has been used to produce medicines and to develop transgenic plants with improved agricultural and nutritional characteristics.

- What are the advantages of using biotechnology for plant breeding compared with traditional methods? **See Figure 18.12**

- What are some of the concerns that people might have about biotechnology? **See pp. 388–389**

?

Are there other uses for microorganisms in environmental cleanup?

ANSWER

Microorganisms currently play a critical role in cleaning up human-made waste products in the environment. For example, composting involves the use of bacteria to break down large molecules, including carbon-rich polymers and proteins in waste products such as wood chips, paper, straw, and kitchen scraps. Other bacteria are used in wastewater treatment to break down human waste and household chemicals in wastewater treatment plants. As we mentioned in the opening story, much research effort has gone into the development of genetically modified bacteria that could be used for environmental cleanup. An interesting example is the radiation-resistant bacterium *Deinococcus radiodurans*, which has been engineered to precipitate heavy metals and to break down components of crude oil. This organism may be useful for bioremediation at radioactively contaminated sites.

CHAPTER**SUMMARY** 18

 18.1 **What Is Recombinant DNA?**

- **Recombinant DNA** is formed by the combination of two DNA sequences from different sources. **Review Figure 18.1**

- Many restriction enzymes make staggered cuts in the two strands of DNA, creating fragments that have **sticky ends** with unpaired bases.

- DNA fragments with sticky ends can be used to create recombinant DNA. DNA molecules from different sources can be cut with the same restriction enzyme and spliced together using **DNA ligase**. **Review Figure 18.2**

18.2 **How Are New Genes Inserted into Cells?**

- One goal of recombinant DNA technology is to **clone** a particular gene, either for analysis or to produce its protein product in quantity.

- Bacteria, yeasts, and cultured plant and animal cells are commonly used as hosts for recombinant DNA. The insertion of foreign DNA into host cells is called **transformation** or (for animal cells) **transfection**. Transformed or transfected cells are called **transgenic** cells.

- Various methods are used to get recombinant DNA into cells. These include chemical or electrical treatment of the cells, the use of viral vectors, and injection. *Agrobacterium tumefaciens* is often used to insert DNA into plant cells.

- To identify host cells that have taken up a foreign gene, the inserted sequence can be tagged with one or more **reporter genes**, which are genetic markers with easily identifiable phenotypes. **Selectable markers** allow for the selective growth of transgenic cells. **Review Figures 18.3, 18.4**

- Replication of the foreign gene in the host cell requires that it become part of a segment of DNA that contains a **replicon** (origin and terminus of replication).

- **Vectors** are DNA sequences that can carry new DNA into host cells. Plasmids and viruses are commonly used as vectors.

18.3 **What Sources of DNA Are Used in Cloning?**

- DNA fragments from a genome can be inserted into host cells to create a **genomic library**. **Review Figure 18.5A**

- The mRNAs produced in a certain tissue at a certain time can be extracted and used to create **complementary DNA (cDNA)** by **reverse transcription**. **Review Figure 18.5B**

- PCR products can be used for cloning. Synthetic DNA containing any desired sequence can be made in the laboratory.

 18.4 **What Other Tools Are Used to Study DNA Function?**

- Homologous recombination can be used to **knock out** a gene in a living organism. **Review Figure 18.6**

- Gene silencing techniques can be used to inactivate the mRNA transcript of a gene, which may provide clues to the gene's function. Artificially created **antisense RNA** or siRNA can be added to a cell to prevent translation of a specific mRNA. **Review Figure 18.7**

- **DNA microarray** technology permits the screening of thousands of cDNA sequences at the same time. **Review Figure 18.8, ANIMATED TUTORIAL 18.1**

18.5 **What Is Biotechnology?**

- **Biotechnology** is the use of living cells to produce materials useful to people. Recombinant DNA technology has resulted in a boom in biotechnology.

- **Expression vectors** allow a transgene to be expressed in a host cell. **Review Figure 18.9, ACTIVITY 18.1**

18.6 **How Is Biotechnology Changing Medicine and Agriculture?**

- Recombinant DNA techniques have been used to make medically useful proteins. **Review Figure 18.10**

- **Pharming** is the use of transgenic plants or animals to produce pharmaceuticals. **Review Figure 18.11**

- Because recombinant DNA technology has several advantages over traditional agricultural biotechnology, it is being extensively applied to agriculture. **Review Figure 18.12**

- Transgenic crop plants can be adapted to their environments, rather than vice versa.

- There is public concern about the application of recombinant DNA technology to food production.

Go to the Interactive Summary to review key figures, Animated Tutorials, and Activities
Life10e.com/is18

CHAPTER**REVIEW**

▦ REMEMBERING

1. Which of the following is used as a reporter gene in recombinant DNA work with bacteria as host cells?
   a. rRNA
   b. Green fluorescent protein
   c. Antibiotic sensitivity
   d. Ability to make ornithine
   e. Vitamin synthesis

2. Which feature is *not* desirable in a vector for gene cloning?
   a. An origin of DNA replication
   b. Genetic markers for the presence of the vector
   c. Many recognition sequences for the restriction enzyme to be used
   d. One recognition sequence each for one or more different restriction enzymes
   e. Genes other than the target gene for transfection

3. RNA interference (RNAi) inhibits
   a. DNA replication.
   b. neither transcription nor translation of specific genes.
   c. recognition of the promoter by RNA polymerase.
   d. transcription of all genes.
   e. translation of specific mRNAs.

4. An expression vector requires all of the following except
   a. genes for ribosomal RNA.
   b. a reporter gene.
   c. a promoter of transcription.
   d. an origin of DNA replication.
   e. restriction enzyme recognition sequences.

5. Pharming is a term that describes
   a. the use of animals in transgenic research.
   b. plants making genetically altered foods.
   c. synthesis of recombinant drugs by bacteria.
   d. large-scale production of cloned animals.
   e. synthesis of a drug by a transgenic plant or animal.

6. Which of the following could *not* be used to test whether expression of a particular gene is necessary for a particular biological function?
   a. RNAi
   b. Knockout technology
   c. Antisense
   d. Mutant tRNA
   e. Transposon mutagenesis

▦ UNDERSTANDING & APPLYING

7. Assume you are using a plasmid vector that contains genes encoding ampicillin resistance and the green fluorescent protein, with a restriction site in the latter gene only. Outline the sequence of steps for inserting a piece of foreign DNA into this plasmid, introducing the recombinant plasmid into bacteria, and verifying that the plasmid and the foreign gene are both present in the bacteria:
   1. Transform host cells.
   2. Select colonies for antibiotic resistance.
   3. Select colonies that do not glow under ultraviolet light.
   4. Digest vector and foreign DNA with a restriction enzyme.
   5. Ligate the digested plasmid together with the foreign DNA.
      a. 4, 5, 1, 3, 2
      b. 4, 5, 1, 2, 3
      c. 1, 3, 4, 2, 5
      d. 3, 2, 1, 4, 5
      e. 1, 3, 2, 5, 4

8. In a genomic library of frog DNA in *E. coli* bacteria,
   a. all bacterial cells have the same sequences of frog DNA.
   b. all bacterial cells have different sequences of frog DNA.
   c. each bacterial cell has a random fragment of frog DNA.
   d. each bacterial cell has many fragments of frog DNA.
   e. the frog DNA is always transcribed into mRNA in the bacterial cells.

9. Compare PCR (see Section 13.5) and cloning as methods to amplify a gene. What are the requirements, benefits, and drawbacks of each method?

10. Compare traditional genetic methods with recombinant DNA methods for producing genetically altered plants. For each case, describe (a) sources of new genes; (b) numbers of genes transferred; and (c) how long the process takes.

▦ ANALYZING & EVALUATING

11. As specifically as you can, outline the steps you would take to (a) insert and express the gene for a new, nutritious seed protein in wheat, and (b) insert and express a gene for a human enzyme in sheep's milk.

12. What are the major public concerns about biotechnology as applied to food? In your locality or country, look up regulations on either producing or labeling foods made using biotechnology. What were the scientific and philosophical bases for these regulations?

Go to BioPortal at *yourBioPortal.com* for Animated Tutorials, Activities, LearningCurve Quizzes, Flashcards, and many other study and review resources.

19 Differential Gene Expression in Development

CHAPTER**OUTLINE**

19.1 What Are the Processes of Development?

19.2 How Is Cell Fate Determined?

19.3 What Is the Role of Gene Expression in Development?

19.4 How Does Gene Expression Determine Pattern Formation?

19.5 Is Cell Differentiation Reversible?

IT WAS A MASTERFUL PERFORMANCE. Baseball star Bartolo Colon pitched a five-hit shutout for the New York Yankees as they defeated the Oakland Athletics. His journey to that day on the mound had been eventful. Growing up poor in the Dominican Republic, he was spotted by baseball talent agents as a teenager, and at age 20 he signed a contract with a U.S. major league team. By age 23 he had won the Cy Young Award as professional baseball's top pitcher. But then Colon partially tore the group of muscles and tendons in the elbow and shoulder of his pitching arm. For several years his performance deteriorated significantly, and he even missed an entire season.

Bartolo Colon   Stem cells helped repair damage to his tendons, and he was able to pitch—and win—again.

Colon came back as good as ever, thanks to stem cell therapy, which he received in the Dominican Republic. A surgeon extracted fat and bone marrow from Colon's body and isolated mesenchymal stem cells from these tissues. Stem cells are actively dividing, unspecialized cells that have the potential to produce different cell types depending on the signals they receive from the body. Mesenchymal stem cells are able to differentiate into various kinds of connective tissue, including bone, cartilage, blood vessels, tendons, and muscle. Colon's stem cells were injected into his elbow and shoulder, and months later he was apparently healed.

Scientists see much promise in the use of stem cell treatments to heal a wide variety of medical conditions, including cancer, Parkinson's disease, brain and spinal cord injuries, heart and other muscular damage, diabetes, blindness—even baldness. The basic idea is to inject stem cells into damaged tissues, where they will differentiate and form new, healthy tissues. Many procedures are being tried experimentally in the U.S., but so far the only kind

of stem cell therapy approved by the U.S. government is hematopoietic stem cell transplantation (see Section 19.5). Other countries, including China, Mexico, Ukraine, and the Dominican Republic, already have clinics that offer various stem cell treatments.

The processes by which an unspecialized stem cell proliferates and forms specialized cells and tissues with distinctive appearances and functions are similar to the developmental processes that occur in the embryo. Much of our knowledge of developmental biology has come from studies on model organisms such as fruit flies, nematodes, frogs, sea urchins, mice, and the small flowering plant *Arabidopsis thaliana*. Eukaryotes share many similar genes, and the cellular and molecular principles underlying their development also turn out to be similar. Thus discoveries from one organism can aid us in understanding other organisms, including ourselves.

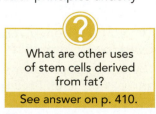

Q What are other uses of stem cells derived from fat?

See answer on p. 410.

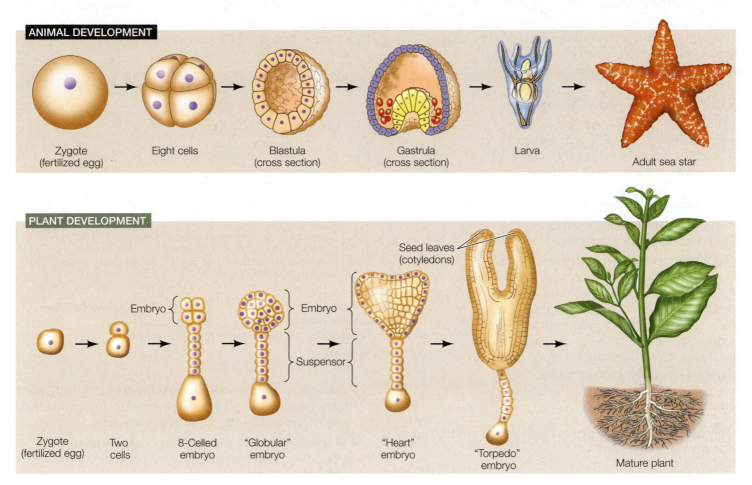

19.1 From Fertilized Egg to Adult The stages of development from zygote to maturity are shown for an animal and for a plant. The blastula is a hollow sphere of cells; the gastrula has three cell layers.

Go to Activity 19.1 **Stages of Development**
Life10e.com/ac19.1

19.1 What Are the Processes of Development?

Development is the process by which a multicellular organism, beginning with a single cell, goes through a series of changes, taking on the successive forms that characterize its life cycle (**Figure 19.1**). After the egg is fertilized it is called a zygote, and in the earliest stages of development a plant or animal is called an **embryo**. Sometimes the embryo is contained within a protective structure such as a seed coat, an eggshell, or a uterus. An embryo does not photosynthesize or feed itself. Instead, it obtains its food from its mother either directly (via the placenta) or indirectly (by way of nutrients stored in a seed or egg). A series of embryonic stages precedes the birth of the new, independent organism. Many organisms continue to develop throughout their life cycles, with development ceasing only at death.

Development involves distinct but overlapping processes

The developmental changes an organism undergoes as it progresses from a fertilized egg to a mature adult involve four processes:

- **Determination** sets the developmental *fate* of a cell—what type of cell it will become—even before any characteristics of that cell type are observable. For example, the mesenchymal stem cells described in the opening story look unspecialized, but their fate to become connective tissue cells has already been determined.

- **Differentiation** is the process by which different types of cells arise, leading to cells with specific structures and functions. For example, mesenchymal stem cells differentiate to become muscle, fat, tendon, or other connective tissue cells.

- **Morphogenesis** (Greek for "origin of form") is the organization and spatial distribution of differentiated cells into the multicellular body and its organs.

- **Growth** is the increase in size of the body and its organs by cell division and cell enlargement.

Determination and differentiation occur largely because of differential gene expression. The cells that arise from repeated mitoses in the early embryo may look the same superficially, but they soon begin to differ in terms of which of the thousands of genes in the genome are expressed.

Morphogenesis involves differential gene expression and the interplay of signals between cells. Morphogenesis can occur in several ways:

- Cell division is important in both plants and animals.
- Cell expansion is especially important in plant development, where a cell's position and shape are constrained by the cell wall.
- Cell movements are very important in animal morphogenesis.
- Apoptosis (programmed cell death) is essential in organ development.

Growth occurs by cell enlargement. In some cases, cell enlargement is coupled to cell division, so the average cell size remains the same as the tissue grows; in other cases (especially in plant tissues), cells enlarge without dividing, so the average cell size increases. Growth continues throughout the individual's life in some organisms, but reaches a more or less stable end point in others.

Cell fates become progressively more restricted during development

During development, each undifferentiated cell will become part of a particular type of tissue—this is referred to as **cell fate**. Cell fate determination occurs as the embryo develops. The timing of this determination varies with the organism, but it is typically quite early. One way to find out the timing is to transplant cells from one embryo to a different region of a recipient embryo (**Figure 19.2**). Do the transplanted cells adopt the differentiation pattern of their new surroundings, or do they continue on their own path, with their fate already sealed?

In amphibian embryos determination happens early. If the donor tissue is from an early-stage embryo (blastula), it adopts the fate of the new surroundings. In this case, cell fate has not been determined and is influenced by the extracellular environment. But if the donor tissue is from an older embryo (gastrula), it continues on its original developmental path. In this case, cell fate has already been determined and is no longer influenced by the extracellular environment.

Cell fate determination is influenced by changes in gene expression as well as the extracellular environment. Determination is not something that is visible under the microscope—cells do not change their appearance when they become determined. Determination is followed by differentiation—the actual changes in biochemistry, structure, and function that result in different cell types. *Determination is a commitment; the final realization of that commitment is differentiation.*

During animal development, cell fate becomes progressively more restricted. This can be thought of in terms of **cell potency**, which is a cell's potential to differentiate into other cell types:

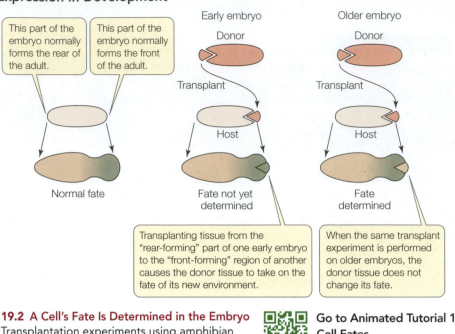

This part of the embryo normally forms the rear of the adult.

This part of the embryo normally forms the front of the adult.

Normal fate

Early embryo
Donor
Transplant
Host
Fate not yet determined

Older embryo
Donor
Transplant
Host
Fate determined

Transplanting tissue from the "rear-forming" part of one early embryo to the "front-forming" region of another causes the donor tissue to take on the fate of its new environment.

When the same transplant experiment is performed on older embryos, the donor tissue does not change its fate.

19.2 A Cell's Fate Is Determined in the Embryo Transplantation experiments using amphibian embryos show that the fate of cells is determined as the early embryo develops.

Go to Animated Tutorial 19.1
Cell Fates
Life10e.com/at19.1

- The cells of an early embryo are **totipotent** (*toti*, "all"; *potent*, "capable"); they have the potential to differentiate into any cell type, including more embryonic cells.

- In later stages of the embryo, many cells are **pluripotent** (*pluri*, "many"); they have the potential to develop into most other cell types, but they cannot form new embryos.

- Through later developmental stages, including adulthood, certain stem cells are **multipotent**; they can differentiate into several different, related cell types. Mesenchymal stem cells (see the opening story) are one kind of multipotent stem cell.

- Many cells in the mature organism are **unipotent**; they can produce only one cell type—their own.

As you will see in Section 19.5, many plant cell types can be manipulated in the laboratory to dedifferentiate, form embryos, and develop into new plants. Much more recently, researchers have found ways to manipulate some mammalian cells to make them dedifferentiate and then redifferentiate into new tissues. So even though cell fate becomes progressively more restricted during normal development, this process can be altered in the laboratory.

■ RECAP 19.1

Development takes place via the processes of determination, differentiation, morphogenesis, and growth. Cells in the very early embryo have not yet had their fates determined; as development proceeds, their potential fates become more and more restricted.

- What are the four processes of development? **See p. 393**

- What did the experiments of the type illustrated in Figure 19.2 tell us about how cell fates become determined? **See p. 394**

- Describe the differences between totipotent, pluripotent, and multipotent cells. **See p. 394**

We have discussed the basic processes that occur during development, and seen that cell fate determination occurs before cells differentiate and become specialized. We will now turn to the mechanisms of cell fate determination.

19.2 How Is Cell Fate Determined?

The fertilized egg undergoes many cell divisions to produce the many differentiated cells in the body (such as liver, muscle, and nerve cells). How can one cell produce so many different cell types? There are two ways that determination occurs:

- **Cytoplasmic segregation** (unequal cytokinesis). A factor within an egg, zygote, or precursor cell may be unequally distributed in the cytoplasm. After cell division, the factor ends up in some daughter cells or regions of cells, but not others.

- **Induction** (cell-to-cell communication). A factor is actively produced and secreted by certain cells to induce other cells to become determined.

Cytoplasmic segregation can determine polarity and cell fate

Some differences in gene expression patterns are the result of cytoplasmic differences among cells. One such cytoplasmic difference is the emergence of distinct "top" and "bottom" ends of an organism or structure; such a difference is called **polarity**. Many examples of polarity are observed as development proceeds. Our heads are distinct from our rear ends, and the distal (far) ends of our arms and legs (wrists, ankles, fingers, toes) differ from the proximal (near) ends (shoulders and hips). Polarity may develop early; even within the fertilized egg, the yolk and other factors are often distributed asymmetrically. During early development in animals, polarity is specified by an animal pole at the top of the zygote and a vegetal pole at the bottom. This polarity can lead to determination of cell fates at a very early stage of development. For example, sea urchin embryos can be bisected at the eight-cell stage in two ways:

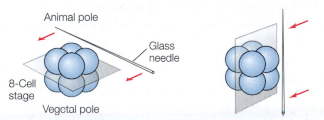

If the two halves of these embryos (each with four cells) are allowed to develop, the results are dramatically different for the two different cuts:

- If the embryo is cut into a top half and a bottom half (left, above), the bottom half develops into a small sea urchin and the top half does not develop at all.

- If the embryo is cut into two side halves (right, above), both halves develop into small sea urchins.

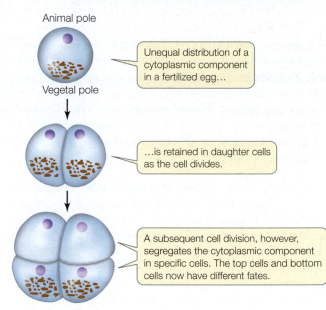

19.3 The Principle of Cytoplasmic Segregation The unequal distribution of some component in the cytoplasm of a cell may determine the fates of its descendants.

Go to Animated Tutorial 19.2
Early Asymmetry in the Embryo
Life10e.com/at19.2

These results indicate that the top and bottom halves of an eight-celled sea urchin embryo have already developed distinct fates. Such observations led to the model of cytoplasmic segregation shown in **Figure 19.3**. The model states that certain materials called **cytoplasmic determinants** are distributed unequally in the egg cytoplasm. Cytoplasmic determinants include specific proteins, small regulatory RNAs, and mRNAs, and they play roles in directing the embryonic development of many organisms. What accounts for the unequal distribution of these determinants?

The cytoskeleton contributes to the asymmetric distribution of cytoplasmic determinants in the egg. Recall from Section 5.3 that an important function of the microtubules and microfilaments in the cytoskeleton is to help move materials around in the cell. Two properties allow these structures to accomplish this:

- Microtubules and microfilaments have polarity—they grow by adding subunits to the plus end.

- The cytoskeletan can bind motor proteins that are used to transport the cytoplasmic determinants.

For example, in the sea urchin egg there is a protein that binds to both the growing (+) end of a microfilament and to an mRNA encoding a cytoplasmic determinant. As the microfilament grows toward one end of the cell, it carries the mRNA along with it. The asymmetrical distribution of the mRNA leads to a similar distribution of the protein it encodes.

Inducers passing from one cell to another can determine cell fates

The term "induction" has different meanings in different contexts. In biology it can be used broadly to refer to the initiation of, or cause of, a change or process. But in the context of cellular differentiation, it refers to the signaling events by which cells in a developing organism communicate and influence one another's developmental fate. Induction involves chemical signals and signal transduction mechanisms. We will describe two examples of this form of induction: one in the developing vertebrate eye, and the other in a developing reproductive structure of the nematode *Caenorhabditis elegans*.

LENS DETERMINATION IN THE VERTEBRATE EYE The development of the lens in the vertebrate eye is a classic example of induction. In a frog embryo, the developing forebrain bulges out at both sides to form the optic vesicles, which expand until they come into contact with the cells at the surface of the head (**Figure 19.4**). The surface tissue in the region of contact thickens, forming a lens placode—tissue that will ultimately form the lens. The lens placode bends inward, folds over on itself, and ultimately detaches from the surface tissue to produce a structure that will develop into the lens. If the growing optic vesicle is cut away before it contacts the surface cells, no lens forms. Placing an impermeable barrier between the optic vesicle and the surface cells also prevents the lens from forming. These observations suggest that the surface tissue begins to develop into a lens when it receives a signal from the optic vesicle. Such a signal is termed an **inducer** (signaling factor).

Inducers trigger sequences of gene expression in the responding cells. How cells switch on different sets of genes that govern development and direct the formation of body plans is of great interest to developmental and evolutionary biologists. They use model organisms to investigate the major principles governing these processes.

VULVAL DETERMINATION IN THE NEMATODE The genome of the nematode *Caenorhabditis elegans* was one of the first eukaryotic genomes to be sequenced (see Section 17.3). It develops from fertilized egg to larva in only about 8 hours, and the worm reaches the adult stage in just 3.5 days. The process is easily observed using a low-magnification dissecting microscope because the body covering is transparent (**Figure 19.5A**).

The adult nematode is hermaphroditic, containing both male and female reproductive organs. It lays eggs through a pore called the vulva on the ventral (lower) surface. During development, a single cell, called the anchor cell, induces the vulva to form from six cells on the worm's ventral surface. In this case there are two molecular signals: the primary (1°) inducer and the secondary (2°) inducer. Each of the six ventral cells has three possible fates: it may become a primary vulval precursor cell, a secondary vulval precursor cell, or simply become part of the worm's skin—an epidermal cell. You can follow the sequence of events in **Figure 19.5B**. The concentration gradient of the primary inducer, LIN-3, is key. (LIN stands for abnormal cell *lin*eage.) The anchor cell produces the LIN-3 protein, which diffuses out of the cell and forms a concentration gradient with respect to adjacent cells. Three cells receive more LIN-3 than the others and become vulval precursor cells; cells farther from the anchor cell receive less LIN-3 and become epidermal cells. The cell closest to the anchor cell receives the most LIN-3—enough LIN-3 to turn on expression of the secondary inducer. The secondary inducer then acts on the two adjacent cells. This second induction event results in the two classes of vulval precursor cells: primary and secondary. Induction leads to the activation or inactivation of specific sets of genes through signal transduction cascades. This differential gene expression leads to cell differentiation.

RECAP **19.2**

Cell fate determination involves cytoplasmic segregation and induction. Cytoplasmic segregation is the unequal distribution of gene products in the egg, zygote, or early embryo. Induction occurs when one cell or tissue sends a chemical signal to another.

- How does cytoplasmic segregation result in polarity in a fertilized egg, and how does polarity affect cell fate determination? **See p. 395 and Figure 19.3**
- Describe how induction influences tissue formation in the vertebrate eye. **See pp. 395–396 and Figure 19.4**
- Describe the process of induction during nematode vulva development. **See p. 396 and Figure 19.5**

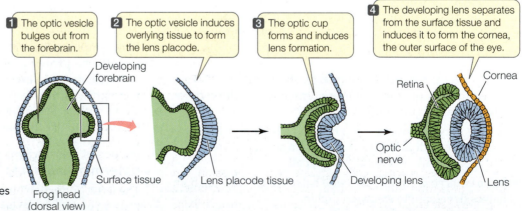

1 The optic vesicle bulges out from the forebrain.

2 The optic vesicle induces overlying tissue to form the lens placode.

3 The optic cup forms and induces lens formation.

4 The developing lens separates from the surface tissue and induces it to form the cornea, the outer surface of the eye.

Developing forebrain

Retina

Cornea

Optic nerve

Developing lens

Lens

Frog head (dorsal view)

Surface tissue

Lens placode tissue

19.4 Embryonic Inducers in Vertebrate Eye Development The eye of a frog develops as different cells induce changes in neighboring cells.

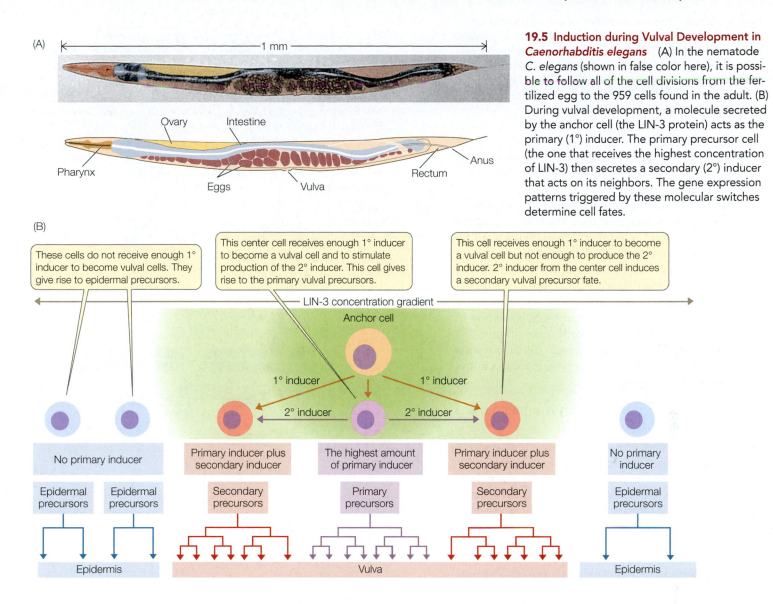

19.5 Induction during Vulval Development in *Caenorhabditis elegans* (A) In the nematode *C. elegans* (shown in false color here), it is possible to follow all of the cell divisions from the fertilized egg to the 959 cells found in the adult. (B) During vulval development, a molecule secreted by the anchor cell (the LIN-3 protein) acts as the primary (1°) inducer. The primary precursor cell (the one that receives the highest concentration of LIN-3) then secretes a secondary (2°) inducer that acts on its neighbors. The gene expression patterns triggered by these molecular switches determine cell fates.

We have seen that cytoplasmic segregation and induction both influence cell fate determination. We have seen two examples of how induction leads to organ formation in developing multicellular organisms. Next we will take a closer look at how gene expression affects cell fate determination and differentiation.

19.3 What Is the Role of Gene Expression in Development?

Although every cell contains all the genes needed to produce every protein encoded by an organism's genome, each cell expresses only selected genes. For example, certain cells in hair follicles produce keratin, the protein that makes up hair, whereas other cell types in the body do not. What determines whether a cell will produce keratin? Chapter 16 described a number of ways in which cells regulate gene expression and protein production—by controlling transcription, translation, and posttranslational protein modifications. The mechanisms that control gene expression during cell fate determination and cell differentiation generally work at the level of transcription.

Cell fate determination involves signal transduction pathways that lead to differential gene expression

As we have seen, cell fate determination can occur by the process of induction. When an inducer molecule binds to its specific receptor on the surface of a cell, a signal transduction pathway leads to the activation of one or more transcription factors. Recall that transcription factors are DNA binding proteins that regulate the expression of specific genes (see Section 16.2). **Figure 19.6** illustrates the induction of one cell (the cell on the left), which is exposed to a high concentration of inducer. This results in the activation of a transcription factor in the cytoplasm, causing it to enter the nucleus and switch on the expression of a specific gene. The cell on the right is exposed to a lower concentration of the inducer, and as a result, gene expression is not activated.

19.6 Induction The concentration of an inducer directly affects the degree to which a transcription factor is activated. The inducer acts by binding to a receptor on the target cell. This binding is followed by signal transduction involving transcription factor activation or translocation from the cytoplasm to the nucleus. In the nucleus it acts to stimulate the expression of genes involved in cell differentiation.

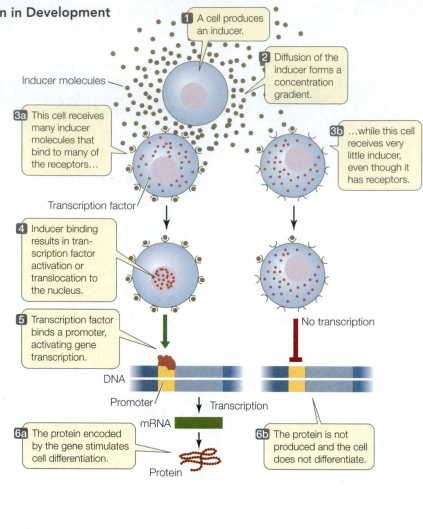

It turns out that development is often controlled by these kinds of molecular switches, which allow a cell to proceed down one of two alternative paths. One challenge for developmental biologists is to find these switches and determine how they work. In the case of vulval development in nematodes (see Figure 19.5), LIN-3 is the primary inducer that determines the fate of vulval precursor cells. LIN-3 is a growth factor that is homologous to a vertebrate growth factor called EGF (*epidermal growth factor*). LIN-3 binds to a receptor on the surfaces of the vulval precursor cells, setting in motion a signal transduction cascade involving the Ras protein and MAP kinases (see Figure 7.10). This results in increased transcription of the genes involved in the differentiation of vulval cells.

Differential gene transcription is a hallmark of cell differentiation

The gene for β-globin, one of the protein components of hemoglobin, is expressed in red blood cells as they form in the bone marrow of mammals. That this same gene is also present—but unexpressed—in neurons in the brain (which do not make hemoglobin) can be demonstrated by nucleic acid hybridization. Recall that in nucleic acid hybridization, a probe made of single-stranded DNA or RNA of known sequence is added to denatured DNA to reveal complementary coding regions in the DNA sample (see Figure 14.7). A probe for the β-globin gene can be applied to DNA from brain cells and immature red blood cells (recall that mature mammalian red blood cells lose their nuclei during development). In both cases, the probe finds its complement, showing that the β-globin gene is present in both types of cells. However, if the β-globin probe is applied to mRNA, rather than DNA, from the two cell types, it finds β-globin mRNA only in the red blood cells, not in the brain cells. This result shows that the gene is expressed in only one of the two cell types.

What leads to this differential gene expression? One well-studied example of cell differentiation is the conversion of undifferentiated muscle precursor cells into cells that are destined to form muscle (**Figure 19.7**). In the vertebrate embryo these cells come from a tissue layer called the mesoderm (see Section 44.3). A key event in the commitment of these cells to become

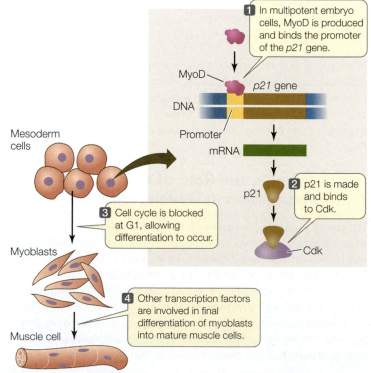

19.7 Transcription and Differentiation in the Formation of Muscle Cells Production of the transcription factor MyoD is important in muscle cell differentiation.

muscle is that they stop dividing. Indeed, in many parts of the embryo, *cell division and cell differentiation are mutually exclusive.* Cell signaling activates the gene for a transcription factor called **MyoD** (for *myo*blast-*d*etermining gene); this in turn activates the gene for p21, which is an inhibitor of the cyclin-dependent kinases (Cdk's) that normally stimulate the cell cycle at G1 (see Figure 11.6). Expression of the *p21* gene causes the cell cycle to stop, and other transcription factors then enter the picture so that differentiation can proceed. Interestingly, *myoD* is also activated in the stem cells that are present in adult muscle, indicating a role of this transcription factor in the repair of muscle tissue as it gets damaged and worn out.

Genes such as *myoD* that direct the most fundamental decisions in development (often by regulating other genes on other chromosomes) usually encode transcription factors. In some cases a single transcription factor can cause a cell to differentiate in a certain way. In others, complex interactions between genes and proteins determine a sequence of transcriptional events that leads to differentiation.

RECAP 19.3

Cell fate determination involves the activation of signal transduction pathways that lead to differential gene expression. Differentiation involves selective gene expression, controlled at the level of transcription by transcription factors.

- How do inducer molecules cause changes in gene expression? **See pp. 397–398 and Figure 19.6**
- What techniques could you use to identify genes expressed during cell differentiation? **See p. 398**
- Describe the roles of transcription factors in controlling differentiation. **See pp. 398–399 and Figure 19.7**

We have seen how cell fate is determined, and we have looked at the roles of gene expression in cell fate determination and differentiation. We will now take a closer look at how gene expression affects differentiation and morphogenesis.

19.4 How Does Gene Expression Determine Pattern Formation?

Pattern formation is the process that results in the spatial organization of a tissue or organism. It is inextricably linked to morphogenesis, the creation of body form. You might expect morphogenesis to involve a lot of cell division, followed by differentiation—and it does. But what you might not expect is the amount of programmed cell death—apoptosis—that occurs during morphogenesis.

Multiple proteins interact to determine developmental programmed cell death

We noted in Section 11.6 that apoptosis is a programmed series of events that leads to cell death. Apoptosis is an integral part of the normal development and life of an organism. For example, in an early human embryo, the hands and feet look like tiny paddles: the tissues that will become fingers and toes are joined by connective tissue. Between days 41 and 56 of development, the cells between the digits die, freeing the individual fingers and toes:

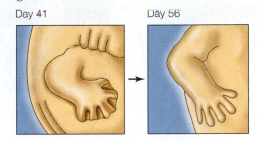

Day 41   Day 56

Many cells and structures form and then disappear during development, in processes involving apoptosis.

Model organisms have been very useful in studying the genes and proteins involved in apoptosis. Mutants with altered cell death phenotypes have been used to identify the genes and proteins involved. For example, the nematode worm *C. elegans* produces precisely 1,090 somatic cells as it develops from a fertilized egg into an adult, but 131 of those cells die, leaving 959 cells in the adult worm. The sequential activation of two proteins called CED-4 and CED-3 (for *cell death*) appears to control this programmed cell death (**Figure 19.8A**). A third protein called CED-9, which is bound to the outside of the mitochondrion, inhibits apoptosis in cells that are not programmed to die. In these cells, CED-9 binds CED-4 and prevents it from activating CED-3. If the cell receives a signal for apoptosis, CED-9 releases CED-4, which then activates CED-3, a protease (an enzyme that breaks down proteins).

A similar system controls apoptosis during human development. The apoptosis pathway in humans involves a class of proteases called caspases (see Figure 11.23). Caspases are similar in amino acid sequence to CED-3 in *C. elegans*. Furthermore, the human proteins Bcl-2 and Apaf1 are similar in

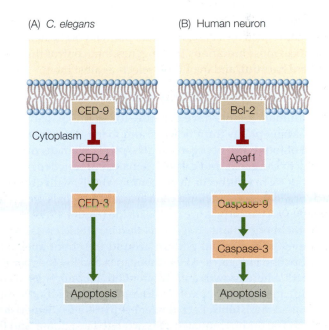

(A) *C. elegans*   (B) Human neuron

19.8 Pathways for Apoptosis In the worm *C. elegans* (A) and humans (B), similar pathways for apoptosis are controlled by genes with similar sequences and functions.

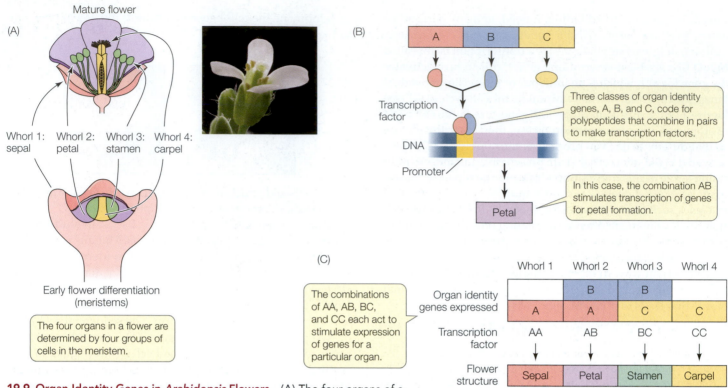

19.9 Organ Identity Genes in *Arabidopsis* Flowers (A) The four organs of a flower—carpels (yellow), stamens (green), petals (purple), and sepals (pink)—grow in whorls that develop from the floral meristem. (B) Floral organs are determined by three classes of organ identity genes whose polypeptide products combine in pairs to form transcription factors. (C) Combinations of polypeptide subunits in transcription factors activate gene expression for specific organs.

structure and function to the *C. elegans* proteins CED-9 and CED-4, respectively (**Figure 19.8B**). So humans and nematodes, two species separated by more than 600 million years of evolutionary history, have similar genes controlling programmed cell death. The conservation of this pathway indicates its importance: most mutations in the genes that control this pathway are harmful, and evolution selects against them.

Plants have organ identity genes

Like animals, plants have organs—for example, leaves and roots. Many plants form flowers, and many flowers are composed of four types of organs: sepals, petals, stamens (male reproductive organs), and carpels (female reproductive organs). These floral organs occur in concentric whorls, with groups of each organ type encircling a central axis. The sepals are on the outside and the carpels are on the inside (**Figure 19.9A**).

In the model plant *Arabidopsis thaliana* (thale cress), flowers develop in a radial pattern around the shoot apex as it develops and elongates. At the shoot apex and in other parts of the plant where growth and differentiation occur (such as the root tip), there are groups of undifferentiated, rapidly dividing cells called **meristems**. Each flower begins as a floral meristem of about 700 undifferentiated cells arranged in a dome, and the four whorls develop from this meristem. How is the identity of a particular whorl determined? Three classes of genes called **organ identity genes** encode proteins that act in combination to produce specific whorl features (**Figure 19.9B and C**):

● Genes in class A are expressed in whorls 1 and 2 (which form sepals and petals, respectively).

● Genes in class B are expressed in whorls 2 and 3 (which form petals and stamens).

● Genes in class C are expressed in whorls 3 and 4 (which form stamens and carpels).

These genes encode transcription factors that are active as dimers, that is, proteins with two polypeptide subunits. The composition of the dimer determines which genes the transcription factor activates. For example, a dimer made up of two class A monomers activates transcription of the genes that make sepals; a dimer made up of A and B monomers results in petals, and so forth. A common feature of the A, B, and C proteins, as well as many other plant transcription factors, is a DNA-binding domain called the **MADS box**. The name "MADS" comes from the initials of four genes encoding proteins with this domain.

Two lines of experimental evidence support this model for floral organ determination:

● *Loss-of-function mutations*: for example, a mutation in a class A gene results in no sepals or petals.

● *Gain-of-function mutations*: for example, a promoter for a class C gene can be artificially coupled to a class A gene. In this case, the class A gene is expressed in all four whorls, resulting in only sepals and petals. In any organism, the replacement of one organ for another is called homeosis, and this type of mutation is a **homeotic mutation**.

Transcription of the floral organ identity genes is controlled by other gene products, including the LEAFY protein. Plants with loss-of-function mutations in the *LEAFY* gene make stems instead of flowers, with increased numbers of modified leaves called bracts. The wild-type LEAFY protein is a transcription factor that stimulates expression of the class A, B, and C genes so that they produce flowers. This finding has practical applications. It usually takes 6 to 20 years for a citrus tree to produce flowers and fruits. Scientists have made transgenic orange trees expressing the *LEAFY* gene coupled to a strongly expressed promoter. These trees flower and fruit years earlier than normal trees.

Morphogen gradients provide positional information

During development, the key cellular question "What will I be?" is often answered in part by "Where am I?" Think of the cells in the developing nematode, which can develop into different parts of the vulva depending on their positions relative to the anchor cell (see Figure 19.5). This spatial "sense" is called **positional information**. Positional information often comes in the form of an inducer called a **morphogen**, which diffuses from one cell or group of cells to surrounding cells, setting up a concentration gradient (as we saw for LIN-3 in *C. elegans* vulval induction). There are two requirements for a signal to be considered a morphogen:

- It must directly affect target cells, rather than triggering a secondary signal that affects target cells.
- Different concentrations of the signal must cause different effects.

Developmental biologist Lewis Wolpert uses the "French flag model" to explain morphogens (**Figure 19.10**). This model can be applied to the differentiation of the vulva in *C. elegans* and to the development of a vertebrate limb.

The vertebrate limb develops from a paddle-shaped limb bud (**Figure 19.11**). The cells that develop into different digits must receive positional information; if they do not, the limb will be totally disorganized—imagine a hand with only thumbs or only little fingers. A group of cells at the posterior base of the limb bud, just where it joins the body wall, is called the zone of polarizing activity (ZPA). The cells of the ZPA secrete a protein morphogen called *Sonic hedgehog* (Shh). Shh forms a gradient that determines the posterior–anterior (little finger to thumb) axis of the developing limb. In humans and other primates, the cells exposed to the highest dose of Shh form the little finger; those that receive the lowest dose develop into the thumb.

A cascade of transcription factors establishes body segmentation in the fruit fly

Perhaps the best-studied example of how morphogens determine cell fate is body segmentation in the fruit fly *Drosophila melanogaster*. The body segments of this model organism are

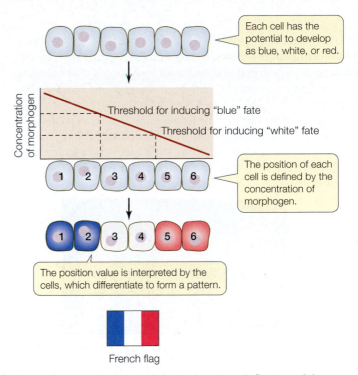

Each cell has the potential to develop as blue, white, or red.

Concentration of morphogen

Threshold for inducing "blue" fate

Threshold for inducing "white" fate

The position of each cell is defined by the concentration of morphogen.

The position value is interpreted by the cells, which differentiate to form a pattern.

French flag

19.10 The French Flag Model In the "French flag" model, a concentration gradient of a diffusible morphogen signals each cell to specify its position.

clearly different from one another. The adult fly has an anterior head (composed of several fused segments), three different thoracic segments, and eight abdominal segments at the posterior end. Each segment develops into different body parts: for example, antennae and eyes develop from head segments, wings from the thorax, and so on.

The life cycle of *Drosophila* from fertilized egg to adult takes about 2 weeks at room temperature. The egg hatches into a larva, which then forms a pupa, which finally is transformed

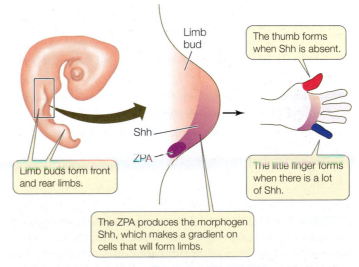

Limb bud

The thumb forms when Shh is absent.

Shh

ZPA

Limb buds form front and rear limbs.

The little finger forms when there is a lot of Shh.

The ZPA produces the morphogen Shh, which makes a gradient on cells that will form limbs.

19.11 Specification of the Vertebrate Limb and the French Flag Model The zone of polarizing activity (ZPA) in the limb bud of the embryo secretes the morphogen *Sonic hedgehog* (Shh). Cells in the bud form different digits depending on the concentration of Shh.

into the adult fly. By the time a larva appears—about 24 hours after fertilization—there are recognizable segments. The thoracic and abdominal segments all look similar, *but the fates of their cells to become different adult segments have already been determined.*

As with other organisms, fertilization in *Drosophila* leads to a rapid series of mitoses. However, the first 12 cycles of nuclear division are not accompanied by cytokinesis. So a multinucleate embryo forms instead of a multicellular embryo (the nuclei are brightly stained in the micrographs below):

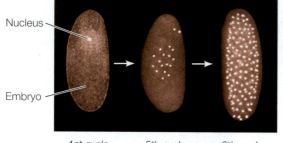

With no cell membranes to cross, morphogens can diffuse easily within the embryo. We focus here on the determination events that occur in the first 24 hours after fertilization.

The events leading to cell fate determination in *Drosophila* were elucidated using experimental genetics:

- First, developmental mutations were identified. For example, a mutant strain might produce larvae with two heads or no segments.

- Then the mutant was compared with wild-type flies, and the gene responsible for the developmental mistake, and the gene's protein product (if appropriate), were isolated.

- Finally, experiments with the gene (making transgenic flies) and protein (injecting the protein into an egg or into an embryo) were done to confirm their roles in the proposed developmental pathway.

These approaches revealed a sequential pattern (cascade) of gene expression that results in the determination of each segment within 24 hours after fertilization. Several classes of genes are involved:

- **Maternal effect genes** set up the major axes (anterior–posterior and dorsal–ventral) of the egg.

- **Segmentation genes** determine the boundaries and polarity of each segment.

- **Hox genes** determine which organ will be made at a given location.

MATERNAL EFFECT GENES Like the eggs and early embryos of sea urchins, *Drosophila* eggs and larvae are characterized by unevenly distributed cytoplasmic determinants (see Figure 19.3). These molecular determinants are the products of specific maternal effect genes. The genes are transcribed in the cells of the mother's ovary, and the mRNAs are passed to the egg

by cytoplasmic bridges. Two maternal effect genes called *bicoid* and *nanos* help determine the anterior–posterior axis of the egg. (The dorsal–ventral axis is determined by other maternal effect genes that will not be described here.)

The mRNAs for *bicoid* and *nanos* diffuse from the mother's cells into what will be the anterior end of the egg. The *bicoid* mRNA is translated into Bicoid protein, a transcription factor that diffuses away from the anterior end, establishing a gradient in the egg cytoplasm (**Figure 19.12A**). Meanwhile, the egg's cytoskeleton transports the *nanos* mRNA from the anterior end of the egg, where it was deposited, to the posterior end, where it is translated (**Figure 19.12B**).

The actions of Bicoid and Nanos establish a gradient of yet another protein, called Hunchback, which determines the anterior and posterior ends of the embryo. Initially, the *hunchback* mRNA is evenly distributed in the embryo, but Nanos inhibits its translation, thus preventing Hunchback protein accumulation at the posterior end of the embryo (**Figure 19.12C**). Meanwhile, at the anterior end of the embryo, Bicoid stimulates increased transcription of the *hunchback* gene, thus increasing the amount of *hunchback* mRNA (and thus Hunchback protein) and further strengthening the Hunchback gradient.

How did biologists elucidate these pathways? Let's look more closely at the experimental approaches used in this case.

- Females that are homozygous for a particular *bicoid* mutation produce larvae with no head and no thorax; thus the Bicoid protein must be needed for the anterior structures to develop.

- If the eggs of these *bicoid* mutants are injected at the anterior end with cytoplasm from the anterior region of a wild-type egg, the injected eggs develop into normal larvae. This experiment also shows that the Bicoid protein is involved in the development of anterior structures.

- If cytoplasm from the anterior region of a wild-type egg is injected into the posterior region of another egg, anterior structures develop there. The degree of induction depends on how much cytoplasm is injected.

- Eggs from homozygous *nanos* mutant females develop into larvae with missing abdominal segments.

- If cytoplasm from the posterior region of a wild-type egg is injected into the posterior region of a *nanos* mutant egg, it will develop normally.

These and other experiments led scientists to understand the cascade of events that determines cell fates.

The events involving *bicoid*, *nanos*, and *hunchback* begin before fertilization and continue after it, during the multinucleate stage, which lasts a few hours. At this stage the embryo looks like a bunch of indistinguishable nuclei under the light microscope. But the cell fates have already begun to be determined. After the anterior and posterior ends have been established, the next step in pattern formation is the determination of segment number and locations.

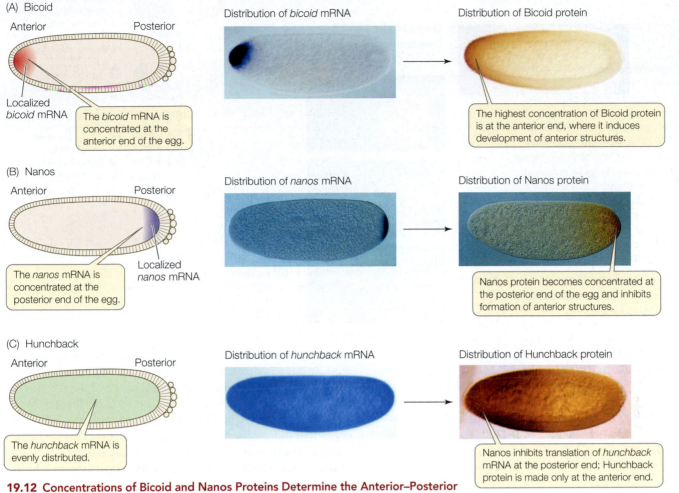

19.12 Concentrations of Bicoid and Nanos Proteins Determine the Anterior–Posterior Axis The anterior–posterior axis of *Drosophila* arises from gradients of the morphogens encoded by (A) *bicoid* and (B) *nanos*. Together, Bicoid and Nanos establish a concentration gradient of Hunchback (C).

SEGMENTATION GENES The number and polarity of the *Drosophila* larval segments are determined by the segmentation genes. These genes are expressed when there are about 6,000 nuclei in the embryo (about 3 hours after fertilization). Three classes of segmentation genes act one after the other to regulate finer and finer details of the segmentation pattern:

- **Gap genes** organize broad areas along the anterior–posterior axis. Mutations in gap genes result in gaps in the body plan—the omission of several consecutive larval segments.

- **Pair rule genes** divide the embryo into units of two segments each. Mutations in pair rule genes result in embryos missing every other segment.

- **Segment polarity genes** determine the boundaries and anterior–posterior organization of the individual segments. Mutations in segment polarity genes can result in segments in which posterior structures are replaced by reversed (mirror-image) anterior structures.

The expression of these genes is sequential (**Figure 19.13**). The products of the gap genes activate pair rule genes, and the pair rule gene products activate segment polarity genes. By the end of this cascade, nuclei throughout the embryo "know" which segment they will be part of in the adult fly.

The next set of genes in the cascade determines the form and function of each segment.

HOX GENES Hox (for "Homeobox") genes encode a family of transcription factors that are expressed in different combinations along the length of the embryo, and help determine cell fate within each segment. Hox gene expression tells the cells of a segment in the head to make eyes, those of a segment in the thorax to make wings, and so on. The *Drosophila* Hox genes occur in two clusters on chromosome 3, in the same order as the segments whose function they determine (**Figure 19.14**). By the time the fruit fly larva hatches, its segments are completely determined. Hox genes are shared by all animals and are homeotic genes—that is, a mutation in a Hox gene can result in one organ being replaced by another.

In *Drosophila*, the maternal effect genes, segmentation genes, and Hox genes interact to "build" a larva step by step, beginning with the unfertilized egg. How do we know that the Hox genes determine segment identity? A clue comes from

19.13 A Gene Cascade Controls Pattern Formation in the *Drosophila* Embryo
(A) Maternal effect genes induce gap, pair rule, and segment polarity genes—collectively referred to as segmentation genes. (B) Expression of two gap genes, *hunchback* (orange) and *Krüppel* (green), overlaps; both genes are transcribed in the yellow area. (C) The pair rule gene *fushi tarazu* is transcribed in the dark blue areas. (D) The segment polarity gene *engrailed* (bright green) is seen here at a slightly more advanced stage than is depicted in (A). By the end of this cascade, a group of nuclei at the anterior of the embryo, for example, is determined to become the first head segment in the adult fly.

Go to Animated Tutorial 19.3 Pattern Formation in the *Drosophila* Embryo
Life10e.com/at19.3

Go to Media Clip 19.1
Spectacular Fly Development in 3D
Life10e.com/mc19.1

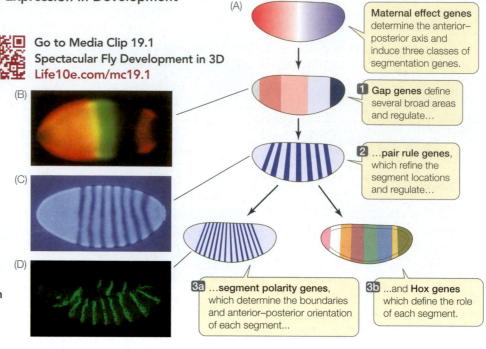

Maternal effect genes determine the anterior–posterior axis and induce three classes of segmentation genes.

1 Gap genes define several broad areas and regulate…

2 …pair rule genes, which refine the segment locations and regulate…

3a …segment polarity genes, which determine the boundaries and anterior–posterior orientation of each segment…

3b …and Hox genes which define the role of each segment.

homeotic mutations. A mutation in the Hox gene *antennapedia* causes legs to grow on the head in place of antennae (**Figure 19.15**). When another Hox gene, *ultrabithorax* is mutated, an extra pair of wings grows in a thoracic segment where wings do not normally occur (see Figure 20.3). So the normal (wild-type) functions of Hox genes must be to "tell" a segment what organ to form.

The *antennapedia* and *ultrabithorax* genes both encode transcription factors and have a common 180 base pair sequence called the **homeobox**. It encodes a 60 amino acid sequence called the **homeodomain**. The homeodomain recognizes and binds to a specific DNA sequence in the promoters of its target genes. This protein domain is found in transcription factors that regulate development in many other animals with an anterior–posterior axis. The evolutionary significance of these common pathways for development will be discussed in Chapter 20.

As we have seen, both plants and animals have homeotic genes that determine organ identity. However, the homeotic genes in plants and animals differ in DNA sequence and encoded protein structure. This is not surprising, given that the

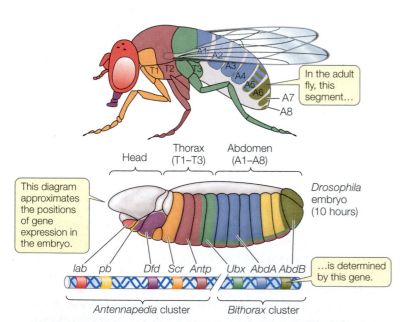

19.14 Hox Genes in *Drosophila* Determine Segment Identity
Two clusters of Hox genes on chromosome 3 (center) determine segment function in the adult fly (top). These genes are expressed in the embryo (bottom) long before the structures of the segments actually appear.

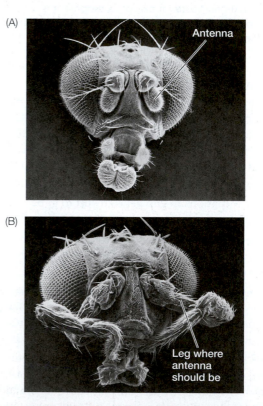

19.15 A Homeotic Mutation in *Drosophila* Mutations of the Hox genes cause body parts to form on inappropriate segments. (A) A wild-type fruit fly. (B) An *antennapedia* mutant fruit fly.

last common ancestor of plants and animals was unicellular, and therefore multicellularity evolved independently in plants and animals.

RECAP 19.4

Cascades of transcription factors govern pattern formation and the subsequent development of animal and plant organs. Often these transcription factors create or respond to morphogen gradients. In plants, cell fate is often determined by MADS box genes; in animal embryos, cell fate is determined in part by Hox genes.

- How is apoptosis crucial in shaping the developing embryo? **See p. 399**
- How do organ identity genes act in *Arabidopsis*? **See pp. 400–401 and Figure 19.9**
- List the key attributes of a morphogen. How does the Bicoid protein fit this definition? **See pp. 401–402 and Figure 19.12**
- How is segment identity established in the *Drosophila* embryo? **Review pp. 401–403 and Figure 19.13**

Is a mesophyll cell in a plant leaf or a liver cell in a human irrevocably committed to that specialization? Under the right experimental circumstances, differentiation is reversible in some cells. The next section will describe how some cells can be manipulated to express different sets of genes used in differentiation.

19.5 Is Cell Differentiation Reversible?

A zygote has the ability to give rise to every type of cell in the organism; in other words, it is totipotent. Its genome contains instructions for all of the structures and functions that will arise throughout the life cycle of the organism. Later in development, the cellular descendants of the zygote lose their totipotency and become determined. These determined cells then differentiate into specialized cells. The human liver cell and the leaf mesophyll cell generally retain their differentiated forms and functions throughout their lives. But this does not necessarily mean that they have irrevocably lost their totipotency. Most of the differentiated cells in an animal or plant have nuclei containing the entire genome of the organism, and they therefore have the genetic capacity for totipotency. We will explore here several examples of how this capacity has been demonstrated experimentally.

Plant cells can be totipotent

A carrot root cell normally faces a dark future. It cannot photosynthesize and generally does not give rise to new carrot plants. However, in 1958 Frederick Steward at Cornell University showed that if he isolated cells from a carrot root and maintained them in a suitable nutrient medium, he could induce them to dedifferentiate—to lose their differentiated characteristics. The cells could divide and give rise to masses of undifferentiated cells called calli (singular "callus"), which could be maintained in culture indefinitely. Furthermore, if they were provided with the right chemical cues, the cells could develop into embryos and eventually into complete new plants (**Figure**

INVESTIGATINGLIFE

19.16 Cloning a Plant   When cells were removed from a plant and put into a medium with nutrients and hormones, they lost many of their specialized features—in other words, they dedifferentiated. Did these cells retain the ability to differentiate again? Frederick Steward found that a cultured carrot cell did indeed retain the ability to develop into an embryo and a new plant.[a]

HYPOTHESIS   Differentiated plant cells can be totipotent and can be induced to generate all types of the plant's cells.

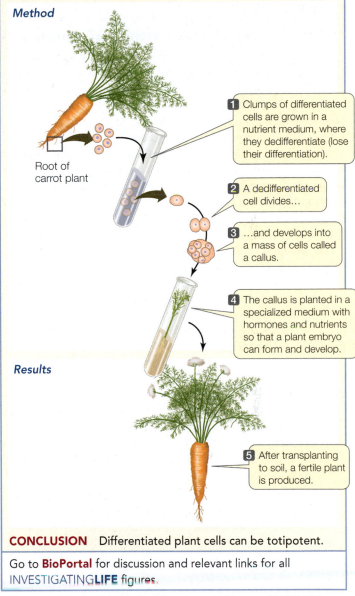

Method

Root of carrot plant

1 Clumps of differentiated cells are grown in a nutrient medium, where they dedifferentiate (lose their differentiation).

2 A dedifferentiated cell divides…

3 …and develops into a mass of cells called a callus.

4 The callus is planted in a specialized medium with hormones and nutrients so that a plant embryo can form and develop.

Results

5 After transplanting to soil, a fertile plant is produced.

CONCLUSION   Differentiated plant cells can be totipotent.

Go to **BioPortal** for discussion and relevant links for all INVESTIGATING**LIFE** figures.

[a]Steward, F. C., et al. 1958. *American Journal of Botany* 45: 705–709.

19.16). Since the new plants were genetically identical to the cells from which they came, they were clones of the original carrot plant.

The ability to clone an entire carrot plant from a differentiated root cell indicated that the cell contained a functional, complete carrot genome, and that under the right conditions, the cell and its descendants could express the appropriate genes in the right sequence to form a new plant. Many types of cells from other plant species show similar behavior in the laboratory. This ability to generate a whole plant from a single cell has been invaluable in agriculture and forestry. For example, trees from planted forests are used in making paper, lumber, and other products. To replace the trees reliably, forestry companies regenerate new trees from the leaves of selected trees with desirable traits. The characteristics of these clones are more uniform and predictable than those of trees grown from seeds.

Nuclear transfer allows the cloning of animals

Animal somatic cells cannot be manipulated as easily as plant cells can. However, experiments such as the one shown on page 395 have demonstrated the totipotency of early embryonic cells from animals. In humans this totipotency permits both genetic screening (see Section 15.4) and certain assisted reproductive technologies (see Section 43.4). A human embryo can be isolated in the laboratory and one or a few cells removed and examined to determine whether a certain genetic condition is present. Because of their totipotency, the remaining cells can develop into a complete embryo, which can be implanted into the mother's uterus, where it develops into a normal fetus and infant.

An isolated animal embryo cell generally won't develop into a complete organism, but the *nucleus* of such a cell has the genetic potential to do so. Nuclear transfer experiments have shown that the genetic information from a single animal cell can be used to create cloned animals. Robert Briggs and Thomas King performed the first such experiments in the 1950s using frog embryos. First they removed the nuclei from unfertilized eggs, forming enucleated eggs. Then, with very fine glass needles, they punctured cells from early embryos and drew up parts of their contents, including the nuclei. Each nucleus was injected into an enucleated egg. They stimulated the eggs to divide, and many went on to form embryos, and eventually frogs, which were clones from the original implanted nuclei. These experiments led to two important conclusions:

- No information is lost from the nuclei of cells as they pass through the early stages of embryonic development. This fundamental principle of developmental biology is known as genomic equivalence.

- The cytoplasmic environment around a cell nucleus can modify its fate.

More recent studies have demonstrated that a cell from a fully developed animal can be induced to dedifferentiate and give rise to an entire new individual. In 1996 Ian Wilmut and his colleagues at the Roslin Institute in Edinburgh cloned the first mammal by somatic cell nuclear transfer. This method involves the fusion of a somatic (nonreproductive) cell from an adult animal, containing the donor nucleus, with an enucleated egg. The fully differentiated donor cells were isolated from a Finn Dorset ewe's udder and starved of nutrients for a week, halting the cells in the G1 phase of the cell cycle. One of these cells was

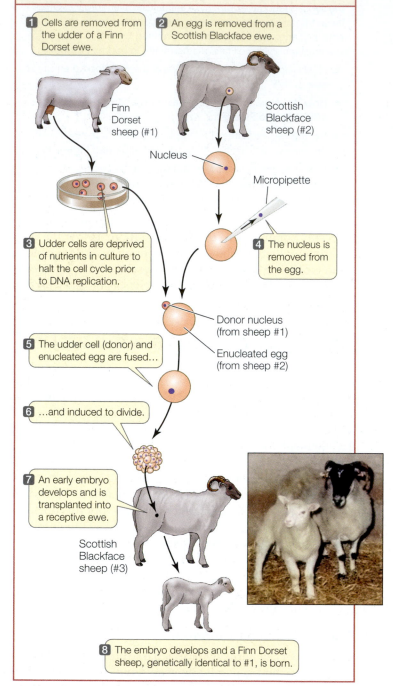

RESEARCHTOOLS

19.17 Cloning a Mammal   The experimental procedure described here produced the first cloned mammal, a Finn Dorset sheep named Dolly (shown on the left in the photo). As an adult, Dolly mated and subsequently gave birth to a normal offspring (the lamb on the right), thus proving the genetic viability of cloned mammals.

1 Cells are removed from the udder of a Finn Dorset ewe.

2 An egg is removed from a Scottish Blackface ewe.

Finn Dorset sheep (#1)

Scottish Blackface sheep (#2)

Nucleus

Micropipette

3 Udder cells are deprived of nutrients in culture to halt the cell cycle prior to DNA replication.

4 The nucleus is removed from the egg.

Donor nucleus (from sheep #1)

5 The udder cell (donor) and enucleated egg are fused…

Enucleated egg (from sheep #2)

6 …and induced to divide.

7 An early embryo develops and is transplanted into a receptive ewe.

Scottish Blackface sheep (#3)

8 The embryo develops and a Finn Dorset sheep, genetically identical to #1, is born.

fused with an enucleated egg from a Scottish Blackface ewe, and this fused cell began to divide. After several cell divisions, the resulting early embryo was transplanted into the womb of a surrogate mother. Eventually, a lamb named Dolly was born (**Figure 19.17**). Dolly showed all the characteristics of a Finn Dorset sheep: she carried the same genetic material as the nuclear donor, and thus was a clone of that donor.

WORKING WITHDATA:

Cloning a Mammal

Original Paper

Wilmut, I., A. E. Schnieke, J. McWhir, A. J. Kind, and K. H. S. Campbell. 1997. Viable offspring derived from fetal and adult mammalian cells. *Nature* 385: 810–813.

Analyze the Data

In 1997 Ian Wilmut and colleagues announced the first successful cloning of a mammal by somatic cell nuclear transfer (SCNT; see Figure 19.17). The team fused mammary epithelium cells from a donor sheep (a Finn Dorset) with enucleated eggs from sheep of a different breed (Scottish Blackface). The eggs were induced to divide, and the resulting embryos were implanted into the uteruses of recipient ewes (surrogate mother sheep; also Scottish Blackface). This resulted in the birth of one live lamb, Dolly, who was genetically identical to the ewe that donated the mammary cells. This work demonstrated that, under appropriate circumstances, animal cells are totipotent.

Aside from several ethical issues surrounding the idea of cloning mammals, the use of the SCNT technique itself raises scientific and medical concerns. One concern is the risk of premature aging. Initially Dolly appeared to be a healthy sheep, but at the age of four she developed severe arthritis, a condition usually associated with older animals. In 1999, research published in the journal Nature suggested that Dolly may have been susceptible to premature aging because of the shortened telomeres in her cells. (See Section 13.3 for a discussion of telomeres and their roles in DNA replication and aging.) Dolly was euthanized at age 6 years, which is approximately half the normal life span of a sheep.

QUESTION 1

In addition to mammary epithelium (ME) cells, Wilmut's team also attempted cloning by nuclear transfer from fetal fibroblasts (FB) and embryo-derived cells (EC). The results are shown in the table. What can you conclude about the efficiency of this cloning process?

QUESTION 2

Compare the efficiencies of cloning using nuclear donors from different sources. What can you conclude about the ability of different nuclei to be reprogrammed?

	Number of attempts that progressed to each stage		
Stage	ME	FB	EC
Egg fusions	277	172	385
Embryos transferred to recipient ewes	29	34	72
Pregnancies	1	4	14
Live lambs	1	2	4

QUESTION 3

Polymorphic DNA markers were used to analyze Dolly's genetic make-up. The data for four short tandem repeat (STR) markers (FCB 11, FCB 304, MAF 33, and MAF 209) are shown in the figure. (See Section 15.3 and Figure 15.14A for an explanation of STR analysis.) In the electrophoresis gels, different genotypes produce DNA bands of different sizes. A sample of Dolly's DNA was compared with samples from her nuclear donor (mammary cells from a Finn Dorset ewe) and from the recipient (her surrogate mother, a Scottish Blackface ewe). Are the DNA bands from Dolly the same sizes as those from her nuclear donor or from her surrogate mother? What does this indicate about Dolly's genetic makeup?

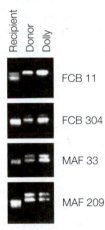

Go to BioPortal for all WORKING WITHDATA exercises

The production of Dolly demonstrated that a fully differentiated cell from a mature organism can revert to a totipotent state, and that this cell can be used to create a new animal. Many other animal species, including cats, dogs, horses, pigs, rabbits, and mice, have since been cloned by nuclear transfer. The cloning of animals has practical uses and has given us important information about developmental biology. There are several reasons to clone animals:

- *Expansion of the numbers of valuable animals*: One goal of Wilmut's experiments was to develop a method for cloning transgenic animals with useful phenotypes. For example, as we mentioned in Section 18.6, a cow was genetically engineered to make human growth hormone in her milk. This animal was then cloned to produce additional cows that do the same thing. Only 15 such cows would supply the world's need for this medication, which is used to treat short stature caused by growth hormone deficiency.

- *Preservation of endangered species*: The banteng, a relative of the cow, was the first endangered animal to be cloned. An enucleated egg from a cow, and a cow surrogate mother were used. Cloning may be the only way to save some endangered species with low rates of natural reproduction, such as the giant panda.

- *Perpetuation of pets*: Many people get great personal benefit from pets, and the death of a pet can be devastating. Companies have been set up to clone cats and dogs from cells provided by their owners. Of course, the behavioral characteristics of the beloved pet, which are certainly derived in part from the environment, may not be the same in the cloned pet as in its genetic parent.

19.18 Stem Cell Transplantation Multipotent stem cells can be used in hematopoietic stem cell transplantation to replace stem cells destroyed by cancer therapy.

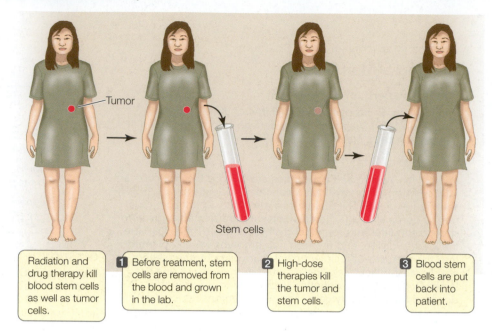

Tumor

Stem cells

Radiation and drug therapy kill blood stem cells as well as tumor cells.

1 Before treatment, stem cells are removed from the blood and grown in the lab.

2 High-dose therapies kill the tumor and stem cells.

3 Blood stem cells are put back into patient.

Multipotent stem cells differentiate in response to environmental signals

Stem cells have been mentioned several times already in this chapter (for example, in the opening story). **Stem cells** are rapidly dividing, undifferentiated cells that can differentiate into diverse cell types.

In plants, stem cells occur in the meristem. In general, plants have far fewer (15–20) broad cell types than animals (as many as 200). In mammals, stem cells are found in adult tissues that need frequent cell replacement, such as the skin, the inner lining of the intestine, and the bone marrow, where blood cells and other types of cells are formed. Canadian cell biologists Ernest McCulloch and James Till discovered mammalian stem cells in the early 1960s when they injected bone marrow cells into adult mice. They noticed that the recipient mice developed small clumps of tissue in their spleens. When they looked more carefully at the clumps, they found that each was composed of undifferentiated stem cells. Before this, stem cells were believed to be present only in animal embryos.

The stem cells found in adult animals are not totipotent, because their ability to differentiate is limited to a relatively few cell types. In other words, they are multipotent (see Section 19.1). For example, there are two types of multipotent stem cells in the bone marrow. Hematopoietic stem cells produce red and white blood cells, whereas mesenchymal stem cells (like those used to treat Bartolo Colon's injury in the opening story) produce bone and connective tissues, including muscle.

The proliferation and differentiation of multipotent stem cells are "on demand." For example, hematopoietic stem cells proliferate in the bone marrow in response to growth factors, and the extra stem cells are released into the blood. This is the basis of an important therapy called hematopoietic stem cell transplantation (**Figure 19.18**). Some cancer treatments kill all dividing cells in the body, so hematopoietic stem cells can be depleted in patients given these treatments. To circumvent this problem, stem cells are harvested from the bone marrow or blood of the patient (prior to cancer treatment) or of a donor, and then the cells are injected back into the patient after cancer treatment.

Signals from adjacent cells can stimulate stem cell differentiation. Many controlled experiments have shown that damaged animal tissues (for example, hearts and tendons) that are injected with stem cells can heal more effectively than tissues that don't receive this treatment. The mechanisms by which this occurs are still not clear. There is some evidence that the injected cells can actually insert themselves into the damaged tissue and differentiate into new cells of that tissue. Alternatively, injected cells may secrete growth factors and other molecules that induce the cells in the surrounding tissue to regenerate into healthy tissue. Whatever the mechanisms by which multipotent stem cells contribute to the healing of damaged tissues, their use in treating diseases is very promising.

Pluripotent stem cells can be obtained in two ways

As stated earlier, totipotent stem cells that can differentiate into any cell type are found only in very early embryos. In both mice and humans, a slightly later embryonic stage is a hollow sphere of cells called a **blastocyst** (see Figure 44.4). A group of cells within the blastocyst is pluripotent: they can differentiate into most cell types but cannot give rise to a complete organism. In mice, these **embryonic stem cells** (**ESCs**) can be removed from the blastocyst and grown in laboratory culture almost indefinitely, if provided with the right conditions. When cultured mouse ESCs are injected back into another mouse blastocyst, the stem cells mix with the resident cells and differentiate to form all the cell types in the mouse. This indicates that the ESCs do not lose any of their developmental potential while growing in the laboratory.

ESCs growing in the laboratory can also be induced to differentiate in a particular way if the right signal is provided (**Figure 19.19A**). For example, treatment of mouse ESCs with a derivative of vitamin A causes them to form neurons, whereas other growth factors induce them to form blood cells. Such experiments demonstrate both the cells' developmental potential and the roles of environmental signals. This finding raises the possibility of using ESC cultures as sources of differentiated cells to repair specific tissues, such as a damaged pancreas in diabetes, or a brain that malfunctions in Parkinson's disease.

ESCs can be harvested from human embryos conceived by in vitro ("under glass"—in the laboratory) fertilization, with the consent of the donors. Since more than one embryo is usually conceived in this procedure, embryos not used for reproduction might be available for embryonic stem cell isolation. These cells could then be grown in the laboratory and used as sources of tissues for transplantation into patients with tissue damage. There are two problems with this approach:

- Some people object to the destruction of human embryos for this purpose.

- The stem cells, and tissues derived from them, would provoke an immune response in a recipient (see Chapter 42).

Shinya Yamanaka and coworkers at Kyoto University in Japan developed another way to produce pluripotent stem cells that gets around these two problems (**Figure 19.19B**). Instead of extracting ESCs from blastocysts, they make pluripotent stem cells from skin cells. They developed this method systematically:

1. First they used microarrays (see Figure 18.8) to compare the genes expressed in ESCs with non-stem cells. They found several genes that were uniquely expressed at high levels in ESCs. These genes were believed to be essential to the undifferentiated state and function of stem cells.

2. Next they isolated the genes, coupled them to highly expressing promoters, and inserted them into skin cells (see Section 18.5). They found that the skin cells now expressed the newly added genes at high levels.

3. Finally, they showed that the altered skin cells were pluripotent and could be induced to differentiate into many tissues. They called these cells **induced pluripotent stem cells (iPS cells)**.

Because the iPS cells can be made from skin cells of the individual who is to be treated, an immune response may be avoided. Such cells have already been used for cell therapy in animals for diseases similar to human Parkinson's disease (a brain disorder), diabetes, and sickle cell anemia. If it can be

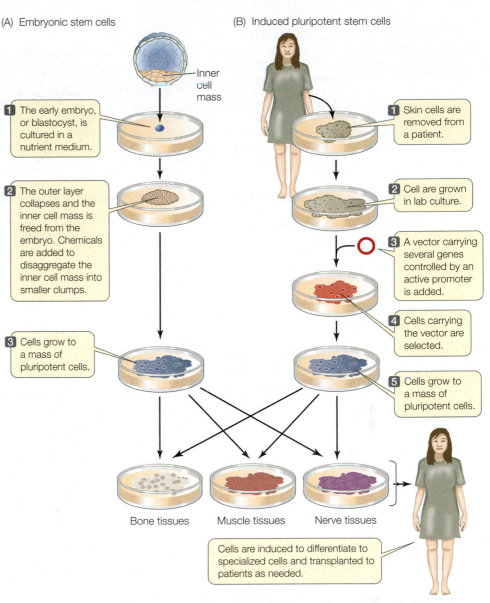

(A) Embryonic stem cells

Inner cell mass

1 The early embryo, or blastocyst, is cultured in a nutrient medium.

2 The outer layer collapses and the inner cell mass is freed from the embryo. Chemicals are added to disaggregate the inner cell mass into smaller clumps.

3 Cells grow to a mass of pluripotent cells.

(B) Induced pluripotent stem cells

1 Skin cells are removed from a patient.

2 Cell are grown in lab culture.

3 A vector carrying several genes controlled by an active promoter is added.

4 Cells carrying the vector are selected.

5 Cells grow to a mass of pluripotent cells.

Bone tissues   Muscle tissues   Nerve tissues

Cells are induced to differentiate to specialized cells and transplanted to patients as needed.

19.19 Two Ways to Obtain Pluripotent Stem Cells Pluripotent stem cells can be obtained either from human embryos (A) or by adding highly expressed genes to skin cells to transform them into stem cells (B).

Go to Animated Tutorial 19.4
Embryonic Stem Cells
Life10e.com/at19.4

shown conclusively that iPS cells have the same properties as ESCs, human uses are sure to follow.

RECAP 19.5

Differentiated cells retain their ability to differentiate into other cell types, given appropriate chemical signals. This has made cloning and stem cell technologies possible.

- How are stem cells found in adult body tissues different from embryonic stem cells? **See p. 408**

- What are the two ways to produce pluripotent stem cells? **See pp. 408–409 and Figure 19.19**

What are other uses of stem cells derived from fat?

ANSWER

In the United States, veterinarians use multipotent stem cells derived from fat to treat injuries and osteoarthritis in animals. In these cases, stem cell injections improve the healing of tendon injuries in horses and ease the symptoms of osteoarthritis in dogs. A procedure has been developed to isolate large quantities of stem cells from human patients in the operating room. These can be used immediately to repair tissues, for example, after surgery for breast cancer.

CHAPTER**SUMMARY** 19

19.1 What Are the Processes of Development?

- A multicellular organism begins its development as an embryo. A series of embryonic stages precedes the birth of an independent organism. **Review Figure 19.1, ACTIVITY 19.1**

- The processes of development are **determination**, **differentiation**, **morphogenesis**, and **growth**.

- Differential gene expression is responsible for the differences among cell types. **Cell fate** is determined by environmental factors, such as the cell's position in the embryo, as well as by intracellular influences. **Review Figure 19.2, ANIMATED TUTORIAL 19.1**

- Determination is followed by differentiation, the actual changes in biochemistry, structure, and function that result in cells of different types. Determination is a commitment; differentiation is the realization of that commitment.

- Over the course of development, embryo cells decrease in **cell potency**. **Totipotent** cells (such as a zygote) are capable of forming every cell type in the adult body. **Pluripotent** cells can give rise to most cell types, **multipotent** cells to several cell types, and **unipotent** cells to only one cell type.

19.2 How Is Cell Fate Determined?

- **Cytoplasmic segregation**—the unequal distribution of **cytoplasmic determinants** in the egg, zygote, or early embryo—can establish **polarity** and lead to cell fate determination. **Review Figure 19.3, ANIMATED TUTORIAL 19.2**

- **Induction** is a process by which embryonic animal tissues direct the development of neighboring cells and tissues by secreting chemical signals, called **inducers**. **Review Figures 19.4, 19.5**

19.3 What Is the Role of Gene Expression in Development?

- Inducers act through signaling pathways to determine cell fate. **Review Figure 19.6**

- Differential gene expression results in cell differentiation. Transcription factors are especially important in regulating gene expression during differentiation.

- Complex interactions of many genes and their products are responsible for differentiation during development. **Review Figure 19.7**

19.4 How Does Gene Expression Determine Pattern Formation?

- **Pattern formation** is the process that results in the spatial organization of a tissue or organism.

- During development, selective elimination of cells by apoptosis results from the expression of specific genes. **Review Figure 19.8**

- Sepals, petals, stamens, and carpels form in plants as a result of combinatorial interactions between transcription factors encoded by **organ identity genes**. **Review Figure 19.9**

- The transcription factors encoded by floral organ identity genes contain an amino acid sequence called the **MADS box** that can bind to DNA.

- Both plants and animals use **positional information** as a basis for pattern formation. Positional information usually comes in the form of a signal called a **morphogen**. Different concentrations of the morphogen cause different effects. **See Figures 19.10, 19.11**

- In the fruit fly *D. melanogaster*, a cascade of transcriptional activation sets up the axes of the embryo, the development of the segments, and finally the determination of cell fate in each segment. The cascade involves the sequential expression of maternal effect genes, gap genes, pair rule genes, segment polarity genes, and Hox genes. **Review Figures 19.13, 19.14, ANIMATED TUTORIAL 19.3**

- Hox genes help determine cell fate in the embryos of all animals. The **homeobox** is a DNA sequence found in Hox genes and other genes that code for transcription factors. The sequence of amino acids encoded by the homeobox is called the **homeodomain**.

19.5 Is Cell Differentiation Reversible?

- The ability to create clones from differentiated cells demonstrates the principle of genomic equivalence. **Review Figures 19.16, 19.17**

- **Stem cells** produce daughter cells that differentiate when provided with appropriate intercellular signals. Some multipotent stem cells in the adult body can differentiate into a limited number of cell types to replace dead cells and maintain tissues. **Review Figure 19.18**

- **Embryonic stem cells** are pluripotent and can be cultured in the laboratory. Under suitable environmental conditions, these cells can differentiate into almost any tissue type. **Induced pluripotent stem cells** have similar characteristics. This has led to technologies designed to replace cells or tissues damaged by injury or disease. **Review Figure 19.19, ANIMATED TUTORIAL 19.4**

 Go to the Interactive Summary to review key figures, Animated Tutorials, and Activities
Life10e.com/is19

CHAPTER**REVIEW**

▬▬ REMEMBERING

1. Which statement about determination is true?
   a. Differentiation precedes determination.
   b. All cells are determined after two cell divisions in most organisms.
   c. A determined cell will keep its determination no matter where it is placed in an embryo.
   d. A cell changes its appearance when it becomes determined.
   e. A differentiated cell has the same pattern of transcription as a determined cell.

2. Cloning experiments on sheep, frogs, and mice have shown that
   a. nuclei of adult cells are not pluripotent.
   b. nuclei of embryonic cells can be totipotent.
   c. nuclei of differentiated cells have different genes than zygote nuclei have.
   d. differentiation is fully reversible in all cells of a frog.
   e. differentiation involves permanent changes in the genome.

3. The term "induction" describes a process in which a cell or cells
   a. influence the development of another cell or group of cells.
   b. trigger cell movements in an embryo.
   c. stimulate the transcription of their own genes.
   d. organize the egg cytoplasm before fertilization.
   e. inhibit the movement of the embryo.

4. Which statement about induction is *not* true?
   a. One group of cells induces adjacent cells to develop in a certain way.
   b. It triggers a sequence of gene expression in target cells.
   c. Single cells cannot form an inducer.
   d. A tissue may be induced as well as make an inducer.
   e. The chemical identification of specific inducers has not been achieved.

5. Homeotic mutations
   a. are often severe and result in structures at inappropriate places.
   b. cause subtle changes in the forms of larvae or adults.
   c. occur only in prokaryotes.
   d. do not affect the animal's DNA.
   e. are confined to the zone of polarizing activity.

6. Which statement about the homeobox is *not* true?
   a. It is transcribed and translated.
   b. It is found only in animals.
   c. Proteins containing the homeodomain bind to DNA.
   d. It is a sequence of DNA shared by more than one gene.
   e. It occurs in Hox genes.

▬▬ UNDERSTANDING & APPLYING

7. In fruit flies, the following genes are used to determine segment polarity: (k) gap genes; (l) Hox genes; (m) maternal effect genes; (n) pair rule genes. In what order are these genes expressed during development?
   a. k, l, m, n
   b. l, k, n, m
   c. m, k, n, l
   d. n, k, m, l
   e. n, m, k, l

8. Molecular biologists can attach genes to active promoters and insert them into cells (see Section 18.5). What would happen if the following were inserted and overexpressed? Explain your answers.
   a. *ced-9* in embryonic neuron precursors of *C. elegans*
   b. *myoD* in undifferentiated myoblasts
   c. the gene for Sonic hedgehog in a chick limb bud
   d. *nanos* at the anterior end of the *Drosophila* embryo

9. A powerful method to test for the function of a gene in development is to generate a "knockout" organism, in which the gene in question is inactivated (see Section 18.4). What do you think would happen in each of the following cases?
   a. a knockout of *ced-9* in *C. elegans*
   b. a knockout of *nanos* in *Drosophila*

10. If you wanted a rose flower with only petals, what kind of homeotic mutation would you seek in the rose genome?

▬▬ ANALYZING & EVALUATING

11. During development, an animal cell's potential for differentiation becomes ever more limited. In the normal course of events, most cells in the adult animal have the potential to be only one or a few cell types. On the basis of what you have learned in this chapter, discuss possible mechanisms for the progressive limitation of the cell's potential.

12. Cloning involves considerable reprogramming of gene expression in a differentiated cell so that it acts like an egg cell. How would you investigate this reprogramming?

Go to BioPortal at **yourBioPortal.com** for Animated Tutorials, Activities, LearningCurve Quizzes, Flashcards, and many other study and review resources.

20 Genes, Development, and Evolution

CHAPTER**OUTLINE**

20.1 How Can Small Genetic Changes Result in Large Changes in Phenotype?

20.2 How Can Mutations with Large Effects Change Only One Part of the Body?

20.3 How Can Developmental Changes Result in Differences among Species?

20.4 How Can the Environment Modulate Development?

20.5 How Do Developmental Genes Constrain Evolution?

A N ICONIC ILLUSTRATION of Charles Darwin's theory of evolution by natural selection is the finches of the Galápagos. Each island in the archipelago is home to a combination of species of these small, dull-colored birds. Their beaks are of different shapes and sizes, ranging from the thick, short, strong beaks of seed-eating species to the thin, long beaks of insect-eating species. Darwin wrote that "Seeing this gradation and diversity of structure in one small intimately related group of birds, one might really fancy that from an original paucity of birds in this archipelago, one species had been taken and modified for different ends."

Darwin had no idea of the genetic basis for such modification. Now we do, but recent discoveries have changed much of our thinking about certain aspects of evolutionary genetics. Based on Mendelian genetics and the central dogma that genes code for proteins, most explanations of evolutionary change have focused on the effects of gene mutations on the structural proteins that make up individual organisms. We now know that mechanisms of evolution can depend as much on changes in the regulatory sequences that control the timing, amount, and location of gene expression as on mutations that change the structural genes themselves.

The beak of the finch develops from tissues at the anterior of the embryo that will form the facial bones. Cell divisions in this embryonic tissue are controlled by signaling proteins, one of which is called bone morphogenetic protein 4 (BMP4); another is the protein calmodulin. If BMP4 is present early and in large amounts, the beak becomes broad and deep. If

Beak Diversity Can Evolve through Changes in Development Genes expressed during embryonic development affect beak length and depth in birds. Changes in the expression patterns of these genes contributed to the evolution of the distinctive beaks of the Australian pelican (*Pelecanus conspicillatus*) and the silver gull (*Chroicocephalus novaehollandiae*).

calmodulin is present early and in large amounts, the beak grows longer and thinner. Thus beak structure is affected by changes in the timing of protein production as well as the amount of protein made.

Beak differences among Galápagos finches are impressive, but consider the diverse beaks seen in other birds—hummingbirds, flamingos, toucans, and pelicans, to name only a few. Could such dramatic differences in beak size and shape also be a result of changes in expression patterns of BMP4 and calmodulin? The realization that major evolutionary change can be the result of subtle changes in spatial, temporal, and quantitative distribution of signaling molecules and changes in the noncoding regions of the DNA that control gene expression has revolutionized the study of evolution, producing the new field of evolutionary developmental biology.

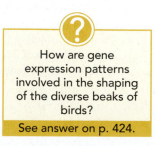

? How are gene expression patterns involved in the shaping of the diverse beaks of birds?

See answer on p. 424.

How Can Small Genetic Changes Result in Large Changes in Phenotype?

Genetic mutations are the source of variation for evolution. But most genetic mutations change only a single nucleotide in a very large genome. How can such small changes lead to the huge diversity we see in living organisms? This question has become acute in recent years with the realization that the actual number of protein-coding genes in many species is many fewer than expected (only about 21,000 in humans, for example). Furthermore, the genomes of related species that appear very different can be very similar, as is the case for the genomes of humans and chimpanzees, which are more than 95 percent identical. Part of the answer to this question has been found in the many and subtle ways that gene expression is controlled in time, space, and amount during development. Study of the relationship between development and evolution has given rise to the new field of **evolutionary developmental biology**, or **evo-devo**. The basic principles of evo-devo are:

- Organisms share similar molecular mechanisms for development that include a "toolkit" of regulatory molecules that control the expression of genes.

- Toolkit regulatory molecules are able to act independently in different tissues and regions of the body, enabling modular evolutionary change.

- Developmental differences can arise from changes in the timing of regulatory molecule action, the location of its action, or the quantity of its action.

- Differences among species can arise from alterations in the expression of developmental genes.

- Developmental changes can arise from environmental influences on developmental processes.

The development of a multicellular organism from a fertilized egg—a single cell—involves an intricate pattern of sequential gene expression. When developmental biologists began to describe the events responsible for the differentiation and controlled proliferation of cells and tissues at the molecular level, they found common regulatory genes and pathways in organisms that don't appear similar at all, such as those that direct eye development in fruit flies and mice.

Developmental genes in distantly related organisms are similar

About a dozen major kinds of eyes are found among the different animals, including the camera-like eyes of humans and the compound eyes of insects (see Section 46.4). Although the eyes of insects and vertebrates evolved independently, a remarkable discovery in the 1990s showed that common developmental pathways were involved in the origin of eyes in both groups. Swiss developmental biologists Rebecca Quiring and Walter Gehring were using lines of mutant fruit flies to identify the transcription factors involved in fly development. One of these mutations, appropriately named *eyeless*, resulted in flies with no eyes. Quiring and Gehring isolated the protein product of the *eyeless* gene, which they determined was a transcription factor that controls the genes responsible for eye development. By making recombinant DNA constructs that allowed the *eyeless* gene to be expressed in different embryonic tissues of transgenic flies, they were able to produce flies with extra eyes on various body parts such as the legs, the antennae, and under the wings.

A big surprise came when database searches revealed that the *eyeless* gene sequence was similar to that of the *Pax6* gene in mice; *Pax6*, when mutated, leads to the development of abnormally small eyes. Could the extremely different eyes of flies and mice be variations on a common developmental theme? To test for functional similarity between the insect and mammalian genes, Gehring and colleagues repeated their experiments on flies, but using the mouse *Pax6* gene instead of the fly *eyeless* gene. Once again, eyes developed at various sites on the transgenic flies. Thus a gene whose expression normally leads to the development of a mammalian eye now led to the development of the very different insect eye.

We have to look very far back in evolutionary time for a common ancestor of fruit flies and mice. Yet the *eyeless* and *Pax6* genes contain sequences that are highly conserved, not only in these two model species but in others as well (**Figure 20.1**). Biologists call such genes **homologous**, meaning that they evolved from a gene present in a common ancestor. In recent years a large number of homologous genes (the regulatory "toolkit") have been shown to control development in distantly related species.

An even more dramatic example of homology in genes that control development in animals is the Hox gene cluster discussed in Chapter 19. This set of genes codes for transcription factors that provide positional information and control the pattern of development in the different body segments of the embryo (see Figure 19.14). The genes in the Hox cluster share a homologous sequence, the homeobox. The similarity among the repeated gene sequences of the homeobox suggests that the Hox genes arose through duplication of an ancestral gene, which then diverged to take on new functions (see Section 24.3). Duplication originally provided protection from loss-of-function mutations in these critical genes. One copy of the gene could provide the critical function even if the other copy acquired mutations that altered its function. New functions in duplicated genes could then be favored through natural selection. Thus duplication and divergence allowed for "evolutionary experiments" using sets of duplicated genes such as the Hox cluster.

The duplication-and-divergence hypothesis derives support from the deep evolutionary history (that is, very early divergences on the tree of life) of the Hox genes. Two clusters of Hox genes are found in cnidarians (jellyfish and their relatives), one expressed in the anterior of the larvae and one expressed in the posterior of the larvae. The number of Hox genes increases among animals that have bilateral symmetry (see Section 31.2) as complexity of the patterning of the body axis increases. The increase in Hox genes reaches a maximum in vertebrates, where there are not only more Hox genes per cluster but there are four clusters, each on a different chromosome. Comparing

Mouse *Pax6* gene:

| DNA | GTATCCAACGGTTGTGTGAGTAAAATTCTGGGCAGGTATTACGAGACTGGCTCCATCAGA |
| Amino acids | V S N G C V S K I L G R Y Y E T G S I R |

Fly *eyeless* gene:

| 77% | GTATCAAATGGATGTGTGAGCAAAATTCTCGGGAGGTATTATGAAACAGGAAGCATACGA |
| 100% | V S N G C V S K I L G R Y Y E T G S I R |

Shark eye control gene:

| 85% | GTGTCCAACGGTTCTGTCAGTAAAATCCTGGGCAGATACTATGAAACAGGATCCATCAGA |
| 100% | V S N G C V S K I L G R Y Y E T G S I R |

Squid eye control gene:

| 78% | GTCTCCAACGGCTGCGTTAGCAAGATTCTCGGACGGTACTATGAGACGGGCTCCATAAGA |
| 100% | V S N G C V S K I L G R Y Y E T G S I R |

20.1 DNA Sequence Similarity in Eye Development Genes Genes controlling eye development contain regions that are highly conserved, even among species with very different eyes. These sequences, from a conserved region of the *Pax6* gene and its homologs in other species, are similar at the DNA level (top sequence in each pair) and identical at the amino acid level (bottom sequence). The percentages beside the sequences represent the percent match with the corresponding DNA and protein sequences in the mouse.

these gene clusters in such distantly related species as *Drosophila* and mice, the genes are arranged in similar clusters and are expressed in similar patterns along the body axes of their embryos (**Figure 20.2**).

The homology of the Hox genes across animal species is remarkable. Over the millions of years that have elapsed since cnidarians, insects, and vertebrates last shared a common ancestor, the genes for these transcription factors have been conserved. This conservation extends beyond similarities of gene sequence and organization to similarities in protein structure and function during development. These similarities led biologists to form one of the fundamental principles of evo-devo: Shared developmental mechanisms controlled by specific DNA sequences comprise a **genetic toolkit** that has been modified and reshuffled to produce the remarkable diversity of plants, animals, and other organisms we know today. Small changes in the application of the genetic toolkit—when, where, and how much the transcription factor genes are expressed—influence the development of the organism and produce the variation upon which natural selection can work.

20.2 Regulatory Genes Show Similar Expression Patterns Homologous genes encoding similar transcription factors are expressed in similar patterns along the anterior–posterior axes of both insects and vertebrates. The mouse (and human) Hox genes are present in multiple copies; this prevents a single mutation from having drastic effects.

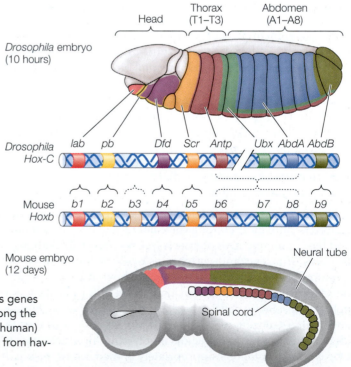

RECAP 20.1

Changes in development underlie many changes in morphology that result in evolution of body form and function. Study of the evolutionary aspects of development has revealed that a genetic toolkit consisting of highly conserved regulatory genes governs pattern formation in multicellular organisms.

- Describe what evolutionary developmental biologists mean by a genetic toolkit. **See pp. 413 and 414**
- How does the story of the eye-determining genes *Pax6* and *eyeless* support the claim that genes controlling development are highly conserved? **See p. 413 and Figure 20.1**
- Using Hox genes as an example, explain how duplication and divergence have enabled the evolution of complexity in body form. **See pp. 413–414 and Figure 20.2**

We saw in Chapter 19 that many developmental mutations in fruit flies result in striking abnormalities that affect only a single structure, segment, or region (e.g., a head segment that forms a leg; see Figure 19.15). The rest of the embryo is often unaffected. How is this possible?

20.2 How Can Mutations with Large Effects Change Only One Part of the Body?

Development involves interactions between gene products that produce a sequence of transcriptional events resulting in differential gene expression. The examples of homeotic mutations revealed that these processes can be localized in different tissues or regions of the embryo. Thus the embryo can be thought of as consisting of **modules** that encompass both the genes and various signaling pathways that determine specific physical structures such as eyes, body segments, wings, or legs. Because developmental genes can be controlled separately in the different modules, structures that arise from different modules can change independently of one another in both developmental and evolutionary time.

 Go to Animated Tutorial 20.1
Modularity
Life10e.com/at20.1

Genetic switches govern how the genetic toolkit is used

Developmental modules based on a common set of genetic instructions can evolve separately within a species because **genetic switches** control how the toolkit is used. These switches include gene promoters and the transcription factors that bind to promoters, as well as the enhancers and repressors that can modulate the interactions of transcription factors and promoters (see Section 16.2). The signal cascades resulting from the interaction of transcription factors, enhancers, and repressors with gene promoters determine where, when, and to what extent genes are turned on and off. Multiple switches control each gene, creating different expression patterns in different

locations. In this way, elements of the genetic toolkit can be involved in multiple developmental processes and still allow individual modules to develop and evolve independently.

Genetic switches integrate positional information in the developing embryo and play key roles in determining the developmental pathways of different modules. In *Drosophila*, for example, each Hox gene codes for a transcription factor that is expressed in a particular segment or appendage of the developing fly. The pattern and function of each segment depends on the unique Hox gene or combination of Hox genes that are expressed in that segment. As an example, let's look at wing development in insects.

Drosophila species are members of the insect group Diptera, which means "two wings"—that is, they have a single pair of wings, whereas most insects have two pairs of wings (i.e., four wings). The single pair of wings of dipterans develops on the second thoracic segment, where the Hox gene *antennapedia* (*Antp*) is expressed. *Antp* is also expressed in the third thoracic segment, but in that segment a pair of balancing organs called halteres develops in dipterans. A critical difference between thoracic segments 2 and 3 is that another Hox gene, *ultrabithorax* (*Ubx*), is expressed along with *Antp* in segment 3. *Ubx* represses the expression of *Antp* in dipterans. If *Ubx* is inactivated by mutation, a second pair of wings forms in thoracic segment 3, as is typical of many other insect groups (**Figure 20.3**). Thus some major morphological differences among groups of animals can result from relatively small changes in gene expression.

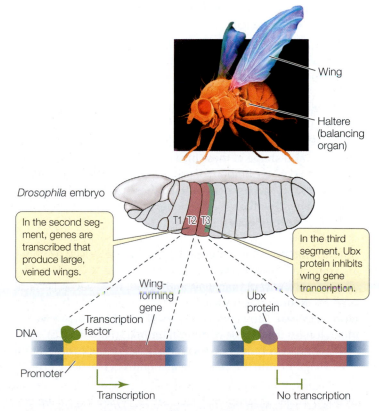

20.3 Segments Differentiate under Control of Genetic Switches The binding of a single protein, Ultrabithorax (Ubx), determines whether a thoracic segment produces full wings or halteres.

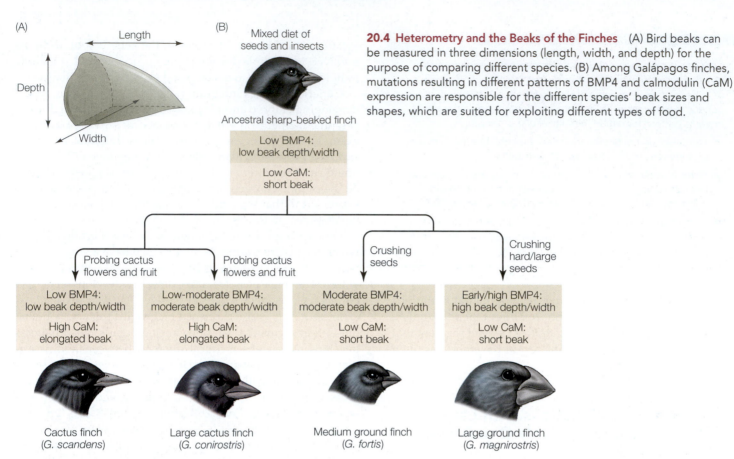

20.4 Heterometry and the Beaks of the Finches (A) Bird beaks can be measured in three dimensions (length, width, and depth) for the purpose of comparing different species. (B) Among Galápagos finches, mutations resulting in different patterns of BMP4 and calmodulin (CaM) expression are responsible for the different species' beak sizes and shapes, which are suited for exploiting different types of food.

Modularity allows for differences in the patterns of gene expression

Because of modularity, transcription factors and other genes of the toolkit can regulate the expression of structural genes in different amounts, at different times, and in different locations.

HETEROMETRY We saw an example of **heterometry** ("different measure") at the opening of this chapter when we described beak development in Galápagos finches. Studies revealed that beak size and shape in these birds is influenced by the level of expression (i.e., the amount of protein produced) of two regulatory genes. The relative amounts of these proteins determines whether an individual's beak is long, thin, and narrow or short, thick, and deep (**Figure 20.4**).

HETEROCHRONY The evolution of the giraffe's neck provides an example of **heterochrony** ("different time"). Giraffes, like all mammals except manatees and sloths, have seven cervical (neck) vertebrae. So giraffes did not get longer necks by adding more vertebrae. However, the cervical vertebrae of giraffes are much longer than those of other mammals (**Figure 20.5**).

Bone growth in mammals is the result of the proliferation of cartilage-producing cells called chondrocytes. Bone growth

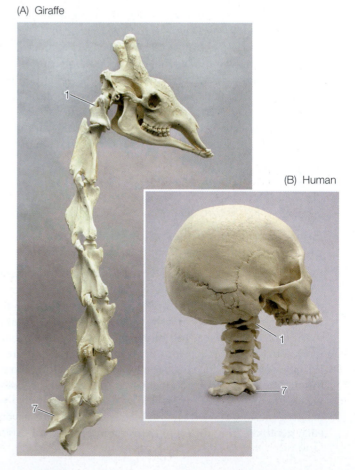

20.5 Heterochrony in the Development of a Longer Neck There are seven vertebrae in the neck of the giraffe (left) and human (right; not to scale). But the vertebrae of the giraffe are much longer (25 cm compared to 1.5 cm) because during development, growth continues for a longer period of time. This timing difference is called heterochrony.

Chick hindlimb   Duck hindlimb

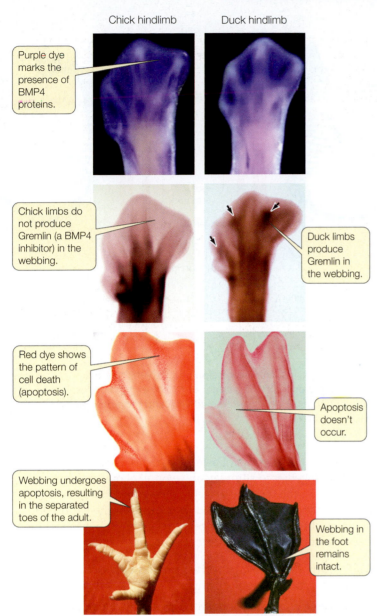

Purple dye marks the presence of BMP4 proteins.

Chick limbs do not produce Gremlin (a BMP4 inhibitor) in the webbing.

Duck limbs produce Gremlin in the webbing.

Red dye shows the pattern of cell death (apoptosis).

Apoptosis doesn't occur.

Webbing undergoes apoptosis, resulting in the separated toes of the adult.

Webbing in the foot remains intact.

20.6 Changes in Gremlin Expression Correlate with Changes in Hindlimb Structure The left column of photos shows foot development in a chicken; the right column shows foot development in a duck. Gremlin protein in the webbing of the duck foot inhibits BMP4 signaling, thus preventing the embryonic webbing from undergoing apoptosis.

is stopped by a genetic signal that results in apoptosis, or cell death, of chondrocytes and calcification of the bone matrix (see Section 48.3). In giraffes this signaling process is delayed in the cervical vertebrae, so that these vertebrae grow longer. Thus the evolution of longer necks resulted from *changes in the timing of expression* of the genes that control bone formation.

HETEROTOPY Spatial differences in the expression of a developmental gene are known as **heterotopy** ("different place"), exemplified by the different development of feet in ducks and chickens. The feet of all bird embryos have webs of skin that connect their toes. This webbing is retained in adult ducks (and other aquatic birds) but not in adult chickens (and other

INVESTIGATINGLIFE

20.7 Changing the Form of an Appendage Ducks have webbed feet and chickens do not—a major difference in the adaptations of these species. Webbing is initially present in the chick embryo, but undergoes apoptosis that is stimulated by the protein BMP4. In ducks another protein, Gremlin, binds to BMP4 and inhibits it, preventing apoptosis and resulting in webbed feet. Juan Hurle and his colleagues at the Universidad de Cantabria in Spain asked what would happen if Gremlin were put onto a developing chick foot.[a]

HYPOTHESIS Adding Gremlin protein (a BMP4 inhibitor) to a developing chicken foot will transform it into a ducklike foot.

Method Open a small window in chick egg shell and carefully add Gremlin-secreting beads to the webbing of embryonic chicken hindlimbs. Add beads that do not contain Gremlin to other hindlimbs (controls). Close the eggs and observe limb development.

Results In the hindlimbs in which Gremlin was secreted, the webbing does not undergo apoptosis, and the hindlimb resembles that of a duck. The control hindlimbs develop the normal chicken form.

Control   Gremlin added

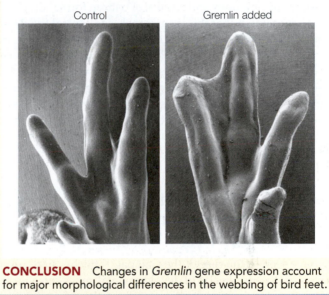

CONCLUSION Changes in *Gremlin* gene expression account for major morphological differences in the webbing of bird feet.

Go to **BioPortal** for discussion and relevant links for all INVESTIGATINGLIFE figures.

[a]Merino, R. et al. 1999. *Development* 126: 5515–5522.

nonaquatic birds). The loss of webbing is controlled by the BMP4 signaling protein, which as we have noted is also involved in beak development (another example of use of a common genetic toolkit to produce different kinds of change).

BMP4 protein instructs the cells that produce webbing to undergo apoptosis and thus eliminates the webbing between the toes. The hindlimbs of both duck and chicken embryos express the *BMP4* gene in the webbing between the toes; however, they differ in expression of the *Gremlin* gene, which encodes a protein that inhibits *BMP4* expression (**Figure 20.6**). In ducks, but not in chickens, *Gremlin* is expressed in the webbing cells and Gremlin protein inhibits *BMP4* expression. With no BMP4 protein to stimulate apoptosis, a webbed foot develops. If chick hindlimbs are experimentally exposed to Gremlin during development, the adult chicken will have ducklike webbed feet (**Figure 20.7**).

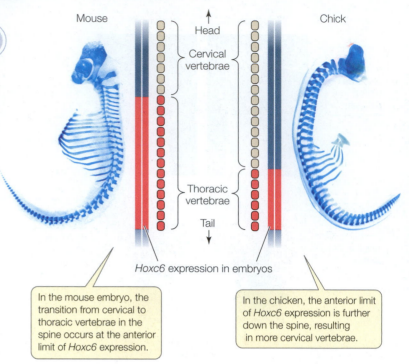

Genetic toolkit genes act independently among modules of a developing embryo. There are various ways in which the expression of these developmental genes can differ among species, which results in major morphological differences. They can differ in amount of expression (heterometry), in the timing of expression (heterochrony), or in the location of expression (heterotopy).

- What are genetic switches, and how can they alter the way a particular gene is expressed? **See p. 415 and Figure 20.3**
- Explain how heterometry and natural selection produced the beak variations we see in today's Galápagos finches. **See p. 416 and Figure 20.4**
- How is heterochrony involved in the evolution of the long neck of giraffes? **See pp. 416–417 and Figure 20.5**
- Explain why the differences in webbing of duck feet and chicken feet are an example of heterotopy. **See p. 417 and Figure 20.6**

Genetic manipulations and studies of pattern formation in embryos have shown that the same signals can control the development of different structures in an individual organism. As we have noted, the protein BMP4 promotes apoptosis between developing digits in bird feet and is also involved in the formation of bird beaks. Perhaps not surprisingly, BMP4—bone morphogenetic protein 4—is also involved in the formation of bone. Might the processes that generate multiple structures *within* an organism also explain how different structures develop in different species?

20.3 How Can Developmental Changes Result in Differences among Species?

We have seen in the case of the Galápagos finches how subtle changes in the expression of developmental genes can alter the morphology of the beak leading to adaptations for exploiting different types of food (see Figure 20.4). Small differences in the expression of developmental genes can produce even more dramatic differences between species.

Differences in Hox gene expression patterns result in major differences in body plans

A diagnostic feature of vertebrates is their vertebral columns, but the vertebral columns exhibit considerable variation across the groups of vertebrates. For example, most mammals have 7 cervical vertebrae, but birds can have many more; chickens have 14 and swans have 25. Mammals have 13 thoracic vertebrae with ribs, birds have fewer, and snakes have many more. These significant differences result from the spatial patterns of Hox gene expression that govern the transitions from one region to another (**Figure 20.8**). The anterior limit of expression of *Hoxc6*, for example, always falls at the boundary between the cervical and thoracic vertebrae in mammals and birds. In addition, the anteriormost segment that expresses *Hoxc6* is the segment where

Hoxc6 expression in embryos

In the mouse embryo, the transition from cervical to thoracic vertebrae in the spine occurs at the anterior limit of *Hoxc6* expression.

In the chicken, the anterior limit of *Hoxc6* expression is further down the spine, resulting in more cervical vertebrae.

20.8 Changes in Gene Expression and Evolution of the Spine Differences in the pattern of *Hoxc6* expression result in a different boundary between the cervical and thoracic vertebrae in mice and chicks.

the forelimbs will develop. Posterior to that segment, *Hoxc6* and *Hoxc8* act together to stimulate development of vertebrae with ribs—that is, the thoracic vertebrae. Over evolutionary time, genetic changes that expanded or contracted the expression domains of the different Hox genes resulted in changes in the characteristic numbers of different vertebrae.

Mutations in developmental genes can produce major morphological changes

Sometimes a major developmental change is due to an alteration in the regulatory molecule itself rather than a change in where, when, or how much it is expressed. This is called **heterotypy** ("different type"). An excellent example of heterotypy is a gene that controls the number of legs in arthropods. Arthropods all have head, thoracic, and abdominal regions with variable numbers of segments. Insects, such as *Drosophila*, have three pairs of legs on their three thoracic segments, whereas centipedes have many legs on both thoracic and abdominal segments. All arthropods express a gene called *Distalless* (*Dll*) that controls segmental leg development. In insects, *Dll* expression is repressed in abdominal segments by the Hox gene *Ubx*. *Ubx* is expressed in the abdominal segments of all arthropods, but it has different effects in different species. In centipedes, Ubx protein activates expression of the *Dll* gene to promote the formation of legs. During the evolution of insects, a change in the *Ubx* gene sequence resulted in a modified Ubx protein that *represses Dll* expression in abdominal segments. A phylogenetic tree of arthropods shows that this change in *Ubx* occurred in the ancestor of insects, at the same time that abdominal legs were lost (**Figure 20.9**).

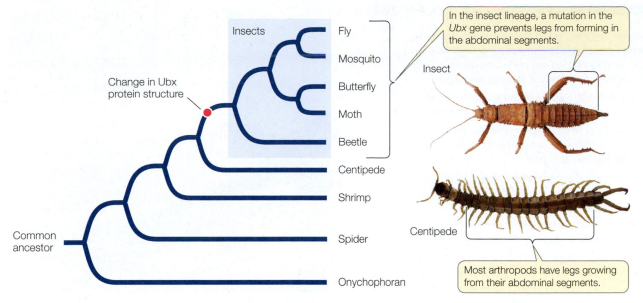

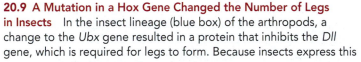

In the insect lineage, a mutation in the *Ubx* gene prevents legs from forming in the abdominal segments.

Most arthropods have legs growing from their abdominal segments.

20.9 A Mutation in a Hox Gene Changed the Number of Legs in Insects In the insect lineage (blue box) of the arthropods, a change to the *Ubx* gene resulted in a protein that inhibits the *Dll* gene, which is required for legs to form. Because insects express this modified *Ubx* gene in their abdominal segments, no legs grow from these segments. Other arthropods, such as centipedes, produce an unmodified Ubx protein and do grow legs from their abdominal segments.

An example of heterotypy that has a huge impact on our food supply is a gene that is partly responsible for the easily accessible edible kernels of domestic corn. The wild relative of domesticated corn—a plant called teosinte—has kernels encased in tough shells (glumes) that develop under control of a gene called *Tga1*. The "liberation" of the edible kernels from these casings that revolutionized this grain's importance to human agriculture is in part based on a mutation in *Tga1* that results in a protein different from the ancestral form by a single amino acid. Genetic experiments have shown that when the domesticated corn *Tga1* gene is inserted into teosinte, teosinte kernels break free of their glumes; conversely, when the teosinte *Tga1* gene is inserted into domestic corn, many of the corn kernels become encased in hard glumes (**Figure 20.10**).

(A)

(B)

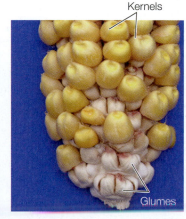

20.10 A Result of Heterotypy The edible kernels of teosinte, the grain from which corn (*Zea mays*) was domesticated, are encased in a hard shell, or glume. Corn kernels have no glumes and thus are easily accessible. *Tga1*, the gene whose product controls glume formation, differs between the two species by one amino acid. (A) If the domestic maize *Tga1* gene is inserted into teosinte, the glumes are incomplete and open to partially expose the kernels. (B) If teosinte *Tga1* is inserted into corn, glumes form over the kernels.

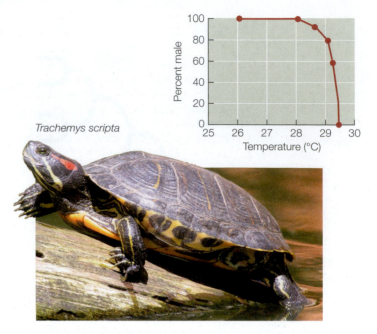

Trachemys scripta

Changes in genetic switches, which determine where, when, and to what extent genes will be expressed, can result in major morphological changes. Natural selection can result in fixation of these differences and subsequent speciation.

- How can Hox gene expression explain the differences in the vertebral columns of different vertebrate groups? **See p. 418**

- How does the activity of *Ubx* explain the fact that insects have three pairs of legs and centipedes have many? **See p. 418 and Figure 20.9**

- Explain how heterotypy accounts for a major morphological difference between corn and its related wild relative, teosinte. **See p. 419 and Figure 20.10**

This chapter has focused on how modular genetic signaling cascades control the development of an organism and how changes in genetic switches can produce differences among species. These processes unfold from the genetic information contained in the fertilized egg, but information from the environment can also influence the genetic signaling cascades and thereby alter the form of the organism.

20.11 Hot Females, Cool Males   The sex of a red-eared slider turtle depends on the temperature at which the egg is incubated. Higher temperatures produce only females and lower temperatures produce only males, apparently due to temperature sensitivity in the synthesis of sex hormones.

20.4 How Can the Environment Modulate Development?

In some cases environmental signals produce developmental changes that affect the morphology (phenotype) of an organism. The ability of an organism to modify its development in response to environmental conditions is called **developmental plasticity** or **phenotypic plasticity**; it means that a single genotype has the capacity to produce two or more different phenotypes.

  Go to Media Clip 20.1
Predator-Induced Development
Life10e.com/mc20.1

Temperature can determine sex

Section 12.4 described how genetic mechanisms can determine sex. In some reptiles, however, sex is determined not by genetic differences between individuals, but by the temperature at which the eggs are incubated. Research in the laboratory of David Crews at the University of Texas has shown that if eggs of the red-eared slider turtle (*Trachemys scripta*) are incubated at temperatures below 28.6°C, all the hatchlings are males, whereas eggs incubated above 29.4°C all produce females. In between these two temperatures—a range of less than 1°C—a clutch of eggs will produce both males and females (**Figure 20.11**). In other species with temperature-dependent sex determination, the incubation temperatures that produce males and females may be different. Thus the effects of temperature on sex determination can vary among species. But how can temperature control developmental plasticity?

In all vertebrate embryos, the development of male and female organs is controlled by the actions of sex steroid hormones. This is the case whether the organism's sex

determination is controlled by its genotype or by temperature. Sex steroid biosynthesis in both males and females begins with cholesterol and goes through many chemical reactions to produce the male sex steroids (androgens) and the female sex steroids (estrogens). In this biosynthetic sequence, the step that produces the first androgen—testosterone—precedes the step that produces estrogens; therefore both males and females produce testosterone.

$$\text{Cholesterol} \rightarrow \rightarrow \rightarrow \text{Testosterone} \begin{array}{c} \xrightarrow{\text{Aromatase}} \text{Estrogen} \\ \xrightarrow{\text{Reductase}} \text{Other androgens} \end{array}$$

In animals with temperature-controlled sex determination, incubation temperature controls the expression of the enzyme aromatase, which converts testosterone to estrogen. If aromatase is abundantly expressed, estrogens are dominant and female organs develop. If aromatase is not expressed, testosterone is dominant and male organs develop. Applying estrogen to eggs results in the development of females, even at the male-inducing temperature.

What is the evolutionary advantage of this sex determination mechanism? Incubation temperature can affect other aspects of the phenotype of the developing organism besides its sex, such as its growth rate and eventual adult body size. For example, red-eared slider females are larger than the males, and the largest females produce many more eggs than smaller females. But even small males can produce enough sperm to fertilize all the eggs of the larger females. For these kinds of reasons, incubation temperature may have a differential effect

INVESTIGATING**LIFE**

20.12 Temperature-Dependent Sex Determination Can Be Associated with Sex-Specific Fitness Differences In some reptiles, sex is determined by the incubation temperature of the developing embryo. This led to the hypothesis that male-inducing temperatures during development result in males with higher reproductive fitness. Daniel Warner and Rick Shine at the University of Sydney, Australia tested this hypothesis by using a drug to block estrogen synthesis, so that males developed instead of females at high and low temperatures. These males had much lower reproductive fitness than males that developed at the normal male-inducing temperature (which is intermediate). In contrast, females showed highest fitness when they developed from eggs incubated at higher temperatures.[a]

HYPOTHESIS Incubation temperature has a differential effect on reproductive success in lizards.

Method Incubate jacky dragon (*Amphibolurus muricatus*) eggs at 23°C, 27°C, and 33°C. Apply aromatase inhibitor to half the eggs. Raise lizards in the natural environment and record the number of offspring produced.

Results Untreated eggs: males at 27°C; females at 23°C and 33°C
Inhibitor-treated eggs: males at 23°C, 27°C, and 33°C

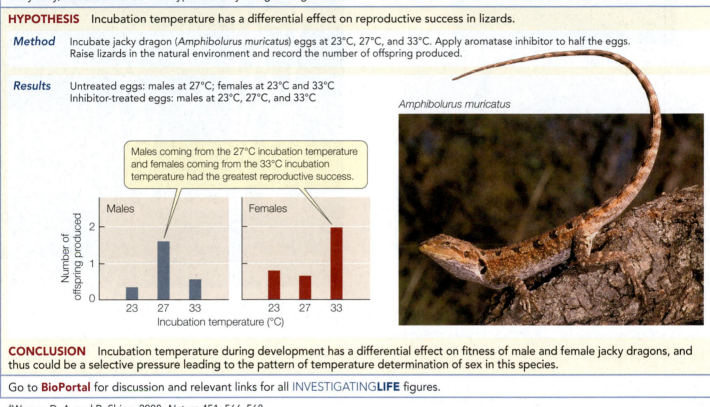

Males coming from the 27°C incubation temperature and females coming from the 33°C incubation temperature had the greatest reproductive success.

Amphibolurus muricatus

CONCLUSION Incubation temperature during development has a differential effect on fitness of male and female jacky dragons, and thus could be a selective pressure leading to the pattern of temperature determination of sex in this species.

Go to **BioPortal** for discussion and relevant links for all INVESTIGATING**LIFE** figures.

[a]Warner, D. A. and R. Shine. 2008. *Nature* 451: 566–568.

on the reproductive success of males and females in a population. **Figure 20.12** describes an experiment that clearly suggests incubation temperature differentially affects the reproductive successes of males and females in one lizard species.

Dietary information can be a predictor of future conditions

In some species with short life spans, an individual may encounter only one of several distinct but predictable seasonal environments. Developmental plasticity allows such individuals to develop the phenotype that enhances their survival in the environment they will encounter as adults. An excellent example is the moth *Nemoria arizonaria*, which produces two generations each year. Caterpillars from eggs that have overwintered and hatch in the spring feed on the oak tree flowers, or catkins, that are present at that time of year. These caterpillars complete their development and transform into adult moths in summer. The summer moths lay their eggs on oak leaves, and the caterpillars that hatch eat the mature leaves. When these caterpillars transform into adult moths, they lay eggs that overwinter and hatch the following spring when the catkins are once again in bloom. Although newly hatched caterpillars all look similar, the body form of spring caterpillars resembles the catkins (**Figure 20.13A**), whereas the body form of summer caterpillars resembles young oak branches (**Figure 20.13B**). Their different larval diets—that is, the different biochemistry of the molecules in oak catkins versus those in mature oak leaves—trigger developmental changes that result in two different phenotypes, each of which is cryptic in the respective environments. This phenotypic plasticity results in lower predation and increased evolutionary fitness.

A variety of environmental signals influence development

In addition to temperature and diet, other environmental signals may initiate developmental change. One ubiquitous source of environmental information is sunlight, which provides predictive information about seasonal changes. Outside the equatorial region, lengthening days herald spring and summer whereas shortening days indicate oncoming winter. Many insects use day length to enter or exit a period of developmental or reproductive arrest called **diapause**, which enables them to better survive harsh conditions. Deer, moose, and elk use day length to time the development and the dropping of antlers,

Catkins

(A)   (B)

Summer adult

Nemoria arizonaria caterpillars

Spring: Caterpillars feed on catkins   Summer: Caterpillars feed on leaves

20.13 Spring and Summer Forms of a Caterpillar
(A) Spring caterpillars of the moth *Nemoria arizonaria* resemble the oak catkins on which they feed. They develop into adults (shown above), which lay eggs on oak leaves. (B) The summer caterpillars feed on oak leaves and are camouflaged by their resemblance to oak twigs.

In the dark, stems elongate and leaf growth is reduced.

In the light, stems are shorter and leaves are bigger.

20.14 Light Seekers   The bean plants on the left were grown under low light levels. The plant's cells have elongated in response to the low light, and the plants have become spindly. The control plants on the right were grown under normal light conditions.

and many organisms use day length to optimize the timing of reproduction or migration. Many plants initiate flowering in response to the length of the night (an absence of light), and in some plant species developmental changes are induced by certain wavelengths of light.

You may wonder why processes such as antler growth, reproductive timing, and seasonal migration are considered in a chapter on development. Development encompasses more than the events that occur before an organism reaches maturity. Development includes changes in body form and function that can occur throughout the life of the organism. For example, giant redwoods that are thousands of years old still have undifferentiated tissues called meristems that produce new differentiated tissues—stems, leaves, reproductive structures, and so on—throughout the life of the tree. These developmental processes are not a simple read-out of a genetic program; they are adjusted to optimize plant form in the environment in which the organism grows. Light, which plants need for photosynthesis, is an important environmental signal in plant development. Dim light stimulates the cells of the stems to elongate, so plants

growing in the shade become tall and spindly (**Figure 20.14**). This developmental plasticity is adaptive because a tall, spindly plant is more likely to reach a patch of brighter light than a plant that remains compact and bushy. A plant in bright light does not need to grow tall and so can put more energy into producing leaves.

RECAP 20.4

Developmental plasticity enables some developing organisms to adjust their forms to fit the environments in which they live. Organisms respond to environmental signals that are accurate predictors of future conditions. Development continues throughout life, and can result in adaptive changes in the forms and functions of adult organisms.

- Describe several examples of how phenotype can be a response to environmental signals. **See pp. 420–422 and Figures 20.11, 20.13, and 20.14**

- How would you determine whether or not an environmental effect on development is adaptive? **See pp. 420–421 and Figure 20.12**

Appropriate responses to new environmental conditions are likely to evolve over time, but what are the limits of such evolution? To what extent do developmental genes dictate the structures and forms that are possible?

20.5 How Do Developmental Genes Constrain Evolution?

Four decades ago, the French geneticist François Jacob made the analogy that evolution works like a tinker, assembling new structures by combining and modifying the available materials, and not like an engineer, who is free to develop dramatically different designs (a jet engine to replace a propeller-driven engine, for example). We have seen that morphological evolution is not usually governed by the acquisition of radically new genes, but proceeds primarily by "tinkering" with the expression patterns of existing genes. Thus developmental genes and their expression constrain evolution in two major ways:

- Nearly all evolutionary innovations are modifications of previously existing structures.
- The basic set of regulatory genes that control development is broadly conserved, changing only slowly over the course of evolution.

Evolution usually proceeds by changing what's already there

The features of organisms usually evolve from preexisting features in their ancestors. New "wing genes" did not suddenly appear in insects, birds, and bats; instead, wings arose as modifications of existing structures. Wings evolved independently in insects and vertebrates—once in insects, and in three independent instances among the vertebrates (**Figure 20.15**). Vertebrate wings are modified forelimbs and share many of the same basic skeletal elements. They have a humerus that connects the wing to the rest of the skeleton; two longer bones (radius and ulna) that project away from the humerus; and metacarpals and phalanges (digits). Although these elements are present in all the various vertebrate wings, the independent origins of vertebrate wings is clear from the different role these skeletal elements play in each wing type. During development these bones grow to different lengths and weights in different organisms, and evolution proceeded in different paths in producing bird, bat, and pterosaur wings.

Developmental regulatory genes also influence how organisms lose structures. The ancestors of present-day snakes lost their forelimbs as a result of changes in the segmental expression of Hox genes. The snake lineage subsequently lost its hindlimbs by the loss of expression of the *Sonic hedgehog* gene in the limb bud tissue. Some snakes, such as the boas and pythons, have rudimentary pelvic and upper leg bones.

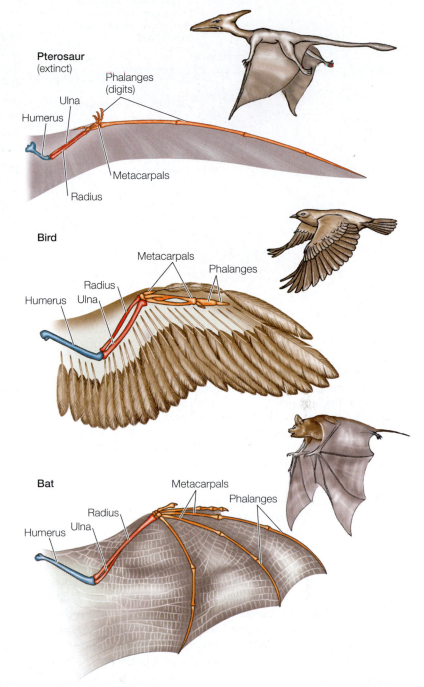

20.15 Wings Evolved Three Times in Vertebrates   The wings of pterosaurs (the earliest flying vertebrates, which lived from 265 to 220 million years ago), birds, and bats are all modified forelimbs and are constructed from the same skeletal components. These components, however, have different forms in the different groups, supporting the independent evolution of wings in each group.

Conserved developmental genes can lead to parallel evolution

The existence of highly conserved developmental genes makes it likely that similar traits will evolve repeatedly, especially among closely related species. This process is known as **parallel evolution**, and a good example is provided by a small fish, the three-spined stickleback (*Gasterosteus aculeatus*).

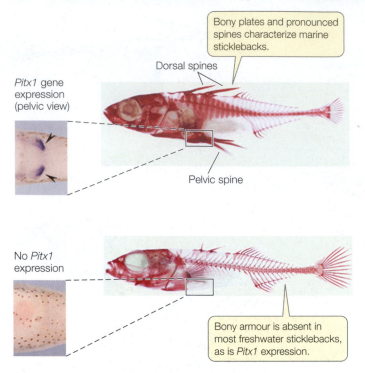

Bony plates and pronounced spines characterize marine sticklebacks.

Dorsal spines

Pitx1 gene expression (pelvic view)

Pelvic spine

No *Pitx1* expression

Bony armour is absent in most freshwater sticklebacks, as is *Pitx1* expression.

20.16 Parallel Phenotypic Evolution in Sticklebacks
A developmental gene, *Pitx1*, encodes a transcription factor that stimulates the production of plates and spines. This gene is active in marine sticklebacks, but mutated and inactive in various freshwater populations of the fish. The fact that this mutation is found in geographically distant and isolated freshwater populations is evidence for parallel evolution.

Sticklebacks are widely distributed throughout the Atlantic and Pacific oceans; they are also found in many freshwater lakes and rivers. Marine sticklebacks spend most of their lives at sea but return to fresh water to breed. Members of freshwater populations that are isolated in lakes never encounter salt water. Genetic evidence shows that freshwater populations have arisen from marine populations many times, and independently. Marine sticklebacks have structures that protect them from predatory marine fish; these are bony plates and well-developed pelvic bones with pelvic spines that lacerate the mouths of predators. Freshwater sticklebacks do not face such predatory dangers; their body armor is greatly reduced, and their dorsal and pelvic spines are shorter or even lacking (**Figure 20.16**).

The differences between marine and freshwater sticklebacks are not induced by environmental conditions. Marine species reared in fresh water still grow armor and spines. The differences are due to the expression of a developmental regulatory gene, *Pitx1*. This gene codes for a transcription factor normally expressed in regions of the developing embryo that in marine sticklebacks form the head, trunk, tail, and pelvis. However, in separate and long-isolated populations of freshwater sticklebacks from Japan, British Columbia, California, and Iceland, the *Pitx1* gene is no longer expressed in the pelvis, and spines do not develop. This same change in regulatory gene expression resulted in similar phenotypic

changes *independently* in several different populations, and is thus a good example of parallel evolution. Is there a common selective mechanism at work in these cases, and what could that mechanism be? One reasonable hypothesis is that decreased predation pressure in the freshwater environment allowed increased reproductive success for sticklebacks that invest less energy in the development of unnecessary protective structures.

RECAP 20.5

Nearly all evolutionary innovations are modifications of previously existing structures. The conservation of many developmental regulatory genes makes it likely that similar traits will evolve repeatedly.

- How have diverse body forms evolved by means of modifications to existing structures? **See p. 423 and Figure 20.15**
- How do the differences between marine and freshwater sticklebacks exemplify parallel evolution via changes in gene regulation? **See p. 424 and Figure 20.16**

Many novel traits have arisen during the course of evolution, but most of them failed to persist beyond even a single generation. Part Six of this book will examine the processes of evolution—the powerful forces that influence the survival and reproductive success of various life forms. We will examine how different adaptations become prevalent in different environments, resulting in the extraordinary diversity of life on Earth today, which we will describe in further detail in Part Seven.

How are gene expression patterns involved in the shaping of the diverse beaks of birds?

ANSWER

As described in Figure 20.4, studies of Galápagos finches indicate that high levels of BMP4 protein expression are associated with the short, robust beaks of the ground finches and that high calmodulin expression is associated with the narrow, sharp beaks of the cactus finches. The timing and level of expression of the genes for these two proteins, under the control of transcription factors and their promotors, enhancers, and repressors—the genetic toolkit—result in morphological modifications of the birds' beaks.

Consider also the difference between the short, narrow, sharp beaks of chickens and the long, thick, broad beaks of ducks. In other studies, analysis of BMP4 in the developing beak tissues of chickens and ducks reveals that its expression is higher in duck embryos. Using molecular genetic means of increasing expression of BMP4 in the developing beak tissue of chicks results in bigger, broader, more robust beaks in the adult chicken. Blocking BMP4 expression in the developing beak tissue of chickens results in adult birds with very small beaks. Thus across these two distantly related groups of birds, the same developmental regulatory genes are involved in shaping the morphology of the beak.

CHAPTER**SUMMARY** 20

20.1 How Can Small Genetic Changes Result in Large Changes in Phenotype?

- **Evolutionary developmental biology**, or **evo-devo**, is the study of the evolutionary aspects of development. This field focuses on the molecular mechanisms that underlie the development of phenotypic diversity.

- Changes in development underlie evolutionary changes in morphology that produce differences in body forms.

- Similarities in the basic mechanisms of development between widely divergent organisms reflect common ancestry. **Review Figure 20.1**

- Evolutionary diversity is produced using a modest number of regulatory genes.

- Genes encoding the transcription factors and other regulatory proteins that govern pattern formation in the developing bodies of multicellular organisms can be thought of as a **genetic toolkit**. These regulatory genes have been highly conserved throughout evolution. **Review Figure 20.2**

20.2 How Can Mutations with Large Effects Change Only One Part of the Body?

- The bodies of developing and mature organisms are organized into self-contained units, or **modules**, that can be modified independently. **See ANIMATED TUTORIAL 20.1**

- The genetic toolkit involves **genetic switches**—promoters, enhancers, and repressors—that can alter the expression of developmental genes in different modules independently of one another.

- Developmental genes can be expressed in a modular fashion in different amounts (**heterometry**), at different times (**heterochrony**), or in different locations (**heterotopy**). **Review Figures 20.3–20.6**

20.3 How Can Developmental Changes Result in Differences among Species?

- Changes in genetic switches that determine where, when, and to what extent a set of genes will be expressed underlie both the

transformation of an individual from egg to adult and the evolution of differences among species.

- Morphological differences among species can result from mutations in the genes that regulate the development of modules such as body segments or wings. **Review Figures 20.8, 20.9**

20.4 How Can the Environment Modulate Development?

- The ability of an organism to modify its development in response to environmental conditions is called **developmental**, or **phenotypic**, **plasticity**.

- In many species of reptiles, sex development is determined by incubation temperature, which acts through genes that control the production, modification, and action of sex hormones. **Review Figure 20.11**

- The adaptive significance of developmental plasticity is not always obvious, but experiments can test for effects on reproductive success. **Review Figure 20.12**

- Some environmental cues, such as those that anticipate seasons, are highly regular and can reliably drive seasonal adaptations in body form and function. **Review Figure 20.13**

- Environmental cues that trigger developmental change are diverse and can act at any stage of the life of an organism.

20.5 How Do Developmental Genes Constrain Evolution?

- Virtually all evolutionary innovations are modifications of preexisting structures. **Review Figure 20.15**

- Because many genes that govern development have been highly conserved, similar traits are likely to evolve repeatedly, especially among closely related species. This process is called **parallel phenotypic evolution**. **Review Figure 20.16**

See ACTIVITY 20.1 for a concept review of this chapter.

 Go to the Interactive Summary to review key figures, Animated Tutorials, and Activities
Life10e.com/is20

CHAPTER**REVIEW**

REMEMBERING

1. Which of the following is *not* one of the principles of evolutionary developmental biology (evo-devo)?
   a. Animal groups share similar molecular mechanisms for morphogenesis.
   b. Changes in the timing of gene expression are important in the evolution of new structures.
   c. Evolution of development is not responsive to the environment.
   d. Changes in the locations of gene expression in the embryo can lead to new structures.
   e. Evolution often occurs by modification of existing developmental genes and pathways.

2. The developmental control pathway that determines the segmental body plan in *Drosophila*
   a. has similar gene sequences and chromosomal arrangements in the mouse.
   b. is unique to insects.
   c. determines only those organs that arise in head segments.
   d. induces the development of wings in each thoracic segment.
   e. arose through new genes that had not existed before in any form.

3. The process whereby changes in the timing of developmental gene expression can change the form of an organism is called
   a. heterochrony.
   b. developmental plasticity.
   c. adaptation.
   d. modularity.
   e. heterometry.

4. Modularity is important for development because it
   a. guarantees that all units of a developing embryo will change in a coordinated way.
   b. coordinates the establishment of the anterior–posterior axis of the developing embryo.
   c. allows changes in developmental genes to change one part of the body without affecting other parts.
   d. guarantees that the timing of gene expression is the same in all parts of a developing embryo.
   e. allows organisms to be built up one module at a time.

5. Organisms often respond to environmental signals that accurately predict future conditions by
   a. stopping development until the signal changes.
   b. altering their development to adapt to the future environment.
   c. altering their development such that the resulting adult can produce offspring adapted to the future environment.
   d. producing new mutants.
   e. developing normally because the predicted conditions may not last long.

UNDERSTANDING & APPLYING

6. What do you predict would happen if you blocked the expression of Gremlin in the feet of duck embryos?

7. If you applied an inhibitor of aromatase to red-eared slider turtle eggs being incubated at warm and at cool temperatures, what would be the outcome?

8. In the mouse, *Hoxc6* is first expressed in the posterior part of the twelfth body segment (somite) and marks the boundary between the cervical and thoracic regions of the vertebral column. *Hoxc8* is expressed a little more posteriorly and is coexpressed with *Hoxc6* for most of the length of the thoracic region (where ribs form). In contrast, in the developing python, *Hoxc6* and *Hoxc8* are expressed together along most of the length of the embryo. What hypotheses can you formulate about the roles of these Hox genes in the development of features of the body axis in mice and snakes?

ANALYZING & EVALUATING

9. In a series of experiments on chick embryos, researchers applied different concentrations of BMP4 to the embryo's beak growth region; they then measured the size of the beak cartilage at a later stage of development. Based on their data in the table below, what would you conclude about the role of BMP4 in beak growth?

Amount of BMP4	Cartilage diameter (mm)
None (control)	0.5
0.1 units	0.7
0.3 units	1.0
1.0 units	1.8

10. *Plasmodium vivax* is a protist that causes a form of malaria. When *P. vivax* enters the blood, it attaches to a glycoprotein on the red blood cells. The same glycoprotein is found on a variety of other cells as well. Some human populations in Africa are immune to *P. vivax* because they lack this particular glycoprotein, but only on their red blood cells. The transcription of this glycoprotein is under the influence of a number of enhancers, and the one that is expressed in red blood cell precursors is mutated in the population that is immune to this form of malaria. What does this case illustrate in evolutionary developmental terms?

Go to BioPortal at **yourBioPortal.com** for Animated Tutorials, Activities, LearningCurve Quizzes, Flashcards, and many other study and review resources.

Appendix A　The Tree of Life

Phylogeny is the organizing principle of modern biological taxonomy. A guiding principle of modern phylogeny is monophyly. A monophyletic group is considered to be one that contains an ancestral lineage and all of its descendants. Any such group can be extracted from a phylogenetic tree with a single cut.

The tree shown here provides a guide to the relationships among the major groups of extant (living) organisms in the tree of life as we have presented them throughout this book. The position of the branching "splits" indicates the relative branching order of the lineages of life, but the time scale is not meant to be uniform. In addition, the groups appearing at the branch tips do not necessarily carry equal phylogenetic "weight." For example, the ginkgo [75] is indeed at the apex of its lineage; this gymnosperm group consists of a single living species. In contrast, a phylogeny of the eudicots [83] could continue on from this point to fill many more trees the size of this one.

The glossary entries that follow are informal descriptions of some major features of the organisms described in Part Seven of this book. Each entry gives the group's common name, followed by the formal scientific name of the group (in parentheses). Numbers in square brackets reference the location of the respective groups on the tree.

It is sometimes convenient to use an informal name to refer to a collection of organisms that are not monophyletic but nonetheless all share (or all lack) some common attribute. We call these "convenience terms"; such groups are indicated in these entries by quotation marks, and we do not give them formal scientific names. Examples include "prokaryotes," "protists," and "algae." Note that these groups cannot be removed with a single cut; they represent a collection of distantly related groups that appear in different parts of the tree. We also use quotation marks here to designate two groups of fungi that are not believed to be monophyletic.

Go to BioPortal at **yourBioPortal.com** for an interactive version of this tree, with links to photos, distribution maps, species lists, and identification keys.

– A –

acorn worms (*Enteropneusta*)　Benthic marine hemichordates [119] with an acorn-shaped proboscis, a short collar (neck), and a long trunk.

"algae"　Convenience term encompassing various distantly related groups of aquatic, photosynthetic eukaryotes [4].

alveolates (*Alveolata*) [5]　Unicellular eukaryotes with a layer of flattened vesicles (alveoli) supporting the plasma membrane. Major groups include the dinoflagellates [51], apicomplexans [50], and ciliates [49].

amborella (*Amborella*) [78]　An understory shrub or small tree found only on the South Pacific island of New Caledonia. Thought to be the sister group of the remaining living angiosperms [15].

ambulacrarians (*Ambulacraria*) [29]　The echinoderms [118] and hemichordates [119].

amniotes (*Amniota*) [36]　Mammals, reptiles, and their extinct close relatives. Characterized by many adaptations to terrestrial life, including an amniotic egg (with a unique set of membranes—the amnion, chorion, and allantois), a water-repellant epidermis (with epidermal scales, hair, or feathers), and, in males, a penis that allows internal fertilization.

amoebozoans (*Amoebozoa*) [84]　A group of eukaryotes [4] that use lobe-shaped pseudopods for locomotion and to engulf food. Major amoebozoan groups include the loboseans, plasmodial slime molds, and cellular slime molds.

amphibians (*Amphibia*) [128]　Tetrapods [35] with glandular skin that lacks epidermal scales, feathers, or hair. Many amphibian species undergo a complete metamorphosis from an aquatic larval form to a terrestrial adult form, although direct development is also common. Major amphibian groups include frogs and toads (anurans), salamanders, and caecilians.

amphipods (*Amphipoda*)　Small crustaceans [116] that are abundant in many marine and freshwater habitats. They are important herbivores, scavengers, and micropredators, and are an important food source for many aquatic organisms.

angiosperms (*Anthophyta* or *Magnoliophyta*) [15]　The flowering plants. Major angiosperm groups include the monocots [82], eudicots [83], and magnoliids [81].

animals (*Animalia* or *Metazoa*) [19]　Multicellular heterotrophic eukaryotes. The majority of animals are bilaterians [22]. Other groups of animals include the sponges [20], ctenophores [95], placozoans [96], and cnidarians [97]. The closest living relatives of the animals are the choanoflagellates [91].

annelids (*Annelida*) [105]　Segmented worms, including earthworms, leeches, and polychaetes. One of the major groups of lophotrochozoans [24].

anthozoans (*Anthozoa*)　One of the major groups of cnidarians [97]. Includes the sea anemones, sea pens, and corals.

anurans (*Anura*)　Comprising the frogs and toads, this is the largest group of living amphibians [128]. They are tail-less, with a shortened vertebral column and elongate hind legs modified for jumping. Many species have an aquatic larval form known as a tadpole.

apicomplexans (*Apicomplexa*) [50]　Parasitic alveolates [5] characterized by the possession of an apical complex at some stage in the life cycle.

arachnids (*Arachnida*)　Chelicerates [114] with a body divided into two parts: a cephalothorax that bears six pairs of appendages (four pairs of which are usually used as legs) and an abdomen that bears the genital opening. Familiar arachnids include spiders, scorpions, mites and ticks, and harvestmen.

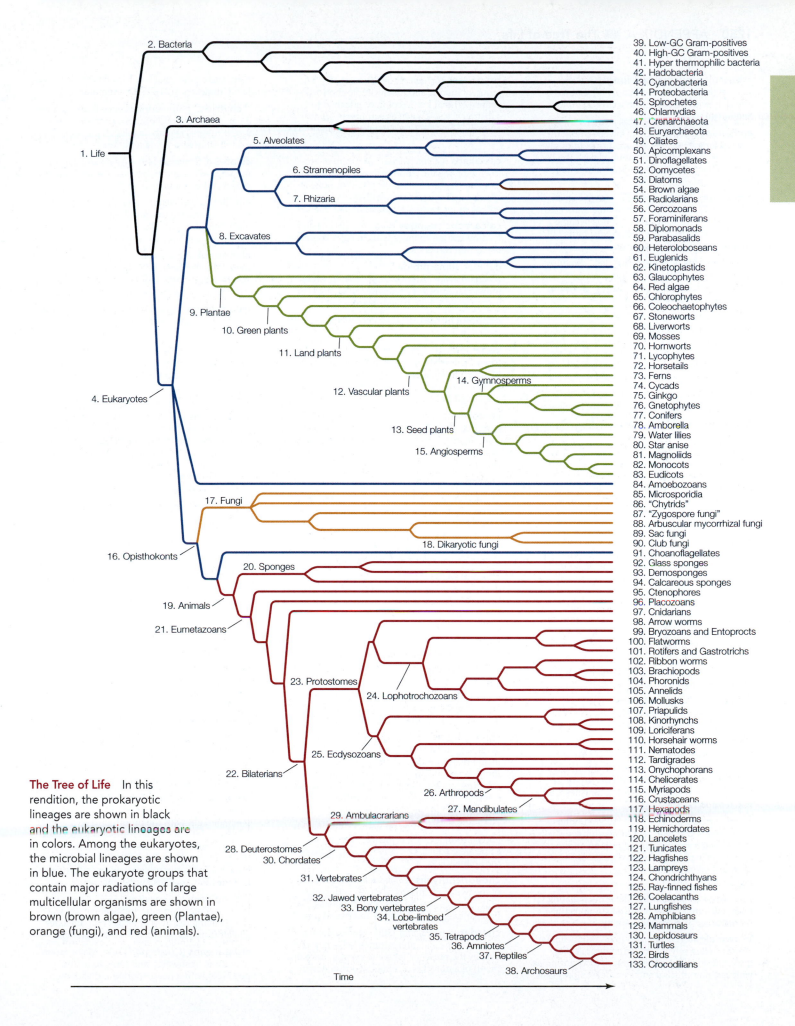

The Tree of Life In this rendition, the prokaryotic lineages are shown in black and the eukaryotic lineages are in colors. Among the eukaryotes, the microbial lineages are shown in blue. The eukaryote groups that contain major radiations of large multicellular organisms are shown in brown (brown algae), green (Plantae), orange (fungi), and red (animals).

1. Life
2. Bacteria
3. Archaea
4. Eukaryotes
5. Alveolates
6. Stramenopiles
7. Rhizaria
8. Excavates
9. Plantae
10. Green plants
11. Land plants
12. Vascular plants
13. Seed plants
14. Gymnosperms
15. Angiosperms
16. Opisthokonts
17. Fungi
18. Dikaryotic fungi
19. Animals
20. Sponges
21. Eumetazoans
22. Bilaterians
23. Protostomes
24. Lophotrochozoans
25. Ecdysozoans
26. Arthropods
27. Mandibulates
28. Deuterostomes
29. Ambulacrarians
30. Chordates
31. Vertebrates
32. Jawed vertebrates
33. Bony vertebrates
34. Lobe-limbed vertebrates
35. Tetrapods
36. Amniotes
37. Reptiles
38. Archosaurs

39. Low-GC Gram-positives
40. High-GC Gram-positives
41. Hyper thermophilic bacteria
42. Hadobacteria
43. Cyanobacteria
44. Proteobacteria
45. Spirochetes
46. Chlamydias
47. Crenarchaeota
48. Euryarchaeota
49. Ciliates
50. Apicomplexans
51. Dinoflagellates
52. Oomycetes
53. Diatoms
54. Brown algae
55. Radiolarians
56. Cercozoans
57. Foraminiferans
58. Diplomonads
59. Parabasalids
60. Heteroloboseans
61. Euglenids
62. Kinetoplastids
63. Glaucophytes
64. Red algae
65. Chlorophytes
66. Coleochaetophytes
67. Stoneworts
68. Liverworts
69. Mosses
70. Hornworts
71. Lycophytes
72. Horsetails
73. Ferns
74. Cycads
75. Ginkgo
76. Gnetophytes
77. Conifers
78. Amborella
79. Water lilies
80. Star anise
81. Magnoliids
82. Monocots
83. Eudicots
84. Amoebozoans
85. Microsporidia
86. "Chytrids"
87. "Zygospore fungi"
88. Arbuscular mycorrhizal fungi
89. Sac fungi
90. Club fungi
91. Choanoflagellates
92. Glass sponges
93. Demosponges
94. Calcareous sponges
95. Ctenophores
96. Placozoans
97. Cnidarians
98. Arrow worms
99. Bryozoans and Entoprocts
100. Flatworms
101. Rotifers and Gastrotrichs
102. Ribbon worms
103. Brachiopods
104. Phoronids
105. Annelids
106. Mollusks
107. Priapulids
108. Kinorhynchs
109. Loriciferans
110. Horsehair worms
111. Nematodes
112. Tardigrades
113. Onychophorans
114. Chelicerates
115. Myriapods
116. Crustaceans
117. Hexapods
118. Echinoderms
119. Hemichordates
120. Lancelets
121. Tunicates
122. Hagfishes
123. Lampreys
124. Chondrichthyans
125. Ray-finned fishes
126. Coelacanths
127. Lungfishes
128. Amphibians
129. Mammals
130. Lepidosaurs
131. Turtles
132. Birds
133. Crocodilians

Time

arbuscular mycorrhizal fungi (*Glomero-mycota*) [88] A group of fungi [17] that associate with plant roots in a close symbiotic relationship.

archaeans (*Archaea*) [3] Unicellular organisms lacking a nucleus and lacking peptidoglycan in the cell wall. Once grouped with the bacteria, archaeans possess distinctive membrane lipids.

archosaurs (*Archosauria*) [38] A group of reptiles [37] that includes dinosaurs and crocodilians [133]. Most dinosaur groups became extinct at the end of the Cretaceous; birds [132] are the only surviving dinosaurs.

arrow worms (*Chaetognatha*) [98] Small planktonic or benthic predatory marine worms with fins and a pair of hooked, prey-grasping spines on each side of the head.

arthropods (*Arthropoda*) The largest group of ecdysozoans [25]. Arthropods are characterized by a stiff exoskeleton, segmented bodies, and jointed appendages. Includes the chelicerates [114], myriapods [115], crustaceans [116], and hexapods (insects and their relatives) [117].

ascidians (*Ascidiacea*) "Sea squirts"; the largest group of tunicates [121]. They are sessile (as adults), marine, saclike filter feeders.

– B –

bacteria (*Eubacteria*) [2] Unicellular organisms lacking a nucleus, possessing distinctive ribosomes and initiator tRNA, and generally containing peptidoglycan in the cell wall. Different bacterial groups are distinguished primarily on nucleotide sequence data.

barnacles (*Cirripedia*) Crustaceans [116] that undergo two metamorphoses—first from a feeding planktonic larva to a nonfeeding swimming larva, and then to a sessile adult that forms a "shell" composed of four to eight plates cemented to a hard substrate.

bilaterians (*Bilateria*) [22] Those animal groups characterized by bilateral symmetry and three distinct tissue types (endoderm, ectoderm, and mesoderm). Includes the protostomes [23] and deuterostomes [28].

birds (*Aves*) [132] Feathered, flying (or secondarily flightless) tetrapods [35].

bivalves (*Bivalvia*) Major mollusk [106] group; clams and mussels. Bivalves typically have two similar hinged shells that are each asymmetrical across the midline.

bony vertebrates (*Osteichthyes*) [33] Vertebrates [31] in which the skeleton is usually ossified to form bone. Includes the ray-finned fishes [125], coelacanths [126], lungfishes [127], and tetrapods [35].

brachiopods (*Brachiopoda*) [103] Lophotrochozoans [24] with two similar hinged shells that are each symmetrical across the midline. Superficially resemble bivalve mollusks, except for the shell symmetry.

brittle stars (*Ophiuroidea*) Echinoderms [118] with five long, whip-like arms radiating from a distinct central disk that contains the reproductive and digestive organs.

brown algae (*Phaeophyta*) [54] Multicellular, almost exclusively marine stramenopiles [6] generally containing the pigment fucoxanthin as well as chlorophylls *a* and *c* in their chloroplasts.

bryozoans (*Ectoprocta* or *Bryozoa*) [99] A group of marine and freshwater lophotrochozoans [24] that live in colonies attached to substrates; also known as ectoprocts or moss animals. They are the sister group of entoprocts.

– C –

caecilians (*Gymnophiona*) A group of burrowing or aquatic amphibians [128]. They are elongate, legless, with a short tail (or none at all), reduced eyes covered with skin or bone, and a pair of sensory tentacles on the head.

calcareous sponges (*Calcarea*) [94] Filter-feeding marine sponges with spicules composed of calcium carbonate.

cellular slime molds (*Dictyostelida*) Amoebozoans [84] in which individual amoebas aggregate under stress to form a multicellular pseudoplasmodium.

cephalochordates (*Cephalochordata*) [120] *See* lancelets.

cephalopods (*Cephalopoda*) Active, predatory mollusks [106] in which the molluscan foot has been modified into muscular hydrostatic arms or tentacles. Includes octopuses, squids, and nautiluses.

cercozoans (*Cercozoa*) [56] Unicellular eukaryotes [4] that feed by means of threadlike pseudopods. Group together with foraminiferans [57] and radiolarians [55] to comprise the rhizaria [7].

charophytes (*Charales*) [67] *See* stoneworts.

chelicerates (*Chelicerata*) [114] A major group of arthropods [26] with pointed appendages (chelicerae) used to grasp food (as opposed to the chewing mandibles of most other arthropods). Includes the arachnids, horseshoe crabs, pycnogonids, and extinct sea scorpions.

chimaeras (*Holocephali*) A group of bottom-dwelling, marine, scaleless chondrichthyan fishes [124] with large, permanent, grinding tooth plates (rather than the replaceable teeth found in other chondrichthyans).

chitons (*Polyplacophora*) Flattened, slow-moving mollusks [106] with a dorsal protective calcareous covering made up of eight articulating plates.

chlamydias (*Chlamydiae*) [46] A group of very small Gram-negative bacteria; they live as intracellular parasites of other organisms.

chlorophytes (*Chlorophyta*) [65] The most abundant and diverse group of green algae, including freshwater, marine, and terrestrial forms; some are unicellular, others colonial, and still others multicellular. Chlorophytes use chlorophylls *a* and *c* in their photosynthesis.

choanoflagellates (*Choanozoa*) [91] Unicellular eukaryotes [4] with a single flagellum surrounded by a collar. Most are sessile, some are colonial. The closest living relatives of the animals [19].

chondrichthyans (*Chondrichthyes*) [124] One of the two main groups of jawed vertebrates [32]; includes sharks, rays, and chimaeras. They have cartilaginous skeletons and paired fins.

chordates (*Chordata*) [30] One of the two major groups of deuterostomes [28], characterized by the presence (at some point in development) of a notochord, a hollow dorsal nerve cord, and a post-anal tail. Includes the lancelets [120], tunicates [121], and vertebrates [31].

"chytrids" [90] Convenience term used for a paraphyletic group of mostly aquatic, microscopic fungi [17] with flagellated gametes. Some exhibit alternation of generations.

ciliates (*Ciliophora*) [49] Alveolates [5] with numerous cilia and two types of nuclei (micronuclei and macronuclei).

clitellates (*Clitellata*) Annelids [105] with gonads contained in a swelling (called a clitellum) toward the head of the animal. Includes earthworms (oligochaetes) and leeches.

club fungi (*Basidiomycota*) [90] Fungi [17] that, if multicellular, bear the products of meiosis on club-shaped basidia and possess a long-lasting dikaryotic stage. Some are unicellular.

club mosses (*Lycopodiophyta*) [71] Vascular plants [12] characterized by microphylls. *See* lycophytes.

cnidarians (*Cnidaria*) [97] Aquatic, mostly marine eumetazoans [21] with specialized stinging organelles (nematocysts) used for prey capture and defense, and a blind gastrovascular cavity. The sister group of the bilaterians [22].

coelacanths (*Actinista*) [126] A group of marine lobe-limbed vertebrates [34] that was diverse from the Middle Devonian to the Cretaceous, but is now known from just two living species. The pectoral and anal fins are on fleshy stalks supported by skeletal elements, so they are also called lobe-finned fishes.

coleochaetophytes (*Coleochaetales*) [66] Multicellular green algae characterized by flattened growth form composed of thin-walled cells. Thought to be the sister-group to the stoneworts [67] plus land plants [11].

conifers (*Pinophyta* or *Coniferophyta*) [77] Cone-bearing, woody seed plants [13].

copepods (*Copepoda*) Small, abundant crustaceans [116] found in marine, freshwater, or wet terrestrial habitats. They have a single eye, long antennae, and a body shaped like a teardrop.

craniates (*Craniata*) Some biologist exclude the hagfishes [122] from the vertebrates [31], and use the term craniates to refer to the two groups combined.

crenarchaeotes (*Crenarchaeota*) [47] A major and diverse group of archaeans [3], defined on the basis of rRNA base sequences. Many are extremophiles (inhabit extreme environments), but the group may also be the most abundant archaeans in the marine environment.

crinoids (*Crinoidea*) Echinoderms [118] with a mouth surrounded by feeding arms, and a U-shaped gut with the mouth next to the anus. They attach to the substratum by a stalk or are free-swimming. Crinoids were abundant in the middle and late Paleozoic, but only a few hundred species have survived to the present. Includes the sea lilies and feather stars.

crocodilians (*Crocodylia*) [133] A group of large, predatory, aquatic archosaurs [38]. The closest living relatives of birds [132]. Includes alligators, caimans, crocodiles, and gharials.

crustaceans (*Crustacea*) [116] Major group of marine, freshwater, and terrestrial arthropods [26] with a head, thorax, and abdomen (although the head and thorax may be fused), covered with a thick exoskeleton, and with two-part appendages. Crustaceans undergo metamorphosis from a nauplius larva. Includes decapods, isopods, krill, barnacles, amphipods, copepods, and ostracods.

ctenophores (*Ctenophora*) [95]   Radially symmetrical, diploblastic marine animals [19], with a complete gut and eight rows of fused plates of cilia (called ctenes).

cyanobacteria (*Cyanobacteria*) [43]   A group of unicellular, colonial, or filamentous bacteria that conduct photosynthesis using chlorophyll *a*.

cycads (*Cycadophyta*) [74]   Palmlike gymnosperms with large, compound leaves.

cyclostomes (*Cyclostomata*)   This term refers to the possibly monophyletic group of lampreys [123] and hagfishes [122]. Molecular data support this group, but morphological data suggest that lampreys are more closely related to jawed vertebrates [32] than to hagfishes.

– D –

decapods (*Decapoda*)   A group of marine, freshwater, and semiterrestrial crustaceans [116] in which five of the eight pairs of thoracic appendages function as legs (the other three pairs, called maxillipeds, function as mouthparts). Includes crabs, lobsters, crayfishes, and shrimps.

demosponges (*Demospongiae*) [93]   The largest of the three groups of sponges [20], accounting for 90 percent of all sponge species. Demosponges have spicules made of silica, spongin fiber (a protein), or both.

deuterostomes (*Deuterostomia*) [28]   One of the two major groups of bilaterians [22], in which the mouth forms at the opposite end of the embryo from the blastopore in early development (contrast with protostomes). Includes the ambulacrarians [29] and chordates [30].

diatoms (*Bacillariophyta*) [53]   Unicellular, photosynthetic stramenopiles [6] with glassy cell walls in two parts.

dikaryotic fungi (*Dikarya*) [18]   A group of fungi [17] in which two genetically different haploid nuclei coexist and divide within the same hypha; includes club fungi [90] and sac fungi [89].

dinoflagellates (*Dinoflagellata*) [51]   A group of alveolates [5] usually possessing two flagella, one in an equatorial groove and the other in a longitudinal groove; many are photosynthetic.

diplomonads (*Diplomonadida*) [58]   A group of eukaryotes [4] lacking mitochondria; most have two nuclei, each with four associated flagella.

– E –

ecdysozoans (*Ecdysozoa*) [25]   One of the two major groups of protostomes [23], characterized by periodic molting of their exoskeletons. Nematodes [111] and arthropods [26] are the largest ecdysozoan groups.

echinoderms (*Echinodermata*) [118]   A major group of marine deuterostomes [28] with fivefold radial symmetry (at some stage of life) and an endoskeleton made of calcified plates and spines. Includes sea stars, crinoids, sea urchins, sea cucumbers, and brittle stars.

elasmobranchs (*Elasmobranchii*)   The largest group of chondrichthyan fishes [124]. Includes sharks, skates, and rays. In contrast to the other group of living chondrichthyans (the chimaeras), they have replaceable teeth.

embryophytes   *See* land plants [11].

entoprocts (*Entoprocta*) [99]   A group of marine and freshwater lophotrochozoans [24] that live as single individuals or in colonies attached to substrates. They are the sister group of bryozoans, from which they differ in having both their mouth and anus inside the lophophore (the anus is outside the lophophore in bryozoans).

eudicots (*Eudicotyledones*) [83]   A group of angiosperms [15] with pollen grains possessing three openings. Typically with two cotyledons, net-veined leaves, taproots, and floral organs typically in multiples of four or five.

euglenids (*Euglenida*) [61]   Flagellate excavates characterized by a pellicle composed of spiraling strips of protein under the plasma membrane; the mitochondria have disk-shaped cristae. Some are photosynthetic.

eukaryotes (*Eukarya*) [4]   Organisms made up of one or more complex cells in which the genetic material is contained in nuclei. Contrast with archaeans [3] and bacteria [2].

eumetazoans (*Eumetazoa*) [21]   Those animals [19] characterized by body symmetry, a gut, a nervous system, specialized types of cell junctions, and well-organized tissues in distinct cell layers (although there have been secondary losses of some or most of these characteristics in a few eumetazoan lineages).

euphyllophytes (*Euphyllophyta*)   The group of vascular plants [12] that is sister to the lycophytes [71] and which includes all plants with megaphylls.

euryarchaeotes (*Euryachaeota*) [48]   A major group of archaeans [3], diagnosed on the basis of rRNA sequences. Includes many methanogens, extreme halophiles, and thermophiles.

eutherians (*Eutheria*)   A group of viviparous mammals [129], eutherians are well developed at birth (contrast to prototherians and marsupials, the other two groups of mammals). Most familiar mammals outside the Australian and South American regions are eutherians (see Table 33.1).

excavates (*Excavata*) [8]   Diverse group of unicellular, flagellate eukaryotes, many of which possess a feeding groove; some lack mitochondria.

– F –

ferns   Vascular plants [12] usually possessing large, frondlike leaves that unfold from a "fiddlehead." Not a monophyletic group, although most fern species are encompassed in a monophyletic clade, the leptosporangiate ferns [73].

flatworms (*Platyhelminthes*) [100]   A group of dorsoventrally flattened and generally elongate soft-bodied lophotrochozoans [24]. May be free-living or parasitic, found in marine, freshwater, or damp terrestrial environments. Major flatworm groups include the tapeworms, flukes, monogeneans, and turbellarians.

flowering plants   *See* angiosperms [15].

flukes (*Trematoda*)   A group of wormlike parasitic flatworms [100] with complex life cycles that involve several different host species. May be paraphyletic with respect to tapeworms.

foraminiferans (*Foraminifera*) [57]   Amoeboid organisms with fine, branched pseudopods that form a food-trapping net. Most produce external shells of calcium carbonate.

fungi (*Fungi*) [17]   Eukaryotic heterotrophs with absorptive nutrition based on extracellular digestion; cell walls contain chitin. Major fungal groups include the microsporidia [85], "chytrids" [86], "zygospore fungi" [87], arbuscular mycorrhizal fungi [88], sac fungi [89], and club fungi [90].

– G –

gastropods (*Gastropoda*)   The largest group of mollusks [106]. Gastropods possess a well-defined head with two or four sensory tentacles (often terminating in eyes) and a ventral foot. Most species have a single coiled or spiraled shell. Common in marine, freshwater, and terrestrial environments.

gastrotrichs (*Gastrotricha*) [101]   Tiny (0.06–3.0 mm), elongate acoelomate lophotrochozoans [24] that are covered in cilia. They live in marine, freshwater, and wet terrestrial habitats. They are simultaneous hermaphrodites.

ginkgo (*Ginkgophyta*) [75]   A gymnosperm [14] group with only one living species. The ginkgo seed is surrounded by a fleshy tissue not derived from an ovary wall and hence not a fruit.

glass sponges (*Hexactinellida*) [92]   Sponges [20] with a skeleton composed of four- and/or six-pointed spicules made of silica.

glaucophytes (*Glaucophyta*) [63]   Unicellular freshwater algae with chloroplasts containing traces of peptidoglycan, the characteristic cell wall material of bacteria.

gnathostomes (*Gnathostomata*)   *See* jawed vertebrates [32].

gnetophytes (*Gnetophyta*) [76]   A gymnosperm [14] group with three very different lineages; all have wood with vessels, unlike other gymnosperms.

green plants (*Viridiplantae*) [10]   Organisms with chlorophylls *a* and *b*, cellulose-containing cell walls, starch as a carbohydrate storage product, and chloroplasts surrounded by two membranes.

gymnosperms (*Gymnospermae*) [14]   Seed plants [13] with seeds "naked" (i.e., not enclosed in carpels). Probably monophyletic, but status still in doubt. Includes the conifers [77], gnetophytes [76], ginkgo [75], and cycads [74].

– H –

hadobacteria (*Hadobacteria*) [42]   A group of extremophilic bacteria [2] that includes the genera *Deinococcus* and *Thermus*.

hagfishes (*Myxini*) [122]   Elongate, slimy-skinned vertebrates [31] with three small accessory hearts, a partial cranium, and no stomach or paired fins. *See also* craniata; cyclostomes.

hemichordates (*Hemichordata*) [119]   One of the two primary groups of ambulacrarians [29];

marine wormlike organisms with a three-part body plan.

heteroloboseans (*Heterolobosea*) [60] Colorless excavates [8] that can transform among amoeboid, flagellate, and encysted stages.

hexapods (*Hexapoda*) [117] Major group of arthropods [26] characterized by a reduction (from the ancestral arthropod condition) to six walking appendages, and the consolidation of three body segments to form a thorax. Includes insects and their relatives (see Table 23.2).

high-GC Gram-positives (*Actinobacteria*) [40] Gram-positive bacteria with a relatively high (G+C)/(A+T) ratio of their DNA, with a filamentous growth habit.

hornworts (*Anthocerophyta*) [70] Nonvascular plants with sporophytes that grow from the base. Cells contain a single large, platelike chloroplast.

horsehair worms (*Nematomorpha*) [110] A group of very thin, elongate, wormlike fresh-water ecdysozoans [25]. Largely nonfeeding as adults, they are parasites of insects and crayfish as larvae.

horseshoe crabs (*Xiphosura*) Marine che-licerates [114] with a large outer shell in three parts: a carapace, an abdomen, and a tail-like telson. There are only five living species, but many additional species are known from fossils.

horsetails (*Sphenophyta* or *Equisetophyta*) [72] Vascular plants [12] with reduced megaphylls in whorls.

hydrozoans (*Hydrozoa*) A group of cnidar-ians [97]. Most species go through both polyp and mesuda stages, although one stage or the other is eliminated in some species.

hyperthermophilic bacteria [41] A group of thermophilic bacteria [2] that live in vol-canic vents, hot springs, and in underground oil reservoirs; includes the genera *Aquifex* and *Thermotoga*.

– I –

insects (*Insecta*) The largest group within the hexapods [117]. Insects are characterized by exposed mouthparts and one pair of an-tennae containing a sensory receptor called a Johnston's organ. Most have two pairs of wings as adults. There are more described species of insects than all other groups of life [1] com-bined, and many species remain to be discov-ered. The major insect groups are described in Table 23.2.

"invertebrates" Convenience term encom-passing any animal [19] that is not a vertebrate [31].

isopods (*Isopoda*) Crustaceans [116] char-acterized by a compact head, unstalked com-pound eyes, and mouthparts consisting of four pairs of appendages. Isopods are abundant and widespread in salt, fresh, and brackish water, although some species (the sow bugs) are terrestrial.

– J –

jawed vertebrates (*Gnathostomata*) [32] A major group of vertebrates [31] with jawed mouths. Includes chondrichthyans [124], ray-finned fishes [125], and lobe-limbed vertebrates [34].

– K –

kinetoplastids (*Kinetoplastida*) [62] Unicellular, flagellate organisms characterized by the presence in their single mitochondrion of a kinetoplast (a structure containing multiple, circular DNA molecules).

kinorhynchs (*Kinorhyncha*) [108] Small (< 1 mm) marine ecdysozoans [25] with bodies in 13 segments and a retractable proboscis.

korarchaeotes (*Korarchaeota*) A group of archaeans [3] known only by evidence from nucleic acids derived from hot springs. Its phy-logenetic relationships within the Archaea are unknown.

krill (*Euphausiacea*) A group of shrimplike marine crustaceans [116] that are important components of the zooplankton.

– L –

lampreys (*Petromyzontiformes*) [123] Elongate, eel-like vertebrates [31] that often have rasping and sucking disks for mouths.

lancelets (*Cephalochordata*) [120] A group of weakly swimming, eel-like benthic marine chordates [30].

land plants (*Embryophyta*) [11] Plants with embryos that develop within protective struc-tures; also called embryophytes. Sporophytes and gametophytes are multicellular. Land plants possess a cuticle. Major groups are the liver-worts [68], mosses [69], hornworts [70], and vascular plants [12].

larvaceans (*Larvacea*) Solitary, planktonic tunicates [121] that retain both notochords and nerve cords throughout their lives.

lepidosaurs (*Lepidosauria*) [130] Reptiles [37] with overlapping scales. Includes tua-taras and squamates (lizards, snakes, and amphisbaenians).

leptosporangiate ferns (*Pteridopsida* or *Polypodiopsida*) [73] Vascular plants [12] usual-ly possessing large, frondlike leaves that unfold from a "fiddlehead," and possessing thin-walled sporangia.

life (*Life*) [1] The monophyletic group that includes all known living organisms. Characterized by a nucleic-acid based genetic system (DNA or RNA), metabolism, and cellular structure. Some parasitic forms, such as viruses, have secondarily lost some of these features and rely on the cellular environment of their host.

liverworts (*Hepatophyta*) [68] Nonvascular plants lacking stomata; stalk of sporophyte elongates along its entire length.

lobe-limbed vertebrates (*Sarcopterygii*) [34] One of the two major groups of bony vertebrates [33], characterized by jointed appendages (paired fins or limbs).

loboseans (*Lobosea*) A group of unicellular amoebozoans [84]; includes the most familiar amoebas (e.g., *Amoeba proteus*).

"lophophorates" Convenience term used to describe several groups of lophotrochozoans [24] that have a feeding structure called a loph-ophore (a circular or U-shaped ridge around the mouth that bears one or two rows of ciliated, hollow tentacles). Not a monophyletic group.

lophotrochozoans (*Lophotrochozoa*) [24] One of the two main groups of protostomes

[23]. This group is morphologically diverse, and is supported primarily on information from gene sequences. Includes bryozoans and entoprocts [99], flatworms [100], rotifers and gastrotrichs [101], ribbon worms [102], brachio-pods [103], phoronids [104], annelids [105], and mollusks [106].

loriciferans (*Loricifera*) [109] Small (< 1 mm) ecdysozoans [25] with bodies in four parts, cov-ered with six plates.

low-GC Gram-positives (*Firmicutes*) [39] A diverse group of bacteria [2] with a relatively low (G+C)/(A+T) ratio of their DNA, often but not always Gram-positive, some producing endospores.

lungfishes (*Dipnoi*) [127] A group of aquatic lobe-limbed vertebrates [34] that are the closest living relatives of the tetrapods [35]. They have a modified swim bladder used to absorb oxygen from air, so some species can survive the tem-porary drying of their habitat.

lycophytes (*Lycopodiophyta*) [71] Vascular plants [12] characterized by microphylls; includes club mosses, spike mosses, and quillworts.

– M –

magnoliids (*Magnoliidae*) [81] A major group of angiosperms [15] possessing two cotyle-dons and pollen grains with a single opening. The group is defined primarily by nucleotide sequence data; it is more closely related to the eudicots and monocots than to three other small angiosperm groups.

mammals (*Mammalia*) [129] A group of tetrapods [35] with hair covering all or part of their skin; -females produce milk to feed their developing young. Includes the prototherians, marsupials, and -eutherians.

mandibulates (*Mandibulata*) [27] Arthropods [26] that include mandibles as mouth parts. Includes myriapods [115], crustaceans [116], and hexapods [117].

marsupials (*Marsupialia*) Mammals [129] in which the female typically has a marsupium (a pouch for rearing young, which are born at an extremely early stage in development). Includes such familiar mammals as opossums, koalas, and kangaroos.

metazoans (*Metazoa*) See animals [19].

microbial eukaryotes See "protists."

microsporidia (*Microsporidia*) [85] A group of parasitic unicellular fungi [17] that lack mito-chondria and have walls that contain chitin.

mollusks (*Mollusca*) [106] One of the major groups of lophotrochozoans [24], mollusks have bodies composed of a foot, a mantle (which often secretes a hard, calcareous shell), and a visceral mass. Includes monoplacophorans, chitons, bivalves, gastropods, and cephalopods.

monilophytes (*Monilophyta*) A group of vascular plants [12], sister to the seed plants [13], characterized by overtopping and posses-sion of megaphylls; includes the horsetails [72] and ferns [73].

monocots (*Monocotyledones*) [82] Angio-sperms [15] characterized by possession of a single cotyledon, usually parallel leaf veins, a fibrous root system, pollen grains with a single opening, and floral organs usually in multiples of three.

monogeneans (*Monogenea*)   A group of ectoparasitic flatworms [100].

monoplacophorans (*Monoplacophora*) Mollusks [106] with segmented body parts and a single, thin, flat, rounded, bilateral shell.

mosses (*Bryophyta*) [69]   Nonvascular plants with true stomata and erect, "leafy" gametophytes; sporophytes elongate by apical cell division.

moss animals   *See* bryozoans [99].

myriapods (*Myriapoda*) [115]   Arthropods [26] characterized by an elongate, segmented trunk with many legs. Includes centipedes and millipedes.

– N –

nanoarchaeotes (*Nanoarchaeota*)   A group of extremely small, thermophilic archaeans [3] with a much-reduced genome. The only described example can survive only when attached to a host organism.

nematodes (*Nematoda*) [111]   A very large group of elongate, unsegmented ecdysozoans [25] with thick, multilayer cuticles. They are among the most abundant and diverse animals, although most species have not yet been described. Include free-living predators and scavengers, as well as parasites of most species of land plants [11] and animals [19].

neognaths (*Neognathae*)   The main group of birds [132], including all living species except the ratites (ostrich, emu, rheas, kiwis, cassowaries) and tinamous. *See* palaeognaths.

– O –

oligochaetes (*Oligochaeta*)   Annelid [105] group whose members lack parapodia, eyes, and anterior tentacles, and have few setae. Earthworms are the most familiar oligochaetes.

onychophorans (*Onychophora*) [113] Elongate, segmented ecdysozoans [25] with many pairs of soft, unjointed, claw-bearing legs. Also known as velvet worms.

oomycetes (*Oomycota*) [52] Water molds and relatives; absorptive heterotrophs with nutrient-absorbing, filamentous hyphae.

opisthokonts (*Opisthokonta*) [16]   A group of eukaryotes [4] in which the flagellum on motile cells, if present, is posterior. The opisthokonts include the fungi [17], animals [19], and choanoflagellates [91].

ostracods (*Ostracoda*)   Marine and freshwater crustaceans [116] that are laterally compressed and protected by two clamlike calcareous or chitinous shells.

– P –

palaeognaths (*Palaeognathae*)   A group of secondarily flightless or weakly flying birds [132]. Includes the flightless ratites (ostrich, emu, rheas, kiwis, cassowaries) and the weakly flying tinamous.

parabasalids (*Parabasalia*) [59]   A group of unicellular eukaryotes [4] that lack mitochondria; they possess flagella in clusters near the anterior of the cell.

phoronids (*Phoronida*) [104]   A small group of sessile, wormlike marine lophotrochozoans [24] that secrete chitinous tubes and feed using a lophophore.

placoderms (*Placodermi*)   An extinct group of jawed vertebrates [32] that lacked teeth. Placoderms were the dominant predators in Devonian oceans.

placozoans (*Placozoa*) [96]   A poorly known group of structurally simple, asymmetrical, flattened, transparent animals found in coastal marine tropical and subtropical seas. Most evidence suggests that placozoans are secondarily simplified eumetazoans [21].

Plantae [9]   The most broadly defined plant group. In most parts of this book, we use the word "plant" as synonymous with "land plant" [11], a more restrictive definition.

plasmodial slime molds (*Myxogastrida*)   Amoebozoans [84] that in their feeding stage consist of a coenocyte called a plasmodium.

pogonophorans (*Pogonophora*) Deep-sea annelids [105] that lack a mouth or digestive tract; they feed by taking up dissolved organic matter, facilitated by endosymbiotic bacteria in a specialized organ (the trophosome).

polychaetes (*Polychaeta*)   A group of mostly marine annelids [105] with one or more pairs of eyes and one or more pairs of feeding tentacles; parapodia and setae extend from most body segments. May be paraphyletic with respect to the clitellates.

priapulids (*Priapulida*) [107]   A small group of cylindrical, unsegmented, wormlike marine ecdysozoans [25] that takes its name from its phallic appearance.

"prokaryotes"   Not a monophyletic group; as commonly used, includes the bacteria [2] and archaeans [3]. A term of convenience encompassing all cellular organisms that are not eukaryotes.

proteobacteria (*Proteobacteria*) [44]   A large and extremely diverse group of Gram-negative bacteria that includes many pathogens, nitrogen fixers, and photosynthesizers. Includes the alpha, beta, gamma, delta, and epsilon proteobacteria.

"protists"   This term of convenience is used to encompass a large number of distinct and distantly related groups of eukaryotes, many but far from all of which are microbial and unicellular. Essentially a "catch-all" term for any eukaryote group not contained within the land plants [11], fungi [17], or animals [19].

protostomes (*Protostomia*) [23]   One of the two major groups of bilaterians [22]. In protostomes, the mouth typically forms from the blastopore (if present) in early development (contrast with deuterostomes). The major protostome groups are the lophotrochozoans [24] and ecdysozoans [25].

prototherians (*Prototheria*)   A mostly extinct group of mammals [129], common during the Cretaceous and early Cenozoic. The five living species—four echidnas and the duck-billed platypus—are the only extant egg-laying mammals.

pterobranchs (*Pterobranchia*)   A small group of sedentary marine hemichordates [119] that live in tubes secreted by the proboscis. They have one to nine pairs of arms, each bearing long tentacles that capture prey and function in gas exchange.

pycnogonids (*Pycnogonida*)   Treated in this book as a group of chelicerates [114], but sometimes considered an independent group of arthropods [26]. Pycnogonids have reduced bodies and very long, slender legs. Also called sea spiders.

– R –

radiolarians (*Radiolaria*) [55]   Amoeboid organisms with needlelike pseudopods supported by microtubules. Most have glassy internal skeletons.

ray-finned fishes (*Actinopterygii*) [125]   A highly diverse group of freshwater and marine bony vertebrates [33]. They have reduced swim bladders that often function as hydrostatic organs and fins supported by soft rays (lepidotrichia). Includes most familiar fishes.

red algae (*Rhodophyta*) [64]   Mostly multicellular, marine and freshwater algae characterized by the presence of phycoerythrin in their chloroplasts.

reptiles (*Reptilia*) [37]   One of the two major groups of extant amniotes [36], supported on the basis of similar skull structure and gene sequences. The term "reptiles" traditionally excluded the birds [132], but the resulting group is then clearly paraphyletic. As used in this book, the reptiles include turtles [131], lepidosaurs [130], birds [132], and crocodilians [133].

rhizaria (*Rhizaria*) [7]   Mostly amoeboid unicellular eukaryotes with pseudopods, many with external or internal shells. Includes the foraminiferans [57], cercozoans [56], and radiolarians [55].

rhyniophytes (*Rhyniophyta*)   A group of early vascular plants [12] that appeared in the Silurian and became extinct in the Devonian. Possessed dichotomously branching stems with terminal sporangia but no true leaves or roots.

ribbon worms (*Nemertea*) [102]   A group of unsegmented lophotrochozoans [24] with an eversible proboscis used to capture prey. Mostly marine, but some species live in fresh water or on land.

rotifers (*Rotifera*) [101]   Tiny (< 0.5 mm) lophotrochozoans [24] with a pseudocoelomic body cavity that functions as a hydrostatic organ, and a ciliated feeding organ called the corona that surrounds the head. Rotifers live in freshwater and wet terrestrial habitats.

roundworms (*Nematoda*) [111]   *See* nematodes.

– S –

sac fungi (*Ascomycota*) [89]   Fungi that bear the products of meiosis within sacs (asci) if the organism is multicellular. Some are unicellular.

salamanders (*Caudata*)   A group of amphibians [128] with distinct tails in both larvae and adults and limbs set at right angles to the body.

salps   *See* thaliaceans.

sarcopterygians (*Sarcopterygii*) [34]   *See* lobe-limbed vertebrates.

scyphozoans (*Scyphozoa*)   Marine cnidarians [97] in which the medusa stage dominates the life cycle. Commonly known as jellyfish.

sea cucumbers (*Holothuroidea*) Echinoderms [118] with an elongate, cucumber-shaped body and leathery skin. They are scavengers on the ocean floor.

sea spiders *See* pycnogonids.

sea squirts *See* ascidians.

sea stars (*Asteroidea*) Echinoderms [118] with five (or more) fleshy "arms" radiating from an indistinct central disk. Also called starfishes.

sea urchins (*Echinoidea*) Echinoderms [118] with a test (shell) that is covered in spines. Most are globular in shape, although some groups (such as the sand dollars) are flattened.

"seed ferns" A paraphyletic group of loosely related, extinct seed plants that flourished in the Devonian and Carboniferous. Characterized by large, frondlike leaves that bore seeds.

seed plants (*Spermatophyta*) [13] Heterosporous vascular plants [12] that produce seeds; most produce wood; branching is axillary (not dichotomous). The major seed plant groups are gymnosperms [14] and angiosperms [15].

sow bugs *See* isopods.

spirochetes (*Spirochaetes*) [45] Motile, Gram-negative bacteria with a helically coiled structure and characterized by axial filaments.

sponges (*Porifera*) [20] A group of relatively asymmetric, filter-feeding animals that lack a gut or nervous system and generally lack differentiated tissues. Includes glass sponges [92], demosponges [93], and calcareous sponges [94].

springtails (*Collembola*) Wingless hexapods [117] with springing structures on the third and fourth segments of their bodies. Springtails are extremely abundant in some environments (especially in soil, leaf litter, and vegetation).

squamates (*Squamata*) The major group of lepidosaurs [130], characterized by the possession of movable quadrate bones (which allow the upper jaw to move independently of the rest of the skull) and hemipenes (a paired set of eversible penises, or penes) in males. Includes the lizards (a paraphyletic group), snakes, and amphisbaenians.

star anise (*Austrobaileyales*) [80] A group of woody angiosperms [15] thought to be the sister-group of the clade of flowering plants that includes eudicots [83], monocots [82], and magnoliids [81].

starfish (*Asteroidea*) *See* sea stars.

stoneworts (*Charales*) [67] Multicellular green algae with branching, apical growth and plasmodesmata between adjacent cells. The closest living relatives of the land plants [11], they retain the egg in the parent organism.

stramenopiles (*Heterokonta* or *Stramenopila*) [6] Organisms having, at some stage in their life cycle, two unequal flagella, the longer possessing rows of tubular hairs. Chloroplasts, when present, surrounded by four membranes. Major stramenopile groups include the brown algae [54], diatoms [53], and oomycetes [52].

– T –

tapeworms (*Cestoda*) Parasitic flatworms [100] that live in the digestive tracts of vertebrates as adults, and usually in various other species of animals as juveniles.

tardigrades (*Tardigrada*) [112] Small (< 0.5 mm) ecdysozoans [25] with fleshy, unjointed legs and no circulatory or gas exchange organs. They live in marine sands, in temporary freshwater pools, and on the water films of plants. Also called water bears.

tetrapods (*Tetrapoda*) [35] The major group of lobe-limbed vertebrates [34]; includes the amphibians [128] and the amniotes [36]. Named for the presence of four jointed limbs (although limbs have been secondarily reduced or lost completely in several tetrapod groups).

thaliaceans (*Thaliacea*) A group of solitary or colonial planktonic marine tunicates [121]. Also called salps.

therians (*Theria*) Mammals [129] characterized by viviparity (live birth). Includes eutherians and marsupials.

theropods (*Theropoda*) Archosaurs [38] with bipedal stance, hollow bones, a furcula ("wishbone"), elongated metatarsals with three-fingered feet, and a pelvis that points backwards. Includes many well-known extinct dinosaurs (such as *Tyrannosaurus rex*), as well as the living birds [132].

tracheophytes *See* vascular plants [12].

trilobites (*Trilobita*) An extinct group of arthropods [26] related to the chelicerates [114]. Trilobites flourished from the Cambrian through the Permian.

tuataras (*Rhyncocephalia*) A group of lepidosaurs [130] known mostly from fossils; there are only two living tuatara species. The quadrate bone of the upper jaw is fixed firmly to the skull. Sister group of the squamates.

tunicates (*Tunicata*) [121] A group of chordates [30] that are mostly saclike filter feeders as adults, with motile larval stages that resemble tadpoles.

turbellarians (*Turbellaria*) A group of free-living, generally carnivorous flatworms [100]. Their monophyly is questionable.

turtles (*Testudines*) [131] A group of reptiles [37] with a bony carapace (upper shell) and plastron (lower shell) that encase the body in a fashion unique among the vertebrates.

– U –

urochordates (*Tunicata*) [121] *See* tunicates.

– V –

vascular plants (*Tracheophyta*) [12] Plants with xylem and phloem. Major groups include the lycophytes [71] and euphyllophytes.

vertebrates (*Vertebrata*) [31] The largest group of chordates [30], characterized by a rigid endoskeleton supported by the vertebral column and an anterior skull encasing a brain. Includes hagfishes [122], lampreys [123], and the jawed vertebrates [32], although some biologists exclude the hagfishes from this group. *See also* craniates.

– W –

water bears *See* tardigrades.

water lilies (*Nymphaeaceae*) [79] A group of aquatic, freshwater angiosperms [15] that are rooted in soil in shallow water, with round floating leaves and flowers that extend above the water's surface. They are the sister-group to most of the remaining flowering plants, with the exception of the genus *Amborella* [78].

– Y –

"yeasts" Convenience term for several distantly related groups of unicellular fungi [17].

– Z –

"zygospore fungi" (*Zygomycota*, if monophyletic) [87] A convenience term for a probably paraphyletic group of fungi [17] in which hyphae of differing mating types conjugate to form a zygosporangium.

Appendix B   Statistics Primer

This appendix is designed to help you conduct simple statistical analyses and understand their application and importance. This introduction will help you complete the Apply the Concept and Analyze the Data problems throughout this book. The formulas for a number of statistical tests are presented here, but the presentation is designed primarily to help you understand the purpose and reasoning of the various tests. Once you understand the basis of the analysis, you may wish to use one of many free, online web sites for conducting the tests and calculating relevant test statistics (such as http://faculty.vassar.edu/lowry/VassarStats.html).

Why Do We Do Statistics?

ALMOST EVERYTHING VARIES   We live in a variable world, but within the variation we see among biological organisms there are predictable patterns. We use statistics to find and analyze these patterns. Consider any group of common things in nature—all women aged 22, all the cells in your liver, or all the blades of grass in your yard. Although they will have many similar characteristics, they will also have important differences. Men aged 22 tend to be taller than women aged 22, but, of course, not every man will be taller than every woman in this age group.

Natural variation can make it difficult to find general patterns. For example, scientists have determined that smoking increases the risk of getting lung cancer. But we know that not all smokers will develop lung cancer and not all nonsmokers will remain cancer-free. If we compare just one smoker to just one nonsmoker, we may end up drawing the wrong conclusion. So how did scientists discover this general pattern? How many smokers and nonsmokers did they examine before they felt confident about the risk of smoking?

Statistics helps us to find general patterns, even when nature does not always follow those patterns.

AVOIDING FALSE POSITIVES AND FALSE NEGATIVES   When a woman takes a pregnancy test, there is some chance that it will be positive even if she is not pregnant, and there is some chance that it will be negative even if she is pregnant. We call these kinds of mistakes *false positives* and *false negatives*.

Doing science is a bit like taking a medical test. We observe patterns in the world, and we try to draw conclusions about how the world works from those observations. Sometimes our observations lead us to draw the wrong conclusions. We might conclude that a phenomenon occurs, when it actually does not; or we might conclude that a phenomenon does not occur, when it actually does.

For example, the planet Earth has been warming over the past century (see Concept 46.4). Ecologists are interested in whether plant and animal populations have been affected by global warming. If we have long-term information about the locations of species and temperatures in certain areas, we can determine whether species movements coincide with temperature changes. Such information can, however, be very complicated. Without proper statistical methods, one may not be able to detect the true impact of temperature or, instead, may think a pattern exists when it does not.

Statistics helps us to avoid drawing the wrong conclusions.

How Does Statistics Help Us Understand the Natural World?

Statistics is essential to scientific discovery. Most biological studies involve five basic steps, each of which requires statistics:

- **Step 1: Experimental Design**
  Clearly define the scientific question and the methods necessary to tackle the question.

- **Step 2: Data Collection**
  Gather information about the natural world through experiments and field studies.

- **Step 3: Organize and Visualize the Data**
  Use tables, graphs, and other useful representations to gain intuition about the data.

- **Step 4: Summarize the Data**
  Summarize the data with a few key statistical calculations.

- **Step 5: Inferential Statistics**
  Use statistical methods to draw general conclusions from the data about the way the world works.

Step 1: Experimental Design

We conduct experiments to gain knowledge about the world. Scientists come up with scientific ideas based on prior research and their own observations. These ideas may take the form of a question like "Does smoking cause cancer?," a hypothesis like "Smoking increases the risk of cancer," or a prediction like "If a person smokes, he/she will increase his/her chances of developing cancer." Experiments allow us to test such scientific ideas, but designing a good experiment can be quite challenging.

We use statistics to guide us in designing experiments so that we end up with the right kinds of data. Before embarking on an experiment, we use statistics to determine how much data will be required to test our idea, and to prevent extraneous factors from misleading us. For example, suppose we want to conduct an experiment on fertilizers to test the hypothesis that nitrogen increases plant growth. If we include too few plants, we will not be able to determine whether or not nitrogen has an effect on growth, and the experiment will be for naught. If we include too many plants, we will waste valuable time and resources. Furthermore, we should design the experiment so that we can detect differences that are actually caused by nitrogen fertilization rather than by variation, for example, in sunlight or precipitation experienced by the plants.

Step 2: Data Collection

TAKING SAMPLES   When biologists gather information about the natural world, they typically collect a few representative pieces of information. For example, when evaluating the efficacy of a candidate drug for medulloblastoma brain cancer, scientists may test the drug on tens or hundreds of patients, and then draw conclusions about its efficacy for all patients with these tumors. Similarly, scientists studying the relationship between body weight and clutch size (number of eggs) for female spiders of a particular species may examine tens to hundreds of spiders to make their conclusions.

We use the expression "sampling from a population" to describe this general method of taking representative pieces of information from the system under investigation (**Figure B1**). The pieces of information in a **sample** are called **observations**. In the cancer therapy example, each observation was the change in a patient's tumor size six months after initiating treatment, and the population of interest was all individuals with medulloblastoma tumors. In the spider example, each observation was a pair of measurements—body size and clutch size—for a single female spider, and the population of interest was all female spiders of this species.

Sampling is a matter of necessity, not laziness. We cannot hope (and would not want) to collect *all* of the female spiders of the species of interest on Earth! Instead, we use statistics to determine how many spiders we must collect in order to confidently infer something about the general population and then use statistics again to make such inferences.

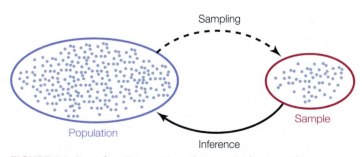

FIGURE B1 Sampling From a Population   Biologists take representative samples from a population, use descriptive statistics to characterize their samples, and then use inferential statistics to draw conclusions about the original population.

TABLE B1
Poinsettia Colors

Color	Frequency	Proportion
Red	108	0.59
Pink	34	0.19
White	40	0.22
Total	182	1.00

DATA COME IN ALL SHAPES AND SIZES   In statistics, we use the word *variable* to mean a measurable characteristic of an individual or a system. Some variables are on a numerical scale, like the daily high temperature (a numerical value constrained by the precision of our thermometer), or the clutch size of a spider (a whole number: 0, 1, 2, 3,…). We call these **quantitative variables**. Quantitative variables that only take on whole number values are called **discrete variables**, whereas variables that can also take on any fractional value are called **continuous variables**.

Other variables take categories as values, like a human blood type (A, B, AB, or O) or an ant caste (queen, worker, or male). We call these **categorical variables**. Categorical variables with a natural ordering, like a final grade in Introductory Biology (A, B, C, D, or F), are called **ordinal variables**.

Each class of variables comes with its own set of statistical methods. We will introduce a few common methods in this Appendix that will help you work on the problems presented in this book, but you should consult a biostatistics textbook for more advanced tests and analyses for other data sets and problems.

Step 3: Organize and Visualize the Data

Tables and graphs can help you gain intuition about your data, design appropriate statistical tests, and anticipate the outcome of your analysis. A **frequency distribution** lists all possible values and the number of occurrences of each value in the sample.

TABLE B2
Fish Weights of *Abramis brama* from Lake Laengelmavesi

Weight (grams)	Frequency	Relative Frequency
201–300	2	0.06
301–400	3	0.09
401–500	8	0.24
501–600	3	0.09
601–700	8	0.24
701–800	3	0.09
801–900	1	0.03
901–1000	6	0.18
Total	34	1.00

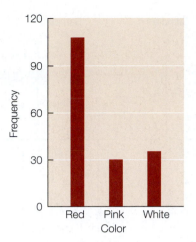

B2 Bar Charts Compare Categorical Data This bar chart shows the frequency of three poinsettia colors that result from an experimental cross.

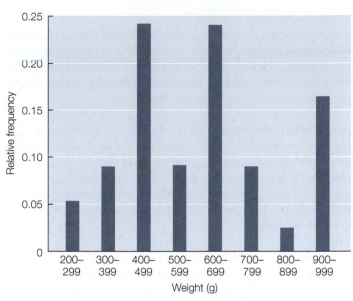

B4 Histograms Depict Frequency Distributions of Quantitative Data This histogram shows the relative frequency of different weight-classes of fish (*Abramis brama*).

Table B1 shows a frequency distribution of the colors of 182 poinsettia plants (red, pink, or white) resulting from an experimental cross between two parent plants. For categorical data like this, we can visualize the frequency distribution by constructing a **bar chart**. The heights of the bars indicate the number of observations in each category (**Figure B2**). Another way to display the same data is in a **pie chart**, which shows the proportion of each category represented like pieces of a pie (**Figure B3**).

For quantitative data, it is often useful to condense your data by grouping (or binning) it into **classes**. In **Table B2**, we see a grouped frequency distribution of fish weights for a sample of 34 fish (*Abramis brama*) caught in Lake Laengelmavesi in Finland. The second column (*Frequency*) gives the number of observations in each class and the third column (*Relative Frequency*) gives the overall proportion of observations falling into each class.

Histograms depict frequency distributions for quantitative data. The histogram in **Figure B4** shows the relative frequencies of each weight class in this study. When grouping quantitative data, it is necessary to decide how many classes to include. It is often useful to look at multiple histograms before deciding which grouping offers the best representation of the data.

Sometimes we wish to compare two quantitative variables. For example, the researchers at Lake Laengelmavesi investigated the relationship between fish weight and length and thus also measured the length of each fish. We can visualize this relationship using a **scatter plot** in which the weight and length of each fish is represented as a single point (**Figure B5**). We say that these two variables have a **linear relationship** since the points in their scatter plot fall roughly on a straight line.

Tables and graphs are critical to interpreting and communicating data, and thus should be as self-contained and comprehensible as possible. Their content should be easily understood simply by looking at them. Axes, captions, and units should be clearly labeled, statistical terms should be defined, and appropriate groupings should be used when tabulating or graphing quantitative data.

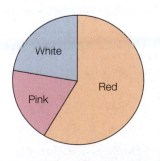

B3 Pie Charts Show Proportions of Categories This pie chart shows the proportions of the three poinsettia colors presented in Table B1.

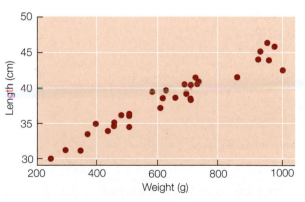
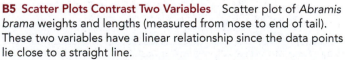

B5 Scatter Plots Contrast Two Variables Scatter plot of *Abramis brama* weights and lengths (measured from nose to end of tail). These two variables have a linear relationship since the data points lie close to a straight line.

Step 4: Summarize the Data

A **statistic** is a numerical quantity calculated from data, while **descriptive statistics** are quantities that describe general patterns in data. Descriptive statistics allow us to make straightforward comparisons between different data sets and concisely communicate basic features of our data.

DESCRIBING CATEGORICAL DATA   For categorical variables, we typically use proportions to describe our data. That is, we construct tables containing the proportions of observations in each category. For example, the third column in Table B1 provides the proportions of poinsettia plants in each color category, and the pie chart in Figure B3 provides a visual representation of those proportions.

DESCRIBING QUANTITATIVE DATA   For quantitative data, we often start by calculating the average value or **mean** of our sample. This familiar quantity is simply the sum of all the values in the sample divided by the number of observations in our sample (**Figure B6**). The mean is only one of several quantities that roughly tell us where the *center* of our data lies. We call these quantities **measures of center**. Other commonly used measures of center are the **median**—the value that literally lies

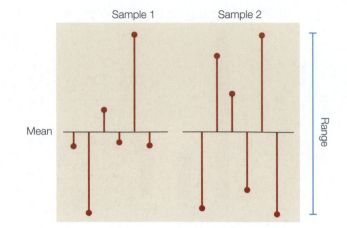

Sample 1   Sample 2

Mean

Range

B7 Measures of Dispersion   Two samples with the same mean (black horizontal lines) and range (blue vertical line). Red lines show the deviations of each observation from the mean. Samples with large deviations have large standard deviations. The left sample has a smaller standard deviation than the right sample.

in the middle of the sample—and the **mode**—the most frequent value in the sample.

It is often just as important to quantify the variation in the data as it is to calculate its center. There are several statistics that tell us how much the values differ from one another. We call these **measures of dispersion**. The easiest to one understand and calculate is the **range**, which is simply the largest value in the sample minus the smallest value. The most commonly used measure of dispersion is the **standard deviation**, which calculates the extent to which the data are spread out from the mean. A deviation is the difference between an observation and the mean of the sample, and the standard deviation is a number that summarizes all of the deviations. Two samples can have the same range, but very different standard deviations if one is clustered closer to the mean than the other. In **Figure B7**, for example, the left sample has a lower standard deviation ($s = 2.6$) than the right sample ($s = 3.6$), even though the two samples have the same means and ranges.

To demonstrate these descriptive statistics, we return to the Lake Laengelmavesi study. The researchers also caught and recorded the weights of six fish in the species *Leusiscus idus*: 270, 270, 306, 540, 800, and 1,000 grams. The mean weight in this sample (equation 1 in Figure B6) is:

$$\bar{x} \frac{\Delta N}{\Delta T} = \frac{(270+270+306+540+800+1000)}{6} = 531$$

Since there is an even number of observations in the sample, then the median weight is the value halfway between the two middle values:

$$\frac{306+540}{2} = 423$$

The mode of the sample is 270, the only value that appears more than once. The standard deviation (equation 2 in Figure B6) is:

RESEARCH**TOOLS**

B6   Descriptive Statistics for Quantitative Data

Below are the equations used to calculate the descriptive statistic we discuss in this appendix. You can calculate these statistics yourself, or use free internet resources to help you make your calculations.

Notation:
$x_1, x_2, x_3, \ldots x_n$ are the n observations of variable X in the sample.

$\sum_{i=1}^{n} x_i = x_1 + x_2 + x_3, \ldots + x_n$ is the sum of all of the observations. (The Greek letter sigma, Σ, is used to denote "sum of.")

In regression, the independent variable is X, and the dependent variable is Y. b_0 is the vertical intercept of a regression line. b_1 is the slope of a regression line.

Equations

1. Mean: $\bar{x} = \dfrac{\sum_{i=1}^{n} x_i}{n}$

2. Standard deviation: $s = \sqrt{\dfrac{\Sigma(x_i - \bar{x})^2}{n-1}}$

3. Correlation coefficient: $r = \dfrac{\Sigma(x_i - \bar{x})(y_i - \bar{y})}{\sqrt{\Sigma(x_i - \bar{x})^2(y_i - \bar{y})^2}}$

4. Least-squares regression line: $Y = b_0 + b_1 X$
   where $b_1 = \dfrac{\Sigma(x_i - \bar{x})(y_i - \bar{y})}{\Sigma(x_i - \bar{x})^2}$ and $b_0 = \bar{y} - b_1\bar{x}$

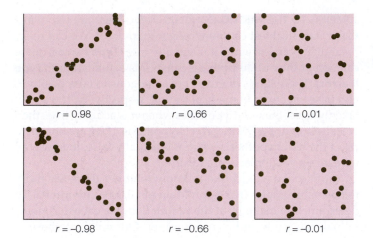

B8 Correlation Coefficients The correlation coefficient (*r*) indicates both the strength and the direction of the relationship.

$$s = \sqrt{\frac{(270-531)^2 + (270-531)^2 + (306-531)^2 + (540-531)^2 + (800-531)^2 + (1000-531)^2}{5}} = 309.6$$

and the range is $1000 - 270 = 730$.

DESCRIBING THE RELATIONSHIP BETWEEN TWO QUANTITATIVE VARIABLES Biologists are often interested in understanding the relationship between two different quantitative variables: How does the height of an organism relate to its weight? How does air pollution relate to the prevalence of asthma? How does lichen abundance relate to levels of air pollution? Recall that scatter plots visually represent such relationships.

We can quantify the strength of the relationship between two quantitative variables using a single value called the Pearson product–moment **correlation coefficient** (equation 3 in Figure B6). This statistic ranges between –1 and 1, and tells us how closely the points in a scatter plot conform to a straight line. A negative correlation coefficient indicates that one variable decreases as the other increases; a positive correlation coefficient indicates that the two variables increase together, and a correlation coefficient of zero indicates that there is no linear relationship between the two variables (**Figure B8**).

One must always keep in mind that *correlation does not mean causation.* Two variables can be closely related without one causing the other. For example, the number of cavities in a child's mouth correlates positively with the size of their feet. Clearly cavities do not enhance foot growth; nor does foot growth cause tooth decay. Instead the correlation exists because both quantities tend to increase with age.

Intuitively, the straight line that tracks the cluster of points on a scatter plot tells us something about the *typical* relationship between the two variables. Statisticians do not, however, simply eyeball the data and draw a line by hand. They often use a method called least-squares **linear regression** to fit a straight line to the data (equation 4 in Figure B6). This method calculates the line that minimizes the overall vertical distances between the points in the scatter plot and the line itself. These distances are called **residuals** (**Figure B9**). Two parameters

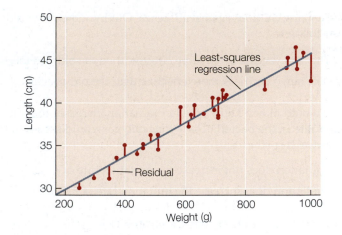

B9 Linear Regression Estimates the Typical Relationship Between Two Variables Linear-least squares regression line for *Abramis brama* weights and lengths (measured from nose to end of tail). The regression line (blue line) is given by the equation $Y = 26.1 + 0.02X$. It is the line that minimizes the sum of the squares of the residuals (red lines).

describe the regression line: b_0 (the vertical intercept of the line, or the expected value of variable Y when $X = 0$), and b_1 (the slope of the line, or how much values of Y are expected to change with changes in values of X).

Step 5: Inferential Statistics

Data analysis often culminates with statistical inference—an attempt to draw general conclusions about the system under investigation. As depicted in Figure B1, the primary reason we collect data is to gain insight into the larger system from which the data are collected. When we test a new medulloblastoma brain cancer drug on ten patients, we do not simply want to know the fate of those ten individuals; rather, we hope to predict its efficacy on the much larger group of all medulloblastoma patients.

STATISTICAL HYPOTHESES When it comes to inferring something about the real world from our data, we often have a *"Whether or not"* question in mind. For example, we would like to know whether or not global warming impacts biodiversity; whether or not the clutch size of a spider increases with body size; or whether or not soil nitrogen increases the growth of a particular plant species.

Before making statistical inferences from data, we must formalize our *"Whether or not"* question into a pair of opposing hypotheses—a **null hypothesis** (denoted H_0) and an **alternative hypothesis** (denoted H_A). The alternative hypothesis is the *"Whether"*—it is formulated to describe the effect that we expect our data to support; the null hypothesis is the *"or not"*—it is formulated to represent the absence of the effect. In other words, we typically conduct our experiment seeking to demonstrate something new (the alternative hypothesis) and thereby reject the idea that it does not occur (the null hypothesis).

Suppose, for example, we would like to know *whether or not* a new vaccine is more effective than an existing vaccine at immunizing children against influenza. Our hypotheses would be as follows:

H_0: The new vaccine is not more effective than the old vaccine.

H_A: The new vaccine is more effective than the old vaccine.

If we would like to know whether radiation increases the mutation rate in the bacteria *Escherichia coli*, we would set up the following hypotheses:

H_0: Radiation does not increase the mutation rate of *E. coli*.

H_A: Radiation does increase the mutation rate of *E. coli*.

STATISTICAL BURDEN OF PROOF In the U.S. justice system, people are innocent until proven guilty. In statistics, the world is *null until proven alternative*. Statistics requires overwhelming proof in favor of the alternative hypothesis before rejecting the null hypothesis. In other words, scientists favor existing ideas and resist adopting new ideas until compelling evidence suggests otherwise. This is based on a philosophy that it is worse to accept new claims when they are false than to miss out on discovering some true facts about world.

When testing a new influenza vaccine, the burden of proof is on the new vaccine. Suppose we were to vaccinate three children with the new vaccine (Group A), three with the old vaccine (Group B) and leave three children unvaccinated (Group C). If no children from Group A, one child from Group B, and one child from Group C became infected, would we have enough evidence to conclude that the new vaccine is superior to the old vaccine? No, we would not. If the study were enlarged, and two out of 100 children in group A, seven out of 100 children in group B, and 22 out of 100 children in group C become infected, would we then have sufficient evidence to choose the new vaccine? Perhaps, but we need to use statistics to be sure.

This is the traditional burden of proof in biology and science in general. As a consequence, scientists are more likely to miss out on discovering something new (and true) about the world than they are to make a false discovery. In recent years, scientists have begun to question this approach and develop an alternative statistical approach, called **Bayesian inference**, which makes it easier to favor new hypotheses. In this primer, we discuss only traditional statistical methods, often called **frequentist statistics**.

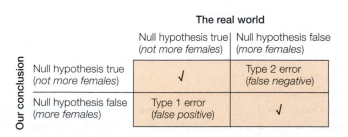

	The real world	
	Null hypothesis true (*not more females*)	Null hypothesis false (*more females*)
Null hypothesis true (*not more females*)	√	Type 2 error (*false negative*)
Null hypothesis false (*more females*)	Type 1 error (*false positive*)	√

Our conclusion (left vertical label)

B10 Two Types of Error Possible outcomes of a statistical test. Statistical inference can result in correct and incorrect conclusions about the population of interest.

JUMPING TO THE WRONG CONCLUSIONS There are two ways that a statistical test can go wrong (**Figure B10**). We can reject the null hypothesis when it is actually true (**Type I error**) or we can accept the null hypothesis when it is actually false (**Type II error**). These kinds of errors are analogous to false positives and false negatives in medical testing, respectively. If we mistakenly reject the null hypothesis when it is actually true, then we falsely endorse the incorrect hypothesis. If we are unable to reject the null hypothesis when it is actually false, then we fail to realize a yet undiscovered truth.

Suppose we would like to know whether there are more females than males in a population of 10,000 individuals. To determine the makeup of the population, we choose 20 individuals randomly and record their sex. Our null hypothesis is that there are *not* more females than males; and our alternative hypothesis is that there are. The following scenarios illustrate the possible mistakes we might make:

- *Scenario 1*: The population actually has 40% females and 60% males. Although our random sample of 20 people is likely to be dominated by males, it is certainly possible that, by chance, we will end up choosing more females than males. If this occurs, and we mistakenly reject the null hypothesis (that there are *not* more females than males), then we make a Type I error.

- *Scenario 2*: The population actually has 60% females and 40% males. If, by chance, we end up with a majority of males in our sample and thus fail reject the null hypothesis, then we make a Type II error.

Fortunately, statistics has been developed precisely to avoid these kinds of errors and inform us about the reliability of our conclusions. The methods are based on calculating the **probabilities** of different possible outcomes. Although you may have heard or even used the word "probability" on multiple occasions, it is important that you understand its mathematical meaning. A probability is a numerical quantity that expresses the likelihood of some event. It ranges between zero and one; zero means that there is no chance the event will occur and one means that the event is guaranteed to occur. This only makes sense if there is an element of chance, that is, if it is possible the event will occur and possible that it will not occur. For example, when we flip a fair coin, it will land on heads with probability 0.5 and land on tails with probability 0.5. When we select individuals randomly from a population with 60% females and 40% males, we will encounter a female with probability 0.6 and a male with probability 0.4.

Probability plays a very important role in statistics. To draw conclusions about the real world (the population) from our sample, we first calculate the probability of obtaining our sample if the null hypothesis is true. Specifically, statistical inference is based on answering the following question:

Suppose the null hypothesis is true. What is the probability that a random sample would, by chance, differ from the null hypothesis as much as our sample differs from the null hypothesis?

If our sample is highly improbable under the null hypothesis, then we rule it out in favor of our alternative hypothesis. If, instead, our sample has a reasonable probability of occurring under the null hypothesis, then we conclude that our data are consistent with the null hypothesis and we do not reject it.

Returning to the sex ratio example, we consider two new scenarios:

- *Scenario 3*: Suppose we want to infer whether or not females constitute the majority of the population (our alternative hypothesis) based on a random sample containing 12 females and eight males. We would calculate the probability that a random sample of 20 people includes at least 12 females assuming that the population, in fact, has a 50:50 sex ratio (our null hypothesis). This probability is 0.13, which is too high to rule out the null hypothesis.

- *Scenario 4*: Suppose now that our sample contains 17 females and three males. If our population is truly evenly divided, then this sample is much less likely than the sample in scenario 3. The probability of such an extreme sample is 0.0002, and would lead us to rule out the null hypothesis and conclude that there are more females than males.

This agrees with our intuition. When choosing 20 people randomly from an evenly divided population, we would be surprised if almost all of them were female, but would not be surprised at all if we ended up with a few more females than males (or a few more males than females). Exactly how many females do we need in our sample before we can confidently infer that they make up the majority of the population? And how confident are we when we reach that conclusion? Statistics allows us to answer these questions precisely.

STATISTICAL SIGNIFICANCE: AVOIDING FALSE POSITIVES Whenever we test hypotheses, we calculate the probability just discussed, and refer to this value as the **P-value** of our test. Specifically, the *P*-value is the probability of getting data as extreme as our data (just by chance) if the null hypothesis is, in fact, true. In other words, it is the likelihood that chance alone would produce data that differ from the null hypothesis as much as our data differ from the null hypothesis. How we measure the difference between our data and the null hypothesis depends on the kind of data in our sample (categorical or quantitative) and the nature of the null hypothesis (assertions about proportions, single variables, multiple variables, differences between variables, correlations between variables, etc.).

For many statistical tests, *P*-values can be calculated mathematically. One option is to quantify the extent to which the data depart from the null hypothesis and then use look-up tables (available in most statistics textbooks, or on the internet) to find the probability that chance alone would produce a difference of that magnitude. Most scientists, however, find *P*-values primarily by using statistical software rather than hand calculations combined with look-up tables. Regardless of the technology, the most important steps of the statistical analysis are still left to the researcher: constructing appropriate null and alternative hypotheses, choosing the correct statistical test, and drawing correct conclusions.

After we calculate a *P*-value from our data, we have to decide whether it is small enough to conclude that our data are inconsistent with the null hypothesis. This is decided by comparing the *P*-value to a threshold called the **significance level**, which is often chosen even before making any calculations. We reject the null hypothesis only when the *P*-value is less than or equal to the significance level, denoted α. This ensures that, if the null hypothesis is true, we have at most a probability α of accidentally rejecting it. Therefore, the lower the value of α, the less likely you are to make a Type I error (lower left cell of Figure B10). The most commonly used significance level is α = 0.05, which limits the probability of a Type I error to 5%.

If our statistical test yields a *P*-value that is less than our significance level α, then we conclude that the effect described by our alternative hypothesis is statistically significant at the level α and we reject the null hypothesis. If our *P*-value is greater than α, then we conclude that we are unable to reject the null hypothesis. In this case, we do not actually reject the alternative hypothesis, rather we conclude that we do not yet have enough evidence to support it.

POWER: AVOIDING FALSE NEGATIVES The **power** of a statistical test is the probability that we will correctly reject the null hypothesis when it is false (lower right cell of Figure B10). Therefore, the higher the power of the test, the less likely we are to make a Type II error (upper right cell of Figure B10). The power of a test can be calculated, and such calculations can be used to improve your methodology. Generally, there are several steps that can be taken to increase power and thereby avoid false negatives:

- **Decrease the significance level**, α. The higher the value of α, the harder it is to reject the null hypothesis, even if it is actually false.

- **Increase the sample size**. The more data one has, the more likely one is to find evidence against the null hypothesis, if it is actually false.

- **Decrease variability in the sample**. The more variation there is in the sample, the harder it is to discern a clear effect (the alternative hypothesis) when it actually exists.

It is always a good idea to design your experiment to reduce any variability that may obscure the pattern you seek to detect. For example, it is possible that the chance of a child contracting influenza varies depending on whether he or she lives in a crowded (e.g., urban) environment or one that is less so (e.g., rural). To reduce variability, a scientist might choose to test a new influenza vaccine only on children from one environment or the other. After you have minimized such extraneous variation, you can use power calculations to choose the right combination of α and sample size to reduce the risks of Type I and Type II errors to desirable levels.

There is a trade-off between Type I and Type II errors: As α increases, the risk of a Type I decreases but the risk of a Type II error increases. As discussed above, scientists tend to be more concerned about Type I errors than Type II errors. That is, they believe that it is worse to mistakenly believe a false hypothesis than it is to fail to make a new discovery. Thus, they prefer to use low values of α. However, there are many real-world scenarios in which it would be worse to make a Type II error than a Type I error. For example, suppose a new cold medication is being tested for dangerous (life-threatening) side effects. The null hypothesis is that there are no such side effects. A Type II error might lead regulatory agencies to approve a harmful medication that could cost human lives. In contrast, a Type I error would simply mean one less cold medication among the many that already line pharmacy shelves. In such cases, policymakers take steps to avoid a Type II error, even if, in doing so, they increase the risk of a Type I error.

STATISTICAL INFERENCE WITH QUANTITATIVE DATA   There are many forms of statistical inference for quantitative data. When measuring a single quantitative variable, like birth weight in lambs, calcium concentration in the blood of pregnant women, or migration rate of birds, we often wish to infer the mean value of the population from which we drew the sample. However, the mean of a randomly chosen sample will not necessarily be the same or even close to the population mean. Suppose we wanted to know the average weight of newborn lambs on a particular farm. By chance, we may end up with a random sample that includes an excess of lightweight lambs and therefore a sample mean that is less than the overall mean in the population.

To infer the population mean from the sample data, we can calculate a **confidence interval for the mean**. This is a statistically derived range of values that is centered on the sample mean and is likely to include the population mean. For example, based on the sample of 34 *Abramis brama* weights from Lake Laengelmavesi (see Table B2; Figure B4), the 95% confidence interval for the mean weight ranges from 554 grams to 698 grams. The true average weight for this species of fish is likely, but not guaranteed, to fall within this range.

Biologists frequently wish to compare the mean values in two or more groups; for example, newborn lamb weights on several different farms, calcium concentration in women in early and late stages of pregnancy, or migration rates in birds of different species. Based on the means and standard deviations calculated for each of the samples, they infer whether or not the means in the different populations are statistically different from one another. There are several statistical methods for this, and the correct method depends on the number of groups, the experimental design, and the nature of the data.

Figure B11 describes the steps of a *t*-test, a simple method for comparing the means in two different groups. To illustrate, we can apply a *t*-test to the Lake Laengelmavesi data to assess whether the two fish species *Abramis brama* and *Leusiscus idus* have significantly different mean weights. We begin by stating our hypotheses and choosing a significance level:

H_0: *Abramis brama* and *Leusiscus idus* have the same mean weight.

H_A: *Abramis brama* and *Leusiscus idus* have different mean weights.

$\alpha = 0.05$

The test statistic is calculated using the means, standard deviations, and sizes of the two samples:

$$t_s = \frac{626-531}{\sqrt{\frac{207^2}{34}+\sqrt{\frac{310^2}{6}}}} = 0.724$$

We can use statistical software or one of the free statistical sites on the internet to find the *P*-value for this result to be $P = 0.497$. Since *P* is considerably greater than α, we fail to reject the null hypothesis and conclude that our study does not provide evidence that the two species have different mean weights.

You may want to consult an introductory statistics textbook to learn more about confidence intervals, *t*-tests, and other basic statistical tests.

STATISTICAL INFERENCE WITH CATEGORICAL DATA   With categorical data, we often wish to infer the distribution of the different categories within the populations from which our samples are drawn. In the simplest case, we have a single categorical variable with two or more categories. If there are just two categories, we can construct a **confidence interval for the proportion** of the population that belongs to one of the two categories. This is a statistically derived range of values that is centered on the sample proportion and is likely to include the population proportion. If there are three or more categories, we can use a **chi-square goodness-of-fit** test to determine whether the distribution of the different categories in the population is consistent with a specific distribution.

Figure B12 outlines the steps of a chi-square goodness-of-fit-test. As an example, consider the data described in Table B1. Many plant species have simple Mendelian genetic systems in which parent plants produce progeny with three different colors of flowers in a ratio of 2:1:1. However, a botanist believes that these particular poinsettia plants have a different genetic system that does not produce a 2:1:1 ratio of red, pink, and

RESEARCH TOOLS

B11   The t-Test

What is the *t*-test? It is a standard method for assessing whether the means of two groups are statistically different from each another.

Step 1: State the null and alternative *hypotheses*:
  H_0: The two populations have the same mean.
  H_A: The two populations have different means.

Step 2: Choose a significance level, α, to limit the risk of a Type 1 error.

Step 3: Calculate the *test statistic*: $t_s = \dfrac{\bar{y}_1 - \bar{y}_2}{\sqrt{\dfrac{s_1^2}{n_1}+\dfrac{s_2^2}{n_2}}}$

  Notation: $\bar{y}_1$ and $\bar{y}_2$ are the sample means; s_1 and s_2 are the sample standard deviations; and n_1 and n_2 are the sample sizes.

Step 4: Use the test statistic to assess whether the data are consistent with the null hypothesis:
  Calculate the *P*-value (P) using statistical software or by hand using statistical tables.

Step 5: Draw conclusions from the test:
  If $P \leq \alpha$, then reject H_0, and conclude that the population distribution is significantly different.
  If $P > \alpha$, then we do not have sufficient evidence to conclude that the means differ.

RESEARCH**TOOLS**

B12 The Chi-Square Goodness-of-Fit Test

What is the chi-square goodness-of-fit test? It is a standard method for assessing whether a sample came from a population with a specific distribution.

Step 1: State the null and alternative *hypotheses*:

H_0: The population has the specified distribution.

H_A: The population does not have the specified distribution.

Step 2: Choose a significance level, α, to limit the risk of a Type 1 error.

Step 3: Determine the *observed frequency* and *expected frequency* for each category:

The observed frequency of a category is simply the number of observations in the sample of that type.

The expected frequency of a category is the probability of the category specified in H_0 multiplied by the overall sample size.

Step 4: Calculate the *test statistic*:
$$\chi_s^2 = \sum_{i=1}^{c} \frac{(O_i - E_i)^2}{E_i}$$

Notation: C is the total number of categories, O_i is the observed frequency of category i, and E_i is the expected frequency of category i.

Step 5: Use the test statistic to assess whether the data are consistent with the null hypothesis:

Calculate the *P-value* (P) using statistical software or by hand using statistical tables.

Step 6: Draw conclusions from the test:

If $P \leq \alpha$, then reject H_0, and conclude that the population distribution is significantly different than the distribution specified by H_0.

If $P > \alpha$, then we do not have sufficient evidence to conclude that population has a different distribution.

We find the *P*-value for this result to be $P = 0.0343$ using statistical software. Since P is less than α, we reject the null hypothesis and conclude that the botanist is correct: The plant color patterns cannot be explained by the simple Mendelian genetic model under consideration.

This introduction is only meant to provide a brief introduction to the concepts of statistical analysis, with a few example tests. **Figure B13** provides a summary of some of the commonly used statistical tests that you may encounter in biological studies.

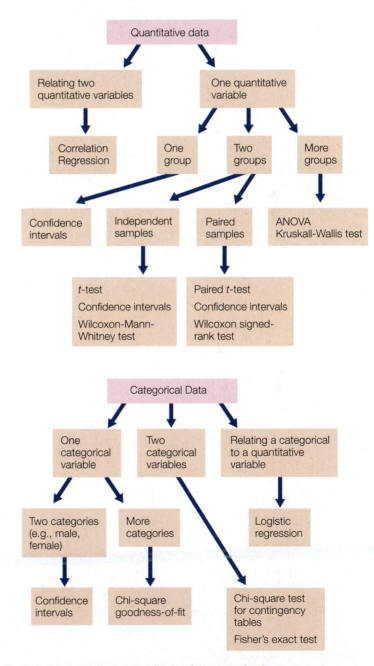

B13 Some Common Methods of Statistical Inference

This flow chart shows some of the commonly used methods of statistical inference for different combinations of data. Detailed descriptions of these methods can be found in most introductory biostatistics textbooks.

---

white plants. A chi-square goodness-of-fit can be used to assess whether or not the data are consistent with this ratio, and thus whether or not this simple genetic explanation is valid. We start by stating our hypotheses and significance level:

H_0: The progeny of this type of cross have the following probabilities of each flower color:

Pr{Red} = .50, Pr{Pink} = .25, Pr{White} = .25

H_A: At least one of the probabilities of H_0 is incorrect.

$\alpha = 0.05$

We next use the probabilities in H_0 and the sample size to calculate the expected frequencies:

	Red	Pink	White
Observed	108	34	40
Expected	(.50)(182) = 91	(.25)(182) = 45.5	(.25)(182) = 45.5

Based on these quantities, we calculate the chi-square test statistic:

$$\chi_s^2 = \sum_{i=1}^{C} \frac{(O_i - E_i)^2}{E_i} = \frac{(108-91)^2}{91} + \frac{(34-45.5)^2}{45.5} + \frac{(40-45.5)^2}{45.5} = 6.747$$

Appendix C Some Measurements Used in Biology

MEASURES OF	UNIT	EQUIVALENTS	METRIC → ENGLISH CONVERSION
Length	meter (m)	base unit	1 m = 39.37 inches = 3.28 feet = 1.196 yards
	kilometer (km)	1 km = 1000 (10^3) m	1 km = 0.62 miles
	centimeter (cm)	1 cm = 0.01 (10^{-2}) m	1 cm = 0.39 inches
	millimeter (mm)	1 mm = 0.1 cm = 10^{-3} m	1 mm = 0.039 inches
	micrometer (µm)	1 µm = 0.001 mm = 10^{-6} m	
	nanometer (nm)	1 nm = 0.001 µm = 10^{-9} m	
Area	square meter (m^2)	base unit	1 m^2 = 1.196 square yards
	hectare (ha)	1 ha = 10,000 m^2	1 ha = 2.47 acres
Volume	liter (L)	base unit	1 L = 1.06 quarts
	milliliter (mL)	1 mL = 0.001 L = 10^{-3} L	1 mL = 0.034 fluid ounces
	microliter (µL)	1 µL = 0.001 mL = 10^{-6} L	
Mass	gram (g)	base unit	1 g = 0.035 ounces
	kilogram (kg)	1 kg = 1000 g	1 kg = 2.20 pounds
	metric ton (mt)	1 mt = 1000 kg	1 mt = 2,200 pounds = 1.10 ton
	milligram (mg)	1 mg = 0.001 g = 10^{-3} g	
	microgram (µg)	1 µg = 0.001 mg = 10^{-6} g	
Temperature	degree Celsius (°C)	base unit	°C = (°F – 32)/1.8
			0°C = 32°F (water freezes)
			100°C = 212°F (water boils)
			20°C = 68°F ("room temperature")
			37°C = 98.6°F (human internal body temperature)
	Kelvin (K)*	K = °C – 273	0 K = –460°F
Energy	joule (J)		1 J ≈ 0.24 calorie = 0.00024 kilocalorie[†]

*0 K (–273°C) is "absolute zero," a temperature at which molecular oscillations approach 0—that is, the point at which motion all but stops.

[†]A *calorie* is the amount of heat necessary to raise the temperature of 1 gram of water 1°C. The *kilocalorie*, or nutritionist's calorie, is what we commonly think of as a calorie in terms of food.

Answers to Chapter Review Questions

CHAPTER 1

1. e   2. b   3. e   4. d   5. e

6. In science, we formulate hypotheses about how the world works, then try to reject those hypotheses with experiments. The experiments must be designed so that we would expect them to uncover problems with our hypothesis. If the experiments are incapable of rejecting a hypothesis, then the experiments are not a rigorous test of the hypothesis.

7. The independent DNA found in mitochondria and chloroplasts is evidence of the origin of these eukaryotic organelles from ancient bacteria that became incorporated in the eukaryotic cell. Since the ancestors of these organelles once existed as independent organisms, they have their own genomes.

8. Controlled experiments, by definition, are able to control many variables in carefully maintained experiments, often in laboratory conditions. Comparative experiments, in contrast, often contain many additional variables that cannot be controlled by the investigator. Comparative experiments often incorporate realistic variation from uncontrolled factors, which accounts for their higher overall variability.

9. If two species share particular changes in the gene we compare, and those changes are not shared by other species we examine, we would expect the two species with the common changes to be more closely related to one another. By comparing many such changes in many genes, we can group species based on their relative evolutionary divergence from one another. For example, we share more changes in our genes with chimpanzees than we do with gorillas. From this, we can deduce that humans and chimpanzees shared a more recent common ancestor than they shared with gorillas.

10. Mitochondrial DNA is often used to follow the history of maternal lineages in a population or species. Nuclear DNA is not used in such cases because it is typically inherited from both parents. This difference can be useful in many circumstances. For example, we might examine a hybrid individual between two species. Equal portions of nuclear DNA from both species could confirm that the individual is a direct hybrid between the two species. If we examine the mitochondrial DNA, however, we can learn which of the two parental species was the female in the cross—and therefore learn by default which was the male.

CHAPTER 2

1. b   2. d   3. c   4. c   5. a   6. d

7.

8. An easy way to answer this question is to make a simple table:

	Covalent H—H	Hydrogen H····O
Electrons	Shared	Remain with H and O
Polarity	Nonpolar	Polar; + at H end
Strength	Stronger	Weaker

9. C—H: nonpolar; hydrophobic
C=O: polar; δ– at O; hydrophilic
O—P: polar; δ– at O; hydrophilic
C—C: nonpolar; hydrophobic

10. This is an example of Van der Waals forces, which act over a short distance and do not involve polarity.

11. The human body has the same elements as Earth's crust but in very different proportions.

CHAPTER 3

1. e   2. e   3. c   4. a   5. c   6. b

7. The observations support explanation "a." Glycine is small and nonpolar. Glutamic acid and arginine are larger and polar (charged). Serine and alanine are small: the protein retains its shape. But serine is polar (it has –OH as its R group), and that does not affect the structure. Valine is larger and nonpolar, and this affects shape. So the issue is size.

8. Mannose and galactose have the same atomic formula, $C_6H_{12}O_6$, but the arrangement of atoms is different: compare carbons 2 and 4. These sugars have the hydroxyl (–OH) functional group. Its polarity helps the sugars dissolve in water. The –OH group also can participate in bonding the sugar to other molecules through condensation reactions (see Figures 3.4 and 3.17).

9. High temperature disrupts weak interactions such as hydrogen bonds. Heat shock proteins might work by stabilizing the protein so that the weak interactions are not necessary to preserve its structure.

10. A change from lysine is a change in primary structure. The change could affect tertiary structure if the protein folds as a result of electrostatic attractions between charged amino acids (+ to –). In this case, the presence of a negatively charged amino acid (aspartic acid) where there should be a positively charged one (lysine) might prevent correct folding if a negatively charged amino acid elsewhere in the polypeptide chain is involved in folding (it is attracted to a + amino acid). The same forces might be at work in the interaction of separate chains for quaternary structure.

11. See Figure 3.10. Heat breaks hydrogen bonds and other weak interactions that maintain protein shape. Disulfide bonds also required for normal protein shape. Styling and perms partially denature keratin, then renature the protein in a new shape. Your investigation might involve measuring keratin protein structure of hair before and after disrupting hydrogen bonds and disulfide bonds.

CHAPTER 4

1. c   2. c   3. c   4. c   5. b   6. b

7. The presence of O_2 in the atmosphere produces an oxidizing condition that prevents the reduction reactions noted by the Miller–Urey experiment.

8. Oligonucleotides of RNA can fold because of hydrogen bonds forming between bases on the single chain and to a lesser extent because of weak interactions of base stacking when bases come near one another. Short strands of about 20 oligonucleotides are enough to produce uniquely folded RNA.

9. Cells provided concentration and compartmentation chemicals for the reactions needed for life, as well as differential permeability to distinguish life's chemical composition from that of the environment.

10. If microbes survived heat, the initial part of Pasteur's experiment might begin with microbes already present. They would grow in both the open and closed flasks. To get the results he did, Pasteur's flasks must not have contained such microbes. An answer for the proposed experiment on heat-stable microbes might be to inactivate them using reagents, such as mercaptoethanol, that destroy proteins.

11. A suggested experiment might be to dry the samples after the Miller–Urey experiment (allowing condensation reactions—polymerization) and then apply energy in the form of heat. This condition might have existed in volcanic rock in early Earth.

CHAPTER 5

1. b   2. d   3. e   4. a   5. d   6. b   7. a

8. Four membranes: two in the chloroplast and two in the mitochondrion
Two membranes: the lysosomal membrane and the plasma membrane (via vesicle; the molecules do not themselves cross any membranes)

No membranes: ribosomes do not have membranes. However, if the ribosomes were associated with the endoplasmic reticulum (ER), the answer would be two membranes: into the ER and out of the ER.

9.

	Animal Cell ECM	Plant Cell Wall
Composition	Collagen fibers in proteoglycan matrix	Cellulose fibers in polysaccharide and protein matrix
Rigidity	Less rigid	More rigid (especially secondary cell walls)
Connections	Some specialized proteins and junctions	Plasmodesmata

10. Microtubules line the long axons of nerve cells, where they act as tracks for vesicles that carry substances down the neuron. Without microtubules, the contents of these vesicles cannot be delivered to their destination, which can result in nerve problems.

   Microtubules are a key part of the mitotic spindle, which is used to move chromosomes during cell division. Depolymerization of microtubules can thus result in loss of dividing cells.

11. For a lysosomal enzyme, the pathway would be ribosome → interior of ER → Golgi → Golgi vesicles → lysosome.

   For an extracellular protein (animal cells), the pathway would be ribosome → interior of ER → Golgi → Golgi vesicles → plasma membrane → extracellular region.

CHAPTER 6

1. c   2. a   3. d   4. c   5. b   6. e   7. c

8. The pumping of Ca^{2+} requires a lipid bilayer membrane to separate compartments, a protein pump in the membrane, and ATP to provide energy for pumping.

9. Diatom wall components move from the Golgi apparatus to the cell wall by exocytosis.

10. Living in a hypotonic environment (cells hypertonic) results in a tendency for water to enter the organism by osmosis, which can cause swelling and dilute cell contents. Some organisms get around this by using reverse pinocytosis (exocytosis) to remove fluid.

11. Experiments might involve the following:

   To measure membrane fluidity, label a small amount of a lipid or protein with a dye and allow it to incorporate into a cell's membrane. This may make a localized labeled spot on the cell. The localized region will be seen to diffuse over the cell over time. In the cancer cells, this rate of diffusion may be faster.

   To measure cell adhesion, dissociate cancer and normal tissue cells. Incubate for a period of time and determine the rate at which cancer cells and normal lung cells bind to cells from the other tissues besides lung. The cancer cells may bind to a greater extent than normal cells.

CHAPTER 7

1. d   2. c   3. d   4. a   5. d   6. a   7. d   8. c

9. Different cells can have different target molecules to which cAMP binds, and these target molecules can have different activities and functions. Binding of cAMP changes the structure (e.g., tertiary structure of a protein) and therefore the function of a target molecule. So cAMP can have many effects.

10. Characteristics of direct communication: the size of signal molecules is limited by the size of openings between cells, it is not specific, it is fast, and there can be cytoplasmic connection between cells.

   Characteristics of receptor-mediated communication: the signal molecules can be larger, it is specific, it is slower, and there is no direct cytoplasmic connection.

   Direct communication is useful for a rapid, coordinated response of many cells.

11. See Figure 7.10. A mutation of the *Raf* gene that activates cell division might involve a protein product that does not need binding of Ras to be active. Cell division would occur without activated Ras, thereby eliminating the need for growth factor binding.

   A mutation of the *MAP kinase* gene would stimulate cell division if the resulting MAP kinase protein did not need to be phosphorylated by MEK to be active. No signaling cascade would be needed for the mutant protein to enter the nucleus and stimulate cell division.

12. Experiments might involve applying a solution containing the antibody to the upper part of the *Hydra* body. The antibody would block diffusion of the signal molecule from the apex to the upper body and—if the hypothesis is correct—would allow a bud to form in the upper body. A sham experiment, in which

the solution without antibody is applied, would be a control. In this case, a bud would not form in the upper body.

CHAPTER 8

1. c   2. e   3. c   4. c   5. d   6. d

7. Endergonic reactions are coupled in time and space with exergonic reactions, which release the energy needed for the endergonic reactions.

8. A cytoplasmic enzyme generally has a globular structure with a hydrophilic exterior and an active site for substrate binding. An ion channel generally has a more linear structure with a hydrophobic membrane-spanning region and no active site.

9.

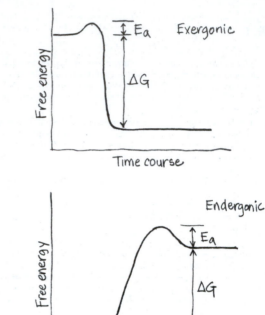

10. (a) The presence of water may prevent O_2 from reaching the enzyme. (b) Boiling denatures proteins, so polyphenol oxidase is irreversibly altered by boiling and its active site is destroyed. (c) Proteins have an optimal pH at which ionized R groups are appropriately charged to give the protein its tertiary structure. A pH 3 of may not be that optimal pH for polyphenol oxidase, so the enzyme is denatured and inactive.

11. See Figure 8.17. A competitive inhibitor binds to the active site of the enzyme and shifts the equilibrium to enzyme molecules in the active form.

12. To determine whether catalase has an allosteric or nonallosteric mechanism, perform an experiment with varying amounts of substrate and plot rate of catalase versus substrate concentration. An S-shaped curve will indicate an allosteric mechanism. A hyperbolic curve indicates a nonallosteric enzyme.

   To determine if a pollutant is a competitive or noncompetitive inhibitor, add the pollutant to the catalase to lower the rate of reaction, then add increasing amounts of substrate. A competitive inhibitor will be removed from the active site and the rate of reaction will increase. A noncompetitive inhibitor will not allow the rate to increase as more substrate is added. (There are more sophisticated kinetic experiments that you will learn in a biochemistry course).

CHAPTER 9

1. d   2. d   3. e   4. c   5. d   6. a

7. If cytochrome *c* remains reduced and cannot accept electrons, the electron transport (respiratory) chain stays reduced and NADH and $FADH_2$ remain reduced. This prevents oxidation reactions in the citric acid cycle and pyruvate oxidation, so pyruvate cannot be converted to acetyl CoA. Instead, pyruvate is converted to lactic acid, regenerating some NAD that can be used so that glycolysis can continue. Because the electron transport chain is not working, there is no proton gradient set up in the mitochondria, and ATP is not made by oxidative phosphorylation.

8. See Figure 9.13. Some amino acids are converted to intermediates of glycolysis. Once they enter glycolysis these intermediates are further metabolized to a

glycolytic intermediate that can be converted to glycerol, which is incorporated into triglycerides. Glycolysis and pyruvate oxidation produce acetyl CoA, which is converted to fatty acids and incorporated into lipids.

Glucose is converted in glycolysis to acetyl CoA, which is then converted to fatty acids as above.

9. (a) Oxidation (removal of H from C2 and C3 of succinate)

(b) Exergonic (because it is an oxidation)

(c) It requires the redox coenzyme NAD or FAD.

(d) The fumarate is converted to other intermediates that regenerate oxaloacetate, the acceptor for the citric acid cycle.

(e) The reduced coenzyme (NADH or FADH2) is reoxidized in the electron transport chain.

10. Anaerobes use alternate electron acceptors to generate energy, such as sulfur, sulfate, and nitrate. Also, they use substrate-level phosphorylation (direct transfer of phosphate to ADP) to make ATP.

11. The proton gradient in the experiment described in Figure 9.9 was generated artificially from the solution and did not require electron transport (a respiratory chain). The presence of antimycin A thus would have no effect on the experiment.

CHAPTER 10

1. e   2. b   3. d   4. d   5. d   6. d

7. In the dark, photosynthetic electron transport stops at photosystem II → reduced PQ (plastoquinone). Initially, the chlorophylls in light-harvesting complexes remain reduced, so reaction-center chlorophylls remain reduced and thus photosystem II remains reduced.

In the dark, the Calvin cycle stops at the reduction phase, which requires NADH. No RuBP is regenerated, so there is no rubisco activity. The initial reactions are no oxidation of photosystem I, and no reduction of NADP to NADPH.

8. These processes can be compared using a table:

	Cyclic Electron Transport	Noncyclic Electron Transport
Products	ATP	ATP, NADPH, O_2
Source of electrons	Electron transport	Electron transport (photosystem I) or water (photosystem II)

9. See Figure 10.18. CO_2 carbons end up in 3PG, which is converted to pyruvate. Pyruvate goes to the citric acid cycle, where some of the intermediates are converted to amino acids, which are incorporated into protein.

In the Calvin cycle, some 3PG is converted to G3P, which can enter glycolysis. Some of the intermediates of glycolysis are converted to amino acids, which are incorporated into protein.

10. (a) O_2

(b) NADPH

(c) 3PG

11. (a) Here is the pathway followed by ^{14}C: $^{14}CO_2$ → cells → photosynthesis → carbohydrate → combustion → $^{14}CO_2$. Release of $^{14}CO_2$ upon combustion would be evidence of photosynthesis (and life).

(b) In this case: $^{14}CO_2$ → heat denatured cells, no photosynthesis. If living things were present, $^{14}CO_2$ would be released in experiment (a), but not in experiment (b).

CHAPTER 11

1. d   2. b   3. d   4. d   5. e   6. d   7. d   8. c

9. See Figure 11.19. In mitotic prophase, there is no pairing of homologous chromosomes, and crossing over is rare. In meiotic prophase I, homologous pairs of chromosomes align, and crossing over is common.

In mitotic anaphase, sister chromatids separate, with one going to each pole. In meiotic anaphase I, sister chromatids do not separate; homologous pairs of chromosomes separate, with one pair going to each pole.

10. Normally, p53 induces expression of p21, which binds to the G1/S Cdk and prevents cyclin from activating it. Without active Cdk, the cell cycle ceases. If p53 is mutated such that it is nonfunctional, p21 is not induced and the cyclin–Cdk complex can form and stimulate the cell cycle at S phase.

11. Cancers often have multiple mutations in different cells of the tumor. If some of these mutations affect different parts of the cell cycle, targeting the different phases may be a useful therapy.

12. Your proposed experiments should involve isolating the synchronous meiotic cells from the lily anthers and establishing them in the lab. As the cells proceed

through the meiotic cell cycle they can be analyzed at different stages for the presence and biochemical activity of various cyclins and Cdks.

CHAPTER 12

1. e   2. a   3. d   4. d   5. d   6. b   7. b   8. d   9. b

10. $BB \times bb$; $bb \times bb$; $Bb \times bb$; $Bb \times Bb$

11. 1/32

12. (a) Autosomal dominant

(b) 1/4

13. (a) Males (XY) contain only one allele and will show only one color, black (X^BY) or yellow (X^bY). Females can be heterozygous (X^BX^b).

(b) X^bY, yellow

14. The body color (G/g) and wing size (A/a) genes are linked; eye color (R/r) is unlinked to the other two genes. The distance between the linked genes is 18.5 units.

15. Yellow, blue, and white in a 1:2:1 ratio.

16. F_1 will all be wild type, $PpSwsw$. F_2 will have phenotypes in the ratio 9:3:3:1; see Figure 12.6 (p. 238) for analogous genotypes.

17. (a) F_1 will all be $PpByby$ and will have wild-type eye color and wings. The ratio of phenotypes in F_2 will be 3:1, $PPByBy$ (wild-type eyes and wings) to $ppbyby$ (pink eyes and blistery wings.

(b) F_1 will all be $PpbyBy$ with wild-type eye color and wings; they will produce just two kinds of gametes (Pby and pBy). Combine them carefully and see the 1:2:1 phenotypic ratio fall out in the F_2: 1 wild-type eyes/blistery wings : 2 wild-type eyes/wild-type wings : 1 pink eyes/wild-type wings.

(c) Pink–blistery

(d) See Figures 11.16 and 11.18 (pp. 220–222). Crossing over took place in F_1.

18. $Rraa$ and $RrAa$

19. (a) $w^+ > w^e > w$

(b) The parents are w^ew and w^+Y. The progeny are w^+w^e, w^+w, w^eY, and wY.

20. (a) BX^a, BY, bX^a, bY

(b) The mother is bbX^AX^a, the father BbX^aY, the son BbX^aY, and the daughter bbX^aX^a.

21. 75 percent

22. Because the gene is carried on mitochondrial DNA, it is passed through the mother only. Thus if the women does not have the disease but her husband does, their child will not be affected. However, if the woman has the disease but her husband does not, their child will have the disease.

23. The cross $RRYY \times rryy$ produces $RrYy$ (round, yellow) F_1 offspring. If the seed shape and seed color genes were linked with no recombination between them, the F_2 would also be all $RrYy$. A distance of 10 map units between two genes means that on average 10% of the F_2 offspring will have recombinant phenotypes, in this case round green (5%) and wrinkled yellow (5%).

The cross in Figure 12.19 is $BbVgvg$ (gray, normal) $\times bbvgvg$ (black, vestigial). If there were no linkage between the genes, then the gray, normal parent would produce four types of gametes: BVg, bVg, Bvg, and bvg. When these combine with the bvg gametes produced by the other parent, four types of offspring in a 1:1:1:1 ratio will result: $BbVgvg$ (gray, normal), $bbVgvg$ (black, normal), $Bbvgvg$ (gray, vestigial), and $bbvgvg$ (black, vestigial).

CHAPTER 13

1. a   2. c   3. b   4. b   5. d   6. c   7. d

8. At 3,000 bp per minute in two directions, each origin grows at 6,000 bp per minute. There are 300 minutes in S phase, so the total bp possible for one origin is $(300 \times 6,000) = 1,800,000$. If there are 120 million bp to replicate, then the total number of origins is 120 million/1.8 million = 66 origins. If there are 3 μm of DNA, this means there are about 22 origins per micrometer of DNA.

9. DNA replication adds new nucleotides to the 3′ end of DNA, where there is an —OH group on the sugar at the 3′ position. If there is no —OH group, there cannot be a condensation reaction and formation of a bond to the next nucleotide, so replication stops.

10. After ten rounds there would still have been some DNA (about 1/512th) as hybrid because the original heavy DNA template strands would still have been there. This tiny amount might not have been detectable in the centrifuge, however.

11. The proposed experiments might use S strain pneumococcus and transform R strain as in Figure 13.1. Incubate separate batches of S strain bacteria in ^{32}P or ^{35}S. Make cell-free extracts of the S strains. Incubate with R cells and look for their transformation to the S phenotype. Then check to see if there is ^{32}P or ^{35}S label in the newly transformed cells. It would be expected that only ^{32}P label (DNA) would enter the cells.

CHAPTER 14

1. b   2. a   3. d   4. b   5. d   6. d   7. d   8. e

9. For 192 amino acids, the triplet genetic code mandates 576 bp of coding sequence. Add the start and stop codons and the total is 582. This is shorter than the actual DNA gene because of promoter and terminator of transcription sequences; introns; and ribosome binding sequences. All except the transcription signals are transcribed into the pre-mRNA. The mature mRNA has the introns removed.

10. Errors in transcription can be tolerated because many copies of each RNA are made; if a few have errors, there are enough perfect ones to overcome any problem. Errors in DNA replication are harmful because DNA is replicated only once in the life of the cell.

11. In the poly CA experiment, threonine is ACA or CAC and histidine is ACA or CAC. In the poly CAA experiment, threonine is CAA, ACA, or AAC. Therefore in the first experiment threonine must be ACA and histidine CAC.

12. Enzymes: $4 \rightarrow 2 \rightarrow 3 \rightarrow 1 \rightarrow 5$
    Compounds: $C \rightarrow F \rightarrow E \rightarrow D \rightarrow G \rightarrow T$

CHAPTER 15

1. a   2. c   3. b   4. b   5. d   6. b

7. (a) In a loss of function mutation, a phenotype is not present; for example, there may be a loss of enzyme activity. In a gain of function mutation, a new phenotype is present; for example, a new signaling protein may be active.

(b) In a missense mutation, a single base pair change results in a codon change and thus an amino acid change in a protein. In a nonsense mutation, a single base pair change results in a codon change to a stop codon and thus premature termination of a protein.

(c) In a spontaneous mutation, DNA changes as a result of unprovoked chemical changes or replication errors. In an induced mutation, DNA changes as a result of outside physical or chemical agents.

8. (a) The mutation that leads to PKU is rare in the human population; most people do not have the harmful allele and the highest probability is that the father is homozygous normal. Because the mother has PKU (she is homozygous mutant), the developing fetus is heterozygous.

(b) High levels of phenylalanine cause brain damage. If the mother's phenylalanine levels were too high, the baby would be born with brain problems.

(c) The woman should be on a phenylalanine-restricted diet.

9. Testing for the cystic fibrosis (CF) allele could be done by allele-specific oligonucleotide hybridization with probes for the normal and CF alleles; see Figure 15.18. Or direct DNA sequencing of the CF gene could be done. A person who is a carrier will test positive for both the normal and the mutant alleles.

To do gene therapy, the normal allele for CF could be inserted into a viral vector that can infect cells in the lung and airway tissues. Then the virus could be sprayed onto these tissues.

10. Early identification of people with multifactorial diseases, even before symptoms appear, could lead to therapeutic interventions to prevent disease development. Ethical issues might include insurability, hiring eligibility, and social stigma.

11. An enzyme test for HEXA would reveal intermediate levels in people who are carriers. This could be done on accessible cells (e.g., blood) if the gene is expressed there. A DNA test could involve sequencing the gene by allele-specific oligonucleotide hybridization (see the answer to Question 9). The advantage of DNA testing is that it can be done on any cells from the body (not just cells that express the enzyme).

Investigation of the stop codon hypothesis would involve isolating the HEXA protein from patients with Tay-Sachs disease and showing that it is shorter in primary structure than the protein encoded by the normal allele.

12. (a) The amino acid sequence would be Leu-Ile-Ser-Ile-Ala. This is a missense mutation.

(b) The mutation replaces proline with serine. Proline is a nonpolar amino acid that is usually part of bends or loops in a protein; serine is a polar amino acid with a smaller side chain. The mutation is likely to affect enzyme activity because it is likely to affect protein structure.

(c) See p. 317. This region of the gene could be amplified by PCR and then digested with *Eco*RV. The mutant DNA will be cut, but the wild-type DNA won't be.

CHAPTER 16

1. b   2. a   3. e   4. b   5. c   6. d

7. The easiest way to answer this question is to construct a simple table:

	Lysogenic Bacteriophage	HIV
(a) Viral entry to host cell	Attachment of viral protein to host cell membrane	Membrane fusion of virus to host cell membrane
(b) Virus release from host	Host cell lysis	Budding and exocytotic release
(c) Viral genome replication	Host DNA polymerase	Virus reverse transcriptase followed by host RNA polymerase
(d) New virus production	Host transcription of virus genes and host-mediated translation of virus proteins	Same as in lysogenic bacteria

8. In a prokaryotic gene, the promoter is a DNA sequence, there are few transcription factors, and there is one RNA polymerase. In a eukaryotic gene, the promoter is a DNA sequence, there are many transcription factors, and there are several RNA polymerases.

9. Here is the structure of the gene:

$$E1 - I1 - E2 - I2 - E3 - I3 - E4$$

(E = exon; I = intron). Assuming that initiation of transcription begins at E1, the possible proteins are composed of exons 1234; 134; 124; and 14.

10. To keep a constant, low-level expression of repressor protein, the regulatory gene would have an inefficient promoter, and synthesis of the repressor would be constitutive.

11. You could sequence the relevant genes of colon cancer cells and look for mutations that lead to aberrant function, then isolate the proteins involved and determine that their functions are indeed abnormal. To show epigenetic silencing, you might sequence the promoters of the genes and look for epigenetic changes (e.g., cytosine methylation, which would be increased if there is transcriptional silencing). Then you could examine the tumor cells to see if the active proteins are there but in small amounts.

CHAPTER 17

1. c   2. b   3. e   4. e   5. b   6. b   7. c   8. c   9. a

10. One gene can produce several proteins by alternative splicing, which makes the proteome highly complex. In addition, many proteins are modified after translation, and this contributes to even more protein diversity. The metabolome is highly variable from cell to cell and from one time to another. It is determined not only genetically but also by responses to environmental conditions.

11. While all of these plants have the same basic genes for "life" as well as "plant" functions (e.g., photosynthesis, cell-wall formation, flowering), there are some genes (and proteins) that are specialized for each plant (e.g., rice genes for growing under water, genes for timing flowering, genes for seed-storage proteins).

12. (a) Extract genomic DNA from the patient's cells and analyze it for SNP polymorphisms. If the SNP that correlates with kidney cancer is present, he has an increased susceptibility.

(b) Isolate both normal and cancerous kidney cells. Do a metabolomic profile on the kidney cancer cells and the normal kidney cells using chemical analyses for small molecules. By comparing the profiles, generate a metabolomic "signature" for the kidney cancer cells. Next, examine the metabolomic profile of kidney tissue from the patient and compare it with the metabolomic signature for kidney cancer cells.

(c) For possible drugs involved in kidney cancer treatment, isolate many cancers (or examine stored tissues) and do a SNP analysis, correlating tumor response to the drug with the SNP polymorphism. Then isolate some of the patient's tumor cells and examine the DNA for SNPs that relate to drug response. Use the drug that the patient's genome indicates will be effective.

CHAPTER 18

1. b   2. c   3. e   4. a   5. e   6. d   7. b   8. c

9. Both PCR and cloning begin with a gene sequence. In PCR, the sequence is amplified in the test tube. In cloning, the sequence is amplified by an organism (typically bacteria). In PCR, amplification is achieved by synthesizing primers that bind to either end of a target DNA sequence and adding nucleotides and DNA polymerase. The doubled DNA is them denatured, and the process is repeated 20 to 40 times.

In cloning, the target DNA is inserted by restriction and ligation into a vector, which has an origin of replication that will function in the organism where amplification will occur. The vector is added to the host cells, which are cultured and divide many times, amplifying the target DNA along with the host chromosome. The vector is then removed from the host cells and cut with a restriction enzyme, releasing amplified, cloned target DNA.

PCR is much simpler and faster but has artifacts where inappropriate fragments of DNA are amplified or sequence errors are introduced by DNA polymerase. Cloning yields the correct DNA without mutations but involves host cell culture and time-consuming DNA purification steps. See Figure 18.12. A simple table can answer this question.

	Conventional	Recombinant DNA
(a) Sources of new genes	Other plants of same species	Any organism or synthetic DNA
(b) Number of genes transferred	Often many	One
(c) How long it takes	At least one growing season, usually many	Weeks

10. (a) The target gene would be inserted into an expression vector with a promoter such that the gene would be expressed in the developing seed. The vector could be added to cultured wheat cells, and those cells carrying the vector selected (the vector could carry a reporter gene for resistance to an antibiotic). The cells could be induced to form a wheat plantlet, which would be transferred to the field and the seeds examined for the new protein.

(b) The target gene could be inserted into a sheep expression vector containing the lactoglobulin promoter so that the gene would be expressed in milk glands. The recombinant vector would be inserted into sheep egg cells. After the female offspring grew up, their milk could be tested for the presence of the human enzyme.

11. Public concerns include the artificiality of unnatural interference with nature, the safety of these foods for human consumption, and environmental dangers if non-host plants receive recombinant genes.

CHAPTER 19

1. c   2. b   3. a   4. e   5. a   6. b   7. c

8. (a) All neuronal precursors might undergo apoptosis and no neurons would form.

(b) The *p21* gene would be activated and the cell cycle would be blocked; in the presence of other factors, muscle cells would form.

(c) There might be no gradient of the protein in the developing limb and therefore no differential development of digits—all the digits would be fingers.

(d) The hunchback protein gradient would not form properly and the embryo would not establish its anterior–posterior axis.

9. (a) No apoptosis would lead to too many cells in developing organs, and the organs would not form properly.

(b) No gradient of hunchback protein would form, and there would be no posterior end determination in the developing fruit fly.

10. A mutation that caused expression of class A genes instead of class C genes. This would lead to an AB combination instead of AC, and petals would develop instead of stamens.

11. Mechanisms might include cell-cycle inhibition as a result of Cdk blocker; induction of transcription of certain genes; and cytoplasmic segregation, so that when a cell divides only one daughter cell gets a factor important in determination.

12. One could analyze mRNA in egg cells, in the parent differentiated cells, and in the reprogrammed cells. This could be done by reverse transcriptase PCR or by gene expression arrays.

CHAPTER 20

1. c   2. a   3. a   4. c   5. c

6. If the expression of Gremlin were blocked, this protein could not inhibit BMP4 signaling. The cells in the webbing of the feet would undergo apoptosis, and the duck would be born with unwebbed feet.

7. All of the hatchlings at any temperature would be expected to develop into males. Aromatase is required to convert testosterone into estrogen, which is required for female development.

8. The coexpression of *Hoxc6* and *Hoxc8* appears to be important in the development of thoracic vertebrae (the vertebrae with ribs). This is a short region in

mice, and mice have only a small number of thoracic vertebrae, and therefore a short body. In snakes, the coexpression of *Hoxc6* and *Hoxc8* along a much greater length of the embryo results in a much larger number of thoracic vertebrae, and therefore a much-elongated body.

9. The results support the conclusion that higher levels of BMP4 expression result in greater cartilage diameter on the beaks of developing chickens.

10. The observations are consistent with the hypothesis that there has been selection in some human populations for mutations on the enhancer that controls expression of the glycoprotein in red blood cells. This genetic change would be expected to have a selective advantage in human populations that are exposed to malaria at high levels, because the mutation confers greater resistance to malaria in humans that carry it.

CHAPTER 21

1. d   2. d   3. d   4. e   5. b

6. Humans select traits in domestic plant and animal populations based on our interest in the trait, rather than on how it affects the natural reproductive rate or survivorship of the organisms. Many of the traits artificially selected by humans would not be advantageous in wild populations. For example, humans have selected many cattle breeds for high body fat and high body weight. These traits result in large calves, which in turn result in calving difficulties for cows. Ranchers often have to assist in the birth of such calves, because the calf (and likely its mother) would often die without such assistance. In a natural population, there would be selection for smaller calf size and birth weight, which would increase the successful reproductive rate and survivorship.

7. Behaviors can respond to environmental cues that are predictive of future conditions, and these behaviors can be selected for if they are under genetic control. For example, day length becomes shorter as we move closer to winter, so individual mammals have a survival advantage if they respond to shortening days by going into hibernation. In this case, the environmental cue (day length) is predictive of future environmental conditions (the cold of winter).

8. Natural selection cannot act when there is no effect on the effective reproductive rate of the organism. Diseases such as Alzheimer's usually occur long after the reproductive years have passed. As long as the disease does not affect the relative likelihood of the survival of the affected person's offspring (as a result of reduced parental care, for example), we would not expect natural selection to lead to any reduction in Alzheimer's disease in human populations.

9. (a) Frequency of allele *a*: 0.60; of allele *A*: 0.40

(b) Frequency of genotype *aa*: 0.40; of genotype *Aa*: 0.40; of genotype *AA*: 0.20

(c) Expected frequency of genotype *aa*: 0.36; of genotype *Aa*: 0.48; of genotype *AA*: 0.16

(d) We would expect some level of deviation because the assumptions of Hardy–Weinberg equilibrium are so restrictive. For example, the finite population size, the presence of mutation, any migration of individuals into or out of the population, gene flow from mating with adjacent populations, nonrandom mating within the population, or selection in the population could all lead to deviations from Hardy–Weinberg expectations.

10. The black mice and white mice are highly unlikely to be mating randomly with each other. The combined population is far from Hardy–Weinberg expectations, with far too few heterozygous (*Aa*) individuals. The much higher frequency of the *a* allele among the black mice, and of the *A* allele among the white mice, suggests that the black and white mice are mostly mating within color types, with few between-color-type matings. Another possibility, though, is that the population consists of two subpopulations (one of mostly black mice, the other of mostly white mice) that have only recently come together in the same location. These two hypotheses could be distinguished by following the mice through another generation. If mating is now occurring at random, we would expect the genotype frequencies to be similar to Hardy–Weinberg expectations after one generation of random mating.

CHAPTER 22

1. e   2. a   3. e   4. a   5. e   6. d

7. The classification is not currently monophyletic. Both genera could be monophyletic if Species 4 were moved from Genus B to Genus A; monophyly could also be achieved if all the species were included in the same genus.

8. Fossils can give us direct evidence of the character states of extinct lineages. For example, all modern birds lack teeth. Is the lack of teeth an ancestral or a derived condition? If we examine extinct species of theropods (the larger group of dinosaurs that includes the living birds), we see that they had teeth. Therefore we know that the lack of teeth is a derived condition in modern birds.

9. The estimated average rate of change is 0.9 amino acid change/500 million years, or 0.0018 amino acid change/million years. If we express this as a percentage rather than as a proportion, we would say there is (on average) 0.18 percent change in amino acid sequences per million years.

10. The West Nile virus in the United States appears to be most closely related to a strain of the virus isolated in Israel. A reasonable hypothesis is that the virus emerged in Africa in the 1930s and subsequently moved into Asia and Europe, probably multiple times. Then in the late 1990s, a strain of the West Nile virus from Israel appears to have been transported to New York, perhaps carried by mosquitoes on an airplane or in a cargo shipment. Once in the United States, the virus spread quickly in native bird populations across North America.

CHAPTER 23

1. e   2. c   3. a   4. e

5. If the only difference between the diverging lineages is at a single locus, then both of the new alleles must be functional when they interact with the products of other gene loci (in both lineages). Any interlocus genetic incompatibility produced by these new alleles would be expected to affect the parental lineages as well. In addition, there are many greater numbers of possible incompatibilities across different gene loci than there are within a single locus. Rather than two deleterious changes at the same locus (one in each lineage), the Dobzhansky–Muller model allows neutral changes at any pair of loci whose products interact. It is the negative interaction of these products in the hybrid between the two lineages that results in genetic compatibility.

6. If two different fusions of chromosomes occur in two different lineages, then the resulting chromosomes cannot pair normally in meiosis in the hybrids. If you attempt to diagram meiosis in the hybrid that would result from a cross of the divergent lineages in Figure 23.4, you will see that homologous pairings require parts of different chromosomes to align with one another. These chromosomes will then be pulled in two different directions as the cell divides in meiosis I, resulting either in a likely failure of the cell to divide, or an uneven distribution of the chromosome arms in the two daughter cells. Production of normal cells with an even distribution of the various chromosomes arms will be limited, so the hybrids will produce few, if any, normal gametes.

7. A likely possibility is that the incompatible alleles have not yet become fixed in the various strains, so only some combinations of crosses result in genetic incompatibility.

8. Species that arise in allopatry initially occur in separate, but usually adjacent, ranges (see Figure 23.6). Therefore we would expect many closely related species to exhibit this same pattern. The ranges of highly mobile species are more likely to change over time, so the pattern should be strongest among relatively sedentary species.

9. There are many possible designs of experiments that might prove informative. Here is an example of one that would examine the effect of flower position on pollinator attraction: Take one species of flower and divide the flowers into two groups. Position each flower to be either upright or pendant, then record the number and type of pollinators that are attracted to flowers in each group. Test to see if the differences between the two groups are statistically significant.

10. (a)

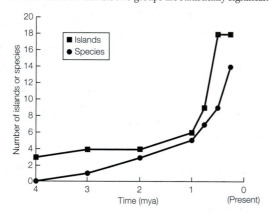

(b) Yes, because the curve for the number of species lags behind the curve for the number of islands, but the two curves exhibit very similar changes in slope through time. As new islands arise, new opportunities for speciation also arise. The number of species at any one time is always just below the number of distinct islands.

(c) There are currently 18 islands in the archipelago and only 14 finch species. This suggests that there are still opportunities for additional speciation by geographic isolation. Based on our graph from Question a, we expect populations of species that occur on two or more islands to diverge into distinct species over time. To test this hypothesis, we could collect samples of each population and examine genetic divergence among the samples. Significant genetic divergence among the populations on different islands suggests that the distance between the islands is a significant barrier to gene flow, so the populations are expected to diverge into distinct species over time.

CHAPTER 24

1. a   2. a   3. a   4. e   5. b   6. e

7. Molecular clocks work best when they are applied within a group of organisms with similar generation times and populations sizes. Population size makes little difference if all or most changes are neutral, but rates of change among deleterious and beneficial changes are affected by population size. In addition, it is important to make comparisons among homologous genes and proteins, since rates of evolution across different genes are likely to vary widely. When molecular clocks are used to make comparisons across species with very different generation times, it is necessary to account for the different generation times.

8. New mutations are introduced into the experiment shown in Figure 24.14 through the errors made in the PCR amplification step. In other words, the mutation rate is a function of the error rate of the DNA polymerase. Using a different DNA polymerase with a higher error rate would increase the overall mutation rate of the experiment, and that would increase the variation in the population of molecules. Another possible answer is to add a mutagen to the PCR amplification step, which would also increase the mutation rate of the experiment. Any process that increases the mutation rate would be expected to increase the genetic variation present in the pool of molecules prior to the next round of selection.

9. This problem can be investigated by sequencing and comparing the genes for opsins in surface-dwelling (eyed) and cave-dwelling (eyeless) crayfishes. If the genes of the eyeless species are no longer under any selection, we would expect to observe a similar rate of synonymous and nonsynonymous substitutions in the genes. If there has been strong selection for a new function (something other than vision), we would expect a higher rate of nonsynonymous substitutions compared with synonymous substitutions (indicating positive selection). We would compare these rates to the rates seen in the surface-dwelling (eyed) species. In the surface-dwelling species, we would expect to see a higher rate of synonymous compared with nonsynonymous substitutions, which is expected under purifying selection.

10. (a) Codon numbers 12, 15, and 61 are likely to be evolving under positive selection for change because these three codons have each experienced a higher rate of nonsynonymous substitutions (which give rise to amino acid replacements) relative to the rate of synonymous substitutions.

(b) Codon numbers 80, 137, 156, and 226 are likely evolving under purifying selection, as the vast majority of changes at these codons are synonymous substitutions, which do not result in amino acid replacements. Substitutions that result in amino acid changes (nonsynonymous substitutions) undoubtedly occur, but they are usually selected against in the population. Codon number 165 has experienced similar numbers of synonymous and nonsynonymous substitutions. However, since there are approximately three times as many possible nonsynonymous substitutions as there are synonymous substitutions, the number of synonymous substitutions is slightly higher than expected if the rates of each type of substitution are equal. Codon 165 may be evolving under weak purifying selection; it is the codon that is closest to neutral among the codons shown in the table.

CHAPTER 25

1. b   2. c   3. a   4. c   5. b   6. c

7. There are many possible answers, but four familiar examples include the study of Earth's past atmosphere by examining the chemical composition of rocks; the study of past climates by examining the growth rings of trees; the study of continental drift by examining the geological record; and the study of the origins of the universe (the "Big Bang") by examining the speed at which galaxies are moving apart.

8. Relative dating provides us an order for events; we learn that Event 1 happened before Event 2. But absolute dating provides us with an estimate of the timing of those events. It is important to know not just that Event 1 occurred before Event 2, but also how much time separated the two events.

9. Multicellular organisms require higher concentrations of oxygen, and the levels of oxygen increased throughout the Precambrian. By the end of the Precambrian, atmospheric oxygen levels were sufficiently high to support a variety of multicellular organisms. In addition, the end of widespread glaciation (the "snowball Earth" period) near the end of the Precambrian probably allowed multicellular organisms to flourish.

10. There are many possible experiments that could be devised. For example, the effects of changing oxygen concentrations on other species (besides flying insects, such as the *Drosophila* used in the described experiment) could be tested. An ideal study organism would have a short generation time (so that many generations could be followed in the course of the experiment) and would be easy to raise in the laboratory. For example, guppies could be raised in elevated and reduced oxygen concentrations, and evolution in the size of the swim bladder (a site of oxygen uptake) could be evaluated as a response.

CHAPTER 26

1. e   2. c   3. e   4. b   5. b   6. d

7. Ribosomal RNA genes are universally present across organisms. They evolve slowly, so they can be compared among the most distantly related species. They are present in multiple copies, so they were relatively easy to isolate and sequence in the earliest days of gene sequencing. Also, since they are required for protein synthesis, and already present in all cellular species, the possibility of lateral gene transfer is greatly reduced. In contrast, different types of metabolism have arisen repeatedly in the history of prokaryotes, so species with similar types of metabolism may not be closely related. Cell structure is useful for identifying some major groups of prokaryotes (Gram-positive versus Gram-negative groups, for example), but the differences are too few to be of great use in classifying most species.

8. A laterally transferred gene does not represent descent from a common ancestor and thus does not reflect a true evolutionary relationship.

Expected tree based on gene x:

Species A
Species B
Species C
Species D

Expected tree based on consensus of non-transferred genes:

Species A
Species B
Species C
Species D

9. Most, but not all, biologists consider viruses to be living organisms. Viruses have their own genomes, and they are composed of proteins much like cellular organisms. They evolved from other living species, and they are clearly a part of life. However, viruses are not composed of cells, and they depend on cellular hosts to carry out many of their biological processes. For these reasons, some biologists consider them to be nonliving components of their cellular hosts rather than distinct living organisms.

10. There are many possible answers, but one widely used approach for detecting new life forms (in any environment, including high-temperature environments) is to directly isolate and amplify conserved gene sequences. The ribosomal RNA genes are often used for such detection because they evolve very slowly and are required for protein synthesis. DNA could be extracted from high-temperature environments, and any ribosomal DNA genes that were present could be amplified and sequenced. The sequences would then be compared with the ribosomal RNA genes of other known species of prokaryotes to classify the organisms living in the extreme environment.

CHAPTER 27

1. e   2. c   3. e

4. (a) Foraminiferans have external shells of calcium carbonate, whereas radiolarians have long, stiff pseudopods and radial symmetry. The external shells of foraminiferans and the internal skeletons of radiolarians are both important components of ocean sediments and sedimentary rocks.

   (b) Ciliates are covered with numerous hairlike cilia, whereas dinoflagellates generally have two flagella (one in an equatorial groove, and the other in longitudinal groove). Both ciliates and dinoflagellates have sacs, called alveoli, just beneath their plasma membranes, which identify them as alveolates.

   (c) Diatoms are unicellular and are typically composed of two nested plates (like a petri dish). Brown algae are large, multicellular organisms composed of branched elements or leaflike growths. Both diatoms and brown algae are photosynthetic.

   (d) The vegetative unit of a plasmodial slime mold is a plasmodium: a wall-less mass of cytoplasm containing numerous diploid nuclei. The vegetative unit of cellular slime molds consists of separate, single amoeboid cells. In both groups, when environmental conditions become unfavorable, the vegetative units form fruiting structures.

5. The independence of sex and reproduction in ciliates suggests that sex has functions apart from reproduction. Sex is important for recombining genes, which is important for several reasons. Sex allows populations of organisms to avoid the accumulation of deleterious alleles, and it allows the formation of new combinations of beneficial alleles. Thus even organisms that reproduce asexually generally have some other means of achieving sexual recombination of their genomes.

6.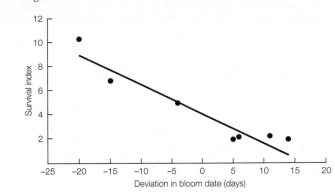

Using the formula for a correlation coefficient shown in Appendix B, $r = -0.948$.

7. The results show that earlier bloom dates are associated with higher survival indices. The relationship between these two measures is very strong and nearly linear, resulting in a correlation coefficient of $r = -0.948$. As noted in the question, larval haddock depend on these blooms for both cover from predation and as a food source. A reasonable hypothesis for this is that earlier blooms provide better cover and more food for the larval haddock, so survivorship of the larval fish is higher in years when the phytoplankton blooms occur earlier. Another (not mutually exclusive) possibility is that the earlier blooms benefit other species that the haddock consume as food, or that the potential predators of haddock target the phytoplankton instead of the haddock.

8. The three rRNA genes of corn are not one another's closest relatives because the nuclear, mitochondrial, and chloroplast genomes have different origins, and the relationships shown in the tree reconstruct the endosymbiotic events that gave rise to mitochondria and chloroplasts.

9. The mitochondrial rRNA gene of corn is more closely related to the rRNA gene of *E. coli* than it is to the nuclear rRNA genes of other eukaryotes because the mitochondria were derived from an endosymbiosis with a proteobacterium. Likewise, the chloroplast rRNA gene of corn is more closely related to the rRNA gene of *Chlorobium* than it is to the nuclear rRNA gene of corn because the chloroplasts were derived from an endosymbiosis with a cyanobacterium.

10. The human and yeast mitochondrial rRNA genes would be expected to cluster on the tree closest to the corn mitochondrial rRNA gene because all of these genes are descended from the same endosymbiotic event (the origin of mitochondria). The human and yeast mitochondrial rRNA genes would be more closely related to each other than either is to the corn mitochondrial rRNA gene because fungi and animals are more closely related to each other than either is to plants (as can be seen in the relationships of the nuclear rRNA genes).

CHAPTER 28

1. c   2. e   3. b   4. b   5. d

6. Microphylls are usually small and typically have a single vascular strand. In contrast, megaphylls are larger and typically contain branched veins. Microphylls may have originated as sterile sporangia. Megaphylls may have originated from flattening of branching stems, between which photosynthetic tissues developed. Among modern plants, microphylls are found in lycophytes, whereas megaphylls are characteristic of the euphyllophytes (such as ferns and seed plants).

7. One advantage of heterospory is that it allows a greater degree of outcrossing, since there are separate male and female gametophytes.

8. Both mosses and ferns are homosporous, and both alternate between a diploid sporophyte and a haploid gametophyte generation. However, the gametophyte generation is the large, dominant portion of the moss life cycle, whereas the sporophyte generation in the large, dominant portion of the fern life cycle. The sporophyte of a moss is completely dependent on the gametophyte, whereas the sporophyte of a fern becomes independent of the gametophyte.

9. Yes. Heterospory is an example of a trait that appears to have evolved multiple times among different groups of vascular plants.

10. One possibility is to examine leaf size as a function of thermal environment among close relatives of living plant species. We might predict that large leaf size is restricted in hot, dry climates but favored in cooler, wetter climates.

CHAPTER 29

1. d   2. a   3. d   4. a   5. a

6. To be functional as a reproductive organ, a flower would need to have at least a carpel or a stamen. The petals function largely for pollinator attraction, and so can easily be lost in wind-pollinated species. The sepals function mostly to protect the flower in bud, and so may be lost in species with simplified flowers.

7. The fossil record is not a complete record of life on Earth. Early angiosperms may have been limited in distribution, or may have lived in environments that were not compatible with easy fossil formation. It is likely that the early angiosperms were not very abundant or widespread. They apparently underwent a rapid radiation in the Cretaceous, where they become common in the fossil record.

8. .

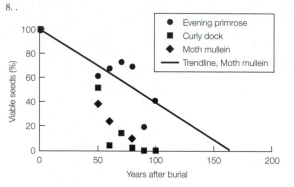

9. One approach to this problem is to calculate a linear trend line for survivorship of moth mullein (*Verbascum blattaria*) seeds by calculating a linear regression line (see Appendix B) and then projecting it forward in time to the point where it intersects zero percent survival. The resulting regression equation is $y = 102.09 - 0.62x$. The graph above shows this approach, which predicts that the last *Verbascum blattaria* seeds would germinate in about Year 165 of the experiment (set y to 0, and solve for x using the regression equation; the result is $x = 164.7$ years). This approach assumes a linear decline in viability of the seeds. It may be more reasonable to assume an exponential decay in seed viability (similar to radioactive decay; see Figure 25.1). If seeds decay exponentially, then we would expect some low level of survivorship of *Verbascum blattaria* seeds well beyond Year 165.

10. At least four factors are related to seed survivorship:
    1.   Size of the seed: larger seeds have more food reserves (endosperm).
    2.   Density of the seed coat: tougher seed coats provide better protection of the seed.
    3.   Level of dormancy of the embryos: deeper dormancy results in longer survivorship.

CHAPTER 30

1. d   2. b   3. c   4. e   5. c

6. If it is a fungus, we would expect to detect chitin in the cells walls, whereas chitin would be absent if it is a plant. We could also examine the specimen for the presence of chloroplasts, which would be expected to produce the green coloration in a plant but not in a fungus. If it is a vascular plant, we would expect to observe vascular tissues in the sample, which would be absent if it is a fungus. We could also sequence a conserved gene from the sample, such as a ribosomal RNA gene, and compare the sequence phylogenetically with other fungi and plants.

7. It is because the nuclei remain separate in dikaryons, even though the two nuclei are contained within a single cell.

8. Fungi play a critical role in decomposition of plants and parts of animals. If there were no more fungi, there would be enormous accumulation of the remains of dead organisms, especially the cellulose and lignin of plants.

9. Site 5 shows the highest diversity and density of lichens and so is probably farthest from the city center. Site 4 is next, followed by Site 1, then Site 3, and finally Site 2. In addition to distance from the city center and prevailing wind direction, other predictive factors could include distance to point-pollution sources (such as factories or power plants) and distance to major highways (a source of pollution from automobile exhaust). Other answers are also possible; it is important for such studies to control for factors such as the species of tree examined and the exposure of the branches to similar light and humidity conditions.

10. One common source of fungal contaminants in plant samples is symbiotic fungi, including endophytic fungi and mycorrhizal fungi. In either case, it is usually possible to collect specific tissues from the plant that exclude these symbionts. If our hypothesis about the source of the fungal genes is correct, then the fungal sequences should be absent from these symbiont-free tissues.

CHAPTER 31

1. d   2. b   3. c   4. d   5. d

6. (a) In radial symmetry, body parts are symmetrical across multiple planes that run through a single axis at the body's center. Animals with radial symmetry have no front or rear ends, and they are often sessile or drift freely with currents. If they move under their own power, they can typically move slowly equally well in any direction. In contrast, bilaterally symmetrical animals have mirror-image right and left halves divided by a single plane that runs along an anterior–posterior midline. They have a front end that usually contains a concentration of sensory systems and nervous tissues in a distinct head. Bilateral animals usually move forward in the direction of the head, so the head encounters new environments first.

   (b) Among the bilaterian animals, there are two distinct forms of gastrulation, or the initial indentation of a hollow sphere of cells early in development to form the blastopore. In protostomes, the blastopore eventually develops into the mouth of the animal. In deuterostomes, the blastopore becomes the anus.

   (c) Diploblastic animals have embryos with two cell layers (an outer ectoderm and an inner endoderm). The embryos of triploblastic animals have an additional cell layer between the ectoderm and the endoderm, known as mesoderm.

   (d) Acoelomate animals lack a body cavity enclosed by mesoderm. Pseudocoelomate animals have a body cavity enclosed in mesoderm; this body cavity contains the gut and internal organs composed of endoderm, but these latter organs are not lined with mesoderm. Coelomate animals have a body cavity that is enclosed with mesoderm, and the internal organs are also lined with mesoderm.

7. The answer to this question will depend on the reader's opinion. However, most biologists would answer that the phylogenetic analyses that result from analysis of animal genomes provide the most definitive evidence of animal monophyly.

8. Bilateral organisms have an anterior and a posterior end. As the animal moves through the environment, the anterior end encounters potential food or predators first. It is therefore advantageous for the sensory organs and central nervous system to be concentrated at the anterior end.

9. A slow metabolic rate requires a low energy budget, and hence a low intake of food.

10. Placement of glass microscope slides (or other smooth substrates for placozoan attachment) in warm tropical waters often results in colonization by placozoans. The glass slides can be suspended in water in survey areas, then later retrieved and examined for the presence of placozoans.

CHAPTER 32

1. e   2. d   3. b   4. d   5. d   6. e

7. Segmentation allows an animal to move different parts of its body independently, which allows for much greater control of movement. However, segmentation tends to constrain the body shape of an organism. Loss of segmentation is often favored in some parasitic and burrowing organisms that live in confined spaces.

8. Several answers are possible, but some examples of key innovations that appear to be associated with major episodes of diversification in protostomes include the evolution of the cuticle in ecdysozoans, the evolution of shells in mollusks, the evolution of jointed limbs in arthropods, and the evolution of wings for flight in insects.

9. Insects have been highly successful in terrestrial environments, in part because flight gives insects greater access to plants. Many insect species are specialists on one or a few plant species, and plant diversity is far greater on land and in freshwater environments than in the oceans. Although some insects live in fresh water for part or all of their life cycles, these freshwater environments are closely associated with surrounding terrestrial environments. Crustaceans have been much more successful in the oceans than have insects, and crustaceans may simply outcompete insects in marine environments.

10. All entomologists agree that many more species of insects remain to be discovered, but many entomologists think that Erwin's estimates were high. Each estimate is highly dependent on how representative *Luehea seemannii* is as a tropical forest tree. If the average tropical forest tree has many fewer host-specific beetle species than does *Luehea seemannii*, then these estimates would be inflated. Likewise, overestimating the number of tropical forest trees, or the percentage of ground-dwelling beetles, or the percentage of all insects that are not beetles, would lead to further inflation of the estimates. In addition, species diversity of beetles may be higher in Panama than in other areas of the tropics. However, any of these estimates could be underestimates as well. Each of Erwin's assumptions is now being tested; these tests require extensive work on additional species of trees, additional groups of insects, and in additional areas of the world.

CHAPTER 33

1. d   2. a   3. d   4. a   5. e   6. b

7. The four appendages common to most vertebrates are the two pectoral appendages and the two pelvic appendages. In most swimming vertebrates, these appendages function as fins. They are commonly used for propulsion (especially the pectoral fins) but are also used for steering, stabilization, and manipulation of the body position in water. Among tetrapods, the appendages are often modified into limbs used for walking, running, jumping, burrowing, climbing, grasping, and manipulating objects. There have been several reversals to fin-like limbs used by aquatic tetrapods (several times among amphibians, turtles, birds, and mammals, for example). There have also been at least three origins of the pectoral limbs of tetrapods into wings for powered flight (among birds, bats, and the extinct pterosaurs). There have also been several other modifications of the limbs for gliding (in fishes, amphibians, reptiles, and mammals). One or both pairs of appendages have been lost (or greatly reduced) in many groups of fishes, amphibians, reptiles (including birds), and mammals. Some well-known examples of limb reduction or loss include the completely legless caecilians and snakes, the loss of external hind limbs in whales and manatees, and the greatly reduced forelimbs of flightless birds.

8. Amphibians exchange gases and fluids through their permeable skins. This makes them highly vulnerable to many environmental toxins. Many species of amphibians have a biphasic life cycle, so they are vulnerable to habitat degradation and loss of both aquatic and terrestrial environments. Most amphibians do not move long distances, so they do not easily move into new habitats when their local environment is destroyed. For these reasons, they are also sensitive to rapid climate changes. Many species of amphibians have highly specialized habitat requirements and live in very restrictive ranges. Habitat loss or changes within these restricted ranges often result in extinction.

9. Fossil remains of extinct theropod dinosaurs shows that many features once thought to be restricted to birds, such as feathers, actually evolved much earlier among the theropods. Other typical "bird" morphological features, such as air-filled bones and a furcula (wishbone), are also typical of the larger group of theropods. Among living reptiles, DNA sequence analyses clearly unite birds with the crocodilians (the other living archosaurs). The combined evidence from many sources that birds are a surviving group of theropod dinosaurs is now overwhelming.

10. Hair evolved in the ancestor of mammals; feathers evolved among theropod dinosaurs (seen today among the birds). Among the living tetrapods, birds and mammals are endothermic. Hair and feathers provide body insulation for mammals and birds, respectively. Without these forms of insulation, the maintenance of metabolic body heat would be difficult. Fossil evidence shows that many extinct theropod dinosaurs also had feathers, so many paleobiologists predict that they were endothermic as well. Endothermy would also be expected in large, active predators—a description that fits our current view of many theropod dinosaurs.

CHAPTER 34

1. b   2. e   3. a   4. b   5. b   6. d   7. c   8. c

9. The cell types can be compared using a table.

Structure/Function	Sclerenchyma	Collenchyma
Cell walls	Secondary, thickened	Primary, thicker at corners
Flexibility	Less flexible	More flexible
Cell conditions	Some dead (apoptosis)	Alive
Presence	Wood, bark	Petioles, growing areas

10. Primary growth involves cell division and cell enlargement, and typically results in growth of an organ in length. Secondary growth involves growth of an organ in thickness, by the addition of more cell layers. Only some angiosperms undergo secondary growth. Herbaceous plants such as peonies have only primary growth. Woody plants such as trees have both primary and secondary growth.

11. The initials are still 1.5 meters above the ground today because the plant grows in height at its apex.

12. Some examples might include a larger root apical meristem to produce thicker carrots and reduced internode growth to produce compact heads of cabbage.

CHAPTER 35

1. c   2. d   3. b   4. b   5. d   6. e

7. Epidermal cells have external walls with a waxy cuticle, which makes them repel water. The epidermal cells of roots might have a thinner (or absent) cuticle, as they take up water; and leaves and stems a thicker cuticle, to conserve water. In addition, the epidermis of leaves, and to a lesser extent stems, has stomata, which regulate gas exchange (including loss of water vapor from the leaf interior).

8. A source is an organ such as a leaf that produces more sugars than it uses. A sink is an organ such as a root that produces less sugars than it needs and so imports sugars from a source. In a deciduous tree, a leaf might be a source in the summer, and roots a sink. But then in spring, the roots might be a source for the buds (newly emerging leaves).

9. To cross the fewest membranes and still get from the soil solution to the atmosphere by way of the stele a water molecule would follow this route: soil to root symplast to stele symplast to stele apoplast to xylem to leaf apoplast to leaf interior air space to stoma to atmosphere. A water molecule could follow this path by crossing as few as two membranes: 1) from soil into a root hair or root cortical cell across a root cell membrane; 2) out of a stele cell into the stele apoplast across a stele cell membrane. Getting from the soil solution to a mesophyll cell in a leaf would require crossing at least three plasma membranes: the two membranes listed for the previous route plus the mesophyll cell membrane (to get from the leaf apoplast into the mesophyll cell).

10. The mutation in the HARDY gene might cause increased expression of a gene that inhibits cation accumulation in stomata, thereby keeping them more closed and conserving water. To test this hypothesis, you could look at the response of stomata to light in the leaves of mutant versus wild-type plants. The stomata of wild-type plants should open rapidly in response to light (see Figure 35.9); the stomata of HARDY mutant plants might open more slowly or less wide.

11. (a) Yes. The difference in water potential between the soil and the leaf (1.7 MPa) is enough to overcome gravity and draw water to the top of the tree.

    (b) No. If the soil water potential decreased to −1.0 MPa, it would be more negative than inside the root cells and water would leave the roots (and enter the soil).

    (c) If all the stomata closed, the leaf water potential would not be as negative. This in turn would make the xylem water potential less negative, and so on down to the roots. This would make the difference between the leaf water potential and root water potential insufficient for water to flow from the roots to the leaves (toward a more negative water potential).

CHAPTER 36

1. d   2. d   3. c   4. a   5. c   6. d

7. The ability of chemists to detect low concentrations of elements is fairly recent. Before then, nutrient solutions thought to be pure often were not.

8. Heavy irrigation after a prolonged dry period may produce runoff of topsoil (the A horizon) and leaching of ions (especially anions) into the subsoil, making fewer nutrients available to plant roots. Converting land use from virgin deciduous forest to crops will change the composition of living organisms in the soil, as many organisms that live in association with tree roots will disappear. The soil structure and texture will also change, because roots will no longer be present to hold the soil together and make air spaces. The soil chemistry will change, because crops take up nutrients from the soils and the nutrients are removed from the system when the crops are harvested.

9. See Figure 36.10, the nitrogen cycle. There are numerous species that fix nitrogen. Loss of one species might allow others to expand and replace it. Loss of all the species would mean that only abiotic methods could be used for nitrogen fixation. This might reduce overall nitrogen in the soil, meaning less would be available for plant growth.

10. The experiment with mutant Arabidopsis suggests that Arabidopsis uses either its own or exogenous strigolactones for growth regulation and has the appropriate receptor and response mechanisms. This reinforces the idea that an ancient mechanism to attract beneficial microbes also is used for modern plant growth regulation. Or the reverse might be true: the original function of strigolactone might have been as a plant hormone and its role in plant–microbe interactions might have evolved later.

11. Because holoparasitic plants can gain reduced carbon through association with hosts, the genes encoding photosynthesis functions are not under selection pressure, because having them would not confer any survival and reproductive advantage for the parasites. So any mutation that renders such a photosynthesis gene nonfunctional will not be deleterious.

CHAPTER 37

1. a   2. d   3. b   4. b   5. a   6. b

7. Fire produces ash, which enriches the soil with plant nutrients. A seed that germinated as a result of fire could have an advantage in such a nutrient-rich soil.

8. If a single species has two mechanisms for breaking seed dormancy, then if environmental conditions for Mechanism A are not present, environmental conditions for Mechanism B might be. This enables the plant to respond to a wider array of environmental conditions. In addition, if the cue for, say, Mechanism A turns out to be misleading (not predictive of favorable conditions) and the

seedling dies, there is still a second seed that can germinate at a different time (by Mechanism B), when conditions might be more favorable.

9. The charcoal in the bag absorbs ethylene gas, which is released by ripening fruits. The lack of ethylene prevents over-ripening and decay.

10. To test for the relationship between corn stunt spiroplasma disease and gibberellins, you could measure gibberellins in plants infected with the bacterium and in normal plants; you might expect the spiroplasma-infected plants to exhibit a reduction in gibberellins. Another approach would be to infect normal plants with the spiroplasma and then spray gibberellins on them; you might expect this to reverse the stunt phenotype.

11. (a) See Figure 37.2 Add a mutagen to hundreds of corn seeds and plant them. In a screen, look for plants that are shorter, and propagate these.

    (b) See Figure 37.11. If the transcription factor in the gibberellin signal transduction pathway is inactivated, the plants will be insensitive to gibberellin and be stunted. A mutation that inactivates the gibberellin receptor would have the same effect.

    (c) Other potential effects might include reduced seed germination and reduced seedling growth due to lack of mobilization of stored reserves in the seed (see Figure 37.5). If the mutant is completely gibberellin-insensitive, these effects will *not* be overcome by adding gibberellin to the seeds as they germinate. If, however, the mutant is a dwarf because of reduced amounts of gibberellin in the plant (because of a mutation that affects gibberellin biosynthesis, for example), the germination effects could be reversed with exogenous gibberellin.

CHAPTER 38

1. b   2. e   3. b   4. e   5. a   6. c

7. In triploid cells undergoing meiosis, there cannot be pairing of homologous chromosomes in meiosis I. So meiosis I is abnormal and functional gametes do not form.

    A fruit is formed from the ovary wall of the flower.

    Seedless grapes are probably propagated by cuttings (vegetative reproduction).

8. Poinsettias are short-day plants; they bloom at a time of year when days are getting shorter (in the Northern Hemisphere).

9. No, it isn't necessary. Just a flash of light during a long night is enough to convert P_r to P_{fr} and to change the photoperiod.

10. (a) The mutation stabilized the CO protein.

    (b) The mutation caused nonfunction of the FD protein.

    (c) The mutation increased expression of the FLC protein.

    (d) The mutation caused constitutive expression of the CO protein.

11. Several approaches might be taken, such as a genetic screen for meiotic cells that do not separate chromosomes at anaphase I, or a search for proteins (and then their genes) that bind to SWII protein.

CHAPTER 39

1. b   2. c   3. a   4. b   5. c   6. c

7. A plant might make a secondary metabolite that kills an insect pest. Plants making this metabolite would be selected for in evolution. However, the insect might develop resistance to the metabolite. Then the insect population would increase while the plant population decreased—until another defense mechanism evolves. This is coevolution. For more examples, see Chapter 56.

8.

Avr2Avr3	Healthy	Healthy	Diseased
Avr1Avr4	Healthy	Healthy	Healthy

9. (a) The effects of reduced rainfall could include dehydration and osmotic stress. Genetic responses might include alterations in leaf anatomy, with a thicker cuticle to reduce evaporation; a more extensive root system to obtain water; and accumulation of solutes in the roots, which would reduce root water potential and result in more water uptake in dry soils.

    (b) Flooding reduces the amount of O_2 available to the plants and results in reduced respiration. Adaptations might include increased production of pneumatophores or aerenchyma to supply air to submerged plant tissues.

    (c) Wheat rust is a fungal pathogen. Plants can adapt by increasing the ability to seal off infected areas and reduce the spread of the fungus within the plant, by developing specific immunity, and by increasing production of phytoalexin and pathogenesis-related proteins that kill the fungus.

10. You could feed one group of hornworms on normal plants and another group on genetically modified plants. The two groups could then be exposed to the parasite. If nicotine is protective, the hornworms that fed on normal plants should have fewer parasites.

CHAPTER 40

1. c   2. c   3. a   4. d   5. b

6. Feedforward information makes it possible to anticipate a physiological challenge to homeostasis and to take preemptive action by changing a set point or the sensitivity of a regulatory system. Feedforward information for the regulation of breathing could be the onset of exercise; for blood pressure it could be the fight-or-flight response to a threat; and for secretion of digestive juices it could be the sight, smell, or expectation of food.

7. In the metabolic rate/environmental temperature curve in Figure 40.17, the equivalent of *HL* would be metabolic rate, as long as the animal's temperature is not rising or falling and the animal is not doing external work. *K* would represent the animal's thermal conductance, or how easily it loses heat; $1/K$ would be a measure of the animal's insulation. The curve projects to 0 at an ambient temperature equal to body temperature because this portion of the curve represents the extra metabolic effort necessary to compensate for heat loss to the environment. If body temperature and environmental temperature were the same, there would be no heat loss to the environment.

8. Biological processes proceed more slowly at lower temperatures. Thus the lower the temperature of the heart or skeletal muscle, the slower will be its ability to generate a contractile force. This could pose a physiological challenge for highly active fish such as great white sharks or giant bluefin tuna that depend on fast swimming and endurance to catch prey. An evolutionary adaptation to this challenge can be seen in these fishes' vascular anatomy: blood from their hearts goes to the gills, where it exchanges respiratory gases but also comes into thermal equilibrium with the cold ocean water. Thus these fishes are sending cold blood to their body tissues.

9. Basal metabolic rate (ml O_2/hr) vs. body mass (kg):

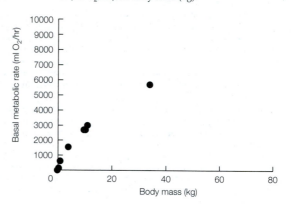

Heart size (g) vs. body mass (kg):

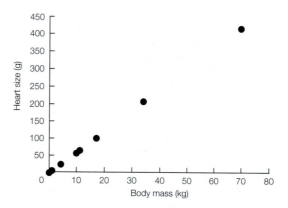

10. The heart has to pump blood against the resistance of the vascular system. Since blood vessels go to all tissues of the body, the total length of blood vessels will be directly proportional to body size, and therefore total peripheral resistance will be directly proportional to body size. The other factor that determines how much blood the heart can pump is the heart rate. The total cardiac output is a function of the size of the heart and the number of times it beats per unit of time.

11. Several factors that determine the heat transfer rate between the iguana and its environment are different in these two conditions. When basking on the lava rocks, the radiation absorbed will be higher and the heat conducted across the skin will be different in comparison to being in the water. Also, when the animal breathes it will be losing heat through evaporation of water in its airways.

    An experiment in which the animal heats up and cools down in the same environment would offer stronger support for the conclusion that blood flow to the skin is a critical factor. The iguana could be placed in water and in air at two different temperatures (e.g., 20°C and 40°C) to compare the rates of heating and cooling.

CHAPTER 41

1. b   2. a   3. d   4. e   5. e

6. The time course of a hormone signaling system depends on several factors, including the rate of release of the hormone, its half-life in the blood, and the interactions it has with its receptors. A hormone signaling system that controls a short-term process such as digestion would be expected to have a rapid (e.g., vesicular) release, a short half-life, and rapid action; these are attributes of a peptide hormone. A hormone signaling system that controls a long-term process such as embryonic development would be expected to have continuous release, a long half-life, and to be slow acting, attributes of a steroid hormone.

7. The high levels of synthetic male steroid hormones exert actions through the testosterone receptors that exist in both males and females. The usual effects of promoting muscle hypertrophy and male secondary sexual characteristics (e.g., hair, a deeper voice) will occur. In addition, the high level of negative feedback on the pituitary gonadotropes and on the production of GnRH by the hypothalamus will reduce testicular and ovarian functions in the males and females. Decrease of circulating estrogens in the females will cause reduction of breast tissue.

8. The most common cause of hypothyroidism is lack of iodine in the diet. Thyroglobulin continues to be produced, but there is a lack of functional T_3 or T_4. As a result, the levels of TRH and TSH rise because of the lack of negative feedback. The elevated TSH induces continued production of thyroglobulin, resulting in goiter. The most common cause of hyperthyroidism is an autoimmune disease called Graves disease. An antibody to the TSH receptor is produced, and the binding of that antibody to the receptor causes activation of the signaling pathway that increases the production of thyroglobulin and the development of goiter. If there is adequate iodine in the diet, however, this condition results in increased secretion of T_3 and T_4, which produces the symptoms of hyperthyroidism.

9. The large size of the larvae together with the lack of adult moths indicates that cycles of growth and molt of the larvae continued without the induction of pupation. The low temperature probably prevented the usual decline in the production and secretion of juvenile hormone by the corpora allata.

10. Insulin controls the entry of glucose into most cells of the body, but not neurons. When insulin levels fall, as happens during the postabsorptive state (i.e., after ingested nutrients have been fully digested), glucose entry into cells slows down and the cells convert to using other sources of energy. Neurons, however, always require glucose; their lack of insulin means that their access to an adequate glucose supply is protected during the postabsorptive state because there is no decrease in their ability to take up glucose from the blood. Also, the decrease in glucose use by other cells of the body preserves the glucose in the blood for use by the nervous tissue.

CHAPTER 42

1. a   2. e   3. e   4. c   5. d   6. a

7. See Figure 42.10. The antigen-binding site of an antibody has heavy and light chains in a unique three-dimensional configuration that binds a particular antigenic determinant. This is similar to an enzyme active site that binds a substrate. In both cases, binding is noncovalent. A major difference is in the result of binding: an antigen does not change its covalent structure when it binds to an antibody, whereas a substrate does change covalently when it binds to an active site.

8. Both immunoglobulins and T cell receptors have constant and variable protein regions, bind antigens, and have great variability in primary structure. T cell receptors are only membrane proteins of T cells. Immunoglobulins can be either membrane proteins of B cells or secreted proteins in the blood.

9. The father's haplotypes are A1B7D11 and A3B5D9; the mother's are A4B6D12 and A2B7D11. These two parents could not have the child with the genotype indicated.

10. There are thousands of different enzymes in an individual but potentially millions of different specific antibodies. Every cell in an animal has the genetic information for all enzymes. Each immunoglobulin, however, is derived from a unique gene (produced by DNA rearrangements) in a B cell or a clone.

11. Experiments might involve testing vaccinated people for neutralizing antibodies against HIV (humoral immunity) and looking for T cell activity against HIV-infected cells (cellular immunity).

CHAPTER 43

1. a   2. d   3. d   4. d   5. d   6. a

7. Leydig and thecal cells have similar functions and characteristics. They are both removed from direct contact with the developing gametes, and they both produce testosterone. Sertoli cells and granulosa cells are both in direct contact with the developing gametes, and they support their development by providing nutrients.

8. Progesterone actions are required to maintain the endometrium in a condition that can support implantation and not degenerate, as occurs during menstruation. By blocking progesterone receptors, RU-486 prevents implantation and the maintenance of the endometrium.

9. Conditions that could favor the evolution of this sexual dimorphism are a sessile existence, dispersed populations, and availability of suitable habitats. If a larva lands on a suitable substrate, it will have high reproductive success if it is a female and can produce lots of eggs; eggs typically have resources that enable them to travel considerable distances if released into the water, and their probability of being fertilized is therefore high. A larva would have a lower probability of success if it developed into a solitary male and produced sperm; sperm have less ability to survive travel over long distances in the water to encounter eggs. However, if a larva lands on a female, it is guaranteed high reproductive success if it can fertilize all of the eggs that female produces. Therefore, by attaching itself to the female and minimizing all of its own physiological processes other than sperm production, a larva can achieve high reproductive success at very little cost.

10. It is likely that the man's offspring would all be daughters. The Y chromosome lacks some essential genes that are on the X chromosome; in the absence of the cytoplasmic bridges, the developing sperm that contain a Y chromosome would lack those gene products. Thus all viable sperm the man produces would contain an X chromosome. This would result in female offspring, since the mother would also contribute an X chromosome.

CHAPTER 44

1. e   2. a   3. d   4. b   5. b

6. You could inject Disheveled protein into the side of the fertilized egg opposite the gray crescent and see if a secondary organizer formed in that region of the resulting blastula. You could also inject the inhibitor of the Disheveled protein into the region of the gray crescent and see if that prevented the formation of the organizer.

7. You would be destroying the cells that normally migrate from the dorsal blastopore lip to form the notochord, and that also produce the signals that determine the anterior–posterior differentiation of the embryo. Thus you might see defects in the development of the nervous system or abnormal segmental development of the body.

8. The flow of fluid over Henson's node may be asymmetrical and thereby create different physical forces on the primary cilia of cells on either side of the node. These differential forces could influence the expression of *Sonic hedgehog*. Experiments to test this hypothesis could include using gene knockouts that eliminate the motile cilia around Henson's node, or experiments in which early embryos are cultured under conditions in which the flow across Henson's node is opposite that of the normal pattern. The prediction for the first experiments would be that the right–left asymmetry of organs in the embryos would be randomized. The prediction for the second experiment would be that the normal asymmetry of organ development would be reversed.

9. A possible mechanism would be cytoplasmic factors that are not distributed randomly or evenly throughout the cytoplasm of the oogonia or the spermatogonia. Thus when they divide by mitosis, factors that control the fates of the daughter cells could be received by one daughter cell but not the other.

10. At an early stage of blastulation (e.g., the 16- or 32-cell stage,) a few blastomeres could be removed from the embryo and cultured separately to produce a population of stem cells. The embryo could go on to develop normally, and the stem cells could be frozen for later use.

CHAPTER 45

1. d   2. b   3. c   4. e   5. c

6. When the stimulus occurs at some point along an axon and an action potential is stimulated, depolarizing current will flow in both directions, bringing adjacent areas of the axon to threshold. However, once an action potential is fired, the Na+ channel inactivation gates close and make that section of the axon refractory to further stimulation until they open again. Thus the action potential cannot reverse its direction of propagation, and if the action potential begins at the axon hillock, it cannot reverse its direction of propagation and is unidirectional.

7. Excitatory synapses cause a depolarization of the neuronal membrane, and inhibitory synapses hyperpolarize it. These two influences are summed by virtue of the resulting membrane potential. If it depolarizes enough to reach threshold, an action potential will be fired at the axon hillock.

8. Because the GABA receptor is inhibitory, benzodiazepines would be expected to slow cognitive processes and make a person more likely to fall asleep.

9. The type of information that an action potential transmits depends on the nature of the sensory cell that generated the action potential and on the nature of the cell that receives input as a result of that action potential. Thus photoreceptors transduce light into action potentials, and those action potentials are interpreted as light in the visual circuits that receive those action potentials. Intensity of the stimulus is coded as the frequency of action potentials. Integration is achieved by the summation of excitatory and inhibitory influences on the target cells.

CHAPTER 46

1. d   2. a   3. e   4. e   5. c

6. Olfactory and taste receptors are both chemosensors that respond to specific molecules in their environment. Olfactory receptors, however, are neurons, whereas taste receptors are epithelial cells that communicate with neurons that are associated with them. Olfactory receptors express a family of genes for olfactory receptor proteins that are then localized on cilia that project out of the olfactory epithelium. All of these olfactory receptor proteins are G protein-linked and are metabotropic. Taste receptors are also located on cilia of the epithelial taste sensor cells. Bitter, sweet, and umami receptors are G protein-linked metabotropic receptors, but salt and sour receptors are ionotropic. The discrimination between an apple and an orange depends on integration of information from both the olfactory and the taste receptors.

7. The sensation of directional motion arises from the vestibular system, which includes the semicircular canals and the vestibule, which contains membranous structures containing a fluid—endolymph. At the base of each semicircular canal is a gelatinous projection, a cupula, that encases a cluster of stereocilia. Movement of the head causes movement of the endolymph, which then exerts force on the cupula and bends the stereocilia, generating action potentials in the vestibular nerves. The vestibule includes two membranous structures called the saccule and the utricle. In these structures, stereocilia tips are in contact with otoliths, which are membranous structures containing crystals of calcium carbonate. When the head is accelerated forward or backward, the momentum of the otoliths causes the stereocilia to bend in a direction that indicates the direction of movement.

8. Underwater, the external ear canals are filled with water. Unlike air, water is not compressible and therefore sound waves are transmitted through water as vibrational movements of the water. These movements exert greater forces on the tympanic membrane than air pressure waves do.

9. As happens in humans in the vestibular system, movements of the fish cause movements of the water in the lateral line canals. The resulting forces are transduced into action potentials by the hair cells and provide information about the movement of the fish through the water. Additionally, vibrations in the water generated by other organisms or physical events will cause movement of the water in the lateral line canal and be transduced into action potentials, providing information to the fish about its environment.

10. The owl depends on auditory stimuli to locate the mouse in total darkness. Directional information comes from the bilateral placement of the ears, which are equally stimulated when the owl is directly facing the source of the sound. The face of the owl is disc-shaped, which helps collect sound waves and direct them to the ears.

CHAPTER 47

1. d   2. d   3. c   4. c   5. a

6. The stab wound must have severed the sympathetic nerves on the left side of the man's neck. Activity in these nerves causes dilation of the pupil. Severing the sympathetic nerves on the left side would remove all sympathetic activity reaching the pupil on the left side, and therefore it would be more constricted. Similarly, sympathetic activity decreases activity in the salivary glands, whereas parasympathetic activity increases salivation. Thus withdrawal of sympathetic input would release the salivary glands from any inhibition that would counteract even low levels of parasympathetic input.

7. Eyes positioned on the sides of the head enable a wider field of vision. Eyes pointing in the same direction create a narrow field of vision but make depth perception possible. You would expect prey species to benefit from wider fields of vision. You would expect predator species to benefit from depth perception, which would facilitate pursuit and capture of prey.

8. Sleepwalking is more likely to occur in non-REM sleep for two reasons: there is motor inhibition in REM sleep, which renders the individual paralyzed; and the nature of sleepwalking activities does not match with the vivid, bizarre content of REM-sleep dreams.

9. We can break this question down into the different observations. First, the loss of motor control of the right leg indicates that motor commands are ipsilateral—they descend on the same side as the limb that is being controlled. The ability to sense painful stimuli applied to the right leg but not to the left foot indicates that the pain pathways cross over to the opposite, or contralateral, side of the spinal cord before they ascend to the brain. The reflex

movement of the right leg to a stimulus applied to the left foot indicates that reflex information is processed at the local level and does not require processing at higher levels of the central nervous system. The conclusion is that motor commands in the spinal cord are ipsilateral to the muscles being controlled but that the pain information ascending to the brain is contralateral. Finally, the different responses to pain and to touch indicate that these two modalities of somatosensory information travel in different tracks: touch ipsilaterally and pain contralaterally.

10. The fact that more slow-wave activity was seen over the right frontal cortex than over the left in response to exercising and training the left hand shows that the right side of the brain controls the left side of the body, and visa versa. This increase in slow-wave activity during sleep following the exercise/training suggests that this sleep slow-wave activity reflects either a restorative process or a learning process.

CHAPTER 48

1. b   2. c   3. b   4. d   5. c

6. One feature is the length of the muscle; half the length of a long muscle is more than half the length of a short muscle. Another feature is the location of insertion of the muscle on the bone; this determines the relative lengths of the effort arm and the load arm of the lever system created by the muscle, bone, and joint. If the ratio of load to effort arm is small, a large movement that can generate only relatively small forces is possible. If the ratio of load to effort arm is large, only small movements that can exert large force are possible.

7. If the break and healing have damaged the epiphyseal plate, and the primary and secondary areas of ossification fuse, the bone can no longer grow at that end.

8. The shoulders will fatigue first because they are not normally responsible for maintaining posture; they are adapted for rapid movements and sudden applications of large force. Thus shoulder muscles have a higher proportion of fast-twitch fibers. The leg muscles are postural muscles and have a higher proportion of slow-twitch fibers.

9. The action potential is conducted throughout the muscle cell by the system of T tubules. In the T tubules, the action potential causes conformational change of the DHP–ryanodine receptor complex. That change opens Ca^{2+} channels in the sarcoplasmic reticulum, and Ca^{2+} diffuses into and throughout the sarcoplasm. Ca^{2+} binds with the troponin units, causing the tropomyosin to expose the actin–myosin binding sites, cross-bridges to form, and the muscle to contract. When the Ca^{2+} concentration in the sarcoplasm falls as a result of being pumped back into the sarcoplasmic reticulum, the process reverses and actin–myosin binding sites are no longer available. The difference in time course of the contraction versus the action potential is due to the time that it takes for the Ca^{2+} to be released, diffuse throughout the sarcoplasm, and then be sequestered back into the sarcoplasmic reticulum.

10. The increased amount and duration of Ca^{2+} in the sarcoplasm causes increased contraction of the muscles and therefore an increase in muscle tension. The increase in muscle tension requires additional expenditure of ATP, raising metabolism and producing more heat. The increased metabolism causes elevated heart rate. This is in addition to the effect of the increased Ca^{2+} in the cardiac muscle itself.

CHAPTER 49

1. d   2. e   3. b   4. c   5. a

6. This fish would not be very active. It would move slowly. It would have a larger heart and larger blood vessels than fishes with hemoglobin, to accommodate a high flow of blood at low pressure. Its gill membranes would be well developed. It would have a high blood volume. Its jaws would show adaptations for sit-and-wait capture rather than pursuit. This fish occurs only in Antarctic waters because Antarctic waters are very cold and therefore the solubility of O_2 in those waters is high.

7. The total surface area for gas exchange is much smaller in a large air cavity than it is in many smaller cavities (alveoli) that add up to the same total volume. If the lung tissue is less elastic, the vital capacity of the lungs will go down, meaning less air can be exchanged during the breathing cycle. The less elastic lung tissue is also less permeable to respiratory gases.

8. Close to the end of inhalation, the pleural cavity pressure is reaching its maximum negative value. At the same time, the alveolar pressure is rising back up to being the same as the atmospheric pressure. Alveolar pressure would be most positive relative to atmospheric pressure at the midpoint of the exhalation phase.

9. Blood cells require energy. When in storage, their initial energy source is the glucose in the blood plasma, but as that supply gets depleted, blood cells also metabolize the intermediates in the glycolytic pathway. That includes 2,3-BPG. As the 2,3-BPG gets metabolized, there is less of it to bind to deoxygenated hemoglobin and therefore the affinity of the hemoglobin for O_2 increases. When

the affinity of the hemoglobin for O_2 gets too high, it can lower the P_{O_2} in the plasma to levels that are below the P_{O_2} in the plasma of a patient.

10. When you go up in altitude, the P_{O_2} in the air you breath goes down, but the P_{CO_2} was already low at low altitude and remains low at high altitude. Thus at higher altitude there is less of a concentration gradient driving diffusion of O_2 into the blood, but no decrease in the concentration gradient driving CO_2 out of the blood. Thus the main stimulus for breathing goes down as the need to increase breathing goes up. As a result, the blood becomes hypoxic and triggers breathing by activating the carotid and aortic chemosensors. The increase in breathing blows off even more CO_2, and breathing slows, which causes another bout of hypoxia and even a rise in blood CO_2. That triggers another bout of rapid breathing, and this cycle repeats.

11. (a) The llama hemoglobin has a higher affinity for O_2.

(b) Llama hemoglobin would be advantageous at high altitudes because it can become 100 percent saturated at the low P_{O_2} of the high-altitude environment. Therefore the hemoglobin can carry a full load of O_2 to the tissues.

(c) Llama hemoglobin allows the transfer of O_2 to occur at lower tissue P_{O_2}s.

CHAPTER 50

1. a   2. c   3. d   4. c   5. e

6. One factor is that at the beginning of a race there is a feedforward signal from the sympathetic nervous system that increases heart rate. Another factor is that the increased heart rate, together with the increased venous return to the heart from the exercising muscles, stretches the ventricles, which then contract with more force as described by the Frank–Starling law. This is due to the fact that a slight stretching of the sarcomeres optimizes the overlap of the actin and myosin fibrils for a maximum contraction. Increased breathing also increases the venous return and induces the Frank–Starling law.

7. There is no time when all four heart valves are open at the same time; if there were, the heart could not pump efficiently. Throughout diastole, the aortic and pulmonary valves are closed and the atrioventricular valves are open. At the beginning of systole, the atrioventricular valves close. There is a brief moment when all four valves are closed, until the aortic and pulmonary valves open, and stay open until the end of systole.

8. There are rapid responses and longer-term responses. A fall in blood pressure lowers the firing rate of baroreceptors in the great arteries. The decrease in baroreceptor input to areas of the brainstem that regulate cardiac function results in increased sympathetic and decreased parasympathetic output to the heart. This increases the heart rate and the force of contraction of the cardiac muscle (the fight-or-flight response). A slower response to blood loss is mediated by the kidney, which responds to the decreased blood pressure by increasing the release of renin, which in turn increases the activation of angiotensin circulating in the blood. Active angiotensin increases blood pressure by constricting peripheral blood vessels and stimulating thirst. Another slow response, stimulated by the fall in baroreceptor activity, is the release of ADH from the posterior pituitary. ADH increases the reabsorption of water by the kidney.

9. The cardiac muscle must be capable of generating and conducting action potentials. This may involve different variants of the ion channels involved in action potential generation and conduction. The cardiac muscle must be able to convert the action potential into the opening of Ca^{2+} channels in the sarcoplasmic reticulum, so there could be adaptive changes in the DHP and ryanodine receptors. Once Ca^{2+} is released into the sarcoplasm, it has to be resequestered into the sarcoplasmic reticulum, and that Ca^{2+} pump is likely to be adapted to operate at lower temperatures in the hibernator.

10. Your graph should look like this:

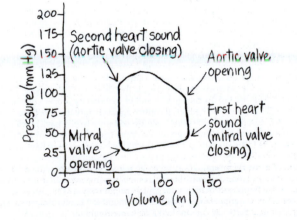

A similar graph for the right ventricle would have the same volumes but lower pressures.

CHAPTER 51

1. e   2. a   3. b   4. d   5. b

6. The rationale for a high-fat and high-protein diet is that it minimizes the secretion of insulin. Insulin promotes the uptake, metabolism, and storage of glucose by various cells of the body; it also inhibits the action of lipase in the adipose tissue. Thus with low insulin, tissues are more likely to metabolize fats and less likely to store them.

7. The following points might be included in your answer.
- Insulin stimulates most cells of the body to take up glucose from the blood by stimulating the insertion of glucose transporters into the plasma membranes of those cells in the absorptive state.
- Insulin inhibits lipase in the adipose tissue, so breakdown of stored lipids is decreased in the absorptive state.
- Insulin stimulates the synthesis of triglycerides in the adipose tissue.
- Insulin activates the enzyme that phosphorylates glucose as it enters cells, thereby preventing it from diffusing back out of the cells. This maximizes the uptake of glucose by cells.
- Insulin activates the liver enzymes that synthesize glycogen.
- The lack of insulin in the postabsorptive state decreases the uptake of glucose by most cells of the body and activates the enzymes of lipolysis and glycogenolysis.

8. Triglyceride in food is emulsified by bile in the duodenum to form micelles, which are broken down by pancreatic lipase into free fatty acids and monoglycerides. Both are absorbed across the plasma membranes of intestinal epithelial cells. In the intestinal epithelial cells, the free fatty acids and monoglycerides are resynthesized into triglycerides and packaged into chylomicrons, which also contain cholesterol and are coated with lipoproteins. The chylomicrons are secreted from the basal ends of the epithelial cells into the center of the intestinal villi, where they enter the lymphatic vessels and circulate through the lymphatic vessels to the thoracic duct, where they enter the blood.

A *direct route* the fatty acid could take would be for the chylomicrons circulating in the blood to come into contact with the damaged endothelium of the coronary arteries and be absorbed into the plaque.

In an *indirect route*, the chylomicrons could be taken up by liver or adipose tissue cells and the triglyceride stored. In the liver, the triglyceride can be repackaged to form low-density lipoproteins or very low-density lipoprotein particles, depending on the amount of cholesterol in the particle. These lipoproteins leave the liver and circulate in the blood. When they come into contact with the damaged endothelial cells of the coronary arteries, they bind to lipoprotein receptors, and their triglyceride and cholesterol are absorbed into the plaque.

9. Carbonic anhydrase catalyzes the hydration of CO_2 to produce carbonic acid that dissociates into H^+ ions and HCO_3^+ ions. In the parietal cells of the stomach, H^+ is secreted into gastric pits and then flows into the stomach lumen. The bicarbonate ions are transported out of the basal side of the cells, where they are absorbed into the blood, raising blood pH. In the ducts of the pancreas, bicarbonate is transported into the lumen of the ducts; H^+ ions are transported out of the basal sides of the cells, where they are absorbed into the blood, lowering its pH.

10. The hypothalamic regulatory pathways control hunger and satiety, not energy use. This tells us that the side of the energy balance equation that is most important in contributing to obesity is Calories in. Of course, obesity can also result in decreased physical activity, which means that decreased Calories out can be a secondary factor in the causation of obesity.

CHAPTER 52

1. d   2. a   3. a   4. b   5. a

6. The glucose contributes to the osmotic concentration of the glomerular filtrate and therefore of the tubular fluid. This results in a greater volume of urine flowing through the collecting ducts and being excreted.

7. ACE inhibitors decrease the production of angiotensin II, the active form, from angiotensin I. Decreasing the level of angiotensin in the blood increases the glomerular filtration rate and therefore the production of urine. Losing more water in the urine lowers the blood volume and therefore the blood pressure. Blocking angiotensin also results in dilation of peripheral blood vessels, which lowers blood pressure, and decreases thirst, which helps maintain a lower vascular volume. Angiotensin stimulates the release of aldosterone, which promotes Na^+ reabsorption and therefore water retention.

8. The rate at which the inulin is filtered is equal to the concentration of inulin in the blood ($[I_b]$) times the glomerular filtration rate (GFR). The rate at which inulin is excreted is equal to the concentration of inulin in the urine ($[I_u]$) times

the urine flow rate (V = 1 ml/min). Since all inulin that is filtered leaves the body in the urine, the rate at which it is filtered must equal the rate at which it is excreted. Therefore $[I_b] \times GFR = [I_u] \times V$, and $GFR = [I_u] \times V/[I_b]$.

9. Any difference in the rate at which the substance (S) is filtered versus the rate at which it is excreted will be due to tubular reabsorption of tubular secretion. If more of S is excreted than filtered, then S must be secreted by the renal tubules. If less of S is excreted than filtered, then S must be reabsorbed by the tubules. Therefore $[S_b] \times GFR - [S_u] \times V$ = the rate of reabsorption (if negative) or secretion (if positive) of S.

10. At high altitude, the concentration difference driving O_2 across the alveolar membranes and into the blood deceases, but the concentration difference driving CO_2 across the alveolar membranes and into the expired air does not change. The increased breathing rate driven by hypoxia will therefore blow off too much CO_2 and the pH of the blood will increase, which will suppress breathing. Bicarbonate is filtered into the glomerular fluid; as that fluid passes through the renal tubules, H^+ ions are excreted, resulting in the formation of H_2CO_3, which dissociates into CO_2 and H_2O. The CO_2 is reabsorbed into the tubule cells, where carbonic anhydrase catalyses its hydration to H^+ ions that are secreted back into the tubular fluid and bicarbonate ions that are secreted back into the extracellular fluid. Blocking carbonic anhydrase results in fewer H^+ ions being secreted into the tubular fluid and more bicarbonate ions remaining in the tubular fluid to be excreted. The retention of H^+ and excretion of bicarbonate lower the blood pH, and that stimulates increased breathing. However, the increased bicarbonate in the tubular fluid raises the osmotic concentration of that fluid, causing more water to be excreted.

CHAPTER 53

1. b   2. e   3. c   4. a   5. d   6. e   7. a   8. e

9. The development of brain circuitry controlling the sexual dimorphism in urination behavior is influenced by the levels of estrogen or testosterone that are circulating in the blood during the early postnatal period. Estrogen must prevent the development of the neuronal patterns of connectivity that are responsible for the male behavior.

10. The major difference between the eusocial insects and vertebrates is the haplodiploid mechanism of sex determination in the insects. This means that a female worker that is a daughter of the queen shares more genes with sisters—the queen's offspring—than she would with her own offspring, which would have a different father. Thus raising a sister contributes more to the female worker's inclusive fitness than raising a daughter would. Haplodiploidy is not found among vertebrates, so this powerful selective force does not operate among them.

11. The variability in the undirected song allows the male to adapt his song to the local variant. The directed song is probably more effective in attracting a female because it more accurately identifies the male as a member of the local population, and is also more effective in competition with neighboring males.

12. Since male cowbirds will not hear the song of his father, his song pattern should be genetically determined. Females probably learn the song of host species and thereby learn to identify and locate potential host nests. You could test the hypothesis about males by raising them in isolation to see what songs develop as they mature. You could test the hypothesis about the females by giving them a choice test such as that done with the zebra finches in testing directed and undirected song preferences (see Figure 53.8). Give each female cowbird a three-chambered cage, and play the songs of host species and non-host species in the opposite end chambers. Place the test bird in the middle chamber. Record the amount of time the female spends in each end chamber as a measure of preference. Another experiment would be to bring a host nest with cowbird eggs into the laboratory and raise the cowbirds in the presence of recordings of another species song. Then repeat the choice experiment with the host's and the other species' songs being played.

13. The classical data suggest that the hygienic behavior is controlled by two genes to give the typical Mendelian ratio of 3 to 1 with 50% hybrid, 25% homozygous dominant, and 25% homozygous recessive. The QTL analysis, however, indicates that more than two genes are involved. The difference in these results could be due to there being two major-effect genes and several modifying genes.

CHAPTER 54

1. d   2. d   3. a   4. c   5. c

6. Based on its location and its weather conditions, the mattoral should have vegetation typical of that found in other areas with Mediterranean climates. Vegetation should be (and is) tough, shrubby and fire-adapted, with slender, leathery leaves and seeds either stored in fire-safe cones or equipped with elaiosomes and dispersed by ants.

7. The fact that species in the genus are known in Australia and in southern Africa, it is likely that the genus originated before the breakup of Gondwanaland, the supercontinent comprising Antarctica, South America, Africa and Australia.

Other species in the genus thus might be found in South America (contemporary conditions are too cold to maintain spider life in Antarctica today but the possibility exists that fossil species in the genus might be found).

8. Examining the x axis of the figure suggests that, if average low temperatures shift upward by five degrees C, the tundra biome would experience a substantial decrease in geographic extent (and may cease to exist altogether); boreal forest might also experience a reduction in geographic extent.

9. The extensive radiation of *Drosophila* species in Hawaii suggests that the genus originated here. Hawaii is an isolated island system and its distance from any continents means that dispersal to such remote islands of ancestral species would likely be very rare. Other lines of evidence to suggest that the genus originated in Hawaii would be whether fossil ancestral taxa are found only in Hawaii or anywhere else in the world. If ancestral taxa are not known from any other continent, it's very likely that the genus originated and subsequently diversified in Hawaii (and then dispersed elsewhere).

CHAPTER 55

1. d   2. c   3. b   4. d   5. a   6. d

7. In both of these cases, human management strategies will be working against the organisms' intrinsic rates of increase. Populations of long-lived organisms with low reproductive rates grow slowly. Such organisms can be categorized as *K*-strategists, which tend to persist at or near the carrying capacity of the environment. They are adapted to predictable environments, and they tend to be more specialized in their resource use, and less tolerant of variation in resource quality, than other organisms. They produce few offspring, but each offspring has a high probability of surviving to adulthood. Recall that the number of births in a population tends to be highest when that population is well below its carrying capacity. For large, long-lived species that we wish to harvest, we should manage the population so that it is far enough below the carrying capacity to have a high birth rate. But some species (such as whales) reproduce so slowly that they cannot sustain any kind of substantial harvest rate. Short-lived organisms with high reproductive rates can be categorized as *r*-strategists. These organisms can generally use a wide variety of resources and tolerate a wide range of conditions, and they can produce large numbers of offspring when conditions are suitable. If we wish to decrease the numbers of a short-lived, rapidly reproducing pest species (such as rats), killing individuals will only increase the birth rate. A better approach is to reduce the species' resources (e.g., clean up garbage) in order to decrease the carrying capacity for the species. In each case, of course, managers need to understand the specific life history and population dynamics of the species they wish to manage.

8. Humans are subject to the same population dynamics that other species are. Resource abundance is a density-dependent population-regulating factor. Like the reindeer population on St. Matthew, the Irish population crashed when its food supply was diminished. Three kinds of changes in the rates of demographic events contributed to the decrease in population size. Recall that $N_1 = N_0 + (B - D) + (I - E)$. First, the emigration rate (E) increased. Second, the age at first reproduction, and thus generation time, increased, so the birth rate (B) decreased. In other words, the population's life history traits changed with changing environmental conditions. Finally, as a direct effect of the food shortage, the death rate (D) increased. A look at the social history of the Irish potato famine might tell us more about the roles of the Irish population's own growth before the famine and whether intraspecific competition (discussed further in Chapter 56) was involved in regulating the food supply—whether other populations monopolized resources and limited the access of the Irish people to those resources.

9. Those who support the view that biological controls should not be used under any circumstances might cite the example of the cane toad in Australia, which not only failed to control the cane beetles it was introduced to control but became a serious pest in its own right. They might note that many species introduced into new regions, where their normal predators and pathogens are absent, reach population densities much higher than those in their native ranges, and they might argue that there is no reason to think this generality would not apply to species introduced as biological control agents. Those who support the view that biological controls can be used safely and effectively might cite the example of the successful control of the cottony-cushion scale in California, which was brought under control within a year by the introduction of a predaceous ladybeetle and a parasitic fly. They might also argue that horror stories like that of the cane toad could be avoided by proper study of the ecology of the proposed biological control agents before they are introduced. Studies in test plots prior to release almost certainly would have revealed that Australian cane beetles stay high on the upper stalks of cane plants, out of reach of the toads, and studies of the toads' life histories might have revealed their generalized and voracious appetites. Strict requirements for extensive testing for specificity and efficacy prior to release can greatly reduce the risk that biological control agents themselves will become pests after they are introduced. But opponents of biological control might respond that, because natural systems are so complex, even careful study might fail to reveal the real risks of introducing a particular species into a new environment.

10. Corridors have to be defined in terms of specific organisms and their dispersal abilities. Corridors consist of habitat between patches through which the organisms of interest can move. An area that serves as a corridor for birds might not be effective as a corridor for small arthropods. On the other hand, the small arthropods would need less habitat area to maintain a viable population. Thus, designing a single study to determine the effects of corridors would be very difficult. A single experiment might be able to determine effects of corridors on animals that are similar in size and mobility but not if organisms differ widely in those attributes. To understand the effects of corridors in fragmented habitats, it is very important to consider multiple organisms because organisms interact within these habitats. Investigators in the Palenque National Park in Mexico discovered that birds are more likely to be recaptured in home forest patches connected by corridors to the patches where they were released than in home forest patches unconnected to the patches where they were released. However, their ability to navigate these corridors successfully depends on the presence of other species, including predators, and their ability to survive in home patches depends on the presence of other species, including prey species, as well. Designing a single study to determine the effects of corridors would be very difficult.

CHAPTER 56

1. a   2. c   3. c   4. d   5. e

6. The interactions among ants, cacti, and pollinators described in the Working With Data exercise represent a diversity of types of interactions. By fending off potential herbivores, the pugnacious ant bodyguards act as mutualists of the cactus, as do the bee species that visit the flowers and serve as pollinators. The five ant species all appear to use the extrafloral nectaries on this plant in similar ways and thus may be competitors for extrafloral nectar. Each type of interaction depends on the relative abundance and activities of the interacting species. Cactus plants that grow where there are no herbivorous insects may have no need of pugnacious bodyguards; under those circumstances, the ants might be considered parasites for removing extrafloral nectar without providing any defensive services. The mutualism between ants and plants could also break down if bee pollinators are scarce and the most aggressive ant bodyguard prevents any bees at all from pollinating the flowers.

7. Which pine trees are susceptible to mountain pine beetle attack could be determined by direct observation and by experimentation. Within infested stands, investigators may be able to identify individual pine trees that do not harbor beetle populations and characterize properties that may make them resistant to the beetles (e.g., ability to produce large quantities of resin). Conversely, trees with especially high beetle populations might have properties that make them particularly susceptible (having a history of surviving fire or lightning strike). Experiments can also be conducted under controlled conditions, in which investigators test various species to determine if beetles display a preference for particular species. The fact that this beetle has a symbiotic partner upon which it depends in order to feed on trees suggests that a novel strategy for managing the outbreak could be to identify a fungicide that kills the symbiont, thereby rendering the beetle incapable of colonizing and killing the trees. Although there are no such programs currently in use today, many researchers are exploring this dimension of interaction ecology to devise novel methods for pest management.

8. (a) By establishing a microbial population that excludes undesirable species, the poultry industry is applying the principle of competitive exclusion. The principle states that two or more species utilizing a limited resource in similar ways cannot coexist. In the broiler chicks given a culture of three species of bacteria, a microbial community was established in which introduced *Salmonella* could not compete.

   (b) Other ecological outcomes this experiment might have produced include ultimate domination of the gut flora by only one species of bacterium or coexistence of all four bacterial species. Whether these species coexist or whether one or more species goes extinct depends on the availability of resources in the chicken intestines.

   (c) The principle of competitive exclusion might be useful in tackling other problems involving a community of organisms growing under confined conditions. This ecological principle provides part of the rationale for the use of probiotics by humans to improve a variety of conditions. Probiotics are live organisms that are consumed in food for health benefits, which are thought to accrue by altering the microbial balance, inhibiting the growth of deleterious species. Probiotics are being investigation for treatment of intestinal inflammatory diseases, pathogen-related diarrhea, and infections of the urogenital tract.

9. If parasites and their hosts coevolve, then the phylogenetic relationships among parasites should reflect the phylogenetic relationships among the hosts. DNA analysis revealed that flamingoes are actually more closely related to grebes than they are to ducks and geese. This relationship leads to the prediction that lice on flamingoes should be more closely related to the lice on grebes than they are to the lice on ducks and geese. Modern methods of molecular analysis that can be used to determine relationships among bird lice and their hosts include constructing a DNA-based phylogeny of multiple species of waterbirds and their lice and then comparing the phylogenies of the hosts and parasites to see if they are congruent (see Chapter 22). In addition to acquiring parasites by shared ancestry between hosts and parasites, bird species may also acquire parasites by virtue of the fact that they share habitats and come in contact with other bird species, each of which has its own parasite fauna. Because flamingoes, ducks, grebes, and geese are all waterbirds, the possibility exists that flamingoes may have acquired some of its louse parasites by this process of host switching.

10. Among the requirements for a mutualistic pollination system is behavior by the pollinator that ensures it will visit more than one individual of the same plant species. Visiting more than one individual provides a pollinator with the opportunity to carry pollen from one plant individual to the receptive stigmatic surface of another individual of the same species. A pollinator that encounters a feeding deterrent that limits the amount of nectar it can imbibe in a single visit is more likely to continue foraging for nectar on another plant individual. The process of taking a larger number of smaller meals by seeking nectar from flowers of different plant individuals increases the likelihood that the pollinator will carry pollen from one individual to another. Too much nicotine in nectar, however, may reduce the likelihood of pollination if it deters future visits to conspecifics altogether or if it impairs the behavior of its pollinator (nicotine is a neurotoxin). Another factor limiting the amount of nicotine is the cost to the plant of biosynthesizing the compound; investing in increased amounts of nicotine may leave fewer resources to invest in producing flowers, seeds and fruits.

CHAPTER 57

1. a   2. a   3. b   4. d   5. a   6. c   7. e

8. The diversity of microbes in the human gut can be compared between individuals or across populations by using the same methods employed for comparing diversity of macroscopic communities. Diversity encompasses both the number of different species present, and richness, or abundances of individuals across species. To determine which microbes might be keystone species, selective antibiotics can be used to eliminate particular species and the effects of that elimination on community composition then monitored. Applying the methods for assessing diversity that were developed for macroscopic communities to assessing diversity for microbial communities is limited, however, by our ability to isolate, identify, and quantify all of the microbial species present. Although molecular methods of identifying microbial species has vastly expanded this capacity, there remain challenges in recognizing and categorizing the full expanse of microbial diversity.

9. According to the theory of island biogeography, the number of species on an island represents a balance between the rate at which species immigrate to and colonize the island and the rate at which resident species go locally extinct. With increasing distance from a source pool, the equilibrium number of species on an island decreases; with increasing size of an island, the species number increases. The pattern of hawk moth diversity documented by Beck and Hitching in the 113 islands of Thailand and mainland Malaysia conforms to several of the predictions of island biogeography. The continental source of colonists includes over 180 species. Borneo, a large island close to Thailand, has a larger number of species (between 113 and 135) than does New Guinea (with 46 to 90 species), which is roughly comparable in size and farther away from Thailand. Generally speaking, too, the prediction that larger islands support higher numbers of species is also upheld; Borneo has a larger number of species than the much smaller Philippines (between 46 and 112 species), even though the two places are about equidistant from Thailand.

10. The pattern documented by Marek Sammul, Lauri Oksanen, and M. Magi—that removal of one perennial species from plant communities resulted in an increase in biomass of its competitors in highly productive communities—has been documented in other communities. One hypothesis postulates that interspecific competition becomes more intense when productivity is very high. Goldenrod (*Solidago virgaurea*) is apparently a superior competitor that, when present in a community, can suppress other species. In less productive communities, competition is less intense, so release from competition with goldenrod does not result in increased growth of any remaining species. The results of this study parallel those of a long-term experiment at the Rothamsted Experiment Station in England, in which fertilizer added regularly to selected plots of land to increase their productivity resulted in a decline in the number of plant species compared with the other plots in the study that were unfertilized; in these less productive, unfertilized plots, species diversity remained essentially the same. An alternative hypothesis could be that goldenrod inhibits the growth of co-occurring plant species (as some colonizing species do in early stages of succession). This hypothesis could be tested directly by extracting root exudates of goldenrod and testing their ability to inhibit germination and growth of other species in the community.

11. Whether lampreys should be eliminated as damaging parasites of game fish, or encouraged as ecosystem engineers that create nesting sites that might increase the reproductive success of game fish depends on many factors. Some of the factors are ecological; it is important to quantify the impact of existing

sea lamprey populations on survivorship of game fish as well as to estimate the lamprey population size that does not influence survivorship. As well, the beneficial impact of nutrient enrichment and provisioning of nesting habitat should be measured. In addition, designing an ecologically sound lamprey management strategy will also require consideration of local cultural values; assessments of the economic value of the sport fishing industry to the local community and the aesthetic and cultural value placed by the local community on maintaining a more natural assemblage of fish species should be made and factored into management plans.

CHAPTER 58

1. e   2. d   3. c   4. c   5. a   6. b

7. The rate of turnover would be important to the recovery rate of a lake, and that would depend on its location. In Lake Washington, which is located in a temperate climate, turnover would occur every spring and fall. When sewage was flowing into the lake, the nutrients it contained would have led to eutrophication and thus to oxygen depletion in the bottom water. Once the flow of sewage stopped, however, biomass production would have decreased. There would have been fewer dead organisms to sink to the lake bottom, less accumulation of nutrients on the lake bottom, and less oxygen-consuming decomposition there. Fall and spring turnover would have brought the accumulated nutrients to the lake surface and oxygen to the bottom, improving conditions for organisms that were typical of the lake's preindustrial community. If the lake had been located in a climate where seasonal temperature changes were not great enough to cause turnover, the excess nutrients that had accumulated would have remained on the lake bottom, and eutrophic conditions would have persisted much longer. It's also possible that other conditions in the area might affect the lake's recovery time. Acid precipitation, for example, can affect the viability of freshwater organisms and, if it were a problem in the region, it might slow the lake community's recovery.

8. A local effect might be nitrogen deposition. Coal—an organic fossil fuel—contains nitrogen, and its combustion would release nitrogen compounds (such as nitrogen dioxide, NO_2, and nitrous oxide, N_2O) through the smokestacks into the atmosphere. Some of this nitrogen would fall back to land in precipitation or as dry particles. The resulting increase of nitrogen in the soil would favor those plant species that are best adapted to take advantage of high nutrient levels, which would then outcompete other species. Thus the composition of the plant community would change and species diversity would be likely to decrease. Nitrogen deposition might also contribute to eutrophication in lakes, and emission of nitrogen into the atmosphere would contribute to smog.

  A regional effect might be acid precipitation. The combustion of fossil fuels releases NO_2 and sulfur dioxide (SO_2) into the atmosphere; both compounds react with water molecules in the atmosphere to form nitric acid (HNO_3) and sulfuric acid (H_2SO_4), respectively. These acids can travel hundreds of kilometers in the atmosphere, so their emission would affect ecosystems far from the smokestacks. Acid precipitation can damage the leaves of plants and reduce their rate of photosynthesis, and it can reduce fish and invertebrate species richness in freshwater lakes.

  A global effect would be climate change. The combustion of fossil fuels releases large amounts of CO_2, as well as lesser amounts of N_2O—both greenhouse gases. The presence of N_2O in the atmosphere also results in the production of trophospheric ozone; it, too, acts as a greenhouse gas as well as contributing to smog. The increasing concentrations of greenhouse gases in the atmosphere are already resulting in global climate warming. This climate change is having a number of worrisome effects on the global ecosystem, such as the shrinking of Arctic sea ice, rising sea level and potential coastal flooding, and profound changes in the abundances and distributions of species.

  The SO_2 released when coal is burned not only produces acid rain, but also contributes to global warming. Scrubbers remove the SO_2 but the scrubbing process contributes to pollution in another form—the process generates solid waste byproducts that contain sulfur, which must be deposited in a landfill, along with other solid waste products generated by burning coal.

9. Iron (Fe) is needed by organisms in only small amounts, but it is nevertheless an essential micronutrient. It is scarce in ocean waters because it is insoluble in oxygenated water, so that iron that enters the oceans sinks rapidly to the seafloor. The experiment described in the text demonstrated that iron is a limiting nutrient in the oceans: when the investigators added dissolved iron to surface waters in the equatorial Pacific Ocean, the large phytoplankton bloom that resulted was accompanied by an increase in the uptake of nitrate and carbon dioxide, showing that these nutrients had been available but underused. This experiment showed that adding iron to ocean waters increased photosynthesis, but to better understand the effects of iron fertilization, more such experiments would have to be carried out, still on an ecosystem scale, but over a longer time span. Investigators would have to observe the effects of the iron increase on the entire food web. The fertilized ecosystem would have to be compared with an unfertilized control ecosystem far enough away from the experimental one that the added iron, and its effects, would not reach it.

10. If the "cap and trade" system worked as intended, it could put the brakes on the ongoing increases quickly, holding emissions to the level that prevailed when the law went into effect. It might also be more acceptable to polluters than an outright ban on or regulation of emissions. And it might encourage some companies to invest more in cleaner technology, since doing so might give them credits to sell, or at least spare them having to buy credits. The drawbacks might include the likelihood that the government would have to set up a system to administer and enforce the law. They also include the fact that it is not easy to pass such a law (the United States, for example, has not succeeded in doing so). The biggest polluters might be reluctant to increase their costs of doing business by paying for credits and might thus be likely to lobby against such a law. From a different viewpoint, environmentalists might argue that a cap and trade system is an inadequate response to global warming—that we must not only stop increases in emissions, but decrease them dramatically. They might also argue that there is a moral hazard in allowing anyone to "pay to pollute"—that it might legitimize pollution.

11. No one hurricane—or even several—can be ascribed to global warming. Remember the difference between weather and climate, described in Chapter 54: "Weather is the short-term state of atmospheric conditions at a particular place and time, whereas climate refers to the average atmospheric conditions, and the extent of their variation, at a particular place over a longer time. In other words, climate is what you expect; weather is what you get." But neither does the observation that hurricanes have occurred for many centuries show that the climate has *not* changed. To address this question, we would have to compile temperature and hurricane data over long periods. First, we would have to show that the average temperatures of ocean waters are increasing over time—which has been done. Second, we would have to show that warmer water is correlated with more or stronger hurricanes—that there have been more and stronger hurricanes during years, or longer periods, when the water was warmer than in periods when it was cooler. Such a correlation would supply evidence that warming of the oceans is increasing hurricane frequency and intensity. It is more difficult to demonstrate that the warming of ocean waters is caused by increasing concentrations of greenhouse gases in the atmosphere, but most scientists believe that the evidence supports that claim.

CHAPTER 59

1. b   2. e   3. e   4. a   5. d   6. c   7. b

8. Conservation biologists and others who wish to preserve biodiversity are usually working with limited resources, so they often face hard choices. They might choose to focus their efforts in biodiversity hotspots and centers of imminent extinction, but within those areas, they would face many other choices. When we discussed the principles of island biogeography, we described the species–area relationship: large islands can support a larger equilibrium number of species than small islands. The same is true of "habitat islands." Protected areas often act as habitat islands, as many are surrounded by habitat that has been made unsuitable for many species by human activities. We must also consider edge effects, keeping in mind that not all of the area we protect will actually remain suitable habitat for communities and species of interest. Therefore, if we wish to preserve natural communities with their full diversity, the larger the preserved area, the better. However, if our concern is focused on one or a few endangered species, we may wish to preserve several separated areas of habitat; that way, if a disturbance or disease should wipe out the population in one area, the entire species will not become extinct. The area or areas we choose, however, must be large enough for the species to maintain a viable population in order to avoid loss of genetic variation. We might favor areas where corridors could be maintained to allow individuals to disperse from the protected area or areas and maintain other populations. But it is rare that a protected area can be designed based on these criteria alone; the plans are also dependent on the willingness of landowners, governments, and area residents to support the preservation of the area.

9. Some might argue that if the sheep constitute only a single population, but there are other populations of pumas, that the puma should be removed from the sheep's range. However, if the puma is a keystone predator in the region, removing it might have unforeseen negative consequences for other species in the community—recall the example of wolves in Yellowstone National Park, described in Chapter 57. If neither the sheep nor the puma is an introduced species—that is, if predator and prey had survived together for a long time before becoming threatened—it might be worth taking a look at what has changed. Have the sheep experienced a loss of habitat or resources, so that their populations are now too small to withstand the rate of predation that they once did? Those observations might suggest an alternative to suppressing the puma population: Could former sheep habitat be restored so that a larger sheep population could be supported, or as a last resort, could the sheep be bred in captivity and then introduced to a new, puma-free area?

10. To some extent, international organizations already have a triage system of sorts in in place. The International Union for the Conservation of Nature (IUCN), e.g., divides species in imminent danger of extinction in all or most of

their range as "endangered" or "critically endangered", differentiating them from those who are less likely to go extinct in the near future (and thus are classified as "vulnerable"). To some degree, this classification system can result in prioritization of rescue efforts. One problem with applying the triage system of World War I to species conservation, however, is that medical science is far more successful at predicting the certainty of death than ecological science is at predicting the certainty of extinction. After all, medical science is focused on one species, which has been the subject of intense scrutiny beginning in the earliest days of scientific research. The cost of erring in assigning certainty to extinction is the loss of an entire species—a unique combination of genes that, at least with current technology, can never be reconstructed.

11. Opinion as to the extent to which ethical and moral arguments should enter into discussions of protecting biodiversity varies widely. Your answer might take into consideration a wide range of cultural, historical, and economic factors, as well as the considerations brought to bear by modern scientific knowledge.

Glossary

A

A horizon   *See* topsoil.

abiotic   (a′ bye ah tick) [Gk. *a*: not + *bios*: life] Nonliving. (Contrast with biotic.)

abomasum   The true stomach of a ruminant.

abortion   Any termination of pregnancy, whether induced or natural (in which case it is called a spontaneous abortion), that occurs after a fertilized egg is successfully implanted in the uterus.

abscisic acid (ABA)   (ab sighs′ ik) A plant growth substance with growth-inhibiting action. Causes stomata to close; involved in a plant's response to salt and drought stress.

abscission   (ab sizh′ un) [L. *abscissio*: break off] The process by which leaves, petals, and fruits separate from a plant.

absorption   (1) Of light: complete retention, without reflection or transmission. (2) Of water or other molecules: soaking up (taking in through pores or by diffusion).

absorption spectrum   A graph of light absorption versus wavelength of light; shows how much light is absorbed at each wavelength.

absorptive heterotrophs   Organisms (primarily fungi) that feed by **absorptive heterotrophy**, i.e., by secreting digestive enzymes into the environment to break down large food molecules, then absorbing the breakdown products.

absorptive state   State in which food is in the gut and nutrients are being absorbed. (Contrast with postabsorptive state.)

abyssal zone   (uh biss′ ul) [Gk. *abyssos*: bottomless] The deepest parts of the ocean.

accessory pigments   Pigments that absorb light and transfer energy to chlorophylls for photosynthesis.

accessory sex organs   Anatomical structures that allow transfer of sperm from male to female for internal fertilization. (contrast with primary sex organs.)

acclimation, acclimatization   Acclimation refers to increased tolerance for environmental extremes (e.g., extreme cold) after prior exposure to them. Acclimatization refers to intrinsic seasonal adjustments in the "set points" of an animal's physiological functioning (e.g., metabolic rate).

acetyl coenzyme A (acetyl CoA)   A compound that reacts with oxaloacetate to produce citrate at the beginning of the citric acid cycle; a key metabolic intermediate in the formation of many compounds.

acetylcholine (ACh)   A neurotransmitter that carries information across vertebrate neuromuscular junctions and some other synapses. It is then broken down by the enzyme acetylcholinesterase (AChE).

acid   [L. *acidus*: sharp, sour] A substance that can release a proton in solution. (Contrast with base.)

acid growth hypothesis   The hypothesis that auxin increases proton pumping, thereby lowering the pH of the cell wall and activating enzymes that loosen polysaccharides. Proposed to explain auxin-induced cell expansion in plants.

acid precipitation   Precipitation that has a lower pH than normal as a result of acid-forming precursor molecules introduced into the atmosphere by human activities.

acidic   Having a pH below 7.0 (i.e., a hydrogen ion concentration greater than 10^{-7} molar). (Contrast with basic.)

acoelomate   An animal that does not have a coelom.

acrosome   (a′ krow soam) [Gk. *akros*: highest + *soma*: body] The structure at the forward tip of an animal sperm which is the first to fuse with the egg membrane and enter the egg cell.

ACTH   *See* corticotropin.

actin   [Gk. *aktis*: ray] A protein that makes up the cytoskeletal microfilaments in eukaryotic cells and is one of the two contractile proteins in muscle. See also myosin.

action potentials   Generated by neurons, these are electrical signals that transmit information via waves of depolarization or hyperpolarization of the cell membrane.

action spectrum   A graph of a biological process versus light wavelength; shows which wavelengths are involved in the process.

activation energy (E_a)   The energy barrier that blocks the tendency for a chemical reaction to occur.

activator   A transcription factor that stimulates transcription when it binds to a gene's promoter. (Contrast with repressor.)

active site   The region on the surface of an enzyme or ribozyme where the substrate binds, and where catalysis occurs.

active transport   The energy-dependent transport of a substance across a biological membrane against a concentration gradient—that is, from a region of low concentration (of that substance) to one of high concentration. (*See also* primary active transport, secondary active transport; contrast with facilitated diffusion, passive transport.)

adaptation   (a dap tay′ shun) (1) In evolutionary biology, a particular structure, physiological process, or behavior that makes an organism better able to survive and reproduce. Also, the evolutionary process that leads to the development or persistence of such a trait. (2) In sensory neurophysiology, a sensory cell's loss of sensitivity as a result of repeated stimulation.

adaptive defenses   One of the two general types of defenses against pathogens. Involves antibody proteins and other proteins that recognize, bind to, and aid in the destruction of specific viruses and bacteria. Present only in vertebrate animals. (Contrast with innate defenses.)

adaptive radiation   A series of evolutionary events that results in an array (radiation) of related species that live in a variety of environments, differing in the characteristics each uses to exploit those environments.

additive growth   Population growth in which a constant number of individuals is added to the population during successive time intervals. (Contrast with multiplicative growth.)

adenine (A)   (a′ den een) A nitrogen-containing base found in nucleic acids, ATP, NAD, and other compounds.

adenosine triphosphate   *See* ATP.

adrenal gland   (a dree′ nal) [L. *ad*: toward + *renes*: kidneys] An endocrine gland located near the kidneys of vertebrates, consisting of two parts, the **adrenal cortex** and **adrenal medulla.**

adrenaline   *See* epinephrine.

adrenergic receptors   G protein-linked receptor proteins that bind to the hormones epinephrine and norepinephrine, triggering specific responses in the target cells.

adrenocorticotropic hormone (ACTH)   *See* corticotropin.

adsorption   Binding of a gas or a solute to the surface of a solid.

adventitious roots (ad ven ti' shus) [L. *adventitius*: arriving from outside] Roots originating from the stem at ground level or below; typical of the fibrous root system of monocots.

aerenchyma In plants, parenchymal tissue containing air spaces.

aerobic (air oh' bic) [Gk. *aer*: air + *bios*: life] In the presence of oxygen; requiring or using oxygen (as in **aerobic metabolism**). (Contrast with anaerobic.)

afferent (af' ur unt) [L. *ad*: toward + *ferre*: to carry] Carrying to, as in neurons that carries impulses to the central nervous system (**afferent neurons**), or a blood vessel that carries blood to a structure. (Contrast with efferent.)

age structure The distribution of the individuals in a population across all age groups.

agonist A chemical substance (e.g., a neurotransmitter) that elicits a specific response in a cell or tissue. (Contrast with antagonist.)

air sacs Structures in the respiratory system of birds that receive inhaled air; they keep fresh air flowing unidirectionally through the lungs, but are not themselves gas exchange surfaces.

alcoholic fermentation *See* fermentation.

aldosterone (al dohs' ter own) A steroid hormone produced in the adrenal cortex of mammals. Promotes secretion of potassium and reabsorption of sodium in the kidney.

aleurone layer In some seeds, a tissue that lies beneath the seed coat and surrounds the endosperm. Secretes digestive enzymes that break down macromolecules stored in the endosperm.

allantoic membrane In animal development, an outgrowth of extraembryonic endoderm plus adjacent mesoderm that forms the allantois, a saclike structure that stores metabolic wastes produced by the embryo.

allantois (al' lun toh is) [Gk. *allant*: sausage] An extraembryonic membrane enclosing a sausage-shaped sac that stores the embryo's nitrogenous wastes.

allele (a leel') [Gk. *allos*: other] The alternate form of a genetic character found at a given locus on a chromosome.

allele frequency The relative proportion of a particular allele in a specific population.

allergic reaction [Ger. *allergie*: altered] An overreaction of the immune system to amounts of an antigen that do not affect most people; often involves IgE antibodies.

allopatric speciation (al' lo pat' rick) [Gk. *allos*: other + *patria*: homeland] The formation of two species from one when reproductive isolation occurs because of the interposition of (or crossing of) a physical geographic barrier such as a river. Also called geographic speciation. (Contrast with sympatric speciation.)

allopolyploidy The possession of more than two chromosome sets that are derived from more than one species.

allosteric regulation (al lo steer' ik) [Gk. *allos*: other + *stereos*: structure] Regulation of the activity of a protein (usually an enzyme) by the binding of an effector molecule to a site other than the active site.

α (alpha) helix A prevalent type of secondary protein structure; a right-handed spiral.

alternation of generations The succession of multicellular haploid and diploid phases in some sexually reproducing organisms, notably plants.

alternative splicing A process for generating different mature mRNAs from a single gene by splicing together different sets of exons during RNA processing.

altruistic Pertaining to behavior that benefits other individuals at a cost to the individual who performs it.

alveolus (al ve' o lus) (plural: alveoli) [L. *alveus*: cavity] A small, baglike cavity, especially the blind sacs of the lung.

amensalism (a men' sul ism) Interaction in which one animal is harmed and the other is unaffected. (Contrast with commensalism, mutualism.)

amine An organic compound containing an amino group (NH_2).

amine hormones Small hormone molecules synthesized from single amino acids (e.g., thyroxine and epinephrine).

amino acid An organic compound containing both NH_2 and COOH groups. Proteins are polymers of amino acids.

amino acid replacement A change in the nucleotide sequence that results in one amino acid being replaced by another.

ammonia NH_3, the most common nitrogenous waste.

ammonotelic (am moan' o teel' ic) [Gk. *telos*: end] Pertaining to an organism in which the final product of breakdown of nitrogen-containing compounds (primarily proteins) is **ammonia**. (Contrast with ureotelic, uricotelic.)

amnion (am' nee on) The fluid-filled sac within which the embryos of reptiles (including birds) and mammals develop.

amniote egg A shelled egg surrounding four extraembryonic membranes and embryo-nourishing yolk. This evolutionary adaptation permitted mammals and reptiles to live and reproduce in drier environments than can most amphibians.

amphipathic (am' fi path' ic) [Gk. *amphi*: both + *pathos*: emotion] Of a molecule, having both hydrophilic and hydrophobic regions.

amplitude The magnitude of change over the course of a regular cycle.

amygdala A component of the limbic system that is involved in fear and the memory of fearful experiences.

amylase (am' ill ase) An enzyme that catalyzes the hydrolysis of starch, usually to maltose or glucose.

anabolic reaction (an uh bah' lik) [Gk. *ana*: upward + *ballein*: to throw] A synthetic reaction in which simple molecules are linked to form more complex ones; requires an input of energy and captures it in the chemical bonds that are formed. (Contrast with catabolic reaction.)

anaerobic (an ur row' bic) [Gk. *an*: not + *aer*: air + *bios*: life] Occurring without the use of molecular oxygen, O_2. (Contrast with aerobic.)

anaphase (an' a phase) [Gk. *ana*: upward] The stage in cell nuclear division at which the first separation of sister chromatids (or, in the first meiotic division, of paired homologs) occurs.

ancestral trait The trait originally present in the ancestor of a given group; may be retained or changed in the descendants of that ancestor.

androgen (an' dro jen) Any of the several male sex steroids (most notably testosterone).

aneuploidy (an' you ploy dee) A condition in which one or more chromosomes or pieces of chromosomes are either lacking or present in excess.

angiosperms Flowering plants; one of the two major groups of living seed plants. (*See also* gymnosperms.)

angiotensin (an' jee oh ten' sin) A peptide hormone that raises blood pressure by causing peripheral vessels to constrict. Also maintains glomerular filtration by constricting efferent vessels and stimulates thirst and the release of aldosterone.

angular gyrus A part of the human brain believed to be essential for integrating spoken and written language.

animal hemisphere The metabolically active upper portion of some animal eggs, zygotes, and embryos; does not contain the dense nutrient yolk. (Contrast with vegetal hemisphere.)

anion (an' eye on) [Gk. *ana*: upward] A negatively charged ion. (Contrast with cation.)

annual A plant whose life cycle is completed in one growing season. (Contrast with biennial, perennial.)

antagonist A biochemical (e.g., a drug) that blocks the normal action of another biochemical substance.

antagonist interactions Interactions between two species in which one species benefits and the other is harmed. Includes predation, herbivory, and parasitism.

antenna system *See* light-harvesting complex.

anterior Toward or pertaining to the tip or headward region of the body axis. (Contrast with posterior.)

anterior pituitary The portion of the vertebrate pituitary gland that derives from gut epithelium. Produces trophic hormones.

anther   (an' thur) [Gk. *anthos*: flower] A pollen-bearing portion of the stamen of a flower.

antheridium   (an' thur id' ee um) [Gk. *antheros*: blooming] The multicellular structure that produces the sperm in nonvascular land plants and ferns.

antibody   One of the myriad proteins produced by the immune system that specifically binds to a foreign substance in blood or other tissue fluids and initiates its removal from the body.

anticodon   The three nucleotides in transfer RNA that pair with a complementary triplet (a codon) in messenger RNA.

antidiuretic hormone (ADH)   *See* vasopressin

antigen   (an' ti jun) Any substance that stimulates the production of an antibody or antibodies in the body of a vertebrate.

antigen-presenting cell   In cellular immunity, a cell that ingests and digests an antigen, and then exposes fragments of that antigen to the outside of the cell, bound to proteins in the cell's plasma membrane.

antigenic determinant   The specific region of an antigen that is recognized and bound by a specific antibody. Also called an epitope.

antiparallel   Pertaining to molecular orientation in which a molecule or parts of a molecule have opposing directions.

antiporter   A membrane transport protein that moves one substance in one direction and another in the opposite direction. (Contrast with symporter, uniporter.)

antisense RNA   A single-stranded RNA molecule complementary to, and thus targeted against, an mRNA of interest to block its translation.

anus   (a' nus) An opening through which solid digestive wastes are expelled, located at the posterior end of a tubular gut.

aorta   (a or' tah) [Gk. *aorte*: aorta] The main trunk of the arteries leading to the systemic (as opposed to the pulmonary) circulation.

aortic body   A chemosensor in the aorta that senses a decrease in blood supply or a dramatic decrease in partial pressure of oxygen in the blood.

aortic valve   A one-way valve between the left ventricle of the heart and the aorta that prevents backflow of blood into the ventricle when it relaxes.

apex   (a' pecks) The tip or highest point of a structure, as of a growing stem or root.

aphasia   a deficit in the ability to use or understand words.

aphotic zone   In bodies of water (lakes and oceans), the region below the reach of light.

apical dominance   In plants, inhibition by the apical bud of the growth of axillary buds.

apical hook   A form taken by the stems of many eudicot seedlings that protects the delicate shoot apex while the stem grows through the soil.

apical meristem   The meristem at the tip of a shoot or root; responsible for a plant's primary growth.

apomixis   (ap oh mix' is) [Gk. *apo*: away from + *mixis*: sexual intercourse] The asexual production of seeds.

apoplast   (ap' oh plast) In plants, the continuous meshwork of cell walls and extracellular spaces through which material can pass without crossing a plasma membrane. (Contrast with symplast.)

apoptosis   (ap uh toh' sis) A series of genetically programmed events leading to cell death.

aposematism   Warning coloration; bright colors or striking patterns of toxic or toxic-mimic species that act as a warning to predators.

appendix   In the human digestive system, the vestigial equivalent of the cecum (blind pouc), which serves no digestive function.

aquaporin   A transport protein in plant and animal cell membranes through which water passes in osmosis.

aquatic   (a kwa' tic) [L. *aqua*: water] Pertaining to or living in water. (Contrast with marine, terrestrial.)

aqueous   (a' kwee us) Pertaining to water or a watery solution.

aquifer   A large pool of groundwater.

archegonium   (ar' ke go' nee um) The multicellular structure that produces eggs in nonvascular land plants, ferns, and gymnosperms.

archenteron   (ark en' ter on) [Gk. *archos*: first + *enteron*: bowel] The earliest primordial animal digestive tract.

area phylogeny   Phylogenetic tree n which the names of the taxa are replaced with the names of the places where those taxa live or lived.

arms race   A series of reciprocal adaptations between species involved in antagonistic interactions, in which adaptations that increase the fitness of a consumer species exert selection pressure on its resource species to counter the consumer's adaptation, and vice versa.

arteriole   A small blood vessel arising from an artery that feeds blood into a capillary bed.

artery   A muscular blood vessel carrying oxygenated blood away from the heart to other parts of the body. (Contrast with vein.)

artificial insemination   An infertility treatment that involves the artificial introduction of sperm into the woman's reproductive tract.

artificial selection   The selection by human plant and animal breeders of individuals with certain desirable traits.

ascus   (ass' cus) (plural: asci) [Gk. *askos*: bladder] In sac fungi, the club-shaped sporangium within which spores (ascospores) are produced by meiosis.

asexual reproduction   Reproduction without sex.

assisted reproductive technologies (ARTs)   Any of several procedures that remove unfertilized eggs from the ovary, combine them with sperm outside the body, and then place fertilized eggs or egg–sperm mixtures in the appropriate location in a female's reproductive tract for development.

association cortex   In the vertebrate brain, the portion of the cortex involved in higher-order information processing, so named because it integrates, or associates, information from different sensory modalities and from memory.

associative learning   A form of learning in which two unrelated stimuli become linked to the same response.

asthenosphere   (ass thenn' o sphere) [Gk. *asthenes*: weak] The viscous, malleable (changeable) layer of Earth's mantle. It is overlain by the solid lithospheric plates.

astrocyte   [Gk. *astron*: star] A type of glial cell that contributes to the blood–brain barrier by surrounding the smallest, most permeable blood vessels in the brain.

atherosclerosis   (ath' er oh sklair oh' sis) [Gk. *athero*: gruel, porridge + *skleros*: hard] A disease of the lining of the arteries characterized by fatty, cholesterol-rich deposits in the walls of the arteries. When fibroblasts infiltrate these deposits and calcium precipitates in them, the disease become arteriosclerosis, or "hardening of the arteries."

atom   [Gk. *atomos*: indivisible] The smallest unit of a chemical element. Consists of a nucleus and one or more electrons.

atomic mass   *See* atomic weight.

atomic number   The number of protons in the nucleus of an atom; also equals the number of electrons around the neutral atom. Determines the chemical properties of the atom.

atomic weight   The average of the mass numbers of a representative sample of atoms of an element, with all the isotopes in their normally occurring proportions. Also called atomic mass.

ATP (adenosine triphosphate)   An energy-storage compound containing adenine, ribose, and three phosphate groups. When it is formed from ADP, useful energy is stored; when it is broken down (to ADP or AMP), energy is released to drive endergonic reactions.

ATP synthase   An integral membrane protein that couples the transport of protons with the formation of ATP.

atrial natriuretic peptide   A hormone released by the atrial muscle fibers of the heart when they are overly stretched, which decreases reabsorption of sodium by the kidney and thus blood volume.

atrioventricular node   A modified node of cardiac muscle that organizes the action potentials that control contraction of the ventricles.

atrium   (a' tree um) [L. *atrium*: central hall] An internal chamber. In the hearts of vertebrates, the thin-walled chamber(s) entered by blood on its way to the ventricle(s). Also, the outer ear.

auditory system A sensory system that uses mechanoreceptors to convert pressure waves into receptor potentials; includes structures that gather sound waves, direct them to a sensory organ, and amplify their effect on the mechanoreceptors.

autocatalysis [Gk. *autos*: self + *kata*: to break down] A positive feedback process in which an activated enzyme acts on other inactive molecules of the same enzyme to activate them.

autocrine A chemical signal that binds to and affects the cell that makes it. (Contrast with paracrine.)

autoimmune diseases Diseases (e.g., rheumatoid arthritis) that result from failure of the immune system to distinguish between self and nonself, causing it to attack tissues in the organism's own body.

autoimmunity An immune response by an organism to its own molecules or cells.

autonomic nervous system (ANS) The portion of the peripheral nervous system that controls such involuntary functions as those of guts and glands. Also called the involuntary nervous system.

autophagy The programmed destruction of a cell's components.

autopolyploidy The possession of more than two entire chromosomes sets that are derived from a single species.

autoregulatory mechanisms In mammalian circulatory systems, local control of blood flow through capillary beds by constriction or dilation of incoming arterioles in response to local metabolite concentrations.

autosome Any chromosome (in a eukaryote) other than a sex chromosome.

autotroph (au′ tow trowf′) [Gk. *autos*: self + *trophe*: food] An organism that is capable of living exclusively on inorganic materials, water, and some energy source such as sunlight (photoautotrophs) or chemically reduced matter (see chemoautotrophs). (Contrast with heterotroph.)

auxin (awk′ sin) [Gk. *auxein*: to grow] In plants, a substance (the most common being indoleacetic acid) that regulates growth and various aspects of development.

avirulence (Avr) genes Genes in a pathogen that may trigger defenses in plants. *See* gene-for-gene resistance.

Avogadro's number The number of atoms or molecules in a mole (weighed out in grams) of a substance, calculated to be 6.023×10^{23}.

axillary bud A bud that forms in the angle (axil) where a leaf meets a stem.

axon [Gk. axle] The process (branching structure) of a neuron that conducts action potentials away from the cell body. *See also* dendrites.

axon hillock The junction between an axon and the neuron's cell body; where action potentials are generated.

axon terminal The end portion of an axon, which passes action potentials to another cell. Axon terminals can form synapses and release neurotransmitter.

B

B cell A type of lymphocyte involved in the humoral immune response of vertebrates. Upon recognizing an antigenic determinant, a B cell develops into a plasma cell, which secretes an antibody. (Contrast with T cell.)

B horizon *See* subsoil.

bacillus (bah sil′ us) [L: little rod] Any of various rod-shaped bacteria.

bacterial conjugation *See* conjugation.

bacteriophage (bak teer′ ee o fayj) [Gk. *bakterion*: little rod + *phagein*: to eat] Any of a group of viruses that infect bacteria. Also called phage.

bacteroids Nitrogen-fixing organelles that develop from endosymbiotic bacteria.

bark All tissues external to the vascular cambium of a plant.

barometric pressure Atmospheric pressure; the total pressure of the gas mixture in air.

baroreceptor [Gk. *baros*: weight] A pressure-sensing cell or organ. Sometimes called a stress receptor.

basal metabolic rate (BMR) The minimum rate of energy turnover in an awake (but resting) bird or mammal that is not expending energy for thermoregulation.

base (1) A substance that can accept a hydrogen ion in solution. (Contrast with acid.) (2) In nucleic acids, the purine or pyrimidine that is attached to each sugar in the sugar–phosphate backbone.

base pair (bp) In double-stranded DNA, a pair of nucleotides formed by the complementary base pairing of a purine on one strand and a pyrimidine on the other. (*See* complementary base pairing.)

basic Having a pH greater than 7.0 (i.e., having a hydrogen ion concentration lower than 10^{-7} molar). (Contrast with acidic.)

basidioma (plural: basiomata) A fruiting structure produced by club fungi.

basidium (bass id′ ee yum) In club fungi, the characteristic sporangium in which four **basidiospores** are formed by meiosis and then borne externally before being shed.

basilar membrane A membrane in the human inner ear whose flexion in response to sound waves activates hair cells; flexes at different locations in response to different pitch.

Batesian mimicry The convergence in appearance of an edible species (mimic) with an unpalatable species (model).

behavioral ecology An evolutionary approach to the study of animal behavior that studies how behaviors are adaptive in different environmental conditions.

behaviorism One of two classical approaches to the study of proximate causes of animal behavior, derived from the discoveries of Ivan Pavlov and focused on laboratory studies. (Compare with ethology.)

benefit Improvement in survival and reproductive success resulting from performing a behavior or having a trait.

benthic zone [Gk. *benthos*: bottom] The bottom of the ocean.

β (beta) pleated sheet A type of protein secondary structure; results from hydrogen bonding between polypeptide regions running antiparallel to each other.

biased gene conversion A mechanism of concerted evolution in which a DNA repair system appears biased in favor of using particular nucleotide sequences as templates for repair, resulting in the rapid spread of the favored sequence across all copies of the gene. (*See* concerted evolution.)

bicarbonate ion Ion (HCO_3^-) resulting from dissociation of carbonic acid in water; important in pH regulation and carbon dioxide (CO_2) transport.

biennial A plant whose life cycle includes vegetative growth in the first year and flowering and senescence in the second year. (Contrast with annual, perennial.)

bilateral symmetry The condition in which only the right and left sides of an organism, divided by a single plane through the midline, are mirror images of each other.

bilayer A structure that is two layers in thickness. In biology, most often refers to the phospholipid bilayer of membranes. (*See* phospholipid bilayer.)

bile A secretion of the liver made up of bile salts synthesized from cholesterol, various phospholipids, and bilirubin (the breakdown product of hemoglobin). Emulsifies fats in the small intestine.

binary fission Reproduction of a prokaryote by division of a cell into two comparable progeny cells.

binocular vision Overlapping visual fields of an animal's two eyes; allows the animal to see in three dimensions.

binomial nomenclature A taxonomic naming system in which each species is given a binomial (Gk.: two names), a genus name followed by a species name.

biodiversity hotspots Regions identified by conservation biologists as being particularly in need of protection because they harbor great species richness and endemism (i.e., large numbers of species, many of which are found nowhere else).

biofilm A community of microorganisms embedded in a polysaccharide matrix, forming a highly resistant coating on almost any moist surface.

biogeochemical cycle Movement of inorganic elements such as nitrogen, phosphorus, and carbon through living organisms and the physical environment.

biogeographic region One of several defined, continental-scale regions of Earth, each of which has a biota distinct from that of the others. (Contrast with biome.)

biogeography   The scientific study of the patterns of distribution of populations, species, and ecological communities across Earth.

bioinformatics   The use of computers and/or mathematics to analyze complex biological information, such as DNA sequences.

biological control   The use of natural enemies (predators, parasites, or pathogens) to reduce the population density of an economically damaging (pest) species.

biological species concept   The definition of a species as a group of actually or potentially interbreeding natural populations that are reproductively isolated from other such groups. (Contrast with lineage species concept; morphological species concept.)

biology   [Gk. *bios*: life + *logos*: study] The scientific study of living things.

bioluminescence   The production of light by biochemical processes in an organism.

biome   (bye' ome) A major division of the ecological communities of Earth, characterized primarily by distinctive vegetation. A given biogeographic region contains many different biomes.

bioremediation   The use by humans of other organisms to remove contaminants from the environment.

biosphere   (bye' oh sphere) All regions of Earth (terrestrial and aquatic) and Earth's atmosphere in which organisms can live.

biota   (bye oh' tah) All of the organisms—animals, plants, fungi, and microorganisms—found in a given area. (Contrast with flora, fauna.)

biotechnology   The use of living cells or organisms to produce materials useful to humans.

biotic   (bye ah' tick) [Gk. *bios*: life] Alive. (Contrast with abiotic.)

biotic interchange   The mixing of biotas previously separated by physical, climatic, or other barriers, for example when two formerly separated land masses fuse.

blastocoel   (blass' toe seal) [Gk. *blastos*: sprout + *koilos*: hollow] The central, hollow cavity of a blastula.

blastocyst   (blass' toe cist) An early embryo formed by the first divisions of the fertilized egg (zygote). In mammals, a hollow ball of cells.

blastodisc   (blass' toe disk) An embryo that forms as a disk of cells on the surface of a large yolk mass; comparable to a blastula, but occurring in animals such as birds and reptiles, in which the massive yolk restricts complete cleavage.

blastomere   Any of the cells produced by the early divisions of a fertilized animal egg.

blastopore   The opening created by the invagination of the vegetal pole during gastrulation of animal embryos.

blastula   (blass' chu luh) An early stage of the animal embryo; in many species, a hollow sphere of cells surrounding a central cavity, the blastocoel. (Contrast with blastodisc.)

block to polyspermy   Any of several responses to entry of a sperm into an egg that prevent more than one sperm from entering the egg.

blood   A fluid connective tissue that is pumped throughout the body. A component of the circulatory system, blood transports gases such as oxygen and carbon dioxide as well as other essential elements.

blood clotting   A cascade of events involving platelets and circulating proteins (clotting factors) that seals damaged blood vessels.

blood–brain barrier   The selective impermeability of blood vessels in the brain that prevents most chemicals from diffusing from the blood into the brain.

blue-light receptors   Pigments in plants that absorb blue light (400–500 nm). These pigments mediate many plant responses including photo-tropism, stomatal movements, and expression of some genes.

body plan   The general structure of an animal, the arrangement of its organ systems, and the integrated functioning of its parts.

Bohr effect   A shift in the O_2 binding curve of hemoglobin in response to excess H^+ ions such that the hemoglobin releases more O_2 in tissues where pH is low.

Bohr model   A model for atomic structure that depicts the atom as largely empty space, with a central nucleus surrounded by electrons in orbits, or electron shells, at various distances from the nucleus.

bond   *See* chemical bond.

bone   A rigid component of vertebrate skeletal systems that contains an extracellular matrix of insoluble calcium phosphate crystals as well as collagen fibers.

Bowman's capsule   An elaboration of the renal tubule, composed of podocytes, that surrounds and collects the filtrate from the glomerulus.

brain   The centralized integrative center of a nervous system.

brainstem   The portion of the vertebrate brain between the spinal cord and the forebrain, made up of the medulla, pons, and midbrain.

brassinosteroids   Plant steroid hormones that mediate light effects promoting the elongation of stems and pollen tubes.

Broca's area   A portion of the human brain essential for speech. Located in the frontal lobe just in front of the primary motor cortex.

bronchioles   The smallest airways in a vertebrate lung, branching off the bronchi.

bronchus   (plural: bronchi) The major airway(s) branching off the trachea into the vertebrate lung.

brown fat   In mammals, fat tissue that is specialized to produce heat. It has many mitochondria and capillaries, and a protein that uncouples oxidative phosphorylation.

budding   Asexual reproduction in which a more or less complete new organism grows from the body of the parent organism, eventually detaching itself.

buffer   A substance that can transiently accept or release hydrogen ions and thereby resist changes in pH.

bulbourethral glands   Secretory structures of the human male reproductive system that produce a small volume of an alkaline, mucoid secretion that helps neutralize acidity in the urethra and lubricate it to facilitate the passage of semen.

bulk flow   The movement of a solution from a region of higher pressure potential to a region of lower pressure potential.

bundle of His   Fibers of modified cardiac muscle that conduct action potentials from the atria to the ventricular muscle mass.

bundle sheath cell   Part of a tissue that surrounds the veins of plants.

C

C horizon   *See* parent rock.

C_3 plants   Plants that produce 3PG as the first stable product of carbon fixation in photosynthesis and use ribulose bisphosphate as a CO_2 receptor.

C_4 plants   Plants that produce oxaloacetate as the first stable product of carbon fixation in photosynthesis and use phosphoenolpyruvate as CO_2 acceptor. C_4 plants also perform the reactions of C_3 photosynthesis.

calcitonin   Hormone produced by the thyroid gland; lowers blood calcium and promotes bone formation. (Contrast with parathyroid hormone.)

calcitriol   A hormone derived from vitamin D whose actions include stimulating the cells of the digestive tract to absorb calcium from ingested food.

calorie   [L. *calor*: heat] The amount of heat required to raise the temperature of 1 gram of water by 1°C. Physiologists commonly use the kilocalorie (kcal) as a unit of measure (1 kcal = 1,000 calories). Nutritionists also use the kilocalorie, but refer to it as the **Calorie** (capital C).

Calvin cycle   The stage of photosynthesis in which CO_2 reacts with RuBP to form 3PG, 3PG is reduced to a sugar, and RuBP is regenerated, while other products are released to the rest of the plant. Also known as the Calvin–Benson cycle.

calyx   (kay' licks) [Gk. *kalyx*: cup] All of the sepals of a flower, collectively.

CAM   *See* crassulacean acid metabolism.

Cambrian explosion   The rapid diversification of multicellular life that took place during the Cambrian period.

cAMP (cyclic AMP)   A compound formed from ATP that acts as a second messenger.

cancellous bone   A type of bone with numerous internal cavities that make it appear spongy, although it is rigid. (Contrast with compact bone.)

canopy   The leaf-bearing part of a tree. Collectively, the aggregate of the leaves and branches of the larger woody plants of an ecological community.

capillaries   [L. *capillaris*: hair] Very small tubes, especially the smallest blood-carrying vessels of animals between the termination of the arteries and the beginnings of the veins. **Capillary beds** are networks of capillaries where materials are exchanged between the blood and the interstitial fluid.

capsid   The outer shell of a virus that encloses its nucleic acid.

carbohydrates   Organic compounds containing carbon, hydrogen, and oxygen in the ratio 1:2:1 (i.e., with the general formula $C_nH_{2n}O_n$). Common examples are sugars, starch, and cellulose.

carbon skeleton   The chains or rings of carbon atoms that form the structural basis of organic molecules. Other atoms or functional groups are attached to the carbon atoms.

carbon-fixation reactions   The phase of photosynthesis in which chemical energy captured in the light reactions is used to drive the reduction of CO_2 to form carbohydrates.

carboxylase   An enzyme that catalyzes the addition of carboxyl functional groups (O=C—OH) to a substrate.

cardiac cycle   Contraction of the two atria of the heart, followed by contraction of the two ventricles and then relaxation.

cardiac muscle   A type of muscle tissue that makes up, and is responsible for the beating of, the heart. Characterized by branching cells with single nuclei and a striated (striped) appearance. (Contrast with smooth muscle, skeletal muscle.)

cardiovascular system   [Gk. *kardia*: heart + L. *vasculum*: small vessel] The heart, blood, and vessels are of a circulatory system.

carnivore   [L. *carn*: flesh + *vorare*: to devour] An organism that eats animal tissues. (Contrast with detritivore, frugivore, herbivore, omnivore.)

carotenoid   (ka rah' tuh noid) A yellow, orange, or red lipid pigment commonly found as an accessory pigment in photosynthesis; also found in fungi.

carotid body   A chemosensor in the carotid artery that senses a decrease in blood supply or a dramatic decrease in partial pressure of oxygen in the blood.

carpel   (kar' pel) [Gk. *karpos*: fruit] The organ of the flower that contains one or more ovules.

carrier   (1) In facilitated diffusion, a membrane protein that binds a specific molecule and transports it through the membrane. (2) In respiratory and photosynthetic electron transport, a participating substance such as NAD that exists in both oxidized and reduced forms. (3) In genetics, a person heterozygous for a recessive trait.

carrying capacity (*K*)   The maximum number of individuals in a population (i.e., maximum population size) that can be supported by the resources present in a given environment.

cartilage   In vertebrates, a tough connective tissue found in joints, the outer ear, and elsewhere. Forms the entire skeleton in some animal groups.

cartilage bone   A type of bone that begins its development as a cartilaginous structure resembling the future mature bone, then gradually hardens into mature bone. (Contrast with membranous bone.)

Casparian strip   A band of cell wall containing suberin and lignin, found in the endodermis. Restricts the movement of water across the endodermis.

caspase   One of a group of proteases that catalyze cleavage of target proteins and are active in apoptosis.

catabolic reaction   (kat uh bah' lik) [Gk. *kata*: to break down + *ballein*: to throw] A synthetic reaction in which complex molecules are broken down into simpler ones and energy is released. (Contrast with anabolic reaction.)

catabolite repression   In the presence of abundant glucose, the diminished synthesis of catabolic enzymes for other energy sources.

catalyst   (kat' a list) [Gk. *kata*: to break down] A chemical substance that accelerates a reaction without itself being consumed in the overall course of the reaction. Catalysts lower the activation energy of a reaction. Enzymes are biological catalysts.

cation   (cat' eye on) An ion with one or more positive charges. (Contrast with anion.)

caudal   [L. *cauda*: tail] Pertaining to the tail, or to the posterior part of the body.

cDNA   *See* complementary DNA.

cDNA library   A collection of complementary DNAs derived from mRNAs of a particular tissue at a particular time in the life cycle of an organism.

cecum   (see' cum) [L. blind] A blind branch off the large intestine. In many nonruminant mammals, the cecum contains a colony of microorganisms that contribute to the digestion of food.

cell   The simplest structural unit of a living organism. In multicellular organisms, many individual cells serve as the building blocks of tissues and organs.

cell adhesion molecules (CAMs)   Molecules on animal cell surfaces that affect the selective association of cells into tissues during development of the embryo. Also a component of desmosomes.

cell cycle   The stages through which a cell passes between one mitotic division and the next. Includes all stages of interphase and mitosis. (*See* mitosis.)

cell cycle checkpoints   Points of transition between different phases of the cell cycle, which are regulated by cyclins and cyclin-dependent kinases (Cdks).

cell division   The reproduction of a cell to produce two new cells. In eukaryotes, this process involves nuclear division (mitosis) and cytoplasmic division (cytokinesis).

cell fate   The type of cell that an undifferentiated cell in an embryo will become in the adult.

cell junctions   Specialized structures associated with the plasma membranes of epithelial cells. Some contribute to cell adhesion, others to intercellular communication.

cell potency   In multicellular organisms, an undifferentiated cell's potential to become a cell of a specific type. (*See* multipotent; pluripotent; totipotent.)

cell recognition   Binding of cells to one another mediated by membrane proteins or carbohydrates.

cell theory   States that cells are the basic structural and physiological units of all living organisms, and that all cells come from preexisting cells.

cell wall   A relatively rigid structure that encloses cells of plants, fungi, many protists, and most prokaryotes, and which gives these cells their shape and limits their expansion in hypotonic media.

cellular immune response   Immune system response mediated by T cells and directed against parasites, fungi, intracellular viruses, and foreign tissues (grafts). (Contrast with humoral immune response.)

cellular respiration   The catabolic pathways by which electrons are removed from various molecules and passed through intermediate electron carriers to O_2, generating H_2O and releasing energy.

cellular specialization   In multicellular organisms, the division of labor such that different cell types become responsible for different functions (e.g., reproduction or digestion) within the organism.

cellulose   (sell' you lowss) A straight-chain polymer of glucose molecules, used by plants as a structural supporting material.

central dogma   The premise that information flows from DNA to RNA to polypeptide (protein).

central nervous system (CNS)   That portion of the nervous system that is the site of most information processing, storage, and retrieval; in vertebrates, the brain and spinal cord. (Contrast with peripheral nervous system.)

central vacuole   In plant cells, a large organelle that stores the waste products of metabolism and maintains turgor.

centrifuge   [L. *centrum*: center + *fugere*: to flee] A laboratory device in which a sample is spun around a central axis at high speed. Used to separate suspended materials of different densities.

centriole   (sen' tree ole) A paired organelle that helps organize the microtubules in animal and protist cells during nuclear division.

centromere (sen' tro meer) [Gk. *centron*: center + *meros*: part] The region where sister chromatids join.

centrosome (sen' tro soam) The major microtubule organizing center of an animal cell.

cephalization (sef ah luh zay' shun) [Gk. *kephale*: head] The evolutionary trend toward increasing concentration of brain and sensory organs at the anterior end of the animal.

cerebellum (sair uh bell' um) [L. diminutive of *cerebrum*, brain] The brain region that controls muscular coordination; located at the anterior end of the hindbrain.

cerebral cortex The thin layer of gray matter (neuronal cell bodies) that overlies the cerebrum.

cerebrum (su ree' brum) [L. brain] The dorsal anterior portion of the forebrain, making up the largest part of the brain of mammals; the chief coordination center of the nervous system and the major information-processing areas of the vertebrate brain consists of two **cerebral hemispheres**.

cervix (sir' vix) [L. neck] The opening of the uterus into the vagina.

cGMP (cyclic guanosine monophosphate) An intracellular messenger that is part of signal transmission pathways involving G proteins. (*See* G protein.)

channel protein An integral membrane protein that forms an aqueous passageway across the membrane in which it is inserted and through which specific solutes may pass.

chaperone A protein that guards other proteins by counteracting molecular interactions that threaten their three-dimensional structure.

character In genetics, an observable feature, such as eye color. (Contrast with trait.)

character displacement An evolutionary phenomenon in which species that compete for the same resources within the same territory tend to diverge in morphology and/or behavior.

chemical bond An attractive force stably linking two atoms.

chemical equilibrium *See* equilibrium

chemical evolution The theory that life originated through the chemical transformation of inanimate substances.

chemical reaction The change in the composition or distribution of atoms of a substance with consequent alterations in properties.

chemical synapse Neural junction at which neurotransmitter molecules released from a presynaptic cell induce changes in a postsynaptic cell. (Contrast with electrical synapse.)

chemically gated channel A type of membrane channel that opens or closes depending on the presence or absence of a specific molecule that binds either to the channel protein itself or to a separate receptor that alters the three-dimensional shape of the channel protein.

chemiosmosis Formation of ATP in mitochondria and chloroplasts, resulting from a pumping of protons across a membrane (against a gradient of electrical charge and of pH), followed by the return of the protons through a protein channel with ATP synthase activity.

chemoautotroph Organisms that obtain energy by oxidizing inorganic substances, using some of that energy to fix carbon. Also known as chemolithotrophs.

chemoheterotroph An organism that must obtain both carbon and energy from organic substances. (Contrast with chemoautotroph, photoautotroph, photoheterotroph.)

chemoreceptor A sensory receptor cell that senses specific molecules (such as odorant molecules or pheromones) in the environment.

chiasma (kie az' muh) (plural: chiasmata) [Gk. cross] An X-shaped connection between paired homologous chromosomes in prophase I of meiosis. A chiasma is the visible manifestation of crossing over between homologous chromosomes.

chief cells One of three types of secretory cell found in the gastric pits of the stomach wall. Chief cells secrete the protein-digesting enzyme pepsin. (*See* mucosal epithelium; parietal cells.)

chitin (kye' tin) [Gk. *kiton*: tunic] The characteristic tough but flexible organic component of the exoskeleton of arthropods, consisting of a complex, nitrogen-containing polysaccharide. Also found in cell walls of fungi.

chlorophyll (klor' o fill) [Gk. *kloros*: green + *phyllon*: leaf] Any of several green pigments associated with chloroplasts or with certain bacterial membranes; responsible for trapping light energy for photosynthesis.

chloroplast [Gk. *kloros*: green + *plast*: a particle] An organelle bounded by a double membrane containing the enzymes and pigments that perform photosynthesis. Chloroplasts occur only in eukaryotes.

choanocyte (ko' an uh site) The collared, flagellated feeding cells of sponges.

cholecystokinin (CCK) (ko' luh sis tuh kai' nin) A hormone produced and released by the lining of the duodenum when it is stimulated by undigested fats and proteins. It stimulates the gallbladder to release bile and slows stomach activity.

chorion (kor' ee on) [Gk. *khorion*: afterbirth] The outermost of the membranes protecting mammal, bird, and reptile embryos; in mammals it forms part of the placenta.

chromatid (kro' ma tid) A newly replicated chromosome, from the time molecular duplication occurs until the time the centromeres separate (during anaphase of mitosis or of meiosis II).

chromatin The nucleic acid–protein complex that makes up eukaryotic chromosomes.

chromatin remodeling A mechanism for epigenetic gene regulation by the alteration of chromatin structure.

chromosomal mutation Loss of or changes in position/direction of a DNA segment on a chromosome.

chromosome (krome' o sowm) [Gk. *kroma*: color + *soma*: body] In bacteria and viruses, the DNA molecule that contains most or all of the genetic information of the cell or virus. In eukaryotes, a structure composed of DNA and proteins that bears part of the genetic information of the cell.

chylomicron (ky low my' cron) Particles of lipid coated with protein, produced in the gut from dietary fats and secreted into the extracellular fluids.

chyme (kime) [Gk. *kymus*: juice] Created in the stomach; a mixture of ingested food with the digestive juices secreted by the salivary glands and the stomach lining.

cilia (sil' ee ah) (singular: cilium) [L. eyelashes] Hairlike organelle used for locomotion by many unicellular organisms and for moving water and mucus by many multicellular organisms. Generally shorter than flagella.

circadian rhythm (sir kade' ee an) [L. *circa*: approximately + *dies*: day] A rhythm of growth or activity that recurs about every 24 hours.

circannual rhythm [L. *circa*: + *annus*: year] A rhythm of growth or activity that recurs on a yearly basis.

circulatory system A physiological system consisting of a muscular pump (heart), a fluid (blood or hemolymph), and a series of conduits (blood vessels) that transports materials around the body.

11-*cis*-retinal The nonprotein, light-absorbing component of the visual pigment rhodopsin. (*See* rhodopsin.)

***cis-trans* isomers** In molecules with a double bond (typically between two carbon items), identifies on which side of the double bond similar atoms or functional groups are found. If they are on the same side, the molecule is a *cis* isomer; in a *trans* isomer, similar atoms are on opposite sides of the double bond. (*See* isomer.)

citric acid cycle In cellular respiration, a set of chemical reactions whereby acetyl CoA is oxidized to carbon dioxide and hydrogen atoms are stored as NADH and $FADH_2$. Also called the Krebs cycle.

clade [Gk. *klados*: branch] A monophyletic group made up of an ancestor and all of its descendants.

class I MHC molecules Cell surface proteins that participate in the cellular immune response directed against virus-infected cells.

class II MHC molecules Cell surface proteins that participate in the cell–cell interactions (of T-helper cells, macrophages, and B cells) of the humoral immune response.

class switching Occurs when a B cell changes the immunoglobulin class it synthesizes (e.g., a B cell making IgM switches to making IgG).

cleavage The first few cell divisions of an animal zygote. *See also* complete cleavage, incomplete cleavage.

climate The long-term average atmospheric conditions (temperature, precipitation, humidity, wind direction and velocity) found in a region. (Contrast with weather.)

climax community The final stage of succession; a community that is capable of perpetuating itself under local climatic and soil conditions and persists for a relatively long time.

clinal variation [Gk. *klinein*: to lean] Gradual change in the phenotype of a species over a geographic gradient.

cloaca The opening through which both urinary wastes and digestive wastes are expelled in most amphibians and in reptiles (including birds).

clonal deletion Inactivation or destruction of lymphocyte clones that would produce immune reactions against the animal's own body.

clonal lineages Asexually reproduced groups of nearly identical organisms.

clonal selection Mechanism by which exposure to antigen results in the activation of selected T- or B-cell clones, resulting in an immune response.

clone [Gk. *klon*: twig, shoot] (1) Genetically identical cells or organisms produced from a common ancestor by asexual means. (2) To produce many identical copies of a DNA sequence by its introduction into, and subsequent asexual reproduction of, a cell or organism.

closed circulatory system Circulatory system in which the circulating fluid is contained within a continuous system of vessels. (Contrast with open circulatory system.)

clumped dispersion pattern *See* dispersion

CO (CONSTANS) Gene coding for a transcription factor that activates the synthesis of florigen (FT); involved in the induction of flowering.

co-repressor In the regulation of bacterial operons, a molecule that binds to the repressor, causing it to change shape and bind to the operator, thereby inhibiting transcription.

coastal zone The marine life zone that extends from the shoreline to the edge of the continental shelf. Characterized by relatively shallow, well-oxygenated water and relatively stable temperatures and salinities.

coccus (kock' us) (plural: cocci) [Gk. *kokkos*: berry, pit] Any of various spherical or spheroidal bacteria.

cochlea (kock' lee uh) [Gk. *kokhlos*: snail] A spiral tube in the inner ear of vertebrates; it contains the sensory cells involved in hearing.

codominance A condition in which two alleles at a locus produce different phenotypic effects and both effects appear in heterozygotes.

codon Three nucleotides in messenger RNA that direct the placement of a particular amino acid into a polypeptide chain. (Contrast with anticodon.)

coelom (see' loam) [Gk. *koiloma*: cavity] An animal body cavity, enclosed by muscular mesoderm and lined with a mesodermal layer called peritoneum that also surrounds the internal organs.

coelomate Possessing a coelom.

coenocytic (seen' a sit ik) [Gk. *koinos*: common + *kytos*: container] Referring to the condition, found in some fungal hyphae, of "cells" containing many nuclei but enclosed by a single plasma membrane. Results from nuclear division without cytokinesis.

coenzyme A nonprotein organic molecule that plays a role in catalysis by an enzyme.

coenzyme A (CoA) A coenzyme used in various biochemical reactions as a carrier of acyl groups.

coevolution Evolutionary processes in which an adaptation in one species leads to the evolution of an adaptation in a species with which it interacts; also known as reciprocal adaptation.

cofactor An inorganic ion that is weakly bound to an enzyme and required for its activity.

cohesin A protein involved in binding chromatids together.

cohesion The tendency of molecules (or any substances) to stick together.

cohort (co' hort) [L. *cohors*: company of soldiers] A group of similar-aged organisms.

cold-hardening A process by which plants can acclimate to cooler temperatures; requires repeated exposure to cool temperatures over many days.

coleoptile A sheath that surrounds and protects the shoot apical meristem and young primary leaves of a grass seedling as they move through the soil.

collagen [Gk. *kolla*: glue] A fibrous protein found extensively in bone and connective tissue.

collecting duct In vertebrates, a tubule that receives urine produced in the nephrons of the kidney and delivers that fluid to the ureter for excretion.

collenchyma (cull eng' kyma) [Gk. *kolla*: glue + *enchyma*: infusion] A type of plant cell, living at functional maturity, which lends flexible support by virtue of primary cell walls thickened at the corners. (Contrast with parenchyma, sclerenchyma.)

colon [Gk. *kolon*] The portion of the gut between the small intestine and the anus. Also called the large intestine.

commensalism [L. *com*: together + *mensa*: table] A type of interaction between species in

which one participant benefits while the other is unaffected.

communication A signal from one organism (or cell) that alters the functioning or behavior of another organism (or cell).

community Any ecologically integrated group of species of microorganisms, plants, and animals inhabiting a given area.

compact bone A type of bone with a solid, hard structure. (Contrast with cancellous bone.)

companion cell In angiosperms, a specialized cell found adjacent to a sieve tube element.

comparative experiment Experimental design in which data from various unmanipulated samples or populations are compared, but in which variables are not controlled or even necessarily identified. (Contrast with controlled experiment.)

comparative genomics Computer-aided comparison of DNA sequences between different organisms to reveal genes with related functions.

competition In ecology, use of the same resource by two or more species when the resource is present in insufficient supply for the combined needs of the species.

competitive exclusion A result of competition between species for resources, in which one species completely eliminates the other from a given habitat.

competitive inhibitor A nonsubstrate that binds to the active site of an enzyme and thereby inhibits binding of its substrate. (Contrast with noncompetitive inhibitor.)

complement system A group of eleven proteins that play a role in some reactions of the immune system. The complement proteins are not immunoglobulins.

complementary base pairing The AT (or AU), TA (or UA), CG, and GC pairing of bases in double-stranded DNA, in transcription, and between tRNA and mRNA.

complementary DNA (cDNA) DNA formed by reverse transcriptase acting with an RNA template; essential intermediate in the reproduction of retroviruses; used as a tool in recombinant DNA technology; lacks introns.

complete cleavage Pattern of cleavage that occurs in eggs that have little yolk. Early cleavage furrows divide the egg completely and the blastomeres are of similar size. (Contrast with incomplete cleavage.)

complete metamorphosis A change of state during the life cycle of an organism in which the body is almost completely rebuilt to produce an individual with a very different body form. Characteristic of insects such as butterflies, moths, beetles, ants, wasps, and flies.

complex ions Groups of covalently bonded atoms that carry an electric charge (e.g., NH_4^+, the ammonium ion).

complex life cycle   In reference to parasitic species, a life cycle that requires more than one host to complete.

composite transposon   Two transposable elements located near one another that transpose together and carry the intervening DNA sequence with them. (*See* transposable element.)

compound   (1) A substance made up of atoms of more than one element. (2) Made up of many units, as in the **compound eyes** of arthropods.

concentration gradient   A difference in concentration of an ion or other chemical substance from one location to another, often across a membrane. (*See* active transport; facilitated diffusion.)

concerted evolution   The common evolution of a family of repeated genes, such that changes in one copy of the gene family are replicated in other copies of the gene family, and thus evolve "in concert." (*See* biased gene conversion; unequal crossing over.)

condensation reaction   A chemical reaction in which two molecules become connected by a covalent bond and a molecule of water is released (AH + BOH → AB + H_2O.) (Contrast with hydrolysis reaction.)

conditional mutation   A mutation that results in a characteristic phenotype only under certain environmental conditions.

conditioned reflex   A form of associative learning first described by Ivan Pavlov, in which a natural response (such as salivation in response to food) becomes associated with a normally unrelated stimulus (such as the sound of a bell).

conduction   The transfer of heat from one object to another through direct contact.

cone   In conifers, a reproductive structure consisting of spore-bearing scales extending from a central axis. (Contrast with strobilus.)

cone cell   In the vertebrate retina, a type of photoreceptor cell responsible for color vision.

conidium   (ko nid' ee um) (plural: conidia) [Gk. *konis*: dust] A type of haploid fungal spore borne at the tips of hyphae, not enclosed in sporangia.

conjugation   (kon ju gay' shun) [L. *conjugare*: yoke together] (1) A process by which DNA is passed from one cell to another through a conjugation tube, as in bacteria. (2) A nonreproductive sexual process by which *Paramecium* and other ciliates exchange genetic material.

connective tissue   A type of tissue that connects or surrounds other tissues; its cells are embedded in a collagen-containing matrix. One of the four major tissue types in multicellular animals, including cartilage, bone, blood, and fat.

connexon   In a gap junction, a protein channel linking adjacent animal cells.

conservation biology   An applied science that carries out investigations with the aim of maintaining the diversity of life on Earth.

conserved   Pertaining to a gene or trait that has evolved very slowly and is similar or even identical in individuals of highly divergent groups.

conspecifics   Individuals of the same species.

constant region   The portion of an immunoglobulin molecule whose amino acid composition determines its class and does not vary among immunoglobulins in that class. (Contrast with variable region.)

constitutive   Always present; produced continually at a constant rate. (Contrast with inducible.)

constitutive genes   Genes that are expressed all the time. (Contrast with inducible genes.)

constitutive proteins   Proteins that an organism produces all the time, and at a relatively constant rate.

consumer   An organism that eats the tissues of some other organism.

consumer–resource interactions   Interactions in which organisms gain their nutrition by eating other living organisms or are eaten themselves.

continental drift   The gradual movements of the world's continents that have occurred over billions of years.

contraception   Birth control methods that prevent fertilization or implantation (conception).

contractile vacuole   (kon trak' tul) A specialized vacuole that collects excess water taken in by osmosis, then contracts to expel the water from the cell.

controlled experiment   An experiment in which a sample is divided into groups whereby experimental groups are exposed to manipulations of an independent variable while one group serves as an untreated control. The data from the various groups are then compared to see if there are changes in a dependent variable as a result of the experimental manipulation. (Contrast with comparative experiment.)

controlled system   A set of components in a physiological system that is controlled by commands from a regulatory system. (Contrast with regulatory system.)

convection   The transfer of heat to or from a surface via a moving stream of air or fluid.

convergent evolution   Independent evolution of similar features from different ancestral traits.

convolutions   Foldings of the vertebrate brain's cerebral cortex into ridges called gyri (sing. **gyrus**) and valleys called sulci (sing. **sulcus**). The level of cortical convolution increases taxonomically and is especially extensive in humans.

copulation   Reproductive behavior that results in a male depositing sperm in the reproductive tract of a female.

cork cambium   [L. *cambiare*: to exchange] In plants, a lateral meristem that produces secondary growth, mainly in the form of waxy-walled protective cells, including some of the cells that become bark.

cork   In plants, a protective outermost tissue layer composed of cells with thick walls waterproofed with suberin.

cornea   The clear, transparent tissue that covers the eye and allows light to pass through to the retina.

corolla   (ko role' lah) [L. *corolla*: a small crown] All of the petals of a flower, collectively.

corpus luteum   (kor' pus loo' tee um) (plural: corpora lutea) [L. yellow body] A structure formed from a follicle after ovulation; produces hormones important to the maintenance of pregnancy.

corridor   A connection between habitat patches through which organisms can disperse; plays a critical role in maintaining subpopulations.

cortex   [L. *cortex*: covering, rind] (1) In plants, the tissue between the epidermis and the vascular tissue of a stem or root. (2) In animals, the outer tissue of certain organs, such as the adrenal gland (adrenal cortex) and the brain (cerebral cortex).

corticosteroids   Steroid hormones produced and released by the cortex of the adrenal gland.

corticotropin   A tropic hormone produced by the anterior pituitary hormone that stimulates cortisol release from the adrenal cortex. Also called adrenocorticotropic hormone (ACTH).

corticotropin-releasing hormone   A hormone produced by the hypothalamus that controls the release of cortisol from the anterior pituitary.

cortisol   A corticosteroid that mediates stress responses.

cost   A decrease in fitness resulting from performing a behavior or having a trait.

cost–benefit approach   An approach to evolutionary studies that assumes an animal has a limited amount of time and energy to devote to each of its activities, and that each activity has fitness costs as well as benefits. (*See also* trade-off.)

cotyledon   (kot' ul lee' dun) [Gk. *kotyledon*: hollow space] A "seed leaf." An embryonic organ that stores and digests reserve materials; may expand when seed germinates.

countercurrent flow   An arrangement that promotes the maximum exchange of heat, or of a diffusible substance, between two fluids by having the fluids flow in opposite directions through parallel vessels close together.

countercurrent heat exchanger   In "hot" fish, an adaptation of the circulatory system such that arterial blood flowing into the muscles is warmed by venous blood flowing out of the muscles, thereby conserving body heat by countercurrent exchange.

countercurrent multiplier   The mechanism that increases the concentration of the interstitial fluid in the mammalian kidney through countercurrent flow in the loops of Henle and selective permeability and active transport of ions by segments of the loops of Henle.

covalent bond   Chemical bond based on the sharing of electrons between two atoms.

CpG islands   DNA regions rich in C residues adjacent to G residues. Especially abundant in promoters, these regions are where methylation of cytosine usually occurs.

crassulacean acid metabolism (CAM)   A metabolic pathway enabling the plants that possess it to store carbon dioxide at night and then perform photosynthesis during the day with stomata closed.

critical night length   In the photoperiodic flowering response of short-day plants, the length of night above which flowering occurs and below which the plant remains vegetative. (The reverse applies in the case of long-day plants.)

critical period   *See* sensitive period.

crop   A simple food storage sac, the first of two stomachlike organs in many animals (including reptiles, earthworms, and various insects). (*See also* gizzard.)

cross section   A section taken perpendicular to the longest axis of a structure. Also called a transverse section.

crossing over   The mechanism by which linked genes undergo recombination. In general, the term refers to the reciprocal exchange of corresponding segments between two homologous chromatids.

crosstalk   Interactions between different signal transduction pathways.

crypsis   [Gk. *kryptos*: hidden] The resemblance of an organism to some part of its environment, which helps it to escape detection by enemies.

cryptochromes   [Gk. *kryptos*: hidden + *kroma*: color] Photoreceptors mediating some blue-light effects in plants and animals.

ctene   (teen) [Gk. *cteis*: comb] In ctenophores, a comblike row of cilia-bearing plates. Ctenophores move by beating the cilia on their eight ctenes.

culture   (1) A laboratory association of organisms under controlled conditions. (2) The collection of knowledge, tools, values, and rules that characterize a human society.

cumulus   A thick gelatinous layer that protects a mammalian ovum.

cupula   Gelatinous swelling in the semicircular canals of the vestibular system. A cupula encloses hair cell stereocilia that react to shifting fluid in the canal ducts.

currents   Circulation patterns in the surface waters of oceans driven by the prevailing winds.

cuticle   (1) In plants, a waxy layer on the outer body surface that retards water loss. (2) In ecdysozoans, an outer body covering that provides protection and support and is periodically molted.

cyclic AMP   *See* cAMP.

cyclic electron transport   In photosynthetic light reactions, the flow of electrons that produces ATP but no NADPH or O_2.

cyclical succession   Pattern of change in community composition (succession) in which the climax community depends on periodic disturbances (e.g., fire) in order to persist. (Contrast with directional succession.)

cyclin   A protein that activates a cyclin-dependent kinase, bringing about transitions in the cell cycle.

cyclin-dependent kinase (Cdk)   A protein kinase whose target proteins are involved in transitions in the cell cycle and which is active only when complexed with additional protein subunits, called cyclins.

cytokine   A regulatory protein made by immune system cells that affects other target cells in the immune system.

cytokinesis   (sy' toe kine ee' sis) [Gk. *kytos*: container + *kinein*: to move] The division of the cytoplasm of a dividing cell. (Contrast with mitosis.)

cytokinin   (sy' toe kine' in) A member of a class of plant growth substances that plays roles in senescence, cell division, and other phenomena.

cytoplasm   The contents of the cell, excluding the nucleus.

cytoplasmic determinants   In animal development, gene products whose spatial distribution may determine such things as embryonic axes.

cytoplasmic segregation   The asymmetrical distribution of cytoplasmic determinants in a developing animal embryo.

cytosine (C)   (site' oh seen) A nitrogen-containing base found in DNA and RNA.

cytoskeleton   The network of microtubules and microfilaments that gives a eukaryotic cell its shape and its capacity to arrange its organelles and to move.

cytosol   The fluid portion of the cytoplasm, excluding organelles and other solids.

cytotoxic T cells (T_C)   Cells of the cellular immune system that recognize and directly eliminate virus-infected cells. (Contrast with T-helper cells.)

D

DAG   *See* diacylglycerol.

data   Quantified observations about a system under study.

daughter chromosomes   During mitosis, the separated chromatids from the beginning of anaphase onward.

dead space   The lung volume that fails to be ventilated with fresh air (because the lungs are never completely emptied during exhalation).

dead zones   Regions in aquatic ecosystems that are devoid of aquatic life because eutrophication has resulted in severe oxygen depletion.

deciduous   [L. *deciduus*: falling off] Pertaining to a woody plant that sheds its leaves but does not die.

declarative memory   Memory of people, places, events, and things that can be consciously recalled and described. (Contrast with procedural memory.)

decomposer   An organism that metabolizes organic compounds in debris and dead organisms, releasing inorganic material; found among the bacteria, protists, and fungi. *See also* detritivore, saprobe.

deductive logic   Logical thought process that starts with a premise believed to be true then predicts what facts would also have to be true to be compatible with that premise. (Contrast with inductive logic.)

defensin   A type of protein made by phagocytes that kills bacteria and enveloped viruses by insertion into their plasma membranes.

deficiency disease   A condition (e.g., scurvy and beriberi) caused by chronic lack of any essential nutrient.

degeneracy   The situation in which a single amino acid may be represented by any of two or more different codons in messenger RNA. Most of the amino acids can be represented by more than one codon.

deletion   A mutation resulting from the loss of a continuous segment of a gene or chromosome. Such mutations almost never revert to wild type. (Contrast with duplication, point mutation.)

demethylase   An enzyme that catalyzes the removal of the methyl group from cytosine, reversing DNA methylation.

demography   The study of population structure and of the processes (**demographic events**, including births and deaths) by which it changes.

denaturation   Loss of activity of an enzyme or nucleic acid molecule as a result of structural changes induced by heat or other means.

dendrites   [Gk. *dendron*: tree] Branching fibers (processes) of a neuron. Dendrites are usually relatively short compared with the axon, and commonly carry information to the neuronal cell body.

denitrification   Metabolic activity by which nitrate and nitrite ions are reduced to form nitrogen gas; carried out by certain soil bacteria.

denitrifiers   Bacteria that release nitrogen to the atmosphere as nitrogen gas (N_2).

density   *See* population density.

density-dependent   Pertaining to an effect on population size that increases in proportion to population density.

density-independent   Pertaining an effect on population size that acts independently of population density.

deoxyribonucleic acid   *See* DNA.

deoxyribonucleoside triphosphates (dNTPs)   The raw materials for DNA synthesis: deoxyadenosine triphosphate (dATP), deoxythymidine triphosphate (dTTP), deoxycytidine triphosphate (dCTP), and deoxyguanosine triphosphate (dGTP). Also called deoxyribonucleotides.

deoxyribose   A five-carbon sugar found in nucleotides and DNA.

dependent variable   In a scientific experiment, the response that is measured and analyzed as the independent variable is manipulated (See independent variable.)

depolarization   A change in the resting potential across a membrane so that the inside of the cell becomes less negative, or even positive, compared with the outside of the cell. (Contrast with hyperpolarization.)

derived trait   A trait that differs from the ancestral trait. (Contrast with synapomorphy.)

dermal tissue system   The outer covering of a plant, consisting of epidermis in the young plant and periderm in a plant with extensive secondary growth. (Contrast with ground tissue system and vascular tissue system.)

descent with modification   Darwin's premise that all species share a common ancestor and have diverged from one another gradually over time.

desmosome   (dez' mo sowm) [Gk. *desmos*: bond + *soma*: body] An adhering junction between animal cells.

desmotubule   A membrane extension connecting the endoplasmic retituclum of two plant cells that traverses the plasmodesma.

determinate growth   A growth pattern in which the growth of an organism or organ ceases when an adult state is reached; characteristic of most animals and some plant organs. (Contrast with indeterminate growth.)

determination   In development, the process whereby the fate of an embryonic cell or group of cells (e.g., to become epidermal cells or neurons) is set (becomes **determined**).

detritivore   (di try' ti vore) [L. *detritus*: worn away + *vorare*: to devour] An organism that obtains its energy from the dead bodies or waste products (**detritus**) of other organisms.

development   The process by which a multicellular organism, beginning with a single cell, goes through a series of changes, taking on the successive forms that characterize its life cycle.

developmental plasticity   The capacity of an organism to alter its pattern of development in response to environmental conditions.

diacylglycerol (DAG)   In hormone action, the second messenger produced by hydrolytic removal of the head group of certain phospholipids.

diapause   A period of developmental or reproductive arrest, entered in response to day length, that enables an organism to better survive.

diaphragm   (dye' uh fram) [Gk. *diaphrassein*: barricade] (1) A sheet of muscle that separates the thoracic and abdominal cavities in mammals; responsible for breathing. (2) A method of birth control in which a sheet of rubber is fitted over the woman's cervix, blocking the entry of sperm.

diastole   (dye ass' toll ee) [Gk. dilation] The portion of the cardiac cycle when the heart muscle relaxes. (Contrast with systole.)

dichotomous   (dye cot' oh mus) [Gk. *dichot*: split in two; *tomia*: removed) A branching pattern in which the shoot divides at the apex producing two equivalent branches that subsequently never overlap.

diencephalon   The portion of the vertebrate forebrain that develops into the thalamus and hypothalamus.

differential gene expression   The hyposthesis that, given that all cells contain all genes, what makes one cell type different from another is the difference in transcription and translation of those genes.

differentiation   The process whereby originally similar cells follow different developmental pathways; the actual expression of determination.

diffuse coevolution   The evolution of similar traits in suites of species experiencing similar selection pressures imposed by other suites of species with which they interact.

diffusion   Random movement of molecules or other particles, resulting in even distribution of the particles when no barriers are present.

digestive vacuole   In protists, an organelle specialized for digesting food ingested by endocytosis.

dihybrid cross   A mating in which the parents differ with respect to the alleles of two loci of interest.

dikaryon   (di care' ee ahn) [Gk. *di*: two + *karyon*: kernel] A cell or organism carrying two genetically distinguishable nuclei. Common in fungi.

dioecious   (die eesh' us) [Gk. *di*: two + *oikos*: house] Pertaining to organisms in which the two sexes are "housed" in two different individuals, so that eggs and sperm are not produced in the same individuals. Examples: humans, fruit flies, date palms. (Contrast with monoecious.)

diploblastic   Having two cell layers. (Contrast with triploblastic.)

diploid   (dip' loid) [Gk. *diplos*: double] Having a chromosome complement consisting of two copies (homologs) of each chromosome. Designated 2*n*. (Contrast with haploid.)

direct development   Pattern of development (notably among insects) in which hatchlings look like miniature versions of adults. (Contrast with metamorphosis.)

direct fitness   That component of fitness resulting from an organism producing its own offspring. (Contrast with inclusive fitness, kin selection.)

directional selection   Selection in which phenotypes at one extreme of the population distribution are favored. (Contrast with disruptive selection, stabilizing selection.)

directional succession   Change in community composition after a disturbance (succession) that is characterized by an orderly progression culminating in a persistent state (the climax community). (Contrast with cyclical succession.)

disaccharide   A carbohydrate made up of two monosaccharides (simple sugars).

discoidal cleavage   In animal development, a type of incomplete cleavage that is common in fishes, reptiles, and birds, the eggs of which contain a dense yolk mass.

dispersal   Movement of organisms away from a parent organism or from an existing population.

dispersion   The distribution of individuals in space within a population. **Clumped dispersion** occurs when individuals tend to occupy the same space; **regular dispersion** is when the presence of one individual decreases the probability of another individual occupying the same space; and **random dispersion** assumes there is equal probability of any individual occupying any given space.

disruptive selection   Selection in which phenotypes at both extremes of the population distribution are favored. (Contrast with directional selection; stabilizing selection.)

distal   Away from the point of attachment or other reference point. (Contrast with proximal.)

distal convoluted tubule   The portion of a renal tubule from where it reaches the renal cortex, just past the loop of Henle to where it joins a collecting duct. (Compare with proximal convoluted tubule.)

disturbance   A short-term event that disrupts populations, communities, or ecosystems by changing the environment.

disulfide bridge   The covalent bond between two sulfur atoms (–S—S–) linking two molecules or remote parts of the same molecule.

DNA (deoxyribonucleic acid)   The fundamental hereditary material of all living organisms. In eukaryotes, stored primarily in the cell nucleus. A nucleic acid using deoxyribose rather than ribose.

DNA fingerprint   An individual's unique pattern of allele sequences, commonly short tandem repeats and single nucleotide polymorphisms.

DNA helicase   An enzyme that unwinds the double helix.

DNA ligase Enzyme that unites broken DNA strands during replication and recombination.

DNA methylation The addition of methyl groups to bases in DNA, usually cytosine or guanine.

DNA methyltransferase An enzyme that catalyzes the methylation of DNA.

DNA microarray A small glass or plastic square onto which thousands of single-stranded DNA sequences are fixed so that hybridization of cell-derived RNA or DNA to the target sequences can be performed.

DNA polymerase Any of a group of enzymes that catalyze the formation of DNA strands from a DNA template.

DNA replication The creation of a new strand of DNA in which DNA polymerase catalyzes the exact reproduction of an existing (template) strand of DNA.

DNA transposons Mobile genetic elements that move without making an RNA intermediate. (Contrast with retrotransposons.)

domain (1) An independent structural element within a protein. Encoded by recognizable nucleotide sequences, a domain often folds separately from the rest of the protein. Similar domains can appear in a variety of different proteins across phylogenetic groups (e.g., "homeobox domain"; "calcium-binding domain"). (2) In phylogenetics, the three monophyletic branches of life (Bacteria, Archaea, and Eukarya).

dominance In genetics, the ability of one allelic form of a gene to determine the phenotype of a heterozygous individual in which the homologous chromosomes carry both it and a different (recessive) allele. (Contrast with recessive.)

dormancy A condition in which normal activity is suspended, as in some spores, seeds, and buds.

dorsal [L. *dorsum*: back] Toward or pertaining to the back or upper surface. (Contrast with ventral.)

dorsal lip In amphibian embryos, the dorsal segment of the blastopore. Also called the "organizer," this region directs the development of nearby embryonic regions.

double fertilization In angiosperms, a process in which the nuclei of two sperm fertilize one egg. One sperm's nucleus combines with the egg nucleus to produce a zygote, while the other combines with the same egg's two polar nuclei to produce the first cell of the triploid endosperm (the tissue that will nourish the growing plant embryo).

double helix Refers to DNA and the (usually right-handed) coil configuration of two complementary, antiparallel strands.

downregulation A negative feedback process in which continuous high concentrations of a hormone can decrease the number of its receptors. (Contrast with upregulation.)

duodenum (do' uh dee' num) The beginning portion of the vertebrate small intestine. (Contrast with ileum, jejunum.)

duplication A mutation in which a segment of a chromosome is duplicated, often by the attachment of a segment lost from its homolog. (Contrast with deletion.)

E

ecdysone (eck die' sone) [Gk. *ek*: out of + *dyo*: to clothe] In insects, a hormone that induces molting.

ecological economics Interdisciplinary field that works to assess the economic value of biodiversity.

ecological efficiency The overall transfer of energy from one trophic level to the next, expressed as the ratio of consumer production to producer production.

ecological survivorship curve *See* survivorship curves

ecological system One or more organisms plus the external environment with which they interact.

ecology [Gk. *oikos*: house] The scientific study of the interaction of organisms with their living (biotic) and nonliving (abiotic) environments.

ecosystem (eek' oh sis tum) The organisms of a particular habitat, such as a pond or forest, together with the physical environment in which they live.

ecosystem engineer An organism that builds structures that alter existing habitats or create new habitats.

ecosystem services Processes by which ecosystems maintain resources that benefit human society.

ecotourism Ecologically responsible travel to natural places.

ectoderm [Gk. *ektos*: outside + *derma*: skin] The outermost of the three embryonic germ layers first delineated during gastrulation. Gives rise to the skin, sense organs, and nervous system.

ectotherm [Gk. *ektos*: outside + *thermos*: heat] An animal that is dependent on external heat sources for regulating its body temperature (Contrast with endotherm.)

edema (i dee' mah) [Gk. *oidema*: swelling] Tissue swelling caused by the accumulation of fluid.

edge effects Changes in ecological processes in a community caused by physical and biological factors originating in an adjacent community.

effector A component of a physiological system that responds to information by *effecting* changes (making change happen) in the internal environment; examples include muscles and the secretory cells of the digestive tract.

effector cells In cellular immunity, B cells and T cells that attack an antigen, either by secreting antibodies that bind to the antigen or by releasing molecules that destroy any cell bearing the antigen.

effector protein In cell signaling, a protein responsible for the cellular reponse to a signal transduction pathway.

efferent (ef' ur unt) [L. *ex*: out + *ferre*: to bear] Carrying outward or away from, as in neurons that carry impulses outward from the central to the peripheral nervous system (**efferent neurons**), or a blood vessel that carries blood away from a structure. (Contrast with afferent.)

egg In all sexually reproducing organisms, the female gamete; in birds, reptiles, and some other vertebrates, a structure within which early embryonic development occurs. *See also* amniote egg, ovum.

electrical synapse A type of synapse at which action potentials spread directly from presynaptic cell to postsynaptic cell. (Contrast with chemical synapse.)

electrocardiogram (ECG or EKG) A graphic recording of electrical potentials from the heart.

electrochemical gradient The concentration gradient of an ion across a membrane plus the voltage difference across that membrane.

electroencephalogram (EEG) A graphic recording of electrical potentials from the brain.

electromagnetic radiation A self-propagating wave that travels though space and has both electrical and magnetic properties.

electron A subatomic particle outside the nucleus carrying a negative charge and very little mass.

electron shell The region surrounding the atomic nucleus at a fixed energy level in which electrons orbit.

electron transport The passage of electrons through a series of proteins with a release of energy which may be captured in a concentration gradient or in chemical form such as NADH or ATP.

electronegativity The tendency of an atom to attract electrons when it occurs as part of a compound.

electrophoresis *See* gel electrophoresis.

element A substance that cannot be converted to a simpler substance by ordinary chemical means.

elongation (1) In molecular biology, the addition of monomers to make a longer RNA or protein during transcription or translation. (2) Growth of a plant axis or cell primarily in the longitudinal direction.

embolus (em' buh lus) [Gk. *embolos*: stopper] A circulating blood clot. Blockage of a blood vessel by an embolus or a bubble of gas is called an **embolism**. (Contrast with thrombus.)

embryo [Gk. *en*: within + *bryein*: to grow] A young animal, or young plant sporophyte, while it is still contained within a protective structure such as a seed, egg, or uterus.

embryo sac In angiosperms, the female gametophyte. Found within the ovule, it consists of eight or fewer cells, membrane bounded, but without cellulose walls between them.

embryonic stem cell (ESC)   A pluripotent cell in the blastocyst.

emergent property   A property of a complex system that is not exhibited by its individual component parts.

emigration   The deliberate and usually oriented departure of an organism from the habitat in which it has been living.

endemic   (en dem' ik) [Gk. *endemos*: native] Confined to a particular region, thus often having a comparatively restricted distribution.

endergonic   A chemical reaction in which the products have higher free energy than the reactants, thereby requiring free energy input to occur. (Contrast with exergonic.)

endocrine cells   Cells that secrete substances into the extracellular fluid. (*See also* endocrine gland.)

endocrine gland   (en' doh krin) [Gk. *endo*: within + *krinein*: to separate] An aggregation of secretory cells that secretes hormones into the blood. The endocrine system consists of all endocrine cells and endocrine glands in the body that produce and release hormones. (Contrast with exocrine gland.)

endocytosis   A process by which liquids or solid particles are taken up by a cell through invagination of the plasma membrane. (Contrast with exocytosis.)

endoderm   [Gk. *endo*: within + *derma*: skin] The innermost of the three embryonic germ layers delineated during gastrulation. Gives rise to the digestive and respiratory tracts and structures associated with them.

endodermis   In plants, a specialized cell layer marking the inside of the cortex in roots and some stems. Frequently a barrier to free diffusion of solutes.

endomembrane system   A system of intracellular membranes that exchange material with one another, consisting of the Golgi apparatus, endoplasmic reticulum, and lysosomes when present.

endometrium   The epithelial lining of the uterus.

endoplasmic reticulum (ER)   [Gk. *endo*: within + L. *reticulum*: net] A system of membranous tubes and flattened sacs found in the cytoplasm of eukaryotes. Exists in two forms: rough ER, studded with ribosomes; and smooth ER, lacking ribosomes.

endorphins   Molecules in the mammalian brain that act as neurotransmitters in pathways that control pain.

endoskeleton   [Gk. *endo*: within + *skleros*: hard] An internal skeleton covered by other, soft body tissues. (Contrast with exoskeleton.)

endosperm   [Gk. *endo*: within + *sperma*: seed] A specialized triploid seed tissue found only in angiosperms; contains stored nutrients for the developing embryo.

endospore   [Gk. *endo*: within + *spora*: to sow] In some bacteria, a resting structure that can survive harsh environmental conditions.

endosymbiosis theory   [Gk. *endo*: within + *sym*: together + *bios*: life] The theory that the eukaryotic cell evolved via the engulfing of one prokaryotic cell by another.

endothelium   The single layer of epithelial cells lining the interior of a blood vessel.

endotherm   [Gk. *endo*: within + *thermos*: heat] An animal that can control its body temperature by the expenditure of its own metabolic energy. (Contrast with ectotherm.)

endotoxin   A lipopolysaccharide that forms part of the outer membrane of certain Gram-negative bacteria that is released when the bacteria grow or lyse. (Contrast with exotoxin.)

energetic cost   The difference between the energy an animal expends in performing a behavior and the energy it would have expended had it rested.

energy   The capacity to do work or move matter against an opposing force. The capacity to accomplish change in physical and chemical systems.

energy budget   A quantitative description of all paths of energy exchange between an animal and its environment.

enhancers   Regulatory DNA sequences that bind transcription factors that either activate or increase the rate of transcription.

enkephalins   Molecules in the mammalian brain that act as neurotransmitters in pathways that control pain.

enteric nervous system   The nerve nets in the submucosa and between the smooth muscle layers of the vertebrate gut.

enthalpy (*H*)   The total energy of a system.

entrain   To advance or delay an organism's circadian clock each day so that it is in phase with the light-dark cycle of the organism's environment.

entropy (*S*)   (en' tro pee) [Gk. *tropein*: to change] A measure of the degree of disorder in any system. Spontaneous reactions in a closed system are always accompanied by an increase in entropy.

enveloped virus   A virus enclosed within a phospholipid membrane derived from its host cell.

environment   Whatever surrounds and interacts with or otherwise affects a population, organism, or cell. May be external or internal.

environmental genomics   Sequencing technique used when biologists are unable to work with the whole genome of a prokaryote species but instead examine individual genes collected from a random sample of the organism's environment.

environmental resistance   Reduction in a population's growth rate caused by preemption of available resources by other individuals in the population.

environmentalism   The use of ecological knowledge, along with economics, ethics, and many other considerations, to inform both personal decisions and public policy relating to stewardship of natural resources and ecosystems.

enzyme   (en' zime) [Gk. *zyme*: to leaven (as in yeast bread)] A catalytic protein that speeds up a biochemical reaction.

enzyme–substrate complex (ES)   An intermediate in an enzyme-catalyzed reaction; consists of the enzyme bound to its substrate(s).

epi-   [Gk. upon, over] A prefix used to designate a structure located on top of another; for example, epidermis, epiphyte.

epiblast   The upper or overlying portion of the avian blastula which is joined to the hypoblast at the margins of the blastodisc.

epiboly   The movement of cells over the surface of the blastula toward the forming blastopore.

epidermis   [Gk. *epi*: over + *derma*: skin] In plants and animals, the outermost cell layers. (Only one cell layer thick in plants.)

epididymis   (epuh did' uh mus) [Gk. *epi*: over + *didymos*: testicle] Coiled tubules in the testes that store sperm and conduct sperm from the seminiferous tubules to the vas deferens.

epigenetics   The scientific study of changes in the expression of a gene or set of genes that occur without change in the DNA sequence.

epinephrine   (ep i nef' rin) [Gk. *epi*: over + *nephros*: kidney] The "fight or flight" hormone produced by the medulla of the adrenal gland; it also functions as a neurotransmitter. (Also known as adrenaline.)

epistasis   Interaction between genes in which the presence of a particular allele of one gene determines whether another gene will be expressed.

epithelial tissue   A type of animal tissue made up of sheets of cells that lines or covers organs, makes up tubules, and covers the surface of the body; one of the four major tissue types in multicellular animals.

epitope   *See* antigenic determinant.

equilibrium   Any state of balanced opposing forces and no net change.

equilibrium potential   The membrane potential at which an ion is at electrochemical equilibrium, i.e., there is no net flux of the ion across the membrane.

ER   *See* endoplasmic reticulum.

error signal   In regulatory systems, any difference between the set point of the system and its current condition.

erythrocyte   (ur rith' row site) [Gk. *erythros*: red + *kytos*: container] A red blood cell.

erythropoietin   A hormone produced by the kidney in response to lack of oxygen that stimulates the production of red blood cells.

esophagus   (i soff' i gus) [Gk. *oisophagos*: gullet] That part of the gut between the pharynx and the stomach.

essential amino acids   Amino acids that an animal cannot synthesize for itself and must obtain from its food.

essential element   A mineral nutrient required for normal growth and reproduction in plants and animals.

essential fatty acids   Fatty acids that an animal cannot synthesize for itself and must obtain from its food.

ester linkage   A condensation (water-releasing) reaction in which the carboxyl group of a fatty acid reacts with the hydroxyl group of an alcohol. Lipids, including most membrane lipids, are formed in this way. (Contrast with ether linkage.)

estivation   (ess tuh vay' shun) [L. *aestivalis*: summer] A state of dormancy and hypometabolism that occurs during the summer; usually a means of surviving drought and/or intense heat. (Contrast with hibernation.)

estrogen   Any of several steroid sex hormones; produced chiefly by the ovaries in mammals.

estrus   (es' trus) [L. *oestrus*: frenzy] The period of heat, or maximum sexual receptivity, in some female mammals. Ordinarily, the estrus is also the time of release of eggs in the female.

estuary   Aquatic biome in which salt water and fresh water mix, as when a river meets the ocean. Includes such ecosystems as salt marshes and mangrove forests.

ether linkage   The linkage of two hydrocarbons by an oxygen atom (HC—O—CH). Ether linkages are characteristic of the membrane lipids of the Archaea. (Contrast with ester linkage.)

ethology   [Gk. *ethos*: character + *logos*: study] An approach to the study of animal behavior that focuses on studying many species in natural environments and addresses questions about the evolution of behavior. (Compare with behaviorism.)

ethylene   One of the plant growth hormones, the gas $H_2C=CH_2$. Involved in fruit ripening and other growth and developmental responses.

euchromatin   Diffuse, uncondensed chromatin. Contains active genes that will be transcribed into mRNA. (Contrast with heterochromatin.)

eudicots   Angiosperms with two embryonic cotyledons. (*See also* monocots.)

Eukarya   One of the three domains of life; organisms made up of one or more eukaryotic cells. (*See also* eukaryotes.)

eukaryotes   (yew car' ree oats) [Gk. *eu*: true + *karyon*: kernel or nucleus] Organisms whose cells contain their genetic material inside a nucleus. Includes all life other than the viruses, archaea, and bacteria. (Contrast with prokaryotes.)

eusocial   Pertaining to a social group that includes nonreproductive individuals, as in honey bees.

eustachian tube   A connection between the middle ear and the throat that allows air pressure to equilibrate between the middle ear and the outside world.

eutrophication   (yoo trofe' ik ay' shun) [Gk. *eu*: truly + *trephein*: to flourish] The addition of nutrient materials to a body of water, resulting in changes in ecological processes and species composition therein.

evaporation   The transition of water from the liquid to the gaseous phase.

evolution   Any gradual change. Most often refers to organic or Darwinian evolution, which is the genetic and resulting phenotypic change in populations of organisms from generation to generation. (*See* macroevolution, microevolution; contrast with speciation.)

evolutionary developmental biology (evo-devo)   The study of the interplay between evolutionary and developmental processes, with a focus on the genetic changes that give rise to novel morphology. Key concepts of evo-devo include modularity, genetic toolkits, genetic switches, and heterochrony.

evolutionary radiation   The proliferation of many species within a single evolutionary lineage.

evolutionary reversal   The reappearance of an ancestral trait in a group that had previously acquired a derived trait.

evolutionary theory   The understanding and application of the mechanisms of evolutionary change to biological problems.

excision repair   DNA repair mechanism that removes damaged DNA and replaces it with the appro-priate nucleotide.

excitable   Capable of generating an action potential.

excitatory   Input from a neuron that causes depolarization of the recipient cell.

excited state   The state of an atom or molecule when, after absorbing energy, it has more energy than in its normal, ground state.

excretion   Release of metabolic wastes by an organism.

excretory systems   In animals, organs that maintain the volume, solute concentration, and composition of the extracellular fluid by excreting water, solutes, and nitrogenous wastes in the form of urine.

exergonic   A chemical reaction in which the products of the reaction have lower free energy than the reactants, resulting in a release of free energy. (Contrast with endergonic.)

exocrine gland   (eks' oh krin) [Gk. *exo*: outside + *krinein*: to separate] Any gland, such as a salivary gland, that secretes to the outside of the body or into the gut. (Contrast with endocrine gland.)

exocytosis   A process by which a vesicle within a cell fuses with the plasma membrane and releases its contents to the outside. (Contrast with endocytosis.)

exon   A portion of a DNA molecule, in eukaryotes, that codes for part of a polypeptide. (Contrast with intron.)

exoskeleton   (eks' oh skel' e ton) [Gk. *exos*: outside + *skleros*: hard] A hard covering on the outside of the body to which muscles are attached. (Contrast with endoskeleton.)

exotoxin   A highly toxic, usually soluble protein released by living, multiplying bacteria. (Contrast with endotoxin.)

expanding triplet repeat   A three-base-pair sequence in a human gene that is unstable and can be repeated a few to hundreds of times. Often, the more the repeats, the less the activity of the gene involved. Expanding triplet repeats occur in some human diseases such as Huntington's disease and fragile-X syndrome.

experiment   A testing process to support or disprove hypotheses and to answer questions. The basis of the scientific method. *See* comparative experiment, controlled experiment.

expiratory reserve volume   The amount of air that can be forcefully exhaled beyond the normal tidal expiration. (Contrast with inspiratory reserve volume, tidal volume, vital capacity.)

exploitation competition   Competition in which individuals reduce the quantities of their shared resources. (Contrast with interference competition.)

exponential growth   Growth, especially in the number of organisms in a population, which is a geometric function of the size of the growing entity: the larger the entity, the faster it grows. (Contrast with logistic growth.)

expression vector   A DNA vector, such as a plasmid, that carries a DNA sequence for the expression of an inserted gene into mRNA and protein in a host cell.

expressivity   The degree to which a genotype is expressed in the phenotype; may be affected by the environment.

extensor   A muscle that extends an appendage. (Contrast with flexor.)

external fertilization   The release of gametes into the environment; typical of aquatic animals. Also called spawning. (Contrast with internal fertilization.)

external gills   Highly branched and folded extensions of the body surface that provide a large surface area for gas exchange with water; typical of larval amphibians and many larval insects.

extinction   The termination of a lineage of organisms.

extracellular matrix   A material of heterogeneous composition surrounding cells and performing many functions including adhesion of cells.

extraembryonic membranes   Four membranes that support but are not part of the developing embryos of reptiles, birds, and mammals, defining these groups phylogenetically as amniotes. (*See* amnion, allantois, chorion, yolk sac.)

extreme halophiles   A group of euryarchaeotes that live exclusively in very salty environments.

extremophiles   Archaea and bacteria that live and thrive under conditions (e.g., extremely high temperatures) that would kill most organisms.

eye cups   Photosensory organs in flatworms; components of one of the simplest visual systems in animals.

F

5′ end   (5 prime) The end of a DNA or RNA strand that has a free phosphate group at the 5′ carbon of the sugar (deoxyribose or ribose).

F₁   The first filial generation; the immediate progeny of a parental (P) mating.

F₂   The second filial generation; the immediate progeny of a mating between members of the F1 generation.

facilitated diffusion   Passive movement through a membrane involving a specific carrier protein; does not proceed against a concentration gradient. (Contrast with active transport, diffusion.)

facilitation   In succession, modification of the environment by a colonizing species in a way that allows colonization by other species. (Contrast with inhibition.)

facultative anaerobe   A prokaryote that can shift its metabolism between anaerobic and aerobic modes depending on the presence or absence of O₂. (Alternatively, facultative aerobe.)

fast-twitch fibers   Skeletal muscle fibers that can generate high tension rapidly, but fatigue rapidly ("sprinter" fibers). Characterized by an abundance of enzymes of glycolysis. (Compare to slow-twitch fibers.)

fat   (1) A triglyceride that is solid at room temperature. (Contrast with oil.) (2) Adipose tissue, one type of connective tissue. (See brown fat, white fat.)

fate map   A diagram of the blastula showing which cells (blastomeres) are "fated" to contribute to specific tissues and organs in the mature body.

fatty acid   A molecule made up of a long nonpolar hydrocarbon chain and a polar carboxyl group. Found in many lipids.

fauna   (faw′ nah) All the animals found in a given area. (Contrast with flora.)

FD (FLOWERING LOCUS D)   Gene coding for a transcription factor in the shoot apical meristem that binds to florigen; involved in the induction of flowering.

feces   [L. *faeces*: dregs] Waste excreted from the digestive system.

fecundity   The average number of offspring produced by each female.

feedback   In regulatory systems, information about the relationship between the set point of the system and its current state (Contrast with feedforward information).

feedback inhibition   A mechanism for regulating a metabolic pathway in which the end product of the pathway can bind to and inhibit the enzyme that catalyzes the first committed step in the pathway. Also called end-product inhibition.

feedforward information   In regulatory systems, information that changes the set point of the system. (Contrast with feedback.)

fermentation   (fur men tay′ shun) [L. *fermentum*: yeast] The anaerobic degradation of a substance such as glucose to smaller molecules such as lactic acid or alcohol with the extraction of energy.

fertilization   Union of gametes. Also known as syngamy.

fertilizer   Any of a number of substances added to soil to improve the soil's capacity to support plant growth. May be organic or inorganic.

fetus   Medical and legal term for the stages of a developing human embryo from about the eighth week of pregnancy (the point at which all major organ systems have formed) to the moment of birth.

fiber   In angiosperms, an elongated, tapering sclerenchyma cell, usually with a thick cell wall, that serves as a support function in xylem. (*See also* muscle fiber.)

fibrin   A protein that polymerizes to form long threads that provide structure to a blood clot.

fibrinogen   A circulating protein that can be stimulated to fall out of solution and provide the structure for a blood clot.

fibrous root system   A root system typical of monocots composed of numerous thin adventitious roots that are all roughly equal in diameter. (Contrast with taproot system.)

Fick's law of diffusion   An equation that describes the factors that determine the rate of diffusion of a molecule from an area of higher concentration to an area of lower concentration.

fight-or-flight response   A rapid physiological response to a sudden threat mediated by the hormone epinephrine.

filament   In flowers, the part of a stamen that supports the anther.

filter feeder   An organism that feeds on organisms much smaller than itself that are suspended in water or air by means of a straining device.

first filial generation   *See* F₁.

first law of thermodynamics   The principle that energy can be neither created nor destroyed.

fission   *See* binary fission.

fitness   The contribution of a genotype or phenotype to the genetic composition of subsequent generations, relative to the contribution of other genotypes or phenotypes. (*See* also inclusive fitness.)

fixed action pattern   In ethology, a genetically determined behavior that is performed without learning, stereotypic (performed the same way each time), and not modifiable by learning.

flagellum   (fla jell′ um) (plural: flagella) [L. *flagellum*: whip] Long, whiplike appendage that propels cells. Prokaryotic flagella differ sharply from those found in eukaryotes.

flexor   A muscle that flexes an appendage. (Contrast with extensor.)

flora   (flore′ ah) All of the plants found in a given area. (Contrast with fauna.)

floral meristem   In angiosperms, a meristem that forms the floral organs (sepals, petals, stamens, and carpels).

floral organ identity genes   In angiosperms, genes that determine the fates of floral meristem cells; their expression is triggered by the products of meristem identity genes.

florigen   A plant hormone involved in the conversion of a vegetative shoot apex to a flower.

flower   The sexual structure of an angiosperm.

fluid feeder   An animal that feeds on fluids it extracts from the bodies of other organisms; examples include nectar-feeding birds and blood-sucking insects.

fluid mosaic model   A molecular model for the structure of biological membranes consisting of a fluid phospholipid bilayer in which suspended proteins are free to move in the plane of the bilayer.

flux   [L.: flow] In ecology, the flow of an element into or out of a compartment of the biosphere.

follicle   [L. *folliculus*: little bag] In female mammals, an immature egg surrounded by nutritive cells.

follicle-stimulating hormone (FSH)   A gonadotropin produced by the anterior pituitary.

food chain   A portion of a food web, most commonly a simple sequence of prey species and the predators that consume them.

food web   The complete set of food links between species in a community; a diagram indicating which ones are the eaters and which are eaten.

forebrain   The region of the vertebrate brain that comprises the cerebrum, thalamus, and hypothalamus.

fossil   Any recognizable structure originating from an organism, or any impression from such a structure, that has been preserved over geological time.

fossil fuels   Fuels, including oil, natural gas, coal, and peat, formed over geologic time from organic material buried in anaerobic sediments.

founder effect   Random changes in allele frequencies resulting from establishment of a population by a very small number of individuals.

fovea [L. *fovea*: a small pit] In the vertebrate retina, the area of most distinct vision.

frame-shift mutation The addition or deletion of a single or two adjacent nucleotides in a gene's sequence. Results in the misreading of mRNA during translation and the production of a nonfunctional protein. (Contrast with missense mutation, nonsense mutation, silent mutation.)

Frank–Starling law The stroke volume of the heart increases with increased return of blood to the heart.

free energy (*G*) Energy that is available for doing useful work, after allowance has been made for the increase or decrease of disorder.

frequency-dependent selection Selection that changes in intensity with the proportion of individuals in a population having the trait.

frontal lobe The largest of the brain lobes in humans; involved with feeling and planning functions; includes the primary motor cortex.

frugivore [L. *frugis*; fruit + *vorare*: to devour] An animal that eats fruit.

fruit In angiosperms, a ripened and mature ovary (or group of ovaries) containing the seeds. Sometimes applied to reproductive structures of other groups of plants.

FT (*FLOWERING LOCUS T*) Gene that codes for florigen, a small, diffusible protein involved in the induction of flowering.

fugitive species A species that leave an otherwise suitable habitat in order to avoid competition with another species.

full census A count of every individual in a population. Can only be achieve if individuals are large and distinct enough to be identifiable by the census taker; population sizes are more usually estimated using sampling methods.

functional genomics The assignment of functional roles to the proteins encoded by genes identified by sequencing entire genomes.

functional group A characteristic combination of atoms that contributes specific properties (such as charge or polarity) when attached to larger molecules (e.g., carboxyl group; amino group).

fundamental niche A species' niche as defined by its physiological capabilities. (Contrast with realized niche.)

G

G protein A membrane protein involved in signal transduction; characterized by binding GDP or GTP.

G protein–linked receptors A class of receptors that change configuration upon ligand binding such that a G protein binding site is exposed on the cytoplasmic domain of the receptor, initiating a signal transduction pathway.

G1 In the cell cycle, the gap between the end of mitosis and the onset of the S phase.

G1-to-S transition In the cell cycle, the point at which G1 ends and the S phase begins.

G2 In the cell cycle, the gap between the S (synthesis) phase and the onset of mitosis.

gain of function mutation A mutation that results in a protein with a new function. (Contrast with loss of function mutation.)

gallbladder In the human digestive system, an organ in which bile is stored.

gametangium (gam uh tan' gee um) (plural: gametangia) [Gk. *gamos*: marriage + *angeion*: vessel] Any plant or fungal structure within which a gamete is formed.

gamete (gam' eet) [Gk. *gamete*/ *gametes*: wife, husband] The mature sexual reproductive cell: the egg or the sperm.

gametogenesis (ga meet' oh jen' e sis) The specialized series of cellular divisions that leads to the production of gametes. (*See also* oogenesis, spermatogenesis.)

gametophyte (ga meet' oh fyte) In plants and photosynthetic protists with alternation of generations, the multicellular haploid phase that produces the gametes. (Contrast with sporophyte.)

ganglion (gang' glee un) (plural: ganglia) [Gk. lump] A cluster of neurons that have similar characteristics or function.

ganglion cells Cells at the front of the human retina that transmit information from the bipolar cells to the brain.

gap genes In *Drosophila* development, segmentation genes that define broad areas along the anterior–posterior axis of the early embryo. Part of a developmental cascade that includes maternal effect genes, pair rule genes, segment polarity genes, and Hox genes.

gap junction A 2.7-nanometer gap between plasma membranes of two animal cells, spanned by protein channels. Gap junctions allow chemical substances or electrical signals to pass from cell to cell.

gastric pits Deep infoldings in the walls of the stomach lined with secretory cells.

gastrin A hormone secreted by cells in the lower region of the stomach that stimulates the secretion of digestive juices as well as movements of the stomach.

gastrovascular cavity Serving for both digestion (gastro) and circulation (vascular); in particular, the central cavity of the body of jellyfish and other cnidarians.

gastrulation Development of a blastula into a gastrula. In embryonic development, the process by which a blastula is transformed by massive movements of cells into a *gastrula*, an embryo with three germ layers and distinct body axes.

gated channel A membrane protein that changes its three-dimensional shape, and therefore its ion conductance, in response to a stimulus. When open, it allows specific ions to move across the membrane.

gel electrophoresis (e lek' tro fo ree' sis) [L. *electrum*: amber + Gk. *phorein*: to bear] A technique for separating molecules (such as DNA fragments) from one another on the basis of their electric charges and molecular weights by applying an electric field to a gel.

gene [Gk. *genes*: to produce] A unit of heredity. Used here as the unit of genetic function which carries the information for a polypeptide or RNA.

gene duplication The generation of extra copies of a gene in a genome over evolutionary time. A mechanism by which genomes can acquire new functions.

gene expression The transcription and translation into a protein of the information (nucleotide sequence) contained in a gene.

gene family A set of similar genes derived from a single parent gene; need not be on the same chromosomes. The vertebrate globin genes constitute a classic example of a gene family.

gene flow Exchange of genes between populations through migration of individuals or movements of gametes.

gene pool All of the different alleles of all of the genes existing in all individuals of a population.

gene therapy Treatment of a genetic disease by providing patients with cells containing functioning alleles of the genes that are nonfunctional in their bodies.

gene tree A graphic representation of the evolutionary relationships of a single gene in different species or of the members of a gene family.

gene-for-gene concept In plants, a mechanism of resistance to pathogens in which resistance is triggered by the specific interaction of the products of a pathogen's *Avr* genes and a plant's *R* genes.

general transcription factors In eukaryotes, transcription factors that bind to the promoters of most protein-coding genes and are required for their expression. Distinct from transcription factors that have specific regulatory effects only at certain promoters or classes of promoters.

genetic code The set of instructions, in the form of nucleotide triplets, that translate a linear sequence of nucleotides in mRNA into a linear sequence of amino acids in a protein.

genetic drift Changes in gene frequencies from generation to generation as a result of random (chance) processes.

genetic linkage Association between genes on the same chromosome such that they do not show random assortment and seldom recombine; the closer the genes, the lower the frequency of recombination.

genetic map The positions of genes along a chromosome as revealed by recombination frequencies.

genetic marker (1) In gene cloning, a gene of identifiable phenotype that indicates the presence of another gene, DNA segment, or chromosome fragment. (2) In general, a DNA sequence such as a single nucleotide polymorphism whose presence is correlated with the presence of other linked genes on that chromosome.

genetic screen   A technique for identifying genes involved in a biological process of interest. Involves creating a large collection of randomly mutated organisms and identifying those individuals that are likely to have a defect in the pathway of interest. The mutated gene(s) in those individuals can then be isolated for further study.

genetic structure   The frequencies of different alleles at each locus and the frequencies of different genotypes in a Mendelian population.

genetic switches   Mechanisms that control how the genetic toolkit is used, such as promoters and the transcription factors that bind them. The signal cascades that converge on and operate these switches determine when and where genes will be turned on and off.

genetic toolkit   A set of developmental genes and proteins that is common to most animals and is hypothesized to be responsible for the evolution of their differing developmental pathways.

genetics   The scientific study of the structure, functioning, and inheritance of genes, the units of hereditary information.

genome   (jee' nome) The complete DNA sequence for a particular organism or individual.

genome sequencing   Determination of the nucleotide base sequence of the entire genome of an organism.

genomic imprinting   The form of a gene's expression is determined by parental source (i.e., whether the gene is inherited from the male or female parent).

genomic library   All of the cloned DNA fragments generated by the breakdown of genomic DNA into smaller segments.

genomics   The scientific study of entire sets of genes and their interactions.

genotype   (jean' oh type) [Gk. *gen*: to produce + *typos*: impression] An exact description of the genetic constitution of an individual, either with respect to a single trait or with respect to a larger set of traits. (Contrast with phenotype.)

genotype frequency   The proportion of a genotype among individuals in a population.

genus   (jean' us) (plural: genera) [Gk. *genos*: stock, kind] A group of related, similar species recognized by taxonomists with a distinct name used in binomial nomenclature.

geographic range   The region within which a species occurs.

germ cell   [L. *germen*: to beget] A reproductive cell or gamete of a multicellular organism. (Contrast with somatic cell.)

germ layers   The three embryonic layers formed during gastrulation (ectoderm, mesoderm, and endoderm). Also called cell layers or tissue layers.

germ line mutation   Mutation in a cell that produces gametes (i.e., a germ line cell). (Contrast with somatic mutation.)

germination   Sprouting of a seed or spore.

gestation   (jes tay' shun) [L. *gestare*: to bear] The period during which the embryo of a mammal develops within the uterus. Also known as pregnancy.

ghrelin   A hormone produced and secreted by cells in the stomach that stimulates appetite.

gibberellin   (jib er el' lin) A class of plant growth hormones playing roles in stem elongation, seed germination, flowering of certain plants, etc.

gill   An organ specialized for gas exchange with water.

gizzard   (giz' erd) [L. *gigeria*: cooked chicken parts] The second of two stomachlike organs in birds, other reptiles, earthworms, and various insects, that grinds up food, sometimes with the aid of fragments of stone. (*See also* crop.)

glia   (glee' uh) [Gk. *glia*: glue] One of the two classes of neural cells (along with neurons, with which glia interact); glia do not typically conduct action potentials. Types of glia include astrocytes, oligodendrocytes, and Schwann cells.

global nitrogen cycle   The movement of nitrogen through the biosphere. Steps in the cycle include the fixation of nitrogen gas (N_2) to ammonia; nitrification of the fixed nitrogen to nitrate by bacteria; nitrate reduction by plants; and denitrification back to N_2 by bacteria.

glomerular filtration rate (GFR)   The rate at which the blood is filtered in the glomeruli of the kidney.

glomerulus   (glo mare' yew lus) [L. *glomus*: ball] Sites in the kidney where blood filtration takes place. Each glomerulus consists of a knot of capillaries served by afferent and efferent arterioles.

glucagon   Hormone produced by alpha cells of the pancreatic islets of Langerhans. Glucagon stimulates the liver to break down glycogen and release glucose into the circulation.

gluconeogenesis   The biochemical synthesis of glucose from other substances, such as amino acids, lactate, and glycerol.

glucose   [Gk. *gleukos*: sugar, sweet wine] The most common monosaccharide; the monomer of the polysaccharides starch, glycogen, and cellulose.

glyceraldehyde 3-phosphate (G3P)   A phosphorylated three-carbon sugar; an intermediate in glycolysis and photosynthetic carbon fixation.

glycerol   (gliss' er ole) A three-carbon alcohol with three hydroxyl groups; a component of phospholipids and triglycerides.

glycogen   (gly' ko jen) [Gk. *glyk*: sweet] An energy storage polysaccharide found in animals and fungi; a branched-chain polymer of glucose, similar to starch.

glycolipid   A lipid to which sugars are attached.

glycolysis   (gly kol' li sis) [Gk. *gleukos*: sugar + *lysis*: break apart] The enzymatic breakdown of glucose to pyruvic acid.

glycoprotein   A protein to which sugars are attached.

glycosidic linkage   Bond between carbohydrate (sugar) molecules through an intervening oxygen atom (–O–).

glycosylation   The addition of carbohydrates to another type of molecule, such as a protein.

glyoxysome   (gly ox' ee soam) An organelle found in plants, in which stored lipids are converted to carbohydrates.

Golgi apparatus   (goal' jee) A system of concentrically folded membranes found in the cytoplasm of eukaryotic cells; functions in secretion from the cell by exocytosis.

Golgi tendon organ   A mechanoreceptor found in tendons and ligaments; provides information about the force generated by a contracting muscle.

gonad   (go' nad) [Gk. *gone*: seed] An organ that produces gametes in animals: either an ovary (female gonad) or testis (male gonad).

gonadotropin   A trophic hormone that stimulates the gonads.

gonadotropin-releasing hormone (GnRH)   Hormone produced by the hypothalamus that stimulates the anterior pituitary to secrete gonadotropins.

Gondwana   The large southern land mass that existed from the Cambrian (540 mya) to the Jurassic (138 mya). Present-day remnants are South America, Africa, India, Australia, and Antarctica.

graded membrane potential   Small local change in membrane potential caused by opening or closing of ion channels.

grafting   Artificial transplantation of tissue from one organism to another. In horticulture, the transfer of a bud or stem segment from one plant onto the root of another as a form of asexual reproduction.

gram stain   A differential purple stain useful in characterizing bacteria. The peptidoglycan-rich cell walls of gram-positive bacteria stain purple; cell walls of gram-negative bacteria generally stain orange.

gravitropism   [L. *gravitas*: weight, force; Gk. *tropos*: to turn] A directed plant growth response to gravity.

gray crescent   In frog development, a band of diffusely pigmented cytoplasm on the side of the egg opposite the site of sperm entry. Arises as a result of cytoplasmic rearrangements that establish the anterior–posterior axis of the zygote.

gray matter   In the nervous system, tissue that is rich in neuronal cell bodies. (Contrast with white matter.)

greenhouse gases Gases in the atmosphere, such as carbon dioxide and methane, that are transparent to sunlight, but trap heat radiating from Earth's surface, causing heat to build up at Earth's surface.

gross primary production The amount of energy captured by the primary producers in a community.

gross primary productivity (GPP) The rate at which the primary producers in a community turn solar energy into stored chemical energy via photosynthesis.

ground meristem That part of an apical meristem that gives rise to the ground tissue system of the primary plant body.

ground tissue system Those parts of the plant body not included in the dermal or vascular tissue systems. Ground tissues function in storage, photosynthesis, and support.

growth An increase in the size of the body and its organs by cell division and cell expansion.

growth factor A chemical signal that stimulates cells to divide.

growth hormone A peptide hormone released by the anterior pituitary that stimulates many anabolic processes.

guanine (G) (gwan' een) A nitrogen-containing base found in DNA, RNA, and GTP.

guard cells In plants, specialized, paired epidermal cells that surround and control the opening of a stoma (pore). *See* stoma.

gustation The sense of taste.

gut An animal's digestive tract.

gymnosperms Seed plants that do not produce flowers or fruits; one of the two major groups of living seed plants. (*See also* angiosperms.)

gyrus (ji' rus) [Gk. *gyros*: spiral] *See* convolutions.

H

habitat The particular environment in which an organism lives.

habitat patches Also called **habitat islands**; areas of suitable habitat for a species that are separated by substantial areas of unsuitable habitat.

Hadley cells Patterns of vertical atmospheric circulation that influence surface winds and precipitation patterns according to latitude.

hair cell A type of mechanoreceptor in animals. Detects sound waves and other forms of motion in air or water.

half-life The time required for half of a sample of a radioactive isotope to decay to its stable, nonradioactive form, or for a drug or other substance to reach half its initial dosage.

halophyte (hal' oh fyte) [Gk. *halos*: salt + *phyton*: plant] A plant that grows in a saline (salty) environment.

Hamilton's rule The principle that, for an apparent altruistic behavior to be adaptive, the fitness benefit of that act to the recipient times the degree of relatedness of the performer and the recipient must be greater than the cost to the performer.

haplodiploidy A sex determination mechanism in which diploid individuals (which develop from fertilized eggs) are female and haploid individuals (which develop from unfertilized eggs) are male; typical of hymenopterans.

haploid (hap' loid) [Gk. *haploeides*: single] Having a chromosome complement consisting of just one copy of each chromosome; designated $1n$ or n. (Contrast with diploid.)

haplotype Linked nucleotide sequences that are usually inherited as a unit (as a "sentence" rather than as individual "words").

Hardy–Weinberg equililbrium In a sexually reproducing population, the allele frequency at a given locus that is not being acted on by agents of evolution; the conditions that would result in no evolution in a population.

haustorium (haw stor' ee um) (plural: haustoria)[L. *haustus*: draw up] A specialized hypha or other structure by which fungi and some parasitic plants draw nutrients from a host plant.

Haversian systems Units of organization in compact bone that reflect the action of intercommunicating osteoblasts.

heart In circulatory systems, a muscular pump that moves extracellular fluid around the body.

heat of vaporization The energy that must be supplied to convert a molecule from a liquid to a gas at its boiling point.

heat shock proteins Chaperone proteins expressed in cells exposed to high or low temperatures or other forms of environmental stress.

helical Shaped like a screw or spring (helix); this shape occurs in DNA and proteins.

helper T cells *See* T-helper cells.

hemiparasite A parasitic plant that can photosynthesize, but derives water and mineral nutrients from the living body of another plant. (Contrast with holoparasite.)

hemizygous (hem' ee zie' gus) [Gk. *hemi*: half + *zygotos*: joined] In a diploid organism, having only one allele for a given trait, typically the case for X-linked genes in male mammals and Z-linked genes in female birds. (Contrast with homozygous, heterozygous.)

hemoglobin (hee' mo glow bin) [Gk. *heaema*: blood + L. *globus*: globe] Oxygen-transporting protein found in the red blood cells of vertebrates (and found in some invertebrates).

Hensen's node In avian embryos, a structure at the anterior end of the primitive groove; determines the fates of cells passing over it during gastrulation.

hepatic (heh pat' ik) [Gk. *hepar*: liver] Pertaining to the liver.

herbivore (ur' bi vore) [L. *herba*: plant + *vorare*: to devour] An animal that eats plant tissues. (Contrast with carnivore, detritivore, omnivore.)

heritable trait A trait that is at least partly determined by genes.

hermaphroditism (her maf' row dite ism) The coexistence of both female and male sex organs in the same organism.

hetero- [Gk.: *heteros*: other, different] A prefix indicating two or more different conditions, structures, or processes. (Contrast with homo-.)

heterochromatin Densely packed, dark-staining chromatin; any genes it contains are usually not transcribed.

heterochrony [Gk: different time] Alteration in the timing of developmental events, contributing to the evolution different phenotypes in the adult.

heterocyst A large, thick-walled cell type in the filaments of certain cyanobacteria that performs nitrogen fixation.

heterometry [Gk: different measure] Alteration in the level of gene expression, and thus in the amount of protein produced, during development, contributing to the evolution of different phenotypes in the adult.

heteromorphic (het' er oh more' fik) [Gk.: different form] Having a different form or appearance, as two heteromorphic life stages of a plant. (Contrast with isomorphic.)

heterosis The superior fitness of heterozygous offspring as compared with that of their dissimilar homozygous parents. Also called hybrid vigor.

heterosporous (het' er os' por us) Producing two types of spores, one of which gives rise to a female megaspore and the other to a male microspore. (Contrast with homosporous.)

heterotherm An animal that regulates its body temperature at a constant level at some times but not others, such as a hibernator.

heterotopy [Gk: different place] Spatial differences in gene expression during development, controlled by developmental regulatory genes and contributing to the evolution of distinctive adult phenotypes.

heterotroph (het' er oh trof) [Gk. *heteros*: different + *trophe*: feed] An organism that requires preformed organic molecules as food. (Contrast with autotroph.)

heterotrophic succession Succession in detritus-based communities, which differs from other types of succession in taking place without the participation of plants.

heterotypy [Gk.: different kind] Alteration in a developmental regulatory gene itself rather than the expression of the genes it controls. (Contraste with heterochrony; heterometry; heterotopy.)

heterozygous (het' er oh zie' gus) [Gk. *heteros*: different + *zygotos*: joined] In diploid organisms, having different alleles of a given gene on the pair of homologs carrying that gene. (Contrast with homozygous.)

heterozygous carrier   An individual that carries a recessive allele for a phenotype of interest (e.g., a genetic disease); the individual does not show the phenotype, but may have progeny with the phenotype if the other parent also carries the recessive allele.

hexose   [Gk. *hex*: six] A sugar containing six carbon atoms.

hibernation   [L. *hibernum*: winter] The state of inactivity of some animals during winter; marked by a drop in body temperature and metabolic rate. (Contrast with estivation)

high-density lipoproteins (HDLs)   Lipoproteins that remove cholesterol from tissues and carry it to the liver; HDLs are the "good" lipoproteins associated with good cardiovascular health.

high-throughput sequencing   Rapid DNA sequencing on a micro scale in which many fragments of DNA are sequenced in parallel.

highly repetitive sequences   Short (less than 100 bp), nontranscribed DNA sequences, repeated thousands of times in tandem arrangements.

hindbrain   The region of the developing vertebrate brain that gives rise to the medulla, pons, and cerebellum.

hippocampus   [Gk. sea horse] A part of the forebrain that takes part in long-term memory formation.

histamine   (hiss' tah meen) A substance released by damaged tissue, or by mast cells in response to allergens. Histamine increases vascular permeability, leading to edema (swelling). (Contrast with histone deacetylase.)

histone   Any one of a group of proteins forming the core of a nucleosome, the structural unit of a eukaryotic chromosome.

histone acetyltransferases   Enzymes involved in chromatin remodeling. Add acetyl groups to the tail regions of histone proteins.

histone deacetylase   In chromatin remodeling, an enzyme that removes acetyl groups from the tails of histone proteins. (Contrast with histone acetyltransferases.)

HIV   Human immunodeficiency virus, the retrovirus that causes acquired immune deficiency syndrome (AIDS).

holoparasite   A fully parasitic plant (i.e., one that does not perform photosynthesis).

homeobox   180-base-pair segment of DNA found in certain homeotic genes. A specific sequence within the homeobox—the **homeodomain**—regulates the expression of other genes and through this regulation controls large-scale developmental processes. (*See* homeotic genes.)

homeostasis   (home' ee o sta' sis) [Gk. *homos*: same + *stasis*: position] The maintenance of a steady state, such as a constant temperature, by means of

physiological or behavioral feedback responses.

homeotic genes   Genes that act during development to determine the formation of an organ from a region of the embryo. (Compare with Hox genes.)

homeotic mutation   Mutation in a homeotic gene that results in the formation of a different organ than that normally made by a region of the embryo.

homing   In animal navigation, the ability to return to a nest site, burrow, or other specific location.

hominid   Lineage that includes all modern and extinct Great Apes (i.e., humans, gorillas, chimpanzees, orangutans, and their ancestors.)

hominin   Lineages that includes modern humans (*Homo sapiens*) and their extinct ancestors (e.g., Australopithecines; *Homo erectus*.)

homo-   [Gk. *homos*: same] A prefix indicating two or more similar conditions, structures, or processes. (Contrast with hetero-.)

homolog   (1) In cytogenetics, one of a pair (or larger set) of chromosomes having the same overall genetic composition and sequence. In diploid organisms, each chromosome inherited from one parent is matched by an identical (except for mutational changes) chromosome—its homolog—from the other parent. (2) In evolutionary biology, one of two or more features in different species that are similar by reason of descent from a common ancestor.

homologous pair   A pair of matching chromosomes made up of a chromosome from each of the two sets of chromosomes in a diploid organism.

homologous recombination   Exchange of segments between two DNA molecules based on sequence similarity between the two molecules. The similar sequences align and crossover. Used to create knockout mutants in mice and other organisms.

homology   (ho mol' o jee) [Gk. *homologia*: of one mind; agreement] A similarity between two or more features that is due to inheritance from a common ancestor. The structures are said to be *homologous*, and each is a *homolog* of the others.

homoplasy   (home' uh play zee) [Gk. *homos*: same + *plastikos*: shape, mold] The presence in multiple groups of a trait that is not inherited from the common ancestor of those groups. Can result from convergent evolution, evolutionary reversal, or parallel evolution.

homosporous   Producing a single type of spore that gives rise to a single type of gametophyte, bearing both female and male reproductive organs. (Contrast with heterosporous.)

homotypic   Pertaining to adhesion of cells of the same type. (Contrast with heterotypic.)

homozygous   (home' oh zie' gus) [Gk. *homos*: same + *zygotos*: joined] In diploid organisms, having identical alleles of a given gene on both homologous chromosomes. An

individual may be a homozygote with respect to one gene and a heterozygote with respect to another. (Contrast with heterozygous.)

horizons   The horizontal layers of a soil profile, including the topsoil (A horizon), subsoil (B horizon) and parent rock or bedrock (C horizon).

hormone   (hore' mone) [Gk. *hormon*: to excite, stimulate] A chemical signal produced in minute amounts at one site in a multicellular organism and transported to another site where it acts on target cells.

host   An organism that harbors a parasite or symbiont and provides it with nourishment.

Hox genes   Conserved homeotic genes found in vertebrates, *Drosophila*, and other animal groups. Hox genes contain the homeobox and specify pattern and axis formation in these animals.

human chorionic gonadotropin (hCG)   A hormone secreted by the placenta which sustains the corpus luteum and helps maintain pregnancy.

Human Genome Project   A publicly and privately funded research effort, successfully completed in 2003, to produce a complete DNA sequence for the entire human genome.

humoral immune response   The response of the immune system mediated by B cells that produces circulating antibodies active against extracellular bacterial and viral infections. (Contrast with cellular immune response.)

humus   (hew' mus) The partly decomposed remains of plants and animals on the surface of a soil.

hybrid   (high' brid) [L. *hybrida*: mongrel] (1) The offspring of genetically dissimilar parents. (2) In molecular biology, a double helix formed of nucleic acids from different sources.

hybrid vigor   *See* heterosis.

hybrid zone   A region of overlap in the ranges of two closely related species where the species may hybridize.

hybridize   (1) In genetics, to combine the genetic material of two distinct species or of two distinguishable populations within a species. (2) In molecular biology, to form a double-stranded nucleic acid in which the two strands originate from different sources.

hydrocarbon   A compound containing only carbon and hydrogen atoms.

hydrogen bond   A weak electrostatic bond which arises from the attraction between the slight positive charge on a hydrogen atom and a slight negative charge on a nearby oxygen or nitrogen atom.

hydrologic cycle   The movement of water from the oceans to the atmosphere, to the soil, and back to the oceans.

hydrolysis reaction   (high drol' uh sis) [Gk. *hydro*: water + *lysis*: break apart] A chemical reaction that breaks a bond by inserting the components of water ($AB + H_2O \rightarrow AH + BOH$). (Contrast with condensation reaction.)

hydrophilic   (high dro fill' ik) [Gk. *hydro*: water + *philia*: love] Having an affinity for water. (Contrast with hydrophobic.)

hydrophobic   (high dro foe' bik) [Gk. *hydro*: water + *phobia*: fear] Having no affinity for water. Uncharged and nonpolar groups of atoms are hydrophobic. (Contrast with hydrophilic.)

hydroponic   Pertaining to a method of growing plants with their roots suspended in nutrient solutions instead of soil.

hydrostatic pressure   Pressure generated by compression of liquid in a confined space. Generated in plants, fungi, and some protists with cell walls by the osmotic uptake of water. Generated in animals with closed circulatory systems by the beating of a heart.

hydrostatic skeleton   A fluid-filled body cavity that transfers forces from one part of the body to another when acted on by surrounding muscles.

hydroxyl group   The —OH group found on alcohols and sugars.

hyper-   [Gk. *hyper*: above, over] Prefix indicating above, higher, more. (Contrast with hypo-.)

hyperaccumulators   Plant species that store large quantities of heavy metals such as arsenic, cadmium, nickel, aluminum, and zinc.

hyperpolarization   A change in the resting potential across a membrane so that the inside of a cell becomes more negative compared with the outside of the cell. (Contrast with depolarization.)

hypersensitive response   A defensive response of plants to microbial infection in which phytoalexins and pathogenesis-related proteins are produced and the infected tissue undergoes apoptosis to isolate the pathogen from the rest of the plant.

hypertonic   Having a greater solute concentration. Said of one solution compared with another. (Contrast with hypotonic, isotonic.)

hypha   (high' fuh) (plural: hyphae) [Gk. *hyphe*: web] In the fungi and oomycetes, any single filament.

hypo-   [Gk. *hypo*: beneath, under] Prefix indicating underneath, below, less. (Contrast with hyper-.)

hypoblast   The lower tissue portion of the avian blastula which is joined to the epiblast at the margins of the blastodisc.

hypothalamus   The part of the brain lying below the thalamus; it coordinates water balance, reproduction, temperature regulation, and metabolism.

hypothermia   Below-normal body temperature.

hypothesis   A tentative answer to a question, from which testable predictions can be generated. (Contrast with theory.)

hypotonic   Having a lesser solute concentration. Said of one solution in comparing it to another. (Contrast with hypertonic, isotonic.)

hypoxia   A deficiency of oxygen.

I

ileum   The final segment of the small intestine. (*See also* duodenum, jejunum.)

imbibition   Water uptake by a seed; first step in germination.

immediate hypersensitivity   A rapid, extensive overreaction of the immune system against an allergen, resuting in the release of large amounts of histamine. (Contrast with delayed hypersensitivity.)

immediate memory   A form of memory for events happening in the present that is almost perfectly photographic, but lasts only seconds.

immunity   [L. *immunis*: exempt from] In animals, the ability to avoid disease when invaded by a pathogen by deploying various defense mechanisms.

immunoassay   The use of antibodies to measure the concentration of an antigen in a sample.

immunoglobulins   A class of proteins containing a tetramer consisting of four polypeptide chains—two identical light chains and two identical heavy chains—held together by disulfide bonds; active as receptors and effectors in the immune system.

immunological memory   The capacity to more rapidly and massively respond to a second exposure to an antigen than occurred on first exposure.

imperfect flower   A flower lacking either functional stamens or functional carpels. (Contrast with perfect flower.)

implantation   The process by which the early mammalian embryo becomes attached to and embedded in the lining of the uterus.

imprinting   In animal behavior, a rapid form of learning in which an animal learns, during a brief critical period, to make a particular response (which is then maintained for life) to some object or other organism. *See also* genomic imprinting.

in vitro   [L.: in glass] A biological process occurring outside of the organism, in the laboratory. (Contrast with in vivo.)

in vitro evolution   A method based on natural molecular evolution that uses artificial selection in the laboratory to rapidly produce molecules with novel enzymatic and binding functions.

in vivo   [L.: in life] A biological process occurring within a living organism or cell. (Contrast with in vitro.)

inclusive fitness   The sum of an individual's genetic contribution to subsequent generations both via production of its own offspring and via its influence on the survival of relatives who are not direct descendants. (Contrast with direct fitness)

incomplete cleavage   A pattern of cleavage that occurs in many eggs that have a lot of yolk, in which the cleavage furrows do not penetrate all of it. (*See also* discoidal cleavage,

superficial cleavage; contrast with complete cleavage.)

incomplete dominance   Condition in which the heterozygous phenotype is intermediate between the two homozygous phenotypes.

incomplete metamorphosis   Insect development in which changes between instars are gradual. (Contrast with direct development; complete metamorphosis.)

independent assortment   During meiosis, the random separation of genes carried on nonhomologous chromosomes into gametes so that inheritance of these genes is random. This principle was articulated by Mendel as his second law.

independent variable   In a scientific experiment, a critical factor that is manipulated while all other factors are held constant. (Contrast with dependent variable.)

indeterminate growth   A open-ended growth pattern in which an organism or organ continues to grow as long as it lives; characteristic of some animals and of plant shoots and roots. (Contrast with determinate growth.)

individual fitness   See direct fitness.

induced fit   A change in the shape of an enzyme caused by binding to its substrate that exposes the active site of the enzyme.

induced mutation   A mutation resulting from exposure to a mutagen from outside the cell. (Contrast with spontaneous mutation.)

induced pluripotent stem cells (iPS cells)   Multipotent or pluripotent animal stem cells produced from differentiated cells in vitro by the addition of several genes that are expressed.

induced responses   Defensive responses that a plant produces only in the presence of a pathogen, in contrast to constitutive defenses, which are always present.

inducer   (1) A compound that stimulates the synthesis of a protein. (2) In embryonic development, a substance that causes a group of target cells to differentiate in a particular way.

inducible genes   Genes that are expressed only when their products—**inducible proteins**—are needed. (Contrast with constitutive genes.)

inducible   Produced only in the presence of a particular compound or under particular circumstances. (Contrast with constitutive.)

induction   In embryonic development, the process by which a factor produced and secreted by certain cells determines the fates other cells.

inductive logic   Involves making observations and then formulating one or more possible scenarios—hypotheses—that might explain those observations. (Contrast with deductive logic.)

inflammation   A nonspecific defense against pathogens; characterized by redness, swelling, pain, and increased temperature.

inflorescence A structure composed of several to many flowers.

inflorescence meristem A meristem that produces floral meristems as well as other small leafy structures (bracts).

ingroup In a phylogenetic study, the group of organisms of primary interest. (Contrast with outgroup.)

inhibitor A substance that blocks a biological process.

inhibitory Input from a neuron that causes hyperpolarization of the recipient cell.

initials Cells that perpetuate plant meristems, comparable to animal stem cells. When an initial divides, one daughter cell develops into another initial, while the other differentiates into a more specialized cell.

initiation complex In protein translation, a combination of a small ribosomal subunit, an mRNA molecule, and the tRNA charged with the first amino acid coded for by the mRNA; formed at the onset of translation.

initiation site The place within a promoter where transcription begins.

innate defenses In animals, one of two general types of defenses against pathogens. Nonspecific and present in most animals. (Contrast with adaptive immunity.)

inner cell mass Derived from the mammalian blastula (bastocyst), the inner cell mass that will give rise to the yolk sac (via hypoblast) and embryo (via epiblast).

inorganic fertilizer A chemical or combination of chemicals applied to soil or plants to make up for a plant nutrient deficiency. Often contains the macronutrients nitrogen, phosphorus, and potassium (N-P-K).

inositol trisphosphate (IP$_3$) An intracellular second messenger derived from membrane phospholipids.

inspiratory reserve volume The amount of air that can be inhaled above the normal tidal inspiration. (Contrast with expiratory reserve volume, tidal volume, vital capacity.)

instar (in' star) An immature stage of an insect between molts.

insula (in' su lah) [L. *insula*: island] An area deep within the forebrain that appears to integrate physiological information from all over the body to create a sensation of how the body "feels" and may be involved in human consciousness. Also called the insular cortex.

insulin (in' su lin) [L. *insula*: island] A hormone synthesized in islet cells of the pancreas that promotes the conversion of glucose into the storage material, glycogen.

integral membrane proteins Proteins that are at least partially embedded in the plasma membrane. (Contrast with peripheral membrane proteins.)

integrin In animals, a transmembrane protein that mediates the attachment of epithelial cells to the extracellular matrix.

integument [L. *integumentum*: covering] A protective surface structure. In gymnosperms and angiosperms, a layer of tissue around the ovule which will become the seed coat.

intercostal muscles Muscles between the ribs that can augment breathing movements by elevating and suppressing the rib cage.

interference competition Competition in which individuals actively interfere with one another's access to resources. (Contrast with exploitation competition.)

interference RNA (RNAi) *See* RNA interference.

interferons Glycoproteins produced by virus-infected animal cells; interferons increase the resistance of neighboring cells to the virus.

internal environment In multicelluar organisms, includes blood plasma and interstitial fluid, i.e., the extracellular fluids that surround the cells.

internal fertilization The release of sperm into the female reproductive tract; typical of most terrestrial animals. (Contrast with external fertilization.)

internal gills Gills enclosed in protective body cavities; typical of mollusks, arthropods, and fishes.**interneuron** A neuron that communicates information between two other neurons.

interneuron A neuron that communicates information between two other neurons.

internode The region between two nodes of a plant stem.

interphase In the cell cycle, the period between successive nuclear divisions during which the chromosomes are diffuse and the nuclear envelope is intact. During interphase the cell is most active in transcribing and translating genetic information.

interspecific competition Competition between members of two or more species. (Contrast with intraspecific competition; see also exploitation competition, interference competition.)

interstitial fluid Extracellular fluid that is not contained in the vessels of a circulatory system.

intertidal zone A nearshore region of oceans that is periodically exposed to the air as the tides rise and fall.

intestine The portion of the gut following the stomach, in which most digestion and absorption occurs.

intraspecific competition Competition among members of the same species. (Contrast with interspecific competition.)

intrinsic rate of increase The rate at which a population is capable of growing when its density is low and environmental conditions are highly favorable.

intron Portion of a of a gene within the coding region that is transcribed into pre-mRNA but is spliced out prior to translation. (Contrast with exon.)

invasive species An exotic species that reproduces rapidly, spreads widely, and has negative effects on the native species of the region to which it has been introduced.

invasiveness The ability of a pathogen to multiply in a host's body. (Contrast with toxigenicity).

inversion A rare 180° reversal of the order of genes within a segment of a chromosome.

involution Cell movements that occur during gastrulation of frog embryos, giving rise to the archenteron.

ion (eye' on) [Gk. *ion*: wanderer] An electrically charged particle that forms when an atom gains or loses one or more electrons.

ion channel An integral membrane protein that allows ions to diffuse across the membrane in which it is embedded.

ion exchange In plants, a process by which protons produced by the plant's root displace mineral cations from clay particles in the surrounding soil.

ionic attraction An electrostatic attraction between positively and negatively charged ions.

ionotropic receptors A receptor that directly alters membrane permeability to a type of ion when it combines with its ligand.

iris (eye' ris) [Gk. *iris*: rainbow] The round, pigmented membrane that surrounds the pupil of the eye and adjusts its aperture to regulate the amount of light entering the eye.

island biogeography A theory proposing that the number of species on an island (or in another geographically defined and isolated area) represents a balance, or equilibrium, between the rate at which species immigrate to the island and the rate at which resident species go extinct.

islets of Langerhans Clusters of hormone-producing cells in the pancreas.

iso- [Gk. *iso*: equal] Prefix used for two separate entities that share some element of identity.

isomers Molecules consisting of the same numbers and kinds of atoms, but differing in the bonding patterns by which the atoms are held together.

isomorphic (eye so more' fik) [Gk. *isos*: equal + *morphe*: form] Having the same form or appearance, as when the haploid and diploid life stages of an organism appear identical. (Contrast with heteromorphic.)

isotonic Having the same solute concentration; said of two solutions. (Contrast with hypertonic, hypotonic.)

isotope (eye' so tope) [Gk. *isos*: equal + *topos*: place] Isotopes of a given chemical element have the same number of protons in their nuclei (and thus are in the same position on the periodic table), but differ in the number of neutrons.

isozymes Enzymes of an organism that have somewhat different amino acid sequences but catalyze the same reaction.

iteroparous [L. itero, to repeat + pario, to beget] Reproducing multiple times in a lifetime. (Contrast with semelparous.)

J

jasmonate Also called jasmonic acid, a plant hormone involved in triggering responses to pathogen attack as well as other processes.

jejunum (jih jew' num) The middle division of the small intestine, where most absorption of nutrients occurs. (*See also* duodenum, ileum.)

joint In skeletal systems, a junction between two or more bones.

juvenile hormone In insects, a hormone maintaining larval growth and preventing maturation or pupation.

K

***K*-strategist** A species whose life history strategy allows it to persist at or near the carrying capacity (*K*) of its environment. (Contrast with *r*-strategist.)

karyogamy The fusion of nuclei of two cells. (Contrast with plasmogamy.)

karyotype The number, forms, and types of chromosomes in a cell.

keystone species Species that have a dominant influence on the composition of a community.

kidneys A pair of excretory organs in vertebrates.

kilocalorie (kcal) *See* Calorie.

kin selection That component of inclusive fitness resulting from helping the survival of relatives containing the same alleles by descent from a common ancestor. (Contrast with direct fitness.)

kinase *See* protein kinase.

kinetic energy (kuh-net' ik) [Gk. *kinetos*: moving] The energy associated with movement. (Contrast with potential energy.)

kinetochore (kuh net' oh core) Specialized structure on a centromere to which microtubules attach.

knockout A molecular genetic method in which a single gene of an organism is permanently inactivated.

Koch's postulates A set of rules for establishing that a particular microorganism causes a particular disease.

Krebs cycle *See* citric acid cycle.

L

lagging strand In DNA replication, the daughter strand that is synthesized in discontinuous stretches. (*See* Okazaki fragments.)

large intestine *See* colon.

larva (plural: larvae) [L. *lares*: guiding spirits] An immature stage of any animal that differs dramatically in appearance from the adult.

lateral [L. *latus*: side] Pertaining to the side.

lateral gene transfer The transfer of genes from one species to another, common among bacteria and archaea.

lateral meristem Either of the two meristems, the vascular cambium and the cork cambium, that give rise to a plant's secondary growth.

lateral root A root extending outward from the taproot in a taproot system; typical of eudicots.

lateralization A phenomenon in humans in which language functions come to reside in one cerebral hemisphere, usually the left.

laticifers (luh tiss' uh furs) In some plants, elongated cells containing secondary plant products such as latex.

Laurasia The northernmost of the two large continents produced by the breakup of Pangaea.

law of independent assortment *See* independent assortment.

law of segregation *See* segregation.

laws of thermodynamics [Gk. *thermos*: heat + *dynamis*: power] Laws derived from studies of the physical properties of energy and the ways energy interacts with matter. (*See also* first law of thermodynamics, second law of thermodynamics.)

leaching In soils, a process by which mineral nutrients in upper soil horizons are dissolved in water and carried to deeper horizons, where they are unavailable to plant roots.

leading strand In DNA replication, the daughter strand that is synthesized continuously. (Contrast with lagging strand.)

leaf (plural: leaves) In plants, the chief organ of photosynthesis.

leaf primordium (plural: primordia) An outgrowth on the side of the shoot apical meristem that will eventually develop into a leaf.

leghemoglobin In nitrogen-fixing plants, an oxygen-carrying protein in the cytoplasm of nodule cells that transports enough oxygen to the nitrogen-fixing bacteria to support their respiration, while keeping free oxygen concentrations low enough to protect nitrogenase.

lek A display ground within which male animals compete for and defend small display areas as a means of demonstrating their territorial prowess and winning opportunities to mate.

lens In the vertebrate eye, a crystalline protein structure that makes fine adjustments in the focus of images falling on the retina.

leptin A hormone produced by fat cells that is believed to provide feedback information to the brain about the status of the body's fat reserves.

leukocyte *See* white blood cell.

lichen (lie' kun) An organism resulting from the symbiotic association of a fungus and either a cyanobacterium or a unicellular alga.

life cycle The entire span of the life of an organism from the moment of fertilization (or asexual generation) to the time it reproduces in turn.

life history strategy The way in which an organism partitions its time and energy among growth, maintenance, and reproduction.

life history The time course of growth and development, reproduction, and death during an average individual organism's life.

life table A summary of information about the progression of individuals in a population through the various stages of their life cycles.

life zones In the aquatic (marine and freshwater) biomes, the regions defined by light penetration and water movement such as wave action. Life zones include, e.g., the intertidal, pelagic (open water) and bethic (bottom) zones.

ligament A band of connective tissue linking two bones in a joint.

ligand (lig' and) Any molecule that binds to a receptor site of another (usually larger) molecule.

light reactions The initial phase of photosynthesis, in which light energy is converted into chemical energy. Followed by the **light-independent reactions** in which the energy captured in the light reactions is used to drive the reduction of CO_2 to form carbohydrates.

light-harvesting complex In photosynthesis, a group of different molecules that cooperate to absorb light energy and transfer it to a reaction center. Also called *antenna system*.

lignin A complex, hydrophobic polyphenolic polymer in plant cell walls that crosslinks other wall polymers, strengthening the walls, especially in wood.

limbic system A group of evolutionarily primitive structures in the vertebrate telencephalon that are involved in emotions, drives, instinctive behaviors, learning, and memory.

limiting resource The required resource whose supply (or lack thereof) most strongly influences the size of a population.

limnetic zone The open-water life zone of a lake

lineage A series of populations, species, or genes descended from a single ancestor over evolutionary time.

lineage species concept The definition of a species as a branch on the tree of life, which has a history that starts at a speciation event and ends either at extinction or at another speciation event. (Contrast with biological species concept; morphological species concept.)

linkage *See* genetic linkage.

lipase (lip' ase; lye' pase) An enzyme that digests fats.

lipid (lip' id) [Gk. *lipos*: fat] Nonpolar, hydrophobic molecules that include fats, oils, waxes, steroids, and the phospholipids that make up biological membranes.

lipid bilayer *See* phospholipid bilayer.

lipoproteins   Lipids packaged inside a covering of protein so that they can be circulated in the blood.

lithoosphere   (lith' o sphere) [Gk. *lithos*: strong] The crust of sold rock plates that overlays the viscous mantle of Earth. The movements of the lithosphere are the source of plate tectonics. (Constrast with asthenosphere.)

littoral zone   The nearshore life zone of a lake that is shallow and is affected by wave action and fluctuations in water level.

liver   A large digestive gland. In vertebrates, it secretes bile and is involved in the formation of blood.

loam   A type of soil consisting of a mixture of sand, silt, clay, and organic matter. One of the best soil types for agriculture.

locus   (low' kus) (plural: loci, low' sigh) In genetics, a specific location on a chromosome. May be considered synonymous with *gene*.

logistic growth   Growth, especially in the size of an organism or in the number of organisms in a population, that slows steadily as the entity approaches its maximum size. (Contrast with multiplicative growth.)

long-day plant (LDP)   A plant that requires long days (actually, short nights) in order to flower. (Compare to short-day plant.)

long-term potentiation (LTP)   A long-lasting increase in the responsiveness of a neuron resulting from a period of intense stimulation.

loop of Henle   (hen' lee) Long, hairpin loop of the mammalian renal tubule that runs from the cortex down into the medulla and back to the cortex; creates a concentration gradient in the interstitial fluids in the medulla.

lophophore   A U-shaped fold of the body wall with hollow, ciliated tentacles that encircles the mouth of animals in several different groups. Used for filtering prey from the surrounding water.

loss of function mutation   A mutation that results in the loss of a functional protein. (Contrast with gain of function mutation.)

low-density lipoproteins (LDLs)   Lipoproteins that transport cholesterol around the body for use in biosynthesis and for storage; LDLs are the "bad" lipoproteins associated with a high risk of cardiovascular disease.

lumen   (loo' men) [L. *lumen*: light] The open cavity inside any tubular organ or structure, such as the gut or a renal tubule.

lung   An internal organ specialized for respiratory gas exchange with air.

luteinizing hormone (LH)   A gonadotropin produced by the anterior pituitary that stimulates the gonads to produce sex hormones.

lymph   [L. *lympha*: liquid] A fluid derived from blood and other tissues that accumulates in intercellular spaces throughout the body and is returned to the blood by the lymphatic system.

lymph node   A specialized structure in the vessels of the lymphatic system. Lymph nodes contain lymphocytes, which encounter and respond to foreign cells and molecules in the lymph as it passes through the vessels.

lymphatic system   A system of vessels that returns interstitial fluid to the blood.

lymphocyte   One of the two major classes of white blood cells; includes T cells, B cells, and other cell types important in the immune system.

lysis   (lie' sis) [Gk. *lysis*: break apart] Bursting of a cell.

lysogeny   A form of viral replication in which the virus becomes incorporated into the host chromosome and remains inactive. Also called a lysogenic cycle. (Contrast with lytic cycle.)

lysosome   (lie' so soam) [Gk. *lysis*: break away + *soma*: body] A membrane-enclosed organelle originating from the Golgi apparatus and containing hydrolytic enzymes. (Contrast with secondary lysosome.)

lysozyme   (lie' so zyme) An enzyme in saliva, tears, and nasal secretions that hydrolyzes bacterial cell walls.

lytic cycle   A viral reproductive cycle in which the virus takes over a host cell's synthetic machinery to replicate itself, then bursts (lyses) the host cell, releasing the new viruses. (Contrast with lysogeny.)

M

M phase   The portion of the cell cycle in which mitosis takes place.

macroevolution   [Gk. *makros*: large] Evolutionary changes occurring over long time spans and usually involving changes in many traits. (Contrast with microevolution.)

macromolecule   A giant (molecular weight > 1,000) polymeric molecule. The macromolecules are the proteins, polysaccharides, and nucleic acids.

macronutrient   In plants, a mineral element required in concentrations of at least 1 milligram per gram of plant dry matter; in animals, a mineral element required in large amounts. (Contrast with micronutrient.)

macrophage   (mac' roh faj) Phagocyte that engulfs pathogens by endocytosis.

MADS box   DNA-binding domain in many plant transcription factors that is active in development.

maintenance methylase   An enzyme that catalyzes the methylation of the new DNA strand when DNA is replicated.

major histocompatibility complex (MHC)   A complex of linked genes, with multiple alleles, that control a number of cell surface antigens that identify self and can lead to graft rejection.

malignant   Pertaining to a tumor that can grow indefinitely and/or spread from the original site of growth to other locations in the body. (Contrast with benign.)

malnutrition   A condition caused by lack of any essential nutrient.

Malpighian tubule   (mal pee' gy un) A type of protonephridium found in insects.

mantle   (1) In mollusks, a fold of tissue that covers the organs of the visceral mass and secretes the hard shell that is typical of many mollusks. (2) In geology, the Earth's crust below the solid lithospheric plates.

map unit   The distance between two genes as calculated from genetic crosses; a recombination frequency.

marine   [L. *mare*: sea, ocean] Pertaining to or living in the ocean. (Contrast with aquatic, terrestrial.)

mark–recapture method   A method of estimating population sizes of mobile organisms by capturing, marking, and releasing a sample of individuals, then capturing another sample at a later time.

mass extinction   A period of evolutionary history during which rates of extinction are much higher than during intervening times.

mass number   The sum of the number of protons and neutrons in an atom's nucleus.

mast cells   Cells, typically found in connective tissue, that release histamine in response to tissue damage.

maternal effect genes   Genes coding for morphogens that determine the polarity of the egg and larva in fruit flies. Part of a developmental cascade that includes gap genes, pair rule genes, segment polarity genes, and Hox genes.

mating type   A particular strain of a species that is incapable of sexual reproduction with another member of the same strain but capable of sexual reproduction with members of other strains of the same species.

maturational survivorship curves   *See* survivorship curves

maximum likelihood   A statistical method of determining which of two or more hypotheses (such as phylogenetic trees) best fit the observed data, given an explicit model of how the data were generated.

mechanically gated channel   A molecular channel that opens or closes in response to mechanical force applied to the plasma membrane in which it is inserted.

mechanoreceptor   A cell that is sensitive to physical movement and generates action potentials in response.

medulla   (meh dull' luh) (1) The inner, core region of an organ, as in the adrenal medulla (adrenal gland) or the renal medulla (kidneys). (2) The portion of the brainstem that connects to the spinal cord.

medusa   (plural: medusae) In cnidarians, a free-swimming, sexual life cycle stage shaped like a bell or an umbrella.

megagametophyte   In heterosporous plants, the female gametophyte; produces eggs. (Contrast with microgametophyte.)

megaphyll   The generally large leaf of a fern, horsetail, or seed plant, with several to many veins. (Contrast with microphyll.)

megaspore [Gk. *megas*: large + *spora*: to sow] In plants, a haploid spore that produces a female gametophyte.

megastrobilus In conifers, the female (seed-bearing) cone. (Contrast with microstrobilus.)

meiosis (my oh' sis) [Gk. *meiosis*: diminution] Division of a diploid nucleus to produce four haploid daughter cells. The process consists of two successive nuclear divisions with only one cycle of chromosome replication. In *meiosis I*, homologous chromosomes separate but retain their chromatids. The second division *meiosis II*, is similar to mitosis, in which chromatids separate.

melatonin A hormone released by the pineal gland. Involved in photoperiodicity and circadian rhythms.

membrane A phospholipid bylayer forming a barrier that separates the internal contents of a cell from the nonbiological environment, or enclosing the organelles within a cell. The membrane regulates the molecular substances entering or leaving a cell or organelle.

membrane potential The difference in electrical charge between the inside and the outside of a cell, caused by a difference in the distribution of ions.

membranous bone A type of bone that develops by forming on a scaffold of connective tissue. (Contrast with cartilage bone.)

memory cells Long-lived lymphocytes produced after exposure to antigen. They persist in the body and are able to mount a rapid response to subsequent exposures to the antigen.

Mendel's laws *See* independent assortment; segregation.

menopause In human females, the end of fertility and menstrual cycling.

menstruation The process by which the endometrium breaks down, and the sloughed-off tissue, including blood, flows from the body.

meristem [Gk. *meristos*: divided] Plant tissue made up of undifferentiated actively dividing cells.

meristem culture A method for the asexual propagation of plants, in which pieces of shoot apical meristem are cultured to produce plantlets.

meristem identity genes In angiosperms, a group of genes whose expression initiates flower formation, probably by switching meristem cells from a vegetative to a reproductive fate.

mesenchyme (mez' en kyme) [Gk. *mesos*: middle + *enchyma*: infusion] Embryonic or unspecialized cells derived from the mesoderm.

mesoderm [Gk. *mesos*: middle + *derma*: skin] The middle of the three embryonic germ layers first delineated during gastrulation. Gives rise to the skeleton, circulatory system, muscles, excretory system, and most of the reproductive system.

mesoglea (mez' uh glee uh) [Gk. *mesos*: middle + *gloia*, glue] A thick, gelatinous noncellular layer that separates the two cellular tissue layers of ctenophores, cnidarians, and scyphozoans.

mesophyll (mez' uh fill) [Gk. *mesos*: middle + *phyllon*: leaf] Chloroplast-containing, photosynthetic cells in the interior of leaves.

messenger RNA (mRNA) Transcript of a region of one of the strands of DNA; carries information (as a sequence of codons) for the synthesis of one or more proteins.

meta- [Gk.: between, along with, beyond] Prefix denoting a change or a shift to a new form or level; for example, as used in metamorphosis.

metabolic pathway A series of enzyme-catalyzed reactions so arranged that the product of one reaction is the substrate of the next.

metabolism (meh tab' a lizm) [Gk. *metabole*: change] The sum total of the chemical reactions that occur in an organism, or some subset of that total (as in respiratory metabolism).

metabolome The quantitative description of all the small molecules in a cell or organism.

metabotropic receptor A receptor that that indirectly alters membrane permeability to a type of ion when it combines with its ligand.

metagenomics The practice of analyzing DNA from environmental samples without isolating intact organisms.

metamorphosis (met' a mor' fo sis) [Gk. *meta*: between + *morphe*: form, shape] A change occurring between one developmental stage and another, as for example from a tadpole to a frog. (*See* complete metamorphosis, incomplete metamorphosis.)

metanephridia The paired excretory organs of annelids.

metaphase (met' a phase) The stage in nuclear division at which the centromeres of the highly supercoiled chromosomes are all lying on a plane (the metaphase plane or plate) perpendicular to a line connecting the division poles.

metapopulation A population divided into subpopulations, among which there are occasional exchanges of individuals.

methylation The addition of a methyl group (—CH$_3$) to a molecule.

MHC *See* major histocompatibility complex.

micelle A particle of lipid covered with bile salts that is produced in the duodenum and facilitates digestion and absorption of lipids.

microbiomes The diverse communities of bacteria that live on or within the body and are essential to bodily function.

microclimate A subset of climatic conditions in a small specific area, which generally differ from those in the environment at large, as in an animal's underground burrow.

microevolution Evolutionary changes below the species level, affecting allele frequencies. (Contrast with macroevolution.)

microfibril Crosslinked cellulose polymers, forming strong aggregates in the plant cell wall.

microfilament In eukaryotic cells, a fibrous structure made up of actin monomers. Microfilaments play roles in the cytoskeleton, in cell movement, and in muscle contraction.

microgametophyte In heterosporous plants, the male gametophyte; produces sperm. (Contrast with megagametophyte.)

microglia Glial cells that act as macrophages and mediators of inflammatory responses in the central nervous system.

micronutrient In plants, a mineral element required in concentrations of less than 100 micrograms per gram of plant dry matter; in animals, a mineral element required in concentrations of less than 100 micrograms per day. (Contrast with macronutrient.)

microphyll A small leaf with a single vein, found in club mosses and their relatives. (Contrast with megaphyll.)

micropyle (mike' roh pile) [Gk. *mikros*: small + *pylon*: gate] Opening in the integument(s) of a seed plant ovule through which pollen grows to reach the female gametophyte within.

microRNA A small, noncoding RNA molecule, typically about 21 bases long, that binds to mRNA to inhibit its translation.

microspore [Gk. *mikros*: small + *spora*: to sow] In plants, a haploid spore that produces a male gametophyte.

microstrobilus In conifers, male pollen-bearing cone. (Contrast with megastrobilus.)

microtubules Tubular structures found in centrioles, spindle apparatus, cilia, flagella, and cytoskeleton of eukaryotic cells. These tubules play roles in the motion and maintenance of shape of eukaryotic cells.

microvilli (sing.: microvillus) Projections of epithelial cells, such as the cells lining the small intestine, that increase their surface area.

midbrain One of the three regions of the vertebrate brain. Part of the brainstem, it serves as a relay station for sensory signals sent to the cerebral hemispheres.

middle lamella (la mell' ah) [L. *lamina*: thin sheet] A layer of polysaccharides that separates plant cells; a shared middle lamella lies outside the primary walls of the two cells.

mineral nutrients Inorganic ions required by organisms for normal growth and reproduction.

mismatch repair A mechanism that scans DNA after it has been replicated and corrects any base-pairing mismatches.

missense mutation A change in a gene's sequence that changes the amino acid at that site in the encoded protein. (Contrast with frame-shift mutation, nonsense mutation, silent mutation.)

mitochondria (my' toe kon' dree uh) (singular: mitochondrion) [Gk. *mitos*: thread + *chondros*: grain] Energy-generating organelles in eukaryotic cells that contain the enzymes of the citric acid cycle, the respiratory chain, and oxidative phosphorylation.

mitochondrial matrix The fluid interior of a mitochondrion, enclosed by the inner mitochondrial membrane.

mitosis (my toe' sis) [Gk. *mitos*: thread] Nuclear division in eukaryotes leading to the formation of two daughter nuclei, each with a chromosome complement identical to that of the original nucleus.

mitosomes Reduced structures derived from mitochondria found in some organisms.

model systems Also known as **model organisms**, these include the small group of species that are the subject of extensive research. They are organisms that adapt well to laboratory situations and findings from experiments on them can apply across a broad range of species. Classic examples include white mice and the fruit fly *Drosophila*.

moderately repetitive sequences DNA sequences repeated 10–1,000 times in the eukaryotic genome. They include the genes that code for rRNAs and tRNAs, as well as the DNA in telomeres.

Modern Synthesis An understanding of evolutionary biology that emerged in the early twentieth century as the principles of evolution were integrated with the principles of modern genetics.

modularity In evolutionary developmental biology, the principle that the molecular pathways that determine different developmental processes operate independently from one another. *See also* developmental module.

mole A quantity of a compound whose weight in grams is numerically equal to its molecular weight expressed in atomic mass units. Avogadro's number of molecules: 6.023×10^{23} molecules.

molecular clock The approximately constant rate of divergence of macromolecules from one another over evolutionary time; used to date past events in evolutionary history.

molecular evolution The scientific study of the mechanisms and consequences of the evolution of macromolecules.

molecular toolkit *See* genetic toolkit.

molecular weight The sum of the atomic weights of the atoms in a molecule.

molecule A chemical substance made up of two or more atoms joined by covalent bonds or ionic attractions.

molting The process of shedding part or all of an outer covering, as the shedding of feathers by birds or of the entire exoskeleton by arthropods.

monoclonal antibody Antibody produced in the laboratory from a clone of hybridoma cells, each of which produces the same specific antibody.

monocots Angiosperms with a single embryonic cotyledon; one of the two largest clades of angiosperms. (*See also* eudicots.)

monoculture In agriculture, a large-scale planting of a single species of domesticated crop plant.

monoecious (mo nee' shus) [Gk. *mono*: one + *oikos*: house] Pertaining to organisms in which both sexes are "housed" in a single individual that produces both eggs and sperm. (In some plants, these are found in different flowers within the same plant.) Examples include corn, peas, earthworms, hydras. (Contrast with dioecious.)

monohybrid cross A mating in which the parents differ with respect to the alleles of only one locus of interest.

monomer [Gk. *mono*: one + *meros*: unit] A small molecule, two or more of which can be combined to form oligomers (consisting of a few monomers) or polymers (consisting of many monomers).

monophyletic (mon' oh fih leht' ik) [Gk. *mono*: one + *phylon*: tribe] Pertaining to a group that consists of an ancestor and all of its descendants. (Contrast with paraphyletic, polyphyletic.)

monosaccharide A simple sugar. Oligosaccharides and polysaccharides are made up of monosaccharides.

monosomic Pertaining to an organism with one less than the normal diploid number of chromosomes.

monosynaptic reflex A neural reflex that begins in a sensory neuron and makes a single synapse before activating a motor neuron.

morphogen A diffusible substance whose concentration gradient determines a developmental pattern in embryonic animals and plants.

morphogenesis (more' fo jen' e sis) [Gk. *morphe*: form + *genesis*: origin] The development of form; the overall consequence of determination, differentiation, and growth.

morphological species concept The definition of a species as a group of individuals that look alike. (Contrast with biological species concept; lineage species concept.)

morphology (more fol' o jee) [Gk. *morphe*: form + *logos*: study, discourse] The scientific study of organic form, including both its development and function.

mortality Death, or the death rate of a population.

mosaic development Pattern of animal embryonic development in which each blastomere contributes a specific part of the adult body. (Contrast with regulative development.)

motif *See* structural motif.

motile (mo' tul) Able to move from one place to another. (Contrast with sessile.)

motor cortex The region of the cerebral cortex that contains motor neurons that directly stimulate specific muscle fibers to contract.

motor end plate The depression in the postsynaptic membrane of the neuromuscular junction where the terminals of the motor neuron sit.

motor neuron A neuron carrying information from the central nervous system to a cell that produces movement.

motor proteins Specialized proteins that use energy to change shape and move cells or structures within cells.

motor unit A motor neuron and the muscle fibers it controls.

mouth An opening through which food is taken in, located at the anterior end of a tubular gut.

mRNA *See* messenger RNA.

mucosal epithelium An epithelial cell layer containing cells that secrete mucus; found in the digestive and respiratory tracts. Also called mucosa.

mucus A viscous substance secreted by mucous membranes (e.g., mucosal epithelium). A barrier defense against pathogens in innate immunity in animals and a protective coating in many animal organ systems.

Muller's ratchet The accumulation—"ratcheting up"—of deleterious mutations in the nonrecombining genomes of asexual species.

Müllerian mimicry Convergence in appearance of two or more unpalatable species.

multifactorial The interaction of many genes and proteins with one or more factors in the environment. For example, cancer is a disease with multifactorial causes.

multipotent Having the ability to differentiate into a limited number of cell types. (Contrast with pluripotent, totipotent.)

muscle fiber A single muscle cell. In the case of skeletal muscle, a syncitial, multinucleate cell.

muscle tissue Excitable tissue that can contract through the interactions of actin and myosin; one of the four major tissue types in multicellular animals. There are three types of muscle tissue: skeletal, smooth, and cardiac.

mutagen (mute' ah jen) [L. *mutare*: change + Gk. *genesis*: source] Any agent (e.g., a chemical, radiation) that increases the mutation rate.

mutation A change in the genetic material not caused by recombination.

mutualism A type of interaction between species that benefits both species.

mycelium (my seel' ee yum) [Gk. *mykes*: fungus] In the fungi, a mass of hyphae.

mycologists Scientists who study fungi.

mycorrhiza (my' ko rye' za) (plural: mycorrhizae) [Gk. *mykes*: fungus + *rhiza*: root] An association of the root of a plant with the mycelium of a fungus.

myelin (my' a lin) Concentric layers of plasma membrane that form a sheath around some axons; myelin provides the axon with electrical insulation and increases the rate of transmission of action potentials.

myocardial infarction (MI) Blockage of an artery that carries blood to the heart muscle; a "heart attack."

MyoD The protein encoded by the *myo*blast *d*etermining gene. A transcription factor involved in the differentiation of myoblasts (muscle precursor cells).

myofibril (my' oh fy' bril) [Gk. *mys*: muscle + L. *fibrilla*: small fiber] A polymeric unit of actin or myosin in a muscle.

myoglobin (my' oh globe' in) [Gk. *mys*: muscle + L. *globus*: sphere] An oxygen-binding molecule found in muscle. Consists of a heme unit and a single globin chain; carries less oxygen than hemoglobin.

myosin One of the two contractile proteins of muscle. See also actin.

N

natural history The characteristics of a group of organisms, such as how the organisms get their food, reproduce, behave, regulate their internal environments (their cells, tissues, and organs), and interact with other organisms.

natural killer cell A type of lymphocyte that attacks virus-infected cells and some tumor cells as well as antibody-labeled target cells.

natural selection The differential contribution of offspring to the next generation by various genetic types belonging to the same population. The mechanism of evolution proposed by Charles Darwin.

nauplius (naw' plee us) [Gk. *nauplios*: shellfish] A bilaterally symmetrical larval form typical of crustaceans.

necrosis (nec roh' sis) [Gk. *nekros*: death] Premature cell death caused by external agents such as toxins.

negative feedback In regulatory systems, information that decreases a regulatory response, returning the system to the set point. (Contrast with positive feedback.)

negative regulation A type of gene regulation in which a gene is normally transcribed, and the binding of a repressor protein to the promoter prevents transcription. (Contrast with positive regulation.)

nematocyst (ne mat' o sist) [Gk. *nema*: thread + *kystis*: cell] An elaborate, threadlike structure produced by cells of jellyfishes and other cnidarians, used chiefly to paralyze and capture prey.

neoteny (knee ot' enny) [Gk. *neo*: new, recent; *tenein*, to extend] The retention of juvenile or larval traits by the fully developed adult organism.

nephron (nef' ron) [Gk. *nephros*: kidney] The functional unit of the kidney, consisting of a structure for receiving a filtrate of blood and a tubule that reabsorbs selected parts of the filtrate.

Nernst equation A mathematical statement that calculates the potential across a membrane permeable to a single type of ion that differs in concentration on the two sides of the membrane.

nerve A structure consisting of many neuronal axons and connective tissue.

nerve nets Diffuse, loosely connected aggregations of nervous tissues in certain non-bilatarian animals such as cnidarians.

nervous tissue Tissue specialized for processing and communicating information; one of the four major tissue types in multicellular animals.

net primary productivity (NPP) The rate at which energy captured by photosynthesis is incorporated into the bodies of primary producers through growth and reproduction.

neural crest cells During vertebrate neurulation, cells that migrate outward from the neural plate and give rise to connections between the central nervous system and the rest of the body.

neural network An organized group of neurons that contains three functional categories of neurons—afferent neurons, interneurons, and efferent neurons—and is capable of processing information.

neural tube An early stage in the development of the vertebrate nervous system consisting of a hollow tube created by two opposing folds of the dorsal ectoderm along the anterior–posterior body axis.

neurohormone A chemical signal produced and released by neurons that subsequently acts as a hormone.

neuromuscular junction Synapse (point of contact) where a motor neuron axon stimulates a muscle fiber cell.

neuron (noor' on) [Gk. *neuron*: nerve] A nervous system cell that can generate and conduct action potentials along an axon to a synapse with another cell.

neurotransmitter A substance produced in and released by a neuron (the presynaptic cell) that diffuses across a synapse and excites or inhibits another cell (the postsynaptic cell).

neurulation Stage in vertebrate development during which the nervous system begins to form.

neutral allele An allele that does not alter the functioning of the proteins for which it codes.

neutral theory A view of molecular evolution that postulates that most mutations do not affect the amino acid being coded for, and that such mutations accumulate in a population at rates driven by genetic drift and mutation rates.

neutron (new' tron) One of the three fundamental particles of matter (along with protons and electrons), with mass slightly larger than that of a proton and no electrical charge.

niche (nitch) [L. *nidus*: nest] The set of physical and biological conditions a species requires to survive, grow, and reproduce.

nitrate reduction The process by which nitrate (NO_3^-) is reduced to ammonia (NH_3).

nitric oxide (NO) An unstable molecule (a gas) that serves as a second messenger causing smooth muscle to relax. In the nervous system it operates as a neurotransmitter.

nitrification The oxidation of ammonia (NH_3) to nitrate (NO_3^-) in soil and seawater, carried out by chemoautotrophic bacteria (nitrifiers).

nitrogen fixation Conversion of atmospheric nitrogen gas (N_2) into a more reactive and biologically useful form (ammonia), which makes nitrogen available to living things. Carried out by **nitrogen fixers**—bacteria, some of them free-living and others living within plant roots.

nitrogenase An enzyme complex found in nitrogen-fixing bacteria that mediates the stepwise reduction of atmospheric N_2 to ammonia and which is strongly inhibited by oxygen.

nitrogenous wastes The potentially toxic nitrogen-containing end products—ammonia, urea, or uric acid—of protein and nucleic acid catabolism in animals. Eliminated from the body by excretion.

node [L. *nodus*: knob, knot] In plants, a (sometimes enlarged) point on a stem where a leaf is or was attached.

node of Ranvier A gap in the myelin sheath covering an axon; the point where the axonal membrane can fire action potentials.

nodule A specialized structure in the roots of nitrogen-fixing plants that houses nitrogen-fixing bacteria, in which oxygen is maintained at a low level by leghemoglobin.

non-REM sleep A state of deep, restorative sleep characterized by high-amplitude slow waves in the EEG. (Contrast with REM sleep.)

noncompetitive inhibitor A nonsubstrate that inhibits the activity of an enzyme by binding to a site other than its active site. (Contrast with competitive inhibitor.)

noncyclic electron transport In photosynthesis, the flow of electrons that forms ATP, NADPH, and O_2.

nondisjunction Failure of sister chromatids to separate in meiosis II or mitosis, or failure of homologous chromosomes to separate in meiosis I. Results in aneuploidy.

nonpolar Having electric charges that are evenly balanced from one end to the other. (Contrast with polar.)

nonrandom mating Selection of mates on the basis of a particular trait or group of traits.

nonsense mutation   Change in a gene's sequence that prematurely terminates translation by changing one of its codons to a stop codon.

nonsynonymous substitution   A change in a gene from one nucleotide to another that changes the amino acid specified by the corresponding codon (i.e., AGC → AGA, or serine → arginine). (Contrast with synonymous substitution.)

norepinephrine   A neurotransmitter found in the central nervous system and also at the postganglionic nerve endings of the sympathetic nervous system. Also called noradrenaline.

normal flora   Microorganisms that normally live and reproduce on or in the body without causing disease, and which form a nonspecific defense against pathogens by competing with them for space and nutrients. See also microbiota.

notochord   (no' tow kord) [Gk. *notos*: back + *chorde*: string] A flexible rod of gelatinous material serving as a support in the embryos of all chordates and in the adults of tunicates and lancelets.

nucleic acid   (new klay' ik) A polymer made up of nucleotides, specialized for the storage, transmission, and expression of genetic information. DNA and RNA are nucleic acids.

nucleic acid hybridization   A technique in which a single-stranded nucleic acid probe is made that is complementary to, and binds to, a target sequence, either DNA or RNA. The resulting double-stranded molecule is a hybrid.

nucleoid   (new' klee oid) The region that harbors the chromosomes of a prokaryotic cell. Unlike the eukaryotic nucleus, it is not bounded by a membrane.

nucleolus   (new klee' oh lus) A small, generally spherical body found within the nucleus of eukaryotic cells. The site of synthesis of ribosomal RNA.

nucleoside   A nucleotide without the phosphate group; a nitrogenous base attached to a sugar.

nucleosome   A portion of a eukaryotic chromosome, consisting of part of the DNA molecule wrapped around a group of histone molecules, and held together by another type of histone molecule. The chromosome is made up of many nucleosomes.

nucleotide   The basic chemical unit in nucleic acids, consisting of a pentose sugar, a phosphate group, and a nitrogen containing base.

nucleotide substitution   A change of one base pair to another in a DNA sequence.

nucleus   (new' klee us) [L. *nux*: kernel or nut] (1) In cells, the centrally located compartment of eukaryotic cells that is bounded by a double membrane and contains the chromosomes. (2) In the brain, an identifiable group of neurons that share common characteristics or functions.

null hypothesis   In statistics, the premise that any differences observed in an experiment are simply the result of random differences that arise from drawing two finite samples from the same population.

nutrient   A food substance; or, in the case of mineral nutrients, an inorganic element required for completion of the life cycle of an organism.

O

obligate anaerobe   An anaerobic prokaryote that cannot survive exposure to O_2.

occipital lobe   One of the four lobes of the brain's cerebral hemisphere; processes visual information.

odorant   A molecule that can bind to an olfactory receptor.

oil   A triglyceride that is liquid at room temperature. (Contrast with fat.)

Okazaki fragments   Newly formed DNA making up the lagging strand in DNA replication. DNA ligase links Okazaki fragments together to give a continuous strand.

olfaction   (ole fak' shun) [L. *olfacere*: to smell] The sense of smell.

olfactory bulb   Structure in the vertebrate forebrain that receives and processes input from olfactory receptor neurons.

olfactory receptor neurons (ORNs)   Neurons with receptors for different odorants.

oligodendrocyte   A type of glial cell that myelinates axons in the central nervous system.

oligophagous   [Gk. *oligo*: few; *phagein*, eat] An animal that feeds on a limited number of foods; generally used of insects that feed on only one or a few plant species.

oligosaccharide   A polymer containing a small number of monosaccharides.

omasum   One of the four chambers of the stomach in ruminants; concentrates food by water absorption before it enters the true stomach (abomasum).

ommatidia   [Gk. *omma*: eye] The units that make up the compound eye of some arthropods.

omnivore   [L. *omnis*: everything + *vorare*: to devour] An organism that eats both animal and plant material. (Contrast with carnivore, detritivore, herbivore.)

oncogene   [Gk. *onkos*: mass, tumor + *genes*: born] A gene that codes for a protein product that stimulates cell proliferation. Mutations in oncogenes that result in excessive cell proliferation can give rise to cancer.

one gene–one polypeptide   The idea, now known to be an oversimplification, that each gene in the genome encodes only a single polypeptide—that there is a one-to-one correspondence between genes and polypeptides.

oocyte   *See* primary oocyte, secondary oocyte.

oogenesis   (oh' eh jen e sis) [Gk. *oon*: egg + *genesis*: source] Gametogenesis leading to production of an ovum.

oogonium   (oh' eh go' nee um) (plural: oogonia) (1) In some algae and fungi, a cell in which an egg is produced. (2) In animals, the diploid progeny of a germ cell in females.

ootid   In oogenesis, the daughter cell of the second meiotic division that differentiates into the mature ovum.

open circulatory system   Circulatory system in which extracellular fluid leaves the vessels of the circulatory system, percolates between cells and through tissues, and then flows back into the circulatory system to be pumped out again. (Contrast with closed circulatory system.)

operator   The region of an operon that acts as the binding site for the repressor.

operon   A genetic unit of transcription, typically consisting of several structural genes that are transcribed together; the operon contains at least two control regions: the promoter and the operator.

opportunity cost   The sum of the benefits an animal forfeits by not being able to perform some other behavior during the time when it is performing a given behavior.

opsin   (op' sin) [Gk. *opsis*: sight] The protein portion of vertebrate visual pigments; associated with the pigment molecule 11-*cis*-retinal. See also rhodopsin.

optic chiasm   [Gk. *chiasma*: cross] Structure on the lower surface of the vertebrate brain where the two optic nerves come together.

optic nerve   The nerve that carries information from the retina of the eye to the brain.

optical isomers   Two molecular isomers that are mirror images of each other.

optimal foraging theory   The application of a cost–benefit approach to feeding behavior to identify the fitness value of feeding choices.

oral   [L. *os*: mouth] Pertaining to the mouth, or that part of the body that contains the mouth.

orbital   A region in space surrounding the atomic nucleus in which the electron is most likely to be found.

organ   [Gk. *organon*: tool] A body part, such as the heart, liver, brain, root, or leaf. Organs are composed of different tissues integrated to perform a distinct function. Organs, in turn, are integrated into organ systems.

organ identity genes   In angiosperms, genes that specify the different organs of the flower. (Compare with homeotic genes.)

organ of Corti   Structure in the inner ear that transforms mechanical forces produced from pressure waves ("sound waves") into action potentials that are sensed as sound.

organ system   An interrelated and integrated group of tissues and organs that work together in a physiological function.

organelle   (or gan el′) Any of the membrane-enclosed structures within a eukaryotic cell. Examples include the nucleus, endoplasmic reticulum, and mitochondria.

organic   (1) Pertaining to any chemical compound that contains carbon. (2) Pertaining to any aspect of living matter, e.g., to its evolution, structure, or chemistry.

organic fertilizers   Substances added to soil to improve the soil's fertility; derived from partially decomposed plant material (compost) or animal waste (manure).

organism   Any living entity.

organizer   Region of the early amphibian embryo that directs early embryonic development. Also known as the primary embryonic organizer.

organogenesis   The formation of organs and organ systems during development.

origin of replication (*ori*)   DNA sequence at which helicase unwinds the DNA double helix and DNA polymerase binds to initiate DNA replication.

orthologs   [Gk. *ortho*: true, direct] Homologous genes whose divergence can be traced to speciation events.

osmoconformer   An aquatic animal that equilibrates the osmolarity of its extracellular fluid to be the same as that of the external environment. (Contrast with osmoregulator.)

osmolarity   The concentration of osmotically active particles in a solution.

osmoregulation   Regulation of the chemical composition of the body fluids of an organism.

osmoregulator   An aquatic animal that actively regulates the osmolarity of its extracellular fluid. (Contrast with osmoconformer.)

osmosis   (oz mo′ sis) [Gk. *osmos*: to push] Movement of water across a differentially permeable membrane, from one region to another region where the water potential is more negative.

ossicle   (oss′ ick ul) [L. *os*: bone] The calcified construction unit of echinoderm skeletons.

osteoblast   (oss′ tee oh blast) [Gk. *osteon*: bone + *blastos*: sprout] A cell that lays down the protein matrix of bone.

osteoclast   (oss′ tee oh clast) [Gk. *osteon*: bone + *klastos*: broken] A cell that dissolves bone.

osteocyte   An osteoblast that has become enclosed in lacunae within the bone it has built.

outgroup   In phylogenetics, a group of organisms used as a point of reference for comparison with the groups of primary interest (the ingroup).

oval window   The flexible membrane that, when moved by the bones of the middle ear, produces pressure waves in the inner ear.

ovarian cycle   In human females, the monthly cycle of events by which eggs and hormones are produced. (Contrast with uterine cycle).

ovary   (oh′ var ee) [L. *ovum*: egg] Any female organ, in plants or animals, that produces an egg.

overtopping   Plant growth pattern in which one branch differentiates from and grows beyond the others.

oviduct   In mammals, the tube serving to transport eggs to the uterus or to the outside of the body.

oviparity   Reproduction in which eggs are released by the female and development is external to the mother's body. (Contrast with viviparity.)

ovoviviparity   Pertaining to reproduction in which fertilized eggs develop and hatch within the mother's body but are not attached to the mother by means of a placenta.

ovulation   Release of an egg from an ovary.

ovule   (oh′ vule) In plants, a structure comprising the megasporangium and the integument, which develops into a seed after fertilization.

ovum   (oh′ vum) (plural: ova) [L. egg] The female gamete.

oxidation   (ox i day′ shun) Relative loss of electrons in a chemical reaction; either outright removal to form an ion, or the sharing of electrons with substances having a greater affinity for them, such as oxygen. Most oxidations, including biological ones, are associated with the liberation of energy. (Contrast with reduction.)

oxidation–reduction (redox) reaction   A reaction in which one substance transfers one or more electrons to another substance. (*See* oxidation; reduction.)

oxidative phosphorylation   ATP formation in the mitochondrion, associated with flow of electrons through the respiratory chain.

oxygenase   An enzyme that catalyzes the addition of oxygen to a substrate from O_2.

oxytocin   A hormone released by the posterior pituitary that promotes social bonding.

ozone layer   A layer of ozone (O_3, a greenhouse gas) in the atmosphere that absorbs a high portion of the sun's potentially mutagenic ultraviology radiation.

P

pacemaker cells   Cardiac cells that can initiate action potentials without stimulation from the nervous system, allowing the heart to initiate its own contractions.

pair rule genes   In *Drosophila* (fruit fly) development, segmentation genes that divide the early embryo into units of two segments each. Part of a developmental cascade that includes maternal effect genes, gap genes, segment polarity genes, and Hox genes.

paleomagnetic dating   A method for determining the age of rocks based on properties relating to changes in the patterns of Earth's magnetism over time.

pancreas   (pan′ cree us) A gland located near the stomach of vertebrates that secretes digestive enzymes into the small intestine and releases insulin into the bloodstream.

Pangaea   (pan jee′ uh) [Gk. *pan*: all, every] The single land mass formed when all the continents came together in the Permian period.

para-   [Gk. *para*: akin to, beside] Prefix indicating association in being along side or accessory to.

parabronchi   Passages in the lungs of birds through which air flows.

paracrine   [Gk. *para*: near] Pertaining to a chemical signal, such as a hormone, that acts locally, near the site of its secretion. (Contrast with autocrine.)

parallel evolution   The repeated evolution of similar traits, especially among closely related species; facilitated by conserved developmental genes.

paralogs   Homologous genes whose divergence can be traced to gene duplication events. (Contrast with orthologs.)

paraphyletic   (par′ a fih leht′ ik) [Gk. *para*: beside + *phylon*: tribe] Pertaining to a group that consists of an ancestor and some, but not all, of its descendants. (Contrast with monophyletic, polyphyletic.)

parasite   An organism that consumes parts of an organism much larger than itself (known as its host). Parasites sometimes, but not always, kill their host.

parasympathetic nervous system   The division of the autonomic nervous system that works in opposition to the sympathetic nervous system. (Contrast with sympathetic nervous system.)

parathyroid glands   Four glands on the posterior surface of the thyroid gland that produce and release parathyroid hormone.

parathyroid hormone (PTH)   A hormone secreted by the parathyroid glands that stimulates osteoclast activity and raises blood calcium levels. Also called parathormone.

parenchyma   (pair eng′ kyma) A plant tissue composed of relatively unspecialized cells without secondary walls.

parent rock   The soil horizon consisting of the rock that is breaking down to form the soil. Also called bedrock, or the C horizon.

parental (P) generation   The individuals that mate in a genetic cross. Their offspring are the first filial (F_1) generation.

parietal cells   One of three types of secretory cell found in the gastric pits of the stomach wall. Parietal cells produce hydrochloric acid (HCl), creating an acidic environment that destroys many of the harmful microorganisms ingested with food. (*See* chief cells; mucosal epithelium.)

parietal lobe   One of four lobes of the cerebral hemisphere; processes complex stimuli and includes the primary somatosensory cortex.

parsimony   Preferring the simplest among a set of plausible explanations of any phenomenon.

parthenocarpy   Formation of fruit from a flower without fertilization.

parthenogenesis   [Gk. *parthenos*: virgin] Production of an organism from an unfertilized egg.

particulate theory   In genetics, the theory that genes are physical entities that retain their identities after fertilization.

passive transport   Diffusion across a membrane; may or may not require a channel or carrier protein. (Contrast with active transport.)

patch clamping   Technique for isolating a tiny patch of membrane to allow the study of ion movement through a particular channel.

pathogen   (path' o jen) [Gk. *pathos*: suffering + *genesis*: source] An organism that causes disease.

pattern formation   In animal embryonic development, the organization of differentiated tissues into specific structures such as wings.

pedigree   The pattern of transmission of a genetic trait within a family.

pelagic zone   [Gk. *pelagos*: deep sea] The open ocean; a marine life zone.

penetrance   The proportion of individuals with a particular genotype that show the expected phenotype.

penis   An accessory sex organ of male animals that enables the male to deposit sperm in the female's reproductive tract.

pentaradial symmetry   Symmetry in five or multiples of five; a feature of adult echinoderms.

pentose   [Gk. *penta*: five] A sugar containing five carbon atoms.

PEP carboxylase   The enzyme that combines carbon dioxide with PEP to form a 4-carbon dicarboxylic acid at the start of C_4 photosynthesis or of crassulacean acid metabolism (CAM).

pepsin   [Gk. *pepsis*: digestion] An enzyme in gastric juice that digests protein.

pepsinogen   Inactive secretory product that is converted into pepsin by low pH or by enzymatic action.

peptide hormones   Relatively large hormone molecules made up of amino acids; encoded by genes and produced by translation.

peptide linkage   The bond between amino acids in a protein; formed between a carboxyl group and amino group (—CO—NH—) with the loss of water molecules.

peptidoglycan   The cell wall material of many bacteria, consisting of a single enormous molecule that surrounds the entire cell.

peptidyl transferase   A catalytic function of the large ribosomal subunit that consists of two reactions: breaking the bond between an amino acid and its tRNA in the P site, and forming a peptide bond between that amino acid and the amino acid attached to the tRNA in the A site.

per capita birth rate (*b*)   In population growth models, the number of offspring that an average individual produces in some time interval.

per capita death rate (*d*)   In population growth models, the average individual's chance of dying in some time interval.

per capita growth rate (*r*)   In population models, the average individual's contribution to total population growth rate.

perennial   (per ren' ee al) [L. *per*: throughout + *annus*: year] A plant that survives from year to year. (Contrast with annual, biennial.)

perfect flower   A flower with both stamens and carpels; a hermaphroditic flower. (Contrast with imperfect flower.)

pericycle   [Gk. *peri*: around + *kyklos*: ring or circle] In plant roots, tissue just within the endodermis, but outside of the root vascular tissue. Meristematic activity of pericycle cells produces lateral root primordia.

periderm   The outer tissue of the secondary plant body, consisting primarily of cork.

period   (1) A category in the geological time scale. (2) The duration of a single cycle in a cyclical event, such as a circadian rhythm.

peripheral membrane proteins   Proteins associated with but not embedded within the plasma membrane. (Contrast with integral membrane proteins.)

peripheral nervous system (PNS)   The portion of the nervous system that transmits information to and from the central nervous system, consisting of neurons that extend or reside outside the brain or spinal cord and their supporting cells. (Contrast with central nervous system.)

peristalsis   (pair' i stall' sis) Wavelike muscular contractions proceeding along a tubular organ, propelling the contents along the tube.

peritoneum   The mesodermal lining of the body cavity in coelomate animals.

peroxisome   An organelle that houses reactions in which toxic peroxides are formed and then converted to water.

petal   [Gk. *petalon*: spread out] In an angiosperm flower, a sterile modified leaf, nonphotosynthetic, frequently brightly colored, and often serving to attract pollinating insects.

petiole   (pet' ee ole) [L. *petiolus*: small foot] The stalk of a leaf.

P$_{fr}$ *See* phytochrome.

pH   The negative logarithm of the hydrogen ion concentration; a measure of the acidity of a solution. A solution with pH = 7 is said to be neutral; pH values higher than 7 characterize basic solutions, while acidic solutions have pH values less than 7.

phage   (fayj) *See* bacteriophage.

phagocyte   [Gk. *phagein*: to eat + *kystos*: sac] One of two major classes of white blood cells; one of the nonspecific defenses of animals; ingests invading microorganisms by **phagocytosis**.

pharmacogenomics   The study of how an individual's genetic makeup affects his or her response to drugs or other agents, with the goal of predicting the effectiveness of different treatment options.

pharming   The use of genetically modified animals to produce medically useful products in their milk.

pharynx   [Gk. throat] The part of the gut between the mouth and the esophagus.

phenotype   (fee' no type) [Gk. *phanein*: to show] The observable properties of an individual resulting from both genetic and environmental factors. (Contrast with genotype.)

phenotypic plasticity   *See* developmental plasticity.

pheromone   (feer' o mone) [Gk. *pheros*: carry + *hormon*: excite, arouse] A chemical substance used in communication between organisms of the same species.

phloem   (flo' um) [Gk. *phloos*: bark] In vascular plants, the vascular tissue that transports sugars and other solutes from sources to sinks.

phosphate group   The functional group —OPO$_3$H$_2$.

phosphodiester linkage   The connection in a nucleic acid strand, formed by linking two nucleotides.

phospholipid   A lipid containing a phosphate group; an important constituent of cellular membranes. (*See* lipid.)

phospholipid bilayer   The basic structural unit of biological membranes; a sheet of phospholipids two molecules thick in which the phospholipids are lined up with their hydrophobic "tails" packed tightly together and their hydrophilic, phosphate-containing "heads" facing outward. Also called lipid bilayer.

phosphorylation   Addition of a phosphate group.

photic zone   The life zone in lakes and oceans that is penetrated by light and therefore supports photosynthetic organisms.

photoautotroph   An organism that obtains energy from light and carbon from carbon dioxide. (Contrast with chemolithotroph, chemoheterotroph, photoheterotroph.)

photoheterotroph   An organism that obtains energy from light but must obtain its carbon from organic compounds. (Contrast with chemoautotroph, chemoheterotroph, photoautotroph.)

photomorphogenesis   In plants, a process by which physiological and developmental events are controlled by light.

photon   (foe' ton) [Gk. *photos*: light] A quantum of visible radiation; a "packet" of light energy.

photoperiodism Control of an organism's physiological or behavioral responses by the length of the day or night (the **photoperiod**).

photophosphorylation Mechanism for ATP formation in chloroplasts in which electron transport is coupled to the transport of hydrogen ions (protons, H⁺) across the thylakoid membrane. Compare with chemiosmosis.

photoreceptors (1) In plants, pigments that trigger a physiological response when they absorb a photon. (2) In animals, the sensory receptor cells that sense and respond to light energy. (See cone cells; rod cells.)

photorespiration Light-driven uptake of oxygen and release of carbon dioxide, the carbon being derived from the early reactions of photosynthesis.

photosynthesis (foe tow sin' the sis) [Gk.: creating from light] Metabolic processes carried out by green plants and some microorganisms by which visible light is trapped and the energy used to synthesize compounds such as ATP and glucose.

photosystem [Gk. *phos*: light + *systema*: assembly] A light-harvesting complex in the chloroplast thylakoid composed of pigments and proteins. **Photosystem I** absorbs light at 700 nm, passing electrons to ferrodoxin and from there to NADPH. **Photosystem II** absorbs light at 680 nm and passes electrons to the electron transport chain in the chloroplast.

phototropism [Gk. *photos*: light + *trope*: turning] A directed plant growth response to light.

phycobilin Photosynthetic pigment that absorbs red, yellow, orange, and green light and is found in cyanobacteria and some red algae.

phylogeny (fy loj' e nee) [Gk. *phylon*: tribe, race + *genesis*: source] The evolutionary history of a particular group of organisms or their genes. A **phylogenetic tree** is a graphic representation of these lines of evolutionary descent.

physiological survivorship curves *See* survivorship curves.

physiology (fiz' ee ol' o jee) [Gk. *physis*: natural form] The scientific study of the functions of living organisms and the individual organs, tissues, and cells of which they are composed.

phytoalexins Substances toxic to pathogens, produced by plants in response to fungal or bacterial infection.

phytochrome (fy' tow krome) [Gk. *phyton*: plant + *chroma*: color] A plant pigment regulating a large number of developmental and other phenomena in plants. It has two isomers: P$_r$, which absorbs red light, and P$_{fr}$, which absorbs far red light. P$_{fr}$ is the active form.

phytomers In plants, the repeating modules that compose a shoot, each consisting of one or more leaves, attached to the stem at a node; an internode; and one or more axillary buds.

phytoplankton Photosynthetic floating organisms. (*See* plankton.)

phytoremediation A form of bioremediation that uses plants to clean up environmental pollution.

pigment A substance that absorbs visible light.

piloting A form of navigation in which an animal finds its way by remembering landmarks in its environment.

pineal gland Gland located between the cerebral hemispheres that secretes melatonin.

pinocytosis Endocytosis by a cell of liquid containing dissolved substances.

pistil [L. *pistillum*: pestle] The structure of an angiosperm flower within which the ovules are borne. May consist of a single carpel, or of several carpels fused into a single structure. Usually differentiated into ovary, style, and stigma.

pith In plants, relatively unspecialized tissue found within a cylinder of vascular tissue.

pituitary gland A small gland attached to the base of the brain in vertebrates. Its hormones control the activities of other glands. Also known as the hypophysis.

placenta (pla sen' ta) The organ in female mammals that provides for the nourishment of the fetus and elimination of the fetal waste products.

plankton Aquatic organisms that float in the water column, dependent on currents and wind for movement. Plankton include many protists, some algae, and larval animals. (See also phytoplankton.)

planula (plan' yew la) [L. *planum*: flat] A free-swimming, ciliated larval form typical of the cnidarians.

plaque (plack) [Fr.: a metal plate or coin] (1) A circular clearing in a layer (lawn) of bacteria growing on the surface of a nutrient agar gel. (2) An accumulation of prokaryotic organisms on tooth enamel. Acids produced by these microorganisms cause tooth decay. (3) A region of arterial wall invaded by fibroblasts and fatty deposits.

plasma (plaz' muh) The liquid portion of blood, in which blood cells and other particulates are suspended.

plasma cell An antibody-secreting cell that develops from a B cell; the effector cell of the humoral immune system.

plasma membrane The membrane that surrounds the cell, regulating the entry and exit of molecules and ions. Every cell has a plasma membrane, and it is often called the cell membrane.

plasmid A DNA molecule distinct from the chromosome(s); that is, an extrachromosomal element; found in many bacteria. May replicate independently of the chromosome.

plasmodesmata (singular: plasmodesma) [Gk. *plassein*: to mold + *desmos*: band] Cytoplasmic strands connecting two adjacent plant cells.

plasmogamy The fusion of the cytoplasm of two cells. (Contrast with karyogamy.)

plastid A class of plant cell organelles that includes the chloroplast, which houses biochemical pathways for photosynthesis.

plate tectonics [Gk. *tekton*: builder] The scientific study of the structure and movements of Earth's lithospheric plates, which are the cause of continental drift.

platelet A membrane-bounded body without a nucleus, arising as a fragment of a cell in the bone marrow of mammals. Important to blood-clotting action.

pleiotropy (plee' a tro pee) [Gk. *pleion*: more] The determination of more than one character by a single gene.

pleural membrane [Gk. *pleuras*: rib, side] The membrane lining the outside of the lungs and the walls of the thoracic cavity. Inflammation of these membranes is a condition known as pleurisy.

pluripotent [L. *pluri*: many + *potens*: powerful] Having the ability to form all of the cells in the body. (Contrast with multipotent, totipotent.)

podocytes Cells of Bowman's capsule of the nephron that cover the capillaries of the glomerulus, forming filtration slits.

point mutation A mutation that results from the gain, loss, or substitution of a single nucleotide.

polar A molecule with separate and opposite electric charges at two ends, or poles; the water molecule (H_2O) is the most prevalent example. (Contrast with nonpolar.)

polar body A nonfunctional nucleus produced by meiosis during oogenesis.

polar covalent bond A covalent bond in which the electrons are drawn to one nucleus more than the other, resulting in an unequal distribution of charge.

polar nuclei In angiosperms, the two nuclei in the central cell of the megagametophyte; following fertilization they give rise to the endosperm.

polarity (1) In chemistry, the property of unequal electron sharing in a covalent bond that defines a polar molecule. (2) In development, the difference between one end of an organism or structure and the other.

pollen [L. *pollin*: fine flour] In seed plants, microscopic grains that contain the male gametophyte (microgametophyte) and gamete (microspore).

pollen tube A structure that develops from a pollen grain through which sperm are released into the megagametophyte.

pollination The process of transferring pollen from an anther to the stigma of a pistil in an angiosperm or from a strobilus to an ovule in a gymnosperm.

poly- [Gk. *poly*: many] A prefix denoting multiple entities.

poly A tail A long sequence of adenine nucleotides (50–250) added after transcription to the 3' end of most eukaryotic mRNAs.

polyandry   Mating system in which one female mates with multiple males.

polygyny   Mating system in which one male mates with multiple females.

polymer   [Gk. *poly*: many + *meros*: unit] A large molecule made up of similar or identical subunits called monomers. (Contrast with monomer.)

polymerase chain reaction (PCR)   An enzymatic technique for the rapid production of millions of copies of a particular stretch of DNA where only a small amount of the parent molecule is available.

polymorphic   (pol' lee mor' fik) [Gk. *poly*: many + *morphe*: form, shape] Coexistence in a population of two or more distinct traits.

polyp   (pah' lip) [Gk. *poly*: many + *pous*: foot] In cnidarians, a sessile, asexual life cycle stage.

polypeptide   A large molecule made up of many amino acids joined by peptide linkages. Large polypeptides are called proteins.

polyphyletic   (pol' lee fih leht' ik) [Gk. *poly*: many + *phylon*: tribe] Pertaining to a group that consists of multiple distantly related organisms, and does not include the common ancestor of the group. (Contrast with monophyletic, paraphyletic.)

polyploid   (pol' lee ploid ee) Possessing more than two entire sets of chromosomes.

polyribosome (polysome)   A complex consisting of a threadlike molecule of messenger RNA and several (or many) ribosomes. The ribosomes move along the mRNA, synthesizing polypeptide chains as they proceed.

polysaccharide   A macromolecule composed of many monosaccharides (simple sugars). Common examples are cellulose and starch.

pons   [L. *pons*: bridge] Region of the brainstem anterior to the medulla.

pool   The total amount of an element in a given compartment of the biosphere.

population   Any group of organisms coexisting at the same time and in the same place and capable of interbreeding with one another.

population bottleneck   A period during which only a few individuals of a normally large population survive.

population density   The number of individuals in a population per unit of area or volume.

population dynamics   The patterns and processes of change in populations.

population genetics   The study of genetic variation and its causes within populations.

population size   The total number of individuals in a population.

positional information   In development, the basis of the spatial sense that induces cells to differentiate as appropriate for their location within the developing organism; often comes in the form of a morphogen gradient.

positive cooperativity   Occurs when a molecule can bind several ligands and each one that binds alters the conformation of the molecule so that it can bind the next ligand more easily. The binding of four molecules of O_2 by hemoglobin is an example of positive cooperativity.

positive feedback   In regulatory systems, information that amplifies a regulatory response, increasing the deviation of the system from the set point. (Contrast with negative feedback.)

positive regulation   A form of gene regulation in which a regulatory macromolecule is needed to turn on the transcription of a structural gene; in its absence, transcription will not occur. (Contrast with negative regulation.)

positive selection   Natural selection that acts to establish a trait that enhances survival in a population. (Contrast with purifying selection.)

post-   [L. *postere*: behind, following after] Prefix denoting something that comes after.

postabsorptive state   State in which no food remains in the gut and thus no nutrients are being absorbed. (Contrast with absorptive state.)

posterior   Toward or pertaining to the rear. (Contrast with anterior.)

posterior pituitary   A portion of the pituitary gland derived from neural tissue; involved in the storage and release of antidiuretic hormone and oxytocin.

postsynaptic cell   The cell that receives information from a neuron at a synapse. (Contrast with presynaptic cell.)

postzygotic isolating mechanisms   Barriers to the reproductive process that occur after the union of the nuclei of two gametes. (Contrast with prezygotic isolating mechanisms.)

potential energy   Energy not doing work but with the potential to do so, such as the energy stored in chemical bonds. (Contrast with kinetic energy.)

P$_r$ *See* phytochrome.

pre-mRNA (precursor mRNA)   Initial gene transcript before it is modified to produce functional mRNA. Also known as the primary transcript.

Precambrian   The first and longest period of geological time, during which life originated.

precapillary sphincter   A cuff of smooth muscle that can shut off the blood flow to a capillary bed.

predator   An organism that kills and eats other organisms.

pressure flow model   An effective model for phloem transport in angiosperms. It holds that sieve element transport is driven by an osmotically generated pressure gradient between source and sink.

pressure potential (Ψ_p)   The hydrostatic pressure of an enclosed solution in excess of the surrounding atmospheric pressure.

(Contrast with solute potential, water potential.)

presynaptic cell   The neuron that transmits information to another cell at a synapse. (Contrast with postsynaptic cell.)

prey   [L. *praeda*: booty] An organism consumed by a predator as an energy source.

prezygotic isolating mechanisms   Barriers to the reproductive process that occur before the union of the nuclei of two gametes (Contrast with postzygotic isolating mechanisms.)

primary active transport   Active transport in which ATP is hydrolyzed, yielding the energy required to transport an ion or molecule against its concentration gradient. (Contrast with secondary active transport.)

primary cell wall   In plant cells, a structure that forms at the middle lamella after cytokinesis, made up of cellulose microfibrils, hemicelluloses, and pectins. (Contrast with secondary cell wall.)

primary consumer   An organism (herbivore) that eats plant tissues.

primary endosymbiosis   The engulfment of a cyanobacterium by a larger eukaryotic cell that gave rise to the first photosynthetic eukaryotes with chloroplasts.

primary growth   In plants, growth that is characterized by the lengthening of roots and shoots and by the proliferation of new roots and shoots through branching. (Contrast with secondary growth.)

primary immune response   The first response of the immune system to an antigen, involving recognition by lymphocytes and the production of effector cells and memory cells. (Contrast with secondary immune response.)

primary meristem   Meristem that produces the tissues of the primary plant body.

primary motor cortex   *See* motor cortex

primary oocyte   (oh' eh site) [Gk. *oon*: egg + *kytos*: container] The diploid progeny of an oogonium. In many species, a primary oocyte enters prophase of the first meiotic division, then remains in developmental arrest for a long time before resuming meiosis to form a secondary oocyte and a polar body.

primary plant body   That part of a plant produced by primary growth. Consists of all the *nonwoody* parts of a plant; many herbaceous plants consist entirely of a primary plant body. (Contrast with secondary plant body.)

primary producer   A photosynthetic or chemosynthetic organism that synthesizes complex organic molecules from simple inorganic ones.

primary sex determination   Genetic determination of gametic sex, male or female. (Contrast with secondary sex determination.)

primary somatosensory cortex   *See* somatosensory cortex.

primary spermatocyte   The diploid progeny of a spermatogonium; undergoes the first meiotic division to form secondary spermatocytes.

primary structure The specific sequence of amino acids in a protein.

primary succession Succession of ecological communities that begins in an area devoid of life, such as on recently exposed glacial till or lava flows. (Contrast with secondary succession.)

primase An enzyme that catalyzes the synthesis of a primer for DNA replication.

primer Strand of nucleic acid, usually RNA, that is the necessary starting material for the synthesis of a new DNA strand, which is synthesized from the 3′ end of the primer.

primordium (plural: primordia) [L. origin] The most rudimentary stage of an organ or other part.

pro- [L.: first, before, favoring] A prefix often used in biology to denote a developmental stage that comes first or an evolutionary form that appeared earlier than another. For example, prokaryote, prophase.

probe A segment of single-stranded nucleic acid used to identify DNA molecules containing the complementary sequence.

procambium Primary meristem that produces the vascular tissue.

procedural memory Memory of motor tasks.; these memories cannot be consciously recalled or described. (Contrast with declarative memory.)

processive Pertaining to an enzyme that catalyzes many reactions each time it binds to a substrate, as DNA polymerase does during DNA replication.

products The molecules that result from the completion of a chemical reation.

progesterone [L. pro: favoring + gestare: to bear] A female sex hormone that maintains pregnancy.

prokaryotes Unicellular organisms that do not have nuclei or other membrane-enclosed organelles. Includes Bacteria and Archaea. (Contrast with eukaryotes.)

prolactin A hormone released by the anterior pituitary, one of whose functions is the stimulation of milk production in female mammals.

prometaphase The phase of nuclear division that begins with the disintegration of the nuclear envelope.

promoter A DNA sequence to which RNA polymerase binds to initiate transcription.

prop roots Adventitious roots in some monocots that function as supports for the shoot.

prophage (pro′ fayj) The noninfectious units that are linked with the chromosomes of the host bacteria and multiply with them but do not cause dissolution of the cell. Prophage can later enter into the lytic phase to complete the virus life cycle.

prophase (pro′ phase) The first stage of nuclear division, during which chromosomes condense from diffuse, threadlike material to discrete, compact bodies.

prostaglandin Any one of a group of specialized lipids with hormone-like functions. It is not clear that they act at any considerable distance from the site of their production.

prostate gland In male humans, surrounds the urethra at its junction with the vas deferens; supplies an acid-neutralizing fluid to the semen.

prosthetic group Any nonprotein portion of an enzyme.

proteases Digestive enzymes that digest proteins.

proteasome In the eukaryotic cytoplasm, a huge protein structure that binds to and digests cellular proteins that have been tagged by ubiquitin.

protein (pro′ teen) [Gk. protos: first] Long-chain polymer of amino acids with twenty different common side chains. Occurs with its polymer chain extended in fibrous proteins, or coiled into a compact macromolecule in enzymes and other globular proteins. The component amino acids are encoded in the triplets of messenger RNA, and proteins are the products of genes.

protein kinase (kye′ nase) An enzyme that catalyzes the addition of a phosphate group from ATP to a target protein.

protein kinase cascade A series of reactions in response to a molecular signal, in which a series of protein kinases activate one another in sequence, amplifying the signal at each step.

proteoglycan A glycoprotein containing a protein core with attached long, linear carbohydrate chains.

proteolysis [protein + Gk. lysis: break apart] An enzymatic digestion of a protein or polypeptide.

proteome The complete set of proteins that can be made by an organism. Because of alternative splicing of pre-mRNA, the number of proteins that can be made is usually much larger than the number of protein-coding genes present in the organism's genome.

prothrombin The inactive form of thrombin, an enzyme involved in blood clotting.

protoderm Primary meristem that gives rise to the plant epidermis.

proton (pro′ ton) [Gk. protos: first, before] (1) A subatomic particle with a single positive charge. The number of protons in the nucleus of an atom determine its element. (2) A hydrogen ion, H+.

proton pump An active transport system that uses ATP energy to move hydrogen ions across a membrane, generating an electric potential.

proton-motive force Force generated across a membrane having two components: a chemical potential (difference in proton concentration) plus an electrical potential due to the electrostatic charge on the proton.

protonephridium The excretory organ of flatworms, made up of a tubule and a flame cell.

protoplast The living contents of a plant cell; the plasma membrane and everything contained within it.

provirus Double-stranded DNA made by a virus that is integrated into the host's chromosome and contains promoters that are recognized by the host cell's transcription apparatus.

proximal convoluted tubule The initial segment of a renal tubule, closest to the glomerulus. (Compare with distal convoluted tubule.)

proximal Near the point of attachment or other reference point. (Contrast with distal.)

proximate causes The immediate genetic, physiological, neurological, and developmental mechanisms responsible for a behavior or morphology. (Contrast with ultimate cause.)

pseudocoelomate (soo′ do see′ low mate) [Gk. pseudes: false + koiloma: cavity] Having a body cavity, called a pseudocoel, consisting of a fluid-filled space in which many of the internal organs are suspended, but which is enclosed by mesoderm only on its outside.

pseudogene [Gk. pseudes: false] A DNA segment that is homologous to a functional gene but is not expressed because of changes to its sequence or changes to its location in the genome.

pseudopod (soo′ do pod) [Gk. pseudes: false + podos: foot] A temporary, soft extension of the cell body that is used in location, attachment to surfaces, or engulfing particles.

pulmonary [L. pulmo: lung] Pertaining to the lungs.

pulmonary circuit The portion of the circulatory system by which blood is pumped from the heart to the lungs or gills for oxygenation and back to the heart for distribution. (Contrast with systemic circuit.)

pulmonary valve A one-way valve between the right ventricle of the heart and the pulmonary artery that prevents backflow of blood into the ventricle when it relaxes.

Punnett square Method of predicting the results of a genetic cross by arranging the gametes of each parent at the edges of a square.

pupa (pew′ pa) [L. pupa: doll, puppet] In certain insects (the Holometabola), the encased developmental stage between the larva and the adult.

pupil The opening in the vertebrate eye through which light passes.

purifying selection The elimination by natural selection of detrimental characters from a population. (Contrast with positive selection.)

purine (pure′ een) One of the two types of nitrogenous bases in nucleic acids. Each of the purines—adenine and guanine—pairs with a specific pyrimidine.

Purkinje fibers Specialized heart muscle cells that conduct excitation throughout the ventricular muscle.

pyrimidine (pe rim′ a deen) One of the two types of nitrogenous bases in nucleic acids. Each of the pyrimidines—cytosine, thymine, and uracil—pairs with a specific purine.

pyrogen [Gk.: *pry, fire;*] Molecule that produces a rise in body temperature (fever); may be produced by an invading pathogen or by cells of the immune system in response to infection.

pyruvate The ionized form of pyruvic acid, a three-carbon acid; the end product of glycolysis and the raw material for the citric acid cycle.

pyruvate oxidation Conversion of pyruvate to acetyl CoA and CO_2 that occurs in the mitochondrial matrix in the presence of O_2.

Q

Q$_{10}$ A value that compares the rate of a biochemical process or reaction over 10°C temperature ranges. A process that is not temperature-sensitive has a Q_{10} of 1; values of 2 or 3 mean the reaction speeds up as temperature increases.

qualitative Based on observation of an unmeasured quality of a trait, as in brown vs. blue.

quantitative Based on numerical values obtained by measurement, as in quantitative data.

quantitative trait loci A set of genes determining a complex character (trait) that exhibits quantitative variation (variation in amount rather than in kind).

quaternary structure The specific three-dimensional arrangement of protein subunits. Contrast with primary, secondary, tertiary structure.

quorum sensing The use of chemical communication signals to trigger density-linked activities such as biofilm formation in prokaryotes.

R

R group The distinguishing group of atoms of a particular amino acid; also known as a side chain.

***r*-strategist** A species whose life history strategy allows for a high intrinsic rate of population increase (*r*). (Contrast with *K*-strategist.)

radial symmetry The condition in which any two halves of a body are mirror images of each other, providing the cut passes through the center; a cylinder cut lengthwise down its center displays this form of symmetry.

radiation The transfer of heat from warmer objects to cooler ones via the exchange of infrared radiation. *See also* electromagnetic radiation; evolutionary radiation.

radicle An embryonic root.

radioisotope A radioactive isotope of an element. Examples are carbon-14 (^{14}C) and hydrogen-3, or tritium (^{3}H).

radiometric dating A method for determining the age of objects such as fossils and rocks based on the decay rates of radioactive isotopes.

rapid eye movement sleep *See* REM sleep.

reactant A chemical substance that enters into a chemical reaction with another substance.

reaction center A group of electron transfer proteins that receive energy from light-absorbing pigments and convert it to chemical energy by redox reactions.

realized niche A species' niche as defined by its interactions with other species. (Contrast with fundamental niche.)

receptive field The area of visual space that activates a particular cell in the visual system.

receptor *See* receptor protein, sensory receptor cell.

receptor potential The change in the resting potential of a sensory cell when it is stimulated.

receptor protein A protein that can bind to a specific molecule, or detect a specific stimulus, within the cell or in the cell's external environment.

receptor-mediated endocytosis Endocytosis initiated by macromolecular binding to a specific membrane receptor.

recessive In genetics, an allele that does not determine phenotype in the presence of a dominant allele. (Contrast with dominance.)

reciprocal crosses A pair of matings in one of which a female of genotype A mates with a male of genotype B and in the other of which a female of genotype B mates with a male of genotype A.

recognition sequence *See* restriction site.

recombinant Pertaining to an individual, meiotic product, or chromosome in which genetic materials originally present in two individuals end up in the same haploid complement of genes.

recombinant DNA A DNA molecule made in the laboratory that is derived from two or more genetic sources.

recombinant frequency The proportion of offspring of a genetic cross that have phenotypes different from the parental phenotypes due to crossing over between linked genes during gamete formation.

recombination frequency The proportion of offspring of a genetic cross that have phenotypes different from the parental phenotypes due to crossing over between linked genes during gamete formation.

reconciliation ecology The practice of making exploited lands more biodiversity-friendly. Compare with restoration ecology.

rectum The terminal portion of the gut, ending at the anus.

redox reaction A chemical reaction in which one reactant is oxidized (loses electrons) and the other is reduced (gains electrons). Short for reduction–oxidation reaction.

reduction Gain of electrons by a chemical reactant. (Contrast with oxidation.)

refractory period The time interval after an action potential during which another action potential cannot be elicited from an excitable membrane.

regeneration The development of a complete individual from a fragment of an organism.

regulative development A pattern of animal embryonic development in which the fates of the first blastomeres are not absolutely fixed. (Contrast with mosaic development.)

regulatory gene A gene that codes for a protein (or RNA) that in turn controls the expression of another gene.

regulatory sequence A DNA sequence to which the protein product of a regulatory gene binds.

regulatory system A system that uses feedback information to maintain a physiological function or parameter at an optimal level. (Contrast with controlled system.)

regulatory T cells (Treg) The class of T cells that mediates tolerance to self antigens.

reinforcement The evolution of enhanced reproductive isolation between populations due to natural selection for greater isolation.

Reissner's membrane *See* tectonic membrane.

releaser Sensory stimulus that triggers performance of a stereotyped behavior pattern.

REM (rapid-eye-movement) sleep A sleep state characterized by vivid dreams, skeletal muscle relaxation, and rapid eye movements. (Contrast with non-REM sleep.)

renal [L. *renes:* kidneys] Relating to the kidneys.

renal tubule A structural unit of the kidney that collects filtrate from the blood, reabsorbs specific ions, nutrients, and water and returns them to the blood, and concentrates excess ions and waste products such as urea for excretion from the body.

renin An enzyme released from the kidneys in response to a drop in the glomerular filtration rate. Together with angiotensin converting enzyme, converts an inactive protein in the blood into angiotensin.

replication The duplication of genetic material.

replication complex The close association of several proteins operating in the replication of DNA.

replication fork A point at which a DNA molecule is replicating. The fork forms by the unwinding of the parent molecule.

replicon   A region of DNA replicated from a single origin of replication.

reporter gene   A genetic marker included in recombinant DNA to indicate the presence of the recombinant DNA in a host cell.

repressor   A protein encoded by a regulatory gene that can bind to a promoter and prevent transcription of the associated gene. (Contrast with activator.)

reproductive isolation   Condition in which two divergent populations are no longer exchanging genes. Can lead to speciation.

rescue effect   The process by which individuals moving between subpopulations of a metapopulation may prevent declining subpopulations from becoming extinct.

residence time   The length of time a chemical element (e.g., carbon or nitrogen) remains in a given compartment of the ecosystem (e.g., in an organic body, in soil, in the atmosphere).

residual volume (RV)   In tidal ventilation, the dead space that remains in the lungs at the end of exhalation.

resistance (R) genes   Plant genes that confer resistance to specific strains of pathogens.

resource   Something in the environment required by an organism for its maintenance and growth that is consumed in the process of being used.

resource partitioning   A situation in which selection pressures resulting from interspecific competition cause changes in the ways in which the competing species use the limiting resource, thereby allowing them to coexist.

respiration   (res pi ra' shun) [L. *spirare*: to breathe] (1) Cellular respiration. (2) Breathing.

respiratory chain   The terminal reactions of cellular respiration, in which electrons are passed from NAD or FAD, through a series of intermediate carriers, to molecular oxygen, with the concomitant production of ATP.

respiratory gases   Oxygen (O_2) and carbon dioxide (CO_2); the gases that an animal must exchange between its internal body fluids and the outside medium (air or water).

resting potential   The membrane potential of a living cell at rest. In cells at rest, the interior is negative to the exterior. (Contrast with action potential.)

restoration ecology   The science and practice of restoring damaged or degraded ecosystems.

restriction enzyme   Any of a type of enzyme that cleaves double-stranded DNA at specific sites; extensively used in recombinant DNA technology. Also called a restriction endonuclease.

restriction fragment length polymorphism   See RFLP.

restriction point (R)   The specific time during G1 of the cell cycle at which the cell becomes committed to undergo the rest of the cell cycle.

restriction site   A specific DNA base sequence that is recognized and acted on by a restriction endonuclease.

reticular activating system   A central region of the vertebrate brainstem that includes complex fiber tracts conveying neural signals between the forebrain and the spinal cord, with collateral fibers to a variety of nuclei that are involved in autonomic functions, including arousal from sleep.

reticulum   One of the four chambers of the ruminant stomach. Along with the rumen, where food is partially digested with the assistance of gut bacteria.

retina   (rett' in uh) [L. *rete*: net] The light-sensitive layer of cells in the vertebrate or cephalopod eye.

retrotransposons   Mobile genetic elements that are reverse transcribed into RNA as part of their transfer mechanism. (Contrast with DNA transposons.)

retrovirus   An RNA virus that contains reverse transcriptase. Its RNA serves as a template for cDNA production, and the cDNA is integrated into a chromosome of the host cell.

reverse genetics   Method of genetic analysis in which a phenotype is first related to a DNA variation, then the protein involved is identified.

reverse transcriptase   An enzyme that catalyzes the production of DNA (cDNA), using RNA as a template; essential to the reproduction of retroviruses.

reversion mutation   A second- or third-round mutation that reverts the DNA to its original sequence or to a new sequence that results in a non-mutant phenotype.

RFLP   Restriction fragment length polymorphism, the coexistence of two or more patterns of restriction fragments resulting from underlying differences in DNA sequence.

rhizoids   (rye' zoids) [Gk. root] Hairlike extensions of cells in mosses, liverworts, and a few vascular plants that serve the same function as roots and root hairs in vascular plants. The term is also applied to branched, rootlike extensions of some fungi and algae.

rhizome   (rye' zome) An underground stem (as opposed to a root) that runs horizontally beneath the ground.

rhodopsin   A vertebrate visual pigment involved in transducing photons of light into changes in the membrane potential of certain photoreceptor cells.

ribonucleic acid   See RNA.

ribose   A five-carbon sugar in nucleotides and RNA.

ribosomal RNA (rRNA)   Several species of RNA that are incorporated into the ribosome. Involved in peptide bond formation.

ribosome   A small particle in the cell that is the site of protein synthesis.

ribozyme   An RNA molecule with catalytic activity.

ribulose bisphosphate carboxylase/oxygenase   See rubisco.

risk cost   The increased chance of being injured or killed as a result of performing a behavior, compared to resting.

RNA (ribonucleic acid)   An often single-stranded nucleic acid whose nucleotides use ribose rather than deoxyribose and in which the base uracil replaces thymine found in DNA. Serves as genome from some viruses. (See ribosomal RNA, transfer RNA, messenger RNA, and ribozyme.)

RNA interference (RNAi)   A mechanism for reducing mRNA translation whereby a double-stranded RNA, made by the cell or synthetically, is processed into a small, single-stranded RNA, whose binding to a target mRNA results in the latter's breakdown.

RNA polymerase   An enzyme that catalyzes the formation of RNA from a DNA template.

RNA splicing   The last stage of RNA processing in eukaryotes, in which the transcripts of introns are excised through the action of small nuclear ribonucleoprotein particles (snRNP).

rod cells   Light-sensitive cells in the vertebrate retina; these sensory receptor cells are sensitive in extremely dim light and are responsible for dim light, black and white vision.

root   The organ responsible for anchoring the plant in the soil, absorbing water and minerals, and producing certain hormones. Some roots are storage organs.

root apical meristem   Undifferentiated tissue at the apex of the root that gives rise to the organs of the root.

root cap   A thimble-shaped mass of cells, produced by the root apical meristem, that protects the meristem; the organ that perceives the gravitational stimulus in root gravitropism.

root hair   A long, thin process from a root epidermal cell that absorbs water and minerals from the soil solution.

root system   The organ system that anchors a plant in place, absorbs water and dissolved minerals, and may store products of photosynthesis from the shoot system.

rough endoplasmic reticulum (RER)   The portion of the endoplasmic reticulum whose outer surface has attached ribosomes. (Contrast with smooth endoplasmic reticulum.)

round window   A flexible membrane at the end of the lower canal of the cochlea in the human ear. (See also oval window.)

rRNA   See ribosomal RNA.

rubisco   Contraction of ribulose bisphosphate carboxylase/oxygenase, the enzyme that combines carbon dioxide or oxygen with ribulose bisphosphate to catalyze the first step of photosynthetic carbon fixation or photorespiration, respectively.

rumen   One of the four chambers of the ruminant stomach. Along with the reticulum, where food is partially digested with the assistance of gut bacteria.

ruminant   Herbivorous, cud-chewing mammals such as cows or sheep, characterized by a stomach that consists of four compartments: the rumen, reticulum, omasum, and abomasum.

S

S phase   In the cell cycle, the stage of interphase during which DNA is replicated. (Contrast with G1 phase, G2 phase, M phase.)

salt glands   Glands on the leaves of some halophytic plants that secrete salt, thereby ridding the plants of excess salt.

saltatory conduction   [L. *saltare*: to jump] The rapid conduction of action potentials in myelinated axons; so called because action potentials appear to "jump" between nodes of Ranvier along the axon.

saprobe   [Gk. *sapros*: rotten] An organism (usually a bacterium or fungus) that obtains its carbon and energy by absorbing nutrients from dead organic matter.

sarcomere   (sark' o meer) [Gk. *sark*: flesh + *meros*: unit] The contractile unit of a skeletal muscle.

sarcoplasm   The cytoplasm of a muscle cell.

sarcoplasmic reticulum   The endoplasmic reticulum of a muscle cell.

saturated fatty acid   A fatty acid in which all the bonds between carbon atoms in the hydrocarbon chain are single bonds—that is, all the bonds are saturated with hydrogen atoms. (Contrast with unsaturated fatty acid.)

Schwann cell   A type of glial cell that myelinates axons in the peripheral nervous system.

scientific method   A means of gaining knowledge about the natural world by making observations, posing hypotheses, and conducting experiments to test those hypotheses.

scion   In horticulture, the bud or stem from one plant that is grafted to a root or root-bearing stem of another plant (the stock).

sclereid   One of the principle types of cells in sclerenchyma.

sclerenchyma   (skler eng' kyma) [Gk. *skleros*: hard + *kymus*: juice] A plant tissue composed of cells with heavily thickened cell walls. The cells are dead at functional maturity. The principal types of sclerenchyma cells are fibers and sclereids.

scrotum   In most mammals, a pouch outside the body cavity that contains the testes.

second filial generation   See F₂.

second law of thermodynamics   The principle that when energy is converted from one form to another, some of that energy becomes unavailable for doing work.

second messenger   A compound, such as cAMP, that is released within a target cell after a hormone (the first messenger) has bound to a surface receptor on a cell; the second messenger triggers further reactions within the cell.

second polar body   In oogenesis, the daughter cell of the second meiotic division that subsequently degenerates. (*See also* ootid.)

secondary active transport   A form of active transport that does not use ATP as an energy source; rather, transport is coupled to ion diffusion down a concentration gradient established by primary active transport.

secondary cell wall   A thick, cellulosic structure internal to the primary cell wall formed in some plant cells after cell expansion stops (Contrast with primary cell wall.)

secondary consumer   An organism that eats primary consumers. (Contrast with primary consumer.)

secondary endosymbiosis   The engulfment of a photosynthetic eukaryote by another eukaryotic cell that gave rise to certain groups of photosynthetic eukaryotes (e.g., euglenids).

secondary growth   In plants, growth that contributes to an increase in girth. (Contrast with primary growth.)

secondary immune response   A rapid and intense response to a second or subsequent exposure to an antigen, initiated by memory cells. (Contrast with primary immune response.)

secondary lysosome   Membrane-enclosed organelle formed by the fusion of a primary lysosome with a phagosome, in which macromolecules taken up by phagocytosis are hydrolyzed into their monomers. (Contrast with lysosome.)

secondary metabolite   A compound synthesized by a plant that is not needed for basic cellular metabolism. Typically has an antiherbivore or antiparasite function.

secondary oocyte   In oogenesis, the daughter cell of the first meiotic division that receives almost all the cytoplasm. (*See also* first polar body.)

secondary plant body   That part of a plant produced by secondary growth; consists of woody tissues. (Contrast with primary plant body.)

secondary sex determination   Formation of secondary sexual characteristics (i.e., those other than gonads), such as external sex organs and body hair. (Contrast with primary sex determination.)

secondary spermatocyte   One of the products of the first meiotic division of a primary spermatocyte.

secondary structure   Of a protein, localized regularities of structure, such as the α helix and the β pleated sheet. (Contrast with primary, tertiary, quarternary structure.)

secondary succession   Succession of ecological communities after a disturbance that did not eliminate all the organisms originally living on the site. (Contrast with primary succession.)

secretin   (si kreet' in) A peptide hormone secreted by the upper region of the small intestine when acidic chyme is present. Stimulates the pancreatic duct to secrete bicarbonate ions.

sedimentary rock   Rock formed by the accumulation of sediment grains on the bottom of a body of water. Often contain stratified fossils that allow geologists and biologist to date evolutionary events relative to each other.

seed   A fertilized, ripened ovule of a gymnosperm or angiosperm. Consists of the embryo, nutritive tissue, and a seed coat.

seedling   A plant that has just completed the process of germination.

segment polarity genes   In *Drosophila* (fruit fly) development, segmentation genes that determine the boundaries and anterior–posterior organization of individual segments. Part of a developmental cascade that includes maternal effect genes, gap genes, pair rule genes, and Hox genes.

segmentation   Division of an animal body into segments.

segmentation genes   Genes that determine the number and polarity of body segments.

segregation   In genetics, the separation of alleles, or of homologous chromosomes, from each other during meiosis so that each of the haploid daughter nuclei produced contains one or the other member of the pair found in the diploid parent cell, but never both. This principle was articulated by Mendel as his first law.

selectable marker   A gene, such as one encoding resistance to an antibiotic, that can be used to identify (select) cells that contain recombinant DNA from among a large population of untransformed cells.

selective permeability   Allowing certain substances to pass through while other substances are excluded; a characteristic of membranes.

self-incompatability   In plants, the possession of mechanisms that prevent self-fertilization.

semelparous   [L. *semel*: once + *pario*: to beget] Reproducing only once in a lifetime. (Contrast with iteroparous.)

semen   (see' men) [L. *semin*: seed] The thick, whitish liquid produced by the male reproductive system in mammals, containing the sperm.

semicircular canals   Three canals in the human inner ear that form part of the vestibular system.

semiconservative replication   The way in which DNA is synthesized. Each of the two partner strands in a double helix acts as a template for a new partner strand. Hence, after replication, each double helix consists of one old and one new strand.

seminiferous tubules   The tubules within the testes within which sperm production occurs.

senescence   [L. *senescere*: to grow old] Aging; deteriorative changes with aging; the increased probability of dying with increasing age.

sensitive period   The life stage during which some particular type of learning must take place, or during which it occurs much more easily than at other times. Typical of song learning among birds. Also known as the critical period.

sensory receptor cell   Cell that is responsive to a particular type of physical or chemical stimulation. Sometimes referred to as a sensor.

sensory system   A set of organs and tissues for detecting a stimulus; consists of sensory cells, the associated structures, and the neural networks that process the information.

sensory transduction   The transformation of environmental stimuli or information into neural signals.

sepal   (see' pul) [L. *sepalum*: covering] One of the outermost structures of the flower, usually protective in function and enclosing the rest of the flower in the bud stage.

septate   [L. wall] Divided, as by walls or partitions.

septum   (plural: septa) (1) A partition or cross-wall appearing in the hyphae of some fungi. (2) The bony structure dividing the nasal passages.

sequence alignment   A method of identifying homologous positions in DNA or amino acid sequences by pinpointing the locations of deletions and insertions that have occurred since two (or more) organisms diverged from a common ancestor.

Sertoli cells   Cells in the seminiferous tubules of the testes that nurture the developing sperm.

sessile   (sess' ul) [L. *sedere*: to sit] Permanently attached; not able to move from one place to another. (Contrast with motile.)

set point   In a regulatory system, the threshold sensitivity to the feedback stimulus.

sex chromosome   In organisms with a chromosomal mechanism of sex determination, one of the chromosomes involved in sex determination (in humans and many other animals, these are the X and Y chromosomes).

sex-linked inheritance   Pattern of inheritance characteristic of genes located on the sex chromosomes of organisms having a chromosomal mechanism for sex determination.

sex pilus   A thin connection between two bacteria through which genetic material passes during conjugation.

sexual reproduction   Reproduction involving the union of gametes.

sexual selection   Selection by one sex of characteristics in individuals of the opposite sex. Also, the favoring of characteristics in

one sex as a result of competition among individuals of that sex for mates.

shared derived trait   *See* synapomorphy.

shoot apical meristem   Undifferentiated tissue at the apex of the shoot that gives rise to the organs of the shoot.

shoot system   In plants, the organ system consisting of the leaves, stem(s), and flowers.

short-day plant (SDP)   A plant that flowers when nights are longer than a critical length specific for that plant's species. (Compare to long-day plant.)

short tandem repeats (STRs)   Short (1–5 base pairs), moderately repetitive sequences of DNA. The number of copies of an STR at a particular location varies between individuals and is inherited.

side chain   *See* R group.

sieve tube element   The characteristic cell of the phloem in angiosperms, which contains cytoplasm but relatively few organelles, and whose end walls (**sieve plates**) contain pores that form connections with neighboring cells.

sigma factor   In prokaryotes, a protein that binds to RNA polymerase, allowing the complex to bind to and stimulate the transcription of a specific class of genes (e.g., those involved in sporulation).

signal sequence   The sequence within a protein that directs the protein to a particular organelle.

signal transduction pathway   The series of biochemical steps whereby a stimulus to a cell (such as a hormone or neurotransmitter binding to a receptor) is translated into a response of the cell.

silencer   A gene sequence binding transcription factors that repress transcription. (Contrast with promoter.)

silent mutation   A change in a gene's sequence that has no effect on the amino acid sequence of a protein either because it occurs in noncoding DNA or because it does not change the amino acid specified by the corresponding codon . (Contrast with frameshift mutation, missense mutation, nonsense mutation.)

silent substitution   *See* synonymous substitution.

similarity matrix   A matrix used to compare the degree of divergence among pairs of objects. For molecular sequences, constructed by summing the number or percentage of nucleotides or amino acids that are identical in each pair of sequences.

simple diffusion   Diffusion that does not involve a direct input of energy or assistance by carrier proteins.

single nucleotide polymorphisms (SNPs)   Inherited variations in a single nucleotide base in DNA that differ between individuals.

single-strand binding protein   In DNA replication, a protein that binds to single strands of DNA after they have been separated from each other, keeping the two strands separate for replication.

sink   In plants, any organ that imports the products of photosynthesis, such as roots, developing fruits, and immature leaves. (Contrast with source.)

sinoatrial node   (sigh' no ay' tree al) [L. *sinus*: curve + *atrium*: chamber] The pacemaker of the mammalian heart.

siRNAs (small interfering RNAs)   Short, double-stranded RNA molecules used in RNA interference.

sister chromatid   Each of a pair of newly replicated chromatids.

sister species   Two species that are each other's closest relatives.

skeletal muscle   A type of muscle tissue characterized by multinucleated cells containing highly ordered arrangements of actin and myosin microfilaments. Also called striated muscle. (Contrast with cardiac muscle, smooth muscle.)

skeletal systems   Organ systems that provide rigid supports—**skeletons**—against which muscles can pull to create directed movements. See also endoskeleton, exoskeleton.

sliding DNA clamp   Protein complex that keeps DNA polymerase bound to DNA during replication.

sliding filament model   Mechanism of muscle contraction based on the formation and breaking of crossbridges between actin and myosin filaments, causing the filaments to slide together.

slow-twitch fibers   Skeletal muscle fibers specialized for sustained aerobic work; contain myoglobin and abundant mitochondria, and are well-supplied with blood vessels. Also called oxidative or red muscle fibers. (Compare to fast-twitch fibers.)

slow-wave sleep   *See* non-REM sleep.

small interfering RNAs   *See* siRNAs.

small intestine   The portion of the gut between the stomach and the colon; consists of the duodenum, the jejunum, and the ileum.

small nuclear ribonucleoprotein particle (snRNP)   A complex of an enzyme and a small nuclear RNA molecule, functioning in RNA splicing.

smooth endoplasmic reticulum (SER)   Portion of the endoplasmic reticulum that lacks ribosomes and has a tubular appearance. (Contrast with rough endoplasmic reticulum.)

smooth muscle   Muscle tissue consisting of sheets of mononucleated cells innervated by the autonomic nervous system. (Contrast with cardiac muscle, skeletal muscle.)

sodium–potassium (Na⁺–K⁺) pump   Antiporter responsible for primary active transport; it pumps sodium ions out of the cell and potassium ions into the cell, both against their concentration gradients. Also called a sodium–potassium ATPase.

soil horizon   *See* horizons.

soil solution   The aqueous portion of soil, from which plants take up dissolved mineral nutrients.

solute   A substance that is dissolved in a liquid (solvent) to form a solution.

solute potential (Ψ_s)   A property of any solution, resulting from its solute contents; it may be zero or have a negative value. The more negative the solute potential, the greater the tendency of the solution to take up water through a differentially permeable membrane. (Contrast with pressure potential, water potential.)

solution   A liquid (the solvent) and its dissolved solutes.

solvent   Liquid in which a substance (solute) is dissolved to form a solution.

somatic cell   [Gk. *soma*: body] All the cells of the body that are not specialized for reproduction. (Contrast with germ cell.)

somatic mutation   Permanent genetic change in a somatic cell (as opposed to a germ cell, the egg or sperm). These mutations affect the individual only; they are not passed on to offspring. (Contrast with germ line mutation.)

somatosensory cortex   An area of the parietal lobe that receives touch and pressure information from mechanoreceptors throughout the body; neurons in this area are arranged according to the parts of the body with which they communicate.

somatostatin   Peptide hormone made in the hypothalamus that inhibits the release of other hormones from the pituitary and intestine.

somite   (so' might) One of the segments into which an embryo becomes divided longitudinally, leading to the eventual segmentation of the animal as illustrated by the spinal column, ribs, and associated muscles.

source   In plants, any organ that exports the products of photosynthesis in excess of its own needs, such as a mature leaf or storage organ. (Contrast with sink.)

spatial summation   In the production or inhibition of action potentials in a postsynaptic cell, the interaction of depolarizations and hyperpolarizations produced at different sites on the postsynaptic cell. (Contrast with temporal summation.)

spawning   *See* external fertilization.

speciation   (spee' see ay' shun) The process of splitting one population into two populations that are reproductively isolated from one another.

species   (spee' sees) [L. kind] The base unit of taxonomic classification, consisting of an ancestor–descendant group of populations of evolutionarily closely related, similar organisms. The more narrowly defined "biological species" consists of individuals capable of interbreeding with each other but not with members of other species.

species composition   The particular mix of species a community contains and the abundances of those species.

species evenness   A measure of species diversity that reflects the distribution of the species' abundances in a community.

species richness   The total number of species living in a region.

species–area relationship   The relationship between the size of an area and the numbers of species it supports.

specific defenses   Defensive reactions of the vertebrate immune system that are based on the reaction of an antibody to a specific antigen. (Contrast with nonspecific defenses.)

specific heat   The amount of energy that must be absorbed by a gram of a substance to raise its temperature by one degree centigrade. By convention, water is assigned a specific heat of one.

sperm   [Gk. *sperma*: seed] The male gamete.

spermatid   One of the products of the second meiotic division of a primary spermatocyte; four haploid spermatids, which remain connected by cytoplasmic bridges, are produced for each primary spermatocyte that enters meiosis.

spermatogenesis   (spur mat' oh jen' e sis) [Gk. *sperma*: seed + *genesis*: source] Gametogenesis leading to the production of sperm.

spermatogonia   In animals, the diploid progeny of a germ cell in males.

spherical symmetry   The simplest form of symmetry, in which body parts radiate out from a central point such that an infinite number of planes passing through that central point can divide the organism into similar halves.

sphincter   (sfink' ter) [Gk. *sphinkter*: something that binds tightly] A ring of muscle that can close an orifice, for example, at the anus.

spicule   [L. arrowhead] A hard, calcareous skeletal element typical of sponges.

spinal cord   Along with the brain, part of the central nervous system; transmits information between the body and the brain and mediates simple reflexes.

spinal reflex   The conversion of afferent to efferent information in the spinal cord without participation of the brain.

spindle apparatus   [O.E. *spindle*, a short stick with tapered ends] Array of microtubules emanating from both poles of a dividing cell during mitosis and playing a role in the movement of chromosomes at nuclear division.

spleen   Organ that serves as a reservoir for venous blood and eliminates old, damaged red blood cells from the circulation.

spliceosome   RNA–protein complex that splices out introns from eukaryotic pre-mRNAs.

splicing   *See* RNA splicing.

spontaneous mutation   A genetic change caused by internal cellular mechanisms, such as an error in DNA replication. (Contrast with induced mutation.)

sporangiophore   A stalked reproductive structure produced by zygospore fungi that extends from a hypha and bears one or many sporangia.

sporangium   (spor an' gee um) (plural: sporangia) [Gk. *spora*: seed + *angeion*: vessel or reservoir] In plants and fungi, any specialized stucture within which one or more spores are formed.

spore   [Gk. *spora*: seed] (1) Any asexual reproductive cell capable of developing into an adult organism without gametic fusion. In plants, haploid spores develop into gametophytes, diploid spores into sporophytes. (2) In prokaryotes, a resistant cell capable of surviving unfavorable periods.

sporocyte   Specialized cells of the diploid sporophyte that will divide by meiosis to produce four haploid spores. Germination of these spores produces the haploid gametophyte.

sporophyte   (spor' o fyte) [Gk. *spora*: seed + *phyton*: plant] In plants and protists with alternation of generations, the diploid phase that produces the spores. (Contrast with gametophyte.)

stabilizing selection   Selection against the extreme phenotypes in a population, so that the intermediate types are favored. (Contrast with disruptive selection.)

stamen   (stay' men) [L. *stamen*: thread] A male (pollen-producing) unit of a flower, usually composed of an anther, which bears the pollen, and a filament, which is a stalk supporting the anther.

starch   [O.E. *stearc*: stiff] A polymer of glucose; used by plants to store energy.

Starling's forces   The two opposing forces responsible for water movement across capillary walls: blood pressure, which squeezes water and small solutes out of the capillaries, and osmotic pressure, which pulls water back into the capillaries.

start codon   The mRNA triplet (AUG) that acts as a signal for the beginning of translation at the ribosome. (Contrast with stop codon.)

stele   (steel) [Gk.: pillar] The central cylinder of vascular tissue in a plant stem.

stem cell   In animals, an undifferentiated cell that is capable of continuous proliferation. A stem cell generates more stem cells and a large clone of differentiated progeny cells. (*See also* embryonic stem cell.)

stem   In plants, the organ that holds leaves and/or flowers and transports and distributes materials among the other organs of the plant.

stereocilia   Fingerlike extensions of hair cell membranes whose bending initiates sound perception. (*See* hair cell.)

steroid   Any of a family of lipids whose multiple rings share carbons. The steroid cholesterol is an important constituent of membranes and is the base of steroid hormones such as testosterone.

sticky ends On a piece of two-stranded DNA, short, complementary, one-stranded regions produced by the action of a restriction endonuclease. Sticky ends facilitate the joining of segments of DNA from different sources.

stigma [L. *stigma*: mark, brand] The part of the pistil at the apex of the style that is receptive to pollen, and on which pollen germinates.

stimulus [L. *stimulare*: to goad] Something causing a response; something in the environment detected by a receptor.

stock In horticulture, the root or root-bearing stem to which a bud or piece of stem from another plant (the scion) is grafted.

stoma (plural: stomata) [Gk. *stoma*: mouth, opening] Small opening in the plant epidermis that permits gas exchange; bounded by a pair of guard cells whose osmotic status regulates the size of the opening.

stomach An organ that physically (and sometimes enzymatically) breaks down food, preparing it for digestion in the midgut.

stomatal crypt In plants, a sunken cavity below the leaf surface in which a stoma is sheltered from the drying effects of air currents.

stop codon Any of the three mRNA codons that signal the end of protein translation at the ribosome: UAG, UGA, UAA.

stratosphere The upper part of Earth's atmosphere, above the troposphere; extends from approximately 18 kilometers upward to approximately 50 kilometers above Earth's surface.

stratum (plural strata) [L. *stratos*: layer] A layer of sedimentary rock laid down at a particular time in the past.

stretch receptor A modified muscle cell embedded in the connective tissue of a muscle that acts as a mechanoreceptor in response to stretching of that muscle.

striated muscle *See* skeletal muscle.

strigolactones Signaling molecules produced by plant roots that attract the hyphae of mycorrhizal fungi.

strobilus (plural: strobili) One of several conelike structures in various groups of plants (including club mosses, horsetails, and conifers) associated with the production and dispersal of reproductive products.

stroke An embolism in an artery in the brain that causes the cells fed by that artery to die. The specific damage, such as memory loss, speech impairment, or paralysis, depends on the location of the blocked artery.

stroma The fluid contents of an organelle such as a chloroplast or mitochondrion.

structural gene A gene that encodes the primary structure of a protein not involved in the regulation of gene expression.

structural isomers Molecules made up of the same kinds and numbers of atoms, in which the atoms are bonded differently.

structural motif A three-dimensional structural element that is part of a larger molecule. For example, there are four common motifs in DNA-binding proteins: helix-turn-helix, zinc finger, leucine zipper, and helix-loop-helix.

style [Gk. *stele*: pillar or column] In the angiosperm flower, a column of tissue extending from the tip of the ovary, and bearing the stigma or receptive surface for pollen at its apex.

sub- [L. under] A prefix used to designate a structure that lies beneath another or is less than another. For example, subcutaneous (beneath the skin); subspecies

subduction In plate tectonics, the movement of one lithospheric plate under another.

suberin A waxlike lipid that is a barrier to water and solute movement across the Casparian strip of the endodermis.

submucosa (sub mew koe' sah) The tissue layer just under the epithelial lining of the lumen of the digestive tract.

subsoil The soil horizon lying below the topsoil and above the parent rock (bedrock); the zone of infiltration and accumulation of materials leached from the topsoil. Also called the B horizon.

substrate (sub' strayte) (1) The molecule or molecules on which an enzyme exerts catalytic action. (2) The base material on which a sessile organism lives.

succession The gradual, sequential series of changes in the species composition of an ecological community following a disturbance. See also cyclical succession, directional selection, heterotrophic succession.

succulence In plants, possession of fleshy, water-storing leaves or stems; an adaptation to dry environments.

sulcus (sul' kus; plural sulci) [Gk.: plowed furrow] *See* convolutions.

superficial cleavage A variation of incomplete cleavage in which cycles of mitosis occur without cell division, producing a syncytium (a single cell with many nuclei).

suprachiasmatic nuclei (SCN) In mammals, two clusters of neurons just above the optic chiasm that act as the master circadian clock.

surface area-to-volume ratio For any cell, organism, or geometrical solid, the ratio of surface area to volume; this is an important factor in setting an upper limit on the size a cell or organism can attain.

surface tension The attractive intermolecular forces at the surface of liquid; an especially important property of water.

surfactant A substance that decreases the surface tension of a liquid. Lung surfactant, secreted by cells of the alveoli, is mostly phospholipid and decreases the amount of work necessary to inflate the lungs.

survivorship The fraction of individuals that survive from birth to a given life stage or age.

survivorship curves Graphic plot of ages at death of a hypothetical cohort, usually of 1,000 individuals, by plotting the numbers of individuals expected to survive to reach each age category. There are three general shapes. Ecological survivorship is linear: individuals face a constant risk of mortality regardless of their age. Physiological survivorship curves are concave: high survivorship through adulthood with steep declines late in life. Maturational survivorship curves are convex, with high mortality early in life but higher survivorship once individuals reach maturity.

suspensor In the embryos of seed plants, the stalk of cells that pushes the embryo into the endosperm and is a source of nutrient transport to the embryo.

sustainable Pertaining to the use and management of ecosystems in such a way that humans benefit over the long term from specific ecosystem goods and services without compromising others.

symbiosis (sim' bee oh' sis) [Gk. *sym*: together + *bios*: living] The living together of two or more species in a prolonged and intimate relationship.

symmetry Pertaining to an attribute of an animal body in which at least one plane can divide the body into similar, mirror-image halves. (*See* bilateral symmetry, radial symmetry.)

sympathetic nervous system The division of the autonomic nervous system that works in opposition to the parasympathetic nervous system. (Contrast with parasympathetic nervous system.)

sympatric speciation (sim pat' rik) [Gk. *sym*: same + *patria*: homeland] Speciation due to reproductive isolation without any physical separation of the subpopulation. (Contrast with allopatric speciation.)

symplast The continuous meshwork of the interiors of living cells in the plant body, resulting from the presence of plasmodesmata. (Contrast with apoplast.)

symporter A membrane transport protein that carries two substances in the same direction. (Contrast with antiporter, uniporter.)

synapomorphy A trait that arose in the ancestor of a phylogenetic group and is present (sometimes in modified form) in all of its members, thus helping to delimit and identify that group. Also called a shared derived trait.

synapse (sin' aps) [Gk. *syn*: together + *haptein*: to fasten] A specialized type of junction where a neuron meets its target cell (which can be another neuron or some other type of cell) and information in the form of neurotransmitter molecules is exchanged across a synaptic cleft.

synapsis (sin ap' sis) The highly specific parallel alignment (pairing) of homologous chromosomes during the first division of meiosis.

synaptic cleft The space between the presynaptic cell and the postsynaptic cell in a chemical synapse.

synergids [Gk. *syn*: together + *ergos*: work] In angiosperms, the two cells accompanying the egg cell at one end of the megagametophyte.

syngamy *See* fertilization.

synonymous (silent) substitution A change of one nucleotide in a sequence to another when that change does not affect the amino acid specified (i.e., UUA → UUG, both specifying leucine). (Contrast with nonsynonymous substitution, missense mutation, nonsense mutation.)

systematics The scientific study of the diversity and relationships among organisms.

systemic acquired resistance A general resistance to many plant pathogens following infection by a single agent.

systemic circuit Portion of the circulatory system by which oxygenated blood from the lungs or gills is distributed throughout the rest of the body and returned to the heart. (Contrast with pulmonary circuit.)

systems biology The scientific study of an organism as an integrated and interacting system of genes, proteins, and biochemical reactions.

systole (sis' tuh lee) [Gk.: contraction] Contraction of a chamber of the heart, driving blood forward in the circulatory system. (Contrast with diastole.)

T

3′ end (3 prime) The end of a DNA or RNA strand that has a free hydroxyl group at the 3′ carbon of the sugar (deoxyribose or ribose).

T cell A type of lymphocyte involved in the cellular immune response. The final stages of its development occur in the thymus gland. (Contrast with B cell; *see also* cytotoxic T cell, T-helper cell.)

T cell receptor A protein on the surface of a T cell that recognizes the antigenic determinant for which the cell is specific.

T tubules A system of tubules that runs throughout the cytoplasm of a muscle fiber, through which action potentials spread.

T-helper (T$_H$) cell Type of T cell that stimulates events in both the cellular and humoral immune responses by binding to the antigen on an antigen-presenting cell; target of the HIV-I virus, the agent of AIDS. (Contrast with cytotoxic T cells.)

taproot system A root system typical of eudicots consisting of a primary root (*taproot*) that extends downward by tip growth and outward by initiating lateral roots. (Contrast with fibrous root system.)

target cell A cell with the appropriate receptors to bind and respond to a particular hormone or other chemical mediator.

taste bud A structure in the epithelium of the tongue that includes a cluster of chemoreceptors innervated by sensory neurons.

TATA box An eight-base-pair sequence, found about 25 base pairs before the starting point for transcription in many eukaryotic promoters, that binds a transcription factor and thus helps initiate transcription.

taxon (plural: taxa) [Gk. *taxis*: put in order] A biological group (typically a species or a clade) that is given a name.

T$_C$ cells *See* cytotoxic T cells.

tectonic membrane One of two membranes (the other is the basilar membrane) that extend along the length of the cochlea in the human ear. Also known as Reissner's membrane.

telencephalon The outer, surrounding structure of the embryonic vertebrate forebrain, which develops into the cerebrum.

telomerase An enzyme that catalyzes the addition of telomeric sequences lost from chromosomes during DNA replication.

telomeres (tee' lo merz) [Gk. *telos*: end + *meros*: units, segments] Repeated DNA sequences at the ends of eukaryotic chromosomes.

telophase (tee' lo phase) [Gk. *telos*: end] The final phase of mitosis or meiosis during which chromosomes become diffuse, nuclear envelopes re-form, and nucleoli begin to reappear in the daughter nuclei.

template A molecule or surface on which another molecule is synthesized in complementary fashion, as in the replication of DNA.

template strand In double-stranded DNA, the strand that is transcribed to create an RNA transcript that will be processed into a protein. Also refers to a strand of RNA that is used to create a complementary RNA.

temporal lobe One of the four lobes of the cerebral hemisphere; receives and processes auditory and visual information; involved in recognizing, identifying, and naming objects.

temporal summation In the production or inhibition of action potentials in a postsynaptic cell, the interaction of depolarizations or hyperpolarizations produced by rapidly repeated stimulation of a single point on the postsynaptic cell. (Contrast with spatial summation.)

tendon A collagen-containing band of tissue that connects a muscle with a bone.

tepal A sterile, modified, nonphotosynthetic leaf of an angiosperm flower that cannot be distinguished as a petal or a sepal.

termination In molecular biology, the end of transcription or translation.

terminator A sequence at the 3′ end of mRNA that causes the RNA strand to be released from the transcription complex.

terrestrial (ter res' tree al) [L. *terra*: earth] Pertaining to or living on land. (Contrast with aquatic, marine.)

territorial behavior Aggressive actions engaged in to defend a habitat or resource such that other animals are denied access.

tertiary consumers Carnivores that consume primary carnivores (secondary consumers).

tertiary endosymbiosis The mechanism by which some eukaryotes acquired the capacity for photosynthesis; for example, a dinoflagellate that apparently lost its chloroplast became photosynthetic by engulfing another protist that had acquired a chloroplast through secondary endosymbiosis.

tertiary structure In reference to a protein, the relative locations in three-dimensional space of all the atoms in the molecule. The overall shape of a protein. (Contrast with primary, secondary, and quaternary structures.)

test cross Mating of a dominant-phenotype individual (who may be either heterozygous or homozygous) with a homozygous-recessive individual.

testis (tes' tis) (plural: testes) [L. *testis*: witness] The male gonad; the organ that produces the male gametes.

tetanus [Gk. *tetanos*: stretched] (1) A state of sustained maximal muscular contraction caused by rapidly repeated stimulation. (2) In medicine, an often fatal disease ("lockjaw") caused by the bacterium *Clostridium tetani*.

tetrad [Gk. *tettares*: four] During prophase I of meiosis, the association of a pair of homologous chromosomes or four chromatids

thalamus [Gk. *thalamos*: chamber] A region of the vertebrate forebrain; involved in integration of sensory input.

theory [Gk. *theoria*: analysis of facts] A far-reaching explanation of observed facts that is supported by such a wide body of evidence, with no significant contradictory evidence, that it is scientifically accepted as a factual framework. Examples are Newton's theory of gravity and Darwin's theory of evolution. (Contrast with hypothesis.)

thermoneutral zone (TNZ) [Gk. *thermos*: temperature] The range of temperatures over which an endotherm does not have to expend extra energy to thermoregulate.

thermophile (ther' muh fyle)[Gk. *thermos*: temperature + *philos*: loving] An organism that lives exclusively in hot environments.

thoracic cavity [Gk. *thorax*: breastplate] The portion of the mammalian body cavity bounded by the ribs, shoulders, and diaphragm. Contains the heart and the lungs.

thoracic duct The connection between the lymphatic system and the circulatory system.

threshold The level of depolarization that causes an electrically excitable membrane to fire an action potential.

thrombin An enzyme involved in blood clotting; cleaves fibrinogen to form fibrin.

thrombus (throm' bus) [Gk. *thrombos*: clot] A blood clot that forms within a blood vessel and remains attached to the wall of the vessel.

thylakoid (thigh' la koid) [Gk. *thylakos*: sack or pouch] A flattened sac within a chloroplast. Thylakoid membranes contain all of the chlorophyll in a plant, in addition to the electron carriers of photophosphorylation. Thylakoids stack to form grana.

thymine (T) Nitrogen-containing base found in DNA.

thymus [Gk. *thymos*: warty] A ductless, glandular lymphoid tissue, involved in development of the immune system of vertebrates. In humans, the thymus degenerates during puberty.

thyroid gland [Gk. *thyreos*: door-shaped] A two-lobed gland in vertebrates. Produces the hormone thyroxine.

thyrotropin Hormone produced by the anterior pituitary that stimulates the thyroid gland to produce and release thyroxine. Also called thyroid-stimulating hormone (TSH).

thyrotropin-releasing hormone (TRH) Hormone produced by the hypothalamus that stimulates the anterior pituitary to release thyrotropin.

thyroxine Hormone produced by the thyroid gland; controls many metabolic processes.

tidal The bidirectional form of ventilation used by all vertebrates except birds; air enters and leaves the lungs by the same route.

tight junction A junction between epithelial cells in which there is no gap between adjacent cells.

tissue A group of similar cells organized into a functional unit; usually integrated with other tissues to form part of an organ.

tissue system In plants, any of three organized groups of tissues—dermal tissue, vascular tissue, and ground tissue—that are established during embryogenesis and have distinct functions.

titin A protein that holds bundles of myosin filaments in a centered position within the sarcomeres of muscle cells. The largest protein in the human body.

tonoplast The membrane of the plant central vacuole.

topsoil The uppermost soil horizon; contains most of the organic matter of soil, but may be depleted of most mineral nutrients by leaching. Also called the A horizon.

totipotent [L. *toto*: whole, entire + *potens*: powerful] Possessing all the genetic information and other capacities necessary to form an entire individual. (Contrast with multipotent, pluripotent.)

toxigenicity The ability of some pathogenic bacteria to produce chemical substances that harm the host.

trachea (tray' kee ah) [Gk. *trakhoia*: tube] A tube that carries air to the bronchi of the lungs of vertebrates. When plural (*tracheae*), refers to the major airways of insects.

tracheary element Either of two types of xylem cells—tracheids and vessel elements—that undergo apoptosis before assuming their transport function.

tracheid (tray' kee id) A type of tracheary element found in the xylem of nearly all vascular plants, characterized by tapering ends and walls that are pitted but not perforated. (Contrast with vessel element.)

trade-off The relationship between the fitness benefits conferred by an adaptation and the fitness costs it imposes. For an adaptation to be favored by natural selection, the benefits must exceed the costs.

trait In genetics, a specific form of a character: eye color is a character; brown eyes and blue eyes are traits. (Contrast with character.)

transcription The synthesis of RNA using one strand of DNA as a template.

transcription factors Proteins that assemble on a eukaryotic chromosome, allowing RNA polymerase II to perform transcription.

transcription initiation site The part of a gene's promoter where synthesis of the gene's RNA transcript begins.

transduction (1) Transfer of genes from one bacterium to another by a bacteriophage. (2) In sensory cells, the transformation of a stimulus (e.g., light energy, sound pressure waves, chemical or electrical stimulants) into action potentials.

transfection Insertion of recombinant DNA into animal cells.

transfer RNA (tRNA) A family of folded RNA molecules. Each tRNA carries a specific amino acid and anticodon that will pair with the complementary codon in mRNA during translation.

transformation (1) A mechanism for transfer of genetic information in bacteria in which pure DNA from a bacterium of one genotype is taken in through the cell surface of a bacterium of a different genotype and incorporated into the chromosome of the recipient cell. (2) Insertion of recombinant DNA into a host cell.

transgenic Containing recombinant DNA incorporated into the genetic material.

transition state In an enzyme-catalyzed reaction, the reactive condition of the substrate after there has been sufficient input of energy (activation energy) to initiate the reaction.

translation The synthesis of a protein (polypeptide). Takes place on ribosomes, using the information encoded in messenger RNA.

translational repressor A protein that blocks translation by binding to mRNAs and preventing their attachment to the ribosome. In mammals, the production of ferritin protein is regulated by a translational repressor.

translocation (1) In genetics, a rare mutational event that moves a portion of a chromosome to a new location, generally on a nonhomologous chromosome. (2) In vascular plants, movement of solutes in the phloem.

transmembrane protein An integral membrane protein that spans the phospholipid bilayer.

transpiration [L. *spirare*: to breathe] The evaporation of water from plant leaves and stem, driven by heat from the sun, and providing the motive force to raise water (plus mineral nutrients) from the roots.

transpiration–cohesion–tension mechanism Theoretical basis for water movement in plants: evaporation of water from cells within leaves (transpiration) causes an increase in surface tension, pulling water up through the xylem. Cohesion of water occurs because of hydrogen bonding.

transposable element (transposon) A segment of DNA that can move to, or give rise to copies at, another locus on the same or a different chromosome.

transversion A mutation that changes a purine to a pyrimidine or vice versa.

tree of life A term that encompasses the evolutionary history of all life, or a graphic representation of that history.

triglyceride A simple lipid in which three fatty acids are combined with one molecule of glycerol.

trimesters The three stages of human pregnancy, approximately 3 months each in length.

tripartite synapse The idea that a synapse includes not only the pre- and postsynaptic neurons involved but also encompasses many connections with glial cells called astrocytes.

triploblastic Having three cell layers.

trisomic Containing three rather than two members of a chromosome pair.

tRNA See transfer RNA.

trochophore (troke' o fore) [Gk. *trochos*: wheel + *phoreus*: bearer] A radially symmetrical larval form typical of annelids and mollusks, distinguished by a wheel-like band of cilia around the middle.

trophic cascade The progression over successively lower trophic levels of the indirect effects of a predator.

trophic interactions The consumer–resource relationships among species in a community.

trophic level [Gk *trophes*: nourishment] A group of organisms united by obtaining their energy from the same part of the food web of a biological community.

trophoblast [Gk *trophes*: nourishment + *blastos*: sprout] At the 32-cell stage of mammalian development, the outer group of cells that will become part of the placenta and thus nourish the growing embryo. (Contrast with inner cell mass.)

tropic hormones Hormones produced by the anterior pituitary that control the secretion of hormones by other endocrine glands.

tropomyosin [troe poe my' oh sin] One of the three protein components of an actin filament; controls the interactions of actin and myosin necessary for muscle contraction.

troponin One of the three components of an actin filament; binds to actin, tropomyosin, and Ca^{2+}.

true-breeding A genetic cross in which the same result occurs every time with respect to the trait(s) under consideration, due to homozygous parents.

trypsin A protein-digesting enzyme. Secreted by the pancreas in its inactive form (trypsinogen), it becomes active in the duodenum of the small intestine.

tube feet A unique feature of echinoderms; extensions of the water vascular system, which functions in gas exchange, locomotion, and feeding.

tubulin A protein that polymerizes to form microtubules.

tumor [L. *tumor*: a swollen mass] A disorganized mass of cells. Malignant tumors spread to other parts of the body.

tumor necrosis factor A family of cytokines (growth factors) that causes cell death and is involved in inflammation.

tumor suppressor A gene that codes for a protein product that inhibits cell proliferation; inactive in cancer cells. (Contrast with oncogene.)

turgor pressure [L. *turgidus*: swollen] *See* pressure potential.

turnover In freshwater ecosystems, vertical movements of water that bring nutrients and dissolved CO_2 to the surface and O_2 to deeper water.

twitch A muscle fiber's minimum unit of contraction, stimulated by a single action potential.

tympanic membrane [Gk. *tympanum*: drum] The eardrum.

U

ubiquitin A small protein that is covalently linked to other cellular proteins identified for breakdown by the proteosome.

ultimate causes In ethology, the evolutionary processes that produce an animal's capacity and tendency to behave in particular ways. (Contrast with proximate causes.)

unequal crossing over When a highly repeated gene sequence becomes displaced in alignment during meiotic crossing over, so that one chromosome receives many copies of the sequence while the second chromosome receives fewer copies. One of the mechanisms of concerted evolution. (See also biased gene conversion.)

uniporter [L. *unus*: one + *portal*: doorway] A membrane transport protein that carries a single substance in one direction. (Contrast with antiporter, symporter.)

unipotent An undifferentiated cell that is capable of becoming only one type of mature cell. (Contrast with totipotent, multipotent, pluripotent.)

unsaturated fatty acid A fatty acid whose hydrocarbon chain contains one or more double bonds. (Contrast with saturated fatty acid.)

upregulation A process by which the abundance of receptors for a hormone increases when hormone secretion is suppressed. (Contrast with downregulation.)

upwelling zones Areas of the ocean where cool, nutrient-rich water from deeper layers rises to the surface.

uracil (U) A pyrimidine base found in nucleotides of RNA.

urea A compound that is the main form of nitrogen excreted by many animals, including mammals.

ureotelic Pertaining to an organism in which the final product of the breakdown of nitrogen-containing compounds (primarily proteins) is urea. (Contrast with ammonotelic, uricotelic.)

ureter (your' uh tur) Long duct leading from the vertebrate kidney to the urinary bladder or the cloaca.

urethra (you ree' thra) In most mammals, the canal through which urine is discharged from the bladder and which serves as the genital duct in males.

uric acid A compound that serves as the main excreted form of nitrogen in some animals, particularly those which must conserve water, such as birds, insects, and reptiles.

uricotelic Pertaining to an organism in which the final product of the breakdown of nitrogen-containing compounds (primarily proteins) is uric acid. (Contrast with ammonotelic, ureotelic.)

urinary bladder A structure in which urine is stored until it can be excreted to the outside of the body.

urine (you' rin) In vertebrates, the fluid waste product containing the toxic nitrogenous by-products of protein and nucleic acid metabolism.

uterine cycle In human females, the monthly cycle of events by which the endometrium is prepared for the arrival of a blastocyst. (Contrast with ovarian cycle).

uterus (yoo' ter us) [L. *utero*: womb] A specialized portion of the female reproductive tract in mammals that receives the fertilized egg and nurtures the embryo in its early development. Also called the womb.

V

vaccination Injection of virus or bacteria or their proteins into the body, to induce immunity. The injected material is usually attenuated (weakened) before injection and is called a *vaccine*.

vacuole (vac' yew ole) Membrane-enclosed organelle in plant cells that can function for storage, water concentration for turgor, or hydrolysis of stored macromolecules.

vagina (vuh jine' uh) [L. sheath] In female animals, the entry to the reproductive tract.

van der Waals forces Weak attractions between atoms resulting from the interaction of the electrons of one atom with the nucleus of another. This type of attraction is about one-fourth as strong as a hydrogen bond.

variable In a controlled experiment, a factor that is manipulated to test its effect on a phenomenon.

variable region The portion of an immunoglobulin molecule or T cell receptor that includes the antigen-binding site and is responsible for its specificity. (Contrast with constant region.)

vas deferens (plural: vasa deferentia) Duct that transfers sperm from the epididymis to the urethra.

vasa recta Blood vessels that parallel the loops of Henle and the collecting ducts in the renal medulla of the kidney.

vascular (vas' kew lar) [L. *vasculum*: a small vessel] Pertaining to organs and tissues that conduct fluid, such as blood vessels in animals and xylem and phloem in plants.

vascular bundle In vascular plants, a strand of vascular tissue, including xylem and phloem as well as thick-walled fibers.

vascular cambium (kam' bee um) [L. *cambiare*: to exchange] In plants, a lateral meristem that gives rise to secondary xylem and phloem.

vascular tissue system The transport system of a vascular plant, consisting primarily of xylem and phloem.

vasopressin A hormone that promotes water reabsorption by the kidney. Produced by neurons in the hypothalamus and released from nerve terminals in the posterior pituitary. Also called antidiuretic hormone or ADH.

vector (1) An agent, such as an insect, that carries a pathogen affecting another species. (2) A plasmid or virus that carries an inserted piece of DNA into a bacterium for cloning purposes in recombinant DNA technology.

vegetal hemisphere The lower portion of some animal eggs, zygotes, and embryos, in which the dense nutrient yolk settles. The *vegetal pole* is to the very bottom of the egg or embryo. (Contrast with animal hemisphere.)

vegetative Nonreproductive, nonflowering, or asexual.

vegetative meristem An apical meristem that produces leaves.

vegetative reproduction Asexual reproduction through the modification of stems, leaves, or roots.

vein [L. *vena*: channel] A blood vessel that returns blood to the heart. (Contrast with artery.)

vena cavae In the circulatory systems of crocodilians, birds, and mammals, large veins that empty into the right atrium of the heart.

ventral [L. *venter*: belly, womb] Toward or pertaining to the belly or lower side. (Contrast with dorsal.)

ventricle A muscular heart chamber that pumps blood through the lungs or through the body.

venule A small blood vessel draining a capillary bed that joins others of its kind to form a vein. (Contrast with arteriole.)

vernalization [L. *vernalis*: spring] Events occurring during a required chilling period, leading eventually to flowering.

vertebral column [L. *vertere*: to turn] The jointed, dorsal column that is the primary support structure of vertebrates.

very low-density lipoproteins (VLDLs) Lipoproteins that consist mainly of triglyceride fats, which they transport to fat cells in adipose tissues throughout the body; associated with excessive fat deposition and high risk for cardiovascular disease.

vesicle Within the cytoplasm, a membrane-enclosed compartment that is associated with other organelles; the Golgi complex is one example.

vessel element A type of tracheary element with perforated end walls; found only in angiosperms. (Contrast with tracheid.)

vestibular system (ves tib' yew lar) [L. *vestibulum*: an enclosed passage] Structures within the inner ear that sense changes in position or momentum of the head, affecting balance and motor skills.

vicariant event (vye care' ee unt) [L. *vicus*: change] The splitting of a taxon's range by the imposition of some barrier to dispersal.

villus (vil' lus) (plural: villi) [L. *villus*: shaggy hair or beard] A hairlike projection from a membrane; for example, from many gut walls.

virion (veer' e on) The virus particle, the minimum unit capable of infecting a cell.

virulence [L. *virus*: poison, slimy liquid] The ability of a pathogen to cause disease and death.

virus Any of a group of ultramicroscopic particles constructed of nucleic acid and protein (and, sometimes, lipid) that require living cells in order to reproduce. Viruses evolved multiple times from different cellular species.

vital capacity (VC) The maximum capacity for air exchange in one breath; the sum of the tidal volume and the inspiratory and expiratory reserve volumes.

vitamin [L. *vita*: life] An organic compound that an organism cannot synthesize, but nevertheless requires in small quantities for normal growth and metabolism.

vitelline envelope The inner, proteinaceous protective layer of a sea urchin egg.

viviparity (vye vi par' uh tee) Reproduction in which fertilization of the egg and development of the embryo occur inside the mother's body. (Contrast with oviparity.)

vivipary Premature germination in plants.

voltage A measure of the difference in electrical charge between two points.

voltage-gated channel A type of gated channel that opens or closes when a certain voltage exists across the membrane in which it is inserted.

vomeronasal organ (VNO) Chemosensory structure embedded in the nasal epithelium of amphibians, reptiles, and many mammals. Often specialized for detecting pheromones.

W

warning coloration *See* aposematism

water potential (psi, Ψ) In osmosis, the tendency for a system (a cell or solution) to take up water from pure water through a differentially permeable membrane. Water flows toward the system with a more negative water potential. (Contrast with solute potential, pressure potential.)

water vascular system In echinoderms, a network of water-filled canals that functions in gas exchange, locomotion, and feeding.

wavelength The distance between successive peaks of a wave train, such as electromagnetic radiation.

weather The state of atmospheric conditions in a particular place at a particular time. (Contrast with climate.)

weathering The mechanical and chemical processes by which rocks are broken down into soil particles.

Wernicke's area A region in the temporal lobe of the human brain that is involved with the sensory aspects of language.

white blood cells Cells in the blood plasma that play defensive roles in the immune system. Also called leukocytes.

white matter In the central nervous system, tissue that is rich in axons. (Contrast with gray matter.)

wild type Geneticists' term for standard or reference type. Deviants from this standard, even if the deviants are found in the wild, are usually referred to as mutant. (Note that this terminology is not usually applied to human genes.)

wood Secondary xylem tissue.

X-Y-Z

xerophyte (zee' row fyte) [Gk. *xerox*: dry + *phyton*: plant] A plant adapted to an environment with limited water supply.

xylem (zy' lum) [Gk. *xylon*: wood] In vascular plants, the tissue that conducts water and minerals; xylem consists, in various plants, of tracheids, vessel elements, fibers, and other highly specialized cells.

yolk [M.E. *yolke*: yellow] The stored food material in animal eggs, rich in protein and lipids.

yolk sac In reptiles, birds, and mammals, the extraembryonic membrane that forms from the endoderm of the hypoblast; it encloses and digests the yolk.

zeaxanthin A blue-light receptor involved in the opening of plant stomata.

zona pellucida A jellylike substance that surrounds the mammalian ovum when it is released from the ovary.

zone of cell division The apical and primary meristems of a plant root; the source of all cells of the root's primary tissues.

zone of cell elongation The part of a plant root, generally above the zone of cell division, where cells are expanding (growing), primarily in the longitudinal direction.

zone of maturation The part of a plant root, generally above the zone of cell elongation, where cells are differentiating.

zoospore (zoe' o spore) [Gk. *zoon*: animal + *spora*: seed] In algae and fungi, any swimming spore. May be diploid or haploid.

zygospore Multinucleate, diploid cell that is a resting stage in the life cycle of zygospore fungi.

zygote (zye' gote) [Gk. *zygotos*: yoked] The cell created by the union of two gametes, in which the gamete nuclei are also fused. The earliest stage of the diploid generation.

zymogen The inactive precursor of a digestive enzyme; secreted into the lumen of the gut, where a protease cleaves it to form the active enzyme.

Illustration Credits

Index

Numbers in *italic* indicate the information will be found in a figure, caption, or table.

A

A band, *987, 988*
A blood group, 243
A horizon, 745
Aardvarks, *698*
Abalones, 662
Abdul-Jabbar, Kareem, 304
Abiotic ecosystem components
  defined, 1122
  factors influencing biomes, 1126–1127
ABO blood types, 53, 243
Abomasum, 1064
Aboral side, 680
Abortion, 897
Abscisic acid (ABA), 734, *759, 762,* 809
Abscission, 765, 769
Abscission zone, 765
Absolute dating, 506–507
Absorption spectrum, of pigments, 189, 190
Absorptive heterotrophs, 556, 609
Absorptive state, 1065
Acacia, 1136
Acacia ants, *1178, 1179*
Acacia cornigera, 1178, 1179
Acanthamoeba polyphaga mimivirus (APMV), *543*
Acanthostega, 690
Accessory fruits, 601
Accessory olfactory bulb, 950
Accessory pigments, 190, 571
Accessory sex organs
  defined, 887, 889
  in vertebrate evolution, 888
Accipiter gentilis, 1116, *1117*
Acclimation, in plants, 806
Acclimatization, in animals, 821
Accommodation, 958
Acetate, 171
Acetic acid, 34
Acetyl coenzyme A (acetyl CoA)
  allosteric regulation of the citric acid cycle, 182
  anabolic interconversions, 180
  β-oxidation, 179–180
  in the citric acid cycle, 170, 171
  in pyruvate oxidation, 170
Acetyl groups, 1051, 1052
Acetylcholine (ACh)
  actions in neuromuscular junctions, 936–938
  in the autonomic nervous system, 975

binding to the acetylcholine receptor, 129
clearing from synapses, 940
effect on gut muscle, 993
effect on heartbeat, 1034
functions of, 939
in skeletal muscle contraction, 989
in smooth muscle relaxation, 135–136
Acetylcholine receptors (AChR)
  at neuromuscular junctions, 936–938
  structure and function, 129
  types of, 939–940
Acetylcholinesterase (AChE)
  function at synapses, *937,* 940
  irreversible inhibitor of, 157, *158*
  isozymes in rainbow trout, 162
N-Acetylgalactosamine, 243
N-Acetylglucosamine, *55*
Acherontia atropos, 1009
Acid–base reactions, 34
Acid growth hypothesis, 766–767
Acid precipitation, 1219–1220
Acid–base balance, regulation by the kidneys, 1084–1085
Acid–base catalysis, 155
Acidification
  of lakes, 1220
  of oceans, 647, 1217
Acidophilic thermophiles, 536
Acids, 34–36, 43
Acinonyx jubatus, 62
Acoelomates, 635
Acoels, *632, 648, 680, 681*
Acorn worms, 682, *683*
Acoustic signals, 1112
Acquired hearing loss, 956
Acquired immune deficiency syndrome (AIDS)
  causal organism, 341
  description of, 876–877
  fungal diseases and, 612
  phylogenetic analyses, 458
  treatments, 301
Acropora millepora, 449
Acrosomal reaction, *885, 886*
Acrosome, 884, *885, 886, 889, 891*
ACTH. *See* Adrenocorticotropic hormone
Actin
  contractile ring, 216
  in microfilaments, 94–95 (*see also* Microfilaments)
  in muscle contraction, *987, 988–990, 991, 993*
  in muscle tissues, 817

Actin filaments
  impact of strength training on, 996
  in muscle fiber contraction, *987, 988–990, 991*
  See also Microfilaments
Actin–myosin bonds
  in insect flight muscle contraction, 998
  in skeletal muscle contraction, 988–990, *991*
  tetanic muscle contractions and, 995
Actinobacteria, 532, 1183
Action potentials
  in cardiac muscle, 991
  defined, 925, 927
  generation at neuromuscular junctions, 938
  generation of, 932–934
  membrane potentials and, 927–932
  principles of bioelectricity, 927
  production by sensory receptor cells, 947–948
  properties of, 934–935
  release of neurotransmitter at neuromuscular junctions, 936, *937*
  saltatory conduction, 935
  in smooth muscle, 993
  speed of transmission, 926
  See also Cardiac action potentials
Action spectra
  of photosynthesis, 189, 190
  of plant phototropism, *771, 772*
Activation energy, 151–152
Activation gate, 933–934
Activator proteins, 329, 332
Active predators, *638*
Active transport
  defined, 113, 118
  directional nature of, 118
  energy sources for, 118–119
Actuarial tables, 1154
Adaptation (sensory), 948
Adaptations
  to climate, 1122, 1125–1126
  defined, 6, 433
  examples in frogs, 7
  to reduce heat loss in endotherms, 828
  trade-offs, 445
Adaptations in plants
  defined, 806
  to growth in saturated soils, 808
  to life on land, 574
  to saline environments, 811

to temperature extremes, 810
to very dry conditions, 806–808
Adaptive defenses, 857, *858*
Adaptive immunity
  cellular immune response, 871–875
  clonal deletion, 865
  clonal selection, 865, *866*
  discovery of, 862, *863,* 864
  humoral immune response, 867–871
  immunological immunity and vaccinations, 866
  immunological memory and secondary immune response, 865–866
  key features, 862–863
  types of responses in, 863–864
Adaptive radiations, 481–482, 489–490
Adenine
  codons and the genetic code, 288–289
  complementary base pairing, 63–65
  in DNA structure, 264, 265, 266, *267*
  poly A tail, 292
  structure, *63*
Adeno-associated virus, 323, 324
Adenosine, 128
Adenosine 2A receptor, 128
Adenosine diphosphate (ADP)
  in ATP hydrolysis, 150
  in ATP synthesis by ATP synthase, *173, 174, 175*
  as a coenzyme, 167
Adenosine monophosphate (AMP), 150
Adenosine triphosphate (ATP)
  in active transport systems, 118–119
  allosteric regulation of glycolysis, 182
  allosteric regulation of the citric acid cycle, 182
  as an "energy currency," 149–151
  in C_4 photosynthesis, 199
  in the Calvin cycle, 194, *196*
  in the charging of tRNAs, 294
  coupled reactions and, 150–151
  effect of ATP supply on muscle performance, 997
  energy released during hydrolysis, 149–150
  functions of, 66, 149
  in protein phosphorylation, 209
  role in catalyzed reactions, *156*

in skeletal muscle contraction, 989, *990*, *991*
slow-twitch fibers and, 995
structure, 149
tetanic muscle contractions and, 995
Adenylyl cyclase, 133
Adiantum, 583
Adipose cells
effect of insulin on, 1066
fat metabolism in the postabsorptive state, 1067
See also Brown fat
Adipose tissues
characteristics of, 819
in the fight-or-flight response, 837
types of, 165
Adrenal cortex, 849–850
Adrenal glands
fight-or-flight response, 837
hormones of, *842*, 849–850
Adrenal medulla, 849, 850
Adrenaline. *See* Epinephrine
Adrenergic receptors, 850, 853
Adrenocorticotropic hormone (ACTH), 849
Adult hemoglobin, 1018
Adventitious roots, 718
Aegilops speltoides, 226
Aegilops tauschii, 226
Aegolius funereus, 1129
Aegopodium podagraria, 597
Aequorea victoria, 378, 449
Aerenchyma, 808
Aerobic exercise
ATP supply and, 997
impact on muscles, 996–997
slow-twitch fibers and, 995
Aerobic metabolism
atmospheric oxygen levels and, 512
cellular respiration, 166–167
energy yield from cellular respiration and fermentation compared, 178
in the evolution of life, 5
pathways of glucose catabolism, 169–171
in prokaryotes, 537
Aerotolerant anaerobes, 537
Afferent blood vessels, 1010
Afferent nervous system, 968
Afferent neurons
defined, 940
spinal reflexes, 942
Afferent renal arterioles
in the autoregulation of the glomerular filtration rate, 1087, *1088*
in the mammalian kidney, *1080,* 1082
in the vertebrate nephron, 1078
Affinity chromatography, 853
Aflatoxins, 310, 624
Africa
ecotourism, 1242, *1243*
ivory trade, 1240–1241
African ass, *1144*
African clawed frog, 612
African elephants, *1135*
fecundity, 1155
full census counting, 1150

population age structure, 1152
See also Elephants
African fish eagle, *218*
African insectivores, *698*
African lions, 1183
African long-tailed widowbird, 435–436
African lungfish, *689,* 1028–1029
African millet beer, 624
African wild dogs, 1116, 1183
Afrosoricida, *698*
After-hyperpolarization, 934
Agalychnis callidryas, 7
AGAMOUS gene, 786
Agarose, 316
Agave, 1132
Agave schottii, 1183
Age-dependent cohort life tables, 1154
Age structure, 1150, 1151–1152, 1165–1166
Aggregate fruit, 601
Aging, telomerase and, 275
Aglaophyton major, 580
Agnosias, 971
Agonists, 940
Agriculture
apomixis studies, 794
applications of biotechnology, 386–388, *389*
biological nitrogen fixation and, 750
dicultures, 1203
effect of levels of atmospheric carbon dioxide on global food production, 202
"Green Revolution," 756
hybrid plants, 778
hybrid vigor, 244–245
impact on ecosystems, 1223
importance of biological research to, 15
important crop plants, 605
improving the nitrogen use efficiency of corn, 740, 753
increasing levels of atmospheric carbon dioxide and, 202
methods to reduce water loss, 738
nitrogen fertilizers and, 740
polyploidy in crops, 225, *226*
problems of monoculture, 1202–1203
secondary succession on agricultural land, 1200
semi-dwarf wheat and rice, 756, 775
soil fertility and fertilizers, 746–747
transgenic rice, 726
use of ethylene in, 769, 770
vegetative reproduction of angiosperms in, 793–794
water demands of, 726
See also Crop plants; Transgenic crops
Agrobacterium tumefaciens, 377, 534
Agropyron repens, 597
AHK protein, 769
AHP protein, 769
AIDS. *See* Acquired immune deficiency syndrome
Air capillaries, 1011

Air quality, lichens as indicators of, 624, *625*
Air sacs, 1011, *1012,* 1013, *1014*
Air temperature
atmospheric circulation patterns and, 1124
causes of seasonal variations, 1123
Airplane metaphor, 1245
Aix sponsa, 468
Alanine, *44*
Alarm calls, 1112, 1116, *1117*
Albumin gene, 335
Alburnus scoranza, 1140
Alcedo atthis, 642
Alces alces, 1129
Alcohol dehydrogenase, 155, *177,* 178
Alcoholic beverages, 178
Alcoholic fermentation, 177, 178
Alcoholism, 1054
Aldehyde group, *40*
Alderflies, *672*
Alders, 1200
Aldosterone
actions of, *842,* 849
regulation of blood pressure and blood volume, 1087, *1088*
structure, *836*
Aleuria aurantia, 622
Aleurone layer, 761, *762*
Alexandrium, 549
Algae
biofuel production and, 585
brown algae, 556
closest relatives of land plants, 572–573
defined, 571
evolution of photosynthesis in, 571–572
red algae, 551, 571–572, 573
See also Green algae
Algal blooms, 1219
Alginic acid, 556
Alkaloids, of endophytic fungi, 615
Alkaptonuria, 282
Allantoic membrane, 918
Allantois, 918, 919
Allele frequency
calculating, 436–437
defined, 432
effect of gene flow on, 433–434
effect of nonrandom mating on, 434–436
Hardy–Weinberg equilibrium, 437–438
microevolutionary change, 446
Allele-specific oligonucleotide hybridization, 321
Alleles
codominance, 243
defined, 236, 431
genetic drift, 434
human pedigree analyses, 240–241
incomplete dominance, 242–243
law of independent assortment, 237–239
law of segregation, 236–237
multiple, 242, 243
mutations, 241–242
neutral, 441
pleiotropic, 243–244

using probability to predict inheritance, 239–240
Allergens, 875–876
Allergic reactions, 875–876
Allergies, 862
Alligators, 694, 1029–1030
Allolactose, 331
Allomyces, 618
Allopatric speciation, 472–474, *475*
Allopolyploidy, 225, *226,* 473, 475
Allosteric enzymes, 159–160
Allosteric regulation
in cell cycle control, 210–211
characteristics of, 159–160
of glucose catabolism, 181–182
of metabolic pathways, 160–161
Allosteric sites, of enzymes, 159
Allostery, 159
Alluaudia procera, 1135
Alnus, 1200
Alpha (α) helix, 45, *46*
Alpha-lactalbumin, 333
Alpine pennycress, 811
Alpine tundra, 1128–1129
Alternation of generations
in ferns, 582
in land plants, 574–575
overview, 218
in protists, 562–563
Alternative splicing, 346–347
Altitude, oxygen availability and, 1007–1008
Altman, Sidney, 72
Altricial young, 642
Altruistic behavior, 1114–1115, 1116
Alveolates, 553–555
Alveoli
in alveolates, 553
of ciliates, 554, *555*
diffusion of carbon dioxide from blood, 1018, *1019*
in humans, 1013, *1014*
lung surfactants, 1015
Alzheimer's disease, 336
Amacrine cells, *962,* 963
Amber, *515*
Amblyrhynchus cristatus, 824
Amborella, 602–603
Ambros, Victor, 347
Ambulacrarians, 680
Ambystoma mavortium, 691
Ambystoma mexicanum, 1007
Amensalism, *1170,* 1171
American bison, *1131*
American chestnut, 622
American elm, 622
American holly, *780*
American Prairie Foundation, 1238
Amine hormones, 836
Amino acid sequences
of cytochrome *c* from different organisms, 488–489
genomic information and, 356
identifying evolutionary changes in, 486
Amino acids
binding of tRNA to, 293, 294
catabolic interconversions, *179,* 180
chemical structure, *44*
codons and the genetic code, 288–289

formed in prebiotic synthesis experiments, 70
found on meteorites, 69
need for in animal diets, 1051–1052
optical isomers, 43
peptide linkages, 43–44, *45*
primary structure of proteins, 45, *46*
properties of, 43
side chains, 43, *44*
Amino group
  of chemically modified carbohydrates, 55
  effect of pH on, 161
  peptide linkages in amino acids, 43–44, *45*
  properties as a base, 34
  properties of, *40*
Amino sugars, 55
Aminoacyl tRNA binding site, 295, 296
Aminoacyl-tRNA synthetases, 294
γ-Aminobutyric acid (GABA), 324, 939
Aminopeptidase, *1062*
Aminopterin, 158
Ammocoetes, 686
Ammonia
  in acid–base regulation by the kidneys, 1085, *1086*
  as a base, 34
  excretion by animals, 1074–1075
  in the global nitrogen cycle, 750, *751*
  inorganic fertilizers, 747
  nitrogen fixation and, 539
  oxidation by nitrifiers, 539
Ammonium
  in the global nitrogen cycle, 750–751
  inorganic fertilizers, 747
  organic fertilizers and, 746
  sources for plants, 200–202
Ammonium transporters, 744
Ammonotelic animals, 1074
Amniocentesis, 321
Amnion, *907*
  in the chicken egg, 918
  origins of, 914
  in placental mammals, 919
Amniote egg, 692, 888
Amniotes
  in the Carboniferous, 520
  evolutionary innovations, 692
  origin of, 690
  phylogeny, *693*
Amniotic cavity, 919
Amniotic fluid, 919
Amoeba proteus, 98
Amoebas
  amoebozoans, 559–561
  heteroloboseans, 558
Amoebozoans, 559–561
AMP-activated protein kinase (AMPK), 1068
Amphibian decline
  atrazine and, 1
  controlled and comparative experiments in, 12, 13
  pathogenic fungi and, 612
Amphibians
  actions of prolactin in, *838*
  anurans, 690, 692

in the Carboniferous, 520
cell fate determination in the embryo, 394
circulatory system, 1029
diversity in, 690, *691*
effects of atrazine on, 1
egg rearrangements following fertilization, 903–904
environmental requirements, 690
gas exchange through the skin, 1029
gastrulation, 909–913
life cycle, *691*
neurulation, 915–916
nitrogenous wastes, 1075
origin of, 690
recent decline in, 692
salamanders, 692
salt and water balance regulation, 1077
sexual reproduction, 888
social behaviors, 692
Amphibolurus muricatus, 421
Amphipathic molecules, 57
Amphipods, 670
Amphiprion, 888, 1171
Amphisbaenians, 693
Ampicillin, 527
Amplitude, of biological cycles, 774
Amygdala, 970, 982
Amylases, 144, 1060
Amylopectin, 53
Amylose, 53
Anabaena, 533, 749
Anabolic interconversions, 180
Anabolic reactions
  endergonic reactions, 147, *148*
  energy changes during, 145
  integration of anabolism and catabolism, 180–181
  linkage to catabolic reactions, 146, 179–180
  repressible systems of regulation in prokaryotes, 332
Anacharis, 189, 190
Anadromous fishes, 689
Anaerobic metabolism
  defined, 5
  energy yield from cellular respiration and fermentation compared, 178
  fermentation pathways, *166, 167,* 177–178
  glycolysis, 166
  non-oxygen electron receptors, 176
  in prokaryotes, 537
Anal fins, 607
Anal sphincter, 1060, 1063
Anaphase (mitosis)
  chromosome separation and movement, 214–216
  comparison between mitosis and meiosis, *223*
  events in mitosis, 212, *217*
Anaphase I (meiosis), 221
Anaphase II (meiosis), *220*
Anaphase-promoting complex (APC), 214, 215, *216*
Ancestral states, reconstruction, 460

Ancestral traits, 452
Anchor cell, 396, *397*
Andansonia, 1136
Andersson, Malte, *435*
Androgens, 850, *851*
  See also Testosterone
Anemia, 292–293, 1053, 1054
Anemonefish, 887, *888,* 1171
Aneuploidy, 222, 224, 313
Anfinsen, Christian, 48, 49
Angel insects, *672*
Angelman syndrome, 345
Angiosperms
  asexual reproduction, 792–794
  coevolution with animals, 588, 598–599, 605
  defining features, 596
  distinguishing characteristics, *574*
  double fertilization, 600–601, *781,* 783
  endosperm, 591
  flowers, 596–598, *599*
  fruit development and dispersal, 784–785
  fruits, 601
  gametophytes, 779–780
  life cycle, 600–601
  monocots and eudicots, 709–710
  monoecious and dioecious, 597
  phylogenetic analysis of fertilization mechanisms, 459–460
  phylogenetic relationships, 601–603
  radiation during the Cretaceous, 521
  root systems, 718
  seed plants, 589
  sexual reproduction, 779–785
  shared derived traits, 596
  strategies for preventing inbreeding, 782, *783*
  in succession, 1201
  transition to the flowering state, 785–792
  vegetative organs, 709
  xylem, 714
Angiotensin, 1044, 1087, *1088*
Angiotensin-converting enzyme, 1087, *1088*
Anglerfishes, *688*
Angraecum sesquipedale, 588
Angular gyrus, 981
Animal behavior
  altruistic behavior, 1114–1115, 1116
  behavioral ecology, 1102
  cost–benefit approach, 1103
  development of, 1098–1102
  evolution of social behavior, 1113–1117
  genetic basis of, 1096–1098
  nest parasitism, 1093
  origins of behavioral biology, 1094–1096
  as a sequence of choices, 1103
  territorial, 1103–1104, *1105*
  underlying physiological mechanisms, 1106–1103
Animal cells
  communication through gap junctions, 139–140
  cytokinesis, 216

extracellular matrix, 100
insertion of genes into, 376
osmosis, 114, *115*
production of metabolic heat and, 822
relationship between cells, tissues, and organs, 817–820
structure of, *86*
  See also Eukaryotic cells
Animal communication, 1110–1113
Animal defense systems. *See* Immunology
Animal development
  activation by fertilization, 903–904
  cleavage, 904–906
  determination of polarity, 395
  determination of the blastomeres, 906–907
  extraembryonic membranes, 918–919
  gastrulation, 908–915
  germ cell lineage, 908
  mosaic and regulative, 907
  organogenesis, 915–918
  overview, *393*
  restriction of cell fate during, 394
  See also Human development
Animal groups
  bilaterians, 643
  cnidarians, 645–648
  ctenophores, 644–645
  eumatozoans, 643
  placozoans, 645
  sister group to all animals, 648
  sponges, 643–644
Animal hemisphere, 903
Animal hormones
  of the adrenal gland, 849–850
  classes of, 836
  compared to plant hormones, *758*
  control of digestion, 1065
  control of insect molting, 839–841
  control of sexual function in human males, 892
  detection and measurement with immunoassays, 852–853
  discovery of, 838–839
  dose–response curves, 853
  factors affecting the action of, 837
  half-life, 852–853
  hormone-mediated signaling cascades, 837
  irisin, 834, 838
  location and function of receptors, 836–837 (*see also* Hormone receptors)
  melatonin, 851
  regulation of blood calcium levels, 847–848
  regulation of blood glucose concentrations, 848–849
  regulation of blood pressure, 1044–1045
  role in determining behavioral potential and timing, 1098–1099
  secondary sex characteristics and, 250
  sex steroids, 850–851
  thyroxine, 845–847

Animal nutrition
  control and regulation of
    nutrients in the body, 1064–
    1068
  macronutrients and
    micronutrients, 1052–1053
  nutrient deficiencies, *1053*, 1054
  vitamins, 1053–1054
  See also Food
Animal pole, 395
Animal-pollinated plants, 480
Animal reproduction
  budding and regeneration, 881
  honey bees, 880
  life cycles, 639–640
  parthenogenic, 881–882
  sexual, 882–889 (*see also* Sexual
    reproduction)
  trade-offs in, 641–642
  See also Human reproduction
Animals
  basic developmental patterns
    in, 633–634
  body plans, 634–637
  cloning, 406–407
  coevolution with angiosperms,
    588, 598–599, 605
  common ancestor of, 631, *633*
  feeding strategies, 637–639
  general characteristics, 630
  genomes, *361* (*see also*
    Eukaryotic genomes)
  life cycles, 639–642
  major living groups, *632*
  monophyly of, 630–631, 633
  opisthokonts, 609
  pharming, 385–386
  phylogenetic tree, *630*
  placozoans, 629
  plant–pollinator mutualisms,
    1180–1181
  pollen transport, 780, *782*
  smallest genome of, 629
  thermoregulation, 822–826
Anions
  defined, 29
  leaching in soils, 746
Annelids
  anatomical characteristics, *652*
  body plan, *660*
  clitellates, 661–662
  closed circulatory system, *1027*
  excretory system, 1075, *1076*
  hydrostatic skeleton, 999
  key features of, 659–660
  major subgroups and number
    of living species, *632*
  polychaetes, 660–661
  segmentation, 636, 659, *660*
  undescribed species, 675
Annona squamosa, 785
Annual rings, 722–723
Annuals, 785
Anomalocaris canadensis, 517
Anopheles, 563, 564
Anoxygenic photosynthesis, 186,
  187
Anser anser, 642, 1099
Ant lions, *672*
Antagonistic interactions
  arms race analogy, 1172
  description of, 1170
  impact on communities,
    1193–1194

predator–prey interactions,
    1172–1175
Antagonistic muscle sets, 942
Antagonists, 128, 940
Antarctica
  adaptations to low
    temperatures, 1125
  lichens, 613
Anteaters, *698*
Antenna systems, 190
  See also Light-harvesting
    complexes
Antennae, 636
Antennapedia (Antp) gene, 404, 415
Anterior, 634, *635*
Anterior pituitary
  control by hypothalamic
    neurohormones, 844
  effect of hypothalamic
    somatostatin on, 849
  hormones produced by, *842,* 843
  in human puberty, 892
  negative feedback loops in
    the regulation of hormone
    secretion, 844–845
  organization and function of,
    842–843
  regulation of human
    spermatogenesis, 892
  regulation of the adrenal cortex,
    849
  regulation of the ovarian and
    uterine cycles, 893, 894
  regulation of thyroxine
    production, 846
Anterior–posterior axis
  body segmentation and Hox
    genes, 916–918
  establishment in the animal
    zygote, 903
Antheridium, 576
Anthers, *591,* 596, 598, 1180
Anthocerophyta, *574*
Anthoceros, 579
Anthocyanin, 93
Anthozoans, 646–647
Anthrax, 531, 542
Anthropoids, 701–702
Anti HIV drugs, 342–343
Anti-inflammatory drugs, 862
Antibiotic resistance
  composite transposons and, 359
  lateral gene transfer and, 496,
    530
  methicillin-resistant *S. aureus,*
    281, 301
  problem of, *626, 627*
Antibiotic resistance genes, 376,
    378, 530
Antibiotics
  derived from fungi, 608
  discovery of penicillin, 608
  effects on bacteria, 281, 295,
    301, 527
  plant phytoalexins, 799–800
  targeting bacterial RNA
    breakdown machinery, 301
Antibodies
  classes of, *868*
  functions of, 858
  in the humoral immune
    response, 864, *865*
  monoclonal, 871

plasma cell production of, 867
  specificity of adaptive
    immunity, 862–863
  See also Immunoglobulins
Anticancer drugs
  antisense drugs, 382
  approaches to designing in
    molecular medicine, 304, 325
  histone deacetylase inhibitors,
    344
  targeting of receptors, 325
  taxol, 604
  treatment of chronic
    myelogenous leukemia, 304
Anticoagulants, 661, 662
Anticodons, 293, 294
Antidepressants, 940
Antidiuretic hormone (ADH)
  actions of, *842,* 843
  regulation of blood pressure
    and blood osmolarity, 1044–
    1045, 1088–1089
Antifreezes, 810, 1125
Antigen-presenting cells, 864, *865,*
    872, 876
Antigenic determinants, 862, 863
  See also Antigens
Antigens
  ABO blood groups and, 243
  in allergic reactions, 875–876
  binding of antibodies, 858
  binding to T cell receptors,
    871–872
  exposure in the adaptive
    immune response, 863–864
  immunoglobulin binding sites,
    867–868
  specificity of adaptive
    immunity, 862–863
  vaccinations and, 866
Antioxidants, 176
Antiparallel strands, in DNA
  structure, 265, 266
Antipodal cells, 779, *781,* 783
Antiporters, 118, 119
Antique bison, *1229*
Antisense drugs, 382
Antisense RNA, 382
Ants
  interference competition, 1183
  fungus farming by, 1169, 1178,
    1185
  mutualisms with plants,
    1178–1179
  survival of seeds in the fynbos
    and, 1121, 1135, 1145, 1146
  See also Hymenoptera
Anurans, 690, 692
Anus, *1058*
  in bryozoans and entoprocts,
    656
  in tubular guts, 1056
Anxiety, 850
Aorta
  aortic bodies, *1021,* 1022, 1045
  in fish, 1028
  in mammals, *1031,* 1032
  in reptiles, 1029, 1030
  stretch receptors in the
    regulation of blood pressure,
    1088–1089
Aortic bodies, *1021,* 1022, 1045
Aortic valve, 1030, 1031, 1032
Apaf1 protein, 399–400

APC gene, *314*
Apes, 701–702
APETALA1 gene, 786, 792
APETALA1 transcription factor,
  786
Aphasia, 981
Aphelocoma coerulescens, 1115
Aphids, *672, 673,* 735, 1153
Aphotic zone, *1139,* 1140
Apical buds, 765
Apical complex, 553
Apical dominance, 765
Apical hook, 770
Apical meristems
  of deciduous trees, 721–722
  origin in plant embryogenesis,
    712
  in plant growth and
    development, 710, 715–716,
    719–720
Apical–basal axis, in plants, 711
Apicomplexans, 553–554, 563
Apis mellifera, 1112, 1150
  See also Honey bees
Apomixis, 778, 794
Apoplast, 729–730, 737
Apoplastic pathway, 736
Apoptosis
  blocking by some cancers, 382
  in bone growth, 417
  cancer and, 228
  in clonal deletion, 865
  hypersensitive response in
    plants, 800–801
  in morphogenesis, 394
  in pattern formation, 399–400
  in plant cells, 226
  reasons for, 225–226
  signals and pathways in, 226,
    227
Aposematism, 662, 1173, *1174*
Appalachian Mountains, 472
Appendages
  of arthropod relatives, 667–668
  of arthropods, 655
  functions in animals, 636–637
  See also Jointed appendages
Appendix, *1058,* 1064
Apple maggot fly, 473
Apples, 601
Aptenodytes forsteri, 1099
Aptenodytes patagonicus, 642
Aquaporins
  actions of antidiuretic hormone
    on, 1088, 1089–1090
  of the collecting duct, 1088,
    1089–1090
  function of, 105, 116
  industrial water purification
    and, 122
  in renal tubules, 1084
  water movement across
    membranes and, 728–729
Aquatic biomes
  estuaries, 1141
  freshwater, 1140
  marine, 1139–1140
Aquatic ecosystems
  consequences of human
    alteration, 1223
  geographic distribution of net
    primary production, 1210
Aquatic species
  gametic isolation and, 477

invertebrate osmoconformers and osmoregulators, 1072–1073
Aqueous solutions
  acids and bases, 34–36
  properties of, 33–34
Aquifers, 1215
Aquifex, 532
Aquilegia, 480
Aquilegia formosa, 477
Aquilegia pubescens, 477
Ara chloropterus, 1106
Arabidopsis thaliana
  apomixis, 794
  camalexin, 801
  cytokinin signaling in, 769
  dwarfed phenotype, 760, *761*
  effect of carbon dioxide on stomatal density, 734
  effect of gibberellins on flowering, 791–792
  embryogenesis, 711–712
  floral organ identity genes, 400–401
  genes involved in the signaling response for flowering, 790
  genetic screens and the identification of signal transduction pathways, 759, 760
  genomic information, *361*
  ion transporters, 744
  meristem identity genes, 786
  metabolone, 370
  as a model organism, 14
  photoperiodism and flowering, 788
  phototropins, 771
  phytochromes, 773, 774
  vernalization studies, 791
  water-use efficiency studies, 726
Arachidonic acid, 1052
Arachnids, 667, 669
Arbuscular mycorrhizae
  description of, 614–615, *616,* 619–620
  expansion of the plant root system, 748–749
  formation, 747, 748
  phylogeny of the fungi, *615*
Arceuthobium americanum, 752
Archaea
  in the evolution of life, 4
  examples of, 2
  extremophiles, 532
  genomes, 356 (*see also* Prokaryotic genomes)
  metabolic pathways, 537–538
  nitrifiers, 539
  nitrogen fixers, 539
  non-oxygen electron receptors, 176
  shapes of, 528
  in the tree of life, 8
  See also Prokaryotes
Archaea (domain)
  distinguishing characteristics, 81, *527*
  habitats, 534
  identification through gene sequencing, 534
  major groups, 535, 536–537
  membrane lipids, 535–536

relationship to Bacteria and Eukarya, *526*
Archaeognatha, *672*
Archaeopteris, 517
Archaeopteryx, 695
Archean eon, 506–*507,* 508
Archegonium, 575–576
Archenteron
  in frog gastrulation, 909, 910
  in sea urchin gastrulation, 909
Archosaurs, 460, *695*
Arctic
  adaptations to low temperatures, 1125
  impact of global climate change on sea ice, 1217, *1218*
  lichens, 613
Arctic ground squirrel, *10*
Arctic hare, *828*
Arctic tundra, 1128–1129
Arctostaphylos montaraensis, 1134
Arcuate nucleus, 1067–1068
Ardipithecines, 702–703
Area phylogenies, 1143, *1144*
Argema mittrei, 674
Argentine ant, 1145, 1146
Arginine
  in histones, 212
  plant canavanine and, 803, 805
  structure, *44*
Argogorytes mystaceus, 1181
Argon, 1211
Argon-40, 507
Argopecten irradians, 663
Argyroxiphium sandwicense, 481
Arils, 594
Ariolimax columbianus, 664
Aristolochia littoralis, 602
Armadillidium vulgare, 670
Armadillos, 698
Armillaria, 623
Arms, of cephalopods, 664
Arms race analogy, 1172, 1175–1176
Aromatase, 420, 850, 896
ARR protein, 769
Arrow worms, *630, 632, 652,* 655–656
Arsenic, 1221
Artemia, 1072, 1073
Artemisia annua, 797
Artemisinin, 797, 812
Arterial blood pressure
  control and regulation of blood flow, 1043–1044
  in fish, 1028
  measuring, 1032, *1033*
  regulation of, 1044–1045
Arteries
  atherosclerosis, 1042–1043
  in blood pressure regulation, 1044
  hardening of, 848
  structure and function, 1027–1028, 1039, *1040*
Arterioles
  in blood pressure regulation, 1043, 1044
  functions in vertebrate circulatory systems, 1028
  regulation of blood flow, 1039, *1040,* 1044
Arthrobotrys dactyloides, 613

Arthropods
  anatomical characteristics, 652
  appendages, 636, 655
  body plan, *655*
  chelicerates, 668–669
  compound eyes, 958
  ectoparasites, 639
  exoskeleton, 999
  hemocoel, 652
  heterotypy and leg number in, 418, *419*
  hormonal control of molting, 839–841
  influence of the exoskeleton on evolution, 655
  insects, 671–673, *674*
  key anatomical features, 667
  major living groups, *632,* 667
  mandibulates, 669–671
  open circulatory system, 1026, *1027*
  relatives, 667–668
  segmentation, 636
  success of, 667
  trilobites, 668
Artificial DNA, 380
Artificial insemination, 897
Artificial life studies, 359, *360*
Artificial ribozymes, 72
Artificial selection
  in agriculture, 15
  Darwin's knowledge of, 6, 433
  description of, 432–433
  plant domestication, 723–724
Artocarpus heterophyllus, 601
Asci, 620, *621,* 622
Ascidians, 684
Asclepias syriaca, 806
Ascoma (ascomata), 620, *621,* 622
Ascomycota, 613, *616,* 620–622
  See also Sac fungi
Ascorbic acid, 1053
Ascospores, 620, *621,* 622
Asexual reproduction
  in angiosperms, 792–794
  in animals, 881–882
  in diatoms, 555
  fungi, 616
  by mitosis, 217
  in protists, 562
Asian apes, 702
Asparagine, 44
Aspartame, 322
Aspartate, 199
Aspartic acid, 44
Aspens, 217, 793, 1193–1194
Aspergillus, 310, 624
Aspergillus nidulans, 624, 626
Aspergillus niger, 624
Aspergillus oryzae, 624
Aspergillus tamarii, 624
Aspicilia, 613
Aspirin, 830, 862
Assisted reproductive technologies (ARTs), 897, 899
Association cortex
  defined, 970
  in human brain evolution, 973
  size in humans, 978
Associative learning, 982
Asterozoans, 681
Asthenosphere, 509
Astrocytes, *819,* 926–927
Astrolithium, 557

Asymmetry, in animals, 634
"Asynchronous" muscle, 998
Atacama Desert, 808, 1127, 1134
Ateles geoffroyi, 702
Atherosclerosis, 1042–1043
Athletes
  effect of enhanced cooling on performance, 831
  heart failure, 1025, 1046
  heat stroke, 815
  replenishment of muscle glycogen and, 997
Athlete's foot, 612
Atlantic croaker, 1207
Atmosphere
  composition and structure, 1211–1211
  global circulation patterns, 1124
  as a medium for respiratory gases, 1007
  methane and, 536
  moderation of Earth's temperature by, 1212
  oxygen in the early atmosphere of Earth, 69
  prevailing winds, 1124, *1125*
Atmospheric carbon dioxide
  biofuels and, 585
  from burning fossil fuels, 1216
  effect of increasing levels on corals, 647
  effect of increasing levels on global food production, 202
  effect of increasing levels on photosynthesis, 185, 202
  effect of oceanic carbon stores on, 1217
  evolution of C_3 and C_4, 200
  evolution of leaves and, 583, 584
  global amounts released by fire, 1214
  in the global carbon cycle, 1216–1218
  global climate change and, 17, 510, 1217–1218
  as a greenhouse gas, 1212
  impact on nitrogen fixation by microorganisms, 1221–1223
  percent of the total atmosphere, 1211
  photosynthetic efficiency of plants and, 201–202
Atmospheric oxygen
  body size in insects and, 505, 513, 514, 522
  during the Cambrian, 516
  during the Carboniferous, 505, 521
  cyanobacteria and, 532, 539
  impact of life on, 511–513
  during the Mesozoic, 521
  percent of the total atmosphere, 1211
  during the Permian, 505, 522
  photosynthesis and the evolution of life, 4–5
  during the Proterozoic, 515
Atmospheric pressure, 1006
Atom bomb, 310
Atomic number, 22, 23
Atomic weight, 23
Atoms
  atomic number, 22

bonding to form molecules, 26–31
in chemical reactions, 31
components of, 22
electric charge, 22
electron orbitals and shells, 24–25
electronegativity, 28
elements and, 22
isotopes, 22–24
mass number, 22
molecules and, 24
reactive and stable, 25
role of electrons in chemical reactions, 24, 25
ATP. *See* Adenosine triphosphate
ATP synthases
in ATP synthesis, 171, *173*, 174
in chloroplast photophosphorylation, 192–193
mechanism of operation, 175, *176*
ATP synthesis
in the citric acid cycle, 171
energy cost of NADH shuttle systems and, 178–179
in fermentation pathways, 177, 178
in glucose catabolism, 168, 169
in glucose oxidation, 166
in glycolysis, 169, 170
mitochondria and, 91
in oxidative phosphorylation, 171–176
in photosynthesis, 188, 191, 192–193
Atrazine
experiments on the effects of, 12, 13
impact on sex development, 1, 18
public policy issues, 17
Atria
in amphibians, 1029
in fish, 1028
in lungfish, 1028
in mammals, 1030–1032
in reptiles, 1029
Atrial natriuretic peptide (ANP), 1090
Atrioventricular node, 1034, *1035*
Atrioventricular valve (AV), 1030, 1031, 1032
Atriplex halimus, 811
Atropine, 604
Atta cephalotes, *1169*, 1185
Atta colombica, 1185
Attention, the parietal lobe and, 973
Attenuation, 866
Auditory canal, 954
Auditory hair cells
hearing loss and, 956
in sound detection, *954, 955*, 956
stereocilia and mechanoreception, 953–954
Auditory processing, 971
Auditory systems
echolocation in bats, 946, 963
flexion of the basilar membrane, 955–956

hair cell stereocilia and mechanoreception, 953–954
hearing loss, 956
structure of the human ear, 954–955
Audounia capitata, 1121
Austin blind salamander, *691*
Australasian region, *1142*
Australia
cane toad populations, 1164, 1235
characteristics of deserts in, 1126
microfossils, 74
Murchison Meteorite, 69
Australian pelican, 412
Australopithecines, 703
Australopithecus afarensis, 703
Austrobaileya, 598
Autism, 540
Autocatalysis, 1061
Autocrine signals, 126, 835
Autoimmunity (autoimmune diseases)
causes and examples of, 876
inflammation and, 862
microbiomes and, 540
multiple sclerosis, 926
Autolysis, 1188
Autonomic nervous system (ANS)
in blood pressure regulation, 1044, *1045*
control of heartbeat, 1034
heart rate and, 992
influence on smooth muscle, 993
structure and function, 974–975
Autophagy, 90–91
Autopolyploidy, 224–225, 473
Autosomes, 249
Autotetraploids, 224–225
Autotriploids, 224–225
Autotrophs
in communities, 1189–1190
defined, 196
as food for heterotophs, 1049
Auxins
discovery of, 760, *761, 762, 763*, 764
ethylene and, 769, 770
gravitropism, 765
interactions with cytokinins, 768
lateral transport, 764–765
molecular mechanisms underlying the activity of, 767
phototropism response and, 762, 763, 764, 765
polar transport, 763–764
structure and typical activities, *759*
Avery, Oswald, 261
Avian malaria, 1236
Avirulence (Avr) genes, 800
Avogadro's number, 33
Axial filaments, of spirochetes, 533
Axillary buds
apical dominance and, 765
of Brussels sprouts, 724
in phytomers, 709
stimulation of growth by cytokinins, 768

Axon hillock, 932–933, 938
Axon terminal
defined, 925
mix of synaptic activity impinging on, 938
neuromuscular junctions, 936
synapses and, 926
Axons
conduction of action potentials, 934–935
functions of, 925
generation of action potentials, 932–934
graded membrane potentials, 932
measurement of the resting potential, 928
myelination, 926
nerves, 968
Azotobacter, 749

B

B blood group, 243
B cells
as antigen-presenting cells, 864, 872
clonal deletion, 865
clonal selection, 865, *866*
cytokines, 859
effector B cells, 865
function of, *858*
generation of diversity in immunoglobulins, 868–869, *870*
in the humoral immune response, 863, 864, *865*, 872, *873*
immunoglobulin class switching, 869–871
maturation in bone marrow, *857*
memory B cells, 865
plasma cells, 867
specificity of adaptive immunity and, 862–863
B horizon, 745
Baboons, *1135*
"Baby boom," 1165
Bacillus, 531, 539
Bacillus (shape), 528, *529*
Bacillus anthracis, 531, 542
Bacillus subtilis, 206
Bacillus thuringiensis toxin, 387–388, 389
Back substitutions, 488
Bacteria
anoxygenic photosynthesis, 186, 187
antibiotic resistance (*see* Antibiotic resistance)
Bacillus thuringiensis toxin in transgenic plants, 387–388
bioluminescence, 525, 534, 546
in bioremediation, 373, 389
cell walls, 527, 528
chlamydias, 533, *534*
conjugation, 253–255
conjugative pili, 84
cyanobacteria, 532, *533*
denitrifiers, 539
DNA replication, 269, 271, 273
effects of antibiotics on, 281, 295, 301, 527
in the evolution of life, 4

experimental studies in molecular evolution, 489–490
expression vectors for, 384
extremophiles, 532
fungal gardens and, 1169
genetic transformation, 260–261
genomes, 356, *361*, 494 (*see also* Prokaryotic genomes)
global nitrogen cycle, 750–751
Gram stains, 527–528
hadobacteria, 532
high-GC Gram-positives, 532
hyperthermophilic, 532
influence on plant uptake of nutrients, 747–748, 748–751
insertion of genes into, 376
lateral gene transfer, 496, 530
low-GC Gram positives, 530–532
metabolic pathways, 537–538
mRNA recycling in, 301
nitrifiers, 539
nitrogen fixers (*see* Nitrogen fixers)
non-oxygen electron receptors, 176
pathogenic, 541–542
proteobacteria, 534
quorum sensing, 525, 539
restriction enzymes, 315
shapes of, 528, *529*
spirochetes, 533
in the tree of life, 8
Bacteria (domain), 81, 526–527
Bacterial conjugation
control of, 254–255
description of, 253–254
in paramecia, 562
Bacteriochlorophyll, 538
Bacteriophage
ecological significance, 546
lysogenic cycle, 340–341
lytic cycle, 339–340, 341
phage λ, 340–341, 377, 379
phage T4, *543*
phage therapy, 545–546
Bacteroids, 748
Bainton, C. R., 1020–1021
Baker's yeast, 620, 623, 624, 626
Balaenoptera musculus, 1164
Bald eagles, *638*, 1228
Baldwin, Ian, 803, 804
Bali island, 1141, *1142*
Ball-and-stick models, *27*
Ballast water regulation, 1241
Bamboos, 792, *793*
Banana slugs, 664
Banded iron formations, 512
Banggai cardinalfish, 1234, *1235*
Banteng, 407
Banting, Frederick, 848
Banyans, 718
Baptista, Luis, 1100
Bar-headed geese, 1010
Bark, 721, 722, 723
Barklice, 672
Barley, "malting," 761
Barn owls, *696*
Barnacles, 670–671, 1184
Baroreceptors, 1044, *1045*
Barr, Murray, 346
Barr bodies, 345–346
Barred tiger salamander, *691*
Barrel cactus, *720*

Barrier methods of contraception, *898*

Basal lamina, *100*

Basal metabolic rate (BMR), *826–827*

Base pairing
  importance to DNA replication and transcription, 65–66, *67*
  in nucleic acids, 63–65

Base pairs
  in DNA structure, 265, 266, *267*
  exposure in DNA grooves, 266, *267*
  "hot spots" for mutations, 310

Bases
  acid–base reactions, 34
  amino acids as, 43
  defined, 34
  in nucleotides, 63
  properties of water as, 34–35

Basidiomata, *621*, 622

Basidiomycota, *615*, *616*
  See also Club fungi

Basidiospores, *622*, 623

Basidium, *621*, 623

Basilar membrane, *954*, 955–956

Basophils, *858*

Batesian mimicry, 1174

Batrachochytrium dendrobatidis, 612

Bats
  echolocation, 946, 963
  excretory physiology of vampire bats, 1071, 1090
  number of species, *698*
  as pollinators, 599, *1181*
  speciation by centric fusion of chromosomes, 470, *471*
  success of, 699–700
  wing evolution, *423*, 452

Bay checkerspot butterfly, 1161–1162

Bayliss, William, 838

Bazzania trilobata, *577*

Bcl-2 protein, 399–400

Bcl2 gene, 382

BCR–ABL oncogene, 304, *305*

BCR–ABL protein, 304

Bdelloid rotifers, 657

Beach grasses, 793

Beadle, George, 282–283

Beagle (HMS), 428, *429*

Beaks
  BMP4 expression in birds and, 412, 424
  heterometry in development, 416

Beauveria bassiana, 627

Beavers, *700*, 1194

Beech–maple forest, 1200

Beef cattle, 245

Beer, 623, 624

Bees
  exploitation competition, 1183
  as pollinators, 599, 605
  See also Honey bees; Hymenoptera

Beetles, *674*
  estimating the number of living species, 651
  fungus farming, 1178
  number of living species, *672*
  in pollination syndromes, *1181*

Begging behavior, 1095

Behavior. *See* Animal behavior; Complex behaviors; Courtship behavior; Sexual behavior; Social behaviors

Behavioral ecology
  defined, 1102
  evolution of behavior, 1102–1105, *1106*

Behavioral epigenetics, 328

Behavioral genetics, 1096–1098

Behavioral isolation, 476–477

Behavioral thermoregulation, 822, *823*, 1126

Behavioral traits, 456

Behaviorism, 1094

Behring, Emil von, 862, *863*, 864

Belding's ground squirrel, 1116, *1117*

Belladonna, *604*

Belly button, 897

Belt, Thomas, 1178

Benecke, Mark, 1188

Bengal tiger, *1136*

Benguela Current, *1125*

Benign tumors, 227

Benson, Andrew, 194, 195

Benthic zone, *1139*, 1140

Benzer, Seymour, 294

Benzopyrene, 309, 311

Beriberi, *1053*, 1054

Bermuda bluegrass, 199

Best, Charles, 848

Beta blockers, 850, 853

Beta-carotene, 58

Beta (β) pleated sheet, 45, *46*

Beta vulgaris, 718

Beutler, Bruce, 860

Beyer, Peter, 388

Biased gene conversion, 498, *499*

Bicarbonate ions
  as a base, 34
  blood transport of carbon dioxide and, 1018, *1019*
  buffering action in the blood, 1084–1085
  in blood, 1041
  in buffers, 35
  in global carbon cycle, 1216
  produced by pancreas, 1062
  production of hydrochloric acid and, *1060*, 1061

Biceps, 995

Bicoid gene, 402, *403*

Bicoid protein, 402, *403*

Bicoordinate navigation, 1109–1110

Biennials, 785

Bilateral symmetry
  in the animal zygote, 903
  in animals, 634–635
  in bilaterians and cnidarians, 643
  in echinoderms and hemichordates, 680, *681*
  in flowers, 597

Bilaterians
  in animal phylogeny, *630*
  functions of the central nervous system, 637
  members of, 634
  monophyly of, 643
  radial cleavage and, 679

Bilayers, lipid, 73–74
  See also Phospholipid bilayers

Bile, 1061–1062

Bile salts, 1061–1062, 1063

Bin Laden, Kahild, 318

Bin Laden, Osama, 318

Binary fission, 206–207

Bindin, 477, *885*, 886

Bindin receptors, *885*, 886

Binocular cells, 978

Binocular vision, 977–978

Binomial nomenclature, 7, 462, 463–464

BioCassava Plus, 724

Biochemical reactions
  energy transformations, 145–149
  role of enzymes in, 151–154

Biodiversity
  economic value, 1241–1243
  human activities threatening, 1232–1237
  importance of understanding and appreciating, 17–18
  meanings of, 1229
  predicting changes in, 1230–1232
  species extinctions and, 1229–1230
  strategies used to protect, 1237–1245
  value to human society, 1230

Biodiversity hotspots, 1237, *1238*

Bioelectric energy, *145*, 927

Biofilms, 281, 539, 540–541

Biofuels, 585

Biogeochemical cycles
  carbon cycle, 1216–1218
  defined, 1215
  hydrologic cycle, 1215
  interaction of, 1221–1223
  iodine, 1221
  iron, 1221
  nitrogen, 1218–1219
  phosphorus, 1220–1221
  sulfur cycle and the burning of fossil fuels, 1219–1220

Biogeographic regions, 1141–1142

Biogeography
  area phylogenies, 1143, *1144*
  biogeographic regions, 1141–1142
  continental drift, 1143
  discontinuous distributions from vicariant or dispersal events, 1143–1145
  human impact on biogeographic patterns, 1145
  origins of, 1141

Bioinformatics, 354

Biological classification, 462–464

Biological clock, 774
  See also Circadian rhythms

Biological control, 626, 627, 1164

Biological hierarchy, 9

Biological information, 5–6

Biological membranes. *See* Membranes

Biological nitrogen fixation. *See* Nitrogen fixation

Biological research
  distinguishing characteristics of science, 14
  generalization of discoveries, 14
  quantification of data, 11
  role of experiments in, 12–13
  role of observation in, 11
  scientific methods, 11–12
  statistics and, 13–14

Biological rhythms
  coordination of animal behavior with environmental cycles, 1106–1108
  melatonin and, 851

Biological species concept, 468–469

Biological warfare agents, 157

Biology
  biological information and genomes, 5–6
  defined, 2
  evolutionary tree of life, 6–9
  impact on public policy, 16–17
  importance for understanding ecosystems, 17
  importance to modern medicine, 15–16
  key characteristics of life and evolution, 2–11
  levels of organization in, *8*
  modern agriculture and, 15
  regulation of the internal environment and, 10
  understanding biodiversity and, 17–18

Bioluminescence
  ATP and, *149*, 150
  by dinoflagellates, 566
  green fluorescent protein, 449
  by *Vibrio*, 525, 534, 546

Biomass
  decrease through time in detritus-based communities, 1202
  distribution in communities, 1190–1192

Biomes
  abiotic factors influencing, 1126–1127
  aquatic, 1139–1141
  boreal and temperate evergreen forests, 1129–1130
  chaparral, 1134–1135
  classification, 1126
  cold deserts, 1133–1134
  defined, 1126
  distribution, 1126, *1127*
  hot deserts, 1132–1133
  temperate deciduous forest, 1130–1131
  temperate grasslands, 1131–1132
  thorn forests and tropical savannas, 1135–1136
  tropical deciduous forest, 1136–1137
  tropical rainforests, 1137–1138
  tundra, 1128–1129

"Bioprospecting," 500–501

"Bioreactors," 385–386

Bioremediation
  microorganisms and, 373, 389
  of oil spills, 373
  phytoremediation, 811, *812*
  use of fungi in, 625

Biosynthesis
  food as a source of carbon skeletons for, 1051–1052
  nutrients as the basis for, 10

Biota, 514

Biotechnology
   agricultural applications,
      386–388
   defined, 383
   expression vectors, 384
   genetically modified organisms
      in bioremediation, 373
   history of, 383–384
   medical applications, 384–386
   patents and, 373
   public concerns, 388–389
   See also Recombinant DNA
      technology
Bioterrorism, 531
Biotic ecosystem components,
   1122
Biotic potential, 1158
Biotin, 156, 1057
Bipedal locomotion, 702–703
Bipolar cells, 925, 962, 963
"Bird flu" virus, 543
Birds
   actions of prolactin in, 838
   advantages of flocking, 1116,
      1117
   altricial and precocial young,
      642
   avian malaria in the Hawaiian
      Islands, 1236
   begging behavior of chicks,
      1095
   BMP4 and beak development,
      412, 424
   circulatory system, 1029, 1030
   daily torpor, 830
   dinosaur origins of, 693, 694
   disruptive selection in bill size,
      440
   extraembryonic membranes,
      918
   fat as stored energy, 1050
   feathers and flight, 695–696
   foveae, 962
   gastrulation, 913–914
   "helping at the nest," 1115
   heterometry in beak
      development, 416
   hindlimb development in ducks
      and chickens, 417
   imprinting, 1099
   monitoring, 1150, 1151
   nest parasitism, 1093, 1102, 1117
   palaeognaths and neognaths,
      694–695
   as pollinators, 599
   salt balance regulation and
      nasal salt glands, 1073
   seed dispersal and, 696
   sex determination in, 249
   sexual reproduction, 888–889
   sexual selection and speciation
      rates, 480–481
   shivering heat production, 827
   unidirectional ventilation in gas
      exchange, 1010–1012
   wing evolution, 423, 452
   See also Hummingbirds;
      Songbirds
Birds of paradise, 480–481
Birdsong
   factors affecting song
      acquisition, 1099–1101
   hormonal control of song
      expression, 1101

Birth control. See Contraception
Birth control pills, 896, 898
Birth process
   effect of oxytocin on, 843
   positive feedback in, 817
   the uterus in, 843, 896–897
Bishop's goutweed, 597
Bison, 1064, 1131, 1239
Bison antiquus, 1229
Bison bison, 1239
1,3-Bisphosphoglycerate, 170
2,3-Bisphosphoglyceric acid
   (BPG), 1018
Bithorax gene, 404
Bitter taste, 951
Bivalents, 219
Bivalves, 662, 663
Black bears, 1130
Black-bellied seedcracker, 440
Black bread mold, 619
Black-eyed Susans, 599
Black lace cactus, 603
Black-legged tick (deer tick), 1151,
   1152
Black rockfish, 1163
Black stem rust, 612
Blackberry, 601
Bladder, 1080
Bladder cancers, 131, 132
Blastocoel
   in the avian blastula, 914
   formation and characteristics
      of, 904, 906, 907
   in frog gastrulation, 910
   in sea urchin gastrulation, 909
Blastocyst
   defined, 906
   implantation in humans, 892,
      893, 907
   pluripotent stem cells, 408–409
   secretion of human chorionic
      gonadotropin, 896
Blastoderm, 905
Blastodisc, 905, 913, 914
Blastomeres
   defined, 904
   determination, 906–907
   in mammalian cleavage, 906
   in sea urchin gastrulation, 909
Blastopore
   defined, 633
   in deuterostomes, 634, 679
   in protostomes, 634, 652
   in sea urchin gastrulation, 909
   See also Dorsal lip of the
      blastopore
Blastula
   defined, 904
   determination of blastomeres
      in, 906–907
   gastrulation, 908–915 (see also
      Gastrulation)
   role of cleavage in forming,
      904–905
Blattodea, 672
Bleak, 1140
"Blebs," 226
Bleeding shiner, 472
Blending inheritance, 233, 234
Blenny, 1174
Block, Barbara, 16, 1036, 1037
Blocks to polyspermy, 885,
   886–887

Blood
   ABO blood groups, 243
   carbonic/bicarbonate buffer
      system, 35, 1084–1085
   composition, 1037, 1038
   as connective tissue, 819
   countercurrent heat exchange
      in endotherms, 828
   countercurrent heat exchange
      in "hot" fish, 825
   filtration in Bowman's capsule,
      1078–1079
   hematocrit, 1037–1038
   hormonal regulation of calcium
      levels, 847–848
   hormonal regulation of
      phosphate levels, 848
   in the mammalian defense
      system, 857
   platelets, 1039
   red blood cells, 1038
   regulation of osmolarity by
      antidiuretic hormone,
      1088–1089
   thermoregulation through the
      skin and, 824–825, 828
   transport of respiratory gases,
      1016–1019
Blood carbon dioxide
   in blood pressure regulation,
      1045
   effect on autoregulation of
      blood flow, 1044
   transport of, 1018–1019
"Blood chamber," 652
Blood clotting, 384–385, 1039
Blood clotting proteins, 323
Blood disorders
   hemoglobin C disease, 312
   sickle-cell disease, 306, 307, 312
Blood factor VIII, 323
Blood flow
   in arteries and arterioles, 1039,
      1040
   control and regulation of,
      1043–1044
   countercurrent flow in fish gills,
      1010, 1011
   through capillary beds, 1040–
      1041
Blood glucose
   hormonal regulation of, 848–
      849, 1066–1067
   maintenance of levels during
      physical exercise, 180–181
   regulatory pathways in glucose
      catabolism, 181–182
Blood groups, 53, 243
Blood meals, vampire bats and,
   1071
Blood pH
   effect on hemoglobin's binding
      of oxygen, 1018
   partial pressure of carbon
      dioxide and, 1021
Blood plasma
   composition of, 1038
   defined, 1026
   extracellular fluid, 816
   filtration in tubule capillaries,
      1072
   loss of proteins during
      starvation, 1050
   oxygen-carrying capacity, 1016

   separation of blood cells from,
      1037
   synthesis of plasma proteins by
      the liver, 1065
   transport of carbon dioxide,
      1018, 1019
Blood pressure
   in capillary beds, 1041
   control and regulation of blood
      flow, 1043–1044
   in fish, 1028
   measuring, 1032, 1033
Blood pressure regulation
   angiotensin and, 1087
   antidiuretic hormone and,
      1088–1089
   atrial natriuretic peptide and,
      1090
   hormonal and neural roles in,
      1044–1045
   vasopressin and, 843
Blood respiratory gases
   in blood pressure regulation,
      1045
   effect on autoregulation of
      blood flow, 1044
   transport, 1016–1019
"Blood sugar." See Blood glucose
Blood vessels
   anatomy, 1040
   arteries and arterioles, 1039,
      1040
   autoregulation of blood flow
      through, 1044
   in blood pressure regulation,
      1044, 1045
   capillary beds, 1039–1041
   carotid and aortic bodies,
      1021–1022
   in closed circulatory systems,
      1026
   clotting and, 1039
   constriction in the fight-or-
      flight response, 837
   dilation in penile erection,
      890–891
   physical stresses in the lungs of
      snorkeling elephants, 1005,
      1022
   smooth muscle in (see Vascular
      smooth muscle)
   veins, 1041–1042
   in vertebrate circulatory
      systems, 1027–1028
Blood–brain barrier, 926, 1041
"Blooms"
   algal, 1219
   diatoms, 556
   red tides, 549, 564
   viral, 546
Blowflies, 674, 1203
Blubber, 697
"Blue" cheese, 622
Blue-green bacteria. See
   Cyanobacteria
Blue-light receptors, 771–772, 788
Blue whales, 637, 1164
Bluebells, 462
Bluebottle flies, 1203
Bluefin tuna, 16–17, 825, 1036, 1037
Bluegill sunfish, 1105, 1106
BMP4. See Bone morphogenetic
   protein 4
BMP4 gene, 417

Boas, 423
Body cavities
  animal movement and, 635–636
  hydrostatic skeletons, 999
  types, 635
Body fat
  buildup during excess food
    consumption, 1051
  See also Adipose tissues
Body plans
  annelids, 660
  appendages, 636–637
  arrow worms, 655
  arthropods, 655
  body cavities and movement,
    635–636
  chelicerates, 668
  crustaceans, 671
  defined, 634
  insects, 671, 672
  key features of, 634
  mollusks, 663
  nematodes, 666
  nervous systems, 637
  of plants, 709–712
  protostomes, 652
  segmentation, 636
  sponges, 644
  symmetry, 634–635
  of vertebrates, 685
Body segmentation
  in animal development, 916, 917
  determination in Drosophila
    melanogaster, 401–405
  regulation by Hox genes,
    916–917
Body size
  brain size and, 973
  chromosome number and, 218
  effect of atmospheric oxygen
    levels on insect body size,
    505, 513, 514, 522
  effect on characteristic species
    density, 1160
  range in birds, 696
  range in mammals, 696–697
  relationship to basal metabolic
    rate, 826–827
  respiratory gas diffusion and,
    1007
  variation between dog breeds,
    352
Body temperature
  acclimatization, 821
  classification system, 822
  impact of changes in, 821
  "point restriction" coat patterns
    and, 245
  regulation (see
    Thermoregulation)
Bogs, 1140
Bohr effect, 1018
Bohr models, 27
Boiga irregularis, 1235
"Bolting," 761
Bombardier beetles, 1173
Bombina bombina, 479
Bombina variegata, 479
Bombus lucorum, 476
Bombycilla cedrorum, 696
Bombyx mori, 950
Bonding behavior
  factors affecting in voles, 125
  oxytocin and, 843

Bone
  as connective tissue, 819
  development, 1001
  hormonal control of turnover,
    847, 848
  interaction with muscle at
    joints, 1001–1002
  osteoporosis and, 1000–1001
  process of growth, 416–417
  structure of, 1000
  of theropod dinosaurs, 695
Bone marrow
  formation of white blood cells
    in, 858
  in the lymphatic system, 857
  red blood cell production, 1038
  types of multipotent stem cells
    in, 408
Bone morphogenetic protein 4
  (BMP4)
  beak development in birds,
    412, 424
  loss of foot webbing in
    nonaquatic birds, 417
Bone morphogenetic proteins
  (BMPs), 913
Bonner, James, 788, 789
Booklice, 672
Boreal forest, 1129–1130
Boreal owl, 1129
Borlaug, Norman, 756
Boron, 741
Borthwick, Harry, 772
Boston Marathon, 815
Bothus lunatus, 445
Botox®, 542, 936
Botrytis fabae, 801
Bottle cells, 909, 910
Botulinum toxin, 936
Botulism, 542
Boulding, Kenneth, 1245
Bowman's capsules, 1078–1079,
  1080, 1081
Boyer, Herbert, 374
Brachiopods
  anatomical characteristics, 652
  description of, 658–659
  major subgroups and number
    of living species, 632
Bracket fungi, 622, 623
Bradypus variegatus, 1137
Brain
  alternative splicing and, 347
  blood–brain barrier, 926, 1041
  complexity in vertebrates, 943
  development in mammals,
    968–969
  diversity in size and
    complexity, 943
  early development, 916
  learning and memory areas,
    981–982
  nervous system information
    flow and, 968
  overinhibition in, 924, 939, 943
  regulation of food intake,
    1067–1068
  sleep and dreaming, 978–980
  in small nervous systems, 941
  structure and function in
    mammals, 969–973
  See also Human brain
"Brain coral," 647

Brain imaging, 970
Brain size
  body size and, 973
  evolution in humans, 704
Brainstem
  changes during sleep, 979
  control of breathing, 1019–1020
  development, 968, 969
  structure and function, 969–970
Branching, in rhyniophytes, 580
Branchiostoma lanceolatum, 683
Branchiopods, 670
Brassica oleracea, 432, 724
Brassinosteroids, 759, 771
Braxton Hicks contractions, 896,
  897
Brazilian mastiff, 352
BRCA1 gene, 245, 246
Bread molds, 282–283, 619
Bread wheat, 225, 226
Breast cancer
  DNA microarray technology
    and diagnosis, 383
  expressivity of the BRCA1 allele
    and, 246
  human SNP scans, 368
  oncogene proteins, 228
  penetrance of the BRCA1 gene
    and, 245
Breast cancer treatment
  drugs targeting the estrogen
    receptor, 325
  Herceptin, 229
  immunotherapy, 871
Breast duct cells, 333
Breast milk, 843
Breathing
  effect on venous blood flow, 1042
  regulation of, 1019–1022
  See also Respiratory gas
    exchange; Ventilation
Breeding seasons, 476
Brenner, Sydney, 362
Brevipalpus phoenicis, 669
Brewer's yeast, 620, 623, 624, 626
Briggs, Robert, 406
Bright-field microscopy, 80
Brine shrimp, 1072, 1073
Brines, 536
Bristlecone pine, 593
Brittle stars, 681, 682
Brno monastery, 233
Broad fish tapeworm, 641
Broca's area, 981
Broccoli, 724
Brock, Thomas, 277–278
Bromelain, 604
Bromeliads, 678
Bronchi, 1013, 1014
Bronchioles, 1013, 1014
Brood parasitism, 1093, 1102, 1117
Brown, Patrick, 742, 743
Brown algae, 556
Brown fat
  in adults, 834
  characteristics of, 819
  conversion of white fat into,
    834, 839
  heat energy from, 165, 174
  in nonshivering heat
    production, 827–828
  UCP1 and weight loss, 182
Brown-headed cowbirds, 1093,
  1233

Brown molds, 624
Brown tree snake, 1235
Brussels sprouts, 724
Bryophyta, 574
Bryozoans
  anatomical characteristics,
    652
  colonies, 641
  description of, 656
  lophophores, 654
  major subgroups and number
    of living species, 632
Bubonic plague, 534, 542
Buckley, Hannah, 1197
Bud scales, 721, 722
Budding, 562, 620, 881
Budding yeast, 362
  See also Saccharomyces
    cerevisiae
Buds
  defined, 709
  of mosses, 576, 577
  of twigs, 721, 722
Buffers
  blood buffer systems, 1084–
    1085
  properties of, 35–36
Bufo marinus, 1164, 1235
Bufo periglenes, 691
Bulbourethral glands, 890
Bulbs, 793
Bulbus arteriosus, 1028
Bulk flow
  defined, 727–728
  in phloem, 728, 735–738
  in xylem, 728
Bull, James, 457
Bull trout, 1228
Bullhorn acacia, 1178, 1179
Bumblebees, 1183
Bundle of His, 1034, 1035
Bundle sheath cells, 198, 199, 721
Burgess Shale, 516
Burs, 784
Bush monkeyflower, 598, 599
Butane, 41
Buteo galapagoensis, 2
Butterflies, 442–443
  See also Lepidopterans; specific
    butterflies

C
C horizon, 745
C ring, 83
C$_3$ plants, 198, 199, 200, 202
C$_4$ plants, 198–200, 202
Cabbage plants, 724, 761
Cactaceae, 1133
Cacti
  adaptations to very dry
    conditions, 807, 808
  number of species, 603
  stomatal function in, 734
Cactus finches, 474, 1154–1155
Caddisflies, 672, 674
Caecilians, 690, 691, 692
Caenorhabditis elegans
  apoptosis studies in
    development, 399
  genomic information, 361, 362,
    363
  microRNA, 347
  as model organism, 666

vulval determination, 396, *397*, 398
See also Nematodes
Caffeine, 128
Caimans, 693–694, *1106*
Calcareous sponges, *630*, 643–644
Calciferol, 848, *1053*, 1054
Calcitonin
actions of, *842*
influence on blood levels of calcium, 847
thyroid source, 845, *846*
Calcitriol, 848
Calcium
in animal nutrition, 1052–1053
in bone, 1000
hormonal regulation of blood calcium, 847–848
in plant nutrition, *741*
Calcium carbonate
in limestone deposits, 557, 565
in oceanic carbon stores, 1217
otoliths, *956*, 957
in the shells of foraminiferans, 557
Calcium-induced calcium release, 992–993
Calcium ion channels
in cardiac muscle contraction, 992–993, 1033, 1034
in cardiac pacemaker cells, 1033, 1034
in the hyperpolarization of neurons at the onset of sleep, 979
in skeletal muscle contraction, 990
See also Voltage-gated calcium channels
Calcium ions
in cardiac muscle contraction, 992–993, 1033, 1034
in cardiac pacemaker cells, 1033, 1034
cycling in cardiac muscle effects heartbeat, 1034, *1036*, 1037
in insect flight muscle contraction, 998
neurotransmitter release at neuromuscular junctions, *936*, 937
roles in animal fertilization, 886, 887
as second messengers, 135, 136
in skeletal muscle contraction, 989–990, *991*
in smooth muscle contraction, 993
in smooth muscle relaxation, 136
tetanic muscle contractions and, 994–995
Calcium phosphate, 21, 1000
Calcium pumps
in sarcoplasmic reticulum of cardiac muscle, 1034
in skeletal muscle contraction, 989–990
California condor, 1244
California plantain, 1161
Callaway, John, *1240*
Calliarthron, *572*
Callose, 799
Callus (calli), 405

Callyspongia plicifera, *1007*
Calmodulin, 412, 424, 993
Caloplaca, 613
Calorie (Cal), 1049
Calories (cal), 147
Calvin, Melvin, 194, 195
Calvin cycle
in C_4 plants, 199
in CAM plants, 200
linkage of photosynthesis and respiration in plants, 201
in photosynthesis, *188*
processes and products, 194—196
radioisotope experiments, 193–194, 195
stimulation by light, 196–197
Calyx, 597
CAM plants, 200
Camalexin, 801
Camarasaurus, 21
Camarhynchus pallidus, *474*
Camarhynchus parvulus, *474*, *1150*
Camarhynchus pauper, *474*
Camarhynchus psittacula, *474*
Cambrian explosion, 516
Cambrian period, *506–507*, 516, *517*
Camembert cheese, 624
cAMP receptor protein (CRP), 332
Campanula rotundifolia, *462*
Campephilus principalis, *1230*, *1231*
Canary Islands, 440
Canarygrass, 762, 763, 764
Canavanine, 803, 805
Cancellous bone, 1001
Cancer drugs
cisplatin, 259, 268, 278
competitive inhibitors of enzymes, 158
disruptors of microtubule dynamics, 96
targeting the cell cycle, 228–229
Cancer treatment
immunotherapy, 871
Ras inhibitors, 131
targeting the cell cycle, 228–229
use of multipotent stem cells, 408
Cancers
affecting Tasmanian devils, 232
blocking of apoptosis, 382
characteristics of cancer cells, 227
chronic myelogenous leukemia, 304, 305
DNA methylation and, 344
HeLa cells, 205
histone deacetylation and, 344
molecular changes in cancer cells, 228
from mutations to somatic cells, 310–311
normal cell death and, 226
prevalence in the United States, 227
RNA retroviruses and, 544
signal transduction pathways and, 131, *132*
somatic mutations and, 314
telomerase and, 275
unregulated cell division and, 227–229

virus-interrupted breakdown of cyclins and, 349
See also specific cancers
Candelabra, *1132*
Candida albicans, 612
Cane toad, 1164, 1235
Canines, 1055
Canis lupus familiaris, 352
See also Dogs
Canis simensis, 1242
5'-Cap, 348
Cap-binding complex, 293
Cape sugarbird, *1134*
Capecchi, Mario, 382
Capillary action, 731
Capillary beds
autoregulation of blood flow, 1044
blood–brain barrier, 1041
effect of autoregulatory actions on mean arterial pressure, 1043
structure and function, 1028, 1039–1041
Capra pyrenaica, *1134*
Capsid, 340
Capsule, of prokaryotes, *82*, 83
Captive breeding programs, 1244
Carapace, 671
"Carbo-loading," 997
Carbohydrases, 1058
Carbohydrates
biochemical roles, 51
breakdown by digestive enzymes, 1057, 1058
catabolic interconversions, 179
categories of, 51
chemically modified, 55
energy yield, 1049, *1050*
general formula, 51
glycosidic linkages, 53
membrane-associated, *106*, 109
monomer components, 40
monosaccharides, 52
polysaccharides, 53–54
production in photosynthesis, 186, 193–197
proportions in living organisms, *41*
as stored energy in the animal body, 1050
Carbon
covalent bonding capability, 27
electronegativity, *28*
isotopes, 23
mass number, 22
octet rule for molecule formation, 25
sources for saprobic fungi, 611
Carbon-12, 507
Carbon-14, 507
Carbon cycling
fungi in, 611
global amounts released by fire, 1214
global cycle, 1216–1218
Carbon dioxide
in the bicarbonate buffer system, 1084, 1085
blood transport, 1018–1019
fixation in photosynthesis (*see* Carbon dioxide fixation)
in the global carbon cycle, 1216–1218

impact on plant stomatal density, 734
produced by yeast in food and drink production, 623
produced during glucose catabolism, 169, 170, 171
respiratory exchange by diffusion, 1008
stomatal function in plant uptake of, 732–734
See also Atmospheric carbon dioxide
Carbon dioxide fixation
in C_3 and C_4 plants, 198–200
in CAM plants, 200
evolution of pathways in, 200
in metabolic interactions in plants, 201
in photosynthesis, 186, 193–197
Carbon isotopes, 74
Carbon monoxide, 1017
Carbon skeletons, 1051–1052
Carbonate ions, 1216
Carbonic acid
bicarbonate ion and, 34
in blood transport of carbon dioxide, 1018, *1019*
carbonic/bicarbonate buffering system, 35
in cation exchange by roots, 746
in ocean acidification, 1217
Carbonic anhydrase, 1018, *1019*, *1060*, 1061
Carboniferous period
atmospheric oxygen levels, 512–513
changes on Earth and major events in life, *506–507*
characteristics of life during, 520
decline of fungi in, 611
evolution of leaves, 583
gigantic insects in, 505
vascular plants in, 580
Carboxyl group
acid properties, 34
of chemically modified carbohydrates, 55
effect of pH on, 161
peptide linkages in amino acids, 43–44, *45*
properties of, *40*
Carboxylic acids, 70
Carcharodon megalodon, 687
Carcinogens, 310, 311
Cardiac action potentials
coordination of muscle contraction, 1034, *1035*
electrocardiograms, 1035–1036
pacemaker cells and cardiac muscle contraction, 1032–1034
Cardiac cycle, 1031–1032
Cardiac muscle
atherosclerosis and, 1043
calcium ion cycling in "hot" fish, *1036*, 1037
contraction, 991–993, 1032–1034, *1035*
Frank–Starling law, 1042
functions of, 817, 987

integration of anabolism and catabolism during exercise, 180–181
pacemaker cells, 1032–1034 (see also Pacemaker cells)
structure, 991
Cardiac output (CO)
mean arterial pressure and, 1043
regulation of, 1044, 1045
Cardiovascular disease, 1042–1043
Cardon, 1132
Caring behavior, 141
Carnivores
defined, 1054
ingestion and digestion of food, 1055
number of species, 698
teeth, 1055
Carnivorous plants, 751–752
β-Carotene
accessory pigment in photosynthesis, 190
in genetically modified plants, 388
Carotenoids
accessory pigments in photosynthesis, 190
in brown algae, 556
in light-harvesting complexes, 191
in red algae, 571
structure and function, 58
Carotid arteries
carotid bodies, 1021–1022, 1045
stretch receptors in the regulation of blood pressure, 1088–1089
Carotid bodies, 1021–1022, 1045
Carpels
anatomy of, 591
evolution of, 598
function of, 779
in the structure of flowers, 596–597
Carpenter bees, 1183
Carpolestes, 701
Carrier-mediated transport, 117
Carrier proteins
defined, 115
in facilitated diffusion, 117
Carrier viruses, 323, 324
Carrion-feeding beetles, 1203
Carrots, 405–406, 718
Carrying capacity, 1158, 1165
Cartilage
as connective tissue, 819
galactosamine in, 55
in vertebrate skeletons, 999–1000
Cartilage bones, 1001
Cartilaginous fishes. See Chondrichthyans
Cartilaginous skeletons, 1000
Cascade mountains, 509
Casparian strip, 730, 744
Caspases, 226, 227, 399–400
Cassava, 708, 724
Cassowaries, 694
Castercantha, 39
Castor canadensis, 700
Catabolic interconversions, 179–180

Catabolic reactions
energy changes during, 146
exergonic reactions, 147, 148
inducible systems of regulation in prokaryotes, 332
integration of anabolism and catabolism, 180–181
linkage to anabolic reactions, 146, 179–180
See also Glucose catabolism
Catabolite repression, 332
Catalase, 156, 176
Catalysts
function of, 151
nonspecific nature of, 152
RNA and the origin of life, 71–72, 73
See also Enzymes
Catasetum, 588, 605
Catasetum macrocarpum, 605
β-Catenine, 904, 911, 912
Caterpillars, 1176
Catfish, 1140
Cations, 29
Cats
point restriction phenotype, 306
retinal ganglion receptive fields, 975, 976
Toxoplasma and, 554
Cattails, 1140
Cattle, 245, 407, 440
Cattle egrets, 1171
Caudal fins, 687
Cauliflower, 724
Cause and effect, 98
Cavanillesia platanifolia, 1136
Cayman crab fly, 1231
CD4 protein, 872
cDNA. See Complementary DNA
Cecal material, 1064
Cech, Thomas, 72
Cecum, 1064
CED-3 protein, 399
CED-4 protein, 399, 400
CED-9 protein, 399, 400
Cedar waxwings, 696
Celestial navigation, 1109–1110, 1111
Celiac ganglion, 974
Cell adhesion
defined, 110
to the extracellular matrix, 111–113
homotypic and heterotypic, 111
roles of proteins and carbohydrates in, 111
significance in multicellular organisms, 110–111
types of cell junctions in, 111, 112
Cell binding. See Cell adhesion
Cell body, of neurons, 925
Cell cycle
abnormal regulation in cancer cells, 228
cancer treatments targeting, 228–229
defined, 208
duration of, 208
gain-of-function mutations in tumor suppressor and, 306
internal signals controlling, 208–211

mitosis, 211–217
summary of events, 217
Cell cycle checkpoints, 210–211, 215
Cell determination. See Determination
Cell differentiation. See Differentiation
Cell division
in asexual reproduction, 217
cancers and, 227–229
cell differentiation in the embryo and, 399
control of, 208–211
in eukaryotes, 207–211
HeLa cells, 205, 229
important consequences of, 206
key events in, 206
mitosis, 211–217 (see also Mitosis)
in morphogenesis, 394
in prokaryotes, 206–207
in sexual reproduction, 217, 218–219 (see also Meiosis)
See also Cytokinesis
Cell expansion
auxin-induced expansion in plants, 766–767
in morphogenesis, 394
Cell fate
defined, 394
processes determining, 395–396, 397 (see also Determination)
restriction during development, 394
Cell fractionation, 84, 85
Cell fusion experiments, 208–209
Cell junctions, 111, 112
Cell movement
cilia and flagella, 96–98
integrins and, 112, 113
microfilaments and, 95, 96, 98, 99
in morphogenesis, 394
Cell plate, 710
Cell potency, 394
Cell recognition
defined, 110
significance in multicellular organisms, 110–111
Cell theory, 78
Cell-to-cell communication
evolution of multicellularity and, 140–141
gap junctions, 139–140
induction in cell fate determination, 395–396, 397
plasmodesmata, 139, 140
See also Intercellular signaling; Signal transduction pathways
Cell walls
of bacteria, 527, 528
of diatoms, 555
loss of in the eukaryotic condition, 551
of plant cells (see Plant cell walls)
of prokaryotes, 82, 83
Cellobiose, 53
Cells
animal cell structure, 86 (see also Animal cells)
cell theory, 78
death of, 225–227
diffusion within, 114

mitosis
effect of osmosis on volume, 1072
effects of signal transduction pathways on, 137–139
experiments on the origin of, 73–74
microscopes, 79, 80–81
plant cell structure, 87 (see also Plant cells)
plasma membrane, 79–81
prokaryotic and eukaryotic, 81 (see also Eukaryotic cells; Prokaryotes)
regulation of protein longevity in, 348–349
regulation of the internal environment, 10
responses to intercellular signaling, 125, 126–127
surface area-to-volume ratio, 78–79
types of work done by, 10
"Cellular drinking." See Pinocytosis
"Cellular eating." See Phagocytosis
Cellular immune response
activation and effector phases, 873, 874
binding of antigens to T cell receptors, 871–872
description of, 863–864, 865
effector T cells in, 865, 866, 871, 872, 873, 874
MHC proteins in, 871, 872, 873, 874
presenting antigens to T cell receptors, 872
suppression by regulatory T cells, 874
in tissue transplants, 874
Cellular respiration
defined, 91
energy yield from cellular respiration and fermentation compared, 178
linkage to photosynthesis in plants, 201
mitochondria and, 91
overview, 166–167, 168
Cellular slime molds, 561
Cellular specialization
in eukaryotes, 4
significance to multicellular life, 9
Cellular therapy, role of transcription factors in, 336, 337, 338
Cellulases, 1063
Cellulose
digestion in herbivores, 1063–1064
in plant cell walls, 710, 711
plant guard cell function and, 733
structure and function, 54
Cenozoic era, 506–507, 521–522
"Centers of imminent extinction," 1237, 1238
Centimorgans (cM), 248
Centipedes, 669–670
"Central dogma," 285
Central nervous system (CNS)
anatomical organization, 968–969

components of, 941
defined, 968
development in humans, 968–969
functions of, 637
glucose as the fuel for, 1067
information flow into and out of, 969
regulation of breathing, 1019–1022
self-perception and, 983
structure and function in mammals, 969–973
Central pattern generators, 969
Central sulcus, 971
Central vacuole, 710
Centric fusion, 470, 471
Centrioles
in animal cells, 86
sperm contribution to the zygote, 903
structure and function in cell division, 212, 215
Centrocercus urophasianus, 1105
Centromeres
alignment during meiotic metaphase II, 220
alignment during mitotic metaphase, 215
comparison between mitosis and meiosis, 223
of sister chromatids in mitosis, 212, 213, 216
Centrosomes
of the animal zygote, 903
structure and function in cell division, 212–213, 214, 215
Century plants, 785
Cephalization, 634–635
Cephalochordates, 684
Cephalopods
description of, 662, 663, 664
image-forming eye, 958, 959
Cephalothorax, 668
Ceramium, 572
Cercozoans, 557
Cerebellum, 968, 969
Cerebral cortex
description of, 969
frontal lobe, 971–972
in human brain evolution, 973
occipital lobe, 973
parietal lobe, 972–973
self-perception and, 983
structure and function, 970–971
temporal lobe, 971
Cerebral hemispheres, 969, 980–981
Cerebrum
cerebral hemispheres, 969
structure and function, 970–973
Certhidea olivacea, 474
Cervical caps, 898
Cervical vertebrae, 418
Cervix
in humans, 892, 893
in labor and childbirth, 896, 897
Cesarean section, 540
Cetaceans
blubber, 697
evolution, 700
number of species, 698
See also Whales
Cetartiodactyla, 698

CFCs. See Chlorinated fluorocarbons
CFTR gene, 318
Chaetopleura angulata, 663
Chagas' disease, 559
Chain, Ernst, 608
Chakrabarty, Ananda, 373
Chalfie, Martin, 449
Challenger Deep, 557
Chaneton, Enrique, 1162
Channel proteins, 115–116
"Chaos amoeba," 559
Chaos carolinensis, 559
Chaparral, 1134–1135, 1146
See also Fynbos
Chaperone proteins, 51, 131
Chaperonins, 810
Chara vulgaris, 573
Character displacement, 1183
Characters
defined, 234
of a phenotype, 431
Charcharodon charcharis, 688
Chargaff, Erwin, 264
Chargaff's rule, 264, 265, 266
Chase, Martha, 261–263
Chasmosaurus belli, 519
Cheese skipper fly, 1188, 1203
Cheeses, molds and, 622, 624
Cheetah Conservation Fund, 75
Cheetahs, 62, 75
Cheirurus ingricus, 668
Chelicerae, 668, 669
Chelicerates, 632, 667, 668–669, 675
Chelonia mydas, 888
Chelonoidis nigra, 2
Chelonoidis nigra abingdonii, 694
Chemical bonds
covalent bonds, 26–28
defined, 26
hydrogen bonds, 30
hydrophobic interactions, 30
ionic attractions, 28–29
phosphoric acid anhydride bond, 150
van der Waals forces, 30
Chemical defenses
of prey, 1173
See also Plant chemical defenses
Chemical energy
in biological systems, 145
released during glucose oxidation, 166–169
Chemical equilibrium
defined, 148
enzymes and, 153
free energy and, 148–149
Chemical evolution
hypotheses on the emergence of polymers, 71
in the origin of life, 3
prebiotic synthesis experiments, 69–71
Chemical fertilizers
ecological impact of, 740, 1207, 1219, 1220–1221
impact on the global nitrogen cycle, 1218–1219
Chemical reactions
defined, 31
energy and, 31
energy transformations, 145–149

law of mass action, 35–36
qualitative and quantitative analyses, 33
reactants and products, 31, 145
reversible, 34, 35–36
role of electrons in, 24, 25
role of enzymes in, 151–154
Chemical signaling systems
receptor proteins, 127–131
"responses" to, 125, 126–127
types and sources of signals, 126
See also Signal transduction pathways
Chemical signals
in animal communication, 1110–1111
in animals, 835–836 (see also Animal hormones)
in slime molds, 838
Chemical synapses, 926, 936–939
See also Synapses
Chemical warfare agents, 940
Chemical weathering, 745–746
Chemically gated ion channels, 930, 937
Chemiosmosis
experimental demonstration of, 174, 175
mechanism for ATP synthesis, 174
in oxidative phosphorylation, 171
photophosphorylation and ATP synthesis, 192–193
proton-motive force and, 173–174
Chemoautotrophs, 538, 1189
Chemoheterotrophs, 538, 539
Chemoreceptors
in blood pressure regulation, 1044, 1045
defined, 949
detection of blood levels of respiratory gases, 1021–1022
detection of pheromones, 950
influence on ion channels, 947, 948
in olfaction, 949–950
in taste buds, 951
in the vomeronasal organ, 950–951
Chemotherapy, 51
Chengjiang, 516
Chenopodium album, 1200
Chernobyl nuclear power plant, 811
Chestnut blight, 622
Chewing, 686
Chiasmata, 219, 220, 222
Chickens
BMP4 and beak development, 424
extraembryonic membranes, 918
gastrulation, 913–914
genomic information, 361
hindlimb development, 417
Chief cells, 1060
Chihuahua, 352
Childbirth, 896–897
Chimaeras, 687, 688
Chimpanzees
comparative genomics, 366

foraging behavior for essential minerals, 1105
origin of, 702
skull, 704
China
artemisinin treatment for malaria, 797
stem cell therapy in, 102
Chiroptera, 698
Chitin
as an elicitor of plant defenses, 799
in exoskeletons, 655, 999
in fungi, 609
structure, 55
Chitinase, 801
Chitons, 662, 663
Chlamydia psittaci, 534
Chlamydias, 533, 534
Chlamydomonas, 140, 141
Chlorella, 14
Chloride ion channels
actions at inhibitory synapses, 938
reduction of overinhibition in the mouse brain, 943
Chloride ions
actions at inhibitory synapses, 938
plant guard cell function and, 733
properties of, 29
transport by nasal salt glands, 1073
uptake by halophytes, 811
Chlorinated fluorocarbons (CFCs), 311, 1211–1212
Chlorine
in animal nutrition, 1052
electronegativity, 28
ionic attraction, 28–29
in plant nutrition, 741
Chlorophyll
absorption of light energy, 188–190
anabolic interconversions and, 180
mutations in plants, 252
Chlorophyll a
in cyanobacteria, 532, 538
molecular structure, 190
photochemical changes following light absorption, 191
in photosystems, 190
in red algae, 571, 572
Chlorophyll b, 190, 191
Chlorophytes, 140–141, 572, 573
Chloroplast DNA (cpDNA), 456
Chloroplasts
carbon dioxide fixation in, 193–197 (see also Carbon dioxide fixation)
in cercozoans, 557
endosymbiotic origins, 102, 550, 551–552
in glaucophytes, 571, 572
of hornworts, 578
photophosphorylation and ATP synthesis, 192–193
in photorespiration, 197, 198
plant cells, 87
sites of photosynthesis in, 188
structure and function, 92, 93

transposition of genes to the nucleus, 366
Choanocytes, *633*, 644
Choanoflagellates, 609, 631, *633*, 644
"Chocolate spot" fungus, *801*
Cholecystokinin (CCK), *1061*, 1065
Cholera, 534, 542, 1063
Cholesterol
  in the absorption of fats in the small intestine, *1062*, 1063
  in atherosclerosis, 1042
  biological membranes and, *106*, 107
  familial hypercholesterolemia, 121
  mechanism of uptake by cells, 121
  structure and function, 58
  synthesis of calciferol from, 848
  in the synthesis of steroid hormones, *836*
Cholinergic neurons, *974*, *975*
Chondrichthyans
  characteristics of, 687, *688*
  excretion of urea, 1074
  regulation of ionic composition of extracellular fluid, 1073
  salt and water balance regulation, *1077*
Chondrocytes, 416–417
Chordamesoderm, *910*, 915–916
Chordates
  in animal phylogeny, *630*
  derived structures in development, 683–684
  evolutionary relationships, *455*
  major clades, 683
  members of, 679
  neurulation, 915–916
  tunicates and lancelets, 683, 684
  See also Vertebrates
Chordin, 916
Chorion
  in the chicken egg, 918
  in placental mammals, 919
Chorionic villus, *907*
Chorionic villus sampling, 321
Chown, S. L., 1199
Christmas tree worm, *638*
Chroicocephalus novaehollandiae, 412
Chromatids
  crossing over and genetic exchange, 247–249, *250*
  differences between chromatids and chromosomes, 215
  events in meiosis, *220–221*
  exchanges during meiosis I, 219–220, 222
  See also Sister chromatids
Chromatin
  defined, 88
  remodeling, 344, 791
  structure, 211–212, *213*
Chromium, *1052*
Chromodoris, *1174*
Chromophores, 773, *774*
Chromoplasts, 92
Chromosomal mutations
  description of, 307–308
  in human genetic diseases, 312–313
Chromosome 1, 366
Chromosome 8, 366

Chromosome 15, 345
Chromosome 19, 366
Chromosome 21, 224, 308, 924
Chromosome number
  haploid and diploid, 218
  reduction in meiosis, 219, 221
  somatic cells, 218
Chromosomes
  defined, 88, 206
  differences between chromatids and chromosomes, 215
  in eukaryotes, 361
  genetic locus, 242
  genomic information, *355*
  global DNA methylation, 345–346
  homologous pairs, 218 (*see also* Homologous chromosomes)
  independent segregation during the formation of gametes, 237, 239
  karyotype, 224, *225*
  linkage of genes, 247, 248, 249–252
  meiotic errors, 222, 224
  numbers in organisms, 218
  origins of replication, 269, 271
  replication in prokaryotes, 206, 207
  separation and movement in anaphase, 214–216
  in somatic cells, 218
  speciation by centric fusion, 470, *471*
  telomeres, 275
Chronic myelogenous leukemia (CML), 304, 305
Chronic obstructive pulmonary disease, 862
Chronic protein deficiency, 1050
Chrysanthemums, 761
Chthamalus stellatus, 1184
Chylomicrons, *1062*, 1063, 1066
Chyme, 1061
Chymotrypsin, *154*, *1062*
Chytrid fungi
  amphibian decline and, 612
  description of, 617, 619
  life cycle, 617–619
  major groups and distinguishing features, *616*
  phylogeny of the fungi, 615, 616
Chytridiomycota, 617
Chytriomyces hyalinus, 617
cI regulatory protein, 340–341
Cicadas, 672, *673*
Cichlids
  prezygotic isolating mechanisms, 477
  speciation in Lake Malawi, 467 (*see also* Haplochromine cichlids)
Cicindela campestris, *1172*
Cigarette smoke
  benzopyrene and induced mutations, 309, 311
  emphysema and, 1013
  public bans on, 311
  smoker's cough, 1015
Cilia
  cilia motion in nodal cells establishes left–right asymmetry, 902, 914–915, 921
  of ciliates, 554, *555*

of ctenophores, 644
of lophophores, 652
structure and function, 96–98
of trophophores, 653
Ciliary muscles, 958–959
Ciliates, 554–555, 562
Cinchona, *604*, *605*, 797
Ciona, 683
Circadian rhythms
  in animal behavior, 1107–1108
  entrainment by light in plants, 774–775
  flowering in angiosperms and, 788
  per gene and, 1096
Circannual rhythms, 1108
Circular chromosomes
  origins of replication, 206, 269, 271
  replication, 206–207
Circulatory systems
  arteries and arterioles, 1039, *1040*
  blood composition, 1037–1039
  blood transport of respiratory gases, 1016–1019
  capillary beds, 1039–141
  components of, 1026
  control and regulation of, 1043–1045
  countercurrent heat exchange, 825, 828
  evolution in vertebrates, 1027–1030
  function of, 1026
  "heat portals" in the skin, 831
  human, *1031*
  lymphatic vessels, 1042
  mammalian heart function, 1030–1037
  open or closed systems, 1026–1027
  perfusion of the lungs, 1016
  sepsis, 862
  thermoregulation through the skin and, 824–825, 828
  vascular disease, 1042–1043
  veins, 1041–1042
Circumcision, 890
Cirrhilabrus jordani, *688*
Cirrhosis, 1039
cis-trans Isomers, 41
Cisplatin, 259, 268, 278
Cisternae, of the Golgi apparatus, 90
Citrate, 170, *1051*
Citrate synthase, 154
Citric acid, 624
Citric acid cycle
  allosteric regulation, 181–182
  anabolic interconversions and, *179*, *180*
  description of, 170–171
  in glucose metabolism, 168
  regulation of, 171
  relationships among metabolic pathways, *179*
Citrus trees, transgenic, 401
Clades, 451, 463
Cladonia subtenuis, 613
Clams, 662, 999
Claspers, 888

Class I MHC proteins, 872, *873*, 874
Class II MHC proteins, 872, *873*
Class switching, 869–871
Classes, 463
Clathrin, 121
Clavelina dellavallei, 684
Clay and clay particles, 71, 745, 746
Clean Air Act of 1990, 1220
Cleaning products, 144
Cleavage
  defined, 633, 904
  in major animal groups, 633
  in mammals, 905–906
  spiral, 653
  types of, 905
Climate
  atmospheric circulation patterns, 1124
  defined, 510, 1122
  evolutionary adaptations in organisms, 1122, 1125–1126
  ocean currents and, 1124–1125
  prevailing winds, 1124, *1125*
  variation of solar radiation over Earth's surface, 1123
  Walter climate diagrams, 1138
Climate change
  detection by isotope analysis of water, 36
  in Earth's history, 510
  photosynthetic efficiency of plants and, 201–202
  species extinctions and, 1236–1237
  See also Global climate change
Climax community, 1200
Clinal variation, 443–444
Clitellates, 661–662
Clitoris, 892, *893*
"Clock genes," 1108
"Clock-shifting" experiments, 1110, *1111*
Clonal deletion, 863, 865, 872
Clonal lineages, 562
Clonal selection, 865, *866*
Clones
  asexual reproduction and, 217, 562
  of cassava, 708
Cloning
  of animals, 406–407
  of plants, 405–406
  See also Molecular cloning
Closed circulatory systems, 1026–1027
Clostridium, 531, 936
Clostridium botulinum, 542
Clostridium difficile, 531
Clostridium tetani, 542
Clothes moths, 1203
Clotting factors, 1039
Clown fish, 887, *888*
Club fungi
  description of, 622
  distinguishing features, *616*
  edible, 624
  fruiting structure, *610*
  life cycle, *621*, 622–623
  phylogeny of the fungi, *615*
  See also Dikarya
Club mosses, 574, 580, 581
Clumped dispersion pattern, 1153

Clunio marinus, 1152
Cnidarians
 in animal phylogeny, *630*
 anthozoans, 646–647
 bilateral symmetry in, 643
 description of, 645–646
 gastrovascular cavity, 1056
 Hox genes, 413
 hydrostatic skeleton, 999
 hydrozoans, 647–648
 life cycle, 645, *646*
 major subgroups and number
   of living species, *632*
 nerve net, 940–941
 radial symmetry, 634
 scyphozoans, 647
Cnidocytes, *646*
Co-repressors, 331, 332
Coal deposits, 580
Coastal redwoods, 593, 1130
Coastal sand dunes, 793
Coastal zone, 1139
Coat color
 epistasis in Labrador retrievers,
   244
 multiple alleles in rabbits, *242*
Coat color patterns, point
   restriction phenotype, 245,
   306
Coat patterns, 245
Coated pits, 121
Coated vesicles, 121
Cobalamin, *1053*
Cobalt, *1052*
Cobalt-60, 24
Coca plants, 626
Cocaine, 626
Coccolithophores, 521
Coccus, 528, *529*
Cochlea
 anatomy of, *954*, 955
 flexion of the basilar
   membrane, 955–956
Cochlear canal, *954*
Cochlear fluid, 955, 956
Cochlear nerve, *954*, 956
Cochons Island, 1199
Cockleburs, 788, 789, 790
Cockroaches, *672*
Coconuts, 784
Cocos Island finch, *474*
Cod-liver oil, 848
Codominance, 243
Codons
 binding of tRNA anticodon to,
   293
 discovery and description of,
   288–289
 frameshift mutations and,
   306–307
 nonsense mutations and, 306,
   307
Coelecanths, 689–690
Coelom, 635, 652
Coelomates, 635
Coelomic fluid, 1075, *1076*
Coelophysis bauri, 518
Coenocyte, 560
Coenocytic hyphae, 609, *610*
Coenzyme A, *156*
Coenzyme Q10. *See* Ubiquinone
Coenzymes
 description of, 155–156

 in oxidation–reduction
   reactions, 167–168
 See also Acetyl coenzyme A
Coevolution
 diffuse coevolution, 1181–1182
 between herbivores and plants,
   1175–1176
 plant–pollinator relationships,
   588, 598–599, 605
 resulting from species
   interactions, 1171–1172
Coffee plantations, 1243
Coffroth, Mary Alice, *565*, 566
Cohen, Stanley, 374
Cohesins
 chromatin structure and,
   211–212
 homologous chromosomes and,
   219
 meiotic errors and, 224
 removal during mitosis, 212,
   213, 214–215, *216*
Cohesion
 of water molecules, 33
 See also Transpiration–
   cohesion–tension mechanism
Cohort life tables, 1154
Cohorts, 1154
Coincident substitutions, 488
Coitus interruptus, 890, *898*
Cold deserts, 1133–1134
"Cold" fish, 825
Cold-hardening, 810
Coleochaetophytes, 572, 573
Coleoptera, *672, 674*
Coleoptiles
 action spectrum of
   phototropism, 771, *772*
 auxin and phototropism, 762,
   763, 764, *765*
 early shoot growth and, 758
Colias, 442–443
Collagen
 in arteries, 1039
 in blood clotting, 1039
 in cartilage, 999–1000
 in connective tissues, 818–819
 in extracellular matrix, 100
"Collapsed lung," 1016
Collard lizard, *1133*
Collared flycatchers, 1103
Collecting duct
 aquaporins and water
   permeability, 1088, 1089–1090
 in the human excretory system,
   1080, 1082
 urine is concentrated in, 1084
Collembola, *672n*
Collenchyma, 713, 719–720
Colon, *1058*, 1063
Colon, Bartolo, 392
Colon cancer, 314
Colonial animals
 bryozoans, *654*, 656
 description of, 640, *641*
 entoprocts, 656
 hydrozoans, 646, 647–648
 sea squirts, 684
Colony-stimulating factor, *385*
Color blindness, *252*, 962
Color vision, 962
Colorectal cancer, 344
Colugos, *698*
Columba palumbus, 1116, 1117

Columbian mammoth, *1229*
Columbines, 477, 480
Columbus, Christopher, 440
Columnar epithelium, *818*
Coma, 970
Comb jellies, 634, 644–645
Combined DNA Index System
   (CODIS) database, 317–318
Comets, 68–69
Commensalism, *1170*, 1171
Common bile duct, 1061
Common carotid artery, *1031*
Common iliac artery and vein,
   1031
Common names, 462
Common wood-pigeons, 1116,
   1117
Communication. *See* Animal
   communication; Cell-to-cell
   communication
Communities
 challenges of identifying
   boundaries, 1189
 characteristics of, 1189
 correlation between
   productivity and species
   diversity, 1192
 defined, 9, 1189
 ecosystems and, 9–10
 energy transfer and biomass
   relationships between trophic
   levels, 1190–1192
 food webs, 1190, *1191*
 impact of disturbances on,
   1199–1202
 impact of species interactions
   on, 1193–1195
 impact of species richness
   on community stability,
   1202–1203
 insect colonization of human
   corpses, 1188
 keystone species, 1194–1195
 microbial, 539, *540*
 patterns of species diversity,
   1195–1198, *1199*
 primary producers and gross
   primary productivity,
   1189–1190
Community composition
 boreal and temperate evergreen
   forest biomes, *1129*
 chapparal biomes, *1134*
 cold desert biomes, *1133*
 hot desert biomes, *1132*
 temperate deciduous forest
   biome, *1130*
 temperate grassland biome,
   1131
 thorn forest and tropical
   savanna biomes, *1135*
 tropical deciduous forest
   biomes, *1136*
 tropical rainforest biomes, *1137*
 tundra biomes, *1128*
Compact bone, 1001
Companion cells, *714*, 715, 735
Comparative experiments, 12, 13
Comparative genomics
 defined, 356
 human genome, 366–367
 myostatin gene, 370
 prokaryotic genomes, 357
"Compass sense," 1109

Competition
 description of, 1170–1171
 exploitation competition,
   1182–1183
 importance in determining a
   species' niche, 1184
 indirect, 1184
 interference competition, 1182,
   1183
 types of, 1182–1183
Competitive exclusion, 1182, 1192
Competitive inhibitors, 158
Complement proteins, 860
Complement system, 860
Complementary base pairing
 in DNA, 266
 importance to DNA replication
   and transcription, 65–66, *67*
 in nucleic acids, 63–65
Complementary diet, 1051
Complementary DNA (cDNA)
 cDNA libraries, 379–380
 creating, 379
 in high-throughput sequencing,
   353
 in HIV infection, 341
 RNA retroviruses and, 544
Complementary mRNA, 544
Complete cleavage, 905
Complete flowers, 779
Complete gut, 644
Complete metamorphosis, 672,
   841
Complex behaviors
 in deuterostomes, 705
 human evolution and, 704, 705
Complex ions, 29
Complex life cycles
 of eukaryotic viruses, 341
 of parasites, 563, *564*, 640, *641*
 in protostome evolution, 674
Composite transposons, 358–359
Compost, 747
Compound eyes, 958
Compound umbel, *597*
Compounds
 covalent bonding, *27*
 defined, 26
Compulsory vaccinations, 856
Concentrated urine, 1071, 1077,
   1082–1084, 1090
Concentration gradients
 effect on diffusion, 114
 passive membrane transport
   and, 113
 secondary active transport and,
   118, 119
Concerted evolution, 498–499
Condensation reactions, 42, 269,
   270
Condensins, 212
Conditional mutations, 306
Conditioned reflexes, 982, 1094
Condoms, *898*
Conducting cells, in cardiac
   muscle, 991–992
Conduction, in heat exchange,
   823, *824*
Conduction deafness, 956
Cone cells, 960, 961–962, 963
Cones, of conifers, 593–594, *595*
Confocal microscopy, 80
Congenital hypothyroidism, 320
Conidia, 616, 622

Coniferophyta, *574*
Conifers
    in boreal and tropical evergreen forests, 1129, 1130
    cones, 593–594, *595*
    distinguishing characteristics, *574*
    fire adaptations, 594, *596*
    life cycle, 594, *595*
    number of species, 593
Conjugation. *See* Bacterial conjugation
Conjugation tube, 253, *254*
Conjugative pili, 84
Connective tissue
    characteristics of, 818–819
    development of bone from, 1001
    endoskeletons, 999–1000
    in the gut, 820
Connell, Joseph, 1184
Connexins, 939
Connexons, 139
Conomyrma bicolor, 1183
Consciousness, 982–983
Consensus sequences, 292, 332–333
Conservation biology
    basic principles of, 1229
    defined, 1229
    goal of protecting and managing biodiversity, 1229–1230
    human activities threatening species persistence, 1232–1237
    prediction of changes in biodiversity, 1230–1232
    risks of deliberate species introductions, 1228, 1245
    strategies used to protect biodiversity, 1237–1245
Conservative replication, 268
Consortium for the Barcode of Life (CBOL), 319
Conspecifics, 1103
CONSTANS (CO) gene, 788, 790
CONSTANS (CO) protein, 788, 790
Constant regions, of immunoglobulins, 867, 869–871
Constipation, 1063
Constitutive plant defenses
    against herbivores, 801, 802–803
    against pathogens, 798, 803
Constitutive proteins, 330
Constricting ring, 612, *613*
Consumers, 1190, *1191*
Continental drift, 509, 1142, 1143
Contraception, 890, 896, 897, *898*
Contractile ring, 216
Contractile vacuoles, 93, 554, *555*
Contractions, in labor and childbirth, 896–897
Contralateral neglect syndrome, 973
"Control" group, 12
Controlled burning, 1240
Controlled experiments, 12–13
Controlled systems, 816
Convection, 823, *824*

Convention on International Trade in Endangered Species (CITES), 1240–1241
Convergent evolution
    defined, 452
    of eyes, *959*
    in lysozyme and foregut fermenters, 492–494
    revealed by phylogenetic analyses, 459–460
Convergent extension, 909–910
Convolutions, 970, 973
Cook Strait, 1145
Copepods, 670
Copper
    in animal nutrition, *1052*
    in plant nutrition, *741*
    role in catalyzed reactions, *156*
Coprophagy, 1064
Copulation, 887, 891
"Copy and paste" transposition, 308, *358, 359,* 365
Coquerelia ventralis, 674
Coral reefs, 1139
Corals
    "bleaching," 565, 566, 1217
    description of, 646–647
    detection of chemical stimuli, 949
    dinoflagellate endosymbionts, 565, 566
    endosymbiotic relationship with dinoflagellates, 647
    environmental threats to, 647
    fluorescent proteins, 449, *464*
Corixidea major, 1230
Cork, 722, *723*
Cork cambium, 715, *716, 721, 722, 723*
Corms, 793
Corn
    domestication, 723–724
    flowers, *780*
    heterotypy and edible kernels, 419
    hybrid vigor, 244, 245
    improving nitrogen use efficiency, 740, 753
    prop roots, 718
    volicitin, 804
Corn oil, 57
Cornea, 958, *959*
Corn–sweet potato dicultures, 1203
Corolla, 597
Corona, of rotifers, 657, *658*
Coronary arteries, 1043
Coronary heart disease, *368*
Coronary thrombosis, 1043
Coronavirus, *543*
Coronosphaera mediterranea, 2
Corpora allata, 841
Corpse communities, 1188, 1203
Corpus callosum, 980
Corpus cardiacum, *841*
Corpus luteum, 893, *894, 895*
Corridors, 1162, 1233–1234
Cortex
    in roots, 717, *718*
    in shoots, 719–720
Cortical granules, *885, 886,* 887
Cortical nephron, *1080*
Corticosteroids, *836*
Corticotropin, *842, 843,* 849

Corticotropin-releasing hormone (CRH), 844, 849
Cortisol
    actions of, *842*
    half-life, 853
    negative feedback signaling to the anterior pituitary and hypothalamus, 844
    stress response and, 849–850
    structure, *836*
Cortisol receptor, 131
Corynebacterium diphtheriae, 542
Cost–benefit analysis
    applied to animal behavior, 1103
    of foraging behavior, 1104–1105, *1106*
    of group living, 1116, *1117*
    of territorial behavior, 1103–1104, *1105*
Cottony-cushion scale, 1164
Coturnix japonica, 1101
Cotyledons
    in angiosperms, *600,* 601
    in embryogenesis, 711–712
    in monocots and eudicots, 602, 709
    in seed germination, 758
Countercurrent flow, in fish gills, 1010, *1011*
Countercurrent heat exchange
    in the appendages of endotherms, 828
    in "hot" fishes, 825
Countercurrent multiplier, 1082–1084
Coupled transporters, 118
Courtship behavior
    experiments on the genetic basis of, 1095
    gene cascades in *Drosophila,* 1097–1098
    impact on speciation, 705
    multiple sensory modalities in *Drosophila,* 1113
    See also Sexual behavior
Covalent bonds
    ester linkages, 56
    formed in condensation reactions, 42
    multiple, 28
    orientation, 27–28
    overview, 26–27
    strength and stability, 27
    unequal sharing of electrons, 28
Covalent catalysis, 155
Covalent disulfide bridges, 47
Covalent modifications, of proteins, 50
Cowbirds, 1093, 1102, 1117, 1233
Cows
    cloning, 407
    See also Cattle
Cozumel thrasher, 1232
CpG islands, 343
Crab lice, 1177
Crabs, 670
Cranberry, 615
Cranial nerves
    autonomic nervous system and, *974,* 975
    development, 968
    relationship to the brainstem, 969

"Craniates," 686
Crassulaceae, 200
Crassulacean acid metabolism (CAM) plants, 200, 734
Crayfish, 670, 947
Creatine phosphate, 997
Crenarchaeota, 535, 536
Cretaceous period
    changes on Earth and major events in life, 506–507
    characteristics of life during, *519,* 521
    mass extinction, 511
Cretinism, 846
Crews, David, 881
Crick, Francis, 264–265, *266,* 430–431
Crickets, *672*
Crinoids, 520, 680–681, *682*
Cristae, of mitochondria, *92*
Critical period, in animal behavior, 1095, 1099
Critical temperature, 827
"Critically endangered" species, 1231
Cro gene, 341
Cro regulatory protein, 340–341
Crocodiles, 693–694, *695,* 1029–1030
Crocodilians, 693–694, *695,* 1029–1030
Crocodylus porosus, 695
Crocuses, 793
Cromileptes altivelis, 688
Crop, of the hoatzin, 494
Crop plants
    applications of biotechnology, 386–388, *389*
    cassava, 708, 724
    domestication, 723–724
    improvement programs, 724
    inoculation of seeds with mycorrhizae, 615
    major species, 605
    pathogenic fungi, 612
    polyploidy in, 225, *226*
    See also Agriculture; Corn; Rice; Transgenic crops; Wheat
Crop rotation, 750
Crossing experiments, 234
Crossing over
    genetic exchange and, 219, 247–249, *250*
    during meiosis, 219, *220, 222*
Crosstalk, between signal transduction pathways, 127, 134
Crotaphytus, 1133
Crown gall, 534
Crustaceans
    description of, 670–671
    larval form, 640
    major subgroups and number of living species, *632*
    undescribed species, 675
Crustose lichens, 613
Crypsis, 1173
Cryptic species, 468, *469*
Crytpochromes, 772
Ctenes, 644
Ctenophores
    description of, 644–645
    major subgroups and number of living species, *632*

in the phylogeny of animals, *630*, *648*
radial symmetry, 634
CTLA4 protein, 876
Cuboidal epithelial cells, *818*
Cuculus canorus, 1093
Cud, 1064
"Cuddle hormone," 843
See also Oxytocin
Culture, humans and, 705
Cumulus, 886
Cup fungi, 620, *622*
Cupula, *956, 957*
Curare plant, *604*
Currents, 1124–1125
Cuscuta, 752
"Cut and paste" transposition, 308, 359
Cut-flower industry, 770
Cuticle
of ecdysozoans, 654, 665
of exoskeletons, 999
of nonvascular land plants, 575
in plant defenses against herbivory, 1175
in plant evolution, 574
of plants, 713, 732
Cuticular plates, 665
Cutin, 713, 798
Cuttings, of stems, 793
Cuttlefish, 662, *663*
Cyanide
in cassava, 708
clinal variation in white clover, 443–444
in the Miller–Urey experiment, 70
protective storage in plants, 805
Cyanobacteria
atmospheric oxygen and, 532, 539
description of, 532, *533*
endosymbiotic origin of eukaryotic chloroplasts and, 102, *550, 551, 552*
in eutrophication, 1219
in lichens, 613
microfossils and, 74
nitrogen fixers, 749
photosynthesis in, 532, 538
stromatolites, *5*, 512, *513*
symbiotic relationship with hornworts, 579
Cyanogenic compounds, 708, 724
Cyathea australis, 582
Cycadophyta, *574*
Cycads, 574, 589, 592, *593*
Cyclic adenosine monophosphate (cAMP)
discovery of, 133, 134
function of, 67
in positive regulation of the *lac* operon, 332, *333*
regulation of, 137
as a second messenger, 133–134
Cyclic electron transport, 192
Cyclic guanosine monophosphate (cGMP)
in penile erection, 890, 891
in smooth muscle relaxation, 136
Cyclical succession, 1201–1202
Cyclin-dependent kinases (Cdk's), 209–211

Cyclin–Cdk complexes, 210–211, 214, *216*
Cyclins, 210–211, 349
Cyclopoid copepod, 670
Cyclosporin, 874
Cynomys ludovicianus, 1239
Cypresses, 808
Cyprinodon, 1160
Cysteine, 43, *44*
Cystic duct, 1061
Cystic fibrosis, *313*, 318, 321
Cytochrome *c*
amino acid sequences from different organisms, *488–489*
in the respiratory chain, 172, *173*
Cytochrome *c* oxidase, 172, *173*
Cytochrome oxidase gene, 319
Cytokines
function of, 859
in the humoral immune response, 872, *873*
in inflammation, 861
interferons, 860
Cytokinesis
in animal cells, 216
in cell division, 206
in eukaryotes, 207
in plant cells, 216–217, 710
in prokaryotes, 207
Cytokinins
discovery of, 768
effects on plant growth, 768
signal transduction pathway, 768–769
structure and typical activities, *759*
Cytoplasm
energy pathways in, *168*
of prokaryotes, 82
Cytoplasmic determinants, 395
Cytoplasmic dynein, 98, 216
Cytoplasmic inheritance, 252–253
Cytoplasmic segregation, 395
Cytoplasmic streaming, 560
Cytosine
codons and the genetic code, 288–289
complementary base pairing, 63–65
deamination, 309, 310
in DNA structure, 264, 265, 266, 267
as a "hot spot" for mutations, 310
methylation, 310, 328, 343 (*see also* DNA methylation)
structure, 63
Cytoskeleton
animal cells, *86*
asymmetric distribution of cytoplasmic determinants and, 395
attachments to membrane proteins, 109
biological membranes and, *106*
in the evolution of the eukaryotic cell, *550, 551*
functions of, 94
intermediate filaments, 95
microfilaments, 94–95, *96*, 98, 99
microtubules, 95–96
of prokaryotes, 84
Cytosol, 82

Cytotoxic T (T_C) cells
in the cellular immune response, 863, *865*, 872, *873*, 874
regulation by Tregs, 874

D

Daddy longlegs, *669*
Daffodils, *603*
Daily torpor, 830
Daltons, 22
Danaus plexippus, 639, 1174
Dandelions, *601, 1153*
Daphnia, 1105, 1106
Dark reactions, 188
See also Light-independent reactions
Darters, *472*
Darwin, Charles
beak diversity in finches, 412
concepts of coevolution, 1171–1172
concepts of evolution and natural selection, 6
description of the focus of ecology, 1122
on divergence of character, 1183
on earthworms, 639
on the evolution of flowers, 605
evolutionary theory, 428–430
experiment on phototropism in coleoptiles, 762, 763, 764
on hybrid vigor, 244
knowledge of artificial selection, 6, 433
on orchids, 588, 605
on sexual selection, 435
Darwin, Francis, 762, 763, 764
Darwin's black spider, 39
Darwin's finches
See Galápagos finches
Darwin's frog, 678
Darwin's rhea, *1131*
Dasyuromorphia, *698*
Data, importance of quantification in science, 11, 14
Date palms, *603*
Dating methods, 506–508
Daucus carota, 718
Daughter chromosomes, 215, 216, *223*
DAX1 gene, 250
DAX1 protein, 250
Day length
impact on development, 421–422
photoperiodic cues in flowering, 787
Day of the Dandelion (Pringle), 778
Db gene, *1067*, 1068
DCC gene, *314*
Dead Sea, 536
Dead space, 1011, 1012, 1013
Dead zones, 740, 1207, 1214, 1219, 1225
Deafness, 956
Deamination, of cytosine, 309, 310
Death Valley, 1160
Decapods, 670
Declarative memory, 982
Decomposers
characteristics of, 639
corpse communities, 1188, 1203

defined, 1054
in food webs, 1190, *1191*
prokaryotes, 538
saprobic fungi, 611
See also Detritivores
Decomposition, 1188
Deductive logic, 12
Deep-sea hydrothermal vent ecosystems
formation of organic polymers and, 71
pogonophorans, 660–661
prokaryotes and, 539
Deepwater Horizon (oil well), 373, 569
Deer tick (black-legged tick), *1151, 1152*
Defecation, 1063
Defensins
in animals, 859
in plants, 801
Deficiency diseases, *1053*, 1054
Deforestation, 1232
Dehydration reactions, 42
Deinococcus, 532
Deinococcus radiodurans, 389
Deiodinase, 845
Delayed hypersensitivity, 876
Deleterious mutations, purging of, 433, 441
Deletions, 308, 312–313
Delonix regia, 603
Demethylase, 343, 345
Demographic events
determination of population size and, 1153–1154
life tables and, 1154–1155
Demography, 1153
Demosponges, *630*, 643
Demyelinating diseases, 926
Denaturation, of proteins, 48, 50
Dendrites, 925–926
Dendritic cells
as antigen-presenting cells, 864, 872
function of, *858*
HIV infection, 876
in innate defenses, 861
pattern recognition receptors, 860
Dendrobates reticulatus, 1174
Dendroctonus frontalis, 1178
Dendroctonus ponderosae, 1202
Dendroica petechia, 1093
Dendrosenecio keniensis, 1128
Denitrification, 750, *751*
Denitrifiers, 539
Density-dependent population regulation, 1159
Density-independent population regulation, 1159
Dental plaque, 539, *540*
Dentine, 1055
Deoxyribonucleic acid (DNA)
artificial, 380
base sequence reveals evolutionary relationships, 66
chromatin, 88, 211–212, *213*
chromosomes, 88
complementary base pairing, 63–65
distinguishing from RNA, 63

double helix structure, 65, *66*, 264–265, 266, *267*
effect of cisplatin on, 259, 268, 278
eukaryotic regulatory sequences, *335*
evidence for being the genetic material, 260–263
exchange in bacteria, 253–255
genetic information and, 5–6
growth of, 63, *64*
hybridization experiments, 290–291
metagenomics, 357–358
methods of study DNA function, 380–383
mutations (*see* Mutations)
noncoding sequences, 290–291, 494–496
normal, daily damage to in humans, 310
PCR amplification, 277–278
recombinant, 374–375 (*see also* Recombinant DNA technology)
relationship of structure to function, 266
repair mechanisms, 275, 276–277
structure and function, 62, 63–67, 264–267
test tube synthesis, 267–268
transcription, 65
transmission of information by, 65–66
Deoxyribonucleoside monophosphates, 268
Deoxyribonucleoside triphosphates (dNTPs)
in DNA replication, 268–269
in high-throughput sequencing, 353, *354*
in the polymerase chain reaction, 277, *278*
Deoxyribose, *52*, 63, 266
Dependent variable, 13
Depolarization
description of, 930–931, *932*
generation of action potentials and, 932–933, 934
Deprivation experiments, 1095
Derived traits, 452, 453, *454*
Dermal tissue system
description of, 712–713
primary meristem giving rise to, 716
Dermaptera, *672*
Dermoptera, *698*
Descent with modification, 428
Deschampsia antarctica, 1125
Desensitization, to allergens, 876
Desert gerbils, 1090
Desert plants
drought avoiders, 806–807
root systems, 808
salt glands, 811
stem modifications, 720
structural adaptations in leaves, 807
succulence, 807
Desert pupfish, 1160
Desert rodents, 1090
Deserts
atmospheric circulation patterns and, 1124

characteristics in Australia, 1126
cold desert biomes, 1133–1134
hot desert biomes, 1132–1133
plant adaptations to, 806–808
Desmodus rotundus, *1071*
Desmosomes, 95, 111, *112*
Desmotubules, *139*, 140
Detergents, 144, 1221
Determinate growth, 715, 720, 786
Determination
during amphibian gastrulation, 910–913
of blastomeres, 907
of cell fate, 394
by cytoplasmic segregation, 395
defined, 393
in development, 710
differential gene expression and, 393, 397–398
by induction, 395–396, *397*
signal transduction pathways and, 397–398
Detritus-based communities, 1202
Detritivores
characteristics of, 639
commensalisms and, 1171
defined, 637, 1054
in food webs, 1190, *1191*
in succession, 1201
See also Decomposers
Detritus, 639, 1190
Deuterium, 22
Deuterostomes
in animal phylogeny, *630*
bilaterians, 634, 643
chordates, 747–679
complex behaviors in, 705
echinoderms and hemichordates, 680–683
fossil ancestors, 679–680
major groups and living species, *632*, 679
pattern of gastrulation in, 634
phylogeny, *679*
shared early developmental patterns, 679
Development
of animal behavior, 1098–1102
defined, 393
developmental constraints on evolution, 444, *445*
DNA methylation and, 344
environmental modulation of, 420–422
evolutionary (*See* Evolutionary developmental biology)
pattern formation, 399–405
phylogenetic analyses and, 455
processes in, 393–394, 710
restriction of cell fates during, 394
role of gene expression in, 397–399
See also Animal development; Human development; Plant growth and development
Developmental genes
differences in expression resulting in differences between species, 418–420
discussion of, 413–414
genetic switches, 415

modularity and differences in the patterns of expression, 416–417
Developmental modules
concept of, 415
differences in the patterns of gene expression, 416–417
genetic switches and the genetic toolkit, 415
Developmental plasticity, 420–422
Devonian period
changes on Earth and major events in life, *506–507*
characteristics of life during, *517, 520*
evolution of leaves, 583
evolution of seed plants, 589
vascular plants in, 580
DHAP. *See* Dihydroxyacetone phosphate
d'Herelle, Felix, 545
DHFR (Dihydrofolate reductase), 158
Diabetes mellitus
among the Pima, 1048
human SNP scans, *368*
as a risk factor for atherosclerosis, 1043
Type I, 848, 876
Type II, 848, 853
Diacylglycerol (DAG), 134–135
Diademed sifaka, *701*
Diadophis punctatus, *694*
Dialysis, 1086, *1087*
Diamondback moth, 1176
Diamondback rattlesnakes, *946*
Diapause, 421
Diaphragm (human respiratory system), *1014*, 1015, 1016, 1019–1020
Diaphragms (in contraception), *898*
Diarrhea, 1063
Diastole, 1031, 1032, 1039
Diastolic pressure, 1032, *1033*
Diatomaceous earth, 565
Diatoms
description of, 555–556
during the Mesozoic, 521
petroleum and natural gas deposits from, 565
as primary producers, 563
red tides and, 564
Dicaeum hirundinaceum, *1182*
Dichotomous branching, 580
Dicksonia antarctica, *581*
Dictyostelium, *98*, 838
Dictyostelium discoideum, *561*
Didelphimorphia, *698*
Didelphis virginiana, *699*, 1173
Didinium nasutum, *554*
Diencephalon, 968–969, *970*
Diet
developmental plasticity and, 421, *422*
effect of specialization on speciation, 480
human birth defects and, 916
impact on replenishment of muscle glycogen, 997
manipulation of epigenetic changes through, 349
Differential gene expression
in cell fate determination, 397–398

in development, 393
in differentiation, 398–399
during induction, 396
Differential interference-contrast microscopy, 80
Differentiation
defined, 393
in development, 710
differential gene expression and, 393
differential gene transcription in, 398–399
distinguished from determination, 394
reversibility, 405–409
role of transcription factors in, 336, *337*, 338
significance to multicellular life, 9
Diffuse coevolution, 1181–1182
Diffusion
across membranes, 114–115
within cells and tissues, 114
discussion of, 113–114
facilitated, 115–117
factors affecting, 114
of gases in water and air, 1007
of membrane proteins, 109, 110
of respiratory gases, 1006, 1008
simple diffusion, 114
through plasmodesmata, 140
Digestion
in ciliates, *555*
digestive enzymes, 1057–1058
external, 1055
gastrovascular cavity, 1056
hormonal control of, 1065
in the small intestine, 1061–1062
"thrifty genes" in humans, 1048
tubular guts, 1056–1057
vertebrate teeth and, 1055–1056
Digestive enzymes, 1057–1058, *1062*
Digestive systems
autonomic influence on smooth muscle, 993
control and regulation of the flow of nutrients, 1064–1068
digestive enzymes, 1057–1058
endocrine cells of, 835
gastrointestinal system in vertebrates, 1058–1064
gastrovascular cavity, 1056
gut microbes and, 539, 540–541
pH in the human stomach, 161
tissue composition, 820
tubular guts, 1056–1057
Digestive vacuoles, 555
Digitalis, 1034
Digitalis purpurea, 1034
Diglycerides, galactose-substituted, 92
Dihybrid crosses, 238, 239–240
Dihydrofolate, 158
Dihydrofolate reductase (DHFR), 158
Dihydropyridine (DHP) receptor, 990, 992
Dihydroxyacetone phosphate (DHAP), 179
Diisopropyl phosphorofluoridate (DIPF), 157, *158*
Dikarya, *615*

club fungi, *621*, 622–623 (*see also* Club fungi)
life cycle, 620, *621*
sac fungi, 620–622 (*see also* Sac fungi)
sister group to arbuscular mycorrhizal fungi, 619
See also Club fungi; Sac fungi
Dikaryon, 620, *621*
Dimethyl sulfide, 1219
Dimethylsulfoniopropionate (DMSP), 1219
Dinoflagellates
beneficial aspects of, 566
bioluminescence, 566
description of, 553, *554*
endosymbionts in corals, 565, 566, 647
endosymbiotic origin of chloroplasts in, 552
during the Mesozoic, 521
red tides and, 549, 564
Dinosaurs
characteristics of, 694
evolution of birds and, 693, 694
evolution of feathers, 695
migration, 21
Dioecious plants
defined, 597, 779
example of, *780*
strategies for preventing self-pollination, 782
Dioecious species, 249, 887
Diomedea melanophris, 1105
Dionaea, 751–752
Dionaea muscipula, 752
Dipeptidase, *1062*
DIPF (diisopropyl phosphorofluoridate), 157, *158*
Diphyllobothrium latum, 641
Diploblastic animals, *630*, 633
Diploids
in alternation of generations life cycle, 562, 563, 574, 575
defined, 218, 236
in sexual reproduction, 218
Diplomonads, 558
Diplontic life cycle, *218*
Diploria labyrinthiformis, 647
Diplura, *672n*
Diprotodonts, *698*
Diptera, 415, *672*, 674
Direct development, 639
Direct fitness, 1114
Directional selection, 439–440
Directional succession, 1199–1201
Disaccharides, 51, 53
Disc flowers, 597
Discoidal cleavage, 905
Disparity, 978
Dispersal
in animal life cycles, 640
discontinuous species distributions and, 1143–1145
dispersal ability and speciation rates, 481
Dispersion patterns, 1150, 1152–1153
Dispersive replication, 268
Disruptive selection, 439, 440, 473
Dissociation constant, 128, 153

Distal convoluted tubule
in the mammalian kidney, *1080*, 1082
in the production of concentrated urine, *1083*, 1084
regulation of the glomerular filtration rate, 1087–1088
Distalless (Dll) gene, 418
Distance-direction navigation, 1109
Disturbances
defined, 1199
impact on species diversity, 1199
restoring disturbance patterns in ecosystems, 1239–1240
species extinctions and, 1232
succession following, 1199–1202
variability in the magnitude of effects from, 1199
Disulfide bonds, 867
Disulfide bridges, 43, *46*, 47, 196–197
Diving mammals, 1017
DNA. *See* Deoxyribonucleic acid
DNA barcode, 319
DNA fingerprinting, 317–318
DNA fragments
arranging, 354, *355*
genomic libraries, 379
DNA helicase, 272, 273
DNA hybridization, 321
DNA libraries, 379–380
DNA ligase, 273, *274*, 276, 374
DNA methylase, 345
DNA methylation
effects of, 343–344
gene regulation and, 328
genomic imprinting and, 344–345
global chromosome changes, 345–346
process of, 343
in promoters, 328
protection from restriction enzymes, 315
DNA methyltransferase, 343
DNA microarrays, 368, 382–383
DNA polymerase I, 273, *274*, 276
DNA polymerase III, 273, *274*, 276
DNA polymerases
in DNA replication, 269, *270*, 271, *272*, 273–274
errors, 276, 309
eukaryotic general transcription factors and, 334
in high-throughput sequencing, 353, *354*
in the PCR reaction, 277–278
processive characteristic, 273
proofreading function, 276–277
sliding DNA clamp, 273–274
structure, 271, 272
in test tube synthesis of DNA, 267
from *Thermus aquaticus*, 532
DNA replication
in cell division, 206
defined, 65
DNA structure and, 266
errors, 309
in eukaryotes, 207

importance of base pairing to, 66, *67*
lagging strand, 272–273, *274*
leading strand, 272–273
Okazaki fragments, 273
possible replication patterns, 267–268
pre-replication complex, 269
primers, 63, 271, *272*, *274*
process of, 268–275
in prokaryotes, 206, *207*
replication complex, 274–275
replication forks, 271, *272*–275
roles of proteins in, *272*, 273
sliding DNA clamp, 273–274
teleomeres, 275
DNA segregation
in cell division, 206
in eukaryotes, 207
in prokaryotes, 207
DNA sequences
alignment, *487*
palindromic, 374
phylogenetic analyses and, 456
recognition by proteins, 335–336
using models to calculate evolutionary divergence, 487–489
DNA sequencing. *See* Genome sequencing
DNA structure
5′ end and 3′ end, 266
antiparallel strands, 265, 266
chemical evidence from base composition, 264
double helix, 65, *66*, 264–265, 266, *267*
importance to DNA function, 266
key features of, 265–266, *267*
major and minor grooves, 265, 266, *267*
physical evidence from X-ray diffraction, 264
Watson and Crick's model, 264–265
DNA technologies
DNA barcode, 319
DNA fingerprinting, 317–318
gel electrophoresis, 316–317
restriction enzymes, 315–316
reverse genetics, 318
DNA templates. *See* Templates
DNA testing, 320–321
DNA transposons, 365
DNA viruses
as carrier viruses, 323, 324
description of, 341, *543*, 544–545
dNTPs. *See* Deoxyribonucleoside triphosphates
Dobsonflies, *672*
Dobzhansky, Theodosius, 470
Dobzhansky–Muller model, 470, *471*
Dodders, 752
Dog Genome Project, 352
Dogs
bleeding and estrus in, 893
conditioned reflexes, 1094
epistasis and coat color, 244
genome, 352
myostatin gene, 370
olfactory sensitivity, 950

partial pressure of carbon dioxide in the regulation of breathing, 1020–1021
variation in body size among breeds, 352
Dolly (cloned sheep), 406–407
Dolphins, 700, 973
Domains, of proteins, 291
Domains (in classification)
common ancestor, 527
distinguishing characteristics of, *527*
lateral gene transfer, 529
overview, 81
relationships among, *526*
shared features of prokaryotes, 526–527
Domestication, impact on plant form, 723–724
Dominance, incomplete, 242–243
Dominance hypothesis, of heterosis, 245
Dominant trait, 235
Dopamine, 940
Dormant seeds. *See* Seed dormancy
Dorsal, 634, *635*
Dorsal aorta, 1028
Dorsal fins, 687
Dorsal horn, 942
Dorsal lip of the blastopore
amphibian neurulation and, 915–916
in embryo formation, 910–911
in gastrulation, 909
organizer cells and, 912
See also Primary embryonic organizer
Dorsal medulla, 1019–1020
Dorsal spines, 424
Dorsal–ventral axis, in animal organogenesis, 917
Dose–response curves, of hormones, 853
Double fertilization, 600–601, *781*, 783
Double helix, of DNA, 65, *66*, 264–265, 266, *267*
Double-stranded DNA viruses, *543*, 544–545
Double-stranded RNA viruses, 544
Doublesex (dsx) gene, 1097–1098
Douglas firs, 1160
Doupe, Allison, 1100–1101
Doushantuo fossils, *516*
Down syndrome, 224, 308, 924, 939
"Down syndrome mouse," 924, 939, 943
Downregulation, of hormone receptors, 853
Downy mildews, 556
Dragline silk, 39, 45
Dragonflies, 505, *672*, *673*, 958
Dreaming, 979
Dreissena polymorpha, 1160, 1235
Drosera, 751
Drosera rotundifolia, 752
Drosophila
allopatric speciation in the Hawaiian Islands, 472, *473*
atmospheric oxygen and body size, 513, 514
complete cleavage, 905

germ cell lineage, 908
long terminal repeats, 495
as a model organism, 282
reproductive isolation
    from increasing genetic
        divergence, *471*
    sex-linked inheritance, 251–252
Drosophila endobranchia, 1231
Drosophila melanogaster
as an *r*-strategist, 1159
artificial selection experiments, 433
determination of body
    segmentation, 401–405
developmental genes in eye
    development, 413, *414*
factors involved in
    characteristic species density, 1160
genetic control of courtship
    behavior, 1097–1098
genomic information, *361, 362*
homeotic mutations, 404
Hox genes, *414*
laboratory experiments on
    speciation, 482
multiple sensory modalities in
    courtship behavior, 1113
wing development, 415
Drought avoiders, 806–807
Drought stress, plant responses
    to, 809
Drugs
agonists and antagonists of
    neurotransmitters, 940
anti-inflammatory, 862
determining dosage level, 128
in HIV treatment, 876–877
metabolic inhibitors, 322–323
pharmacogenomics, 368
psychoactive, 135
reduction of overinhibition in
    the brain, 943
See also Cancer drugs;
    Pharmaceuticals
Drummond, Thomas, 470–471
Dryas octopetala, 1200
Dubautia menziesii, 481
Duchenne muscular dystrophy, 313
Duck-billed platypus, 697, 951
Ducks
BMP4 and beak development, 424
hindlimb development, 417
Duckweed, 1140
Dugongs, 700
Dulbecco, Renato, 353
Dung beetles, 1171
Dunman, Paul, 301
Dunn, Casey, 631
Duodenum, 1061, 1065
Duplication-and-divergence
    hypothesis, 413–414
Duplications, 308
Dutch elm disease, 622, 793
"Dutchman's pipe," 602
Dwarf mistletoe, 752
Dynein, 97, 98
Dyscophus guineti, 7
Dystrophin, 313

E

E7 protein, 228
"Ear stones." *See* Otoliths

Eardrum. *See* Tympanic
    membrane
Earphones, 956
Ears
anatomy of the human ear,
    954–955
See also Auditory systems; Inner
    ear
Earth
atmospheric circulation
    patterns, 1124
magnetic fields and
    paleomagnetic dating, 508
prevailing winds, 1124, *1125*
variation in received solar
    radiation, 1123
Earthworms
closed circulatory system, 1026,
    1027
coelomate, *635*
description of, 661
detritivores, 639
excretory system, 1075, *1076*
hydrostatic skeleton, 999
infolding of the gut in, 1057
neural network, 941
rapid climate change and,
    1236–1237
simultaneous hermaphroditism,
    887
Earwigs, 672, 673
Easterlies, 1124, *1125*
Eastern gray kangaroo, *699*
eBay, 1241
Ecdysone, 840–841
Ecdysozoans
anatomical characteristics, *652*
in animal phylogeny, *630*
arthropods, 667–673, *674*
cleavage pattern, 633
cuticle and molting, 654
horsehair worms, 666, *667*
major subgroups and number
    of living species, *632*
nematodes, 666
priapulids, kinorhynchs, and
    loriciferans, 665–666
Echidnas, 697
Echiniscus, 667
Echinocereus reichenbachii, 603
Echinoderms
in animal phylogeny, *630*
appendages, 636
in deuterostome phylogeny, *679*
features of, 680–682
major groups and living
    species, *632, 679*
radial symmetry, 634
regeneration in, 881
Echinozoans, 681
"Echo generation," 1165
Echolocation, in bats, 946, 963
Ecological communities, 1189–
    1193
See also Communities
Ecological economics, 1241–1243
Ecological efficiency, 1191
Ecological survivorship curves,
    1156
Ecology
defined, 1122
distinguished from
    environmentalism, 1122

factors affecting climate,
    1122–1125
species interactions studied by,
    1170–1172 (*see also* Species
    interactions)
study of the biotic and abiotic
    components of ecosystems,
    1122
terrestrial biomes, 1126–1138
    (*see also* Biomes)
Walter climate diagrams, 1138
EcoRI restriction enzyme, 315–316,
    374
Ecosystem engineers, 1194
Ecosystem services, 1241–1242
Ecosystems
biotic and abiotic components,
    1122
causes and impact of dead
    zones in the Gulf of Mexico,
    1207, 1219, 1225
consequences of human
    alterations, 1223–1224
defined, 9, 1208
examples, 9–10
global climate change and, 17
goods and services provided,
    1223–1224
recent metaphors for, 1244–1245
restoration ecology, 1237–1239,
    1240
restoring disturbance patterns,
    1239–1240
sustainable management, 1224
See also Global ecosystem
Ecotourism, 1242, *1243*
Ectoderm
defined, 908
in diploblastic and triploblastic
    animals, 633
in extraembryonic membranes,
    918
nervous system development in
    amphibians, 912–913
in neurulation, 916
tissues and organs derived
    from, *907,* 908
Ectomycorrhizae, 614, 615, 622
Ectoparasites
features of, 639, 1176–1177
leeches, 661
monogeneans, 657
Ectopic pregnancy, 906
Ectopistes migratorius, 1230
Ectotherms
control of blood flow to the
    skin, 824–825
defined, 822
differences from endotherms,
    822
energy budget and, 823–824
metabolic heat production,
    825–826
reptile circulatory system,
    1029–1030
response to changes in
    environmental temperature,
    822, *823*
Edema, 1041, 1050
Edge effects, 1233
Ediacaran fossils, *516*
Eelgrass, 1141
Eels, *688*
Effector B cells. *See* Plasma cells

Effector cells, 865–866
Effector proteins, 130
Effector T cells, 865, 866, 871, 872,
    873, 874
Effector-triggered immunity (ETI),
    799
    See also Specific plant immunity
Effectors
in allosteric regulation, 159
in physiological systems, 816
plant specific immunity and,
    799, 800
Efferent blood vessels, in fish gills,
    1010
Efferent nervous system, 968
Efferent neurons
defined, 940
spinal reflexes, 942
Efferent pathways, of the
    autonomic nervous system,
    975
Efferent renal arterioles
in the autoregulation of the
    glomerular filtration rate,
    1087, *1088*
in the mammalian kidney, *1080,*
    1082
in the vertebrate nephron, 1078
"Efficiency genes," 1048
EGF (epidermal growth factor),
    398
Egg cell, 779, *781, 783*
Egg cytoplasm
components of, 903
rearrangement following
    fertilization, 903–904
Eggs
amniote, 692 (*see also* Amniote
    egg)
of conifers, *595*
cytoplasmic segregation, 395
fertilization in animals,
    884–887 (*see also* Sperm–egg
    interactions)
fertilization in humans, 892
gametic isolation and, 477
genomic imprinting in
    mammals, 344–345
ovulation in humans, 892
parthenogenic reproduction,
    881–882
production in animals, 882–883,
    884
release in spawning, 887
reproductive technologies in
    humans, 897, 898
reproductive trade-offs in
    animals, 642
of seed plants, 592
Ehrlich, Anne, 1245
Ehrlich, Paul, 1161–1162, 1175,
    1245
Einkorn wheat, *226*
Ejaculation, 891
Ejaculatory duct, 890
EKG. *See* Electrocardiogram
Elaiosomes, 1135, 1146
Elastin, 819, 1039, *1040*
Elbert, Thomas, 328
Electric charge, of atoms, 22
Electric currents
creation of, 927
ionic, 931–932
Electric eels, 485

Electric organs, 485
Electric signals, employed by fish, 485
Electrical synapses, 926, 936, 939
Electrical work, 10
Electricity. *See* Bioelectricity/ Bioelectric energy
Electrocardiogram (ECG), 1035–1036
Electrochemical gradients
    membrane potentials and, 929
    root uptake of mineral ions and, 729
Electrodes, measuring membrane potentials with, 928
Electroencephalogram (EEG), 978–979
Electromagnetic radiation
    photobiology of light, 189–190
    photochemistry of light, 188–189
Electromagnetic spectrum, *189*
Electromyogram (EMG), 978
Electron acceptors
    in denitrifiers, 539
    non-oxygen acceptors, 176
Electron carriers
    in the citric acid cycle, 170–171
    in oxidation–reduction reactions, 167–168
    reoxidation during glucose catabolism, 171
    in the respiratory chain, 172, *173*
Electron donors
    in anoxygenic photosynthesis, 187
    in oxygenic photosynthesis, 187
Electron microscopes, 79, 81, 108
Electron shells, 24–25
Electron transport
    in glucose metabolism, 168
    light-induced, 196–197
    with non-oxygen electron acceptors, 176
    in oxidative phosphorylation, 171
    in photosynthesis, 191–192
    relationships among metabolic pathways, *179*
    toxic intermediates, 175–176
Electronegativity, 28
Electrons
    in atoms, 22
    chemical bonding and, 24, 25
    covalent bonds, 26–28
    shells and orbitals, 24–25
    transport in the respiratory chain, 172, *173*
Electroocculogram (EOG), 978
Electrosensors, 947, *948*
Elegant madtom, *472*
Element cycling
    prokaryotes and, 538–539
    See also Biogeochemical cycles
Elementary bodies, 533, *534*
Elements
    atomic number, 22
    atomic weight, 23
    defined, 22
    isotopes, 22–24
    periodic table, 22, *23*
Elephant-nosed fish, *485*
Elephant seals, 1104, *1105*, 1114

Elephant shrews, *698*
Elephantiasis, 666
Elephants, *1135*
    brain size–body size relationship, 973
    fecundity, 1155
    full census counting, 1150
    impact of human overexploitation, 1234
    importance of ending international ivory trade, 1240–1241
    number of species, *698*
    physical challenges of snorkeling, 1005, 1022
    population age structure, 1152
    sound communication, 946
Elicitors
    activation of plant defenses to herbivory, 803–804
    defined, 798
    plant responses to, 798–801
Elk, 1193–1194, 1239
Elliott's milkpea, 1221–1223
Elongation, in transcription, *287*, 288
Embioptera, *672*
Embolism, 1043
Embolus, 1043
Embryo sac, 779, *781*
Embryogenesis
    basic patterns in animals, 633–634
    DNA methylation and, 344
    in mammals, 697
    in plants, 711–712
Embryonic stem cells (ESCs)
    culturing of, 408–409
    in homologous recombination and knockout mice, 381
    in the mammalian embryo, 906
Embryophytes, 573
    See also Land plants
Embryos
    of angiosperms, *600, 601, 781, 783*
    cell fate determination during development, 394
    cleavage of the zygote in animals, 904–906
    of conifers, 594, *595*
    defined, 393
    development in humans, 920
    development in mammals, 697
    development in plants, *393*, 711–712
    developmental modules, 415–418
    features of, 393
    of land plants, 575
    patterns of care and nurture in animals, 889
    in plant evolution, 574
    from in vitro fertilization, 899
Emerging diseases, 501
Emission, 891
Emlen, Stephen, 1110
Emmenanthe penduliflora, 757
Emmer wheat, *226*
Emperor penguins, 1099
Emphysema, 1013
Emu, 694
Emulsifiers, 556, 1062
Enamel, 21, 1055

Enceladus (moon of Saturn), 70
Encephalartos, 593
End-product inhibition, of metabolic pathways, 160–161
Endangered species
    captive breeding programs, 1244
    cloning, 407
    defined, 1231
Endangered Species Act, 17
Endeis, 668
Endemism, 1237
Endergonic reactions, 147, *148*, 150–151
Endocrine cells, 835–836
Endocrine glands
    adrenal gland, 849–850
    anterior pituitary, 842–843, 844–845 (*see also* Anterior pituitary)
    defined, 835
    gonads, 850
    in humans, *842*
    pancreas, 848–849
    parathyroid glands, 847–848
    pineal gland, 851
    posterior pituitary, 842, 843
    thyroid gland, 845–847
Endocrine signaling, 835–836
    See also Animal hormones
Endocrine system
    in humans, *842*
    interactions with the nervous system, 842–845
    major glands and hormones, 845–852
    types of chemical signaling in, 835–836
    See also Animal hormones
Endocytosis
    in ciliates, 555
    defined, 120
    receptor-mediated, 120, *121*
    types of, 120, *122*
Endoderm
    in avian gastrulation, 914
    defined, 908
    in diploblastic and triploblastic animals, 633
    in frog gastrulation, 909, 910
    in sea urchin gastrulation, 909
    tissues and organs derived from, *907*, 908
Endodermis, 717, *718*, 729–730
Endogenous retroviruses, 308
Endomembrane system, 88–91
    See also Endoplasmic reticulum; Golgi apparatus; Lysosomes
Endometrium
    implantation of the blastocyst, 892, 906, *907*, 919
    in the uterine cycle, 893, *894*
Endoparasites
    features of, 639, 1176
    flatworms, 657
    horsehair worm larvae, 666, *667*
Endophytic fungi, 615
Endoplasmic reticulum (ER)
    evolution in eukaryotic cells, 551
    glycosylation of proteins, 301
    structure and function, 88–90
    See also Rough endoplasmic reticulum; Sarcoplasmic

reticulum; Smooth endoplasmic reticulum
Endorphins, *842*, 843, 940
Endoskeletons
    connective tissue in, 999–1000
    of humans, 999, *1000*
    interactions with skeletal muscle, 999, 1001–1002
    of radiolarians, *557, 558*
    See also Internal skeletons
Endosperm
    in angiosperm seeds, 591, *600*, 601
    formation in angiosperms, 601, *781*, 783
    in seed germination, 758
Endospores, 531
Endosymbionts, 564–565, 566
Endosymbiosis
    chloroplasts and photosynthesis in eukaryotes, 102, *550*, 551–552, 570
    overview, *101*, 102
    primary, 570
    transposons and, 366
Endosymbiotic bacteria
    colonization of the intestines, 1056–1057
    in pogonophorans, 660
Endothermic reactions. *See* Endergonic reactions
Endotherms
    adaptations to cold, 828, 1125
    basal metabolic rate and body size, 826–827
    control of blood flow to the skin, 824–825, 828
    defined, 822
    differences from ectotherms, 822
    dissipation of heat with water and evaporation, 829
    energy budget and, 823–824
    fevers, 830
    heat production in, 822, 827–828
    hibernation, 830, *831*
    hypothermia, 830
    response to changes in environmental temperature, 822, *823*
Endotoxins, 542
Endurance, impact of exercise on, 996–997
Endymion non-scriptus, 462
Energetic costs, in animal behavior, 1103
Energy
    activation energy, 151–152
    ATP and, 149–151
    in chemical reactions, 31
    defined, 31
    from food, 1049–1050
    laws of thermodynamics, 146–147
    measures of, 1049
    nutrients as sources of, 10
    storage in the animal body, 1050
    transfer between trophic levels in communities, 1190–1192
    types of, 145
Energy budgets, 823–824, 1050
Energy maximization hypothesis, of foraging behavior, 1105

Energy transformation, organelles involved in, 91–93
English elm, 793
Engrailed gene family, 499–500
Enhancers
    as genetic switches, 415
    of transcription factors, 335
Enkephalins, *842, 843,* 940
Enteric nervous system, 968, 1059
Enterococcus, 4
Enterokinase, 1062, *1062*
Enthalpy, 146, 147
Entomology, forensic, 1188
Entoprocts, *632, 652, 656*
Entrainment, 1107–1108
Entropy, 146, 147
Enveloped virus, 341
Environment
    carrying capacity, 1158
    gene–environment interactions, 245–246
    modulation of development, 420–422
Environmental cleanup, 373, 389
    See also Bioremediation
Environmental cycles, coordination of animal behavior with, 1106–1108
Environmental genomics, 530
Environmental Protection Agency (EPA), 17
Environmental resistance, 1158
Environmentalism, 1122
Enzyme-catalyzed reactions
    effect of pH on, 161
    effect of substrate concentration on reaction rate, 156
    effect of temperature on, 161–162
    lowering of the energy barrier in, 151–154
    mechanisms of enzyme function, 154–156
Enzyme-substrate complex (ES), 152–153
Enzymes
    binding to substrates, 152–153
    as biological catalysts, 151
    chemical equilibrium, 153
    in cleaning aids, 144
    commercial applications, 144, 162
    effect of pH on, 161
    effect of temperature on, 161–162
    effect on the rate of reaction, 153–154
    functions, 42
    induced fit, 155
    interactions with substrates, 154–155
    isozymes, 162
    loss of function mutations, 311–312
    lowering of the energy barrier in biochemical reactions, 151–154
    mechanisms in catalyzing reactions, 154–156
    modification during signal transduction, 138
    naming convention, 152
    nonprotein chemical "partners," 155–156
    one-gene, one-enzyme hypothesis, 282–283, 284

regulation of, 156–162
regulation of signal transduction, 136–137
relationship of molecular structure to function, 155
in signal transduction pathways, 126–127
Eons, *506–507,* 508
Eosinophils, *858*
Ephedra, 604
Ephedrine, *604,* 802
Ephemeroptera, *672*
Ephestia kuehniella, 1184
Epiblast, *907,* 913–914, 919
Epidemics, influenza, 427
Epidermal growth factor (EGF), 398
Epidermal growth factor receptor, 325
Epidermis
    of leaves, 720–721
    of plants, 712–713
    of roots, 717, *718*
Epididymis, 890, *891*
Epigenetics
    changes induced by the environment, 344
    defined, 328, 343
    DNA methylation, 343–344 (*see also* DNA methylation)
    effects of "royal jelly" on honey bee development, 899
    FLC gene expression in angiosperms, 791
    gene regulation and, 343–346
    histone modifications, 344
    manipulation of epigenetic changes through diet, 349
Epigenomes, 344
Epiglottis, 1059
Epilepsy, 982
Epiloby, 910
Epinephrine (adrenaline)
    actions of, *842,* 850
    activation of glycogen phosphorylase, 132–133
    in blood pressure regulation, 1044, *1045*
    endocrine source, 849
    fight-or-flight response, 837
    half-life, 852–853
    regulation of glucose metabolism in liver cells, 138
    structure, *836*
Epiphytes, 718
Epistasis, 244
Epithelial cells, 112
Epithelial tissues
    characteristics of, 817
    in the gut, 820
    types of, 817–819
Epitopes, 862, 864, 868
    See also Antigenic determinants
Epochs, 521, 522
Epstein–Barr virus, 876
Equatorial Countercurrent, 1124, *1125*
Equatorial plate
    in meiosis, *220, 221*
    in mitosis, 215
Equidae, 1143, *1144*
Equilibrium, organs of, *956*
    see also Vestibular system
Equisetum, 581

Equisetum pratense, 581
Equus, 1144
Equus occidentalis, 1229
Erectile dysfunction (ED), 891
Eremias lugubris, 1174
Erinaceomorpha, *698*
Erlotinib, 325
Erosion, 1213
Error signals, 816
Erwin, Terry, 651, 673
Erwinia uredovora, 388
Erythrocytes. *See* Red blood cells
Erythropoietin, 211, *385,* 1038
Escherichia, 542
Escherichia coli
    cell division, 206
    circular chromosome, 206
    comparative genomics, 357
    conjugation in, 253–254
    DNA replication, 273
    functional genomics, 357
    gene cloning, 375–376
    genomic comparison to *S. cerevisiae, 362*
    genomic information, *361*
    illustration of, *2*
    lac operon in, 330, *331*
    LexA protein, 341
    Meselson–Stahl experiment on DNA replication, 268, *269,* 270
    as a model organism, 282
    negative and positive regulation of the *lac* operon, 330–332, *333*
    origins of replication, 269, 271
    pollution of lakes, 1221
    proteobacteria, 534
    recombinant plasmids, 374
    regulation of lactose metabolism, 329–330
    reporter genes, 378
    sigma factors, 333
    strain O157:H7, 357
    uses of genomic information from, 357
    viral infection experiments on DNA, 261–263
Eschrichtius robustus, 1109
Escovopsis, 1169
Esophageal sphincter, 1060
Esophagitis, 612
Esophagus, 1059–1060
Essential amino acids, 1051–1052
Essential elements, in plant nutrition, 741–743
Essential fatty acids, 1052
Ester linkages
    in lipids, 535
    in phospholipids, 57, *58*
    in triglycerides, 56
    in waxes, 59
Estivation, 1077
Estradiol, 850
Estrogen receptor, 129, 325
Estrogens
    actions of, *842*
    defined, 850
    in follicle selection for ovulation, 896
    in human pregnancy, 896
    in human puberty, 894
    in labor and childbirth, 896

in parthenogenic whiptail lizards, 882
production in human ovaries, 893, 894
in regulation of the ovarian and uterine cycles, *894, 895*
structure, *836*
temperature-dependent sex determination and, 420
    See also Sex steroids
Estrus, 893
Estuaries, 1141
Ethanol
    from alcoholic fermentation, *177, 178*
    produced by yeast, 623
Etheostoma tetrazonum, 472
Etheostoma variatum, 472
Ether linkages, 535
Ethiopian region, *1142*
Ethiopian wolf, 1242
Ethnobotany, 605
Ethology, 1094–1096
Ethylene
    effects on plants, *759,* 769–770
    genetic screen in *Arabidopsis,* 760
    signal transduction pathway in plants, 770–771
    structure, *759*
Etiolated seedlings, 772
Eublepharis macularius, 694
Eucalyptus leaves, 1055
Euchromatin, 345
Eudicots
    characteristics of, 601, 602, 709–710
    early shoot development, 758
    examples of, *603*
    leaf anatomy, 720–721
    leaf veins, 720
    root anatomy, *717*
    root systems, *718*
    shoot anatomy, 719–720
Eudocimus ruber, 696
Eudorina, 140, 141
Euglena, 558, 559
Euglenids, 551–552, 558–559
Eukarya
    distinguishing characteristics, 81, *527*
    relationship to prokaryotes, *526,* 527
    in the tree of life, 8–9
Eukaryotes
    appearance in the Proterozoic, 515–516
    cellular specialization, 4
    defined, 81
    in the evolution of life, 4
    evolution of multicellularity in, 552–553
    origin of, 550–552
    Precambrian divergence of major groups, 552, *553*
    proteomes, 369, *370*
    protists, 550, 552–561
    relationship to prokaryotes, 527
    shared features with prokaryotes, 526
Eukaryotic cells
    animal cell structure, *86* (*see also* Animal cells)

atmospheric oxygen levels and, 512
cell division, 207–211 (*see also* Meiosis; Mitosis)
cellular locations of energy pathways, *168*
characteristics of gene expression in, *291*
compartmentalization in, 84
cytoskeleton, 94–98
endomembrane system, 88–91
endosymbiosis and, *550*, 551–552
evolution of, 550–552
extracellular structures, 99–100
genes and gene transcription, 290–293
genetic transformation, 263, 376–377
location and functions of RNAs in, *286*
methods of studying and analyzing organelles, 84, 85
nucleus, 85, *86, 87*, 88
organelles that transform energy, 91–93
origin of, 101–102
other types of organelles, 93–94
plant cell structure, *87* (see also Plant cells)
posttranscriptional gene regulation, 346–349
ribosomes, 84–85, 88
transcriptional gene regulation, 333–338
Eukaryotic genomes
Caenorhabditis elegans, *362, 363*
Drosophila melanogaster, 362
features of, 361
gene families, 363–364
plants, 363
repetitive sequences, 364–366
yeast, 361–362
Eukaryotic viruses, 341
Eumetazoans, *630*, 643
Eupholus magnificus, *674*
Euphorbia, 801
Euphydryas editha bayensis, 1161–1162
Euphyllophytes, 583
Euplectella aspergillum, *643*
Euplectes progne, *435*
Euplotes, *554*
Euprymna scolopes, 546
Europa (moon of Jupiter), 70
European bee-eaters, *888*
European common cuckoo, 1093
Euryarchaeota, 535, 536–537
Eurycea waterlooensis, *691*
Eurylepta californica, *657*
Eusociality, 1115–1116
Eustachian tube, 954–955
Eusthenopteron, *690*
Eusthenopteron foordi, 517
Eutherians
cleavage in, 905–906
evolutionary relationships in, 697–698, *699*
herbivores, 700
key features, 697
major living groups and number of species, *698*
primates, 701–705

return to an aquatic environment, 700
rodents and bats, 699–700
Eutrophication, *533*, 1219, 1221
Evans, Martin, 382
Evaporation
in the global hydrologic cycle, 1215
heat dissipation in endotherms and, 829
in heat exchange between animals and their environment, 823, *824*
Evapotranspiration, 1192
Even-toed hoofed mammals, *698*
Evo-devo, 413
See also Evolutionary developmental biology
Evolution
adaptations, 6, 7
biological classification and, 463
concerted evolution, 498–499
constraints on, 423–424, 444–446
defined, 6, 428, 432
emergence and impact of photosynthesis, 4–5
evolutionary history and phylogenies, 451
evolutionary tree of life, 6–9
genetic basis of, 431
in genome function, 496–499
in genome size, 494–496
maintenance of genetic variation in populations, 441–444
mechanisms in, 432–436
methods of measuring, 436–440
"modern synthesis" with genetics, 430–431
by natural selection, 428–430
"opportunistic," 753
parallel, 423–424
of populations, 6
relationship between fact and theory in, 428–431
role of membranes in, 3–4
short-term and long-term outcomes, 446
use of genomes in the study of, 486–491
using molecular clocks to date evolutionary events, 461–462
in viruses, 427
in vitro evolution, 500–501
See also Molecular evolution
Evolutionary developmental biology
basic principles of, 413
beak diversity in birds, 412, 424
developmental genes, 413–414, 423–424
developmental modules, 415–418
differences in gene expression resulting in differences between species, 418–420
Evolutionary radiations, 481–482
Evolutionary reversal, 452
Evolutionary theory
Darwin's and Wallace's concepts, 428–430
defined, 428
development following Darwin, 430–431

practical applications of, 427, 428, 446
Evolutionary tree of life, 6–9
Evolutionary trends, 440
Ex vivo gene therapy, 323
Excavates, 558–559
Excision repair, 276, 277
Excitable cells, 925
Excitatory synapses, 938
Excretory systems
function of, 1071
of invertebrates, 1075–1077
in mammals, 1079–1086
mechanisms to maintain homeostasis, 1072–1074
nitrogen excretion, 1074–1075
in vertebrates, 1077–1079
See also Kidneys; Nephrons
Excurrent siphon, 662, *663, 664*
Exercise
impact on muscle strength and endurance, 996–997
integration of anabolism and catabolism during, 180–181
Exergonic reactions
activation energy, 151–152
ATP and energy-coupling, 150–151
characteristics of, 147, *148*
Exit site, 295, 296
Exocrine cells, 333
Exocrine glands, 835, 1062
Exocytosis
description of, *120*, 122
sweating and, 105
Exons
alternative splicing, 346–347
description of, *290, 291*
Exoskeletons
arthropod evolution and, 655
of arthropods, 667
of ecdysozoans, 655
features of, 999
in protostome evolution, 674–675
Exothermic reactions. *See* Exergonic reactions
Exotoxins, 542
Expanding triplet repeats, 313–314
Expansins, 710
"Experimental" group, 12
Experiments, types of, 12–13
Expiratory reserve volume (ERV), *1012, 1013*
Exploitation competition, 1182–1183
Exponential population growth, 1157–1158, 1164–1166
Expression vectors, 384
Expressivity, 246
Extensor muscle, 1002
External anal sphincter, 1063
External digestion, 1055
External fertilization, 887, 888
External gills, 1008, *1009*
Extinction. *See* Mass extinctions; Species extinctions
Extinction rates, in island biogeography theory, 1196–1198, *1199*
Extracellular fluid
in closed circulatory systems, 1026

in the internal environment of multicellular animals, 816
in open circulatory systems, 1026
regulation of homeostasis by excretory systems, 1071, 1072–1074 (see also Excretory systems)
Extracellular matrix
biological membranes and, *106*
cell adhesion to, 111–113
collagen fibers in, 818–819
of ctenophores, 644
of sponges, 644
structure and function, 100
Extraembryonic membranes
in the amniote egg, 692
in the chicken egg, 918
functions of, 918
origins of, 914
in placental mammals, 919
Extreme halophiles, 536–537
Extremophiles, 532
Exxon Valdez (oil tanker), 373
Eye color
determination in humans, 246
sex-linked inheritance in *Drosophila*, 251–252
Eye cups, 958
Eye impairments, brain development in humans and, 921
Eye infections, 533
Eyeless gene, 413, *414*
Eyes
of cephalopods, 664
compound, 958
developmental genes and, 413, *414*
image-forming, 958–959
lens determination in vertebrates, 396
structure and function of the retina, 959–963
"Eyes," of potatoes, *720*

F
F-box-containing proteins, 767
Fabaceae, 747–748
"Face neurons," *971*
Facial recognition, 971
Facilitated diffusion
aquaporins, 116
carrier proteins and, 117
channel proteins and, 115–116
characteristics of, *118*
saturation of, 117
Facilitation, 1201
Factor VIII, *385*
Facultative anaerobes, 537
Facultative parasites, 611
Fallopian tubes, 892, *893*
Familial hypercholesterolemia, 121, *313*
Family (taxonomic category), 462–463
"Fan worms," *661*
Fanged striped blenny, *1174*
Far-red light
photomorphogenesis and phytochromes in plants, 772–774
in photoperiodic control of flowering, 788

Farber, Sidney, 158
"Farming behavior," of cowbirds, 1093
Fas protein, 873, 874
Fast block to polyspermy, 885, 886
Fast-twitch fibers, 995
Fat metabolism
   control by the liver, 1065–1066
   in the postabsorptive state, 1067
Fat-soluble vitamins, 1053–1054
Fat tissues. See Adipose tissues
Fate maps, of blastomeres, 907
Fats
   absorption in the small
      intestine, 1062, 1063
   breakdown by digestive
      enzymes, 1057, 1058
   digestion in the small intestine,
      1061–1062
   energy yield, 1049, 1050
   storage in the liver, 1065
   as stored energy in the animal
      body, 1050
   structure and function, 56–57
   triglycerides of animal fats, 57
Fatty acid synthase, 182
Fatty acids
   anabolic interconversions, 180
   catabolic interconversions,
      179–180
   formed in prebiotic synthesis
      experiments, 70
   need for in the human diet,
      1051
   in phospholipids, 57, 58, 107
   significance in the evolution of
      life, 3–4
   structure and function, 56–57
   in waxes, 59
Fauna, 514
FD gene, 790
FD protein, 791
Fear and fear memory, 970, 982
"Feather duster worms," 661
Feather stars, 520, 680–681
Feathers
   anatomy of, 696
   evolution of, 695
   flight in birds and, 695–696
   as thermal insulation, 828
Fecal matter, 1063
Feces, 1056, 1063, 1064
Fecundity, 1154–1155, 1156
Federal Bureau of Investigation
   (FBI), 317–318
Feedback
   in the mechanisms regulating
      breathing, 1020–1022
   in regulatory systems, 816
Feedback inhibition, of metabolic
   pathways, 160–161
Feedforward information, 817
Feeding anthers, 1180
Feeding strategies
   detritivores, 639
   filter feeders, 637, 638
   herbivores, 637–638
   overview, 637
   parasites, 638–639
   predators and omnivores, 638
Feeding structures
   in protostome evolution, 674
   See also Teeth
Feigin, Andrew, 323, 324

Female athlete triad, 1000–1001
Female flowers, 779, 780
Female reproductive system
   childbirth, 896–897
   components and function of,
      892, 893
   embryo–mother connection
      in, 889
   follicle selection for ovulation,
      895–896
   ovarian and uterine cycles,
      893–895
   pregnancy, 896
Females
   genomic imprinting in
      mammals, 344–345
   length of meiosis in, 220
Feminization, atrazine and, 1, 18
Fenestrations, 1041
Feral cattle, 440
Fermentation
   in beer and wine production,
      623–624
   energy yield from cellular
      respiration and fermentation
      compared, 178
   in glucose metabolism, 168
   overview, 166, 167
   pathways in, 177–178
Ferns
   in the Devonian, 520
   distinguishing characteristics,
      574, 582
   life cycle, 582
   monilophytes, 581
   number of chromosomes in, 218
   sexual life cycle, 218
Ferocactus wislizeni, 1179
Ferredoxin, 196, 197
Ferritin, 348.23
Ferrous ion, 187
Fertile Crescent, 388
Fertilization
   activation of development in
      animals, 903–904
   in animals, 884–887
   defined, 218, 882
   double fertilization in
      angiosperms, 600–601, 781,
      783
   fruit development in
      angiosperms and, 785
   in humans, 892
   internal and external, 887
   in mosses, 576
   phylogenetic analysis of
      mechanisms in angiosperms,
      459–460
   in seed plants, 590, 592
Fertilization anthers, 1180
Fertilization cone, 885, 886
Fertilizers
   inorganic, 747
   organic, 746–747
   to treat plant deficiencies, 742
   See also Chemical fertilizers;
      Nitrogen fertilizers
Fetal hemoglobin, 1018
Fetscher, Elizabeth, 599
Fetus
   birth of, 896–897
   defined, 920
   development of, 920
Feulgen, Robert, 260

Fevers, 830, 861–862
Fibers, in plants, 596, 713, 714
Fibrin, 1039
Fibrinogen, 890, 1039
Fibrinolysin, 890
Fibroblasts, 336, 337, 338
Fibrous root systems, 718
Ficedula albicollis, 1103
Ficke, Henry, 21
Fick's law of diffusion, 1006
Fig trees, 1194–1195
Fight-or-flight response, 837, 850,
   852–853, 974
Figs, 601
Filamentous sac fungi, 620, 622
Filaments (of anthers), 591, 596,
   598
Filopodia, 909
Filter feeders, 637, 638, 1054
Filtration
   of blood plasma in tubule
      capillaries, 1072
   in the vertebrate nephron,
      1078–1079
Fimbriae (fimbria), 84, 892, 893
Finches, 440
   See also Darwin's finches;
      Galápagos finches
Fins, 686–687, 688
Fire
   Australian deserts and, 1126
   controlled burning, 1240
   the fynbos and, 1121, 1135, 1145,
      1146
   movement of elements through
      ecosystem compartments
      and, 1214
Fire adaptations, in pines, 594, 596
Fire-bellied toad, 479
Fireflies
   bioluminescence, 149, 150
   visual signaling, 1111
Firmicutes, 530–532
First filial generation (F_1), 234, 235
First law of thermodynamics, 146
First polar body, 883, 884
Fishes
   acclimatization to seasonal
      temperatures, 821
   actions of prolactin in, 838
   calcium ion cycling in the
      hearts of "hot" fish, 1036,
      1037
   circulatory system, 1027, 1028
   countercurrent heat exchange,
      825
   developmental constraints on
      evolution, 444, 445
   evolution of electric organs, 485
   external fertilization in, 887, 888
   jawed, 686–689
   jawless, 685–686, 687
   overharvesting, 16–17, 1163–
      1164, 1235
   parallel evolution in, 423–424
   respiratory gas exchange in
      gills, 1009–1010, 1011
   salt and water balance
      regulation in marine fish,
      1077
   sequential hermaphroditism,
      887, 888
   taste buds, 951

   See also Chondrichthyans;
      Haplochromine cichlids
Fishflies, 672
Fishhook barrel cactus, 1179
Fishing industry, 16–17, 1163–
   1164, 1235
Fission yeast, 624, 626
Fitness
   altruistic behavior and, 1114–
      1115
   defined, 438
   direct, 1114
   inclusive, 1115
   indirect, 1115
   maximization by mating
      systems, 1113–1114
   sex-specific differences in
      lizards, 421
Fitzpatrick, John, 1115
5' Cap, 291–292, 293, 296
Fixed action patterns, 1095
Fixed populations, 437
Flabellina iodinea, 664
Flagella
   of chytrids, 617
   of dinoflagellates, 553, 554
   of euglenids, 558, 559
   of opisthokonts, 609
   of prokaryotes, 82, 83–84
   of stramenopiles, 555
   structure and function, 96–98
Flagellin, 799
Flame cells, 1075
Flame fairy wrasse, 688
"Flame tree," 603
Flamingos, 637, 638
Flathead Lake, 1228
Flatworms, 675
   acoelomate, 635
   anatomical characteristics, 652
   description of, 656–657
   excretory system, 1075
   eye cups, 958
   major subgroups and number
      of living species, 632
Flavin adenine dinucleotide
   (FAD)
   in the citric acid cycle, 170, 171
   in glucose metabolism, 168
   in NADH shuttle systems, 179
   in the respiratory chain, 172,
      173
   role in catalyzed reactions, 155,
      156
Flavonoids, 747, 748
Flax, 800
FLC gene, 791
FLC protein, 792
Fleas, 639, 672, 986
Flegal, A. Russell, 625
Fleming, Alexander, 383, 608
Flesh flies, 1181, 1203
Flexors, 942, 1002
Flight
   bird physiology and, 695–696
   contraction in insect flight
      muscles, 997–998
Flight feathers, 695, 696
Flightless birds, 694, 695
Flightless weevils, 1144–1145
Flocking, 1116, 1117
Flooding, 1140
Flor, Harold Henry, 800
Flora, 514

Floral meristems, 400, 715, 785–786
Floral organ identity genes, 400–401, 786
Florey, Howard, 608
Florida scrub-jays, 1115
Florigen, 790
    See also FT–FD florigen pathway
Flounders, 444, 445
Flowering
    categories of, 785
    cues from an "internal clock," 792
    floral organ identity genes, 400–401, 786
    florigen, 790
    the flowering stimulus originates in leaves, 788–790
    induction by gibberellins, 791–792
    induction by temperature, 791
    meristem identity genes, 786
    photoperiodic cues, 787–788
    transition of shoot apical meristems to inflorescence meristems, 785–786
Flowering plants. See Angiosperms
Flowers
    anatomy, 591, 596–597
    in angiosperm sexual reproduction, 779, 781
    coevolution with animals and, 599
    Darwin on the evolution of, 605
    evolution of, 597–598, 599
    forms, 596, 597
    in monocots and eudicots, 710
    monoecious and dioecious plants, 597
    organ determination, 400–401
    parts of, 779, 781
    perfect and imperfect, 597, 598
    pollination (see Pollination)
    pollination syndromes, 1180–1181
    symmetry, 597
    types of, 779
Flu epidemics, 427
Flu vaccines, 427, 446
Fluid feeders, 1054
Fluid mosaic model, 106
Flukes, 657, 675
Fluorescence microscopy, 80
Fluorescent dyes, 353, 354
Fluorescent proteins
    discovery of green fluorescent protein, 449
    evolution in corals, 464
Fluorine, 1052
5-Fluorouracil, 228, 229
Fluxes, 1211
Flying insects. See Pterygotes
Flying squirrels, 962
FMR1 gene, 313–314
FMR1 protein, 314
Fog-basking beetle, 1132
Folate, 916
Folic acid, 349, 916, 1053
Foliose lichens, 613
Follicle cells, 895–896
Follicle-stimulating hormone (FSH)
    actions of, 842

endocrine source, 842, 843
    in follicle selection for ovulation, 896
    in puberty, 850–851, 894
    in regulation of human spermatogenesis, 892
    in regulation of the ovarian and uterine cycles, 893–894
Follicles, of the thyroid, 845, 846
Follicles (human ovaries)
    selection for ovulation, 895–896
    structure and function, 893, 894
    in the uterine cycle, 894–895
Fontanelle, 1001
Food
    as an energy source, 1049–1050
    brain areas controlling intake, 1067–1068
    effect of food supply on population growth, 1159
    ingestion and digestion in animals, 1054–1058
    movement through the vertebrate gut, 1059–1060
    as a source of carbon skeletons for biosynthesis, 1051–1052
Food acquisition. See Foraging
Food allergies, 876
Food chains, 1190
Food webs, 1190, 1191
Foot
    development in ducks and chickens, 417
    molluscan, 662, 663
Foraging
    costs of foraging in groups, 1116
    optimal foraging theory, 1104–1105, 1106
Foraminiferans, 557, 565
Forebrain
    in human brain evolution, 973
    insular cortex, 983
    structure and function, 968–969, 970
Foregut fermenters, 492–494
Forelimbs, loss in snakes, 423
Forensics
    DNA fingerprinting, 317–318
    forensic entomology, 1188
    use of phylogenetic analyses in, 458, 459
Foreskin, 890
Forest corridors, 1162
Forest fires
    conifer adaptations, 594, 596
    cyclical succession and, 1201–1202
    See also Fire
Forests. See individual forest types
Forewings, 672
Forgetting, 980
Formation of Vegetable Mould Through the Action of Worms, The (Darwin), 639
Fossil fuel burning
    acid precipitation and, 1219–1220
    global climate change and, 17, 510
    impact on the global carbon cycle, 1216
Fossil fuels
    formation of, 1211
    phytoplankton and, 569

Fossil record
    Cambrian fossil beds, 516
    dating fossils, 506–508
    microfossils, 74
    number of species identified, 514
    phylogenetic analyses and, 455
    reasons for the paucity of, 514–515
    See also Paleontology
Founder effect, 434
Founder events, in speciation, 467, 472, 473
Four-chambered heart
    in birds, 1029, 1030
    in crocodilians, 1029
    in fishes, 1028
    in mammals, 697, 1030
Four o'clock plants, 252
Fovea, 959, 962
Foxglove, 604
FOXP2 gene, 367
Fraenkel, Gottfried, 1175
Fragile-X syndrome, 313–314
Fragmented habitats, 1233–1234
Frame-shift mutations, 306–307
Frankenia palmeri, 811
Franklin, Rosalind, 264
Frank–Starling law, 1042
Free-air concentration enrichment (FACE), 185
Free energy
    ATP and, 149, 150
    chemical equilibrium and, 148–149
    defined, 146–147
    enzymes and, 153
    harvested during glucose oxidation, 166–167
Free radicals, 309
Free-running circadian clock, 1107–1108
Freeze-fracture microscopy, 81, 108
Freeze-tolerant plants, 810
"French flag model," of morphogens, 401
Frequency-dependent selection, 441–442
Freshwater biomes, 1140
Freshwater ecosystems
    consequences of human alteration, 1223
    impact of pollution on, 1233
    transport of elements through ecosystem compartments, 1213–1214
Freshwater protists, 93
Frog eggs
    aquaporins, 116
    complete cleavage, 905
    rearrangements following fertilization, 903–904
Frogs
    adaptations in, 7
    experiments on the effects of atrazine, 12, 13
    fate map of the blastula, 907
    features of, 690–691
    feminization induced by atrazine, 1
    gastrulation, 909–910
    germ cell lineage, 908
    jumping ability, 986

lens determination in the embryo, 396
    mating calls, 705
    nitrogenous wastes excreted, 1075
    nuclear transfer experiments with the embryo, 406
    parental care, 678
    prezygotic isolating mechanisms, 476, 477
    salt and water balance regulation, 1077
    sexual signals, 435
Fromia, 881
Frontal lobe, 971–972
Fructose
    absorption in the small intestine, 1063
    production in photosynthesis, 195
    in seminal fluid, 890
    structure, 52
Fructose 1,6-bisphosphate, 55, 169
Frugivores, 1181–1182
Fruit flies
    compound eye, 958
    genomic information, 361, 362
    as a model organism, 282
    sex determination in, 249
    sex-linked inheritance, 251–252
    See also Drosophila
Fruiting structures
    of fungi, 609–610, 620, 621, 622
    of slime molds, 561
Fruitless (fru) gene, 1097–1098
Fruits
    auxins in fruit development, 766
    defined, 784
    dispersal, 784
    ethylene and fruit ripening, 769–770
    functions of, 784
    gibberellins in fruit growth, 761
    parthenocarpic, 766
    plant–frugivore mutualisms in seed dispersal, 1181–1182
    relationship between seed development and fruit development, 784–785
    seedless, 784–785
    a synapomorphy of angiosperms, 596
Fruticose lichens, 613
FSH receptors, 895–896
FT gene, 790
FT protein, 790, 791
FT–FD florigen pathway, 790, 791, 792
Fucoxanthin, 556
Fugitive species, 1183
Fulcrum, 1002
Full census, 1150
Functional genomics, 355–356, 357
Functional groups
    of chemically modified carbohydrates, 55
    of macromolecules, 40
    properties of, 40
Functional residual volume (FRV), 1012, 1013
Fundamental niche, 1184
Fungal mutalisms
    discussion of, 612–615

evolution of land plants and, 574
fungus farming, 1169, 1178, 1185
with nonvascular land plants, 575
See also Mycorrhizae
Fungi
    antibiotics derived from, 608
    arbuscular mycorrhizal fungi, 619–620
    in bioremediation, 625
    chytrids, 617–619
    classification of major groups, 616
    Dikarya, 620–623
    distinguishing characteristics, 609
    endophytic, 615
    in food and drink production, 623–624
    global carbon cycle and, 611
    hyphae, mycelium, and fruiting structures, 609–610
    life cycles, 616, 618, 621
    microsporidia, 617
    microtubule organizing centers, 212
    as model organisms in lab studies, 625, 626
    parasitic, 611–612
    pathogenic, 612
    phylogenetic relationships, 615–616
    predatory, 612, 613
    in reforestation efforts, 626
    sexual and asexual reproduction, 616–617
    sexual life cycle, 218
    surface area-to-volume ratio of mycelium, 610
    tolerance for hypertonic environments, 610–611
    tolerance of temperature extremes, 611
    unicellular yeasts, 609
    used to control diseases and pests, 626, 627
    used to study environmental contamination, 624, 625
    zygosopore fungi, 618, 619
    See also Fungal mutualisms
Fungus farming, 1169, 1178, 1185
Funk, Casimir, 1054
Fur, 697, 825, 828
Furcula, 694
Fusarium oxysporum, 626
Fynbos
    characteristics of, 1121, 1134–1135
    economic benefits of, 1242–1243
    fire and, 1121, 1135, 1145, 1146
    introduced Argentine ant, 1145
    Mediterranean climate, 1126

G

G protein-linked receptors
    in the IP$_3$/DAG pathway, 134, 135
    structure and function, 129–130
G1 phase (of mitosis), 208, 217
G1-to-S transition
    description of, 208

internal signals controlling, 210–211
G2 phase (of mitosis), 208, 211–212, 217
GABA. *See* γ-Aminobutyric acid
GABA receptors, 943
Gage, Phineas, 972
Gain-of-function mutations, 305, 306, 400–401
Galactia elliottii, 1221–1223
Galactosamine, 55
Galactose, 52
β-Galactosidase, 329
β-Galactosidase gene, 378
β-Galactoside, 329
β-Galactoside permease, 329
β-Galactoside transacetylase, 329
Galagos, 701
Galápagos finches
    adaptive radiation, 481
    allopatric speciation, 473, 474
    beak diversity, 412, 424
    heterometry in beak development, 416
    See also Darwin's finches
Galápagos hawk, 2
Galápagos Islands
    Darwin's visit to, 428, 429
    exploitation competition, 1183
Galápagos tortoises, 694
Gallbladder, 1058, 1061, 1065
Gallus gallus, 361
Gametangia
    of mosses, 575, 576
    in plant evolution, 574
    of zygospore fungi, 618, 619
Gametes
    in alternation of generations life cycle, 562, 563
    genomic imprinting in mammals, 344–345
    Mendelian law of segregation and, 236–237, 239
    number of chromosomes in, 218
    production in animals, 882–884
    in sexual reproduction, 218, 219
Gametic isolation, 477
Gametogenesis
    in animals, 882–884
    defined, 882
Gametophytes
    of angiosperms, 596, 600, 779–780
    of conifers, 594, 595
    of ferns, 582
    in homospory, 584
    of hornworts, 578, 579
    of land plants, 575
    of liverworts, 577
    of mosses, 577–578
    of nonvascular land plants, 575–576
    relationship to sporophytes in plant evolution, 590
    of seed plants, 589, 590, 591, 592
γ-Aminobutyric acid (GABA), 324, 939
Ganglia
    in annelids, 659–660, 941
    of the autonomic nervous system, 974, 975
Ganglioside, 91
Ganoderma applanatum, 623
Gap genes, 403, 404

Gap junctions
    in cardiac muscle, 991
    structure and function, 111, 112, 139–140
Garrod, Archibald, 282
Garter snakes, 445, 889
Gas exchange systems
    components of, 1006
    fish gills, 1009–1010, 1011
    human lungs, 1013–1016
    in insects, 671, 1009
    maximization of gas exchange surface area, 1008–1009
    maximization of partial pressure gradients, 1009
    physical factors affecting, 1006–1008
    snorkeling elephants, 1005, 1022
    unidirectional ventilation in birds, 1010–1012
Gases
    of the atmosphere, 1211
    partial pressures and diffusion, 1006
    solubility in liquids, 1006
    See also Carbon dioxide; Hydrogen; Nitrogen; Oxygen; Respiratory gases
Gasterosteus aculeatus, 423–424
Gastric brooding, 678
Gastric mucosa, 1060, 1061
Gastric pits, 1060–1061
Gastrin, 1065
Gastritis, 1054
Gastrointestinal disease, 534
Gastrointestinal disorders, autism and, 540
Gastrointestinal system
    absorption of nutrients by the liver, 1063
    absorption of nutrients in the small intestine, 1062, 1063
    chemical digestion in the mouth and stomach, 1060–1061
    concentric tissue layers in, 1058–1059
    control and regulation of the flow of nutrients, 1064–1068
    digestion in the small intestine, 1061–1062
    digestion of cellulose in herbivores, 1063–1064
    enteric nervous system, 1059
    large intestine, 1063
    movement of food through, 1059–1060
    peritoneum, 1059
Gastrophryne carolinensis, 477
Gastrophryne olivacea, 477
Gastropods, 662, 663, 664
Gastrotrichs, 632, 652, 656, 658
Gastrovascular cavity, 645, 648, 1056
Gastrulation
    in amphibians, 909–913
    basic patterns, 633–634
    defined, 908
    germ layers formed during, 908
    in mammals, 914–915
    in reptiles and birds, 913–914
    in sea urchins, 908–909
Gated ion channels
    chemically gated channels, 930, 937

mechanically gated channels, 930
membrane potentials and, 930–932
    as receptors, 129
    structure and function, 115, 116
    types of, 930
    See also Voltage-gated ion channels
Ge Hong, 797
Gehring, Walter, 413
Geiger counter, 24
Gel electrophoresis, 316–317
Gemmae, 577
Gemmae cups, 577
Gemsbok, 1132
Gender identification, 1096–1097
Gene cascades, in control of animal behavior, 1097–1098
Gene duplications
    in the evolution of electric organs, 485
    genetic diversity and, 310
    as a source of new genome functions, 496–498
    transposons and, 365
Gene expression
    blocked by antibiotics in bacteria, 281
    "central dogma" of molecular biology, 285
    changes in plants in response to pathogens, 799, 800
    defined, 65
    differences between prokaryotes and eukaryotes, 291
    differences in expression resulting in differences between species, 418–420
    effect on noncoding sequences on, 494
    evidence for proteins as major products of, 282–284
    genome characteristics and, 486
    genomic information and, 355
    methods of studying, 380–383
    overview, 284–285
    pattern formation and, 399–405
    primary embryonic organizer and, 911–912
    in RNA genomes, 285
    role in development, 6, 397–399
    sequential pattern during determination of fruit fly body segmentation, 401–405
    signal sequences and polypeptide movement within the cell, 298–300
    transcription, 286–293
    translation, 293–297, 298
    visualizing with fluorescent proteins, 449
Gene families
    concept of, 497, 498
    concerted evolution and, 498–499
    engrailed gene family, 499–500
    in eukaryotes, 363–364
Gene flow
    barriers leading to speciation, 472–475
    effect on allele frequencies, 433–434

Gene-for-gene concept, 800
Gene-for-gene resistance, 800
"Gene guns," 376
Gene mutations
  effect on phenotype, 305–306
  molecular medicine and, 304
  point mutations, 306–307
  reversal of, 306
  types in multicellular
    organisms, 305
  See also Mutations
Gene pool, 431
Gene regulation
  DNA methylation and, 328
  effect of stress on during
    prenatal development, 328
  epigenetic changes, 343–346
  posttranscriptional mechanisms
    in eukaryotes, 346–349
  potential points for in
    eukaryotes, 334
  in prokaryotes, 329–333
  transcriptional regulation in
    eukaryotes, 333–338
  in viruses, 339–343
Gene regulatory proteins, 42
Gene sequences
  environmental genomics, 530
  evolutionary relationships in
    prokaryotes and, 528–529
  identification of the Archaea
    and, 534
Gene-silencing mechanisms,
  347–348
Gene therapy, 323–324
Gene trees
  description of, 497, 498
  of the engrailed gene family,
    499–500
  lateral gene transfer in
    prokaryotes and, 529–530
General plant immunity, 799–800,
  801
General transcription factors,
  333–334, 335, 336
Generative cell, 780, 783
Genes
  alleles, 431
  animal behavior and, 1096–1098
  biological information and, 5, 6
  cloning, 375–376 (see also
    Molecular cloning)
  comparing through sequence
    alignment, 486–487
  defined, 66
  duplication (see Gene
    duplications)
  epistasis, 244
  evidence for being DNA,
    260–263
  gene trees, 499–500
  gene–environment interactions,
    245–246
  genetic markers, 318–319
  genome characteristics and, 486
    (see also Genomes)
  genotype and phenotype, 431
  homologous, 413–414, 499–500
  hybrid vigor, 244–245
  inactivation by homologous
    recombination, 381–382
  incompatibilities and
    reproductive isolation, 470,
    471

inheritance of organelle genes,
  252–253
inserting in cells, 376–377 (see
  also Recombinant DNA
  technology)
largest human gene, 291
lateral gene transfer, 496,
  529–530
law of independent assortment,
  237–239
law of segregation, 236–237, 239
linkage, 247, 248, 249–252
in Mendelian inheritance, 236
methods of studying gene
  expression, 380–383
methylated, 343–344 (see also
  DNA methylation)
multiple alleles, 242
mutations (see Gene mutations;
  Mutations)
numbers in the human genome,
  366
one-gene, one-enzyme
  hypothesis, 282–283, 284
one-gene, one-polypeptide
  relationship, 283–284
phenotype and, 237
promoters, 286
pseudogenes, 491, 494–495
quantitative trait loci, 246
recombinant frequencies, 248,
  250
recombination and mapping,
  247–249, 250
reporter genes, 377–378, 379
sizes in the human genome, 366
therapeutic, 323–324
transfer in prokaryotes, 253–255
Genetic code
  commonality of, 289
  description of, 288–289
  missense mutations, 306, 307
  redundancy, 289
"Genetic determinism," 245
Genetic diversity
  gene duplications and, 310
  generated by chromatid
    exchanges during meiosis I,
    219–220, 222
  generated by independent
    assortment of homologous
    chromosomes, 222
  from meiosis and sexual
    reproduction, 218–219, 245,
    882
  mutations and, 310
  See also Genetic variation
Genetic drift
  evolution and, 6
  fixation of neutral mutations,
    492
  impact on small populations,
    434
Genetic markers
  in transformation experiments,
    263
  uses of, 318–319, 352
Genetic recombination
  in ciliates, 562
  genetic mapping and, 248–249
  with homologous
    chromosomes, 247–248, 250
  inactivation of genes by,
    381–382

Genetic screening
  allele-specific oligonucleotide
    hybridization, 321
  DNA testing, 320–321
  identification of plant signal
    transduction pathways, 759,
    760
  purposes, 320
  screening for disease
    phenotypes, 320
Genetic structure, 437
Genetic switches, 415
Genetic toolkit, 414, 415
Genetic transformation, 260–261,
  263, 376–377
  See also Recombinant DNA
  technology
Genetic variation
  generation by mutation, 432
  geographically distinct
    populations within species,
    443–444
  maintenance by frequency-
    dependent selection, 441–442
  mechanisms maintaining in
    populations, 441–444
  phenotypic variation and, 431
  selection on leads to new
    phenotypes, 432–433
  transposons and, 365–366
  See also Genetic diversity
Genetically modified organisms
  (GMOs)
  in agriculture, 386–388, 389
  patenting, 373
  public concerns, 388–389
Genetics
  behavioral, 1096–1098
  codominance, 243
  evidence for DNA being the
    genetic material, 260–263
  incomplete dominance, 242–243
  mechanisms of gene interaction,
    244–246
  Mendelian laws of inheritance,
    233–241
  model organisms, 282
  "modern synthesis" with
    evolution, 430–431
  monohybrid crosses, 234–236
  multiple alleles, 242
  mutations, 241–242
  pleiotropy, 243–244
  probability calculations,
    239–240
  test crosses, 237, 238
Genitalia, 887
Genome sequencing
  applications, 353
  defined, 353
  "genetic determinism," 245
  information yield, 355–356
  methods in, 353–354, 355
  prokaryotes, 356
Genomes
  accumulation of deleterious
    mutations, 441
  biological information and, 5–6
  characteristics of, 486
  of cheetahs, 62
  defined, 66, 486
  detecting positive and
    purifying selection in,
    492–494

differential gene expression, 6
  of dogs, 352
  endogenous retroviruses in
    vertebrate genomes, 545
  environmental genomics, 530
  of eukaryotes, 361–366
  evolution in size, 494–496
  gain of new functions, 496–499
  gene–environment interactions
    and, 245
  "genetic determinism," 245
  of humans, 366–369 (see also
    Human genome)
  "junk" sequences, 347
  microRNA, 347, 348
  of mimiviruses, 545
  minimal genome studies, 359,
    360
  mutations and, 6
  phylogenetic analyses and, 456
  positive-sense, 544
  of prokaryotes, 356–360
  representative organisms, 361
  small interfering RNAs,
    347–348
  smallest animal genome, 629
  in the study of evolution,
    486–491
  of Thermoplasma, 537
  variation in Tasmanian devils,
    232
Genomic imprinting, 344–345
Genomic libraries, 379
Genotype
  defined, 431
  gene–environment interactions,
    245–246
  in Mendelian genetics, 236
  relationship of phenotype to, 431
  sexual recombination amplifies
    the possible number of, 441
Genotype frequency
  calculating, 436–437
  defined, 432
  effect of nonrandom mating on,
    434–436
  Hardy–Weinberg equilibrium,
    437–438
Genotyping technology, 368
Genus (genera), 7, 462
Geographical Distribution of
  Animals, The (Wallace), 1141
Geology
  geological time scale, 506–507,
    508
  influence on biomes, 1126–1127
Georges Bank, 1163–1164
Geospiza conirostris, 474
Geospiza difficilis, 474
Geospiza fortis, 474
Geospiza fuliginosa, 474, 1183
Geospiza magnirostris, 474
Geospiza scandens, 474, 1154–1155
Germ cell plasm, 908
Germ cells
  in animal gametogenesis,
    882–883
  in development, 908
  epigenetic changes and, 344
  in human males, 891
  mutations and, 310
Germ layers
  formed during gastrulation, 908
  in frog gastrulation, 909–910

in sea urchin gastrulation, 908–909
Germ line gene therapy, 323
Germ line mutations, 305
Germination
    of angiosperm pollen, 780, 782
    of seeds (*see* Seed germination)
Gestation, 919–920
    See also Pregnancy
Gey, George and Margaret, 205
GFP. *See* Green fluorescent protein
Gharials, 693–694
Ghost bat, *700*
Ghrelin, 1067, 1068
Giant bluefin tuna, 825
Giant groundsel, *1128*
Giant kelp, 556, *1139*
Giant petrel, *1073*
Giant redwoods, 422
Giant sequoia, 1130, 1160
Giant tortoise, *2*
Giardia lamblia, 558
Giardiasis, 558
Gibberellic acid, *759*, 760
Gibberellins
    activated by phytochromes in seed germination, 774
    discovery of, 760, *761*
    effects on plant growth and development, *759*, 760–762
    induction of flowering and, 791–792
    molecular mechanisms underlying the activity of, 767
    semi-dwarf plants and, 775
    structure, *759*
Gibbons, 702
Gigantactis vanhoeffeni, *688*
Gigantism, 843
Gill arches
    development from pharyngeal arches, 684
    in the evolution of jaws, *687*
    of fish gills, 1009–1010
    in lungfish, 1028
Gill arteries, 1028
Gill filaments, 1010
Gills
    countercurrent heat exchange in fish, 825
    in mollusks, 662, *663*
    respiratory gas exchange, *1007*, 1009–1010, *1011*
    surface area maximization, 1008, *1009*
Gills (of mushrooms), 623
Ginkgo biloba, 592, *593*
Ginkgophyta, *574*
Ginkgos, *574*, 589, 592, *593*
Giraffes, 416–417
Girdle, of chitons, 662
Gizzard, 1056
Glacial moraines, 1200
Glaciation
    allopatric speciation and, 472
    during the Carboniferous, 521
    climate change through time and, 510
    during the Proterozoic, 515–516
    in the Quaternary, 522
    sea level and, 509, *510*
Glacier Bay, 1200

Glaciers, impact of climate warming on, *17*
Gladiators, *672*
Gladioli, 793
Glans penis, 890
Glass sponges, *630*, 643
Glaucophytes, 551, *552*, 571, *572*, *573*
Gleevec, 304
Glial cells (glia)
    characteristics of, 820
    neural networks, 940–943
    types and functions of, 926–927
Global Biodiversity Outlook 3, 1233
Global climate change
    atmospheric carbon dioxide levels and, *17*, 510, 1217–1218
    greenhouse gases and, 1212
    methane cycling studies, 357
Global ecosystem
    biogeochemical cycles, 1214–1223
    energy flow through, 1208–1210
    movement of elements through, 1210–1214
Global nitrogen cycle, 750–751
"Globe crab," *670*
α-Globin gene cluster, 364
β-Globin
    gene cluster, 364
    hemoglobin C disease, 312
    missense mutation and sickle-cell disease, 306, 312
    nonsense mutation and thalassemia, 306
    transcriptional regulation red blood cells, 335
β-Globin gene
    differential transcription, 398
    transcriptional regulation red blood cells, 335
γ-Globin, 364
Globin gene family, 364, 497, *498*
"Globular" embryo, *393*
Glomeromycota, 615, *616*, 619–620
    See also Arbuscular mycorrhizae
Glomerular capillaries, 1078–1079, 1082
Glomerular filtration rate (GFR), 1087–1088
Glomeruli, olfactory, *949*, 950
Glomerulus
    blood flow to, 1082
    function in the vertebrate nephron, 1078–1079
    of the mammalian kidney, *1080*, *1081*
Glomus mosseae, *614*
Glucagon, *842*, 848, 849, 1067
β-1,3-Glucan, 799
Glucoamylase, 162
Glucocorticoids, 849
Gluconeogenesis, 180, 1065, 1067
Glucosamine, *55*
Glucose
    carrier-mediated transport, 117
    catabolic interconversions, 179
    energy released during oxidation, 166–169
    formation of glucose 6-phosphate, 150, *151*
    forms of, 52
    as the fuel for the nervous system, 1067

gluconeogenesis, 180
    in glycogen, 53, 54
    in positive regulation of the *lac* operon, 332
    production in photosynthesis, 195
    secondary active transport, 119
    in starches, 53, *54*
Glucose 1-phosphate, 148, 149
Glucose 6-phosphate, 148, 149, 150, *151*
Glucose catabolism
    aerobic pathways, 169–171
    allosteric regulation, 181–182
    under anaerobic conditions, 177–179
    citric acid cycle, 170–171
    energy cost of NADH shuttle systems, 178–179
    energy released during glucose oxidation, 166–169
    energy yield from cellular respiration and fermentation compared, 178
    glycolysis, 169–170
    oxidative phosphorylation and ATP synthesis, 171–176
    pyruvate oxidation, 170
    regulation of pyruvate oxidation and the citric acid cycle, 171
Glucose metabolism, liver regulation of, 138, 1065, 1066–1067
Glucose transporters, 117, 1066
Glucuronic acid, 55
Glumes, 419
Glutamate
    anabolic interconversions and, 180
    catabolic interconversions, 180
    as a neurotransmitter, 939
Glutamate decarboxylase, 323, 324
Glutamic acid, 44
Glutamine, 44
Glutamine synthetase (GS), 753
Glyceraldehyde, 52
Glyceraldehyde 3-phosphate (G3P), *169*, 170, 194, 195, *196*, 201
Glyceraldehyde 3-phosphate dehydrogenase, 170
Glycerate, 197, *198*
Glycerol
    catabolic interconversions, 179
    in phospholipids, 57, *58*
    in triglycerides, 56
Glycerol 3-phosphate, 179
Glycine
    as a neurotransmitter, 939
    in photorespiration, 197, *198*
    side group properties, 43
    structure, *44*
Glycogen
    buildup during excess food consumption, 1050–1051
    in exercising muscles, 997
    glucose levels during exercise and, 181
    in glucose metabolism, 1066, 1067
    regulation by a protein kinase cascade, 138

regulation of blood glucose, 848, 849
    storage in the liver, 1065
    as stored energy, 1050
    structure and function, 53–54
Glycogen phosphorylase, 132–133
Glycogen synthase, 138
Glycogen synthase kinase-3 (GSK-3), 904
Glycolate, 197, *198*
Glycolipids, 109, 111
Glycolysis
    allosteric regulation, 181–182
    description of, 169–170
    energy yield from cellular respiration and fermentation compared, 178
    in glucose catabolism, 166, 168
    relationships among metabolic pathways, *179*
Glycolytic muscle, 995
Glycolytic system, in skeletal muscle, 997
Glycoproteins
    cell adhesion and, 111
    formation of, 301
    structure and function, 109
Glycosidic linkages, 53, 54
Glycosylation, *300*, 301
Glyoxysomes, 93
Glyphosate, 160, 388, 389
Glyphosate resistance, 389
Gnathonemus petersi, *485*
Gnathostomes, 686–689
Gnetophytes, *574*, 593
Goats, 1099
Gobi Desert, *1192*
Goiter, 846–847, 1054
Golden feather star, *682*
Golden lion tamarin, 1114, *1137*
Golden toads, *691*
Goldenrod, 752, *1201*
Goldman equation, 929, 931
Golgi, Camillo, 90, 953
Golgi apparatus
    in animal cells, *86*
    evolution in eukaryotic cells, 551
    glycosylation of proteins, 301
    Inclusion-cell disease, 300
    plant cell plate and, 710
    in plant cells, *87*
    processing of newly translated polypeptides, *299*, 300
    structure and function, *89*, 90
Golgi tendon organs, 953
Gonadotropin-releasing hormone (GnRH)
    control of gonadotropins, 851
    discovery of, 844
    in puberty, 892, 894
    in regulation of the ovarian and uterine cycles, *894*, 895
Gonadotropins
    defined, 850–851
    in pregnancy, 896
    in puberty, 894
    regulation of the ovarian and uterine cycles, 893–894
Gonads
    establishment of the germ cells, 908
    gametogenesis in, 882–884
    hormones of, *842*

phenotypic determination, 850, *851*
sex steroids produced by, 850–851
Gondwana, 521, 1143
Gonionemus vertens, *647*
Gonium, 140–141
Gonzales, Andrew, *1162*
Goosecoid gene, 911, 912
Goosecoid transcription factor, 911, 912
Goshawks, 1116, *1117*
Gout, 1074
Graded membrane potential, 932, 938, 947
Grafting, 793, *794*
Grains, genetically modified, 388
Gram-negative bacteria, 528, 542
Gram-positive bacteria, 527–528
Gram stains, 527–528
Grant, Peter and Rosemary, 1154–1155
Granulocytes, 858
Granulosa cells, 895, 896
Granum, 92, *93*
Grapevines, 761
Grasses
    defenses against herbivory, 1175
    endophytic fungi and, 615
    spikes, *597*
Grasshoppers, *672*, *674*
Grasslands
    restoration projects, 1237–1239, *1240*
    temperate grassland biome, 1131–1132
Grave's disease, 846–847
Gravitropism, 765
Gravity, vestibular detection of, 956–957
Gray crescent, 903–904, 909–910
Gray-headed albatross, 1109–1110
Gray matter, 941
Gray tree-frogs, *469*, *1173*
Gray whales, 1109
Gray wolves, 1193–1194
Graylag geese, 1099
Grazers
    trophic cascades in savanna communities, 1194
    See also Herbivores/Herbivory
Great apes, 983
Great Barrier Reef, 646–647
Great Lakes
    eutrophication in, 1221
    invasive zebra mussels, 1160
Great Plains, 1131
Great Rift Valley, 509
Great white sharks, 825
Greater bilby, *699*
Greater flamingo, *638*
Greater prairie-chicken, 434
Greater sage-grouse, *1105*
Green algae
    biofuel production and, 585
    closest relatives of land plants, 572–573

endosymbiotic origin of chloroplasts in, 551
in lichens, 613
"Volvocine line," 140–141
Green fluorescent protein (GFP), 378, *379*, 449
Green molds, 624, 1169
Green plants, 572, 573
"Green revolution," 747
Green sea turtle, *888*
Green sulfur bacteria, 187
Green tiger beetle, *1172*
Greenbottle flies, 1203
Greenhouse gases, 185, 1212
Gremlin gene, 417
Gremlin protein, 417
Grévy's zebra, *1144*
Grey goose, *642*
Grooming behavior, 1177
Gross primary production (GPP), 1189–1190
Ground finch, 1183
Ground ivy, *720*
Ground meristem
    ground tissue system and, 716
    in root growth, 716–717
    in shoot growth, 719
Ground squirrels, *831*
Ground tissue system
    description of, *712*, 713–714
    primary meristem, 716
    of shoots, 720
Groundwater, 1213, 1215
Groundwater depletion, 1215
Group living, 1116, *1117*
Growing season, 1138
Growth
    defined, 393
    determinate, 715, 720, 786
    in development, 710
    indeterminate, 715–720, 786
    processes contributing to, 394
    See also Plant growth and development
Growth factors
    induction of cell division, 211
    primary embryonic organizer and, 911, 912
    signal transduction pathways and, 131
Growth hormone (GH)
    actions of, *842*, 843
    production through biotechnology, 385
Growth hormone deficiency, 386
Growth hormone release-inhibiting hormone, 844
Growth hormone-releasing hormone, 844
Grylloblattodea, *672*
Guam, 1235
Guanine
    codons and the genetic code, 288–289
    complementary base pairing, 63–65
    in DNA structure, 264, 265, 266, *267*
    induced mutations from cigarette smoke, 309
    mutagens and, 310
    structure, *63*

Guanosine diphosphate (GDP)
    in the citric acid cycle, 170, 171
    G protein-linked receptors and, 130
Guanosine triphosphate (GTP)
    5' cap, 291–292
    in the citric acid cycle, 171
    function of, 67
    G protein-linked receptors and, 130
Guanylyl cyclase, 136
Guard cells, 712, 721, 732, 733–734
"Guide proteins," 559
Guillain–Barre syndrome, 926
Guillemin, Roger, 844
Gulf of Mexico
    causes and impact of dead zones, 1207, 1214, 1219, 1225
    oil spills, 373, 569
    red tides, 549
Gulf Stream, 1124–1125
Gulls, begging behavior of chicks, 1095
Guppies, 889, 1157
Gustation, 951
Gut
    defined, 630
    of flatworms, 656, *657*
    of herbivores, 638
    of phoronids, 659, *660*
    tissue composition, 820
    tubular, 1056–1057
    See also Gastrointestinal system
Gut hormones, 838
Gut microbes, benefits of, 539, 540–541
Gut muscle
    function of, 820
    influence of the autonomic nervous system on, 993
    layers of, 1058, *1059*
Gymnogyps californianus, *1244*
Gymnosperms
    conifers, 593–596
    major groups and distinguishing characteristics, *574*, 592–593
    seed development, *592*
    swimming sperm, 589
Gymnothorax meleagris, *688*
Gypsy moth, 1110
Gyres, 1124
Gyri, 970

H

H zone, *987*, 988
H1N1 influenza virus, *427*, 877
H5N1 influenza virus, *543*
Haber process, 750
Habitat corridors, 1233–1234
Habitat fragmentation, impact on biodiversity, 1233–1234
Habitat islands, 1198
Habitat isolation, 477, 482
Habitat loss
    impact on biodiversity, 1233–1234
    species extinctions and, 1232
Habitat patches, 1161–1162, 1233
Habitats
    animal selection of, 1103
    defined, 1103
    effect on population dynamics, 1161–1163

impact of interference competition on habitat use, 1183
Haddock, 1163–1164
Hadean eon, *506–507*, 508
Hadobacteria, 532
Hadrurus arizonensis, *636*
Haeckel, Ernst, 1122
Haemophilus influenzae
    comparative genomics, 357
    functional genomics, 356, *357*
    genome sequencing, 356
    genomic information, 361
Hagfish, 685–686, 1073, 1074
Hair
    a distinguishing feature of mammals, 697
    pigmentation and the *MCIR* gene, 367
Hair cells
    mechanism of mechanoreception, 953–954
    of the vestibular system, 953–954, *956*, 957
    See also Auditory hair cells
Hair follicles, 397
Hairgrass, 1125
"Hairy backs." *See* Gastrotrichs
Haldane, J. B. S., 1115
Half-cell reactions, 187
Half-life
    of hormones, 852–853
    of radioisotopes, 507
Haliaeetus leucocephalus, *638*, *1228*
Haliaeetus vocifer, *218*
Hall, Donald, 1105, *1106*
Halophiles, 536–537
Haltares, 672
Hamilton, W. D., 1115
Hamilton's rule, 1115
Hamner, Karl, 788, 789
Haplochromine cichlids
    prezygotic isolating mechanisms, 476, 477, 482
    sexual selection and rates of speciation, 480
    speciation, 467
Haplodiploidy, 1115, *1116*
Haploids
    in alternation of generations life cycle, 562, 563, 574, 575
    defined, 218, 236
    generation during meiosis, *220–221*, *223* (see also Meiosis)
    in sexual reproduction, 218, 219
Haplontic life cycle, *218*
Haplotype mapping, 367–368
Haplotypes, 367
Hardy, Godfrey, 437
HARDY gene, 726
Hardy–Weinberg equilibrium, 437–438
Hartwell, Leland, 210
Harvestmen, 669
Hashimoto's thyroiditis, 876
Hatena, 102
Haustoria, 611–612, 752
Haversian bone, 1001
Haversian systems, 1001
Hawaiian bobtail squid, 546
Hawaiian Islands
    adaptive radiations, 481–482
    allopatric speciation in *Drosophila*, 472, 473

avian malaria, 1236
distribution of long-horned beetles, 1145
effect of dispersal ability on speciation rates, 481
Hawks, fovea, 962
Hayes, Tyrone, 1, 12, 13
Hazel, *590*
Head (inflorescence), *597*
Hearing. *See* Auditory systems
Hearing loss, 956
Heart attack
  atherosclerosis and, 1042–1043
  human SNP scans, *368*
  from mutation-based wall thickening, 1025, 1046
  treatment with TPA, 385
Heart disease, stem cell therapy, 102
"Heart" embryo, *393*
Heart muscle. *See* Cardiac muscle
Heart pacemaker. *See* Pacemaker cells
Heart stage, 711
Heartbeat
  autonomic nervous system control of, 1034
  in blood pressure regulation, 1044, *1045*
  effect of calcium ion cycling on, 1034, *1036, 1037*
  in the fight-or-flight response, 837
  Frank–Starling law, 1042
  pacemaker cells and cardiac muscle contraction, 1032–1034
Hearts
  in amphibians, 1029
  calcium ion cycling in "hot" fish, *1036, 1037*
  in crocodilians and birds, 1029
  in fish, 1028
  four-chambered heart of mammals, 697
  in lungfish, 1028
  mammalian heart function, 1030–1037
  mutation-based wall thickening, 1025, 1046
  in open circulatory systems, 1026, *1027*
  production of atrial natriuretic peptide in blood pressure regulation, 1090
Heat
  in biological systems, *145*
  from brown fat, 165, 174
  muscle fatigue and, 998
Heat of vaporization, 33
Heat-resistant DNA polymerases, 277–278
Heat shock proteins, *51*, 810
Heat shock response, 810
Heavy metal tolerance, in plants, 811–812
Heavy nitrogen, 268, *269*, 270
Hebert, Paul, 319
Hedgehogs, *698*
Heidmann, Thierry, 545
Height, quantitative variation in humans, 246
HeLa cells, 205, 208–209, 229

Helices
  alpha helix, 45, *46*
  double helix structure of DNA, 65, *66, 264–265, 266, 267*
  left- and right-handed, *47*
Helicobacter pylori, 542
Heliconius, *1174*, 1175
Helium, 22, *22*
Helix pomatia, *663*
Helix-turn-helix motif, 335–336
"Helping at the nest," 1115
Hematocrit, 1037–1038
Hematopoietic stem cells, 408
Heme, *156*
Hemichordates
  in animal phylogeny, *630*
  in deuterostome phylogeny, *679*
  features of, 680, 682–683
  major groups, *632*, 679
Hemiparasites, 752
Hemipenes, 888
Hemipterans, 480, *672, 673, 674*
Hemitrichia serpula, *560*
Hemizygous, 251
Hemocoel, 652, 662
Hemoglobin
  binding of carbon dioxide, *1019*
  binding of carbon monoxide, 1017
  binding of oxygen, 1016–1017
  factors affecting oxygen affinity, 1017–1018
  globin gene family and, 364, *497, 498*
  hemoglobin C disease, 312
  missense mutation and sickle-cell disease, 306, *307*, 312
  in pogonophorans, 660
  polymorphism in, *312*
  quaternary structure, 48, *49*
  in red blood cells, 1038
  structure, 1016
  β-thalassemia, 292–293, 306
Hemoglobin C disease, 312
Hemolymph, 1026, *1027*
Hemophilia, *313, 323*, 365, 1039
Hendricks, Sterling, 788
Hensen's node, 914
Henslow, John, 428
Hepatic duct, 1061
Hepatic portal vein, 1063
Hepatic portal vessel, *1031*
Hepatic veins, *1031*
Hepatitis, 1039
Hepatophyta, *574*
HER2 receptor, 228, 229
Herbicide resistance, 388, 389
Herbicides, 1, 160
Herbivores/Herbivory
  circumvention of plant defenses, 806
  defined, 1054, 1170
  description of, 1175
  digestion of cellulose in, 1063–1064
  eutherian, 700
  feeding strategy, 637–638
  ingestion and digestion of food, 1054–1055
  plant defenses against, 801–805, 1175
  reciprocal interactions between herbivores and plants, 1175–1176

salt balance in, 1073
  teeth, *1055*
  trophic cascades in savanna communities, 1194
  See also Primary consumers
Herbivorous bugs, 480
Herceptin, 229, 871
"Herd immunity," 856
Herelle, Felix d', 545
Heritable traits, 431
Hermaphroditic flowers, 597, 779
Hermaphroditism, in animals, 887–888
Hermodice carunculata, *636*
Herpes viruses, 341, *543*
Herpesviridae, *543*
Hershey, Alfred, 261–263
Hershey–Chase experiment, 261–263, 339
Heterocephalus glaber, *1116*
Heterochromatin, 345, 346, 364
Heterochrony, 416–417
Heterocysts, 532, *533*
Heteroloboseans, 558
Heterometry, 416
Heteromorphic alternation of generations, 563
Heterophrynus batesii, *654*
Heteropods, 662
Heterosis, 244–245
Heterospory
  in angiosperms, 600–601
  appearance in vascular plants, 584–585
  in seed plants, 589–590
Heterotherms, 822
Heterotopy, 417
Heterotrophic succession, 1202
Heterotrophs
  absorptive heterotrophy, 609
  classified by acquisition of nutrition, 1054
  in communities, 1189, 1190
  defined, 196
  facilitation of succession by, 1201
  requirements from food, 1049–1054
Heterotypic cell binding, 111
Heterotypy, 418–419
Heterozygote advantage, 442–443
Heterozygotes
  defined, 236
  incomplete dominance, 242–243
Hexacontium, *557*
Hexapods, 667, 671, 673
  See also Insects
Hexokinase, *151*, 155
Hexose phosphates, 201
Hexoses, 52, 195
Hibernation/Hibernators, 830, *831*, 1108
Hide beetles, 1188, 1203
High-density lipoproteins (HDLs), 1066
High-GC Gram-positives, 532
High-throughput sequencing, 353–354
Highly repetitive sequences, *355*, 364, 366
Hillis, David, 457
Himalayan mountains, 509
Himanthalia elongata, *556*
Hindbrain, 968, *969*

Hindlimbs
  development in ducks and chickens, 417
  loss in snakes, 423
Hippocampus
  functions of, 970
  of London taxi drivers, 967
  memory and, 967, 982, 983
  "place cells" in rats, 967
  regulation of the stress response and, 850
Hirudo, 1056
Hirudo medicinalis, 661–662
Hispaniola, 440
Histamine, 861, 1041
Histidine, *44*
Histone acetyltransferases, 344
Histone deacetylase inhibitors, 344
Histone deacetylases, 344
Histone methylation, 344
Histone phosphorylation, 344
Histone tails, 344
Histones
  in chromatin structure, 212, *213*
  modifications in epigenetic gene regulation, 344
HIV-1, 458, 461–462
HIV-2, 458
HIV protease, 301
HMGA2 gene, 246
Hoatzin, 494
Hodgkin, A. L., 929
Holdfast, 556
Holocene epoch, 522
Holometabolous insects, *672, 673*
Holoparasites, 752, 753
Homeobox, 404, 413, 917
"Homeobox" genes. *See* Hox genes
Homeodomain, 404
Homeosis, 401
Homeostasis
  defined, 10
  of the internal environment, 816
  plasma membrane and, 80
  regulation by physiological systems, 816–817
  regulation of enzymes and, 156–157
Homeotherms, 822
Homeotic genes
  in fruit fly body segmentation, 403–405
  regulation of body segmentation, 916–918
Homeotic mutations, 401, 404
Homing, 1109
Hominins
  ancestors of humans, 702–705
  bipedal locomotion, 702–703
  diet of, 1056
Hominoids, 522
Homo, early members of, 703–704
Homo erectus, 703, 704
Homo ergaster, *703*
Homo floresiensis, *703*, 704
Homo habilis, 703
Homo neanderthalensis, 7, *703*, 704
Homo sapiens
  evolution of, 522, *703*, 704–705
  genome size, 494
  meaning of, 7
  See also Humans

Homogentisic acid, 282
Homogentisic acid oxidase, 282
Homologous chromosomes
  characteristics of, 218
  duplications, 308
  events in meiosis, 219, 220–222, 223
  genetic recombination, 219, 220, 222, 247–249, 250
  meiotic errors, 222, 224
Homologous features, 452
Homologous genes, 413–414, 499–500
Homologous pairs, 218
  See also Homologous chromosomes
Homologous recombination, 381–382
Homologs, 218
Homology
  in genes, 413–414, 499–500
  between macromolecules, 486
Homoplasies, 452
Homospory, 584
Homotypic cell binding, 111
Homozygotes, 236
Honest signals, 435
Honey bees
  epigenetic effects of "royal jelly," 899
  exploitation competition, 1183
  metabolic heat production of colonies, 826
  monitoring, 1150, 1151
  as pollinators, 1181, 1182
  sex and reproduction in, 880
  waggle dance, 1112
Honeypot ant, 1183
Hoofed mammals, 698
Hooke, Robert, 78
Hoover, Jeffrey, 1117
Hoplostethus atlanticus, 1235
Horizons, in soils, 745
Horizontal cells, 962, 963
Horizontal stems, 720
Hormone-based contraceptives, 898
Hormone receptors
  location and function, 836–837
  multiple receptors for a single hormone, 853
  upregulation and downregulation, 853
  See also individual hormone receptors
Hormones
  affinity chromatography, 853
  comparison of animal and plant hormones, 758
  criteria for defining a molecule as, 853
  defined, 126, 758, 835
  detection and measurement with immunoassays, 852–853
  dose–response curves, 853
  half-life, 852–853
  See also Animal hormones; Plant hormones
Hornworts, 573, 574, 578–579
Horowitz, Norman, 283, 284
Horse family, 1143, 1144
Horsehair worms, 632, 652, 666, 667
Horseshoe crabs, 668–669

Horsetails, 520, 574, 580, 581
Hosts
  lytic viral reproductive cycle in, 339–340
  parasites and, 638–639, 1170
Hot deserts, 1132–1133
"Hot" fish, 825, 1036, 1037
Hot sulfur springs, 536
House flies, 1203
"Housekeeping genes," 332–333
Hox genes
  differences in expression and spine evolution, 418
  duplication-and-divergence hypothesis, 413–414
  in ecdysozoans, 654
  in fruit fly body segmentation, 401, 403–405
  heterotypy and leg number in insects, 418, 419
  loss of limbs in snakes and, 423
  regulation of body segmentation, 916–918
  wing development in insects, 415
Hoxc6 gene, 418
Hoxc8 gene, 418
HSP60, 51
Hubel, David, 977
Human activities
  impact on ecosystems, 1223–1224
  impact on energy flow through the global ecosystem, 1210
  impact on the global nitrogen cycle, 1218–1219
  impact on the global phosphorus cycle, 1220–1221
  overexploitation of species, 1234–1235
  predicting the effects of humans on biodiversity, 1231–1232
  species extinctions and, 1229–1230
  threatening species persistence, 1232–1237
  See also Fossil fuel burning
Human birth defects
  spina bifida, 916
  from thalidomide, 920
Human brain
  brainstem structure and function, 969–970
  complexity of neural networking in, 943
  consciousness, 982–983
  development, 920–921, 968–969
  forebrain structure and function, 970
  language areas, 980–981
  learning and memory areas, 981–982
  size and evolution, 704, 973
  sleep and dreaming, 978–980
  telencephalon structure and function, 970–973
Human chorionic gonadotropin (hCG), 871, 896
Human development
  apoptosis in, 399–400
  left–right asymmetry of internal organs, 902
  regulative, 907
  stages in, 919–921

Human diseases
  fungal, 612
  pathogenic bacteria, 542
  pathogenic protists, 563, 564
  pathogenic trypanosomes, 559
  viral agents, 544–545
Human epidermal growth factor (HER2) receptor, 228, 229
Human genetic diseases
  abnormal hemoglobin, 312
  cancer and somatic mutations, 314
  evolutionary studies of sodium channel genes and, 502
  examples of, 313
  expanding triplet repeats, 313–314
  Inclusion-cell disease, 300
  IPEX, 874
  knockout mouse models, 382
  loss of enzyme function, 311–312
  multifactorial nature of, 314–315
  point mutations, 312
  prevalence, 315
  strategies for treating, 322–325
  β-thalassemia, 292–293
  transposons and, 365
Human genome
  alternative splicing, 346–347
  characteristics of, 366
  comparative genomics, 366–367
  endogenous retroviruses in, 545
  gene duplication in, 497
  genomic library of, 379
  Human Genome Project, 353
  key parameters, 361
  largest gene, 291
  medical benefits from studying, 367–368
  microRNA, 347
  normal damage to DNA, 310
Human Genome Project, 353
Human growth hormone (hGH), 385–386
Human immunodeficiency virus (HIV)
  as an RNA retrovirus, 543, 544
  course of infection, 876, 877
  gene expression in, 285
  gene regulation at the level of transcription elongation, 341–343
  HIV protease, 301
  molecular clock dating of the origin of HIV-1 in humans, 461–462
  phylogenetic analyses, 458, 459
  treatment, 342–343, 876–877
Human lungs
  anatomy of, 1013, 1014
  diseases of, 1013
  inhalation and exhalation, 1015–1016
  perfusion by the circulatory system, 1016
  respiratory tract secretions, 1013, 1015
  tidal ventilation, 1012–1013
Human nervous system
  functional and anatomical organization, 968–969

left–right crossover between brain and body, 970
  See also Human brain
Human papillomavirus (HPV), 228, 229, 349
Human reproduction
  contraception, 897, 898
  female reproductive system, 892–897
  implantation, 906, 907
  male reproductive system, 889–892
  reproductive technologies to solve infertility, 897, 899
  twins, 907
Humans
  abnormal sex chromosome arrangements, 250
  ABO blood groups, 243
  basal energy expenditure, 1049
  brown fat, 165, 828
  chromosome number, 218
  circulatory system, 1031
  decomposition and corpse communities, 1188, 1203
  development of language and culture, 705
  digestive enzymes, 1062
  digestive system, 1058
  Down syndrome, 224
  ear structure, 954–955
  effects of prenatal stress on child behavior, 328
  embryonic stem cells, 409
  endocrine system overview, 842
  endosymbiotic bacteria in the intestines, 1057
  essential amino acids, 1051–1052
  evolutionary responses to ectoparasites, 1177
  excretory system, 1080–1086
  exponential population growth, 1164–1166
  extracellular fluid in, 816
  eye, 959, 962–963
  fevers, 830
  founder effect in, 434
  gene flow in, 434
  genomic imprinting, 345
  global climate change and, 17
  growth hormone deficiency in children, 386
  heart attack from mutation-based wall thickening, 1025, 1046
  heart function, 1030–1037
  heat stroke, 815
  jumping ability, 986
  karyotypes, 224, 225
  lymphatic system, 857, 858
  metabolomes, 370
  microbiomes and human health, 539–541
  muscular segmentation, 636
  myostatin gene, 370
  Neanderthal ancestors, 704
  nitrogenous wastes excreted, 1074–1075
  origin of, 702
  overnutrition and obesity, 165, 1048, 1051, 1068
  oxytocin and, 141
  pedigrees, 240–241

primate ancestors, 702–704
proteomes, 369
puberty, 850–851, 894
quantitative variation in, 246
regulation of breathing, 1019–1022
sex-linked inheritance, 252
sex steroids and phenotypic sex determination, 850, *851*
skeleton, 999, *1000*
sociobiology and, 1116–1117
stabilizing selection on birth weight, 439
tactile receptors of the skin, 952
taste buds and taste, 951
trisomies and monosomies, 224
value of biodiversity to, 1230
vitamin requirements, 1053–1054
See also Infants
Humata tyermanii, 218
Humboldt, Alexander von, 1196
Humerus, 423
Hummingbirds
    daily torpor, 830
    feeding on nectar, *477*
    "nectar corridor" migration, 1137
    as pollinators, 598, *599*, *1136*, 1181, *1182*
    weight of, 696
Humoral immune response
    activation and effector phases, 872, *873*
    description of, 864, *865*
    generation of immunoglobulin diversity, 868–869, *870*
    immunoglobulin class switching, 869–871
    immunoglobulin classes, 868
    immunoglobulin structure and function, 867–868
    monoclonal antibodies, 871
    overview, 863–864
    plasma cells, 867
    suppression by regulatory T cells, 874
Humpback whales, 1112
Humus, 745, 746
Hunchback gene, 402, *403*
Hunchback protein, 402, *403*
Hungate, Bruce, 1221–1223
Huntington's disease, 314
Hurricane Katrina, 1223–1224
Hussein, Saddam, *317*, 318
Huxley, A. F., 929
Huxley, Thomas, 605
Hyacinths, *603*
Hyalophora cecropia, 841
Hybrid seeds, 778
Hybrid vigor, 244–245
Hybrid zones, 478–479
Hybridization
    allopolyploidy in wheat and, 225, 226
    lateral gene transfer and, 496
    mechanisms preventing, 475–478
Hybrids
    hybrid plants in agriculture, 778
    postzygotic isolating mechanisms, 478, 479
Hydra, 881
Hydrocarbon molecules, 30

Hydrochloric acid
    production by the stomach, 1058, *1060*, 1061
    properties of, 34
Hydrogen
    atomic stability, 25
    covalent bonding capability, 27
    electronegativity, 28
    isotopes, 22
Hydrogen bonds
    in alpha helices, 45, *46*
    in beta pleated sheets, 45, *46*
    description of, 30
    in DNA, *66*, 266
    features of, *26*
    in protein binding, 50
    in protein quaternary structure, 48
    in protein tertiary structure, 47
Hydrogen ions
    acids and, 34
    bases and, 34
    pH of solutions, 35
    transfer during oxidation–reduction reactions, 167, 168
Hydrogen peroxide, 93, 175
Hydrogen sulfide
    as an electron donor in anoxygenic photosynthesis, 187
    as an electron donor in photoautrophs, 538
    in the global sulfur cycle, 1219
    in pogonophoran metabolism, 660, 661
Hydroid, 578
Hydrolagus colliei, 688
Hydrologic cycle, 1215
Hydrolysis
    of ATP, 149–150
    hydrolysis reactions, 42
Hydrolytic enzymes, 161
Hydronium ion, 34
Hydrophilic molecules, 30
Hydrophilic regions
    of integral membrane proteins, 108
    of phospholipids, 106
Hydrophobic interactions
    description of, 30
    features of, *26*
    in protein binding, 50
    in protein quaternary structure, 48
Hydrophobic molecules, 30
Hydrophobic regions
    of integral membrane proteins, 108
    of phospholipids, 106
Hydroponics, 742–743
Hydrostatic skeletons, 635–636, 999
Hydrothermal vents. *See* Deep-sea hydrothermal vent ecosystems
Hydroxide ion, 34, 175
Hydroxyl group, *40*
Hydroxyl radical, 175, 176
Hydrozoans, 646, 647–648
Hyla chrysoscelis, 469
Hyla versicolor, 469, 1173
Hylobates lar, 702
Hymen, 892

Hymenoptera
    haplodiploidy and eusociality, 1115, *1116*
    number of living species, *672*
Hyperaccumulators, 811–812
Hyperemia, 1044
Hypericum perforatum, 1175
Hyperosmotic regulation, 1073
Hyperpolarization
    description of, 931, *932*
    at inhibitory synapses, 938
    of neurons at the onset of sleep, 979
    of rod cells, 960–961
Hypersensitive response, 226, 799, 800–801
Hypersensitivity, allergic reactions and, 875–876
Hypertension, atherosclerosis and, 1043
Hyperthermophilic bacteria, 532
Hyperthyroidism, 846–847
Hypertonic environments, fungal tolerance of, 610–611
Hypertonic solutions, 114, *115*
Hyphae
    characteristics of, 609–610
    in lichens, 613
    in mycorrhizae, 614
    of parasitic fungi, 611–612
    of predatory fungi, 612, *613*
Hypoblast
    of the human blastocyst, *907*
    in the origin of extraembryonic membranes, 918, 919
    in yolky eggs, 913, 914
Hypocotyls, 770
Hypoosmotic regulation, 1073
Hypothalamus
    body temperature regulation in mammals, 829–830
    functions of, 969, 970
    hormones of, *842*
    in human puberty, 851, 892, 894
    interconnections with the pituitary gland, 842, 843
    in negative feedback regulation of hormone secretion, 844, 845
    neurohormones of, 843, 844
    regulation of blood pressure and blood osmolarity, 1088, 1089
    regulation of cortisol release in the stress response, 849–850
    regulation of food intake, 1067–1068
    regulation of the ovarian and uterine cycles, 894
    regulation of thyroxine production, 846
    somatostatin production, 849
Hypotheses
    defined, 12
    hypothesis–prediction methodology, 11–12
    significance to scientific inquiry, 14
Hypothyroidism, 846, 847, 1054
Hypotonic solutions, 114, *115*
Hypoxia
    dead zones in the Gulf of Mexico, 1207, 1225

release of erythropoietin by the kidneys in response to, 1038
Hypoxia-inducible factor 1 (HIF-1), 1038
Hypoxic conditions, insect body size and, 522
Hyracoidea, 698
Hyracotherium leporinum, 519
Hyraxes, 698

I

I band, *987*, 988
Ice
    effect of ice crystals on plant cells, 810
    properties, 32
"Ice ages," 522
Ice caps, impact of global climate change on, 1217, *1218*
Ice-crawlers, *672*
Identical twins, 907
IgA, *868*
IgD, *868*
IgE, *868*, 875, 876
IgG, *868*, 876
IgM, *868*, 870
Igneous rocks, 507–508
Ignicoccus, 537
Ileocaecal sphincter, 1060
Ileum, 1061
Ilex opaca, 780
Illicium floridanum, 602
Image-forming eyes
    anatomy of, 958, *959*
    focusing, 958–959
    structure and function of the retina, 959–963
Imatinib, 304
Imbibition, 758
Immediate hypersensitivity, 875–876
Immediate memory, 982
Immigration, in island biogeography theory, 1196–1198, *1199*
Immune system
    adaptive immunity (*see* Adaptive immunity)
    chemical signaling in, 835
    discovery of immunity, 862, *863*, 864
    inhibition by cortisol, 849
    innate immunity (*see* Innate immunity)
    "plant immune system," 799–801
    See also Immunology
Immune system proteins, types and functions of, 858–859
Immunization, 866
Immunoassays, 852–853, 871
Immunodeficiency viruses, 458
    See also Human immunodeficiency virus
Immunoglobulin genes, 868–869, *870*
Immunoglobulins
    class switching, 869–871
    classes of, 868
    constant regions, 867, 869–871
    generation of diversity in, 868–869, *870*
    structure and function, 867–868
    variable regions, 867, *869*, *870*
    See also Antibodies

Immunological memory
  secondary immune response and, 865–866
  vaccinations and, 863, 866
Immunology
  adaptive immunity, 862–867
  cellular immune response, 871–875
  characteristics of innate defenses, 859–862
  humoral immune response, 867–871
  immune system malfunctions, 875–877
  phases in the defensive response, 857
  role of immune system proteins, 858–859
  roles of blood and lymph in, 857–858
  roles of white blood cells in, 858
  types of defense systems, 857, 858
  vaccinations, 856
Immunotherapy, 871
Imperfect flowers, 597, 598, 779, 780
Implantation, 906, 907, 919, 920
Implantation blockers, 898
Impotence, 891
Imprinting, 1099
In vitro evolution, 500–501
In vitro fertilization (IVF), 899, 907
In vivo gene therapy, 323
Inactivation gate, 933–934
Inbreeding
  inbreeding depression, 244–245
  self-incompatibility studies in plants, 380–381
  strategies for preventing in angiosperms, 782, 783
Inbreeding depression, 244–245
Incisors, 1055
Inclusion-cell disease, 300
Inclusive fitness, 1115
Incomplete cleavage, 905
Incomplete dominance, 242–243
Incomplete metamorphosis, 672, 841
Incubation temperature, impact on sex determination, 420–421
Incus, 954, 955
Independent assortment
  Mendelian law of, 237–239
  separation of homologous chromosomes during meiosis, 220–222
Independent variable, 13
Indeterminate growth, 715–720, 786
Indianmeal moth, 1184
Indirect competition, 1184
Indirect fitness, 1115
Indole-3-acetic acid, 759, 760, 762, 763
  See also Auxins
Induced fit, 155
Induced mutations, 309
Induced plant defenses
  against herbivores, 803–805
  against pathogens, 798–801
Induced pluripotent stem cells (iPS cells), 409

Inducers
  in cell fate determination, 395–396, 397
  defined, 330
  in negative regulation of the lac operon, 331, 332
Inducible promoters, 384
Inducible proteins, 330
Inducible systems, in transcriptional regulation of operons, 330–331, 332
Induction
  in cell fate determination, 395–396, 397
  defined, 395
Inductive logic, 12
Industrial nitrogen fixation, 749, 750
Infants
  brown fat in, 828
  essential amino acids, 1051
  genetic screening, 320, 321
  microbiomes and, 540
  respiratory distress syndrome in premature babies, 1015
Infection thread, 748
Inferior mesenteric ganglion, 974
Inferior vena cava, 1030, 1031
Infertility, in humans, 897, 899
Inflammation
  anti-inflammatory drugs, 862
  description of, 861–862
  medical problems, 862
  necrosis and, 225
Inflammatory disease therapy, 158
Inflammatory response, 860, 861–862
Inflorescence meristems, 715, 785–786
Inflorescences, 596, 597, 786
Influenza virus
  as an RNA virus, 341
  epidemics, 427
  strain H5N1, 543
  vaccines, 427, 446
Infrared perception, in rattlesnakes, 946
Ingroup, 453
Inheritance
  blending inheritance, 233, 234
  human pedigrees, 240–241
  Mendelian laws of, 233–241
  of organelle genes, 252–253
  particulate theory of, 233–234, 236
  probability calculations, 239–240
  sex-linked, 249, 251–252
Inhibin, 892, 896
Inhibition
  in demonstrating cause and effect, 98
  of enzymes, 157–159
Inhibitors, 128
Inhibitory synapses, 938
Initials, 715
Initiation complex, in translation, 295, 296
Initiation site, in transcription, 286, 287
Innate immunity
  barrier and local agents, 859–860
  cell signaling pathways, 860

inflammation, 861–862
  overview, 857, 858, 859
  specialized proteins and cells, 860–861
Inner cell mass, 906, 907, 914
Inner ear
  anatomy of, 954, 955
  flexion of the basilar membrane, 955–956
  vestibular system, 954, 955, 956–957
Inorganic cofactors, of enzymes, 155, 156
Inorganic fertilizers, 747
Inorganic ions
  absorption in the large intestine, 1063
  absorption in the small intestine, 1063
Inosine (I), 294
Inositol trisphosphate (IP_3), 134–135, 136
Inouye, Isao, 102
Insect wings
  development, 415
  evolution of, 673, 675
  first appearance of, 520
  of pterygotes, 671, 672
  success of the insects and, 673
Insecticides
  inhibition of acetylcholinesterase, 940
  irreversible inhibitors of enzymes, 157
  produced by genetically modified plants, 386–388
Insects
  appearance of flight in, 520
  atmospheric oxygen and body size, 505, 513, 514, 522
  chemical defenses, 1173
  colonization of human corpses, 1188, 1203
  contraction in flight muscles, 997–998
  diapause, 421
  estimating the number of living species, 651, 673
  excretory system, 1076–1077
  fossils in amber, 515
  gas exchange in, 671, 1009
  gigantic, 505
  herbivorous, 1175
  heterotypy and leg number in, 418, 419
  hormonal control of molting, 839–841
  human impact on the distribution of, 1145
  internal fertilization, 887
  key features and body plan, 671, 672
  major groups, 671–673, 674
  metabolic heat production, 825–826
  as pollinators, 599
  superficial cleavage, 905
  undescribed species, 675
  wingless relatives, 671
Insomnia, 978
Inspiratory reserve volume (IRV), 1012, 1013
Instars, 671
Insula, 983

Insular cortex, 983
Insulin
  actions of, 842
  blood glucose regulation, 848–849, 1066–1067
  control of food intake, 1067, 1068
  diabetes in the Pima and, 1048
  production through biotechnology, 385
Insulin-dependent diabetes mellitus, 848, 876
Insulin-like growth factor 1 (IGF-1) gene, 352
Insulin receptors, 129, 853
"Insurance population," 255
Integral membrane proteins, 106, 107–108, 121
Integrase, 544
Integrase inhibitors, 342
Integrin
  in cell attachment to the extracellular matrix, 112, 113
  in cell movement, 112, 113
Integument
  in angiosperms, 597
  in conifers, 594, 595
  development of seed coat from, 591, 592, 784
  in seed plants, 590
Intercalated discs, 991
Intercellular signaling
  bonding behavior in voles and, 125
  cell responses to, 125, 126–127
  effects on cell function, 137–139
  evolution of multicellularity and, 140–141
  second messengers, 131–137
  signal receptors, 127–131
  types and sources of signals, 126
  See also Signal transduction pathways
Intercostal muscles, 1014, 1016, 1020
Interference competition, 1182, 1183
Interference RNA. See RNA interference
Interferons, 860
Interglacial intervals, 510
Interleukins, 211, 874
Intermediate filaments, 95
Intermediate muscle fibers, 996
Internal anal sphincter, 1063
Internal environment
  homeostatic regulation, 816–817
  importance of self-regulation, 10
  importance to multicellular animals, 816
  plasma membrane and, 80
Internal fertilization, 888–889
Internal gills
  respiratory gas exchange in fishes, 1009–1010, 1011
  surface area maximization, 1008, 1009
Internal jugular vein, 1031
Internal membranes
  origins of, 101–102
  of prokaryotes, 83
Internal shells, 664

Internal skeletons
of echinoderms, 680, *681*
of glass sponges and demosponges, 643
of humans, 999, *1000*
of vertebrates, 685, 687
See also Endoskeletons; Skeletal systems
International animal trade
consequences of, 1234, *1235*
ending, 1240–1241
International Fund for Animal Welfare, 1241
International System of Units (ISU), 1049
International Union for the Conservation of Nature (IUCN), 1231
Interneurons
defined, 940
in the retina, 963
in spinal reflexes, 942–943
Internodes, 709, 719
Interphase
defined, 208
in meiosis, 219
in mitosis, *217*
subphases, 208
Interspecific competition
community productivity and, 1192
defined, 1182
impact on life history traits, 1157
Interstitial fluid, 816, 1026, 1042
Intertidal zone
defined, 1139
impact of competition on barnacle niche determination, 1184
inhibition of succession by green algae, 1201
keystone species, 1194
Intestines
colonization by endosymbiotic bacteria, 1056–1057
digestion in, 1056
epithelial absorption of nutrients and inorganic ions, 1063
infolding of the walls, 1057
See also Large intestine; Small intestine
Intracellular receptors, 129, 130–131
Intracytoplasmic sperm injection (ICSI), 899
Intraspecific competition, 1182
Intrauterine device (IUD), *898*
Intrinsic factor, 1054, 1058
Intrinsic rates of increase, 1156, 1159, 1163
Introduced species
high population densities and, 1149, 1160
impact on the fynbos, 1145
population dynamics of reindeer, 1149, 1158, 1166
See also Invasive species; Non-native species
Introns
in alternative splicing, 346
in archaea, 356
description of, *290*, 291
splicing to remove, *291*, 292–293

Inuit peoples, 1054
Invasive species
controlling or preventing, 1241, *1242*
"decision tree" for evaluating invasive plants, 1241, *1242*
negative impact of, 1235–1236
See also Introduced species; Non-native species
Invasiveness, of pathogens, 542
Inversions, 308
Invertase, 736
Invertebrates
body size and respiratory gas diffusion, 1007
excretory systems, 1075–1077
visual systems, 958
Involution, 909, *910*
Iodine/Iodide, 845, *846*, 847, *1052*, 1221
Iodine/Iodide deficiency, 847, 1054
Ion channel receptors, 129
Ion channels
actions of sensory receptor proteins on, 947, *948*
activation in stereocilia, 954
in cardiac pacemaker cells and heart contraction, 1033–1034
in the generation of membrane potentials, 928, 929
ionotropic receptors, 938–939
opening in response to signals, 137–138
patch clamp studies, 929, *931*
as receptors, 129
root uptake of mineral ions and, 729
specificity of, 115–116
structure and function, 115, *116*
See also Gated ion channels; *specific types of ion channels*
Ion pumps, root uptake of mineral ions and, 729
Ion transporters, in the generation of membrane potentials, 928, 929
Ionic attractions
description of, 28–29
features of, *26*
in protein quaternary structure, 48
Ionic electric current, 931–932
Ionic interactions
in protein binding, 50
in protein tertiary structure, 47
Ionic regulators, 1073
Ionization, of water, 34
Ionizing radiation, 309
Ionotropic receptors, 938–939
Ionotropic sensory receptor proteins, 947, *948*
Ions
complex, 29
description of, 29
electrochemical gradients, 929
See also specific ions
IP₃/DAG pathway, 134–135
IPEX, 874
Ipomoea batatas, 718
Iridium, 511
Iris (eye), 958, *959*
Iris (Greek god), 834
Irisin, 834, 838

Iron
in animal nutrition, *1052*, 1053
in catalyzed reactions, *156*
global cycle, 1221
impact on nitrogen fixation, 1222–1223
in plant nutrition, *741*, 742
Iron deficiency, 1053, 1054
Iron oxide, 512
Irreversible inhibition, of enzymes, 157, *158*
Island biogeography theory, 1196–1198, *1199*
Islets of Langerhans, 848–849
Isobutane, *41*
Isocitrate dehydrogenase, 182
Isoleucine, *44*
Isomers, 41
Isomorphic alternation of generations, 563
Isopods, 670
Isoptera, 672
Isosmotic reabsorption, 1082
Isotonic solutions, 114, *115*
Isotope analysis, of water to detect climate change, 36
Isotopes
description of, 22–24
of oxygen, 21
Isozymes, 162
Istiophorus albicans, 1139
Iteroparous species, 1156
Ivory-billed woodpecker, 1230, *1231*
Ivory trade, 1240–1241
Ixodes scapularis (= *Ixodes dammini*), *1151*, 1152

J

Jackfruit, *601*
Jacky dragons, 421
Jacob, François, 331
Jacobson, Henning, 856
Janzen, Daniel, 1178, *1179*
Japanese macaque, 1177
Japanese mint, *604*
Japanese quail, 1101
Jasmonic acid (jasmonate), 759, 799, 804, 805
Jasper Ridge Biological Preserve, 1161
Jaw worms, 657–658
Jawed fishes, 686–689
Jawless fishes, 685–686, *687*
Jaws
development in vertebrates, 684
as joints, 1002
JAZ protein, 805
Jejunum, 1061
Jellyfish
characteristics of, 646–647
gastrovascular cavity, 1056
green fluorescent protein, 449
life cycle, 646, *647*
See also Cnidarians
Johnny jump-ups, *597*
Johnson, R. T., 209
Johnston's organ, 671
Jointed appendages
in animal evolution, 636
arthropods, 667
crustaceans, 671
trilobites, 668
Joints, 1001–1002

Joules (J), 147
Joyner-Kersee, Jackie, 986
Juan de Fuca oceanic plate, 509
Jumping
diversity of ability in animals, 986
efficiency of kangaroos in, 1003
Jumping bristletails, 671, *672*
Junipers, 594, *1133*
Juniperus, *1133*
Jurassic period, *506–507*, 521
Juvenile hormone, 841
Juxtaglomerular cells, 1088
Juxtacrine signals, 126

K

K-strategists, 1159, 1201
Kalanchoe, 793
Kamen, Martin, 186, 187
Kangaroos
jumping efficiency, 986, 1003
as marsupials, 697, *698*, 699
Kaposi's sarcoma, 876
Karenia, 549
Kartagener's syndrome, 902
Karyogamy, *618*, 619, 620, *621*
Karyotype, 224, *225*
Kashefi, Kazem, 535
Katydid, 1173
Katz, Lawrence, 951
Keeley, John, 757
Kentucky bluegrass, 199
Kenward, R. E., 1116
Keratin, 95, 397
Keratin genes, 66
Keto group, *40*
α-Ketoglutarate, 180
Keystone species, 1194–1195
Kidney stones, 848
Kidney transplants, 1085–1086
Kidneys
in amniote evolution, 692
in blood pressure regulation, 1044–1045
blood vessels of, *1080*, 1082
effect of aldosterone on, 849
effect of antidiuretic hormone on, 843
effect of parathyroid hormone on, 848
in the human excretory system, *1080*
nephron structure in mammals, *1080*, 1081–1082
production of concentrated urine in mammals, 1071, 1077, 1082–1084, 1090
regulation of, 1087–1090
regulation of acid–base balance, 1084–1085
secretion of erythropoietin, 1038
structure and function, 1077, 1078–1079, *1080*
treatments for renal failure, 1085–1086, *1087*
Kilocalorie (kcal), 1049
Kimura, Motoo, 492
Kin selection, 1115–1116
Kinesins
in cilia and flagella, 97–98
in kinetochores and cytokinesis, 216
Kinetic energy, 145
Kinetin, 768

Kinetochore microtubules, 214–216
Kinetochores, 213, 214, *215*, 216
Kinetoplast, 559
Kinetoplastids, 558, 559
King, Thomas, 406
King penguin, *642*
Kingdoms (taxonomic category), 463
Kingfishers, *642*
Kinorhynchs, *632*, 665
"Kiss and run" process, 122
Kitasato, Shibasaburo, 862, *863*, 864
Kiwis, 694
Klebsiella, 1169
Klein, David, 1149
Klinefelter syndrome, 250
Klok, C. Jaco, 513
Knee-jerk reflex, 969, 1002
Knee joint, 1002
Knockout model experiments, 381–382, 1096–1097
Koalas, 1055, 1175
Koch, Robert, 541
Koch's postulates, 541–542
Kokanee salmon, 1228
Komodo dragon, 693
Korarchaeota, 535, 537
Krakatau, 510
Krebs cycle. *See* Citric acid cycle
Kruger National Park, 1242
Kuffler, Stephen, 975, 976
Kuroshio Current, *1125*
Kuwait, 373
Kwashiorkor, 1050

L

L ring, *83*
La Selva Biological Station, 1237
Labia majora, 892, *893*
Labia minora, 892, *893*
Labor, in childbirth, 896–897
Labrador Current, *1125*
Labrador retrievers, 244
lac Operon
    description of, 330, *331*
    negative and positive regulation of, 330–332, *333*
lac Repressor, 331–332, *333*, 335
Lace lichen, 625
Lacewings, *672*
Lacks, Henrietta, 205
Lactases, *1062*, 1063
Lactate, 177
Lactate dehydrogenase, 177
Lacteals, 1063
Lactic acid
    accumulation in muscle, 997
    formed in prebiotic synthesis experiments, 70
Lactic acid fermentation, 177
Lactoglobulin, 385
Lactose
    in negative regulation of the *lac* operon, 331, *333*
    in the small intestine, 1063
Lactose metabolism, regulation in *E. coli*, 329–332, *333*
LacZ gene, 378
Laetiporus sulphureus, 623
Lagging strand, in DNA replication, 272–273, *274*
Lagomorpha, *698*

Lagopus lagopus, 1128
Lake Erie, 1221
Lake Malawi, 467, 509
Lake Malawi cichlids. *See* Haplochromine cichlids
Lake St. Clair, 1160
Lake Superior, 1189, 1215
Lake trout, 1228
Lakes
    acidification, 1220
    characteristics of, 1140
    effect of nitrate pollution on arsenic levels, 1221
    eutrophication, 1219, 1221
    thermocline, 1214
    turnover, 1213–1214
Lamar Valley, 1193–1194
Lamb's-quarter, 1200
Lamellae, 1010
Lamin proteins, 95
Lampreys, 686
Lampshells, 659
    See also Brachiopods
Lancelets, *632*, 679, 683, 684
Land plants
    adaptations to life on land, 574
    alternation of generations, 574–575
    classification, *574*
    closest relatives, 572–573
    colonization of the land, 574–579
    endosymbiotic origin of chloroplasts in, 551
    evolution of, *570*, 571
    major clades, 573, *574*
    nonvascular, 575–579
    vascular, 579–585
Landsteiner, Karl, 243
Language, humans and, 705
Language areas, of the human brain, 980–981
Langurs, 492–494
Laqueus, 659
Large cactus finch, *474*
Large ground finch, *474*
Large intestine
    in humans, *1058*
    in ruminants, 1064
Large tree finch, *474*
Larva
    complex life cycles of parasites, 640, *641*
    defined, 639
    dispersal, 640
    in metamorphosis, 639
Larvaceans, 684
Larynx, 1013, *1014*, 1059
Lateral gene transfer, 496, 529–530
Lateral meristems, 715, *716*
Lateral roots, 718, *718*
Lateralization, of language functions, 980–981
Latex, 801, 806
Laticifer-cutting beetles, 1175
Laticifers, 806
Latimeria chalumnae, 689
Latimeria menadoensis, 689
Latitudinal gradients, in species diversity, 1196, 1197
Laupala, 495
Laurasia, 521, 1143
Law of independent assortment, 237–239

Law of mass action, 35–36
Law of segregation, 236–237, 239
Laws of thermodynamics, 146–147
LDL. *See* Low-density lipoproteins
LDL receptors, 121
LEA proteins, 809
Leaching, 745, 746
Leading strand, in DNA replication, 272–273
Leafcutter ants, 1169, 1178, 1185
Leaf primordia, 719, 720
Leafhoppers, *672*, 673, 1175, 1184
LEAFY gene, 401, 792
LEAFY transcription factor, 786
Learning
    cellular basis of, 982
    defined, 981
    human capacity for, 981–982
    sleep and, 980
Leaves
    abscission, 765, 769
    adaptations to very dry conditions, 807
    aerenchyma, 808
    anatomy, 720–721
    anatomy in C_3 and C_4 pants, *198, 199*
    of carnivorous plants, 751–752
    defenses against herbivory, 1175
    development, 719, 720
    evolution of, 583, *584*
    function, 709
    hairs, 713, 1175
    invasion by fungal hyphae, *612*
    origin of the flowering stimulus in, 788–790
    salt glands, 811
    senescence, 768, 769–770
    stomatal control, 732–734 (*see also* Stomata)
    transpiration, 732
    trichomes, 713
    vegetative reproduction in angiosperms, 793
    veins, 710, 720, *721*
Lecithin, *58*
Leeches, 661–662, 1056
Left-handed helices, 47
Left subclavian artery, 858
Leg muscles
    impact on venous blood flow, 1042
    integration of anabolism and catabolism during exercise, 180–181
Leghemoglobin, 750
Legs. *See* Jointed appendages; Limbs
Legumes
    in crop rotation, 750
    cyanide production, 805
    evolution of nitrogen-fixing symbiosis, 521
    root nodule formation, 747–748
Leiobunum rotundum, 669
Leishmania major, 559
Leishmaniasis, *559*
Leks and lekking, 481, 1104, *1105*
Lemurs, *10*, 701
Lens
    determination in vertebrates, 396
    of the image-forming eye, 958–959
    in ommatidia, 958

Lens placode, 396
Lenticels, 722
Leontopithecus rosalia, 1114, *1137*
Leopard frogs
    experiments on the effects of atrazine, 12
    jumping ability, *986*
    temporal isolation, 476
Leopard gecko, *694*
Lepas pectinata, 670
Lepidodermella, 658
Lepidopterans, *672*, 674, 675
Lepidosaurs, 693
Lepomis macrochirus, 1105, *1106*
Leptin, 1067, 1068
Leptin receptor, *1067*, 1068
Leptosiphon, 460
Leptosiphon bicolor, 460
Leptosiphon liniflorus, 460
Lepus alleni, 828
Lepus arcticus, 828
Lesser long-nosed bats, *1137*
Lettuce seeds, 772
Leucine, *44*
Leucoplasts, 92
Leucospermum cordifolium, 1134
Leukemia therapy, 158
Leukocytes. *See* White blood cells
Levers, 1002
Levin, Donald, 478, 479
Lewis, Reggie, 1025
Lews, Cynthia, 565, 566
LexA protein, 341
Leydig cells, 889, *891*, 892
Libellula quadrimaculata, 674
Lice, *672*, 1177
Lichens
    description of, 613–614
    edible, 624
    indicators of air pollution, 624, *625*
    in succession on glacial moraines, 1200
Life
    common ancestry of, 3
    common characteristics of, 2
    elements of living organisms, 22
    evolution of, 3–5 (*see also* Evolution)
    origin of, 3 (*see also* Origin of life)
    scale of, *78*
    timeline of, *3, 515*
Life cycles
    amphibians, *691*
    angiosperms, 600–601
    animals, 639–642
    cnidarians, 645, *646*
    complex, 563, *564*
    ferns, 582
    fungi, 616, 618, 621
    mosses, 575–577
    pine, 594, *595*
    seed plants, 589–590
    of viruses, 339–341
    See also Alternation of generations
Life history strategies, 1156–1157, 1159, 1163
Life tables, 1154–1155
Life zones
    in freshwater biomes, 1140
    in marine biomes, 1139–1140

Lifestyle, obesity and, 1048
Ligaments, 953, 1002
Ligand-gated ion channels, 115, 116
Ligand–receptor complexes, 128
Ligands
  in binding to receptor proteins, 127–128
  chemoreceptors and, 949
  defined, 115, 127
Light
  absorption by pigments, 190–191
  aspects plants are responsive to, 771
  detection by animal sensory systems, 957–963
  entrainment of circadian rhythms in plants, 774–775
  light energy in biological systems, 145
  photobiology, 189–190
  photochemistry, 188–189
  photomorphogenesis in plants, 772
Light-harvesting complexes, 190–191
Light-independent reactions, 188
Light-induced electron transport, 196–197
Light microscopes, 79, 80
Light reactions, 188
Lignin, 711, 798
Lilies, 598, 780
Lilium, 780
Lima (British merchant vessel), 525
Limbic system, 970, 982
Limbs
  evolution in lobe-limbed vertebrates, 690
  evolution of the insect wing, 673, 675
  heterotypy and leg number in insects, 418, 419
  loss of forelimbs in snakes, 423
  morphogens and positional information in vertebrate development, 401
  See also Hindlimbs; Jointed appendages
Limestone deposits, 557, 565
Limnephilus, 674
Limpets, 662
Limulus polyphemus, 668
LIN-3 protein, 396, 397, 398
Lin-14 mutations, 347
Lind, James, 1053
Lindley, John, 588
Lineage species concept, 469
Linnaean classification, 462–463
Linnaeus, Carolus, 468
Linnean Society of London, 429
Linoleic acid, 57, 1052
Linum usitatissimum, 800
Lions, exploitation competition, 1183
Lipases, 144, 1058, 1062
Lipid bilayers, 73–74
Lipid-derived second messengers, 134–135
Lipid monolayers, 535–536
Lipid-soluble hormones, 837
Lipids
  of archaea, 535–536

in atherosclerosis, 1042
  in biological membranes, 106–107
  carotenoids, 58
  catabolic interconversions, 179–180
  membrane fluidity and, 107
  monomer components, 40
  phospholipids, 57, 58 (see also Phospholipids)
  properties of, 56
  proportions in living organisms, 41
  steroids, 58
  thylakoid lipids, 92
  triglycerides, 56–57
  types, 56
  vitamins, 58
  waxes, 58–59
Lipoproteins, 1065–1066
Lithium, 135
Lithosphere, 509
Lithotrophs, 538
Littoral zone, 1139
Liver
  absorption of nutrients by, 1063
  activation of glycogen phosphorylase, 132–133
  control of fat metabolism, 1065–1066
  control of glucose metabolism, 1065, 1066–1067
  effect of insulin on, 1066
  familial hypercholesterolemia, 121
  gluconeogenesis in, 1065, 1067
  in humans, 1058
  maintenance of blood glucose levels during exercise and, 181
  production of clotting factors, 1039
  reabsorption of bile salts, 1063
  role in digestion, 1061–1062
  uptake of low-density lipoproteins, 121
Liver cells
  in the fight-or-flight response, 837
  regulation of glucose metabolism, 138
Liver diseases, 1039, 1041
Liverworts, 573, 574, 577
Lizards
  behavioral thermoregulation, 822, 823
  evolution and characteristics of, 693, 694
  heat exchange through the skin, 824
  hemipenes, 888
  parthenogenic reproduction, 881–882
  temperature-dependent sex determination and sex-specific fitness differences, 421
  territorial behavior, 1103, 1104
Loading, of phloem sieve tubes, 736
Loams, 745
Lobe-limbed vertebrates, 687, 689–690
Loboseans, 560

Lobsters, 670
Locomotion
  body cavities and, 635–636
  in ciliates, 554
  in plasmodial slime molds, 560
  in protostome evolution, 675
Locus, 242, 431
Locusta migratoria, 802
Lodgepole pines, 594, 596, 1201–1202, 1236
Lofenelac, 322
Logic, inductive and deductive, 12
Logistic population growth, 1158
Lombok island, 1141, 1142
London taxi drivers, 967
Long bones, development, 1001
Long-day plants (LDPs), 787, 788
Long-distance athletes, 995
Long-horned beetles, 1145
Long interspersed elements (LINEs), 364–365
Long-tailed widowbird, 435–436
Long-term depression (LTD), 982
Long-term memory, 982, 983
Long-term potentiation (LTP), 982
Long terminal repeats (LTRs), 364–365, 495
Longhorn cattle, 440
Longwing butterflies, 1174, 1175, 1176
Loops of Henle
  aquaporins, 1084
  countercurrent multiplier, 1082–1084
  in desert rodents, 1090
  organization of, 1080, 1081–1082
  regulation of the glomerular filtration rate, 1088
"Lophophorates," 652
Lophophores
  in brachiopods and phoronids, 658, 659, 660
  in bryozoans, 654, 656
  description of, 652
  in entoprocts, 656
Lophotrochozoans
  anatomical characteristics, 652
  in animal phylogeny, 630
  annelids, 659–662
  brachiopods and phoronids, 658–659, 660
  bryozoans and entoprocts, 656
  diversity in, 656
  flatworms, rotifers, and gastrotrichs, 656–658
  lophophores and trochophores, 652–653, 654
  major subgroups and number of living species, 632
  mollusks, 662–664
  ribbon worms, 658, 659
  spiral cleavage, 633, 653
  wormlike body forms, 653–654
Lordosis, 1099
Lorenz, Konrad, 1095, 1099
Loriciferans, 632, 665–666
Lorikeets, 696
Lorises, 701
Loss-of-function mutations, 305, 311–312, 400, 401
Lovley, Derek, 535
Low-density lipoproteins (LDLs), 121, 1066

Low-GC Gram positives, 530–532
Low temperatures, animal adaptations to, 1125
Lowland gorillas, 702
Loxodonta africana, 1135, 1150
Luciferase, 150
Luciferin, 149, 150
Lucilia caesar, 674
"Lucy," 703
Luehea seemannii, 651
Lumbricus terrestris, 661
Lumen, of the vertebrate gut, 1058
Lung cancer cells, 227
Lung surfactants, 1015
Lungfish, 689, 690, 1028–1029
Lungs
  diffusion of carbon dioxide from blood, 1018, 1019
  evolution in lungfish, 1028–1029
  human, 1013–1016
  perfusion by the circulatory system, 1016
  physical stresses in snorkeling elephants, 1005, 1022
  surface area maximization, 1008, 1009
  tidal ventilation, 1012–1013
  unidirectional ventilation in birds, 1010–1012
Luteal phase, 893
Luteinizing hormone (LH)
  actions of, 842
  endocrine source, 842, 843
  in follicle selection for ovulation, 895–896
  in puberty, 850–851, 892, 894
  regulation of the ovarian and uterine cycles, 893–894
Luxilus coccogenis, 472
Luxilus zonatus, 472
Lycaon pictus, 1183, 1242, 1243
Lycoperdon perlatum, 610
Lycophytes, 574, 580, 581, 582–583
Lycopodiophyta, 574
Lycopodium, 583
Lycopodium annotinum, 581
Lyell, Charles, 428
Lyme disease, 533, 1152
Lymph, 857, 858, 1042
Lymph capillaries, 858
Lymph ducts, 857
Lymph nodes, 857, 858, 1042
Lymphatic system, 857, 858, 1042
Lymphatic vessels, 1042
Lymphocytes
  cell membranes, 4
  clonal deletion, 865
  clonal selection, 865, 866
  diversity of adaptive immunity and, 863
  effector cells and memory cells, 865–866
  specificity of adaptive immunity and, 862–863
  types and functions of, 858
Lymphoma tumors, 876
Lyperobius huttoni, 1144–1145
Lysine, 44, 212
Lysogenic cycle, of bacteriophage, 340–341
Lysosomal storage diseases, 91
Lysosomes
  evolution in eukaryotic cells, 551

Inclusion-cell disease, 300
structure and function, 90–91
Lysozyme
convergent molecular evolution
in foregut fermenters,
492–494
in innate defenses, 859
interactions with substrate, 153,
154
molecular models of, 47
turnover number, 156
Lytic cycle, of bacteriophage,
339–340, 341

M

M band, 987, 988
Macaca fuscata, 1177
MacArthur, Robert, 1197
MacKinnon, Roderick, 115
Macrocystis, 556, 1139
Macroderma gigas, 700
Macroevolutionary change, 446
Macromolecules
characteristics of, 39
condensation and hydrolysis
reactions, 42
defined, 40
endocytosis and exocytosis,
120–122
functional groups, 40
isomers, 41
molecular evolution, 486
proportions in living
organisms, 41
relationship of structure to
function, 41–42
types found in living things, 40
See also Carbohydrates; Lipids;
Nucleic acids; Proteins
Macronectes giganteus, 1073
Macronucleus, 555, 562
Macronutrients
in animal nutrition, 1052–1053
in plant nutrition, 741, 742
Macroparasites, 1176–1177
Macroperipatus torquatus, 667
Macrophages
as antigen-presenting cells, 872
cytokines, 859
degradation of old red blood
cells, 1038
digestion of pus, 862
function of, 858
HIV infection, 876
pattern recognition receptors,
860
Macropus giganteus, 699
Macroscelidea, 698
Macrotis lagotis, 699
Macula densa, 1088
Macular degeneration, 382
Madagascan shield bug, 674
Madagascar ocotillo, 1135
Madia sativa, 481, 482
Madreporite, 680, 681
MADS box, 400
"Mafia behavior," 1093, 1102, 1117
Magnesium
in animal nutrition, 1052
in chlorophyll a, 190
in plant nutrition, 741
Magnetic fields
animal navigation and, 1109
paleomagnetic dating and, 508

Magnetic resonance imaging
(MRI), 970
Magnolia, 598
Magnolia, 602
Magnoliids, 602
Maguire, Eleanor, 967
Mahadevan, Lakshminarayanan,
752
Maidenhair tree, 592
See also Ginkgos
Maintenance methylase, 343
Major histocompatibility complex
(MHC) proteins
antigen-presenting function,
872, 873
in the cellular immune
response, 871, 872, 873, 874
functions of, 858–859
organ transplant surgery and,
874
role in T cell selection, 872
Malaria
biological control, 627
causative agent, 553, 564
cause of primary symptoms
in, 563
treatments, 605, 797, 812
Malate, 171, 199
Malate dehydrogenase, 171
Malathion, 157, 940
Malawi, 1240–1241
Malay Archipelago, 1141, 1142
Male flowers, 779, 780
Male reproductive system
components of semen, 889, 890
emission and ejaculation, 891
erectile dysfunction, 891
hormonal control of, 892
penile erection, 890–891
spermatogenesis in, 889–890,
891
Malignant tumors, 227
Maller, James, 209
Malleus, 954, 955
Malnutrition, 1054
Malpighi, Marcello, 734–735
Malpighian tubules, 1076–1077
Maltase, 1062
Malthus, Thomas Robert, 1164
"Malting," 761
Maltose, 53
Mammalian heart
anatomy of, 1030, 1031
blood flow through, 1030–1031
cardiac cycle, 1031–1032
coordination of muscle
contraction, 1034, 1035
electrocardiograms, 1035–1036
pacemaker cells and cardiac
muscle contraction, 1032–
1034
sustained contraction of
ventricular muscles, 1034,
1035, 1036
Mammalian nervous system
brainstem structure and
function, 969–970
forebrain structure and
function, 969–970
functional and anatomical
organization, 968–969
higher functions in cellular
terms, 978–983

information processing by
neural networks, 973–978
spinal cord functions, 969
Mammalian thermostat, 829–830
Mammals
actions of prolactin in, 838
basal metabolic rate and body
size, 826–827
blocks to polyspermy, 886–887
bone growth in, 416–417
circulatory system, 1030
cleavage in, 905–906
convergent molecular evolution
in foregut fermenters,
492–494
defense systems, 857–859 (see
also Immunology)
dissipation of heat with water
and evaporation, 829
distinguishing features, 697
egg-laying, 889
eutherians, 697–705
evolutionary radiation, 696
excretory system, 1079–1086
extraembryonic membranes,
919
gastrulation, 914–915
genomic imprinting during
gamete formation, 344–345
heart function, 1030–1037
heat production in, 827–828
heat stroke, 815
hibernation, 830, 831
Hox genes and body
segmentation, 917
hypothalamus as the
"thermostat" of, 829–830
length of gestation in, 920
lens of the eye, 958–959
major endocrine glands and
hormones, 845–852
major living groups and
number of species, 698
master circadian "clock," 1108
multipotent stem cells, 408
origin of, 692, 693
production of concentrated
urine, 1071, 1077, 1082–1084,
1090
prototherians, 697, 698
range in body size, 696–697
sex determination in, 249–250
teeth, 697, 1055–1056
therians, 697–700
thermoregulation through the
skin, 825
viviparity in, 889
vomeronasal organ, 950, 951
Mammary glands, 697, 843
Mammuthus columbi, 1229
Manatees, 700
Mandibles, 669, 1056
Mandibulates, 667, 669–671
Mandrills, 702
Mandrillus sphinx, 702
Manduca sexta, 802
Manganese
in animal nutrition, 1052
in plant nutrition, 741
Mangold, Hilde, 911, 912
Mangrove forests, 1141
Mangrove island experiment,
1198, 1199
Mangroves, 808, 811, 1141

Manihot esculenta, 708
Manihot glaziovii, 724
Mannose, 52
Mantidflies, 672
Mantids, 672, 673
Mantle, 662, 663, 664
Mantodea, 672
Mantophasmatodea, 672
Manucodes, 480, 481
Manucodia comrii, 480
Manure, 747
MAP kinase, 132
"Map sense," 1109
MAPK (mitogen-activated protein
kinase), 139
Mapping, genetic, 248–249
Maquis, 1134
Marathon runners, 815
Marchantia, 577
Marchantia polymorpha, 577
Mariana Islands, 509
Mariana Trench, 509
Marine animals
larval forms and dispersal, 640
undescribed species of
annelids, 675
Marine biome, 1139–1140
Marine ecosystems, negative
impact of invasive species,
1235
Marine fireworms, 636
Marine flatworms, 1007
Marine iguanas, 824
Marine mollusks, 999
Marion Island, 1199
Marker genes, selectable, 376, 378
Mark–recapture method, 1151
Marler, Catherine, 1103, 1104
Marler, Peter, 1100
Marrella splendens, 517
Mars, 68, 69, 70
Marshall, Barry, 542
Marshes, 1140
Marsilea, 581
Marsupial moles, 698
Marsupials, 697, 698, 699
Marthasterias glacialis, 682
Maryland Mammoth tobacco, 787
Mass, defined, 22
Mass extinctions
Carboniferous, 521
Cretaceous, 521
Devonian, 520
meteorite-caused, 511, 520, 521
periodic nature of, 508
Permian, 522
sea level drops and, 510
Triassic, 521
See also Species extinctions
Mass spectrometry, 369
Mast cells, 858, 861, 875
Mastax, 657, 658
Maternal diet, human birth
defects and, 916
Maternal effect genes, 402, 403
Mating
costs and risks, 882
effect on genotype or allele
frequencies, 434–436
heterozygote advantage and
polymorphic loci, 442–443
impact of signaling systems on
speciation, 705

maximization of the fitness of both partners, 1113–1114
olfactory cues and, 705
types of, 1113
variety in deuterostomes, 705
Mating calls, 476, 477, 705
Mating seasons, 476
Mating types, of fungi, 617
Matter
  atomic structure, 22–25
  in chemical reactions, 31
Matthaei, J. H., 288
Maturation promoting factor, 209, 210
Maturational survivorship curves, 1156
Matz, Mikhail, 449, 464
Maxillipeds, 671
Maximum likelihood methods, 456
Mayflies, 672, 673
Mayr, Ernst, 468–469
McCulloch, Ernest, 408
McFall-Ngai, Margaret, 546
MCIR gene, 367
Mean arterial pressure (MAP), 1043–1044
Measles, 877
Mechanical energy, 145
Mechanical isolation, 476
Mechanical weathering, 745
Mechanical work, 10
Mechanically gated ion channels, 930
Mechanoreceptors
  auditory and vestibular hair cells, 953–954
  in auditory systems, 954–956
  influence on ion channels, 947, 948
  in muscles, tendons, and ligaments, 952–953
  response to physical forces, 952
  tactile, 952
  vestibular system, 956–957
Mechanosensory signals, in animal communication, 1112
Mecoptera, 672
Medawar, Sir Peter, 339
Medicinal leeches, 661–662
Medicinal plants, 604–605
Medicine
  benefits of human genomics, 367–368
  biotechnology and, 384–386
  discovery of penicillin, 608
  importance of biological research to, 15–16
  Koch's postulates, 541–542
  medicinal plants, 604–605
  phage therapy, 545–546
  stem cells and, 77, 392, 408, 409, 410
  use of leeches in, 661–662
  uses of molecular evolution, 501–502
  See also Cancer drugs; Cancer treatment; Molecular medicine
Mediterranean climate, 1121, 1126, 1134, 1146
Mediterranean flour moth, 1184
Medium ground finch, 474
Medium tree finch, 474

Medulla
  in blood pressure regulation, 1044, 1045
  in control of breathing, 1019–1020
  development in the human brain, 968, 969
  sensitivity to the partial pressure of carbon dioxide in blood, 1021
Medusa
  of cnidarians, 645, 646
  of scyphozoans, 647, 648
Megafauna, 1229
Megagametophyte
  in angiosperms, 600, 779, 781
  in seed plants, 590, 591, 592
  in vascular plants, 584
Megakaryocyte, 1039
Megaloptera, 672
Meganeuropsis permiana, 505
Megaphylls, 583, 584
Megasporangia
  in angiosperms, 596–597, 600
  in conifers, 594, 595
  in seed plants, 590, 591, 592
  in vascular plants, 584, 585
Megaspore
  in angiosperms, 600, 779, 781
  in conifers, 595
  in seed plants, 590, 592
  in vascular plants, 584
Megasporocyte, 595, 600
Megastrobilus, 594, 595
Megatypus schucherti, 518
Meiacanthus grammistes, 1174
Meiosis
  chromatid exchanges during, 219–220, 222
  comparison with mitosis, 223
  defined, 207
  errors leading to abnormal chromosome structures and numbers, 222, 224
  final products, 219, 221, 222, 223
  during gametogenesis, 883–884
  length of, 219–220
  nondisjunctions, 309
  overall function of, 219
  reduction of chromosome number in, 219
  segregation of alleles, 237
  separation of homologous chromosomes by independent assortment, 220–222
  in sexual life cycles, 217, 218–219
Meiosis I
  errors leading to abnormal chromosome structures and numbers, 222, 224
  events of, 220–221
  unique features of, 219
Meiosis II
  events of, 219, 220–221
  separation of sister chromatids, 222
Meissner's corpuscles, 952
MEK, 132
Melampsora lini, 800
Melanin, 246
Melanocyte-stimulating hormone (MSH), 842

Melatonin, 842, 851
Membrane-associated carbohydrates, 106, 109, 111
Membrane currents, in rod cells, 961
Membrane lipids, of archaea, 535–536
Membrane potential
  of cardiac pacemaker cells, 1032–1034
  defined, 927
  gated ion channels and, 930–932
  generation of, 928–929, 930
  generation of action potentials and, 932–934
  graded changes, 932
  measuring with electrodes, 928
Membrane proteins
  cell adhesion and, 111
  rapid diffusion of, 109, 110
  types of, 106, 107–109
Membrane receptors, 129
Membrane transport
  active, 118–120
  function of transporters, 42
  mechanisms in, 118
  passive, 113–117
Membranes
  aquaporins and permeability, 116
  of archaea, 535–536
  diffusion across, 114
  dynamic nature of, 109
  factors affecting fluidity, 107
  fluid mosaic model, 106
  fluidity of, 107
  membrane-associated carbohydrates, 109
  osmosis, 114–115
  passive transport, 113–117
  pleural membranes, 1014, 1015–1016
  protocells, 73–74
  significance in the evolution of life, 3–4
  structure of, 106–110
  thickness of, 107
  See also Extraembryonic membranes; Internal membranes; Plasma membranes
Membranous bones, 1001
Memory
  cellular basis of, 982
  defined, 981
  emotional content, 982
  fear memory, 970, 982
  hippocampus and, 967, 982, 983
  human capacity for, 981–982
  sleep and, 980
  types of, 982
Memory cells, 865–866
Menadione, 1053
Mendel, Gregor
  inheritance experiments with garden peas, 233–236
  law of independent assortment, 237–239
  law of segregation, 236–237, 239
  rediscovery of, 428
  test crosses, 237, 238
Menopause, 893
Menstrual cycle, 892, 893–894
  See also Uterine cycle

Menthol, 604
Meristem culture, 793
Meristem identity genes, 786
Meristems
  floral, 400, 715, 785–786
  hierarchy in plant growth and development, 715, 716
  indeterminate primary growth, 715–720
  initials, 715
  lateral, 715, 716
  origin in plant embryogenesis, 712
  in plant development, 710
  secondary, 715, 716
  stem cells in, 408
  See also Apical meristems; Ground meristem; Primary meristems
Merkel's discs, 952
Merops apiaster, 888
Merozoites, 564
Mertensia virginica, 462
Meselson, Matthew, 268, 269, 270
Meselson–Stahl experiment, 268, 269, 270
Mesenchymal stem cells, 392, 394, 408
Mesenchyme
  in acoelomates, 635
  in frog gastrulation, 910
  in sea urchin gastrulation, 908, 909
Mesoderm
  in avian gastrulation, 914
  body cavity types and, 635
  body segmentation in vertebrates, 916–918
  defined, 908
  differentiation of muscle precursor cells, 398–399
  in extraembryonic membranes, 918
  in frog gastrulation, 909, 910
  in protostomes, 652
  in sea urchin gastrulation, 908, 909
  tissues and organs derived from, 907, 908
  in triploblastic animals, 633
Mesoglea, 644, 646, 647
Mesophyll
  in C_3 and C_4 plants, 198, 199
  in eudicot leaves, 720, 721
  response to water stress, 734
Mesozoic era, 506–507, 521, 593
Mesquite, 808
Messenger RNA (mRNA)
  alternative splicing, 346–347
  blocking translation to study gene expression, 382
  cDNA libraries and, 379–380
  codons and the genetic code, 288–289
  DNA microarray technology, 382–383
  location and role in eukaryotic cells, 286
  modification of pre-mRNA, 291–293
  movement out of the nucleus, 293
  nucleic acid hybridization, 290–291

produced by transcription, 286–288
recycling in bacteria, 301
relation to protein abundance, 348
role in gene expression, 285
transcriptional regulation in prokaryotes, 329–330
translation, 293–297, *298*
Metabolic heat, 822, 825
Metabolic inhibitors, 322–323
Metabolic pathways
    allosteric regulation, 160–161
    cellular locations of energy pathways, *168*
    glucose oxidation, 166–169
    governing principles, 166
    linkages between catabolism and anabolism, 179–180
    in prokaryotes, 166, 168
    as regulated systems, 181–182
    systems biology and, 157
    transcriptional regulation in prokaryotes, 329–330
Metabolic pool, 180
Metabolic rate, basal, 826–827
Metabolism
    defined, 145
    in early prokaryotes, 4
    energy transformations, 145–149
    linkage of anabolic and catabolic reactions, 146
    types of, 145–146
Metabolites, 370
Metabolomes, 370
Metabolomics, 370
Metabotropic receptors, 939
Metabotropic sensory receptor proteins, 947, *948*
Metacarpals, 423
Metagenomics, 357–358
Metal ion catalysis, 155
Metamorphosis
    defined, 672
    description of, 639–640
    holometabolous insects, 673
    incomplete, 841
    juvenile hormone and, 841
    in lampreys, 686
    types of, 672
Metanephridia, 1075, *1076*
Metaphase
    comparison between mitosis and meiosis, *223*
    determination of karyotype during, 224, *225*
    in meiosis, *220, 221*
    in mitosis, 212, 214, 215, *217*
Metaphase plate
    in meiosis, *220, 221*
    in mitosis, *215*
Metapopulations, 1161–1162
Metarhizium anisopliae, 627
Metastasis, 227
Meteorites
    meteorite-caused mass extinctions, 511, 520, 521
    origin of life and, 69
Methane
    bacteria in methane cycling, 357
    covalent bonds in, 27
    as a greenhouse gas, 1212

in pogonophoran metabolism, 660, 661
production by methanogens, 536
Methanococcus, 357
Methanogens, 536
Methicillin-resistant *S. aureus* (MRSA), 281, 301
Methionine, *44*, 296
Methotrexate, 158
Methyl bromide, 1198
Methylation
    of cytosine, 310
    of histone, 344
1-Methylcyclopropene, 770
5'-Methylcytosine, 310, 328, 343, 345
Methylglucosinolide, *802*
Methylococcus, 357
Meyer, Axel, 328
Mice
    albumin gene promoter, 335
    "Down syndrome mouse," 924
    embryonic stem cells, 408–409
    exercise-induced irisin production, 834
    homologous recombination and knockout mice, 381–382
    Hox genes, *414*
    immunoglobulin genes, 868
    inversus viscerum mutant, 915
    knockout experiments in behavior, 1096–1097
    vomeronasal organ, 951
Micelles, 1062
Microbial communities, 539, *540*
Microbial eukaryotes, 552
    See also Protists
Microbial rhodopsin, 537
Microbiomes, 539–541
Microbiothere, *698*
Microclimates, 1126
Microevolutionary change, 446
Microfilaments
    asymmetric distribution of cytoplasmic determinants and, 395
    cell movement and, 95, *96*, 98, 99
    contractile ring, 216
    structure and function, 94–95
Microfossils, 74
Microgametophyte
    in angiosperms, 600, 779
    in seed plants, 589–590
    in vascular plants, *584, 585*
    See also Pollen grains
Microglia, 927
Micromolar solutions, 34
Micronucleus, *555*, 562
Micronutrients
    in animal nutrition, 1052, 1053
    in plant nutrition, 741–743
Microorganisms
    commercial production of proteins, 383–384
    digestion of cellulose in herbivores and, 1063–1064
    interference competition, 1183
Microparasites, 1176
Microphylls, 581, 583
Micropogonias undulatus, 1207
Micropterus dolomieu, 1160
Micropyle, *592, 594, 595*

in pogonophoran metabolism, 660, 661
Microraptor gui, 695
MicroRNA (miRNA), *286*, 347, 348, 382
Microscopes, 79, 80–81, 84
Microsporangium
    in angiosperms, 597
    in conifers, *595*
    in seed plants, 590, 591
    in vascular plants, *584, 585*
Microspores
    in angiosperms, *600, 780, 781*
    in conifers, *595*
    in seed plants, 590
    in vascular plants, *584, 585*
Microsporidia, *615, 616*, 617
Microsporocytes, *600, 781*
Microstrobilus, 594, *595*
Microtubule organizing centers, 212
Microtubules
    asymmetric distribution of cytoplasmic determinants and, 395
    cilia and flagella, 96–98
    in the evolution of the eukaryotic cell, *550, 551*
    in plant cell cytokinesis, 216–217
    in rearrangements of egg cytoplasm following fertilization, 903–904
    spindle apparatus, 212, 213–214, *215*
    structure and function, 95–96
Microtus montanus, 125, 1113
Microtus ochrogaster, 125, 1113
Microvilli
    in the intestines, 1057, 1063
    microfilaments in, 95, *96*
    in taste buds, 951
Midbrain, 968, *969*
Middle ear, 954–955, *956*
Middle lamella, 713
Mifepristone (RU-486), *898*
Migration
    adaptive value of, 1126
    of dinosaurs, 21
    fat as stored energy in birds, 1050
    navigation over great distance, 1109–1110
Milk
    breast milk, 843
    of prototherians, 697
Milkweed grasshopper, 2
Milkweeds, 806, 1176, 1200
Miller, Stanley, 70, 71
Millet beer, 624
Millimolar solutions, 34
Millipedes, 669, 670
Mimetica, 1173
Mimicry
    mechanical reproduction isolation and, *476*
    mimicry systems, 1173–1175
Mimiviruses, *543, 544*, 545
Mimulus aurantiacus, 598, 599
Mineral nutrients
    plant requirements, 741
    transport in xylem, 730–732
    uptake by plants, 727–730
Mineralcorticoids, 849
Minty taste, 951
miRNA. *See* MicroRNA

Mirounga angustirostris, 1105
Mismatch repair, 276, 277, 310
Missense mutations, 306, *307*, 311, 312
Missense substitutions, 491
    See also Nonsynonymous substitutions
Missouri saddled darter, *472*
Mistletoebird, 1182
Mistletoes, 752, 1182
Mites, 668, 669
Mitochondria
    absence in microsporidia, 617
    absence in some excavates, 558
    in animal cells, *86*
    in animal fertilization, *885, 886*
    apoptosis and, 226
    β-oxidation in, 179–180
    in developing human sperm, 889, *891*
    endosymbiotic origin in eukaryotic cells, *550, 551*
    energy pathways in, *168*
    in exercised muscle fibers, 997
    generation of heat in brown fat and, 165
    of kinetoplastids, 559
    membrane impermeability to NADH, 178–179
    oxidative phosphorylation and ATP synthesis, 171–176
    in photorespiration, 197, *198*
    in plant cells, *87*
    structure and function, 91–92
    transposition of genes to the nucleus, 366
Mitochondrial DNA (mtDNA)
    origins of, 92
    phylogenetic analyses and, 456
Mitochondrial genes
    inheritance of, 252–253
    mutations, 252
Mitochondrial matrix
    contents of, 92
    energy pathways in, *168*
    pyruvate oxidation in, 170
Mitogen-activated protein kinase (MAPK), 139
Mitogens, 139
Mitosis
    in asexual reproduction, 217
    in the cell cycle, 208
    centromeres and the plane of cell division, 212–213
    chromatin structure, 211–212, *213*
    chromosome separation and movement, 207, 214–216
    comparison with meiosis, *223*
    cytokinesis, 216–217
    determination of karyotype during, 224, *225*
    overview of events and phases, 211, 212, *214–215, 217*
    spindle apparatus formation, 213–214
Mitosomes, 617
Mitotic center, 213
Mitotic spindles, 905
    See also Spindle apparatus
Mitter, Charles, 480
Mnemiopsis, 644
Mobley, Cuttino, 1025

Model organisms
  apoptosis studies in development, 399
  Caenorhabditis elegans, 666
  eukaryotic genome studies, 361–363
  in genetics, 282
  importance in biological research, 14
Moderately repetitive sequences, *355*, 364–366
Modules, developmental, 415–418
Molar solutions, 34
Molars (teeth), 1055
Molds
  in biological control, 626
  bread molds, 282–283, 619
  brown molds, 624
  description of, 622
  green molds, 624, 1169
  water molds, 556, *557*
  See also Slime molds
Mole (chemistry), 33–34
Mole (mammal), *698*
Molecular biology
  "central dogma," 285
  using data in phylogenetic analyses, 456
Molecular chaperones, 51
Molecular clocks
  concept of, 492
  defined, 461
  using to date evolutionary events, 461–462
Molecular cloning
  commercial production of proteins, 383–384
  with recombinant DNA technology, 375–379
  sources of DNA used in, 379–380
Molecular evolution
  comparing genes and proteins through sequence alignment, 486–487
  description of, 486
  detecting positive and purifying selection in the genome, 492–494
  experimental studies, 489–491
  in genome size, 494–496
  neutral theory of, 492
  practical applications, 499–502
  synonymous and nonsynonymous substitutions, 491, 492–494
  using models of sequence evolution to calculate evolutionary divergence, 487–489
Molecular medicine
  approaches to gene mutations, 304
  chronic myelogenous leukemia, 304
  DNA microarray technology, 383
  genetic screening, 320–322
  halplotype mapping, 367–368
  knockout mouse models, 382
  reverse genetics, 318
  RNAi-based therapy, 382
  strategies for treating genetic diseases, 322–325

using genetic markers to find disease-causing genes, 318–319
Molecular mimicry, 876
Molecular weight, 26–27
Molecules
  amphipathic, 57
  chemical bonds, 26–31
  defined, 24
  hydrophilic and hydrophobic, 30
  models for representing, *27*
  molecular weight, 26–27
  octet rule, 25
  See also Macromolecules
Mollusks
  anatomical characteristics, 652, *652*
  body plans, *663*
  cephalopod image-forming eye, *958*, *959*
  chemical defenses, 1173
  larval form, 640
  major body components, 662, *663*
  major groups, *632*, 662, 664
  monoplacophorans, 662
  open circulatory system, 1026, *1027*
  red tides and, 549
  spiral cleavage, 905
  undescribed species, 675
Molothrus ater, 1093, 1233
Molting
  in ecdysozoans, 654
  exoskeletons and, 999
  hormonal control in insects, 839–841
Molybdenum
  impact on nitrogen fixation, 1222–1223
  in plant nutrition, *741*
  in sediments, 69
Momentum, vestibular detection of, 956–957
Monachanthus, 588, 605
Monarch butterfly, *639*, *1174*, 1176
Monilophytes, *574*, 581–582, *582*–583
Monito del monte, 697
Monkeys, 701, *702*
Monoamines, 939–940
Monoclonal antibodies, 871
Monocots
  characteristics of, 601–602, 709–710
  diversity in, 603
  early shoot development, 758
  guard cells, 733
  leaf veins, 720
  root anatomy, *717*
  root systems, 718
  thickening of stems, 723
  vascular bundles, 719
Monoculture, 1202–1203
Monocytes, *858*
Monod, Jacques, 331
Monoecious plants
  defined, 597, 779
  example of, *780*
  strategies for preventing self-pollination, 782
Monoecious species
  defined, 249, 887

hermaphroditism in animals, 887–888
Monogeneans, 657
Monohybrid crosses, 234–236, 239
Monomers, 40
  of carbohydrates, 51
  condensation and hydrolysis reactions, 42
Monomorphic populations, 437
Monophyletic groups, 463
Monoplacophorans, 662
Monosaccharides, 51–53
  membrane-associated, 109
Monosodium glutamate (MSG), 951
Monosomics, 224
Monosynaptic reflexes, 942
Monotremes, *698*, 889
Monozygotic twins, 344, 907
Montane voles, 125, 1113
Montreal Protocol, 311
Moore, Michael, 1103, *1104*
Moose, *1129*
Moraines, 1200
Moray eels, *688*
Morchella esculenta, 622
Morels, 620, 622
Morgan, Thomas Hunt, 247, 248, 251, 430
Morphine, *604*, 940
Morphogenesis
  defined, 393
  in development, 710
  impact of mutations in developmental genes, 418–419
  pattern formation, 399–405
  in plants, 710–712
  processes contributing to, 394
Morphogens
  in fruit fly body segmentation, 401–405
  positional information in development and, 401
Morphological species concept, 468, 469
Morphology
  adaptations to low temperature, 1125
  phylogenetic analyses and, 455
Mortality, 1154
Morus bassanus, *1153*
Mosaic development, 907
Mosaic viruses, 544
Mosquitoes
  biological control, 627
  malaria and, 563, *564*
Mosses
  description of, 577–578
  distinguishing characteristics, *574*
  life cycle, 575–577
  as nonvascular land plants, 573
  in succession on glacial moraines, 1200
Moths
  impact of diet on developmental plasticity, 421, *422*
  in pollination syndromes, 1180–1181
  See also Lepidopterans; *individual moths*
Motile animals, 637

Motor end plate, 936, 937, 989, *990*
Motor neurons
  induction by Sonic hedgehog, 917
  neuromuscular junctions, 936–938
  number of muscle fiber innervated, 994
  in skeletal muscle contraction, 989, *990*
  spinal reflexes, 942–943
Motor programs, 969
Motor proteins
  in cilia and flagella, 97–98
  functions of, 94
  microtubules and, 96
  See also Myosin
Motor units, 989, 995
Mottle viruses, 544
Mount Everest, 1010
Mount Pinatubo, 510
Mountain avens, 1200
Mountain climbers, 1008
Mountain pine beetle, 1202
Mountain zebra, *1144*
Mountains, rain shadows and, 1126–1127
Mourning cloak butterfly, *1130*
Mouth
  chemical digestion in, 1060
  chewing of food, 1059
  in tubular guts, 1056
Mouthparts
  of chelicerates, 668
  of mandibulates, 669
"Movement proteins," 140
MRSA. *See* Methicillin-resistant *S. aureus*
Mucopolysaccharides, 716
Mucosa, 820, 1058
Mucosal epithelium, 1058
Mucus
  of the human respiratory tract, 1013, 1015
  in innate defenses, 859
  of the nasal epithelium, 949
  in seminal fluid, 890
Mucus escalator, 1013, 1015
Muir Glacier, 17
Muller, Hermann Joseph, 441, 470
Müllerian mimicry, *1174*, 1175
Muller's ratchet, 441
Mullis, Kary, 278
Multicellular animals
  internal environment and homeostasis, 816–817
  physiological systems (*see* Physiological systems)
  relationship between cells, tissues, and organs, 817–820
Multicellular organisms
  cell communication in, 139–141
  genome and gene expression, 6
  importance of cell adhesion and cell recognition in, 110–111
  importance of cellular specialization and differentiation, 9
  intercellular communication and the evolution of, 140–141
  types of mutations in, 305
Multicellularity
  appearance in the Proterozoic, 515–516

atmospheric oxygen levels and, 512
evolution in eukaryotes, 552–553
nematode genome, 362, *363*
through geological time, 515–522
Multifactorial phenotypes, 314–315
Multiple alleles, 242, 243
Multiple covalent bonds, 28
Multiple fruits, 601
Multiple genes, coordinated regulation, 336–337
Multiple sclerosis, 926
Multiple substitutions, 488, 489
Multipotency, 394
Multipotent stem cells, 408, 410
Multisubunit allosteric enzymes, 159–160
Murchison Meteorite, 69
Muscari armeniacum, 603
Muscarinic ACh receptors, 940
Muscle
   contraction (*see* Muscle contraction)
   differentiation of precursor cells, 398–399
   exercise-induced irisin production, 834
   integration of anabolism and catabolism during exercise, 180–181
   interactions with exoskeletons, 999
   interactions with hydrostatic skeletons, 999
   jumping ability in animals, 986
   myoglobin, 1017
   types of, 987
   See also Cardiac muscle; Skeletal muscle; Smooth muscle
Muscle contraction
   in cardiac muscle, 991–993, 1032–1034, *1035*
   factors affecting the strength of, 994–995
   impact on venous blood flow, 1041–1042
   in insect flight muscle, 997–998
   sliding filament model, 987–990, *991*
   in smooth muscle, *992, 993*
Muscle fatigue
   effect of enhanced cooling on, 831
   heat and, 998
Muscle fibers
   impact of strength training on, 996
   intermediate, 996
   optimal length for generating maximum tension, 996
   role of muscle fiber types in strength and endurance, 995–996
   in skeletal muscle, 987–988
   sliding filament contractile mechanism, 987–990, *991*
   twitches and tetanus, 994–995
Muscle spindles, 952–953
Muscle tissues
   characteristics of, 817

lactic acid fermentation in, 177
   types of, 817
Muscle tone, 995
Muscular dystrophy, 365, 370
Muscular segmentation, 636
Musculoskeletal systems
   factors affecting muscle performance, 994–998
   interactions of muscles and skeletal systems, 999–1003
   muscle contraction, 987–994
Mushrooms, 609–610, 622, 623, 624
Mussels, 662, 1194
   See also Zebra mussels
Mustard, 724
Mustard oil glycosides, 1176
Mutagens
   defined, 283
   induced mutations, 309
   natural or artificial, 310
   public policy goals regarding, 311
   use of transposons in minimal genome studies, 359, 360
Mutant alleles, 241–242
Mutation rates, 432
Mutations
   base pairs that are "hot spots," 310
   benefits and costs, 310–311
   caused by retroviruses and transposons, 308
   chromosomal, 307–308
   creation of new alleles, 241–242
   defined, 283, 305, 432
   in demonstrating cause and effect, 98
   of DNA, 266
   effect on phenotype, 305–306
   evolution and, 6
   generation of genetic variation, 432
   in genomes, 6
   morphological impact of mutations in developmental genes, 418–419
   Muller's ratchet, 441
   in organelle genes, 252
   point mutations, 306–307
   purging of deleterious mutations, 433, 441
   reversal of, 306
   spontaneous or induced, 308–309
   that lead to human genetic diseases, 311–315
   transposons and, 365
   types and effects of mutagens, 309, 310
   types in multicellular organisms, 305
   using to study gene function, 381
   See also Gene mutations; *specific types of mutations*
Mutualisms
   characteristics of, 1177–1178
   defined, 612, 1170
   food exchange for care or transport, 1178
   food exchange for housing or defense, 1178–1179
   food exchange for seed transport, 1181–1182

pollination syndromes, 1180–1181
   See also Fungal mutualisms; Mycorrhizae
Myanthus, 588, 605
Mycelium, 609, 610
Mycobacterium tuberculosis, 532
Mycologists, 611
Mycoplasma capricolum, 359
Mycoplasma genitalium
   comparative genomics, *357*
   functional genomics, *357*
   genomic information, *361*
   minimal genome studies, 359, 360
Mycoplasma mycoides, 359
Mycoplasma mycoides JCV1-syn.1.0., 359, 360
Mycoplasmas, 531–532
Mycorrhizae
   arbuscular mycorrhizae, 619–620, 749 (*see also* Arbuscular mycorrhizae)
   description of, 614–615
   ectomycorrhizae, 614, 615, 622
   expansion of the plant root system, 748–749
   formation, 747, 748
   in plant evolution, 574
   in reforestation efforts, 626
Myelin, 926
Myelinated axons, conduction of action potentials, 935
Myliobatis australis, 688
Myocardial infarction, 1043
MyoD gene, 399
MyoD transcription factors, 399
Myofibrils, *987, 988*
Myogenic heartbeat, 992
Myoglobin
   binding of oxygen, 1017
   in exercised muscle cells, 996–997
   globin gene family and, 497, *498*
   in slow-twitch fibers, 995
Myosin
   contractile ring, 216
   functions of, 94
   in muscle tissues, 817
   in skeletal muscle contraction, *987, 988–990, 991*
   in smooth muscle contraction, 993
Myosin filaments
   impact of strength training on, 996
   in muscle fiber contraction, *987, 988–990, 991*
Myosin kinase, 993
Myosin phosphatase, 993
Myostatin, 370
Myostatin gene, 370
Myotonic dystrophy, 314
Myriapods, *632, 667*, 669–670, 675
Myrmecocystus mexicanus, 1183
Mysis diluviana, 1228
Mytilus californianus, 1194
Myxamoebas, 561
Myxomyosin, 560
Myxozoans, 646

N

NADH-Q reductase, 172
Naegleria, 558

Naked mole-rats, 1115–1116
Nanaloricus mysticus, 665
Nanoarchaeota, 535, 537
Nanos gene, 402, *403*
Nanos protein, 402, *403*
Narcis, 603
Nasal epithelium, 949, 950
Nasal salt glands, 1073
National Wildlife Federation, 1244
Natural gas deposits, 565
Natural history, 18
Natural killer cells
   function of, *858*
   in innate defenses, 861
   pattern recognition receptors, 860
Natural Resources Defense Council, 17
Natural selection
   categories of, 433
   Darwin's theory of, 6
   direct action on phenotypes, 438
   generation of new phenotypes, 433
   origin of the concept of, 428–430
   possible effects on populations, 439–440
Naupilus, 640
Nautilus, 664
Nautiluses, 662, 664
Navigation, 1109–1110, *1111*
Neanderthals, 7, 367, 434, *703, 704*
Nearctic region, *1142*
Nebela collaris, 560
Neck, evolution in the giraffe, 416–417
Necrosis, 225
Nectar, 599, 1180
"Nectar corridor," 1137
"Nectar thieves," 1180
Negative feedback
   in physiological systems, 816
   regulation of hormone secretion, 844–845
Negative gravitropism, 765
Negative regulation
   defined, 329
   of the *E. coli lac* operon, 330–332, *333*
   in eukaryotes, 333
   in virus reproductive cycles, 340
Negative selection, 863, 872, 876
Negative-sense single-stranded RNA viruses, *543*, 544
Neher, E., 929
Nematodes
   anatomical characteristics, *652*
   description of, 666
   genomic information, *361, 362, 363*
   major subgroups and number of living species, *632*
   predatory fungi, 612, *613*
   undescribed species, 675
   vulval determination, 396, *397*, 398
   See also Caenorhabditis elegans
Nemerteans, 658, *659*
Nemoria arizonaria, 421, 422
Nemtaocysts, 645, *646*
Neognaths, 694–695

Neopterans, 672–673
Neoteny, 692, 704
Neotropical region, *1142*
Nephridiopore, 1075, *1076*
Nephrons
  production of concentrated
    urine in mammals, 1082–1084
  structure and function in
    vertebrates, 1077, 1078–1079
  structure in mammals, *1080,
    1081*–1082
Nephroselmis, 102
Nephrostome, 1075, *1076*
Neptune's grass, *603*
Nernst equation, 929, *930*
Neruoptera, *672*
Nerve cells. *See* Neurons
Nerve cord, 659–660, 683
Nerve deafness, 956
Nerve gases, 157, 940
Nerve nets, 637, 940–941
Nerves, 968
Nervous system development
  initiation in amphibians,
    912–913
  neurulation, 915–916
Nervous systems
  cell types, 925–927
  chemical signaling in, 835, 836
  electric signaling in, 927–936
    (*see also* Action potentials)
  effect of blood calcium levels
    on, 847
  gray matter and white matter,
    941
  interactions with endocrine
    system, 842–845
  neural networks, 940–943
  neurotransmitters, 936–940
  overinhibition, 924, 939, 943
  synapses, 936–940
  types and functions, 637
  See also Mammalian nervous
    system
Nervous tissues, 819–820
Nest parasitism, 1093, 1102, 1117
Net primary productivity (NPP),
  1190, 1208–1210
Neural crest cells, 916
Neural networks
  functional categories of neurons
    in, 940
  information processing in the
    mammalian nervous system,
    973–978
  range of complexity in, 940–941
  spinal reflexes, 941–943
  vertebrate brain, 943
Neural plate, 916
Neural tube, 916, 917, 968
Neuroeconomics, 141
Neurohormones
  defined, 836
  hypothalamic, 843, 844, 849
Neuromuscular junctions
  actions at, 936–938
  in skeletal muscle contraction,
    989, *990*
Neurons
  changes during sleep, 979
  characteristics of, 819–820, 925
  communication via
    neurotransmitters, 836
  components of, *819*

electroencephalograms, 978
functional categories, 940
generated from fibroblasts, 336,
  337, 338
generation and transmission of
  electric signals, 927–936 (*see
  also* Action potentials)
measurement of the resting
  potential, 928
neural networks, 940–943
number in the human brain,
  943
structure and function, 925–926
summation of synaptic input by
  the postsynaptic cell, 938
synapses and
  neurotransmitters, 936–940
varied morphologies of, *925*
Neurospora, 282–283
Neurospora crassa, 624, 626
Neurotoxins, 500
Neurotransmitter receptors
  agonists and antagonists, 940
  ionotropic and metabotropic,
    938–939
  multiple types for each
    neurotransmitter, 940
Neurotransmitters
  of the autonomic nervous
    system, *975*
  clearing from the synapse, 940
  functions of, 836, 926
  multiple receptor types for each
    neurotransmitter, 940
  at neuromuscular junctions,
    936–938
  release from the presynaptic
    membrane, 936, *937*
  response of the postsynaptic
    membrane to, 936–938
  types and properties of,
    939–940
Neurulation
  in amphibians, 915–916
  body segmentation during,
    916–918
Neutral alleles
  accumulation in populations,
    441
  genetic drift, 434
Neutral mutations
  accumulation in populations,
    441
  fixation by genetic drift, 492
Neutral theory, of molecular
  evolution, 492
Neutrons
  defined, 22
  isotopes and, 22–24
  mass number and, 22
Neutrophils, *858, 862*
"New" diseases, 501
New Orleans (LA), 1223–1224
New World monkeys, 701, *702*
New World opossums, 697, *698,
  699*
New Zealand, 1144–1145
Newborns. *See* Infants
"Newly rare" species, 1231
Nexin, 97
NF-κB transcription factor, 860
Niacin, *1053*
Niche, 1184
Nickel, *741, 742, 743*

Nicotinamide, 70
Nicotinamide adenine
    dinucleotide (NAD$^+$/NADH)
  in alcoholic fermentation, 177,
    178
  allosteric regulation of the citric
    acid cycle, 182
  in catalyzed reactions, *156*
  in the citric acid cycle, 170, 171
  in glucose catabolism, 169
  in glycolysis, 169, 170
  impermeability of the
    mitochondrial membrane to,
    178–179
  in oxidation–reduction
    reactions, 167–168
  in pyruvate oxidation, 170
  in the respiratory chain, 172, 173
Nicotinamide adenine
    dinucleotide phosphate
    (NADP$^+$)
  in the Calvin cycle, 194, *196*
  in photosynthesis, 188, 191, 192
Nicotine, 803, 804
Nicotinic ACh receptors, 940
Night blindness, 58, *1053*
Night length
  melatonin release and, 851
  photoperiodic cues in
    flowering, 788
Night vision, 962
"9 + 2" Array, 96, *97*
Nirenberg, Marshall W., 288
Nitrate
  as an electron acceptor in
    denitrifiers, 539
  in the global nitrogen cycle,
    750, 751
  impact on the cycling of
    arsenic, 1221
  inorganic fertilizers and, 747
  leaching in soils, 746
  organic fertilizers and, 746
  production by nitrifiers, 539
Nitrate transporters, 744
Nitric acid, in acid precipitation,
  1219–1220
Nitric oxide (NO)
  hypersensitive response in
    plants and, 800–801
  as a neurotransmitter, 940
  in penile erection, 890, 891
  in plant responses to
    pathogens, 799
  as a second messenger, 135–136
Nitrification, 750, *751*
Nitrifiers
  in the global nitrogen cycle,
    750, *751*
  oxidation of ammonia by, 539
Nitrite
  in the global nitrogen cycle, 751
  as a mutagen, 310
  production by nitrifiers, 539
Nitrobacter, 539
Nitrogen
  in the atmosphere, 1211
  covalent bonding capability, *27*
  electronegativity, *28*
  heavy nitrogen in the
    Meselson–Stahl experiment,
    268, *269*, 270
  octet rule for molecule
    formation, 25

in plant nutrition, *741*
  sources for fungi, 611
Nitrogen-14, 507
Nitrogen cycling
  global nitrogen cycle, 750–751,
    1218–1219
  prokaryotes and, 539
Nitrogen deposition, 1219
Nitrogen dioxide, 1219
Nitrogen excretion
  forms of nitrogenous waste
    excreted, 1074–1075
  in invertebrates, 1075–1077
  See also Excretory systems
Nitrogen fertilizers
  energy costs, 740, 750
  environmental costs, 740
  See also Chemical fertilizers
Nitrogen fixation
  by heterocysts of cyanobacteria,
    532, *533*
  by humans, 1218–1219
  impact of atmospheric carbon
    dioxide on, 1221–1223
  prokaryotes and, 539
  by soil bacteria, 749–750
Nitrogen fixers
  evolution of symbiosis with
    legumes, 521
  formation of root nodules,
    747–748
  nitrogen fixation by, 539,
    749–750
Nitrogen runoff, 740, 1207, 1219
Nitrogen use efficiency, 740, 753
Nitrogenase, 750
Nitrogenous waste
  forms excreted by animals,
    1074–1075
  See also Excretory systems
Nitrosamines, 310
Nitrosococcus, 539
Nitrosomonas, 539
Nitrous acid, 309
Nitrous oxide
  as a greenhouse gas, 1212
  increased atmospheric levels
    from human activity, 1219
NO. *See* Nitric oxide
NO synthase, 136
Noctiluca, 549
Nod factors, 747
Nod genes, 747
Nodal cells, 902, 914–915, 921
Node, of the mammalian embryo,
  902, 914–915, 921
Nodes
  of phylogenetic trees, 450
  of plants, 709
Nodes of Ranvier, 935
Nodule meristem, 747, *748*
Noggin, 916
Noller, Harry, 296
Non-identical twins, 907
Non-native species
  invasives, 1235–1236
  regulating the importation of
    non-native plants, 1241
  risks of deliberate
    introductions, 1228, 1245
  See also Introduced species;
    Invasive species
Non-REM sleep, 978–979

Noncoding DNA sequences
  in eukaryotic genes, 290–291, 361
  genomic information and, 356
  significance in genomes,
    494–496
Noncompetitive inhibitors, of
  enzymes, 158, 159
Noncovalent interactions, in
  protein binding, 50
Noncyclic electron transport,
  191–192
Nondisjunction
  abnormal sex chromosome
    arrangements, 250
  in the formation of
    allopolyploids, 225, 226
  as a mutation, 309
  production of aneuploid cells
    and, 222, 224
Nonpolar covalent bonds, 28
Nonpolar substances, effect on
  protein structure, 50
Nonrandom mating, 434–436
Nonself, distinguishing from self,
  243, 863, 874
Nonsense mutations, 306, 307
Nonshivering heat production,
  827–828
Nonspecific defenses. See Innate
  immunity
Nonsynonymous substitutions,
  491, 492–494
Nonvascular land plants
  defined, 573
  distinguishing characteristics,
    575
  hornworts, 578–579
  liverworts, 577
  members and distinguishing
    characteristics, 574
  mosses, 577–578
  sporophyte and gametophyte
    generations, 575–577
Noradrenaline. See
  Norepinephrine
Noradrenergic neurons, 974, 975
Norepinephrine (noradrenaline)
  actions of, 842, 850
  in the autonomic nervous
    system, 975
  in blood pressure regulation,
    1044, 1045
  effect on heartbeat, 1034
  effects on gut muscle, 993
  endocrine source, 849
  as a neurotransmitter, 940
North Atlantic Drift, 1125
North Pacific Drift, 1125
Nose
  nasal epithelium, 949, 950
  olfaction, 949–950
  vomeronasal organ, 950–951
Nostoc, 749
Nostoc punctiforme, 533
Nothofagus, 1129, 1130, 1143
Notochord
  in chordates, 683–684
  development in amphibians,
    915–916
  dorsal–ventral signaling during
    organogenesis, 917
  evolutionary relationships
    among chordates and, 455
  in tunicates and lancelets, 684

Notoryctemorphia, 698
Notropis, 472
Nottebohm, Fernando, 1101
Noturus, 472
Nuclear bombs, 310
Nuclear envelope
  breakdown in meiosis I, 220
  breakdown in mitosis, 214
  in the evolution of the
    eukaryotic cell, 551
  formation during telophase, 216
  function of, 85
  origins of, 101–102
Nuclear gene sequences,
    phylogenetic analyses and,
    456
Nuclear lamina, 85, 88
Nuclear localization signal (NLS),
  299, 300
Nuclear pores, 85, 88, 293
Nuclear reactors, 310
Nuclear transfer experiments,
  406–407
Nucleases, 1058, 1062
Nucleic acid hybridization, 290–
  291, 321, 398
Nucleic acids
  anabolic interconversions and,
    179, 180
  growth of, 63, 64
  molecular evolution, 486
  monomer components, 40
  in the origin of life, 3
  proportions in living
    organisms, 41
  in protocells, 74
  structure and function, 63–67
  types of, 63
  See also Deoxyribonucleic acid;
    Ribonucleic acid
Nucleoid, 82
Nucleolus, 85, 87
Nucleosides, 63
Nucleosomes, 212, 213
Nucleotide bases
  Chargaff's rule, 264, 265, 266
  in DNA structure, 264–267
  exposure in the grooves of
    DNA, 266, 267
  formed in prebiotic synthesis
    experiments, 70
  induced mutations, 309
  point mutations, 306–307
  spontaneous mutations,
    308–309
Nucleotide sequences
  evolutionary relationships in
    prokaryotes and, 528–529
  identifying evolutionary
    changes in, 486
  phylogenetic analyses and, 456
  point mutations, 306–307
Nucleotide substitutions
  in gene evolution, 486
  synonymous and
    nonsynonymous
    substitutions, 491, 492–494
  using models of to calculate
    evolutionary divergence,
    487–489
Nucleotides
  components of, 63
  in DNA replication, 268–269,
    270

genetic information and, 5, 6
  other functions of, 66–67
  phosphodiester linkages, 63, 64
Nucleus (atomic), 22
Nucleus (cell)
  in animal cells, 86
  in ciliates, 554, 555, 562
  defined, 81
  in eukaryotic cells, 85, 88
  movement of mature mRNA
    out of, 293
  in plant cells, 87
  transposition of organelle genes
    to, 366
Nucleus (group of neurons),
  969–970
Nudibranchs, 662, 664, 1174
Null hypothesis, 13
Nutrient deficiencies, 1053, 1054
  See also Deficiency diseases
Nutrients
  as the basis of cellular
    biosynthesis, 10
  cycling through ecosystems,
    1208, 1214–1223
  importance of viruses to
    nutrient cycling, 546
  as sources of energy, 10
  See also Animal nutrition;
    Macronutrients;
    Micronutrients; Mineral
    nutrients; Plant nutrition
Nutritional categories, of
  prokaryotes, 537–538
Nymphaea, 597
Nymphalis antiopa, 1130

O

O blood group, 243
Ob gene, 1067, 1068
Obelia, 648
Obesity, 1048, 1050–1051, 1068
  human SNP scans, 368
  recent epidemic in, 165
  as a risk factor for
    atherosclerosis, 1043
  single-gene mutations affecting,
    1067, 1068
  UCP1 protein and, 165, 182
Obligate aerobes/anaerobes, 536,
  537
Obligate parasites, 553–554, 611
Oblimersen, 382
"Observational" learning, 982
Observations, in science, 11–12
Occam's razor, 454
Occipital lobe, 973
Ocean currents, 1124–1125
Oceans
  acidification, 647, 1217
  biomass distribution, 1191–1192
  importance of viruses to ocean
    ecology, 546
  iron in sediments, 1221
  marine biome, 1139–1140
  osmolarity, 1073
  transport of elements through
    ecosystem compartments,
    1214
  uptake and release of carbon
    dioxide, 1217
  upwelling zone, 1214
  See also Sea level
Ochre sea star, 1194

Octet rule, 25
Octopus macropus, 664
Octopuses, 662, 664, 959
Odd-toed hoofed mammals, 698
Odonata, 672
Odorant receptors, 138
Odorants, 138, 949, 950
Off-center receptive fields, 975
Oil spills, 569
  bioremediation, 373
Oils, 56–57
Okamoto, Noriko, 102
Okazaki fragments, 273
Olcese, James, 895
Old World monkeys, 701, 702
Olduvai Gorge, 703
Olfaction
  description of, 949–950
  signal transduction in, 137–138
Olfactory bulb, 949–950
Olfactory cilia, 949
Olfactory cues, in mate selection,
  705
Olfactory receptor neurons
  (ORNs), 949–950
Olfactory receptor proteins,
  949–950
Oligochaetes, 661
Oligodendrocytes, 926
Oligonucleotide primers, 380
Oligonucleotide probes, 321
Oligonucleotides, 63, 353, 383
Oligophagous herbivores, 1175,
  1176
Oligosaccharides
  as antigens on red blood cells,
    243
  defined, 51
  in glycoproteins, 109
  glycosidic linkages, 53
Omasum, 1064
Ommatidia, 958
Omnivores
  defined, 1054
  feeding strategy, 638
  in food webs, 1190, 1191
  teeth, 1055
On-center receptive fields, 975
On the Origin of Species (Darwin),
  6, 428, 429, 430, 454, 1122,
  1171–1172, 1183
Onager, 1144
Oncogene proteins, 228
Oncogenes
  causing chronic myelogenous
    leukemia, 304
  DNA methylation and, 344
  gain of function mutations in
    tumor suppressor and, 306
  somatic mutations and cancers,
    314
Oncorhynchus clarkii lewisi, 1228
Oncorhynchus mykiss, 1140
Oncorhynchus nerka, 1228
One-gene, one-enzyme
  hypothesis, 282–283, 284
One-gene, one-polypeptide
  relationship, 283–284
"One-hour midge," 1152
Onychophorans, 632, 667, 668
Onymacris unguicularis, 1132
Oocytes
  in human reproduction, 893, 894
  in oogenesis, 883, 884

Oogenesis, *883*, 884
Oogonia, *883*, 884, 908
Oogpister beetle, 1174
Oomycetes, 556, *557*
Ootid, *883*, 884
Open circulatory systems, 1026, 1027
Open reading frames, 355–356
Operant conditioning, 1094
Operators, 330
Opercular cavity, 1010
Opercular flaps, 1009–1010
Operculum, 687
Operons
  description of, 330, *331*
  negative and positive regulation of, 330–332, *333*
Ophioglossum reticulatum, 218
Ophiopholis aculeata, *682*
Ophrys apifera, *476*
Ophrys insectifera, *1181*
Opisthokonts, 559, 609
Opium poppy, *604*
Opossum shrimp, 1228
Opossums, 697, *698*, *699*, 1173, 1190
"Opportunistic evolution," 753
Opportunity costs, in animal behavior, 1103
Opsins, 460, *957*, 958
Optic chiasm, 977–978
Optic nerve, *962*, 963, 977–978
Optic vesicles, 396
Optical isomers, 41, *43*, 52
Optimal foraging theory, 1104–1105, *1106*
Optix gene, 1175
Oral hormone contraceptives, 896, *898*
Oral side, 680
Oral–aboral body orientation, 680, *681*
Orange roughy, 1235
Orange trees, transgenic, 401
Orangutans, 366, 702
Orbitals, 24–25
Orcas, *1153*
Orchids
  Darwin and, 588, 605
  mechanical isolation, 476
  mycorrhizae and seed germination, 615
  plant–pollinator mutualisms, 1181
  pollinators, 588, 605
  types of flowers, 605
Orcinus orca, *1153*
Orders, 463
Ordovician period, *506–507*, 516, 520
Organ identity genes, 400–401
Organ of Corti, *954*, *955*, 956
Organ systems
  biological hierarchy concept, 9
  defined, 820
  development in animals, 915–918
  development in humans, 920
Organ transplant surgery
  kidney transplants, 1085–1086
  MHC proteins and, 874
Organelle genes, 252–253
Organelles
  defined, 81

endomembrane systems, 88–91
endosymbiotic origin in eukaryotic cells, *550*, 551–552
in energy transformation, 91–93
methods of studying and analyzing, 84, 85
origin of, 101–102
other types, 93–94
transposition of genes to the nucleus, 366
See also individual organelles
Organic fertilizers, 746–747
Organismal trees, 529–530
Organizer, 911–913
Organogenesis
  in animal development, 915–918
  in human development, 920
Organophosphates, 387
Organs
  biological hierarchy concept, 9
  development in animals, 915–918
  development in humans, 920
  left–right asymmetry, 902, 914–915, 921
  organization in systems, 820
  in physiological systems, 817
  tissue composition of, 820
Orgasm, 891
Orgyia antiqua, *1157*
ori. See Origin of replication
Oriental region, 1142
Origin of life
  chemical evolution, 3
  extraterrestrial hypothesis, 69
  importance of water to, 68–69
  prebiotic synthesis experiments, 69–71
  protocell experiments, 73–74
  RNA's catalytic properties and, 71–72, *73*
  spontaneous generation concept, 67–68
  timeline, *74*
Origin of replication (*ori*), 206, 269, 271
Ornithorhynchus anatinus, 697
Orotidine monophosphate decarboxylase, 154
Orthologs, 499–500
Orthonectids, *632*, 648
Orthopterans, *672*, 674
Oryx gazella, *1132*
Oryza sativa, 361, *362*, 605
  See also Rice
Osculum, *633*
Osmoconformers, 1072–1073
Osmolarity, 1072
Osmoreceptors, of the kidney, 1088
Osmoregulators, *1072*, 1073
Osmosis
  effect on cell volume, 1072
  movement of water into root xylem, 730
  osmolarity and, 1072
  water potential and, 727
Osmotic pressure, of blood in capillary beds, 1041
Osprey, *1140*
Ossicles, *954*, *955*, 956
Ossification, 1001
Osteoarthritis, 410

Osteoblasts, 847, 1000
Osteoclasts, 847, 1000
Osteocytes, 1000
Osteoporosis, 1000–1001
Ostia, 1026
Ostracods, 670
Ostrander, Elaine, 352
Ostrich, 694, *695*, 696
Otoliths, *956*, 957
Otters, 700
Ottoia, 517
Ouachita Mountains, 472
Outer ear, 954
Outer membrane, of prokaryotes, 83
Outgroup, 453
Ova
  fertilization in animals, 884–887 (*see also* Sperm–egg interactions)
  production in animals, 882–883, 884
  See also Eggs
Oval window, *954*, 955
Ovarian cycle
  defined, 892
  description of, 893, *894*
  hormonal regulation of, 894–895
Ovaries (in animals)
  hormones of, *842*
  production of eggs in, 882–883, 884
  structure and function in humans, 892, *893*, *894*
Ovaries (in plants)
  angiosperms, 597, *600*
  development into fruit, 784
  evolution in flowers, 598
  seed plants, *591*
Ovary wall, in angiosperms, 784
Overdominance hypothesis, of heterosis, 245
Overharvesting
  of fishes, 16–17, 1163–1164, 1235
  species extinctions and, 1234–1235
Overtopping growth, 583
Oviducts
  ectopic pregnancy, 906
  in humans, 892, *893*
Oviparity, 889
Ovoviviparity, 889
Ovulation, in humans, 892, 893, *894*, 895–896
Ovule
  in angiosperms, 597, *600*, 601
  in conifers, 594, *595*
  in seed plants, 590, *591*, *592*
Oxaloacetate, 170, 171, 180, 198, *199*
Oxidation
  beta-oxidation, 179–180
  defined, 167
  of glucose, 166–169
  of pyruvate, 170, 171
Oxidation–reduction reactions
  coenzyme NAD$^+$ in, 167–168
  glucose oxidation, 166–169
  in glycolysis, 170
  importance of enzymes to, 155
  in photosynthesis, 187
  transfer of electrons during, 167
Oxidative muscle, 995

Oxidative phosphorylation, 171–176
Oxidative system, in skeletal muscle, 997
Oxycomanthus bennetti, *682*
Oxygen
  as an oxidizing agent, 167
  atomic number, 22
  binding to hemoglobin, 1016–1017
  binding to myoglobin, 1017
  blood transport of, 1016–1018
  chemical reaction with propane, 31
  covalent bonding capability, 27
  diffusion in water, 1007
  in early atmosphere, 69
  effect of altitude, 1007–1008
  electronegativity, *28*
  factors affecting hemoglobin's affinity for, 1017–1018
  inhibition of nitrogenase, 750
  isotopes, 21
  mass number, 22
  in photorespiration, 197–198
  production during photosynthesis, 186–187
  See also Atmospheric oxygen
Oxygenic photosynthesis, 186–188
  See also Photosynthesis
Oxytocin
  actions of, *842*, 843
  bonding behavior in voles and, 125
  in labor and childbirth, 896, 897
  sensitive period in parent–offspring recognition and, 1099
Oxytocin receptors, 125, 129
Oysters, 662, 887
Ozark madtom, 472
Ozark minnow, 472
Ozark Mountains, 472
Ozone layer, 5, 311, 1211–1212

P

P orbital, 24
P ring, 83
p21 gene, 399
p21 protein, 210–211, 399
p53 gene, *314*
p53 protein, 349
p53 transcription factor, 228
"P$_{680}$" chlorophyll, 191, *192*
"P$_{700}$" chlorophyll, 191, *192*
Pääbo, Svante, 367
Pace, Norman, 357–358
Pacemaker cells
  autonomic nervous system and, 975
  in blood pressure regulation, 1044
  in cardiac muscle contraction, 991–992, 1032–1034, *1035*
Pacific barrel sponges, *643*
Pacific yew, *604*
Pacinian corpuscles, 952
Paclitaxel, 228, *229*
PAH gene, 311
Paine, Robert, 1194
Pair bonding, 843
Pair rule genes, 403, *404*
Pairwise sequence comparison, *487*

Palaeognaths, 694
Palearctic region, *1142*
Palenque National Park, 1162
Paleomagnetic dating, 508
Paleontology
    dating fossils and rocks,
        506–508
    phylogenetic analyses and, 455
    See also Fossil record
Paleozoic era, *506–507*, 579
Palindromic DNA sequences, 374
Palisade mesophyll, 720, *721*
Palmitic acid, *57*, *1051*
Palms, *603*, 723
PAMP-triggered immunity (PTI),
    799–800
Pampas, 1131
PAMPs, 799
Pan troglodytes, 702
Pancreas
    blood glucose regulation,
        848–849
    hormonal control of, 1065
    hormones produced by, *842*
    in humans, *1058*
    procarboxypeptidase A of
        exocrine cells, 333
    role in digestion, 1062
    secretion of insulin, 1066
Pancreatic amylase, *1062*
Pancreatic duct, 1061, 1062
Pancreatitis, 1062
Pandanus trees, 718
Pandemics, 427
Pandion haliaetus, 1140
Pandorina, 140, 141
Pangaea, 520, 521, 1143
Pangolins, *698*
Panther groupers, *688*
Panthera leo, 1183
Panthera tigris, *1172*, 1235
Panthera tigris tigris, *1136*
Panting, 829
Pantothenic acid, 70, *1053*
    See also Vitamin B$_6$
Paper wasps, 674
Papillomaviruses, 341
Papio, 1135
Parabasalids, 558
Parabronchi, 1011, *1012*
Paracrine signals, 126, 835
Paradisaea minor, 480
Paragordius tricuspidatus, 667
Parallel evolution, 423–424
Parallel substitutions, 488
Paralogs, 499–500
Paralysis, 970
    during REM sleep, 979
Paralytic shellfish poisoning, 549
Paramecium, 554, *555*, 562
Paraminohippuric acid (PAH),
    1079
Paranthropus aethiopicus, 703
Paranthropus boisei, 703, 1056
Paranthropus robustus, 703
Paraphyletic groups, 463
Parapodia, 660
Parasite–host interactions,
    pathogenic, 1176–1177
Parasites
    ampicomplexans, 563
    chlamydias, 533, *534*
    complex life cycles, 563, *564*
    defined, 637

feeding strategy, 638–639
flatworms, 657
fungi, 611–612
kinetoplastids, 559
life cycles, 640, *641*
malaria and, 563, *564*
myxozoans, 646
nanoarchaeota, 537
nematodes, 666
orthonectids, 648
parasitic fungi, 611–612, 617
parasitic plants, 752–753, 753
rhombozoans, 648
Parasitism
    defined, 1170
    pathogenic parasite–host
        interactions, 1176–1177
    in protostome evolution, 674
Parasitoid wasps
    in indirect competition, 1184
    in sweet potato–corn dicultures,
        1203
"Parasol effect," 510
Parasympathetic division
    effect on heartbeat, 1034
    influence on smooth muscle,
        993
    structure and function, 974–975
Parathormone, 848
Parathyroid glands, *842*, 847–848
Parathyroid hormone (PTH), *842*,
    847–848
Parenchyma, 713, 730
Parent rock, 745
Parental care, in frogs, 678
Parental generation (P), 234
Parent–offspring recognition, 1099
Parietal cells, 1060, 1061
Parietal lobe, 972–973
Parkinson's disease, 323–324
Parmotrema, 613
Parotid salivary gland, *1058*
Parsimony principle, 454, 1144–
    1145
Parthenocarpy, 766
Parthenogenesis, 881–882
Partial pressure gradients,
    maximization in respiratory
    gas exchange, 1009
Partial pressure of carbon dioxide
    blood transport of carbon
        dioxide and, 1018
    regulation of breathing, 1020–
        1021
    respiratory gas exchange and,
        1008
Partial pressure of oxygen
    altitude and, 1008
    binding of oxygen to
        hemoglobin and, 1016–1017
    binding of oxygen to
        myoglobin and, 1017
    detection of blood levels by
        aortic and carotid bodies,
        1021–1022
    diffusion of oxygen and, 1006
    gas exchange in birds and, 1012
    gas exchange in fish gills and,
        1010
    in lungs with tidal ventilation,
        1013
Partial pressures, of gases, 1006
Particulate inheritance, 233–234,
    236

Passenger pigeon, 1230
Passeriform birds, *696*
Passiflora, 2, *2*, *603*, *1174*, 1175, 1176
Passionflower, 2, *603*, *1174*, 1175,
    1176
Passive transport (diffusion),
    113–117
    See also Diffusion; Facilitated
        diffusion
Pasteur, Louis, 68, 259
Patch clamping, 929, *931*
Patents, 373
Pathogen associated molecular
    patterns (PAMPs), 799, 860,
    861
Pathogenesis-related (PR) genes,
    799, 800
Pathogenic fungi, 612
Pathogens
    animal defense systems and,
        857
    bacterial, 357, 541–542
    effect on population growth,
        1159
    macroparasites, 1176–1177
    microparasites, 1176
    negative impact as invasives,
        1236
Pattern formation
    apoptosis in, 399–400
    body segmentation in
        Drosophila melanogaster,
        401–405
    defined, 399
    morphogen gradients and
        positional information, 401
    plant organ identity genes,
        400–401
Pattern recognition receptors
    (PRRs), 799, 860
Paucituberculata, *698*
Pauling, Linus, 461
Pavlov, Ivan, 838, 982, 1094
Pax6 gene, 413, *414*
pBR322 plasmid, 377
Pdm gene, 675
Peanut butter, 624
Pears, 601, 714
Peas
    loss of function mutations, 305
    Mendel's experiments with,
        233–236
    as a model organism, 282
Peat, 578, 611
Peatlands, 578
Pectin, 713
Pectinase, 162
Pectoral fins, 686–687, *688*
Pedigrees, 240–241
Peforin, *873*, 874
Pegea, 684
Pelagic zone, 1139–1140
Pelecanus conspicillatus, 412
Pellagra, *1053*
Pellicle, 554, *555*
Pelvic fins, 686–687, *688*
Pelvic spines, 424
Pelvis, 1082
Penetrance, 245
Penicillin, 527, 608
Penicillin resistance, 626
Penicillium, 383, 608, 622, 624, 626
Penis
    in birds, 888

erection, 136, 890–891
role in internal fertilization, 887
Pentaradial symmetry, 680, *681*
Pentoses, 52, 63
PEP carboxylase, 199, 200
Pepsin, 1060, 1061, *1062*
Pepsinogen, *1060*, 1061
Peptidases, 300, 1058
Peptide bonds. *See* Peptide
    linkages
Peptide hormones, 836
Peptide linkages, 43–44, *45*, 296,
    297
Peptide neurotransmitters, 936,
    940
Peptidoglycans, 527, *528*, 551, *552*,
    571
Peptidyl transferase, 296, *297*
Peptidyl tRNA binding site, 295,
    296
Per capita growth rate, 1156
Per genes, 1096, 1108
Peramelemorphia, *698*
Perching birds, *696*
Pereiopods, *671*
Pereira, Andrew, 726
Perennials, 785
Perfect flowers, 597, 598, 779, *780*
Perfusion
    in the human lung, 1016
    partial pressure gradients, 1009
Perianth, 597
Periarbuscular membrane, 747
Pericycle, 717
Periderm, 712, 722
Peridinium, 554
Period, of biological cycles, 774
Periodic table, 22, *23*
Peripheral membrane proteins,
    106, 107, 108
Peripheral nerves, 968
Peripheral nervous system (PNS)
    afferent and efferent portion,
        968
    brainstem components, 969
    components of, 941
Periplasmic space, 528
Perissodactyla, *698*
Perissodus microlepis, 441–442
Peristalsis, 1059–1060
Peritoneum, 635, 1059
Peritubular capillaries, 1078, *1080*,
    1082, 1084
Periwinkle, *604*
Permafrost, 1128
Permian period
    atmospheric oxygen levels,
        512–513, 522
    changes on Earth and major
        events in life, *506–507*
    characteristics of life during,
        518, 520–521
    fungi in, 611
    gigantic insects in, 505
    gymnosperms in, 593
    mass extinction, 522
    vascular plants in, 580
    volcanic activity during, 511
Pernicious anemia, *1053*, 1054
Peroxidase, 443
Peroxides, 93
Peroxisome disorders, 93
Peroxisomes
    in animal cells, *86*

conversion of peroxide to water, 176
in photorespiration, 197, *198*
in plant cells, *87*
structure and function, 93
Personal genomics, 368
Personality, the frontal lobe and, 971–972
Petals, *591, 597*
Petroleum deposits, 565, 569
Petroscirtes breviceps, 1174
Pets
cloning, 407
See also International animal trade
Pfiesteria piscicida, 553
pH
buffers, 35–36
concept of, 35
effect on enzymes, 161
effect on protein structure, 50
generation of low pH in the stomach, 1060–1061
optimal soil pH for plants, 746
regulation by the kidneys, 1084–1085
See also Blood pH
Phaeolus schweinitzii, 1202
Phage therapy, 545–546
Phagocytes
defensins, 859
functions of, 90, *858*
in inflammation, 861
in innate defenses, 860–861
types of, *858*
Phagocytosis
defined, 120, *122*
in the evolution of the eukaryotic cell, 551
lysosomes and, 90, *91*
Phagosomes, 90, *91*
Phalanges, 423
Phalaris canariensis, 762, 763, 764
Phallus indusiatus, 2
Phanerozoic eon, 508
Pharmaceuticals
medicinal plants and, 604–605
pharming, 385–386
See also Drugs
Pharmacogenomics, 368
Pharming, 385–386
Pharyngeal arches, 684
Pharyngeal basket, 684
Pharyngeal slits, 684
Pharyngotympanic tube, 954
Pharynx
in ecdysozoans, 655
in humans, 1013, *1014,* 1059
in tunicates and lancelets, 684
Phase-contrast microscopy, 80
Phasmida, *672*
Phelloderm, 722
Phenolics, *802*
Phenotype
action of natural selection on, 438
defined, 431
DNA structure and, 266
effects of mutations on, 305–306
epistasis and, 244
fitness and, 438
gene–environment interactions and, 245–246

generation of new phenotypes through selection, 432–433
genes and, 237
incomplete dominance and, 242–243
in Mendelian genetics, 236
modifying in the treatment of genetic diseases, 322–323
multifactorial, 314–315
pleiotropy, 243–244
qualitative and quantitative variation, 246
relationship of genotype to, 431
Phenotypic plasticity, 420–422
Phenotypic sex determination, 850, *851*
Phenotypic variation, genetic variation and, 431
Phenyl butyrate, 899
Phenylalanine
in phenylketonuria, 311, 320, 322, 1051
structure, *44*
Phenylalanine hydroxylase (PAH), 311, 320, 322
Phenylketonuria (PKU)
genetic screening for, 320
knockout mouse model, 382
mutations causing, 311, *312*
pleiotropy in, 243–244
prevalence, *312*
treatment for, 322, 1051
Phenylpyruvic acid, 311
Pheromones
animal communication and, 1110–1111
detection of, 950
Philodina, 658
Philodina roseola, 658
Phlebopteris smithii, 518
Phloem
in angiosperms, 596
bulk flow, 728
function of, 579
in leaves, *721*
in roots, 717
secondary, 721, 722, *723*
in shoots, 719
structure and function, 714–715
translocation in, 734–738
Phloem sap, 734–738
Phlox cuspidata, 478, 479
Phlox drummondii, 470–471, 478, 479
Phoenicopterus ruber, 638
Phoenix dactylifera, 603
"Phoenix" virus, 545
Pholidota, *698*
Phoronids, *632,* 652, 658, 659, *660*
Phoronis australis, 660
Phosphatases, 837
Phosphate-based detergents, 1221
Phosphate group
of chemically modified carbohydrates, 55
free energy released by ATP and, 150
in nucleotides, 63
properties of, *40*
protein phosphorylation, *300,* 301
Phosphate ions
hormonal regulation of blood phosphate levels, 848
inorganic fertilizers, 747

Phosphate transporters, 744
Phosphatidylcholine, *58*
Phosphatidylinositol bisphosphate (PIP$_2$), 134, *135*
Phosphodiester linkages
formation during DNA replication, 269, *270,* 273
between nucleotides, 63, *64*
Phosphodiesterase (PDE), 137, 961
Phosphoenolpyruvate (PEP), 199
Phosphofructokinase, 182
Phosphoglucose isomerase (PGI), 442–443
3-Phosphoglycerate (3PG), 170, 194, 197, *198*
Phosphoglycerate kinase, 170
Phosphoglycolate, 197, *198*
Phospholipases, 134, 135
Phospholipid bilayers
simple diffusion across, 114
structure, 57, *58*
See also Membranes
Phospholipids
in the absorption of fats in the small intestine, *1062, 1063*
in biological membranes, 106–107
fatty acid chain characteristics, 107
hydrophilic and hydrophobic regions, 106
lipid-derived second messengers, 134–135
structure and function, 57, *58*
Phosphoric acid anhydride bond, 150
Phosphorus
in animal nutrition, *1052*
covalent bonding capability, 27
ecological impact of soil accumulation and runoff, 1207, 1220–1221
electronegativity, *28*
global cycle, 1220–1221
in plant nutrition, *741*
radioactive isotope in the Hershey–Chase experiment, 262–263
Phosphorylase kinase, 138
Phosphorylation
of histones, 344
oxidative, 171–176
photophosphorylation, 192–193
of proteins, 209, *300,* 301
reversible, 161
substrate-level, 170
Photic zone, 1139, 1140
Photoautrophs, 538
Photochemistry, 188–189
Photoheterotrophs, 538
Photomorphogenesis, *772*
Photons
absorption by pigments, 188–189
defined, 188
Photoperiod
as an indicator of seasonal change, 1108
flowering and, 787–788
melatonin and, 851
Photophosphorylation, 192–193
Photoreceptor cells (animal)
function of, 957

influence on ion channels, 947, *948*
in ommatidia, 958
receptive fields of ganglion cells and, 963, 975–977
of the retina, 959
rhodopsin and the response to light, 957–958
rod cells and cone cells, 960–962, 963
Photoreceptors (plant), 759, 771–775
Photorespiration, 197–198
Photosensitivity, of visual pigments, 957–958
Photosynthates, phloem translocation, 734–738
Photosynthesis
action spectrum, 189, 190
atmospheric oxygen and, 4–5
in C$_3$ and C$_4$ plants, 198–200
in CAM plants, 200
conversion of light energy into chemical energy, 188–193
in cyanobacteria, 532, 538
defined, 186
effect of increasing levels of atmospheric carbon dioxide on, 185, 202
evolution in plants, 570–573
in the evolution of life, 4–5
general equation for, 186
global consumption of carbon dioxide, 1217
impact on atmospheric oxygen levels, 511–513
interactions with other pathways, 200–202
overview of pathways in, 188
photophosphorylation and ATP synthesis, 192–193
photorespiration and, 197–200
photosynthetic efficiency, 201–202
source of oxygen produced by, 186–187
synthesis of carbohydrates, 193–197
Photosynthetic autotrophs, 1189–1190
Photosynthetic bacteria
impact on atmospheric oxygen levels, 511–512
stromatolites, 512, *513*
Photosynthetic endosymbionts, 553
Photosynthetic lamellae, 532
Photosynthetic pigments
in red algae, 571, 572
See also Carotenoids; Chlorophyll
Photosynthetic protists, 252
Photosystem I, 191, 192, *193*
Photosystem II, 191–192, *193*
Photosystems
description of, 191–192
organization of, 190
in photophosphorylation, *193*
Phototropins, 771–772
Phototropism
action spectrum, 771, *772*
coleoptile experiments and the role of auxin in, 762, *763,* 764, *765*

Phthiraptera, *672*
Phycobilins, 190
Phycocyanin, 571
Phycoerythrin, 571, 572
Phyla, 463
Phyllactinia guttata, 612
PhyloCode, 464
Phylogenetic trees
  area phylogenies and, 1143, *1144*
  evolutionary history and, 6–9, 451
  gene trees and, 497, 499
  how to read, 450–451
  identifying neutral, purifying, or positive selection in, 492
  lateral gene transfer events and reticulations, 496
  maximum likelihood methods, 456
  methods in construction, 452–458
  monophyletic groups, 463
  parsimony principle, 454
  sources of data for, 454–456
  testing the accuracy of, 456, 457–458
  uses and components of, 450, 458–462
Phylogeny
  defined, 449, 450
  evolutionary history and, 451
  evolutionary perspective on comparing species, 451–452
  evolutionary possibilities of traits, 452
  homologous features, 452
  maximum likelihood methods, 456
  parsimony principle, 454
  relationship to biological classification, 462–464
  sources of data for, 454–456
  testing the accuracy of, 456, 457–458
  using to reconstruct protein sequences from extinct organisms, 464
Phylogeography, 1143, *1144*
Phymateus morbillosus, 2, 674
Physalia physalis, 646
Physiological survivorship curves, 1155
Physiological systems
  effect of temperature on, 820–821
  heat stroke in mammals, 815
  maintenance of the internal environment, 816
  regulation of homeostasis, 816–817
  relationship between cells, tissues, and organs, 817–820
  thermoregulation (*see* Thermoregulation)
Phytoalexins, 799–800
Phytochromes
  entrainment of circadian rhythms in plants, 774–775
  nuclear localization sequence and protein kinase domain, 774
  in photoperiodic control of flowering, 788

properties of, 772–773
  stimulation of gene transcription, 773–774
Phytomers, 709, 719
Phytomining, 812
Phytoplankton
  biomass distribution in open oceans and, 1191–1192
  during the Mesozoic, 521
  petroleum deposits and, 569
  as primary producers, 563
Phytoremediation, 811, *812*
Pierid butterflies, *1106*
Pigeons
  artificial selection, 6
  homing, 1109
  time-compensated solar compass in, *1111*
Pigmented epithelium, 959
Pigments
  absorption of photons, 188–189
  absorption spectrum, 189
  accessory pigments, 190, 571
  defined, 189
  photochemical changes, 190–191
  in plant evolution, 574
  in red algae, 571, 572
  in vacuoles, 93
  visual pigments, 957–958
  See also Carotenoids; Chlorophyll; Skin pigmentation
Pigs, allantoic sac, 919
Pigweed, 1200
Pijio tree, *1136*
Pili, *82, 84*
Pilobolus, 619
Pilosa, 698
Piloting, 1109
Pima peoples, 1048
Pinaroloxias inornata, 474
Pincushion protea, *1134*
Pine bark beetles, 1217–1218
Pineal gland, *842,* 851
Pineapples, *604*
Pines
  cones, 594, *595*
  fire adaptations, 594, *596*
  life cycle, 594, *595*
  pine bark beetle infestations and climate change, 1217–1218
  response to rapid climate change, 1236
Pinnae, 954
Pinocytosis, 120, *122*
Pinus, 592
Pinus contorta, 594, 1201–1202, *1236*
Pinus longaeva, 593
Pinus ponderosa, 443, 1241
Pioneer Hi-Bred seed company, 753
Pioneer species, 793, 1200, 1201
Piophila casei, 1188
Pisaster ochraceus, 1194
Pisolithus tinctorius, 614
Pistils, 597, 598
Pisum sativum, 282
  See also Peas
Pit organs, 946
Pitcher plants, 751, 1197
Pith, 718, 719, 720

Pith rays, 719, 722
Pits, in xylem, 714
Pituitary gland, 842–845
  See also Anterior pituitary; Posterior pituitary
Pitx1 gene, 424
Pivot, 1002
"Place cells," 967
Placenta
  in childbirth, 897
  development in mammals, 906, 907
  functions of, 697, 889, 892
  fusion of cells in the outer layer, 545
  hormones produced by, 896
  origin and development, 919
Placental mammals
  characteristics of, 697–700
  cleavage in, 905–906
  extraembryonic membranes, 919
Placoderms, *687*
Placozoans
  in animal phylogeny, *630*
  asymmetry in, 634
  description of, 645
  major subgroups and number of living species, *632*
  structural simplicity of, 629, 648
  Trichoplax, 629
Plaice, 444
Plains zebra, *1144*
Plan B®, *898*
Planaria, 1075
Plankton, 515
Plant biotechnology, 386–388, *389*
Plant cell walls, *87*
  auxin-induced cell expansion and, 766–767
  in plant development, 710–711
  plasmodesmata, 100
  primary and secondary, 710–711
  responses to pathogenic invasions, 798
  structure and function, 99–100
Plant cells
  apoptosis, 226
  auxin-induced expansion and, 766–767
  cold-hardening and, 810
  communication through plasmodesmata, *139,* 140
  cytokinesis, 216–217, 710
  expansion, 710
  ice crystals and, 810
  methods of transformation, 376, 377
  microtubule organizing centers, 212
  osmosis, 114–115
  structure, *87*
  totipotency and cloning, 405–406
  turgor pressure, 114–115, 727
  vacuoles, 93
  See also Eukaryotic cells
Plant chemical defenses
  artemisinin, 797, 812
  constitutive, 798
  lignin, 798
  plant self-protection from, 805
  secondary metabolites, 802–803, 804

Plant defenses
  defensins, 859
  against herbivores, 801–806, 1175
  to pathogens, 798–801
  See also Constitutive plant defenses; Induced plant defenses; Specific plant immunity
Plant diseases
  bacterial, 534
  club fungi, 622
  molds, 622
  pathogenic fungi, 612
  powdery mildews, 622
  viral agents, 544
Plant genomes
  features of, 362
  key parameters, *361*
Plant growth and development
  developmental plasticity in response to light, 422
  effect of increasing levels of atmospheric carbon dioxide on, 185
  effects of gibberellins on, 760–762
  embryogenesis, 711–712
  leaf development, 719, 720
  organ identity genes, 400–401
  overview, 393
  primary and secondary growth defined, 715
  primary indeterminate growth, 715–720
  processes in, 710
  properties affecting, 710–711
  role of auxin in, 762–767
  secondary growth, 721–723
  seed germination and seedling growth, 757–758
Plant growth regulation
  auxin and, 762–767
  brassinosteroids and, 771
  cytokinins and, 768–769
  ethylene and, 769–771
  genetic screens and the identification of signal transduction pathways, 759, 760
  gibberellins and, 760–762
  hormones and photoreceptors in, 758–759, 771–775
  key factors in, 757
Plant hormones
  auxins, 760, 762–767
  brassinosteroids, 771
  compared to animal hormones, 758
  cytokinins, 768–769
  ethylene, 769–771
  gibberellins, 760–762, 767
  in plant defenses against herbivory, 804, *805*
  in plant growth and development, 758–759
  production in response to pathogens, 799
  structures of, *759*
"Plant immune system," 799–801
"Plant kingdom," 573
Plant mutualisms
  food exchange for housing or defense, 1178–1179

food exchange for seed transport, 1181–1182
pollination syndromes, 1180–1181
See also Mycorrhizae

Plant nutrition
carnivorous plants, 751–752
deficiency symptoms, 742
essential macronutrients and micronutrients, 741–743
hydroponic experiments, 742–743
impact of soil on nutrient availability, 744–747
improving nitrogen use efficiency, 740
influence of fungi and bacteria on root uptake of nutrients, 747–751
parasitic plants, 752–753
plant acquisition of nutrients, 743–744
plant regulation of nutrient uptake and assimilation, 744

Plant pathogens
plant defenses against, 798–801
plant–pathogen signaling, *799*

Plant physiology, 708

Plant reproduction
asexual reproduction in angiosperms, 792–794
sexual reproduction in angiosperms, 779–785

Plant signal transduction pathways
activation in response to pathogens, 799, 800
ethylene pathway, 770–771
involving auxins and gibberellins, 767
involving cytokinins, 768–769
in plant defenses against herbivory, 804, *805*

Plant tissue culture, 768

Plant tissue systems
dermal tissue, 712–713
ground tissue, *712*, 713–714
primary meristems in the origin of, 716
vascular tissue, *712*, 714–715

Plantae, 570–571

Plantlets, 793

Plant–pollinator relationship
coevolution in, 588, 598–599, 605
pollination syndromes, 1180–1181

Plants
adaptations (*see* Adaptations in plants)
aerenchyma, 808
agricultural applications of biotechnology, 386–388, *389*
aspects of light responsive to, 771
"bolting," 761
carnivorous, 751–752
challenges of saline environments, 810–811
cloning, 405–406
colonization of the land, 574–579
cuticle, 732
cytoplasmic inheritance, 252

development in (*see* Plant growth and development)
effect of pollination strategies on speciation rates, 480
effects of domestication on, 723–724
endophytic fungi and, 615
entrainment of circadian rhythms, 774–775
environmental stresses on, *806*
evolution of, *570*
evolution of photosynthesis in, 570–573
global nitrogen cycle, 750–751
heavy metal tolerance, 811–812
hypersensitive response, 226
impact of soil structure on, 744–747
improving nitrogen use efficiency, 740
interactions with the external environment, 9
lateral gene transfer, 496
metabolic interactions involving photosynthesis, 200–202
metabolomes, 370
morphogenesis, 710–712
mycorrhizae and, 614–615
nitrogen composition, 751
nutrients and nutrition (*see* Plant nutrition)
optimal soil pH, 746
organ identity genes, 400–401
parasitic, 752–753
partial reproductive isolation in, 470–471
photomorphogenesis, 772
photosynthetic efficiency, 201–202
phytomining, 812
phytoremediation, 811, *812*
plastid gene mutations, 252
plastid structure and function, 92–93
polyploidy and agriculture, 225, *226*
postzygotic isolating mechanisms, 478, *479*
prezygotic isolating mechanisms, 476, *477*
reciprocal interactions with herbivores, 1175–1176
responses to drought and water stress, 734, 809
root and shoot systems, 709
self-incompatibility studies, 380–381
self-protection from chemical defenses, 805
starches and starch grains, 53
stem cells, 408
stomatal control of water loss and carbon dioxide uptake, 732–734
stress response, 337
sympatric speciation through polyploidy, 475
translocation in phloem, 734–738
transport systems for ions, 744
triglycerides of, 57
uptake of water and solutes, 727–730

water-use efficiency, 726
waxes, 59
xylem transport of water and minerals, 730–732
See also Angiosperms; Crop plants; Gymnosperms; Land plants; Nonvascular land plants; Seed plants; Vascular plants

Planula
of cnidarians, 645, *646*
of scyphozoans, 647, *648*

Plaque, in atherosclerosis, 1042

Plasma. *See* Blood plasma

Plasma cells
development of B cells into, 867
function of, 865
in the humoral immune response, 872, *873*

Plasma membranes
in animal cells, *86*
in cell adhesion and cell recognition, 110–113
depolarization and hyperpolarization, 930–931, *932*
endocytosis, 120–121
energy pathways on, *168*
membrane-associated carbohydrates, 109
membrane proteins, *106, 107*–109
origin of eukaryotic organelles and, 101–102
in plant cells, *87*
principles of bioelectricity, 927
properties and characteristics of membrane potentials, 927–932
structure and function, 79–81

Plasmid pBR322, 377

Plasmids
in bacterial conjugation, 254–255
in prokaryotes, 356
recombinant, 374, 376
reporter genes, 378
as vectors in transformation, 377

Plasmin, 385

Plasminogen, 385

Plasmodesmata
blocking in response to pathogenic invasions, 798, *799*
in phloem, 714, 715
in plant cells, *87*, 710
structure and function, 100, *139*, 140

Plasmodial slime molds, 560–561

Plasmodium, 553, 554, 563, 564

Plasmodium falciparum, 564

Plasmogamy, *618, 619, 620, 621*

Plastid genes, inheritance of, 252–253

Plastids
endosymbiosis theory of, 102
structure and function, 92–93

Plate tectonics, 509

Platelet-derived growth factor, 211, *385*

Platelets
in atherosclerosis, 1042
in blood clotting, *1038*, 1039

platelet-derived growth factor, 211
Platinum, 259, 278
Platyspiza crassirostris, 474
"Playing possum," 1173
Plecoptera, 672
Pleiotropic alleles, 243–244
Pleistocene epoch, 472, 522
Pleodorina, 140, 141
Pleopods, *671*
Plesiadapis fodinatus, 519
Pleural membranes, *1014*, 1015–1016
Plimsoll line metaphor, 1244–1245
Plodia interpunctella, 1184
Plumatella repens, 641, 654
Pluripotency, 394
Pluripotent stem cells, 408–409
Pneumatophores, 808
Pneumococcus, 260–261
Pneumocystis jirovecii, 612, 876
Pneumonia, 533
Podocytes, 1078, 1079
Poecilia reticulata, 1157
Poecilotheria metallica, 669
Pogonophorans, 660–661
Poikilotherms, 822
Point mutations, 306–307, 312, 486
Point restriction phenotype, 245, 306
Poison dart frogs, *1174*
Polar auxin transport, 763–764
Polar bodies, *883*, 884
Polar bonds, 28
Polar covalent bonds, 28
Polar microtubules, 214, *215*
Polar molecules, 30
Polar nuclei, *600*, 779, *781, 783*
Polar substances, effect on protein structure, 50
Polar tube, 617
Polarity
defined, 395
determination by cytoplasmic segregation, 395
establishment in the animal zygote, 903
Pole plasm, 908
Polistes nympha, 674
Pollen grains
in angiosperms, 600, 779, 780–782, 783
in conifers, 594, *595*
mechanisms of transport, 780, *782*
in monocots and eudicots, 710
in seed plants, 590, *591*
Pollen tubes
in angiosperms, 779, 780–782
in conifers, 594
in seed plants, 590, *591*
Pollination
in angiosperms, 600, 780–782
evolution of flowers and, 598, *599*
influence on speciation rates, 480
in orchids, 588, 605
in seed plants, 590, *591*
strategies for preventing inbreeding in angiosperms, 782, *783*
wind-pollination, 480, 780

See also Plant–pollinator relationship; Pollinators
Pollination syndromes, 1180–1181
Pollinators
  economic benefits to coffee plantations, 1243
  hummingbirds, 598, *599*
  of orchids, 588, 605
  pollination syndromes, 1180–1181
  prezygotic isolating mechanisms in plants, 476, 477
  See also Plant–pollinator relationship
Pollinia, 605
Poll's stellate barnacle, 1184
Pollution
  impact on habitats and biodiversity, 1233
  lichens as indicators of air quality, 624, 625
  use of fungi to study environmental contamination, 624, 625
Poly A sequence, in expression vectors, 384
Poly A tail, 292
Polyacrylamide, 316
Polyadenylation sequence, 292
Polyandrous mating systems, 1113, 1114
Polychaetes, *638*, 640, 660–661
Polygynous mating systems, 480–481, 1113–1114
Polymerase chain reaction (PCR)
  description of, 277–278
  in DNA fingerprinting, 317
  in DNA testing, 321
  in high-throughput sequencing, 353, *354*
  in metagenomics, 357–358
  RT-PCR, 380
  using to create synthetic DNA, 380
  in vitro evolution studies, 501
Polymerization, origin of life and, 71
Polymers
  condensation and hydrolysis reactions, 42
  defined, 40
  origin of life and, 71
Polymorphic loci, 242, 442–443
Polymorphisms, 317
Polymorphus marilis, 658
Polynucleotides, 63, 74
Polyorchis penicillatus, 647
Polypeptide chains, 43, 45, *46*
Polypeptides
  modification after translation, 300–301
  signal sequences and movement within the cell, 298–300
  synthesis in translation, 293–297, *298*
Polyphagous herbivores, 1175–1176
Polyphyletic groups, 463
Polyploidy
  description of, 224–225, *226*
  gene duplication, 497–498

sympatric speciation and, 473, 475
Polyps
  of anthozoans, 646–647
  of cnidarians, 645, *646*
  of hydrozoans, 647–648
  of scyphozoans, 647
Polyribosomes, 297, *298*
Polysaccharides
  catabolic interconversions, 179
  cellulose, 54
  defined, 51
  features of, 53
  glycogen, 53–54
  glycosidic linkages, 53
  starch, 53, *54*
Polysomes, 297, *298*
Polyspermy blocks, *885*, 886–887
Polysynaptic reflexes, 942–943
Polytrichum, 578
Polyubiquitin, 349
Polyubiquitination, 767
Pombe, 624
"Pond scum," *533*
Ponderosa pine, 443, *1241*
Ponds, 1140
Pongo pygmaeus, 702
Pons, 968, *969*, 1020
Poplar trees, 362
Population bottleneck, 434
Population density
  defined, 1150
  estimating, 1151
  factors limiting, 1157–1161
Population dynamics
  defined, 1150
  demographic events determine population size, 1153–1154
  effect of habitat variation on, 1161–1163
  influence on population management, 1163–1164
  introduced reindeer populations, 1149, 1158, 1166
  life tables, 1154–1155
  survivorship curves, 1155–1156
Population growth
  effects of density-dependent or density-independent factors on, 1159
  exponential, 1157–1158
  human population growth, 1164–1166
  logistic, 1158
  per capita growth rate, 1156
Population management, 1163–1164
Population size, noncoding DNA in the genome and, 496
Populations
  age structure, 1150, 1151–1152, 1165–1166
  biotic potential, 1158
  contribution of genetic variation to phenotypic variation, 431
  defined, 6, 432, 1150
  dispersion patterns, 1150, 1152–1153
  effect of environmental conditions on life histories, 1156–1157
  evolution of, 6
  fixed, 437

genetic structure, 437
genetic variation in geographically distinct populations, 443–444
Hardy–Weinberg equilibrium, 437–438
interactions between individuals, 9
intrinsic rate of increase, 1156
maintenance of genetic variation by frequency-dependent selection, 441–442
measuring or counting, 1150–1151, 1152
mechanisms of evolution in, 432–436
mechanisms of maintaining genetic variation in, 441–444
possible effects of natural selection on, 439–440
properties of, 1150
r-strategists and *K*-strategists, 1159
speciation and, 6
using ecological principles to manage, 1163–1164
Populus trichocarpa, 362
Portal blood vessels, 844
Portuguese man-of-war, 640, 646
Portuguese water dogs, 352
Posidonia oceanica, *603*
Positional information, 401
Positive cooperativity, 1017
Positive feedback, in physiological systems, 816–817
Positive gravitropism, 765
Positive regulation
  defined, 329
  of the *E. coli lac* operon, 332, *333*
  in eukaryotes, 333
  in virus reproductive cycles, 339–340
Positive selection, 433, 492–494
Positive-sense genomes, 544
Positive-sense single-stranded RNA viruses, *543*, 544
Positron emission tomography (PET), 970, *981*
Possessions Island, 1199
Post-traumatic stress disorder (PTSD), 970
Postabsorptive state, 1065, 1067
Postelsia palmiformis, *556*
Posterior, 634, *635*
Posterior hippocampus, 967
Posterior pituitary, 842, 843, 896–897
Posterior–anterior axis, determination in vertebrate limb development, 401
Postganglionic neurons, *974*, 975
Postsynaptic cells
  defined, 925
  overinhibition of neurons in the brain, 943
  summation of excitatory and inhibitory input, 938
Postsynaptic membrane
  at electrical synapses, 939
  responses to neurotransmitter, 936–938
Posttranscriptional gene regulation, 346–349
Postural muscles, 995

Postzygotic isolating mechanisms, 475, 478, 479
Potassium
  in animal nutrition, 1052
  electronegativity, 28
  in plant nutrition, 741
Potassium-40, 507
Potassium equilibrium potential (E_K), 929
Potassium ion channels
  in action potentials, 932, 933, 934
  in cardiac pacemaker cells and heart contraction, 1033, 1034
  in the hyperpolarization of neurons at the onset of sleep, 979
  membrane hyperpolarization and, 931, *932*
  membrane potentials and, 929
  specificity of, 115–116
Potassium ions
  in cardiac pacemaker cells and heart contraction, 1033, 1034
  generation of action potentials and, 933, 934
  inorganic fertilizers, 747
  membrane potential and, 927
  plant guard cell function and, 733
  reabsorption in the kidney, 1084
Potato beetles, 1184
Potatoes
  "eyes," 720, 792
  sink strength of tubers, 736–737
  tubers, 720
  vegetative reproduction, 792
Potential energy, 145
Potrykus, Ingo, 388
Powdery mildews, 622
PR genes. *See* Pathogenesis-related genes
PR proteins, 801
Prader-Willi syndrome, 345
Prairie dogs, *1239*
Prairie voles, 125, 1113
Prairies
  restoration projects, 1237–1239, *1240*
  See also Grasslands
Pre-mRNA. *See* Precursor mRNA
Pre-replication complex, 269
Prebiotic synthesis experiments, 69–71
Precambrian, *506–507*, 508, 515–516
Precapillary sphincters, 1044
Precipitation
  acid precipitation, 1219–1220
  atmospheric circulation patterns and, 1124
  boreal and temperate evergreen forest biomes, *1129*
  chaparral biomes, *1134*
  cold desert biomes, *1133*
  effect on terrestrial biomes, 1126–1127
  hot desert biomes, *1132*
  rain shadows, 1126–1127
  temperate deciduous forest biome, *1130*
  temperate grassland biome, *1131*

thorn forest and tropical savanna biomes, *1135*
tropical deciduous forest biomes, *1136*
tropical rainforest biomes, *1137*
tundra biomes, *1128*
Walter climate diagrams, 1138
Precocial young, 642
Precursor mRNA (pre-mRNA)
  alternative splicing, 346–347
  hybridization experiments, 291
  processing before translation, 291–293
Predation
  defined, 1170
  impact on life history traits, 1157
Predation hypothesis, of latitudinal gradients in diversity, 1196
Predator–prey interactions
  overview, 1172–1173
  prey defenses, 1173–1175
Predators
  defined, 637
  effect on population growth, 1159
  feeding strategy, 638
  types of, 1054
Predatory fireflies, 1111
Predatory fungi, 612, *613*
Preganglionic neurons, *974*, *975*
Pregnancy
  consequences of stress during, 328
  effect of exposure to environment factors, 920
  in humans, 896
  length of, 919
  methods of preventing, 897, *898* (see also Contraception)
  stages of development during, 920
Pregnancy tests, 871, 896
Preimplantation genetic diagnosis (PGD), 899
Preimplantation screening, 321
Premature babies, respiratory distress, 1015
Premolars, 1055
Prenatal screening, 320, 321
Prepenetration apparatus (PPA), *747*, *748*
Pressure, detection of, 952
Pressure chambers, 732
Pressure flow model, 735–738
Pressure potential
  defined, 727
  movement of water and solutes in plants and, 727–728
  pressure flow model of phloem transport, 735–738
Presynaptic cell, 925
Presynaptic membrane, 939
"Pretzel mold," *560*
Prevailing winds
  description of, 1124, *1125*
  ocean currents and, 1124–1125
Prey
  defenses, 1173–1175
  defined, 638
  See also Predator–prey interactions
Prezwalski's horse, *1144*

Prezygotic isolating mechanisms, 475, 476–477, 482
Priapulids, *632*, 665
Priapulus caudatus, 665
Priapus, 665
Primary active transport, 118–119
Primary bronchi, 1011, *1012*
Primary cell wall, 710
Primary consumers, 1190, *1191*
Primary embryonic organizer, 911–913
Primary endosymbiosis, 551, *552*, 570
Primary growth
  defined, 715
  role of apical meristems in, 715–716
  in roots, 716–718
  in shoots, 719–720
Primary immune response, 866
Primary lysosomes, 90, *91*
Primary meristems
  origins and types of, 715
  in root development, 716–718
  in shoot growth, 719–720
  tissues produced by, 716
Primary metabolites, 370, 802
Primary motor cortex
  location, 971
  mapping of the body in, 971, *972*
Primary nodule meristem, 747, *748*
Primary oocytes, *883*, *884*, *893*, *894*
Primary producers, 563, 1189–1190
Primary sex determination, 250
Primary sex organs, 887, 889
  See also Ovaries; Testes
Primary somatosensory cortex, *972*–973
Primary spermatocytes, 883–884, *891*
Primary structure
  of proteins, 45, *46*
  specifies protein tertiary structure, 48, 49
Primary succession, 1200
Primase, 271, 273, *274*
Primates
  anthropoids, 701–702
  bipedal locomotion, 702–703
  brain size–body size relationship, 973
  comparative genomics, 366–367
  fossil record, 701
  grooming behavior, 1177
  hominins, 702–705
  number of species, *698*
  phylogeny, *701*
  prosimians, 701
  See also Chimpanzees
Primers
  in creating synthetic DNA, 380
  in DNA replication, 63, 271, 272, *274*
  in the PCR reaction, 277, *278*
Primitive groove, 914
Primitive gut
  formation in sea urchins, 909
  in frog gastrulation, 909
Primitive streak, 914
"Primordial soup" hypothesis, 71
Probability, 13–14, 239–240
Probes, 290

Proboscidea, *698*
Proboscis
  of ecdysozoans, 665
  of hemichordates, 682, 683
  of ribbon worms, 658, *659*
Procambium
  in root growth, 716–718
  in shoot growth, 719
  vascular tissue system and, 716
Procarboxypeptidase A, 333
Procedural memory, 982
Productivity
  defined, 1189–1190
  impact of species richness on, 1202, *1203*
  species diversity and, 1192
Products, in chemical reactions, 31, 145
Progesterone
  actions of, *842*
  in human pregnancy, 896
  in labor and childbirth, 896
  in parthenogenic whiptail lizards, 882
  produced in the ovaries, 850, *893*, *894*
  in regulation of the ovarian and uterine cycles, *894*, 895
Progestin-only pill (Plan B®), 898
Programmed cell death. *See* Apoptosis
Progymnosperms, 589, 591
Prokaryotes
  archaea, 534–537 (*see also* Archaea)
  atmospheric oxygen levels and, 511–512
  beneficial relationships with eukaryotes, 539
  cell division, 206–207
  cellular locations of energy pathways, *168*
  characteristic features of, 82–84, 526–527
  characteristics of gene expression in, *291*
  complex communities, 539, *540*
  defined, 81
  discordant gene trees, 529–530
  in element cycling, 538–539
  environmental genomics, 530
  in the evolution of life, 4
  evolutionary relationships from nucleotide sequences, 528–529
  gene regulation in, 329–333
  gene transfer in, 253–255, 529–530
  insertion of genes into, 376–377
  major bacterial groups, 530–534
  metabolism and metabolic pathways, 4, 166, 168, 537–538
  microbiomes and human health, 539–541
  nucleic acid hybridization, 291
  origins of eukaryotic cells and, 101–102
  origins of photosynthesis, 4, *5*
  pathogenic, 541–542
  phenotypic characteristics used in classification, 527–528, *529*
  shared and unique features, 526
  success of, 530

Prokaryotic genomes
  artificial life studies, 359, *360*
  benefits of sequencing, 357
  comparative genomics, 357
  features of, 356
  functional genomics, 356, 357
  metagenomics, 357–358
  minimal genome studies, 359, 360
  sequencing, 356
  transposons, 358–359
Prolactin, *838*, 842
Prolactin-inhibiting hormone, 844
Prolactin-releasing hormone, 844
Proline, 43, *44*
Promerops cafer, *1134*
Prometaphase
  in meiosis, *220*
  in mitosis, 212, 214, *216*, *217*
Promoters
  binding of transcription factors to, 328
  consensus sequences, 292, 332–333
  eukaryotic general transcription factors and, 333–334, *335*, *336*
  in expression vectors, 384
  as genetic switches, 415
  in transcription, 286, *287*
  in the viral lytic reproductive cycle, 340
  viral regulatory proteins and, 340–341
Proofreading, 276–277
Prop roots, 718
Propane, 31
Prophage, 340–341
Prophase
  centrosome separation, 212
  comparison between mitosis and meiosis, 223
  events in mitosis, 212, 214, 217
  spindle apparatus formation, 213–214
Prophase I (meiosis), 219, 220
Prophase II (meiosis), 220
Propithecus diadema, *701*
Propithecus verreauxi, *10*
Propranolol, 940
Prosimians, 701
Prosopis, 808
Prostaglandins
  in inflammation, 861, 862
  production by the prostate gland, 890
Prostate cancer, *368*
Prostate fluid, 890
Prostate gland, 890
Prosthetic groups, of enzymes, 155, *156*
Protease inhibitors, 342, 805
Proteases
  function of, 1057–1058
  produced by the pancreas, 1062
  in proteolysis, 300–301
Proteasomes, 349
Protected areas, 1237, *1238*
Protein hormones, 836
Protein kinase C (PKC), 134–135
Protein kinase cascades, 131–132, 138
Protein kinase receptors, 129, *130*
Protein kinases
  in cell cycle control, 209–211

in hormone-mediated signaling cascades, 837
in protein phosphorylation, 301
receptors, 129, *130*
regulation of glucose metabolism in liver cells, 138
Protein phosphatase, 161
Protein starvation, 1041
Protein synthesis
  inducers, 330
  modifications after translation, 300–301
  polysomes, 297, *298*
  signal sequences and polypeptide movement within the cell, 298–300
  steps in gene expression, 284–285
  transcription, 286–293
  transcriptional regulation in prokaryotes, 329–333
  translation, 293–297, *298*
Proteinases, 300
Proteins
  in animal "self-consumption," 1050
  biological information and, 5, 6
  breakdown by digestive enzymes, 1057–1058
  catabolic interconversions, *179*, 180
  commercial production, 383–384
  comparing through sequence alignment, 486–487
  denaturation, 48, 50
  digestion in animals to constituent amino acids, 1051–1052
  domains, 291
  dysfunctional proteins and human genetic diseases, 311–312
  effect of temperature on, 820
  energy yield, 1049, *1050*
  environmental effects on structure, 50
  fluorescent, 449, 464
  functions, *42*
  gain-of-function mutations, 305
  genomic information and, *355*, 356
  identifying homologous parts, 486
  inducible and constitutive, 330
  loss-of-function mutations, 305, 311–312
  modification after translation, 300–301
  molecular chaperones, 51
  molecular evolution, 486
  monomer components, 40
  in the origin of life, 3
  peptide linkages, 43–44, *45*
  phosphorylation, 209
  point mutations and, 306–307
  primary structure, 45, *46*
  primary structure specifies tertiary structure, 48, *49*
  production of medically useful proteins through biotechnology, 384–386
  in prokaryotic gene regulation, 329

proportions in living organisms, *41*
protein–DNA interactions, 266, *267*
proteomes, 369, *370*
quaternary structure, *46, 48, 49*
reconstruction of sequences from extinct organisms, 464
regulation of longevity in the cell, 348–349
relation of mRNA abundance to cell protein abundance, 348
roles in DNA replication, 272, *273*
secondary structure, 45, *46*
shape modifications, 50
specificity of binding, 48, 50
spider silk, 39
structural characteristics, 43
structural motifs and binding to DNA, 335–336
tertiary structure, 46–48
use of gene evolution to study protein function, 500
Proteobacteria, 534, 538, *550*, 551
Proteoglycans, 100, 109, 111
Proteolysis, 300–301
Proteomes, 369, *370*
Proteomics, 369, *370*
Proterozoic eon, *506–507*, 508, 515–516
Prothoracicotropic hormone (PTTH), 840, *841*
Prothrombin, 1039
Protists
  alveolates, 553–555
  amoebozoans, 559–561
  ancestor to fungi and animals, 609, 631, 633
  approaches to classifying, 553
  defined, 550
  endosymbionts, 564–565, 566
  evolution of multicellularity and, 552–553
  excavates, 558–559
  pathogenic, 563, *564*
  primary producers, 563
  rhizaria, 557–558
  sex and reproduction in, 562–563
  stramenopiles, 555–556, *557*
Protocells, 73–74
Protoderm
  dermal tissue system and, 716
  in root growth, 716–717
  in shoot growth, 719
Protohominids, 702–703
Proton gradients, photophosphorylation and ATP synthesis, 192–193
Proton-motive force, 173–174
Proton pumps
  active transport in plants and, 729
  in auxin-induced cell expansion, 766–767
  in polar auxin transport, 763
  in the uptake of ions by roots, 746
Protonema, *576, 577*
Protonephridium, 1075
Protons
  atomic number and, 22
  defined, 22

mass number and, 22
proton-motive force and ATP synthesis, 173–176
transport in the respiratory chain, 172–173
Protopterus annectens, 689
Protostomes
  anatomical characteristics, 652
  in animal phylogeny, *630*
  arrow worms, 655–656
  bilaterians, 634, 643
  defined, 652
  ecdysozoans, 654–655, 665–673, *674*
  key aspects of evolution in, 673–675
  lophotrochozoans, 652–654, 656–664
  major derived traits, 652
  major groups, *632*, 652
  pattern of gastrulation in, 634
  phylogenetic tree, *653*
  undescribed species, 675
Prototherians, 697, *698*, 889
Proturans, 671, *672n*
Province Islands, 1199
Provirus, 341–342, 544
Proximal convoluted tubule (PCT), *1080, 1081, 1082, 1083,* 1084
Proximate causes, of animal behavior, 1096
Prozac, 940
Przewalski's horse, *1131*
Pseudobiceros, 1007
Pseudocoel, 635, 652
Pseudocoelomates, 635
Pseudogenes, 364, 491, 494–495, 497
Pseudomonas, 373, 539
Pseudomonas aeruginosa, 82, 207
Pseudomonas fluorescens, 489–490
Pseudomyrmex, 1178, 1179
Pseudonocardia, 1169
Pseudoplasmodium, 561
Pseudopodia, 94
Pseudopods
  of amoebozoans, 559
  of foraminiferans, 557
  of radiolarians, 557
Pseudotsuga menziesii, 1160
Pseudouroctonus minimus, 669
Psocoptera, *672*
Psoriasis, 158
Psychoactive drugs, 135
Pterapogon kauderni, 1234, 1235
Pterobranchs, 682–683
Pteroeides, 647
Pterosaur, *423*
Pterygotes
  diversity in, *674*
  instars, 671
  major groups and number of living species, *672*
  mayflies and dragonflies, 672
  metamorphosis, 672
  neopterans, 672–673
  wings and flight, 671, 672, 673
Pthirus pubis, 1177
PTTH. *See* Prothoracicotropic hormone
Puberty
  in females, 894
  in males, 892
  overview, 850–851

Public policy
  importance of biological research to, 16–17
  to reduce the effects of mutagens on human health, 311
Puccinia graminis, 612
Puffballs, *610*
Pufferfish, *361*, 500
Pulmonary arteriole, *1014*
Pulmonary artery, 1031, 1032
Pulmonary circuit
  in amphibians, 1029
  in birds and mammals, 1030
  blood vessels of, 1027–1028
  defined, 1027
  in lungfish, 1028
  in reptiles, 1029–1030
Pulmonary valve, 1030, *1031, 1032*
Pulmonary veins, 1031
Pulmonary venule, *1014*
Pulp cavity, 1055
Punnett, Reginald, 236
Punnett square, 236–237
Pupa, *639*
Pupil, 958, *959*
Purifying selection, 433, 492–494
Purines
  anabolic interconversions and, *179*, 180
  found on meteorites, 69
  structure, 63
  transition and transversion mutations, 306
Purkinje cells, *925*
Purkinje fibers, 1034, *1035*
Purple foxglove, 1034
Purple owl's clover, 1161
Purple pitcher plant, 1189
Purple sand crab, *670*
Purple sulfur bacteria, 187, *538*
Pus, 862
Putrefaction, 1188
Pycnogonids, 668
Pycnophyes kielensis, 665
Pygmy tarsiers, 1230–1231
Pyloric sphincter, *1059, 1060, 1061*
Pyramid diagrams, 1191, *1192*
Pyramidal cells, *925*
Pyrenestes ostrinus, 440
Pyrethrin, *802*
Pyridoxine, *1053*
Pyrimidines
  anabolic interconversions and, *179*, 180
  found on meteorites, 69
  structure, 63
  transition and transversion mutations, 306
Pyrogens, 830
Pyrophosphate, 150, 269, *270*
Pyruvate
  in alcoholic fermentation, 177, 178
  anabolism in the liver during exercise, 181
  in C_4 photosynthesis, 199
  in cellular respiration, 166–167
  in glucose catabolism, 168, 169
  in glycolysis, 166
  in lactic acid fermentation, 177
  in metabolic interactions in plants, 201

produced during glycolysis, 169–170
Pyruvate decarboxylase, *177*, 178
Pyruvate dehydrogenase, 171
Pyruvate oxidation
    in glucose catabolism, 168, 170
    regulation of, 171
    relationships among metabolic pathways, *179*
Pyruvic acid. *See* Pyruvate
Pythons, 423

Q

Q. *See* Ubiquinone
Q_{10}, 821
Qiu, Yin-Long, 571
QTL analysis, 1096
Quack grass, *597*
Quadrats, 1151
Quadriceps, 1002
Qualitative analysis, 33
Qualitative traits, 439
Qualitative variation, 246
Quantifiable data, 11, 14
Quantitative analysis, 33
Quantitative trait analysis, 1096
Quantitative trait loci, 246
Quantitative traits
    defined, 439
        possible actions of natural selection on, 439–440
Quantitative variation, 246
Quaternary period, *506–507*, 510, 522
Quaternary structure, of proteins, *46*, 48, *49*
Queen honey bees, 880, 899
Quiescent center, 716, *717*
Quill, *696*
Quinine, *604*, *605*, 797
Quinine-resistant malaria, 797
Quiring, Rebecca, 413
Quorum sensing, 525, 539

R

R genes, 800
R groups
    of amino acids, 43
    in an alpha helix, 45
    of histones, 212
    in polypeptide chains, 44
    in protein binding, 50
    in protein shape changes, 50
    in protein tertiary structure, 47
    See also Side chains
R proteins, 799, 800
r-Strategists, 1159, 1201
RAAS system, 1087, *1088*
Rabbits
    multiple alleles for coat color, 242
    number of species, *698*
    point restriction phenotype, 245, 306
Rachis, *696*
Radcliffe, Paula, 815
Radial axis, in plants, 711
Radial cleavage, 633, 679, 905
Radial symmetry
    in animals, 634, *635*
    in echinoderms and hemichordates, 680, *681*
    in flowers, 597

Radiation
    in heat exchange between animals and their environment, 823, *824*
    human-made or natural, 310
    mutagenic effects, 309
Radiation treatments, 229
Radicle, 718, 758
Radio frequency identification (RFID), 1151
Radioactive contamination, bioremediation of, 389
Radioactive decay, 23
Radioimmunoassays, 852
Radioisotopes
    in experiments revealing the Calvin cycle, 193–194, 195
    Hershey–Chase experiment on DNA, 262–263
    as mutagens, 310
    properties of, 23–24
    radiometric dating, 507–508
Radiolarians, 557–558, *564*
Radiometric dating, 507–508
Radius, 423
Radula, 662, *663*, 1056
Rafflesia arnoldi, *603*
Ragweed, 1200
Rain shadows, 1126–1127
Rainbow trout, 162, 1140
Rainey, Paul, 489–490
Ralph, Martin, 1108
Ramalina menziesii, 625
Rana berlandeieri, *476*
Rana blairi, *476*
Rana pipiens, 12
Rana sphenocephala, *476*
Rana sylvatica, 642, *1125*, *1126*
Randallia ornata, *670*
Random dispersion pattern, 1153
Rangifer tarandus, *700*, 1149
Rao, P. N., 209
Raphidoptera, *672*
"Rapid-cooling" technology, 831
Rapid eye movement (REM) sleep, 978, *979*
Ras protein, 131, *132*
Ras signaling pathway, 139
Rashes, 876
Ras–MAP signal transduction pathway, 398
Raspberry, 601
Rats
    hormonal control of sexual behavior, 1098–1099
    mutation affecting obesity, *1067*, 1068
    "place cells" of the hippocampus, 967
    regulation of food intake by the hypothalamus, 1067
    Toxoplasma and, 554
Rattlesnakes, 946
Raven, Peter, 1175
Ray-finned fishes, *685*, 687–689
Ray flowers, *597*
Rays
    cartilaginous skeleton, 1000
    claspers in sexual reproduction, 888
    evolution in body morphology, 444, *445*
    excretion of urea, 1074

features of and diversity in, 687, *688*
    regulation of ionic composition of extracellular fluid, 1073
    salt and water balance regulation, 1077
Reabsorption, isosmotic, 1082
Reactants, in chemical reactions, 31, 145
Reaction center, of photosystems, 190, 191, 192
Reactive atoms, 25
Reactive oxygen
    hypersensitive response in plants and, 800–801
    in plant defenses against herbivory, 804
    in plant responses to pathogens, 799
Realized niche, 1184
Recent, the. *See* Holocene epoch
Receptacle (floral), *591*
Receptive fields
    of neurons in the visual cortex, 977
    of photoreceptors, 963
    of retinal ganglion cells, 975–977
Receptor cells
    in sensory transduction, 947–948
    See also Sensory receptor cells
Receptor-mediated endocytosis, 120, 121, *122*
Receptor potential, 947
Receptor proteins
    binding of signal ligand to, 127–128
    classification by function, 129–130
    classification by location, 128–129
    functions, *42*
    intracellular receptors, 130–131
    in receptor-mediated endocytosis, 121
    in sensory transduction, 947, *948*
    specificity, 127
Recessive traits, 235
Recognition sequence, 315, 316
Recombinant chromatids, 219
Recombinant DNA, 374–375
Recombinant DNA technology
    agricultural applications, 386–388, *389*
    cloning genes, 375–376
    methods of creating recombinant DNA, 374–375
    methods of transformation, 376–377
    origin of, 374
    production of medically useful proteins, 385–386
    public concerns, 388–389
    reporter genes, 377–378, *379*
    sources of DNA used in cloning, 379–380
    using to produce vaccines, 866
    using to study gene function, 381
Recombinant frequencies, 248, *250*
Reconciliation ecology, 1243–1244
Rectum, 1056, *1058*, 1063

Red algae, 551, 571–572, 573
Red-and-green macaws, *1106*
Red blood cells
    ABO blood groups, 243
    β-globin gene expression in, 398
    induction of cell division in, 211
    malaria, 563, *564*
    pernicious anemia, 1054
    production and elimination, 1038
    sickle-cell disease, 306, *307*, 312
    transcriptional regulation of β-globin, 335
Red-eared slider turtle, 420
Red fluorescent pigments, 449, 464
Red-green color blindness, *252*
Red light
    photomorphogenesis and phytochromes in plants, 772–774
    in photoperiodic control of flowering, 788
    See also Far-red light
Red mangrove, 1198, *1199*
Red muscle, 995
Red tides, 549, *564*
Red-winged blackbirds, 1114
Redi, Francesco, 67–68
Redox reactions. *See* Oxidation–reduction reactions
Reduction, 167
Reefs
    byrozoan, 656
    corals and, 646–647
Reflexes. *See* Spinal reflexes
Reforestation, mycorrhizal fungi and, 626
Refractory period, of voltage-gated sodium channels, 933–934
Regeneration, 881
Regular dispersion pattern, 1153
Regulative development, 907
Regulatory sequences
    in eukaryotic genomes, 361
    genomic information and, 356
    of operons, 330
Regulatory subunits, of enzymes, 159
Regulatory systems
    components and functions of, 816–817
    maintenance of stability in the internal environment, 10
Regulatory T cells (Tregs), 863, 874
Reindeer, *700*, 1149, 1158, 1166
Reindeer moss, 613
Reinforcement, 475, 478
Relative atomic mass, 23
Release factor, *297*, *298*
Releasers, 1095
Religion, 14
REM sleep, 978, *979*
Remediation. *See* Bioremediation
Renal artery, *1080*, 1081, 1082
Renal cortex, *1080*, 1081
Renal dialysis, 1086, *1087*
Renal failure, 1085–1086, *1087*
Renal medulla, *1080*, 1081, 1082–1084
Renal pyramids, *1080*, 1081, 1090
Renal tubules
    conversion of glomerular filtrate to urine, 1079

in the mammalian kidney, 1081
reabsorption in, 1078, 1082
water channels, 1084
Renal vein, *1080*, 1081
Renal venule, *1078*
Renin, 1087, *1088*
Reoxidation reactions, 171
Repetitive sequences, in
    eukaryotic genomes, 364–366
Replication. *See* DNA replication
Replication complex, 206, 274–275
Replication forks, 271, 272–275
Replicons, 377
Reporter genes, in recombinant
    DNA technology, 377–378,
    379
Repressible systems, in
    transcriptional regulation of
    operons, 330, 331–332
Repressor proteins
    as genetic switches, 415
    helix-turn-helix motif, 336
    in negative regulation, 329
    operator–repressor interactions
        controlling operon
        transcription, 330–332, *333*
    strategies in the repression of
        transcription, 336
Reproduction. *See* Animal
    reproduction; Asexual
    reproduction; Human
    reproduction; Plant
    reproduction; Sexual
    reproduction
Reproductive capacity, estimating,
    1154–1155
Reproductive isolation
    biological species concept and,
        468–469
    defined, 468
    hybrid zones, 478–479
    importance to speciation, 469
    from incompatibilities between
        genes, 470, *471*
    from increasing genetic
        divergence, 470–471
    mechanisms preventing
        hybridization, 475–478
Reproductive signal
    in cell division, 206
    in eukaryotic cell division, 207
    in prokaryotic cell division, 206
Reproductive success, evolution
    by natural selection and, 6
Reproductive technologies, 897,
    899
Reptiles
    circulatory systems, 1029–1030
    crocodilians and birds, 693–695
    evolution of the amniote egg,
        888
    gastrulation, 913–914
    incomplete cleavage, 633
    lepidosaurs, 693
    origin of, 692, *693*
    radiation during the Triassic,
        521
    salt and water balance
        regulation, 1078
    temperature-dependent sex
        determination and sex-
        specific fitness differences,
        421
    turtles, 693, *694*

Rescue effect, 1161
Residence time, 1211, 1215
Residual volume (RV), *1012*, 1013
Resistance *(R)* genes, 800
Resolution, of microscopes, 79,
    80, 81
Resource partitioning, 1182
Respiration. *See* Cellular
    respiration
Respiratory chain
    allosteric regulation, 181
    controlled release of energy
        by, 172
    description of, 172–173
    in glucose metabolism, 168
    in oxidative phosphorylation,
        171
Respiratory distress syndrome,
    1015
Respiratory gas exchange
    in amphibians, 1029
    fish gills, 1009–1010, *1011*
    fully separated pulmonary and
        systemic circuits, 1030
    human lungs, 1013–1016
    partial pressure gradients, 1009
    physical factors governing,
        1006–1008
    regulation of breathing, 1019–
        1022
    snorkeling elephants, 1005,
        1022
    surface area of respiratory
        organs, 1008–1009
    unidirectional ventilation in
        birds, 1010–1012
Respiratory gases
    air and water as media for, 1007
    blood transport, 1016–1019
    defined, 1006
    diffusion, 1006
    maximization of partial
        pressure gradients in gas
        exchange, 1009
Respiratory organs
    bird lungs, 1010–1012
    fish gills, 1009–1010, *1011*
    human lungs, 1013–1016
    surface area maximization,
        1008–1009
Respiratory tract, anatomy of,
    1013, *1014*
Resting membrane potential
    of cardiac pacemaker cells,
        1032–1033
    defined, 927
    measurement of, 928
Restoration ecology, 1237–1239,
    1240
Restriction digestion, 315
Restriction endonucleases,
    315–316
Restriction enzymes, 315–316,
    317, 374
Restriction (R) point, 210
Restriction site, 315, 316
Reticular activating system, 970
Reticular formation, 979
Reticulate bodies, 533, *534*
Reticulations (on phylogenetic
    trees), 496
Reticulum, 1064
Retina
    cone cells, 960, 961–962, 963

in the image-forming eye, 958,
    959
information flow in, 962–963
inputs to the visual cortex,
    977–978
receptive fields of ganglion
    cells, 975–977
rod cells, 960–961, *962*, 963
structure of, 959
Retinal
    changes with the absorption of
        light, 957, 958
    in cone cells, 962
    role in catalyzed reactions, *156*
    vitamin A and, 58
Retinal ganglion cells
    information flow through the
        retina, *962*, 963
    receptive fields, 975–977
Retinoblastoma, 349
Retinoblastoma (RB) protein, 210,
    228
Retinol, *1053*
Retrotransposons, 364–365, 495
Retroviruses
    description of, 341, *543*, 544
    endogenous retroviruses in the
        vertebrate genome, 545
    mutations caused by, 308
    reverse transcription, 285
Reverse genetics, 318
Reverse transcriptase
    in the production of cDNA,
        379–380
    in retrovirus infections, 341
    synthesis of DNA from RNA, 72
Reverse transcriptase inhibitors,
    342
Reverse transcription, 285
Reversible chemical reactions, 34,
    35–36
Reversible inhibition, of enzymes,
    157–159
Reversible phosphorylation, 161
Reversion mutations, 306
Reversions, 488
Reznick, David, 1157
Rhacophorus nigropalmatus, 7
Rhagoletis pomonella, 473
Rhcg protein, *1086*
Rhea pennata, *1131*
Rheas, 694
Rhenium, 69
Rheobatrachus silus, *678*
Rheumatoid arthritis, 158, 862, 876
Rhinoceroses, 1234
Rhizaria, 557–558
Rhizobia, formation of root
    nodules, 747–748
Rhizobium, 357, 534
Rhizoids
    of fungi, 609
    of nonvascular plants, *576*, 577
    of rhyniophytes, 580
Rhizomes, 580, 792, *793*
Rhizophora, 1141
Rhizophora mangle, *1198*, *1199*
Rhizopus oligosporus, 218
Rhizopus stolonifer, *618*, 619
Rhizosphere, interference
    competition in, 1183
Rhodnius prolixus, 839–840, 841
Rhodopsin, 957–958, 960, 961
Rhogeessa tumida, 470, *471*

Rhombozoans, *632*, 648
Rhyniophytes, 580
Rhynocoel, 658, *659*
Rhythm method of contraception,
    898
Ribbon model, of protein tertiary
    structure, *47*, 48
Ribbon worms, *632*, *652*, 658, *659*
Riboflavin, *1053*
Ribonucleic acid (RNA)
    antisense RNA, 382
    complementary base pairing,
        63–64, *65*
    distinguishing from DNA, 63
    DNA transcription and, 65
    growth of, 63, *64*
    origin of life and, 71–72, *73*
    as a primer in DNA replication,
        271, *272*, 274
    reverse transcription, 285
    roles in gene expression, 285
    structure and function, 63–67
    translation, 65
    types of RNAs produced by
        transcription, 286
    in vitro evolution, 501
Ribonucleoside triphosphates,
    287, 288
Ribose, *52*, 63
Ribosomal RNA (rRNA)
    in eukaryotic ribosomes, 84–85
    location and role in eukaryotic
        cells, *286*
    produced by transcription, 286
    role in translation, 285
    See also Ribosomes
Ribosomal RNA genes
    concerted evolution in, 498
    evolutionary relationships in
        prokaryotes and, 528
Ribosomes
    action of antibiotics on
        prokaryotic ribosomes, 295
    in animal cells, *86*
    functions in translation,
        294–295
    interaction with tRNAs in
        translation, 294
    in plant cells, *87*
    polyribosomes, 297, *298*
    in the process of translation,
        295–297, *298*
    of prokaryotes, *82*, 83
    structure and function in
        eukaryotes, 84–85, 88
    subunits, 295, 364
Riboswitch, 348
Ribozymes
    as biological catalysts, 72, *73*,
        151
    lowering of the energy barrier
        in biochemical reactions,
        151–154
Ribulose 1,5-bisphosphate (RuBP),
    194, 195, *196*, 197–198
Ribulose bisphosphate
    carboxylase/oxygenase
    (rubisco)
    in C_3 plants, 198
    in C_4 plants, 199
    in the Calvin cycle, 194
    photorespiration and, 197–198
Ribulose monophosphate (RuMP),
    194, *196*

Rice
  genetically modified, 388
  genome, *361, 362*
  as a primary human food
    source, 605
  quantitative variation in grain
    production, 246
  sake, 624
  semi-dwarf, 756
  transgenic improvement of
    water-use efficiency, 726
  water demands of, 726
Rice, William, 482
Rice "paddies," 605
Richardson, A., *300*
Rickets, *1053*
Ricketts, Taylor, 1243
Ridley–Tree Condor Preservation
    Act, 1244
Riftia, 661
Riggs Glacier, *17*
Right-handed β-spirals, *46*
Right-handed helices, *47, 265*
Rigor mortis, 989
Ring canal, 680, *681*
Ringneck snake, *694*
Ringworm, 612
Risk costs, in animal behavior,
    1103
Rivers
  freshwater biomes, 1140
  in the global hydrologic cycle,
    1215
  nitrogen runoff and "dead
    zones," 740, 1207, 1219
RNA-dependent RNA
    polymerase, 544
RNA genes
  genomic information and, *355,
    356*
  moderately repetitive
    sequences, 364, *365*
RNA interference (RNAi), 346, 382
RNA polymerases
  eukaryotic general transcription
    factors and, 334, *335, 336*
  roles in transcription, 286, *287,*
    288
  sigma factors in prokaryotes,
    333
  structure and function, 286
  in transcriptional regulation in
    prokaryotes, 332–333
  in the viral lytic reproductive
    cycle, 340
RNA retroviruses, *543, 544*, 545
RNA splicing, *291*, 292–293
RNA viruses
  description of, 341
  plant systemic acquired
    resistance, 801
  types of, *543, 544*
  See also RNA retroviruses
"RNA world," 71–72, *73*
RNAi. *See* RNA interference
Roaches, 673
Robinson, Scott, 1117
Rock barnacle, 1184
Rocks
  dating, 506–508
  weathering of, 745–746, 1213
Rod cells, 960–961, *962, 963*
Rodents, *698, 699*
Rohm, Otto, 144

Rooibos, 1242
Root apical meristem, 712,
    715–716
Root cap, 716, *717*
Root hairs, *713, 717*
Root nodules
  formation, 747–748
  nitrogen fixation in, 750
Root systems
  adaptations to saturated soils,
    808
  adaptations to very dry
    conditions, 808
  in monocots and eudicots, 710
  primary indeterminate growth,
    716–718
  root apical meristems, 712,
    715–716
  secondary growth, 721–723
  structure and function, 709
  types of, 718
  uptake of water and minerals,
    718
Roots
  auxins in the initiation of, 765
  cation exchange with the soil
    solution, 746
  endodermis, 729–730
  ethylene and, 770
  evolution in vascular plants,
    582–583
  gravitropism, 765
  movement of water and
    mineral ions across the
    plasma membrane, 728–729
  mycorrhizae, 747, 748–749
  of phylogenetic trees, 450
  root nodule formation, 747–748
  uptake and transport of water
    and mineral ions, 728–730
  vegetative reproduction in
    angiosperms and, 793
Roquefort cheese, *622*, 624
Rosenberg, Barnett, 259, 278
Rotational cleavage, 906
Rothamsted Experiment Station,
    1192
Rotifers
  anatomical characteristics, *652*
  description of, 656, 657–658
  major subgroups and number
    of living species, *632*
Rough endoplasmic reticulum
    (RER)
  in animal cells, *86*
  in plant cells, *87*
  processing of newly translated
    polypeptides, *299, 300*
  structure and function, 88–89
Rough-skinned newt, 445
Round window, *954, 955*
Roundup, 160
Roundworms, 635, 666
  See also Nematodes
Rove beetles, 1203
"Royal jelly," 899
Royal poinciana, *603*
RPL21 gene, 497
rRNA. *See* Ribosomal RNA
rRNA genes, 364, *365*
RT-PCR, 380
RU-486 (mifepristone), *898*
Ruben, Samuel, 186, 187

Rubisco. *See* Ribulose
    bisphosphate carboxylase/
    oxygenase
Rudbeckia fulgida, 599
Ruffini endings, 952
Rufous hummingbirds, 1137
Rumen, 1064
Ruminants, 492–494, 1064
Runners, 720, 792
Ruppell's griffon, 1010
Rushes, 793
Russian steppe, 1131
Rust fungi, 622, 800
Rusty tussock moth, *1157*
Ryanodine receptor, 990, 992
Rye, 718
Rye mosaic virus, 669

S

S-adenosyl methinione (SAM-e),
    349
S genes, 381, 782
S orbital, 24
S phase
  cell fusion experiments on cell
    cycle control, 208–209
  centrosomes in, 212
  description of, 208
  in meiosis, 219
  in mitosis, *214, 217*
Sac fungi
  distinguishing features, *616*
  edible, 624
  filamentous, 620, 622
  in lichens, 613
  life cycle, 620, *621*
  as model organisms, 625, 626
  phylogeny of the fungi, *615*
  yeasts, 620 (*see also* Yeasts)
  See also Dikarya
Saccharomyces cerevisiae, 609
  characteristics of, 620
  genomic information, *361, 362*
  insertion of genes into, 376
  as a model organism, 624
  in the production of food and
    drink, 623
Saccharum, 603
Saccoglossus kowalevskii, 683
Saccule, 956, 957
Sagartia modesta, 647
Sailfish, *1139*
Sake, 624
Sakmann, B., 929
Salamanders
  experiments on embryo
    formation, 910–911, *912*
  features of, 690, *691*
  gills and respiratory gas
    exchange, *1007*
  neoteny in, 692
Salamone, Daniel, 386
Salicylic acid, 759, 799, 801
Saline environments
  challenges to plants, 810–811
  plant adaptations to, 811
Saliva, 1059
Salivary amylase, *1062*
Salivary glands, *1058, 1060, 1062*
Salivation, 1064–1065
Salmon, 689, 887, 1141
Salmonella, 357, 542
Salmonella typhimurium, 534
Salps, 684

Salt, George, 482
Salt and water balance regulation
  in aquatic invertebrates, 1072–
    1073
  in invertebrates, 1075–1077
  by the mammalian kidney,
    1082–1084
  in vampire bats, 1071
  in vertebrates, 1073, 1077–1079
Salt bridges, 47
Salt glands
  nasal, 1073
  in plants, 811
Salt marshes, 1141
Salt tolerance
  in genetically modified plants,
    388, *389*
  See also Halophiles
Saltatory conduction, 935
Salty taste, 951
Salvelinus confluentus, 1228
Salvelinus namaycush, 1228
Salvinia, 581
SAM-e (*S*-adenosyl methinione),
    349
San Andreas Fault, 509
Sand verbena, 793
Sandy soils, 745
Sanger, Frederick, 353
Sapindopsis belviderensis, 519
Saprobes, 556, 611, 1054
Saprolegnia, 556, 557
Saprotrophs. *See* Saprobes
Sarcolemma, *987*
Sarcomeres, *987, 988, 996*
Sarcophilus harrisii, 232
Sarcoplasm, 989, 990
Sarcoplasmic reticulum
  effect of calcium ion cycling on
    cardiac muscle contraction,
    1034, *1036, 1037*
  in skeletal muscle contraction,
    989–990, *991*
Sardinella aurita, 1139
Sardines, *1139*
Sargasso Sea, 556
Sargassum, 556
Sarin, 157
Sarracenia, 751
Sarracenia purpurea, 1189, 1197
Satiety factors, 1067–1068
Saturated fatty acids
  phospholipids and, 107
  structure and function, 56, *57*
Saturated soils, plant adaptations
    to, 808
Saturation, of facilitated diffusion,
    117
Savannas, 1135–1136, 1194
SBE1 gene, 305
SBE1 protein, 305
Scale of life, *78*
Scales
  of lepidosaurs, 693
  of ray-finned fishes, 687
Scallops, 662
Scandentia, *698*
Scanning electron microscopy, 81
Scarab beetles, 826
Scarlet ibis, *696*
Sceloporus jarrovii, 1103, 1104
Schally, Andrew, 844
Schindler, David, 1220
Schistosomiasis, 657

Schizosaccharomyces pombe, 361,
    624, 626
Schomburgk, Robert, 588
Schopf, J. William, 74
Schulze, Franz, 629
Schwann cells, 926
Scientific methods
    distinguishing characteristics
        of, 14
    experiments, 12–13
    key features of, 11–12
    statistics and, 13–14
    See also Biological research
Scientific names, 7, 462, 463–464
Scion, 793, *794*
Sclera, 958, *959*
Sclereids, *713*, 714
Sclerenchyma, 713–714
Scleria goossensii, 440
Scleria verrucosa, 440
Sclerotium, 560
Scolopendra hardwicki, 670
Scorpionflies, *672*
Scorpions, *636,* 669
Scottish deerhound, 352
"Scouring rushes," 581
Scrotum, 889
Scurvy, 1053
Scyphozoans, 647
Sea anemones, 646, *647,* 1171
Sea butterflies, 662
Sea cucumbers, 681, 682
Sea grass, 1141
Sea level
    effect of plate tectonics on, 509
    glaciation and, 509, *510*
    mass extinctions and, 510
Sea lilies, 520, 680–681, *682*
Sea lions, 700
Sea palms, *556*
Sea pens, 646, *647*
Sea slugs, 662, *664,* 1173, *1174*
Sea spiders, 668
Sea squirts, *455,* 683, *684*
Sea stars, 681, 682, 881
    See also Starfish
Sea turtles, 693
Sea urchin eggs
    fertilization, 884–886
    maturation promoting factor,
        209, 210
Sea urchins
    blastopore, *634*
    determination of polarity in the
        embryo, 395
    gametic isolation, 477
    gastrulation, 908–909
    key features of, 681, *682*
Seabirds
    dispersion patterns, *1153*
    territory of, 1104, *1105*
Seagrasses, *603*
Seals, 700, 1017
Seasonal temperatures,
    acclimatization in animals,
    821
Seasons
    causes of, 1123
    See also Environmental cycles
Seawater evaporating ponds, *536*
Sebastes melanops, 1163
Secale cereale, 718

Second filial generation (F$_2$),
    234–236
Second law of thermodynamics,
    146–147
Second messengers
    calcium ions, 135
    defined, 131
    discovery of, 132–133, 134
    in hormone-mediated signaling
        cascades, 837
    lipid-derived, 134–135
    nitric oxide, 135–136
Second polar body, *883,* 884
Secondary active transport, 118,
    119
Secondary bronchi, 1011, *1012*
Secondary cell wall, 711
Secondary consumers, 1190, *1191*
Secondary endosymbiosis,
    551–552
Secondary growth, 591–592, 715
Secondary immune response, 866
Secondary lysosomes, 90, *91*
Secondary meristems, 715, *716*
Secondary metabolites
    of the metabolome, 370
    in plant defenses against
        herbivores, 802–803, 804,
        1175
Secondary oocyte, *883,* 884
Secondary phloem, 16, *17,* 721,
    722, *723*
Secondary sex characteristics, 250,
    889
Secondary spermatocytes, *883,*
    884, *891*
Secondary structure, of proteins,
    45, *46*
Secondary succession, 1200, *1201*
Secondary xylem, 591–592, 721,
    722–723
    See also Wood
Secretin, 838–839, 1065
Sedges, 440
Sedimentary rocks, 506, 507, 565
Seed coat, 591, *592,* 784
Seed companies, 753
Seed dispersal
    birds and, 696
    conifers, 594
    fruits and, 601
    plant–frugivore mutualisms in,
        1181–1182
    seed plants, 591
    strategies in, 784
Seed dormancy, 591, 757, 762, 784
Seed ferns, 589, 591
Seed germination
    cytokinins and, 768
    effects of light on, 772
    overview, 758
    phytochromes and, 774
    role of abscisic acid in, 762
    role of gibberellins in, 761–762
    smoke-induced, 1121
    vacuoles and, 93
"Seed leaves." *See* Cotyledons
Seed plants
    angiosperms, 596–604
    euphyllophytes, 583
    evolution of, 589
    gymnosperms, 592–596
    important crop plants, 605

life cycle, 589–590
major groups, *574,* 589
medicinal, 604–605
pollination, 590, *591*
secondary growth, 591–592
seeds, 590, 591, *592*
Seedless fruits, 784–785
Seedless grapes, 761
Seedless watermelon, 225
Seedlings
    ethylene and the apical hook, 770
    etiolation, 772
    growth of, 758
Seeds
    aleurone layer, 761, *762*
    in angiosperm sexual
        reproduction, *781*
    conifers, 594, *595*
    development, 590, *592*
    dispersal (*see* Seed dispersal)
    dormancy, 591, 757, 762
    fruit development and, 784–785
    germination (*see* Seed
        germination)
    hybrid, 778
    inoculation with mycorrhizae,
        615
    quiescence, 757
    tissues in, 591
Segment polarity genes, 403, *404*
Segmentation
    in animals, 636
    in annelids, 659, *660*
    in arthropods, 667
    in protostome evolution, 673
    in trilobites, 668
Segmentation genes, 402, 403, *404*
Segmentation movements, 1060,
    1063
Segregation
    Mendelian law of, 236–237, 239
    See also DNA segregation
Selectable marker genes, 376, 378
Selection. *See* Artificial selection;
    Natural selection; Sexual
    selection
Selenium, *1052*
Self, distinguishing from nonself,
    863, 874
Self antigens, 874, 876
Self-compatibility, 460
Self-incompatibility
    in angiosperms, 782, *783*
    phylogenetic analysis, 459–460
    studies in plants, 380–381
Self-perception, 983
Self-pollination, 598, 782, *783*
Semelparous species, 1156, *1157*
Semen
    components of, 889, 890
    ejaculation, 891
Semi-dwarf grains, 756, 775
Semibalanus balanoides, 1184
Semicircular canals, *954,* 956–957
Semicircular ducts, 956, 957
Semiconservative replication, 268,
    269, 270
    See also DNA replication
Seminal fluid, 890
Seminal vesicles, 890
Seminiferous tubules, 889, *891*
Senescence, 768, 769–770

Senescence hormone, 769
    See also Ethylene
Sensations
    activation of neurons by action
        potentials, 947–948
    intensity, 948
Sensitive period, in animal
    behavior, 1099
Sensors
    in regulatory systems, 816
    in sensory transduction,
        947–948
    See also Sensory receptor cells
Sensory organs, 948
Sensory receptor cells
    adaptation, 948
    chemoreceptors, 949–952
    conversion of stimuli into
        action potentials, 947–948
    in sensory transduction,
        947–948
Sensory receptor proteins, 947, *948*
Sensory systems
    conversion of stimuli into
        action potentials by sensory
        receptor cells, 947–948
    defined, 948
    detection of chemical stimuli,
        949–952
    detection of light, 957–963
    detection of mechanical forces,
        952–957
    functions of, 637
    infrared perception in
        rattlesnakes, 946
Sensory transduction, 947–948
Sepals, *591,* 597
Separase, 215, *216*
Sepia, 663
Sepsis, 862
Septa, 609, *610*
Septate hyphae, 609, *610*
Sequence alignment, 486–487
Sequential hermaphroditism, 887,
    888
Sequoiadendron giganteum, 1160
Serine
    in photorespiration, 197, *198*
    structure, *44*
Serotonin, 940
Sertoli cells, 889, *891, 892*
Sessile animals
    dispersal, 640
    filter feeding, 637, *638*
    radial symmetry in, 634
Set point, 816
Setae, *636,* 660
Seven-transmembrane domain
    receptors. *See* G protein-
    linked receptors
Severe acute respiratory
    syndrome (SARS), 357, *543*
Sex
    advantages and disadvantages
        of, 441
    in protists, 562
Sex chromosomes
    abnormal arrangements, 250
    of ginkgos, 592
    sex determination by, 249–250
    sex-determining gene, 250
    sex-linked inheritance, 249,
        251–252

Sex determination
    in animal groups, *249*
    haplodiploidy, 1115, *1116*
    primary, 250
    secondary sex characteristics, 250
    by sex chromosomes, 249–250
    sex steroids and phenotypic determination, 850, *851*
    temperature-dependent, 420–421
Sex development, impact of atrazine on, 1, 18
Sex-linked genetic diseases, 312–313
Sex-linked inheritance, 249, 251–252
Sex pheromones, 1110
Sex pili, 84, 253, *254*
Sex steroids
    endocrine sources, 850
    functions of, 849
    in phenotypic sex determination, 850, *851*
    in puberty, 850–851, 894
    in regulation of the ovarian and uterine cycles, 894, 895
    sexual behavior in rats and, 1099
    structures, *836*
    temperature-dependent sex determination and, 420
    types of, 850
    See also Estrogens; Testosterone
Sexual behavior
    hormonal control in rats, 1098–1099
    in parthenogenic whiptail lizards, 882
    See also Courtship behavior
Sexual dimorphisms, 480–481
Sexual life cycles
    meiosis and, 217, 218–219 (see *also* Meiosis)
    types of, 218–219
Sexual reproduction
    costs and risks, 882
    evolution in vertebrates, 888–889
    fertilization, 884–887
    in flowering plants, 779–785
    fundamental steps in, 882
    fungi, 616–617
    genetic diversity and, 218–219, 245, 882
    hermaphroditism, 887–888
    honey bees, 880
    patterns of embryo care and nurture, 889
    in protists, 562–563
    spawning, 887
    See also Human reproduction
Sexual selection
    description of, 435–436
    in evolution, 6
    speciation rates and, 480–481
Sexual stimulation
    engorgement of the labia minora and clitoris, 892
    oxytocin and, 141
    penile erection, 890–891
Sexually selected traits, 435–436, 459

Sexually transmitted diseases, 533, 558
Shark Bay, *513*
Sharks
    cartilaginous skeleton, 1000
    claspers in sexual reproduction, 888
    excretion of urea, 1074
    eye control gene, *414*
    features of, 687, *688*
    motor programs of the spinal cord, 969
    regulation of ionic composition of extracellular fluid, 1073
    salt and water balance regulation, 1077
    spiral valve of the intestines, 1057
Sharp-billed ground finch, *474*
Sheep, cloning, 406–407
Shellfish industry, 549
Shells
    of bivalves, 662, *663*
    of brachiopods, 659
    of cephalopods, 664
    as exoskeletons, 999
    of gastropods, 662, *663*
    in protostome evolution, 674–675
    of turtles, 693
Sherman, Paul, 1116
Shigella, 357
Shimomura, Osamu, 449
Shindagger agave, 1183
Shine, Rick, 421
Shine–Dalgarno sequence, 295
Shiners, *472*
Shivering heat production, 827
Shoot apical meristem
    indeterminate growth, 715, *716*, 786
    origin in plant embryogenesis, 712
    primary shoot growth, 719–720
    transition to inflorescence meristems, 785–786
Shoot system
    primary indeterminate growth, 719–720
    secondary growth, 721–723
    shoot apical meristems, 715, *716*
    structure and function, 709
Shoots
    apical dominance, 765
    auxin-induced root formation with cuttings, 765
    early development, 758
    gravitropism, 765
    vegetative reproduction in angiosperms, *792*
Short-beaked echidna, *697*
Short-day plants (SDPs), 787, 788
Short interspersed elements (SINEs), 364–365
Short-tailed shrew, 1172–1173
Short-tandem repeats (STRs), 317–318, 364
Short-term memory, 982
Short-term work, fast-twitch fibers and, 995
Shrew opossums, 697, *698*
Shrews, *698*
Shrimps, 670
Shull, George, 244

Siamese cats, 245, 306
Siamois gene, 911–912
Siamois transcription factor, 911–912
Siberian hamsters, *851*
Sickle-cell disease
    DNA testing by allele-specific oligonucleotide hybridization, *321*
    missense mutation causing, 306, *307*, 312
    prevalence among African-Americans, 312
    use of reverse genetics to discover the DNA mutation in, 318
Side chains, of amino acids, 43, *44*
Siegelman, William, 788
Sierra Madre Occidental, 1137
Sieve plates, *714*, 715, 735
Sieve tube elements
    companion cells, 735
    pressure flow model of translocation, 735–738
    structure and function, 714–715, 735
Sifaka, *10*
Sigma factors, 286, 333
Sigmoria trimaculata, 670
Signal amplification, 133–134, 138
Signal peptides, 298–300
Signal sequences, 298–300, 384
Signal transduction pathways
    cancers and, 131, *132*
    in cell fate determination, 397–398
    crosstalk, 127, 134
    defined, 126
    effects on cell function, 137–139
    elements of, 126–127
    initiation of DNA transcription, 139
    in innate defenses, 860
    protein kinase cascades, 131–132
    receptor proteins, 127–131
    regulation of, 136–137
    second messengers, 131, 132–136
    signal amplification, 133–134, 138
    types and sources of signals, 126
    See also Plant signal transduction pathways
Signals
    acoustic signals, 1112
    autocrine signals, 126, 835
    cell responses to signal molecules, 125
    effects on cell function, 137–139
    electric signals, 485
    error signals, 816
    functions of signal proteins, *42*
    honest signals, 435
    initiation of DNA transcription, 139
    juxtracrine signals, 126
    mechanosensory signals, 1112
    paracrine signals, 126, 835
    types and sources of intercellular signals, 126
    visual signals, 1111

    See also Chemical signals; Reproductive signal
Sildenafil (Viagra), 136, 137
Silencers, of transcription factors, 335
Silent mutations, 305, 306, *306*
Silent substitutions, 491
    See also Synonymous substitutions
Silica, 555, 1175
Silicates, 71
Silicon dioxide, 643
Silkworm moth, 59, 841, 950
Silurian period, *506–507*, 520
Silver gull, 412
Silver salts, 770
Silverfish, 671, *672*
Silverswords, 481–482
Simberloff, Daniel, 1198, *1199*
Similarity matrix, 487
Simple diffusion, 114, *118*
Simple fruit, 601
Simple sugars. *See* Monosaccharides
Simultaneous hermaphroditism, 887–888
Single nucleotide polymorphisms (SNPs)
    description of, 317
    haplotype mapping, 367–368
    in the human genome, 366
    human genome scans and diseases, *368*
    pharmacogenomics and, 368
    using to find disease-causing genes, 318, *319*
Single-strand binding proteins, 272
Sink strength, 736–737
Sinks, in phloem translocation, 714, 734, 736–737
Sinoatrial node, 1032–1033, 1034, *1035*
Sinus venosus, 1028
Siphonaptera, *672*
Siphonops annulatus, 691
Siphons, of cephalopods, 662, *663*, 664
Sirenians, *698*
Sirius Passet, 516
siRNAs. *See* Small interfering RNAs
Sister chromatids
    centrosomes, 212–213
    chromatin structure, 211–212, *213*
    comparison between mitosis and meiosis, 223
    defined, 207
    events in meiosis, 220–221
    separation in meiosis II, 219, 222
    separation in mitosis, 214–215
Sister clades, 451
Sister species, 451, 472
Sit-and-wait predators, *638*
Situs inversus, 902, 915
Skates, 444, 687, *688*
Skeletal muscle
    antagonistic sets, 942
    effect of ATP supply on performance, 997
    effect of heat on fatigue, 998

factors affecting the strength of muscle contraction, 994–995
functions of, 817, 987
impact of exercise on strength and endurance, 996–997
interaction with bone at joints, 1001–1002
jumping ability in animals, 986
in the knee-jerk reflex, 942
motor units, 989
muscle spindles, 952–953
neuromuscular junctions, 936–938
optimal length for generating maximum tension, 996
paralysis during REM sleep, 979
role of muscle fiber types in strength and endurance, 995–996
sliding filament contractile mechanism, 987–990, *991*
structure of, 987–988
Skeletal systems
interactions with muscle, 999, 1001–1002
types of, 999–1001
See also Endoskeletons; Exoskeletons; Internal skeletons
Skin
in amniote evolution, 692
blood flow and thermoregulation, 824–825, 828
epithelial tissues of, 817
gas exchange in amphibians, 1029
"heat portals," 831
of lepidosaurs, 693
tactile receptors, 952
Skin beetles, 1203
Skin cancer, 277, 382
Skin cells, induced pluripotent stem cells from, 409
Skin pigmentation
MC1R gene and, 367
vitamin D and, 1054
Skinner, B. F., 1094
Skoog, Folke, 768
Skull cap, 1001
Skulls, in humans and chimpanzees, 704
Sleep, 978–980
Sleeping sickness, *559*
Sliding DNA clamp, 273–274
Sliding filament model, 987–990, *991*
Slime, of roots, 716
Slime molds
cellular, 561
chemical signaling in, 838
plasmodial, 560–561
Sloths, *698*
Slow block to polyspermy, *885, 886*
Slow-twitch fibers, 995
Slow-wave sleep, 978
Slug (cellular slime molds), 561
Slugs, 662
Small ground finch, *474*
Small interfering RNAs (siRNAs)
description of, 347–348

location and role in eukaryotic cells, *286*
in plant systemic acquired resistance to RNA viruses, 801
in RNA interference, 382
Small intestine
absorption of nutrients in, *1062, 1063*
digestion in, 1061–1062
digestive enzymes of, *1062*
in humans, *1058*
movement of stomach contents into, 1061
production of secretin, 1065
sections of, 1061
segmentation movements, 1060
Small nuclear ribonucleoprotein particles (snRNPs), 292
Small nuclear RNA (snRNA), *286, 292*
Small populations
impact of genetic drift on, 434
noncoding DNA in the genome and, 496
Small tree finch, *474*
Smallmouth bass, 1160
Smallpox virus, 856
Smell
description of, 949–950
signal transduction in, 137–138
Smith, Hamilton, 356
Smithies, Oliver, 382
Smoke
breaking of seed dormancy, 757, 1121
See also Cigarette smoke
Smoker's cough, 1015
Smooth endoplasmic reticulum (SER)
in animal cells, *86*
conversion of nitrites to nitrosamines, 310
in plant cells, *87*
in plasmodesmata, *139, 140*
structure and function, 89–90
Smooth muscle
acetylcholine-stimulated relaxation, 135–136
contraction, *992, 993*
functions of, 817, 987, 993
in the gut, 820, 993, 1058, *1059*
peristalsis, 1059–1060
structure, 993
vascular (*see* Vascular smooth muscle)
Smut fungi, 622
Snails, 662
Snakeflies, *672*
Snakes
features of, 693, *694*
hemipenes, 888
loss of limbs, 423
ovoviviparity, 889
pit organs and infrared perception, 946
vomeronasal organ, 950–951
"Snowball Earth" hypothesis, 515–516
SNPs. *See* Single nucleotide polymorphisms
snRNA. *See* Small nuclear RNA
snRNPs. *See* Small nuclear ribonucleoprotein particles

Soap, 144
Social behaviors
in amphibians, 692
evolution of, 1113–1117
Social organization, effect on characteristic species density, 1160
Sociobiology, 1116–1117
Sociobiology (Wilson), 431
Socratea exorrhiza, 1194
Sodium
in animal nutrition, *1052*
electronegativity, 28
ionic attraction, 28–29
uptake by halophytes, 811
Sodium bicarbonate, 36
Sodium channel genes
evolutionary studies, 485, 500, 502
gene duplication, 498
Sodium chloride, 29
See also Salt and water balance regulation
Sodium hydroxide, 34
Sodium ion channels
in action potentials, 932–935
blocking by TTX, 500
in cardiac pacemaker cells and heart contraction, 1033–1034
evolutionary studies, 485, 500, 502
membrane depolarization and, 930–931, *932*
in rod cells, 961
in taste bud sensory cells, 951
TTX resistant, 445, 500
See also Acetylcholine receptors; Voltage-gated sodium channels
Sodium ion transporters, 1063
Sodium ions
absorption in the small intestine, 1063
in cardiac pacemaker cells and heart contraction, 1033–1034
generation of action potentials and, 932–934
ionic electric current and, 931–932
properties of, 29
salt tolerance in plants, 388
Sodium tripolyphosphate (STPP), 1221
Sodium–potassium pump
action potentials and, 933
in the generation of membrane potentials, 928, 929
in primary active transport, 118–119
Soft-shelled crabs, 999
Soil bacteria
global nitrogen cycle, 750–751
influence on plant uptake of nutrients, 747–748, 748–751
interference competition, 1183
nitrogen fixation, 749–750
Soil fertility
defined, 745
factors determining, 746
Soil fungi, 612, *613*
Soil solution, 741
Soils
adding fertilizers to, 746–747

availability of nutrients to plants, 746
formation of, 745–746
heavy metals and plants, 811–812
leaching, 745, 746
optimal pH for plants, 746
saline, 810–811
structure, 745
of tundra, 1128
Solar radiation
atmospheric circulation patterns and, 1124
geographic distribution and ecosystems, 1208–1210
impact on community productivity, 1192
impact on development, 421–422
photosynthetic efficiency of plants, 201–202
variation in input across Earth's surface, 1123
Sole, 444
Soleus, 995
Solidago, 752
Solute potential, 727, *728*, 736
Solutes
accumulation in xerophytes, 808
defined, 33
uptake by plants, 727–730
Solutions
buffers, 35–36
concentration of gases in, 1006
osmolarity, 1072
pH, 35
properties of aqueous solutions, 33–34
water potential, 727, *728*
Solvents, 33
Somatic cell gene therapy, 323
Somatic cell nuclear transfer experiments, 406–407
Somatic cells
chromosomes in, 218
cloning animals from, 406–407
harmful consequences of mutations in, 310–311
Somatic mutations, 305, 310–311, 314
Somatosensory cortex, 983
Somatostatin, *842, 844*, 848, 849
Somites, 916, *917*
Songbirds
factors affecting song acquisition, 1099–1101
hormonal control of song expression, 1101
Sonic hedgehog (*Shh*) gene, 423
Sonic hedgehog (Shh) protein, 401, 917
Soredia, 613–614
Sorghum, 805
Sori, 582
Soricomorpha, *698*
Sound
in animal communication, 946, 1112
definition of, 946
perception of (*see* Auditory systems)
Sour taste, 951

Sources, in phloem translocation, 714, 734, 736
South Africa. *See* Fynbos
South Georgia Island, 1149, 1166
Southern beeches, *1129*, 1130, 1143
Southern pine bark beetle, 1178
Sow bugs, 670
Soy sauce, 624
Soybeans, 624
Space-filling models, *27, 47*
Spaceship Earth, 1245
Spanish ibex, *1134*
Spatial heterogeneity hypothesis, of latitudinal gradients in diversity, 1196
Spatial summation, 938
Spawning, 882, 887
Specialization hypothesis, of latitudinal gradients in diversity, 1196
Speciation
   from barriers to gene flow, 472–475
   defined, 468
   divergence of populations, 6
   diversification of mating behaviors and, 705
   factors affecting rates of speciation, 480–482
   genetic basis of, 470–471
   hybrid zones, 478–479
   laboratory experiments with *Drosophila*, 482
   Lake Malawi cichlids, 467
   lineage species concept, 469
   mechanisms preventing hybridization, 475–478
   reproductive isolation and, 468–469
Species
   characteristic densities, 1159–1160
   concepts of, 468–469
   defined, 468
   estimated number of living species, 514, 1230
   evolution of populations, 6
   evolutionary perspective on comparing, 451–452
   evolutionary tree of life, 6–9
   genetic variation in geographically distinct populations, 443–444
   making comparisons between, 7
   mechanisms preventing hybridization, 475–478
Species abundance, role of evolutionary history in, 1160
Species concepts, 468–469
Species diversity
   community productivity and, 1192
   contributions of species richness and species evenness to, 1195–1196
   decrease through time in detritus-based communities, 1202
   impact of disturbances on, 1199
   island biogeography theory, 1196–1198, *1199*
   latitudinal gradient in, 1196, 1197

Species evenness, 1195
Species extinctions
   biodiversity loss and, 1229–1230
   "centers of imminent extinction," 1237, *1238*
   from human activity, 1229–1230
   from invasives, 1235–1236
   from overexploitation, 1234–1235
   predictors of, 1231–1232
   from rapid climate change, 1236–1237
   See also Mass extinctions
Species immigration, in island biogeography theory, 1196–1198, *1199*
Species interactions
   categories of, 1170–1171
   coevolution and, 1171–1172
   competition, 1182–1185
   evolution of antagonistic interactions, 1172–1177
   existing as a continuum, 1171
   herbivory, 1175–1176
   impact on communities, 1193–1195
   mutualisms, 117–1182
Species names, 462
Species pool, 1197, 1198
Species richness
   defined, 1195
   habitat loss and, 1232
   impact on community stability, 1202–1203
   species diversity and, 1195–1196
   using as a criterion for protected areas, 1237
   wetlands restorations and, 1239, *1240*
Species trees, 529–530
Species–area relationship, 1197
Specific heat, 32
Specific plant immunity
   defined and described, 799
   gene-for-gene resistance, 800
   hypersensitive response, 800–801
   phytoalexins, 799–800
   systemic acquired resistance, 801
Spemann, Hans, 910–911, 912
Sperm
   of conifers, 594, *595*
   contributions to the zygote, 903
   double fertilization in angiosperms, 600–601, *781, 783*
   fertilization in animals, 884–887
   fertilization in humans, 892
   gametic isolation and, 477
   genomic imprinting in mammals, 344–345
   of mosses, 576
   production in animals, 882–884
   production in humans, 889–890, *891*
   release in spawning, 887
   reproductive technologies in humans, 897, 898
   of seed plants, 589
   in semen, 889
   transfer in internal fertilization, 887
   See also Sperm–egg interactions

Sperm cells, in angiosperm sexual reproduction, 780, *781, 783*
Spermatids, *883, 884, 889, 891*
Spermatocytes, *883–884, 889, 891*
Spermatogenesis
   in animals, 883–884
   hormonal regulation in humans, 892
   in humans, 889–890, *891*
Spermatogonia, 883, 889, *891*, 908
Spermatophores, 887
Spermatozoa, 889, *891*
Sperm–egg interactions
   activation of development in animals, 903–904
   blocks to polyspermy, *885*, 886–887
   specificity in, 884–886
Spermicidal jellies, *898*
Spermophilus beldingi, *1117*
Spermophilus parryii, *10*
Sperry, Roger, 980
Sphaerechinus granularis, *682*
Sphagnum, 578
Sphenodon punctatus, *694*
Spherical symmetry, 634
Sphincter muscles, *1059*, 1060, 1061, 1080
Sphinx moth, *588*
Sphygmomanometer, 1032, *1033*
Spicules, *633*, 643, 644
Spicy/hot taste, 951
Spider beetles, 1201, 1203
Spider monkeys, *702*
Spider silk
   bioengineering of, 59
   properties of, 39
   protein structure, 39, 45, 46
Spiders
   deprivation experiments, 1095
   external digestion of food, 1055
   nervous system, 941
   sperm transfer in spermatophores, 887
   webs, 669
Spiegelman, Bruce, 839
Spike mosses, 581
Spikelets, *597*
Spikes (inflorescence), *597*
Spina bifida, 916
Spinal cord
   development, 968
   early development, 916
   functions of, 969
   nerves of the autonomic nervous system, *974, 975*
   reflexes, 942–943
   structure and function, 941–942
Spinal cord injuries, 970
Spinal cord transection, 969
Spinal nerves, 942, 968
Spinal reflexes, 942–943, 969, 1080
Spindle apparatus, 212, 213–214, *215*
   See also Mitotic spindles
Spindle assembly checkpoint, 215
Spindle cells, of the insular cortex, 983
Spines
   of cacti, 807
   in plant defenses against herbivory, 1175
   of sea urchins, 681
Spinner dolphins, *700*

Spiny-headed worms, 657–658
Spiracles, 671, *672*, 1009
Spiral cleavage, 633, 653, 905
Spiral valve, 1057
Spiralians, 633, 653
   See also Lophotrochozoans
Spirillum, 528, *529*
Spirobranchus, *638*
Spirochetes, 533
Spirographis spallanzanii, *661*
Spirometer, 1012
Spleen, *857*, 1038
Spliceosomes, 292
"Split-brain" studies, 980–981
Sponges
   in animal phylogeny, *630*
   asymmetry in, 634
   cell adhesion and cell recognition in, 110, 111
   description of, 643–644
   digestion in, 1056
   filter feeding, 637
   major subgroups and number of living species, 632
   respiratory gas exchange, *1007*
   similarity of choanoflagellates to, 631, *633*, 644
   as the sister group to all animals, 648
Spongy mesophyll, 720, *721*
Spontaneous abortion, 897
Spontaneous generation, 67–68
Spontaneous mutations, 308–309
Sporangia
   of club mosses, 581
   of ferns, 582
   of land plants, 575
   of liverworts, 577
   of slime molds, 561
   of zygospore fungi, *618*, 619
Sporangiophores, *618*, 619
Spore walls, in plant evolution, 574
Spores
   in alternation of generations, 575
   of cyanobacteria, 532, *533*
   dispersal in liverworts, 577
   in heterospory, 584–585
   in homospory, 584
   of seed plants, 590
   of sporocytes, 563
   of zygospore fungi, *618*, 619
Sporocytes, 563
Sporophytes
   of angiosperms, 596, 600
   of conifers, *595*
   of ferns, 582
   in homospory, 584
   of hornworts, 578, *579*
   of land plants, 575
   of liverworts, 577
   of mosses, 578
   of nonvascular land plants, 575, 576–577
   relationship to gametophytes in plant evolution, *590*
   of seed plants, 590, 591
   of vascular plants, 579
Sporopollenin, 590
Sporozoites, *564*
Sporulation, 562
Spring wheat, 791
Springtails, 671, *672n*

Spruces, 1200
Squamates, 693
Squamous cells, *818*
Squid giant axons, 929, *930*
Squids
  eye control gene, *414*
  features of, 662–663
  sperm transfer in
    spermatophores, 887
Srb, Adrian, 283, 284
SRY gene, 250
SRY protein, 250
St. Johnswort, 1175
St. Matthew Island, 1149, 1166
Stabilizing selection, 439
Stable atoms, 25
Stage-dependent cohort life tables,
  1154
Stahl, Franklin, 268, *269*, 270
Stained bright-field microscopy,
  80
Stamen, *591*, 596, 598, 779
Standard free energy
  from glucose oxidation, 149, 166
  from the oxidation of NADH,
    168
Stapes, *954*, 955
Staphylococcus, 531, 608
Staphylococcus aureus, 281, 301, 531
Star anise, 602
Starch branching enzyme 1
  (SBE1), 237
Starch grains, 53
Starches
  in cassava, 708
  conversion to ethanol by yeast,
    623
  production in photosynthesis,
    195–196
  structure and function, 53, *54*
Starfish, 681, 682
  See also Sea stars
Starling, Ernest, 838, 1041
Starling's forces, 1041
Stars, navigation by, 1109–1110
Start codon, 289, 295, 296
Starvation, 1050
Statistics, 13–14
Stele, 717, *718*
Stellate barnacles, 1184
Stem cell therapy, 102, 392
Stem cells
  defined, 381, 408
  mesenchymal, 392, 394, 408
  multipotent, 408, 410
  pluripotent, 408–409
  potential medical uses, 77
  See also Embryonic stem cells
Stem elongation
  gibberellins and, 761
  inhibition by cytokinins, 768
Stems
  aerenchyma, 808
  annual rings, 722–723
  apical dominance, 765
  cuttings, 793
  effect of ethylene on growth,
    770
  function, 709, 720
  modified, 720
  primary indeterminate growth,
    719–720
  secondary growth, 721–723
  thickening in monocots, 723

vegetative reproduction in
    angiosperms, 792–793
Stenella longirostris, *700*
Steno, Nicolaus, 506
Steppe, 1131
Steppuhn, Anke, 804
Stereocilia, 953–954, 955
Sterilization, as a method of
    contraception, *898*
Sternum, 695
Steroid hormones
  of the adrenal cortex, 849–850
  characteristics of, 836
  ecdysone, 840–841
  in plants, 771
  structure and function, 58
  synthesis pathway, *836*
  See also Sex steroids
Steward, Frederick, 405
Stewart, Caro-Beth, 492–493, 494
Stick insects, *672*, 673
Stick model, of protein tertiary
    structure, 47–48
Sticky ends, 374
Stigmas
  in flower structure, *591*, 597
  germination of pollen grains,
    780, 782
  retraction response in bush
    monkeyflowers, 598, *599*
Stilt palm, 1194
Stingrays, *445*, 687
Stinkhorn mushrooms, 2
Stock, in grafting, 793, *794*
Stolons, 792
Stomach
  buffering of acid, 36
  chemical digestion in, 1060–
    1061
  digestive enzymes, *1062*
  function of mucosal epithelial
    cells, 1058
  in humans, *1058*
  pH in humans, 161
  production of gastrin, 1065
  production of ghrelin, 1067
  release of chyme into the small
    intestine, 1061
  of ruminants, 1064
  segmentation movements, 1060
  in tubular guts, 1056
Stomata
  in CAM plants, 734
  closure in response to drought
    stress, 809
  control of water loss and carbon
    dioxide uptake, 732–734
  dermal tissue system origin, 712
  functions of, 198, 721
  in mosses, 577
  in plant evolution, 574
  plant regulation of number and
    function, 734
  in xerophytes, 807
Stomatal crypts, 807
Stone cells, 714
Stoneflies, *672*, 673
Stoneworts, 572–573
Stop codons
  in the genetic code, 289
  nonsense mutations, 306, *307*
  in translation, 296–297, *298*
Storage proteins, 42
Strabismus, 921

Stramatolites, 539
Stramenopiles, 555–556, *557*
Strata, 506
Stratified epithelium, *818*
Stratigraphy, 506
Stratosphere, 1211, *1212*
Strawberries, 601
Strawberry plants, 720, 792
Streams, 1140
Strength, impact of exercise on,
    996
Strength training, impact on
    muscle, 996
Strepsipterans, *672*
Streptococcus pneumoniae, 260–261
Streptokinase, 385
Streptomyces, 532
Streptophytes, 572, 573
Stress, during pregnancy,
    implications of, 328
Stress response
  cortisol and, 849–850
  in plants, 337
Stress response element (SRE), 337
Stretch receptors
  in blood pressure regulation,
    1044, *1045*
  function in crayfish, 947
  in the knee-jerk reflex, 942
  regulation of blood pressure,
    1088–1089
Striated muscle. *See* Skeletal
    muscle
Striga, 626, 753
Strigolactones, 747, *748*, 753
Strobili, 581
Strokes
  atherosclerosis and, 1042–1043
  treatment with TPA, 385
Stroma
  carbohydrate synthesis in,
    193–197
  light-independent reactions,
    188
  light-induced pH changes, 196
  structure and function, 92, *93*
Stromatolites, *5*, 512, *513*
STRs. *See* Short-tandem repeats
Structural genes, regulation in
    prokaryotes, 330–332, *333*
Structural isomers, 41
  of hexoses, 52
Structural motifs, 335–336
Structural proteins, *42*
Struthio camelus, *695*
Styles, *591*, 597, 782
Subclavian artery, *1031*
Subclavian vein, *1031*
Subduction, 509
Suberin, 717, 798
Sublingual salivary gland, *1058*
Submandibular salivary gland,
    1058
Submucosa, 1058, 1059
Subpopulations, 1161, 1162
Subsoil, 745
Substance P, 940
Substrate-level phosphorylation,
    170
Substrates
  effect on reaction rate, 156
  in enzyme-catalyzed reactions,
    152–153

enzyme interactions with,
    154–155
Succession, 1201–1203
Succinate dehydrogenase, 172
Succinic acid, 70
Succulence, 807
Succulents
  absence in Australia, 1126
  adaptations to very dry
    conditions, 807
  crassulacean acid metabolism
    in, 200
  in hot desert biomes, *1132*, 1133
  vegetative reproduction, 793
Suckers (of leeches), 661
Suckers (of plants), 793
"Sucking chest wound," 1016
Sucrases, 152, *1062*
Sucrose
  phloem translocation, 734–735
  production in plants, 195, 201
  structure, *53*
Sudden Acute Respiratory
    Syndrome (SARS), 501
Sugar apples, 785
Sugar beets, 718
Sugar phosphates, *55*
Sugarcane, *603*
Sugars
  formation of glycoproteins, 301
  formed in prebiotic synthesis
    experiments, 70
  found on meteorites, 69
  phloem translocation, 734–738
Sulci, 970
Sulfate ions
  inorganic fertilizers, 747
  leaching in soils, 746
Sulfated polysaccharides, 111
Sulfhydryl group, *40*
Sulfolipids, 92
Sulfolobus, 2, 536
Sulfur
  in animal nutrition, *1052*
  covalent bonding capability, 27
  in plant nutrition, *741*
  produced by photoautrophic
    bacteria, 538
  radioactive isotope in the
    Hershey–Chase experiment,
    262–263
Sulfur cycling
  global cycle, 1219–1220
  prokaryotes and, 539
Sulfur dioxide
  in the global sulfur cycle, 1219
  from volcanoes, 510
Sulfuric acid
  acid precipitation and, 1219–
    1220
  properties of, 34
Sun, Yuxiang, 182
Sundews, 751, *752*
Sunflowers, 811
Sunlight. *See* Solar radiation
"Superbugs," 357, 373
Superficial cleavage, 905
Superior vena cava, 1030, *1031*
Superoxide, 175, 176
Superoxide dismutase, 176
Suprachiasmatic nuclei (SCN),
    1108
Surface area-to-volume ratio, of
    cells, 78–79

Surface runoff
  ecological impact of chemical fertilizers, 740, 1207, 1219, 1221
  movement of elements and, 1213
Surface tension
  lung surfactants and, 1015
  of water, 33
Surfactants, 1015
Survival, evolution by natural selection and, 6
Survivorship, 1154, 1156
Survivorship curves, 1155–1156
Suspension feeders, 637, *638*
Suspensor, 711
Sustainable management, 1224
Sutherland, Earl, 132–133
Sutterella, 540
Svedberg unit, 364
Swallowing, 1059
Swamps, 1140
Swarm cells, 561
Sweat glands, 105, 697
Sweating, 105, 829
Sweet potato, 718
Sweet potato–corn dicultures, 1203
Sweet taste, 951
Sweet wormwood, 797
SWII gene, 794
Swim bladders, 687
Swine flu, 877
Swordtails, 459
Sycon, 643
Symbiotic interactions
  defined, 102, 612
  dinoflagellate endosymbionts in corals, 565, 566
  evolution of nitrogen-fixation in legumes, 521
  hornworts and cyanobacteria, 579
  between plants and soil bacteria, 747–748
  Vibrio and Hawaiian bobtail squid, 546
Symmetry
  in animal body plans, 634–635
  pentaradial, 680, *681*
  See also Bilateral symmetry; Radial symmetry
Sympathetic division
  in blood pressure regulation, 1044, *1045*
  effect on heartbeat, 1034
  influence on smooth muscle, 993
  structure and function, 974–975
Sympatric speciation, 473, 475
Symplast, 729, 730, 711
Symplastic pathway, 736
Symporters, 118, 1063
Synapomorphies, 452
Synapses
  clearing of neurotransmitter, 940
  function of, 926
  functions of astrocytes at, 926–927
  inhibitory or excitatory, 938
  ionotropic and metabotropic, 938–939

long-term potentiation and long-term depression, 982
neuromuscular junctions, 936–938
number in the human brain, 943
summation of synaptic input by the postsynaptic cell, 938
tripartite, 927
types of, 926, 936
Synapsis, 219, *220*
Synaptic cleft, 936, *937*
Synaptula, 682
"Synchronous" muscle, 998
Syncytium, 905
Synergids
  in angiosperm sexual reproduction, 779, *781*
  degeneration of, 783
  pollen tube growth and, 782
Synonymous substitutions
  defined, 491
  effect of modes of selection on substitution rates, 492–494
Synthetic cells, 359, *360*
Synthetic DNA, 380
Synthetic hormones, in birth control pills, 896
Syphilis, 533
Systematics, 451
Systemic acquired resistance, in plants, 801
Systemic circuit
  in amphibians, 1029
  in birds and mammals, 1030
  blood vessels of, 1027–1028
  defined, 1027
  in lungfish, 1028
  in reptiles, 1029–1030
Systemic lupus erythematosis (SLE), 876
Systems biology, 157, 181
Systole, 1031–1032, 1039
Systolic pressure, 1032, *1033*
Szostak, Jack, 73

T

T cell receptors
  binding of antigens to, 871–872
  in the cellular immune response, 864, *865*
  function of, 859
  in the humoral immune response, 864, *865*
  specificity of adaptive immunity and, 862–863
  structure, *871*
T cells
  as antigen-presenting cells, 864
  binding of T cell receptors to antigens, 871–872
  in the cellular immune response, 863, 864, *865*
  clonal deletion, 865
  clonal selection, 865, *866*
  cytokines, 859
  in delayed hypersensitivity, 876
  effector T cells, 865, 866, 871, 872, 873, 874
  function of, *858*
  interaction with antigen-presenting cells, 872
  maturation in the thymus, *857*

memory T cells, 865
selection in the thymus, 872
T DNA, 377
T-helper (T_H) cells
  in the cellular immune response, 864, *865*, 872, *873*
  in class switching, 870–871
  in delayed hypersensitivity, 876
  development of plasma cells and, 867
  in HIV infections, 876, *877*
  in the humoral immune response, 864, *865*, 872, *873*
  regulation by Tregs, 874
T tubules, 989, 990, 992
T2 phage, 261–263
T4 phage, *543*
T7 bacteriophage, 316
Tachyglossus aculeatus, *697*
Tadpole shrimp, *670*
Tadpoles, 678
Taeniopygia guttata, 1100–1101
Taeniura lymma, 445
Taiga, 1129–1130
Tamoxifen, 325
Tapeworms, 638, *641*, 657, 675, 888, 1176
Taproots, 710, 718, 808
Tarantulas, *669*
Taraxacum officinale, 1153
Tardigrades, 632, 667–668
Taricha granulosa, 445
Tarsius pumilus, 1230–1231
Tarweeds, 481–482
Tasmanian Devil Genome Project, 255
Tasmanian devils, 232, 245, 255
Taste, 951
Taste buds, 951
Taste pore, *951*
Tat protein, 342
TATA box, 333–334, *335*
Tatum, Edward, 282–283
Taxi drivers, 967
Taxol, 96, 604
Taxon (taxa)
  biological nomenclature, 462, 463–464
  defined, 451
  Linnaean classification, 462–463
  monophyletic, 463
Taxonomy, 462–464
Taxus brevifolia, 604
Tay-Sachs disease, 91
Tcf-3 transcription factor, 912
Tectorial membrane, *954*
Teeth
  enamel, 21
  in mammals, 697
  plaque, 539, *540*
  structure and function, 1055–1056
Tegeticula yuccasella, 1181
Telencephalon, 968, 969, 970–973
Teleomeres, 275
Telomerase, 229, 275
Telophase (mitosis), 212, *215*, 216, *217*
Telophase I (meiosis), *221*, 223
Telophase II (meiosis), *221*
Teloschistes exilis, 613
Temperate forests
  deciduous, 1126, 1130–1131

global climate change and, 1217–1218
Temperate grasslands, 1131–1132
Temperature
  acclimatization to, 821
  boreal and temperate evergreen forest biomes, *1129*
  chaparral biomes, *1134*
  cold desert biomes, *1133*
  effect on diffusion, 114
  effect on enzymes, 161–162
  effect on living organisms, 820–821
  effect on protein structure, 50
  effect on terrestrial biomes, 1126
  highest temperature compatible with life, 535
  hot desert biomes, *1132*
  impact of solubility of gases in liquids, 1006
  impact on respiratory gas exchange for aquatic animals, 1007, *1008*
  induction of flowering and, 791
  influence on sex determination, 420–421
  membrane fluidity and, 107
  plant adaptations and responses to temperature extremes, 810
  temperate deciduous forest biome, *1130*
  temperate grassland biome, *1131*
  thorn forest and tropical savanna biomes, *1135*
  tolerance of extremes in fungi, 611
  tropical deciduous forest biomes, *1136*
  tropical rainforest biomes, *1137*
  tundra biomes, *1128*
  Walter climate diagrams, 1138
Temperature-dependent sex determination, 420–421
Temperature-sensitive mutations, 306
Temperature sensitivity (Q_{10}), 821
Templates
  in DNA replication, 268, 272
  in the PCR reaction, 277, *278*
  telomeric, 275
  in test tube synthesis of DNA, 267
Temporal isolation, 476
Temporal lobe, 971
Temporal summation, 938
Tendons, 953, 1002, 1003
Tennessee shiner, *472*
Tension
  generated by skeletal muscle fibers, 996
  See also Transpiration–cohesion–tension mechanism
Tentacles
  of cephalopods, 664
  of lophophores, 652
Teosinte, 419, 723–724
Teosinte branched 1 (tb1) gene, 723–724
Ter site, 206
Terminal buds, 709

Termination, of transcription, *287*, 288
Termites, *672, 673, 1135*
Terpenes, *802*
Terrestrial biomes. *See* Biomes
Territorial behavior, 1103–1104, *1105*
Territorial calls, 1112
Tertiary consumers, 1190, *1191*
Tertiary endosymbiosis, 552
Tertiary period, *506–507, 519,* 522
Tertiary protein structure, 46–49
Test crosses, 237, *238*
Testate amoebas, 560
Testes
  hormone of, *842*
  in humans, 889, *891*
  spermatogenesis in, 882–884, 889–890, *891*
Testicular cancer, 259
Testosterone
  actions of, *842*
  as an androgen, 850
  in control of song expression in songbirds, 1101
  in follicle selection for ovulation, 895–896
  in human puberty, 892
  sexual behavior in rats and, 1099
  spermatogenesis and, 889, 892
  structure, *836*
  temperature-dependent sex determination and, 420
  territorial behavior in lizards and, 1103, *1104*
  See also Sex steroids
Tetanus (disease), 542
Tetanus (muscle contraction), 994–995
Tetanus toxin, 936
Tetracycline, 281, 295
Tetrads, 219, 247–249, *250*
Tetrahydrofolate, 158
Tetraiodothyronine (T_4), 845, 846, 847
Tetraodon nigroviridis, 361
Tetraploids, 224–225, *226*
Tetrapods
  limb evolution, 690
  modifications to pharyngeal slits, 684
Tetrodotoxin (TTX), 445, 500
Texas Longhorn cattle, 440
TFIIB protein, 334
TFIID protein, 334, *335*
TFIIE protein, 334
TFIIF protein, 334
TFIIH protein, 334
Tga1 gene, 419
Thalamus
  functions of, 969, 970
  during sleep, 979
  in visual processing, 975, 977
Thalassarche chrysostoma, 1109–1110
β-Thalassemia, 292–293
Thale cress. *See Arabidopsis thaliana*
Thaliaceans, 684
Thalidomide, 920
Thalloid liverworts, 577
Thallus, 613
Thamnophis sirtalis, 445
Thecal cells, 895–896
Therapeutic abortion, 897

Therapeutic genes, 323–324
Therians, 697–700, 889
Thermal insulation, 825, 828
Thermal limits, 820
Thermocline, 1214
Thermodynamics, laws of, 146–147
Thermogenin, 827–828
Thermoneutral zone, 826
Thermophiles, 532, 536
Thermoplasma, 537
Thermoreceptors, 947, *948*
Thermoregulation
  behavioral, 822, *823*
  conservation of metabolic heat in "hot" fish, 825
  control of blood flow to the skin, 824–825, 828
  energy budgets and, 823–824
  metabolic heat production in ectotherms, 825–826
  production of metabolic heat, 822
  role of the hypothalamus in mammals, 829–830
  strategies in ectotherms, *824*
  strategies in endotherms, *824,* 826–831
Thermotoga, 529, 532
Thermus aquaticus, 277–278, 532
Theropods, 694, 695
Thiamin, *1053,* 1054
Thioredoxin, 196–197
Thistles, 784, 1201
Thlaspi caerulescens, 811
Thoracic cavity, *1014,* 1015–1016
Thoracic ducts, *857, 858,* 1042, 1063
Thoracic vertebrae, 418
Thorn forest, 1135–1136
Thorns, in plant defenses against herbivory, 1175
"Threatened" species, 1231
Three-chambered hearts, 1029
Three-dimensional vision, 977–978
3′ End, 266
Three-spined sticklebacks, 423–424
Three-toed sloth, *1137*
Threonine, *44*
Threshold membrane potential, 933
"Thrifty genes," 1048
Thrips, *672, 673*
Thrombin, 1039
Thrombus, 1042, 1043
Thylakoids
  electron transport systems, 191–192
  light reactions, *188*
  photophosphorylation and ATP synthesis, 192–193
  photosystems, 190
  structure and function, 92, *93*
Thymine
  complementary base pairing, 63–65
  dimerization by UV light, 277
  in DNA structure, 264, 265, 266, *267*
  effect of ultraviolet radiation on, 309
  formed from cytosine deamination, 310
  structure, *63*

Thymine dimers, 277
Thymosin, *842*
Thymus
  clonal deletion in, 865, 872
  hormones of, *842*
  in the lymphatic system, *857*
  maturation of regulatory T cells in, 874
  T cells selection in, 872
Thyroglobulin, 845, *846*
Thyroid gland, 320, *842,* 845–847
Thyroid-stimulating hormone (TSH), 846, 847, 1101
Thyrotropin, *842, 843,* 846
  See also Thyroid-stimulating hormone
Thyrotropin-releasing hormone (TRH), 844, 846
Thyroxine
  actions of, *842*
  goiter, 846–847
  half-life, 853
  production and regulation, 845–846
  structure, *836*
Thysanoptera, *672*
Thysanura, *672*
Ti plasmid, 377
Ticks, 638, 639, 668, 669
Tidal ventilation, 1012–1013
Tigers, *1172, 1234, 1235*
Tight junctions, 111, *112*
Tijuana Estuary, 1239, *1240*
Tiktaalik roseae, 689, 690
Till, James, 408
Tilman, David, 1202
Time-compensated solar compass, *1111*
Time hypothesis, of latitudinal gradients in diversity, 1196
Tinamous, 694
Tinbergen, Niko, 1095–1096
Tissue plasminogen activator (TPA), 384–385
Tissue-specific promoters, 384
Tissue systems. *See* Plant tissue systems
Tissue transplants, 874
Tissues
  biological hierarchy concept, 9
  diffusion within, 114
  relationship between cells, tissues, and organs, 817–820
  See also specific tissue types
Titin, 291, 988, 996
Toads
  characteristics of, 690–691
  hybrid zones, 479
  nitrogenous wastes excreted, 1075
Tobacco
  flowering cues from an "internal clock," 792
  Maryland Mammoth, 787
  nicotine in flower nectar, 1181
Tobacco hornworm moth, 348
Tobacco mosaic virus, 285
Tocopherol, *1053*
Toll-like receptors, 860
Tomatoes
  diseases of, *798*
  dwarfed phenotype, 760, *761*
  genetically modified for salt tolerance, 388, *389*

Tomocerus minor, 671
Tongue, taste buds in humans, 951
Tonoplast, 710
Topography, impact on biomes, 1126–1127
Topsoil, 745, 1132
"Torpedo" embryo, *393*
Total peripheral resistance (TPR), 1043
Totipotency
  animal cloning, 406–407
  defined, 394
  plant cloning, 405–406
  in plant development, 710
Touch, 952
Toxigenicity, 542
Toxins
  Bacillus thuringiensis toxin in transgenic plants, 387–388, 389
  bacterial, 542
  of dinoflagellates, 549
  impact on synaptic proteins, 936
  in plant nectar, 1181
Toxoplasma, 554
Toxostoma guttatum, 1232
TP53 gene, 306
TP53 protein, 306
TPA. *See* Tissue plasminogen activator
Trachea
  in birds, 1011, *1012*
  in humans, 1013, *1014*
Tracheae, 671, 1009
Tracheal system, 1009
Tracheary elements, 714
Tracheids
  description of, 714
  in the evolution of vascular plants, 579
  in gymnosperms, 593
  in vascular plants, 573
Trachemys scripta, 420
Tracheophytes, 573
  See also Vascular plants
Trade-offs
  in animal reproduction, 641–642
  constraints on evolution, 445
Trade winds, 1124, *1125*
Traits
  adaptation, 433
  ancestral and derived, 452
  defined, 234, 431
  dominant and recessive, 235
  in Mendel's monohybrid crosses, 234–236
  qualitative and quantitative, 439
  in sexual selection, 435–436
  sources of data for phylogenetic analyses, 454–456
  using to construct phylogenetic trees, 453, *454*
Transcription
  compared in prokaryotes and eukaryotes, 334
  components needed for, 286
  differential gene transcription in differentiation, 398–399
  of DNA, 65
  effect of histone modifications on, 344
  error rates, 288

in gene expression, 284–285
gene expression in RNA genomes, 285
genetic code, 288–289
importance of base pairing to, 65, 67
initiation by signal transduction, 139
initiation in eukaryotes, 335, 336
noncoding sequences, 290–291
processing of gene transcripts before translation, 291–293
regulation in prokaryotes, 329–333
RNAs produced by, 286
role of mRNA in, 285
signals that start and stop, 297
steps in, 286–288
stimulation by phytochromes in plants, 773–774
structure and function of RNA polymerases, 286
in the viral lytic reproductive cycle, 340
of X chromosome genes, 345–346
Transcription elongation, HIV gene regulation and, 341–343
Transcription factors
binding to promoters, 328
cell differentiation and, 336, 337, 338, 399
in cell fate determination, 397–398
coordinated regulation of sets of genes, 336–337
determination of fruit fly body segmentation and, 401–405
enhancers and silencers, 335
as genetic switches, 415
as intracellular receptors, 130–131
plant organ identity genes and, 400
primary embryonic organizer and, 911–912
role in transcription, 286
in signal transduction pathways, 126–127
structural motifs and binding to DNA, 335–336
transcriptional regulation in eukaryotes and, 333–334, 335
Transcriptional regulation
in eukaryotes, 333–338
in prokaryotes, 329–333
Transducin, 961
Transects, 1151
Transfection, 263, 376
See also Transformation
Transfer RNA (tRNA)
binding sites on ribosomes, 295
charging with an amino acid, 293, 294
location and role in eukaryotic cells, 286
produced by transcription, 286
role in translation, 285, 293–294, 295–297, 298
specificity in binding to an amino acid, 294
wobble, 294
Transformation
defined, 376

methods, 376–377
See also Genetic transformation
Transforming growth factor-β (TGF-β), 912
Transfusions, 243
Transgenic animals
cloning, 407
pharming, 385–386
Transgenic cells, 376
Transgenic crops
water-use efficiency, 726
overview, 386–388, 389
public concerns, 388–389
Transition mutations, 306
Transition state, 151–152
Transition-state intermediates, 152
Translation
blocking to study gene expression, 382
elongation, 296, 297
in gene expression, 284–285
initiation, 295–296
overview, 293
polyribosomes, 297, 298
regulation in eukaryotes, 348.23
of RNA, 65
role of ribosomes in, 285, 294–295
role of tRNAs in, 285, 293–294, 295–297, 298
signals that start and stop, 297
termination, 296–297, 298
in the viral lytic reproductive cycle, 340
Translocations, 224, 304, 308
Transmembrane domains, 108
Transmembrane proteins, 108, 112, 113
Transmission electron microscopy, 81
Transpiration, 732
Transpiration–cohesion–tension mechanism, 731–732
Transport proteins
functions, 42
in the small intestine, 1063
Transposons (transposable elements)
in eukaryotic genomes, 364–366
in the human genome, 366
as mutagens in minimal genome studies, 359, 360
mutations caused by, 308
as noncoding DNA, 495
in prokaryotic genomes, 358–359
small interfering RNAs and, 348
Transverse tubules. See T tubules
Transversion mutations, 306
Trastuzumab, 871
Travisano, Michael, 489–490
Tree ferns, 520, 580, 581
Tree of life, 6–9, 451, 522
Tree shrews, 698
Treg cells, 863, 874
Treponema pallidum, 533
Triassic period, 506–507, 518, 521
Tricarboxylic acid cycle. See Citric acid cycle
Trichinella spiralis, 666
Trichinosis, 666
Trichocysts, 554, 555
Trichoglossus haematodus, 696

Trichomes, 713, 807
Trichomonas vaginalis, 558
Trichoplax adhaerens, 629
Trichoptera, 672
Trifolium repens, 443–444
Trigger hairs, 751–752
Triglycerides
absorption in the small intestine, 1062, 1063
structure and function, 56–57
synthesis, 56
Triiodothyronine (T$_3$), 845, 846, 847
Trilobites, 668
Trimesters, of pregnancy, 919–920
Trimethylamine oxide (TMAO), 1073
Triops longicaudatus, 670
Triose phosphates, 195
Tripartite synapse, 927
Triploblastic animals, 633
Triploids, 224–225, 473, 475, 783
Trisomics, 224
Triticum aestivum, 226
Triticum monococcum, 226
Triticum turgidum, 226
Tritium, 22
tRNA. See Transfer RNA
tRNA genes, 364
Trochophores, 640, 652, 653
Trophic cascades, 1193–1194
Trophic levels
in communities, 1190, 1191
energy transfer between, 1190–1192
Trophoblast, 906, 907, 919
Trophosome, 660
Tropic hormones, 843
Tropical alpine tundra, 1128
Tropical forests
deciduous, 1136–1137
evergreen, 1129, 1130
keystone species, 1194–1195
Tropical rainforests
current rate of loss, 1232
description of, 1137–1138
estimating the number of insect species in, 651
fragmentation and habitat corridors, 1233–1234
Tropical savannas, 1135–1136
Tropics
atmospheric circulation patterns and, 1124
influence of the dry season on, 1126
Tropidolaemus wagleri, 638
Tropomyosin, 988, 990, 991
Troponin, 988, 990, 991, 998, 1034
Troposphere, 1211, 1212
Trp operon, 331
True bugs (hemipterans), 480, 672, 673, 674
True flies, 672, 672, 674
True navigation, 1109–1110
Truffles, 620
Trypanosoma brucei, 559
Trypanosoma cruzi, 559
Trypanosomes, 559
Trypsin, 144, 1062
Trypsinogen, 1062
Tryptophan, 44, 331
TSH. See Thyroid-stimulating hormone

TSH receptors, 846–847
Tsien, Roger, 449
Tsunami of 2004, 1223
TTX-resistant sodium channels, 499, 500
Tuataras, 693, 694
Tubal ligation, 898
Tubal pregnancy, 906
Tube cell, 780
Tube feet, 680, 681, 682
Tuberculosis, 532
Tubers, 720, 792
Tubocurarine, 604
Tubulanus sexlineatus, 659
Tubular guts, 1056–1057
Tubular heart, 1027
Tubular reabsorption, in the vertebrate nephron, 1078, 1079
Tubular secretion, in the vertebrate nephron, 1078, 1079
Tubulidentata, 698
Tubulin
in microtubules, 95–96 (see also Microtubules)
spindle apparatus, 213–214
Tubulinosema ratisbonensis, 617
Tulips, 603
Tumor necrosis factor, 861
Tumor suppressor genes
DNA methylation and, 344
gain of function mutation and, 306
mutations in colon cancer, 314
Tumors
benign and malignant, 227
chaperone proteins and, 51
treatments targeting the cell cycle, 228–229
Tundra, 1128–1129
Tunic, 684
Tunicates, 632, 679, 683, 684
Turbellarians, 657
Turgor pressure
guard cell function and, 733
mechanism generating, 727
osmosis and, 114–115
in plant cell expansion, 766
in plant growth, 710
in the pressure flow model of phloem transport, 736
vacuoles and, 93
Turner syndrome, 250
Turnover, in lake water, 1213–1214
Turtles, 420, 693, 694
Twigs, 721–722
Twin studies, on epigenetic changes, 344
Twins, 907
Twitches, 994–995
Twitters, 1183
Two-dimensional gel electrophoresis, 369
"Two-point spatial discrimination test," 952
Two-pronged bristletails, 671, 672
Tympanic canal, 954, 955
Tympanic membrane, 954, 955, 956
Tympanuchus cupido, 434
Type I diabetes, 848, 876
Type II diabetes, 848, 853
Typhlosole, 1057

Tyrosine, *44*
Tyto alba, 696

U

Ubiquinone (coenzyme Q$_{10}$), 172, *173*
Ubiquitin, 349
UCP1 (uncoupling 1) protein, 165, 174, 182
Ulmus procera, 793
Ulna, 423
Ultimate causes, of animal behavior, 1096
Ultrabithorax (Ubx) gene, 415, 418, *419*
Ultrabithorax (Ubx) protein, 418
Ultraviolet (UV) radiation
    evolution of life and, 5
    mutagenic effects, 309, 310
    ozone layer and, 1212
    thymine dimers and skin cancer, 277
Ulva, 538, 572, 1201
Ulva rigida, 573
Umami, 951
Umbelliferone, *802*
Umbels, *597*
Umbilical cord, 897, 906, 919
Umbilicus, 897
Uncompetitive inhibitors, of enzymes, 158
Underground stems, 792, 793
Undershoot, 934
Unequal crossing over, 498, *499*
Unicellular yeasts, 609
Unidirectional ventilation, 1010–1012
Uniporters, 118
Unipotency, 394
United States
    population age structure, 1165–1166
    population growth, 1165
Unloading, of phloem sieve tubes, 736–737
Unsaturated fatty acids
    phospholipids and, 107
    structure and function, 56, *57*
Upregulation, of hormone receptors, 853
Upwelling zone, 1214
Uracil
    codons and the genetic code, 288–289
    formed by cytosine deamination, 309
    in RNA, 63, 64, *65*
    structure, *63*
Uranium-234, *507*
Uranium-235, *507*
Urea
    effect on protein structure, 50
    excretion as a nitrogenous waste, 1074
    in the extracellular fluid of cartilaginous fish, 1073
    inorganic fertilizers, 747
Ureotelic animals, 1074
Ureter, 1080, 1082
Urethra
    excretion through, 1080
    in the female reproductive system, 892, *893*

in the male reproductive system, 890, 891
Urey, Harold, 70, 71
Uric acid, 1074, 1076–1077
Uricotelic animals, 1074
Urinary bladder, 1080
Urine
    defined, 1072
    formation in the vertebrate kidney, 1078–1079
    glucose levels with diabetes, 848
    production of concentrated urine in mammals, 1071, 1077, 1082–1084, 1090
    See also Excretory systems
Ursus americanus, 1130
U.S. Coast Guard, 1241
Uterine cycle
    defined, 892
    description of, 893
    hormonal regulation of, 894–895
Uterus
    function of, 889
    in humans, 892, *893*
    in labor and childbirth, 843, 896–897
    origin and development of the placenta, 919
Utricle, 956, 957

V

Vaccinations
    "drive-through," *16*
    eradication of smallpox, 856
    evolution of viruses and, 427, 446
    "herd immunity," 856
    immunological memory and, 863, 866
    reasons people resist, 877
Vaccine proteins, production through biotechnology, *385*
Vacuoles
    in ciliates, 554, *555*
    functions, 93
    in plant cell expansion, 766
    in plant cells, *87*
    in plant development, 710
    in plant self-protection from chemical defenses, 805
Vagina
    in childbirth, 897
    in humans, 892, *893*
    role in internal fertilization, 887
Vagus nerve (cranial nerve X), 969
Valence shell, 25
Valine, *44*
Vampire bats, 1071, 1090
Van der Waals interactions
    description of, 30
    in DNA, 266
    features of, *26*
    in protein quaternary structure, 48
    in protein tertiary structure, 47
Variable regions, of immunoglobulins, 867, *869,* 870
Variables, 12, 13
Variegated darter, *472*
Vas deferens, 890, 891
Vasa recta, *1080, 1082, 1084*

Vascular bundles, 710, 719
Vascular cambium
    in grafting, 793, *794*
    secondary plant growth and, 715, *716,* 721, 722
Vascular disease, 1042–1043
Vascular plants
    ancient forests, 580
    branching, independent sporophyte, 579
    distinguishing characteristics, 573
    evolution of leaves, 583, *584*
    evolution of roots, 582–583
    evolutionary significance of vascular tissue, 579
    heterospory, 584–585
    horsetails and ferns, 581–582
    lycophytes, 581
    major groups and distinguishing characteristics, *574*
    rhyniophyte relatives, 580
Vascular rays, 722
Vascular smooth muscle
    in arteries and arterioles, 1039, *1040*
    in autoregulation of blood flow, 1044
    calcium ions in the relaxation of, 136
    control of blood distribution in the body, 993
Vascular tissue
    in angiosperms, 596
    evolution of land plants and, 579
    in gymnosperms, 593
    See also Phloem; Xylem
Vascular tissue system
    description of, *712,* 714–715
    in leaves, 720, *721*
    primary meristem giving rise to, 716
    in roots, 717, *718*
    in shoots, 719
Vasectomy, *898*
Vasopressin
    actions of, 843
    in blood pressure regulation, 1044–1045
    bonding behavior in voles and, 125
    See also Antidiuretic hormone
Vasopressin receptors, 125, 129
Vectors, in transformation, 377
Vegetal hemisphere, 903–904
Vegetal pole, 395
Vegetarian diet, 1051, 1054
Vegetarian finch, *474*
Vegetative cells, of cyanobacteria, 532, *533*
Vegetative meristems, 715
    See also Shoot apical meristem
Vegetative reproduction
    in agriculture, 793–794
    disadvantages, 793
    forms of, 792–793
    See also Asexual reproduction
Veins (blood vessels)
    anatomy, *1040*
    blood flow through, 1041–1042
    function in vertebrate circulatory systems, 1028

Veins (of leaves), 710, 720, *721*
Veldt, 1131
Velvet worms, *667,* 668
Venter, Craig, 356, 359, 360
Ventilation
    in human lungs, 1015–1016
    maximization of partial pressure gradients, 1009
    respiratory tract secretions and, 1013, 1015
    tidal, 1012–1013
    unidirectional ventilation in birds, 1010–1012
Ventral, 634, *635*
Ventral horn, 942
Ventral medulla, 1020
Ventricles
    in amphibians, 1029
    in crocodilians and birds, 1029
    in fish, 1028
    in lungfish, 1028
    mammalian heart, 1030–1032
    mutation-based wall thickening in humans, 1025, 1046
    in reptiles, 1030
    in three-chambered hearts, 1029
Venturia canescens, 1184
Venules, 1028
Venus flytraps, 751–752
Vernalization, 791
Vernonia amygdalina, 1105
Vernonioside B1, 1105
Vertebrae, 684
Vertebral column
    characteristic of vertebrates, 684
    evolution of the giraffe neck, 416–417
    evolutionary impact of differences in Hox gene expression, 418
Vertebrate genomes
    endogenous retroviruses in, 545
    gene duplication in, 497–498
    Hox genes, 413–414
Vertebrates
    amniotes, 692, *693*
    amphibians, 690–692
    appendages, 636
    body plan, *685*
    brain, 943
    central nervous system, 941
    characteristic features, 684–685
    circulatory systems, 1027–1030
    in deuterostome phylogeny, *679*
    endoskeleton, 999–1002
    evolution of, 685
    gastrointestinal system, 1058–1064
    Hox gene expression and spine evolution, 418
    jawed fishes, 686–689
    jawless fishes and, 685–686, *687*
    lens determination, 396
    lobe-limbed, 689–690
    major subgroups and number of living species, *632*
    mammals, 696–700
    muscle types, 987
    neurulation and body segmentation, 916–918
    osmoregulation and ionic regulation in, 1073
    phylogeny of living vertebrates, *685*

regulative development in, 907
reptiles, 693–696
salt and water balance regulation, 1077–1079
teeth, 1055–1056
wing evolution in, 423
Vertical life tables, 1155
Very low-density lipoproteins (VLDLs), 1066
Vesicles
  in endocytosis, 121
  in exocytosis, 122
Vessel elements, 596, 714
Vestibular canal, 955
Vestibular hair cells, 953–954, *956*, *957*
Vestibular membrane, *954*, 955
Vestibular nerve, *954*
Vestibular system
  anatomy of, *954*, 955
  detection of gravity and momentum, 956–957
  hair cell stereocilia and mechanoreception, 953–954
Vestibule, 956
Vestibulocochlear nerve, *954*, 956
Veterinary science, multipotent stem cells and, 410
Viagra, 136
Vibrio
  bioluminescence, 525, 534, 546
  symbiotic relationship with Hawaiian bobtail squid, 546
Vibrio cholerae, 534, 542
Vibrio fischeri, 546
Vicariant events, 1143–1145
Victoria amazonica, *602*
Vicugna vicugna, *1133*
Vicuña, *1133*
Vietnam War, 797
Villi, 1057
Vincristine, 96, *604*
Viola tricolor, *597*
Violets, *597*
Viral blooms, 546
Viral genomes, 357
  See also Prokaryotic genomes
Virginia opossum, 697
Virions, 339
Viruses
  bacteriophage and phage therapy, 545–546
  characteristics and diversity, 543
  defined, 339
  DNA viruses, *543*, 544–545
  ecological significance, 546
  emerging diseases and, 501
  endogenous retroviruses in vertebrate genomes, 545
  enveloped, 341
  evolution in, 427
  flu epidemics, 427
  forensic investigations into transmission, 458, *459*
  gene expression in RNA genomes, 285
  gene regulation, 339–343
  Hershey–Chase experiment on DNA, 261–263
  interferons and, 860
  lateral gene transfer, 496
  life cycles, 339–341
  as living organisms, 3

movement through plasmodesmata, 140
phylogenetic challenges, 543–544
RNA viruses, *543*, 544, 545
vaccines and, 427, 446
as vectors in transformations, 377
Visceral mass, 662
Vision
  disparity, 978
  See also Visual systems
Visual cortex
  inputs from the eyes, 975
  organization of, 978
  receptive fields of neurons, 977
Visual pigments, 957–958
Visual signals, in animal communication, 1111
Visual systems
  image-forming eyes, 958–959
  of invertebrates, 958
  photosensitivity and photoreceptors cells, 957
  retinal receptive fields, 975–977
  structure and function of the retina, 959–963
  three-dimensional vision, 977–978
Vital capacity (VC), *1012*, 1013
Vitamin A, 58, 388, *1053*
Vitamin A deficiency, 58
Vitamin B$_1$, *1053*, 1054
Vitamin B$_2$, *1053*
Vitamin B$_6$, 70, *1053*
  See also Pantothenic acid
Vitamin B$_{12}$, 540, *1053*, 1058
Vitamin C, 1053
Vitamin D, 848, *1053*, 1054
Vitamin E, 176, *1053*
Vitamin K, 540, 1057
Vitamins
  in animal nutrition, 1053–1054
  antioxidants, 176
  coenzymes and, 156
  deficiency in humans, 58, *1053*, 1054
  structure and function, 58
Vitelline envelope, 884, *885*
Vitreous humor, *959*
Viviparity, 889
Vocal cords, 1013
Vocalization, *FOXP2* gene and, 367
Voice box. *See* Larynx
Volcanoes, 510–511
Voles, bonding behavior, 125
Volicitin, 804
Voltage, 927
Voltage-gated calcium channels
  actions at the neuromuscular junction, 936, *937*
  in cardiac pacemaker cells and heart contraction, 1033, 1034
  in the sustained contraction of ventricular muscles, 1034, *1035*
Voltage-gated ion channels
  in action potentials, 932–935
  in cardiac pacemaker cells and heart contraction, 1033–1034
  defined, 115, 930
  at neuromuscular junctions, 936, 937, 938

at nodes of Ranvier, 935
  in the sustained contraction of ventricular muscles, 1034, *1035*
Voltage-gated potassium channels
  in action potentials, 932, 933, 934
  in cardiac pacemaker cells and heart contraction, 1033, 1034
Voltage-gated sodium channels
  in action potentials, 932–935
  of axon hillocks, 938
  in cardiac pacemaker cells and heart contraction, 1033–1034
  electric organs, 485
  evolutionary studies of, 485
  at neuromuscular junctions, 937, 938
  refractory period, 933–934
  in skeletal muscle contraction, 989, *990*
  in the sustained contraction of ventricular muscles, 1034, *1035*
"Volvocine line," 140–141
Volvox, *140*, 141, 572, *573*
Vomeronasal organ (VNO), 950–951, 1096–1097
von Frisch, Karl, 1095, *1112*
von Humboldt, Alexander, 1196
"Vulnerable" species, 1231
Vulva, determination in nematodes, 396, *397*, 398
Vulval precursor cells, 396, *397*, 398

W

Waggle dance, 1112
Wagler's pit viper, *638*
Walchia piniformis, *518*
Walker, M. P., 980
Wallace, Alfred Russel, 429, 1141
"Wallace's line," 1141, *1142*
Walruses, 700
Walter, Heinrich, 1138
Walter climate diagrams, 1138
Wamionoa, *681*
Warbler finch, *474*
Warner, Daniel, 421
Warning coloration, 1173, *1174*
Warpaint shiner, *472*
Warren, Robert, *542*
Warts, 341
Wasps, 1181
  See also Hymenoptera
Wasser, Samuel, 1240
Waste treatment, 389, 1221
Water
  absorption in the large intestine, 1063
  absorption in the small intestine, 1063
  dissolving of ionic solids in, 29
  fern life cycle and, 582
  global hydrologic cycle, 1215
  hydrogen bonding and, 30
  isotope analysis to detect climate change, 36
  as a medium for respiratory gases, 1007
  molecular weight, 27
  moss life cycle and, 576
  movement in capillary beds, 1040–1041

movement through aquaporins, 116
  origin of life and, 68–69
  osmosis, 114–115
  polar covalent bonding in, 28
  seed imbibition, 758
  as the source of oxygen produced during photosynthesis, 186–187
  structure and properties, 32–33
  transport in xylem, 730–732
  transport of elements through ecosystem compartments, 1213–1214
  uptake by plants, 727–730
Water balance. *See* Salt and water balance regulation
Water bears, 667–668
  See also Tardigrades
Water-breathers
  diffusion of oxygen in water, 1007
  effect of high temperature on, 1007, *1008*
  respiratory exchange of carbon dioxide, 1008
Water ferns, *581*
Water fleas, 1105, *1106*
Water lilies, *597*, 602
Water loss
  methods of reducing, 738
  stomatal control of, 732–734
  in trees, 731
Water molds, 556, *557*
Water potential
  components of, 727–728
  defined, 727
  of imbibing seeds, 758
Water purification, 122
Water-soluble hormone receptors, 836–837
Water-soluble vitamins, 1053
Water stress
  effect on plants, 809
  plant response to, 734
Water-use efficiency, 726
Water vapor, in the atmosphere, 1211
Water vascular system, of echinoderms, 680, *681*
Watermelons, seedless, 225
Watson, James D., 264–265, 266, 430–431
Wavelength, light, 188
Waxes
  in constitutive plant defenses, 798
  structure and function, 58–59
Weather, 510, 1122
Weathering
  release of nitrogen, 1218
  of rock, 745–746, 1213
Webbing, in avian feet, 417
Webspinners, *672*
Weeds, 1235–1236
Weevils, *674*
Wegener, Alfred, 509
Weight loss, 165, 182
Weight training
  fast-twitch fibers and, 995
  impact of muscle, 996
  impact on osteoporosis, 1000
Weinberg, Wilhelm, 437
Welwitschia, 593

Welwitschia mirabilis, 593
Werner, Earl, 1105, *1106*
Wernicke's area, 981
Wernig, Marius, 336, *337*, 338
West African finches, 440
West Wind Drift, *1125*
Westerlies, 1124, *1125*
Western horses, *1229*
Westslope cutthroat trout, 1228
Wetlands
  characteristics of, 1140
  consequences of human
    alteration, 1223
  restorations, 1239, *1240*
Whalers, 1164
Whales
  evolution, 700
  overhunting, 1164
  piloting, 1109
  songs of, 1112
  See also Cetaceans
Wheat
  allopolyploidy in, 225, *226*
  effect of temperature on
    flowering, 791
  semi-dwarf, 756, 775
Wheat mosaic virus, 669
Whelks, 662
Whip scorpion, *654*
Whippets, *370*
Whiptail lizards, 881–882
White adipose tissue ("white fat")
  browning of, 834, 839
  characteristics of, *819*
  function of, 165
  See also Adipose tissues
White blood cells, *1038*
  chronic myelogenous leukemia,
    304, 305
  defensive roles of, 858
  induction of cell division in, 211
  types, *858*
White clover, 443–444
White-crowned sparrow, 1100
White light, absorption by
  pigments, 189
White matter, 941
White muscle, 995
White oaks, *534*
White-winged doves, 1137
Widowbirds, 435–436
Wiesel, Torsten, 977
Wigglesworth, Sir Vincent,
  839–840
Wild dogs, ecotourism in Africa
  and, 1242, *1243*
Wild mustard, 432–433, 724
Wild type, 241
Wildfire. *See* Fire
Wilkesia gymnoxiphium, 481
Wilkins, Maurice, 264
Willmitzer, Lothar, 736–737
Willow ptarmigan, *1128*
Wilmut, Ian, 406, 407

Wilson, E. O., 431, 1113, 1116, 1197,
  1198, *1199*, 1229
Wilson, M. A., 967, 983
Wind
  prevailing winds, 1124–1125
  pollination and, 480, 780
Wine, 623–624
Wings
  evolution in vertebrates, 423
  evolutionary comparisons, 452
  multiple occurrences in
    animals, 636
  See also Insect wings
Winter wheat, 791
Wishart, David, 370
Witchweed, 626, 753
Withdrawal reflex, 942–943, 969
Wobble, 294
Wolpert, Lewis, 401
Wolves, 1193–1194, 1239
Womb. *See* Uterus
Wood
  production by the vascular
    cambium, 721, 722–723
  secondary growth in seed
    plants, 591–592
  tracheary elements in, 714
Wood ducks, *468*
Wood frogs, *642*, 1125, *1126*
Wood pigeons, 1116, *1117*
Woodpecker finch, *474*
Woods Hole Oceanographic
  Institution (WHOI), 549
Woolfenden, Glen, 1115
Woolley, Sarah, 1100–1101
Work, done by cells, 10
Worker honey bees, 880
World Wildlife Fund, 1238
Wrasses, *688*
Wu, Ray, 809

X

X chromosomes
  in ginkgos, 592
  importance to spermatocyte
    development, 884
  inactivation, 345–346
  in mammals, 249–250
  mutations leading to human
    genetic diseases, 312–313
  sex-linked inheritance in fruit
    flies, 251–252
X-Gal, 378
X-linked diseases, IPEX, 874
X-linked genes, transcription of,
  345–346
X-ray crystallography, 264
X-ray diffraction, 264
Xanthopan morgani, 588
Xenopus laevis, 7, 612
Xenoturbellids, *632*, 648, *680*
Xeroderma pigmentosum, 277
Xerophytes, 807–808, *809*, 811
Xestospongia testudinaria, 643

Xiphophorus, 459
Xist gene, 346
XO genotype, 250
XXX genotype, 346
XXXX genotype, 346
XXY genotype, 250, 346
Xylem
  in angiosperms, 596
  bulk flow, 728
  defined, 730
  functions of, 579
  in gymnosperms, 593
  in leaves, *721*
  mechanisms of sap flow, *736*
  movement of water and ions to
    in the root, 729–730
  in roots, 717
  secondary, 721, 722–723
  secondary growth and, 591–592
  in shoots, 719
  structure and function, 714
  transport of water and
    minerals, 730–732
Xylem sap
  formation of, 730
  measuring the pressure of, 732
  mechanisms of flow, *736*
Xylocopa darwinii, 1183

Y

Y chromosomes
  evolution of, 345
  in ginkgos, 592
  in mammals, 250
  number of genes in humans,
    366
  sex-linked inheritance in fruit
    flies, 251–252
Yalow, Rosalyn, 844, 852
Yamanaka, Shinya, 409
Yarrow's spiny lizards, 1103, *1104*
Yeasts
  cell cycle control, 210
  features of, 609
  in food and drink production,
    623–624
  genome, 361–362
  insertion of genes into, 376
  as model organisms in lab
    studies, 625, 626
  pathogenic, 612
  relation of mRNA abundance to
    protein abundance, 348
  sac fungi, 620
Yellow-bellied toad, *479*
Yellow warbler, *1093*
Yellowstone National Park, 594,
  596, 1193–1194
Yellowstone to Yukon
  Conservation Initiative, 1234
Yersinia pestis, 534, 542
Yolk, 692, 918
Yolk plug, 910
Yolks sac, 918, 919

Yolky eggs
  extraembryonic membranes,
    918
  gastrulation, 913–914
Yoshida, Masasuke, 175
Yoshimura, Takashi, 1101
Yucca, 599, 1181–1182
Yucca filamentosa, 1181
Yucca moths, 599, 1181–1182
Yunnanozoans, 679, *680*

Z

Z lines, *987*, *988*, 996
Zambia, 1240–1241
Zea mays, 419
  See also Corn
Zeatin, 768
Zeaxanthin, 772
Zebra finches, 1100–1101
Zebra mussels, 1160, 1235
Zebrafish, *905*
Zebras, *1144*
Zellweger syndrome, 93
Zinc
  in animal nutrition, *1052*
  in catalyzed reactions, *156*
  in plant nutrition, *741*
  structural motifs and, 335
  uptake by hyperaccumulators,
    811
Zinnia elegans, 597
Zona pellucida, 886, 906
Zone of cell division, 716, 717
Zone of cell elongation, 716, 717
Zone of cell maturation, 716, 717
Zone of polarizing activity (ZPA),
  401
Zonotrichia leucophrys, 1100
Zoonotic diseases, 458
Zoraptera, 672
Zostera, 1141
Zuckerkandl, Emile, 461
Zygomycota, *616*
Zygosporangium, *618*, 619
Zygospore, *618*, 619
Zygospore fungi, 615, 616, *618*,
  619
Zygote
  in alternation of generations life
    cycle, 563
  of angiosperms, 600, 601, *781*,
    783
  cleavage in animals, 904–906
  of conifers, 594, *595*
  defined, 218
  determination of polarity, 395
  formation in humans, 892
  of land plants, 575
  of mosses, 576–577
  production in animals, 884–887
Zymogens, 1061, 1062